Proceedings of the Eighteenth Lunar and Planetary Science Conference

Proceedings of the Eighteenth Lunar and Planetary Science Conference

Graham Ryder, editor

Cambridge University Press
Lunar and Planetary Institute

Published by the Press Syndicate of the University of Cambridge
The Pitt Building, Trumpington Street, Cambridge, CB2 1RP
32 East 57th Street, New York, NY 10022, USA
10 Stamford Road, Oakleigh, Melbourne 3166, Australia

and

The Lunar and Planetary Institute, Houston, TX 77058

Typeset in the United States of America by the Lunar and Planetary Institute
Printed in the United States of America

© *Lunar and Planetary Institute 1988*

Papers prepared by U.S. government employees as part of their official duties are in the public domain and are not available for copyright protection.

LIBRARY OF CONGRESS CATALOGING IN PUBLICATION DATA

Lunar and Planetary Science Conference (18th : 1987 : Houston, Tex.)
    Proceedings of the Eighteenth Lunar and Planetary Science Conference / Graham Ryder, editor.
        p.   cm.
    Includes bibliographies and indexes.
    ISBN 0-521-35090-5
    1. Lunar geology—Congresses. 2. Planets—Geology—Congresses. 3. Cosmochemistry—Congresses.
I. Ryder, Graham. II. Lunar and Planetary Institute. III. Title.
QB592.L84 1987
523.3—dc 19   87-33290
                                                                                        CIP
ISBN 0-521-35090-5

EDITOR

Graham Ryder

BOARD OF ASSOCIATE EDITORS

Steven K. Croft
Ronald Greeley
Charles M. Hohenberg
Lon L. Hood
Fred Hörz
John H. Jones

James J. Papike
Graham Ryder
Edward R. D. Scott
John Shervais
Paul D. Spudis
Michael Zolensky

LPI PUBLICATIONS OFFICE

Lisa Bowman
Renee Dotson
Pat Pleacher

Assisted by: Penny Auman, Barbara Carter, Mildred Dickey, Carolyn Kohring, Jackie Lyon, and Joan Shack

LPI PRODUCTION SERVICES

Deborah Barron
Shirley Brune
Donna Chady
Carl Grossman
Pam Thompson

Preface

The year 1987 marked the 30th anniversary of mankind's first successful launching of an artificial satellite into earth orbit. Since then many spacecraft have provided us with information about many of the planets and satellites in the solar system, with flybys, orbiters, and landers. These spacecraft have taken photographs and made *in situ* measurements of their environment, and some have brought samples back from the Moon. Meteorite falls such as Allende and the wealth of finds in Antarctica have added to the previously known samples. Over the last 30 years, a rapid and broad growth in planetary studies and our understanding of the solar system have been witnessed. These studies have caused us to look at the Earth itself with a different perspective. The program for the Eighteenth Lunar and Planetary Science Conference, held in Houston, March 16-20, 1987, reflects this wide variety of planetary studies, with subjects ranging from planetary accretion to space utilization.

Since the *Proceedings of the Apollo 11 Lunar Science Conference* (1970), the publication of papers delivered at the Conference has changed with the Conference itself. The present Proceedings volume is being published in a different way from that of the last few years, with a new publisher and returning to the hardcover book format. All papers are published simultaneously, rather than in two separate publications, and the Proceedings returns to having its own, not borrowed, identity. I have chosen to retain the page size and page formats of the last few years, both for continuity and to allow ease of reading. The publication change follows much debate and consideration of options, and I believe it to be in the interests of most of those who contribute to and extract information from the Proceedings.

Both the number of papers submitted and the number published are very similar to the corresponding numbers for the *Proceedings of the Seventeenth Lunar and Planetary Science Conference,* despite the fact that the new procedures required much tighter schedules and deadlines for review, revision, and publication preparation. The Editor's task was eased by a capable and supportive Board of Associate Editors and by the extraordinary efforts and dedication of the staff of the Publications Office and the excellent work from the staff of the Production Services Office at the Lunar and Planetary Institute.

The Eighteenth Lunar and Planetary Science Conference was sponsored by the Lunar and Planetary Institute, the Lyndon B. Johnson Space Center, the American Association of Petroleum Geologists, the American Geophysical Union, the Division for Planetary Sciences of the American Astronomical Society, the Geological Society of America, the International Union of Geological Sciences, and the Meteoritical Society. This constitutes an impressive list of interested organizations. The Conference was organized into many concurrent technical sessions, including poster and oral presentations, as well as public sessions. This volume is also organized into topics, but these do not correspond particularly with the organization of the Conference. Compared with the Conference program, the Proceedings are biased toward lunar studies, both geological and petrochemical, and to planetary geology topics. Other fields are more poorly represented, in part because the papers are published elsewhere, including publications of some of the other sponsoring organizations. Any paper that is considered by the Program Committee for the Lunar and Planetary Science Conference to be suitable for presentation is considered by the Editor to be suitable subject matter for submission to the Proceedings and is then judged only on its scholarship and interest. Papers from all fields and disciplines of planetary science are solicited; indeed, it is not necessary that the paper be presented at the Conference for it to be submitted to the Proceedings. The Proceedings is an excellent vehicle for the publication of papers in, for example, meteoritics, planetary physics, impact mechanics, and isotopic studies, because it is seen by an interested audience.

—*Graham Ryder*

TABLE OF CONTENTS

Petrogenesis and Chemistry of Lunar Samples

On Identifying Parent Plutonic Rocks from Lunar Breccia and Soil Fragments — 1
L. A. Haskin and D. J. Lindstrom

Highland Materials at Apollo 17: Contributions from 72275 — 11
P. A. Salpas, M. M. Lindstrom, and L. A. Taylor

Petrology of Brecciated Ferroan Noritic Anorthosite 67215 — 21
J. J. McGee

Mineralogical Studies of Clasts in Lunar Highland Regolith Breccia 60019 — 33
H. Takeda, M. Miyamoto, H. Mori, and T. Tagai

Olivine Vitrophyres: A Nonpristine High-Mg Component in Lunar Breccia 14321 — 45
J. W. Shervais, L. A. Taylor, and M. Lindstrom

Apollo 14 Regolith Breccias: Different Glass Populations and Their Potential for Charting Space/Time Variations — 59
J. W. Delano

Glasses in Ancient and Young Apollo 16 Regolith Breccias: Populations and Ultra Mg' Glass — 67
S. J. Wentworth and D. S. McKay

^{10}Be Profiles in Lunar Surface Rock 68815 — 79
K. Nishiizumi, M. Imamura, C. P. Kohl, H. Nagai, K. Kobayashi, K. Yoshida, H. Yamashita, R. C. Reedy, M. Honda, and J. R. Arnold

Solar Wind Record in the Lunar Regolith: Nitrogen and Noble Gases — 87
U. Frick, R. H. Becker, and R. O. Pepin

Importance of Lunar Granite and KREEP in Very High Postassium (VHK) Basalt Petrogenesis — 121
C. R. Neal, L. A. Taylor, and M. M. Lindstrom

Apollo 14 Mare Basalt Petrogenesis: Assimilation of KREEP-like Components by a Fractionating Magma — 139
C. R. Neal, L. A. Taylor, and M. M. Lindstrom

Geology and Petrogenesis of the Apollo 15 Landing Site

Materials and Formation of the Imbrium Basin — 155
P. D. Spudis, B. R. Hawke, and P. G. Lucey

Apennine Front Revisited: Diversity of Apollo 15 Highland Rock Types — 169
M. M. Lindstrom, U. B. Marvin, S. K. Vetter, and J. W. Shervais

Chemistry and Petrology of the Apennine Front, Apollo 15, Part I: KREEP Basalts and Plutonic Rocks — 187
S. B. Simon, J. J. Papike, and J. C. Laul

Chemistry and Petrology of the Apennine Front, Apollo 15, Part II: Impact Melt Rocks 203
J. C. Laul, S. B. Simon, and J. J. Papike

The Reliability of Macroscopic Identification of Lunar Coarse Fines Particles and the
Petrogenesis of 2-4 mm Particles in Apennine Front Sample 15243 219
G. Ryder, M. M. Lindstrom, and K. Willis

The Origin of Pristine KREEP: Effects of Mixing Between urKREEP and the Magmas
Parental to the Mg-rich Cumulates 233
P. H. Warren

The Formation of Hadley Rille and Implications for the Geology of the Apollo 15 Region 243
P. D. Spudis, G. A. Swann, and R. Greeley

Petrology and Geochemistry of Olivine-Normative and Quartz-Normative Basalts from
Regolith Breccia 15498: New Diversity in Apollo 15 Mare Basalts 255
S. K. Vetter, J. W. Shervais, and M. M. Lindstrom

Chemical Dispersion Among Apollo 15 Olivine-Normative Mare Basalts 273
G. Ryder and A. Steele

Petrology and Provenance of Apollo 15 Drive Tube 15007/8 283
A. Basu, D. S. McKay, T. Gerke

Laser Probe ^{40}Ar-^{39}Ar Dating of Impact Melt Glasses in Lunar Breccia 15466 299
M. A. Laurenzi, G. Turner, and P. McConville

Lunar Geology and Applications

Lunar Mare Ridges: Analysis of Ridge-Crater Intersections and Implications for the
Tectonic Origin of Mare Ridges 307
V. L. Sharpton and J. W. Head III

Highland Contamination and Minimum Basalt Thickness in Northern Mare Fecunditatis 319
W. H. Farrand

Thickness and Volume of Mare Deposits in Tsiolkovsky, Lunar Farside 331
R. A. Craddock and R. Greeley

Geologic and Remote Sensing Studies of Rima Mozart 339
C. R. Coombs, B. R. Hawke, and L. Wilson

A Remote Mineralogic Perspective on Gabbroic Units in the Lunar Highlands 355
P. G. Lucey and B. R. Hawke

Electromagnetic Energy Applications in Lunar Resource Mining and Construction 365
D. P. Lindroth and E. R. Podnieks

Cratering Records and Cratering Effects

Detecting a Periodic Signal in the Terrrestrial Cratering Record 375
R. A. F. Grieve, V. L. Sharpton, J. D. Rupert, and A. K. Goodacre

Crater Identification and Resolution of Lunar Radar Images — 383
H. J. Moore and T. W. Thompson

A Search for Water on the Moon at the Reiner Gamma Formation, A Possible Site of Cometary Coma Impact — 397
T. L. Roush and P. G. Lucey

Water Content of Tektites and Impact Glasses and Related Chemical Studies — 403
C. Koeberl and A. Beran

The Effects of Impact Velocity on the Evolution of Experimental Regoliths — 409
M. J. Cintala and F. Hörz

Formation of Agglutinate-like Particles in an Experimental Regolith — 423
T. H. See and F. Hörz

Dehydration Kinetics of Shocked Serpentine — 435
J. A. Tyburczy and T. J. Ahrens

Shock-enhanced Dissolution of Silicate Minerals and Chemical Weathering on Planetary Surfaces — 443
M. B. Boslough and R. T. Cygan

Microkrystites: A New Term for Impact-produced Glassy Spherules Containing Primary Crystallites — 455
B. P. Glass and C. A. Burns

Differentiated Meteorites and Related Studies

Phase Equilibrium Constraints on the Howardite-Eucrite-Diogenite Association — 459
J. Longhi and V. Pan

Partitioning of Siderophile and Chalcophile Elements Between Sulfide, Olivine, and Glass in a Naturally Reduced Basalt from Disko Island, Greenland — 471
W. Klöck and H. Palme

Metal with Anomalously Low Ni and Ge Concentrations in the Allan Hills A77081 Winonaite — 485
A. Kracher

Formation of the Lamellar Structure in Group IA and IIID Iron Meteorites — 493
J. A. Kowalik, D. B. Williams, and J. I. Goldstein

Bidirectional Reflectance Properties of Iron-Nickel Meteorites — 503
D. T. Britt and C. M. Pieters

Chondritic Meteorites and Asteroids

Nature and Origin of C-rich Ordinary Chondrites and Chondritic Clasts — 513
E. R. D. Scott, A. J. Brearley, K. Keil, M. M. Grady, C. T. Pillinger, R. N. Clayton, T. K. Mayeda, R. Wieler, and P. Signer

Solar, Planetary, and Other Inert Gases in Two Sieve Fractions of a Disaggregated Allende Sample: A Study by Stepwise Heating Extraction ... 525
   R. L. Palma and D. Heymann

An Assemblage of a Metallic Particle with a CAI in the Efremovka CV Chondrite ... 537
   A. V. Fisenko, K. I. Ignatenko, A. Yu. Ljul, and A. K. Lavrukhina

Origin of Fragmental and Regolith Meteorite Breccias—Evidence from the Kendleton L Chondrite Breccia ... 545
   A. J. Ehlmann, E. R. D. Scott, K. Keil, T. K. Mayeda, R. N. Clayton, H. W. Weber, and L. Schultz

Initial $^{87}Sr/^{86}Sr$ and Sm-Nd Chronology of Chondritic Meteorites ... 555
   J. C. Brannon and F. A. Podosek, and G. W. Lugmair

Cosmic-ray-produced Kr in St. Severin Core AIII ... 565
   B. Lavielle and K. Marti

Spectral Alteration Effects in Chondritic Gas-rich Breccias: Implications for S-Class and Q-Class Asteroids ... 573
   J. F. Bell and K. Keil

Interactions of Light with Rough Dielectric Surfaces: Spectral Reflectance and Polarimetric Properties ... 581
   S. A. Yon and C. M. Pieters

Extraterrestrial Grains: Observations and Theories

Characteristics and Origin of Greenland Fe/Ni Cosmic Grains ... 593
   E. Robin, C. Jéhanno, and M. Maurette

The Search for Refractory Interplanetary Dust Particles from Preindustrial-Aged Antarctic Ice ... 599
   M. E. Zolensky, S. J. Webb, and K. Thomas

Stratospheric Particles: Synchrotron X-Ray Fluorescence Determination of Trace Element Contents ... 607
   S. R. Sutton and G. J. Flynn

Analytical Electron Microscopy of a Hydrated Interplanetary Dust Particle ... 615
   D. F. Blake, A. J. Mardinly, C. J. Echer, and T. E. Bunch

Identification of Two Populations of Extraterrestrial Particles in a Jurassic Hardground of the Southern Alps ... 623
   C. Jéhanno, D. Boclet, Ph. Bonté, A. Castellarin, and R. Rocchia

Trapping Ne, Ar, Kr, and Xe in Si_2O_3 Smokes ... 631
   J. A. Nuth III, C. Olinger, D. Garrison, C. Hohenberg, and B. Donn

Stochastic Histories of Refractory Interstellar Dust ... 637
   K. Liffman and D. D. Clayton

Venus, Mars, and Icy Satellites

Venus Lives!   659
C. A. Wood and P. W. Francis

The Resurfacing History of Mars: A Synthesis of Digitized, Viking-based Geology   665
K. L. Tanaka, N. K. Isbell, D. H. Scott, R. Greeley, and J. E. Guest

A Widespread Common Age Resurfacing Event in the Highland-Lowland Transition Zone in Eastern Mars   679
H. Frey, A. M. Semeniuk, J. A. Semeniuk, and S. Tokarcik

Ages of Fracturing and Resurfacing in the Amenthes Region, Mars   701
T. A. Maxwell and G. A. McGill

Gossans on Mars   713
R. G. Burns

Martian Mantle Primary Melts: An Experimental Study of Iron-rich Garnet Lherzolite Minimum Melt Composition   723
C. M. Bertka and J. R. Holloway

A Model of the Porous Structure of Icy Satellites   741
J. Eluszkiewicz and J. Leliwa-Kopystynski

Author Index   749

Subject Index   750

Sample Index   752

Meteorite Index   753

On Identifying Parent Plutonic Rocks from Lunar Breccia and Soil Fragments

Larry A. Haskin and David J. Lindstrom*

Department of Earth and Planetary Sciences and McDonnell Center for the Space Sciences, Washington University, St. Louis, MO 63130

*Now at Lockheed EMSCO, 2400 NASA Road One, Houston, TX 77058

Conditions of lunar crustal formation that would produce a preponderance of nearly monomineralic anorthosite (e.g., a magma ocean or variant thereof) may be more global and restricted than conditions that would produce a spectrum of rocks ranging from smaller proportions of anorthosite through troctolitic and noritic variants to norite and troctolite. Compositions of most lunar soils and breccias cannot be made by simple mixing of "pristine" lunar nonmare rocks. As a test of our ability to identify parental rock types from examination of breccia and soil fragments, we have modeled breccia fragments expected from a well-studied boulder of Stillwater anorthosite. Depending on their size (number of mineral grains), the boulder fragments give distributions that suggest mixtures of rock types. These include monomineralic anorthosite with subordinate amounts of more gabbroic anorthosite, anorthosite, and gabbro for small fragments. They evolve to two distinct groups of anorthositic gabbro and gabbroic anorthosite for larger fragments, and produce gabbroic anorthosites of approximately the average composition of the boulder for very large fragments. The distribution of FeO in samples of lunar ferroan anorthosite (FAN) indicates that FAN has a heterogeneous distribution of mafic minerals like the boulder, but a smaller proportion of mafic minerals. However, the "pristine" samples include fragments of ferroan noritic and troctolitic rocks that are, by definition, excluded from classification as FAN but that might be parts of the same rocks or formations as FAN.

INTRODUCTION

The lunar crust is enriched in plagioclase, relative to any plausible planetary starting material. This conclusion stems from compositions of lunar materials acquired during the Apollo and Luna missions and from remote sensing of lunar surface soils. Some investigators regard the early lunar crust as consisting mainly of nearly pure plagioclase anorthosite, especially the variety known as ferroan anorthosite (FAN), (e.g., *Warren*, 1985; *Shirley*, 1983). The presence of accumulated plagioclase to form FAN is a major argument for the magma ocean hypothesis of lunar igneous differentiation. This argument is based in part on a presumed positive Eu anomaly for the lunar crust (e.g., *Warren*, 1985).

The highlands crustal surface does not have the bulk composition of anorthosite, but the composition of anorthositic norite (e.g., *Korotev et al.*, 1980; *Taylor*, 1982), with substantially higher proportions of mafic minerals and trace elements than are found in FAN. This conclusion stems from the overall compositions of acquired highlands soils and breccias and from data taken by remote sensing. Examination of data on lunar highlands samples in light of Th concentrations observed by remote sensing indicates that there is little if any Eu anomaly for the lunar highlands crust; i.e., while the lunar crust is clearly deficient in mafic minerals relative to the composition of the bulk Moon, it may not be deficient in trace-element-rich residual liquid from crystallization of crustal plagioclase and its complementary mafic minerals, as would be expected if FAN had floated high above its supporting liquid (*Korotev and Haskin*, 1987).

Understanding what kinds of lunar rocks make up the bulk of the lunar highlands is important to the development and testing of the magma ocean hypothesis. The increasing variety of plutonic rock types inferred by continuing studies of lunar breccias is difficult to explain on the basis of a simple concept for a magma ocean. If there was a magma ocean, then our knowledge of rock types is crucial to helping us define how such a large scale differentiating system operates. It is important to know whether the soil compositions were derived from mixing of a very early, nearly pure-plagioclase anorthosite with noritic and troctolitic rocks that intruded later, or whether they were derived from plagioclase-rich rocks with higher proportions of mafic minerals than found in FAN. Our understanding of early lunar differentiation has substantial implications for our ideas about the differentiation of other planetary bodies; i.e., the Moon serves automatically as an important model for the evolution of other small planets because we have some knowledge of the products and time scale of its igneous differentiation.

The most vigorous attempt to determine what igneous rocks are indigenous to the lunar highlands has centered around the "pristine" suite of specimens found in our samples of lunar materials (e.g., *Warren and Wasson*, 1977; *Warren et al.*, 1987 and intermediate papers in the series). The suite of "pristine" lunar rocks is, at most, a limited subset of the principal types of highlands igneous rocks, however. The compositions of the sampled highlands soils and most breccias do not correspond to those obtainable by mixing together common types of "pristine" lunar rocks. In particular, these soils and breccias require addition of highlands components with higher proportions of ferroan mafic minerals to plagioclase than are found in FAN (e.g., *Korotev et al.*, 1983; *Lindstrom and Salpas*, 1983). These components must have a lower mg' value [mols MgO/(mols MgO + mols FeO)] than found in the mafic rocks of the "magnesian pristine suite." They must either be strongly ferroan, to offset the presence of the magnesian rocks when those are present, or they may be more intermediate in mg' value and be present instead of materials of the magnesian suite. The mg' value for sampled highlands soils varies only within the narrow range of 0.62 to 0.70. The range of variation in mg' for the lunar highlands surface overall is unknown; presently available data from remote sensing are too imprecise to allow its determination.

The soils and breccias themselves might seem to be the logical materials from which to identify their precursor igneous rocks. This is difficult to do, owing to the relatively small sizes of most grains in soils, to the conversion of a high fraction of the material to glass, and to difficulty in making the correct assignments of the rock types from which individual lithic and mineral fragments derive. The most obvious means of estimating precursor rock type from mineral grains is to assign each grain to the pristine rock type in which grains with similar characteristics most commonly occur. This procedure must be incorrect, however, since common pristine rocks cannot be combined to produce materials with the compositions of the soils. Similar problems arise in assigning parent rock types to many materials taken from breccias. Clasts of substantial size can be classified, with caveats that will be evident from the body of this paper, but the bulk of most breccias consists of fine-grained, and in some cases metamorphosed, matrix whose relationship to the larger clasts is ambiguous. Nevertheless, we must learn how to extract from soils and breccias the identities of their precursor rocks.

Hoping to obtain a better idea of what igneous precursors were mixed to produce highlands soils, we are analyzing, chemically and petrographically, large numbers of 2-4 mm fragments from Apollos 15 and 16. As an early step in considering how best to interpret the anticipated data, we have modeled the mineral modes of randomly selected fragments from a plutonic rock whose composition and mineral distribution we have determined, namely, a boulder of anorthosite from the Stillwater Intrusion. We believe this boulder is a pertinent analog to the precursors of lunar soils and breccias in the sense that it is a plutonic rock of reasonable grain size (~3 mm dia) and has the composition of gabbroic anorthosite. The results of this exercise are instructive and bear on the interpretation of fragments reported from petrographic analyses of lunar soil grains.

EXPERIMENTAL

The boulder, from the AN-II anorthosite layer of the Stillwater Middle Banded Zone, was collected, sliced, polished, mapped for pyroxene and plagioclase, extensively sampled, and analyzed by Salpas and coworkers (*Salpas et al.*, 1987) (Fig. 1). The distribution of pyroxene in the boulder is clearly heterogeneous; there are regions of nearly pure plagioclase and regions of plagioclase-laden pyroxene oikocrysts. Even in the oikocrysts, the ratio of plagioclase to pyroxene is high, because the pyroxene formed around an extensive framework of pre-existing feldspar. The boulder contains oikocrysts of both ortho- and clinopyroxene, and these are compositionally zoned (*Gitlin et al.*, 1985). In this preliminary modeling, we ignore the mineralogical differences and the zoning, as well as minor amounts of magnetite.

To generate modeled breccia fragments from this boulder, we first constructed a digital map of the 13 × 13 cm region of the boulder face marked in Fig. 1. Each axis was divided into hundredths, resulting in 10,000 1.3 mm × 1.3 mm pixels. Each pixel in the map is designated as plagioclase or as pyroxene, according to which mineral dominated it. We then wrote a computer program to select randomly a value for x and a value for y and to draw a square of requested dimension with those coordinates as the lower left-hand corner. Next, we tallied the number of pixels of plagioclase and the (complementary) number of pixels of pyroxene within that randomly chosen square. We repeated this process 10,000 times for each selected size of square (only 3,000 times for squares containing 8,100 pixels) to obtain a statistically representative group of fragments. The distributions of mineral modes of the modeled fragments are given in Fig. 2 as frequency diagrams of the percent of fragments with a particular proportion of pyroxene versus percent of pyroxene per fragment.

The overall mineral mode for the mapped portion of the boulder is 88.1% plagioclase and 11.9% pyroxene. When the selected number of pixels per fragment is only one (Fig. 2), the composition of each fragment is either pure plagioclase or pure pyroxene. The frequency diagram has two peaks, one for pure plagioclase at (0,88.1) and another (not shown) for pyroxene at (100,11.9). At the other extreme, 10,000-pixel fragments, there is but a single peak, at (11.9,100), because each fragment represents the entire map. (Whenever the selected square passed over a boundary of the map, the

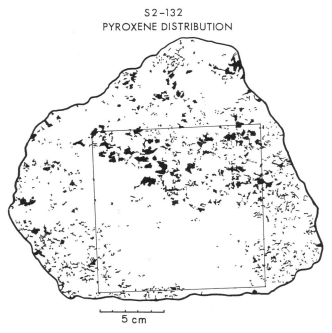

Fig. 1. This is a map of pyroxene distribution (dark regions) of a slice taken through a boulder of Stillwater anorthosite (*Salpas et al.*, 1984, 1987). Light regions are plagioclase. The experiments were done on a computerized map of the 13 cm × 13 cm region marked by the square.

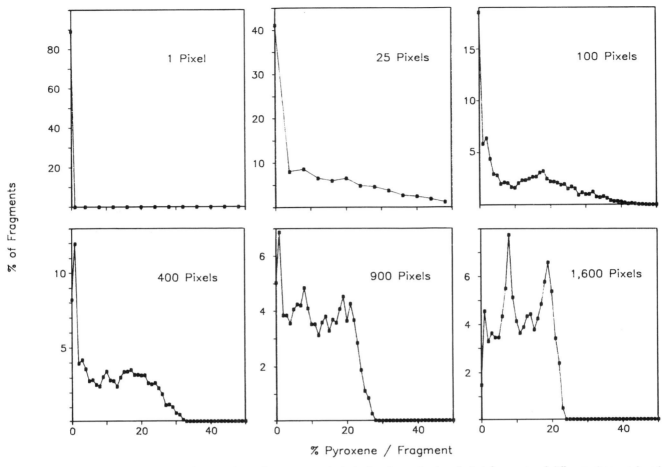

Fig. 2. This shows the distributions of proportions of pyroxene and plagioclase in randomly selected fragments of different sizes produced from the map. The percent of all fragments with a particular range of pyroxene concentration is plotted along the ordinate versus percent pyroxene per fragment along the abscissa.

computer program merely continued the square by folding its remaining portion over to the opposite side of the map. This is a limitation of the model, because it assumes that the boulder looks like a vinyl tile floor with an exactly repeating pattern in all directions.)

With increasing fragment size between 1 and 10,000 pixels the frequency diagrams must show a transition from two peaks to a single one. The peak for pure pyroxene disappears by the time the fragment size reaches 9 pixels, because all of the pyroxene is dispersed. At 25 pixels per fragment, some 40% of the fragments still consist of pure plagioclase; that mineral is clustered in a manner that leaves substantial portions of the map without pyroxene. The distribution of fragments containing pyroxene is broad, tapering off to zero at about 50% pyroxene; i.e., gabbroic fragments are still present. Might we be tempted to interpret such an array of fragments from a lunar soil or breccia as representing a mixture of anorthosite and norite?

By 100 pixels per fragment, fewer than 20% of the fragments are pure plagioclase, and a small peak of anorthosite with some 2% pyroxene has developed. Fragments with increasing pyroxene modes define a broad, but skewed, peak centered at about 19% pyroxene. This broad peak represents the typical proportions of plagioclase and pyroxene in the more pyroxene-rich regions of the map; observed out of the known context of the boulder, this peak might seem to imply brecciation of a rock containing about 19% pyroxene, nearly twice the actual pyroxene mode. (The exact position of the peak, at 19%, and the small bump there, while reproducible, are artifacts of the particular pattern of the map and its endless repetition; a more random selection of fragments this size from the entire boulder or its outcrop would show the same general features but not these specific ones.)

By 400 pixels per fragment, there are more plagioclase-rich fragments that contain some pyroxene than plagioclase-rich fragments that do not. The shoulder on the peak for plagioclase-rich fragments has grown; many of the fragments contributing to it come from regions consisting mainly of plagioclase but bordering on more pyroxene-rich regions. The broad peak of more pyroxene-rich fragments has narrowed and risen, and a small peak at about 10% pyroxene has begun to develop.

By 900 pixels per fragment, the distribution of fragments is trimodal, with a sharp peak at very low pyroxene mode, a broader peak at 7%-8% pyroxene, and a still broader peak

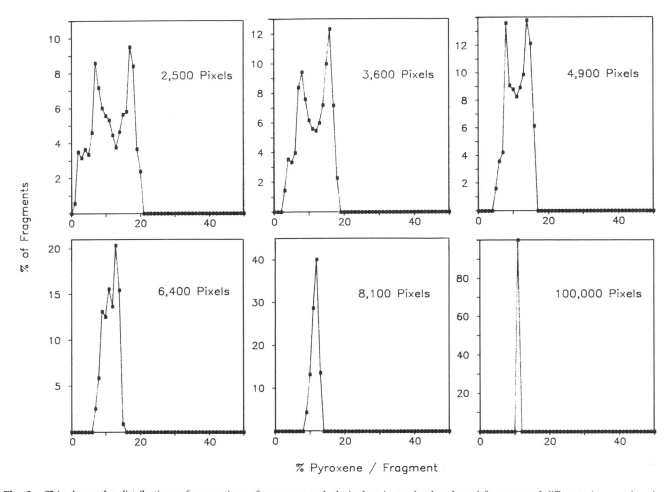

Fig. 2. This shows the distributions of proportions of pyroxene and plagioclase in randomly selected fragments of different sizes produced from the map. The percent of all fragments with a particular range of pyroxene concentration is plotted along the ordinate versus percent pyroxene per fragment along the abscissa.

at 19%-20% pyroxene. The intermediate peak does not represent the 11.9% average pyroxene mode for the map; it falls below that value, and its exact position is another artifact of the specific distribution of pyroxene in this map. The high background beneath the peaks stems mainly from the continuum of modes for fragments in regions of sparse pyroxene and in regions of overlap of the boundary between pyroxene-poor and pyroxene-rich materials.

Not until a fragment size of 1,600 pixels is reached does the proportion of fragments consisting of nearly pure plagioclase become subordinate to those of fragments richer in pyroxene. The distribution of pyroxene modes has two predominant peaks. The left-hand peak represents more plagioclase-rich regions, including those that border on pyroxene-rich regions, and the right-hand peak represents the pyroxene-rich, plagioclase-laden oikocrysts. Surely we would be tempted to interpret such a strongly bimodal distribution from a lunar soil as this one as representing two distinct rock compositions or populations. This shows the importance of determining mineral compositions and values for other parameters in addition to determining mineral modes, to verify the need for more than one parental rock type or composition. We note that the sharpness of the bimodality may have been enhanced somewhat by repeated use of the map in Fig. 1 rather than a broader sampling of the outcrop.

By 2,500 pixels per fragment, the majority of the fragments fall beneath the two main peaks. By 3,600 pixels per fragment, the peaks are closing together, and plagioclase-rich fragments are rare, but still present. This closure of the distance between the peaks continues through 4,900 and 6,400 pixels per fragment until by 8,100 pixels per fragment there is only a single peak, which represents the average pyroxene mode of the map with some residual spread.

In a more realistic model, the fragment distributions would have somewhat different specific characteristics, and the widths of the peaks in the range of a few thousand pixels per fragment and above would be greater, because we would not be repeating the same pattern over and over as in this exercise. At only 10,000 pixels per fragment there would still be significant width to the distribution because not all 10,000 pixel fragments would be alike, as in the present case. Nevertheless, the same general types of characteristics would be observed, because they stem from the bimodality in pyroxene distribution of the Stillwater anorthosites. This exercise demonstrates the complexity and the systematics of fragment modes that can reasonably be produced from impacts into plutonic rocks. It illustrates the caution that must be exercised in interpreting petrographic data on lithic and mineral fragments in terms of the number and type of precursor rocks in a soil or breccia. The problem stems in part from the heterogeneity of distribution of pyroxene in the boulder, but lunar plutonic rocks are also heterogeneous in their mineral distributions, as discussed below.

AVERAGING

We can improve the estimate of mineral mode and parental rock type by averaging the modes of several fragments, provided that we can convince ourselves that fragments of widely dispersed modes might be portions of the same source rock. This is illustrated in Fig. 3 for groups of 10 fragments each, for fragments consisting of 25, 100, and 400 pixels each. For this experiment 1,000 fragments of each size range were averaged in groups of 10, beginning with the first fragment generated by the program and averaging it with the 9 successive fragments (numbers 2 through 10), then proceeding to the second fragment generated and averaging it with the next 9 fragments (numbers 3 through 11), and so on.

The diagrams in Fig. 3 represent the probability of observation of an average with a particular pyroxene mode. For fragments as small as 25 pixels each, the most probable averages would indicate a rock with some 6% to 16% pyroxene. There is little probability that the average would indicate a rock consisting mainly of plagioclase, although some 40% of the individual fragments consist mainly of that mineral. There is a substantial chance that the average would suggest a rock of more than 16% pyroxene.

The most probable average for 100-pixel fragments is in the 9% to 11% pyroxene range and there is a shoulder of probability in the 13% to 16% pyroxene range. The probability of obtaining an average outside the range of 4% to 18% pyroxene is small, and there would be no suggestion of a FAN-type precursor. The most probable distribution of ten 400-pixel fragments is in the 10% to 14% pyroxene range; there remains a significant probability for averages as low as 6% pyroxene and as high as 17%, but most averages would yield a decent definition of the actual rock type. However, 400 grains, or even 100 grains, constitute a rather large sample of many lunar plutonic rocks.

The averages for groups of 10 fragments do not yield the same distributions as found for individual fragments with 10 times the number of pixels. This is because each individual fragment samples a contiguous region while the averages sample a variety of regions of the map. Thus, the averages are less restricted to a single type of material, (plagioclase-rich regions, plagioclase-laden pyroxenes, or interfaces between the two).

Since averages offer better estimates than do distributions of individual fragments alone, it will be important to develop carefully criteria for deciding when averaging is appropriate. Such criteria are likely to include mineral compositions, not just mineral identification. Fragments with common mg' values for the mafic components and common An contents of feldspars are the most obvious candidates for averaging.

Those particular criteria would be too severe in the case of the actual Stillwater boulder, however. In that rock, the oikocrysts are zoned over distances of mm (*Gitlin et al.,* 1985), so some breccia fragments with core values of mg' and some with various lower values of mg' (all associated with rather irregularly zoned feldspar) would result. Even if the feldspar were recognized to be the same from fragment to fragment,

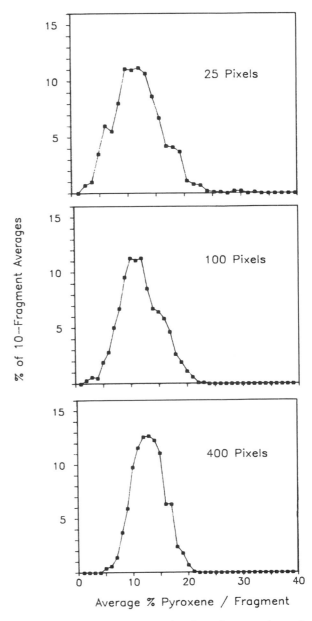

Fig. 3. This shows the distribution, for three fragment sizes, of proportions of pyroxene estimated for the rock if averages are taken of ten fragments.

grains of pyroxene would not be, and might not show zoning, because the gradient for zoning is not very steep. That is, some fragments from the zoned parts of oikocrysts might appear not to be zoned, but to be fragments of unzoned crystals of some composition other than the principal, or core, composition. This would likely lead us to decide that the fragments were part of a polymict breccia. Recognition that the fragments all came from a single rock would be further complicated because the rock contains both orthopyroxene oikocrysts and clinopyroxene oikocrysts; this alone might lead us to conclude that the host breccia was polymict.

Furthermore, we know that the 10 m × 10 m outcrop from which the boulder was taken consists of meter-sized regions, each relatively constant in its proportions of plagioclase and pyroxene, but each somewhat different in its composition and mineral mode from other regions; some of these regions have less than 5% pyroxene (Salpas et al., 1984). This complex pattern is typical of other outcrops of the thick Stillwater anorthosites. It would be very difficult to determine whether a soil or a breccia containing a limited number of lithic fragments of Stillwater anorthosite stemmed from a single rock, or formation, or represented more than one formation.

LUNAR ROCK SAMPLING

We do not have enough large samples of intact lunar plutonic rocks to establish how compositionally heterogeneous lunar host rocks or formations are. The typical grain size of the Stillwater boulder is 0.2 to 0.4 cm; this corresponds to ~2 to 9 pixels per mineral grain. A 100-pixel fragment corresponds to some 10 to 50 grains, for an approximate mass of 8 g. The typical size of analyzed lunar rock samples is less than that. Some FAN samples showing relict igneous textures have grain sizes for plagioclase of the order of 1 cm. The largest lunar sample collected, breccia 61016 with a mass of 11,745 g, contains only some 4,000 grains, if that grain size is applicable. A 100-grain fragment would weigh some 280 g, substantially more than any analyzed sample. Among the "pristine" FAN samples obtained by the Apollo missions (*Ryder and Norman*, 1978) are only four specimens weighing > 500g, five in the range 100-500 g, six in the range 1-100 g, and six < 1 g. The majority of grains and their distributions in each lunar rock have not been observed or studied, and rightly so, as only a fraction of each rock has been allocated for petrographic examination.

Consider the first hand-specimen of anorthosite collected, 15415. The mass of 15415 was 269 g, corresponding to roughly 80 grains. In this context, the entire specimen is just a single, small breccia fragment, equivalent to some 200 to 800 pixels of the Stillwater boulder; that number of pixels is in the range of substantial breadth of distribution (Fig. 2). In fact, all of our samples of highland plutonic rocks are breccias or breccia fragments. Perhaps our sampling is better with a rock such as 60015, which has a mass of 5,574; it is cataclastic and may contain fragments of very many grains drawn from a broad region of parent rock. Our sampling of rock formations may be better for samples that are "slightly polymict," i.e., samples consisting mainly of a single rock type but containing a few percent of another, than it is for "pristine" rocks; presence of some material exotic to the bulk of the breccia suggests that at least some large-scale mixing has occurred.

The distribution of mineral grains in lunar rocks is observedly heterogeneous for specimens of the sizes we have collected. FAN sample 60025 was initially identified as gabbroic because the first fragment of it examined was relatively rich in mafic minerals; subsequent analyses confirm its heterogeneity (*Dixon and Papike*, 1975; *Warren and Wasson*, 1980). Troctolite 76535 has a mass of < 60 g and, at a grain size of 0.3 cm, contains only 440 grains; it is heterogeneous, with clots of plagioclase up to 1 cm and of olivine up to 0.8 cm in diameter (*Gooley et al.*, 1974). Norite 78535 has a mass of 199 g; at a grain diameter of 0.3 cm, it contains only 2,500 grains. Are we certain that these rocks such as these are not merely mafic-mineral-rich chunks from more plagioclase-rich formations? Sample 61016 consists of about one-third anorthosite set in a troctolitic matrix. Is it really polymict, or might its anorthositic and troctolitic parts derive from the same rock or formation? Should lunar plutonic rocks be any more homogeneous than terrestrial ones? Such questions are not easily answered.

The sizes of most "pristine" lunar rocks are such that, in terms of numbers of grains, most of those rocks correspond to relatively small breccia fragments. Samples analyzed chemically and petrographically are far smaller yet. Also, the "pristine" rocks constitute a small fraction of the mass of highlands materials sampled.

Perhaps we owe more attention to the rest of the highlands samples. The compositions of the highlands soils tell us that other rock types are quantitatively important to the lunar crust. It seems risky to presume that the only quantitatively important classes of highland rocks are nearly pure plagioclase anorthosites, norites, and troctolites, just because most "pristine" rocks are anorthosites, norites, or troctolites. Unless conditions of formation of the lunar crust were very restricted, compared with those under which terrestrial plutonic rocks form, we should expect more of a range of rock types, one that includes noritic and troctolitic anorthosites and anorthositic troctolites and norites as well as the more extreme types mentioned above. It is precisely the possibility that lunar conditions were very restricted that makes the question of types of highlands plutonic rocks so important.

The Apollo highlands collection contains large, "nonpristine" rocks with compositions closer to those of soils and to the typical highlands composition as determined from remote sensing data. It is surely too harsh a criterion to declare that any rock that is demonstrably polymict should not be considered as possibly representing mainly an originally igneous composition rather than presuming that such a rock must be a mechanical mixture of nearly monomineralic anorthosite plus norite or troctolite. The presence of a small amount of meteoritic material in a breccia does not much alter the major-element composition of the breccia from that of its precursor igneous rock; neither does the presence of a small amount of most other types of lunar rock. Some samples of clearly "nonpristine" rocks may nearly represent monomict rocks; an example is granulitic breccia 67215 (*McGee*, 1987), with a mass of only 276 g but a bulk composition corresponding to ferroan anorthositic norite, the type of material needed to account for the compositions of soils. Rock 67215 is clearly polymict, as indicated by the presence of mafic minerals of different compositions. However, the majority of the mafic grains are within a narrow compositional range, and it can be argued that the composition of the breccia may sensibly represent that of an anorthositic-norite precursor rock, as opposed to a mixture of anorthosite and norite. Some granulites may represent rock types rather than heterogeneous mixtures (*Lindstrom and Lindstrom,* 1986).

EFFECTS ON CHEMICAL COMPOSITION

Determining chemical compositions of soil and breccia fragments is an important technique for discovering new rock types and new compositional variants of known rock types. What compositional spread might we expect for breccia fragments from a lunar boulder or its parent formation? Modeling of most terrestrial plutonic rocks is a fairly complex task. For example, describing the chemical compositions of hand-specimen-sized samples of Stillwater anorthosite requires four separate mixing components (*Haskin and Salpas,* 1985), and making a map of the final distribution of those components after they have interacted to produce zoned oikocrysts and regions of trace-element-rich residues of crystallization of trapped liquid would require more than four compositions. Detailed modeling is beyond the scope of this paper, but to illustrate the way in which the distribution of chemical compositions are related to breccia fragment size, we present a model for FeO compositions based on the boulder map. We will use this model to test the proposition that lunar FAN might have a more mafic-mineral-rich composition than we believe, owing to heterogeneity in the distribution of mafic minerals as in the Stillwater boulder.

Most petrological studies of FAN have concluded that FAN plagioclase modes are ≥95%, and that many are ≥98% (e.g., *James,* 1980). Thus, we generally regard FAN as being a nearly pure plagioclase rock. Might our sampling have misled us? Might our FAN samples only be plagioclase-rich portions of somewhat more mafic rocks? Conceivably, plagioclase-rich portions might survive brecciation better than mixed-mineral portions of more mafic rocks. Also, we may not recognize small, somewhat mafic samples as belonging to the same class of rocks we designate as FAN. Finally, we have analyzed very small fractions of the specimens of FAN we have acquired, and may not have accounted properly, at least in our thinking, for the amounts of mafic minerals that they actually contain. Suppose we compare our present information of the distribution of mafic minerals in FAN with possible distributions as estimated from the boulder map. One way to do this is to compare the distribution of FeO concentrations in analyzed samples of FAN with those estimated for the boulder.

First, we must estimate the distribution of FeO concentrations in lunar fragments. To obtain just a first-order idea of the distribution of FeO, we will treat every analyzed sample of FAN as if it were a fragment from a single, precursor source of all FAN. Since the samples of FAN were obtained on different missions and from widely varying locations, this assumption is an oversimplification.

Of the 57 samples of FAN for which FeO has been analyzed, some 31 have <0.5% FeO (consistent with nearly pure plagioclase), 6 have between 0.5 and 1% FeO (requiring some mafic mineral), and 20 have >1% FeO (requiring a significant amount of mafic mineral to be present). Clearly, lunar FAN has mafic-mineral-poor and mafic-mineral-rich regions. Fig. 4 shows the distribution of FeO concentrations in these samples as a histogram of percent of samples within a particular range of FeO concentration versus range of FeO concentration.

We estimate the principal effects of pyroxene distribution in the boulder on FeO concentrations of hypothetical breccia fragments by assigning FeO concentrations to plagioclase and pyroxene, then combining the contributions from these minerals to each fragment. These estimates ignore pyroxene zoning and other effects of trapped liquid and ignore the presence of minor magnetite. In a real rock, these effects are clearly important, which can be demonstrated as follows for the Stillwater boulder: If we assign for plagioclase the measured value of 0.58% FeO observed for the boulder plagioclase and assign for pyroxene the estimated value of 12.6% (the value for the adcumulus portion of the pyroxene, *Haskin and Salpas,* 1985), the estimated FeO concentration for the mapped portion of the boulder is 2.0%. Typical concentrations for samples from the region of the outcrop from which the boulder came, however, are in the range 2.5% to 3.2% FeO. The differences stem from the more ferroan, wide rims of the oikocrysts and

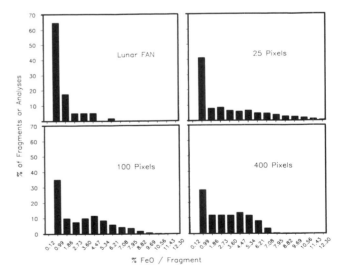

Fig. 4. The histogram for lunar FAN was developed by regarding all 57 analyzed samples (data compiled in *Ryder and Norman,* 1978 and *Haskin et al.,* 1981) as equal and independent fragments belonging to a single population. The histograms of hypothetical FeO concentrations represent those expected for breccia fragments of rocks in which minerals of lunar highlands compositions are distributed in the same amount and manner as they are distributed in the Stillwater boulder. The pixel ranges were chosen to match the relative grain sizes of FAN and the Stillwater minerals.

the minor magnetite component. Accurate modeling of the composition of the actual boulder would require a much more detailed map than that used here.

We could expect to model the distribution of FeO concentrations in a FAN boulder by this technique to a better approximation than we obtained for the Stillwater boulder. This is because lunar anorthositic rocks have low proportions of mafic minerals and little if any trapped liquid. Thus, their compositions, especially for major elements, tend toward matching the theoretically expected values based on simple mixtures of unzoned plagioclase and mafic minerals. To compare the amount and distribution of mafic minerals in a hypothetical FAN with a distribution of mafic minerals like that of the Stillwater boulder, we use the plagioclase and pyroxene distributions of the Stillwater boulder map, but model the distributions using lunar values of FeO concentrations for plagioclase (0.25%) and pyroxene (22%). We ignore the difference between pyroxene and the olivine that is present in some FAN samples. This results in a mean value for the FeO concentration of the pseudo-lunar boulder of 2.6%. How would this FeO be distributed among boulder fragments if lunar FAN had the same proportions and distribution of plagioclase and mafic minerals as found in the boulder?

To compare the lunar distribution to the estimates from the boulder map, we must take grain size into account. How many grains of FAN are present in each sample analyzed? Most FAN samples that have been analyzed for FeO concentration were in the size range of 0.5 to 3 g. At a grain dimension of 1 cm, this sample size would correspond to about 1 grain. We use a more realistic estimate of 0.3 cm for the effective grain size. This gives a range of about 6 to 40 grains analyzed per sample. Grain dimensions of plagioclase in the Stillwater boulder are 0.2-0.4 cm, corresponding to about 3 to 9 pixels per grain. By scaling the 6 to 40 lunar grains to the same number of pixels per grain as for the Stillwater plagioclase, we determine that we need to compare the distribution of observed lunar FeO concentrations with the estimated distributions for as few as 20 pixels per fragment and as many as 360. Fig. 4 shows the estimates for 25, 100, and 400 pixels, along with the observed lunar distribution.

The observed lunar distribution has features in common with those estimated from the boulder distributions. The lunar distribution has a relatively flat region from about 2.7% to 4.5% FeO, which demonstrates that FAN also has mafic-mineral-rich and mafic-mineral-poor regions. However, the highest concentration observed for a FAN sample is 6.6%, well below the maxima of 12.5%, 10.2%, and 7.7% calculated for the 25-, 100-, and 400-pixel fragments. Also, there is no peak corresponding to the equivalent of the plagioclase-laden oikocrysts of the boulder, which is seen for the 100- and 400-pixel fragments at about 4.5% FeO, as a small peak lying on a high background. About 84% of the lunar fragments have FeO concentrations <2%. Only 58%, 46%, and 41% of the modeled 25-, 100-, and 400-pixel boulder fragments have <2% FeO. The mean concentration for the FAN samples is 1.07%; that for the boulder fragments is 2.6%. The greater frequency of fragments with FeO concentrations <2% and the shorter tail for the FAN distribution indicate that the proportion of mafic minerals in analyzed FAN samples must be less than half that for the boulder value of 11.9%, consistent with petrographic observations of FAN.

Unfortunately, this exercise does not fully resolve the nature of the lunar plagioclase-rich rocks. In fact, it was inevitable that the result would indicate a higher proportion of plagioclase in FAN than in the boulder. The FeO distribution for FAN was based on a definition of FAN that excludes any small, mafic-mineral-rich fragments from that category. Thus, the population in the FAN histogram may be biased. This ambiguity does not mean that there really are "missing," more mafic fragments of FAN; however, we should be mindful of their possibility as we consider the samples collection. For example, at least six fragments of "pristine" troctolitic anorthosite (15437, 15455, 62237, 76335) and noritic anorthosite or anorthositic norite (62236, 15455) have been characterized (*Ryder and Norman,* 1979), of which three are ferroan. There are in addition small samples of "pristine" ferroan troctolite and norite. Addition of just a few samples such as these to the pool of FAN fragments would bring the distribution of FeO concentrations of FAN into line with that observed for the Stillwater boulder. There appear to be enough such samples in the Apollo collection to do this, if such samples belong with FAN.

There may well be massive amounts of FAN as we presently envision it in the lunar crust, and noritic and troctolitic varieties may be rare. Nevertheless, as pointed out earlier, mixtures of FAN with most "pristine" norites and troctolites do not explain the compositions of the highlands soils and many breccias, and we must still determine what the missing igneous rock types are. The answer surely lies within the soils and breccias we have sampled. Considerations of compositional and mineral distributions of fragments in various size ranges should help us find it.

CONCLUSIONS

This work is based on the proposition that it is important to understand the nature of lunar highlands plutonic rocks to test the viability of a magma ocean model and to provide meaningful detail to such a model, because the Moon is an important analog to the differentiation of other silicate bodies. We begin with the following observations: (1) Highlands soils and many breccias do not have chemical compositions that can be produced by mixing together common "pristine" lunar rocks. (2) The average composition of the surface highlands crust is that of anorthositic norite. (3) Although the crustal surface is depleted in mafic minerals relative to plagioclase, compared with any reasonable bulk Moon composition, it does not appear to be depleted in products of residual liquid crystallization, as evidenced by the significant proportions of mafic minerals and the absence of a significant Eu anomaly. (4) To the extent that the processes that produce lunar plutonic rocks are similar to those that produce terrestrial plutonic rocks, we might expect that nearly pure plagioclase anorthosite would be rarer than noritic anorthosite and anorthositic norite. (5) We have only soils and breccia fragments from which to deduce the nature of the Moon's earliest crust, and the subset of breccia fragments which make up the "pristine" suite of lunar rocks is a very biased sample of those materials from a compositional point of view.

To examine the distribution of products we might expect from impact brecciation of an anorthositic plutonic rock, we have modeled the distributions of mineral proportions of plagioclase and pyroxene for randomly chosen fragments from a mapped area on the surface of a boulder of Stillwater anorthosite. The resulting distribution of fragment compositions is complex, even with the simple two-phase map used, because the pyroxene is heterogeneously distributed. The map shows regions of pure or nearly pure plagioclase and regions of plagioclase-laden pyroxene oikocrysts. The distribution of fragment compositions evolves with increasing fragment size from one dominated by nearly purely plagioclase anorthosite plus lesser amounts of gabbroic anorthosite, anorthositic gabbro, and gabbro, through distributions with peaks corresponding to anorthosite and strongly gabbroic anorthosite, through a bimodal distribution of gabbroic anorthosite and anorthositic gabbro, to samples large enough to be representative of the map.

Sizes of lunar "pristine" breccia fragments correspond to rather poor sampling of lunar rocks or formations, when grain size is taken into account. Most entire samples fall in the range of fragment size for which the distributions for the Stillwater boulder show nonrepresentative peaks. This makes it hard to identify the parent rock types of the breccia fragments. When iron concentrations of lunar rocks with distributions of plagioclase and pyroxene like those for Stillwater boulder fragments of comparable size are compared with those for FAN (treating all analyzed FAN samples as part of a single population), it is evident that FAN, like the boulder, has a heterogeneous distribution of mafic minerals within a framework of plagioclase. The proportion of mafic minerals to plagioclase in FAN is less than half the 11.9% found in the boulder. However, the FAN "population" may be biased because observed ferroan anorthositic norite, anorthositic troctolite, norite, and troctolite samples found in the sample collection have been excluded from consideration owing to the definition of FAN.

It is not clear that the Apollo and Luna rock samples support the notion of a magma ocean and its variants in which a thick crust of nearly pure plagioclase built up and was later intruded or stirred by deep impacts to produce the present surface composition of the Moon.

Acknowledgments. We thank Randy Korotev for helpful comments on the manuscript. We acknowledge gratefully the partial support of this work by the National Aeronautics and Space Administration through grant NAG-956.

REFERENCES

Dixon J. R. and Papike J. J. (1975) Petrology of anorthosites from the Descartes region of the Moon: Apollo 16. *Proc. Lunar Sci. Conf. 6th*, pp. 263-291. Gitlin E. C., Salpas P. A., McCallum I. S., and Haskin, L. A. (1985) Small scale compositional heterogeneities in Stillwater anorthosite (abstract). In *Lunar and Planetary Science XVI*, pp. 272-273, Lunar and Planetary Institute, Houston.

Gooley R., Brett R., Warner J., and Smyth J. R. (1974) A lunar rock of deep crustal origin: sample 76535. *Geochim. Cosmochim. Acta*, 38, 1329-1339.

Haskin L. A. and Salpas P. A. (1985) Petrogenesis of Stillwater anorthosites of the Middle Banded Zone: Causes of geochemical trends (abstract). In *Lunar and Planetary Science XVI*, pp. 325-326.

Haskin L. A., Lindstrom M. M., Salpas P. A., and Lindstrom D. J. (1981) On compositional variations among lunar anorthosites. *Proc. Lunar Planet. Sci. 12B*, pp. 41-66.

James O. B. (1980) Rocks of the early lunar crust. *Proc. Lunar Planet. Sci. Conf. 11th*, pp. 432-436.

Korotev R. L. (1983) Compositional relationships of the pristine nonmare rocks to the highlands soils and breccias. *Proceedings of the Workshop on Pristine Highlands Rocks*, (J. Longhi and G. Ryder, eds.), pp. 52-55. LPI Tech. Rept. 83-02, Lunar and Planetary Institute, Houston.

Korotev R. L. and Haskin L. A. (1987) Does the lunar crust have a Europium anomaly? (abstract). In *Lunar and Planetary Science XVIII*, pp. 507-508. Lunar and Planetary Institute, Houston.

Korotev R. L., Haskin L. A., and Lindstrom M. M. (1980) A synthesis of lunar highlands data. *Proc. Lunar Planet Sci. Conf. 11th*, pp. 395-429.

Korotev R. L., Lindstrom M. M., Lindstrom D. J., and Haskin L. A. (1983) Antarctic meteorite ALHA81005—not just another lunar anorthositic norite. *Geophys. Res. Lett.*, 10, 829-832.

Lindstrom M. M. and Lindstrom D. J. (1986) Lunar granulites and their precursor anorthositic norites of the early lunar crust. *Proc. Lunar Planet. Sci. Conf. 16th*, in *J. Geophys. Res.*, 91, D263-D276.

Lindstrom M. M. and Salpas P. A. (1983) Geochemical studies of feldspathic fragmental breccias and the nature of North Ray Crater ejecta. *Proc. Lunar Planet. Sci. Conf. 13th*, in *J. Geophys. Res.*, 88, A671-A683.

McGee J. J. (1987) Petrologic evaluation of the components of granulitic breccia 67215 (abstract) In *Lunar and Planetary Science XVIII*, pp. 618-619. Lunar and Planetary Institute, Houston.

Ryder G. and Norman M. (1978) *Catalog of pristine non-mare materials. Part 2. Anorthosites.* NASA Johnson Space Center Rept. JSC 14603.

Ryder G. and Norman M. (1979) *Catalog of pristine non-mare materials. Part. 1. Non-anorthosites* (revised). JSC Publ. No. 14565. NASA Johnson Space Center, Houston.

Salpas P. A., Haskin L. A., and McCallum I. S. (1984) The scale of compositional heterogeneities in Stillwater anorthosites AN-I and AN-II (abstract). In *Lunar and Planetary Science XV*, pp. 713-714. Lunar and Planetary Institute, Houston.

Salpas P. A., Haskin L. A., and McCallum I. S. (1987) Trace-element distributions among subunits of a Stillwater anorthosite boulder (abstract). In *Lunar and Planetary Science XVIII*, pp. 872-873. Lunar and Planetary Institute, Houston.

Shirley D. N. (1983) A partially molten magma ocean model. *Proc. Lunar Planet. Sci. Conf. 13th*, in *J. Geophys. Res.*, 88, A519-A527.

Taylor S. R. (1982) Planetary Science: A Lunar Perspective. Lunar and Planetary Institute, Houston.

Warren P. H., Jerde E. A., and Kallemeyn G. W. (1987) Pristine Moon rocks: A "large" felsite and a metal-rich ferroan anorthosite. *Proc. Lunar Planet. Sci. Conf. 17th*, in *J. Geophys. Res.*, 92, E303-E313.

Warren P. H. (1985) The magma ocean concept and lunar evolution. *Annu. Rev. Earth Planet. Sci.*, 13, 201-240.

Warren P. H. and Wasson J. T. (1977) Pristine non-mare rocks and the nature of the lunar crust. *Proc. Lunar Sci. Conf. 8th*, pp. 2215-2235.

Warren P. H. and Wasson J. T. (1980) Further foraging for pristine nonmare rocks: Correlations between geochemistry and longitude. *Proc. Lunar Planet. Sci. Conf. 11th*, pp. 431-470.

Highland Materials at Apollo 17: Contributions from 72275

Peter A. Salpas[1]

Department of Geological Sciences, University of Tennessee, Knoxville, TN 37996

Marilyn M. Lindstrom[2]

Department of Earth and Planetary Sciences, Washington University, St. Louis, MO 63130

Lawrence A. Taylor

Department of Geological Sciences, University of Tennessee, Knoxville, TN 37996

[1] *Now at Department of Geology, 210 Petrie Hall, Auburn University, AL 36849*
[2] *Now at Planetary Materials Branch, Code SN2, NASA/Johnson Space Center, Houston, TX 77058*

A suite of Apollo 17 feldspathic highland clasts from polymict breccia 72275 have been studied. The clasts include six granulites and a rare Apollo 17 ferroan anorthosite (FAN), 72275,350/9018. This anorthosite possesses elevated incompatible trace element compositions (e.g., La = 1.7× chondritic), as does the only other possible sample so far described from the site. Such compositions suggest three possibilities regarding the petrogenesis of Apollo 17 ferroan anorthosites: (1) These rocks are products of late-stage crystallization from their parent magma(s); (2) they solidified with a high amount of trapped liquid relative to ferroan anorthosites from other Apollo sites; or (3) the highlands rocks that formed at this site crystallized from a parent magma(s) with a composition distinct from that of other lunar highlands sites. The granulites have bulk compositions with 10-15× chondritic abundances of the REE and ~26 wt % Al_2O_3, generally intermediate between FAN and Mg-suite rocks. However, two of the granulites have mineral compositions that plot at the extremes for granulites on the diagram of mafic mineral Mg# versus coexisting plagioclase An content—one is Fe-rich, the other Mg-rich. The remaining granulites plot within the "gap" on the mineral composition diagram. Compositionally, the 72275 granulites appear to be significant components of the soil at Apollo 17, in sharp contrast to the Apollo 17 KREEPy basalts, found only in 72275, whose presence had led to the interpretation that the materials that make up 72275 were not locally derived.

INTRODUCTION

The region of the Moon sampled during the Apollo 17 mission includes highland rocks that form the massifs and mare basalts that fill the valley floors. The massifs are part of the third ring of the lower Serenitatis basin, which is thought to define the limit of the transient cavity of the basin-forming impact (*Wolfe*, 1975). The highland samples should therefore provide a window into the deep crust at Apollo 17.

Polymict breccia 72275 was collected from Boulder 1, Station 2 at Apollo 17 as a representative sample of boulder matrix (e.g., *Marvin*, 1975). The sample contains an assortment of clasts and fine-grained material, some of which have compositions that are somewhat anomalous at the Apollo 17 site. In particular, the Apollo 17 KREEPy basalt samples (*Ryder et al.*, 1975, 1977; *Salpas et al.*, 1987) represent a distinct basalt type that is not found at any other sampled lunar site or in any other Apollo 17 sample. Furthermore, *Blanchard et al.* (1975a,b) demonstrated that the fine-grained matrix material of 72275 is not a significant component of the soil at Apollo 17. *Salpas et al.* (1987) showed that the fine-grained portion of 72275 is dominated by the Apollo 17 KREEPy basalts and inferred from this that the boulder was assembled during comminution of a flow(s) of these basalts.

Other matter was brought in and added to the material from which Boulder 1 was finally formed. Among this material were three "large" clasts that were collected as samples 72215, 72235, and 72255. There were also numerous small clasts that were retrieved from 72275. These include melt breccias [e.g., the marble cake clast of *Marvin* (1975) and *Ryder et al.* (1975)], other microbreccias, granites, and a suite of granulitic breccias of feldspathic compositions [which were originally identified as anorthositic breccias by *Ryder et al.* (1975)] that are reported here. Also found within this diverse suite of lunar rock types is a ferroan anorthosite clast 72275,350/9018 (INAA/PM#), also reported here. Although abundant at many other Apollo sites, ferroan anorthosite is scarce at Apollo 17. In addition to the one here, only a few other possible ferroan anorthosites have been reported. Like ,350/9018, these are all small (4-10 mm). They include a clast each from 77539 (*R. D. Warner et al.*, 1977) and 73217,35a (*Warren et al.*, 1983), the unusually sodic 76504,18 (*Warren et al.*, 1986), and 71504,12, 72464,7, 76034,10, and 76504,13 (*Warren et al.*, 1987).

The paucity of FAN at Apollo 17 relative to other highland sites (e.g., Apollo 16) must be accounted for in an interpretation of the geological history of the site (e.g., *Warren et al.*, 1986). As highland material, the granulite samples will help clarify the nature of the highland component at Apollo 17. By investigating samples from 72275 we can also address the "exotic" nature of this rock. That is, how well do the compositions of the highland clasts in 72275 correspond to those of other highland material at the site? We also present these granulite samples as part of an effort to better define the compositional and textural characteristics of this lunar rock type, and to augment similar work that has been done in recent years (e.g., *Lindstrom and Salpas*, 1981, 1983; *Lindstrom*, 1984; *Lindstrom and Lindstrom*, 1986).

The samples discussed in this paper are from clasts removed from breccia 72275 for INAA and microprobe analyses. The

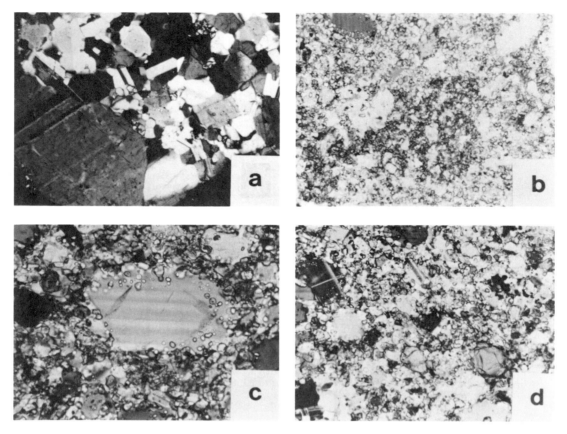

Fig. 1. Photomicrographs of highland clasts from 72275. Field of view in the longest dimension is 0.95 mm except for (c), which is 0.47 mm. (a) Ferroan anorthosite ,9018. (b) Granulite ,480. (c) Plagioclase fragment in ,480 with necklace of ovoid olivines. (d) Granulite ,502. (e) Granulite ,9019. Large grain in lower center is olivine. (f) Granulite ,9021. On the left is an aggregate of polygonal plagioclase grains. (g) Granulite ,495. (h) Granulite ,493 displaying a dominant cataclastic texture.

clasts were selected as part of an effort to characterize highland samples at Apollo 17. During the course of clast selection, 72275 was mapped and a guidebook was made (*Salpas et al.,* 1985).

PETROGRAPHY AND CHEMISTRY

Petrographic and electron microprobe studies were performed on probe mounts from the selected clasts. Photomicrographs are presented in Fig. 1. Microprobe analyses were conducted with an automated MAC 400S using the standard operating procedures for this lab (e.g., *Shervais et al.,* 1985). Whole-rock analyses for trace elements and some major elements were performed by instrumental neutron activation analysis (INAA). Procedures were similar to those described in *Korotev* (1987). An additional irradiation for short-lived nuclides (Mg, Al, Ti, Mn) was carried out on the granulite samples. Analytical uncertainties based on one sigma counting statistics are: 1-2% Al_2O_3, FeO, Na_2O, Sc, Cr, Mn, Co, La, Sm, Eu; 3-5% MgO, CaO, Ce, Tb, Yb, Lu, Hf, Ta, Th; 5-10% Sr, Cs, Ba, Nd, U; 15-25% TiO_2, Ni, Rb, Zr, Ir, Au. Data are presented in Table 1.

Ferroan Anorthosite

Clast 350/9018 (INAA#/PM#) occurred as a 3 × 4 × 5 mm clast in breccia 72275. PM 9018 is a cataclasized, monomict anorthosite composed of approximately 95% plagioclase (An 95.1-97.1) with about 5% pyroxene consisting of augite (Mg# 54-57) and pigeonite (Mg# 42-48) in approximately equal proportions. Mineral compositions plot within and below the field for FAN on the mafic Mg# versus plagioclase An diagram (Fig. 2). Two distinct textural domains are observed; one is a granular mosaic texture of plagioclase grains ranging from 30 to 50 μm, the other is a single large, broken grain (500 × 850 μm). The plagioclase is twinned and displays undulose extinction. Pyroxene occurs as small (40-90 μm) grains interstitial to the granular plagioclase. The large compositional differences between coexisting pyroxenes (i.e., representing the "miscibility gap") (Fig. 3) and the calculated temperature of equilibration (approximately 850°C; *Lindsley and Anderson,* 1983) attest to their formation under plutonic conditions.

Whole-rock analysis of ,350 by INAA reveals a composition that is similar to but in some ways not typical of ferroan anorthosites (Table 1), with high CaO (19.2 wt %) and low

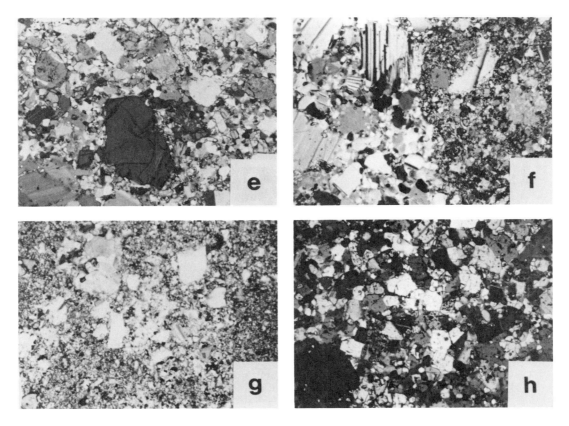

FeO, Sc, Cr, and Co (0.49 wt %, 1.1, 46.6, and 0.44 ppm, respectively) concentrations compared to more mafic rocks. However, the concentrations of the transition metals are somewhat elevated compared to other ferroan anorthosites. The concentrations of the REE are also somewhat elevated (e.g., La = 0.57 ppm) with respect to their concentrations in other ferroan anorthosites, but the shape of the REE pattern (Fig. 4) is similar to other ferroan anorthosites (e.g., 15415; *Haskin et al.*, 1981), displaying LREE enrichment (chondrite normalized La/Yb = 2.75) and a large, positive Eu anomaly (chondrite normalized Sm/Eu = 0.094). Concentrations of Ir (<2 ppb) and Ni (<7 ppm) are interpreted as evidence for a lack of meteoritic contamination of the analyzed sample. On the basis of compositional and textural criteria, we interpret 72275,350 to be a "pristine" igneous rock.

Granulites

Textures, mineralogy, and mineral compositions of the six granulite samples are summarized in Table 2. Bulk compositions are presented in Table 1. In general, the granulites are composed of rounded to angular fragments of plagioclase and

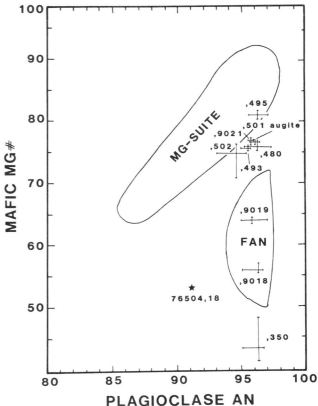

Fig. 2. Mg# [molar Mg/(Mg + Fe) × 100] in mafic silicates versus An content in coexisting plagioclase for the 72275 granulites and ferroan anorthosite. Two probe mounts of the ferroan anorthosite were analyzed, ,9018 and ,350. The mafic minerals are olivine and orthopyroxene except for augite in granulite ,501 and in ferroan anorthosite ,350. 76504,18 is the sample analyzed by *Warren et al.* (1986). See text for explanation of remaining sample numbers.

TABLE 1. 72275 highland rocks.

	Granulites						FAN
	351A	351B	355A	397	433	439	350
	Major Elements (wt %)						
TiO_2	0.31	0.32	0.29	0.22	0.15	0.32	na
Al_2O_3	22.1	23.1	27.2	26.2	24.6	26.3	na
FeO	8.87	7.83	4.85	5.71	5.10	4.95	0.485
MgO	11.5	9.9	7.6	7.9	8.0	9.7	na
CaO	11.9	12.6	14.8	14.8	14.2	14.5	19.2
Na_2O	0.307	0.316	0.349	0.353	0.362	0.350	0.456
	Trace Elements (ppm)						
Sc	14.97	12.92	8.13	7.81	8.24	7.12	1.12
V	69	65	19	20	24	25	na
Cr	2414	1646	810	842	881	846	46.6
Mn	934	792	489	499	481	462	na
Co	35.1	34.1	27.0	39.3	30.6	52.0	0.440
Ni	250	290	340	455	422	540	< 7
Sr	124	129	157	160	160	163	205
Cs	0.124	0.118	0.164	0.19	0.23	0.10	0.016
Ba	58	55	70	72	87	62	40
La	4.86	3.56	4.04	3.66	4.72	3.76	0.567
Ce	10.9	8.62	9.87	10.1	12.6	10.5	1.48
Nd	7.0	5.5	5.6	5.7	6.2	5.0	< 2.5
Sm	2.04	1.60	1.82	1.56	1.93	1.67	0.228
Eu	0.698	0.713	0.864	0.835	0.860	0.870	0.928
Tb	0.473	0.410	0.456	0.375	0.49	0.381	0.045
Yb	2.05	1.66	1.67	1.69	2.06	1.55	0.125
Lu	0.302	0.242	0.251	0.238	0.292	0.230	0.020
Hf	1.67	1.24	1.46	1.46	1.98	1.22	0.133
Ta	0.266	0.199	0.233	0.302	0.309	0.190	0.015
Th	1.81	1.38	1.18	1.17	2.06	1.02	0.047
U	0.39	0.27	0.30	0.34	0.37	0.19	0.020
Ir (ppb)	9.6	11.3	13.0	16.4	14.0	22.2	nd
Au (ppb)	3.4	3.6	5.0	6.8	6.5	4.3	< 0.8

na = not analyzed.
nd = not detected (Ir detection limit = 2 ppb).

olivine in granoblastic or poikiloblastic matrices composed mainly of plagioclase and pyroxene. Also contained within the matrices are small grains of FeNi metal. Due to their small size, analyses of these metal grains are available for only one sample, probe mount ,480. Compositions are "meteoritic" with 0.40-0.58 wt % Co and 3.27-6.69 wt % Ni.

Plagioclase clasts in probe mounts ,502, ,9019, ,9021, and ,480 contain necklaces of ovoid olivine inclusions near their rims. Similar inclusions have been observed in plagioclase fragments in other granulites (e.g., *Chao*, 1973; *Ashwal*, 1975; *Bickel et al.*, 1976) and probably represent the products of annealing of the edges of plagioclase grains. Edges containing

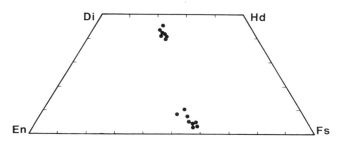

Fig. 3. Pyroxene quadrilateral showing compositions of pyroxenes in Apollo 17 ferroan anorthosite 72275,350/9018.

these inclusions tend to have 2-4 mole % less An than cores of the same grain, similar to the observation by *Ashwal* (1975). In one of these 72275 samples (,9021), an olivine ovoid that was large enough for analysis had a composition indistinguishable from that of the fragmental olivines in the clast (Table 2). Another probe mount (,493) has ovoid olivine inclusions that are restricted to plagioclase fragments of intermediate grain size (about 150 μm). These do not occur as necklaces; rather, they are homogeneously distributed through the fragments. The sample represented by this probe mount has a dominant cataclastic texture, different from the typical metamorphic textures in most of the other probe mounts.

Aggregates of polygonal plagioclase were observed in three probe mount samples: ,495, ,9021, and ,480. Compositions of plagioclase polygons are the same as those of the fragments that make up the rest of the clasts (Table 2). One such aggregate in PM ,9021 also contains minor olivine (<5%) with compositions identical to the larger olivine fragments in the clast.

Probe mount ,501 reveals an anorthosite fragment composed of a single, euhedral grain of plagioclase (2 × 3.5 mm; An 95.8-96.4) and is not strictly a granulite, although portions of the grain are cataclasized. Minor (<5%), small (<100 μm) inclusions of augite (Mg# 76-77) are present. PM ,501 was extracted from 72275 to be representative of INAA sample

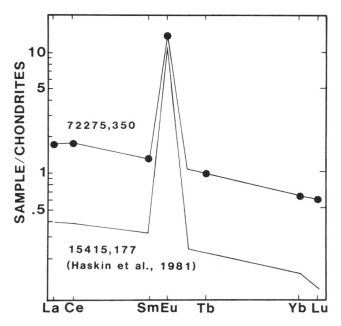

Fig. 4. REE composition of Apollo 17 anorthosite 72275,350 normalized to the chondritic abundances of *Haskin et al.* (1968). Also shown is the pattern of ferroan anorthosite 15415 (*Haskin et al.*, 1981) for comparison.

,351. However, the compositional nature of ,351 does not indicate a sample of nearly pure plagioclase. Rather, its low CaO and Al_2O_3 and high FeO, MgO, and Sc concentrations (see Table 1) indicate a significant mafic component not observed in PM ,501. Furthermore, high Ir and Ni concentrations (see Table 1) in ,351 indicate meteorite contamination that is not consistent with the monomict nature of PM ,501. Due to the unusual texture of PM ,501, probe mount ,9019 was made from chip ,351 subsequent to its analysis by INAA. As described in Table 2, this probe mount displays the metamorphic textures expected of a granulite and we conclude that PM ,501 represents a clast different from sample ,351.

The bulk compositions of the six granulites (Table 1) are typical of granulites in general with compositions intermediate between ferroan anorthosites and Mg-suite troctolites and norites. For instance, CaO and Al_2O_3 (which have ranges of 11.9-14.8 and 22.1-27.2 wt %, respectively) are high for Mg-suite rocks but low for anorthosites, and FeO (4.85-8.87 wt %), MgO (7.6-11.5 wt %), and Sc (8.13-14.97 ppm) also have intermediate concentrations. The REE occur in concentrations between about 10 and 15× chondritic with flat patterns and little or no Eu anomalies (Fig. 5).

Meteoritic contamination of the granulites is indicated by the Ni and Co concentrations in the FeNi metal of PM ,480 as described above, and by high whole rock concentrations of Ir (9.6-22.2 ppb) and Ni (250-540 ppm). These elements, along with the other siderophile element Co (Table 1), distinguish the 72275 granulites from Apollo 16 granulites, which have lower Co, Ni, and Ir concentrations (e.g., *Palme*, 1980; *Lindstrom and Lindstrom*, 1986; *Ostertag et al.*, 1987).

DISCUSSION

Highlands Component in 72275

The compositions of the highlands samples extracted from breccia 72275 are consistent with a local derivation. This is in contrast to the KREEPy basalt clasts also found in 72275 whose compositions appear to be exotic to the Apollo 17 site (e.g., *Blanchard et al.*, 1975a,b; *Salpas et al.*, 1987). On the basis of comparisons with chemical trends found within Apollo 17 soils, *Blanchard et al.* (1975a,b) demonstrated that the Apollo 17 KREEPy basalts and the fine-grained material that makes up the matrix of 72275, and all of Boulder 1 as well, are exotic to Apollo 17. *Salpas et al.* (1987) demonstrated that the composition of the fine-grained material is dominated by the KREEPy basalts, thereby providing its exotic nature.

TABLE 2. Textures, mineralogy, and mineral compositions of 72275 granulites.

INAA/PM:	439/495	355/502	351/9019	397/9021	433/493	none/480
Texture	c-g	g-c	c-g	g-c	c	g-c
Plagioclase						
Mode	46%	66%	59%	57%	76%	52%
Size	< 10-350 μm	< 10-400 μm	< 10-650 μm	< 10-500 μm	< 10-340 μm	< 10-500 μm
An	95.9-97.1	94.1-95.3	95.8-96.4	95.4-96.3	94.8-95.8	95.2-97.3
Pyroxene						
Mode	50%	31%	35%	40%	20%	44%
Size	< 10 μm	< 10 μm	< 10 μm	< 10 μm	< 10 μm	< 10 μm
	na	na	na	na	na	na
Olivine						
Mode	1%	3%	5%	1%	2%	2%
Size	40-175 μm	50-150 μm	80-450 μm	100-150 μm	30-45 μm	45-500 μm
Fo	80-82	71-77	63-64	76-77	75-76	75-77
Fe metal						
Mode	3%	< 1%	1%	1%	2%	2%
Size	< 10 μm	< 10 μm	< 10 μm	< 10 μm	< 10 μm	< 10-20 μm
	na	na	na	na	na	see text

c = cataclastic; g = granulitic; na = not analyzed. Analyzed compositions are for mineral fragments and do not include groundmass minerals that were generally too small for accurate analysis.

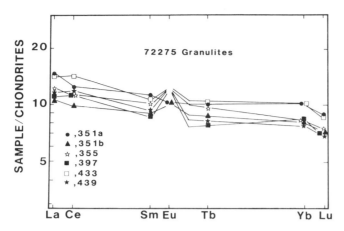

Fig. 5. REE compositions of the 72275 granulites normalized to the chondritic abundances of *Haskin et al.* (1968).

However, highlands clasts, in the form of the six granulites reported here, also plucked from the fine-grained matrix of 72275, have compositions that can be related to the soils at Apollo 17 (Fig. 6). Sample 72275,350, the ferroan anorthosite, does not plot within the Apollo 17 soil fields on Fig. 5 because its concentrations of FeO, Sc, and Sm are too low. Given the scarcity of ferroan anorthosite samples in the Apollo 17 collection, we do not expect them to be a significant constituent of the soils at the site.

Apollo 17 Ferroan Anorthosites

The sample of FAN recovered from 72275 (,350/,9018) has a composition that is enriched in the rare earth and other incompatible elements relative to ferroan anorthosites from other Apollo sites [e.g., La in ,350 = 5× La in 15415 (*Haskin et al.*, 1981)]. Except for sample 77539 (*R. D. Warner et al.* 1977), for which the data are not available, the other possible samples of FAN from Apollo 17 [73217,35a (*Warren et al.*, 1983), 76504,18 (*Warren et al.*, 1986), and 71504,11, 72464,6, 76034,9, and 76504,12 (*Warren et al.*, 1987)] also have relatively elevated incompatible element concentrations. For example, the concentration of La in 73217,35a is 29 times that in the sample of 15415 analyzed by *Haskin et al.* (1981). For the other Apollo 17 samples, these values are as follows: 76504,18 = 17×; 71504,12 = 24×; 72464,7 = 3×; 76034,10 = 15×; 76504,13 = 13×. However, of these samples, only 76504,18 is thought to be pristine (*Warren et al.*, 1987), so it cannot be said with certainty that the incompatible trace element compositions of the other samples are indigenous. The plagioclase compositions of 76504,18 are also unusually sodic for a ferroan anorthosite (An 91) and, on the basis of these compositional characteristics, *Warren et al.* (1986) interpreted this rock to be a sample of evolved FAN. If the lunar anorthositic crust formed as flotation cumulates then the presence of such an evolved sample of FAN is consistent with deep excavation of the crust, which is thought to have been responsible for the highlands rocks at Apollo 17 (e.g., *Warren et al.*, 1986). Although ferroan anorthosite 72275,350 is also characterized by elevated incompatible trace element concentrations, the compositions of its plagioclase grains fall within the range of compositions more generally associated with FAN, i.e., An 95-98.

On the basis of only two possible pristine Apollo 17 ferroan anorthosites, 72275,350 and 76504,18, we make the following speculations regarding the compositional nature of the lunar highland crust at Apollo 17. FAN from Apollo 17 is characterized by high incompatible trace element compositions. In this regard, all samples collected at Apollo 17 might represent evolved ferroan anorthosites, as suggested for 76504,18 by *Warren et al.* (1986), an interpretation consistent with their presumed deep-seated origin. However, of the two samples of Apollo 17 FAN, only one, 76504,18, has evolved plagioclase compositions. It seems unusual from the perspective of terrestrial geochemistry and petrology that evolved mineral and whole-rock compositions should not go hand in hand. But the compositional characteristics of all ferroan anorthosites are unusual by terrestrial standards. The mechanism(s) by which the compositions of so much plagioclase should be buffered in order to produce such a narrow range has not been established. Based on simple mass balance considerations, we

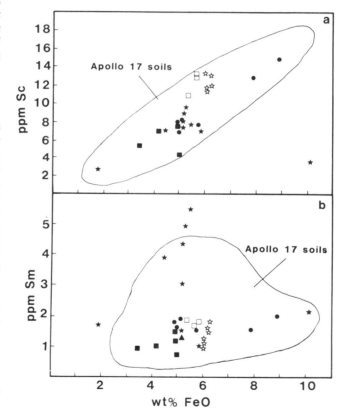

Fig. 6. Plots of FeO against **(a)** Sc and **(b)** Sm for some Apollo 17 granulites. The Apollo 17 soil fields are from *Blanchard et al.* (1975b). The granulites are: filled circle: 72275 (this work); filled stars: 73215 (*Blanchard et al.*, 1977); filled triangle: 76230 (*Hubbard et al.*, 1974); open star: 77017 (*Hubbard et al.*, 1974; *Laul et al.*, 1974; *Lindstrom and Lindstrom*, 1986); open box: 78155 (*Laul and Schmitt*, 1973; *Hubbard et al.*, 1974; *Wanke et al.*, 1976; *Lindstrom and Lindstrom*, 1986); filled box: 79215 (*Blanchard et al.*, 1977).

presume that the mechanism would eventually have had to break down during the later stages of formation of the anorthositic crust. What would be the characteristics of the products of the breakdown of this mechanism? Should we expect that all anorthosites that formed in the late stages of crystallization of a lunar magma ocean would contain evolved mineral and whole-rock compositions? Or would breakdown of the mechanism produce an inhomogeneous distribution of evolved plagioclase compositions?

Alternatively, we consider the possibility that the elevated incompatible trace element composition of 72275,350 does not signify a rock that formed during the later stages of crystallization of a magma ocean. This sample could represent a cumulate that formed relatively early, and its enriched incompatible trace element composition might reflect a higher proportion of trapped liquid than is found in other ferroan anorthosites. The low incompatible trace element compositions that are generally associated with FAN have been recognized as necessitating unusual circumstances for the petrogenesis of these lunar rocks (e.g., *Haskin et al.*, 1981; *Morse*, 1982). Thus ferroan anorthosites that possess distinctly higher concentrations of incompatible trace elements would seem to make more sense from a petrologic point of view than do the incompatible-trace-element-depleted samples of FAN that are so familiar. In this case, the REE compositions of 72275,350 and 76504,18 can be related if the liquid in equilibrium with the former underwent about 60% crystallization of plagioclase + opx + cpx = 55:20:25, roughly cotectic proportions, to produce the liquid from which the latter rock could have crystallized (e.g., *Warren et al.*, 1986). This statement is only semiquantitative because the calculations are highly model dependent and involve making a number of assumptions regarding: (1) the cumulate origin of 72275,350 and 76504,18; (2) the mineralogies of the crystallized assemblages; and (3) above all, the manner in which the lunar crust became so enriched in plagioclase. Thus nearly any solution can be achieved depending upon the assumptions that are made.

The degree to which the plagioclase compositions of 76504,18 have evolved from the supposed precursor rock 72275,350 is small relative to the 60% crystallization necessary to go from the equilibrium liquid of the latter to that of the former. Assuming that the two samples really are related, then the manner in which this small change in plagioclase composition was effected probably lies hidden in the mechanism by which most other ferroan anorthosites crystallized with nearly constant plagioclase compositions. In fact, that a sample of ferroan anorthosite should display evolved plagioclase compositions is quite remarkable and should serve as a reminder that the relatively constant compositions of plagioclase observed in the suite of ferroan anorthosites is not an inviolable rule. The major argument that these compositions should be constant is based on empirical data—a circular process of reasoning. Other than this, a range in compositions of plagioclase would seem to make better petrologic sense.

Finally, we suggest that the compositions of Apollo 17 ferroan anorthosites may be reflections of a difference in composition between the parent magma of these rocks and those that crystallized FAN elsewhere on the Moon.

The speculative nature of this discussion rests in its derivation from data on only two samples. In this regard, we also recognize the possibility that neither 72275,350 nor 76504,18 may be representative of the Apollo 17 site. More pristine ferroan anorthosites from Apollo 17 with whole-rock and mineral chemical data are necessary before statistically meaningful hypotheses can be formed regarding the compositional nature of FAN at Apollo 17.

The 72275 Granulitic Breccias

The most common type of feldspathic highlands rock in breccia 72275 is granulite. These lunar rocks in general are described as metamorphosed, polymict breccias composed of mixtures of pristine material from Mg-suite rocks and ferroan anorthosites (e.g., *Warner et al.*, 1976; *James*, 1980). They typically display metamorphic textures, and their major and trace element compositions are usually intermediate between rocks from the two pristine suites. Mineral compositions are also of an intermediate nature. Granulites are usually "gap-fillers" on the plagioclase An content versus coexisting mafic mineral Mg# diagram (see Fig. 2). High siderophile element concentrations are characteristic of granulites and are taken as evidence for meteorite contamination (e.g., *Warren*, 1985). Granulites probably represent metamorphically reequilibrated impact mixtures of early crustal rocks that have undergone multiple episodes of brecciation and reheating to temperatures approaching partial melting (e.g., *J. L. Warner et al.*, 1977; *Bickel and Warner*, 1978; *James*, 1980).

Within the 72275 granulites, plagioclase compositions cluster around An 95, while mafic minerals have Mg# ranging from 63 to 82. Thus, on a plot of major mineral compositions, all but two of the granulites fall in the gap between Mg-suite rocks and ferroan anorthosites (Fig. 2). Bulk compositions are those of anorthositic norites or anorthositic troctolites with around 26% Al_2O_3. Major and trace element concentrations are intermediate between Mg-suite rocks and ferroan anorthosites and are similar to the estimated average highlands composition of *Taylor* (1982). High concentrations of siderophile elements (Ir 10-22 ppb, Au 3-7 ppb) demonstrate meteoritic contamination. The compositional and textural characteristics discussed above are typical of granulitic breccias found throughout the lunar highlands.

Interpretations of granulite textures and compositions generally focus on the similarities among samples instead of the differences. However, the 72275 granulites exhibit significant variations in both composition and texture. For example, sample 72275,433/493 is more cataclastic and less metamorphosed than the others. Mineral compositions are sufficiently homogeneous in most samples that a 3 mole % difference in Mg# reflects a real difference between samples (Mg# 71, 74, 78 are all different). However, one sample (72275,355/502) has less homogeneous mineral compositions that cover the entire range of most granulites and span the whole Mg# gap of the pristine rocks. This is the most obvious example of a mixture of the two pristine suites. The small ranges in mineral compositions observed in the others suggest that if these clasts represent mixtures then they must have become metamorphosed under extreme conditions that would have completely homogenized the compositions of their minerals. As described, the textures of at least one of these granulites (PM ,493) does not allow such an interpretation.

Although most samples have gap-filling mineral compositions, two do not. One (72275,351/9019) is ferroan (Mg# 63), and the other (72275,439/495) is magnesian (Mg# 82). *Lindstrom and Lindstrom* (1986) noted similar differences among large samples of granulitic breccias. Some of these have relatively simple cataclastic textures with clasts of annealed plutonic rocks. Some have compositions that fall well within the fields of either ferroan anorthosites or Mg-suite rocks. They concluded that the most ferroan granulites (e.g., 67215) are not mixtures of the two common types of pristine rocks, but instead represent the most mafic members of the ferroan suite. Likewise, the most magnesian granulites (e.g., 67415), which have simple textures, probably represent feldspathic members of the Mg-suite rather than mixtures of ferroan and magnesian components. Thus, while some of the granulites with gap-filling compositions are clearly polymict, some of those with extreme compositions appear to be monomict. Both the similarities and differences among the granulitic breccias must be considered in evaluating their origins. The precursors of the ferroan and magnesian granulites in 72275 may be quite different, while they share similar histories of brecciation and metamorphism.

CONCLUSIONS

Based on the compositional natures of the highlands samples from breccia 72275 we conclude that the rock, and therefore Boulder 1, Station 2 whose matrix it represents, does not consist of material that is entirely foreign to the Apollo 17 site. The suite of samples that make up the Apollo 17 KREEPy basalts are certainly unique to the site but could have been derived locally without being a significant component in the Apollo 17 soils if they originated as a small flow of limited areal extent.

Both possible samples of pristine ferroan anorthosite at Apollo 17 have relatively high concentrations of incompatible trace elements. This may be a reflection of either: (1) a fundamentally different compositional nature of the lunar crust at Apollo 17; (2) the fact that ferroan anorthosites from Apollo 17 were derived from deep within the crust and represent samples that crystallized during the later stages of the magma ocean; or (3) the present collection of Apollo 17 ferroan anorthosites contain a higher proportion of trapped liquid than do samples of FAN collected elsewhere.

The 72275 granulites are representative of an important rock type that constitutes the bulk of lunar highland samples at Apollo 17. The variety of the compositions and textures of the phases of these rocks and the polymict nature of these samples have contributed to an enhanced understanding of the petrogenesis of the highland rocks at the Apollo 17 site.

Acknowledgments. This paper has benefited from reviews by G. Kurat, P. Warren, and an anonymous reviewer. Logistical support by the Planetary Material Curatorial Staff (and especially K. Willis) is greatly appreciated. This study was supported by NASA grants NAG 9-62 (to L. A. Taylor) and NSG 9-56 (to L. A. Haskin).

REFERENCES

Ashwal L. D. (1975) Petrologic evidence for a plutonic igneous origin of anorthositic norite clasts in 67955 and 77019. *Proc. Lunar Sci. Conf. 6th*, pp. 221-230.

Blanchard D. P., Korotev R. L., Brannon J. C., Jacobs J. W., Haskin L. A., Reid A. M., Donaldson C. H., and Brown R. W. (1975a) A geochemical and petrographic study of 1-2 mm fines from Apollo 17. *Proc. Lunar Sci. Conf. 6th*, pp. 2321-2341.

Blanchard D. P., Haskin L. A., Jacobs J. W., Brannon J. C., and Korotev R. L. (1975b) Major and trace element chemistry of boulder 1 at station 2, Apollo 17. *The Moon, 14*, 359-371.

Blanchard D. P., Jacobs J. W., and Brannon J. C. (1977) Chemistry of ANT-suite and felsite clasts from consortium breccia 73215 and of gabbroic anorthosite 79215. *Proc. Lunar Sci. Conf. 8th*, pp. 2507-2525.

Bickel C. E. and Warner J. L. (1978) Survey of lunar plutonic and granulitic lithic fragments. *Proc. Lunar Planet. Sci. Conf. 9th*, pp. 629-652.

Bickel C. E., Warner J. L., and Phinney W. C. (1976) Petrology of 79215: Brecciation of a lunar cumulate. *Proc. Lunar Sci. Conf. 7th*, pp. 1793-1819.

Chao E. C. T. (1973) The petrology of 76055,10, a thermally metamorphosed fragment-laden olivine micronorite hornfels. *Proc. Lunar Sci. Conf. 4th*, pp. 719-732.

Haskin L. A., Haskin M. A., Frey F. A., and Wildeman T. R. (1968) Relative and absolute abundances of the rare earths. In *Origin and Distribution of the Elements* (L. H. Ahrens, ed.), pp. 889-912. Pergamon, New York.

Haskin L. A., Lindstrom M. M., Salpas P. A., and Lindstrom D. J. (1981) On compositional variations among lunar anorthosites. *Proc. Lunar Planet. Sci. 12B*, pp. 41-66.

Hubbard N. J., Rhodes J. M., Wiesmann H., Shih C.-Y., and Bansal B. M. (1974) The chemical definition and interpretation of rock types returned from non-mare regions of the moon. *Proc. Lunar Sci. Conf. 5th*, pp. 1227-1246.

James O. B. (1980) Rocks of the early lunar crust. *Proc. Lunar Planet. Sci. Conf. 11th*, pp. 365-393.

Korotev R. L. (1987) National Bureau of Standards coal flyash (SRM 1633a) as a multielement standard for instrumental neutron activation analysis. *J. Radioanal. Nuc. Chem.*, in press.

Laul J. C. and Schmitt R. A. (1973) Chemical compositions of Apollo 15, 16, and 17 samples. *Proc. Lunar Sci. Conf. 4th*, pp. 1349-1367.

Laul J. C., Hill D. W., and Schmitt R. A. (1974) Chemical studies of Apollo 16 and 17 samples. *Proc. Lunar Sci. Conf. 5th*, pp. 1047-1066.

Lindsley D. H. and Andersen D. J. (1983) A two-pyroxene thermometer. *Proc. Lunar Planet. Sci. Conf. 13th*, in *J. Geophys. Res., 88*, A887-A906.

Lindstrom M. M. (1984) Alkali gabbronorite, ultra-KREEPy melt rock and the diverse suite of clasts in North Ray Crater feldspathic fragmental breccia 67975. *Proc. Lunar Planet. Sci. Conf. 15th*, in *J. Geophys. Res., 89*, C50-C62.

Lindstrom M. M. and Lindstrom D. J. (1986) Lunar granulites and their precursor anorthositic norites and the early lunar crust. *Proc. Lunar Planet. Sci. Conf. 16th*, in *J. Geophys. Res., 91*, D263-D276.

Lindstrom M. M. and Salpas P. A. (1981) Geochemical studies of rocks from North Ray Crater, Apollo 16. *Proc. Lunar Planet. Sci. 12B*, pp. 305-322.

Lindstrom M. M. and Salpas P. A. (1983) Geochemical studies of feldspathic fragmental breccias and the nature of North Ray Crater ejecta. *Proc. Lunar Planet. Sci. Conf. 13th*, in *J. Geophys. Res., 88*, A671-A683.

Marvin U. B. (1975) The boulder. *The Moon, 14*, 315-326.

Morse S. A. (1982) Adcumulus growth at the base of the lunar crust. *Proc. Lunar and Planet. Sci. Conf. 13th*, in *J. Geophys. Res., 92*, A10-A18.

Ostertag R., Stoffler D., Borchardt R., Palme H., Spettel B., and Wanke H. (1987) Precursor lithologies and metamorphic history of granulitic breccias from North Ray crater, Station 11, Apollo 16. *Geochim. Cosmochim. Acta, 51*, 131-142.

Palme H. (1980) The meteoritic contamination of terrestrial and lunar impact melts and the problem of indigenous siderophiles in the lunar highland, *Proc. Lunar Planet. Sci. Conf. 11th*, pp. 481-506.

Ryder G., Stoeser D. B., Marvin U. B., Bower J. F., and Wood J. A. (1975) Boulder 1, station 2, Apollo 17: Petrology and petrogenesis. *The Moon, 14*, 327-357.

Ryder G., Stoeser D. B., and Wood J. A. (1977) Apollo 17 KREEPy basalt: A rock type intermediate between mare and KREEP basalts. *Earth Planet. Sci. Lett., 35*, 1-13.

Salpas P. A., Willis K. J., and Taylor L. A. (1985) *Breccia Guidebook No. 8: 72275*. JSC Publ. 20419, NASA Johnson Space Center, Houston, Texas. 44 pp.

Salpas P. A., Taylor L. A., and Lindstrom M. M. (1987) Apollo 17 KREEPy basalts: Evidence for non-uniformity of KREEP. *Proc. Lunar Planet. Sci. Conf. 17th*, in *J. Geophys. Res., 92*, E340-E348.

Shervais J. W., Taylor L. A., and Lindstrom M. M. (1985) Apollo 14 mare basalts: Petrology and geochemistry of clasts from consortium breccia 14321. *Proc. Lunar Planet. Sci. Conf. 15th*, in *J. Geophys. Res., 90*, C375-C395.

Taylor S. R. (1982) *Planetary Science: A Lunar Perspective*. Lunar and Planetary Institute, Houston. 481 pp.

Wanke H., Palme H., Kruse H., Baddenhausen H., Cendales M., Dreibus G., Hofmeister H., Jagoutz E., Palme C., Spettel B., and Thacker R. (1976) Chemistry of lunar highland rocks: A refined evaluation of the composition of the primary matter, *Proc. Lunar Sci. Conf. 7th*, pp. 3479-3499.

Warner J. L., Simonds C. H., and Phinney W. C. (1976) Genetic distinction between anorthosites and Mg-rich plutonic rocks: New data from 76255 (abstract). In *Lunar and Planetary Science XII*, pp. 915-917. Lunar and Planetary Institue, Houston.

Warner J. L., Phinney W. C., Bickel C. E., and Simonds C. H. (1977) Feldspathic granulitic impactites and pre-final bombardment lunar evolution. *Proc. Lunar Sci. Conf. 8th*, pp. 2051-2066.

Warner R. D., Taylor G. J., and Keil K. (1977) Petrology of crystalline matrix breccias from Apollo 17 rake samples. *Proc. Lunar Sci. Conf. 8th*, pp. 1987-2006.

Warren P. H. (1985) The magma ocean concept and lunar evolution. *Ann. Rev. Earth Planet. Sci., 13*, 201-240.

Warren P. H., Taylor G. J., Keil K., Kallemeyn G. W., Rosener P. S., and Wasson J. T. (1983) Sixth foray for pristine nonmare rocks and an assessment of the diversity of lunar anorthosites. *Proc. Lunar Planet. Sci. Conf. 13th*, in *J. Geophys. Res., 88*, A615-A630.

Warren P. H., Shirley D. N., and Kallemeyn G. W. (1986) A potpourri of pristine moon rocks, including a VHK mare basalt and a unique, augite-rich Apollo 17 anorthosite. *Proc. Lunar Planet. Sci. Conf. 16th*, in *J. Geophys. Res., 91*, D319-D330.

Warren P. H., Jerde E. A., and Kallemeyn G. W. (1987) Pristine moon rocks: A "large" felsite and a metal-rich ferroan anorthosite. *Proc. Lunar Planet. Sci. Conf. 17th*, in *J. Geophys. Res., 92*, E303-E313.

Wolfe E. W. (1975) Geologic setting of Boulder 1, Station 2. Apollo 17 landing site. *The Moon, 14*, 307-314.

Petrology of Brecciated Ferroan Noritic Anorthosite 67215

James J. McGee

U. S. Geological Survey, 959 National Center, Reston, VA 22092

A petrologic study of breccia 67215 demonstrates that the rock, which has the bulk composition of a ferroan noritic anorthosite (bulk Mg/(Mg+Fe) = 0.60), is a polymict breccia containing several lithic clast types within a crushed, cataclastic matrix. Lithic clasts include: (1) igneous and metamorphic-textured noritic anorthosite clasts, containing approximately 80% plagioclase and varied proportions of low- and high-Ca pyroxenes and olivine; (2) granulated and sheared clasts, most of which have mineralogy similar to that of the igneous and metamorphic clasts; (3) coarse-grained anorthosite, with coarsely exsolved, relatively Fe-rich augite; (4) aphanitic, feldspathic microporphyritic melt breccias; and (5) an impact-melt rock with strongly zoned, relatively Mg-rich pyroxene. The igneous and metamorphic noritic anorthosite clasts are the dominant clast types in the breccia, and the breccia matrix is composed predominantly of mineral fragments from parent rocks similar to these noritic anorthosite clasts. The granulated/sheared clasts were deformed during the 67215 breccia-forming event. The coarse-grained anorthosite clast is more ferroan than the noritic-anorthosite lithologies, but the mineral chemistry suggests that the anorthosite and noritic anorthosite clasts crystallized from the same parent magma, possibly by adcumulus growth during varied crystallization and cooling conditions at the base of the lunar crust. The feldspathic melt breccias and magnesian impact-melt rock are lithic clasts that are derived from different precursors than the noritic anorthosite clasts and possibly contain some non-ferroan-anorthosite suite components. Numerous clasts with ferroan noritic anorthosite composition similar to 67215 have been identified among the samples from North Ray Crater, suggesting that this rock type is relatively common in the highlands regolith excavated by the crater.

INTRODUCTION

The pristine lunar highlands rocks that have been identified in the returned sample collection consist predominantly of ferroan anorthosites and more mafic, Mg-suite plutonic rocks such as norite and troctolite. During the past few years, many new types of pristine rocks have been identified. Considerable diversity now exists among the Mg-suite rocks, and significant, additional pristine rock types such as alkali anorthosite, alkali norite, and alkali gabbronorite have been identified (*Warren et al.*, 1983a,b; *James*, 1980, 1983; *James and Flohr*, 1983; *James et al.*, 1987b; *Shervais et al.*, 1983, 1984). Still absent in the collection of identified pristine rocks are ferroan-anorthosite-suite rocks that have relatively high mafic mineral contents. Various mixing models have invoked the presence of a mafic ferroan anorthositic component within the lunar crust to model the chemical constituents of polymict highlands breccias (*Lindstrom and Salpas*, 1981, 1983; *Stöffler et al.*, 1985). Studies of Apollo 16 feldspathic fragmental breccias, particularly those from the North Ray Crater area, have shown that a possible candidate for this component may be a ferroan anorthositic norite or noritic anorthosite (*Taylor*, 1982; *Lindstrom and Salpas*, 1983; *Lindstrom*, 1983). Small lithic clasts of ferroan anorthositic norite and noritic anorthosite have been found in 67015 (*Marvin and Lindstrom*, 1983; *Marvin et al.*, 1987), 67016 (*Norman*, 1981; *Lindstrom and Salpas*, 1983), 67455 (*Minkin et al.*, 1977), and 67975 (*McGee*, 1987a), but no large pristine samples have been found.

Apollo 16 breccia 67215, the subject of the present study, is the best hand-specimen of a ferroan noritic anorthosite. Previous studies (*Taylor*, 1982; *Warren et al.*, 1983a; *Lindstrom and Salpas*, 1983) suggested that 67215 might be a pristine rock; however, the low but measurable content of siderophile trace elements reported by *Lindstrom and Lindstrom* (1986) and *Warren et al.* (1986) indicates that 67215 is not pristine. *Lindstrom and Lindstrom* (1986) also contended that, despite the siderophile-element enrichment which was presumably caused by meteoritic contamination, the silicate-mineral fraction of the breccia appeared to have a single anorthositic norite precursor. If this is the case, then the mineralogy and major-element composition of 67215 may be characteristic of an important, relatively mafic, ferroan-anorthositic component of the ancient lunar crust (*Warren et al.*, 1983a; *Lindstrom and Salpas*, 1983; *Lindstrom and Lindstrom*, 1986). The detailed petrologic study of 67215 presented here was undertaken to characterize the breccia's mineralogy and texture, to determine if the rock's silicate fraction is monomict, and to present a detailed petrologic description of this apparently rare but significant mafic, ferroan-anorthosite suite lithology.

Breccia 67215 was collected at Apollo 16, Station 11, on the south rim of North Ray Crater in the white breccia boulder area. The sample was placed in a padded bag to protect its supposedly unabraded surface (*Ulrich et al.*, 1981), and was first examined and described by *Taylor* (1982). The bulk composition of the breccia is that of a ferroan noritic anorthosite with a bulk Mg' [100 × molar Mg/(Mg+Fe)] of 60, making it one of the more mafic ferroan-anorthosite suite rocks returned from the lunar highlands (*Lindstrom and Salpas*, 1983; *Lindstrom and Lindstrom*, 1986; *Warren et al.*, 1983a). The siderophile element data of *Lindstrom and Lindstrom* (1986) and *Warren et al.* (1986) indicate that the breccia contains minor meteoritic contamination. Concentrations of the rare-earth elements in the breccia (*Lindstrom and Lindstrom*, 1986) are low, ranging from 2-5× chondrites, and the sample has a positive Eu anomaly (~10× chondrites). ^{40}Ar-^{39}Ar

stepwise-heating data were reported for 67215 by *Marvin et al.* (1987), but a well-defined plateau was not obtained and no reliable age determination is available.

SAMPLING AND METHODS

In hand specimen, 67215 is a cataclastic breccia consisting of a crushed light gray matrix and numerous coherent lithic clasts. Two types of lithic clast were extracted from the breccia at the Lunar Curatorial Lab and thin-sectioned for additional study (*McGee,* 1987b). One type is a dark aphanitic rock. Such clasts are abundant on one side of the hand sample and are also present in the chips and fines. Thin sections of aphanitic clasts from both the hand sample (67215,34) and the chips and fines (67215,33) were studied. The other clast type that was extracted is a coarse-grained anorthosite. The clast of this rock was approxiately 5 mm long and had two dark half-millimeter-sized mineral grains that appeared to be pyroxene along its edge. Thin section 67215,35 was prepared from this clast. Three other thin sections (67215,6, ,9, ,11) that are representative of the 67215 bulk breccia were also studied.

All six thin sections were studied petrographically with optical and scanning electron microscopes; backscattered electron images were obtained with an accelerating voltage of 20 kV on the scanning microscope. Mineral composition data were obtained with an automated ARL-SEMQ electron microprobe. Operating conditions were 15 kV accelerating voltage (25 kV for metal analyses), 0.1 microamp beam current, and 40-second counting time. Analyses were corrected for instrumental drift, deadtime, and background. Natural and synthetic mineral and glass standards were used, and matrix effects were corrected using the method of *Bence and Albee*

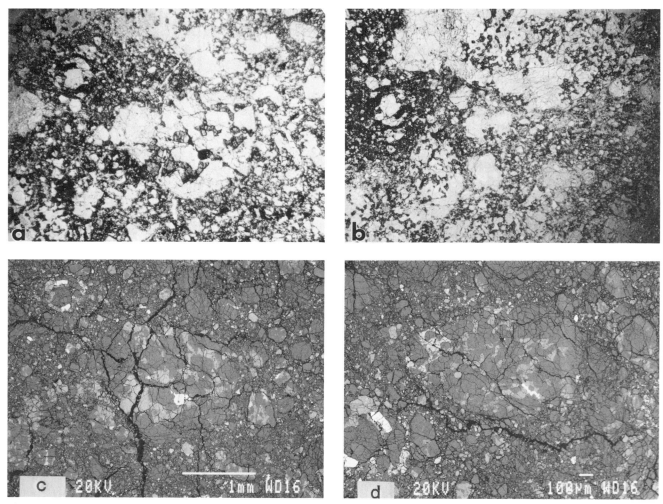

Fig. 1. Transmitted light (TL) photomicrographs and backscattered electron (BSE) images (obtained at 20 kV) of 67215. (**a**) Lithic clast (center) with igneous (subophitic) texture surrounded by granulated matrix; TL, field of view (FOV) = 3 × 4 mm. (**b**) Lithic clasts (top center and lower left) with metamorphic (granoblastic) texture included in matrix; TL, FOV = 3 × 4 mm. (**c**) BSE image showing same igneous clast (center) as in (**a**). Dark gray = plagioclase, light gray = pyroxene. White, blocky grain in lower center of clast is ilmenite. A smaller igneous clast containing two ilmenite grains is visible at top left of image. (**d**) BSE image of metamorphic clast (center) and smaller, ilmenite-bearing igneous clast (lower left). Irregular white, fracture-filling phase (right) in metamorphic clast is iron-metal.

TABLE 1. Compositions of plagioclase and olivine in 67215.

	Plagioclase (Mol % Anorthite)			Olivine (Mol % Forsterite)		
	No. Analyses	Average An	Range An	No. Analyses	Average Fo	Range Fo
Matrix	39	96.3	95-97	13	51.2	48-52
Igneous clasts	22	96.2	94-97	11	51.0	49-51
Metamorphic clasts	32	96.1	95-97	14	51.8	48-53
Granulated/sheared clasts	11	96.4	95-97	12	55.9	51-60
Impact-melt rock	14	97.5	96-98 *	5	55.5	54-56
Anorthosite	19	96.4	94-97	-	-	-
FMMBs[†]	90	93.5	84-98	15	58.5	57-60
Lithic clasts in FMMBs	17	94.6	88-97	9	56.5	51-60

* Plus two inclusions (An_{90} and An_{91}) in pyroxene.
[†] FMMBs = feldspathic microporphyritic melt breccias.

TABLE 2. Compositions of pyroxenes in 67215.

	Low-Ca Pyroxene					Average (wt %)				High-Ca Pyroxene					Average (wt %)			
	No. Analyses	Wo	En	Fs	Mg'	Al_2O_3	TiO_2	Cr_2O_3	MnO	No. Analyses	Wo	En	Fs	Mg'	Al_2O_3	TiO_2	Cr_2O_3	MnO
Matrix	26	6.4	58.0	35.7	59-65	0.64	0.41	0.17	0.40	13	40.3	42.7	17.0	69-73	1.69	1.16	0.38	0.23
Igneous clasts	18	5.4	58.0	36.6	59-65	0.59	0.39	0.17	0.39	7	41.1	42.0	16.9	70-73	1.77	0.98	0.38	0.21
Metamorphic clasts	20	3.4	59.0	37.5	58-64	0.52	0.35	0.15	0.39	19	40.6	42.0	17.4	68-73	1.71	1.06	0.42	0.21
Granulated/sheared clasts	21	4.6	57.7	37.7	50-70	0.67	0.40	0.19	0.42	6	41.5	40.9	17.6	62-78	2.00	1.36	0.46	0.23
Impact-melt rock	16	10.6	64.7	24.7	58-80	2.41	0.33	0.48	0.29	4	32.8	44.5	22.7	58-72	3.07	0.89	0.51	0.26
Anorthosite	4	4.2	44.7	51.1	46-47	0.52	0.34	0.14	0.52	4	41.9	34.4	23.7	58-61	1.59	0.85	0.36	0.24
FMMBs	18	19.7	51.1	29.2	55-69	3.05	1.11	0.35	0.32	-	-	-	-	-	-	-	-	-
Lithic clasts in FMMBs	4	12.4	55.9	31.7	62-66	1.56	1.09	0.22	0.36	2	38.9	44.7	16.4	72-74	3.51	2.24	0.75	0.25

(1968), with the correction factors of *Albee and Ray* (1970). Approximately 600 quantitative mineral analyses were obtained for the breccia; complete analyses are available from the author on request.

TEXTURE AND MINERALOGY

Matrix

The bulk of the breccia consists of a granulated matrix of angular mineral fragments with seriate grain size, ranging from less than 10 μm to as much as 500 μm across (Fig. 1). Most grains in the matrix are less than 50 μm across, but mineral fragments in the 100-150 μm size range are also abundant. The matrix shows evidence of varied, but generally weak, shock deformation: Grains are fractured and sheared, some plagioclase grains show undulatory extinction, and some pyroxene grains have shock-induced twinning. Plagioclase makes up approximately 80% of the matrix and low-Ca pyroxene is the dominant mafic mineral, followed by olivine and high-Ca pyroxene. Average compositions of the major minerals are given in Tables 1 and 2 and compositional ranges are shown in Figs. 3-5. Minor phases in the matrix include ilmenite with 1.5% MgO, iron-metal with 3.5% Ni and 1.2% Co, and troilite.

Lithic Clasts

Approximately 20% (by volume) of the breccia consists of lithic clasts; these clasts commonly grade into the matrix. The clasts range in size from 0.5 to 2mm; most have dimensions on the order of a millimeter. The lithic clasts are modally and texturally varied, generally containing between 10% and 25% mafic minerals and having either igneous or metamorphic textures (Fig. 1). Several clasts are granulated and sheared to such an extent that their primary textures are difficult to recognize. Traces of both igneous and metamorphic textures are visible in these "granulated/sheared" clasts but are not preserved well enough to identify their primary textures unequivocally (Fig. 2a).

Twenty-one lithic clasts were studied in detail. Of these, approximately one-half have metamorphic texture, one-quarter have igneous texture, and one-quarter have granulated/sheared texture. One large clast has intergranular texture like that observed in impact-melt rocks.

Igneous clasts. The igneous clasts have subophitic to intergranular to poikilitic textures. Lathy or subhedral plagioclase grains range in size between 0.2-0.3 mm (rarely, as much as 0.5 mm). Plagioclase compositions are similar to those in the matrix (Table 1; Fig. 3). Low- and high-Ca pyroxene

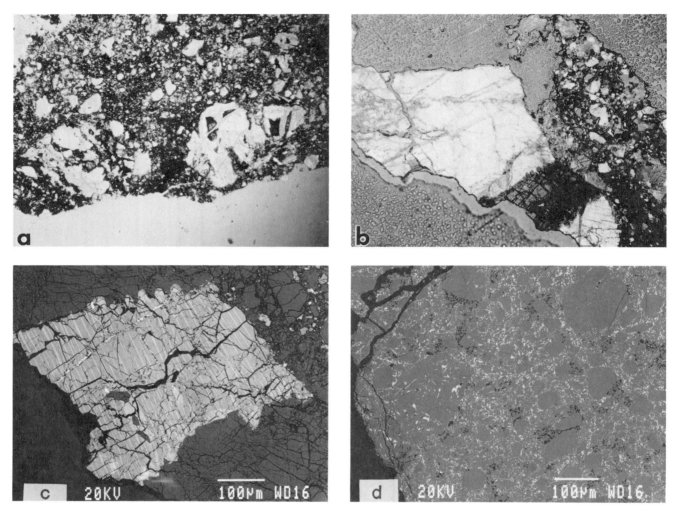

Fig. 2. Transmitted light (TL) photomicrographs and backscattered electron (BSE) images (obtained at 20 kV) of impact-melt rock, coarse-grained anorthosite, and feldspathic microporphyritic melt breccia (FMMB) clasts in 67215. **(a)** Impact-melt rock clast (lower right) and irregular granulated/sheared clast (lower left) within matrix; TL, FOV = 3 × 4 mm. **(b)** Part of coarse-grained anorthosite clast with augite (dark grain in lower center) and adjacent matrix; TL, FOV = 1.5 × 2 mm. **(c)** BSE image of coarse-grained anorthosite clast. Plagioclase (dark gray) forms large grains surrounding pyroxene and small inclusions in pyroxene. The large augite grain (medium gray) contains abundant exsolution lamellae of low-Ca pyroxene (light gray). Note the corrugated border (top) containing relatively coarse, granular low-Ca pyroxene exsolution. Fracture crossing left side of augite contains abundant troilite blebs (white). **(d)** BSE image of FMMB clast. Note network of poikilitic mafic minerals (bright phases) and abundant plagioclase fragments (dark gray). Slightly coarser grained, lithic clast with intergranular texture is visible in left center of field.

and olivine are intergranular, with grain sizes slightly smaller than those of the plagioclase, or are poikilitic and extend for as much as 0.5 mm (Fig. 1). The pyroxenes contain varied amounts of thin (micrometer-sized) exsolution lamellae. Average pyroxene compositions are given in Table 2 and compositional ranges are shown in Fig. 5; average olivine compositions are given in Table 1 and ranges are shown in Fig. 4. The clasts also contain sparse, large blocky or lathy grains of ilmenite, as much as 0.2 mm long, with 1.6% MgO. In some clasts the modal proportion of ilmenite is as much as 5-10 vol %. No iron-metal was found in any of the igneous clasts.

Metamorphic clasts. The metamorphic-textured clasts have polygonal plagioclase and granoblastic or poikiloblastic high- and low-Ca pyroxene and olivine (Fig. 1). Grain sizes are generally in the 40-250 μm range, similar to grain sizes in the matrix. Pyroxenes in the metamorphic clasts have fine exsolution lamellae which are more common and slightly coarser than the lamellae in the pyroxenes in the igneous clasts. Modally, the metamorphic clasts are similar to the igneous clasts and the matrix, containing approximately 15-20 vol % mafic minerals dominated by low-Ca pyroxene. However, within a few metamorphic clasts olivine or high-Ca pyroxene is the dominant mafic mineral instead of low-Ca pyroxene.

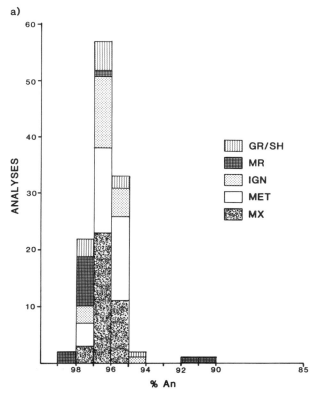
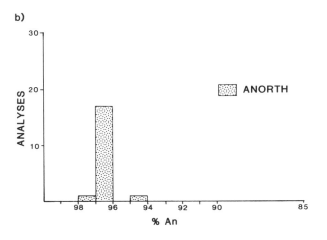
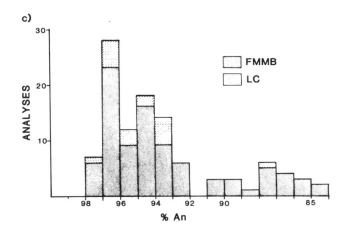

Fig. 3. Histograms of plagioclase compositions, expressed as anorthite [An = Ca/(Ca+Na+K)] contents, in 67215 lithic clasts and matrix. (a) Plagioclase compositions in granulated/sheared clasts (GR/SH), impact-melt rock (MR), igneous clasts (IGN), metamorphic clasts (MET), and matrix (MX). (b) Plagioclase compositions in coarse-grained anorthosite clast (ANORTH). (c) Plagioclase compositions in feldspathic microporphyritic melt breccia clasts (FMMB) and in the lithic clasts (LC) within the FMMB.

Compositions of plagioclase, olivine, high-Ca pyroxene, and low-Ca pyroxene are similar, both in averages and in total ranges, to those in the igneous clasts and in the matrix (Tables 1 and 2; Figs. 3-5). Trace chromite is present in the metamorphic clasts and has a uniform composition of 46% Cr_2O_3, 33% FeO, 15% Al_2O_3, 3% TiO_2, and 2% MgO. Iron-metal is present both as irregular grains and as vein fillings as much as 0.2 mm long. This metal contains 3.5% Ni and 1.2% Co, the same Ni and Co concentrations as those that are observed in the matrix metal grains.

Granulated/sheared clasts. Most of the granulated/sheared clasts have modes and mineralogies similar to those of the igneous and metamorphic clasts, but one clast adjacent to the impact-melt rock clast described below appears to be a sheared and deformed fragment of that rock (Fig. 2a). Although these clasts have been granulated or sheared, they retain vestiges of textures like those in the metamorphic clasts; plagioclase usually forms a mosaic and mafic minerals display traces of poikilitic or poikiloblastic textures. Plagioclase compositions (Table 1; Fig. 3) in most of the clasts are like those in the matrix, igneous clasts, and metamorphic clasts, but the mafic minerals differ slightly (Tables 1 and 2; Figs. 4 and 5). Although the average pyroxene compositions do not differ greatly from those of the pyroxenes in the matrix and igneous/metamorphic clasts, the range of pyroxene compositions is much greater in the granulated/sheared clasts (Fig. 5). Also, minor-element concentrations in the pyroxenes in the granulated/sheared clasts are generally higher than in the matrix and igneous/metamorphic clasts (Table 2). Olivines in the granulated/sheared clasts are on average significantly more magnesian than those in the matrix and igneous/metamorphic clasts (Table 1; Fig. 4).

Impact-melt rock. The impact-melt rock has intergranular texture and consists of approximately 90% plagioclase and 10% pyroxene, mostly pigeonite (Fig. 2a). The plagioclase is blocky or lathy with patches of mosaic intergrowths and displays minor undulatory extinction, indicating that the clast was weakly shocked. The pigeonite is subhedral to intergranular and is zoned and/or contains scattered micrometer-sized blebs of high-Ca pyroxene and olivine. High-Ca pyroxene also occurs as rims or overgrowths on the low-Ca pyroxene. Traces of ilmenite, chromite, troilite, and felsic glass are also present.

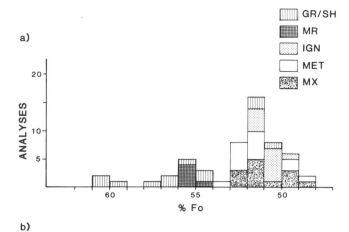

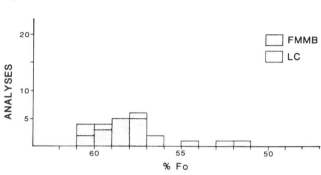

Fig. 4. Histograms of olivine compositions, expressed as forsterite [Fo = Mg/(Mg+Fe)] contents, in 67215 lithic clasts and matrix. (a) Olivine compositions in granulated/sheared clasts (GR/SH), impact-melt rock (MR), igneous clasts (IGN), metamorphic clasts (MET), and matrix (MX). (b) Olivine compositions in feldspathic microporphyritic melt breccia clasts (FMMB) and in the lithic clasts (LC) within the FMMB. No olivine is present in the coarse-grained anorthosite clast.

Mineral compositions in the impact-melt rock are significantly different from those in the breccia matrix and other lithic clasts. Plagioclase in the impact-melt rock is slightly more calcic than in the rest of the breccia (Table 1; Fig. 3), but locally small grains of more sodic plagioclase, An_{90}, are included in pyroxene. The pyroxene is significantly more magnesian and more highly zoned than the pyroxene in the other types of lithic clasts (Table 2; Fig. 5). Minor-element contents of pyroxenes are higher than those of pyroxenes in the rest of the breccia (Table 2). The low-Ca pyroxenes have total $Al_2O_3+TiO_2+Cr_2O_3$ contents nearly a factor of three higher than those of the low-Ca pyroxenes in the breccia matrix, igneous/metamorphic, and granulated/sheared clasts. Olivine in the impact-melt rock is on average more magnesian than olivine in the igneous and metamorphic clasts and breccia matrix (Table 1; Fig. 4).

Coarse-grained Anorthosite Clast

The anorthosite clast extracted from the breccia is nearly monomineralic plagioclase except for one 0.4 mm grain of augite (Fig. 2b). The plagioclase is coarse grained; one grain is nearly 4 mm in length. The augite grain has relatively coarse (2-10 μm) exsolution lamellae of low-Ca pyroxene; some of the exsolution lamellae are bent and one edge of the grain has coarser, granular exsolution forming a corrugated border (Fig. 2c). The exsolution lamellae terminate on either side of irregular veins of low-Ca pyroxene and troilite that crosscut the augite. Traces of chromite and ilmenite and a few plagioclase grains are also included in the augite.

Plagioclase compositions within the anorthosite clast are similar to those in the other clasts (Table 1; Fig. 3). However, the pyroxenes in the clast are significantly more ferroan than any observed in the other lithic clasts in 67215 (Table 2; Fig. 5). A few of the pyroxene fragments in the matrix that surrounds the anorthosite clast contain abundant exsolution lamellae and have compositions similar to those observed within the anorthosite (Fig. 5). The minor-element concentrations in the anorthosite pyroxenes, however, do not differ significantly from those in the pyroxenes in the breccia matrix and in the igneous and metamorphic clasts (Table 2).

An ilmenite inclusion in the pyroxene contains 1.6% MgO and a chromite inclusion has a composition of 47% Cr_2O_3, 34% FeO, 12% Al_2O_3, 3.5% TiO_2, and 1.5% MgO. These compositions are the same as those found in the oxide minerals in the matrix and in the igneous and metamorphic clasts.

Aphanitic Clasts (Feldspathic Microporphyritic Melt Breccias)

The two clasts of feldspathic microporphyritic melt breccia (FMMB) consist of an aggregate of mineral fragments and lithic clasts set in a very fine grained (<5-10 μm) matrix of plagioclase, pyroxene, olivine, and trace ilmenite and iron metal (Fig. 2d). The FMMBs contain approximately 80-90 vol % plagioclase and the two clasts studied have slightly different amounts of mafic minerals; sample ,33 is more mafic (15-20 vol %) than sample ,34 (5-10 vol %). The mineral fragments within these clasts are angular plagioclase grains, with sizes between 50-100 μm and rarely as large as 0.5 mm. The mafic minerals (mostly low-Ca pyroxene) form spray-like poikilitic grains that extend for as much as 0.5 mm across the clast.

Plagioclase within the FMMB clasts shows a much greater spread of compositions and ranges to much more sodic compositions than does plagioclase in the other clast types (Table 1; Fig. 3). Approximately 15% of the analyzed plagioclase grains have compositions more sodic than An_{91}; these sodic compositions occur in both small and large grains, some of which have overgrowth rims with more calcic compositions typical of the rest of the grains. Matrix pyroxene is mostly pigeonite and is compositionally heterogeneous (Table 2; Fig. 5). Matrix olivine averages more magnesian than olivine in the other 67215 lithologies (Table 1; Fig. 4). Ilmenite is also more magnesian (3.3% MgO) than in the rest of the breccia. Iron-metal compositions in the FMMBs are also distinct from those in the rest of the breccia, with 5.5-11.5% Ni and 0.7-1.4% Co.

The lithic clasts within the FMMBs are 0.2-0.5 mm across and have an intergranular texture that is slightly coarser grained

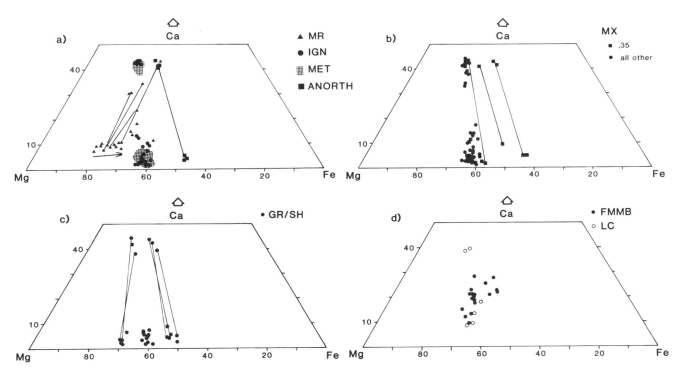

Fig. 5. Compositions of 67215 pyroxenes plotted in the pyroxene quadrilateral. (**a**) Impact-melt rock (MR), igneous (IGN), metamorphic (MET) and coarse-grained anorthosite (ANORTH) clasts. Arrow shows direction of core-rim zoning in impact-melt rock pyroxenes. Tie-lines connecting impact-melt rock data points show compositions of zoned pyroxenes or coexisting low- and high-Ca pyroxenes. Compositions of high- and low-Ca pyroxene in the exsolved pyroxene of the coarse-grained anorthosite are also connected by a tie-line. (**b**) Matrix pyroxenes. Pyroxene compositions in the matrix surrounding the coarse-grained anorthosite clast in thin section ,35 are shown as squares; tie-lines connect compositions of host and lamellae in exsolved pyroxene fragments in this matrix. Compositions of matrix pyroxene in all other thin sections of the breccia are shown as circles. (**c**) Granulated/sheared clasts. Tie-lines connect compositions of coexisting pyroxenes in the same clasts. (**d**) Feldspathic microporphyritic melt breccia (FMMB) clasts and lithic clasts (LC) within the FMMB clasts.

(5-30 μm) than the FMMB matrix. These clasts consist of lathy plagioclase and minor (~5%) intergranular pyroxene (both high- and low-Ca) and olivine, and they are slightly more feldspathic than the bulk FMMBs. Minor numbers of 50-80 μm-sized plagioclase mineral fragments are also present in the lithic clasts; one of these grains has a relatively sodic ($An_{87.5}$) core and a more calcic rim, like that found in some of the grains in the FMMB matrix. However, the plagioclase in the lithic clasts generally has more restricted and more calcic compositions than the plagioclase in the FMMB matrix (Table 1; Fig. 3). The pigeonites in the lithic clasts have the same average Mg' as those in the FMMB matrix but they also average slightly lower in wollastonite content than those in the matrix (Table 2; Fig. 5). Olivine in the lithic clasts has a greater compositional range than olivine in the FMMB matrix; this range is similar to that of olivine in the bulk 67215 matrix and other lithic clasts (Table 1; Fig. 4).

DISCUSSION

Breccia Components and Precursors

Previous studies which concluded that 67215 is a monomict, possibly pristine rock were based on limited textural and mineral compositional data (*Ryder and Norman*, 1980; *Taylor*, 1982; *Lindstrom and Salpas*, 1983; *Warren et al.*, 1983a; *Lindstrom and Lindstrom*, 1986). *Lindstrom and Lindstrom* (1986) and *Warren et al.* (1986) demonstrated that 67215 is not pristine, based upon siderophile-element enrichment, but *Lindstrom and Lindstrom* (1986) contended that there was no evidence of more than one precursor for the silicate fraction of the breccia. The present study demonstrates that 67215 is polymict; the breccia contains a variety of lithologies that display different textures, mineral modes, and mineral compositions.

The igneous and metamorphic lithic clasts all have similar mineralogical assemblages and modes; their modal compositions are representative of an anorthositic norite or noritic anorthosite, consistent with the classification of the breccia based upon bulk chemical data (*Lindstrom and Lindstrom*, 1986). The nearly identical mineral compositions also suggest that both the igneous and metamorphic clasts were derived from the same noritic anorthosite precursor, which was partly recrystallized in order to produce the annealed rocks now represented by the metamorphic clasts.

The textures and mineralogies indicate that the breccia matrix was primarily derived from the same precursor as the igneous and metamorphic noritic-anorthosite lithic clasts;

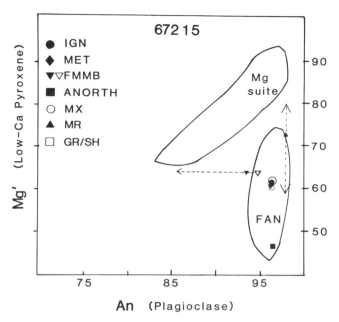

Fig. 6. Anorthite (An) content of plagioclase vs. Mg' of low-Ca pyroxenes in 67215 clasts and matrix. IGN = igneous clasts, MET = metamorphic clasts, FMMB = feldspathic microporphyritic melt breccias (open inverted triangle is for lithic clasts within FMMB), ANORTH = coarse-grained anorthosite, MX = matrix, MR = impact-melt rock clast, GR/SH = granulated/sheared clasts. Fields for pristine Mg-suite rocks and ferroan-anorthosite suite (FAN) rocks are shown as defined by *McGee* (1987a). Vertical dashed line with arrows shows range of pigeonite Mg' in the impact-melt rock clast. Horizontal dashed line with arrow shows extent of relatively sodic plagioclase compositions in the FMMB clasts.

evidence of this relation includes the gradational textures between the matrix and lithic clasts and the similarity of mineral modes and compositions among the matrix, igneous clasts, and metamorphic clasts. On the familiar plot of An [Ca/(Ca+Na+K)] content of plagioclase versus Mg' of low-Ca pyroxene (Fig. 6), compositions of the plagioclase and low-Ca pyroxene in the matrix plot in the same position within the ferroan-anorthosite (FAN) suite field as the compositions of the plagioclase and low-Ca pyroxene in the igneous and metamorphic clasts. In addition, the pyroxenes in the matrix and in the igneous and metamorphic clasts all have similar concentrations of minor elements. In the remaining discussion, reference to the noritic anorthosite lithology or precursor pertains to that of the igneous and metamorphic clasts, and the breccia matrix that presumably was derived from these clasts.

Several mineralogical features suggest that the precursor of the coarse-grained anorthosite is related to the precursor of the noritic anorthosite. Although the mafic minerals in the anorthosite clast are distinctly more ferroan than are those in the noritic anorthosite, the plagioclases in the anorthosite have approximately the same An contents as do those in the noritic anorthosite. These mineral compositional relations result in formation of a near-vertical trend between the anorthosite and the noritic anorthosite, as plotted on the Mg' versus An diagram (Fig. 6). This near-vertical trend of mineral compositions is common in rocks of the ferroan-anorthosite suite and is consistent with that which would be produced initially by strong fractional crystallization of a cotectic magma with intermediate bulk Mg' and very low alkali concentration (*Longhi*, 1982; *Ryder*, 1982). The anorthosite also has pyroxene minor-element concentrations and ilmenite and chromite compositions nearly the same as those of the noritic anorthosite. These data suggest that both the anorthosite and the noritic anorthosite crystallized from the same parent magma.

Most of the granulated/sheared clasts are samples of the same noritic anorthosite precursor as represented by the matrix and igneous and metamorphic clasts. Other granulated/sheared clasts appear to be deformed and slightly more magnesian equivalents of the impact-melt rock, a lithology that is distinctly more magnesian than the rest of the breccia. Because there is more than one precursor for these granulated/sheared clasts, their deformation most likely took place during the 67215 breccia-forming event, either during initial mixing of the various breccia components or during subsequent burial and compaction.

The mineral compositions in the impact-melt rock indicate that it was derived from a more magnesian precursor than the noritic anorthosite. This impact-melt rock also had a different thermal history, and possibly a slightly more intense shock history, than did the noritic anorthosite. The texture and mineralogy of this impact-melt rock are apparently unusual among the returned Apollo samples. A review of the catalogs of lunar rocks (*Ryder and Norman*, 1980) revealed no other rocks exactly like this relatively magnesian, relatively feldspathic melt rock, with the possible exception of impact melt 68415, which has some mineralogical and textural similarities to the melt-rock clast (*Gancarz et al.*, 1972; *Helz and Appleman*, 1973; *Walker et al.*, 1973).

The FMMB clasts clearly had different histories and appear to have been derived from different precursors than the breccia matrix and the other 67215 lithic clasts. The FMMBs formed by rapid cooling and crystallization of fragment-laden impact melts. Compositions of the iron-metal grains in the FMMBs fall within the range defined for polymict highlands rocks by *Ryder et al.* (1980). A significant proportion of the plagioclase in the mineral fragment population in the FMMBs is relatively sodic and approaches compositions typical of alkali anorthosites or relatively iron-rich members of the Mg-suite. In contrast, plagioclase and pyroxene compositions in the lithic clasts in the FMMBs plot within the ferroan-anorthosite-suite field (Fig. 6). These lithic clasts appear to have a different source, with more calcic plagioclase and pyroxene, than the mineral fragments.

Origin of Lithologies and Breccia Formation

Although the anorthosites and noritic anorthosites in 67215 presumably were derived from the same parent magma, the different textures and mineralogies of these two lithologies indicate that they crystallized under different conditions. The anorthosite has a much coarser grain size than the noritic anorthosites and contains mafic minerals that are much more ferroan than those of the noritic anorthosites. The coarseness of the exsolution in the anorthosite's pyroxene indicates that the anorthosite also had a different post-crystallization thermal history (slower cooling or more prolonged subsolidus reequilibration) than the noritic anorthosite.

If the anorthosite and noritic anorthosite crystallized from the same parent magma, then the crystallization environment must have been able to produce rocks of varied grain size and modal composition. The relatively coarse grain size and homogeneous plagioclase composition in the anorthosite are consistent with an origin by adcumulus growth at a magma/roof interface, following nucleation in a zone within supercooled magma (*Morse,* 1982; *Ryder,* 1982). Under these conditions a supercooled magma is remotely cooled, possibly at upwellings generated by impact events, and is the principal heat sink during crystallization and growth (*Morse,* 1982). Following plagioclase crystallization, mafic minerals would crystallize as the magma composition oscillates about the cotectic composition (*Morse,* 1979, 1982); however, these mafic minerals would sink, forming complementary, coarse-grained orthocumulates unlike the mafic (noritic anorthosite) lithology in 67215. Finer-grained noritic anorthosite like that observed in 67215 may have been produced by varied extents of adcumulus growth at the magma/roof interface. Variable crystallization conditions, such as changing accumulation rates or degree of supercooling of the magma, can affect the relative positions of the accumulation and solidification interfaces or the location of the nucleation zone at or near the magma/roof interface, as discussed by *Morse* (1982). These crystallization conditions could vary enough to allow nucleation and growth of mafic minerals or entrapment of liquid during plagioclase accumulation, resulting in noncotectic proportions of plagioclase and mafic minerals (*Longhi,* 1982; *Morse,* 1982). Thus, both the anorthosite and noritic anorthosite lithologies in 67215 may have been produced from the same parent magma by these varied conditions during adcumulus growth. Intermittent availability of impact-generated upwelling sites, and thus of remote cooling sites, could have been a factor contributing to these variable crystallization conditions.

Two 67215 breccia clast types appear to be derived from sources other than the noritic anorthosite/anorthosite precursor. The impact-melt rock is texturally and compositionally distinct from the rest of the breccia. This clast's average pigeonite composition plots at an Mg' above the field of ferroan anorthosites and some of the pigeonite compositions extend up to an Mg' of nearly 80 (Fig. 6). The FMMBs also have mineral compositions that lie outside the field of ferroan anorthosites. The average plagioclase composition plots at a slightly lower An content than in ferroan anorthosites (Fig. 6), and about a third of the plagioclase fragments in the FMMBs have compositions that are more sodic than those generally found in ferroan anorthosite. The precursors of the impact-melt rock and the FMMBs cannot be determined from the present data. However, the relatively Mg-rich compositions of the pyroxenes in the impact-melt rock and the relatively sodic compositions of many of the plagioclase fragments in the FMMBs suggest that some components other than typical ferroan-anorthosite-suite rocks were present among the precursors of these clasts.

Formation of a breccia such as 67215, which consists predominantly of granulated noritic anorthosite mixed with minor amounts of other lithologies, may have occurred as follows. Adcumulus growth of anorthosite and noritic anorthosite lithologies in 67215 may have occurred during varying crystallization and cooling conditions, as discussed above. The lithologies formed a layered pluton or rockberg as a result of these varying conditions; metamorphism of part of this pluton may have occurred during later reheating following excavation and burial in a hot ejecta blanket. The noritic anorthosite was subsequently crushed, mixed with minor amounts of other lithologies, and reburied by impact processes; some of the admixed lithologies, such as the coarse-grained anorthosite, were apparently derived from the same pluton or parent body as the noritic anorthosite whereas others, such as the FMMBs and the impact-melt rock, were derived from other sources. Burial and crushing of the breccia lithologies produced the granulated matrix and the sheared and granulated clasts; the breccia was lithified at about this time with little or no annealing or reequilibration following lithification. The breccia presumably was excavated by the North Ray Crater-forming impact.

Several other samples of the noritic anorthosite precursor may be preserved in the rocks collected at Station 11 in the vicinity of North Ray Crater. Clasts in several other breccias from this area of the Apollo 16 site have ferroan noritic anorthosite or anorthositic norite compositions similar to 67215. These clasts are from samples 67455 (*Minkin et al.,* 1977), 67016 (*Norman,* 1981; *Lindstrom and Salpas,* 1983), and 67015 (*Marvin et al.,* 1987). Other North Ray Crater samples, such as anorthosite 67075 (*McCallum et al.,* 1975) and a troctolitic anorthosite clast in 67915 (*Roedder and Weiblen,* 1974), have relatively high concentrations of Fe-rich mafic minerals, suggesting that these clasts may be related to the other ferroan noritic anorthosites. The relative abundance of these mafic ferroan samples at the North Ray Crater station indicates that there may be a significant amount of noritic anorthosite buried in the Apollo 16 regolith, some of which may have been sampled by the North Ray cratering event. *James et al.* (1987b) suggested that these ferroan, mafic-rich anorthositic lithologies form one of four distinctive subgroups of the ferroan-anorthosite-suite rocks. If this subgroup is truly distinct from other ferroan anorthosites, then the data on 67215 and the other ferroan, mafic-rich anorthosites will provide important evidence as to the nature and evolution of the lunar crust and the Moon's primordial magma ocean.

CONCLUSIONS

Sample 67215 is a polymict breccia that consists of clasts of several lithologies, including igneous and metamorphic-textured noritic anorthosite, coarse-grained anorthosite, a relatively magnesian impact-melt rock, and feldspathic microporphyritic melt breccias that are themselves polymict. The breccia matrix appears to be composed predominantly of granulated noritic anorthosites like those preserved as metamorphic- and igneous-textured lithic clasts. The granulated/sheared clasts consist of more than one lithology; both igneous and metamorphic noritic anorthosite clasts, and minor lithologies such as the impact-melt rock, are represented in the granulated/sheared clast population that was produced during the breccia-forming event. The coarse-grained anorthosite clast is mineralogically and texturally distinct, yet the similarities of pyroxene minor-element concentrations and of trace phase compositions among this clast and the noritic anorthosite clasts, and the near-vertical trend of mineral compositions on the An versus Mg′ diagram, suggests that these lithologies may be samples of a layered ferroan-anorthositic series produced by adcumulus growth at the base of the lunar curst during varying crystallization and cooling conditions. Breccia 67215 is the largest sample of a group of relatively mafic, noritic anorthosites from the North Ray Crater area that are among the most ferroan, mafic-rich, ferroan-anorthosite-suite rocks in the lunar highlands samples. Additional studies of these rocks, particularly isotopic analysis, would provide evidence concerning the relation of these rocks to each other and to the other members of the ferroan-anorthosite suite.

Acknowledgments. I thank A. Mosie and the NASA/JSC Lunar Curatorial Staff for assistance in obtaining the samples, and O. B. James, P. A. Salpas, M. K. Flohr, and R. D. Warner for helpful reviews of the manuscript. This work was done on behalf of the National Aeronautics and Space Administration under contract T3040C (O. B. James, Principal Investigator).

REFERENCES

Albee A. L. and Ray L. (1970) Correction factors for electron probe microanalysis of silicates, oxides, carbonates, phosphates, and sufates. *Anal. Chem.*, 42, 1408-1414.

Bence A. E. and Albee A. L. (1968) Empirical correction factors for the electron microanalysis of silicates and oxides. *J. Geol.*, 76, 382-403.

Gancarz A. J., Albee A. L., and Chodos A. A. (1972) Comparative petrology of Apollo 16 sample 68415 and Apollo 14 samples 14276 and 14310. *Earth Planet. Sci. Lett.*, 16, 307-330.

Helz R. T. and Appleman D. E. (1973) Mineralogy, petrology, and crystallization history of Apollo 16 rock 68415. *Proc. Lunar Sci. Conf. 4th*, 643-659.

James O. B. (1980) Rocks of the early lunar crust. *Proc. Lunar Planet. Sci. Conf. 11th*, 365-393.

James O. B. (1983) Mineralogy and petrology of the pristine rocks. In *Workshop on Pristine Highlands Rocks and the Early History of the Moon* (J. Longhi and G. Ryder, eds.), pp. 44-51. LPI Tech. Rpt. 83-02, Lunar and Planetary Institute, Houston.

James O. B. and Flohr M. K. (1983) Subdivision of the Mg-suite noritic rocks into mg-gabbronorites and Mg-norites. *Proc. Lunar Planet. Sci. Conf. 13th*, in *J. Geophys. Res.*, 88, A603-A614.

James O. B., Lindstrom M. M., and Flohr M. K. (1987a) Petrology and geochemistry of alkali gabbronorites from lunar breccia 67975. *Proc. Lunar Planet. Sci. Conf. 17th*, in *J. Geophys. Res.*, 92, E314-E330.

James O. B., Lindstrom M. M., and Flohr M. K. (1987b) Ferroan anorthosite from lunar breccia 64435: Implications for the origin of granulitic breccias and the ferroan-anorthosite suite. In preparation.

Lindstrom M. M. (1983) Ferroan anorthositic norite: An important highland cumulate. In *Workshop on Pristine Highlands Rocks and the Early History of the Moon* (J. Longhi and G. Ryder, eds.), pp. 56-57. LPI Tech Rpt. 83-02. Lunar and Planetary Institute, Houston.

Lindstrom M. M. and Lindstrom D. J. (1986) Lunar granulites and the role of anorthositic norites in the early lunar crust. *Proc. Lunar Planet. Sci. Conf. 16th*, in *J. Geophys. Res.*, 91, D263-D276.

Lindstrom M. M. and Salpas P. A. (1981) Geochemical studies of rocks from North Ray Crater, Apollo 16. *Proc. Lunar Planet. Sci. 12B*, 305-322.

Lindstrom M. M. and Salpas P. A. (1983) Geochemical studies of feldspathic fragmental breccias and the nature of North Ray Crater ejecta. *Proc. Lunar Planet. Sci. Conf. 13th*, in *J. Geophys. Res.*, 88, A671-A683.

Longhi J. (1982) Effects of fractional crystallization and cumulus processes on mineral composition trends of some lunar and terrestrial rock series. *Proc. Lunar Planet. Sci. Conf. 13th*, in *J. Geophys. Res.*, 87, A54-A64.

Marvin U. B. and Lindstrom M. M. (1983) Rock 67015: A feldspathic fragmental breccia with KREEP-rich melt clasts. *Proc. Lunar Planet. Sci. Conf. 13th*, in *J. Geophys. Res.*, 88, A659-A670.

Marvin U. B., Lindstrom M. M., Bernatowicz T. J., and Podosek F. A. (1987) The composition and history of breccia 67015 from North Ray Crater. *Proc. Lunar Planet. Sci. Conf. 17th*, in *J. Geophys. Res.*, 92, E471-E490.

MaCallum I. S., Okamura F. P., and Ghose S. (1975) Mineralogy and petrology of sample 67075 and the origin of lunar anorthosites. *Earth Planet. Sci. Lett.*, 26, 36-53.

McGee J. J. (1987a) Petrology and precursors of lithic clasts from feldspathic fragmental breccia 67975. *Proc. Lunar Planet. Sci. Conf. 17th*, in *J. Geophys. Res.*, 92, E513-E525.

McGee J. J. (1987b) Petrologic evaluation of the components of granulitic breccia 67215 (abstract). In *Lunar and Planetary Science XVIII*, pp. 618-619. Lunar and Planetary Institute, Houston.

Minkin J. A., Thompson C. L., and Chao E. C. T. (1977) Apollo 16 white boulder consortium samples 67455 and 67475: Petrologic investigation. *Proc. Lunar Sci. Conf. 8th*, 1967-1986.

Morse S. A. (1979) Kiglapait geochemistry I: Systematics, sampling, and density. *J. Petrol.*, 20, 555-590.

Morse S. A. (1982) Adcumulus growth of anorthosite at the base of the lunar crust. *Proc. Lunar Planet. Sci. Conf. 13th*, in *J. Geophys. Res.*, 87, A10-A18.

Norman M. D. (1981) Petrology of suevitic lunar breccia 67016. *Proc. Lunar Planet. Sci. 12B*, 235-252.

Roedder E. and Weiblen P. W. (1974) Petrology of clasts in lunar breccia 67915. *Proc. Lunar Sci. Conf. 5th*, 303-318.

Ryder G. (1982) Lunar anorthosite 60025, the petrogenesis of lunar anorthosites, and the composition of the Moon. *Geochim. Cosmochim. Acta*, 46, 1591-1601.

Ryder G. and Norman M. D. (1980) *Catalog of Apollo 16 Rocks.* NASA JSC Curatorial Branch Publication 52, JSC 16904, 1144 pp.

Ryder G., Norman M. D., and Score R. A. (1980) The distinction of pristine from meteorite-contaminated highlands rocks using metal compositions. *Proc. Lunar Planet. Sci. Conf. 11th*, 471-479.

Shervais J. W., Taylor L. A., and Laul J. C. (1983) Ancient crustal components in the Fra Mauro breccias. *Proc. Lunar Planet. Sci. Conf. 14th*, in *J. Geophys. Res.*, 88, B177-B192.

Shervais J. W., Taylor L. A., Laul J. C., and Smith M. R. (1984) Pristine highland clasts in consortium breccia 14305: Petrology and geochemistry. *Proc. Lunar Planet. Sci. Conf. 15th*, in *J. Geophys. Res.*, 89, C25-C40.

Stöffler D., Bischoff A., Borchardt R., Burghele A., Deutsch A., Jessberger E. K., Ostertag R., Palme H., Spettel B., Reimold W. V., Wacker K., and Wänke H. (1985) Composition and evolution of the lunar crust in the Descartes Highlands, Apollo 16. *Proc. Lunar Planet. Sci. Conf. 15th*, in *J. Geophys. Res., 89*, C449-C506.

Taylor G. J. (1982) A possibly pristine, ferroan anorthosite (abstract). In *Lunar and Planetary Science XIII*, p. 798. Lunar and Planetary Institute, Houston.

Ulrich G. E., Hodges C. A., and Muehlberger W. R., eds. (1981) *Geology of the Apollo 16 Area, Central Lunar Highlands*. U.S.G.S. Professional Paper 1048, 539 pp.

Walker D., Longhi J., Grove T. L., Stolper E., and Hays J. F. (1973) Experimental petrology and origin of rocks from the Descartes Highlands. *Proc. Lunar Sci. Conf. 4th*, 1013-1032.

Warren P. H., Taylor G. J., Keil K., Kallemeyn G. W., Rosener P. S., and Wasson J. T. (1983a) Sixth foray for pristine nonmare rocks and an assessment of the diversity of lunar anorthosites. *Proc. Lunar Planet. Sci. Conf. 13th*, in *J. Geophys. Res., 88*, A615-A630.

Warren P. H., Taylor G. J., Keil K., Kallemeyn G. W., Shirley D. N., and Wasson J. T. (1983b) Seventh foray: Whitlockite-rich lithologies, a diopside-bearing troctolitic anorthosite, ferroan anorthosites, and KREEP. *Proc. Lunar Planet. Sci. Conf. 14th*, in *J. Geophys. Res., 88*, B151-B164.

Warren P. H., Kallemeyn G. W., and Jerde E. A. (1986) Siderophile elements in some peculiar Moon rocks and eucrites (abstract). In *Meteoritics, 21*, 531.

Mineralogical Studies of Clasts in Lunar Highland Regolith Breccia 60019 and in Lunar Meteorite Y82192

Hiroshi Takeda
Mineralogical Institute, Faculty of Science, University of Tokyo, Hongo, Tokyo 113, Japan

M. Miyamoto
College of Arts and Sciences, University of Tokyo, Komaba, Meguro-ku, Tokyo 153, Japan

H. Mori and T. Tagai
Mineralogical Institute, Faculty of Science, University of Tokyo, Hongo, Tokyo 113, Japan

Large (>8 mm) clasts found on the cut surfaces of new slabs of Apollo 16 regolith breccia 60019 have been studied by microprobe and by analytical transmission electron microscope and were compared with lunar meteorites. Three large clasts exposed on the large cut surface of 60019, WF-1 (,205), WF-4 (,208), and WC-1 (,207), have different plagioclase types set in poikilitic matrices. WF-1 consists of fine white angular plagioclase fragments in a fine-grained matrix. WF-4 is a poikilitic breccia with a few small subrounded plagioclase grains in a poikilitic matrix. WC-1 includes coarse clasts of plagioclase, olivine, and granulite. Pyroxene composition trends within the poikilitic matrices are similar to the Apollo 16 poikilitic breccias. The ranges of the An contents of plagioclase differ from one clast to the others. The plagioclase fragments in WF-1 show a narrow range and those in WF-4 show the widest range. These clast types are also dominant in small clasts of the matrix portion. The most noteworthy difference in terms of lithic types in the Y82192 lunar meteorite and in 60019 is that dominant clasts in 60019 appear to be mostly poikilitic, whereas the lunar meteorite contains abundant granulitic breccia clasts. This difference may be explained by the derivation of poikilitic rock components from large circular basin-forming impact events. The low abundance of such clasts and high abundance of the granulitic clasts in Y82192 and other lunar meteorites are consistent with a view that the lunar meteorites may not have originated from the near side.

INTRODUCTION

All four lunar meteorites, Allan Hills 81005 (e.g., *Warren et al.*, 1983a), Yamato 791197 (*Yanai and Kojima*, 1984), and Yamato 82192 and 82193 (*Yanai et al.*, 1984) are lunar highland regolith breccias. Very few samples of their lunar analogs have been characterized mineralogically for comparison with the lunar meteorites (*Takeda et al.*, 1979, 1985). Recently, Apollo 16 regolith breccias have been characterized in terms of their petrography, chemistry, and noble gas contents by *McKay et al.*, (1986). We have recently compared the mineralogy of pyroxenes in the lunar meteorites with those in regolith breccias 60016 and 60019 (*Takeda et al.*, 1985, 1986), and found that ALH81005 and Y791197 bear a strong resemblance to one another and contain extremely Mg-poor pyroxenes from mare basalts, whereas 60016, 60019, Y82192, and Y82193 appear to have significantly fewer pyroxenes with extremely low Mg contents (*Warren et al.*, 1983a; *Takeda et al.*, 1987a). This difference suggests that the ALH81005 and Y791197 lunar meteorites came from regions close to mare basalt flows, whereas the Yamato-82 meteorites either came from a region far from mare basalt flows such as the Apollo 16 site, or the mare basalts they contain may not have low-Mg pyroxenes. *Takeda et al.* (1987b) noted that a mare basalt clast in 60019 (Ba-2) contains only small amounts of Fe-rich pyroxene.

During the course of mapping a large sawn piece of 60019, *Galindo* (1985) found many large clasts up to 20 × 20 mm in size. This paper discusses mineralogical analyses of three dominant clasts and of matrix samples of 60019. The results are compared with those from lunar meteorite Y82192. We also report analytical transmission electron microscope (AEM) data for the matrix and clasts of 60019 and the lunar meteorites. Major goals of this study are to characterize the dominant lithologies in the Apollo 16 regolith breccias and in the Y-82 lunar meteorites and to use these data to constrain the origins of the lunar meteorites. Comparison of the Apollo 16 regolith breccia 60019 and the Yamato lunar meteorites revealed that the dominant clast-type in 60019 is poikilitic whereas granulitic breccias and glassy matrix breccias are more abundant in Y82192. This difference may be attributed to the abundance of poikilitic rocks resulting from large circular basin-forming events on the near side versus an inferred abundance of granulitic compositions on the far side (*Lindstrom and Lindstrom*, 1986). Further studies of Apollo regolith breccias and granulitic breccias may give a better understanding of the impact sites and the origins of the lunar meteorites.

SAMPLES AND EXPERIMENTAL TECHNIQUES

The new slabs of sample 60019 have been described by *Galindo* (1985) and the Ba-2 clast it contains was investigated by us (*Takeda et al.*, 1987b). A greyish white fine-grained clast (WF-4) was found at the end of ,127. Slab ,124 consists mostly of a white coarse-grained clast (WC-1). The largest clast found in the second slab (,135) is WF-1. A general

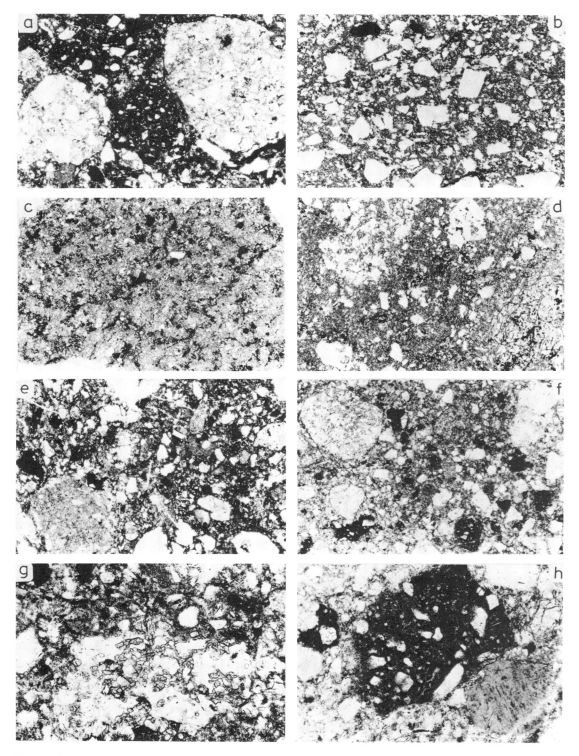

Fig. 1. Photomicrographs of clasts and matrices in regolith breccia 60019 and lunar meteorites Y82192; unpolarized light. (a) A portion of PTS 60019,91. Two poikilitic clasts are shown. Width = 3.3 mm. (b) Plagioclase fragments in 60019,205. Width = 3.3 mm. (c) Poikilitic clast 60019,208. Width = 3.3 mm. (d) Granulitic clast (upper right) and olivine fragments (lower left) in 60019,207. Width = 3.3 mm. (e) Typical matrix portion of 60019,206. Width = 3.3 mm. (f) Overall view of Y82192. Width = 1.3 mm. (g) Granulitic clast GR1 of Y82192. Width = 1.3 mm. (h) Glassy-matrix breccia in Y82191. Width = 1.3 mm.

description of the clast types found on the two slabs is given by *Galindo* (1985). The WF and WC clasts are the most abundant clasts observed on the cut surfaces. Other clasts include anorthosites and very rare basalts. The generic relationship of the samples used is given in Table 1. A new PTS of lunar meteorite Y82192,50-4, was also examined for comparison with 60019.

Chemical analyses were made with a JEOL 733 Super Probe at the Ocean Research Institute of the University of Tokyo, employing the same parameters and method used by *Nakamura and Kushiro* (1970). The bulk compositions of the glassy breccia matrices and the matrices of glassy clasts of 60019 and of lithic clasts in Y82192 were determined by broad-beam (40 micrometers) microprobe analyses (average of 5-10 spots).

We also investigated the glassy matrices of small chips of 60019,205 and ,206, ALH81005, Y791197, and Y82192 with a Hitachi H-600 analytical transmission electron microscope (AEM) equipped with a Kevex 8000Q system, which is capable of analyzing the chemical composition, texture, and atomic arrangements of a region as small as 800 Å. Chips were mounted in resin, sliced, and polished to about 10 micrometers thickness. These samples were then glued to a 3-mm molybdenum AEM grid for support, and thinned in a GATAN ion-thinning machine until perforation occurred.

RESULTS

General Description of 60019

Part of the description of 60019 has been given in our previous paper (*Takeda et al.*, 1987b). *McKay et al.* (1986) reported that the porosity of 60019 is intermediate, glass spheres are rare, agglutinates are absent, and shock features are common. Sample 60019,91 is representative of the matrix and clasts of 60019. This PTS includes six large clasts including four poikilitic matrix breccias with plagioclase fragments (Fig. 1a), one noritic anorthosite, and one large exsolved augite fragment. The largest poikilitic breccia clast (2.6 × 1.1 mm) is similar to one in PTS ,208 that includes very few plagioclase fragments. The three other poikilitic clasts about 1 mm in diameter include plagioclase fragments in micropoikilitic matrices. One large crystalline clast (1.4 × 0.7 mm) is mostly plagioclase with minor pyroxene and olivine. An unusually large augite fragment (2.1 × 0.75 mm) has exsolved low-Ca pyroxene.

PTS ,206 is a representative portion of the matrix without large clasts as observed on the slab surface. Angular to subangular lithic and mineral clasts smaller than about 1 mm in diameter are set in a dark aphanitic glassy groundmass with many glass-lined fractures and glass inclusions. Major classes of lithic fragments include anorthosites, norites, and troctolites; subophitic to intergranular varieties of impact melt rocks; and poikilitic melt rocks as observed in other Apollo 16 regolith breccias (*McKay et al.*, 1986). Breccia fragments are almost entirely of the poikilitic, melt matrix, and granulitic varieties. These breccias represent the WF and WC clast-types.

A granulitic spinel-plagioclase clast (1.9 × 1.3 mm) was found in 60019,206. The plagioclase in this clast is calcic and ranges in composition from An_{98} to An_{95}. The atomic Cr/(Cr+Ti+Al)

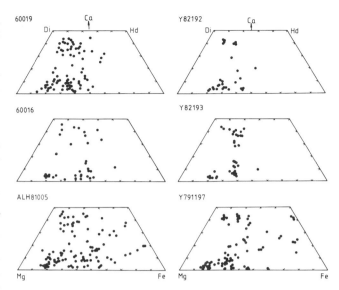

Fig. 2. Pyroxene quadrilaterals for Y791197, ALH81005, Y82192, Y82193, 60016, and 60019. Some data from *Yanai and Kojima* (1984), *Ryder and Ostertag* (1983), and *Takeda et al.* (1979, 1985, 1986, 1987a).

ratios of spinel in the clast vary from 0.033 to 0.10 and the mg number [Mg/(Mg+Fe)] ranges from 0.78 to 0.65. A small grain of olivine with Fa_{21} is present. A similar spinel-plagioclase clast in Y791197 (*Takeda et al.*, 1986) is comparable in mineral composition.

The coarser mineral fragments in ,206 are almost entirely plagioclase except for a single fragment of augite in 60019,91 and one olivine crystal with Fa_{22} in 60019,207. Olivines and

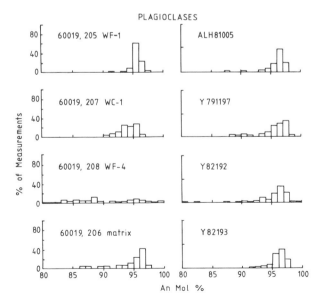

Fig. 3. Histograms of distributions of the An mol % of plagioclase in 60019 and in lunar meteorites.

TABLE 1. Sample numbers used in this study and their generic relations (*Galindo*, 1985).

PTS No.	Sample Name	Chip No. Parent	Chip No. Daughter	Clast-type	Slab No.
205	Feldspathic clastic breccia	181	180	WF-1	135
206	Matrix without large clasts	183	182	matrix WF-1	135
207	Poikilitic breccia	187	186	WC-1	124
208	Poikilitic breccia	191	190	WF-4	127
91	Representative matrix and clasts	-	-	-	-

pyroxenes larger than 0.5 mm are rare in the thin sections examined. The major component in all size ranges is monomineralic plagioclase. Olivines and pyroxenes are the next most abundant minerals. The chemical data on 20-200 micrometer plagioclase and pyroxenes are most representative and are summarized in Figs. 2 and 3. Olivines vary in composition from Fa_{11} to Fa_{35}, with most compositions between Fa_{15} and Fa_{25}.

Poikilitic Clasts in 60019

The most common large (>1 mm) lithic clasts observed in hand specimen on the two new slabs of 60019 are milky-white to greyish-white coarse-grained clasts (WC) and greyish-white fine-grained clasts (WF). Other common clasts are milky white to translucent, fine-grained pure plagioclase clasts and milky-white to translucent, fine-grained, sugary plagioclase clasts spotted with round to oblong, yellow to translucent mafic mineral grains. Clasts WC-1, WF-1, and WF-4 (Table 1) are the three largest clasts found in the new slabs (*Galindo*, 1985). Microscopic textures of these clasts can be related to that of Apollo 16 poikilitic breccias such as 65015. The poikilitic texture implies an optically continuous grain that encloses other grains, but the poikilitic textures in these clasts are not very well developed in some cases. In this paper the term implies that the matrix is recrystallized and that grain orientations tend to be the same in a small region.

WF-1 (,181) is 10 × 15 mm in size. The whole clast is subangular and whitish grey. It is a mixture of 80% sugary or milky-white plagioclase and pyroxene with 20% shiny black opaques (mostly ilmenite) occurring in chains. A few plagioclase inclusions occur along with one metal bleb (0.6 × 0.6 mm) and several smaller "rusty" metal blebs were found.

The PTS 60019,205 of WF-1 shows that the texture of this clast is that of a feldspathic clastic breccia with poorly developed micropoikilitic matrix. It shows a unique texture, shown in Fig. 1b. Very angular plagioclase fragments are abundant and are set in a fine-grained micropoikilitic matrix of orthopyroxene, plagioclase, and olivine (Fig. 1b). The plagioclase sizes range between 0.1 and 0.8 mm in the longest dimension. The range of chemical variation among the plagioclase fragments (Fig. 3) is smaller than that of other fragments in the entire matrix. A few large blebs of metal are present.

The micropoikilitic texture of the matrix in PTS 60019,205 is not as pronounced as that of the other WF and WC clasts. The matrix consists of fine-grained pyroxene, olivine, and plagioclase. The pyroxene compositions are distributed in a manner similar to Apollo 16 poikilitic breccias such as 65015 (Fig. 4). Mineral sizes in the matrix of this clast are larger than in the typical breccia matrix as observed by AEM. Matrix minerals in the clast are up to a few microns in diameter; most are plagioclase. Matrix plagioclase grains are heavily shocked, and some are maskelynitized.

WF-4 (,190) is one of the large WF clasts in the slab (8 × 17 mm). It is an elongated, whitish-grey coherent clast with a distinct boundary. It consists of 70% plagioclase and pyroxene, with 30% opaque chains (mostly ilmenite) making mafic-opaque zones distinguishable. Abundant "rusty" metal blebs and rarer plagioclase inclusions are found. This clast also contains glass-lined zap pits along its west edge (*Galindo*, 1985).

The texture of PTS 60019,208 of WF-4 resembles that of typical low-K Fra Mauro poikilitic breccia 65015 (Fig. 1c). It contains a very few fine-grained subrounded plagioclase fragments ranging from 0.03 to 0.2 mm in diameter. The largest plagioclase fragment is 0.14 × 0.25 mm in size. The matrix has moderately developed poikilitic textures and is rich in opaque minerals (ilmenite and metal). The matrix consists of very fine-grained aggregates of pyroxene, plagioclase, and olivine. The distribution of matrix pyroxene compositions in the pyroxene quadrilateral is similar to that of ,205 (Fig. 4). The variation of the matrix plagioclase compositions is greater than in the other clasts (Figs. 3 and 5); the most An-poor plagioclase is $An_{80}Ab_{19}Or_2$. Olivine compositions are relatively uniform, and range between Fa_{15} and Fa_{26}.

WC-1 (,186) is a large (20 × 20 mm) subrounded, whitish, coherent clast with a sharp boundary. It consists of 80% milky-white crushed sugary plagioclase and some translucent

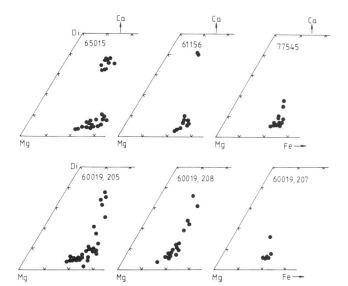

Fig. 4. Pyroxene quadrilaterals of poikilitic clasts in 60019 and other poikilitic rocks (*Warner et al.*, 1979) from Apollo 16 and Apollo 17. The data plotted are compositions of individual pyroxenes of the poikilitic matrices.

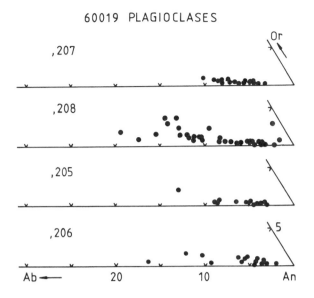

Fig. 5. Plagioclase compositions in 60019. The data plotted are point analyses of plagioclase fragments in the matrix.

Granulitic Clasts in Yamato Lunar Meteorites

The PTS of Y82192 that we examined for the consortium study (*Takeda et al.*, 1987a) included a breccia within a breccia occupying almost half of the PTS, and was therefore not a representative PTS of Y82192. The new PTS for this study (Y82192,50-4, 9.1 × 8.5 mm in size) is texturally similar to 60019,206 (Fig. 1e) and has an even distribution of various sized clasts of different kinds (Fig. 1f). Among clasts larger than 0.5 mm, the most prominent lithologies are a devitrified glassy-matrix breccia (GL1: 1.29 × 0.98 mm), three shocked anorthosites (AN1 to AN3, up to 1.4 × 0.8 mm), three granulitic breccias (GR1 to GR3, Fig. 1g, up to 0.95 × 0.75 mm) and three devitrified glassy-matrix breccia clasts with plagioclase fragments (VT1 to VT3, Fig. 1h, up to 1.3 × 0.6 mm). More detailed statistics of the clast types have been given by *Bischoff et al.* (1986). It is noteworthy that the poikilitic clasts dominant in 60019 are virtually absent. The bulk chemical compositions obtained by microprobe analyses for each of these clasts are given in Table 2 and are plotted in an olivine-silica-anorthosite pseudo-ternary diagram in Fig. 6.

The largest granulitic clast, GR1, commonly includes connected chains of rounded mafic silicate grains up to 0.06 mm in diameter. Mafic silicates are less abundant in GR2 than in GR1, and individual crystals are isolated in a granoblastic plagioclase matrix. Mafic silicate grains in GR3 up to 0.05 mm in diameter are connected to each other, forming complex chain-like textures commonly observed in granulitic breccias. The olivine compositions of GR1, GR2, and GR3 are Fa_{27}, Fa_{29}, and Fa_{18}, and the pyroxene compositions of GR2 and GR3 are $Ca_3Mg_{60}Fe_{37}$ and $Ca_3Mg_{81}Fe_{16}$, respectively. The plagioclase in these clasts ranges in composition from An_{95} to An_{97}. The chemical variations of pyroxenes and plagioclases are included in Figs. 2 and 3. The An versus mg number diagram for these mineral compositions is shown in Fig. 7. A rare lithic clast includes olivine with Fa_{22}, pyroxene with $Ca_4Mg_{77}Fe_{19}$, and plagioclase with An_{97}. These mineral compositions lie in the range of the Mg-suite rocks, but they may be the Mg-rich extension of the ferroan anorthosite trend as was found in Y79117 (*Takeda et al.*, 1986).

pyroxene showing few crystal faces, with about 20% shiny black opaques, predominantly ilmenite, distributed throughout the clast in chains. Less than 1% metal blebs up to 0.25 mm across are scattered throughout the clast; some exhibit rust and rust-staining along plagioclase-pyroxene borders. Also within the clast are plagioclase inclusions less than 0.5 mm across.

The PTS 60019,207 of WC-1 includes large fragments of plagioclase (up to 0.63 × 0.80 mm) and one large olivine fragment (0.92 × 1.11 mm). The matrix is poikilitic and the grain boundaries of the mineral fragments are not sharp, indicating thermal annealing. One granulitic breccia clast contains fine-grained mafic minerals almost identical to the grain-sizes in the matrix, and the bordering minerals are continuous with those in the matrix (Fig. 1d). The pyroxene and olivine compositions in this clast are also similar to those in the matrix.

TABLE 2. Bulk chemical compositions (wt %) of Apollo granulitic breccias and granulitic clasts in Y82192.

	Y82192			67415	67955	76230	79215
	GR1*	GR2*	GR3*	3	56	4	38.1
SiO_2	43.1	41.0	41.25	-	-	-	-
TiO_2	0.26	0.04	0.07	0.32	0.27	0.20	0.3
Al_2O_3	23.7	29.5	29.02	26.0	27.7	27.01	27.9
FeO	6.48	3.26	3.56	4.64	3.84	5.14	3.40
MgO	9.76	4.17	8.12	7.77	7.69	7.63	6.18
CaO	10.27	15.79	14.76	15.1	15.5	15.17	16.0
Na_2O	0.33	0.58	0.28	0.40	0.35	0.616	-
K_2O	0.03	0.02	0.02	0.05	0.06	0.128	-

Data of Apollo granulites are after *Lindstrom and Lindstrom* (1986). Symbol (-) indicates elements that were not analyzed for a particular sample.

*GR1-3 obtained by average of 5-10 broad-beam electron microprobe analyses.

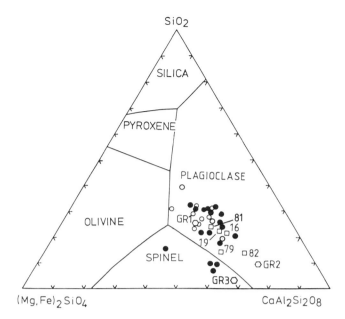

Fig. 6. Compositions of the matrix glasses in glassy-matrix breccias of Y82192 (solid circles) and of the fine-grained matrices of poikilitic clasts in 60019 (open circles) plotted in the silica-olivine plagioclase pseudoternary system. Liquidus phase boundaries are from *Walker et al.* (1972). Each symbol represents an average value of 5-10 individual broad-beam analyses. The bulk chemical compositions of granulitic clasts in Y82192 (open hexagon = GR1, GR2, and GR3) and the other compositions from Table 3 (open square) are included. The first two digits of the lunar meteorite numbers (81, 82, 79) and the last two digits of lunar sample numbers (16, 19) are given for bulk analyses of ALH81005, Y82192, Y791197, 60016, and 60019.

The compositional distributions of the pyroxene fragments in matrices of Y82192,50-4 is almost identical to the old PTS from this sample and to Y82193 (*Takeda et al.,* 1987a). This similarity between the Y82 samples is consistent with the fact that they have the same terrestrial ages (*Eugster,* 1987). The pyroxenes are compared with those of other lunar meteorites and with 60019 in Fig. 2. Histograms of the An contents of plagioclase are also compared in Fig. 3. The lunar meteorites all show similar trends with maximum between An_{96} and An_{98}. Many shocked plagioclases show relatively higher concentrations of FeO and MgO. Molar Fe/(Fe+Mg) of plagioclase is plotted against An content in Fig. 8 to display the contribution of mare plagioclase in the matrices.

Plagioclase grains in 60019 show two chemical trends. One trend, which is identical to others with the constant An value and variable mg number, can be ascribed to the highland plagioclase. The other trend with decreasing An values has been attributed to the mare component (*Takeda et al.,* 1987b). The other plot of (Mg+Fe) versus (Si+Al) indicates that the amount of Mg and Fe increases with increasing (Si+Al) (*Takeda et al.,* 1987a, Fig. 9), and corresponds with shock effects in the plagioclase structure, because it implies formation of fine shock melt glass droplets.

Although the statistics of clast types in other lunar meteorites are given in *Bischoff et al.* (1986), reexamination of the new Y791197 PTS shows that recrystallized matrix breccias containing plagioclase fragments are the other abundant clasts in Y791197. The poikilitic matrix texture is not as well developed as that observed in 60019, and it is very fine-grained. *Warren and Kallemeyn* (1986) designated them as micropoikilitic breccias. In hand specimen, this is grey in color and can be distinguished from the sugary-white anorthositic fragments and from the mafic-mineral containing clasts, which have yellowish spots in their white matrices. Hand specimen observations by the consortium group (*Takeda et al.,* 1986) show that this grey type of clast is another abundant type in Y791197.

Matrix glass compositions of the glassy matrix breccias are plotted in an olivine-silica-anorthite pseudoternary diagram in Fig. 6. The compositions shown in Fig. 6 are distributed in a manner similar to ALH81005, Y791197 (*Takeda et al.,* 1986, Fig. 4) and 60019. The matrix glasses scatter around the bulk-rock compositions and are similar to 60019 and 60016 glassy matrices and to matrix compositions of the poikilitic clasts in 60019.

Microtextures of the Breccia Matrices

AEM studies show that the matrix materials of 60019, Y791197, Y82192 and ALH81005 are mineralogically very similar. These matrices consist of angular mineral fragments

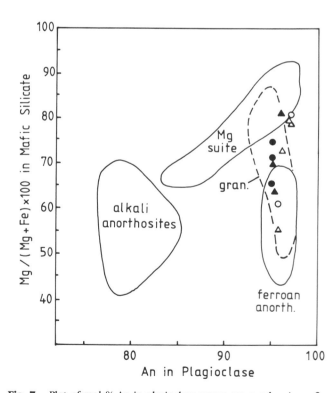

Fig. 7. Plot of mol % An in plagioclase versus mg number in mafic silicates of crystalline clasts in Y82192 (open symbols) and Y791197 (solid symbols). Both dashed (granulites) and solid fields for highland rock types are after *Lindstrom and Lindstrom* (1986). Square = spinel; triangle = olivine; and circle = pyroxene.

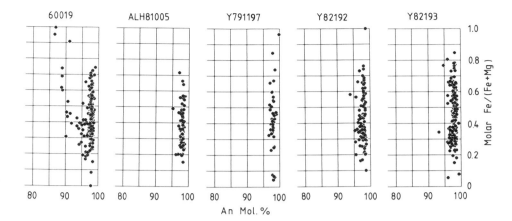

Fig. 8. Variation of molar Fe/(Fe+Mg) in cation numbers per eight oxygens versus An mol % of plagioclase in 60019 and in lunar meteorites.

with interstitial glass. The mineral fragments are micrometer-sized and are composed primarily of plagioclase with small amounts of orthopyroxene and olivine (Fig. 9). Very minor amounts of Fe-Ni metal, troilite, ilmenite, chromite, and whitlockite also occur. All of the interstitial glasses in 60019, Y82192, and ALH81005 are devitrified to submicron-sized fine crystals of plagioclase with small amounts of orthopyroxene and other accessory minerals (Fe-Ni metal, troilite, and ilmenite). In Y791197, however, the interstitial glass is not devitrified. This observation indicates that Y791197 had a metamorphic annealing history different from the other lunar meteorites and 60019. The matrix of lunar breccia 60019,206 contains abundant heavily-shocked plagioclase and orthopyroxene fragments. These fragments show extensive fracturing; some of the plagioclase crystals are maskelynitized. The lunar meteorites also have weak shock deformation, but the degree of shock deformation is less extensive than in 60019,206.

DISCUSSION

Sample 60019 is the largest regolith breccia with large sawn pieces available for study. There are other regolith breccias from other landing sites, but the Apollo 16 breccias include the highland components most comparable to the lunar meteorites.

Because many of the Apollo 16 regolith breccias are compositionally different from the Apollo 16 soils in that they lack an important mafic component present in the soils, *McKay et al.* (1986) concluded that most of the Apollo 16 regolith breccias were not formed from any known Apollo 16 soil.

All four lunar meteorites are small highland regolith breccias ranging from 52 to 27 g (*Yanai et al.*, 1984). For lunar comparisons, it is important to study the variation of clast types within a large specimen. Similarity between 60019 and Y791197 has been suggested on the basis of their glassy matrices and on their content of well-comminuted materials containing ancient regolith developed during the late-stage heavy bombardment of the Moon (*McKay et al.*, 1986; *Takeda et al.*, 1986). Our previous comparisons of the pyroxene compositions recognized both similarities and differences between the lunar meteorites and the Apollo 16 regolith breccias (*Takeda et al.*, 1985, 1986, 1987a,b). On the basis of new data from the dominant clasts and from the AEM observation of breccia matrices of lunar meteorites and 60019, we will discuss similarities and differences between 60019 and the lunar meteorites.

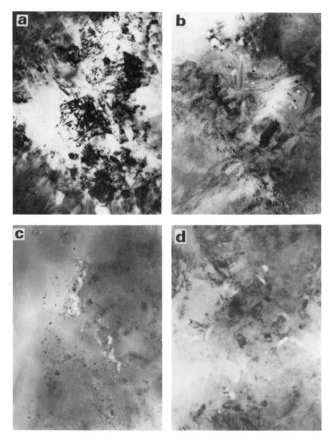

Fig. 9. Transmission electron micrographs of very fine-grained matrices in three lunar meteorites, ALH81005, Y791197, Y82192, and lunar rock 60019,206. Bright field (a) Recrystallized matrix of 60019,206. Width = 1.2 microns. (b) ALH81005 matrix. Width = 2.0 microns. (c) Glassy matrix of Y791197. Width = 1.1 microns. (d) Y82192 matrix. Width = 1.6 microns.

Similarities Between the Anorthositic Highland Regolith Breccias and the Lunar Meteorites

Instead of discussing microscopic similarity of these samples, we stress mineralogical similarity between the breccia components. Fragments of minerals in the breccias are known to represent the minerals in representative lithologies widely distributed around the sampling site. The mineralogic characterstics of pyroxenes in highland regolith breccias have been detailed in our study of 60016 (*Takeda et al.,* 1979). Pyroxene quadrilaterals for 60016, 60019, and the lunar meteorites (Fig. 2) indicate compositions distributed over a wide range, covering almost all known pyroxenes in nonmare pristine rocks (*Warren et al.,* 1983b; *Ryder and Norman,* 1978a,b), except for the Fe-rich pyroxenes of Y791197 and ALH81005.

The distributions of plagioclase compositions among fragments in the matrices of 60019 and in the lunar meteorites (Fig. 3) also show similar histograms. This similarity is not surprising considering the small ranges in An content of all highland rocks. However, the An contents of plagioclases in the poikilitic clasts of 60019,208 and 205 are distinct. The narrow range of the ,205 clast suggests brecciation of a local rock, and the Na-rich histogram of the ,208 clast indicates recrystallization of fine plagioclase laths from the matrix. Olivine compositions among fragments in the matrices of 60019 and in the lunar meteorites also represent the range of known nonmare rocks.

Variations in mg number [Mg/(Mg+Fe)] in mafic minerals versus An in plagioclase for the lunar meteorites are important for comparison with lunar highland rock types. Figure 7, with data from ALH81005 (e.g., *Treiman and Drake,* 1983; *Goodrich et al.,* 1984), indicates that all lunar meteorites show a similar trend intermediate between that of the Mg-suites of the lunar highland crust and that of ferroan anorthosites. Although it is possible that the intermediate clasts are melted mixtures of these two suites, a mixture of the Mg-suite rocks with ferroan anorthosites would lead to plagioclases with distinctly lower An contents than are observed. Lunar granulites described by *Lindstrom and Lindstrom* (1986) also show the same trend as the lunar meteorites.

Bulk chemical compositions of 60016, 60019, and the lunar meteorites given in Table 3 show that they are similar to each other except for FeO. The FeO value ranges from 7 to 3.5 wt %. The data plotted in an olivine-silica-anorthite pseudoternary diagram (Fig. 6) are located close to each other and span the same compositional range as the glassy-matrix breccia clasts, the matrix of the poikilitic clasts in 60019, and the bulk compositions of the granulitic clasts in Y82192. This observation is consistent with a complex history of comminution and recrystallization events. *Lindstrom and Lindstrom* (1986) pointed out that the compositions of granulites and of the lunar meteorite ALH1005 more closely resemble the average composition of lunar highland (*Taylor,* 1982, p.230), as determined by remote sensing, than do those of any other lunar samples. Our results on the Y82192 granulitic clasts are in agreement with their interpretation.

TABLE 3. Bulk chemical compositions of three lunar meteorites and their lunar analogs (wt %).

	Lunar Meteorite			Lunar Regolith Breccias			
	Y791197[*] Bulk rock	ALH81005[†] Bulk rock	Y82192[*] Bulk rock	60016[‡] Bulk	60019[§] Matrix	60019 ,207	60019 ,208
SiO_2	43.14	46.46	43.05	45.5	45.3	43.8	48.4
TiO_2	0.35	0.23	0.22	0.29	0.35	0.35	0.41
Al_2O_3	26.01	25.32	27.78	27.4	26.3	25.9	20.1
Cr_2O_3	0.13	0.12	0.10	0.07	0.10	0.10	0.28
Fe_2O_3	0.04	-	1.47				
FeO	7.02	5.40	3.69	4.8	5.3	3.98	5.15
MnO	0.08	0.076	0.08	0.057	0.06	0.04	0.07
MgO	6.22	7.98	5.64	6.2	6.7	7.18	11.09
CaO	15.33	15.11	16.28	15.2	14.9	15.07	10.94
Na_2O	0.33	0.31	0.45	0.48	0.46	0.58	0.74
K_2O	0.02	0.029	0.02	0.10	0.14	0.18	0.35
$H_2O(-)$	0.10		0.10				
$H_2O(+)$	0.48		0.60				
P_2O_5	0.31		0.51		0.19		
Ni	180	186	130	300	795		
Co	30	20.2	30	27	49		
S	4100		1900		920		

60019,207 and ,208 are analyses of matrices of poikilitic clasts.
[*]Wet chemical analysis by H. Haramura (Geol. Inst., Univ. of Tokyo).
[†]*Palme et al.* (1983).
[‡]*Wänke et al.* (1975).
[§]*Rose et al.* (1975).

Microtextures of the breccia matrices in 60019 and in the lunar meteorites studied by the AEM technique are also mineralogically very similar, except that the Y791197 interstitial glass is not devitrified. This difference, however, may indicate only a different metamorphic annealing history within local ejecta. In the lunar meteorites we could not detect any evidence of shock effects recorded when they were ejected from the Moon. Heavy impact signatures such as maskelynitization of plagioclase were observed only in the matrix of the 60019,205 clast by AEM.

Differences Between 60019 and Lunar Meteorites

Yamato 791197 and ALH81005 (e.g., *Ryder and Ostertag,* 1983; *Warren et al.,* 1983a; *Treiman and Drake,* 1983) both contain extremely Mg-poor pyroxenes, whereas 60016 (*Takeda et al.,* 1979) and 60019 both appear to have significantly fewer pyroxenes with extremely low Mg contents. Our new data from 60019 and Y82192 also show an absence of Fe-rich pyroxene. This difference can be accounted for if (1) Y791197 and ALH81005 came from a region close to mare basalt flows whereas Y82192, Y82193, as well as 60016 and 60019 came from a region far from mare flows, or (2) the minor mare basalt in the latter four samples contains few Fe-rich pyroxenes (*Takeda et al.,* 1986, 1987a).

The possible presence of mare basalt clasts similar to Luna 16 in 60019, and of clasts similar to VLT basalts and a Luna 24 ferrogabbro (24077,13) described by *Schaeffer et al.* (1978) in the lunar meteorites ALH81005 (*Ryder and Ostertag,* 1983; *Treiman and Drake,* 1983; *Warren et al.,* 1983a) and Y791197 (*Lindstrom et al.,* 1986; *Takeda et al.,* 1986), accounts for some of the differences in the distribution of Fe-rich pyroxene composition between 60019 and the lunar meteorites, as discussed by these authors. Although 60019 contains a mare basalt component, the zoning of the pyroxene is so minor that it does not show up in the entire trend. It is of interest to speculate what this difference will tell us about the source basalt flows and the impact sites of the lunar meteorites. The possible absence of pyroxenes from VLT basalts in Y82192 and Y82193 (*Takeda et al.,* 1987a) suggests an impact site for the Yamato-82 lunar meteorites different from the other lunar meteorites. VLT basalt flows near Eimmart at the northern margin of Mare Crisium have been proposed from the source of the VLT clast in ALH81005 (*Treiman and Drake,* 1983).

Granulitic breccias are the dominant clast types in the lunar meteorites ALH81005, Y791197, and Y82192/193 (*Ryder and Ostertag,* 1983; *Ostertag et al.,* 1986; *Bischoff et al.,* 1986). Although our previous PTS of Y82192 did not contain large granulitic clasts, our observations from the new PTS of Y82192 support the above statement. Among the four largest lithic clasts, three were granulitic.

Lunar granulites are brecciaas with metamorphic textures. They are not pristine, having been contaminated with meteoritic siderophiles. Nonetheless, they constitute an abundant class of highlands rocks that are important because they are both ancient and KREEP-free. *Lindstrom and Lindstrom* (1986) suggested that granulites may represent our best samples of the early lunar crust. Lunar granulites can be divided into ferroan and magnesian groups, and each group can be distinguished on the basis of mineral compositions and REE concentrations. Textural and compositional data show that some granulites (especially 67215 and 67415) may be derived from distinct anorthositic norite precursors, while some other granulites (79215, 78155) are clearly polymict (*Lindstrom and Lindstrom,* 1986). Even the polymict granulites probably had anorthositic norites as their dominant precursors, and not anorthosites, norites, or troctolites.

The chemical characteristics of granulites indicate that they are possibly derived from a KREEP-free or KREEP-poor region of the lunar crust. The lack of KREEP in granulites has a very important bearing on the regional evolution of the primordial lunar crust. If the intermediate members observed in Fig. 7 (dotted field) were not mixtures between that of the Mg-rich suites of the lunar highland crust and the ferroan anorthosite, then the Mg-rich extension of the trend may represent early-crystallized anorthositic "rockbergs" in a magma ocean which was crystallizing the Mg-rich portion of the Mg-suite.

Microscopic observation and microprobe analyses of three thin sections (,205, ,207, ,208) of the WC and WF clasts show that they are crystalline matrix breccias containing plagioclase fragments. The main difference between the WF and WC clasts lies in the size of the plagioclase fragments. The matrices have poikiloblastic textures that consist of very fine-grained pyroxenes with minor plagioclase and olivines. The chemical compositions of the pyroxenes are similar to those in large rocks among the Apollo 16 samples. These crystalline-matrix breccias are rather rare in Y791197, and otherwise similar breccias do not have well-developed poikilitic textures (*Takeda et al.,* 1986).

Poikilitic impact-melt breccias represented by the WF and WC clasts in 60019 are among the abundant clast types at the Apollo 16 site (*Ryder and Seymour,* 1982; *Stöffler et al.,* 1981). They may represent a local impact melt sheet, possibly from Nectaris as suggested by *Spudis* (1984). The absence of such clasts in the Y82 meteorites suggests that the Y82 source may be far from any large circular mare basins. This conclusion is in agreement with the proposed farside origin of the lunar meteorites. The predominance of the granulitic breccia clasts in the Y82 meteorites is in agreement with the proposal that the crustal composition of the far side may be rich in granulitic rock types (*Lindstrom and Lindstrom,* 1986).

The compositional similarity of clasts in the lunar polymict meteorites to monomict granulites suggests that anorthositic norites similar to the plutonic precursors of the granulites are the dominant source rocks of the lunar meteorites. As suggested by *Lindstrom and Lindstrom* (1986), plutonic anorthositic norites may be more abundant in the early lunar crust, especially of the far side, than is implied by the Apollo collection. The evidence that the dominant rock types of the lunar far side may be different from the near side, as revealed by studies of the lunar meteorites, suggests that exploration of the far side by remote-sensing satellites will very likely alter our understanding of the entire Moon.

CONCLUSIONS

(1) The absence in lunar meteorites of poikilitic clasts that are dominant in the Apollo 16 breccias, which are proposed to have been derived from large circular basin-forming impacts, suggests that the lunar meteorites may have originated far from any large circular mare basins. (2) This conclusion is in general agreement with the proposed farside origin for the lunar meteorites, more specifically because large circular mare basins are less numerous on the far side. (3) The predominance of granulitic breccia clasts in Y82192 supports the proposal that the crustal composition of the far side may be rich in the granulitic rock types (*Lindstrom and Lindstrom,* 1986).

Acknowledgments. We are indebted to the U.S. Meteorite Working Group, Dr. K. Yanai, and Mr. H. Kojima, and the National Institute of Polar Research (NIPR) for samples, and to the Ministry of Education for the financial support for this study. We thank D. S. McKay, P. H. Warren, G. Ryder, D. Bogard, M. Lindstrom, T. Ishii, and C. Galindo for helpful suggestions and discussions, and O. Tachikawa, M. Hatano, and M. Sugawara for technical assistance. D. S. McKay kindly suggested to us the similarity of 60019 to lunar meteorites. This work was also supported in part by funds from Cooperative Program (No. 84134) provided by Ocean Research Institute, University of Tokyo. We thank S. B. Simon, D. T. Vaniman, and D. S. McKay for critical reading of the manuscript.

REFERENCES

Bischoff A., Palme H., Spettel B., Stöffler D., Wänke H., and Ostertag R. (1986) Yamato 82192 and 82193: Two other meteorites of lunar origin (abstract). In *Eleventh Symp. Antarctic Meteorites,* pp. 34-36. Natl. Inst. Polar Res., Tokyo.

Eugster O. (1987) Lunar meteorites Y-82192 and Y-82193: Identical Cosmic-ray exposure history and terrestrial age (abstract). In *Lunar and Planetary Science XVIII,* pp. 273-274. Lunar and Planetary Institute, Houston.

Galindo C. (1985) Regolith breccia 60019. *Lunar Sample Newsletter No. 43,* pp. 23-37. NASA/JSC, Houston.

Goodrich C. A., Taylor G. J., Keil K., Boynton W. V., and Hill D. H. (1984) Petrology and chemistry of hyperferroan anorthosites and other clasts from lunar meteorite ALHA81005. *Proc. Lunar Planet. Sci. Conf. 15th,* in *J. Geophys. Res., 89,* C87-C94.

Lindstrom M. M., Lindstrom D. J., Korotev R. L., and Haskin L. A. (1986) Lunar meteorites Yamato 791197: A polymict anorthositic norite breccia. *Mem. Natl. Inst. Polar Res., Spec. Issue 41,* pp. 58-75. Natl. Inst. Polar Res., Tokyo.

Lindstrom M. M. and Lindstrom D. J. (1986) Lunar granulites and their precursor anorthositic norites of the early lunar crust. *Proc. Lunar Planet. Sci. Conf. 16th,* in *J. Geophys. Res., 91,* D263-D276.

McKay D. S., Bogard D. D., Morris R. V., Korotev R. L., Johnson P., and Wentworth S. J. (1986) Apollo 16 regolith breccias: Characterization and evidence for early formation in the megaregolith. *Proc. Lunar Planet. Sci. Conf. 16th,* in *J. Geophys. Res., 91,* D277-D303.

Nakamura Y. and Kushiro I. (1970) Composition relations of coexisting orthopyroxene, pigeonite and augite in a tholeiitic andesite from Hakone volcano. *Contrib. Mineral. Petrol., 26,* 265-275.

Ostertag R., Stöffler D., Bischoff A., Palme H., Schultz L., Spettel B., Weber H., Weckwerth G., and Wänke H. (1986) Lunar meteorite Yamato-791197: Petrography, shock history and chemical composition. *Mem. Natl. Inst. Polar Res., Spec. Issue 41,* pp. 17-44. Natl. Inst. Polar Res., Tokyo.

Palme H., Spettel B., Weckwerth G., and Wänke H. (1983) Antarctic meteorite ALHA81005, a piece from the ancient lunar crust. *Geophys. Res. Lett, 10,* 817-820.

Rose H. J., Jr., Baedecker P. A., Berman S., Christian R. P., Dwornik E. J., Finkelman R. B., and Schnepfe M. M. (1975) Chemical composition of rocks and soils returned by the Apollo 15, 16, and 17 missions. *Proc. Lunar Sci. Conf. 6th,* 1363-1373.

Ryder G. and Norman M. (1978a) *Catalog of Pristine Non-mare Materials, Part 1, Non-anorthosites.* JSC 14565, NASA Johnson Space Center, Houston. 146 pp.

Ryder G. and Norman M. (1978b) *Catalog of Pristine Non-mare Materials, part 2, Anorthosites.* JSC 14603, NASA Johnson Space Center, Houston. 86 pp.

Ryder G. and Ostertag R. (1983) ALHA 81005: Moon, Mars, petrography, and Giordano Bruno. *Geophys. Res. Lett., 10,* 791-794.

Ryder G. and Seymour R. (1982) Chemistry of Apollo 16 impact melts: Numerous melt sheets, lunar cratering history, and the Cayley-Descartes distinction (abstract). In *Lunar and Planetary Science XIII,* pp. 673-674. Lunar and Planetary Institute, Houston.

Schaeffer O. A., Bence A. E., Eichhorn G., Papike J. J., and Vaniman D. T. (1978) ^{39}Ar-^{40}Ar and petrologic study of Luna 24 sample 24077,13 and 24077,63. *Proc. Lunar Planet. Sci. Conf. 9th,* 2363-2373.

Spudis P. D. (1984) Apollo 16 site geology and impact melts: Implications for geologic history of the lunar highlands. *Proc. Lunar Planet. Sci. Conf. 15th,* in *J. Geophys. Res., 89,* C95-C107.

Stöffler D., Ostertag R., Reimold W. U., Borchardt R., Malley J., and Rehfeldt A. (1981) Distribution and provenance of lunar highlands rock types at North Ray Crater, Apollo 16. *Proc. Lunar Planet. Sci. 12B,* 185-207.

Takeda H., Miyamoto M., and Ishii T. (1979) Pyroxenes in early crustal cumulates found in achondrites and lunar highland rocks. *Proc. Lunar Planet. Sci. Conf. 10th,* 1095-1107.

Takeda H., Miyamoto M., and Ishii T. (1985) Mineralogy of lunar glassy matrix breccia 60019, a lunar analog of lunar meteorites. *The 18th ISAS Lunar Planet. Symp.,* pp. 38-39. Inst. Space Aeron. Sci., Tokyo.

Takeda H., Mori H., and Tagai T. (1986) Mineralogy of Antarctic lunar meteorites and differentiated products of the lunar crust. *Mem. Natl. Inst. Polar Res., Spec. Issue 41,* pp. 45-57. Natl. Inst. Polar Res., Tokyo.

Takeda H., Mori H., and Tagai T. (1987a) Mineralogy of lunar meteorites, Yamato-82192 and -82193 with reference to breccias in a breccia. *Mem. Natl. Inst. Polar Res., Spec. Issue 46,* pp. 43-55. Natl. Inst. Polar Res., Tokyo.

Takeda H., Miyamoto M., Galindo C., and Ishii T. (1987b) Mineralogy of basaltic clast in lunar highland regolith breccia 60019. *Proc. Lunar Planet. Sci. Conf. 16th,* in *J. Geophys. Res., 92,* E462-E470.

Taylor S. R. (1982) *Planetary Science: A Lunar Perspective.* Lunar and Planetary Institute, Houston. 481 pp.

Treiman A. H. and Drake M. J. (1983) Origin of lunar meteorite ALHA 81005: Clues from the presence of terrae clasts and a very low-titanium mare basalt clast. *Geophys. Res., Lett., 10,* 783-786.

Wänke H., Palme H., Baddenhausen H., Dreibus G., Jagoutz E., Kruse H., Palme C., Spettel B., Teschke F., and Thacker R. (1975) New data on the chemistry of lunar samples: primary matter in the lunar highlands and the bulk composition of the moon. *Proc. Lunar Sci. Conf. 6th,* 1313-1340.

Walker D., Longhi J., and Hays J. F. (1972) Experimental petrology and origin of Fra Mauro rocks and soil. *Proc. Lunar Sci. Conf. 3rd,* 797-818.

Warner R. D., Taylor G. J., and Keil K. (1979) Petrology of crystalline matrix breccias from Apollo 17 rake samples. *Proc. Lunar Sci. Conf. 8th,* 1987-2006.

Warren P. H. and Kallemeyn G. W. (1986) Geochemistry of lunar meteorite Yamato-791197: comparison with ALHA81005 and other lunar samples. *Mem. Natl. Inst. Polar Res., Spec. Issue 41*, pp. 3-16. Natl. Inst. Polar Res., Tokyo.

Warren P. H., Taylor G. J., and Keil K. (1983a) Regolith breccia Allan Hills A81005: Evidence of lunar origin and petrography of pristine and nonpristine clasts. *Geophys. Res. Lett., 10*, 779-782.

Warren P. H., Taylor G. J., and Keil K. (1983b) Seventh foray: Whitlockite-rich lithologies, a diopside-bearing troctolitic anorthosite, ferroan anorthosite, and KREEP. *Proc. Lunar Planet. Sci. Conf. 14th*, in *J. Geophys. Res., 88*, B151-B164.

Yanai K. and Kojima H. (1984) Yamato-791197: A lunar meteorite in the Japanese collection of Antarctic meteorites. *Mem. Natl. Inst. Polar Res., Spec. Issue 35*, pp. 18-34. Natl. Inst. Polar Res., Tokyo.

Yanai K., Kojima H., and Katsushima T. (1984) Lunar meteorites in Japanese collection of the Yamato meteorites (abstract). *Meteoritics, 19*, 342-343.

Olivine Vitrophyres: A Nonpristine High-Mg Component in Lunar Breccia 14321

John W. Shervais
Department of Geological Sciences, University of Tennessee, Knoxville, TN 37996 and Department of Geology, University of South Carolina, Columbia, SC 29280

Lawrence A. Taylor
Department of Geological Sciences, University of Tennessee, Knoxville, TN 37996

Marilyn M. Lindstrom*
Department of Earth and Planetary Sciences, Washington University, St. Louis, MO 63130

*Now at NASA Johnson Space Center, Code SN2, Houston, TX 77058

Geochemical modeling of the highlands crust has established the need for three principal components: anorthosite, KREEP, and a high-Mg component. The high-Mg component is most elusive because its presence can only be established indirectly; no pristine rocks of suitable compositions are known. Olivine vitrophyre clasts from breccia 14321 have high bulk MgO (~20 wt %) and 100 × Mg/(Mg+Fe) = 78, which is about the same as the cryptic Mg-component at Apollo 16. The 14321 olivine vitrophyres also have high incompatible element concentrations (~0.5 × high-K KREEP), with REE ratios similar to KREEP (e.g., La/Lu = 22) and CaO/Al$_2$O$_3$ ratios similar to highland melt rocks. The high siderophile element concentrations of these rocks show that they are impact-generated melt rocks. Because impact melts form by total or near total melting of the target, their compositions represent the average composition of the crust, or of some portion of the crust. The high modal olivine content of the vitrophyres is not the result of accumulation, as almost all of the modal olivine has skeletal morphologies indicating rapid crystallization. Thus, the crust represented by this melt sheet is extremely mafic compared to the crust sampled by lunar surface geochemistry. Least squares modeling of the mafic component in this crust shows that it is not the same as that found at Apollo 16 (MAF, SCCRV, Primary matter), but must have a higher Mg/Fe ratio (Mg# = 84-85) and a high MgO concentration (~42 wt %). We suggest that this component is not an isolated rock type, but represents the mafic portion of some prevalent lunar rock type, e.g., Mg-rich troctolite. The mixing calculations show that the Mg-suite rocks comprise about 30% of the crust; alkali-rich lithologies are dominant (~60% of crust) in the form of KREEPy norites, alkali anorthosites, and granites. Olivine vitrophyres are higher in MgO than supposed Imbrium basin impact melts (15445 and 15455) and cannot represent melt rocks formed during this event. *Pieters* (1982) has shown that the central uplift in Copernicus (95 km diameter) consists largely of olivine derived from less than 10 km depth. Pre-Imbrium craters of similar size are common in the Fra Mauro region. Two of these craters ("northwest" and Fra Mauro) are less than one km from the Apollo 14 landing site, and may be the source area of the 14321 olivine vitrophyres.

INTRODUCTION

Geochemical modeling has shown that soils and breccias in the lunar highlands consist of three principal components: anorthosite, KREEP, and a high-Mg component (e.g., *Wänke et al.*, 1976; *Wasson et al.*, 1977; *Korotev et al.*, 1980). In addition, many also contain small contributions from meteorites and mare basalts. The anorthosite component used in the modeling is usually taken to be the ferroan anorthosite that characterizes the Apollo 16 site. While plagioclase-rich rocks are the dominant pristine rock type at the other highland sites, these vary from anorthosite to troctolite, and at the western sites the anorthosites are sodic. KREEP is a cryptic component that dominates incompatible element concentrations in soils and breccias and only rarely occurs as a pristine rock. Major element concentrations for KREEP are poorly constrained, but are generally thought to be noritic (*Warren and Wasson*, 1979).

The high-Mg component has been referred to variously as "primary matter" (*Wänke et al.*, 1976, 1977), SCCRV (*Boynton et al.*, 1975; *Wasson et al.*, 1977), and MAF (*Korotev et al.*, 1980). Like KREEP, the high-Mg component is also cryptic: no pristine rock types with sufficiently high Mg are known except dunite and lherzolite. However, these rocks are rare in the returned sample suites and they have Mg/Fe ratios that are too high by a factor of 2 (*Korotev et al.*, 1980).

We report here on a nonpristine rock type that may represent this illusory Mg-rich component—the olivine vitrophyres of lunar breccia 14321. These rocks, which have been described previously from thin sections (*Allen et al.*, 1979), occur as large (6.5 × 4.5 cm) clasts in lunar breccia 14321 (*Shervais and Taylor*, 1984; *Shervais et al.*, 1984a, 1985a). We present the first bulk-rock major and trace element analyses of this rock type by INAA, with mineral and glass compositions by electron microprobe.

These data show that the 14321 olivine vitrophyres are impact melt rocks, as inferred by *Allen et al.* (1979), and that they probably formed by the total or near total melting of mafic-rich crustal lithologies in the Fra Mauro region. These lithologies comprise the mafic-rich portions of both Mg-rich suite and alkali anorthosite suite plutons. The 14321 olivine vitrophyres represent the average composition of this crustal material and include much of the "anorthosite" component isolated by previous models.

PREVIOUS WORK

Lunar Breccia 14321

Lunar breccia 14321 (*alias* Big Bertha) is the largest sample returned from the Apollo 14 site and, at 9 kg, the third largest sample returned from the Moon. It was collected from the rim of Cone Crater and represents Fra Mauro formation excavated from a depth of about 60 m below the surface (*Swann et al.*, 1977). The Fra Mauro formation was originally thought to consist of primary ejecta from the Imbrium basin, but based on comparative studies of terrestrial impact events (e.g., *Hörz et al.*, 1983), it now seems more likely that the Fra Mauro breccias consist dominantly of locally derived material entrained by secondary impacts of the primary Imbrium ejecta (*Hawke and Head*, 1977). This conclusion has important implications for the composition of the lunar crust at the Apollo 14 site.

A detailed petrographic study of breccia 14321 by *Grieve et al.* (1975) identified three principal lithologic components: (1) rounded fragments of pre-existing microbreccia, comprising mineral and rock fragments in a dark, fine-grained matrix; (2) igneous rock fragments of basaltic composition, ranging in texture from glassy to ophitic; and (3) a light-colored matrix that binds the clastic components.

Grieve et al. (1975) found that mineral fragments in the matrix have compositons similar to minerals in the mare basalts, suggesting that the matrix is dominated by comminuted basalt fragments, with a smaller component of dissaggregated microbreccia. This interpretation is supported in general by chemical mixing models (*Lindstrom et al.*, 1972; *Duncan et al.*, 1975).

Olivine Vitrophyres

Olivine vitrophyres were first discovered and named by *Allen et al.* (1979) during a petrographic survey of 16 thin sections from breccia 14321. They discovered a total of 21 olivine vitrophyre clasts ranging from 0.1-14 square mm (*Allen et al.*, 1979). Bulk compositions of olivine vitrophyres were determined by modal reconstruction using average mineral and glass compositions from electron microprobe analyses. Olivine vitrophyres are distinguished from mare basalt clasts by their high modal olivine, high MgO concentrations, and low CaO/Al_2O_3 ratios. *Allen et al.* (1979) concluded that the olivine vitrophyres were impact melts of Mg-enriched highland crust and attributed their high apparent Mg/Fe partition coefficients (Kd ≅ 0.5) to olivine accumulation in the melt.

Grover et al. (1980) performed piston cylinder experiments at high pressures on an average olivine vitrophyre bulk composition to determine whether or not it could be derived as an endogenous melt of the lunar interior. They found that the average olivine vitrophyre did not become multiply saturated until pressures of 25-30 kb (470-570 km). *Grover et al.* (1980) concluded that olivine vitrophyres probably did not represent endogenous melts but formed near the surface as suggested by *Allen et al.* (1979). *Grover et al.* (1980) also determined equilibrium olivine/liquid partition coefficients experimentally for an olivine vitrophyre bulk composition. They found K_d = 0.30-0.35 and agreed with *Allen et al.* (1979) that the high apparent K_d's for Fe/Mg partitioning resulted from olivine accumulation.

METHODS

In 1983, while the senior author was a research associate at the University of Tennessee, a new consortium study of breccia 14321 was instituted (*Shervais and Taylor*, 1984). This study was to characterize the diverse assemblage of clasts present and to place these clasts within the context of our current petrologic and geochemical understanding of lunar rocks. A breccia guidebook to sample 14321 was published that documents the size and distribution of clasts (*Shervais et al.*, 1984). We have already published our results on pristine highland lithologies (*Lindstrom et al.*, 1984) and mare basalts (*Shervais et al.*, 1985b) that occur as clasts in breccia 14321. Further work on the microbreccia clasts and matrix is in progress (*Shervais et al.*, 1985c; *Knapp*, 1986).

Previous studies of breccia 14321 were limited to thin sections or small slabs that did not represent the bulk sample (e.g., *Grieve et al.*, 1975; *Duncan et al.*, 1975; *Allen et al.*, 1979). As part of our consortium effort, all remaining subsamples of breccia 14321 greater than 10 g were examined and mapped, including two new faces formed by a saw cut through the largest subsample (*Shervais et al.*, 1984a). Three of the largest clasts exposed on these faces are olivine vitrophyres that range up to 6.5 × 4.5 cm. Additional smaller clasts of olivine vitrophyre are distributed throughout the matrix (*Shervais et al.*, 1984a).

Representative chips of the three largest olivine vitrophyre clasts exposed on the new west face of 14321 ,37 (DA-1, DA-3, and DA-4) were extracted and split into two portions: one for whole-rock analysis and the other to make polished probe mounts (*Shervais and Taylor*, 1984). Most of these clasts remain in place and are available for further study. In addition, mirror images of these clasts may be found on the east faces of 14321 ,1082 ,1083 and ,1084 (*Shervais et al.*, 1984a).

Whole-rock samples were analyzed for major and trace elements by INAA at Washington University based on the procedures of *Lindstrom* (1984), and data were reduced using the TEABAGS program of *Lindstrom and Korotev* (1982). Analytical uncertainties based on counting rate at the 1-σ level are as follows: 1-2% Al, Fe, Na, Sc, Cr, Mn, Co, La, Sm, Eu; 3-5% Ca, Ce, Tb, Yb, Lu, Hf, Ta, Th; 5-15% Ti, Mg, Sr, Cs, Ba, Nd, U; 15-25% K, V, Ni, Rb, Zr, Ir. The detection limit for Ir is 2 ppb.

Mineral analyses were performed using the automated MAC 400S electron microprobe at the University of Tennessee. Standards, operating procedures, and data reduction techniques are those normally used in this lab (*Shervais et al.*, 1984b). Major element analyses of residual glass in the vitrophyres were made by broad beam (~ 20 microns) electron microprobe analysis with a 30 nA beam current to minimize sodium loss.

Fig. 1. Mug shot of 14321 ,37 (south face), showing new surface produced by slabbing in November, 1983. The three largest clasts exposed on this face are olivine vitrophyres, labeled DA-1, DA-3, and DA-4. Note the small light-colored clast near the pointed end of DA-3.

PETROGRAPHY AND PHASE CHEMISTRY

The distribution of olivine vitrophyre clasts in breccia 14321 is shown in Fig. 1, a photograph of the west face of sample ,37. These clasts are mapped as Dark Aphanitic (DA) because in hand-sample the olivine vitrophyres have no discernable texture and are dark greenish-brown or olive-drab in color. Three clasts from this face (DA-1, DA-3, and DA-4) were sampled (Fig. 1). DA-3 is the only true vitrophyre, consisting of about 30% olivine microphenocrysts set in a dark reddish-brown glass with traces of Fe metal (Fig. 2a). Sample DA-1 and DA-4 have similar olivine microphenocryts set in a glassy to variolitic matrix of plagioclase, low-Ca pyroxene, augite, ilmenite, Fe-metal, and glass (Figs. 2b,c). The variolitic textures imply that these clasts experienced a slower crystallization rate than DA-3 after an intitial quenching. Clast DA-1 also contains large, partially resorbed xenocrysts of olivine, plagioclase, and maskelynite (Fig. 2b), and DA-3 contains rare interclasts of anorthosite or troctolite (seen in hand-sample).

Mineral Chemistry

Olivine crystals are divided into three groups based on size and morphology: (1) microphenocrysts, (2) phenocrysts, and (3) xenocrysts. Quenching of all three samples is indicated by the size of the olivine microphenocrysts (20-100 microns) and their characteristic skeletal morphologies (Fig. 3a). Olivine microphenocrysts range in composition from Fo_{65} to Fo_{87}, but most cluster in the narrow range of Fo_{79} to Fo_{85} (Table 1). The average composition of olivine microphenocryst in DA-3, based on 50 random analyses, is Fo_{83}. The average compositions of olivine microphenocrysts in DA-1 and DA-4 are slightly lower (Fo_{81} in DA-1; Fo_{80} in DA-4), and have even more restricted compositional ranges ($Fo_{77.5}$ to Fo_{83}).

Larger euhedral to subhedral olivine grains (100-200 microns), some with internal skeletal growth structures, have cores as magnesian as Fo_{89} and zone outward to rims of Fo_{86} to Fo_{84} (Fig. 3a). We interpret these larger grains to be early-

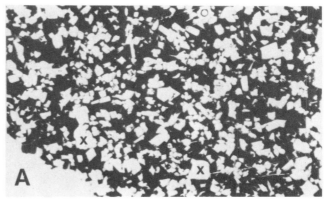

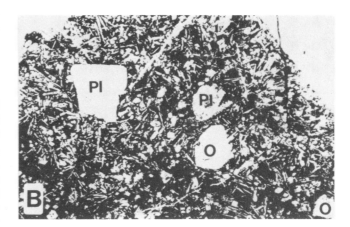

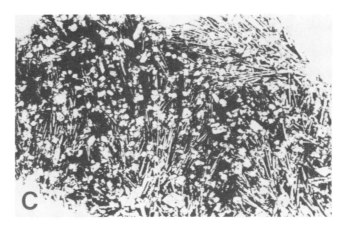

Fig. 2. Photomicrographs of olivine vitrophyre clasts from breccia 14321, all at same scale (FOV = 2.8 mm). **(a)** DA-3: olivine phenocrysts and microphenocrysts in glass; **(b)** DA-1: olivine microphenocrysts in quenched groundmass of plagioclase, pyroxene, and glass; large rounded crystals are xenocrysts of olivine (o) and feldspar (p); **(c)** DA-4: olivine microphenocrysts in quenched groundmass of plagioclase, pyroxene, and glass; note variolitic texture.

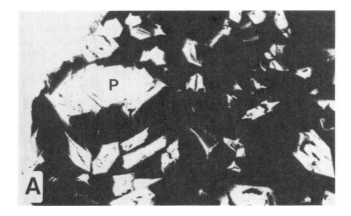

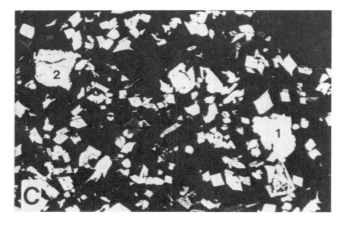

Fig. 3. Photomicrographs of olivine crystals in vitrophyres DA-3 and DA-1. (a) Olivine phenocryst (b) and microphenocrysts in glass, DA-3. Note the small size and delicate, skeletal morphologies of the olivine microphenocrysts. The slightly larger phenocryst has morphology similar to the microphenocrysts. FOV = 0.7 mm. (b) Olivine xenocrysts (X1, X2) surrounded by smaller microphenocrysts and phenocrysts in DA-3. Xenocryst X1 has an Fo_{87} core, Fo_{85} rim; xenocryst X2 has an Fo_{66} core, Fo_{73} rim, suggesting back-reaction with a more magnesian melt. FOV = 1.4 mm. (c) Large xenocryst of magnesian olivine with an Fo_{89} core, Fo_{84} rim in DA-1. FOV = 0.7 mm.

formed phenocrysts that did not re-equilibrate with the evolving residual melt, but zoned outwards to more Fe-rich compositions. The presence of large, magnesian phenocrysts in addition to the more common microphenocrysts implies a two-stage crystallization history.

Olivine also occurs as xenocrysts, along with plagioclase and maskelynite (Figs. 3b,c). The xenocrysts range in size from 0.2 to 0.6 mm and are generally subhedral with resorbed grain boundaries. Plagioclase and maskelynite are commonly rounded or embayed, while olivine xenocrysts have subhedral or irregular grain shapes with frittered grain boundaries. Olivine xenocrysts generally have homogenous core compositions of Fo_{87} to Fo_{90} and thin rims Fo_{84} to Fo_{86} in composition. We found one olivine xenocryst, however, with a core composition of Fo_{66} and a rim of Fo_{73}. This suggests back-reaction of an Fe-rich xenocryst with a more magnesian melt. Plagioclase and maskelynite xenocrysts range in composition from An_{83} to An_{95} and are not zoned (Table 2).

The variolitic matrices of DA-1 and DA-4 consist of sheaf-like arrays of slender plagioclase crystals (An_{79}, Ab_{19}, Or_2) that engulf the olivine phenocrysts and microphenocrysts, and are intergrown with granular aggregates of low-Ca pyroxene ($Wo_{3.5-7.5}$, En_{72}), augite (Wo_{24}, En_{44}), ilmenite, and Fe-metal. Minor residual glass is found between plagioclase laths and in nonvariolitic patches.

Residual Glass Chemistry

Residual glass was analyzed in DA-3, the only true vitrophyre. Twenty random spots were analyzed with a 20 micron beam and the results averaged (Table 3). The residual glass is alkali rich, with $Na_2O \cong 1\%$ and $K_2O \cong 0.4\%$. The Mg# [=100 · Mg/(Mg+Fe)] is 64 (higher than most mare basalts) and the CaO/Al_2O_3 ratio = 0.69, which is typical of highland melt rocks (*Vaniman and Papike,* 1980). The DA-3 glass composition is compared, in Table 3, to alkali-rich mare vitrophyres found in breccia 14321. The mare vitrophyres have SiO_2, TiO_2, MgO, and Na_2O in the same range as the olivine vitrophyres, but they are lower in Al_2O_3 and K_2O, and higher in FeO. The mare vitrophyres have Mg#s (43-45) and CaO/Al_2O_3 ratios (0.86) that are typical of the aluminous basalts found at the Apollo 14 site. Thus, the major element chemistry of the residual glass in DA-3 clearly distinguishes it from mare vitrophyres such as those described by *Shervais et al.* (1985b) and confirm its highland affinities.

As noted by *Allen et al.* (1979), residual glass in the olivine vitrophyres is compositionally variable due to the quench growth of olivine. This is shown in Table 3 by representative glass analyses that span the range of observed compositions. When plotted in the Ol-An-Q ternary of *Walker et al.* (1973), the residual glass data form a field that extends from near the Ol-Px-Plg reaction point into the plagioclase stability field and away from the olivine apex (Fig. 4). Because plagioclase is not observed as a liquidus or near liquidus phase in DA-3, these data are consistent with metastable expansion of the olivine phase volume due to suppression of plagioclase nucleation during rapid crystallization, as shown by *Taylor and Nabelek* (1979), and *Nabelek et al.* (1978).

TABLE 1. Olivines in olivine vitrophyre clast DA-3 from breccia 14321 analyzed by electron microprobe.

	1	2	3	4	5	6	7	8
SiO_2	37.67	38.84	38.84	39.67	39.14	39.29	36.36	37.89
FeO	24.64	19.73	16.14	14.25	12.80	14.04	33.12	25.55
MnO	0.20	0.13	0.11	0.09	0.05	0.17	0.26	0.21
MgO	37.87	42.20	44.24	46.53	48.49	47.09	30.68	36.61
CaO	0.16	0.09	0.15	0.11	0.07	0.12	0.13	0.23
Sum	100.54	100.99	99.47	100.65	100.55	100.71	100.55	100.57
Cations per four oxygens								
Si	0.986	0.987	0.987	0.987	0.970	0.977	0.990	0.991
Fe	0.539	0.419	0.342	0.296	0.265	0.292	0.754	0.559
Mn	0.003	0.002	0.002	0.001	0.000	0.002	0.005	0.002
Mg	1.477	1.599	1.675	1.725	1.791	1.746	1.245	1.427
Ca	0.003	0.002	0.003	0.002	0.001	0.002	0.003	0.006
Sum	3.009	3.009	3.009	3.011	3.027	3.019	3.001	2.996
% Fo	73.3	79.2	83.0	85.3	87.1	85.6	62.3	71.9

Columns 1-4 are quench olivines with skeletal habits; columns 5-6 are the core and rim of a large relict xenocryst; columns 7-8 are the core and rim of a smaller relict xenocryst that is more Fe-rich than the quench olivines.

WHOLE-ROCK GEOCHEMISTRY

Whole-rock major and trace element analyses of DA-1, DA-3, and DA-4 by INAA are presented in Table 4. The major element data are complete only for DA-3; however, for the elements analyzed, all three are virtually identical. Also shown for comparison are major element analyses of DA-3 by modal reconstruction and the Average Olivine Vitrophyre (AOV) of *Allen et al.* (1979).

The major element data for DA-3 confirm the highland affinities of the olivine vitrophyres, with high Mg# (78) and low CaO/Al_2O_3 (0.69). Mass balance calculations using an average olivine microphenocryst and the average residual glass show that the whole-rock composition may be reconstructed if we assume that olivine constitutes about 30% of the mode by weight. The best fit is obtained if we assume that part of this olivine has high Fo content (similar to the early phenocrysts and the xenocrysts) and that the rest is similar to the average microphenocryst composition (Table 4, column 5).

The AOV composition of *Allen et al.* (1979) has MgO = 24%, which is considerably higher than the concentration in DA-3 (19.2%). *Allen et al.* (1979) base their AOV composition on the four most magnesian vitrophyres they found in thin section. Since these particular clasts are all small (< 6.0 square mm), their modes may reflect nonmodal enrichment in olivine or overcounting of olivine. *Allen et al.* (1979) report six other modal reconstructions of olivine vitrophyre clasts that average around 20% MgO, and are almost identical to our analysis of DA-3. We suggest that the MgO concentration used by *Allen et al.* (1979) for their average olivine vitrophyre is too high, and recommend our own "preferred average olivine vitrophyre" composition, based on the average of our whole rock data (Table 4).

The olivine vitrophyres studied here have nearly identical trace element concentrations (Table 4). All are greatly enriched in incompatible trace elements relative to mare basalts and

TABLE 2. Plagioclase in olivine vitrophyre clast DA-1 from breccia 14321.

	1	2	3
SiO_2	45.18	45.29	47.41
Al_2O_3	35.84	35.87	33.84
FeO	0.22	0.28	0.23
CaO	19.05	19.00	17.48
Na_2O	0.60	0.42	1.23
K_2O	0.10	0.08	0.09
Sum	100.98	100.94	100.28
Cations per eight oxygens			
Si	2.064	2.068	2.169
Al	1.930	1.931	1.824
Fe	0.008	0.009	0.008
Ca	0.932	0.930	0.856
Na	0.052	0.036	0.108
K	0.005	0.004	0.004
Sum	4.991	4.978	4.970
% An	94.2	95.9	88.4

TABLE 3. Residual glass in olivine vitrophyre clast DA-3 analyzed by electron microprobe with mare basalt vitrophyres from breccia 14321 (*Shervais et al.*, 1985) shown for comparison.

	1	2	3	4	5	6	7	8
SiO_2	49.52	49.36	49.73	49.23	49.49	0.96	47.49	47.04
TiO_2	1.55	2.04	2.00	1.79	1.87	0.22	2.22	2.30
Al_2O_3	15.98	18.22	17.75	16.99	17.64	0.77	13.28	13.33
FeO	10.18	9.25	7.59	9.52	8.43	1.06	16.74	17.17
MnO	0.20	0.13	0.15	0.14	0.13	0.04	0.27	0.27
MgO	11.96	8.78	8.09	7.56	8.34	1.18	7.63	7.16
CaO	8.92	10.68	13.05	12.09	12.18	1.45	11.41	11.41
Na_2O	0.75	0.95	0.98	0.86	0.99	0.18	0.81	0.92
K_2O	0.27	0.41	0.34	0.85	0.42	0.13	0.14	0.15
Cr_2O_3	0.16	0.18	0.24	0.26	0.23	0.02	0.29	0.30
Sum	99.50	100.00	99.92	99.28	99.73	S.D.	100.28	100.05
Mg#	67.7	62.8	65.6	58.7	63.9		44.8	42.6
CaO/Al_2O_3	0.558	0.586	0.735	0.712	0.690		0.859	0.856

Columns 1-4 are representative residual glass analyses from olivine vitrophyre DA-3; column 5 is the mean of 20 glass analyses from DA-3, and column 6 is one standard deviation of this average. Column 7 is the mean of 30 glass analyses of mare vitrophyre DV-1 from breccia 14321, and column 8 is the mean of 25 glass analyses from mare vitrophyre DA-6 from breccia 14321 (*Shervais et al.*, 1985).

pristine highlands rocks. The rare earth elements are KREEPy, with La = 170 × chondrite, Yb = 78 × chondrite, and La/Lu = 22, which is the same ratio as KREEP (Fig. 5). The other KREEPy incompatible elements (K, Rb, Hf, Th, U) show similar enrichments. Unlike KREEP, however, the olivine vitrophyres are also enriched in the compatible transition metals Sc, Co, Cr, and V. These elements are more commonly identified with either mare basalts or the cryptic mafic component of the lunar highlands.

One of the more important results of our trace element analyses is the demonstration of high siderophile element concentrations in the olivine vitrophyres. The high concentration of Ir and Ni in all three olivine vitrophyres studied here shows that these rocks represent impact melts, and they are not pristine volcanic rocks.

DISCUSSION

The data presented above confirm the hypothesis presented by *Allen et al.* (1979) that olivine vitrophyre clasts in lunar breccia 14321 represent impact melts of lunar highlands material. Their highlands affinity is shown by high Mg#s, low Ca/Al ratios, and KREEPy incompatible element enrichments. Their origin by impact melting is shown by high siderophile concentrations, which implies the addition of a meteoritic component to the target. As noted by *Allen et al.* (1979) their compositions do not match any known lunar soil or regolith breccia, and cannot be modeled using common pristine rock types.

We must now face the dilemma of how to account for the mafic-rich composition of these vitrophyres. There are two possible solutions. The enrichment in mafic components may result from (1) the accumulation of olivine in a more normal highland impact melt, or (2) from the melting of a protolith rich in the mafic component. We will address first the question of olivine accumulation, which hinges on the partitioning of Mg and Fe between olivine and the melt.

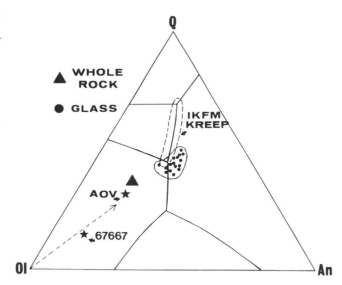

Fig. 4. Residual glass compositions from DA-3 plotted on the Ol-An-Q ternary of *Walker et al.* (1973). Shown for comparison are the average olivine vitrophyre of *Allen et al.* (1979), our preferred olivine vitrophyre whole-rock analysis, feldspathic peridotite 67667, and the field of IKFM KREEP. The residual glass data form a field that extends away from the Ol-An-Px reaction point into the plagioclase phase volume. Since plagioclase is not present in these glasses, this trend into the plagioclase phase volume suggests suppression of plagioclase nucleation due to rapid cooling rates (*Nabelek et al.*, 1978).

TABLE 4. Whole rock analyses of olivine vitrophyres from breccia 14321.

	DA-4 ,1159	DA-1 ,1180	DA-3 ,1158	MR-3	AOV*	Preferred 14321 AOV
SiO$_2$	na	na	48.3	46.6	45.7	46.5
TiO$_2$	na	na	1.3	1.31	1.03	1.30
Al$_2$O$_3$	na	na	12.4	12.35	10.3	12.4
FeO	9.85	9.84	9.88	10.57	10.8	9.86
MnO	na	na	0.1312	0.12	0.12	0.13
MgO	na	na	19.2	19.51	24.3	19.2
CaO	8.5	7.9	7.4	8.56	6.27	7.9
Na$_2$O	0.801	0.806	0.796	0.69	0.78	0.80
K$_2$O	0.57	0.41	0.56	0.29	0.46	0.51
Cr$_2$O$_3$	0.227	0.235	0.225	0.16	0.17	0.229
Sum	none	none	100.0*	100.0	99.93	98.83

Trace Elements

ppm						
V			41			41
Sc	17.87	17.62	17.93			17.80
Cr	1555	1605	1537			1566
Co	34.88	38.2	33.4			35.5
Ni	310	340	240			297
Rb	18	27	20			22
Sr	134	176	157			156
Cs	0.72	0.66	0.7			0.69
Ba	790	730	780			767
La	58.9	57.1	58.7			58.2
Ce	159	156	160			158
Nd	94	92	93			93
Sm	25.4	24.	26.5			25.3
Eu	2.05	1.96	2.01			2.01
Tb	6.05	5.92	6.17			6.05
Yb	19.7	19.5	20.1			19.8
Lu	2.63	2.62	2.68			2.64
Hf	21.9	21.8	22.4			22.0
Zr	800	810	830			815
Ta	2.55	2.53	2.61			2.56
Th	11.71	11.19	11.7			11.53
U	2.82	2.92	3.08			2.94
ppb						
Ir	7.2	6.7	5.8			6.57

Major and trace elements in DA-1, DA-3, and DA-4 by INAA; major elements in MR-3 and the Average Olivine Vitrophyre (AOV*) of *Allen et al.* (1979) by modal reconstruction from electron microprobe analyses of glass and olivine. Preferred AOV is our best estimate of major and trace elements in an average olivine vitrophyre, based on averages of DA-1, DA-3, and DA-4 (except silica from MR-3).

Partitioning of Fe and Mg Between Olivine and Glass

A major problem in deciphering the origin of the 14321 olivine vitrophyres is the apparent disequilibrium in Fe/Mg partitioning between olivine and a liquid having the bulk rock composition. The bulk-rock Mg# (= 77.6) implies equilibrium with Fo$_{91}$ olivine, using the experimentally determined maximum K_d = 0.35 of *Grover et al.* (1980), or Fo$_{92}$ for K_d = 0.30. The most magnesian olivine phenocryst observed, however, is Fo$_{89}$, which implies that the K_d is equal to 0.5. Alternatively, this high apparent K_d may result from olivine accumulation or kinetic factors (e.g., super-cooling of the melt; *Bianco and Taylor*, 1976).

Both *Allen et al.* (1979) and *Grover et al.* (1980) favor olivine accumulation to explain this discrepancy between the calculated and observed olivine composition. *Allen et al.* (1979) calculate that 25% olivine accumulation is required for equilibrium between their calculated average olivine vitrophyre and the average microphenocryst composition; less accumulation is needed if our lower MgO "preferred average olivine vitrophyre" composition is used. However, this conclusion is not consistent with the observed petrography. The vast majority

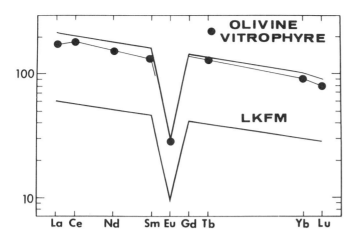

Fig. 5. Chondrite-normalized REE concentrations of our preferred average olivine vitrophyre, compared to the range of low-K Fra Mauro (LKFM) melt rocks, after *Vaniman and Papike* (1980). Olivine vitrophyres have REE concentrations at the upper extreme of the LKFM range, and are only 0.5 × high-K KREEP.

of modal olivine formed by quench crystallization during rapid cooling within the melt sheet. These olivines are small and have delicate skeletal morphologies. Accumulation of such olivine crystals by gravitational settling is not possible under these circumstances.

One explanation for this discrepancy is that the apparent accumulation of olivine results from the incomplete digestion of xenocrystic olivine in the melt sheet. In this scenario, the liquid composition was never equivalent to the whole rock, which contains clasts of material entrained during flow. The major problem with this explanation is that the observed content of xenocrystic olivine is too low (< 1% modally) to account for the apparent enrichment, which requires at least 10% olivine accumulation (based on an equilibrium olivine of Fo_{89} and our preferred AOV composition). Thus, while the incomplete digestion of xenocrystic olivine is a factor in this problem, it is not the complete solution.

Relationships between observed olivine compositions and calculated equilibrium values in DA-3 are shown in Fig. 6. The average olivine microphenocryst ($Fo_{83.1}$) is virtually identical to the calculated olivine in equilibrium with the average residual glass ($Fo_{83.5}$). Further, the range in observed olivine microphenocryst compositions is similar to the range in equilibrium olivine compositions calculated from the range in residual glass analyses (Fig. 6). The rims on most large olivines also fall within this same range. The implication is that conditions approaching equilibrium were maintained between the olivine microphenocrysts and residual liquid on a local scale near individual grains, but not on a large scale. This suggests that kinetic factors may be important in the origin of these rocks.

There are at least two possible explanations for the discrepancy between observed and calculated olivine compositions that focus on kinetic considerations: (1) the rapid quenching to form olivine microphenocrysts may have been preceeded by a period of slower crystallization, during which the larger, euhedral olivine phenocrysts formed, or (2) the earliest olivine formed by "disequilibrium" crystallization, during which the effective K_d was close to 0.5. *Donaldson et al.* (1975) and *Bianco and Taylor* (1976) have shown that effective K_d values will approach 0.5 at high cooling rates (48° to 400°C/hour) because cations are not able to diffuse through the melt fast enough to maintain bulk equilibrium at the crystal-liquid interface. This explanation implies that cooling rates were highest during initial crystallization and slowed as crystallization proceeded, consistent with the inferred cooling history of impact melt sheets (*Onorato et al.*, 1976, 1978).

It is difficult to distinguish between these hypotheses with the data presented here. The large size of the olivine phenocrysts relative to the microphenocrysts, and the presence of thin rims on the phenocrysts that are more or less in equilibrium with the adjacent glass, supports the first explanation—early crystallization at rates slow enough for the crystals to maintain bulk equilibrium with the melt, followed by more rapid crystallization. On the other hand, disequilibrium crystallization (i.e., crystallization where the effective K_d is ~0.5) is supported by the metastable expansion of the olivine phase volume into the plagioclase phase volume (Fig. 4). In any event, it is not necessary to invoke olivine accumulation to account for the apparent discrepancy in Fe-Mg partitioning. Olivine accumulation is not consistent with the observed textures, which must have formed by rapid "quench" crystallization.

In summary, we conclude that the high MgO concentrations observed in olivine vitrophyre bulk-rock samples is not the result of gravitational accumulation of olivine; the high MgO content of the bulk rock must be a primary characteristic of the original melt.

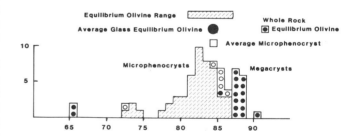

Fig. 6. Compositional data for olivine microphenocrysts (shaded), phenocrysts (open circles), and xenocrysts (closed circles) in vitrophyre DA-3, compared to the range of olivines in equilibrium with the residual glass (shaded bar), the olivine in equilibrium with the average residual glass (large filled circle), the average olivine microphenocryst (open square), and the olivine in equilibrium with the whole rock (filled circle in square). Note that the range of olivine in equilibrium with residual glass and the olivine in equilibrium with the average glass (based on $K_d = 0.33$) correspond to the observed range in olivine microphenocryst compositions and the average microphenocryst composition, respectively. This implies that equilibrium was maintained on a local scale, but not throughout the rock.

Olivine Vitrophyres and the Highland Mafic Component

All geochemical models of the soils and breccias in the lunar highlands have demonstrated the need for a mafic component that is not represented by any of the common lunar rock types. Mare basalts cannot fulfill this role because, although they are rich in mafic elements, their Mg/Fe ratio is too low. Dunite and lherzolite are more magnesian, but they are exceedingly rare and their Mg/Fe ratios are too high by a factor of 2 (*Korotev et al.*, 1980). The most recent suggestion is the postulate of *Ringwood et al.* (1987) that the mafic component at Apollo 16 derives from ancient komatiite lava flows.

The Mg content of the 14321 olivine vitrophyres is much higher than "normal" lunar soils or regolith breccias (e.g., *Laul et al.*, 1982; *Jerde et al.*, 1987) and presumably contains a higher proportion of the elusive mafic component. Because of their high MgO concentrations and intermediate Mg/Fe ratios, the 14321 olivine vitrophyres cannot be modeled as simple mixtures of the more common, felsic highland lithologies (anorthosite, norite, troctolite).

Least squares mixing models using both major and trace elements show that our "preferred" average olivine vitrophyre can be modeled as mixtures of various cryptic mafic components (e.g., SCCRV, MAF) plus KREEP, anorthosite, and an "extra lunar" (meteorite) component. Most models require ~30-35% mafic component, 50-55% KREEP, 8-10% anorthosite, and about 1.5% meteorite (Table 5). The fits are poor, however, and cannot be improved by adding other components such as mare basalt. The problem is exacerbated if we use a mafic component derived from Apollo 14 regolith breccias ("SCCRV-14"; *Jerde et al.*, 1987) because the MgO content of this component (~20% MgO) is the same as the 14321 vitrophyres. This leaves no room for the KREEP component, which must be over 50% to account for the high incompatible trace element concentrations.

A major problem with all of the mafic components calculated previously is that their Mg/Fe ratios and MgO concentrations are too low to reproduce the 14321 olivine vitrophyre composition when mixed with the large proportion of KREEP required to balance the trace element data. One solution to this problem is to calculate a mafic component that is tailored to the 14321 vitrophyres, using the same iterative technique as *Wasson et al.* (1977) and *Korotev et al.* (1980). The results of this exercise are shown in Table 7 as "MAF-14."

The major differences between MAF-14 and the other mafic components are its MG# (= 84) and its MgO concentration (~42 wt %), both of which are significantly higher than SCCRV, MAF, and Primary matter. Our preferred AOV comprises about 30% MAF-14, 53% KREEP, 13% alkali anorthosite, and 2% extra lunar component (Table 5). The Mg-rich nature of MAF-14 compared to other calculated mafic components must represent a real difference in the composition of mafic rocks that contribute to this component. This difference reflects the Mg-rich nature of the western highlands crust compared to the Apollo 16 highlands. In the next section, we attempt to relate this cryptic Mg-rich component to pristine rock types that constitute the western highlands crust.

Olivine Vitrophyres and the Pristine Highland Rocks

Previous studies of Apollo 14 breccias have established that Mg-suite troctolites and anorthosites, alkali anorthosites and norites, and mare basalts are the most common pristine rock types found at that site (*Hunter and Taylor*, 1983; *Warren et al.*, 1983a,b; *Shervais et al.*, 1983, 1984b, 1985b; *Lindstrom et al.*, 1984). Less common, but still important from a petrogenetic standpoint, are granite and ilmenite norites/gabbros, both of which may be associated with the alkali suite (*Shervais and Taylor*, 1986; *James et al.*, 1987). Dunite, pyroxenite and ferroan anorthosite are rare at the Apollo 14 site.

TABLE 5. Mixing results for Preferred 14321 AOV with various mafic components derived from the literature and by unmixing of our preferred AOV composition.

Component	Mix 1	Mix 2	Mix 3	Mix 4	Mix 5	Mix 6
SCCRV	31.1%	35.0%	-	-	-	-
MAF	-	-	34.7%	33.4%	-	-
MAF-14	-	-	-	-	29.4%	30.1%
KREEP	46.4%	38.9%	42.9%	45.1%	53.7%	52.1%
Meteorite	1.65%	1.68%	1.41%	1.42%	1.71%	1.70%
Granite	3.14%	2.98%	2.77%	1.51%	1.49%	0.76%
FAN	11.6%	-	11.9%	-	10.9%	-
ALKAN	-	17.0%	-	12.2%	-	13.5%
Sum	94.0%	95.5%	93.6%	93.6%	97.2%	98.1%
Chi-sq	23.5	19.4	26.1	27.4	5.6	4.3

Mafic components are SCCRV (*Wasson et al.*, 1977), MAF (*Korotev et al.*, 1980), and MAF-14 (derived here). Results for the Primary matter of *Wänke et al.* (1977) are similar to those of SCCRV. Other components are KREEP (*Warren and Wasson*, 1979), FAN (*Wasson et al.*, 1977), ALKAN (14305,400; *Shervais et al.*, 1984b), meteorite (*Boynton et al.*, 1975), and granite (14321,1198; unpublished data). MAF-14 derived by successive approximations to provide reasonable match to bulk MgO and Mg#. KREEP adjusted to higher SiO_2 (49.5%) and TiO_2 (2.0%) to provide better fits.

The composition of olivine and plagioclase xenocrysts in the 14321 vitrophyres provide important clues concerning the constitution of the source rocks. Olivine xenocrysts are mostly magnesian (Fo_{87} to Fo_{90}), but one ferroan olivine is also found (Fo_{66}). Plagioclase/maskelynite xenocrysts span a wide range of compositions, from $An_{83.6}$ to $An_{94.6}$ (Table 2). These xenocrysts reflect a range in parent rock types that corresponds to the common pristine rock types at the Apollo 14 site: Mg-suite troctolites (calcic plagioclase, magnesian olivine), alkali anorthosites or gabbronorites (sodic plagioclase), and possibly mare basalt (ferroan olivine). Ferroan anorthosite (calcic plagioclase, ferroan olivine) may be present also, but is not required. The correspondence between the inferred parent lithologies and rock types common in the Apollo 14 sample suite supports the idea that local crust was sampled by the impact.

Geochemical studies of Apollo 14 pristine rocks have shown that these lithologies are generally enriched in incompatible trace elements relative to similar rocks from other sites (*Warren et al.*, 1983a,b; *Shervais et al.*, 1983, 1984b, 1985b; *Lindstrom et al.*, 1984). Least squares mixing models using pristine rock compositions from the Apollo 14 site show that the preferred average olivine vitrophyre may be modeled with moderate success as a mix of KREEP (~50-60%), dunite (~25-30%), mare basalt (~10-12%), granite (~0-2%), anorthosite (8-12%), and an extra lunar component (~2%). The exact proportions vary somewhat, depending on the type of anorthosite used (Table 6). Mare basalt is required to account for the high Sc, V, Fe, and Ti; small amounts of granite help raise the Si content of the mix.

The high concentration of KREEP derived from the mixing models is required by the incompatible trace element data. Olivine vitrophyres have K, REE, P, and Hf concentrations 10× to 200× higher than most pristine highland rocks, and about half that of high-K KREEP. Because Apollo 14 alkali anorthosite has higher REE, K, and Na concentrations than FAN, slightly lower KREEP contents are necessary when alkali anorthosite is used in place of FAN as the anorthosite component (Table 6). Regardless of which anorthosite is used, however, the overall chemistry of the mix is strongly dominated by KREEP.

We suggest that the mixing components used here do not necessarily represent distinct components in the lunar crust. As discussed by *Korotev et al.* (1980), the cryptic mafic component (dunite in this case) may represent the mafic portion of some prevalent lunar rock type. If we mix dunite with anorthosite, the result is an Mg-rich troctolite (Table 7) that probably represents the average composition of Mg-suite rocks melted in the impact event. Alkali-rich lithologies are most likely represented by the KREEP component, which has affinities to alkali anorthosites, norites, and ilmenite gabbronorites. These rocks have an average major element composition that is roughly equivalent to norite and similar to high-K KREEP, and their trace elements are dominated by a KREEPy overprint, so they may easily be represented in the mixing models by a single component.

Mixing models that use only these two components (Mg-rich troctolite and KREEP) plus an extra lunar component provide good solutions (Table 6) that suggest the proportions of Mg-suite and alkali suite rocks in the Apollo 14 crust. Alkali suite rocks predominate by a two-to-one ratio over Mg-suite rocks, approximately the inverse found for pristine rocks. The difference may be explained by the common occurrence of KREEP as an extrusive rock (*Shirley and Wasson*, 1981), or from the exclusion of the nonpristine KREEPy granulitic norites, which are probably the most common alkali suite lithology.

Implications for the Composition of the Western Lunar Crust

Studies of terrestrial impact melt sheets have shown that they have extremely homogeneous compositions due to turbulent mixing of the superheated melt phase with entrained clasts (*Simonds et al.*, 1976, 1978). Further studies have shown that most of the melt sheet is confined to the crater cavity,

TABLE 6. Mixing results for Preferred 14321 olivine vitrophyre using pristine highland lithologies from Apollo 14.

Component	Mix 1	Mix 2	Mix 3	Mix 4	Mix 5	Mix 6
Dunite	23.7%	23.7%	24.4%	24.3%	-	-
MGTROC	-	-	-	-	32.9%	33.7%
FAN	8.2%	-	-	-	0.33%	-
Magan	-	9.9%	-	-	-	-
ALKAN	-	-	10.8%	10.6%	-	-
KREEP	53.2%	51.0%	48.9%	51.0%	51.7%	61.2%
Meteorite	1.83%	1.80%	1.83%	1.83%	1.80%	1.80%
Granite	1.65%	1.79%	1.41%	-	1.72%	-
Mare	8.2%	8.6%	10.0%	9.3%	8.3%	-
Sum	96.8%	96.7%	97.4%	97.0%	96.7%	96.7%
Chi-sq	4.3	3.7	2.5	2.45	3.7	4.3

MGTROC derived from mix of dunite (72%) and Mg-suite anorthosite (28%). Components are same as Table 5, except dunite and Magan from *Lindstrom et al.* (1984) and mare from *Shervais et al.* (1985b).

although ejected melt may be transported as far as one crater diameter (*Howard and Wilshire*, 1975; *Hörz et al.*, 1983). This implies a relatively local origin for melt rocks. Melt rocks that remain within the crater cavity may be redistributed by later impacts.

The 14321 olivine vitrophyres are anomalously mafic compared to average lunar crust calculated from surface chemistry (e.g., *Adler and Trombka*, 1977; *Taylor and Jakes*, 1974; *Taylor*, 1982) and compared to the Apollo 14 regolith, which averages about 9.5% MgO (*Laul et al.*, 1982). Since surface geochemistry is skewed towards surface lithologies (e.g., *Ryder and Wood*, 1977), these data imply that the mafic

TABLE 7. Cryptic components derived from 14321 vitrophyres.

	MAF-14	MGTROC
Wt %		
SiO_2	42.0	41.4
TiO_2	0.4	0.116
Al_2O_3	1.20	9.195
FeO	15.0	8.30
MgO	41.0	36.78
CaO	0.8	5.42
Na_2O	0.25	0.165
K_2O	0.05	0.035
ppm		
Sc	30	4.35
Cr	2800	434
Co	40	45.5
Ni	120	64.4
La	0.2	9.72
Ce	0.54	24.7
Sm	0.12	4.16
Eu	0.10	0.75
Yb	0.13	2.73
Lu	0.02	0.472
Hf	0.10	1.745

component of lunar crust must reside at deeper crustal levels. The outstanding question is, how deep?

The generation of mafic-rich impact melts is commonly thought to require a basin-forming impact large enough to sample the deeper crustal levels where mafic-rich protoliths are thought to dominate (e.g., *Ryder and Bower*, 1977). The only basin close enough to the Apollo 14 site to be a reasonable candidate, however, is the Imbrium basin. Olivine vitrophyres are too high in MgO, however, to represent Imbrium basin impact melt; the best-documented samples of Imbrium basin impact melt (15445 and 15455) contain only 13% to 14% MgO (*Ryder and Bower*, 1977).

Pieters (1982) has reported spectral reflectance data that show that the central peak of Copernicus is dominated by dunite, which we have shown may represent the mafic component in the 14321 olivine vitrophyres. Extrapolation of data on cratering dynamics to a crater the size of Copernicus (~96 km) suggests that the central peak material was excavated from depths of up to 10 km (*Pieters*, 1982; *Pieters and Wilhelms*, 1985). Thus, the mafic component is closer to the surface in some areas than one would assume from the surface chemistry data. *Pieters and Wilhelms* (1985) attribute this near-surface enrichment in mafic material to removal of the early ferroan crust by ballistic erosion during basin-forming impacts (e.g., the Procellarum basin event).

Hawke and Head (1977) made an extensive survey of pre-Imbrium craters in the Fra Mauro region. They found four craters ("northwest," "central," "site," and Fra Mauro) that are close enough to the Apollo 14 landing site to deposit signicant amounts of impact melt at the site. Only two of these ("northwest" and Fra Mauro) are large enough (95 to 136 km in diameter) to generate impact melt deep enough to involve mafic rocks like those inferred to underlie Copernicus (95 km diameter) by *Pieters* (1982). There are many other potential source craters that are farther away, but the lack of shock features in the olivine vitrophyre clasts argues against emplacement by later impacts into an already solid melt sheet.

The enrichment in KREEP observed in Apollo 14 soils, breccias, and orbital geochemistry must extend to at least 12-17 km depth in the Apollo 14 crust, based on the estimated maximum depths of excavation calculated by *Hawke and Head* (1977) for Fra Mauro and "northwest" craters. KREEPy element concentrations are as high in the 14321 olivine vitrophyres as they are in the Apollo 14 regolith, despite the high concentration of mafic components in the vitrophyres. The implication here is that KREEP is an important component in the intermediate crust, where it is part of the indigenous rock suite. This deeper crustal KREEP may exist in the form of KREEPy norite plutons (e.g., *Ryder and Wood*, 1977; *Ryder*, 1976; *Warren et al.*, 1983a,b; *Lindstrom et al.*, 1984; *Shervais and Taylor*, 1986).

The mixing models calculated here for the olivine vitrophyres show that granite remains a minor but important mixing component at the Apollo 14 site, consistent with its observed modal abundance in the A-14 pristine rock suite, in the A-14 regolith, and in A-14 regolith breccias (*Warren et al.*, 1983; *Simon et al.*, 1982; *Jerde et al.*, 1987). The inferred abundance of granite in A-14 regolith breccias (*Jerde et al.*, 1987) and in the 14321 vitrophyres (1-2 wt %) is much higher than that estimated for A-14 regolith (0.3-0.5 wt %; *Shervais and Taylor*, 1983; *Jerde et al.*, 1987). This suggests that granite may also be more common deeper in the crust, probably in association with noritic plutons of the alkali-rich suite (e.g., *Ryder*, 1976; *James*, 1980).

CONCLUSIONS

The data presented here show that the olivine vitrophyres of lunar breccia 14321 are impact melt rocks, as proposed by *Allen et al.* (1979). These rocks are unusual because they are exceptionally rich in both KREEP and a high-Mg mafic component. KREEPy trace elements are present in concentrations approximately 0.5 times the high-K KREEP component of *Warren and Wasson* (1979). The high MgO concentration of the vitrophyres is manifest by skeletal quench crystals of olivine that constitute approximately 30% of the mode.

Mixing calculations show that the mafic component of the 14321 olivine vitrophyres is more magnesian than cryptic mafic components isolated at the Apollo 16 site (SCCRV, MAF, Primary matter) and cannot be modeled by any mixture that uses these cryptic components. Mixing calculations based on pristine highland lithologies require about 25% dunite and 10% anorthosite, along with KREEP, granite, and extra lunar material. Since dunite is not a common rock type, we suggest that its components occur combined with "anorthosite" as Mg-suite troctolite. Alkali-rich lithologies are represented by KREEP and minor granite.

The absence of ferroan anorthosite from this assemblage is consistent with a crust that is locally dominated by later Mg-suite and alkali-suite plutons. Since dense Mg-suite magmas are not likely to be emplaced at shallow levels in a ferroan anorthosite crust, the emplacement of Mg-rich material close enough to the surface to be sampled by 100-km size craters probably requires that the early ferroan crust was thinned or removed by ballistic erosion during basin-forming impacts (e.g., *Shervais and Taylor,* 1986; *Pieters and Wilhelms,* 1985).

The importance of impact melt rocks in deciphering lunar crustal origins should not be overlooked in the current fervor for pristine rocks. Many important rock types may not be sampled in their pristine state because they formed on the lunar surface during the later stages of accretion, or because they occur deep within the crust and are only sampled by large impacts which destroy their pristinity. The merits of nonpristine rocks for understanding the origin of the lunar crust are aptly demonstrated by the data presented here on the 14321 olivine vitrophyres.

Acknowledgments. This paper benefited from useful discussions with J. W. Delano. Logistical support was provided by the Planetary Materials Curatorial Staff, and especially Rene Martinez. This work was supported by NASA grants NAG 9-62 (L. A. Taylor), NSG 9-56 (L. A. Haskin), and NAG 9-169 (J. Shervais).

REFERENCES

Adler I. and Trombka J. (1977) Orbital chemistry—lunar surface analysis from the X-ray and gamma ray remote sensing experiments. In *Chemistry of the Moon, Phys. Chem. Earth, 10,* 17-43.

Allen F. M., Bence A. E., Grove T. L. (1979) Olivine vitrophyres in Apollo 14 breccia 14321: Samples of the high-Mg component of the lunar highlands. *Proc. Lunar Planet. Sci. Conf. 10th,* 695-712.

Bianco A. and Taylor L. A. (1976) Cooling rate experimentation on olivine normative mare basalts: effects of FeO content on mineral texture and chemistry. *Geol. Soc. Am., Abstracts with Programs, 8,* 778.

Boynton W. V., Baedecker P. A. Chou C.-L., Robinson K. L., and Wasson J. T. (1975) Mixing and transport of lunar surface materials: Evidence obtained by the determination of lithophile, siderophile, and volatile elements. *Lunar Sci. Conf. 6th,* 2241-2259.

Donaldson C., Usselman T., Williams R., and Lofgren G. (1975) Experimental modeling of the cooling history of Apollo 12 olivine basalts. *Proc. Lunar Planet. Sci. Conf. 6th,* 843-869.

Duncan A. R., McKay S. M., Stoesser J. W., Lindstrom M. M., Lindstrom D. J., Fruchter J. S., and Goles G. A. (1975) Lunar polymict breccia 14321: a compositional study of its principal components. *Geochim. Cosmochim. Acta, 39,* 247-260.

Grieve R. A. F., McKay G. A., Smith H. D., and Weill D. F. (1975) Lunar polymict breccia 14321: a petrographic study. *Geochim. Cosmochim. Acta, 39,* 229-246.

Grover J. E., Lindsley D. H., and Bence A. E. (1980) Experimental phase relations of olivine vitrophyres from breccia 14321: the temperature and pressure dependence of Fe-Mg partitioning for olivine and liquid in a highlands melt-rock. *Proc. Lunar Planet. Sci. Conf 11th,* 179-196.

Hawke R. and Head J. W. (1977) Pre-Imbrium history of the Fra Mauro region and Apollo 14 sample provenance. *Proc. Lunar Sci. Conf. 8th,* 2741-2761.

Hörz F., Ostertag R. and Rainey D. (1983) Bunte breccia of the Ries: continuous deposits of large impact craters. *Rev. Geophys. Space Phys., 21,* 1667-1725.

Howard K. A. and Wilshire H. G. (1975) Flows of impact melt at lunar craters. *J. Res. U.S. Geological Survey, 3,* 237-251.

Hunter R. H. and Taylor L. A. (1983) The magma ocean from the Fra Mauro shoreline: An overview of the Apollo 14 crust. *Proc. Lunar Planet. Sci. Conf. 13th,* in *J. Geophys. Res., 88,* A591-A602.

James O. B. (1980) Rocks of the early lunar crust. *Proc. Lunar Planet. Sci. Conf 11th,* 365-393.

James O. B., Lindstrom M. M., and Flohr M. K. (1987) Petrology and geochemistry of alkali gabbronorites from lunar breccia 67975. *Proc. Lunar Planet. Sci. Conf. 17th,* in *J. Geophys. Res., 92,* E314-E330.

Jerde E., Warren P. H., Morris R. V., Heiken G., and Vaniman D. T. (1987) A potpourri of regolith breccias: "New" samples from the Apollo 14, 16, and 17 landing sites. *Proc. Lunar and Planet. Sci. Conf. 17th,* in *J. Geophys. Res., 92,* E526-E536.

Knapp S. A. (1986) *Petrogenesis of Apollo 14 Lunar Breccia 14321.* M.S. Thesis, University of Tennessee. 118 pp.

Korotev R. L., Haskin L. A., and Lindstrom M. M. (1980) A synthesis of lunar highlands compositional data. *Proc. Lunar Planet. Sci. Conf 11th,* 395-429.

Laul J. C., Papike J. J. and Simon S. B. (1982) The Apollo 14 regolith: chemistry of cores 14210/14211 and 14220 and soils 14141, 14148, 14149. *Proc. Lunar Planet. Sci. Conf. 13th,* in *J. Geophys. Res., 87,* A247-A259.

Lindstrom D. J. and Korotev R. L. (1982) TEABAG: computer programs for instrumental neutron activation analysis. *J. Radioanal. Chem., 70,* 439-458.

Lindstrom M. M. (1984) Alkali gabbronoritic, ultra-KREEPy melt rock and the diverse suite of clasts in North Ray Crater feldspathic fragmental breccia 67975. *Proc. Lunar Planet. Sci. Conf. 15th,* in *J. Geophys. Res., 90,* C50-C62.

Lindstrom M. M., Duncan A. R., Fruchter J. S., McKay S. M., Stoeser J. W., Gales G. G., and Lindstrom D. J. (1972) Compositional characteristics of some Apollo 14 clastic materials. *Proc. Lunar Sci. Conf. 3rd,* 1201-1214.

Lindstrom M. M., Knapp S. A., Shervais J. W., and Taylor L. A. (1984) Magnesian anorthosites and associated troctolites and dunite in Apollo 14 breccias. *Proc. Lunar Planet. Sci. Conf. 15th,* in *J. Geophys. Res., 89,* C41-C49.

Nabelek P. I., Taylor L. A. and Lofgren G. E. (1978) Nucleation and growth of plagioclase and development of texture in a high-alumina basaltic melt. *Proc. Lunar Planet. Sci. Conf. 9th,* 725-741.

Onarato P., Uhlmann D., and Simonds C. (1976) Heat flow in impact melts: Apollo 17 station 6 boulder and some applications to other breccias and xenolith laden melts. *Proc. Lunar Planet. Sci. Conf 7th,* 2449-2467.

Onorato P., Uhlman D., and Simonds C. (1978) The thermal history of the Manicougan impact melt sheet, Quebec. *J. Geophys. Res., 83,* 2789-2798.

Pieters C. M. (1982) Copernicus crater central peak: lunar mountain of unique composition. *Science, 215,* 59-61.

Pieters C. M. and Wilhelms D. E. (1985) Origin of olivine at Copernicus. *Proc. Lunar Planet. Sci. Conf. 15th,* in *J. Geophys. Res., 90,* C415-C420.

Ringwood A. E., Seifert S., and Wänke H. (1987) A komatiite component in Apollo 16 highland breccias: implications for the nickel-cobalt systematics and bulk composition of the moon. *Earth Planet. Sci. Lett.*, 81, 105-117.

Ryder G. (1976) Lunar sample 15405: remnant of a KREEP basalt-granite granite differentiated pluton. *Earth Planet. Sci. Lett.*, 29, 255-268.

Ryder G. and Bower J. F. (1977) Petrology of Apollo 15 black-and-white white rocks 15445 and 15455—fragments of the Imbrium melt sheet? *Proc. Lunar Planet. Sci. Conf. 8th*, 1895-1923.

Ryder G. and Wood J. A. (1977) Serenitatis and Imbrium impact melts: implications for large scale layering in the lunar crust. *Proc. Lunar Planet. Sci. Conf. 8th*, 655-668.

Shervais J. W. and Taylor L. A. (1983) Micrographic granite: More from Apollo 14 (abstract). *Lunar and Planetary Science XIV*, pp. 696-697. Lunar and Planetary Institute, Houston.

Shervais J. W. and Taylor L. A. (1984) Consortium breccia 14321: Petrology of mare basalts and olivine vitrophyres (abstract). *Lunar and Planetary Science XV*, pp. 766-767. Lunar and Planetary Institute, Houston.

Shervais J. W. and Taylor L. A. (1986) Petrologic constraints on the origin of the Moon. In *Origin of the Moon* (W. K. Hartman, R. J. Phillips, and G. J. Taylor, eds.), pp. 173-202. Lunar and Planetary Institute, Houston.

Shervais J. W., Taylor L. A., and Laul J. C. (1983) Ancient crustal components in the Fra Mauro breccias: implications for igneous processes. *Proc. Lunar Planet. Sci. Conf. 14th*, in *J. Geophys. Res.*, 88, B177-B192.

Shervais J. W., Knapp S., and Taylor L. A. (1984a) *Breccia guidebook No. 7: 14321*. NASA-JSC Publication #69, Planetary Materials Branch, 57 pp.

Shervais J. W., Taylor L. A., Laul J. C., and Smith M. R. (1984b) Pristine highland clasts in consortium breccia 14305: Petrology and geochemistry. *Proc. Lunar Planet. Sci. Conf. 15th*, in *J. Geophys. Res.*, 89, C25-C40.

Shervais J. W., Taylor L. A., and Lindstrom M. M. (1985a) Olivine vitrophyres—a nonpristine high-Mg component in lunar breccia 14321 (abstract). In *Lunar and Planetary Science XVI*, pp. 771-772. Lunar and Planetary Institute, Houston.

Shervais J. W., Taylor L. A., and Lindstrom M. M. (1985b) Apollo 14 mare basalts: petrology and geochemistry of clasts from consortium breccia 14321. *Proc. Lunar Planet. Sci. Conf. 15th*, in *J. Geophys. Res.*, 90, C375-C395.

Shervais J. W., Knapp S., Taylor L. A., and Lindstrom M. M. (1985c) Petrology and geochemistry of microbreccia clasts and matrix in lunar breccia 14321 (abstract). In *Lunar and Planetary Science XVI*, pp. 767-768. Lunar and Planetary Institute, Houston.

Shirley D. and Wasson J. T. (1981) A mechanism for the extrusion of KREEP. *Proc. Lunar Planet. Sci. 12B*, 965-978.

Simon S. B., Papike J. J., Laul J. C. (1982) The Apollo 14 regolith: Petrology of cores 14210/14211 and 14220 and soils 14141, 14148, and 14149. *Proc. Lunar Planet. Sci. Conf. 13th*, in *J. Geophys. Res.*, 87, A232-A246.

Simonds C., Warner J., Phinney W., and McGee P. (1976) Thermal model for impact breccia lithification: Manicougan and the moon. *Proc. Lunar Planet. Sci. Conf. 7th*, 2509-2528.

Simonds C., Floran R., McGee P., Phinney W., and Warner J. (1978) Petrogenesis of melt rocks, Manicougan impact structure, Quebec. *J. Geophys. Res.*, 83, 2773-2788.

Swann G. A., Bailey N. G., Batson R. M., Eggleton R. E., Hait M. H., Holt H. E., Larson K. B., Reed V. S., Schaber G. G., Sutton R. L., Trask N. J., Ulrich G. E., and Wilshire H. G. (1977) Geology of the Apollo 14 landing site in the Fra Mauro highlands. *U.S. Geological Survey Prof. Pap.*, 880, 103 pp.

Taylor S. R. (1982) Planetary Science—a lunar perspective. Lunar and Planetary Institute, Houston. 481 pp.

Taylor S. R. and Jakes P. (1974) The geochemical evolution of the Moon. *Proc. Lunar Sci. Conf. 5th*, 1287-1305.

Taylor L. A. and Nabelek P. I. (1979) Effects of kinetics on the crystallization sequences in basalts (abstracts). In *Lunar and Planetary Science Conference X*, pp. 1212-1213. Lunar and Planetary Institute, Houston.

Vaniman D. T. and Papike J. J. (1980) Lunar highland melt rocks: Chemistry, petrology and silicate mineralogy. *Proc. Conf. Lunar Highlands Crust* (J. J. Papike and R. B. Merrill, eds.), pp. 271-337.

Walker D. (1983) Lunar and terrestrial crust formation. *Proc. Lunar Planet. Sci. Conf. 14th*, in *J. Geophys. Res.*, 88, B17-B25.

Walker D., Grove T. L., Longhi J., Stolper E. M., and Hays J. F. (1973) Origin of lunar feldspathic rocks. *Earth Planet. Sci. Lett.*, 20, 325-336.

Wänke H., Baddenhausen H., Blum K., Cendales M., Dreibus G., Hofmeister H., Kruse H., Jagoutz E., Palme C., Spettel B., Thacker R., and Vilcsek E. (1977) On the chemistry of lunar samples and achondrites: Primary matter in the lunar highlands, a re-evaluation. *Proc. Lunar Planet. Sci. Conf. 8th*, 2191-2213.

Wänke H., Palme H., Kruse H., Baddenhausen H., Blum K., Cendales M., Dreibus G., Hofmeister H., Jagoutz E., Palme C. Spettel B., Thacker R. (1976) Chemistry of lunar highland rocks: a refined evolution of the composition of the primary matter. *Proc. Lunar Sci. Conf. 7th*, 3479-3499.

Warren P. H. and Wasson J. T. (1979) The origin of KREEP. *Rev. Geophys. Space Phys.*, 17, 73-88.

Warren P. H., Taylor G. J., Keil K., Kallemeyn G. W., Rosener P. S., and Wasson J. T. (1983a) Sixth foray for pristine nonmare rocks and an assessment of the diversity of lunar anorthosites. *Proc. Lunar Planet. Sci. Conf. 13th*, in *J. Geophys. Res.*, 88, A615-A630.

Warren P. H., Taylor G. J., Keil K., Kallemeyn G. W., Shirley D., and Wasson J. T. (1983b) Seventh foray: whitlockite-rich lithologies, a diopside bearing troctolitic anorthosite, ferroan anorthosites, and KREEP, *Proc. Lunar Planet. Sci. Conf. 13th*, in *J. Geophys. Res.*, 88, A615-A630.

Wasson J. T., Warren P. H., Kallemyn G. W., McEwing C. E., Mittlefehldt D. W., and Boynton W. V. (1977) SCCRV, a major component of highland rocks. *Proc. Lunar Planet. Sci. Conf. 8th*, 2237-2252.

Apollo 14 Regolith Breccias: Different Glass Populations and Their Potential for Charting Space/Time Variations

John W. Delano
Department of Geological Sciences, State University of New York, Albany, NY 12222

Apollo 14 regolith breccias (14313, 14307, 14301, 14049, 14047) have been found to have different populations of nonagglutinitic, mare-derived glasses. These variations appear to not only reflect different source regoliths but also different closure ages for these breccias. Based upon these different glass populations, 14301 is inferred to have a closure age sometime *during* the epoch of mare volcanism. All of the other four breccias were formed *after* the termination of mare volcanism with a possible age sequence from old to young of the following: 14307, 14313, 14049, 14047. Due to the relative simplicity of acquiring high-quality chemical data on large numbers of glasses by electron microprobe, mare glass populations allow (1) classification of regolith breccias with respect to provenance and (2) estimation of their relative and absolute closure ages. The determination of ^{40}Ar-^{39}Ar ages on individual glass spherules within breccias using the laser probe should in the future prove to be a promising extension of the present study.

INTRODUCTION

During a search for pristine (i.e., volcanic) glasses contained within five Apollo 14 regolith breccias, the author analyzed nearly 1200 glasses by electron microprobe, as shown below. The following polished thin-sections were studied: *14313*, 41; *14307*, 3, 4, 36, 45, 48, 49; *14303*, 9, 11, 13, 15, 16, 17, 84; *14049*, 38; *14047*, 106.

Sample #	14313	14307	14301	14049	14047
Number of glasses	297	437	215	123	125

It was found that these five regolith breccias possessed different populations of mare-derived glass. Following the first report of this phenomenon at Apollo 14 (*Delano*, 1986a), intrasite variations in the glass populations contained within other regolith breccias have also been described (*Jerde et al.*, 1987; *Wentworth and McKay*, 1987a,b).

These five regolith breccias were collected at four locations (Fig. 1) by the Apollo 14 astronauts during their second EVA along the Fra Mauro Formation. The most thorough and detailed petrographic examination of the Apollo 14 regolith breccias was carried out by *Simonds et al.* (1977), who convincingly demonstrated that (1) these breccias were chemically and petrographically distinct from the metamorphosed Fra Mauro breccias (Fig. 2) and (2) they must have been derived from different sources than the Fra Mauro breccias. These regolith breccias, although compositionally indistinguishable from the local Apollo 14 regolith, were assembled from comparatively immature regoliths (*Simonds et al.*, 1977).

The specific varieties of glass that are discussed in this paper are schematically illustrated in Fig. 3. The *pristine* glasses (indicated by "p" in Fig. 3) belong to six chemically distinct groups (Table 1): VLT; Green A; Green B; Yellow; Orange; Red/Black. The chemistries of these Apollo 14 pristine glasses have been previously discussed in detail by the author (e.g., *Delano*, 1981a,b, 1986b). The *impact* glasses occur as three groups in Fig. 3: Fra Mauro Basalt; Highland Basalt; Mare group "M." The high-Al_2O_3 impact glasses in Fig. 3 known as "Fra Mauro basalt" and "highland basalt" were identified by the *Apollo Soil Survey* (1971) and *Reid et al.* (1972a,b) through studies of glass contained in the Apollo 14 regolith. Those investigators noted that nearly 90% of the glasses in the present regolith belonged to these two groups. "Fra Mauro basalt" glasses are chemically similar to the local Apollo 14 regolith and regolith breccias (Table 2), whereas "highland basalt" glasses compositionally resemble Apollo 16 regolith.

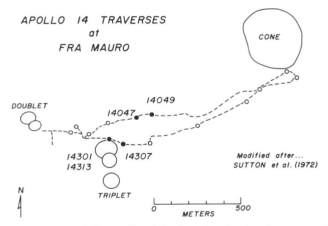

Fig. 1. Map of the Apollo 14 landing site showing the routes of the two EVAs (dashed lines) and the sampling locations (circles). The principal craters in the area are labeled. The sampling locations for the five regolith breccias are as follows: 14047 from station B; 14049 from about 80 m east of station B; 14307 from station G; 14301 and 14313 from station G1. Diagram modified after *Sutton et al.* (1972). Breccia 14318, which is discussed in the text, was collected at station H about 100 m north of the lunar module.

RESULTS

The populations of nonagglutinitic glass observed in these Apollo 14 regolith breccias are illustrated in Fig. 4. By the very nature of formation of these breccias by shock compaction

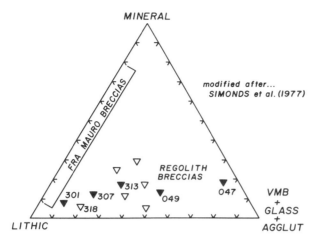

Fig. 2. Ternary diagram (modified after *Simonds et al.*, 1977) with the following petrographically identified components: (1) lithic clasts; (2) mineral fragments; and (3) clasts of regolith breccias + glass + agglutinates. The range observed among the Apollo 14 regolith breccias (triangles) by *Simonds et al.* (1977) is substantial. Those samples analyzed in the present study are identified by sample numbers and *solid* triangles. Note that these five samples span most of the range observed for all Apollo 14 regolith breccias. The field for the recrystallized Fra Mauro Breccias is shown for comparison.

of regolith (e.g., *Kieffer*, 1975; *McKay et al.*, 1986; *Stoffler et al.*, 1979), these glass populations must have been present in the source regoliths. Although differences are evident in Fig. 4, the similarities will be mentioned initially. First, all of these regolith breccias contain *some* pristine mare glasses. Second, all of these breccias contain a wide assortment of mare-derived impact glasses containing $\lesssim 12$ wt % Al_2O_3 and $\gtrsim 2$ wt % TiO_2. Third, "Fra Mauro basalt" and "highland basalt" glasses are abundant components in all of these breccias.

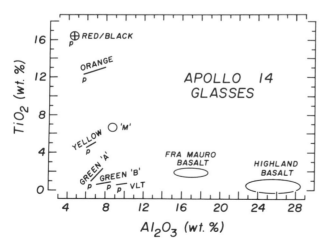

Fig. 3. The groups of glass observed in the Apollo 14 regolith breccias during the present study are shown on this diagram of Al_2O_3 (wt %) versus TiO_2 (wt %). The pristine (i.e., volcanic) glasses are designated by "p." Those groups *without* the pristine designation are believed to be of impact origin.

TABLE 1. The six varieties of pristine glass observed in Apollo 14 regolith breccias.

	Green A	Green B	VLT	Yellow	Orange	Red/Black
SiO_2	44.1	44.8	46.0	40.8	37.2	34.0
TiO_2	0.97	0.45	0.55	4.58	12.5	16.4
Al_2O_3	6.71	7.14	9.30	6.16	5.69	4.6
Cr_2O_3	0.56	0.54	0.58	0.41	0.86	0.92
FeO	23.1	19.8	18.2	24.7	22.2	24.5
MnO	0.28	0.24	0.21	0.30	0.31	0.31
MgO	16.6	19.1	15.9	14.8	14.5	13.3
CaO	7.94	8.03	9.24	7.74	7.04	6.9
Na_2O	n.d.	0.06	0.11	0.42	0.28	0.23
K_2O	n.d.	0.03	0.07	0.10	0.29	0.16

Refer to *Delano* (1986b) for discussion of petrogenesis and for complete data compilation.

TABLE 2. Major-element compositions of Apollo 14 regolith breccias, "Fra Mauro Basalt" glass, and Apollo 14 regolith.

	14313	14307	14301	14049	14047	"Fra Mauro Basalt" (glass)	Apollo 14 Regolith
SiO_2	n.a.	46.9	47.6	48.3	47.2	48.0	47.9
TiO_2	n.a.	1.84	1.77	1.75	1.75	2.1	1.74
Al_2O_3	17.4	16.5	15.9	16.2	18.2	17.0	17.6
Cr_2O_3	0.20	0.23	0.20	0.20	0.15	n.a.	0.25
FeO	10.7	12.3	11.9	10.8	10.5	10.9	10.4
MnO	0.13	0.14	0.14	0.14	0.14	n.a.	0.14
MgO	n.a.	9.76	10.4	10.0	8.89	8.7	9.24
CaO	10.6	10.8	10.1	10.6	11.5	10.7	11.2
Na_2O	0.77	0.70	0.74	0.74	0.68	0.7	0.68
K_2O	0.72	0.55	0.69	0.60	0.48	0.54	0.55
P_2O_5	n.a.	0.51	0.58	0.44	0.50	n.a.	0.53
Ref.	1,2	3	3	4	5	6	7

References:
(1) *Boynton et al.* (1975); (2) *Alexander and Kahl* (1974); (3) *Hubbard et al.* (1972); (4) *Philpotts et al.* (1972); (5) *Taylor et al.* (1972); (6) *Reid et al.* (1972b); (7) *Taylor* (1975).

TABLE 3. From the nearly 1200 nonagglutinate glasses analyzed, the following proportions (%) of pristine (i.e., volcanic) glass were observed in five Apollo 14 regolith breccias.

	14307	14313	14301	14047	14049
14 Green A	15.6	6.1	n.d.*	n.d.†	4.1
14 Green B	7.6	2.4	n.d.*	2.4	4.1
14 VLT	18.8	3.4	n.d.*	3.2	2.4
14 Yellow	1.6	1.0	n.d.*	n.d.†	0.8
14 Orange	0.7	0.3	18.6	1.6	n.d.†
14 Red/Black	2.5	1.3	23.3	n.d.†	n.d.†
Total % Pristine	46.8	14.5	42	8	12

*n.d. = not detected < 0.5%.
†n.d. = not detected < 0.8%.

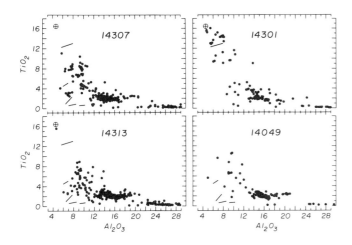

Fig. 4. The populations of glasses having ≤ 29 wt % Al_2O_3 observed in four Apollo 14 regolith breccias are shown. Compare these populations with the complete suite of glasses shown in Fig. 3. Breccia 14047 (not shown) has a population nearly identical to that of breccia 14049.

TABLE 4. Petrographic comparisons (volume %) among six Apollo 14 regolith breccias (after *Simonds et al.*, 1977).

	14307	14313	14047	14049	14301	14318
Matrix (< 39 μm)	65	72	85.5	78	46	42.5
Crystalline Matrix Breccias	6.5	5	1	2.5	28.5	27
Mare Basalt Clasts	6.5	0.5	<0.5	0.5	3.5	4

TABLE 5. Major-element composition (wt %; ± 1σ) of the group "M" mare glasses observed in Apollo 14 regolith breccias.

Oxide	Wt %	Oxide	Wt %
SiO_2	39.6 ± 0.2	MnO	0.31 ± 0.01
TiO_2	6.74 ± 0.07	MgO	8.89 ± 0.19
Al_2O_3	8.84 ± 0.15	CaO	9.68 ± 0.21
Cr_2O_3	0.35 ± 0.02	Na_2O	0.58 ± 0.03
FeO	24.9 ± 0.2	K_2O	0.11 ± 0.01

Several differences are apparent in Fig. 4. First, 14307 and 14313 possess generally similar populations of pristine glass, whereas 14301 contains only the high-Ti varieties. Second, the absolute proportions of pristine glass in these breccias are highly variable (Table 3), ranging from nearly 50% to less than 10%. Third, based in part on the petrographic data from *Simonds et al.* (1977), it appears that 14301 and 14318 are fundamentally distinct from the other Apollo 14 regolith breccias (Fig. 5; Table 4). Fourth, 14301 does not contain any pieces of the mare group "M" glass, whereas the other breccias all contain this variety of impact glass (Table 5).

DISCUSSION

Breccia Taxonomy

The Apollo 14 regolith breccias display significant differences in their abundances and proportions of pristine mare glass. Those data (Figs. 4 and 5; Tables 3 and 4) suggest that 14301 was formed from a regolith that was distinct from the other Apollo 14 regolith breccias. If 14318 is related to 14301, as implied by Fig. 5a (*Simonds et al.*, 1977), and suggested earlier by *Dence and Plant* (1972), then its population of pristine glass should also be observed in future studies to have similar characteristics (Figs. 4 and 5; Table 4). At present the only indication that this may indeed be the case for 14318 is the following qualitative statement made by *Swindle et al.* (1985, p. C520): "The small mare component...is mainly in the form of orange glass." As seen in Table 3 and Fig. 5b, 14301 is indeed exclusively dominated by the high-Ti pristine glasses (orange to red), whereas the other breccias are dominated by the low-Ti pristine glasses (green). Detailed chemical analysis of the glasses contained in breccia 14318 needs to be performed in order to confirm this potential affinity with 14301.

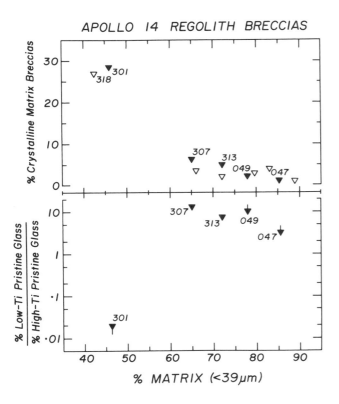

Fig. 5. The volume % of matrix (< 39 μ; *Simonds et al.*, 1977) in the Apollo 14 regolith breccias is shown plotted against: **(a)** percentage of crystalline matrix breccia clasts from *Simonds et al.* (1977) and **(b)** the frequency-ratio of low-Ti pristine glass to high-Ti pristine glass from the present study. Note that 14301 and 14318 are clearly distinct from the other Apollo 14 regolith breccias. The "tails" on three symbols signifies that they are upper or lower limits.

Figure 2 demonstrates that the remaining Apollo 14 regolith breccias form a wide petrographic range (*Simonds et al.*, 1977). A similar feature is evident in 14307, 14313, 14049, and 14047 by the factor-of-6 range in the total percentage of pristine glasses present (Table 3). The sequence among these breccias of increasing proportions of pristine glasses is 14047 to 14049 to 14313 to 14307. This same petrographic sequence is evident in Fig. 2 using other independent petrographic parameters (*Simonds et al.*, 1977).

Although Fig. 4 suggests that breccia 14049 has a distinctive population of pristine glasses in that it does not appear to have the high-Ti varieties (i.e., inverse of 14301 population), it appears likely that this is only an artifact of sampling. Note in Table 3 that the best-analyzed breccias 14313 and 14307 have ratios of pristine low-Ti to pristine high-Ti glasses of about 10/1 (Fig. 5b). Since only 123 glasses were analyzed for breccia 14049, these comparatively rare high-Ti pristine glasses could have been missed in a sample containing a total of only 12% pristine glasses (i.e., only 15 pristine glasses). Similar arguments could also be made for breccia 14047 where only 125 glasses were analyzed. Consequently, those rare high-Ti pristine glasses are believed to be present in both 14047 and 14049 but are at or below the detection limits (< 0.8%) of the present study. In support of this notion, Table 3 shows that one piece of pristine high-Ti orange glass was observed in 14047.

In summary, based on present data, the author concludes that 14047 and 14049 belong to the same general suite of breccias as 14313 and 14307. Breccia 14301, and perhaps 14318, formed from a distinctly different regolith than the other Apollo 14 breccias and merit being considered a separate suite. This view is compatible with the trace-element data on different size-fractions of 14301 and 14318 reported by *Swindle et al.* (1985) and with the petrography of *Dence and Plant* (1972).

Provenance of Regolith Breccias

The Apollo 14 landing site is situated about 30 km from maria to the east and west. Rima Parry, a graben known to have pyroclastics (i.e., pristine glasses) outcropping sporadically along its several hundred kilometer length, is located only about 15 km east of the Apollo 14 site (*Kosofsky and El-Baz*, 1970). Unfortunately, descriptions of the Apollo 14 geology have not discussed the potential sources for the signficant mare component occurring in the Apollo 14 samples (e.g., *Hawke and Head*, 1977; *Head and Hawke*, 1975; *Sutton et al.*, 1972; *Swann et al.*, 1977; *Wilshire and Jackson*, 1972). Consequently, the task of locating the volcanic vents of these Apollo 14 pristine glasses (Table 1) awaits (1) detailed, ground-based, multispectral mapping comparable to that already reported for the Apollo 15 site by *Hawke et al.* (1979) and (2) high-resolution orbital geochemistry by the Lunar Geoscience Orbiter (LGO). Since the lateral transport of lunar surface materials has generally been small during the postmare epoch (e.g., *Rhodes*, 1977), it seems likely that the mare components at the Apollo 14 landing site have been derived from distances of less than a few tens of kilometers.

Closure Ages

Regolith breccias contain the components that were present in their source regoliths at the time of their shock lithification. The differences noted with respect to the Apollo 14 regolith breccias require that different source regoliths were involved. Figure 6 schematically illustrates the possible significance of this observation in relation to the closure ages of the breccias. Note in Fig. 6 that three hypothetical varieties of pristine mare glass (X, Y, Z) are produced by pyroclastic volcanism at 3.2 Gy, 3.5 Gy, and 3.9 Gy, respectively. Within specific time intervals, the regolith will therefore contain diagnostic characteristics with respect to its mare components. For example, since the regolith prior to 3.9 Gy would contain no pristine glasses, breccias formed during that period would share that feature. Following the eruption of glass "Z" onto the lunar surface at 3.9 Gy, but prior to 3.5 Gy, the regolith would contain only pristine glass "Z." Breccias with closure ages of < 3.9 Gy but > 3.5 Gy would possess only pristine glass "Z." It should be evident from Fig. 6 that during the epoch of mare volcanism when new magmatic components are entering the regolith, *the characteristic assemblages in the regolith change markedly with time.* From 3.2 Gy onward, a "lithologic stasis" occurs wherein the volcanic assemblage remains constant.

Based on Fig. 6, the abbreviated assemblage of pristine glasses in 14301 (Table 3; Fig. 4) suggests that its closure age is sometime *during* the period of mare volcanism in the region of the Apollo 14 landing site. In addition, it also suggests that the high-Ti red/black and orange pristine glasses at Apollo 14 are older than the other four varieties of pristine glass (Table 1). This latter inference can be tested by performing ^{40}Ar-^{39}Ar age-determinations on these pristine glasses (refer to section entitled "Future Work").

As shown schematically in Fig. 6, the *post*mare epoch of lithologic stasis is about a factor of 4 longer than the mare epoch. Consequently, for a near-constant cratering rate during the last 3.3 Gy (e.g., *Guinness and Arvidson*, 1977), most

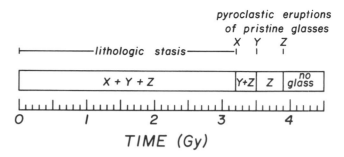

Fig. 6. Hypothetical scenario to illustrate how the volcanic assemblage in lunar regolith changes through time. "X", "Y", and "Z" are chemically distinct, hypothetical pristine glasses erupted onto the lunar surface by pyroclastic volcanism at 3.2 Gy, 3.5 Gy, and 3.9 Gy, respectively. The horizontal bar shows the assemblage of pristine glasses occurring in the regolith at various times.

Apollo 14 regolith breccias should (1) have closure ages in the *post*mare epoch and (2) possess a full complement of pristine glasses. Indeed, four out of the five breccias studied in this work (14313, 14307, and probably 14047, 14049) appear to have closure ages lying within the *post*mare epoch of lithologic stasis.

Table 3 shows that the total abundance of pristine glasses within these Apollo 14 breccias varies from 47% to 8% of the total glass population. These differences may also be a function of closure age. For instance, following the emplacement of pristine glasses onto the lunar surface by pyroclastic volcanism as a concentrated deposit, impact gardening will gradually stir and disperse that component. With time, the volcanics should become thoroughly mixed into the ambient regolith. This progressive dilution with time would suggest that breccia 14307 with 47% pristine glass (i.e., least diluted) has the oldest *post*mare closure age, whereas breccia 14047 with 8% pristine glass (most diluted) has the youngest. Of course, as discussed previously, breccia 14301 has a closure age *within* the mare epoch and is the oldest of the five breccias studied.

While it is tempting to interpret the total abundances of pristine glasses in these breccias as being dominantly a function of closure age, other factors such as the spatial proximity of the source regoliths to the pyroclastic vents may be more important. Under those circumstances, the relative closure ages of these breccias should not follow the old-to-young sequence shown here: 14301, 14307, 14313, 14049, 14047. A possible test of this hypothesis is discussed in the section entitled "Future Work."

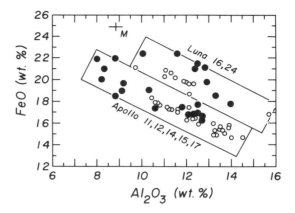

Fig. 7. On a plot of Al_2O_3 (wt %) versus FeO (wt %), the mare basalts (solid symbols) and mare regoliths (open symbols) sampled by the Apollo missions are distinct from those sampled by the Luna missions. The Apollo 14 mare group "M" glass ($\pm 2\sigma$) seems related to the type of *ferrobasalts* sampled by the Luna 16 and Luna 24 missions in the eastern maria, Crisium and Fecunditatis. The presence of mare group "M" glass at Apollo 14 in four of the five regolith breccias suggests that ferrobasalts were also extruded in the western maria of the Moon. A crystalline clast of Apollo 14 ferrobasalt has been reported by *Shervais et al.* (1985). Chemical data are from *Albee et al.* (1972), *Blanchard et al.* (1978), *Compston et al.* (1971), *Dickinson et al.* (1985), *Duncan et al.* (1975), *Gillum et al.* (1972), *Laul et al.* (1978), *Ma et al.* (1978), *Rhodes et al.* (1974), *Rose et al.* (1974), and *Willis et al.* (1972).

Mare Group "M" Glass

A rare variety of mare glass (known as group "M") occurs in breccias 14313, 14307, 14049, and 14047 but *not* in 14301, as shown below:

Sample #	14313	14307	14301	14049	14047
% of "M" glasses	1%	2%	<0.5%	1%	1%

Compared to the pristine mare glasses of *Delano* (1986b), this variety is low in MgO and has a concomitantly low Mg/Al ratio (Table 5). Although this variety is believed to be of impact origin, additional study is required to check its characteristics against the seven criteria for pristinity (*Delano*, 1986b). Figure 7 shows that this unusual group of glasses (1) has a lower abundance of Al_2O_3 than any mare regolith sampled by Apollo and Luna missions and (2) has a higher abundance of FeO than any mare basalt or mare regolith previously sampled. Group "M" Apollo 14 glass is interpreted as having a ferrobasalt composition with apparently little or no contamination by regolith. Consequently, Group "M" glass may be an impact melt formed by a meteorite hitting a ferrobasalt flow-unit somewhere in the general region of the Apollo 14 landing site. The occurrence of this glass suggests that ferrobasalts, which appear to be common in the eastern maria visited by Luna 16 and 24, were also extruded onto the western maria. *Shervais et al.* (1985) and *Dasch et al.* (1986) have also described rare clasts of ferrobasalt from breccia 14321 that support this view.

The lack of group "M" glass in breccia 14301 suggests that this breccia was produced prior to the formation and ballistic transport of these glasses into the vicinity of the Apollo 14 landing site. In addition, it once again implies that breccia 14301 has an older closure age than any of the other four regolith breccias.

Future Work

Although this study has described the value of using different populations of pristine mare glass to infer *relative* closure ages of five Apollo 14 regolith breccias, the application of a laser probe (*Eichhorn et al.*, 1978; *Laurenzi and Turner*, 1987; *Megrue*, 1973; *Muller et al.*, 1977; *Plieninger and Schaeffer*, 1976) provides a great potential for quantifying this further. Specifically, *the age of the youngest component provides an upper limit on the closure age of the breccia.* This would be a logical extension of the pioneering work by *McKay et al.* (1986) using bulk $^{40}Ar/^{36}Ar$ ratios in breccias for inferring relative closure ages.

CONCLUSIONS

Based upon the results of this investigation on the glass populations contained within five Apollo 14 regolith breccias, the following conclusions can be made:

(1) Apollo 14 regolith breccias contain different populations of pristine mare glass.

(2) The relative closure ages among five breccias appear to be the following: (oldest) 14301, 14307, 14313, 14049, 14047 (youngest).

(3) Breccia 14301 was formed *during* the mare epoch but *before* the eruption of the lower-Ti magmas.

(4) The Apollo 14 high-Ti red/black and orange pristine glasses are older than the lower-Ti pristine glasses.

(5) ^{40}Ar-^{39}Ar dating of distinct components within lunar breccias using laser probes has great potential for inferring the closure ages of breccias and for acquiring age information on rare lunar components.

Acknowledgments. The author wishes to thank E. Jerde, J. C. Laul, and J. W. Shervais for their insightful comments during the review process. This research was supported by NASA grant NAG 9-78.

REFERENCES

Albee A. L., Chodos A. A., Gancarz A. J., Haines E. L., Papanastassiou D. A., Ray L., Tera F., Wasserburg G. J., and Wen T. (1972) Mineralogy, petrology, and chemistry of a Luna 16 basaltic fragment, sample B-1. *Earth Planet. Sci. Lett., 13*, 353-367.

Alexander E. C., Jr. and Kahl S. B. (1974) ^{40}Ar-^{39}Ar studies of lunar breccias. *Proc. Lunar Sci. Conf. 5th*, 1353-1373.

Apollo Soil Survey (1971) Apollo 14: Nature and origin of rock types in soil from the Fra Mauro formation. *Earth Planet. Sci. Lett., 12*, 49-54.

Blanchard D. P., Brannon J. C., Aaboe E., and Budahn J. R. (1978) Major and trace element chemistry of Luna 24 samples from Mare Crisium. In *Mare Crisium: The View from Luna 24*, (J. J. Papike and R. B. Merrill, eds.), pp. 613-630. Pergamon, New York.

Boynton W. V., Baedecker P. A., Chou C.-L., Robinson K. L., and Wasson J. T. (1975) Mixing and transport of lunar surface materials: Evidence obtained by the determination of lithophile, siderophile, and volatile elements. *Proc. Lunar Sci. Conf. 6th*, 2241-2259.

Compston W., Berry H., Vernon M. J., Chappell B. W., and Kaye M. J. (1971) Rubidium-strontium chronology and chemistry of lunar material from the Ocean of Storms. *Proc. Lunar Sci. Conf. 2nd*, 1471-1485.

Dasch E. J., Shih C.-Y., Bansal B. M., Wiesmann H., and Nyquist L. E. (1986) Isotopic provenance of aluminous mare basalts from the Fra Mauro Formation (abstract). In *Lunar and Planetary Science XVII*, pp. 150-151. Lunar and Planetary Institute, Houston.

Delano J. W. (1981a) Major-element composition of volcanic green glasses from Apollo 14 (abstract). In *Lunar and Planetary Science XII*, pp. 217-219. Lunar and Planetary Institute, Houston.

Delano J. W. (1981b) Major-element compositions of high-titanium volcanic glasses from Apollo 14 (abstract). In *Lunar and Planetary Science XII*, pp. 220-222. Lunar and Planetary Institute, Houston.

Delano J. W. (1986a) A new aspect of Apollo 14 regolith breccias: Different glass populations and their potential for charting space/time variations (abstract). In *Lunar and Planetary Science XVII*, pp. 168-169. Lunar and Planetary Institute, Houston.

Delano J. W. (1986b) Pristine lunar glasses: Criteria, data, and implications. *Proc. Lunar Planet. Sci. Conf. 16th*, in *J. Geophys. Res., 91*, D201-D213.

Dence M. R. and Plant A. G. (1972) Analysis of Fra Mauro samples and the origin of the Imbrium Basin. *Proc. Lunar Sci. Conf. 3rd*, 379-399.

Dickinson T., Taylor G. J., Keil K., Schmitt R. A., Hughes S. S., and Smith M. R. (1985) Apollo 14 aluminous mare basalts and their possible relationship to KREEP. *Proc. Lunar Planet. Sci. Conf. 15th*, in *J. Geophys. Res., 90*, C365-C374.

Duncan A. R., Sher M. K., Abraham Y. C., Erlank A. J., Willis J. P., and Ahrens L. H. (1975) Interpretation of the compositional variability of Apollo 15 soils. *Proc. Lunar Sci. Conf. 6th*, 2309-2320.

Eichhorn G., James O. B., Schaeffer O. A., and Muller H. W. (1978) Laser ^{39}Ar-^{40}Ar dating of two clasts from consortium breccia 73215. *Proc. Lunar Planet. Sci. Conf. 9th*, 855-876.

Gillum D. E., Ehmann W. D., Wakita H., and Schmitt R. A. (1972) Bulk and rare earth abundances in the Luna-16 soil levels A and D. *Earth Planet. Sci. Lett., 13*, 444-449.

Guinness E. A. and Arvidson R. E. (1977) On the constancy of the lunar cratering flux over the past 3.3×10^9 yr. *Proc. Lunar Sci. Conf. 8th*, 3475-3494.

Hawke B. R. and Head J. W. (1977) Pre-Imbrian history of the Fra Mauro region and Apollo 14 sample provenance. *Proc. Lunar Sci. Conf. 8th*, 2741-2761.

Hawke B. R., MacLaskey D., McCord T. B., Adams J. B., Head J. W., Pieters C. M., and Zisk S. H. (1979) Multi-spectral mapping of the Apollo 15-Apennine region: The identification and distribution of regional pyroclastics. *Proc. Lunar Planet. Sci. Conf. 10th*, 2995-3015.

Head J. W. and Hawke B. R. (1975) Geology of the Apollo 14 region (Fra Mauro): Stratigraphic history and sample provenance. *Proc. Lunar Sci. Conf. 6th*, 2483-2501.

Hubbard N. J., Gast P. W., Rhodes J. M., Bansal B. M., Wiesmann H., and Church S. E. (1972) Nonmare basalts: Part II. *Proc. Lunar Sci. Conf. 3rd*, 1161-1179.

Jerde E. A., Warren P. H., Morris R. V., Heiken G. H., and Vaniman D. T. (1987) A potpourri of regolith breccias: "New" samples from the Apollo 14, 16, and 17 landing sites. *Proc. Lunar Planet. Sci. Conf. 17th*, in *J. Geophys. Res., 92*, E526-E536.

Kieffer S. W. (1975) From regolith to rock by shock. *The Moon, 13*, 301-320.

Kosofsky L. J. and El-Baz F. (1970) *The Moon as Viewed by Lunar Orbiter*. NASA SP-20, NASA, Washington, D.C., 152 pp.

Laul J. C., Vaniman D. T., and Papike J. J. (1978) Chemistry, mineralogy and petrology of seven > 1 mm fragments from Mare Crisium. In *Mare Crisium: The View from Luna 24* (J. J. Papike and R. B. Merrill, eds.), pp. 537-568. Pergamon, New York.

Laurenzi M. A. and Turner G. (1987) Laser probe ^{39}Ar-^{40}Ar dating of impact glasses in lunar breccia 15466 (abstract). In *Lunar and Planetary Science XVIII*, pp. 537-538. Lunar and Planetary Institute, Houston.

Ma M.-S., Schmitt R. A., Taylor G. J., Warner R. D., Lange D. E., and Keil K. (1978) Chemistry and petrology of Luna 24 lithic fragments and < 250 μm soils: Constraints on the origin of VLT mare basalts. In *Mare Crisium: The View from Luna 24* (J. J. Papike and R. B. Merrill, eds.), pp. 569-592. Pergamon, New York.

McKay D. S., Bogard D. D., Morris R. V., Korotev R. L., Johnson P., and Wentworth S. J. (1986) Apollo 16 regolith breccias: Characterization and evidence for early formation in the megaregolith. *Proc. Lunar Planet. Sci. Conf. 16*, in *J. Geophys. Res., 91*, D277-D303.

Megrue G. H. (1973) Spatial distribution of $^{40}Ar/^{39}Ar$ ages in lunar breccia 14301. *J. Geophys. Res.,* 78, 3216-3221.

Muller H. W., Plieninger T., James O. B., and Schaeffer O. A. (1977) Laser probe ^{39}Ar-^{40}Ar dating of materials from consortium breccia 73215. *Proc. Lunar Sci. Conf. 8th,* 2551-2565.

Philpotts J. A., Schnetzler C. C., Nava D. F., Bottino M. L., Fullagar P. D., Thomas H. H., Schuhmann S., and Kouns C. W. (1972) Apollo 14: Some geochemical aspects. *Proc. Lunar Sci. Conf. 3rd,* 1293-1305.

Plieninger T. and Schaeffer O. A. (1976) Laser probe ^{39}Ar-^{40}Ar ages of individual mineral grains in lunar basalt 15607 and lunar breccia 15465. *Proc. Lunar Sci. Conf. 7th,* 2055-2066.

Reid A. M., Ridley W. I., Harmon R. S., Warner J., Brett R., Jakes P., and Brown R. W. (1972a) Highly aluminous glasses in lunar soils and the nature of the lunar highlands. *Geochim. Cosmochim. Acta,* 36, 903-912.

Reid A. M., Warner J., Ridley W. I., Johnston D. A., Harmon R. S., Jakes P., and Brown R. W. (1972b) The major element compositions of lunar rocks as inferred from glass compositions in the lunar soils. *Proc. Lunar Sci. Conf. 3rd,* 363-378.

Rhodes J. M. (1977) Some compositional aspects of lunar regolith evolution. *Phil. Trans. Roy. Soc. London,* A285, 293-301.

Rhodes J. M., Rodgers K. V., Shih C., Bansal B. M., Nyquist L. E., Wiesmann H., and Hubbard N. J. (1974) The relationship between geology and soil chemistry at the Apollo 17 landing site. *Proc. Lunar Sci. Conf. 5th,* 1097-1117.

Rose H. J., Jr., Cuttitta F., Berman S., Brown F. W., Carron M. K., Christian R. P., Dwornik E. J., and Greenland L. P. (1974) Chemical composition of rocks and soils at Taurus-Littrow. *Proc. Lunar Sci. Conf. 5th,* 1119-1133.

Shervais J. W., Taylor L. A., and Lindstrom M. M. (1985) Apollo 14 mare basalts: Petrology and geochemistry of clasts from consortium breccia 14321. *Proc. Lunar Planet. Sci. Conf. 15th,* in *J. Geophys. Res.,* 90, C375-C395.

Simonds C. H., Phinney W. C., Warner J. L., McGee P. E., Geeslin J., Brown R. W., and Rhodes J. M. (1977) Apollo 14 revisited, or breccias aren't so bad after all. *Proc. Lunar Sci. Conf. 8th,* 1869-1893.

Stoffler D., Knoll H.-D., and Maerz U. (1979) Terrestrial and lunar impact breccias and the classification of lunar highland rocks. *Proc. Lunar Planet. Sci. Conf. 10th,* 639-675.

Sutton R. L., Hait M. H., and Swann G. A. (1972) Geology of the Apollo 14 landing site. *Proc. Lunar Sci. Conf. 3rd,* 27-38.

Swann G. A., Bailey N. G., Batson R. M., Eggleton R. E., Hait M. H., Holt H. E., Larson K. B., Reed V. S., Schaber G. G., Sutton R. L., Trask N. J., Ulrich G. E., and Wilshire H. G. (1977) Geology of the Apollo 14 landing site in the Fra Mauro highlands. *U. S. Geol. Survey Prof. Paper 880,* 103 pp.

Swindle T. D., Caffee M. W., Hohenberg C. M., Hudson G. B., Laul J. C., Simon S. B., and Papike J. J. (1985) Noble gas component organization in Apollo 14 breccia 14318: ^{129}I and ^{244}Pu regolith chronology. *Proc. Lunar Planet. Sci. Conf. 15th,* in *J. Geophys. Res.,* 90, C517-C539.

Taylor S. R. (1975) *Lunar Science: A Post-Apollo View.* Pergamon, New York. 372 pp.

Taylor S. R., Kaye M., Muir P., Nance W., Rudowski R., and Ware N. (1972) Composition of the lunar uplands: Chemistry of Apollo 14 samples from Fra Mauro. *Proc. Lunar Sci. Conf. 3rd,* 1231-1249.

Wentworth S. J. and McKay D. S. (1987a) Ancient Apollo 16 regolith breccias: Glass populations and high Mg glass (abstract). In *Lunar and Planetary Science XVIII,* pp. 1072-1073. Lunar and Planetary Institute, Houston.

Wentworth S. J. and McKay D. S. (1987b) Glasses in Apollo 15 regolith breccias (abstract). In *Lunar and Planetary Science XVIII,* pp. 1074-1075. Lunar and Planetary Institute, Houston.

Willis J. P., Erlank A. J., Gurney J. J., Theil R. H., and Ahrens L. H. (1972) Major, minor, and trace element data for some Apollo 11, 12, 14, and 15 samples. *Proc. Lunar Sci. Conf. 3rd,* 1269-1273.

Wilshire H. G. and Jackson E. D. (1972) petrology and stratigraphy of the Fra Mauro Formation at the Apollo 14 site. *U.S.G.S. Prof. Paper 785,* 26 pp.

Glasses in Ancient and Young Apollo 16 Regolith Breccias: Populations and Ultra Mg' Glass

Susan J. Wentworth
Lockheed/EMSCO, 2400 NASA Road One, Houston, TX 77058

David S. McKay
Mail Code SN4, NASA Johnson Space Center, Houston, TX 77058

Major element compositions of glass spheres and fragments in seven possibly ancient (~4 Gy) and three young (<4 Gy) Apollo 16 regolith breccias were determined by electron microprobe analysis for comparison to soil glass compositions. Only two glasses of mare origin were identified in the ancient breccias; both are low TiO_2 glasses. The scarcity of mare glasses supports the possible ancient age of these regolith breccias because it is consistent with an age that predates the most active stages of mare volcanism. In contrast, the young regolith breccia group contains abundant mare glasses, which have a wide variety of compositions. The young regolith breccia mare glass component is similar to that of present-day Apollo 16 soil. The regolith breccias contain several compositional types of highland glass. KREEP, LKFM, and highland basalt glasses are common in all sample types (ancient breccias, young breccias, and soils), but proportions of these glass types in the ancient breccias are different from those in the young breccias and soils, indicating that the ancient regolith was enriched in KREEP and LKFM glasses. The ancient breccias are also distinct from the young breccias and soils in that they do not contain silica-poor gabbroic anorthosite glasses. These silica-poor glasses, which are not as volatile-depleted as HASP (high alumina-silica poor), are common in the young breccias and soils. An unusual type of glass, with atomic Mg/Mg + Fe (Mg') ≥ 0.90, was identified. These ultra Mg' glasses (seven in ancient breccias and one in a young breccia) have Mg' values higher than any known lunar rocks and are unique to the Apollo 16 regolith breccias. The ultra Mg' glasses have a wide range of compositions, and include troctolitic-noritic, quartz normative, KREEPy, and silica-poor glasses. The origins of the high Mg' glasses are not clear, but there are several possibilities; the most exciting is that they represent previously unsampled lunar rock type(s); e.g., a lunar komatiite.

INTRODUCTION

On the basis of unusually high trapped $^{40}Ar/^{36}Ar$ ratios (~8-12), several Apollo 16 regolith breccias may represent regolith that formed ~4 Gy ago (*McKay et al.,* 1986). These are termed ancient regolith breccias, although their ages have not been determined by other means. A few Apollo 16 regolith breccias have lower $^{40}Ar/^{36}Ar$ ratios, ranging from <1 to 4.25 (i.e., like present-day soils to possibly ~3 Gy; *McKay et al.,* 1986), and can be termed young regolith breccias. The Apollo 16 regolith breccias, particularly the ancient ones, have very low surface exposure maturities, which are indicated by low agglutinate contents, trapped gas abundances, and I_s/FeO ratios. Ancient regolith breccia maturities are much lower than those of Apollo 16 soils or regolith breccias from other Apollo sites. Although the breccias are immature, they are relatively fine-grained and their grain size distributions are similar to those of submature and mature lunar soils. The ancient breccias also resemble soils in that they contain abundant (10-14 modal %) nonagglutinitic glass clasts, including homogeneous, clast-bearing, and quench-crystallized fragments and spheres.

The unusual properties of the ancient breccias, i.e., low maturities combined with fine grain sizes and abundant glass clasts, suggest that the regolith precursor of the ancient breccias formed during a time when the effects of large impacts dominated over those of small impacts. *McKay et al.* (1986) proposed that the ancient regolith breccias represent megaregolith formed during the late-stage heavy bombardment associated with large basin formation. If so, this is the first recognized set of samples from the megaregolith, although many other rocks and clasts were undoubtedly megaregolith components during part of their history. The ancient Apollo 16 regolith breccias therefore may be of key importance in gaining an understanding of early lunar surface conditions and processes.

The ancient regolith breccias may contain unusual components such as glass fragments, lithic fragments, and fine-grained material, which reflect regolith conditions at a much earlier time. Glasses in the ancient regolith breccias are important for several reasons. First, the glasses may contain some evidence of an ancient age; for example, proportions of glass types may be unlike those in present-day Apollo 16 soils and young regolith breccias. Second, if the breccias really are ancient, they may contain unusual types of glass that either are not present or are extremely rare in soils because the soils were reworked and diluted by more recent material after the ancient breccias became closed systems. Third, the ancient regolith breccias may lack certain components, such as younger volcanic or impact glasses, so it might be possible to determine lower age limits for the breccias. And finally, the glasses may contain important clues to megaregolith compositions and early surface conditions.

METHODS

Glasses were identified by means of optical microscopy. Because of the heterogeneity of some glasses, a JEOL 35 CF scanning electron microscope was used for backscattered

TABLE 1. Apollo 16 regolith breccias studied, with bulk trapped ^{40}Ar/^{36}Ar ratios (*McKay et al.*, 1986).

Ancient Regolith Breccias		Young Regolith Breccias	
Sample	^{40}Ar/^{36}Ar	Sample	^{40}Ar/^{36}Ar
60016	12.2	61175	4.25
61516	9.5	63507	0.55
65095	-	63588	3.3
65715	11.3		
66035	10.5		
66036	10.4		
66075	11.7		

TABLE 2. Apollo 16 glass classification.

Glass Type	Criteria
Mare	CaO/Al$_2$O$_3$ > 0.75* and FeO > 13 wt %†
Highland	CaO/Al$_2$O$_3$ < 0.75* and FeO < 13 wt %†
Ultra Mg'	Mg/Mg + Fe (atomic) ≥ 0.90‡
KREEP	K$_2$O ≥ 0.25 wt %†
Highland basalt	Al$_2$O$_3$ ≥ 23 wt %§
LKFM	K$_2$O < 0.25 wt % and Al$_2$O$_3$ <23 wt %
Silica-poor gabbroic anorthosite	Insufficient SiO$_2$ for norm†,¶

*E.g., *Papike et al.* (1982).
†*Warner et al.* (1979a).
‡*This study.*
§*Reid et al.* (1972).
¶*Naney et al.* (1976).

imaging of glass textures. The purpose of the textural study was to ensure representative microprobe analyses.

Quantitative major element glass compositions were obtained with an automated Cameca Camebax Microbeam electron microprobe. Analyses were performed at 15 KV accelerating potential and 30 nA beam current, using standard Cameca instrument control and ZAF data reduction software. Extreme care was taken to determine an average composition for each glass: A large (~10 μm) rastered beam was used and, if permitted by grain size, several analyses were done of each glass. Obvious clasts were avoided.

A large portion of nonagglutinitic glasses in the Apollo 16 regolith breccias are quench-crystallized or clast-bearing. Therefore, in order to create the largest possible sampling base, quench-crystallized and clast-bearing glasses were included among the analyzed glasses in addition to optically homogeneous glasses. Many of the quench-crystallized and clast-bearing varieties were spheres or sphere fragments and were obviously originally melt droplets. Quench-crystallized, clast-bearing, and homogeneous glasses in soil core thin section 60010,6037 were selected and analyzed by the methods used for the regolith breccia glasses. Core glass compositions and proportions of compositional types were found to be typical of previously documented Apollo 16 soil glasses (e.g., *Ridley et al.*, 1973), indicating that the procedures used in this study are valid.

Glasses were analyzed in polished thin sections of the seven ancient regolith breccias listed in Table 1. Sample 65095 contains no trapped gases but is assumed to be ancient because it is indistinguishable from the ancient breccias with respect to other characteristics, such as maturity parameters and petrography. Glasses from three young Apollo 16 regolith breccias were also analyzed (Table 1) for comparison to the ancient breccia glasses. Soil glass analyses used for comparison include new data for soil core 60010 and previously published soil glass data (*Glass*, 1976; *Ridley et al.*, 1973). The glass classification (Table 2) is based strictly on composition; no direct relationships between glass types and lunar rocks with similar names are implied. Most of these glass categories were originally defined by the Soil Survey (*Reid et al.*, 1972; *Ridley et al.*, 1973) and we have used them here so that we could easily compare our results with that large data base. Whether any of these glass types actually reflect real lunar rock types is a complex issue that is beyond the scope of this paper. The primary concern of the study was to compare glasses in the ancient breccias, young breccias, and soils, emphasizing the three groups of samples rather than individual samples.

RESULTS

Mare Glasses

Although the Apollo 16 site is in the highlands, mare glasses of exotic origin are commonly present in Apollo 16 soils. Mare glasses are characterized by having high CaO/Al$_2$O$_3$ ratios compared to highland glasses (*Wood*, 1975; *Naney et al.*, 1976). They also contain high FeO and low Al$_2$O$_3$ abundances relative to highland glasses. *Warner et al.* (1979a) defined glasses with FeO > 15 wt % and CaO/Al$_2$O$_3$ > 0.8 as true mare glasses; mare glass criteria used in this study (Table 2) include mixtures of mare and highland material according to the *Warner et al.* (1979a) classification.

TABLE 3. Compositions of mare glasses in ancient Apollo 16 regolith breccias with average soil glass shown for comparison.

Wt %	60016,165-1	66036,13-1	Average Mare 1*
SiO$_2$	48.30	45.47	46.21 (1.43)
TiO$_2$	0.73	2.45	1.76 (1.53)
Al$_2$O$_3$	12.69	13.26	14.34 (3.88)
Cr$_2$O$_3$	0.44	0.37	0.29 (0.10)
FeO	16.37	16.95	14.86 (3.35)
MnO	0.27	0.21	—
MgO	7.97	11.18	10.31 (1.84)
CaO	11.96	10.20	11.00 (1.43)
Na$_2$O	0.10	0.10	0.36 (0.21)
K$_2$O	0.04	0.04	0.10 (0.09)
P$_2$O$_5$	0.04	0.03	—
Total	98.91	100.26	99.23
CaO/Al$_2$O$_3$	0.94	0.77	0.77
Mg'	0.46	0.54	0.58

Numbers in parentheses are standard deviations.
Ridley et al. (1973).

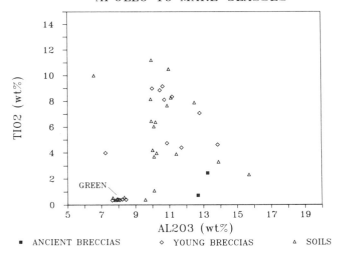

Fig. 1. TiO$_2$ versus Al$_2$O$_3$ for mare glasses in ancient regolith breccias, young breccias, and soils; soil points include new 60010 analyses and data from *Glass* (1976).

Using these criteria, mare glasses are extremely rare in the ancient regolith breccias; of the few glasses that have CaO/Al$_2$O$_3$ ratios > 0.75, only two have FeO contents in the mare range. Thus, mare glasses make up <1% of the 241 glasses analyzed in the ancient regolith breccias. Compositions of the two glasses, which are from samples 60016 and 66036, are given in Table 3. Both glasses are relatively low in TiO$_2$ and similar to the Mare 1 glass defined by *Ridley et al.* (1973) for Apollo 16 soils (Table 3). In contrast to the ancient breccias, the young regolith breccias, as a group, contain significant amounts of mare glass, as do Apollo 16 soils. The range of mare glass compositions in the young regolith breccias is like that in the soils (Fig. 1): In addition to low TiO$_2$ glasses similar to those in the ancient breccias, the young breccias and soils also contain high TiO$_2$ and green glasses.

Highland Glasses

Classification. Highland glasses, with CaO/Al$_2$O$_3$ ratios <0.75, are the dominant glasses at the Apollo 16 site and have a wide range in composition. *Ridley et al.* (1973) defined seven types of highland glass, including two kinds of plagioclase glass, highland basalt, gabbroic anorthosite, LKFM (low-K Fra Mauro), KREEP, and high Mg. The classification scheme used in this study (Table 2) is a modified version of the *Ridley et al.* (1973) scheme. Plagioclase glass is not included as part of the glass population. We define ultra Mg′ glass to be any glass with an atomic Mg/Mg + Fe ratio (Mg number or Mg′ value) ≥ 0.90. Ultra Mg′ glass is not the same as the high Mg glass of *Ridley et al.* (1973), which was defined on the basis of high MgO contents rather than high Mg numbers. Highland basalt glasses are feldspathic (≥23 wt % Al$_2$O$_3$), with compositions similar to those of anorthosites and related crustal rocks. They are equivalent to *Ridley et al.'s* (1973) highland basalt and gabbroic anorthosite glasses, except that gabbroic anorthosite glasses with insufficient SiO$_2$ for a standard norm calculation are considered separately in this study. KREEP glass

TABLE 4. Average compositions of common highland glasses in ancient Apollo 16 regolith breccias, young Apollo 16 regolith breccias, and Apollo 16 soils.

Wt %	Ancient Regolith Breccia Glasses*			Young Regolith Breccia Glasses†			Soil Highland Glasses‡,§		
	KREEP	LKFM	Highland Basalt	KREEP	LKFM	Highland Basalt	KREEP	LKFM	Highland Basalt
SiO$_2$	48.74 (2.22)	47.67 (2.21)	45.11 (0.97)	49.93 (2.50)	45.12 (2.91)	44.71 (1.23)	49.70 (3.36)	46.65 (2.44)	44.63 (1.58)
TiO$_2$	1.59 (0.64)	1.31 (0.64)	0.36 (0.25)	1.90 (1.03)	1.49 (1.19)	0.42 (0.25)	2.19 (1.35)	1.28 (1.23)	0.34 (0.18)
Al$_2$O$_3$	19.29 (3.00)	19.03 (2.55)	27.19 (2.14)	17.32 (3.38)	19.50 (2.24)	27.51 (1.98)	17.60 (5.06)	18.76 (2.86)	27.81 (2.68)
Cr$_2$O$_3$	0.20 (0.06)	0.19 (0.06)	0.09 (0.04)	0.19 (0.04)	0.16 (0.06)	0.09 (0.04)	—	—	—
FeO	7.07 (1.87)	7.12 (1.60)	4.70 (0.97)	8.37 (2.74)	8.74 (1.80)	4.83 (1.38)	9.97 (2.96)	10.09 (2.64)	4.99 (1.54)
MnO	0.11 (0.03)	0.11 (0.03)	0.06 (0.02)	0.13 (0.04)	0.13 (0.04)	0.06 (0.02)	0.14 (0.03)	0.14 (0.03)	0.09 (0.03)
MgO	9.89 (3.63)	12.29 (3.24)	5.93 (1.85)	9.05 (3.27)	11.84 (2.88)	5.72 (1.62)	6.96 (2.21)	9.86 (1.67)	6.13 (2.03)
CaO	11.18 (1.55)	11.38 (1.34)	15.92 (1.18)	10.50 (1.55)	11.96 (1.57)	15.88 (1.02)	10.36 (2.38)	11.55 (1.13)	15.25 (1.32)
Na$_2$O	0.99 (0.29)	0.47 (0.31)	0.44 (0.18)	0.89 (0.34)	0.38 (0.25)	0.45 (0.24)	1.51 (0.48)	0.65 (0.48)	0.49 (0.36)
K$_2$O	0.66 (0.46)	0.10 (0.07)	0.05 (0.05)	0.58 (0.34)	0.10 (0.08)	0.06 (0.05)	0.70 (0.56)	0.10 (0.08)	0.10 (0.08)
P$_2$O$_5$	0.32 (0.36)	0.18 (0.21)	0.06 (0.05)	0.51 (0.45)	0.17 (0.17)	0.14 (0.18)	—	—	—
Total	100.04	99.85	99.90	99.36	99.59	99.75	99.12	99.07	99.84
CaO/Al$_2$O$_3$	0.58	0.60	0.59	0.61	0.61	0.58	0.59	0.62	0.55
Mg′	0.70	0.75	0.69	0.64	0.71	0.67	0.55	0.64	0.66
No. of Glasses	79	58	95	19	11	56	24	21	106
Highland Percent	33.1	24.3	39.7	20.0	11.6	58.9	13.9	12.1	61.3

Numbers in parentheses are standard deviations.
*Highland glasses also include 2.9% ultra Mg′ glass.
†Highland glasses also include 8.4% silica-poor and 1.1% ultra Mg′ glass.
‡Highland glasses also include 2.9% HASP and 9.8% silica-poor gabbroic anorthosite glass.
§*Glass* (1976) and new 60010 data.

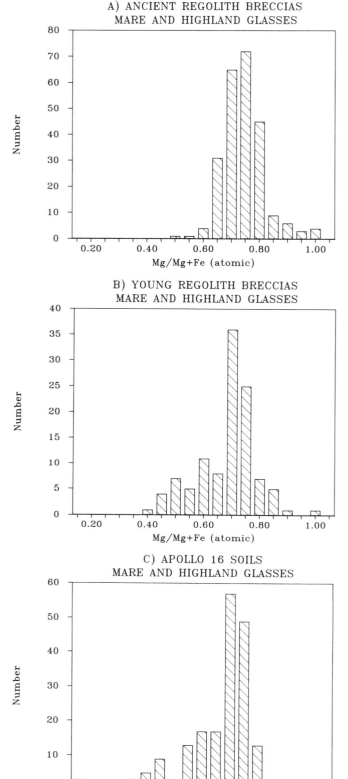

contains ≥0.25 wt % K_2O; LKFM glass is similar to KREEP with respect to most major elements but contains <0.25 wt % K_2O.

Common highland glass types. KREEP, LKFM, and highland basalt glasses are common in all Apollo 16 regolith breccias and soils. Average compositions for the common highland glass types in the ancient regolith breccias, young breccias, and soils are very much alike (Table 4). Averages for each glass type are generally well within one standard deviation of the average for the equivalent glass type in all sample groups. The averages for the ancient breccias, young breccias, and soils are also nearly identical to those originally determined by *Ridley et al.* (1973) for Apollo 16 soil glasses. The one major difference is that Mg' values are high in ancient regolith breccia glasses, especially for the KREEP and LKFM categories, compared to soil and young breccia glasses (Table 4). The differences in Mg' values among the sample groups reflect a slight decrease in MgO and an increase in FeO in going from ancient regolith breccia glasses to young regolith breccia glasses to soil glasses. Figure 2 shows the distribution of Mg' values for all glasses in the three sample groups. It is clear that the ancient regolith breccia glasses are significantly different from the soil glasses because the breccias glasses extend to more magnesian values and lack the iron-rich varieties present in the soils, while the glass distribution for the young regolith breccias is intermediate between ancient breccia and soil glass distributions.

Silica-poor gabbroic anorthosite glass. Silica-poor gabbroic anorthosite glasses are present in the young regolith breccias and soils but not in the ancient regolith breccias. Average compositions of these glasses in the young regolith breccias and soils are very much alike (Table 5). These glasses are distinct from highland basalt glasses in that they contain insufficient SiO_2 to permit a standard norm calculation. They are enriched in refractory oxides, such as Al_2O_3 and CaO, and depleted in volatiles. Similar silica-poor glasses were noted by *Naney et al.* (1976) and were included as part of their gabbroic anorthosite glass group. *Naney et al.* (1976) concluded that these glasses are compositionally intermediate between highland basalt and HASP (high alumina-silica poor) glasses. They are not HASP glasses, according to *Naney et al.'s* (1976) original definition, because SiO_2 is slightly higher and CaO and Al_2O_3 are slightly lower than abundances defined for HASP (Table 5). Glass compositions are within the ranges described by *Warner et al.* (1979a) for Apollo 17 silica-alkali-poor glasses, which also have Al_2O_3 contents below HASP limits. Average compositions for the Apollo 16 and 17 silica-poor glasses are much different (Table 5), however; the Apollo 17 glasses contain a mare component (*Warner et al.*, 1979a) that is not present in the Apollo 16 glasses. No true HASP glasses were found in the samples of the present study, although a few were found in Apollo 16 regolith breccias by Heavilon and Basu (personal communication, 1987).

Fig. 2. Histograms of atomic Mg/Mg + Fe ratios (Mg') for mare and highland glasses in (**a**) ancient regolith breccias, (**b**) young regolith breccias, and (**c**) Apollo 16 soils.

TABLE 5. Average compositions of silica-poor gabbroic anorthosite and HASP glasses in young Apollo 16 regolith breccias and Apollo 16 and Apollo 17 soils.

Wt %	Si-Poor Gabbroic Anorthosites		HASP [†]	Si-Poor [‡]
	Young Breccias	Soils[*]	Soil	Apollo 17
SiO_2	37.16 (2.54)	39.00 (1.98)	31.91 (1.36)	36.27 (2.64)
TiO_2	0.60 (0.22)	0.49 (0.27)	0.82 (0.14)	4.18 (2.26)
Al_2O_3	31.87 (1.41)	29.94 (1.62)	34.16 (1.30)	20.26 (3.21)
Cr_2O_3	0.05 (0.02)	—	0.03 (0.02)	0.19 (0.07)
FeO	4.43 (1.16)	5.80 (1.97)	4.98 (0.81)	12.27 (3.28)
MnO	0.05 (0.02)	0.08 (0.03)	—	0.15 (0.08)
MgO	6.85 (0.92)	6.94 (1.64)	8.48 (1.00)	12.16 (1.57)
CaO	18.20 (0.67)	16.92 (0.86)	19.52 (0.54)	13.75 (1.13)
Na_2O	0.02 (0.02)	0.07 (0.06)	0.13 (0.01)	0.00 (0.00)
K_2O	0.01 (0.01)	0.10 (0.08)	tr	0.02 (0.01)
P_2O_5	0.17 (0.17)	—	—	0.10 (0.01)
Total	99.40	99.33	100.04	99.25
CaO/Al_2O_3	0.57	0.57	0.58	0.68
Mg'	0.74	0.68	0.75	0.64
No. of Glasses	8	17	5	20

Numbers in parentheses are standard deviations.
[*]Glass (1976) and new 60010 data.
[†]Naney et al. (1976).
[‡]Warner et al. (1979b).

Ultra Mg' glass. The Apollo 16 glasses have a wide range in Mg' values (Figs. 2 and 3). Glasses with extremely high Mg' values (0.90-0.98) are present in four ancient regolith breccias (60016, 65095, 65715, and 66035) and one young breccia (61175), but none were found in the soils. These ultra Mg' glasses are very unusual because even the most magnesian lunar rocks, such as pristine peridotites and troctolites, have Mg' values up to only ~0.92 (*Warren*, 1986). Mg' values for most glasses in the ancient breccias, young breccias, and soils, are between 0.60 and 0.80 (Figs. 2 and 3).

Most of the ultra Mg' glasses in the ancient breccias are spheres, ranging from 30-50 μm in diameter; they exhibit a variety of textures, ranging from nearly holohyaline to extensively quench-crystallized (Fig. 4). Quench crystals include olivine and pyroxene (determined by qualitative SEM energy dispersive analysis). Iron metal and troilite droplets are also present in minute amounts in some of the glasses. One glass, 65095,53-22 (Fig. 4), contains small (<1 μm) patches of SiO_2 that may be remnant clasts; no other possible remnant clasts were identified.

The ultra Mg' glasses have a wide range in composition, indicated by the scatter of SiO_2 contents for glasses with Mg' values ≥0.90 in Fig. 3. One glass (65715,5-4) is troctolitic-noritic and characterized by high normative contents of olivine and hypersthene (Table 6). This composition resembles that of the ancient lunar komatiite proposed by *Ringwood et al.* (1987), especially for SiO_2, MgO, and CaO. The glass, however, has a much lower FeO content, so its Mg' value is proportionally higher (Table 6). The glass is also enriched in Al_2O_3 relative to the proposed komatiite; in fact, the Al_2O_3 enrichment in 65715,5-4 is large enough that corundum appears as a normative mineral. The glass is similar in composition to the high Mg glasses defined by *Ridley et al.* (1973) for Apollo 16 soils but the Mg' values of the soil glasses (Table 6) are below the lower limit for ultra Mg' glass; the soil glasses fall into the LKFM category because Al_2O_3 and K_2O contents are, respectively, less than 23 wt % and 0.25 wt %.

Compositions of the other ultra Mg' glasses are even more unusual. Three (60016,171-2; 65095,53-22; 66035,14-12) are quartz normative; two of these have Mg' values of 0.98, the highest of all the glasses (Table 6). The quartz normative glasses contain low abundances of volatile elements, such as Na, K, and P. Two other ultra Mg' glasses (65095,53-13 and 66035,14-14), however, are KREEPy but not quartz normative. The KREEPy glasses are characterized by high volatile abundances (Table 6), which are similar to volatile contents of normal KREEP glasses.

Two ultra Mg' glasses, one from an ancient breccia (60016,165-269) and the only ultra Mg' glass from the young regolith breccias (61175,108-49), are silica-poor (Table 6). These glasses have some similarities to the silica-poor gabbroic anorthosite glasses, especially with respect to SiO_2, Al_2O_3, and CaO abundances (Tables 4 and 6). The silica-poor ultra Mg' glasses, however, contain very little FeO, so their Mg' values are very high.

DISCUSSION

Glass Distributions

The relative abundance of mare glass in the ancient regolith breccias is distinctly lower than mare glass abundances in young regolith breccias and soils (Fig. 5, Table 7). The scarcity of

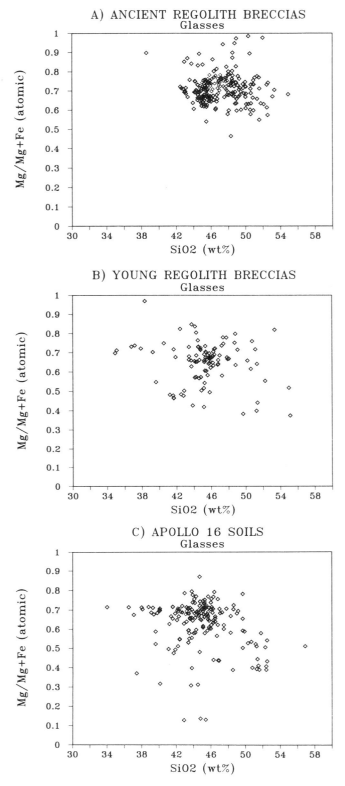

Fig. 3. Atomic Mg/Mg + Fe(Mg′) versus SiO$_2$ for Apollo 16 highland and mare glasses in (**a**) ancient regolith breccias, (**b**) young regolith breccias, and (**c**) soils.

mare glasses in the ancient regolith breccias supports the idea that the breccias, or their regolith precursors, formed prior to the extrusion of most mare basalts. If the ancient breccias represent ~4-Gy-old regolith, then their mare glasses sample very early mare volcanism. Mare basalts with ages >4 Gy have been identified (*Taylor et al.*, 1983; *Dasch et al.*, 1987), so it is possible that related volcanic or impact glasses also exist.

The young regolith breccias, as a group, are quite similar in mare glass abundances to Apollo 16 soils (Fig. 5, Table 7). The high abundances relative to mare glasses in the ancient regolith breccias are consistent with a more recent age for the young regolith breccias, because of the increase in mare volcanic activity with time. Individual young breccias, however, have a wide range in mare glass abundances. There is no clear relationship between mare glass percentages and ^{40}Ar/^{36}Ar ratios for individual regolith breccias of the present study (Fig. 6). Breccia 61175, which has the highest ^{40}Ar/^{36}Ar ratio (4.25) of the young breccias, contains the most mare glasses, even more than typical Apollo 16 soils. A possible explanation for this high mare glass content is that the 61175 regolith may have formed during the most active stages of lunar mare volcanism, since the ^{40}Ar/^{36}Ar ratio suggests that it was exposed at the surface ~3 Gy ago. Breccia 63588, however, has a slightly lower ^{40}Ar/^{36}Ar ratio (3.3) than 61175 and contains no mare glasses, while 63507, with a very soil-like ^{40}Ar/^{36}Ar ratio (0.55), contains less mare glass than do present-day soils.

Relative percentages of highland glasses in the ancient regolith breccias are also unlike those in the young breccias and soils (Fig. 7). Of the common glass types, KREEP and LKFM glasses are enriched and highland basalt glass is depleted in the ancient breccias relative to the young breccias and soils. In addition, the ancient regolith breccias do not contain silica-poor gabbroic anorthosite glass, which is present in the young breccias and soils, suggesting that these glasses were added to the regolith after the ancient breccias were removed from the surface environment. It is likely that the silica-poor gabbroic anorthosite glasses are a less extreme form of HASP glass, and they probably underwent some vapor fractionation and loss of silica. The absence of these silica-depleted glasses in the ancient regolith breccia suite suggests that they may have formed by small-scale impact events and not by the larger impacts that may have dominated the history of the ancient regolith breccias (*McKay et al.*, 1986). Another distinction is that the ancient breccias contain a small but significant amount of ultra Mg′ glass, which is not found in the soils, and which is present in the young breccias only in trace amounts. The unusual highland glass population is further proof that the ancient regolith breccia group is different from the other breccias and soils, and supports the idea that the group represents ancient regolith.

The young regolith breccia group has highland glass proportions similar to those in soils (Fig. 7); i.e., the highland glass component is dominated by highland basalt glass and the content of silica-poor gabbroic anorthosite glass is significant, but ultra Mg′ glass is nearly absent. Compared to the soils, KREEP glass is slightly enriched in the young breccias, indicating that the glass population is intermediate between

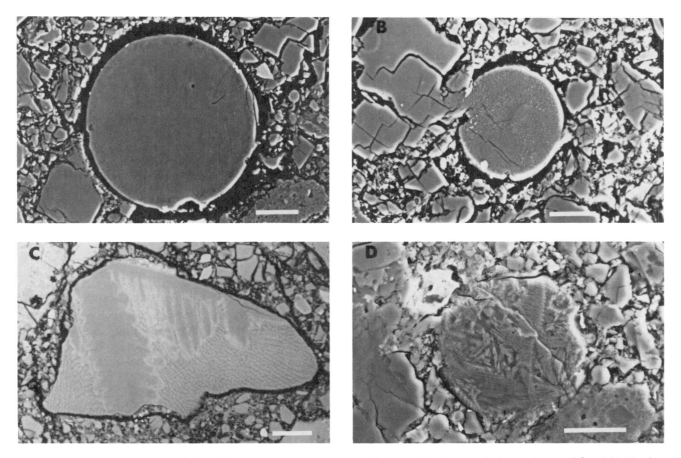

Fig. 4. Photomicrographs of typical ultra Mg' glasses; scale bar on each is 20 μm. (a) Backscattered electron image of 65095,53-22 sphere with quench-crystallized rim and possible remnant SiO$_2$ clasts (small dark patches near rim), (b) backscattered electron image of cryptocrystalline sphere 66035,14-12, (c) backscattered electron image of partly quench-crystallized fragment 65715,5-4, (d) scanning electron image of extensively quench-crystallized sphere 65095,53-13.

soil and ancient breccia populations, consistent with their possible age relationships.

There are differences in highland glass proportions among individual regolith breccias within both the ancient and the young groups. Although percentages of glass types are somewhat scattered relative to the $^{40}Ar/^{36}Ar$ ratios, the general trends found in the ancient breccia, young breccia, and soil groups are also present in the individual samples: With decreasing $^{40}Ar/^{36}Ar$, highland basalt and silica-poor gabbroic anorthosite glass generally increase, while KREEP, LKFM, and ultra Mg' glass decrease. The decrease in KREEP with decreasing $^{40}Ar/^{36}Ar$ is consistent with the data of *Heavilon and Basu* (1987), who found a decrease in maximum glass K$_2$O content with decreasing $^{40}Ar/^{36}Ar$ in Apollo 16 regolith breccias. The individual regolith breccias therefore seem to reflect a change in highland glass population with time (i.e., with decreasing $^{40}Ar/^{36}Ar$ ratio) from ancient breccia-like to soil-like.

The enrichment of KREEP and LKFM glasses in ancient regolith breccias relative to present-day soils (by a factor of about two) is distinct but puzzling. A possible explanation is that these glasses came from different depths in the crust and that the relative contributions to the regolith from various depths has changed over time. Another possibility is that the KREEP and LKFM glasses came from ancient surface basalts and pyroclastics that were mostly destroyed by subsequent bombardment and diluted in younger highland regolith. In any case, these differences in glass populations are real and may provide important clues to early lunar history.

Origin of Ultra Mg' Glasses

The ultra Mg' glasses could have formed either by impact or by volcanic processes. *Delano and Livi* (1981) defined several criteria for distinguishing volcanic glasses from impact glasses; these include (1) absence of schlieren and exotic inclusions, (2) intraglass chemical uniformity, (3) clustering of analyses, (4) high Mg/Al ratios, (5) uniform Ni abundances, and (6) surface-correlated volatiles. The quench-crystallized textures of some ultra Mg' glasses suggest a possible volcanic origin (e.g., Fig. 4d). Likely remnant clasts, other than rare metallic Fe and troilite droplets, are present in only one glass (65095,53-22; Fig. 4a). The glasses that are not quench-

TABLE 6. Compositions of ultra Mg' highland glasses in Apollo 16 regolith breccias.

Wt %	Proposed Komatiite*	Soil Avg. High Mg'†	65715 ,5-4	60016 ,171-2	65095 ,53-22	66035 ,14-12	65095 ,53-13	66035 ,14-14	60016 ,165-269	61175 ,108-49
SiO₂	45.16	45.36	47.40	51.98	50.33	48.49	48.74	48.91	38.57	37.52
TiO₂	0.54	0.19	1.33	1.33	1.61	0.70	1.51	1.56	0.63	0.06
Al₂O₃	6.61	15.41	12.58	16.76	18.41	23.46	17.61	18.81	31.45	32.21
Cr₂O₃	0.40	0.25	0.24	0.27	0.16	0.13	0.20	0.16	0.09	0.03
FeO	14.77	8.46	3.15	0.76	0.47	2.10	1.69	0.52	1.75	0.82
MnO	0.19	—	0.03	0.00	0.08	0.09	0.12	0.17	0.02	0.02
MgO	26.92	21.10	28.67	17.84	16.88	10.44	17.29	10.58	8.55	13.50
CaO	5.31	9.55	5.86	10.16	11.07	14.83	9.30	16.72	18.94	16.17
Na₂O	0.10	0.17	0.08	0.23	0.31	0.24	1.96	1.26	0.01	0.01
K₂O	—	0.01	0.20	0.03	0.08	0.05	1.00	0.50	0.01	0.03
P₂O₅	—	—	0.07	0.02	0.02	0.01	0.40	0.00	0.01	0.03
Total	100.00	100.50	99.60	99.38	99.42	100.54	99.81	99.19	100.03	100.37
CaO/Al₂O₃	0.80	0.62	0.47	0.61	0.60	0.63	0.53	0.89	0.60	0.50
Mg'	0.76	0.82	0.90	0.98	0.98	0.90	0.95	0.97	0.90	0.97
Normative Minerals (wt %)										
Q	0.0	0.0	0.0	3.4	0.7	0.6	0.0	0.0		
C	0.0	0.0	1.7	0.0	0.0	0.0	0.0	0.0		
or	0.0	0.1	1.2	0.2	0.5	0.3	5.9	3.0		
ab	0.8	1.4	0.7	2.0	2.6	2.0	16.6	10.7		
an	17.6	41.2	28.7	44.6	48.6	62.8	36.3	44.2		
di	7.0	4.9	0.0	2.7	1.9	8.4	5.8	27.8		
hy	29.5	19.2	39.3	43.2	41.2	24.8	8.0	3.8		
ol	43.4	33.0	25.0	0.0	0.0	0.0	23.2	6.8		
cm	0.6	0.4	0.4	0.4	0.2	0.2	0.3	0.2		
il	1.0	0.4	2.5	1.3	1.0	1.3	2.9	1.3		
ap	0.0	0.0	0.2	0.0	0.0	0.0	0.9	0.0		

*Ringwood et al. (1987).
†Ridley et al. (1973).

crystallized are compositionally homogeneous. There are not enough glasses to determine whether or not they fall into compositional clusters. The glasses do have high Mg/Al ratios; this criterion is for volcanic mare glasses, however, and may or may not hold true for highland volcanic glasses. There are no data available for Ni or surface-correlated volatiles. It is not clear how closely the different types of ultra Mg' glasses are related to each other, so it is even possible that some are volcanic and others are impact-derived.

There are several possible origins for the ultra Mg' glasses. The first possibility is that some process removed Fe and left Mg in the glasses; the most obvious process is impact-induced differential volatilization because Fe is more volatile than Mg. For the silica-poor ultra Mg' glasses, which are enriched in refractory oxides (e.g., Ca, Al, and Mg) and depleted in volatiles (e.g., Na, K, P, and even Si), some differential volatilization seems likely. Volatilization does not completely account for the high Mg' values, however. Evidence for this is that HASP glasses (*Naney et al.*, 1976), which are even more depleted in volatiles, have much lower Mg' values than do the ultra Mg' glasses. The other ultra Mg' glasses are not depleted in SiO₂ and other volatiles, so volatilization is not likely for them. Another possible process for removing iron from the glasses is by reduction of iron to metal and physical separation during an impact event. Carbon from carbon-rich meteorites, or carbon and hydrogen from implanted solar wind, may have caused reduction of FeO to Fe metal in an impact melt; subsequent centrifugal separation in impact droplets (or gravity separation in large melt sheets) might have resulted in an FeO-depleted melt. This process, however, has not been found to be significant in the many previously studied examples of lunar impact glasses and impact melt rocks. While reduction of iron occurs during agglutinate production, nearly all of the reduced iron remains in the agglutinate in the form of tiny droplets, which contribute to the ferromagnetic resonance signature used to determine the relative maturity of soils.

Another possible explanation for the ultra Mg' glasses is that they might be mixtures of ultra Mg' meteoritic material and lower Mg' (i.e., normal) lunar material. An objection to this is that meteoritic materials with ultra Mg' values are rare; e.g., ordinary and carbonaceous chondrites generally have bulk Mg' values below 0.90. Although some meteoritic olivines, enstatites, and chondrules have Mg' values above 0.90 (e.g., *Wlotzka*, 1983), these phases are found as inclusions in meteorites with much lower bulk Mg' values.

The last possibility is that the ultra Mg' glasses represent previously unidentified lunar rock types. They may have been derived by impact from deep-seated rocks with very high Mg' values (e.g., troctolites or peridotites), some of which may have combined with SiO₂- and volatile-rich rocks to produce

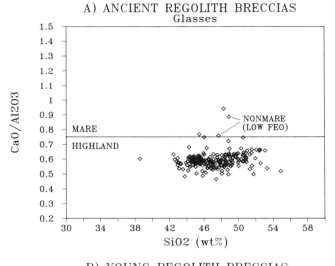

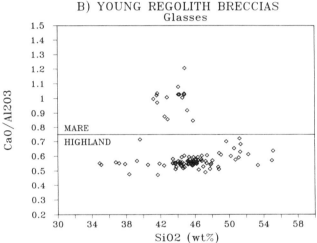

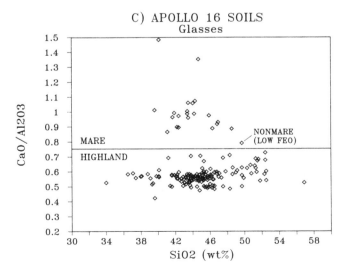

Fig. 5. CaO/Al$_2$O$_3$ versus SiO$_2$ for Apollo 16 mare and highland glasses in (a) ancient regolith breccias, (b) young regolith breccias, and (c) soils.

TABLE 7. Relative percentages of Apollo 16 glass types.

Type of Sample	No. of Glasses	Mare Glass	Highland Glass
Regolith Breccia			
Ancient	241	0.8%	99.2%
Young	111	14.4	85.6
Soil*	190	11.6	88.4

*Glass (1976) and new 60010 data.

the quartz normative and KREEPy glasses. An extrusive volcanic origin is also feasible. The troctolitic-noritic glass, in particular, could be a komatiite similar to the ancient lunar komatiite (Table 6) proposed by *Ringwood et al.* (1987). There are some problems in comparing any of the ultra Mg' glasses to the Ringwood komatiite, however. The Mg' values of the Ringwood komatiite are much lower than those in the glasses. Additionally, the proposed composition of the komatiite was determined by extrapolation of Apollo 16 breccia compositions to a primitive chondritic composition with a Ca/Al ratio of 1.08 (CaO/Al$_2$O$_3$ ratio of 0.80; *Ringwood et al.*, 1987), a much different ratio than is found in most ultra Mg' glasses (Table 6). Although the troctolitic-noritic glass may be komatiitic, the quartz normative and KREEPy ultra Mg' glasses are more difficult to explain, since high Mg' values and volatile enrichments are on the opposite ends of magmatic fractionation trends. Perhaps they formed as a result of mixing, by either impact or magmatic processes, of very mafic mantle material with highly evolved SiO$_2$-rich or KREEP rocks. One possible scenario is that a very hot komatiitic magma, rising through conduits in the upper crust, melted and assimilated the low-melting fractions of the surrounding country rock. It also incorporated xenoliths, the remnants of which may be the silica-rich clasts observed in one of the ultra Mg' droplets (Fig. 4). The komatiite magma

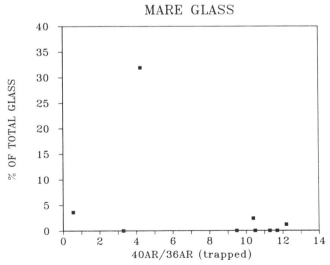

Fig. 6. Mare glass (percent of total glass) versus bulk trapped ^{40}Ar/^{36}Ar in Apollo 16 regolith breccias; possible age increases from left to right.

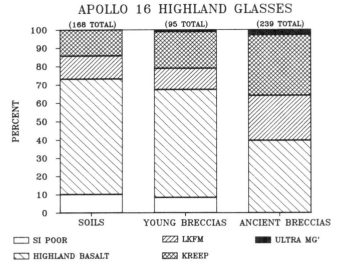

Fig. 7. Relative proportions of Apollo 16 highland glass types in soils, young regolith breccias, and ancient regolith breccias.

could have formed fire fountains similar to those postulated for the orange and black volcanic glass at the Apollo 17 site (*Heiken et al.*, 1974). Thus, lunar komatiites might be preserved only as glass droplets and not as extrusive flows.

CONCLUSIONS

1. Glass populations in the possibly ancient Apollo 16 regolith breccia group are unlike those in young regolith breccia and present-day soil groups, increasing the evidence that regolith breccias with unusually high trapped $^{40}Ar/^{36}Ar$ ratios represent ancient regolith.

2. Mare glasses are extremely rare in the ancient regolith breccias, constituting <1% of all glasses. The relative lack of mare glasses supports the idea that the ancient breccias represent regolith that formed prior to the extrusion of most mare basalts.

3. Highland glass populations indicate that the ancient regolith was enriched in KREEP and LKFM glasses by a factor of two. This enrichment constitutes an important fundamental property of the ancient regolith breccias.

4. Silica-poor gabbroic anorthosite and HASP glasses are not present in the ancient breccias; this suggests that silica-poor glasses are created mainly in small-scale impact events.

5. Individual regolith breccias generally reflect a change in glass population with time (indicated by decreasing $^{40}Ar/^{36}Ar$ ratios), which includes the addition of mare and silica-poor glasses and possibly highland basalt glass.

6. Ultra Mg' glasses, with Mg' values ≥0.90, are unique to the Apollo 16 regolith breccias and are the most common in the ancient breccias. They may be of great importance to lunar petrogenetic studies because their Mg' values range higher than those of any known lunar samples. While only eight ultra Mg' glasses have been identified, they have diverse compositions, including troctolitic-noritic, quartz normative, KREEPy, and silica-poor varieties. Because of the relatively few samples and their wide compositional ranges, relationships among the ultra Mg' glass types and relationships to other highland glasses are not clear.

7. The origin of the ultra Mg' glasses remains puzzling, but there are several possibilities. The ultra Mg' glasses could indicate the presence of previously unsampled crustal or volcanic rock types on the Moon. An exciting possibility is that some may represent lunar komatiites similar to the ancient lunar komatiite proposed by *Ringwood et al.* (1987).

Acknowledgments. We thank J. J. Papike (associate editor), S. B. Simon, and L. A. Taylor for useful reviews, and A. Basu for helpful comments.

REFERENCES

Dasch E. J., Shih C.-Y., Bansal B. M., Wiesmann H., and Nyquist L. E. (1987) Isotopic analysis of basaltic clasts from lunar breccia 14321: Chronology and petrogenesis of pre-Imbrium mare volcanism. *Geochim. Cosmochim. Acta*, in press.

Delano J. W. and Livi K. (1981) Lunar volcanic glasses and their constraints on mare petrogenesis. *Geochim. Cosmochim. Acta, 45*, 2137-2149.

Glass B. P. (1976) Major element composition of glasses from Apollo 11, 16 and 17 soil samples. *Proc. Lunar Sci. Conf. 7th*, 679-693.

Heavilon C. and Basu A. (1987) Compositions of glass fragments in Apollo 16 regolith breccias (abstract). In *Lunar and Planetary Science XVIII*, pp. 413-414. Lunar and Planetary Institute, Houston.

Heiken G. H., McKay D. S. and Brown R. W. (1974) Lunar deposits of possible pyroclastic origin. *Geochim. Cosmochim. Acta, 38*, 1703-1718.

McKay D. S., Bogard D. D., Morris R. V., Korotev R. L., Johnson P., and Wentworth S. J. (1986) Apollo 16 regolith breccias: Characterization and evidence for early formation in the megaregolith. *Proc. Lunar Planet. Sci. Conf. 16th*, in *J. Geophys. Res., 91*, D277-D303.

Naney M. T., Crowl D. M., and Papike J. J (1976) The Apollo 16 drill core: Statistical analysis of glass chemistry and the characterization of a high alumina-silica poor (HASP) glass. *Proc. Lunar Sci. Conf. 7th*, 155-184.

Papike J. J., Simon S. B., and Laul J. C. (1982) The lunar regolith: Chemistry, mineralogy, and petrology. *Rev. Geophys. Space Phys., 20*, 761-826.

Reid A. M., Warner J., Ridley W. I., and Brown R. W. (1972) Major element composition of glasses in three Apollo 15 soils. *Meteoritics, 7*, 395-415.

Ridley W. I., Reid A. M., Warner J. L., Brown R. W., Gooley R., and Donaldson C. (1973) Glass compositions in Apollo 16 soils 60501 and 61221. *Proc. Lunar Sci. Conf. 4th*, 309-321.

Ringwood A. E., Seifert S., and Wänke H. (1987) A komatiite component in Apollo 16 highland breccias: Implications for the nickel-cobalt systematics and bulk composition of the moon. *Earth Planet. Sci. Lett., 81*, 105-117.

Taylor L. A., Shervais J. W., Hunter R. H., Shih C.-Y., Nyquist L. E., Bansal B. M., Wooden J., and Laul J. C. (1983) Pre-4.2 AE marebasalt volcanism in the lunar highlands. *Earth Planet. Sci. Lett., 66*, 33-47.

Warner R. D., Taylor G. J., and Keil K. (1979a) Composition of glasses in Apollo 17 samples and their relation to known lunar rock types. *Proc. Lunar Planet. Sci. Conf. 10th*, 1437-1456.

Warner R. D., Taylor G. J., Wentworth S. J., Huss G. R., Mansker W. L., Planner H. N., Sayeed U. A., and Keil K. (1979b) Electron microprobe analyses of glasses from Apollo 17 rake sample breccias and Apollo 17 drill core. *Inst. Meteoritics Spec. Publ. No. 20,* University of New Mexico, Albuquerque. 20 pp.

Warren P. H. (1986) The bulk-moon MgO/FeO ratio: A highlands perspective. In *Origin of the Moon* (W. K. Hartmann, R. J. Phillips, and G. J. Taylor, eds.), pp. 279-310. Lunar and Planetary Institute, Houston.

Wlotzka F. (1983) Compositions of chondrules, fragments and matrix in unequilibrated ordinary chondrites Tieschitz and Sharps. In *Chondrules and Their Origins* (E. A. King, ed.), pp. 296-318. Lunar and Planetary Institute, Houston.

Wood J. A. (1975) Lunar petrogenesis in a well-stirred magma ocean. *Proc. Lunar Sci. Conf. 6th,* 1087-1102.

^{10}Be Profiles in Lunar Surface Rock 68815

K. Nishiizumi[1], M. Imamura[2], C. P. Kohl[1], H. Nagai[3], K. Kobayashi[4], K. Yoshida[5], H. Yamashita[6], R. C. Reedy[7], M. Honda[3], and J. R. Arnold[1]

—Author affiliations follow References section—

Cosmic-ray-produced ^{10}Be ($t_{1/2} = 1.6 \times 10^6$ years) activities have been measured in 14 carefully ground samples of lunar surface rock 68815. The ^{10}Be profiles from 0 to 4 mm are nearly flat for all three surface angles measured and show a very slight increase with depth from the surface to a depth of 1.5 cm. These depth profiles are in contrast to the SCR (solar cosmic ray)-produced ^{26}Al and ^{53}Mn profiles measured from these same samples. There is no sign of SCR-produced ^{10}Be in this rock. The discrepancy between the data and the Reedy-Arnold theoretical calculation for SCR ^{10}Be production (about 2 atoms/min/kg at the surface) can be explained in two ways: (1) The low-energy cross sections for proton-induced ^{10}Be production from oxygen are lower than those used in the calculations, or (2) compared to the reported fits for ^{26}Al and ^{53}Mn, the solar proton spectral shape is actually softer (exponential rigidity parameter R_0 less than 100 MV), the omnidirectional flux above 10 MeV is higher (more than 70 protons/cm^2 s), and the erosion rate is higher (greater than 1.3 mm/m.y.). ^{10}Be, as a medium- to high-energy product, is a very useful nuclide for determining the SCR spectral shape in the past.

INTRODUCTION

Variations in the flux and spectrum of solar cosmic ray (SCR) particles are related to the variations of solar activity. Knowledge of the history of solar activity is extremely important not only to understanding solar physics but also quite possibly the climatic history of the Earth (for example, the glaciation cycle). Although direct SCR measurements by satellites have been performed only for the last few decades, we do have a good way to study the past record. Cosmic rays produce radioactive and stable nuclides during interactions with lunar surface materials and meteorites. The concentrations of these cosmogenic nuclides are directly related to the average cosmic-ray intensity in the past. The nuclides of interest are produced not only by SCR but also by galactic cosmic rays (GCR). The GCR do not have a solar origin although their spectrum and flux are modulated to some extent by solar activity. The much lower energy (but higher flux) of the SCR means that their effects can only be seen in the top few millimeters of lunar materials while GCR-produced nuclides dominate at greater depths. Except for a few cases (e.g., *Nishiizumi et al.*, 1986; *Evans et al.*, 1987), the record of SCR effects is erased in meteorites during passage through the Earth's atmosphere. Details concerning these two types of cosmic rays and their interactions are given in several articles (e.g., *Reedy and Arnold*, 1972; *Reedy*, 1980; *Reedy et al.*, 1983).

Previously, we have studied the depth profiles of SCR-produced radionuclides in the upper 2 cm of lunar rocks and determined the average SCR parameters, flux, and energy spectrum over a time scale from a few months to 10 m.y. The comparison of ^{26}Al ($t_{1/2} = 7.05 \times 10^5$ years) and ^{53}Mn ($t_{1/2} = 3.7 \times 10^6$ years) depth profiles in the surface of three lunar rocks, 12002 (*Finkel et al.*, 1971), 14321 (*Wahlen et al.*, 1972), and 68815 (*Kohl et al.*, 1978), with the theoretical SCR production profiles (*Reedy and Arnold*, 1972) indicates that the flux of solar protons over the past 5 to 10 m.y. was similar to that during the past million years and that the average SCR spectrum and flux were characterized by an exponential rigidity with a spectral shape parameter $R_0 = 100$ MV (cf. *Reedy and Arnold*, 1972) and a flux $J = 70$ protons/cm^{2}s (E > 10 MeV, 4 π). These calculations assume 0.5-2.2 mm/m.y. erosion rate for the three rocks (*Kohl et al.*, 1978; *Russ and Emerson*, 1980). It is possible to fit the data also with R_0 in the range 70-150 MV (a higher value of R_0 corresponds to more higher-energy protons per lower-energy proton) with appropriate adjustments of flux J and erosion rate. [The excitation functions for producing ^{26}Al and ^{53}Mn by proton-induced reactions are quite similar (*Reedy and Arnold*, 1972)]. In this present work, we measured ^{10}Be ($t_{1/2} = 1.6 \times 10^6$ years) in rock 68815 by accelerator mass spectrometry (AMS) to investigate the SCR production of this nuclide and to verify the SCR parameters. Rock 68815 is a breccia that was collected from the top of a meter-high boulder. The ^{81}Kr-Kr exposure age of this rock is 2.04 ± 0.08 m.y. (*Drozd et al.*, 1974), and 68815 is thought to have been ejected by the South Ray crater event from a depth that cosmic-ray particles could not reach.

EXPERIMENTAL METHODS AND RESULTS

Fourteen samples were separated from aliquant samples that we had previously dissolved and used for ^{26}Al and ^{53}Mn measurements (*Kohl et al.*, 1978). The samples measured were from three different zenith angles [A 48 ± 16°, B 41 ± 17°, and C 29 ± 15° (*Russ and Emerson*, 1980)] and four different nominal depths (0-0.5, 0.5-1.0, 1.0-2.0, and 2.0-4.0 mm). A 4-8 mm and a 10-15 mm layer were also obtained from near the bottom of our specimen from beneath face A and face C. The details of the grinding procedures were described by *Kohl et al.* (1978). The sample sizes ranged from 0.6 to 2.9 g. About 700 µg of Be carrier was added to each sample dissolved. Beryllium was separated from other elements and purified by anion exchange, cation exchange, and Be-acetylacetone extraction. Finally, Be(OH)$_2$ was precipitated

TABLE 1. ^{10}Be in 68815 (dpm ^{10}Be/kg).

Depth (g/cm^2)*	Face A	Face B	Face C
0 -0.13	6.19 ± 0.21	7.28 ± 0.37	6.92 ± 0.24
0.13-0.28	6.75 ± 0.23	7.22 ± 0.65	6.47 ± 0.30
0.28-0.57	6.61 ± 0.18	7.24 ± 0.66	6.81 ± 0.20
0.57-1.03	6.81 ± 0.21	7.21 ± 0.32	7.07 ± 0.29
1.03-2.19†	7.22 ± 0.21		
2.8 -4.2†,‡	7.43 ± 0.53		

*These depths are averages over the three faces according to calculations by *Russ and Emerson* (1980). The depths of the individual faces differ no more than 15% from the averages. Values in the figures were plotted at the actual depth for each face. A density of 2.8 g/cm^3 was used in the depth calculations.

†The two deepest samples were ground from areas under both faces A and C.

‡The depth for this layer is from *Kohl et al.* (1978).

with water containing about 2% of ^{17}O so that ^{9}Be^{17}O would pass into the accelerator with ^{10}Be^{16}O (*Imamura et al.*, 1984).

The ^{10}Be measurements were carried out at the University of Tokyo's tandem Van de Graaff accelerator. The apparatus and method used for the accelerator mass spectrometry were essentially those described previously (*Imamura et al.*, 1984). We selected a 3.5 MV terminal voltage for the ^{10}Be measurements. We measured ^{10}Be/^{9}Be ratios in the range 1-5 × 10^{-10} with experimental errors of 3-9% (±1 σ). The ^{10}Be/^{9}Be measured values were normalized to ICN-UCSD ^{10}Be standard (*Nishiizumi et al.*, 1984). The ^{10}Be activities obtained from 68815 are given in Table 1.

DISCUSSION

The ^{10}Be activity depth profiles in the three faces A, B, and C of 68815 are shown in Fig. 1a. The ^{10}Be results were adjusted to saturation using the ^{81}Kr-Kr exposure age of 2.04 ± 0.08 m.y. (*Drozd et al.*, 1974). The saturation activities are used for all the following discussions. The ^{26}Al and ^{53}Mn depth profiles in the same samples are also shown in Figs. 1b and 1c. The curves shown in Figs. 1b and 1c are the Reedy-Arnold theoretical profiles for the sum of SCR and GCR production rates for each nuclide (*Reedy and Arnold*, 1972). The curves are slightly modified from the previous paper (*Kohl et al.*, 1978) by calculating the production rate on a point by point basis (*Russ and Emerson*, 1980). SCR parameters R$_0$ = 100 MV and J = 70 p/cm^2 s, a 2.0 m.y. exposure age, and 0.0 and 1.0 mm/m.y. erosion rates are adopted for these calculations. The ^{10}Be profiles are essentially the same for all three faces and are nearly flat except for a slight increase with increasing depth. This profile is in remarkable contrast to the ^{26}Al and ^{53}Mn profiles, which show sharp increases in activity toward the surface due to SCR production of these nuclides.

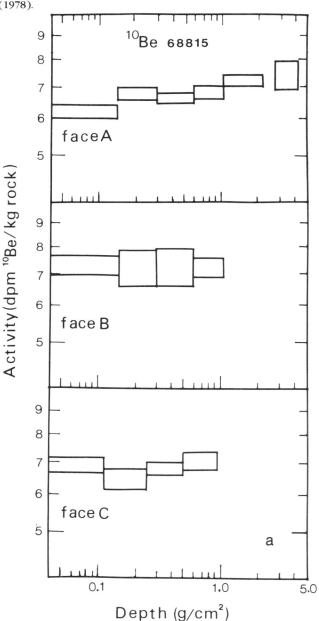

Fig. 1. Measured (**a**) ^{10}Be, (**b**) ^{26}Al, and (**c**) ^{53}Mn activity depth profiles in the three faces of 68815. The width of the error bars indicates the average depth interval sampled as determined from the maps made during grinding (*Russ and Emerson*, 1980). The curves shown for ^{26}Al and ^{53}Mn are the theoretical profiles of the SCR plus GCR production calculated by *Russ and Emerson* (1980) on a point by point basis for 68815 using a 2 m.y. exposure age and the Reedy-Arnold model. Figures 1b and 1c are from *Murrell* (1980).

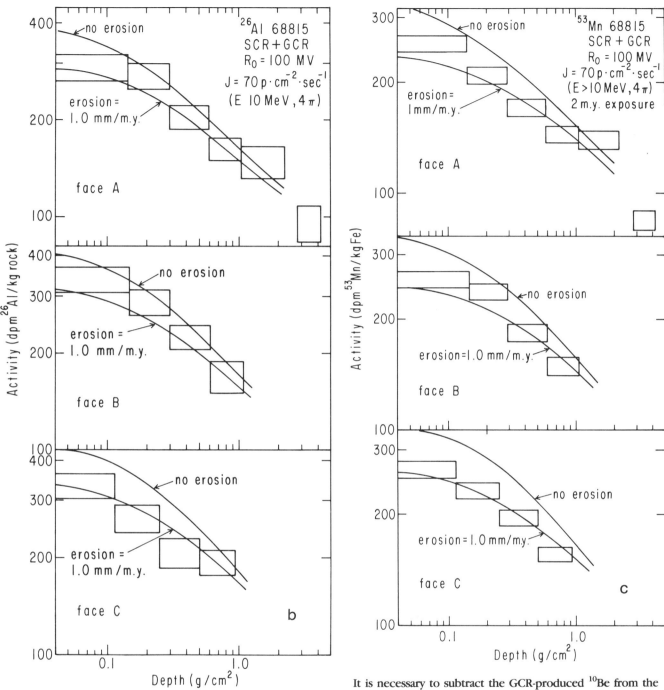

Russ and Emerson (1980) recalculated ^{26}Al and ^{53}Mn depth profiles in 68815 using point by point mapping of all grinding faces. Even though their detailed calculation shows that the average angles of the faces from horizontal are substantially different from those used by *Kohl et al.* (1978), they obtained essentially the same conclusions with regard to the SCR parameters, R_0 and J. The more detailed calculations of *Russ and Emerson* (1980) showed no evidence of SCR anisotropy or of differential erosion for the three surfaces.

It is necessary to subtract the GCR-produced ^{10}Be from the observed ^{10}Be to see the SCR component. The expected GCR production profile using the chemical composition of 68815 was calculated based on the Reedy-Arnold model (*Reedy and Arnold*, 1972) and is shown in Figs. 2a,b. The measured ^{10}Be profile for face B is essentially the same as for face A and C, but the data contain somewhat larger errors. The original model (*Reedy and Arnold*, 1972) and the evaluated cross sections of *Tuniz et al.* (1984) were used for both GCR and SCR calculations. The Reedy-Arnold GCR profile fits the 68815 data well without any of the normalization that was required for both the ^{26}Al and ^{53}Mn GCR production profiles (*Nishiizumi et al.*, 1983). The Reedy-Arnold GCR ^{10}Be profile using the new cross sections also fits the ^{10}Be results for the Apollo 15 drill core (*Nishiizumi et al.*, 1984). However, the Reedy-Arnold GCR profile for ^{10}Be appears to increase with depth slower than the measured data, suggesting that the Reedy-Arnold GCR model might be slightly inaccurate for the production rate versus depth profile near the surface, at least for high-energy products. As pure GCR production profiles are hard to find (almost all nuclides have significant SCR components near the surface), it is difficult to test the Reedy-Arnold model at such shallow depths.

We would expect to see excess ^{10}Be due to SCR production, if present, in near-surface samples. Figures 2a and 2b, however, show no indication of SCR-produced ^{10}Be in this rock. The Reedy-Arnold SCR model predicts a ^{10}Be SCR production rate of about 2 atoms/min/kg in the surface layer using a SCR flux with $R_0 = 100$ MV and $J(>10$ MeV$) = 70$ p/cm^2 s (see Fig. 3a), the parameters obtained from ^{26}Al and ^{53}Mn profiles in this and other lunar surface rocks (*Kohl et al.*, 1978). Taking values at the limits of the errors, minimum values for the deeper samples and maximum for the near surface samples, we obtained a value for SCR-produced ^{10}Be of less than 1 dpm/kg.

The use of an exponential-rigidity spectral shape for the solar protons needs to be justified here much more than in previous works because much of the SCR production of ^{10}Be occurs for proton energies above 100 MeV, whereas most of the production of ^{53}Mn and ^{26}Al occurs at much lower energies. For example, only about 4% of ^{53}Mn or ^{26}Al but 55% of ^{10}Be

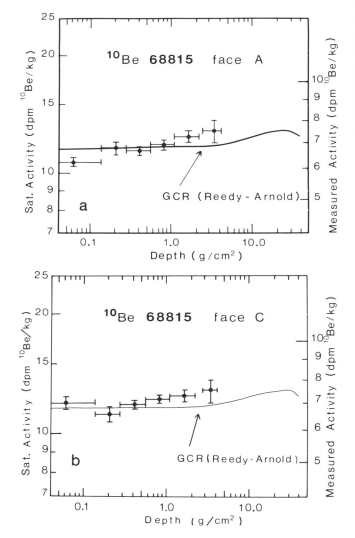

Fig. 2. ^{10}Be activity depth profiles in (a) face A and (b) face C of 68815. Values are plotted as measured activities on the right-hand scale and as saturation activities on the left-hand scale. The ^{81}Kr-Kr exposure age of 2.04 ± 0.08 m.y. (*Drozd et al.*, 1974) was used to calculate the saturation activities. The curves are the unnormalized GCR production profiles calculated using the Reedy-Arnold model (*Reedy and Arnold*, 1972) and the revised cross sections (*Tuniz et al.*, 1984). They include no corrections for erosion or surface inclination. The two deepest points were ground from under both faces A and C and are shown in both (a) and (b).

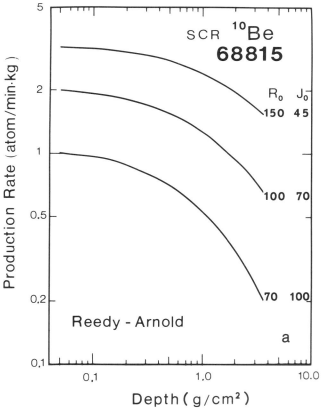

Fig. 3. SCR production profiles for (a) ^{10}Be, (b) ^{26}Al, and (c) ^{53}Mn, calculated using 68815 chemical composition. The depth profiles were calculated using the Reedy-Arnold model (*Reedy and Arnold*, 1972) and the cross sections of *Tuniz et al.* (1984) for ^{10}Be. They show expected saturation levels for each nuclide for three sets of SCR parameters ($R_0 = 150$ MV, $J = 45$ p/cm^2 s, $R_0 = 100$, $J = 70$, and $R_0 = 70$, $J = 100$). Erosion was assumed to be 0 for these calculations.

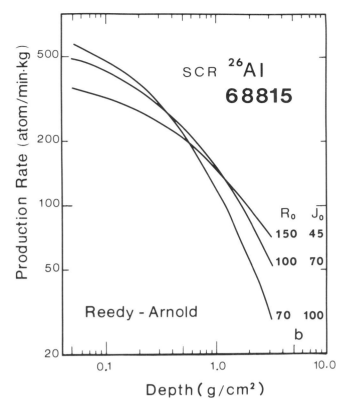

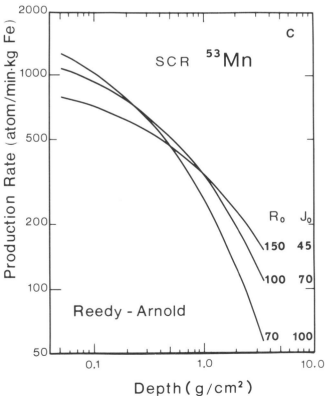

are made at energies above 100 MeV by primary protons with a spectral shape of $R_0 = 100$ MV. Below the lunar surface, these fractions for production by protons with E >100 MeV increase as the lower energy protons are preferentially stopped. While some investigators use a power law in energy for the spectral shape of solar protons over limited energy intervals below 100 MeV, all spectral observations for proton energies above 100 MeV show a curvature down relative to a pure power law (see, e.g., *McGuire and von Rosenvinge*, 1984). *Freier and Webber* (1963) used several techniques to measure the fluxes of solar protons up to energies of a few GeV and concluded that an exponential rigidity was a good fit to solar proton spectra at high energies. However, most of their spectra were based on indirect observations using balloons. *McGuire and von Rosenvinge* (1984) used instruments on the IMP-7 and IMP-8 satellites to measure proton fluxes for many energies up to 400 MeV during several solar flares. While they preferred a modified Bessel function of the second kind to fit the proton's energy spectra, in the figures of their paper an exponential rigidity function appears to give equally good fits for energies from 10 to 400 MeV. As less than 1% of the ^{10}Be is made by protons with E >400 MeV, the study by *McGuire and von Rosenvinge* (1984) confirms the validity of our using an exponential rigidity for the spectral shape of solar protons for calculating and comparing the SCR productions of ^{10}Be, ^{26}Al, and ^{53}Mn.

The Reedy-Arnold model is a well-developed method for calculating cosmic-ray interactions in bodies of various sizes for both SCR and GCR. However, the model contains some uncertainties with regard to estimating the GCR flux at near-surface depths. If the calculated Reedy-Arnold GCR ^{10}Be production rates are overestimated near the surface, the flat profile we measured may be due to a real SCR component in the surface layers. However, SCR ^{10}Be production rates decrease drastically with increasing depth below a few g/cm^2 regardless of the SCR parameters (see Fig. 3a). The observed ^{10}Be activities at a few g/cm^2 and below are almost entirely produced by GCR interactions and the measured values at these depths are in agreement with the theoretical values and the Apollo 15 drill cores. The GCR production rate would have to decrease 15-20% from 1 g/cm^2 to the surface to match the measured profile if the SCR production rate is 2 atoms ^{10}Be/min/kg in this region. There is no theoretical or experimental support for such an abrupt change for the penetrating GCR particles, and this explanation is unlikely. The discrepancy between the Reedy-Arnold model for SCR ^{10}Be production and the data can be explained as discussed below.

1. The ^{10}Be proton-induced cross sections that are used for Reedy-Arnold SCR calculations may be too high, especially for the low-energy region. Although the SCR spectrum varies from flare to flare, SCR particle intensity decreases exponentially with increasing energy (*Reedy and Arnold*, 1972). The low-energy proton-induced cross sections, especially below 100 MeV, are therefore very important for total SCR production. However, there are no cross section measurements for ^{10}Be production below 135 MeV for protons on any target element. *Reedy and Arnold* (1972) estimated the ^{10}Be cross sections from nuclear systematics and comparisons with the measured

^{7}Be cross sections at lower energies. The original Reedy-Arnold calculation for R_0 = 100 MV, J = 70 protons/cm^2 s, predicted about 4 dpm ^{10}Be/kg produced by SCR in the surface layer. The new calculation, which uses new and lower-proton cross sections (*Tuniz et al.*, 1984), predicts 2 dpm ^{10}Be/kg at the top layer of 68815, still more than we find experimentally. The target element responsible for the majority of ^{10}Be produced by SCR protons is oxygen. The elemental abundance of oxygen in 68815 is 44.8% (*Apollo 16 Preliminary Science Report*, 1972; *Wänke et al.*, 1974). ^{10}Be is also produced by proton interactions with Mg (3.85% in 68815), Al (14.2%), and Si (21.8%). However, the threshold energies for these nuclear reactions are higher than those interactions of O, and the elemental abundances of Mg, Al, and Si are also lower than oxygen. The threshold energy for the ^{10}Be-producing reaction with O is 34 MeV. There are only two cross section measurements for ^{10}Be production from O available below 500 MeV proton energy. *Yiou et al.* (1969) reported the cross section to be 0.37 ± 0.12 mb at 135 MeV. *Amin et al.* (1972) reported the cross section to be 0.59 ± 0.04 mb at 135 MeV. [The result by *Amin et al.* (1972) has been corrected for the revised half-life of ^{10}Be.] At 135 MeV, the higher cross section was used for the Reedy-Arnold calculations because, as noted in *Tuniz et al.* (1984), the *Yiou et al.* (1969) cross sections are consistently lower than other measurements at 600 MeV and higher energies. There are no other cross section measurements below 500 MeV proton energy except on boron and carbon, which are not abundant elements in lunar rocks. This matter could be resolved by AMS measurements of the low-energy cross sections below 100 MeV for ^{10}Be production from O and other elements. Lower cross sections, especially below ~300 MeV, could decrease the calculated SCR production rates by factors of 2 or more.

2. The second possibility is that the average SCR flux and mean rigidity over the last two million years differed from the adopted parameters R_0 = 100 MV and J(>10 MeV) = 70 p/cm^2 s. The Reedy-Arnold SCR production rates of ^{10}Be, ^{26}Al, and ^{53}Mn with three different rigidities (R_0 = 70, 100, and 150 MV) are shown in Figs. 3a-c. Although the depth profiles of both ^{26}Al and ^{53}Mn at near-surface depths are very insensitive to changes in R_0, different SCR fluxes and erosion rates would be required to fit those profiles to the data. To fit the observed ^{26}Al and ^{53}Mn profiles in lunar surface rocks, different proton fluxes and rock erosion rates must be chosen for each rigidity. If we use a lower R_0, a higher SCR flux and larger erosion rate would be required. It is known that R_0 = 100 MV, J = 70 p/cm^2 s, and an erosion rate of 1.3 mm/My (*Kohl et al.*, 1978) are not unique parameters to fit the ^{26}Al and ^{53}Mn profiles in 68815 (*Russ and Emerson*, 1980). For example, R_0 = 70 MV, J(>10 MeV) = 150 p/cm^2 s, and an erosion rate of 3 mm/m.y. (which was the erosion rate reported from track data for 68815 by *Blanford et al.*, 1975) can also fit the measured ^{26}Al and ^{53}Mn activities in 68815.

The SCR ^{10}Be depth profile is very different from both the ^{26}Al and ^{53}Mn profiles. As shown in Fig. 3a, the ^{10}Be production rates change from 1-3 dpm/kg at near-surface depths depending on the rigidity used. The production rate for ^{10}Be in the top of rock 68815 using R_0 = 70 MV and J(>10 MeV) = 150 p/cm^2 s is 1.0, only 60% of that calculated for the other set of spectral and flux parameters. Even though the SCR production rate of ^{10}Be is lower than the GCR production, the amount of ^{10}Be activity produced by SCR is very sensitive to changes in R_0. The very low SCR production of ^{10}Be observed in 68815 could indicate that the mean SCR rigidity over the last two million years was lower than the 100 MV adopted by *Kohl et al.* (1978). However, lowering the R_0 conflicts with the conclusion by *Bhandari et al.* (1976). They proposed a higher rigidity (R_0 = 150 MV) and a higher flux [J(> 10 MeV) = 140 p/cm^2 s] based on their nondestructive ^{26}Al measurements in Apollo 16 rocks. Their SCR parameters do not fit the observed ^{10}Be depth profiles in 68815 nor the ^{26}Al and ^{53}Mn profiles in 68815 and the other lunar surface rocks. A higher R_0, such as 150 MV, is most unlikely unless the ^{10}Be cross sections are more than factor of 5 smaller than the values adopted by *Reedy and Arnold* (1972) for their calculations. It should be noted that measurements of ^{26}Al in five pieces from the top 4.4 cm of lunar rock 74275 by *Fruchter et al.* (1982) gave results in good agreement with those reported in *Kohl et al.* (1978) and not with those of *Bhandari et al.* (1976). *Reedy* (1980), using ^{81}Kr data in 12002 (*Yaniv et al.*, 1980), found a somewhat higher R_0 for the period of 3 × 10^5 years, but this is not necessarily a contradiction because the main reactions producing ^{81}Kr have threshold energies above 60 MeV and the chemical abundances of the target elements were not well measured in the sample. Also, ^{81}Kr, because of its half-life of 2.1 × 10^5 years, integrates solar protons for a much shorter period than the other radionuclides. Also, unpublished ^{81}Kr measurements (K. Marti, personal communication) in 68815 support a lower R_0.

In general, SCR production rates of high-energy products such as ^{10}Be and ^{36}Cl are potentially very useful for obtaining the SCR spectrum. ^{10}Be has a distinct advantage over ^{36}Cl since ^{36}Cl is produced in both high-energy and low-energy reactions. The comparison of ^{53}Mn and ^{26}Al profiles made by *Kohl et al.* (1978) is very useful for detecting variations in cosmic rays with time, since the half-lives differ by a factor of ~5 while the excitation functions are similar. Since both ^{53}Mn and ^{26}Al are low-energy products and their profiles are steep near the surface, they are also sensitive to the erosion rate. The comparison of these two nuclides with ^{10}Be is useful in a complementary way. Because the excitation functions are quite different, only a narrow range of R_0 values can satisfy the constraints imposed by the three profiles, even though the ^{10}Be production by SCR can only be given as an upper limit. If we accept the published proton cross sections as representative, this fixes R_0 close to 70 MV. New low-energy cross section data would be most desirable, but unless they are lowered by a factor of 2 or more this conclusion will remain valid.

Previous ^{10}Be measurements, which used decay counting techniques (*Finkel et al.*, 1971; *Wahlen et al.*, 1972) to study lunar surface rocks 12002, 14310, and 14321, give results that are in good agreement with the ^{10}Be activity found in 68815 by AMS measurements. The ^{10}Be activities in the above rocks also show no increase of ^{10}Be at the surface and therefore no evidence of SCR production. Since substantially all the ^{10}Be in these rocks was produced by GCR and since they have different exposure ages, we conclude that no significant changes in the GCR flux were observed during at least the last few million years.

SUMMARY

Cosmogenic ^{10}Be activities were measured in lunar surface rock 68815. Four different depths were sampled for three different angles. The ^{10}Be profiles are flat or increase slightly with depth for all three faces and show no sign of SCR-produced ^{10}Be. The extremely low SCR production of ^{10}Be compared to the calculations of the Reedy-Arnold model suggests that either (1) low-energy proton-induced cross sections for ^{10}Be production are lower than expected, or (2) the SCR rigidity R_0 is lower than 100 MV averaged over the last few million years.

Acknowledgments. We thank T. Inoue for his help with the experiment and F. Kirchner for typing the manuscript. This work was supported by NASA grant NAG 9-33 and NASA order T-294M.

REFERENCES

Amin B. S., Biswas S., Lal D., and Somayajulu B. L. K. (1972) Radiochemical measurements of ^{10}Be and ^{7}Be formation cross-sections in oxygen by 135 and 550 MeV protons. *Nucl. Phys.* A195, 311-320.

Apollo 16 Preliminary Science Report (1972) *NASA SP-315.* NASA, Washington, DC.

Bhandari N., Bhattacharya S. K., and Padia J. T. (1976) Solar proton fluxes during the last million years. *Proc. Lunar Sci. Conf. 7th,* 513-523.

Blanford G. E., Fruland R. M., and Morrison P. A. (1975) Long-term differential energy spectrum for solar-flare iron-group particles. *Proc. Lunar Sci. Conf. 6th,* 3557-3576.

Drozd R. J., Hohenberg C. M., Morgan C. J., and Ralston C. E. (1974) Cosmic-ray exposure history at the Apollo 16 and other lunar sites: lunar surface dynamics. *Geochim. Cosmochim. Acta,* 38, 1625-1642.

Evans J. C., Reeves J. H., and Reedy R. C. (1987) Solar cosmic ray produced radionuclides in the Salem meteorite (abstract). In *Lunar and Planetary Science XVIII,* pp. 271-272. Lunar and Planetary Institute, Houston.

Finkel R. C., Arnold J. R., Imamura M., Reedy R. C., Fruchter J. S., Loosli H. H., Evans J. C., Delany A. C., and Shedlovsky J. P. (1971) Depth variation of cosmogenic nuclides in a lunar surface rock and lunar soil. *Proc. Lunar Sci. Conf. 2nd,* 1773-1789.

Freier P. S. and Webber W. R. (1963) Exponential rigidity spectrums for solar-flare cosmic rays. *J. Geophys. Res.,* 68, 1605-1629.

Fruchter J. S., Evans J. C., Reeves J. H., and Perkins R. W. (1982) Measurement of ^{26}Al in Apollo 15 core 15008 and ^{22}Na in Apollo 17 rock 74275 (abstract). In *Lunar and Planetary Science XIII,* pp. 243-244.

Imamura M., Hashimoto Y., Yoshida K., Yamane I., Yamashita H., Inoue T., Tanaka T., Nagai H., Honda M., Kobayashi K., Takaoka N., and Ohba Y. (1984) Tandem accelerator mass spectrometry of ^{10}Be/^{9}Be with internal beam monitor method. *Nucl. Inst. Methods,* 233, 211-216.

Kohl C. P., Murrell M. T., Russ G. P., III, and Arnold J. R. (1978) Evidence for the constancy of the solar cosmic ray flux over the past ten million years: ^{53}Mn and ^{26}Al measurements. *Proc. Lunar Planet. Sci. Conf. 9th,* 2299-2310.

McGuire R. E. and von Rosenvinge T. T. (1984) The energy spectra of solar energetic particles. *Adv. Space Res. (COSPAR),* 4, 117-125.

Murrell M. T. (1980) Cosmic ray produced radionuclides in extraterrestrial material. Ph.D. thesis.

Nishiizumi K., Murrell M. T., and Arnold J. R. (1983) ^{53}Mn profiles in four Apollo surface cores. *Proc. Lunar Planet. Sci. Conf. 14th,* in *J. Geophys. Res.,* 88, B211-B219.

Nishiizumi K., Elmore E., Ma X. Z., and Arnold J. R. (1984) ^{10}Be and ^{36}Cl depth profiles in an Apollo 15 drill core. *Earth Planet. Sci. Lett.,* 70, 157-163.

Nishiizumi K., Arnold J. R., Goswami J. N., Klein J., and Middleton R. (1986) Solar cosmic ray effects in Allan Hills 77005. *Meteoritics,* 21, 472-473.

Reedy R. C. (1980) Lunar radionuclide records of average solar-cosmic-ray fluxes over the last ten million years. In *The Ancient Sun* (R. O. Pepin, J. A. Eddy, and R. B Merrill, eds.), pp. 365-386. Pergamon, New York.

Reedy R. C. and Arnold J. R. (1972) Interaction of solar and galactic cosmic-ray particles with the moon. *J. Geophys. Res.,* 77, 537-555.

Reedy R. C., Arnold J. R., and Lal D. (1983) Cosmic-ray record in solar system matter. *Ann. Rev. Nucl. Part. Sci.,* 33, 505-537.

Russ G. P., III and Emerson M. T. (1980) ^{53}Mn and ^{26}Al evidence for solar cosmic ray constancy—an improved model for interpretation. In *The Ancient Sun* (R. O. Pepin, J. A. Eddy, and R. B. Merrill, eds.), pp. 387-399. Pergamon, New York.

Tuniz C., Smith C. M., Moniot R. K., Kruse T. H., Savin W., Pal D. K., Herzog G. F., and Reedy R. C. (1984) Beryllium-10 contents of core samples from the St. Severin meteorite. *Geochim. Cosmochim. Acta,* 48, 1867-1872.

Wahlen M., Honda M., Imamura M., Fruchter J. S., Finkel R. C., Kohl C. P., Arnold J. R., and Reedy R.C. (1972) Cosmogenic nuclides in football-sized rocks. *Proc. Lunar Sci. Conf. 3rd,* 1719-1732.

Wänke H., Palme H., Baddenhausen H., Dreibus G., Jagoutz E., Kruse H., Spettel B., Teschke F., and Thacker R. (1974) Chemistry of the Apollo 16 and 17 samples: Bulk composition, late stage accumulation and early differentiation of the moon. *Proc. Lunar Sci. Conf. 5th,* 2231-2247.

Yaniv A., Marti K., and Reedy R. C. (1980) The solar cosmic-ray flux during the last two million years (abstract). In *Lunar and Planetary Science XI,* pp. 1291-1293. Lunar and Planetary Institute, Houston.

Yiou F., Seide C., and Bernas R. (1969) Formation cross sections of lithium, beryllium, and boron isotopes produced by spallation of oxygen by high energy protons. *J. Geophys. Res.,* 74, 2447-2448.

[1]Department of Chemistry, B-017, University of California, San Diego, La Jolla, CA 92093
[2]Institute for Nuclear Study, University of Tokyo, Midori-cho, Tanashi-shi, Tokyo, Japan
[3]Department of Chemistry, College of Humanities and Sciences, Nihon University, Setagaya-ku, Tokyo, Japan
[4]Research Center for Nuclear Science and Technology, University of Tokyo, Bunkyo-ku, Tokyo, Japan
[5]Department of Chemistry, Faculty of Science, University of Tokyo, Bunkyo-ku, Tokyo, Japan
[6]Department of Physics, Faculty of Science, University of Tokyo, Bunkyo-ku, Tokyo, Japan
[7]Earth and Space Sciences Division, Los Alamos National Laboratory, Mail Stop D-438, Los Alamos, NM 87545

Solar Wind Record in the Lunar Regolith: Nitrogen and Noble Gases

Urs Frick*, Richard H. Becker, and Robert O. Pepin

School of Physics and Astronomy, University of Minnesota, Minneapolis, MN 55455

*Now at NAGRA, Parkstrasse 23, CH-5401 Baden, Switzerland

We have measured elemental and isotopic abundances of noble gases and nitrogen in five different kinds of lunar regolith material: a size separate of a young Apollo 16 soil, 67701, from the rim of North Ray Crater; an apparently old soil breccia, 79035, from Van Serg Crater at Apollo 17; and three mineral separates (plagioclase, pyroxene, and ilmenite) from the Apollo 17 soil 71501. Analyses were by stepwise heating either in vacuum or in the presence of oxygen, with the exception of a room-temperature acid etch of the 67701 sample. The data reveal the presence of two distinct noble gas reservoirs in the samples. The first exhibits noble gas elemental ratios that closely match solar abundances, and helium and neon isotopic ratios that agree with directly measured values in the solar wind. Isotopic data on heavier gases from this reservoir indicate that $^{36}Ar/^{38}Ar$ in the solar wind is 5.6 ± 0.1, and that solar wind krypton is isotopically lighter than krypton in the terrestrial atmosphere. The second reservoir appears to be a fractionated residue of the first. Its noble gases display fractionation patterns that to first approximation are mineral-independent for argon, krypton, and xenon. The solar-composition reservoir was sampled by low-temperature oxidation of ilmenite and by the acid etch of the 67701 soil, and is thus presumed to exist in grain surfaces. Gases from the fractionated reservoir are released at higher temperatures, and are assumed to have diffused to some depth below the grain surfaces in which they were originally implanted. The presence of the two reservoirs is consistent with regolith dynamics models in which grains experience repeated brief exposures to the solar wind at the lunar surface, interrupted by relatively long periods of burial in the shallow subsurface. In the context of such models, the contents of the grain-surface reservoir, acquired by implantation during the surface exposure, diffuse both outward (and are lost) and inward to grain interiors (to constitute the fractionated reservoir) during the episodes of shallow burial. Solar reservoirs survive in grains that are deeply buried in the cold lower regolith shortly after their last surface exposure. In contrast to the noble gases, nitrogen is not present in either reservoir in its expected solar abundance, but instead is enhanced in both by a factor of ~13 relative to argon. Although there is no entirely satisfactory explanation for this nitrogen excess, it seems most likely that it represents nitrogen that was preferentially retained, relative to noble gases, during diffusive loss. The high ratio of nitrogen to noble gases, as well as the evidence for a fractionated noble gas reservoir, lead to the conclusion that significant loss of all noble gases from lunar samples has occurred. Minimum loss estimates are on the order of 70% for Xe and correspondingly larger fractions for the lighter gases.

INTRODUCTION

Long-term secular variations in elemental and isotopic ratios of solar wind elements, which relate to possible changes over the lifetime of the sun in physical mechanisms operating at the solar surface, have been extensively investigated through measurements of solar-wind-derived gases trapped in lunar regolith materials and gas-rich meteorites. These measurements indicate an increase in $^4He/^3He$ into the past (*Eberhardt et al.*, 1972) and provide strong evidence for a $^{15}N/^{14}N$ ratio in the ancient solar wind more than 30% below that at present (*Kerridge*, 1980a,b; *Clayton and Thiemens*, 1980; *Pepin*, 1980; *Kerridge*, 1982), although this interpretation of the observed variation of $^{15}N/^{14}N$ in lunar samples has been challenged (*Geiss and Bochsler*, 1982). A lower $^{20}Ne/^{22}Ne$ ratio in the past has also been suggested (*Pepin*, 1980), although newer evidence (*Wieler et al.*, 1983) indicates little if any secular change in the solar-wind Ne isotopic composition. Secular isotopic variations in the heavier noble gases have not been demonstrated. For the most part, nonradiogenic trapped Ar compositions are comparatively uniform in solar system reservoirs, although solar wind $^{36}Ar/^{38}Ar$ implanted in metal grains in the Weston meteorite (*Becker et al.*, 1986) is ~6% higher than in bulk lunar soils (*Eberhardt et al.*, 1972) or the terrestrial atmosphere (*Nier*, 1950). A notable exception to this uniformity appears in gases trapped in the glassy lithology of the antarctic shergottite EETA 79001, in which $^{36}Ar/^{38}Ar$ is nearly 25% lower than other solar system reservoirs (*Wiens et al.*, 1986). Isotopic variations in solar wind Kr present in both lunar and meteoritic matter are known to occur (*Basford et al.*, 1973; *Bogard et al.*, 1974; *Marti*, 1980), but appear to have been generated by mass fractionation and thus do not require compositional changes in the primary solar wind. Lunar Xe is variable and complex, but a first attempt to unravel the components and processes contributing to its isotopic composition suggests that the composition of solar wind Xe has changed little if at all during the past 4 Gyr or so (*Pepin and Phinney*, unpublished data, 1978).

The ability to measure simultaneously N and noble gases in small samples led us to readdress the problem of deducing the variation in time of the composition of solar wind by analyzing an old lunar breccia (79035) and a young lunar soil (67701). It became apparent that a more appropriate choice of samples might exist when a violation of sample protocol resulted in a purely accidental *in vacuo* acid etch of soil 67701 (see *Frick and Pepin*, 1981c, and Experimental section below). Light noble gases liberated by this etch displayed elemental and isotopic abundances very similar to those in modern solar wind. This raised the question of what kinds of regolith samples may best have recorded the composition of the primary solar wind and preserved it without significant alteration. Soil 67701 is rich in feldspar mineral grains. *Signer et al.* (1977) and *Wieler et al.* (1980) had concluded from studies of He, Ne, Ar, and from particle track densities in separated regolith components that pristine mineral grains were the most reliable recorders of the solar flux. Their results indicated that noble gas losses from lunar minerals, particularly pyroxenes and ilmenites, are small. Since solar wind nitrogen appeared to be better retained in the regolith than even heavy noble gases (*DesMarais et al.*, 1974; *Jull and Pillinger*, 1977), this led to an expectation that mineral separates would be the best candidates for accurate detectors of the past solar wind. We therefore included in this study a suite of small, pure samples of plagioclase, pyroxene, and ilmenite from Apollo 17 soil 71501, in addition to our original choices of bulk lunar samples.

EXPERIMENTAL

Samples

Breccia sample 79035, from the Van Serg ejecta blanket, was chosen for analysis because of the presumed "antiquity" of its solar wind irradiation. *Kerridge* (1975) and *Clayton and Thiemens* (1980) have argued for a roughly linear increase in solar wind $\delta^{15}N$ over time, and 79035 has extremely light, and thus by hypothesis extremely old, nitrogen (*Clayton and Thiemens*, 1980; *Becker and Epstein*, 1981; *Carr et al.*, 1985). Its high spallogenic gas contents (*Hintenberger et al.*, 1974; *Clayton and Thiemens*, 1980) imply a long residence time in the regolith and thus also contribute to the idea that it may have received its solar wind content in the distant past. Although labeled a breccia, 79035 resembles a rather loosely agglomerated soil.

Soil 67701, collected from the rim of North Ray Crater, which was excavated about 50 m.y. ago (*Behrmann et al.*, 1973), is rather coarse, feldspar-rich, and submature (*Morris*, 1978). It was selected for this study because it was expected to contain recently implanted solar wind. There is, however, strong evidence that the fine-grained fraction of this soil contains significant material whose cosmic ray exposure—and thus perhaps solar wind exposure as well—predates the young crater (*Alexander et al.*, 1976; *Eberhardt et al.*, 1976). Grain-size separates from soil 67701,22 were previously prepared and analyzed for noble gases in unirradiated and neutron-irradiated aliquots in this laboratory, although only K/Ar and cosmic-ray exposure ages have been published (*Alexander et al.*, 1976, 1977). We reanalyzed a <4 μm fraction, obtained by sieving in acetone.

The pyroxene, plagioclase, and ilmenite samples from submature (*Morris*, 1978) soil 71501,38 were provided by R. Wieler and P. Signer of the Federal Institute of Technology at Zürich, Switzerland. The starting material was a 100-150 μm size fraction, obtained by sieving the bulk soil in acetone. Ilmenite WS104.5 was obtained by density separation in heavy liquids and further purified by handpicking. Pyroxene WS104.3 and plagioclase WS104.1 were carefully handpicked from low-magnetic and non-magnetic (Frantz separator) fractions, respectively. Optical inspection revealed homogeneous samples without noticeable evidence of shock features, glass, or agglutinates, but one cannot rule out microscopic "contamination" of the grains by glass splashes.

All samples were wrapped in Pt or Ag (for 67701) foils for analysis. Sample weights were 9.213 ± 0.010 mg for 79035, 3.40 ± 0.10 mg for 67701, 2.88 ± 0.02 mg for the pyroxene, 2.06 ± 0.02 mg for the plagioclase, and 29.12 ± 0.02 mg for the ilmenite.

Gas Extraction and Processing Procedures

The principal procedures for static isotopic analysis of noble gases plus nitrogen are described in detail in *Frick and Pepin* (1981a,b). A revised version of the sample processing system is presented in Fig. 1. Modifications include (1) a major rearrangement of the extraction area and (2) replacement of charcoal and zeolite traps for gas transfer by metal frits. The latter consist of sintered stainless steel filter elements (NUPRO SS-8FE-7) welded into metal fingers. These traps are bakeable to 900°-1000°C in vacuum and are thus extremely clean. Thermal response is fast, and adsorption and desorption for a given gas occur over a smaller temperature interval than for zeolite or charcoal, permitting better gas separation by differential release. Capacity is more than adequate; at liquid nitrogen (LN) temperature two of these filter elements retain more than 0.001 ccSTP of O_2. Typical amounts of calibration argon (~10^{-6} ccSTP) adsorbed on the frit at LN temperature yield lower equilibrium pressures (<<10^{-10} Torr) than when adsorbed on 10 g of Linde 5A zeolite.

Blank levels have been substantially reduced by changing the protocol for producing O_2. The CuO finger (20 g of CuO wrapped in Pt foil) is heated to 750°C, and the O_2 produced is continuously frozen into a trap containing ~15 g of zeolite for 10-20 minutes. The CuO is then isolated and the zeolite warmed until the desired amount of O_2 is released. The decreased duration of heating and lower CuO temperature reduce the blank markedly. Uptake of O_2 during combustion can also now be monitored, using a GRANVILLE-PHILLIPS Convectron vacuum gauge (1 mTorr to 1000 Torr).

Samples were heated by an external resistance heater if temperatures were not to exceed 1200°-1260°C. For more complete gas release from the ilmenite and pyroxene separates, samples were raised to a temperature estimated to be >1500°C by RF heating of the Pt wrappings in several pulses of a few minutes each. This procedure minimizes recrystallization and prevents collapse of the quartz finger, since only the quartz immediately adjacent to the Pt is heated. Total pressure during RF heating of the pyroxene was < 1 mTorr. The ilmenite, however, released oxygen, probably due to reduction of Fe^{+3}. Total pressure was kept below 3 mTorr, to avoid RF-discharge, by exposure of the released gases to CuO at ~600°C.

Gas clean-up and mass spectrometry procedures (*Frick and Pepin*, 1981a,b) have remained basically unchanged for the noble gases, except that the standard splitting throughout these runs was ~50% each for the noble gas and nitrogen fractions. In a few cases, where little or no N or heavy noble gases were expected, the extracted gas was not split. Instead, He and Ne were analyzed while the heavier gases, including N_2, were frozen on metal frit 1. Argon and krypton were then transferred quantitatively from frit 1 at about -100°C to the main separation charcoal finger. The desorption characteristics of Xe and N from the metal frit are virtually identical, so separation of these two gases was impossible; Xe was omitted in favor of N_2 analyses in these cases.

We now detail where procedures deviated from those originally used (*Frick and Pepin*, 1981a,b). In combustion runs (7-11 Torr O_2), samples were held at the nominal temperatures for 45 minutes. Gases were then exposed to cooling CuO to remove excess O_2. Splitting into noble gas and nitrogen fractions was done with the CuO below 300°C. In pyrolyses, samples were heated in vacuum for 45 minutes at the extraction temperature, then cooled. Noble gas fractions were split off and the nitrogen fractions exposed to CuO at ~700°C for 30 minutes to completely oxidize volatiles. In both combustions and pyrolyses the CuO was slowly cooled to <100°C, and condensible gases (CO_2, H_2O, SO_2, etc.) were frozen onto Pyrex capillary 1 with liquid nitrogen. This clean-up stage for N lasted at least one hour, while noble gases were being processed. Nitrogen was then exposed to Pyrex capillary 2 in liquid nitrogen for a few minutes before transfer into metal frit 1. The extraction area was isolated, and capillary 2 defrosted and pumped. When the last noble gas was in the mass spectrometer, frit 1 was warmed to -70° to -60°C in an ethanol bath and the N_2 frozen into frit 2. Since Ar accompanied the N_2, the Ar was then desorbed from frit 2 at -120° to -100°C using ethanol ice sludge, and stored elsewhere for clean-up and analysis. Separation of Ar is necessary, since modest amounts of Ar in the N_2 were found to cause a reproducible shift of a few per mil in $\delta^{15}N$ values, an effect also observed in dynamic mass spectrometry (*Mariotti*, 1984). The duplicate analysis of Ar provided an internal check of the split sizes and the reproducibility of the Ar isotopic composition. Nitrogen was then analyzed as discussed in *Frick and Pepin* (1981b). Preexposure to the hydrogen getter was for 2 minutes rather than 1, and 12 ratios

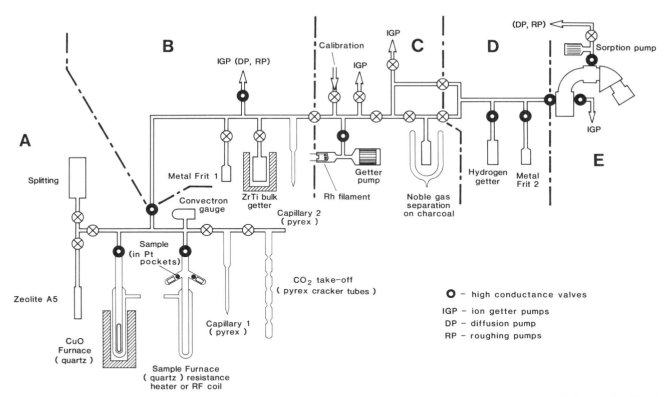

Fig. 1. System schematic for nitrogen and noble gas extraction and processing. The tasks of the various sectors (nitrogen handling in parentheses) are as follows: (A) sample pyrolysis or combustion, splitting into noble gas and nitrogen fractions (N: N_2-conversion stage I clean-up on Pyrex capillary 1, CO_2, SO_2, and H_2O take-off); (B) stage 1 noble gas clean-up on ZrTi bulk getter (N: transfer into metal frit 1 and stage II clean-up on capillary 2); (C) manifold and calibration gas inlet, stage II noble gas clean-up with "cracking" Rh filament and Westinghouse WL-32742 getter pump equipped with an SAES St. 101 alloy cartridge, separation of noble gases by desorption from charcoal (N: getter and charcoal valved off, volumes used for further splitting of the nitrogen fraction); (D) transfer area for separated Ar, Kr, and Xe (N: transfer, Ar-separation while N is adsorbed on frit 2, hydrogen gettering on cold ZrTi alloy); (E) sequential isotopic analysis of noble gases and nitrogen in a statically operating, 6" double focusing Nier-type mass spectrometer (N: sorption pump SAES NP10 with St. 101 alloy is valved off during nitrogen analysis).

were taken instead of 16-25. When N_2 amounts were small, ratio splits were combined rather than run as replicates.

Soil 67701 (<4 μm) was initially etched at room temperature in a side arm of the extraction finger during analysis of a terrestrial chlorite sample. Gases evolved from the chlorite apparently produced sulfur-containing acids, as liquid condensed around the lunar sample and blackened the silver foil wrapper. As this was the first lunar sample in the extraction system, there is no possibility of the acid attack having been on a residue of a previous analysis. Thermal releases from the etched sample up to 800°C were first done by pyrolysis and then repeated in the presence of ~10 Torr O_2, to explore possible preferential oxidation of gas-rich surface sites. Higher temperature steps were by combustion only.

Sample 79035 was open to the CuO at 700°C (20-50 mTorr O_2) for all steps. Gases in the 71501 minerals were extracted by combustion only at temperatures below 650°-750°C. Although combustion is a more time-consuming procedure and yields larger blanks, it is more effective in removing contaminant nitrogen at low temperatures. While some contamination might be tolerable for gas-rich samples such as the fines from 67701 or the breccia, far too much was seen in the relatively gas-poor plagioclase from 71501 (five 600°C clean-up steps were required). We eventually realized that overnight baking of the walls of the extraction system with heating tapes had transferred substantial nitrogen (presumably in the form of organic matter present on the walls) onto the samples, which were stored at <50°C during the bakes. We then minimized these bakes and, in addition, exposed the extraction area to O_2 from the CuO while heating with the tapes for several hours, in order to remove the organic contaminants. In later analyses each sample was pumped for a few hours while held at 140°-150°C, with the quartz above the sample kept at higher temperature to avoid condensation of N-containing species onto areas near the samples. This drastically reduced the "contaminant" N in the 71501 ilmenite and pyroxene separates. The remaining small amounts of nitrogen of terrestrial isotopic composition released at 300°C may well have been N_2 adsorbed on the pristine surfaces of these grains.

Typical blanks for the combustion procedures were 3×10^{-9} ccSTP ^{4}He, 1×10^{-13} ccSTP ^{3}He, 3×10^{-11} ccSTP ^{20}Ne, 2×10^{-9} ccSTP ^{40}Ar, 1×10^{-11} ccSTP ^{36}Ar, 1×10^{-13} ccSTP ^{84}Kr, 1×10^{-14} ccSTP ^{132}Xe, and about 2.5×10^{-7} ccSTP N (~300 picograms). The pyrolysis procedures yielded substantially smaller amounts of Ar and N, particularly when utilizing the RF coil, as that minimized heating of the quartz. We reached N blank levels of 25 pg using RF heating. Hydrogen and hydrocarbon interferences (e.g., on ^{3}He and ^{78}Kr, respectively) were negligible, due to the high mass resolution ($M/\Delta M$ ~800) of the spectrometer. Hydride formation with noble gases was not observed. Interference corrections for Ne($H_2^{18}O^+$ and $^{40}Ar^{+2}$ on mass 20, $H_3^{18}O^+$ on mass 21, and CO_2^{+2} on mass 22) were determined empirically and applied where necessary. These corrections were usually smaller than the quoted errors, except for steps having very little Ne.

TABLE 1a. Nitrogen, helium, neon, and argon yields and isotopic compositions from the stepwise extraction of 3.40 mg of soil 67701,22 (<4 μm).

Temp. (°C)	^{14}N (ppm)	δ^{15}N (‰)	^{4}He (10^{-4}ccSTP/g)	$\frac{^4He}{^3He}$	^{22}Ne (10^{-6}ccSTP/g)	$\frac{^{20}Ne}{^{22}Ne}$	$\frac{^{21}Ne}{^{22}Ne}$	^{36}Ar (10^{-6}ccSTP/g)	$\frac{^{36}Ar}{^{38}Ar}$	$\frac{^{40}Ar}{^{36}Ar}$
25*	-	-	359	2681	22.99	13.56	0.0332	6.53	5.650	<7.57
etch			14	43	0.99	0.15	0.0002	0.26	0.012	0.08
400	17.7	15.8	173	3110	15.08	12.86	0.0323	6.96	5.622	4.210
pyrol	0.7	2.6	8	53	0.54	0.07	0.0002	0.28	0.010	0.012
400	28.9	17.4	105.3	2775	12.71	13.20	0.0323	5.22	5.626	2.33
comb	1.2	2.8	4.4	47	0.46	0.06	0.0002	0.20	0.011	0.14
600	72.4	33.4	411	3192	40.9	12.73	0.0317	97.4	5.562	1.458
pyrol	3.0	3.1	18	54	1.5	0.06	0.0002	3.9	0.009	0.005
600	30.4	65.6	46.3	3919	13.05	12.76	0.0320	63.6	5.513	0.983
comb	1.3	3.3	2.0	70	0.48	0.06	0.0002	2.5	0.009	0.013
800	124.7	61.9	64.6	3750	48.4	12.70	0.0317	488	5.288	0.454
pyrol	5.2	5.6	2.8	73	1.8	0.06	0.0002	26	0.008	0.002
800	2.0	-71.7	2.40	3995	10.26	12.56	0.0317	7.21	5.234	0.568
comb	0.2	8.4	0.10	85	0.38	0.06	0.0002	0.29	0.009	0.091
850	3.6	-83	2.51	3979	3.95	12.39	0.0317	5.51	5.176	1.07
comb	0.2	10	0.11	90	0.14	0.06	0.0002	0.23	0.014	0.13
950	25.5	-4.2	3.73	4151	5.85	12.33	0.0314	14.1	5.243	0.944
comb	1.1	2.9	0.16	86	0.21	0.06	0.0002	0.3	0.009	0.052
1200	96.5	10.0	0.41	3930	4.29	12.10	0.0313	71.8	5.161	0.946
comb	4.1	2.7	0.03	180	0.15	0.06	0.0002	2.9	0.008	0.009
1350	4.9	-13.2	0	-	0.115	12.37	0.0328	1.78	5.064	5.53
comb	0.3	6.2			0.004	0.06	0.0004	0.08	0.016	0.41
Total	407	32.6	1169	3052	177.5	12.83	0.0320	768	5.355	0.747
	17	3.9	49	52	6.4	0.06	0.0002	39	0.009	0.009
Total†	-	-	979	3170	166.2	12.75	0.0321	829	5.328	0.729
melt			28	180	6.6	0.05	0.0002	14	0.011	0.003
Isotopic‡		0.2		1301.5		9.647	0.02874		5.355	2.1161
standards		0.4		6.6		0.042	0.00015		0.008	0.0021

RESULTS

Nitrogen and noble gas data for soil 67701,22 (<4 μm) are set out in Table 1. Data from breccia 79035,24 are in Table 2. For soil 71501,38 (100-150 μm), plagioclase and pyroxene data are listed in Table 3, ilmenite data in Table 4. Krypton and xenon data from the earlier analyses of 67701 grain-size separates (*Alexander et al.*, 1976, 1977) are given in Table 5 for reference purposes. Data are blank corrected except as noted. Tabulated errors are 1σ only, and include statistical errors of peak height measurements (electrometer noise and ion statistics) and errors arising from extrapolation of peak heights and ratios to "zero-time" in the spectrometer, reproducibility of calibrations with a known gas mixture, blank and interference corrections, and uncertainties in nonlinear pressure-dependent sensitivities. An additional error of 5-10% should be added to abundances due to uncertainty in the gas pipette volume. Brief discussions of some of the data have been published previously (*Frick and Pepin*, 1981c,d, 1982; *Pepin and Frick*, 1981; *Frick et al.*, 1982, 1986).

Comparison with Previous Measurements

Noble gases from related samples have been analyzed in several laboratories. For example, *Walton et al.* (1974) measured separates from 67701,20; their finest grain size (11 μm) contained, as expected, one-third to one-half the gas in our <4 μm fraction. Breccia 79035 was analyzed by *Hintenberger et al.* (1974) and by *Wieler et al.* (1983); our total isotopic abundances are in excellent agreement with these earlier data. *Signer et al.* (1977) measured He, Ne, and Ar from melt runs of 71501,38 mineral separates in the size range 150-200 μm. Our plagioclase concentrations are a factor of ~2 lower for all gases, pyroxene concentrations from both laboratories are in excellent agreement, and ilmenite concentrations are about twice those of the Zürich analysis. Such inconsistency could arise both from the relatively small number of grains in many of the separates and from sampling bias in handpicking larger grains from sieved fractions. Measurements of He, Ne, and Ar in an individual 71501,23 ilmenite grain have been carried out by *Warbaut et al.* (1979).

TABLE 1b. Krypton yields and isotopic compositions from 67701,22 (<4 μm).

Temp. (°C)	^{84}Kr (10^{-10}ccSTP/g)	$\frac{^{78}Kr}{^{84}Kr}$	$\frac{^{80}Kr}{^{84}Kr}$	$\frac{^{82}Kr}{^{84}Kr}$	$\frac{^{83}Kr}{^{84}Kr}$	$\frac{^{86}Kr}{^{84}Kr}$
25* etch	-	-	-	-	-	-
400 pyrol	19.72	0.00801	0.04210	0.2067	0.2023	0.2985
	0.56	0.00027	0.00091	0.0011	0.0013	0.0019
400 comb	13.31	0.00855	0.0431	0.2069	0.2050	0.2934
	0.47	0.00052	0.0039	0.0036	0.0031	0.0049
600 pyrol	144.8	0.00693	0.04201	0.2094	0.2044	0.2971
	3.9	0.00012	0.00025	0.0007	0.0006	0.0011
600 comb	44.4	0.00742	0.04236	0.2065	0.2029	0.3019
	1.3	0.00023	0.00053	0.0011	0.0016	0.0016
800 pyrol	3030	0.00623	0.03995	0.2031	0.2019	0.3040
	82	0.00003	0.00010	0.0003	0.0002	0.0005
800 comb	113.6	0.00571	0.03883	0.1991	0.1997	0.3098
	3.4	0.00023	0.00034	0.0006	0.0006	0.0011
850 comb	199.1	0.00569	0.03881	0.1986	0.1997	0.3108
	5.4	0.00009	0.00028	0.0007	0.0006	0.0007
950 comb	233.8	0.00580	0.03827	0.1993	0.1993	0.3092
	9.5	0.00049	0.00027	0.0007	0.0009	0.0010
1200 comb	763	0.00586	0.03902	0.2003	0.2004	0.3078
	22	0.00017	0.00009	0.0004	0.0004	0.0007
1350 comb	35.8	0.00636	0.03913	0.2001	0.1989	0.3107
	1.4	0.00052	0.00069	0.0012	0.0014	0.0017
Total	4598	0.00616	0.03974	0.2024	0.2015	0.3051
	86	0.00009	0.00015	0.0004	0.0004	0.0006
Total† melt	4680	0.00612	0.03979	0.2019	0.2018	0.3056
	210	0.00002	0.00007	0.0001	0.0002	0.0003
Isotopic‡ standards		0.006087	0.039599	0.20217	0.20136	0.30524
		0.000020	0.000020	0.00004	0.00021	0.00015

Elemental and isotopic ratios agree within ±2σ with the total ilmenite data in Table 4a, except for higher $^{20}Ne/^{22}Ne$ (13.7 versus 13.06). We estimate our ^{4}He content to be ~10^{-3} ccSTP/cm^2 in Warhaut et al.'s units, roughly 2.5 times their surface concentration value.

Direct comparison for nitrogen is possible only for breccia 79035, analyzed by *Clayton and Thiemens* (1980) and *Becker and Epstein* (1981); our yield (114 ppm) is somewhat higher than seen previously (93, 74, 73 ppm). The yield of 407 ppm N from 67701,22 (<4 μm) greatly exceeds a bulk 67701 value of 47 ppm (*Kerridge et al.,* 1975), as would be expected. The analyses of N in mineral separates show much lower abundances (2.6-5.8 ppm N) than in typical bulk soils, similar to the situation for noble gases (*Hübner et al.,* 1975; *Signer et al.,* 1977; *Wieler et al.,* 1980).

Elemental Release Patterns

In contrast to total gas amounts, one must consider the effects of different experimental protocols when comparing gas release patterns as a function of temperature. The use of different temperature intervals from sample to sample, or the presence or absence of O$_2$, could introduce variations in evolution profiles that might otherwise not be seen and that can complicate sample intercomparisons. The effects of varying temperature intervals should be small, since linear heatings of lunar samples at various heating rates (100° to 300°C per hour) yield almost identical release patterns with only minor shifts in the temperatures of maximum release (unpublished data of the Zürich laboratory). One expects, however, that the presence of O$_2$ would enhance gas evolution from oxidizable phases, such as iron metal or ilmenite. With such considerations in mind, we now consider the gas release patterns.

Gas releases as a function of temperature for He, Ne, and Ar from 67701 and from the plagioclase and pyroxene from 71501 are virtually identical, within the resolution of the temperature intervals, to those established earlier by linear heatings in the Zürich laboratory (*Baur et al.,* 1972; *Frick et al.,* 1973a,b). Releases for the ilmenite and for 79035, where

TABLE 1c. Xenon yields and isotopic compositions from soil 67701,22 (<4 μm).

Temp. (°C)	^{132}Xe (10^{-12}ccSTP/g)	$\frac{^{124}Xe}{^{132}Xe}$	$\frac{^{126}Xe}{^{132}Xe}$	$\frac{^{128}Xe}{^{132}Xe}$	$\frac{^{129}Xe}{^{132}Xe}$	$\frac{^{130}Xe}{^{132}Xe}$	$\frac{^{131}Xe}{^{132}Xe}$	$\frac{^{134}Xe}{^{132}Xe}$	$\frac{^{136}Xe}{^{132}Xe}$
25* etch	-	-	-	-	-	-	-	-	-
400 pyrol	98.3	0.0034	0.0035	0.0750	1.116	0.1588	0.814	0.389	0.316
	9.3	0.0022	0.0016	0.0045	0.028	0.0032	0.016	0.027	0.019
400 comb	57.5	0.0059	0.0021	0.0833	1.078	0.1639	0.807	0.387	0.317
	9.3	0.0018	0.0035	0.0067	0.043	0.0098	0.032	0.012	0.013
600 pyrol	208.1	0.0044	0.0028	0.0799	1.045	0.1583	0.8174	0.3769	0.3130
	9.3	0.0031	0.0019	0.0038	0.030	0.0044	0.0098	0.0072	0.0078
600 comb	54.7	0.0034	0.0036	0.0880	1.026	0.1594	0.8123	0.3564	0.3231
	9.3	0.0011	0.0013	0.0038	0.030	0.0044	0.0098	0.0072	0.0078
800 pyrol	19090	0.00463	0.00411	0.08412	1.0579	0.1660	0.8238	0.3644	0.2942
	570	0.00013	0.00009	0.00061	0.0045	0.0013	0.0021	0.0022	0.0015
800 comb	392	0.00288	0.00398	0.0842	1.029	0.1632	0.8181	0.3893	0.2895
	15	0.00055	0.00076	0.0024	0.011	0.0029	0.0090	0.0074	0.0061
850 comb	1105	0.00392	0.00404	0.0800	1.0356	0.1617	0.8139	0.3606	0.2954
	34	0.00063	0.00049	0.0026	0.0099	0.0016	0.0077	0.0040	0.0032
950 comb	4890	0.00428	0.00405	0.0844	1.0327	0.1634	0.8200	0.3685	0.2989
	150	0.00024	0.00026	0.0012	0.0062	0.0015	0.0057	0.0033	0.0033
1200 comb	22400	0.00448	0.00451	0.08245	1.0270	0.16310	0.8142	0.3695	0.3045
	670	0.00009	0.00009	0.00048	0.0024	0.00059	0.0028	0.0018	0.0018
1350 comb	782	0.00435	0.00510	0.0772	1.014	0.1608	0.8169	0.3750	0.3133
	25	0.00045	0.00064	0.0030	0.011	0.0018	0.0074	0.0041	0.0034
Total	49100	0.00449	0.00429	0.08315	1.0399	0.1641	0.8186	0.3675	0.2998
	1500	0.00018	0.00015	0.00074	0.0042	0.0011	0.0032	0.0024	0.0020
Total† melt	49500	0.00479	0.00437	0.0845	1.0423	0.1649	0.8210	0.3689	0.3000
	1700	0.00004	0.00005	0.0006	0.0020	0.0006	0.0013	0.0009	0.0008
Isotopic‡ standards		0.003537	0.003300	0.07136	0.9832	0.15136	0.7890	0.3879	0.3294
		0.000011	0.000017	0.00009	0.0012	0.00012	0.0011	0.0006	0.0004

Analytical errors are indicated below the measured values in the table.
*Data for N, ^{40}Ar, Kr, and Xe dominated by terrestrial gases in etch step (see text).
†Replicate analysis of <4 μm sample (from Table 5).
‡Nitrogen cross-calibrated against air (*Frick and Pepin*, 1981b); MLSSA standard (*Pepin et al.*, 1975) for the noble gases with air Kr and Xe compositions from *Basford et al.* (1973). Errors of standards not included in error propagation.

no comparison data are available, are shown in Figs. 2a and 2b, respectively, It should be noted in the case of the ilmenite that substantial amounts of Ne and Ar are retained to temperatures >1000°C while significant Kr and Xe loss occurs below 600°C. In stepwise pyrolyses of lunar samples, fractions of the total gases evolved are generally in the order He>Ne>Ar>Kr>Xe at low temperature, reversing as the temperature increases, consistent with their expected mobilities as a function of mass and size. Stepwise combustion of the ilmenite differs, however, in that the fractions of low-temperature (300°-630°C) N, Ar, Kr, and Xe liberated at 400°C, 500°C, and 600°C (temperatures of major gas evolution) are roughly mass-independent. Gas release appears to be due to chemical attack on the ilmenite, rather than mass-dependent diffusion.

Nitrogen profiles show no consistent pattern, and no consistent relationship to those of the noble gases. Nitrogen and argon releases are roughly congruent for 67701 and 79035 and particularly so for the 71501 ilmenite, but the two remaining mineral separates are very different from these and from each other. For 71501 pyroxene, 70% of the N is outgassed prior to the maximum Ar release at 900°C. In contrast, N is more retentively sited than Ar in the plagioclase. Although some N was undoubtedly lost in the multiple 600°C heatings of the plagioclase, at least 70% of the recovered nitrogen was held until 1200°C (possibly more, since extraction may have been incomplete), whereas only ~2% of the Ar was held to 1200°C (Table 3a). Apart from the plagioclase, however, it appears that under laboratory heating conditions N_2 is lost from lunar regolith materials about as easily as Ar.

TABLE 2a. Nitrogen, helium, neon, and argon yields and isotopic compositions from the stepwise extraction of 9.213 mg of breccia 79035,24.

Temp. (°C)	^{14}N (ppm)	$\delta^{15}N$ (‰)	^{4}He (10^{-4}ccSTP/g)	$\frac{^{4}He}{^{3}He}$	^{22}Ne (10^{-6}ccSTP/g)	$\frac{^{20}Ne}{^{22}Ne}$	$\frac{^{21}Ne}{^{22}Ne}$	^{36}Ar (10^{-6}ccSTP/g)	$\frac{^{36}Ar}{^{38}Ar}$	$\frac{^{40}Ar}{^{36}Ar}$
310	43.3*	2.2	87.5	2320	1.74	13.82	0.03334	0.131	5.53	14.82
comb	3.1	0.9	2.2	100	0.03	0.05	0.00005	0.003	0.15	0.10
600/1	123*	2.1	767	2112	181.2	12.86	0.03312	23.7	5.509	6.357
comb	14	1.0	19	95	9.8	0.04	0.00008	0.5	0.007	0.013
600/2	5.06	-22.1	38.7	2830	8.69	12.47	0.03348	4.93	5.480	3.110
comb	0.36	1.2	1.0	130	0.47	0.04	0.00008	0.99	0.008	0.008
600/3	5.70	-14.8	29.1	2880	5.09	12.32	0.03387	-†	-	-
comb	0.28	1.1	0.7	130	0.28	0.03	0.00006			
700	11.52	-92.8	160.9	2830	20.0	12.23	0.03454	31.10	5.406	2.370
comb	0.82	1.0	4.0	130	1.1	0.04	0.00010	0.62	0.007	0.005
750	14.4	-99.6	119.7	3050	9.13	12.22	0.03668	38.2	5.332	1.967
comb	1.0	1.2	3.0	140	0.49	0.03	0.00007	0.8	0.007	0.004
800	7.95	-182.4	103.5	3350	5.48	12.29	0.03880	33.7	5.262	1.717
comb	0.56	3.5	2.6	150	0.30	0.03	0.00007	0.7	0.007	0.004
850	11.02	-200.4	101.3	3540	4.44	12.52	0.04070	48.1	5.251	1.505
comb	0.78	1.1	2.5	160	0.25	0.03	0.00005	1.0	0.007	0.010
900	9.85	-170.0	107.1	4110	4.70	12.66	0.04117	4.40	5.234	1.656
comb	0.70	1.4	2.7	190	0.26	0.04	0.00006	0.88	0.007	0.005
1000	9.85	-148.8	97.5	3710	11.89	12.38	0.03833	28.90	5.205	1.828
comb	0.66	1.4	2.4	170	0.67	0.05	0.00011	0.58	0.007	0.007
1100	9.82	-124.2	1.33	3210	8.24	11.83	0.04129	24.84	5.160	1.925
comb	0.70	1.0	0.03	150	0.44	0.03	0.00006	0.50	0.007	0.004
1200	20.3	-90.0	0.68	11800	2.78	11.28	0.05670	38.81	5.067	2.161
comb	1.4	1.0	0.02	600	0.16	0.03	0.00008	0.78	0.007	0.012
1260	5.68	-4.6	0.64	256000	0.42	10.02	0.1013	3.98	4.621	2.130
comb	0.40	1.1	0.02	28000	0.02	0.03	0.0002	0.08	0.011	0.007
1300‡	3.24	22.0	0.26	200000	0.10	9.84	0.1213	0.69	4.334	2.230
comb	0.23	1.6	0.01	52000	0.01	0.04	0.0004	0.02	0.038	0.098
Total	113.8	-109.1	1615	2550	263.9	12.68	0.03465	321.1	5.250	2.225
	8.1	1.3	21	120	9.9	0.04	0.00008	2.1	0.007	0.007

Elemental Abundance Ratios

Elemental abundance ratios calculated with respect to both ^{36}Ar and ^{132}Xe and normalized to *Cameron's* (1982) "solar system" values are presented in Table 6 for our summed samples and some subsets of the data. Cameron's compilation uses no solar wind data from lunar regolith samples, and thus provides an independent estimate of relative elemental abundances in the solar wind with which to compare our results. Partly for this reason, but primarily because they yield smoother relationships when our data are plotted relative to them (Fig. 3), we have chosen to use Cameron's values rather than those of *Anders and Ebihara* (1982) for normalization.

For all five samples, integrated noble gas abundances relative to Xe are fractionated with respect to solar values, as shown in Fig. 3a. Depletions increase smoothly with decreasing mass for the 71501 ilmenite, as well as for four comparison ilmenites (*Eberhardt et al.,* 1970, 1972; *Hübner et al.,* 1975). The other 71501 minerals and the 67701 (<4 μm) sample show essentially the same depletions for Ar and Kr as the ilmenites, while 79035 appears somewhat more fractionated. The well-known ease of He and Ne loss from feldspar, and He from pyroxene, is reflected by much greater depletions of these gases in the plagioclase and pyroxene separates, and in the feldspar-rich 67701 soil, than in the ilmenites.

The two release fractions shown in Fig. 3b, normalized here to Ar rather than Xe, define a noble gas pattern with respect to solar abundances strikingly different from that in Fig. 3a. One is the 25°C acid-etch fraction from 67701 (for which only the He, Ne, and Ar were free of contamination), the other

TABLE 2b. Krypton yields and isotopic compositions from breccia 79035,24.

Temp. (°C)	^{84}Kr (10^{-10}ccSTP/g)	$\frac{^{78}Kr}{^{84}Kr}$	$\frac{^{80}Kr}{^{84}Kr}$	$\frac{^{82}Kr}{^{84}Kr}$	$\frac{^{83}Kr}{^{84}Kr}$	$\frac{^{86}Kr}{^{84}Kr}$
310	2.47	0.00775	0.0417	0.2057	0.2046	0.3020
comb	0.08	0.00091	0.0012	0.0027	0.0029	0.0039
600/1	40.2	0.00658	0.04191	0.2046	0.2024	0.3016
comb	1.2	0.00018	0.00043	0.0009	0.0007	0.0010
600/2	3.28	0.00818	0.0402	0.2097	0.2060	0.2982
comb	0.10	0.00061	0.0011	0.0022	0.0023	0.0023
600/3	-†	-	-	-	-	-
comb						
700	20.01	0.00673	0.04212	0.2063	0.2049	0.3024
comb	0.60	0.00032	0.00039	0.0011	0.0011	0.0011
750	74.0	0.00693	0.04254	0.2059	0.2040	0.2996
comb	2.2	0.00017	0.00033	0.0006	0.0007	0.0008
800	116.3	0.00662	0.04196	0.2058	0.2043	0.2997
comb	3.5	0.00012	0.00023	0.0006	0.0004	0.0009
850	213.3	0.00650	0.04147	0.2043	0.2039	0.3017
comb	6.4	0.00011	0.00017	0.0005	0.0005	0.0006
900	225.3	0.00643	0.04038	0.2036	0.2028	0.3039
comb	6.8	0.00010	0.00018	0.0005	0.0004	0.0008
1000	187.2	0.00670	0.04147	0.2043	0.2048	0.3038
comb	5.6	0.00011	0.00018	0.0005	0.0005	0.0005
1100	206.7	0.00676	0.04186	0.2054	0.2059	0.3038
comb	6.2	0.00012	0.00017	0.0006	0.0005	0.0005
1200	391	0.00748	0.04324	0.2073	0.2092	0.3045
comb	12	0.00006	0.00013	0.0005	0.0003	0.0006
1260	51.7	0.00903	0.04745	0.2143	0.2188	0.3066
comb	1.6	0.00009	0.00025	0.0009	0.0007	0.0013
1300‡	-†	-	-	-	-	-
comb						
Total	1531	0.00693	0.04212	0.2057	0.2060	0.3032
	18	0.00011	0.00019	0.0006	0.0005	0.0007

the integrated release from the oxidation steps (300°-630°C) of the 71501 ilmenite. Relative noble gas abundances are close to the *Cameron* (1982) values, except for a He depletion of ~73% in 67701, likely due to mineral-specific He loss from this fine-grained, feldspar-rich soil. The only direct determination of solar wind elemental abundances, for He, Ne, and Ar from the Apollo Solar Wind Composition (SWC) experiments (*Geiss et al.*, 1972; *Bochsler and Geiss*, 1977), is also shown. Relative to the SWC, He/Ar is low by ~40% in the ilmenite and by a factor of ~4.7 in 67701, while ^{22}Ne/^{36}Ar in both samples is within 7% of the SWC value.

As an aside, we note that there is actually a discrepancy between the SWC Ne/Ar ratio and that of *Cameron* (1982). Cameron isotopically partitions total Ne using a ^{20}Ne/^{22}Ne of 8.2 (corresponding to the Ne-A of *Pepin*, 1967), which he considers to be the most representative solar system value (and which lies within the range allowed for solar flare Ne, according to *Mewaldt et al.*, 1984). With this value and his total Ne abundance, Cameron's ^{22}Ne/^{36}Ar ratio is very close to that of the SWC experiment. However, were this total Ne repartitioned using the SWC ^{20}Ne/^{22}Ne ratio of 13.7, the ^{22}Ne/^{36}Ar ratio would only be ~60% of the SWC value. By presenting our abundances relative to Ar in plots such as Fig. 3b, we can relate Kr and Xe fractionations to Camerons's values while at the same time comparing our He and Ne data to the SWC results (which are, after all, the relevant values).

Except for He and Ne, the gas reservoirs represented in Fig. 3b contain only minor fractions of the total gas in these samples. For example, two-thirds of the total He in the ilmenite was released between 300° and 630°C, but only ~7.5% of the Xe. The 67701 etch released ~30% of the total He; taking into account the inferred loss factor of ~4.7 relative to SWC noted above, it would have again contributed ~two-thirds of the total He. Retention of He in the etched reservoir, poor as it is, is dramatically greater than in the sample as a whole, where the He/Ar ratio is more than 35 times lower.

The abundance of N in the <630°C fraction of the ilmenite (Fig. 3b) is distinctly at odds with the approximately solar pattern displayed by the noble gases, being in excess by more

TABLE 2c. Xenon yields and isotopic compositions from breccia 79035,24.

Temp. (°C)	^{132}Xe (10^{-12}ccSTP/g)	$\frac{^{124}Xe}{^{132}Xe}$	$\frac{^{126}Xe}{^{132}Xe}$	$\frac{^{128}Xe}{^{132}Xe}$	$\frac{^{129}Xe}{^{132}Xe}$	$\frac{^{130}Xe}{^{132}Xe}$	$\frac{^{131}Xe}{^{132}Xe}$	$\frac{^{134}Xe}{^{132}Xe}$	$\frac{^{136}Xe}{^{132}Xe}$
310	29.0	-0.0016	0.0022	0.0791	1.040	0.1596	0.844	0.377	0.333
comb	0.9	0.0020	0.0015	0.0053	0.047	0.0076	0.019	0.013	0.010
600/1	394	0.00432	0.00401	0.08065	1.076	0.1619	0.8216	0.3723	0.3011
comb	9	0.00036	0.00032	0.00016	0.021	0.0021	0.0065	0.0038	0.0032
600/2	17.4	0.0023	0.0035	0.0799	1.018	0.1553	0.805	0.3880	0.310
comb	0.7	0.0018	0.0024	0.0037	0.011	0.0055	0.012	0.0059	0.010
600/3	30.0	0.0032	0.0069	0.0710	0.975	0.1508	0.774	0.390	0.326
comb	9.9	0.0030	0.0022	0.0031	0.023	0.0092	0.017	0.013	0.011
700	81.7	0.00432	0.0046	0.0774	1.020	0.1603	0.8218	0.3756	0.3095
comb	2.2	0.00056	0.0011	0.0034	0.013	0.0050	0.0099	0.0083	0.0059
750	171.9	0.0051	0.00521	0.0780	1.054	0.1579	0.807	0.3787	0.3183
comb	4.9	0.0011	0.00086	0.0014	0.024	0.0043	0.011	0.0039	0.0067
800	742	0.00454	0.00448	0.0858	1.076	0.1642	0.8289	0.3641	0.2938
comb	17	0.00025	0.00029	0.0015	0.015	0.0023	0.0050	0.0031	0.0048
850	2333	0.00464	0.00466	0.08228	1.0679	0.1630	0.8226	0.3692	0.2994
comb	54	0.00012	0.00014	0.00094	0.0043	0.0013	0.0042	0.0017	0.0022
900	4720	0.00463	0.00433	0.08421	1.0632	0.16533	0.8205	0.3703	0.3029
comb	110	0.00012	0.00010	0.00066	0.0031	0.00093	0.0029	0.0017	0.0019
1000	5420	0.00481	0.00477	0.08331	1.0640	0.16449	0.8219	0.3719	0.3018
comb	130	0.00008	0.00011	0.00046	0.0030	0.00076	0.0026	0.0009	0.0008
1100	8230	0.00506	0.00537	0.08430	1.0630	0.16441	0.8228	0.3722	0.3043
comb	190	0.00007	0.00009	0.00039	0.0026	0.00091	0.0023	0.0011	0.0008
1200	13470	0.00574	0.00664	0.08584	1.0607	0.1648	0.8322	0.3718	0.3035
comb	310	0.00012	0.00015	0.00050	0.0024	0.0010	0.0027	0.0009	0.0007
1260	2060	0.00632	0.00809	0.08755	1.0522	0.16626	0.8336	0.3740	0.3043
comb	47	0.00038	0.00020	0.00081	0.0043	0.00094	0.0033	0.0012	0.0014
1300‡	289	0.00751	0.01062	0.0919	1.0399	0.1639	0.8300	0.3743	0.3036
comb	7	0.00066	0.00094	0.0015	0.0058	0.0015	0.0044	0.0027	0.0024
Total	37980	0.00524	0.00571	0.08474	1.0620	0.1646	0.8264	0.3716	0.3031
	410	0.00013	0.00014	0.00059	0.0035	0.0010	0.0029	0.0012	0.0012

Analytical errors are indicated below the measured values in the table.
*Low-temperature steps with $\delta^{15}N \cong 0 \pm 3$‰ considered to be contaminated; not included in nitrogen totals.
†Not measured.
‡Heater burn-out after only ~10 minutes at 1300°C.

than an order of magnitude. For N/Ar the enrichment is 12.7 times Cameron's solar ratio (N was not measured in the SWC experiment), which is identical to that shown by the total ilmenite in Fig. 3a, and close to the ~15-fold excess seen in bulk lunar soils (*DesMarais et al.*, 1974; *Jull and Pillinger*, 1977). A similar result has also been obtained for an ilmenite separate from 79035 analyzed in the same way (*Becker et al.*, 1984). Severe N contamination prevented a determination of N/Ar in the 67701 etch.

Isotopic Release Patterns

Helium, neon, and argon. Isotopic compositions of the light noble gases as a function of temperature roughly follow trends commonly seen in linear heatings (e.g., *Baur et al.*, 1972) for all five samples (see Fig. 4). Initial releases are near solar wind values, followed by isotopically heavier fractions at higher temperatures and excursions toward spallogenic compositions at the highest temperatures. The heavier fractions are probably mass-fractionated residuals of reservoirs that initially had solar wind isotopic compositions, but only Ne, with three nonradiogenic isotopes, allows a direct evaluation of a mass-fractionation relationship between the low and high temperature compositions. In the middle panel inset of Fig. 4a, Ne isotopic variations in 67701 (<4 μm), for which spallation effects are small, are seen to fall closely along a straight line. In 79035 spallogenic Ne is more important, and deviations to the right of this line occur that are probably due to spallation, although we cannot rule out a trapped component lying somewhere to the right of the line. The line itself could arise from mixing of a solar wind-produced reservoir with another trapped component, such as the SEP-Ne of *Wieler et al.* (1986, 1987), rather than from fractionation, but as gas release from 67701 and 79035 in our analyses was primarily

TABLE 3a. Nitrogen, helium, neon, and argon yields and isotopic compositions from stepwise extractions of plagioclase and pyroxene separates from soil 71501,38 (100-150 μm).

Temp. (°C)	^{14}N (ppm)	$\delta^{15}N$ (‰)	^{4}He (10^{-4}ccSTP/g)	$\frac{^{4}He}{^{3}He}$	^{22}Ne (10^{-6}ccSTP/g)	$\frac{^{20}Ne}{^{22}Ne}$	$\frac{^{21}Ne}{^{22}Ne}$	^{36}Ar (10^{-6}ccSTP/g)	$\frac{^{36}Ar}{^{38}Ar}$	$\frac{^{40}Ar}{^{36}Ar}$
Plagioclase (2.06 mg)										
600/1	149.5*	2.2	13.69	2380	0.937	12.90	0.05386	4.19	5.661	2.45
comb	6.3	1.0	0.71	76	0.051	0.05	0.00020	0.42	0.078	0.06
600/2	7.0*	1.1	0.878	2630	0.0558	12.87	0.0510	1.20	5.415	1.48
comb	0.3	1.1	0.047	110	0.0042	0.22	0.0017	0.12	0.049	0.23
600/3	7.5*	1.9	0.472	3030	0.0230	13.77	0.05242	0.776	5.33	0.74
comb	0.3	1.2	0.026	170	0.0015	0.11	0.00016	0.078	0.12	0.24
600/5‡	0.28*	-1.0	-†	-	-†	-	-	0.232	4.42	2.7
comb	0.01	9.0						0.024	0.30	1.5
800	0.247	29	2.07	3230	0.1240	13.38	0.04083	3.77	5.181	0.583
pyrol	0.014	23	0.11	130	0.0047	0.06	0.00049	0.38	0.083	0.091
1000	1.27	21.6	1.306	4010	0.2317	13.08	0.03701	1.27	4.10	8.73
pyrol	0.05	3.6	0.068	200	0.0088	0.04	0.00015	0.13	0.13	0.24
1200/1	2.17	26.5	0.020	-	0.0289	12.58	0.03551	0.173	2.268	32.7
pyrol	0.11	3.4	0.006		0.0011	0.15	0.00093	0.018	0.065	0.3
1200/2	1.08	75	0.0026	-	0.0029	13.6	0.044	0.0251	2.89	37.8
pyrol	0.09	12	0.0027		0.0005	2.3	0.010	0.0032	0.40	1.9
Total	4.77	36.3	18.43	2567	1.404	12.98	0.04939	11.63	5.125	2.85
	0.27	6.0	0.72	79	0.052	0.06	0.00031	0.60	0.098	0.16
Pyroxene (2.88 mg)										
300	0.908*	2.3	6.72	2170	0.149	14.62	0.03491	0.0149	5.76	1.1
comb	0.055	2.0	0.41	70	0.009	0.34	0.00082	0.0021	0.62	5.0
500	1.144	29.6	40.3	2608	9.17	13.20	0.03231	0.734	5.70	2.11
comb	0.069	2.0	2.5	84	0.49	0.04	0.00006	0.074	0.21	0.33
700	0.702	26.3	11.65	904	7.91	12.59	0.03151	5.30	5.443	0.434
comb	0.042	3.0	0.72	30	0.43	0.04	0.00005	0.53	0.019	0.047
900	0.499	-54.0	3.84	2330	3.05	11.84	0.03431	5.72	5.381	0.277
pyrol	0.030	9.0	0.24	80	0.16	0.03	0.00006	0.58	0.032	0.043
1050	0.163	38	0.363	1810	0.556	11.15	0.09259	0.633	5.23	0.64
pyrol	0.010	10	0.024	100	0.030	0.03	0.00025	0.064	0.10	0.29
1200	0.091	-	0.0166	450	0.210	8.78	0.2476	1.73	4.969	0.409
pyrol	0.005		0.0019	110	0.011	0.03	0.0012	0.15	0.013	0.010
1500/1§	0.030	470	0.0185	7200	0.138	3.02	0.700	1.43	3.584	0.425
pyrol	0.003	50	0.0011	5200	0.008	0.08	0.013	0.15	0.008	0.055
1500/2§ pyrol	<0.001	-	-†	-	-†	-	-	<0.001	-	-
Total	2.629	14.4	62.7	1887	21.18	12.62	0.04039	15.55	5.125	0.460
	0.087	4.4	2.6	62	0.67	0.04	0.00016	0.82	0.032	0.070

diffusive, some fractionation is to be expected in any case.

Spallation also affects the 71501 mineral separates (Fig. 4b), but is seen predominantly at high temperatures. The prominent dip at 700°C for ^{4}He/^{3}He in the pyroxene appears to be an exception. This is near the temperature of maximum release of spallogenic He from meteoritic pyroxene (*Huneke et al.*, 1969), however, and the amount of spallogenic He implied is not unreasonable, so this "anomaly" in the ^{4}He/^{3}He pattern can be understood.

Helium and Ne isotopic compositions in the chemically-released gas fractions (inset to Fig. 3b and Table 6) are in excellent agreement with modern solar wind values determined from SWC foils (*Bochsler and Geiss*, 1977) or, for He in 67701, can be related to these values by simple diffusive fractionation (*Frick and Pepin*, 1981c). Argon isotopic ratios were poorly determined in the foils; the estimate of 5.3 ± 0.3 (*Bochsler and Geiss*, 1977) is nondiagnostic, since it spans virtually the entire range observed for nonspallogenic Ar in solar system

TABLE 3b. Krypton yields and isotopic compositions for plagioclase and pyroxene separates from 71501,38 (100-150 μm).

Temp. (°C)	^{84}Kr (10^{-10}ccSTP/g)	$\frac{^{78}Kr}{^{84}Kr}$	$\frac{^{80}Kr}{^{84}Kr}$	$\frac{^{82}Kr}{^{84}Kr}$	$\frac{^{83}Kr}{^{84}Kr}$	$\frac{^{86}Kr}{^{84}Kr}$
Plagioclase (2.06 mg)						
600/1¶	2.90	0.0045	0.0422	0.2184	0.1976	0.3158
comb	0.19	0.0050	0.0015	0.0071	0.0055	0.0062
800	9.76	0.0047	0.0374	0.2063	0.2045	0.3024
pyrol	0.65	0.0021	0.0018	0.0041	0.0040	0.0038
1000	51.5	0.00650	0.04039	0.2029	0.2055	0.3065
pyrol	3.3	0.00032	0.00059	0.0014	0.0015	0.0018
1200/1	2.18	0.0137	0.0724	0.256	0.314	0.312
pyrol	0.15	0.0054	0.0066	0.014	0.017	0.013
1200/2	0.86	0.022	0.0704	0.270	0.269	0.264
pyrol	0.06	0.017	0.0093	0.018	0.018	0.019
Total	67.2	0.0066	0.0415	0.2067	0.2093	0.3059
	3.4	0.0012	0.0011	0.0027	0.0026	0.0029
Pyroxene (2.88 mg)						
300	0.62	-0.0021	0.0357	0.203	0.2005	0.3163
comb	0.06	0.0090	0.0077	0.013	0.0061	0.0095
500	0.66	0.0027	0.0383	0.199	0.206	0.286
comb	0.12	0.0041	0.0052	0.018	0.015	0.015
700	6.40	0.0072	0.0441	0.2082	0.2161	0.3024
comb	0.41	0.0012	0.0010	0.0049	0.0067	0.0034
900	46.4	0.00676	0.0390	0.2034	0.2030	0.3077
pyrol	3.0	0.00053	0.0012	0.0013	0.0009	0.0019
1050	7.22	0.0080	0.0467	0.2033	0.2049	0.3189
pyrol	0.47	0.0049	0.0056	0.0037	0.0096	0.0057
1200	14.3	0.0079	0.0393	0.2104	0.2131	0.3192
pyrol	0.9	0.0011	0.0009	0.0033	0.0058	0.0067
1500/1§	12.8	0.01202	0.0478	0.2106	0.2260	0.3088
pyrol	0.8	0.00097	0.0018	0.0047	0.0044	0.0049
1500/2§ pyrol	-†	-	-	-	-	-
Total	88.4	0.00762	0.0413	0.2059	0.2090	0.3100
	3.3	0.00090	0.0016	0.0028	0.0035	0.0037

samples. The two ratios shown in the Fig. 3b inset suggest that ^{36}Ar/^{38}Ar in the solar wind lies in the range 5.6 ± 0.1, considerably higher than a previous estimate based on lunar ilmenite separates of 5.33 ± 0.03 (*Eberhardt et al.*, 1972), but within error of the values of 5.4 ± 0.2 for Ar trapped in Surveyor 3 surfaces (*Warasila and Schaeffer*, 1974) and 5.53 ± 0.03 calculated by extrapolation from the compositions of surface-correlated Ar measured in lunar feldspar separates (*Frick et al.*, 1975). Results from surface oxidations of a metal separate from the Weston meteorite (*Becker et al.*, 1986) indicate a similar range of 5.5-5.75 for the solar wind ^{36}Ar/^{38}Ar ratio.

Nitrogen. No consistent patterns emerge in the nitrogen isotopic profiles shown in Fig. 5, other than a general increase in δ^{15}N at high temperature attributable to the release of spallogenic N (*Becker and Clayton*, 1977). Nitrogen in the 71501 plagioclase (rightmost panel of Fig. 5) has been discussed previously (*Frick and Pepin*, 1982). Not much can

TABLE 3c. Xenon yields and isotopic compositions for plagioclase and pyroxene separates from 71501,38 (100-150 μm).

Temp. (°C)	^{132}Xe (10^{-12}ccSTP/g)	$\frac{^{124}Xe}{^{132}Xe}$	$\frac{^{126}Xe}{^{132}Xe}$	$\frac{^{128}Xe}{^{132}Xe}$	$\frac{^{129}Xe}{^{132}Xe}$	$\frac{^{130}Xe}{^{132}Xe}$	$\frac{^{131}Xe}{^{132}Xe}$	$\frac{^{134}Xe}{^{132}Xe}$	$\frac{^{136}Xe}{^{132}Xe}$
				Plagioclase (2.06 mg)					
600/1¶	44.0**	0.0063	0.0039	0.0704	1.002	0.1603	0.810	0.385	0.309
comb	2.9	0.0026	0.0020	0.0052	0.025	0.0059	0.025	0.011	0.011
800	17.9**	0.0100	-0.0020	0.101	1.016	0.128	0.800	0.374	0.368
pyrol	1.7	0.0069	0.0058	0.010	0.029	0.098	0.026	0.017	0.013
1000	519	0.0046	0.0049	0.0756	1.037	0.1557	0.817	0.376	0.296
pyrol	31	0.0017	0.0024	0.0040	0.009	0.0046	0.016	0.006	0.007
1200/1	124.5	0.0079	0.011	0.085	1.028	0.1652	0.835	0.403	0.316
pyrol	7.9	0.0093	0.011	0.016	0.050	0.0025	0.029	0.027	0.034
1200/2	17.2**	0.002	-0.001	0.090	0.944	0.156	0.798	0.333	0.340
pyrol	1.3	0.011	0.011	0.016	0.028	0.015	0.030	0.024	0.017
Total	722	0.0054	0.0055	0.0779	1.031	0.1569	0.819	0.380	0.303
	33	0.0034	0.0034	0.0065	0.018	0.0085	0.019	0.010	0.012
				Pyroxene (2.88 mg)					
300	22.2**	-0.006	0.0026	0.0851	0.967	0.139	0.803	0.354	0.312
comb	3.9	0.011	0.0064	0.0040	0.038	0.013	0.040	0.024	0.016
500	11.3**	-0.0012	0.0094	0.080	1.001	0.173	0.817	0.365	0.361
comb	1.8	0.0067	0.0063	0.011	0.063	0.024	0.043	0.024	0.025
700	35.3	0.0059	0.0105	0.0764	1.004	0.1404	0.772	0.380	0.309
comb	2.8	0.0033	0.0032	0.0047	0.033	0.0098	0.023	0.014	0.014
900	401	0.00501	0.0030	0.0791	1.070	0.1741	0.807	0.362	0.291
pyrol	24	0.00081	0.0012	0.0049	0.020	0.0081	0.011	0.012	0.011
1050	177	0.0022	0.0085	0.0789	1.034	0.170	0.817	0.342	0.279
pyrol	11	0.0015	0.0034	0.0046	0.020	0.014	0.034	0.018	0.020
1200	248	0.0063	0.0069	0.0796	1.049	0.1661	0.820	0.382	0.289
pyrol	15	0.0019	0.0013	0.0043	0.040	0.0081	0.029	0.014	0.009
1500/1§	172	0.0032	0.0074	0.0811	1.044	0.1524	0.813	0.364	0.314
pyrol	11	0.0026	0.0024	0.0067	0.024	0.0066	0.023	0.009	0.010
1500/2§	-†	-	-	-	-	-	-	-	-
pyrol									
Total	1066	0.0043	0.0044	0.0795	1.050	0.1662	0.811	0.364	0.294
	33	0.0014	0.0020	0.0050	0.026	0.0092	0.022	0.013	0.012

Analytical errors are indicated below the measured values in the table.
*,†See footnotes to Table 2.
‡All gases from the 4th extraction step at 600°C were discarded without measurement.
§Sample foil wrapping inductively heated by RF; temperature uncertain.
¶Kr and Xe not measured in the repeat extractions at 600°C.
**Xe close to blank level; no blank correction applied.

be said about solar wind N in this sample. Some was probably lost in the multiple steps at 600°C, and the 800°C extraction yielded only 0.5 ng N, with poor isotopic precision. Although the measured $\delta^{15}N$ values are similar to those for the pyroxene, excluding the 900°C pyroxene step, the extreme retentivity of nitrogen in the plagioclase led *Frick and Pepin* (1982) to conclude that it might contain lunar indigenous nitrogen. The $\delta^{15}N$ value measured is nondiagnostic in this regard. The ilmenite shows a clear discontinuity between low and high temperature fractions, with high $\delta^{15}N$ values for most of the combustion steps and negative $\delta^{15}N$ for pyrolyses prior to the appearance of spallogenic N above 1100°C. Weighted averages of $\delta^{15}N$ are +57‰ for the 400°-600°C fraction and -7‰ for the 800°-1150°C fraction.

The $\delta^{15}N$ profile for 67701 is shown in the left panel of Fig. 5. It differs from a comparison profile for bulk 67601 (*Becker and Clayton*, 1977), a similar North Ray soil, for several reasons. As a <4 μm separate, 67701's solar-wind content is much higher, so spallogenic contributions are negligible at all temperatures. It is also much more susceptible to "contamination" by fine regolith particles transported from elsewhere at the Apollo 16 site (*Eberhardt et al.*, 1976; *Alexander et al.*, 1976, 1977). Such particles, irradiated on average >>50 m.y. ago, presumably are responsible for the strikingly light

TABLE 4a. Nitrogen, helium, neon, and argon yields and isotopic compositions from the stepwise extraction of 29.12 mg of ilmenite from soil 71501,38 (100-150 μm).

Temp. (°C)	^{14}N (ppm)	δ^{15}N (‰)	^{4}He (10^{-4}ccSTP/g)	$\frac{^4\text{He}}{^3\text{He}}$	^{22}Ne (10^{-6}ccSTP/g)	$\frac{^{20}\text{Ne}}{^{22}\text{Ne}}$	$\frac{^{21}\text{Ne}}{^{22}\text{Ne}}$	^{36}Ar (10^{-6}ccSTP/g)	$\frac{^{36}\text{Ar}}{^{38}\text{Ar}}$	$\frac{^{40}\text{Ar}}{^{36}\text{Ar}}$
300	0.504*	16.7	69.7	2162	0.602	14.39	0.03454	0.0149	5.31	11.94
comb	0.025	3.2	4.9	69	0.023	0.21	0.00054	0.0015	0.19	0.92
400	0.222	68.6	384	2387	7.05	13.62	0.03288	0.507	5.495	1.331
comb	0.011	3.0	40	76	0.16	0.05	0.00012	0.051	0.019	0.095
500	0.249	73.1	173	2695	4.86	13.31	0.03257	1.32	5.456	0.518
comb	0.012	4.0	12	87	0.11	0.05	0.00011	0.13	0.012	0.085
600	0.773	48.7	242	2662	8.05	13.38	0.03242	3.69	5.514	0.850
comb	0.039	2.0	16	86	0.18	0.04	0.00016	0.30	0.009	0.036
630	0.090	12.0	68.3	2825	1.33	13.15	0.03242	0.929	5.460	0.964
comb	0.005	8.0	4.6	91	0.03	0.08	0.00041	0.075	0.027	0.046
Sum	1.45*	55	937	2516	21.90	13.46	0.03266	6.46	5.492	0.862
≤630	0.05	5	45	81	0.26	0.05	0.00016	0.34	0.013	0.054
680	0.101	12	114.1	3090	1.86	13.48	0.03302	1.53	5.411	1.003
pyrol	0.005	14	7.7	100	0.04	0.07	0.00029	0.12	0.010	0.022
800	0.756	6.0	241	3530	5.82	13.44	0.03273	4.44	5.316	0.669
pyrol	0.038	2.5	14	110	0.13	0.05	0.00010	0.32	0.009	0.027
910	0.918	-11.0	148.1	3860	5.13	12.73	0.03239	4.11	5.243	0.743
pyrol	0.046	1.5	8.6	130	0.11	0.04	0.00018	0.29	0.009	0.029
1000/1	0.383	-11.0	23.9	3390	2.63	12.36	0.03298	2.16	5.289	0.705
pyrol	0.019	2.0	0.8	110	0.06	0.04	0.00006	0.15	0.009	0.048
1100	0.373	-22.0	3.87	1592	3.19	12.36	0.03305	2.32	5.256	0.523
pyrol	0.019	2.0	0.09	51	0.07	0.04	0.00009	0.17	0.011	0.027
1150	0.167	-5.2	0.947	610	0.940	12.34	0.03384	0.677	5.140	0.623
pyrol	0.008	2.0	0.031	20	0.021	0.04	0.00008	0.055	0.037	0.067
1200	0.404	26.0	0.706	412	1.52	12.38	0.03456	1.87	5.065	0.58
pyrol	0.020	2.0	0.030	13	0.03	0.04	0.00006	0.15	0.008	0.15
1260/1	0.633	95.5	0.453	393	4.98	12.39	0.03352	3.15	5.018	0.334
pyrol	0.032	2.0	0.024	13	0.11	0.04	0.00005	0.16	0.008	0.039
1000/2	0.018	113	0.040	362	0.062	12.40	0.03838	0.0227	4.47	0.20
comb	0.002	17	0.002	14	0.003	0.05	0.00017	0.0023	0.11	0.66
1260/2	0.074	179.0	0.026	328	0.817	12.41	0.03503	0.265	4.754	0.27
comb	0.006	9.0	0.002	13	0.044	0.04	0.00005	0.027	0.060	0.12
1500/1†	0.511	248.0	0	-	0.472	12.15	0.04477	0.628	4.062	0.60
pyrol	0.026	2.5			0.025	0.04	0.00015	0.051	0.012	0.11
1500/2†	0.005	-	0	-	0	-	-	0	-	-
pyrol	0.005									
Total	5.79*	47.7	1470	2773	49.3	13.06	0.03303	27.6	5.242	0.683
	0.29	2.3	49	89	0.4	0.05	0.00013	0.7	0.011	0.049

nitrogen at 800°C-850°C, and probably reduce δ^{15}N values at other temperatures as well (*Frick and Pepin*, 1981c). Most important, 67701 was chemically etched. As discussed above, this removed light noble gases of roughly solar composition, and presumably removed a high-δ^{15}N component as well (*Becker et al.*, 1976). As the etching took place in the presence of large amounts of terrestrial nitrogen, it may also have led to contamination of the sample, so that part of the N subsequently released could be terrestrial (*Frick and Pepin*, 1981c). For this reason we regard the high N/Ar ratio in 67701 (~33 times solar) as an upper limit to the true ratio.

Our δ^{15}N profile for breccia 79035 is shown in the center panel of Fig. 5, along with the original results of *Clayton and Thiemens* (1980), which gave a minimum δ^{15}N value of -210‰ at 900°C. The bulk δ^{15}N value obtained by Clayton and Thiemens was confirmed by *Becker and Epstein* (1981); their minimum δ^{15}N was -197‰ in the 800°-1000°C fraction. Our lowest value of -200‰ at 850°C is in excellent agreement with these, but our profile is quite different, as is the integrated δ^{15}N, indicating that the light component constitutes a much smaller fraction of the total N in our sample. Some of the N included in Fig. 5 below 850°C may be contamination, as

TABLE 4b. Krypton yields and isotopic compositions for ilmenite from 71501,38 (100-150 μm).

Temp. (°C)	^{84}Kr (10^{-10}ccSTP/g)	$\frac{^{78}Kr}{^{84}Kr}$	$\frac{^{80}Kr}{^{84}Kr}$	$\frac{^{82}Kr}{^{84}Kr}$	$\frac{^{83}Kr}{^{84}Kr}$	$\frac{^{86}Kr}{^{84}Kr}$
300	-†	-	-	-	-	-
comb						
400	1.94	0.00507	0.04019	0.2039	0.2014	0.3027
comb	0.12	0.00071	0.00078	0.0017	0.0015	0.0021
500	5.10	0.00572	0.0383	0.2027	0.2012	0.3048
comb	0.33	0.00038	0.0016	0.0014	0.0012	0.0013
600	8.26	0.00637	0.04101	0.2061	0.2038	0.3019
comb	0.53	0.00021	0.00040	0.0009	0.0012	0.0015
630	1.25	0.00604	0.0405	0.1986	0.2019	0.3090
comb	0.08	0.00084	0.0029	0.0062	0.0048	0.0073
Sum	16.54	0.00599	0.04003	0.2042	0.2026	0.3034
≤630	0.64	0.00037	0.00099	0.0016	0.0015	0.0022
680	1.87	0.00646	0.0395	0.2029	0.2024	0.3028
pyrol	0.08	0.00062	0.0011	0.0029	0.0023	0.0028
800	8.79	0.00636	0.04114	0.2026	0.2015	0.3038
pyrol	0.35	0.00025	0.00044	0.0011	0.0011	0.0014
910	26.7	0.00596	0.03912	0.2005	0.2020	0.3077
pyrol	1.1	0.00014	0.00059	0.0014	0.0011	0.0012
1000/1	15.10	0.00658	0.03999	0.2052	0.2049	0.3051
pyrol	0.97	0.00024	0.00076	0.0010	0.0012	0.0013
1100	15.48	0.00676	0.04108	0.2052	0.2060	0.3068
pyrol	0.99	0.00028	0.00033	0.0011	0.0012	0.0008
1150	5.70	0.00885	0.04327	0.2073	0.2098	0.3083
pyrol	0.37	0.00044	0.00068	0.0032	0.0039	0.0046
1200	18.0	0.00878	0.04479	0.2092	0.2122	0.3100
pyrol	1.2	0.00014	0.00042	0.0010	0.0009	0.0022
1260/1	21.3	0.01140	0.05193	0.2177	0.2266	0.3072
pyrol	1.4	0.00015	0.00059	0.0020	0.0012	0.0021
1000/2	0.133	0.033	0.0592	0.2238	0.267	0.3000
comb	0.010	0.010	0.0094	0.0079	0.013	0.0077
1260/2	1.68	0.0217	0.0659	0.2408	0.2564	0.2973
comb	0.11	0.0011	0.0022	0.0051	0.0043	0.0036
1500/1†	4.44	0.0570	0.1465	0.3510	0.3960	0.2937
pyrol	0.29	0.0017	0.0027	0.0035	0.0060	0.0025
1500/2†	0.065	0.0179	0.0593	0.2351	0.267	0.273
pyrol	0.005	0.0013	0.0027	0.0042	0.011	0.014
Total	135.8	0.00940	0.04650	0.2118	0.2155	0.3061
	2.7	0.00030	0.00071	0.0016	0.0016	0.0018

there was significant nitrogen with $\delta^{15}N$ of +2‰ produced through the first 600°C step. Assuming as a limit that the $\delta^{15}N$ of solar nitrogen in all steps below 850°C is actually -200‰, we derive a total of ~24 ppm of +2‰ contaminant in these steps; the corrected bulk $\delta^{15}N$ is then -139‰ and the corrected abundance 90 ppm. Even after this correction there remains a discrepancy between the 79035 profiles, apparently due to a heavy component mixing in increasing proportion with the -200‰ nitrogen at increasing temperatures. Spallogenic N accounts for only part of this effect; the amount of such nitrogen required to produce the integrated increase in $\delta^{15}N$ above 850°C is more than twice that calculated for the Chicago sample above 910°C (*Clayton and Thiemens*, 1980). It should be noted that our sample was much smaller than the others (35-70 times smaller than those of Becker and Epstein, for example), and thus more subject to sampling

TABLE 4c. Xenon yields and isotopic compositions for ilmenite from 71501,38 (100-150 μm).

Temp. (°C)	^{132}Xe (10^{-12}ccSTP/g)	$\frac{^{124}Xe}{^{132}Xe}$	$\frac{^{126}Xe}{^{132}Xe}$	$\frac{^{128}Xe}{^{132}Xe}$	$\frac{^{129}Xe}{^{132}Xe}$	$\frac{^{130}Xe}{^{132}Xe}$	$\frac{^{131}Xe}{^{132}Xe}$	$\frac{^{134}Xe}{^{132}Xe}$	$\frac{^{136}Xe}{^{132}Xe}$
300 comb	-‡	-	-	-	-	-	-	-	-
400	17.4	0.0053	0.0031	0.0765	1.011	0.1679	0.806	0.349	0.3134
comb	1.2	0.0020	0.0024	0.0042	0.017	0.0062	0.016	0.011	0.0061
500	43.8	0.0033	0.0035	0.0815	1.069	0.1651	0.805	0.3759	0.3130
comb	2.7	0.0012	0.0026	0.0033	0.026	0.0034	0.011	0.0070	0.0044
600	58.4	0.00495	0.00324	0.0842	1.027	0.1655	0.807	0.3730	0.3096
comb	3.5	0.00066	0.00077	0.0020	0.012	0.0030	0.011	0.0063	0.0043
630	1.7	0.0082	0.017	0.065	1.057	0.165	0.799	0.397	0.346
comb	0.3	0.0094	0.011	0.016	0.061	0.019	0.043	0.035	0.022
Sum	121.3	0.0044	0.0035	0.0819	1.041	0.1657	0.806	0.3709	0.3119
≤630	4.6	0.0012	0.0018	0.0030	0.018	0.0038	0.012	0.0076	0.0048
680	3.5	-0.0108	0.0046	0.0833	1.039	0.164	0.814	0.424	0.336
pyrol	0.3	0.0085	0.0081	0.0081	0.046	0.011	0.032	0.025	0.023
800	16.0	0.0025	0.0041	0.0833	1.111	0.1597	0.833	0.368	0.308
pyrol	1.0	0.0027	0.0024	0.0041	0.041	0.0067	0.015	0.013	0.009
910	246	0.00494	0.00374	0.0863	1.056	0.1669	0.8266	0.3728	0.2884
pyrol	15	0.00044	0.00018	0.0017	0.017	0.0025	0.0055	0.0082	0.0070
1000/1	209	0.00522	0.00494	0.0835	1.0424	0.1619	0.8221	0.3751	0.3035
pyrol	13	0.00051	0.00053	0.0016	0.0081	0.0039	0.0065	0.0080	0.0054
1100	288	0.00556	0.00595	0.0862	1.0229	0.1631	0.8340	0.3768	0.3097
pyrol	17	0.00034	0.00023	0.0013	0.0050	0.0012	0.0043	0.0028	0.0028
1150	100.0	0.00832	0.0123	0.0922	1.0282	0.1673	0.828	0.3702	0.2944
pyrol	6.0	0.00095	0.0012	0.0038	0.0091	0.0026	0.011	0.0092	0.0057
1200	280	0.01310	0.01792	0.1040	1.0280	0.1716	0.8658	0.3723	0.3020
pyrol	17	0.00052	0.00049	0.0015	0.0059	0.0013	0.0065	0.0031	0.0033
1260/1	271	0.00831	0.00955	0.0927	1.026	0.1668	0.831	0.3639	0.3003
pyrol	16	0.00060	0.00078	0.0029	0.011	0.0030	0.010	0.0048	0.0032
1000/2	3.8	0.009	-0.0010	0.0641	1.084	0.234	0.829	0.390	0.276
comb	0.4	0.014	0.0086	0.0082	0.059	0.028	0.046	0.045	0.038
1260/2	16.5	0.0018	0.0045	0.083	1.022	0.162	0.820	0.384	0.332
comb	1.3	0.0052	0.0056	0.020	0.041	0.028	0.027	0.028	0.035
1500/1†	58.6	0.0082	0.0169	0.0992	0.992	0.1624	0.8241	0.3781	0.303
pyrol	3.5	0.0012	0.0034	0.0051	0.020	0.0029	0.0082	0.0060	0.010
1500/2† pyrol	-‡	-	-	-	-	-	-	-	-
Total	1613	0.00725	0.00872	0.0902	1.034	0.1662	0.8331	0.3723	0.3019
	36	0.00067	0.00077	0.0024	0.011	0.0028	0.0078	0.0061	0.0050

Analytical errors are indicated below the measured values in the table.

*300°C fraction assumed to be a mixture of +2‰ contaminant N (78%) with ~+70‰ N such as was released in the next two fractions (22%). The total N includes 22% of the 300°C nitrogen.

†See Table 3, footnote (§).

‡Not measured.

bias or to contamination by small amounts of recently irradiated soil or agglutinate fragments during collection or subsequent processing.

Krypton. During the stepwise heating of 67701 (<4 μm), the ^{86}Kr/^{84}Kr ratio varied by ~10%. Isotopic trends in ^{86}Kr/^{82}Kr versus ^{84}Kr/^{82}Kr, shown in Fig. 6, are strikingly consistent with mass fractionation as the responsible mechanism; the data fall with great precision along a simple fractionation curve through the integrated composition. Earlier measurements in this laboratory on grain-size separates from 67701 (Table 5) are in good accord with simple two-component mixing of spallogenic Kr with a trapped composition lying on the fractionation curve close to the position of the integrated composition for the new <4 μm sample (see inset to Fig. 6).

Also shown in the inset are temperature fractions from Pesyanoe (*Marti,* 1980), which are consistent with a spallogenic component superimposed on variably fractionated trapped Kr.

TABLE 5a. Krypton yields and isotopic compositions from grain-size separates of soil 67701,22.

Grain-size (μm)	^{84}Kr (10^{-10}ccSTP/g)	$\frac{^{78}Kr}{^{84}Kr}$	$\frac{^{80}Kr}{^{84}Kr}$	$\frac{^{82}Kr}{^{84}Kr}$	$\frac{^{83}Kr}{^{84}Kr}$	$\frac{^{86}Kr}{^{84}Kr}$
<4	4680	0.00612	0.03979	0.2019	0.2018	0.3056
(4.434 mg)	210	0.00002	0.00007	0.0001	0.0002	0.0003
4-16	2210	0.00616	0.03980	0.2026	0.2020	0.3054
(3.777 mg)	90	0.00002	0.00011	0.0003	0.0003	0.0005
37-74	653	0.00634	0.04053	0.2028	0.2029	0.3054
(21.49 mg)	30	0.00002	0.00008	0.0002	0.0002	0.0003
74-250	317	0.00650	0.04076	0.2037	0.2037	0.3055
(45.18 mg)	15	0.00002	0.00007	0.0002	0.0002	0.0003
>250	150	0.00674	0.04177	0.2050	0.2056	0.3051
(47.65 mg)	6	0.00002	0.00013	0.0004	0.0003	0.0004

TABLE 5b. Xenon yields and isotopic compositions from grain-size separates of 67701,22.

Grain-size (μm)	^{132}Xe (10^{-12}ccSTP/g)	$\frac{^{124}Xe}{^{132}Xe}$	$\frac{^{126}Xe}{^{132}Xe}$	$\frac{^{128}Xe}{^{132}Xe}$	$\frac{^{129}Xe}{^{132}Xe}$	$\frac{^{130}Xe}{^{132}Xe}$	$\frac{^{131}Xe}{^{132}Xe}$	$\frac{^{134}Xe}{^{132}Xe}$	$\frac{^{136}Xe}{^{132}Xe}$
<4	49500	0.00479	0.00437	0.0845	1.0423	0.1649	0.8210	0.3689	0.3000
(4.434 mg)	1700	0.00004	0.00005	0.0006	0.0020	0.0006	0.0013	0.0009	0.0008
4-16	20700	0.00499	0.00474	0.08280	1.0406	0.1649	0.8229	0.3698	0.3015
(3.777 mg)	700	0.00010	0.00011	0.00062	0.0034	0.0010	0.0024	0.0012	0.0008
37-74	6980	0.00558	0.00567	0.08544	1.0362	0.1654	0.8278	0.3689	0.3013
(21.49 mg)	240	0.00007	0.00009	0.00062	0.0023	0.0007	0.0018	0.0015	0.0011
74-250	3900	0.00597	0.00661	0.08617	1.0408	0.1663	0.8301	0.3698	0.3011
(45.18 mg)	140	0.00006	0.00009	0.00033	0.0027	0.0009	0.0018	0.0013	0.0010
>250	1580	0.00671	0.00775	0.08805	1.0384	0.1661	0.8317	0.3677	0.3009
(47.65 mg)	60	0.00013	0.00011	0.00077	0.0025	0.0011	0.0026	0.0020	0.0019

Data are from the work of *Alexander et al.* (1976, 1977). The 16-37 μm separate has been excluded because of problems with the spectrometer electronics during the analysis. Analytical errors are indicated below the measured values in the table.

Figures 7 and 8 show compositions of Kr extracted from 79035 and from the 71501 minerals on this same three-isotope plot. Spallogenic Kr is a relatively major constituent in these samples compared to the fine 67701 separate, and at high temperatures trends toward the spallation end point are evident. However, the low temperature fractions fall, within the precision of the data, along the same fractionation curve defined by 67701.

Only three of the six Kr isotopes are shown in Figs. 6-8. Mass fractionation requires that all isotopes are affected in a coherent way. In a representation of isotopic mass M versus per mil deviation of (M/82) from the composition of air Kr (Fig. 9a), one sees that this is indeed the case for 67701, where spallation contributions are small even for M = 78. Variations in all isotopes fall smoothly along fractionation curves for total Kr and for Kr in each of the three principal release intervals. Uncertainties in spallation corrections, particularly at the lightest isotopes, compromise a similar demonstration for 79035 and Pesyanoe. The general trends shown in Fig. 9b are, however, qualitatively consistent with fractionation.

Gases in individual temperature steps are subject to isotopic fractionation by the release process itself. No such problem exists for total gas contents. Compositional variations occur in total trapped Kr in bulk regolith samples, as indicated in Fig. 6. Variations in ^{86}Kr/^{82}Kr span ~4%, from SUCOR to the lightest Minnesota sample, 14149, and fall more or less within error along the 67701 fractionation curve. Only Luna 16 deviates significantly, and this was the one sample for which no spallation target element chemistry was available for inferring Kr compositions (*Basford et al.,* 1973). Two other important solar system Kr reservoirs, the terrestrial atmosphere and average carbonaceous chondrites, fit nicely the fractionation trend followed by the lunar data (Figs. 6 and 9b). Krypton in the lunar regolith derives from solar wind Kr and must be closely related to it, almost certainly through mass fractionation. Thus solar, terrestrial, lunar, and meteoritic trapped Kr are probably all simply mass-fractionated versions of a single solar system composition. Only components produced in nuclear processes—spallation, fission, or nucleosynthesis (e.g., the s-process Kr found by *Srinivasan and Anders,* 1978, and the anomalous Kr discovered by *Frick,* 1977, 1980)—appear to be isotopically independent. The mass-fractionation relationship for the various trapped Kr

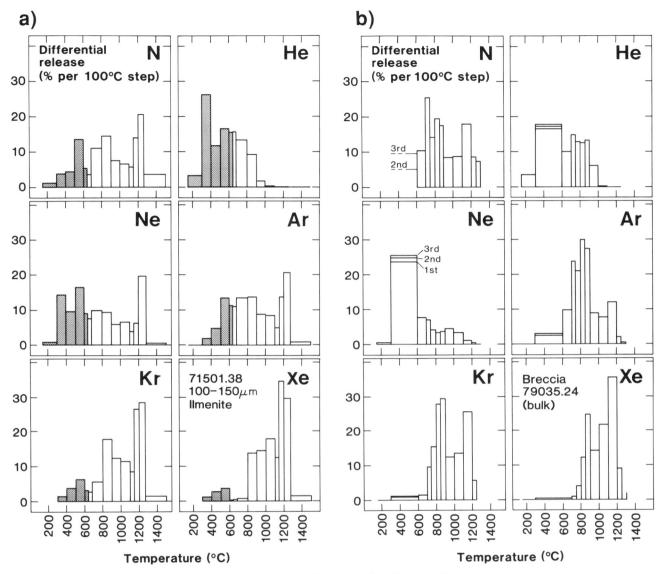

Fig. 2. Temperature release of nitrogen and the noble gases (^{4}He, ^{22}Ne, ^{36}Ar, ^{84}Kr, and ^{132}Xe) during stepwise extraction of (a) 71501 ilmenite and (b) breccia 79035. The area of each column of the histogram is proportional to the fraction released and the total areas of the histograms (100%) are the same in each plot. For 71501 ilmenite, the shaded areas indicate combustion steps (~10 Torr O$_2$). All steps for 79035 were combustions (20-50 mTorr O$_2$).

components has been previously pointed out (*Basford et al.,* 1973; *Bogard et al.,* 1974). It is of interest to note that all these trapped Kr compositions are heavier than the low-temperature integrated composition for 71501 ilmenite.

Xenon. Xenon compositions measured in the stepwise heating of 67701 are plotted in Fig. 10. As with Kr, Xe isotopic variations are in complete accord with an origin by mass fractionation. This is also the case for Xe from Pesyanoe (*Marti,* 1969), shown in the inset to Fig. 10. Total trapped Xe compositions from a number of other lunar regolith samples, however, lie below the fractionation curve through the 67701 data. This is most clear for 14149, a sample that is known to contain parentless fission Xe from extinct ^{244}Pu (*Drozd et al.,* 1975; Pepin and Phinney, unpublished data, 1978). For Xe, therefore, simple relationships among compositions from different samples, or between individual temperature fractions in a given sample, are perturbed to a significantly greater degree than Kr by the presence of nuclear components. A three-isotope Xe plot shown in Fig. 11 illustrates this point with regard to spallogenic Xe in 79035. This sample has such a high cosmic-ray exposure age (*Frick et al.,* 1982) that the bulk composition is significantly displaced from the combined mass-fractionation, fission, and air-contamination trends followed by most other samples, including 14149.

Xenon compositions from the 71501 minerals are displayed in Fig. 12. Small gas amounts lead to correspondingly large

TABLE 6. Elemental ratios relative to ^{36}Ar and ^{132}Xe, normalized to solar abundances.

Reservoir	^{4}He	^{14}N	^{22}Ne	^{36}Ar	^{84}Kr	^{132}Xe	$\frac{^4He}{^3He}$	$\frac{^{20}Ne}{^{22}Ne}$	$\frac{^{21}Ne}{^{22}Ne}$	$\frac{^{36}Ar}{^{38}Ar}$
Solar*	2.02 (+4)	2.59 (+1)	3.16 (+0)	≡1	2.63 (−4)	1.70 (−5)	−	8.2†	−	−
	1.19 (+9)	1.52 (+6)	1.86 (+5)	5.88 (+4)	1.55 (+1)	≡1				
SWC-foil‡	1.267	−	1.038	≡1	−	−	2350 ±120	13.7 ±0.3	0.0323 ±.0042	5.40§ ±.20
71501 ilmenite <630°C	0.718	12.7	1.073	≡1	0.973	1.106	2516 ±81	13.46 ±0.05	0.0327 ±.0002	5.492 ±.013
	0.649	11.5	0.970	0.904	0.880	≡1				
71501 ilmenite 630°-1500°C	0.125	12.7	0.410	≡1	2.14	4.15				
	0.030	3.06	0.099	0.241	0.516	≡1				
71501 ilmenite total	0.263	12.7	0.564	≡1	1.87	3.44				
	0.0766	3.69	0.164	0.291	0.544	≡1				
71501 pyroxene total	0.0223	11.6	0.478	≡1	2.41	4.50				
	0.0049	2.58	0.106	0.222	0.534	≡1				
71501 plagioclase total	0.0078	25.4	0.0383	≡1	2.20	3.65				
	0.00215	6.95	0.0105	0.274	0.602	≡1				
79035 breccia	0.0248	21.9	0.260	≡1	1.81	6.94				
	0.0036	3.15	0.0374	0.144	0.261	≡1				
67701 (<4 μm)	0.0075	32.7	0.0731	≡1	2.28	3.76				
total	0.0020	8.70	0.0195	0.266	0.606	≡1				
67701 (<4 μm)¶ 25°C etch	0.272	−	1.11	≡1	−	−	2681 ±43	13.56 ±0.15	0.0332 ±.0002	5.650 ±.012

*From *Cameron* (1982). Values in parentheses are exponents of ten by which numbers should be multiplied.

†Value chosen by *Cameron* (1982) because of its prevalence in primitive chondrites ("planetary" neon); the ratio lies within the range of solar flare Ne values (7.4-11.1) as determined by *Mewaldt et al.* (1984).

‡Solar Wind Composition experiment (*Bochsler and Geiss*, 1977).

§Solar wind Ar as analyzed in recovered Surveyor 3 material (*Warasila and Schaeffer*, 1974).

¶N, Kr, and Xe contaminated with terrestrial gases.

uncertainties, and spallation contributes to some fractions. Despite these ambiguities the data seem consistent with mass fractionation. In comparison with 67701, the magnitude of the fractionation appears to be small in plagioclase, comparable in ilmenite and significantly larger in pyroxene.

The component structure of solar system Xe is much more complex than that of Kr. Although nonradiogenic Xe in the terrestrial atmosphere derives from primitive solar system Xe by mass fractionation, this primitive U-Xe may be isotopically distinct from the solar wind composition to which the lunar regolith Xe is related (Pepin and Phinney, unpublished data, 1978). For this reason, and because air Xe and carbonaceous chondrite Xe (the range of which is shown in Fig. 10) contain varying proportions of several independent components (Pepin and Phinney, unpublished data, 1978), these solar system reservoirs are not simply related to the solar-lunar compositions by fractionation alone, in the way that the Kr reservoirs appear to be.

DISCUSSION

The most interesting result of our analyses is the observation of two distinct elemental abundance patterns for the noble gases, one seemingly "solar" (Fig. 3b) and the other "fractionated" with respect to the first and to measured (*Bochsler and Geiss*, 1977) or estimated (*Cameron*, 1982) solar abundance ratios. Both patterns are present in the 71501 ilmenite, in 67701, in ilmenite from 79035 (*Becker et al.*, 1984), and presumably in most lunar samples. In addition, we observe that noble gas isotopic compositions within and across samples tend to be interrelated, when nucleogenic components can be accounted for, by simple mass fractionation of a single component (Figs. 4, 6-10, and 12). Nitrogen, in contrast to the noble gases, appears to be anomalous. The results have important consequences for our understanding of solar wind elemental and isotopic abundances and of the processes occurring at the lunar surface.

Identification of Components

One may ask which, if either, of the abundance patterns observed actually represents the incoming solar wind. *Signer et al.* (1977) and *Wieler et al.* (1980) concluded that lunar mineral grains, particularly ilmenites and pyroxenes, have quantitatively retained implanted solar wind Ar, Kr, and Xe. The fact that Ar:Kr:Xe ratios in these minerals were apparently fractionated with respect to the solar wind was mentioned by *Wieler et al.* (1980), with the implication that Cameron's solar estimates could be systematically in error. Considering the consistent, smooth patterns seen in all mineral separates as well as many bulk soil samples, and the fact that Cameron's Ar, Kr, and Xe abundances were derived by interpolation and

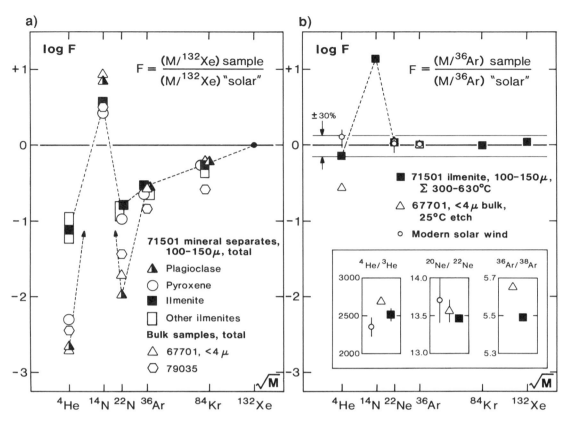

Fig. 3. Nitrogen and noble gas fractionation factors F versus square root of mass M. Gas concentrations are normalized to "solar" abundances (*Cameron*, 1982). (a) Integrated gas concentrations for 67701, 79035, and the 71501 mineral separates, normalized to Xe. For comparison with 71501 ilmenite the range of ilmenite noble gas data from the literature is shown (10084: *Eberhardt et al.*, 1970; 12001 and 10046: *Eberhardt et al.*, 1972; 74241: *Hübner et al.*, 1975). (b) Gases released by oxidation of 71501 ilmenite below 630°C and He, Ne, and Ar from the etch of 67701, normalized to ^{36}Ar. The horizontal band depicts ±30% offsets from *Cameron's* (1982) estimates of relative solar abundances. The relative He, Ne, and Ar abundances in modern solar wind as determined by the Solar Wind Composition (SWC) experiment (*Bochsler and Geiss*, 1977) are also shown. The inset presents isotopic data for He, Ne, and Ar.

not from direct measurements, their point of view was not unreasonable. However, one of their major arguments for quantitative Ar retention, namely the absence of significant depletions in ^{36}Ar/^{38}Ar ratios, is not valid, as it appears to have been based on gas loss from a well-mixed reservoir rather than by Fick's Law diffusion from a solid. In view of this, and in light of our low-temperature ilmenite data, the *Cameron* (1982) Ar:Kr:Xe ratios need not be in error, and indeed now seem to have some basis in measurement.

One might argue that our "solar-type" pattern does not derive from an independent reservoir with abundance ratios like those of *Cameron* (1982), but instead is an artifact of diffusive release from sites populated by the "fractionated" component. Such an argument would require that outgassing of the ilmenite during combustion (300°-630°C) occurred by diffusion. We find this hard to accept for several reasons. First, not only are the integrated gas amounts up to 630°C approximately solar, but so are those for the individual steps. One would normally expect Ar release to precede that of Xe, as seen in the 680°-800°C steps or in 79035 (Fig. 2b). Instead, Xe is slightly enriched relative to Ar and Kr in the first steps. Second, if the gases were diffusing from a single, initially uniform reservoir in the ilmenite, it would be difficult to account for the hiatus in Xe release at the end of the combustion steps (Fig. 2a). Third, the acid etch of 67701, at 25°C, could not have involved diffusion at all, yet the ^{22}Ne/^{36}Ar and ^{20}Ne/^{22}Ne ratios in the etched fraction and in the low-temperature ilmenite component are identical, as illustrated in Fig. 3b.

We therefore conclude that the solar-type pattern in Fig. 3b is not an artifact of the stepwise heating but instead reflects the presence in the grains of near-surface reservoirs whose noble gas elemental ratios are, except perhaps for He, close to those predicted by *Cameron* (1982). These reservoirs can be sampled without fractionation by chemical attack, as in the acid etch of 67701 and the controlled etching experiments carried out by *Wieler et al.* (1986), or by oxidation of iron

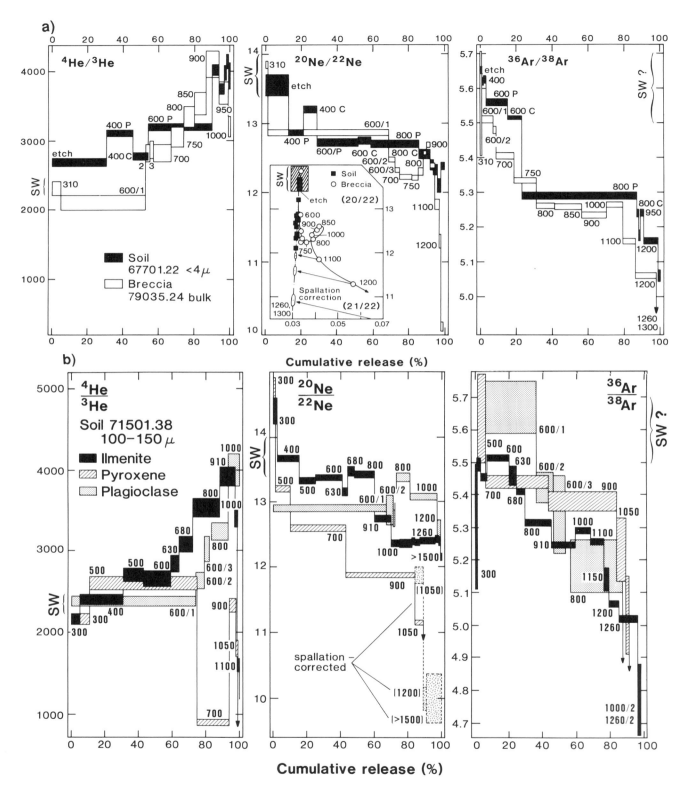

Fig. 4. Isotopic composition of He, Ne, and Ar from stepwise degassing of (a) soil 67701 and breccia 79035 and (b) the mineral separates from soil 71501. Ratios are plotted versus cumulative release of ^{4}He, ^{22}Ne, and ^{36}Ar. The isotopic compositions of He and Ne in the solar wind (*Bochsler and Geiss*, 1977) are indicated by the brackets labeled SW. The composition of solar wind Ar is not yet well determined; its suggested range is indicated by the "SW?" bracket on the right side of the diagrams (see text). The inset in the Ne graph in (a) illustrates the data from soil 67701 and breccia 79035 on a three-isotope diagram, with the hatched field representing solar wind neon.

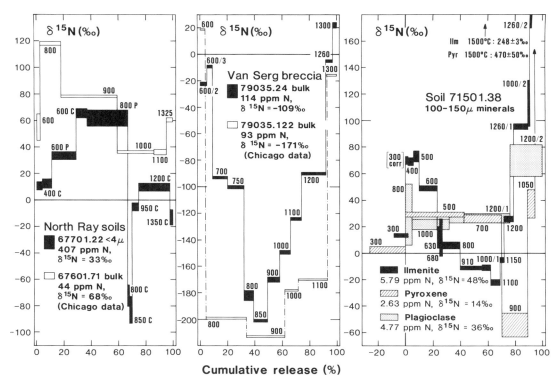

Fig. 5. Nitrogen isotopic composition versus cumulative release from stepwise degassing of 67701 (left), 79035 (center), and 71501 mineral separates (right). Data from 67701 are compared with the isotopic release pattern of a contiguous soil, 67601 (*Becker and Clayton, 1977*). Both are presumed to represent typical young soil samples with rather recent acquisition of solar gases. Data from soil breccia 79035 are depicted together with the analysis performed earlier by *Clayton and Thiemens* (1980). Nitrogen data from the ilmenite and pyroxene separates of soil 71501 are shown with the cumulative releases extended to the left, illustrating outgassing of contaminant nitrogen with terrestrial isotopic composition. Some low-temperature fractions with large amounts of contaminant nitrogen (300°C, 600°C/1 from 79035, and all 600°C steps from 71501 plagioclase) have been omitted from the plots.

grains (*Becker et al.*, 1986) and ilmenite. In the case of ilmenite, gas release presumably results from structural transformations induced by the oxidation (*Rao and Rigaud*, 1975).

If the noble gas spectrum in Fig. 3b reflects the composition of ions implanted in grain surfaces by the solar wind, the relative abundances in Fig. 3a require a different origin. The pattern of depletions suggests mass fractionation, as might be produced by diffusive loss from the solar composition reservoir. Such loss would be accompanied by transfer of residual gases in the grains to depths below the original implantation zone. Support for the idea of inward diffusion can be found in the etching experiments of *Eberhardt et al.* (1970), which indicate that about half the trapped noble gases in a 41 μm ilmenite grain-size separate are sited below ~1400 Å, well below the expected implantation depth. The depth to which etching occurred is indirectly calculated, however, and thus is subject to certain assumptions. More direct evidence for inward diffusion comes from investigations by *Warhaut et al.* (1979) of the concentrations and depth distribution profiles of He and Ne in lunar mineral grains using the Gas Ion Probe (GIP). In the majority of ilmenite profiles, the gases are sited in the outermost ~300 Å, comparable to solar wind ion implantation depths. However, about 15% of the ilmenites show deep profiles in which He peaks near ~2500 Å, with little or no gas in the implantation zone. In the one of these cases where Ne could also be detected, a deep Ne profile and a very low He/Ne ratio were observed. These observations are consistent with diffusive transfer of the gases inwards from the grain surfaces. (They may also be consistent with the presence of a solar-energetic-particle (SEP) component proposed by *Wieler et al.*, 1986, which we discuss below.)

Warhaut et al. (1979) interpreted their deep profiles as the products of long exposure times to the solar wind, during which sufficient radiation damage occurred that the implantation zones became amorphous and could no longer retain gases. In the context of a different model, involving bubble diffusion, *Tamhane and Agrawal* (1979) also suggested that the implantation zones of grains would become incapable of gas retention after some length of exposure. These suggestions are consistent with the apparent absence of the fractionated component in the noble gases released up to 600°C from the ilmenite, despite the fact that, except for He and Ne, this component dominates the total noble gas inventory. The fractionated component appears to be "shielded" from

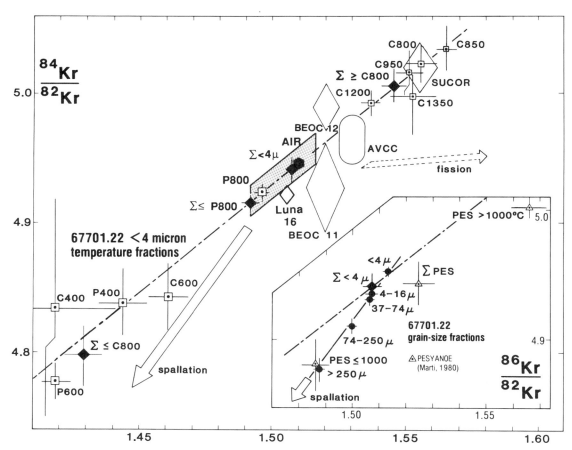

Fig. 6. Three-isotope representation of Kr isotopic compositions from stepwise extraction of soil 67701. The open squares symbolize temperature release data, the filled diamonds integrated fractions. The letters P and C associated with temperatures stand for "pyrolysis" or "combustion" steps, respectively. A mass fractionation line has been arbitrarily drawn through the P800 datum. The shaded area depicts the range of surface-correlated Kr compositions from various soils (10084, 14149, 14259, and 15531) previously analyzed at Minnesota by *Basford et al.* (1973). Other lunar Kr compositions shown are from the literature, as compiled by *Basford et al.* (1973); AVCC is from *Eugster et al.* (1967). The lower right inset, on the same scale, shows data from 67701 grain-size fractions (filled circles) measured earlier in this laboratory (Table 5). For comparison, Kr compositions from the solar wind-rich aubrite Pesyanoe (*Marti*, 1980) are plotted as open triangles, and the integrated isotopic composition from our stepwise extraction of 67701 as a filled diamond. All the literature data have been adjusted slightly to conform with the atmospheric standard values used in this work (Table 1b).

release by low-temperature oxidation, presumably by some thickness of surficial material devoid of this component.

In the 71501 ilmenite the fractionated reservoir contains about 77%, 88%, and 92% of the total Ar, Kr, and Xe, respectively. This might seem inconsistent with the apparent rarity of deep profiles in the *Warhaut et al.* (1979) GIP measurements. There are two possible reasons for this difference. The fractionated component may be confined to a small subset of the grain population, or it may be generally present but undetectable by the GIP technique. Severe diffusive migration will produce flat, low-concentration profiles that extend far into grains, where they could either appear as the deep tailing seen on some surface profiles by Warhaut et al., or escape observation altogether. The fractionated component could thus be present in many more grains than were recognized by *Warhaut et al.* (1979). This view is supported by their mass-spectrometric analysis of He, Ne, and Ar in an individual 71501 ilmenite grain that had revealed only a surface He profile with the GIP. Ratios of ^{4}He/^{22}Ne and ^{22}Ne/^{36}Ar were similar to those in our total ilmenite, which is clearly fractionated (Fig. 3a). Thus, this particle at least contained a fractionated component but no identifiable deep He profile.

We should point out that one might expect to find near-surface abundance ratios that are solar, and abundances at depth showing depletions of light relative to heavy gases, in the case of equilibrium between sputtering and ion implantation, even with no diffusive redistribution (see Fig. 2 of *Tamhane and Agrawal*, 1979). It appears, however, that sputter equilibrium

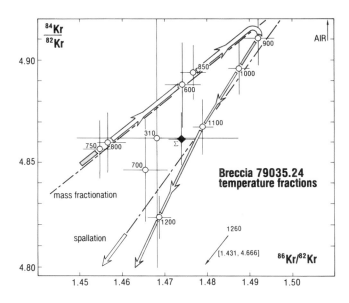

Fig. 7. Krypton isotopic trends during stepwise degassing of soil breccia 79035. The isotopic range includes only a portion of that in Fig. 6. Ignoring the uncertain low-temperature steps (310°C, 600°C, and 700°C) the data follow a mass-fractionation trend (line based on 67701 data, adopted from Fig. 6.) before deviating toward a more spallogenic composition at 1000°C and above. The spallation trend line has been drawn arbitrarily through the 1100°C datum. Temperature fractions are plotted as open circles, the integrated composition as a filled diamond.

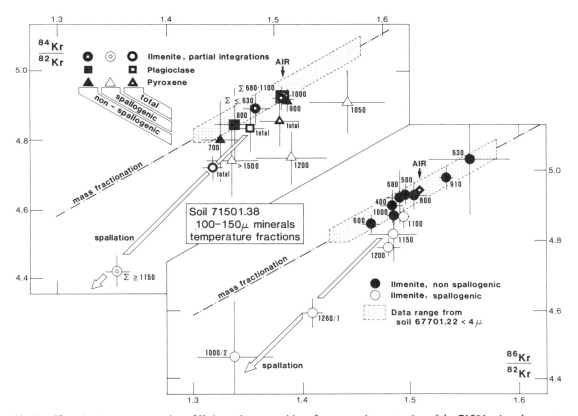

Fig. 8. Three-isotope representation of Kr isotopic compositions from stepwise extraction of the 71501 mineral separates. Temperature data from the ilmenite are plotted in the lower right panel; data from the pyroxene and the plagioclase samples are shown in the upper left panel together with the integrated fractions from the ilmenite. Some high-temperature data from the ilmenite (1260°C/2, 1500°C/1, 1500°C/2) and plagioclase (1200°C/1, 1200°C/2) plot outside the frame to the lower left, within error on the spallation trend lines. The lightly shaded areas depict the data range obtained from soil 67701, as shown in Fig. 6. The mass-fractionation line is also taken from that figure.

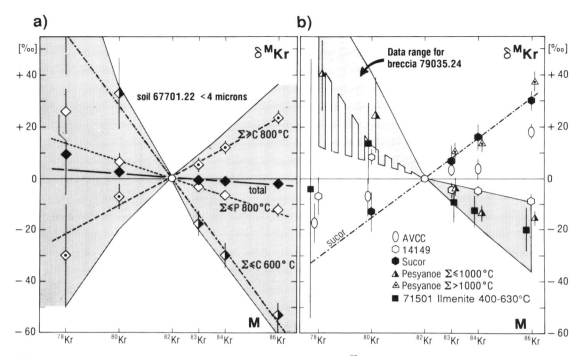

Fig. 9. Krypton isotopic compositions versus mass M, normalized to ^{82}Kr and plotted as per mil deviations from the composition of atmospheric Kr. **(a)** Various integrated Kr compositions from soil 67701. The entire 67701 data range is covered by the shading. **(b)** The data range obtained from stepwise extraction of 79035, depicted by shading; spallation-free Kr from this sample would plot within the broken shading. Low-temperature Kr from 71501 ilmenite, our putative solar wind composition, is represented by the solid squares. For comparison a number of other lunar and meteoritic Kr compositions are also plotted (for references, consult caption of Fig. 6).

was less important than diffusion in establishing the depleted gas pattern (Kerridge and Kaplan, 1978). Concentrations of solar wind Ar in mineral separates do not corresponds to the predictions of sputter-induced equilibrium (*Wieler et al.,* 1980). Depletion factors relative to Xe predicted by the model of *Tamhane and Agrawal* (1979) for the fractionated reservoir in silicates (0.86, 0.68, 0.21, and 0.03 for Kr, Ar, Ne, and He, respectively) do not accord well with the observed depletions in Table 6. Finally, the time scale for establishment of equilibrium depends on the sputter rate, and the rate of 0.2 Å/yr used by Tamhane and Agrawal is at the high end of a wide range of estimates (see, for example, *Zinner et al.,* 1977; *Borg et al.,* 1983). A choice of a lower rate leads to unreasonably long times for the establishment of sputter equilibrium, compared to the turnover time of grains at the lunar surface.

Diffusive Fractionation

To test the assumption that diffusion is the main mechanism for redistributing implanted solar wind gases in lunar grains, and therefore produces the fractionated component, we have explored some of the properties of Fick's Law diffusion in a simple system (spherical particles of 70 μm radius into which thin gas shells with gaussian radial abundance profiles have been implanted). Ratios of diffusion coefficients were assumed to be constant; that is, no allowance was made for diffusion occurring over a range of temperatures, with different species having different temperature dependencies. This is probably a safe assumption for isotopes of a single element, but may fail when intercomparing elements. We further assumed that diffusion coefficients were constant throughout the grains. This is probably the major defect in the calculations. While the assumption may hold for short exposure times, it presumably breaks down as radiation damage alters grain surfaces, since outward diffusion should then be favored over inward diffusion.

Calculations were carried out for diffusion occurring after the entire complement of gas was implanted, and for diffusion during implantation. It is not clear which is the appropriate case. The existence of unfractionated reservoirs suggests that diffusive loss does not occur instantaneously, but the presence of an unfractionated reservoir in ilmenite from 79035 (*Becker et al.,* 1984), a breccia presumably closed to solar wind implantation for $>10^9$ years, implies that diffusion can be very slow at average lunar regolith temperatures. A localized heat source, such as energy deposited by the solar wind during ion implantation, may be required as a driving force. Alternatively, significant diffusion may occur only at shallow depths in the regolith where heating by insolation is effective, with 79035 residing at greater depths for most of its history. Results are similar in the two cases, with diffusion during implantation yielding somewhat smaller depletions of light gases relative to heavy and concentration profiles peaked closer to the surface for a given set of starting conditions. Since, under natural conditions, diffusion will probably continue for a while after implantation stops in any case, the true situation will probably lie somewhere between our two extremes.

Calculated fractionations between species in diffusive loss depend on the ratio of the diffusion coefficients, as would be expected, but

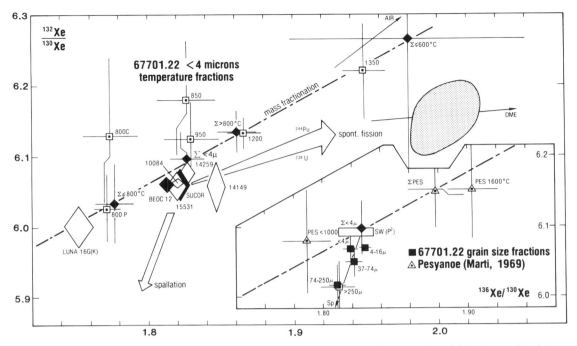

Fig. 10. Three-isotope correlation plot of Xe isotopic compositions from stepwise extraction of 67701. Symbol definitions as in Fig. 6. The LUNA 16G(K) datum (*Kaiser*, 1972) represents an unusually light isotopic composition for lunar surface-correlated Xe (see *Basford et al.*, 1973). The mass-fractionation line is arbitrarily plotted through the $\Sigma \leq 800°C$ Xe point. Directions of excursions resulting from addition of spontaneous fission Xe and spallation Xe are indicated by the arrows. The shaded oval depicts typical meteoritic (Cold Bokkeveld, Renazzo, Novo Urei, Orgueil, and Murray) Xe data, as compiled from stepwise heating of bulk samples (Pepin and Phinney, unpublished data, 1978). Acid resistant residues from the carbonaceous chondrites Murchison, Cold Bokkeveld, and Murray (*Srinivasan et al.*, 1977; *Reynolds et al.*, 1978) also plot within the oval. The arrow labeled DME points to "anomalous" Xe as revealed by partially oxidized residues of carbonaceous chondrites (Pepin and Phinney, unpublished data, 1978). As in Fig. 6, data from grain-size fractions of 67701 and from the Pesyanoe meteorite (*Marti*, 1969) are shown in the inset. The rectangle labeled SW(P^2) depicts a model-derived solar wind composition from Pepin and Phinney (unpublished data, 1978). All literature data are slightly adjusted to conform with the atmospheric standard values used in this work (Table 1c).

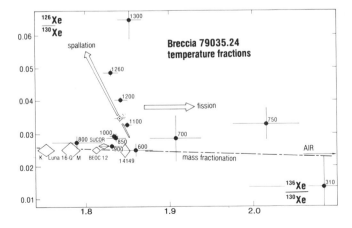

Fig. 11. Three-isotope correlation plot of Xe isotopic compositions from stepwise degassing of soil breccia 79035. Temperature fractions are plotted as filled circles, the integrated composition as an open diamond. Surface-correlated lunar Xe compositions BEOC 12, SUCOR, LUNA 16-G, and 14149 are selected from the literature [compiled in *Basford et al.* (1973); for LUNA 16, K refers to Kaiser (1972) data and M to Minneapolis data]. The mass-fractionation line is obtained by plotting our nonspallogenic 67701 data (not shown) on the same diagram. The arrow pointing toward the spallation Xe composition is drawn arbitrarily through the integrated 79035 Xe composition.

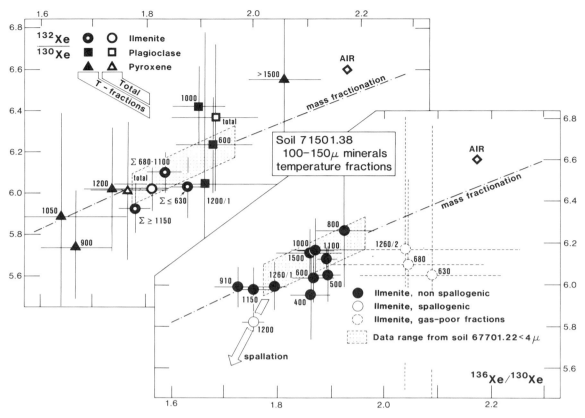

Fig. 12. Three-isotope correlation plot of Xe isotopic compositions from the stepwise extraction of the 71501 mineral separates. Symbol definitions as in Fig. 8. The data range for 67701 (light shading) and the mass-fractionation line are taken from Fig. 10. Ilmenite data from the 1000°C/2 fraction fall outside the frame of the lower right panel on the spallation trend line. Several gas-poor fractions with very large uncertainties (800°C and 1200°C/2 from plagioclase and 300°-700°C from pyroxene) are omitted from the upper left panel.

also on the relative implantation depths assumed. This can be illustrated for Ne by taking diffusion coefficients related as $(22/20)^{1/2}$ and calculating $^{20}Ne/^{22}Ne$ fractionations both for the case of equal implantation depths for the two isotopes and for implantation of ^{20}Ne ~6.5% (~12 Å) shallower than ^{22}Ne (*Huneke*, 1973). Fractionation factors average 0.977 in the first case and 0.915 in the second, with less than 1% variation from these values over a range of Ne losses from ~60% to 99.9%, assuming diffusion after implantation. *Huneke* (1973) obtained the same values for the same relative implantation depths. Solar wind implantation depths can be constrained using literature data. We chose the projected ranges and range straggling values for silicates as given by *Tambane and Agrawal* (1979), and scaled these values for ilmenite using a projected range for Kr of ~260 Å with ~100 Å of range straggling (*Jull et al.*, 1980). Results are nearly independent of absolute implantation depths, if relative depths remain the same.

One of the most interesting results of the calculations is that, relative to a reference species, depletions or enrichments of other species are essentially constant for diffusive losses of the reference species ranging between about 20% and 98% assuming diffusion after implantation, and between about 40% and 98% assuming diffusion during implantation. This could explain the observation (*Wieler et al.*, 1980; Fig. 3a) that many lunar regolith materials show similar depletion patterns, since it obviates the need to assume that all these samples underwent the same degree of lunar processing. Mineral-specific differences in He and Ne diffusion coefficients do exist, so larger He and Ne depletions in certain minerals (e.g., feldspars) are understandable, and bulk sample-to-sample variations can be attributed, at least in part, to mineralogical differences. A consequence of this result is that one cannot use isotopic fractionations in the noble gases to assess the extent of diffusive losses that might have occurred, because the fractionations are nearly independent of loss for a large range of loss factors.

We have applied these diffusion calculations specifically to data from the 71501 mineral separates. As the extent of loss from each sample is unknown *per se*, we examined three representative cases: minor Xe loss (~8%), severe loss (98%), and an intermediate case (~60-70% loss). The first case may serve for illustrative purposes only, since observed Xe/N ratios in lunar samples imply a minimum Xe loss corresponding to the intermediate case. [Note that this constraint is removed if the initial N/Xe ratio is the implanted reservoir was higher than the *Cameron* (1982) value, a possibility discussed below.] Relative diffusion coefficients D(M)/D(Xe) that reproduce the depletions observed in ilmenite above 630°C, in total pyroxene, and in total plagioclase (Table 6) were calculated and are

TABLE 7. Diffusion coefficients, relative to D(^{132}Xe), required to produce the observed depletions of noble gases in the 71501 mineral separates by Fick's Law diffusion from an initial implanted solar-composition reservoir.

Mineral	D(^{4}He)	D(^{22}Ne)	D(^{36}Ar)	D(^{84}Kr)	D(^{132}Xe)
Ilmenite ($\geq 680°$C)					
8% Xe loss	–	100	30	6.2	≡1
60% Xe loss	240	45	12.5	3.2	≡1
98% Xe loss	110	35	11.0	3.0	≡1
Pyroxene					
8% Xe loss	–	89	28	5.8	≡1
60% Xe loss	11500	38	11.5	3.0	≡1
98% Xe loss	920	31	10.0	2.8	≡1
Plagioclase					
8% Xe loss	–	8400	22	4.6	≡1
60% Xe loss	45000	3300	9.5	2.3	≡1
98% Xe loss	1450	600	8.0	2.2	≡1
[M/132]$^{-\frac{1}{2}}$		5.74	2.45	1.91	1.25	≡1

Implantation depths and their spreads, with gaussian distributions assumed, were taken to be: for ilmenite, ^{4}He = 130 ± 150 Å, ^{22}Ne = 185 ± 150 Å, ^{36}Ar = 240 ± 150 Å, ^{84}Kr = 260 ± 150 Å, ^{132}Xe = 290 ± 150 Å; for pyroxene and plagioclase, ^{4}He = 180 ± 195 Å, ^{22}Ne = 257 ± 195 Å, ^{36}Ar = 330 ± 195 Å, ^{84}Kr = 364 ± 195 Å, ^{132}Xe = 403 ± 195 Å (*Tambane and Agrawal*, 1979). The observed depletions are given in Table 6.

presented in Table 7. Diffusion after implantation was assumed. Except for the lightest gases, D(M)/D(Xe) does not vary much as a function of Xe loss, consistent with what was said above. Because of the large depletions of He, D(He)/D(Xe) values are sensitive to the assumed Xe loss.

We emphasize that the ratios given in Table 7 have been chosen to fit our data, and thus only indicate what values are required if the fractionated component in the minerals arose by diffusion. To the extent that the lunar minerals actually resemble homogeneous media in their diffusive properties, the D(M)/D(Xe) values in Table 7 stand as a testable consequence of the diffusion model for the origin of the fractionated component. The predicted values clearly do not follow the simple mass dependence frequently obeyed by isotopes of a single element (*Huneke*, 1973), as shown in the bottom line of Table 7. Literature data on diffusion of two or more noble gases in these minerals are sparse. One relevant measurement, based on thermal release of spallogenic He and Ne from pyroxene in the Mócs meteorite, gives D(He)/D(Ne) at 30°C of ~500 with large uncertainties (*Huneke et al.*, 1969); the value for pyroxene in Table 7 is ~300 (60% Xe loss).

Our calculations show that, as gases diffuse from the implantation zone, the peak of the residual gas profile shifts to greater depths and the radial profile becomes very elongated. For moderate Xe loss, however, a significant fraction of the residual Xe remains within 500 Å of the grain surface: about 30% for the case of 60% Xe loss and ~10% even with 80% loss. If 10% of the Xe from the fractionated reservoir were released in the ilmenite combustion, Xe/Ar would have been about twice the *Cameron* (1982) value. Since we do not observe excess Xe in the Fig. 3b pattern, either 90% or more of the original Xe implanted was diffusively lost, or, as mentioned above, a mechanism existed for removing the near-surface portion of the diffused reservoir. In the latter case, two types of surface regimes would be necessary in lunar grains, one consisting of heavily damaged areas where near-surface gases could not be retained but diffused gases existed at some depth, and the other consisting of areas where exposures were short enough that implanted gases could be retained.

Irradiation History and Surface Dynamics

The possibility that lunar mineral grains have sustained large losses of Ar, Kr, and Xe can be explored by comparing measured gas contents to predictions from lunar regolith dynamic studies. Specifically, do the amounts of Ar measured in the mineral separates accord with what might have been implanted, or are they in fact significantly lower? Regolith dynamic models suggest that solar wind reservoirs in soil particles are populated in a number of individual episodes. A grain cycling n times to the regolith surface, with a fraction f of its surface area exposed to the solar wind flux F for t years in each episode, should accumulate $Fftn$ atoms/cm^2, assuming no losses by diffusion or sputtering. Concentrations per gram will then be $6Fftn/\rho l$ assuming cubic crystals with density ρ and edge length l, or $3Fftn/\rho r$ assuming spherical grains of radius r.

The average proton flux incident on the Apollo 17 site is 6.5×10^7 cm^{-2} sec^{-1}, calculated from the average solar wind flux of 2.7×10^8 cm^{-2} sec^{-1} at 1 AU, a geometric factor of 0.30 at 20° lunar latitude, and a mean reduction of ~20% due to shielding by the earth's magnetospheric tail (*Etique*, 1981). The surface flux F of ^{36}Ar is about 3.1×10^9 cm^{-2} yr^{-1}, assuming solar wind ratios for ^{4}He/H of 0.04 and for ^{36}Ar/^{4}He of 3.8×10^{-5} (*Bochsler and Geiss*, 1977). Ilmenite densities lie in the range 4.7-4.8 g/cm^3, and feldspar around 2.75 g/cm^3. The grains are 100-150 µm across.

The Monte Carlo soil mixing model (Solmix) developed by the Orsay group (*Duraud et al.*, 1975; *Bibring et al.*, 1975; *Borg et al.*, 1980) predicts an average surface residence of ~5000 years ± 50% each time a grain is cycled to the top of the regolith and, for 50-100 µm radius grains, about 30 such exposures during the ~25 m.y. lifetime of a few centimeters thick surface stratum if the grains are incorporated into the upper few millimeters of the stratum. Using the values given here, the expected content of ^{36}Ar in the ilmenite would be about $5 \times 10^{16} f$ atoms/g, as compared to the measured content of 7.4×10^{14} atoms/g (Table 4). For the feldspar sample, we expect about $8 \times 10^{16} f$ atoms/g and observe 3.1×10^{14} atoms/g. Nominal lower-limit values for f would then be about 0.015 and 0.004 for the ilmenite and feldspar, for the case where implanted Ar has not been subsequently depleted.

Whether the fraction f of the grain surface exposed to solar wind when a grain is at the regolith surface is as low as the values indicated above is a question we cannot answer. There is a geometric upper limit to f of 0.5, but the actual value depends on the extent to which small particles shield a larger grain while at the lunar surface. If shielding is limited to 50-75% of a grain's upper surface, then Ar must have been severely depleted by some process after implantation. If, however, shielding averages 95% or more, then one does not require large losses of implanted Ar. The Solmix model does at least allow for enough solar wind implantation to accommodate the diffusive losses that would accompany migration of significant amounts of heavy noble gases to sites below their initial implantation depths.

It might be argued that the grains in soil 71501 have not had the total length of exposure suggested by the Solmix model of 150,000 yr on average because, for example, some grains may have been added recently and did not experience the complete history of the soil. Data of *Wieler et al.* (1980) on solar flare tracks in feldspar grains from this soil indicate, however, that 100% of the feldspars are track-rich by their definition, and all members of a suite of feldspar grains measured individually for solar wind gases were exposed at the lunar surface at least once. Soil 71501 shows no signs of being anomalous, and thus, to the extent that the Solmix model is valid, the parameters derived from that model should be applicable to this soil.

The ilmenite surface reservoir can also be examined in the context of the Solmix model. The amount of ^{36}Ar obtained through 630°C corresponds to a concentration of 1.74×10^{14} atoms/g. This yields a value of 555 years for the product of f, t, and n, which is consistent with the surface reservoir representing a single exposure for roughly 5000 yr with an average f value of ~0.1. Since we would expect that at any given time some of the grains in a soil have already undergone diffusive loss from their last surface exposure, and indeed data of *Warhaut et al.* (1979) suggest that only about two-thirds of the ilmenites in a given soil have surface profiles, we would infer that f was actually closer to 0.2. If such a value of f can be applied to the bulk mineral data, then major depletions of ^{36}Ar must have occurred after implantation.

The rapid attenuation of the lunar surface thermal cycle with depth constrains the most favorable environment for diffusion out of grains to the upper centimeter or so of the regolith. Solmix calculations indicate that a 100 μm grain initially in the ~0.5 cm lunar "skin" tends to remain near the surface for the ~25 m.y. lifetime of a surface stratum prior to burial (*Duraud et al.*, 1975). We therefore propose, in the context of this model of regolith dynamics, that gases acquired by the average grain in each of its ~30 exposures at the top of the regolith are depleted by diffusion during the long intervals (~10^6 yr) of shallow subsurface residence between surface episodes. The last loading of the grain surface reservoir survives when the gases are frozen in by deep cycling or stratum burial shortly after implantation.

As an aside, we note that the essentially solar composition of noble gases in ilmenite grain surfaces indicates that the ilmenite sputtering rate of 0.05 Å/yr given by *Bertaut et al.* (1975) may be incorrect. In ~5000 yr, removal of only about 250 Å would lead to noble gas ratios in sputtered surfaces distinct from solar values (~80% of solar for Ne/Ar), contrary to observation (Fig. 3b). At least twice as much sputtering would be necessary to establish sputter equilibrium. On the other hand, preservation of the initially implanted gases in their solar ratios (excluding He) implies sputtering rates at least 3-4 times lower than 0.05 Å/yr.

Consistency of Noble Gas Isotopes

If we are correct in concluding that unfractionated and fractionated solar wind reservoirs exist in these samples, then there are certain expectations as to the relationship of the isotopic compositions of the two reservoirs; in particular, that the fractionated reservoir should be isotopically heavier. As is seen in Fig. 4, this is indeed the case for the nonspallogenic light noble gases. Whether the agreement is quantitative as well as qualitative is harder to evaluate. Consider Ne in 71501 ilmenite. Above 630°C, ^{20}Ne/^{22}Ne ratios range from 13.5 to 12.1 with an integrated value of 12.7. As indicated earlier, depending on the assumed relative implantation depths for the Ne isotopes, fractionations between 0.915 and 0.977 can be expected for this ratio. For an initial solar ratio of 13.5 (Fig. 3b), fractionated values of 12.4 to 13.2 are obtained, which bracket the ilmenite value. A similar situation exists for Ar, assuming the solar ratio is 5.6 ± 0.1.

The situation for Ne must be considered in more detail, because it may include the proposed SEP-Ne component with ^{20}Ne/^{22}Ne of ~11.3 (*Wieler et al.*, 1986, 1987). It is clear from our calculations that such a component cannot be derived by diffusion from implanted solar wind for any reasonable assumptions about relative implantation depths of the Ne isotopes, and it therefore constitutes a true third component to be considered in addition to unfractionated and fractionated solar wind. An SEP component would not be observable in our data if present at a level of ~10% of the total Ne or less, because of the uncertainty in the expected diffusive fractionation and because mixing of components could not be distinguished from laboratory-induced fractionation in the high-temperature release steps. We therefore have no problem in accepting the existence of SEP-Ne at this level. But if the SEP component amounts to 20% or more of total Ne (*Wieler et al.*, 1986), difficulties do arise, because the "fractionated" component would then have to be either a mixture of SEP and solar wind components or essentially an SEP component alone.

The two possibilities arise for the following reason. One can consider the SEP-Ne as a discrete component with a particular implantation energy range and a ^{20}Ne/^{22}Ne ratio of ~11.3, or as a continuum with a wide range of implantation energies above the solar wind value, in which case it is not clear that a single ^{20}Ne/^{22}Ne ratio should characterize the entire spectrum of energies. *Wieler et al.* (1986) take the position that SEP and solar flare Ne have essentially the same ^{20}Ne/^{22}Ne ratio, distinct from the solar wind, while *Black* (1983) allows that there might be an energy dependence of the isotopic ratio. In the first case, our "fractionated" component would consist at least in part of a mixture of two end-members, SEP and solar wind. In the second case, the fractionated component might be more or less completely SEP, with internal variations in isotopic structure reflecting variations in primary isotopic composition as a function of ion energy. We do not think that either of these possibilities could account for the "fractionated" component, because one needs to explain the elemental abundance pattern as well as the isotopic ratios. In essence, there is no evident reason for either a pure SEP component or a mixture of SEP and solar wind to show the light element depletions necessary to account for the Fig. 3a pattern. Therefore, while the evidence for an 11.3 component in the data of *Wieler et al.* (1986, 1987) is strong, it appears that this component contributes less than 10% of the total Ne in our samples.

Turning to Kr, one can again ask whether the "solar" ilmenite reservoir is isotopically consistent with other observed Kr reservoirs. For bulk lunar soils, this appears to be the case, as all the examples in Fig. 6 are similar to or isotopically heavier than Kr from the low-temperature ilmenite reservoir. Note that SUCOR-Kr (*Podosek et al.*, 1971) is heavier than Kr in the bulk soils, and would therefore seem not to be the candidate for primary solar wind that it was suggested to be, at least if one accepts that mass fractionation has been operating at the lunar surface. Integrated Kr isotopic compositions from high-temperature reservoirs in 67701 and 79035, as well as the ilmenite, are heavier than that in the "solar" reservoir. This is not the case, however, for the low-temperature data. In particular, the integrated Kr composition for the first four temperature-release steps from 67701, plotted in Fig. 9a as $\Sigma \leq C600°C$, is significantly lighter than the low-temperature ilmenite Kr reservoir (Fig. 9b). One must account in some way for differences in this direction if the ilmenite reservoir is presumed to contain unaltered solar wind.

Release from 67701 was primarily by diffusion, and gases evolved in early stages of diffusive loss can be isotopically lighter than their source reservoir. For a reservoir where gas concentration profiles extend to grain surfaces, as might be expected here since the sample was etched, Fick's Law predicts enhancements of light (l) relative to heavy (h) species in first releases by a factor $[D_l/D_h]^{1/2}$, which for isotopic species should be $\alpha^{1/4}$ where $\alpha = [M_h/M_l]$. Comparing the data in Tables 1b and 4b, which are relative to ^{84}Kr rather than renormalized to ^{82}Kr as in Fig. 9, the 1σ uncertainties in the integrated low-temperature Kr compositions from 67701 and the 71501 ilmenite overlap, except for ^{78}Kr, when the $\alpha^{1/4}$ factor is taken into account. Thus a solar wind composition somewhere between these two might be consistent with both data sets. However, the pattern of fractionation appears too uniform to arise from random scatter of errors for a single reservoir composition, so 67701 Kr probably is isotopically lighter than the ilmenite Kr, even with the $\alpha^{1/4}$ correction. Light isotope enhancements larger than $\alpha^{1/4}$ are possible when the initial profiles of the diffusing species do not extend to the grain surface, because diffusive separation occurs during transport to the surface. If a thin layer of precipitated material through which the noble gases subsequently had to diffuse were left on grain surfaces when the etching acids were pumped away, then it is possible for the low-temperature Kr from 67701 to represent diffusively fractionated Kr from a reservoir that closely resembles the low-temperature ilmenite Kr. It is even possible to account for the step-to-step variations in the first four steps for 67701 in this fashion. However, this scenario becomes much more difficult to defend when extended to include Ar as well as Kr: Predicted Ar fractionations are such that the ^{36}Ar/^{38}Ar ratio in the reservoir being sampled at 400° and 600°C would be ~5.3, rather than the value of 5.6 ± 0.1 characterizing the 67701 etch and the low-temperature ilmenite reservoir.

The case of breccia 79035 (Fig. 7) is similar to 67701. The low-temperature data are enriched in the light isotopes relative to the ilmenite surface reservoir. Since this is a sample in which lunar processes such as agglutinate formation and glass splashing may have led to deposition of gas-free material on top of most of the implanted gases, diffusive separation can again be invoked. The point of this discussion is to emphasize that it is possible, with the aid of some *ad hoc* but reasonable assumptions, to account for our Kr data in the context of an implanted solar-composition reservoir plus diffusive fractionation. More precise data are clearly needed, but at the moment we would argue that the isotopic composition of solar wind Kr resembles that in the low-temperature 71501 ilmenite reservoir, perhaps falling near the isotopically light ends of the 71501 error bars in Fig. 9.

Interpretation of the Xe isotopic data is intrinsically more difficult, because of the possible presence of components such as reimplanted fission Xe, as well as significant spallogenic contributions. The data do not point unambiguously toward a particular composition for unaltered solar wind Xe. The 71501 mineral data in Fig. 12 illustrate the difficulties. The 900°-1200°C pyroxene fractions, while not very precise, are clearly isotopically light. This cannot be simply a diffusive enhancement in first release, because these fractions together constitute ~80% of the total Xe in the sample. Moreover, as this Xe is presumably dominated by the fractionated component, it should be isotopically heavy compared to the primary solar wind. The pyroxene data thus point to a solar wind composition falling to the left of their intergrated value in Fig. 12. This is, however, inconsistent with the supposition that the low-temperature ilmenite data represent the primary solar wind, since the $\Sigma \leq 630$ ilmenite point lies well to the *right* of the integrated pyroxene point. In addition, the composition of total ilmenite Xe appears to be similar to or even lighter than that in the low-temperature fractions, even when spallation is taken into account, despite the fact that ~90% of the total Xe belongs to the fractionated component and should therefore be heavier. These observations may be evidence for reimplanted fission Xe in the ilmenite surface reservoir, with the true solar wind composition lying somewhere to the left of the $\Sigma \leq 630$ point. The light Xe isotopes can in principle be used to check this hypothesis, since they are shielded from fission contributions. Light isotope abundances relative to ^{130}Xe should be lower in pyroxene and total ilmenite than in the putatively unfractionated ilmenite surface reservoir. This is in fact the case for ^{128}Xe/^{130}Xe in the 900°-1200°C steps of the pyroxene: The average of this ratio, even ignoring possible spallation contributions, is significantly less than that in the $\Sigma \leq 630$ for ilmenite. Measurement uncertainties do not allow the corresponding comparisons to be made for ^{124}Xe, ^{126}Xe, and

^{129}Xe. For the total ilmenite, spallation contributions at the light isotopes are too large, as is clear from Table 4c, to permit accurate derivation of the nonspallogenic composition for comparison with the isotopic ratios in the ilmenite surface reservoir.

Finally, the same kinds of uncertainties hamper interpretation of the 67701 data in Fig. 10. One might suppose, by the arguments concerning diffusive fractionation in first releases used for Kr, that the solar wind composition should be isotopically heavier than the low-temperature Xe composition. But the $\Sigma \leq 600°C$ composition in Fig. 10, though imprecise, is much too heavy to support this view: Fission Xe contributions or (in this case) air contamination must be present. The pyroxene data in Fig. 12 also suggest that the solar wind composition derived by Pepin and Phinney (unpublished data, 1978), and apparently supported by the grain-size data shown in the Fig. 10 inset, is isotopically too heavy for solar wind, in all likelihood because these are bulk soil samples dominated by the fractionated Xe component.

Nitrogen

Nitrogen in our samples stands out as an apparent anomaly above the noble gas patterns defined in Fig. 3. Bulk soils have long been known to contain excess nitrogen relative to noble gases, with the excess attributed to preferential loss of noble gases during regolith processing (*DesMarais et al.,* 1974; *Jull and Pillinger,* 1977). The overabundance of N in the unfractionated ilmenite reservoir (Fig. 3b) complicates the situation, since we argue that this reservoir has not lost noble gases. Several possible explanations for the excess can be imagined, though none is completely satisfactory.

One possibility is surficial contamination, resulting in excess N in the combustion steps. Contamination from the extraction system was low for ilmenite and pyroxene when compared to other samples, as judged by yields of N and by the absence of δ^{15}N values near +2‰ at low temperatures (compare Tables 2a and 3a with Table 4a). The 300°C ilmenite nitrogen has already been excluded in the sums given in Table 6 and Fig. 3, as its δ^{15}N suggests it might be compromised. An upper limit for contaminants can be obtained by assuming that solar wind N in ilmenite grain surfaces is characterized by δ^{15}N around +156‰, the highest lunar value reported (*Norris et al.,* 1983). Subtracting this maximum contamination leaves a total of 0.49 ppm N in the ilmenite combustion steps, and a N/Ar ratio that is still 4.7 times the *Cameron* (1982) value.

To reduce the N/Ar ratio in the 400°-630°C steps to the solar value (assuming the 300°C step represents laboratory contamination) requires subtraction of a hypothetical nitrogen component with δ^{15}N around +45‰. Such a component could not be a system contaminant, but must have as a source the sample itself. An adsorbed atmospheric component could be proposed, but the δ^{15}N of +45‰ cited above requires an enormous isotopic fractionation. This seems unlikely since xenon adsorbed on active surfaces is not isotopically fractionated (*Niemeyer and Leich,* 1976). In addition, the isotopic ratio of the 600°C step implies that nearly all the nitrogen in this step consisted of the adsorbed component, requiring that this component be retained to higher temperatures than implanted N. In view of these considerations, it is difficult to assign a substantial fraction of the low-temperature nitrogen in the ilmenite to contamination.

Indigenous nitrogen might be proposed to account for the excess above the solar ratio. This raises several problems, however. First, the enrichment of N relative to Ar is the same in both gas reservoirs, even though the low-temperature reservoir gases presumably come only from the outermost thin shell of the ilmenite grains. This would require the *ad hoc* assumption that indigenous nitrogen is concentrated near the outside of grains. Second, as noted above, the excess component of nitrogen in grain surfaces must have a δ^{15}N value of +45‰, if the solar wind is +156‰. If the excess N released at higher temperatures has the same δ^{15}N, as it must for an indigenous component, a simple mass balance requires that the high-temperature solar component have a δ^{15}N around -600‰. Evidence for such light nitrogen has not been seen in other lunar materials. Finally, ilmenite from 79035 shows similar nitrogen excesses but very different δ^{15}N values (*Becker et al.,* 1984), a result that cannot be reconciled with the proposition that nitrogen in the ilmenite is dominated by an indigenous lunar component.

One could argue that the excess nitrogen in the ilmenite surface reservoir is a residue from the fractionated component. That is to say, when noble gases diffused out of the implantation zone, nitrogen, because of its chemical activity, was retained to some degree. This would correspond to the generally held view that nitrogen is better retained than the noble gases in lunar regolith materials as a whole. In the extreme, one could argue that no N has been lost from lunar soils, and the excess N in the ilmenite combustion steps thus represents all the nitrogen originally implanted. This requires nitrogen in the higher temperature steps to be indigenous. The isotopic restriction noted above for an indigenous component no longer applies, but one still must explain the remarkable coincidence of N/Ar ratios in both 71501 reservoirs and in the 79035 ilmenite (*Becker at al.,* 1984). A less extreme view is that some nitrogen diffusion and loss occurred, but much less than for the noble gases. If nitrogen were initially present in the *Cameron* (1982) abundance relative to Xe, implanted at a depth of 160 Å (between He and Ne), approximately 80% N loss would yield a residual profile in which ~25% of the retained N is sited in the surface reservoir (above ~500 Å), corresponding to the observation for ilmenite. In addition, the near-surface 25% of N would be enriched in δ^{15}N by 40-50‰ relative to the rest of the residual N, in rough accord with the actual measurements (Fig. 5). Note that this particular explanation requires nitrogen to be retained in radiation-damaged portions of grain surfaces, even if noble gases are not. Otherwise, no enrichment of nitrogen would appear in the solar component.

A consequence of this scenario for N is the severe losses it implies for the noble gases. If nitrogen has suffered an 80% loss and the original N/Xe ratio in the ilmenite were solar, then the current N/Xe (3.9 times solar in the fractionated component) requires a total Xe depletion of about a factor of 20 and correspondingly greater depletions for the other gases (see Table 6). Absolute retentions for 80% N loss would be: ^{4}He = 0.15%, ^{22}Ne = 0.50%, ^{36}Ar = 1.2%, ^{84}K = 2.6%, ^{132}Xe = 5.0%, and ^{14}N = 20%. Such low retention factors are consistent with the earlier argument that large loss factors would be required for the noble gases if the fractionated component presently lies predominantly below the original implantation depth.

There is one serious problem with the proposition that excess N in ilmenite can be attributed to a relatively smaller diffusive loss: In the laboratory, nitrogen and argon diffuse from ilmenite at essentially the same rate (see Fig. 2a). This is true for the other samples as well, excepting the 71501 plagioclase. Unless one invokes a very large difference in the behaviors of Ar and N under lunar conditions (see below), this apparently satisfactory explanation fails.

One must finally consider the possibility that the high nitrogen to noble gas ratios seen in ilmenite, and indeed in all lunar samples, reflects an equally high ratio in the implanted gases; that is, either in the solar wind itself or in a combination of solar wind plus reimplanted lunar atmospheric species (*Heymann and Yaniv*, 1970; *Manka and Michel*, 1970, 1971). *Ray and Heymann* (1982) have argued, for example, that the observed secular variation in δ^{15}N in lunar soils is due to a significant reimplanted lunar wind component. We note at least two problems in invoking reimplanted species. First, the excess N is superimposed on a normal solar pattern for the noble gases, requiring that there is a selective enrichment process for N in the lunar atmosphere. Second, lunar wind species are implanted more shallowly than solar wind species, and this initial difference will survive diffusion to some degree, yielding higher ratios of lunar to solar species near-surface than at depth. This occurs in ^{40}Ar/^{36}Ar profiles, but not in N/Ar. The alternative, in this scenario of excess N in the irradiating ion flux, is that the solar wind itself is enhanced in nitrogen by some process relative to the *Cameron* (1982) abundance ratios. However, satellite measurements on carbon in solar wind (*Bochsler et al.*, 1986), which must also be enriched severalfold in order to account for observed lunar C/N ratios (*Kerridge and Kaplan*, 1978), do not support this possibility.

The above discussion has been confined essentially to data from the 71501 ilmenite. The much smaller plagioclase and pyroxene samples were measured with less temperature resolution and lower isotopic precision. Effects of combustion were not recognizable in the gas release patterns, so both minerals can only be presumed by analogy with the 67701 etch to contain a surface reservoir of solar-type gases. Both are dominated by the fractionated component, as shown in Fig. 3a by Ar:Kr:Xe ratios similar to bulk ilmenite. Nitrogen to argon ratios differ from that in the ilmenite, in particular for plagioclase, but are clearly high relative to the solar ratio. *Frick and Pepin* (1982) concluded that the retentivity of N in the laboratory heating of the plagioclase suggested a nonsolar wind origin for this N. The high N/Ar ratio in the feldspar-rich 67701 sample (Table 6), which has a much larger solar component but still has a large excess of N, if not due to contamination during the etch, might indicate instead that the plagioclase has an enhanced mineral-specific retentivity for solar wind N, rather than a substantial content of indigenous N.

This possibility can be tested by further feldspar analyses. In either case, however, the pyroxene, which releases nitrogen in the laboratory at least as readily as Ar, would seem to support the idea that we are not dealing with a diffusive depletion of Ar relative to N of a factor of 12 or more in the ilmenite, but that some other process must be responsible for the high N/Ar ratios.

We are left with the situation that the nitrogen data, as well as our interpretation of the fractionated noble gas pattern, indicate that lunar grains have lost massive amounts of their solar wind gases, amounting to 70% or more of the Xe and correspondingly larger fractions for the lighter noble gases. The exact extent of the losses depends on how the nitrogen to noble gas ratios are interpreted. Concerning the nitrogen excesses, we find no completely satisfactory explanation. It may be that the laboratory data on relative diffusion rates of N and Ar are misleading us. Diffusion in the lunar regolith takes place over long times at relatively low temperatures, as compared to the elevated temperatures used in the lab. Nitrogen is implanted as N atoms, and may be more chemically active in this form than as N_2 molecules. If there is a threshold temperature at which association of atoms to N_2 molecules occurs, then diffusion of N on the Moon may be retarded relative to noble gases, and preferential retention of N may be the preferred explanation for the excesses despite our laboratory observations. If this is not the case, then the implanted gases may have to have a nitrogen to argon ratio more than an order of magnitude greater than the *Cameron* (1982) value.

SUMMARY AND CONCLUSIONS

The studies reported here and by the Zürich laboratory (*Signer et al.*, 1977; *Wieler et al.*, 1980, 1983, 1986) illustrate the complexity of extracting quantitative information about the present and past solar wind from the lunar regolith: The record has been overprinted to varying degrees by a variety of processes acting on regolith components in the lunar surface environment. These studies demonstrate, however, that in selected samples the record may be coherent and subject to self-consistent interpretation. Virtually pristine recordings of the solar wind extending back to the early history of the sun almost certainly exist on the lunar surface. The experimental challenge lies in finding and reading them.

It would be a long extrapolation of our current data base—two samples, of which one was incompletely analyzed—to propose that all grain surfaces contain unaltered solar wind noble gases. There are some conclusions, however, that one may draw from the present work:

1. Noble gases liberated by chemical attack on grain surfaces from two quite different kinds of lunar samples are strikingly solar. Surface-sited Ne/Ar ratios in fines from an anorthositic soil, and Ne/Ar, Kr/Ar, and Xe/Ar ratios in an ilmenite separate from another soil, agree with estimates of solar elemental ratios to within 10%; the ilmenite He/Ar ratio is solar to within 30%. Grain surfaces of feldspar do not retain implanted solar wind He quantitatively.

2. *Cameron's* (1982) solar system abundances describe very well the elemental composition of the noble gases in the solar wind.

3. The solar wind $^{36}Ar/^{38}Ar$ ratio is 5.6 ± 0.1. The $^{20}Ne/^{22}Ne$ ratio in the two grain surface reservoirs is 13.5-13.6, within error of the average value of 13.7 ± 0.3 measured in the Apollo solar wind collection foils (*Bochsler and Geiss*, 1977).

4. Noble gases in grain interiors have suffered severe mass fractionation, presumably as a result of the process acting to populate the grain interiors, which we believe was diffusive redistribution from the grain surfaces. The fractionation appears to have been both mass and size dependent, since relative magnitudes of the diffusion coefficients derived from the residual noble gas ratios are not directly related to the relative masses.

5. The fractions of the initially implanted gases now present in the minerals cannot be determined from the available data. Estimates of 70-95% loss of Xe and correspondingly larger losses of the light noble gases may be derived, however, for the ilmenite. The lower limit assumes an initial Xe/N as given by *Cameron* (1982) and no loss of N. The higher loss estimate derives from the observation that little of the fractionated noble gas component now resides in the original implantation zone, and is also a consequence of one of the possible scenarios for explaining nitrogen abundances and distributions.

6. In sharp contrast to the noble gases, surface-sited N/Ar in the ilmenite exceeds the predicted solar ratio by more than a factor of 10. This may represent a separate source of nitrogen, a depletion of the noble gases, or an incorrect choice for the solar wind ratio. The siting of the nitrogen does not appear to be compatible with an excess component added by either low-energy lunar surface mechanisms (e.g., atmospheric ion retrapping or adsorption) or by postlunar contamination.

7. Nitrogen to argon ratios in the interior of ilmenite grains are similar to those in the surface reservoir. Under laboratory conditions, N and Ar show similar diffusive releases. Since Ar in the interior reservoir has been depleted, one could assume that N has been as well, implying that N is not fully retained in lunar samples, and that implanted N/Ar ratios are much larger that the *Cameron* (1982) value. If, however, N can be shown to behave at lunar surface temperatures differently from Ar, then N/Ar ratios implanted into lunar grains can have been "solar." Some diffusive loss of N would still be implied by the present distribution of N in the ilmenite grains.

Acknowledgments. We thank Elizabeth Nilles for word processing assistance. This research was supported by grants NGL-24-004-225 and NAG 9-60 from the National Aeronautics and Space Administration.

REFERENCES

Alexander E. C., Jr., Bates A., Coscio M. R., Jr., Dragon J. C., Murthy V. R., Pepin R. O., and Venkatesan T. R. (1976) K/Ar dating of lunar soils II. *Proc. Lunar Sci. Conf. 7th*, pp. 625-648.

Alexander E. C., Jr., Coscio M. R., Jr., Dragon J. C., Pepin R. O., and Saito K. (1977) K/Ar dating of lunar soils III: Comparison of ^{39}Ar-^{40}Ar and conventional techniques; 12032 and the age of Copernicus. *Proc. Lunar Sci. Conf. 8th*, pp. 2725-2740.

Anders E. and Ebihara M. (1982) Solar-system abundances of the elements. *Geochim. Cosmochim. Acta, 46*, 2363-2380.

Basford J. R., Dragon J. C., Pepin R. O., Coscio M. R., Jr., and Murthy V. R. (1973) Krypton and xenon in lunar fines. *Proc. Lunar Sci. Conf. 4th*, pp. 1915-1955.

Baur H., Frick U., Funk H., Schultz L., and Signer P. (1972) Thermal release of helium, neon, and argon from lunar fines and minerals. *Proc. Lunar Sci. Conf. 3rd*, pp. 1947-1966.

Becker R. H. and Clayton R. N. (1977) Nitrogen isotopes in lunar soils as a measure of cosmic-ray exposure and regolith history. *Proc. Lunar Sci. Conf. 8th*, pp. 3685-3704.

Becker R. H. and Epstein S. (1981) Carbon isotopic ratios in some low-$\delta^{15}N$ lunar breccias. *Proc. Lunar Planet. Sci. 12B*, pp. 289-293.

Becker R. H., Clayton R. N., and Mayeda T. K. (1976) Characterization of lunar nitrogen components. *Proc. Lunar Sci. Conf. 7th*, pp. 441-458.

Becker R. H., Wiens R. C., and Pepin R. O. (1984) Comparison of solar wind gases in two lunar ilmenites of different antiquities (abstract). In *Lunar and Planetary Science XV*, pp. 42-43. Lunar and Planetary Institute, Houston.

Becker R. H., Pepin R. O., Rajan R. S., and Rambaldi E. R. (1986) Light noble gases in Weston metal grain surfaces (abstract). *Meteoritics, 21*, 331-332.

Behrmann C., Crozaz G., Drozd R., Hohenberg C., Ralston C., Walker R., and Yuhas D. (1973) Cosmic-ray exposure history of North Ray and South Ray material. *Proc. Lunar Sci. Conf. 4th*, pp. 1957-1974.

Bertaut J. L., Bibring J. P., Borg J., Burlingame A. L., Langevin Y., Maurette M., and Walls F. C. (1975) The maturation of lunar dust grains (abstract). In *Lunar Science VI*, pp. 42-44. The Lunar Science Institute, Houston.

Bibring J. P., Borg J., Burlingame A. L., Langevin Y., Maurette M., and Vassent B. (1975) Solar-wind and solar-flare maturation of the lunar regolith. *Proc. Lunar Sci. Conf. 6th*, pp. 3471-3493.

Black D. C. (1983) The isotopic composition of solar flare noble gases. *Astrophys. J., 266*, 889-894.

Bochsler P. and Geiss J. (1977) Elemental abundances in the solar wind. *Trans. Int. Astron. Union, XVIB*, 120-123.

Bochsler P., Geiss J., and Kunz S. (1986) Abundances of carbon, oxygen, and neon in the solar wind during the period from August 1978 to June 1982. *Solar Physics, 103*, 177-201.

Bogard D. D., Hirsch W. C., and Nyquist L. E. (1974) Noble gases in Apollo 17 fines: Mass fractionation effects in trapped Xe and Kr. *Proc. Lunar Sci. Conf. 5th*, pp. 1975-2003.

Borg J., Chaumont J., Jouret C., Langevin Y., and Maurette M. (1980) Solar wind radiation damage in lunar dust grains and the characteristics of the ancient solar wind. In *The Ancient Sun* (R. O. Pepin, J. A. Eddy, and R. B. Merrill, eds.), pp. 431-461. Pergamon, New York.

Borg J., Bibring J. P., Cowsik G., Langevin Y., and Maurette M. (1983) A model for the accumulation of solar wind radiation damage effects in lunar dust grains, based on recent results concerning implantation and erosion effects. *Proc. Lunar Planet. Sci. Conf. 13th*, in *J. Geophys. Res., 88*, A725-A730.

Cameron A. G. W. (1982) Elemental and nuclidic abundances in the solar system. In *Essays in Nuclear Astrophysics* (C. A. Barnes, D. D. Clayton and D. N. Schramm, eds.), pp. 23-43. Cambridge University Press, Cambridge.

Carr L. P., Wright I. P., and Pillinger C. T. (1985) Nitrogen content and isotopic composition of lunar breccia 79035: a high resolution study (abstract). *Meteoritics, 20*, 622-623.

Clayton R. N. and Thiemens M. H. (1980) Lunar nitrogen: Evidence for secular change in the solar wind. In *The Ancient Sun* (R. O. Pepin, J. A. Eddy, and R. B. Merrill, eds.), pp. 463-473. Pergamon, New York.

DesMarais D. J., Hayes J. M., and Meinschein W. G. (1974) Retention of solar wind-implanted elements in lunar soils (abstract). In *Lunar Science V,* pp. 168-170. The Lunar Science Institute, Houston.

Drozd R., Hohenberg C., and Morgan C. (1975) Krypton and xenon in Apollo 14 samples: Fission and neutron capture effects in gas-rich samples. *Proc. Lunar Sci. Conf. 6th,* pp. 1857-1877.

Duraud J. P., Langevin Y., Maurette M., Comstock G., and Burlingame A. L. (1975) The simulated depth history of dust grains in the lunar regolith. *Proc. Lunar Sci. Conf. 6th,* pp. 2397-2415.

Eberhardt P., Geiss J., Graf H., Grögler N., Krähenbühl U., Schwaller H., Schwarzmüller J., and Stettler A. (1970) Trapped solar wind noble gases, exposure age and K/Ar-age in Apollo 11 lunar fine material. *Proc. Apollo 11 Lunar Sci. Conf.,* pp. 1037-1070.

Eberhardt P., Geiss J., Graf H., Grögler N., Mendia M. D., Mörgeli M., Schwaller H., Stettler A., Krähenbühl U., and von Gunten H. R. (1972) Trapped solar wind noble gases in Apollo 12 lunar fines 12001 and Apollo 11 breccia 10046. *Proc. Lunar Sci. Conf. 3rd,* pp. 1821-1856.

Eberhardt P., Eugster O., Geiss J., Grögler N., Guggisberg S., and Mörgeli M. (1976) Noble gases in the Apollo 16 special soils from the East-West split and the permanently shadowed area. *Proc. Lunar Sci. Conf. 7th,* pp. 563-585.

Etique Ph. (1981) L'utilisation des plagioclases du régolithe lunaire comme détecteurs des gaz rares provenant des rayonnements corpusculaires solaires. Ph.D. thesis, No. 6924, ETH Zürich, Switzerland. 246 pp.

Eugster O., Eberhardt P., and Geiss J. (1967) Krypton and xenon isotopic composition in three carbonaceous chondrites. *Earth Planet. Sci. Lett., 3,* 249-257.

Frick U. (1977) Anomalous krypton in the Allende meteorite. *Proc. Lunar Sci. Conf. 8th,* pp. 273-292.

Frick U. (1980) Nucleosynthetic origin of anomalous krypton: test of a simple model (abstract). *Meteoritics, 15,* 292-293.

Frick U. and Pepin R. O. (1981a) On the distribution of noble gases in Allende: a differential oxidation study. *Earth Planet. Sci. Lett., 56,* 45-63.

Frick U. and Pepin R. O. (1981b) Microanalysis of nitrogen isotope abundances: association of nitrogen with noble gas carriers in Allende. *Earth Planet. Sci. Lett., 56,* 64-81.

Frick U. and Pepin R. O. (1981c) Study of solar wind gases in a young lunar soil (abstract). In *Lunar and Planetary Science XII,* pp. 303-305. Lunar and Planetary Institute, Houston.

Frick U. and Pepin R. O. (1981d) Fractionation in solar system krypton and xenon, and their isotopic compositions in the solar wind (abstract). *Meteoritics, 16,* 316-317.

Frick U. and Pepin R. O. (1982) The solar wind record in lunar regolith mineral grains: Noble gases and nitrogen (abstract). In *Lunar and Planetary Science XIII,* pp. 241-242. Lunar and Planetary Institute, Houston.

Frick U., Baur H., Funk H., Phinney D., Schäfer Chr., Schultz L., and Signer P. (1973a) Diffusion properties of light noble gases in lunar fines. *Proc. Lunar Sci. Conf. 4th,* pp. 1987-2002.

Frick U., Baur H., Funk H., Phinney D., Schultz L., and Signer P. (1973b) Diffusional gas release patterns of trapped gases from lunar soil separates (abstract). *Meteoritics, 8,* 367-369.

Frick U., Baur H., Ducati H., Funk H., Phinney D., and Signer P. (1975) On the origin of helium, neon, and argon isotopes in sieved mineral separates from an Apollo 15 soil. *Proc. Lunar Sci. Conf. 6th,* pp. 2097-2129.

Frick U., Kenny T. W., and Pepin R. O. (1982) The irradiation history of lunar breccia 79035 (abstract). *Meteoritics, 17,* 216-217.

Frick U., Becker R. H., and Pepin R. O. (1986) Release of noble gases and nitrogen from grain-surface sites in lunar ilmenite by closed-system oxidation (abstract). In *Workshop on Past and Present Solar Radiation: The Record in Meteoritic and Lunar Regolith Material* (R. O. Pepin and D. S. McKay, eds.), pp. 22-23. LPI Tech. Rpt. 86-02, Lunar and Planetary Institute, Houston.

Geiss J. and Bochsler P. (1982) Nitrogen isotopes in the solar system. *Geochim. Cosmochim. Acta, 46,* 529-548.

Geiss J., Buehler F., Cerutti H., Eberhardt P., and Filleux Ch. (1972) Solar wind composition experiment. In *Apollo 16 Prelim. Sci. Report,* pp. 14-1 to 14-10. NASA SP-315, Washington, D.C.

Heymann D. and Yaniv A. (1970) Ar^{40} anomaly in lunar samples from Apollo 11. *Proc. Apollo 11 Lunar Sci. Conf.,* pp. 1261-1267.

Hintenberger H., Weber H. W., and Schultz L. (1974) Solar, spallogenic, and radiogenic rare gases in Apollo 17 soils and breccias. *Proc. Lunar Sci. Conf. 5th,* pp. 2005-2022.

Hübner W., Kirsten T., and Kiko J. (1975) Rare gases in Apollo 17 soils with emphasis on analysis of size and mineral fractions of soil 74241. *Proc. Lunar Sci. Conf. 6th,* pp. 2009-2026.

Huneke J. C. (1973) Diffusive fractionation of surface implanted gases. *Earth Planet. Sci. Lett., 21,* 35-44.

Huneke J. C., Nyquist L. E., Funk H., Köppel V., and Signer P. (1969) The thermal release of rare gases from separated minerals of the Mócs meteorite. In *Meteorite Research* (P. M. Millman, ed.), pp. 901-921. Reidel, Dordrecht.

Jull A. J. T. and Pillinger C. T. (1977) Effects of sputtering on solar wind element accumulation. *Proc. Lunar Sci. Conf. 8th,* pp. 3817-3833.

Jull A. J. T., Grant W. A., Christodoulides C., and Pillinger C. T. (1980) Experimental ranges of the solar wind ions krypton and xenon in minerals (abstract). In *Lunar and Planetary Science XI,* pp. 523-525. Lunar and Planetary Institute, Houston.

Kaiser W. A. (1972) Rare gas studies in Luna-16-G-7 fines by stepwise heating technique. A low fission solar wind Xe. *Earth Planet. Sci. Lett., 13,* 387-399.

Kerridge J. F. (1975) Solar nitrogen: Evidence for a secular increase in the ratio of nitrogen-15 to nitrogen-14. *Science, 188,* 162-164.

Kerridge J. F. (1980a) Accretion of nitrogen during the growth of planets. *Nature, 283,* 183-184.

Kerridge J. F. (1980b) Secular variations in composition of the solar wind: Evidence and causes. In *The Ancient Sun* (R. O. Pepin, J. A. Eddy and R. B. Merrill, eds.), pp. 475-489. Pergamon, New York.

Kerridge J. F. (1982) Whence so much ^{15}N? *Nature, 295,* 643-644.

Kerridge J. F. and Kaplan I. R. (1978) Sputtering: Its relationship to isotopic fractionation on the lunar surface. *Proc. Lunar Planet. Sci. Conf. 9th,* pp. 1687-1709.

Kerridge J. F., Kaplan I. R., and Petrowski C. (1975) Light element geochemistry of the Apollo 16 site. *Geochim. Cosmochim. Acta, 39,* 137-162.

Manka R. H. and Michel F. C. (1970) Lunar atmosphere as a source of argon-40 and other lunar surface elements. *Science, 169,* 278-280.

Manka R. H. and Michel F. C. (1971) Lunar atmosphere as a source of lunar surface elements. *Proc. Lunar Sci. Conf. 2nd,* pp. 1717-1728.

Mariotti A. (1984) Natural ^{15}N abundance measurements and atmospheric nitrogen standard calibration. *Nature, 311,* 251-252.

Marti K. (1969) Solar-type xenon: a new isotopic composition of xenon in the Pesyanoe meteorite. *Science, 166,* 1263-1265.

Marti K. (1980) On krypton isotopic abundances in the sun and in the solar wind. In *The Ancient Sun* (R. O. Pepin, J. A. Eddy and R. B. Merrill, eds.), pp. 423-429. Pergamon, New York.

Mewaldt R. A., Spalding J. D., and Stone E. C. (1984) A high-resolution study of the isotopes of solar flare nuclei. *Astrophys. J., 280,* 892-901.

Morris R. V. (1978) The surface exposure (maturity) of lunar soils: Some concepts and I_s/FeO compilation. *Proc. Lunar Planet. Sci. Conf. 9th,* pp. 2287-2297.

Niemeyer S. and Leich D. A. (1976) Atmospheric rare gases in lunar rock 60015. *Proc. Lunar Sci. Conf. 7th*, pp. 587-597.

Nier A. O. (1950) A redetermination of the relative abundances of the isotopes of carbon, nitrogen, oxygen, argon, and potassium. *Phys. Rev.*, 77, 789-793.

Norris S. J., Swart P. K., Wright I. P., Grady M. M., and Pillinger C. T. (1983) A search for correlatable, isotopically light carbon and nitrogen components in lunar soils and breccias. *Proc. Lunar Planet. Sci. Conf. 14th*, in *J. Geophys. Res.*, 88, B200-B210.

Pepin R. O. (1967) Trapped neon in meteorites. *Earth Planet. Sci. Lett.*, 2, 13-18.

Pepin R. O. (1980) Rare gases in the past and present solar wind. In *The Ancient Sun*. (R. O. Pepin, J. A. Eddy and R. B. Merrill, eds.), pp. 411-421. Pergamon, New York.

Pepin R. O. and Frick U. (1981) Noble gases and nitrogen in lunar breccia 79035 (abstract). *Meteoritics*, 16, 376-377.

Pepin R. O., Dragon J. C., Johnson N. L., Bates A., Coscio M. R., Jr., and Murthy V. R. (1975) Rare gases and Ca, Sr, and Ba in Apollo 17 drill-core fines. *Proc. Lunar Sci. Conf. 6th*, pp. 2027-2055.

Podosek F. A., Huneke J. C., Burnett D. S., and Wasserburg G. J. (1971) Isotopic composition of xenon and krypton in the lunar soil and in the solar wind. *Earth Planet. Sci. Lett.*, 10, 199-216.

Rao D. B. and Rigaud M. (1975) Kinetics of the oxidation of ilmenite. *Oxid. of Metals*, 9, 99-116.

Ray J. and Heymann D. (1982) Long-term solar wind activity as inferred from lunar regolith samples (abstract). In *Lunar and Planetary Science XIII*, pp. 640-641. Lunar and Planetary Institute, Houston.

Reynolds J. H., Frick U., Neil J. M., and Phinney D. L. (1978) Rare-gas-rich separates from carbonaceous chondrites. *Geochim. Cosmochim. Acta*, 42, 1775-1797.

Signer P., Baur H., Derksen U., Etique Ph., Funk H., Horn P., and Wieler R. (1977) Helium, neon, and argon records of lunar soil evolution. *Proc. Lunar Sci. Conf. 8th*, pp. 3657-3683.

Srinivasan B. and Anders E. (1978) Noble gases in the Murchison meteorite: Possible relics of s-process nucleosynthesis. *Science*, 201, 51-56.

Srinivasan B., Gros J., and Anders E. (1977) Noble gases in separated meteoritic minerals: Murchison (C2), Ornans (C3), Karoonda (C5), and Abee (E4). *J. Geophys. Res.*, 82, 762-778.

Tamhane A. S. and Agrawal J. K. (1979) Diffusion of rare gases of solar wind origin from lunar fines as bubbles. *Earth Planet. Sci. Lett.*, 42, 243-250.

Walton J. R., Heymann D., Jordan J. L., and Yaniv A. (1974) Evidence for solar cosmic ray proton-produced neon in fines 67701 from the rim of North Ray Crater. *Proc. Lunar Sci. Conf. 5th*, pp. 2045-2060.

Warasila R. L. and Schaeffer O. A. (1974) Trapped solar wind He, Ne, and Ar and energetic He in Surveyor 3. *Earth Planet. Sci. Lett.*, 24, 71-77.

Warhaut M., Kiko J., and Kirsten T. (1979) Microdistribution patterns of implanted rare gases in a large number of individual lunar soil particles. *Proc. Lunar Planet. Sci. Conf. 10th*, 1531-1546.

Wieler R., Etique Ph., Signer P., and Poupeau G. (1980) Record of the solar corpuscular radiation in minerals from lunar soils: A comparative study of noble gases and tracks. *Proc. Lunar Planet. Sci. Conf. 11th*, pp. 1369-1393.

Wieler R., Etique Ph., Signer P., and Poupeau G. (1983) Decrease of the solar flare/solar wind flux ratio in the past several aeons deduced from solar neon and tracks in lunar soil plagioclases. *Proc. Lunar Planet. Sci. Conf. 13th*, in *J. Geophys. Res.*, 88, A713-A724.

Wieler R., Baur H., and Signer P. (1986) Noble gases from solar energetic particles revealed by closed system stepwise etching of lunar soil minerals. *Geochim. Cosmochim. Acta*, 50, 1997-2017.

Wieler R., Baur H., Benkert J. P., Pedroni A., and Signer P. (1987) Noble gases in the meteorite Fayetteville and in lunar ilmenite originating from solar energetic particles (abstract). In *Lunar and Planetary Science XVIII*, pp. 1080-1081. Lunar and Planetary Institute, Houston.

Wiens R. C., Becker R. H., and Pepin R. O. (1986) The case for a martian origin of the shergottites, II. Trapped and indigenous gas components in EETA 79001 glass. *Earth Planet. Sci. Lett.*, 77, 149-158.

Zinner E., Walker R. M., Chaumont J., and Dran J. C. (1977) Ion probe surface concentration measurements of Mg and Fe and microcraters in crystals from lunar rock and soil samples. *Proc. Lunar Sci. Conf. 8th*, pp. 3859-3883.

Importance of Lunar Granite and KREEP in Very High Potassium (VHK) Basalt Petrogenesis

Clive R. Neal and Lawrence A. Taylor
Department of Geological Sciences, University of Tennessee, Knoxville, TN 37996

Marilyn M. Lindstrom*
Department of Earth and Planetary Sciences, Washington University, St. Louis, MO 63130

*Now at NASA Johnson Space Center, Code SN2, Houston, TX 77058

Five new Very High Potassium (VHK) basalts, from Apollo 14 breccia 14303, extend the compositional ranges of previously reported VHK basalts. It is demonstrated that although lunar granite is essential in achieving VHK compositions, a KREEP component is also present in the 14303 VHK basalts. An Assimilation and Fractional Crystallization model (AFC) is presented whereby a primitive magma [represented by a high-Al (HA) mare basalt composition] undergoing AFC with KREEP comes into contact with granite during migration. It is envisaged that VHK parental magmas lie along the mare basalt evolution path (*Neal et al.*, 1987a). The VHK compositions diverge from this trend because of the effect of granite assimilation. This process stresses the significance of granite in the lunar crust. Proportions of fractionating phases vary according to the position along the mare basalt evolutionary path prior to granite assimilation and can be constrained by major element modeling. The presence of VHK basalts containing only a granite signature (i.e., representing two-component AFC) and those with both granite and KREEP signatures (three-component AFC) argues for at least two different VHK basalt flows at the Apollo 14 site. An impact melt of similar composition to the VHK basalts (14303,277) is also reported and can only be distinguished on the basis of whole-rock chemistry.

INTRODUCTION

The study of mare basalts is critical in determining the nature of the lunar interior. Calculations of source compositions for these basalts and comparisons with other lunar rock types allow the extent of igneous processes, such as partial melting, fractional crystallization, and assimilation, to be evaluated with regard to the evolution of the Moon (*Nyquist et al.*, 1979; *Ringwood and Kesson*, 1976; *Shervais et al*, 1985a). For example, assimilation of a KREEP component has been invoked to explain certain anorthosite compositions (*Warren et al*, 1983a) as well as mare basalt evolution (*Binder*, 1982, 1985; *Dickinson et al.*, 1985; *Shervais et al.*, 1985a; *Neal et al.*, 1987a).

The recognition of Very High Potassium (VHK) mare basalts in Apollo 14 breccias (*Shervais et al.*, 1985b) created problems in achieving such compositions by partial melting and/or fractional crystallization (*Shih et al.*, 1986, 1987). VHK basalts would seem to require a unique source region that, as implied by the rarity of VHK basalts, was highly localized. Major element compositions are remarkably similar to High Aluminum (HA) mare basalts except for the characteristically elevated K_2O contents (>0.5 wt %). Consequently, VHK basalts are also characterized by K_2O/Na_2O ratios of >1 and K/La ratios between 500 and 1500 (*Shervais et al.*, 1985b). In addition, Rb and Ba are enriched relative to other HA mare basalts. In an isotopic study of two VHK basalt clasts, *Shih et al.* (1986) reported the most radiogenic $^{87}Sr/^{86}Sr$ values to date from the Moon, but with initial ratios indistinguishable from other analyzed basalts. More recent studies by *Shih et al.* (1987) of 14304 VHK basalts reported similar conclusions. This indicates an approximate 15-fold enrichment of Rb in the VHK magma relative to HA mare basalts at the time of crystallization.

The above observations have led to the formulation of two hypotheses to explain the production of VHK basalts: (1) a metasomatized source for the VHK basalts that is similar to that of other HA basalts but is enriched in K, Rb, and Ba (e.g., *Goodrich et al.*, 1986); and (2) assimilation of lunar granite by an HA basaltic magma (e.g., *Shervais et al.*, 1985b). However, both hypotheses have serious shortcomings. For instance, it is unlikely that a metasomatizing fluid enriched in volatiles would also be enriched in Ba; enrichment in other volatiles besides K and Rb would also be expected (see *Shervais et al.*, 1985b). The high U and Th contents of most lunar granites pose a mass balance problem for any simple granite assimilation model (*Shervais et al.*, 1985b; *Warren et al.*, 1986), reflected in the normally low U and Th contents of VHK basalts. This has led to the explanation by *Shih et al.* (1987) that granitic feldspars are preferentially assimilated during VHK basalt petrogenesis. There exists yet a third possibility to explain the VHK petrogenesis wherein these basalts are the products of meteorite impact melts (G. Ryder and R. Schmitt, personal communication, 1987). This hypothesis is based upon Ni/Co ratios, Au and Ir abundances, and geochemical comparisons with the breccia matrix. These considerations will be discussed in detail within the chemistry and modeling sections below.

In this paper, we report the results of a study of five new VHK basalts from Apollo 14 breccia 14303 that form part of an ongoing investigation into VHK basalt petrogenesis (e.g., *Neal et al.*, 1987b). A sixth clast is very similar, but evidence presented below suggests it is an impact melt. The new data are compared and contrasted with previously reported VHK compositions, and the role of KREEP and lunar granite in generating all reported VHK compositions is evaluated. The ultimate aim of this research is to clarify the processes involved in VHK petrogenesis.

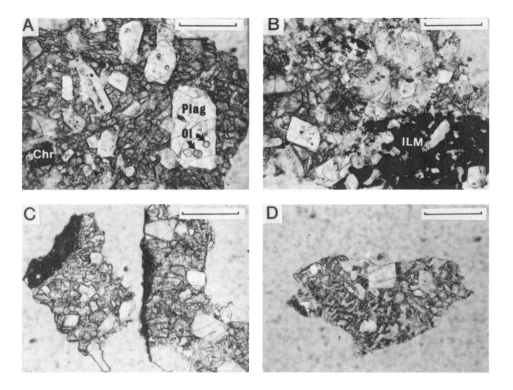

Fig. 1. Photomicrographs of the 14303 VHK basalts. The scale bar represents 0.5 mm. **(a)** 14303,245 contains plagioclase phenocrysts with rounded olivine inclusions, suggesting a precursory cotectic magma. One grain of chromite is seen. **(b)** Interstitial ilmenite in 14303,245 containing plagioclase phenocrysts. **(c)** Texture of 14303,246 with plagioclase phenocrysts and interstitial pyroxene. **(d)** Texture of 14303,249 also exhibiting plagioclase phenocrysts and interstitial pyroxene.

PETROGRAPHY AND MINERAL CHEMISTRY

Six VHK basalt clasts were extracted from Apollo 14 breccia 14303. They were small (0.25-1 cm diameter) with grain sizes reaching a maximum of 0.4 mm. Probe mounts (PM) were made for ,245 (INAA = ,244) ,246 (INAA = ,247) and ,249; in addition, probe mounts were made from "hot" INAA samples for ,266 ,275 and ,277. Mineral chemistry was determined using an automated MAC 400S electron microprobe at the University of Tennessee. Standards, operating procedures, and techniques of data reduction have been reported elsewhere (e.g., *Shervais et al.*, 1984).

The VHK clasts from breccias 14303 are fine- to medium-grained basalts with a general subophitic texture (Fig. 1). Estimated modal abundances were determined by careful examination of each probe mount (Table 1); small sample size renders detailed point counting less than realistic. The mineralogy in ,245 is especially dominated by feldspar ($\cong$ 60%) with relatively little pyroxene ($\cong$ 22%) and olivine ($\cong$ 10%).

TABLE 1. Estimated modal proportions for the 14303 VHK basalts.

.PM No.	Feldspar	Olivine	Pyroxene	Oxide	Glass
,245	60	3	21	10	6
,246	50	0	42	3	5
,249	60	0	30	4	6
,266	40	10	30	12	8
,275	50	0	43	4	3
,277	35	15	30	15	5

Plagioclase is also the dominant phase in all other samples (except ,277), although the proportion of mafic minerals increases. The proportions of interstitial glass shows little variation throughout the suite ($\cong$ 3-8%).

Large plagioclase phenocrysts ($\cong$ 0.4 mm diameter) occur in four of our samples (,245 ,246 ,249 ,275) with core compositions of An96-97 (Table 2), but they become more sodic toward the rims (An84-87). There is also a smaller, compositionally indistinguishable, groundmass plagioclase generation displaying similar zonation. The remaining two samples (,266 and ,277) have no plagioclase phenocrysts. K-feldspar is present in all samples (Table 3), except ,245, as small crystals (<0.1 mm) in interstitial K-rich glass and also as thin overgrowths on plagioclase (Or 77-93 in ,246; 82-94 in ,249; 88-92 in ,266; 86-96 in ,275; 85-95 in ,277). These K-feldspars are generally rich in Ba (0.12-5.11 wt % BaO). The petrographic relations indicate K-feldspar is a late-stage crystallization product in these basalts. Interstitial K-rich glass containing Ba is present in all six 14303 VHK basalts (Table 4). The composition of this interstitial glass in ,245 is distinct from other samples in that it has elevated FeO contents (7.09 wt %). This glass composition may be due to the lack of substantial crystallization of late-stage fayalitic olivines, possibly implying faster cooling rates for this sample. The larger grain size of the plagioclase phenocrysts in ,245, coupled with the lack of K-feldspar, again suggests that this sample has undergone greater pre-eruption fractionation and cooled more quickly after eruption, thus inhibiting K-feldspar and fayalite crystallization from interstitial liquid.

TABLE 2. Representative plagioclase analyses from 14303 VHK basalts.

PM No	,245	,246	,249	,266	2,75	,277
INAA NO.	,244	,247	-	,266	,275	,277
SiO_2	45.4	45.0	44.8	45.2	46.5	45.6
Al_2O_3	34.8	35.3	35.5	34.8	34.6	34.2
FeO	0.32	0.06	0.14	0.70	0.04	0.79
CaO	18.7	18.5	18.9	17.7	17.8	18.2
Na_2O	0.93	0.96	0.82	0.99	1.29	1.03
K_2O	0.14	0.11	0.10	0.14	0.11	0.15
Total	100.29	99.93	100.26	99.53	100.34	99.97
Formula moles based on eight oxygens						
Si	2.091	2.075	2.065	2.097	2.129	2.109
Al	1.887	1.919	1.928	1.898	1.869	1.863
Fe	0.012	0.001	0.005	0.027	0.001	0.030
Ca	0.923	0.917	0.931	0.880	0.871	0.902
Na	0.082	0.085	0.071	0.088	0.114	0.092
K	0.006	0.005	0.005	0.006	0.005	0.008
Total	5.000	5.002	5.006	4.997	4.991	5.004
An	91.3	91.1	92.4	90.3	88.0	90.0
Ab	8.1	8.4	7.1	9.1	11.5	9.2
Or	0.6	0.5	0.5	0.6	0.5	0.8

Olivine is present only in ,245 ,266 and ,277 with two generations present, Mg- and Fe-rich, as noted by *Shervais et al.* (1985b) from 14305 VHK basalts. In ,245 olivine occurs as two corroded ($\cong 0.1$ mm) grains (Fo51-52 and Fo70) (Table 5). Olivine is more common in 266 (Fo34-39 and Fo65-69) and 277 (Fo37-43 and Fo55-65), occurring as corroded phenocrysts ($\cong 0.25$ mm). However, also in sample ,245, some of the larger plagioclase phenocrysts contain inclusions of olivine (Figs. 1a,b and Table 5) that are quite variable in Fo content (55-73) and span the compositional gap reported by *Shervais et al.* (1985b). The rimming of olivine by pigeonite in all three samples, where olivine is not present as inclusions in plagioclase, indicates a reaction between olivine and melt. No pigeonite rims are present on the olivine inclusions in plagioclase of ,245.

Pyroxene compositions range from orthopyroxene to augite (Table 6 and Fig. 2), and although there is some evidence of augite rimming pigeonite, the small grain size of these samples makes the exact relationship between these phases unclear. Generally there are two pyroxene populations: Ca- and Mg-rich. The ranges in composition are displayed in Figs. 2a-f and 3a-f. In Figs. 3d and 3f (samples ,266 and ,277), the Al/Ti ratio of the pyroxenes decreases with Mg#, generally between Mg and Ca pyroxenes. This suggests plagioclase crystallization after subcalcic augite (*Bence and Papike*, 1972), similar to pyroxenes from 14305 VHK basalts. In Fig. 3e (14303,245), the Al/Ti ratio remains constant at approximately

TABLE 3. Representative orthoclase analyses from the 14303 VHK basalts.

PM No.	,246	,249	,266	,275	,277
INAA No.	,247	-	,266	,275	,277
SiO_2	61.9	62.8	63.5	61.2	63.6
Al_2O_3	19.5	19.0	19.2	19.9	18.8
FeO	-	0.12	0.36	0.32	0.82
CaO	0.43	0.38	0.33	0.56	0.10
BaO	3.34	1.88	0.36	3.69	0.94
Na_2O	0.67	0.52	0.78	0.80	0.53
K_2O	14.6	15.0	15.4	13.5	15.2
Total	100.44	99.70	99.93	99.97	99.99
Formula moles based on eight oxygens					
Si	2.913	2.944	2.947	2.892	2.961
Al	1.079	1.050	1.051	1.107	1.032
Fe	-	0.004	0.013	0.012	0.031
Ca	0.021	0.018	0.015	0.027	0.004
Ba	0.061	0.034	0.005	0.068	0.016
Na	0.061	0.046	0.069	0.072	0.047
K	0.875	0.897	0.912	0.816	0.898
Total	5.008	4.994	5.012	4.994	4.990
An	2.2	1.9	1.5	3.0	0.4
Ab	6.4	4.8	6.9	7.9	5.0
Or	91.4	93.3	91.6	89.1	94.6

TABLE 4. Representative K-rich glass analyses from the 14303 VHK basalts.

PM No.	,245	,246	,249	,266	,275	,277
INAA No.	,244	,247	-	,266	,275	,277
SiO_2	58.0	76.9	68.9	65.8	71.4	66.9
Al_2O_3	19.0	13.6	15.6	18.4	14.4	17.9
FeO	7.09	0.69	0.53	0.36	0.49	0.25
CaO	1.08	0.77	0.46	0.52	0.58	0.02
BaO	1.76	0.30	0.90	0.73	0.64	0.12
Na_2O	0.89	0.45	0.60	0.66	0.57	0.45
K_2O	12.2	7.10	12.3	13.1	11.3	15.1
Total	100.02	99.81	99.29	99.57	99.38	100.74

TABLE 5. Representative olivine analyses from the 14303 basalts.

PM No.	,245				,266		,277	
INAA No.	,244				,266		,277	
	Mg-Rich	Fe-Rich	Inclusions in Plagioclase		Mg-Rich	Fe-Rich	Mg-Rich	Fe-Rich
SiO_2	37.1	35.1	35.9	37.8	36.0	33.2	36.6	33.5
FeO	27.2	40.8	35.9	25.0	28.9	51.1	31.7	49.2
MnO	0.23	0.47	0.32	0.24	0.25	0.74	0.41	0.71
MgO	35.3	24.1	28.1	37.0	35.5	15.0	30.9	16.5
Total	99.83	100.47	100.22	100.04	100.65	100.04	99.61	99.91
	Formula moles based on four oxygens							
Si	0.990	0.997	0.997	0.994	0.964	1.003	1.002	1.004
Fe	0.606	0.969	0.832	0.551	0.647	1.294	0.725	1.236
Mn	0.004	0.011	0.007	0.004	0.005	0.019	0.009	0.018
Mg	1.405	1.022	1.163	1.453	1.415	0.677	1.260	0.736
Total	3.006	2.999	2.999	3.002	3.032	2.993	2.995	2.994
Fo	69.9	51.3	58.3	72.5	68.6	34.3	63.5	37.3

2 while Mg# decreases from 79-60. At Mg# 60-56, the Al/Ti ratio decreases to 1. This suggests plagioclase crystallization occurred after the Mg# became lower than 60. Sample 14303,246 (Fig. 3b) contains pyroxenes exhibiting a small range in Mg# (67.5-70) or Al/Ti ratio (1.25-2.25). There is a possible positive correlation suggesting co-precipitation of pyroxene and plagioclase. Pyroxene compositions in the remaining two VHK basalts have constant Al/Ti ratios of approximately 2 (Figs. 3a and 3c), while Mg# decreases. This indicates possible plagioclase crystallization prior to pyroxene, as Al is reduced in pyroxene relative to other 14303 VHK clasts. Plagioclase phenocrysts are present in these samples.

Ilmenite is an interstitial phase that can form large (up to 0.75 mm in length), homogeneous masses (Table 7), some poikilitically enclosing plagioclase phenocrysts, as in ,245 (Fig. 1c). Also in ,245, ilmenite is associated with titaniferous chromite subsolidus reduction, and has a distinct composition (Mg# = 11.3) from the interstitial variety in this specimen (average Mg# = 17.7). Ilmenite also exhibits small ranges in composition in the other samples (Mg#: ,246 = 18.9-20.4; ,249 = 16.4-18.6; ,266 = 5.6-14.4; ,277 = 7.0-14.4). No ilmenite is present in ,275.

Members of the chromite-ulvöspinel series (Table 8) are present in ,245 ,266 and ,277. In ,245 they occur as two grains,

TABLE 6. Representative pyroxene analyses from the 14303 basalts.

PM No.	,245			,246		,249		
INAA No.	,244			,247		-		
	En	Pig	Aug	En	Aug	En	Pig	Aug
SiO_2	52.8	51.6	50.2	53.1	51.4	51.8	51.8	50.7
TiO_2	0.77	0.67	1.09	0.81	1.58	0.89	0.88	1.46
Al_2O_3	1.73	0.80	1.63	0.85	1.60	1.06	0.81	1.70
Cr_2O_3	0.64	0.24	0.48	0.30	0.31	0.41	0.32	0.48
FeO	15.3	23.1	13.4	19.2	10.1	21.3	20.4	10.4
MnO	0.20	0.34	0.25	0.25	0.19	0.35	0.32	0.21
MgO	25.9	18.9	14.2	24.1	15.2	22.5	20.1	15.6
CaO	2.31	4.64	18.2	1.72	20.0	1.70	6.01	19.6
Total	99.65	100.29	99.45	100.33	100.38	100.01	100.64	100.15
	Formula moles based on six oxygens							
Si	1.923	1.951	1.909	1.952	1.913	1.933	1.937	1.900
Ti	0.020	0.019	0.030	0.021	0.043	0.024	0.024	0.042
Al	0.073	0.035	0.072	0.037	0.070	0.046	0.035	0.075
Cr	0.018	0.006	0.014	0.008	0.008	0.012	0.008	0.018
Fe	0.465	0.730	0.424	0.589	0.316	0.664	0.636	0.387
Mn	0.005	0.010	0.008	0.007	0.005	0.011	0.010	0.006
Mg	1.404	1.062	0.803	1.319	0.845	1.252	1.120	0.848
Ca	0.089	0.187	0.424	0.066	0.798	0.066	0.240	0.729
Total	3.999	4.000	4.003	4.000	3.999	4.008	4.010	4.004
En	71.8	53.7	40.8	66.8	43.2	63.2	56.1	43.2
Wo	4.5	9.4	37.7	3.2	40.7	3.3	12.0	37.1
Fs	23.7	36.9	21.5	29.8	16.1	33.5	31.9	19.7

TABLE 6. (continued)

PM No.	,266			,275		,277	
INAA No.	,266			,275		,277	
	Pig	Aug	Ferro-Aug	Pig	Aug	Pig	Aug
SiO$_2$	50.3	49.4	48.8	50.8	50.5	50.6	50.3
TiO$_2$	0.71	1.36	1.46	0.85	1.33	0.89	1.48
Al$_2$O$_3$	2.49	1.79	1.55	0.90	1.72	1.52	2.01
Cr$_2$O$_3$	1.03	0.33	0.24	0.33	0.61	0.72	0.47
FeO	17.5	16.5	21.1	23.2	11.2	22.2	12.1
MnO	0.41	0.29	0.47	0.39	0.24	0.43	0.14
MgO	20.6	11.0	9.22	18.2	14.6	18.9	14.4
CaO	6.55	19.2	17.7	5.34	20.5	4.78	18.4
Total	99.59	99.87	100.54	99.98	100.70	100.04	99.30
	Formula moles based on six oxygens						
Si	1.889	1.907	1.907	1.935	1.891	1.918	1.906
Ti	0.020	0.038	0.042	0.023	0.037	0.024	0.041
Al	0.108	0.080	0.071	0.040	0.076	0.068	0.089
Cr	0.029	0.009	0.007	0.009	0.018	0.021	0.013
Fe	0.545	0.532	0.688	0.740	0.351	0.703	0.383
Mn	0.012	0.009	0.007	0.009	0.018	0.021	0.013
Mg	1.148	0.633	0.536	1.032	0.816	1.066	0.814
Ca	0.262	0.796	0.741	0.218	0.821	0.193	0.745
Total	4.013	4.003	4.006	4.010	4.016	4.007	3.996
En	58.7	32.3	27.3	51.8	41.0	54.4	41.9
Wo	13.4	40.6	37.7	11.0	41.3	9.8	38.4
Fs	27.9	27.1	35.0	37.2	17.7	35.8	19.7

a euhedral 0.2×0.2 mm and a <0.1 mm diameter grain (see Fig. 1b). The larger grain is homogeneous [Mg# = 8.4-9.1;Cr/(Cr+Al) = 0.73-0.75] and is associated with ilmenite (see above) and native Fe, an assemblage typically indicative of a reduction reaction (*El Goresy et al.,* 1972). Only one analysis was possible from the smaller grain [Mg# = 16.1; Cr/(Cr+Al) = 0.67]. TiO$_2$ contents are higher in the larger grain. In ,266 and ,277, chromite-ulvöspinels are associated with pyroxene and olivine. Compositional heterogeneity is observed in the rare titaniferous chromites from ,266 [Mg# = 2.6-5.5; Cr/(Cr+Al) = 0.69-0.75; 27-35 wt % TiO$_2$] and in chromites from ,277 [Mg# = 9.0-23.6; Cr/(Cr+Al) = 0.60-0.79]. No members of the chromite-ulvöspinel series are present in ,275, possibly a sampling problem. Zonation of these opaques to higher values for Ti, and lower Mg# and Cr/(Cr+Al) ratio toward the rims, is observed only in interstitial types and those associated with pyroxene. Chromites associated with olivine are relatively homogeneous. Such chromites formed early and

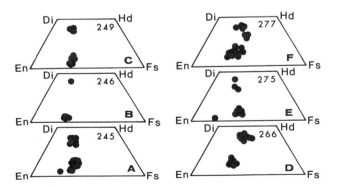

Fig. 2. Pyroxene compositions of the 14303 VHK basalts represented on a pyroxene quadrilateral. Generally there are two populations: Ca-poor and Ca-rich.

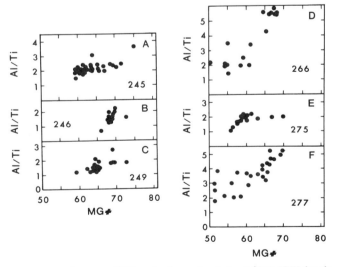

Fig. 3. Mg# versus Al/Ti ratio for each of the 14303 VHK basalt pyroxene populations. This figure serves to illustrate the effect of plagioclase crystallization (Al/Ti ratio) during the evolution of the VHK magma (decreasing Mg#). See text for discussion.

TABLE 7. Representative ilmenite analyses from the 14303 VHK basalts.

PM No.	,245	,246	,249	,266	,277
INAA No.	,244	,247	-	,266	,277
TiO_2	55.6	53.8	56.3	55.3	53.6	53.3
Al_2O_3	0.18	0.16	0.18	0.13	0.25	0.16
Cr_2O_3	0.44	0.46	0.43	0.50	0.57	0.36
FeO	38.7	42.0	38.6	39.5	43.7	43.6
MgO	5.22	3.06	5.07	4.82	1.99	2.83
Total	100.14	99.48	100.58	100.25	100.11	100.25
Formula moles based on three oxygens						
Ti	1.005	1.000	1.013	1.004	0.997	0.988
Al	0.005	0.004	0.005	0.003	0.006	0.004
Cr	0.007	0.008	0.007	0.009	0.011	0.006
Fe	0.779	0.868	0.771	0.796	0.904	0.899
Mg	0.187	0.112	0.180	0.173	0.073	0.104
Total	1.983	1.992	1.976	1.985	1.991	2.001
Mg#	19.4	11.3	18.9	17.9	7.5	10.4

were encased in and protected by olivine, ensuring no further reaction with the remaining magma. As pyroxene crystallized after olivine and in some cases plagioclase, chromites associated with these minerals display zonation of the type described above. These chromites continued to experience further interaction with the remaining magma (*El Goresy et al.*, 1972).

Native Fe grains occur either enclosed in or interstitial to the other mineral phases. Grain sizes can reach up to 0.1 mm, but are usually <0.05 mm. These metal particles may be associated with ilmenite, probably the result of the subsolidus reduction of ulvöspinel (*El Goresy et al.*, 1972). Native Fe compositions show a wide variation in Ni and Co contents (Table 9), but do not plot within the "meteorite field" (*Goldstein and Yakowitz*, 1971). Interstitial iron sulfide (troilite) is present in all 14303 VHK basalts and exhibits no significant compositional variation. Grain size is similar to the native Fe grains.

TABLE 8. Representative chromite-ulvöspinel analyses from the 14303 VHK basalts.

PM No.	,245		,266		,277
INAA No.	,244		,266		,277
TiO_2	24.7	11.2	26.7	4.95	22.3
Al_2O_3	4.89	12.0	3.90	15.9	5.25
Cr_2O_3	19.9	35.5	14.1	42.3	23.7
FeO	47.4	35.8	53.5	32.5	46.2
MgO	2.55	3.85	1.46	4.35	2.59
Total	100.46	98.65	99.66	100.00	99.95
Formula moles based on four oxygens					
Ti	0.648	0.291	0.725	0.124	0.590
Al	0.201	0.472	0.166	0.626	0.218
Cr	0.548	0.963	0.401	1.121	0.659
Fe	1.382	1.030	1.618	0.912	1.361
Mg	0.132	0.197	0.078	0.216	0.135
Total	2.934	2.965	2.987	2.999	2.964
Cr/(Cr+Al)	73.2	67.1	70.7	64.2	78.7
Mg#	8.7	16.1	4.6	19.1	9.0

TABLE 9. Ranges of Co and Ni in native Fe from the 14303 VHK basalts.

PM No./INAA No.	Co	Ni	Ni/Co
,245/,244	0.33-1.26	1.52-9.71	1.85-8.5
,246/,247	0.61	8.39	13.8
,249/ ---	0.31-0.51	3.16-4.83	7.94-11.1
,266/,266	0.01-0.10	0-0.95	0-39
,275/,275	0.12-0.80	0.71-6.85	2.25-10.3
,277/,277	0.03-1.27	0.07-2.84	0.11-6.07

WHOLE ROCK CHEMISTRY

Analytical Techniques

Compositions of five 14303 VHK basalts were determined by INAA at Washington University using their standard procedure (*Lindstrom*, 1984). One clast (,249) was too small for INA analysis. All data were reduced employing the TEABAGS program of *Lindstrom and Korotev* (1982). Relative uncertainties based on counting statistics at the one sigma level are: 1-2% for Fe, Na, Sc, Cr, Co, La, Sm, and Eu; 3-5% for Ca, Ce, Tb, Yb, Lu, Hf, Ta, and Th; 5-15% for Cs, Ba, Nd, and U; 15-25% for K, Sr, Ni, Rb, and Zr. All samples were analyzed for Ir and Au. Four samples contained Ir and Au abundances below the detection limit of 2 ppb, supporting a pristine origin for these basalts (*Warren and Wasson*, 1979; *Ryder et al.*, 1980; *Warren et al.*, 1983b). However, ,277 has a high Ni content (210 ppm) with Au and Ir abundances slightly higher than 2 ppb (3.9 and 2.7 ppb, respectively), indicative of meteoritic contamination.

Two VHK samples (,244 and ,247) were crushed and fused on a molybdenum strip in an atmosphere of argon. This permitted the major element composition (Table 10) of the whole rock to be determined on the quench glass by electron microprobe.

TABLE 10. Whole-rock compositions of the 14303 VHK basalts measured by INAA.

PM No.	,245	,246	,266	,275,	277
INAA No.	,244	,247	,266	,275	,277
Weight (mg)	24.4	13.6	15.2	20.9	22.4
	P	P	I	I	I
SiO$_2$	48.7	48.9	na	na	na
TiO$_2$	1.61	1.70	na	na	na
Al$_2$O$_3$	18.2	13.4	na	na	na
FeO	9.55	13.6	16.8	9.90	12.9
MgO	8.27	9.68	na	na	na
CaO	12.3	10.4	10.5	11.0	9.9
Na$_2$O	0.45	0.55	0.49	0.72	0.64
K$_2$O	<0.75*	1.05	1.30	1.10	0.86
Sc	30.0	46.0	57.8	20.8	37.9
Cr	1777	2595	3350	1014	2720
Co	19.4	31.8	31.6	15.3	34.9
Ni	50	150	110	65	210
Rb	10	39	37	38	23
Sr	155	125	<100	160	92
Cs	0.20	1.41	0.89	1.75	0.99
Ba	300	800	587	930	2720
La	18.1	41.3	8.51	37.1	52.0
Ce	47.2	109	22.7	94.0	127
Nd	27.0	60.0	12.0	54.0	79.0
Sm	9.26	19.3	4.82	16.3	24.4
Eu	1.62	1.56	0.94	1.96	1.90
Tb	2.13	4.14	1.12	3.65	4.83
Yb	6.91	13.7	4.49	16.3	17.5
Lu	1.05	2.09	0.71	2.60	2.45
Zr	245	500	nd	550	700
Hf	6.00	13.0	2.82	14.0	17.7
Ta	0.73	1.81	0.59	1.65	2.08
Th	2.20	7.97	1.15	8.27	9.76
U	0.65	2.37	0.55	2.40	3.35
Ir(ppb)	nd	nd	nd	nd	3.9
Au(ppb)	<3	<3	<5	<4	2.7

na = not analyzed.
nd = not detected.
P = major elements by probe analysis of fused glass.
I = all elements by INAA.
*Analysis from INAA.

Results

Whole rock compositions are presented in Table 10. The major element composition of ,247 (PM = ,246) determined by the fusion method is comparable with those major elements analyzed by INAA. However, although ,244 has similar results from INAA and the fusion method for Ca and Na, the absolute determination of K by the fusion method (0.23 wt % K$_2$O) is within the mare basalt range, rather than VHK. We consider 0.23 wt % K$_2$O too low to account for the interstitial high-K glass (Tables 1 and 4) in this sample. This highlights the problem of sample size and representative sampling, especially as the INAA sample masses are particularly low (e.g., 13.6-24.4 mg). It would seem apparent that 14303,244 used for the INA and EMP fusion analyses did not sufficiently sample the high-K glass witnessed in the companion probe mount (,245).

An approximate K$_2$O value may be determined by modal reconstruction, using the probe mount ,245. By employing this method, 0.75 wt % may be taken as the maximum K$_2$O abundance in PM ,245 (,244). Abundances of Na and Ca are similar to those in HA basalts (*Shervais et al.*, 1985b; *Dickinson et al.*, 1985; *Neal et al.*, 1987a) of 0.48-0.72 and 9.9-12.3 wt %, respectively. Iron concentrations in the 14303 VHK basalts (9.9-16.8 wt % FeO) overlap and are lower than in mare basalts. The low Fe and high Al in ,244 indicate that this sample is plagioclase-rich relative to other VHK basalts. There is a much higher concentration of K in the 14303 samples (0.75-1.30 wt %) than in mare basalts. These major element analyses are similar to other previously reported VHK basalts (*Shervais et al.*, 1985b; *Goodrich et al.*, 1986; *Warren et al.*, 1986), except for the high Al content of ,244 (18.2 wt %). The high K$_2$O contents of the parental magmas account for the presence in these basalts of K-rich glass in all specimens

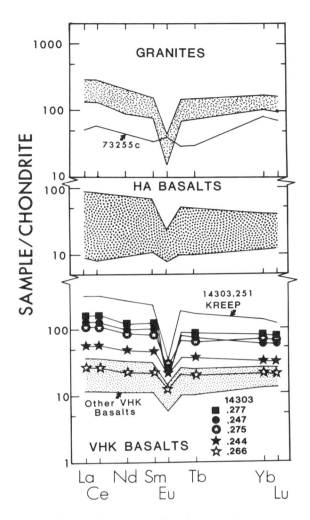

Fig. 4. The REE profiles of the 14303 VHK basalts are compared to others previously reported (stippled area: data of *Shervais et al.*, 1985b; *Goodrich et al.*, 1986; *Warren et al.*, 1986) and a KREEP basalt also from breccia 14303. Shown for comparison are the entire range of Apollo 14 HA mare basalts (*Dickinson et al.*, 1985; *Shervais et al.*, 1985a; *Neal et al.*, 1987a) and granite compositions (*Blanchard and Budahn*, 1979; *Warren et al.*, 1983c; *Salpas et al.*, 1985).

and K-feldspar in all but ,244 (,245). This sample [,244 (,245)] has lower K_2O than other 14303 VHK samples reflecting the lack of K-feldspar.

The trace element abundances presented here extend the ranges previously reported for VHK basalts. These elements are sensitive to the many igneous processes that operate during magma evolution. For ease of presentation, the trace elements will be discussed as groups of elements with similar physicochemical affinities. The 14303 samples will be compared and contrasted with those previously reported (*Shervais et al.*, 1985b; *Goodrich et al.*, 1986; *Warren et al.*, 1986), as well as with ordinary Apollo 14 mare basalts (*Dickinson et al.*, 1985; *Neal et al.*, 1987a; *Shervais et al.*, 1985a).

Rare earth elements (REE). The REE contents of the 14303 VHK basalts exhibit a wide range of abundances (Table 6) that overlap those of mare and other VHK basalts and extend to much higher values (Fig. 4). Samples containing the greatest REE contents are LREE enriched, showing a similarity to KREEP (Fig. 4), being distinguished from KREEP on the basis of K/La ratio (except ,277). All 14303 samples have the characteristic negative Eu anomaly and there is a slight enrichment of the LREE over the HREE. REE abundances are within the ranges reported for Apollo 14 HA basalts (Fig. 4).

Compatible elements: Sc, Cr, Co, Ni. The five 14303 VHK basalts have a wide range in compatible element abundances. Abundances of Sc (20.8-57.8 ppm), Cr (1014-3350 ppm), and Co (15.3-34.9 ppm) overlap but are generally lower than abundances in mare basalts and previously reported VHK basalts. However, nickel abundances (50-210 ppm) overlap and are higher than mare basalts and previously reported VHK compositions. If ,277 is omitted, the range is reduced to 50-150 ppm Ni, which is comparable with other VHK basalts. In fact, the Ni/Co ratio of ,277 is the highest in any reported VHK basalt (6.0) and is similar to the Ni/Co ratios of Apollo 15 impact melts reported by *Ryder and Spudis* (1987). However, ,247 and ,275 also fall within this range (4.7 and 4.2, respectively). Therefore, it is pertinent to compare the abundances of Ir and Au in these samples to determine whether any significant meteorite component is present (i.e., origin as an impact melt). Indeed, sample ,277 contains the highest amounts of Ir (3.9 ppb) and Au (2.7 ppb). Therefore, we conclude that *14303,277 contains a meteoritic component and is probably the crystallized product of an impact melt rather than a pristine VHK basalt.* This will be highlighted in the discussion below.

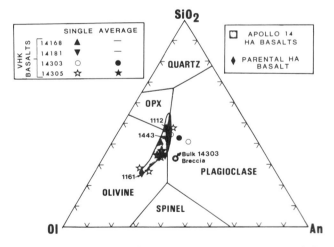

Fig. 5. Representation of VHK basalt data on a SiO_2-Ol-Plag pseudoternary (after *Walker et al.*, 1972). These are compared with all Apollo 14 HA mare basalts (*Dickinson et al.*, 1985; *Shervais et al.*, 1985a; *Neal et al.*, 1987a). Generally, the VHK basalts fall along the inferred crystallization path of the HA compositions. Phase boundaries are calculated for an Fe/(Fe+Mg) ratio of 0.6, the magma composition upon just encountering the Ol-Plag cotectic (composition calculated using the method of *Longhi*, 1977). Note that the two 14303 basalts plotted bear no similarity to the bulk rock 14303 breccia composition (data of *Brunfelt et al.*, 1972). The three HA magma compositions, taken as parental to VHK basalts, are also shown.

TABLE 11a. Crystal/liquid partition coefficients.

	Olivine	Pyroxene	Plagioclase	Chromite	Ilmenite
K	0.0068	0.014	0.17	0	0
Ba	0.03	0.013	0.686	0.005	0.005
Hf	0.04	0.063	0.05	0.38	1.817
Th	0.03	0.13	0.05	0.55	0.55
La	(0.0001)	(0.012)	0.051	0.029	0.029
Ce	0.0001	0.038	0.037	0.038	0.038
Sm	0.0006	0.054	0.022	0.053	0.053
Eu	0.0004	0.1	1.22	0.02	0.02
Yb	0.02	0.671	0.012	0.39	0.39
Lu	(0.02)	0.838	0.011	0.47	0.47

Brackets indicate estimated partition coefficient.
Chromite REE partition coefficients taken as those for ilmenite.
References: *Arth and Hanson* (1975); *Binder* (1982); *Drake and Weill* (1975); *Haskin and Korotev* (1977); *Irving and Frey (1984)*; *McKay et al.* (1986); *Schnetzler and Philpotts* (1970); *Villemant et al.* (1981).

TABLE 11b. Modeling parameters

	Parental Magmas			Assimilated Granite	
	,1161	,1443	,1112	73255c	Average Granite
K	564	747	1826	62665	58864
Ba	32	130	216	5470	4490
Hf	1.78	5.43	11.5	16.0	15.4
Th	0.20	1.50	2.90	9.50	42.3
La	2.89	13.5	30.6	20.3	50.7
Ce	8.40	35.0	82.9	50.0	123
Sm	2.11	7.59	15.3	6.74	16.7
Eu	0.56	1.12	1.67	2.71	2.52
Yb	2.89	6.33	9.72	10.3	24.7
Lu	0.47	0.91	1.33	1.50	3.71

References: *Shervais et al.* (1985a); *Neal et al.* (1987a); *Blanchard and Budahn* (1979); *Warren et al.* (1983c); *Salpas et al.* (1985).

Large ion lithophile (LIL) elements: Rb, Sr, Cs, Ba. The LIL elements exhibit a wide range of abundances (Rb = 10–39 ppm; Sr = 92–160 ppm; Cs = 0.2–1.75 ppm; Ba = 300–2720 ppm). All extend the upper ranges of previously reported VHK compositions to higher values. Sample 14303,277 contains anomalously high Ba abundances (2720 ppm) that, if omitted, reduce the Ba range to 300–930 ppm, comparable with other VHK basalts. Rubidium and Cs abundances in the 14303 samples are higher than and overlap those of the mare basalts, whereas Sr is within the mare basalt range. However, Ba is distinctly higher than any mare basalt composition, even if ,277 is omitted.

High field strength (HFS) elements: Zr, Hf, Ta, Th, U. As with the element groups discussed above, the HFS element abundances exhibit wide variances (Zr = 245–700 ppm; Hf = 2.82–17.7 ppm; Ta = 0.59–2.08 ppm; Th = 1.15–9.76 ppm; U = 0.55–3.35 ppm). Sample 14303,277 contains the greatest abundances of these elements, which are similar to abundances present in KREEP (e.g., *Warren et al.*, 1983d). However, even if sample ,277 is omitted, the 14303 VHK basalts still have HFS abundances that are higher than and overlap the compositions of mare and previously reported VHK basalts.

MODELING

Introduction

The analysis and recognition of very high potassium (VHK) basalts (*Warner et al.*, 1980; *Shervais et al.*, 1985b) presented a highly significant rock type whose origin will be indicative of petrologic processes that have operated at the Apollo 14 site. Several authors (e.g., *Shervais et al.*, 1985b; *Warren et al.*, 1986) have favored the theory of granite assimilation versus a metasomatized source for VHK basalt petrogenesis. However, it is recognized that this model is not complete; the high K/Th and K/U ratios of the VHK basalts cannot be generated by simple mass balance granite assimilation (e.g., *Shervais et al.*, 1985b; *Warren et al.*, 1986; *Shih et al.*, 1987).

In this paper, we present a combined Assimilation and Fractional Crystallization (AFC) model for VHK basalt petrogenesis. Our model is designed to encompass all reported VHK compositions into one dynamic process that can explain both the wide range of elemental abundances and the rarity of VHK compositions at the lunar surface.

TABLE 12. Modeling results (ppm).

Parent/Granite	,1161/73255c				.1161/Av. Granite			
F Value	0.95	0.90	0.85	0.80	0.95	0.90	0.85	0.80
K	3890	7584	11709	16347	3690	7162	11039	15398
Ba	321	641	998	1398	312	614	898	1298
Hf	2.70	3.71	4.83	6.08	2.67	3.64	4.73	5.94
Th	0.71	1.26	1.88	2.57	2.43	4.88	7.59	10.6
La	4.11	5.46	6.98	8.68	5.71	8.84	12.3	16.3
Ce	11.8	14.9	18.7	23.0	15.3	23.0	31.6	41.3
Sm	2.57	3.09	3.67	4.31	3.10	4.20	5.42	6.80
Eu	0.73	0.92	1.14	1.38	0.72	0.90	1.10	1.33
Yb	3.57	4.31	5.14	6.06	4.32	5.90	7.66	9.62
Lu	0.57	0.68	0.80	0.94	0.69	0.92	1.19	1.48
Parent/Granite	,1443/73255c				.1443/Av. Granite			
F Value	0.95	0.90	0.85	0.80	0.95	0.90	0.85	0.80
K	4063	7561	11316	15476	3863	7154	10684	14595
Ba	415	711	1027	1371	364	608	868	1151
Hf	6.51	7.57	8.65	9.82	6.48	7.51	8.55	9.68
Th	2.06	2.64	3.25	3.93	3.78	6.16	8.67	11.4
La	15.2	17.0	19.0	21.2	16.8	20.3	24.1	28.3
Ce	39.4	44.2	49.1	54.6	43.2	52.1	61.4	71.8
Sm	8.33	9.10	9.92	10.8	8.85	10.2	11.6	13.2
Eu	1.25	1.39	1.54	1.71	1.24	1.36	1.51	1.66
Yb	7.17	7.87	8.46	9.09	7.92	9.39	10.8	12.2
Lu	1.03	1.12	1.19	1.27	1.15	1.36	1.55	1.74
Parent/Granite	,1112/73255c				.1112/Av. Granite			
F Value	0.95	0.90	0.85	0.80	0.95	0.90	0.85	0.80
K	5198	8920	13054	17671	4999	8503	12392	16738
Ba	507	823	1169	1547	456	717	1002	1314
Hf	12.7	13.9	15.3	16.7	12.6	13.8	15.2	16.6
Th	3.50	4.16	4.89	5.69	5.22	7.75	10.5	13.6
La	33.2	36.1	39.3	42.9	34.8	39.4	44.6	50.4
Ce	89.6	96.9	105	114	93.4	105	118	132
Sm	16.4	17.6	18.9	20.4	16.9	18.7	20.7	22.9
Eu	1.82	1.98	2.16	2.34	1.81	1.96	2.12	2.30
Yb	10.3	10.9	11.6	12.3	11.0	12.5	13.9	15.5
Lu	1.40	1.47	1.55	1.63	1.51	1.70	1.91	2.12

Modeling Systematics

In the proposed model, granite is inferred to be an assimilated component by an Apollo 14 high-Al (HA) mare basalt magma. Although treatment of assimilation as a mass balance or bulk mixing problem may be a good first approximation, any assimilation will be accompanied by some magma crystallization (*DePaolo*, 1981). This is a more plausible approach than simple mixing of granite and HA basalt (*Shih et al.*, 1987). The composition of the crystallizing phases will evolve as this process proceeds. We have attempted, where possible, to account for these evolving compositions within our major element modeling in order to represent a realistic situation. The proportions and nature of fractionating phases can be calculated from phase equilibria. These results can then be applied to the trace elements. Such an approach is critical for consistency between major and trace element modeling and is essential in the development of any model for magma petrogenesis.

Major elements. A major constraint in any petrologic modeling is the phase equilibria applicable to that particular melt, and numerous experimental investigations of mare basalt liquids have been performed (e.g., *Walker et al.*, 1972; *Kesson*, 1975). It was determined that the basaltic phase relations could most readily be explained by consideration of the SiO_2-Ol-An pseudoternary system (often referred to as the "Walker Diagram"). We feel this is a feasible approach as the diopside component is low (approximately 10% in each case), which will impart insignificant errors in the overall modeling. All reported major element compositions of VHK basalts are represented on such a Walker diagram in Fig. 5 along with the range in Apollo 14 HA basalts. The petrogenesis of the

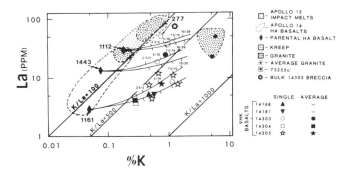

Fig. 6. K% versus La (ppm) to demonstrate the role of granite assimilation in the petrogenesis of VHK basalts. AFC paths are calculated from the three HA magmas (these span the range of Apollo 14 HA basalts), with the upper and lower paths in each case representing average granite and 73255c as the respective assimilants. Tick marks represent increments of 5% fractional crystallization (2.5% granite assimilation). These same increments will be repeated, but without numbers, in subsequent diagrams. Note that 14303,277 cannot be generated by any of these paths and has similar K/La ratios to the bulk rock 14303 breccia.

HA major element compositions by fractional crystallization (*Neal et al.*, 1987a) can be extended to the VHK basalts because of the similarity of their major element abundances. The phase boundaries (e.g., cotectics) in this system have been shown to be particularly sensitive to the Fe/(Fe+Mg) ratio of the system (*Walker et al.*, 1973; *Longhi*, 1977). As such, the phase boundaries are drawn for the Fe/(Fe+Mg) of the system just as the evolving melt encounters the Ol-Plag cotectic (Fe/(Fe+Mg) ≅ 0.6). The major elements are modeled by fractional crystallization only, as the major elements are effectively controlled by the fractionating phases (*DePaolo*, 1981). However, with a high proportion of granite assimilation, this may have an effect on the major element modeling, probably slightly increasing the proportion of plagioclase fractionation. We feel that this will not significantly affect the overall results of our fractional crystallization modeling.

Trace elements. The process of assimilation will have a profound effect on the trace elements, and thus these elements cannot be modeled by fractional crystallization alone. We have modeled the trace elements by combined assimilation and fractional crystallization, using equation (6a) of *DePaolo* (1981)

$$C_m/C_m^o = F^{-z} + \left(\frac{r}{r-1}\right)\left(\frac{C_a}{zC_m^o}\right)\left(1 - F^{-z}\right)$$

where r = mass assimilated/mass crystallized; z = r+D-1/r-1; D = bulk distribution coefficient; C_m^o = concentration of element in the parental magma; C_m = concentration of element in the residual magma; C_a = concentration of element in the assimilated component; F = proportion of liquid remaining.

The inferred ratio of the mass assimilated to the mass crystallized (r value) is relatively high (0.5). This value represents a realistic figure of a high-temperature basaltic magma assimilating a granitic component of lower melting point (e.g., *Watson*, 1982).

Components. As has been reported by *Neal et al.* (1987a), HA mare basalts show a wide range of compositions at the Apollo 14 site. This compositional range is due to evolution of a primitive HA magma by combined assimilation and fractional crystallization with KREEP. Therefore, in order to represent this wide range of HA basalts, we have chosen three compositions as our parental VHK magmas (14321,1161 of *Shervais et al.*, 1985a; 14321,1443 and 14321,1112 of *Neal et al.*, 1987a). These compositions span the range of Apollo 14 HA mare basalts (see Fig. 5). Sample ,1161 plots within the olivine field; ,1443 falls on the Ol-Plag cotectic; and ,1112 falls on the Ol-Opx-Plag peritectic reaction point.

The proportions of fractionating phases for the three HA parental magmas have been calculated by the Lever Rule from Fig. 5. If 14321,1161 produced all reported VHK compositions, it would have to undergo approximately 60% fractional crystallization. During the first 25% crystallization, olivine (90%) + chromite (10%) are the only phases to fractionate (*Walker et al.*, 1972), at which point the Ol-Plag cotectic is encountered. During 25% to 45% crystallization, plagioclase (50%) + olivine (40%) + chromite (10%) are fractionated, after which the Ol-Opx-Plag peritectic is encountered. It is envisaged that all fractionating phases are effectively removed from the system; therefore, olivine does not undergo any back-reaction. Between 45% and 60% crystallization of ,1161, orthopyroxene (60%) + plagioclase (30%) + ilmenite (10%) are fractionated.

If composition 14321,1443 is taken as parental to the VHK basalts, it would have to undergo approximately 25% fractional crystallization. During the first 7% crystallization, plagioclase (50%) + olivine (40%) + chromite (10%) would fractionate. At this point, the Ol-Opx-Plag peritectic is encountered, and orthopyroxene replaces olivine on the liquidus. From 7% to 25% crystallization, orthopyroxene (60%) + plagioclase (30%) + ilmenite (10%) are fractionated.

If composition 14321,1112 is taken as parental to the VHK basalts, it would only be required to undergo between 10-15% fractional crystallization. This composition cannot be parental to all VHK basalts. As ,1112 lies approximately on the Ol-Opx-Plag peritectic, orthopyroxene (60%) + plagioclase (30%) + ilmenite (10%) would be the only fractionating phases.

The proportions of fractionating mineral phases are the same as in Apollo 14 mare basalt evolution by AFC with KREEP (*Neal et al.*, 1987a), indicating the similarity in petrogeneses.

Reported granite compositions demonstrate that although these tend to be heavily enriched in the LIL and HFS elements, a wide range in compositions exists (e.g., *Warren et al.*, 1983c). For our modeling purposes, we have taken: (1) an average of seven granite compositions (five from *Warren et al.*, 1983c and two "new" samples from 14305 of *Salpas et al.*, 1985); and (2) the composition of 73255c (*Blanchard and Budahn*, 1979; *Warren et al.*, 1983c). The majority of granite compositions approximate that of the average, except for 73225c. This sample generally has the lowest granite trace element abundances. A peculiar feature of granite 73255c is the positive Eu anomaly rather than the usual negative one. Therefore, use of an average composition will best represent all other compositions, apart from 73255c. Incorporation of these two granite compositions in our modeling should be more representative than a single, skewed average composition.

Effects of short-range unmixing. The small sample sizes analyzed will undoubtedly induce errors due to unrepresentative sampling. The effects of these errors are attributable to short-range unmixing (*Haskin and Korotev,* 1977; *Lindstrom and Haskin,* 1978) of a solidifying magma. This process is similar to fractional crystallization, except there is no large-scale separation of crystallized products and residual liquid. Short-range unmixing deals with centimeter-sized inhomogeneties in materials that may be homogeneous on a meter scale (*Lindstrom and Haskin,* 1978). This may explain the plagioclase-rich nature of 14303,245 relative to 14303,246. In order to minimize the possible effects of short-range unmixing, average compositions have been calculated in Fig. 5 (and succeeding figures) for VHK compositions from the 14303, 14304, and 14305 breccias. Single analyses for VHK basalts from 14168 and 14181 are also shown. However, short-range unmixing alone cannot account for the wide range in major and especially trace element abundances observed in the VHK basalts.

Major element modeling. We have not presented quantitative modeling on Fig. 5, as it is envisaged that *one parental HA magma is not responsible for the array of VHK compositions*. However, it is demonstrable that a large proportion of olivine fractionation is required in order to generate these basalts if 14321,1161 is taken as the parental composition [e.g., 14303,246 may be generated from ,1161 by 25% crystallization of olivine (90%) + chromite (10%), followed by 20% crystallization of olivine (40%) + plagioclase (50%) + chromite (10%)]. In order to recreate realistic crystallizing conditions, the Fo content of fractionating olivine was calculated iteratively in crystallization increments of 2% using the Mg/Fe ratio of the resultant liquid to calculate the Fo content of the crystallizing olivine (similar to *Longhi,* 1977). With ,1161 as the parental composition, olivine remains on the liquidus approximately until 45% is crystallized. This yields a range in the *Fo content from 80 to 60* and may explain the upper range observed in VHK basalts. But there are basalts with olivines of various Fo contents. The lower range of olivine Fo contents is considered to be a late-stage phenomenon (*Shervais et al.,* 1985b). The simplified assumption of complete olivine removal (i.e., no back-reaction) during crystallization may not be totally correct.

Comparison of the major element data (where available) on a Walker diagram (Fig. 5) illustrates that, in general, the VHK basalts have similar compositions to the Apollo 14 mare basalts. Only 14303,247 plots in the Plag field, reflecting the feldspar-rich nature of this sample. The VHK basalt major element compositions display a similar crystallization path to that of Apollo 14 HA mare basalts. Two 14305 VHK compositions reported by *Shervais et al.* (1985b) are primitive (14305,304 and ,384), similar to the parental composition 14321,1161. Modeling of the major elements in this way is feasible, even though we also envisage assimilation of a granitic component (see section on Modeling Systematics).

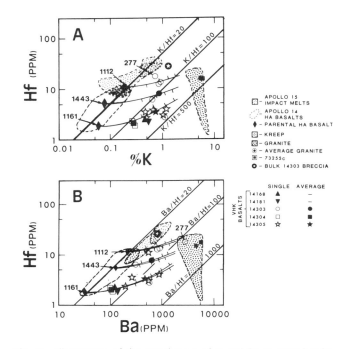

Fig. 7. Illustrations of the enrichment of K and Ba in VHK basalts due to the effect of granite assimilation. (**a**) K% versus Hf (ppm); AFC paths are calculated from the three parental HA magmas. As average granite and 73255c have similar K and Hf abundances, the calculated paths plot on top of each other. Note that 14303,277 cannot be generated by these AFC paths and has similar K/Hf ratios to the bulk rock 14303 breccia. (**b**) Ba (ppm) versus Hf (ppm); VHK basalts generated by ,1161 and 73255c can be generated in the same order as in (**a**) along calculated AFC paths, but amounts are reduced. This is due to the high Ba abundance in 73255c. The upper and lower paths in each case represent 73255c and average granite as the respective assimilants. Note that 14303,277 falls within the granite field on this diagram, due to elevated Ba abundance.

Trace Element Modeling

We have attempted to model the critical elements involved in VHK basalt petrogenesis, namely K, Ba, Hf, Th, and the REE. We have not modeled U because of the incomplete data set for U in the VHK basalt and granite suites and the lack of accurately determined crystal-liquid partition coefficients. Three parental magma compositions are again taken as 14321,1161 ,1443 and ,1112. Crystal-liquid partition coefficients and parameters used are presented in Table 11. Results obtained are presented in Table 12. In Figs. 6-8, all VHK basalt data are plotted with (1) granite (*Warren et al.,* 1983c; *Salpas et al.,* 1985), (2) KREEP (*Vaniman and Papike,* 1980; *Warren et al.,* 1983d), (3) Apollo 14 HA mare basalts (*Neal et al.,* 1987a; *Dickinson et al.,* 1985; *Shervais et al.,* 1985a), (4) the bulk 14303 breccia composition (*Brunfelt et al.,* 1972), and (5) Apollo 15 impact melt compositions (*Ryder and Spudis,* 1987). This approach is designed to demonstrate the petrogenesis of VHK basalts from assimilation of lunar granite by a HA mare basalt, rather than from an impact melt.

Figure 6 illustrates the familiar plot of K versus La for lunar granite, KREEP, and HA basalts. This diagram demonstrates that KREEP and Apollo 14 mare basalts have similar K/La ratios (50-100), but VHK basalts diverge from this trend, removed somewhat toward granite (K/La = 500-750). The Apollo 15 impact melts have similar K/La ratios to mare basalts and KREEP, unlike the VHK basalts, with the possible exception of 14303,277. AFC paths have been calculated from the three parental magmas defined above, using average granite and 73255c as assimilants. It is obvious that it is not possible to generate all VHK data using one parent and one granite composition.

In Fig. 6, we have averaged the VHK data according to the breccia in which they were identified (cf. Fig. 5). As depicted in Fig. 6, the 14304 VHK basalts can be generated from the parental basalt 14321,1161 by approximately *4% fractional crystallization and 2% assimilation of granite 73255c*. The 14305 VHK basalts can be generated from the parental basalt 14321,1161 by *10% fractional crystallization and 5% assimilation of the average granite composition*. It is possible that both of these averaged compositions can be generated by either the average granite or 73255c assimilants. *Shih et al.* (1986, 1987) performed mass balance mixing calculations and arrived at the same amounts of granite assimilation for 14304 and 14305 VHK basalts as calculated by our AFC considerations. When the amounts of assimilation and fractional crystallization are small, the AFC process can be approximated by more simple mass balance calculations. However, it is important to consider crystallization from the melt concommitant with assimilation of the granite. We insist that these AFC calculations represent a more realistic approach to the natural situation.

The VHK compositions described by *Warren et al.* (1986) and *Shervais et al.* (1985b) from breccias 14181 and 14168 also fall along the AFC path between 14321,1161 and 73255c.

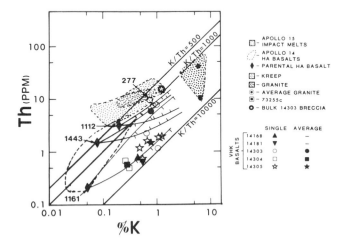

Fig. 8. K% versus Th (ppm); AFC paths are calculated as in Figs. 6 and 7. The proportions of fractional crystallization and granite assimilation are the same as in Figs. 6 and 7. The upper and lower paths in each case represent average granite and 73255c as the respective assimilants. Note that 14303,277 falls within the impact melt field on this diagram and has a similar K/Th ratio to the bulk rock 14303 breccia.

14168 is generated after *6% fractional crystallization (3% granite assimilation)* and 14181 is generated after *7% fractional crystallization (3.5% granite assimilation)* There is a progression of VHK compositions along this AFC path (i.e., increasing granite component) from 14304, through 14168 and 14181, to 14305.

The average 14303 VHK composition cannot be generated by parental basalt 14321,1161 (recall that the average was calculated without including ,277 since this is considered to be an impact melt). The average 14303 VHK basalt can be generated by approximately *11.5% fractional crystallization of parental basalt 14321,1443, with 5.75% assimilation of average granite*. Those individual 14303 VHK basalts with lower K/La ratios can only be generated by AFC between parental basalt 14321,1112 and average granite. Note that 14303,277 cannot be generated by this model.

Figure 7a presents K versus Hf, and Fig. 7b depicts Ba versus Hf. These plots serve to illustrate the similarity between the behaviour of K and Ba in VHK basalt petrogenesis. Mare basalts, KREEP, and impact melts have similar K/Hf and Ba/Hf ratios (10-20 and 20-50, respectively), with VHK compositions removed toward granite, which has elevated K/Hf and Ba/Hf ratios (200-1000 and 200-1200, respectively). The proportions of granite assimilation and fractional crystallization are the same as in Fig. 6 in order to generate the average VHK basalt compositions in Fig. 7a (K versus Hf) for 14168, 14181, 14303, 14304, and 14305. For Ba against Hf (Fig. 7b), these proportions are reduced for VHK basalts generated by parent magma ,1161 and granite 73255c. This is necessary because although most trace elements are less abundant in 73255c relative to average granite, Ba abundance is higher (5470 ppm). The average 14303 VHK basalt is generated by the same proportion of fractional crystallization of ,1443 and assimilation of average granite as in Fig. 6. However, the same general progression along the AFC path between 14321,1161 and 73255c is maintained, described in Fig. 6 for 14304, 14168, 14181, and 14305 compositions. In both Figs. 7a and 7b, those individual 14303 VHK basalts with lower K/Hf and Ba/Hf can only be generated by AFC between 14321,1112 and average granite. Note here again, as in Fig. 6 14303,277 cannot be generated on either plot by our model, which suggests an origin different from that proposed for other VHK compositions.

Elements that are problematic from a mass balance point of view in VHK basalt petrogenesis are U and Th (see above). We have not modeled U due to the lack of published crystal/liquid partition coefficients. However, it is considered that U and Th should behave in a similar manner, implying that the AFC paths for U will approximate the results obtained for Th. Partition coefficients for Th are sparse, and we have conducted our modeling using the few we could find in the literature.

It has been considered that K/Th ratios in VHK basalts are too high to be the result of bulk granite assimilation (*Warren et al.*, 1986), but may be generated by selective assimilation of granitic feldspars only (*Shervais et al.*, 1985b; *Shih et al.*, 1987). In Fig. 8, K is plotted against Th, demonstrating that Apollo 14 HA mare basalts, KREEP, and Apollo 15 impact melts have similar K/Th ratios of 500-1000. The majority of individual VHK samples and the average 14304 and 14305 compositions can be generated by AFC between parental basalt 14321,1161

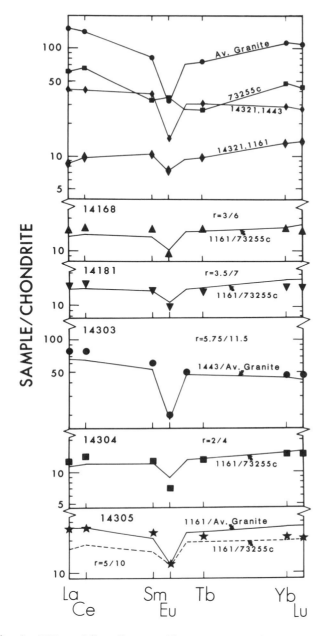

Fig. 9. REE modeling diagrams. The proportions of fractional crystallization and granite assimilation, as well as the composition of the parental HA magma and assimilated granite, have been determined in Figs. 5-8. Average compositions are modeled for 14303, 14304, and 14305 VHK basalts, and the REE profiles of the components used are shown. Symbols represent the actual reported compositions, and the solid lines represent the modeling results. The "r" values quoted represent the "mass assimilated/mass crystallized."

and the granite 73255c. The average 14304 VHK basalt is generated by *4% fractional crystallization (2% granite assimilation)*. 14181 and 14168 also fall on this AFC path, being generated by *10% fractional crystallization (5% granite assimilation)*, whereas the average 14305 VHK basalt is generated by *7% and 6% fractional crystallization (3.5% and 3% granite assimilation)*, respectively. Likewise, the average 14303 VHK basalt is generated by *11.5% fractional crystallization of 14321,1443 (5.75% assimilation of average granite)*. These values were also obtained in Fig. 6 and 7a. Our results demonstrate that the main problem in VHK basalt petrogenesis, of generating K/Th (and K/U) ratios, is overcome by our realistic AFC modeling approach and by the more detailed examination of published granite compositions. Note once again that individual 14303 samples with reduced K/Th ratios are generated by AFC between 14321,1112 and average granite. Also, attention is called to the fact that 14303,277 falls in the Apollo 15 impact melt field.

The average REE patterns from 14303, 14304, and 14305, and those of 14168 and 14181 are presented in Fig. 9. We have attempted to model the REE profiles using the components and proportions of AFC determined in Figs. 5-8. There is reasonable agreement between measured and modeled REE profiles for 14168, 14181, 14303, and 14304. However, the average profile of the 14305 VHK basalts cannot be generated accurately by AFC between parental magma 14321,1161 and granite 73255c. The LREE of this profile are accurately modeled by using an average granite composition (as suggested in Fig. 6), whereas the HREE are best generated by using 73255c. It may be that the actual granite assimilant has a composition intermediate between these two, or the REE profile of the parental basalt (14321,1661) is not representative of the 14305 VHK parental magma.

IMPACT MELT ORIGIN FOR VHK BASALTS

The identification of 14303,277 as a probable impact melt highlights the importance of detailed geochemical analyses of the VHK basalts in order to determine the pristinity of these samples. It is important to note that only this sample has been defined as an impact melt, primarily because of the Ni/Co ratio and Au and Ir abundances. In Figs. 5-8, the bulk rock analysis of breccia 14303 (*Brunfelt et al.,* 1972) is plotted in order to determine whether the 14303 VHK basalts are crystalline products of remelted (via meteorite impact) matrix material. On Fig. 5, only two 14303 VHK basalts are plotted (,244 and ,247), but neither of these has major elements composition similar to the bulk 14303 breccia. However, throughout Figs. 6-8, K/La, K/Hf, and K/Th ratios of 14303,277 and the bulk rock 14303 breccia support the contention that ,277 is an impact melt. In fact, as shown in Fig. 8, this sample falls within the Apollo 15 impact melt field on a plot of K versus Th. No other reported VHK compositions can be regarded as impact melts, although Au and Ir abundances for the 14305 and 14168 VHK basalts are not reported.

DISCUSSION

The above modeling demonstrates that VHK basalt compositions can be generated by a combined assimilation and fractional crystallization process between a HA mare basalt parent magma and granite. The major element compositions of VHK basalts are generally very similar to the Apollo 14

HA mare basalts, consistent with the crystallization path described by *Neal et al.* (1987a). Only elevated K₂O contents give an indication as the the petrogenesis of these VHK basalts. *Watson* (1982) and *Watson and Jurewicz* (1984) have demonstrated that during granite assimilation by basaltic materials, the basalt becomes highly enriched in K with some depletion in Na. No real differences can be defined in Na abundance between VHK and HA mare basalts, but enrichment of K in the former appears to be indicative of granite assimilation.

In a study of HA mare basalt petrogenesis at the Apollo 14 site, *Neal et al.* (1987a) have proposed that the range of observed compositions is produced by and AFC process between a primitive magma and KREEP (r = 0.22). This requires that 14321,1443 and 14321,1112 compositions, taken as parental for the 14303 VHK basalts, already contain a KREEP component. Basalt ,1443 has undergone 30% fractional crystallization (6.6% KREEP assimilation), whereas ,1112 has undergone approximately 60% fractional crystallization (13.2% KREEP assimilation).

It is evident that VHK and HA mare basalts are genetically related at the Apollo 14 site. A primitive mare basalt magma evolves beneath the Apollo 14 site by combined AFC with KREEP to produce the array of HA mare basalt compositions. This primitive magma type may also undergo AFC with a granite component, as VHK compositions 14168, 14181, 14303, and 14305 can be generated without any incorporated KREEP component. However, this primitive magma cannot produce the 14303 VHK compositions. These require a more evolved basalt parent (i.e., mare basalts with an assimilated KREEP component) and a different assimilated granite composition, corresponding to an average lunar granite. It is envisaged that in the dynamic process of AFC, KREEP is replaced by granite as the assimilated component during magma migration within the lunar crust. This produces the 14303 VHK basalts.

The rarity of both granite and VHK basalts in the lunar crust suggest that VHK basalt production is a highly localized phenomenon. Identification of both KREEP and granitic components in the 14303 VHK basalts indicates that *there must be at least two VHK basalt flows at the Apollo 14 site.*

The results of our research indicate that bulk granite assimilation can produce VHK basalts just as well as preferential incorporation of granitic feldspars. This observation is not at variance with the isotopic data reported by *Shih et al.* (1986, 1987). These authors conclude that at the time of crystallization, there is a dramatice increase in Rb/Sr ratio with little change in Sm/Nd ratio, attributing this to preferential assimilation of granitic feldspars. However, bulk granite has a relatively chondritic to slightly enriched Sm/Nd ratio (Fig. 4), similar to that of the parental basalts (especially ,1443). Therefore, if bulk granite is combined with parental basalts used in this study, little alteration in Sm-Nd systematics will occur.

CONCLUSIONS

The above discussion outlines the importance of lunar granite in VHK basalt petrogenesis. However, unlike similar models, we have demonstrated that KREEP is also involved, and this has major implications not only for VHK basalts, but also for KREEP-granite relationships (see #3 below). The main conclusions resulting from this study are as follows:

1. Major element compositions of VHK basalts fall along the general crystallization path proposed for Apollo 14 high Al (HA) mare basalts. Potassium is the only major element that deviates from a normal HA mare basalt composition. As the major elements in an AFC process can be effectively modeled by fractional crystallization alone, this suggests similar petrogeneses for VHK and HA basalts.

2. VHK basalts from breccias 14168, 14181, 14304, and 14305 are produced by combined assimilation and fractional crystallization of a primitive HA mare basalt magma (composition of 14321,1161) and granite (composition of 73255c).

3. VHK basalts from breccia 14303 are produced from a more evolved HA mare basalt magma (compositions of 14321,1443 and 14321,1161) and an average granite composition. This argues for these magmas to have resulted from normal Apollo 14 mare basalt evolution by AFC with KREEP. At a certain point along this evolutionary path, the KREEP assimilant is replaced by granite in the AFC process in order to produce VHK compositions. The actual AFC path taken is controlled not only by the composition of the parental magma, but also by the composition of the assimilated granite.

4. The different AFC components for 14168, 14181, 14304, and 14305 versus 14303 argue for at least two VHK basalt flows at the Apollo 14 site.

5. All reported VHK basalts are produced by granite assimilation. In this petrogenesis, the amounts of granite assimilation are small-i.e., maximum of 5.75% for 14303 VHK compositions. This process highlights the significance of the rare granite occurrences in the lunar crust.

6. The identification of 14303,277 as an impact melt highlights the importance of detailed geochemical analysis in determining the pristinity of VHK basalts.

Acknowledgments. This study would not have been possible without the capable assistance of the Planetary Materials Curatorial Staff at Johnson Space Center, especially Ms. Kim Willis. The sawing of the breccias, the elaborate documentation, and the actual "plucking" of the individual clasts was performed by Kim. We thank Randy Korotev at Washington University for assistance in the INA analyses, and Tom See at JSC for assistance in the fusion process for major element analyses. Acknowledgments are also due to John Shervais, Roman Schmitt, Gordon McKay, Graham Ryder, and John Jones for critical reviews and valuable discussions. The research reported in this paper was supported by NASA grants NGR 9-62 to L. A. Taylor and NAG 9-56 to L. A. Haskin.

REFERENCES

Arth J. G. and Hanson G. N. (1975) Geochemistry and origin of the early Pre-Cambrian crust of north-eastern Minnesota. *Geochim. Cosmochim. Acta 39,* 325-362.

Bence A. E. and Papike J. J. (1972) Pyroxenes as recorders of lunar basalt petrogenesis: Chemical trends due to crystal-liquid interactions. *Proc. Lunar Sci. Conf. 3rd,* 431-469.

Binder A. B. (1982) The mare basalt magma source region and mare basalt magma genesis. *Proc. Lunar Planet. Sci. Conf. 13th*, in *J. Geophys. Res., 87*, A37-A53.

Binder A. B. (1985) Mare basalt genesis: Modeling trace elements and isotopic ratios. *Proc. Lunar Planet. Sci. Conf. 16th*, in *J. Geophys. Res., 90*, D19-D30.

Blanchard D. P. and Budahn J. R. (1979) Remnants from the ancient lunar crust: clasts from consortium breccia 73255. *Proc. Lunar Planet. Sci. Conf. 10th*, 803-816.

Brunfelt A. O., Heier K. S., Nilssen B., and Sundvoll B. (1972) Distribution of elements between different phases of Apollo 14 rocks and soils. *Proc. Lunar Sci. Conf. 3rd*, 1133-1147.

DePaolo D. J. (1981) Trace element and isotopic effects of combined wallrock assimilation and fractional crystallization. *Earth Planet. Sci. Lett., 53*, 189-202.

Dickinson T., Taylor G. J., Keil K., Schmitt R. A., Hughes S. S., and Smith M. R. (1985) Apollo 14 aluminous mare basalts and their possible relationship to KREEP. *Proc. Lunar Planet. Sci. Conf. 16th*, in *J. Geophys. Res., 90*, C365-C374.

Drake M. J. and Weill D. F. (1975) Partition of Sr, Ba, Ca, Y, Eu^{2+} and Eu^{3+}, and other REE between plagioclase feldspar and magmatic liquid: an experimental study. *Geochim. Cosmochim. Acta, 39*, 689-712.

El Goresy A., Taylor L. A., and Ramdohr P. (1972) Fra Mauro crystalline rocks: Mineralogy, geochemistry, and subsolidus reduction of opaque minerals. *Proc. Lunar Sci. Conf. 3rd*, 333-348.

Goldstein J. I. and Yakowitz H. (1971) Metallic inclusions and metal particles in the Apollo 12 lunar soil. *Proc. Lunar Sci. Conf. 2nd*, 177-191.

Goodrich C. A., Taylor G. J., Keil K., Kallemeyn G. W., and Warren P. H. (1986) Alkali norite, troctolites, and VHK mare basalts from breccia 14304. *Proc. Lunar Planet. Sci. Conf. 16th*, in *J. Geophys. Res., 91*, D305-D318.

Haskin L. A. and Korotev R. L. (1977) Test of a model for trace element partition during closed-system solidification of a silicate liquid. *Geochim. Cosmochim. Acta, 41*, 921-939.

Irving A. J. and Frey F. A. (1984) Trace element abundances in megacrysts and their host basalts: Constraints on partition coefficients and megacryst genesis. *Geochim. Cosmochim. Acta, 48*, 1201-1221.

Kesson S. E. (1975) Mare basalts: melting experiments and petrogenetic interpretations. *Proc. Lunar Sci. Conf. 6th*, 921-944.

Lindstrom M. M. (1984) Alkali gabbronorite, ultra-KREEPy melt rock, and the diverse suite of clasts in North Ray Crater feldspathic fragmental breccia 67975. *Proc. Lunar Planet. Sci. Conf. 15th*, in *J. Geophys. Res., 89*, C50-C62.

Lindstrom M. M. and Haskin L. A. (1978) Causes of compositional variations within mare basalt suites. *Proc. Lunar Planet. Sci. Conf. 9th*, 465-486.

Lindstrom D. J. and Korotev R. L. (1982) TEABAGS: Computer programs for instrumental neutron activation analysis. *J. Radioanal. Chem. 70*, 439-458.

Longhi J. (1977) Magma oceanography 2: Chemical evolution and crustal formation. *Proc. Lunar Sci. Conf. 8th*, 601-621.

McKay G. A., Wagstaff J., and Yang S.-R. (1986) Zr, Hf, and REE partition coefficients for ilmenite and other minerals in high-Ti lunar mare basalts: an experimental study. *Proc. Lunar Planet. Sci. Conf. 16th*, in *J. Geophys. Res., 91*, D229-D237.

Neal C. R., Taylor L. A., and Lindstrom M. M. (1987a) Mare basalt petrogenesis: Assimilation of KREEP-like components by a fractionating magma. *Proc. Lunar Planet. Sci. Conf. 18th*, this volume.

Neal C. R., Taylor L. A., and Lindstrom M. M. (1987b) Very High Potassium (VHK) basalt petrogenesis: The role of granite and KREEP components (abstract). In *Lunar and Planetary Science XVIII*, pp. 710-711. Lunar and Planetary Institute, Houston.

Nyquist L. E., Shih C.-Y., Wooden J. L., Bansal B. M., and Weismann H. (1979) The Sr and Nd isotopic record of Apollo 12 basalts: Implications for lunar geochemical evolution. *Proc. Lunar Planet. Sci. Conf. 10th*, 77-114.

Ringwood A. E. and Kesson S. E. (1976) A dynamic model for mare basalt petrogenesis. *Proc. Lunar Sci. Conf. 7th*, 1697-1722.

Ryder G. and Spudis P. (1987) Chemical composition and origin of Apollo 15 impact melts. *Proc. Lunar Planet. Sci. Conf. 17th*, in *J. Geophys. Res., 92*, E432-E446.

Ryder G., Norman M. D., and Score R. A. (1980) The distinction of pristine from meteorite-contaminated highland rocks using metal compositions. *Proc. Lunar Planet. Sci. Conf. 11th*, 471-480.

Salpas P. A., Shervais J. W., Knapp S. A., and Taylor L. A. (1985) Petrogenesis of lunar granites: The results of apatite fractionation (abstract). In *Lunar and Planetary Science XVI*, pp 726-727. Lunar and Planetary Institute, Houston.

Schnetzler C. C. and Philpotts J. A. (1970) Partition coefficients of REE between igneous matrix material and rock-forming mineral phenocrysts - II. *Geochim. Cosmochim. Acta, 34*, 331-340.

Shervais J. W., Taylor L. A., Laul J. C., and Smith M. R. (1984) Pristine highland clasts in consortium breccia 14305: Petrology and geochemistry. *Proc. Lunar Planet. Sci. Conf. 15th*, in *J. Geophys. Res., 89*, C25-C40.

Shervais J. W., Taylor L. A., and Lindstrom M. M. (1985a) Apollo 14 mare basalts: Petrology and geochemistry of clasts from consortium breccia 14321. *Proc. Lunar Planet. Sci. Conf. 15th*, in *J. Geophys. Res., 90*, C375-C395.

Shervais J. W., Taylor L. A., Laul J. C., Shih C.-Y., and Nyquist L. E. (1985b) Very High Potassium (VHK) basalt: Complications in mare basalt petrogenesis. *Proc. Lunar Planet. Sci. Conf. 16th*, in *J. Geophys. Res., 90*, D3-D18.

Shih C.-Y., Nyquist L. E., Bogard D. D., Bansal B. M., Weismann H., Johnson P., Shervais J. W., and Taylor L. A. (1986) Geochronology and petrogenesis of Apollo 14 Very High Potassium mare basalts. *Proc. Lunar Planet. Sci. Conf. 16th*, in *J. Geophys. Res., 91*, D214-D228.

Shih C.-Y., Nyquist L. E., Bogard D. D., Dasch E. J., Bansal B. M., and Weismann H. (1987) Geochronology of high-K aluminous mare basalt clasts from Apollo 14 breccia 14303. *Geochim. Cosmochim. Acta*, in press.

Vaniman D. T. and Papike J. J. (1980) Lunar highland melt rocks: Chemistry, petrology, and silicate mineralogy. *Proc. Conf. Lunar Highlands Crust (J. J. Papike and R. B. Merrill, eds.)*, pp. 271-337. Pergamon, New York.

Villemant B., Jaffrezic H., Joron J. L., and Treuil M. (1981) Distribution coefficients of major and trace elements: fractional crystallization in the alkali basalt series of Chaine des Puys (Massif Central, France). *Geochim. Cosmochim. Acta, 45*, 1997-2016.

Walker D., Longhi J., and Hays J. R. (1972) Experimental petrology and origin of Fra Mauro rocks and soil. *Proc. Lunar Sci. Conf. 3rd*, 797-817.

Walker D., Longhi J., Grove T. L., Stolper E., and Hays J. F. (1973) Experimental petrology and origin of rocks from the Descartes Highlands. *Proc. Lunar Sci. Conf. 4th*, 1013-1032.

Warner R. D., Taylor G. J., Keil K., Ma M.-S., and Schmitt R. A. (1980) Aluminous mare basalts: New data from Apollo 14 coarse fines. *Proc. Lunar Planet. Sci. Conf. 11th*, 87-104.

Warren P. H. and Wasson J. T. (1979) Effects of pressure on the crystallization of a "chondritic" magma ocean and implications for the bulk composition of the Moon. *Proc. Lunar Planet. Sci. Conf. 10th*, 2051-2083.

Warren P. H., Taylor G. J., Keil K., Kallemeyn G. W., Shirley D. N., and Wasson J. T. (1983a) Seventh foray: Whitlockite-rich lithologies, a diopside-bearing troctolitic anorthosite, ferroan anorthosites, and KREEP. *Proc. Lunar Planet. Sci. Conf. 14th*, in *J. Geophys. Res., 88*, B151-B164.

Warren P. H., Taylor G. J., and Keil K. (1983b) Regolith breccia Allan Hills A81005: Evidence of lunar origin, and petrography of pristine and non-pristine clasts. *Geophys. Res. Lett., 10,* 779-782.

Warren P. H., Taylor G. J., Keil K., Shirley D. N., and Wasson J. T. (1983c) Petrology and chemistry of two "large" granite clasts from the Moon. *Earth Planet. Sci. Lett., 64,* 175-185.

Warren P. H., Taylor G. J., Keil K., Kallemeyn G. W., Shirley D. N., and Wasson J. T. (1983d) Seventh foray: Whitlockite-rich lithologies, a diopside-bearing troctolitic anorthosite, ferroan anorthosites and KREEP. *Proc. Lunar Planet. Sci. Conf. 14th*, in *J. Geophys. Res., 88*, B151-B164.

Warren P. H., Shirley D. N., and Kallemeyn G. W. (1986) A potpourri of pristine moon rocks, including a VHK mare basalt and a unique, augite-rich Apollo 17 anorthosite. *Proc. Lunar Planet. Sci. Conf. 16th*, in *J. Geophys. Res., 91,* D319-D330.

Watson E. B. (1982) Basalt contamination by continental crust: some experiments and models. *Contrib. Mineral. Petrol., 80,* 73-87.

Watson E. B. and Jurewicz S. R. (1984) Behaviour of alkalies during diffusive interaction of granitic xenoliths with basaltic magma. *J. Geol., 92,* 121-131.

Apollo 14 Mare Basalt Petrogenesis: Assimilation of KREEP-Like Components by a Fractionating Magma

Clive R. Neal and Lawrence A. Taylor
Department of Geological Sciences, University of Tennessee, Knoxville, TN 37996

Marilyn M. Lindstrom[*]
Department of Earth and Planetary Sciences, Washington University, St. Louis, MO 63130

*Now at NASA Johnson Space Center, Code SN2, Houston, TX 77058

Whole rock and mineral data from 22 new 14321 high-Al (HA) mare basalts demonstrate a complicated petrogenesis for these rocks. The trace element characteristics of these basalts exhibit considerable variation that cannot be related to short-range unmixing alone. This variation is facilitated by the combined effects of KREEP assimilation and fractional crystallization of liquidus phases (AFC). Such a process is reflected in the REE profiles, which range from flat, primitive types to LREE-enriched (KREEP-like), evolved types. Earlier work on Apollo 14 HA mare basalts had defined five groups. We suggest this is an artifact of sampling; when the new data are added, a continuum of compositions is observed. Detailed trace element modeling demonstrates that a primitive parental magma (taken as 14321,1422: La = 3.19 ppm; La/Lu = 7.2; SiO_2 = 46%; MgO = 11%) evolves by AFC with KREEP. The ratio of mass assimilated to mass crystallized ("r" value) is 0.22. Proportions of crystallizing phases vary as fractional crystallization proceeds. Major element modeling indicates that only olivine (90%) and chromite (10%) are precipitated during the first 14% of crystallization. Plagioclase, olivine, and chromite (50:40:10) fractionate as crystallization proceeds, and finally opx, plagioclase, and ilmenite (60:30:10) fractionate together after 21% crystallization. It is envisaged that this AFC process occurs at depth within the lunar crust and the fractionating phases are effectively removed from the system. This model places the Apollo 14 HA mare basalts within the confines of a single dynamic process which can explain both major and trace element compositions.

INTRODUCTION

The petrogenesis of basaltic samples returned by the Apollo missions is critical in understanding the composition and development of the Moon's interior and the evolution of the lunar crust. Paramount in the study of mare basalts have been the "pull-apart" efforts on lunar breccias, which have greatly increased the basalt data base. In particular, investigations of basalt clasts from Apollo 14 breccias have been extremely rewarding. At the Apollo 14 site, the mare basalts form a minor, yet highly significant, component of the lunar crust. For example, the oldest mare basalts have been described from this area (*Taylor et al.*, 1983), as well as the important new VHK (very high potassium) rock-type (*Shervais et al.*, 1984a, 1985).

As a result of such research, three general models have been proposed for basalt petrogenesis: (1) cumulate melting (*Smith et al.*, 1970; *Taylor and Jakes*, 1974); (2) melting of a primitive source (e.g., *Green et al.*, 1971); and (3) assimilation of crustal components by a primitive magma (e.g., *Ringwood and Kesson*, 1976). However, none of these models can, on its own, account for the range in compositions exhibited by high-Al (HA) basalt samples from the Apollo 14 site, which show restricted major element variation, but considerable range in trace element abundances (*Shervais et al.*, 1985).

Earlier work on the Apollo 14 HA mare basalts has described 5 basalt groups (*Dickinson et al.*, 1985; *Shervais et al.*, 1985). The latter authors described a tridymite ferrobasalt/gabbro and an ilmenite ferrobasalt (Group 5) not noted in the work by *Dickinson et al.* (1985). The groupings are based on rare earth and high field strength (HFS) element abundances. These two papers concluded that simple fractional crystallization of a parent magma cannot be responsible for the range displayed in trace element compositions, although the major elements can be generated by such a process (*Dickinson et al.*, 1985). *Shervais et al.* (1985) argued for three distinct source regions below the Apollo 14 site. These authors assumed that the more evolved HA basalts were derived from sources comparatively richer in cpx, containing more trapped liquid, and crystallizing when the magma ocean had greater incompatible element abundances. Both this and a similar approach by *Dickinson et al.* [using a source region proposed for 12038 by *Nyquist et al.* (1981)] cannot generate the observed trace element abundances in all their proposed groups. These authors concluded that variable degrees of cumulate melting of a single source is not responsible for Apollo 14 mare basalt generation.

The process of assimilation has also been addressed. *Shervais et al.* (1985) demonstrated that their 14321-type basalts are generated by assimilation of an "IKFM" KREEP component (intermediate potassium Fra Mauro basalt 15386; *Rhodes and Hubbard*, 1973; *Vaniman and Papike*, 1980) by their 14072-type basalt. *Dickinson et al.* (1985) calculated the effects of assimilation of Apollo 14 KREEP, as well as 15386, by their Group 5 basalts to generate Groups 1-4. They demonstrated that there is better agreement between calculated and observed abundances using 15386 IKFM rather than Apollo 14 KREEP. The Apollo 14 HA mare basalts were generated by 4%-27% assimilation of this composition. However, both *Shervais et al.* (1985) and *Dickinson et al.* (1985) have treated assimilation as a mass balance problem, although the latter authors have

TABLE 1. Numbers, texture, and compositional variation within analyzed mare basalts.

T.S./INAA No.	Texture	Plagioclase			Pyroxene			Olivine
		An	Ab	Or	Fs	Wo	En	Fo
,1115/,1106	O	69-92	7-27	0-4	27-32	6-31	41-67	72
,1118/,1112	G	81-91	9-15	0-1	25-52	6-25	23-68	-
- /,1338	-	-			-			-
- /,1344	-	-			-			-
- /,1345	-	-			-			-
- /,1346	-	-			-			-
,1376/,1318	SO	79-91	9-18	0-2	27-48	7-34	22-63	61-74
,1378/,1329	SO	74-93	7-23	0-3	27-62	7-33	10-66	-
,1382/,1353	O	84-95	5-14	0-2	24-42	9-32	30-62	64-71
,9087/,1347	O	85-94	6-14	0-1	27-34	9-19	42-58	67-70
,9088/,1349	O	87-91	8-12	0-1	28-52	10-26	26-61	-
,9089/,1350	O	77-95	5-20	0-3	25-49	10-28	29-62	62
,9090/,1365	O	80-94	6-18	0-2	27-36	9-33	34-55	65-71
,1471/ -	V	83-94	6-16	0-1	-			68-88
,1473/,1422	V	81-95	4-17	0-2		-		60-71
,1474/,1425	O	80-93	6-19	0-2	25-49	11-36	29-51	58-69
,1475/,1429	O	76-93	7-21	0-3	14-47	8-28	25-61	60-65
,1476/,1432	O	82-94	6-16	0-2	25-51	9-37	23-61	63-68
,1477/,1435	O	78-93	7-18	0-4	24-67	9-36	7-61	60-75
,1478/,1437	O	79-96	4-19	0-3	27-62	9-34	9-64	55-62
,1479/,1439	O	77-94	6-21	0-2	24-51	11-32	22-53	63-72
,1480/,1441	O	85-96	4-14	0-1	25-43	8-34	26-58	56-74
,1481/,1443	O	83-93	7-15	0-2	26-52	9-35	20-63	59-65
,1482/ -	O	75-95	5-23	0-3	25-59	9-34	19-64	58-72
,1483/ -	O	79-93	7-19	0-3	23-55	9-37	8-56	71-86
,1485/ -	O	82-97	3-16	0-3	24-73	11-41	2-61	-
,1486/ -	V	76-95	5-20	0-4	-			71-86

G = granular; SO = sub-ophitic; O = ophitic; V = vitrophyre.

admitted that the magma will also crystallize during assimilation.

Binder (1982, 1985) suggested remelting a layered cumulate in the lunar mantle, followed by storage in the lower crust and 30% fractional crystallization of olivine ± pyroxene as a possible petrogenesis for mare basalts (*Binder*, 1982, Fig. 18, 1985, Fig. 1). This crystallization was coupled with assimilation of residues of urKREEP (*Warren and Wasson*, 1979a) which produced the trace element variability seen in mare basalts.

The present paper refines such models by combining previous Apollo 14 HA mare basalt data with that from 22 new 14321 samples. We demonstrate that "groups" of basalts defined by previous authors are an artifact of sampling and are part of a continuum of compositions generated by a combined assimilation and fractional crystallization (AFC) process between primitive basalt and a KREEP component.

PETROGRAPHY AND MINERAL CHEMISTRY

Freshly sawn surfaces of breccia 14321 were prepared and examined in the facilities at Johnson Space Center. Numerous clasts were extracted for examination. Probe mounts (PM) of 20 mare basalts were prepared and studied using standard petrographic techniques. Mineral chemistry was determined using an automated MAC 400S electron microprobe at the University of Tennessee. Standards, operating procedures, and methods of data reduction have been reported elsewhere (e.g., *Shervais et al.*, 1984b).

Table 1 summarizes the textures and mineral compositions of the basalts studied. These samples are generally ophitic to sub-ophitic basalts dominated by plagioclase and pyroxene, with subordinate olivine phenocrysts. Textures are variable as highlighted in Fig. 1. Sample ,1118 has been granulated (Fig. 1a) and includes two areas of glass (≃ 0.4 mm diameter) containing small (< 0.1 mm) plagioclase phenocrysts. Figure 1b displays the typical texture of the majority of the analyzed basalts. Olivine phenocrysts (see Table 1) can reach > 1 mm diameter and are always rimmed by pigeonitic pyroxene; in ,1118, ,1485, and ,9088 olivine appears to have been totally replaced. In Fig. 1c, ,1476 displays a general ophitic (0.5-0.7 mm), but contains a greater proportion of interstitial glass, whereas ,1474 (Fig. 1d) is much finer grained (0.1-0.3 mm), but still maintains a general ophitic texture.

Euhedral Ti-chromite grains (up to 0.2 mm) are included in olivine, where they are relatively honmogeneous (Fig. 2a,b), and also in pigeonite, where TiO_2 typically increases toward the rims. This zonation is considered to be due to continued growth of earlier-formed chromites in the evolving magma, before inclusion in pyroxene, which crystallizes after olivine. In ,1477, reduction of the late-formed Cr-ulvöspinel is evidenced by the association of native Fe and ilmenite with the spinel phase (cf., *El Goresy et al.*, 1972).

Pyroxenes (up to 1 mm) show partial zonation from pigeonite/enstatite cores to augite/ferroaugite rims (Fig. 3). These pyroxenes are interstitial to An-rich plagioclase laths, generally 0.3-0.7 mm in length, which also display a partial zonation, becoming more Ab- and Or-rich toward the rims (Fig. 5). Plagioclase crystallization in all basalts occurred during pyroxene fractionation, as Al/Ti ratios decrease in more Fe-rich pyroxenes (Figs. 3 and 4). This ratio can be seen to decrease from core to rim of the pyroxene.

Cristobalite is an interstitial ($\simeq$ 0.15 mm) phase in some of the basalts studied (e.g., ,1481 and ,9088), where it is associated with opaque glass. Euhedral ilmenites (up to 0.3 mm long), usually found within the opaque glass, are homogeneous (Table 2). However, some variation occurs in MgO from grain to grain within a given sample, reflecting the chemically isolated nature of the residual glass pockets from which ilmenite crystallized.

Three vitrophyres also have been studied (,1471, ,1473, and ,1486); each contains euhedral plagioclase and olivine phenocrysts in an opaque glass. Figure 1e is of vitrophyre ,1471 which contains two populations of olivine phenocrysts. The smaller (0.2 mm diameter), euhedral, and more abundant olivines have a lower Fo content (Fo $\simeq$ 70) than the larger (0.5 mm diameter), subhedral, and less abundant type (Fo $\simeq$ 85). Small (0.25 mm long), rare plagioclase laths are present. Figure 1f is of ,1473 which has more abundant plagioclase laths (up to 0.5 mm long) and large olivine phenocrysts (generally 0.6 mm diameter) containing chromite-ulvöspinel. Glass compositions of the vitrophyres have contrasting compositions: (a) high Al/low Mg (11-31 wt % Al_2O_3; 0.5-7.6 wt % MgO); (b) low Al/high Mg types (2-5 wt % Al_2O_3; 17-23 wt % MgO). Ilmenite and pyroxene are not present in the vitrophyres.

Native Fe is present in all samples, usually < 0.2 mm in diameter, either enclosed in the mineral phases or in the interstices, sometimes too small for analysis. Most show low Ni and Co abundances similar to Apollo 15 mare basalts and

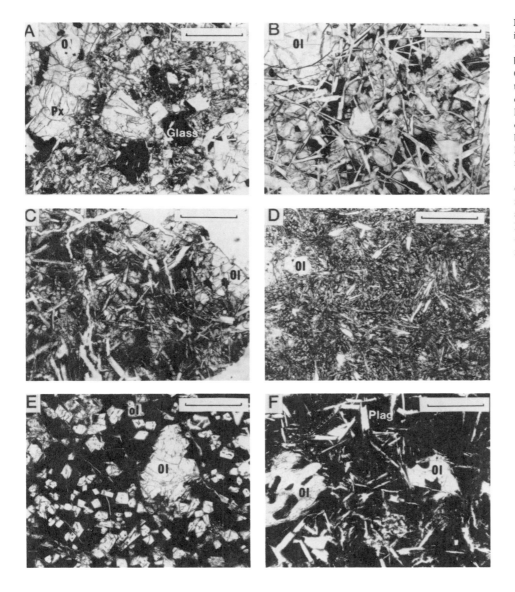

Fig. 1. Photomicrographs depicting the main textural features of the 14321 HA mare basalts. The scale bar represents 1 mm in A-D, and 0.5 mm in E and F. (**a**) Granulated texture of 14321,1118; small areas of glass contain euhedral plagioclase laths. (**b**) General ophitic texture of 14321,1477 which is exhibited by the majority of the 14321 basalts; large olivine phenocrysts (> 1 mm) may be present. (**c**) Sample 14321,1476 also has a general ophitic texture, but contains a much greater proportion of interstitial glass. (**d**) Sample 14321,1474 has a general ophitic texture with olivine phenocrysts, but is much finer grained. (**e**) Olivine vitrophyre 14321,1471 contains two populations of olivine: A large subhedral type (0.5 mm) and a smaller, euhedral type (0.2 mm) set in an opaque glass. (**f**) Olivine vitrophyre 14321,1475 consists of olivine phenocrysts containing chromite, as well as plagioclase laths set in an opaque glass.

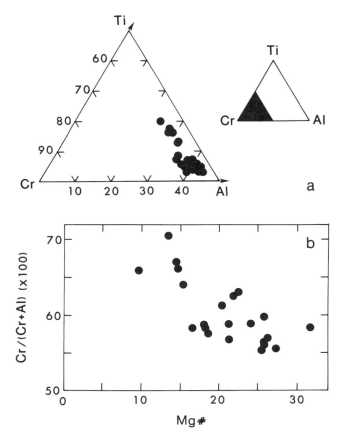

Fig. 2. Graphical representations of chromite-ulvöspinel compositions: **(a)** the Cr, Al, Ti ternary diagram; **(b)** Mg# versus 100[Cr/(Cr + Al)].

to previously studied Apollo 14 basalts (*Shervais et al.*, 1983). Basalts which contain native Fe grains with > 1 wt % Ni fall within the field of Apollo 14 mare basalt cumulates (*Shervais et al.*, 1983) or above the field indicating "meteorite contamination." However, four samples (,1115, ,1118, ,1473, and ,1475) have metal compositions which plot above the field of Apollo 14 and Apollo 16 polymict rocks, and meteorite contamination (*Ryder et al.*, 1980).

WHOLE ROCK GEOCHEMISTRY

Analytical Techniques

INAA of 22 new 14321 HA basalt clasts were conducted at Washington University using the procedure of *Lindstrom* (1984). All data were reduced employing the TEABAGS program of *Lindstrom and Korotev* (1982). Uncertainties based on counting rate at the one sigma level are: 1%-2% for Fe, Na, Sc, Cr, Co, La, Sm, and Eu; 3%-5% for Ca, Ce, Tb, Yb, Lu, Hf, Ta, and Th; 5%-15% for Cs, Ba, Nd, and U; 15%-25% for K, Sr, Rb, Ni, and Zr. All samples were analyzed for Ir and Au, which occur below the detection limit of 2-3 ppb, supporting a pristine (i.e., non-impact) origin for these basalts (*Warren and Wasson*, 1979b; *Ryder et al.*, 1980; *Warren et al.*, 1983). Major element compositions were obtained either by INAA (where SiO_2 is calculated by difference) or by broad beam electron microprobe analysis. The latter method involved crushing of INAA samples and fusing on a molybdenum strip in an argon atmosphere followed by quenching. This procedure was carried out at the Johnson Space Center. The elements Ca, Fe, and Na have also been measured by INAA, serving as a check for the fused glass method. In general, there is good agreement, although Fe is sometimes low, possibly due to small Fe-metal grains present in the quench glass.

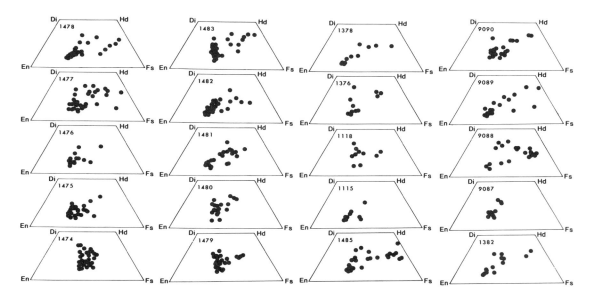

Fig. 3. Pyroxene compositions from all analyzed 14321 HA mare basalts represented on a pyroxene quadrilateral. Note that in practically all cases, there is a general trend toward more Ca- and Fe-rich compositions.

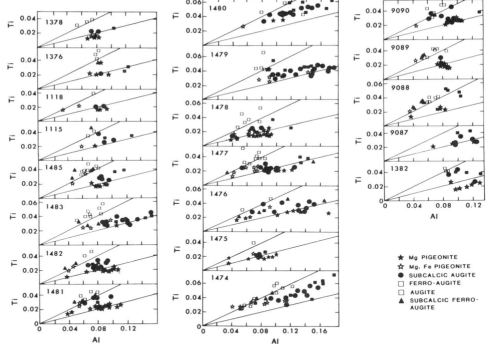

Fig. 4. Pyroxene compositions from all analyzed 14321 HA mare basalts represented on Al versus Ti diagrams. A general decrease in Al/Ti ratio is observed from Ca- and Mg-rich compositions to Fe-rich pyroxenes. This is indicative of plagioclase fractionation during pyroxene crystallization (*Bence and Papike*, 1972).

Results

Whole rock compositions are presented in Table 3. The mare basalts fall within the ranges of low-Ti, high-Al mare basalts reported by *Dickinson et al.* (1985) and *Shervais et al.* (1985), and the Apollo 11 low-K group of *Beaty et al.* (1979). The range of SiO_2 (46.0-50.1) forms a broad negative correlation with MgO (7.00-11.0), indicating these rocks have experienced some degree of olivine and/or pyroxene fractionation. Positive correlations of SiO_2, and negative correlations of Mg#, with the incompatible trace elements (La and Hf) again suggest such a hypothesis.

Large variations are apparent in comparisons of the trace element contents of these basalts. These elements are very sensitive to the many igneous processes that operate during magmatic evolution. The extent to which trace elements are incorporated into different mineral phases varies by several orders of magnitude, according to the respective trace element and the mineral phase considered. These factors all enhance the use of trace elements as petrologic guides of magma evolution, and the trace element behavior is the main subject of the following discussion. For ease of presentation, the trace elements are divided into groups of similar physico-chemical affinity.

Rare earth elements (REE). The REE profiles (Fig. 6) display considerable variation from flat, primitive patterns (at approximately 10 × chondritic abundances) with a small negative Eu anomaly, to evolved LREE-enriched patterns (LREE almost 100 × chondrite abundance), with a large negative Eu anomaly, which mirror KREEP (Fig. 6). Increase in REE abundance (Table 3) is accompanied by an increase in the La/Lu ratio. The new basalt data from this study conveniently fall into three groups on the basis of REE patterns; however, if compared with previous Apollo 14 HA mare basalt REE data (Fig. 6) a continuum of compositions is seen. Therefore, we do not classify the basalts into specific groups but describe them in terms of their La/Lu ratio: Those with a low La/Lu ratio are described as "primitive" and those with a high La/Lu ratio are termed "evolved" types. In addition, the ratio Sm/Eu (a measure of the negative Eu anomaly) increases from the primitive to evolved types.

Compatible elements: Ni, Cr, Sc, and Co. Within the HA basalts analyzed, there is a moderate variation in the

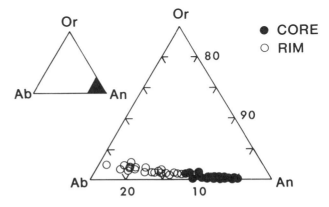

Fig. 5. Core and rim plagioclase compositions represented on an An-Ab-Or ternary diagram, demonstrating an increase in Ab and Or, and decrease in An from core to rim.

TABLE 2. Representative ilmenite analyses from the 14321 basalts.

	,1474	,1475	,1476	,1477	,1478	,1479	,1480	,1481	,1482	,1483	,1485	,1115	,1118	,1376	,1378	,1382	,9087	,9088	,9089	,9090
TiO_2	53.4	53.2	53.4	53.4	54.1	54.4	52.8	52.7	53.1	52.9	53.3	53.0	54.2	53.3	53.2	53.2	54.3	52.8	53.7	52.7
Al_2O_3	0.20	0.25	0.21	0.20	0.12	0.48	0.14	0.23	0.21	0.17	0.21	0.26	0.30	0.39	0.27	0.15	0.59	0.30	0.21	0.23
Cr_2O_3	0.43	0.41	0.39	0.30	0.42	0.31	0.45	0.36	0.21	0.17	0.43	0.41	0.30	0.39	0.27	0.15	0.59	0.30	0.46	0.21
FeO	44.9	44.0	44.5	45.3	44.2	44.3	44.1	45.9	46.4	46.2	44.7	45.0	42.3	46.5	44.5	46.3	42.2	46.8	42.8	46.1
MgO	1.13	1.88	1.42	1.22	1.91	1.04	1.86	0.75	0.63	0.78	1.22	1.19	3.04	0.27	1.17	0.66	2.98	0.59	3.20	0.77
Total	100.06	99.74	99.92	100.42	100.75	100.53	99.35	99.94	100.85	100.22	99.86	99.86	100.14	100.59	99.46	100.62	100.26	100.70	100.37	100.01
Formula on the basis of 3 oxygens																				
Ti	1.000	0.995	0.999	0.998	1.000	1.009	0.993	0.993	0.996	0.995	0.999	0.996	1.000	1.000	1.000	0.996	1.000	0.991	0.991	0.993
Al	0.005	0.007	0.005	0.005	0.002	0.014	0.003	0.006	0.006	0.004	0.006	0.007	0.008	0.003	0.009	0.008	0.005	0.005	0.005	0.006
Cr	0.008	0.008	0.007	0.005	0.008	0.005	0.009	0.006	0.003	0.003	0.008	0.008	0.005	0.007	0.005	0.002	0.011	0.005	0.008	0.003
Fe	0.936	0.914	0.926	0.941	0.908	0.913	0.921	0.961	0.968	0.967	0.934	0.939	0.866	0.970	0.929	0.964	0.864	0.977	0.878	0.966
Mg	0.042	0.069	0.052	0.045	0.070	0.038	0.069	0.028	0.023	0.028	0.044	0.044	0.111	0.009	0.043	0.024	0.108	0.021	0.117	0.028
Total	1.989	1.992	1.990	1.995	1.989	1.978	1.996	1.995	1.997	1.997	1.991	1.994	1.990	1.990	1.986	1.994	1.988	2.000	2.000	1.997
Mg#	4.3	7.0	5.3	4.6	7.2	4.0	7.0	2.8	2.3	2.8	4.5	4.5	11.4	0.9	4.4	2.4	11.1	2.1	11.8	2.8

TABLE 3. Whole rock compositions of 14321 basalts.

INAA #	,1106	,1112	,1318	,1329	,1338	,1344	,1345	,1346	,1347	,1349	,1350	,1353	,1365	,1422	,1425	,1429	,1432	,1435	,1437	,1439	,1441	,1443
PM #	,1115	,1118	,1376	,1378	-	-	-	-	,9087	,9088	,9089	,1382	,9090	,1473	,1474	,1475	,1476	,1477	,1478	,1479	,1480	,1481
Wt (mg)	13.2	11.9	106	72.1	57.3	13.6	21.8	16.1	39.2	36.0	25.6	92.2	83.4	46.0	47.0	46.1	49.4	48.8	32.2	36.9	27.6	52.6
	I	I	P	I	P	P	P	I	P	I	I	I	P	P	P	P	P	P	P	P	P	P
SiO_2	48.7*	49.3*	48.1	47.9*	48.0	48.9	48.3	50.1*	49.4	46.3*	47.1*	48.1*	49.4	46.0	50.2	49.0	48.1	48.9	47.6	48.0	46.7	47.4
TiO_2	1.28	1.91	1.96	2.72	2.06	2.18	2.07	1.66	2.14	2.90	2.22	2.85	2.19	2.72	2.16	2.23	2.65	1.88	2.22	2.07	2.64	2.61
Al_2O_3	11.5	14.5	12.1	12.4	12.0	11.5	12.3	13.1	12.7	13.1	13.4	12.5	12.2	11.3	11.4	11.6	12.1	10.9	13.1	12.6	11.6	11.5
FeO	17.1	14.8	17.0	16.1	16.0	15.9	15.3	14.4	16.0	15.7	15.3	16.5	16.0	17.9	16.7	16.1	16.4	16.9	15.9	15.9	17.2	16.6
MnO	nd	nd	0.20	nd	0.25	0.16	0.16	nd	0.25	nd	nd	nd	0.16	0.23	0.21	0.24	0.33	0.24	0.40	0.22	0.36	0.33
MgO	10.7	7.00	10.2	9.60	9.60	9.49	8.91	9.20	8.05	9.20	8.10	8.40	8.60	11.0	8.80	9.37	8.48	9.28	8.89	8.78	10.7	9.56
CaO	9.90	11.7	10.6	10.4	10.7	10.4	12.0	10.8	11.0	13.3	12.5	10.7	10.5	10.6	10.1	10.6	10.9	10.4	11.3	10.8	10.3	10.5
Na_2O	0.43	0.52	0.50	0.45	0.65	0.53	0.47	0.30	0.58	0.46	0.46	0.55	0.65	0.48	0.48	0.51	0.49	0.51	0.27	0.47	0.41	0.45
K_2O	nd	0.22	0.10	0.16	0.29	0.19	0.20	nd	0.23	nd	nd	nd	0.16	0.03	0.14	0.13	0.05	0.11	0.05	0.14	0.03	0.09
Total	99.61	99.73	100.76	99.73	99.55	99.25	99.71	99.56	100.35	100.96	99.08	99.56	99.86	100.34	100.19	99.78	99.50	99.12	99.73	99.45	99.94	99.04
Sc	57.3	55.0	59.7	56.9	58.2	59.7	61.6	59.8	60.4	60.2	58.7	59.9	59.4	64.1	58.1	61.3	62.0	58.4	58.7	59.3	63.5	58.4
Cr	4190	2906	3819	3230	3188	3386	3264	3284	2870	2734	2980	3144	3470	3770	3070	2975	2920	3150	3315	3220	4000	3485
Co	38.7	40.0	33.4	34.2	34.0	34.6	33.9	33.9	31.8	28.3	32.1	31.2	34.7	32.0	29.9	37.4	28.0	31.9	43.4	33.7	32.6	34.8
Ni	120	80	<100	90	115	150	120	<140	72	60	90	<100	125	<100	<80	<70	<110	49	80	<100	58	<100
Rb	<27	<26	<9	<8	<13	<13	<9	<9	<9	<16	<15	<100	<10	<10	<10	<10	<8	<8	<20	<9	<8	<10
Sr	135	110	72	62	160	150	100	<200	85	100	90	121	100	<100	70	84	60	80	80	55	nd	57
Cs	0.11	0.19	0.18	0.22	0.30	0.26	0.29	0.21	0.26	0.19	0.20	0.20	0.24	<0.20	0.11	0.13	0.15	0.14	0.11	0.15	<0.20	0.18
Ba	150	216	160	170	200	180	200	81	190	155	205	175	170	28	190	190	110	155	83	160	33	130
La	21.1	30.6	20.8	20.3	22.8	21.0	22.7	6.74	24.9	13.2	27.6	20.8	20.5	3.19	23.7	23.3	11.9	21.8	8.35	20.5	3.41	13.5
Ce	59.7	82.9	59.7	56.0	63.8	58.4	63.2	21.0	67.9	37.2	75.2	57.8	57.1	9.3	60.9	62.3	31.2	58.1	22.3	54.1	9.7	35.2
Nd	40	50	30	28	40	35	42	10	40	20	44	33	33	4.3	39	41	18	37	11	30	11	20
Sm	12.1	15.3	10.4	10.5	10.8	10.7	11.6	3.85	11.9	6.76	14.3	10.5	10.5	2.10	12.0	12.1	6.44	11.12	4.75	10.67	2.35	7.59
Eu	1.27	1.67	1.27	1.28	1.33	1.24	1.37	0.672	1.43	1.39	1.53	1.30	1.26	0.62	1.37	1.35	1.03	1.32	0.72	1.30	0.64	1.12
Tb	2.43	3.32	2.40	2.39	2.52	2.49	2.67	1.09	2.71	1.71	3.10	2.40	2.32	0.68	2.63	2.52	1.48	2.46	1.23	2.35	0.75	1.79
Yb	6.72	9.72	7.27	7.20	7.40	6.97	7.61	3.85	8.01	6.33	8.29	7.38	7.27	3.02	7.88	7.93	5.70	7.48	4.49	6.77	2.95	6.33
Lu	1.02	1.33	0.98	0.99	1.00	1.07	1.09	0.57	1.09	0.90	1.24	0.98	0.97	0.44	1.09	1.11	0.82	1.03	0.64	0.98	0.47	0.91
Zr	305	395	280	320	285	390	380	250	400	240	450	330	300	nd	370	400	300	370	230	290	nd	200
Hf	8.14	11.5	8.15	7.94	8.43	8.06	8.58	3.20	9.00	5.75	10.4	8.17	7.95	1.70	8.64	8.62	5.02	8.21	3.62	7.50	1.87	5.43
Ta	0.88	1.31	1.08	1.05	1.11	1.03	1.05	0.42	1.20	0.99	1.22	1.02	0.99	0.45	1.15	1.15	0.89	1.10	0.59	1.06	0.43	0.98
Th	2.07	2.90	1.73	1.84	1.98	2.10	1.95	0.70	2.33	1.59	2.38	1.74	1.79	0.20	2.04	2.40	1.46	1.95	1.35	1.60	0.23	1.50
U	0.32	0.70	0.44	0.42	0.70	0.39	0.57	0.24	0.58	0.50	0.90	0.55	0.42	0.20	0.62	0.58	0.32	0.56	0.23	0.83	0.07	0.48
Ir (ppb)	nd	nd	nd	nd	nd	<5	<5	<3	nd	nd	<7	<3	<3	nd	<2	<3	nd	nd	nd	<2	nd	nd
Au (ppb)	<6	<4	<4	<2	nd	<5	<5	<3	nd	nd	<3	<3	<3	<3	<2	<3	nd	<2	<2	<2	5	<2

* SiO_2 calculated by difference; I = analysis by INA; P = analysis by electron microprobe of fused INAA sample; na = not analyzed; nd = not detected.

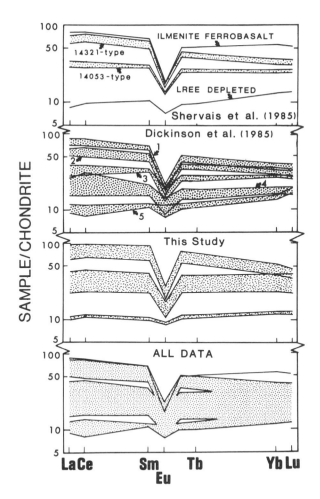

Fig. 6. Rare earth element profiles of all analyzed 14321 HA mare basalts. Each data set (upper 3 diagrams) conveniently falls into groupings, but when all compositions are plotted together, these disappear. Instead, a broad continuum is established (bottom diagram). Although minor gaps still exist, it is evident that 5 basalt groups are not present at the Apollo 14 site. It is expected that these remaining gaps will be filed in as more data is collected.

compatible elements (Ni = 40-180 ppm; Cr = 2734-4190 ppm; Sc = 55.0-64.1 ppm; Co = 28.0-43.4 ppm: Table 3) which classifies these as high-Al basalts corresponding with the major elements described above. Nickel analyses have uncertainties that are too large to be of any constraint and will not be considered further. The variation in Cr and Co cannot be correlated with that in the REE (La/Lu, Sm/Eu), but Sc generally decreases with increasing La/Lu and Sm/Eu.

Large ion lithophile (LIL) elements: K, Rb, Sr, Ba, and Cs. Abundances of K range from 0.03 to 0.25 wt % (Table 3). The high uncertainties associated with Rb and Sr analyses do not allow them to be of any constraint and will not be considered further. Ba displays a considerable range within the 22 analyzed HA basalts (28-216 ppm). K and Ba increase from the primitive basalts to the evolved types, demonstrated by a positive correlation with La/Lu and Sm/Eu. Cs shows moderate variation (0.11-0.30 ppm).

High field strength (HFS) elements: Zr, Hf, Ta, U, and Th. Moderate to large variations in concentration exist within the HFS group. Zr ranges from 200-400 ppm, and an approximate positive correlation between Zr and La/Lu, and Zr and Sm/Eu is noted. However, even better positive correlations are seen with Hf (1.70-11.5 ppm), Ta (0.42-1.31 ppm), U (0.20-0.90 ppm), and Th (0.20-2.90 ppm). These correlations (e.g., Fig. 11e-f) demonstrate concentration of the HFS elements in the basaltic magma from primitive to evolved types.

MODELING

Introduction

It has been previously demonstrated that Apollo 14 HA mare basalt compositions cannot be generated by partial melting of a single source and/or fractional crystallization of a common parent magma (*Dickinson et al.*, 1985; *Shervais et al.*, 1985). A more complicated model for mare basalt petrogenesis by assimilation of urKREEP residues (*Binder*, 1982, 1985) and KREEP (*Dickinson et al.*, 1985; *Shervais et al.*, 1985) has been invoked in order to explain the large variation in trace element contents with limited differences in the major elements. *Binder* (1982, 1985) and *Shervais et al.* (1985) have suggested possible anatexis of KREEPy material in the lower crust by a primitive basaltic magma (e.g., the composition represented by 14321,1422). *Dickinson et al.* (1985) postulated a source for this uncontaminated basaltic magma (Group 5 of their classification) as being an "olivine-orthopyroxene cumulate that crystallized from a magma having chondritic relative abundances." Such a cumulate source would have an inherent

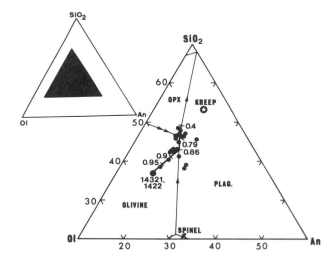

Fig. 7. The major element compositions of the 14321 HA mare basalts from this study represented on a Walker diagram. Phase boundaries are drawn for a Fe/(Fe + Mg) ratio of 0.6, the composition of the evolving parental magma just prior to encountering the plagioclase-olivine cotectic. The fractional amount of melt remaining is shown at various points along the liquidus path. The details of the fractionation modeling are given in the text.

negative Eu anomaly, accounting for the small negative Eu anomaly in the primitive basalts (*Warren*, 1985). *Binder* (1982) suggested a cumulate source consisting of two or more density-graded rhythmic bands whose compositions grade from that of the very low TiO_2 magma source regions (< 0.2% ilmenite) to that of the very high TiO_2 magma source regions (9% ilmenite). However, our research indicates that assimilation alone (i.e., bulk mixing), and even the proposed AFC scheme of *Binder* (1982, 1985), cannot account for all the chemical variations exhibited by the Apollo 14 HA mare basalts (e.g., the Ba and Hf abundances). On the basis of this rationale and the results of our research, we have formulated a combined AFC model for Apollo 14 HA mare basalt petrogenesis.

Modeling Systematics

In the proposed model, KREEP is inferred as an assimilated contaminant by a primitive HA basaltic magma. We have taken the composition of 15386 "IKFM" (*Vaniman and Papike*, 1980) as our KREEP component, so as to be consistent with

TABLE 4. Crystal/liquid partition coefficients used in AFC modeling.

	Pyroxene	Plagioclase	Olivine	Ilmenite	Chromite
La	(0.012)*	0.051	(0.0001)	0.029	0.029
Ce	0.038	0.037	0.0001	0.038	0.038
Sm	0.054	0.022	0.0006	0.053	0.053
Eu	0.100	1.22	0.0004	0.02	0.02
Yb	0.671	0.012	0.02	0.39	0.39
Lu	0.838	0.011	(0.02)	0.47	0.47
Sc	1.6	0.065	0.4	1.5	1.5
Ba	0.013	0.686	0.03	0.005	0.005
Ta	(0.21)	0.04	0.03	0.53	0.53
Hf	0.063	0.05	0.04	1.817	0.38
Th	0.13	0.05	0.03	0.55	0.55

RESULTS
Proportion of Liquid Remaining

	Parent	0.95	0.9	0.86	0.79	0.7	0.6	0.5	0.4	KREEP
SiO_2	46.1	46.8	47.6	48.3	49.1	49.2	49.4	49.6	49.9	50.8
TiO_2	2.73	2.59	2.44	2.31	2.09	2.14	2.21	2.30	2.42	2.23
Al_2O_3	11.3	11.9	12.5	13.1	12.8	12.8	12.8	12.9	12.9	14.8
FeO	17.9	17.4	16.9	16.4	16.5	16.4	16.3	16.1	16.0	10.6
MnO	0.23	0.23	0.22	0.22	0.23	0.24	0.26	0.28	0.30	0.16
MgO	11.1	9.86	8.52	7.34	6.75	6.67	6.58	6.45	6.29	8.17
CaO	10.1	10.7	11.2	11.8	11.9	11.9	11.9	11.8	11.8	9.71
Na_2O	0.48	0.51	0.53	0.56	0.57	0.54	0.52	0.49	0.45	0.73
La	3.19	4.60	6.16	7.54	9.86	13.5	18.1	25.0	31.7	83.5
Ce	9.30	12.9	16.9	20.5	26.4	35.8	47.4	64.6	81.8	211
Sm	2.10	2.77	3.51	4.16	5.26	6.96	9.06	12.2	15.2	37.5
Eu	0.62	0.69	0.77	0.84	0.91	1.04	1.19	1.37	1.60	2.72
Yb	3.84	3.53	4.09	4.59	5.43	6.42	7.59	8.98	10.7	24.4
Lu	0.44	0.51	0.59	0.66	0.78	0.91	1.05	1.32	1.43	3.40
Sc	64.1	65.6	67.2	68.6	71.9	69.3	66.5	63.5	60.2	23.6
Ba	28	42	57	71	91	124	164	214	278	837
Ta	0.45	0.54	0.63	0.71	0.85	1.06	1.31	1.63	2.03	4.60
Hf	1.7	2.25	2.86	3.39	4.29	5.59	7.17	10.3	11.6	31.6
Th	0.2	0.36	0.53	0.68	0.94	1.33	1.82	2.42	3.21	10.0
% crystallized		0-14	15-21	22-60						
Olivine		90	40	0						
Plagioclase		0	50	30						
Orthopyroxene		0	0	60						
Ilmenite		0	0	10						
Chromite		10	10	0						

References: *Binder*, 1982; *Haskin and Korotev* (1977); *Irving and Frey* (1984); *Arth and Hanson* (1975); *McKay et al.* (1986); *Drake and Weill* (1975); *Schnetzler and Philpotts* (1970); *Villemant et al.* (1981).

*Brackets indicate estimated partition coefficient. Chromite REE partition coefficients taken as those for ilmenite.

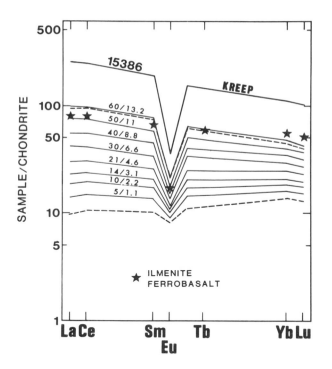

Fig. 8. The range of REE patterns for all Apollo 14 HA mare basalts are modeled using equation (6a) of *DePaolo* (1981) and the phase proportions deduced from major element modeling. Partition coefficients are presented in Table 4. Note that the range of REE patterns can be generated after 60% fractional crystallization of a parental composition represented by 14321,1422. An r value of 0.22 is required, and the numbers on each of the generated patterns represents the percentage crystallized/percentage assimilated. The KREEP composition of *Vaniman and Papike* (1980) was used in these calculations. The ilmenite ferrobasalt of *Shervais et al.* (1985) cannot be effectively generated using this model.

the previous mass balance models of *Dickinson et al.* (1985) and *Shervais et al.* (1985). As the primitive HA magma, the composition of 14321,1422 is taken from our new data as it has the lowest trace element abundances, lowest SiO_2 (46.1 wt %), highest MgO contents (11 wt %), and among the highest Mg# (52.4).

Although the treatment of assimilation as a mass balance problem is a good first approximation, realistically any assimilation will be accompanied by some magma crystallization (*DePaolo*, 1981). The composition of the crystallizing phases will evolve as this process proceeds. We have attempted, where possible, to account for these evolving compositions within our major element modeling, in order to represent a realistic situation. As we envisage a low r value of 0.22 (r = mass assimilated/mass crystallized) for our AFC scheme, the major elements can be effectively modeled by fractional crystallization alone (*DePaolo*, 1981). This may well explain the observation of *Dickinson et al.* (1985) that the major elements may be produced by fractional crystallization alone of a common parent magma. Of course, this conclusion is at variance with the trace element data. The proportions and nature of the fractionating phases can be calculated from phase equilibria (i.e., by representation of data on the Ol-An-SiO_2 pseudoternary). These are applied to the trace elements, upon which assimilation has a profound effect. Such an approach is critical for consistency between major and trace element modeling. This is essential in the development of any model of magma petrogenesis.

Major Element Modeling

A major constraint in any petrologic modeling is the phase equilibria applicable to that particular melt, and numerous experimental investigations of mare basalt liquids have been performed (e.g., *Walker et al.*, 1972; *Kesson*, 1975; *Usselman et al.*, 1975). It was determined that the phase relations could most readily be explained by consideration of the Ol-An-SiO_2 pseudoternary system (often referred to as the "Walker Diagram"), which is extremely useful in representing basaltic compositions. Our new 14321 HA mare basalt data are represented on such a Walker diagram in Fig. 7. The phase boundaries (e.g., cotectics) have been shown to be particularly sensitive to the Fe/(Fe + Mg) ratio of the system (*Walker et al.*, 1973; *Longhi*, 1977). As such, the phase boundaries are drawn for the Fe/(Fe + Mg) of the evolving system just before the evolving melt composition encounters the cotectic (Fe/(Fe + Mg) ≃ 0.6). As stated above, we have modeled the major elements by fractional crystallization only of parent magma 14321,1422, as minor assimilation will have a negligible effect on the major elements (*DePaolo*, 1981).

A least squares regression program was used to calculate the major element compositions of the evolving magma. Olivine and chromite are the first phases to crystallize (*Walker et al.*, 1972). Olivine (90%) and chromite (10%) only fractionate during the first 14% crystallization (Fig. 7). After 14% crystallization, the olivine-plagioclase cotectic is reached and plagioclase (50%), olivine (40%), and chromite (10%) fractionate until 21% of the parental liquid has crystallized. At this point, the olivine-opx-plagioclase peritectic is reached, and olivine fractionation ceases with orthopyroxene becoming the dominant liquidus phase. In our model, we envisage that all olivine has been effectively removed, and no appreciable back reaction occurs at the onset of pyroxene fractionation. Plagioclase (30%), orthopyroxene (60%), and ilmenite (10%) are the fractionating phases until 60% of the parental liquid is solidified. In order to recreate realistic crystallizing conditions, the Fo content of fractionating olivine was calculated iteratively in crystallization increments of 2% using the Mg/Fe ratio of the resultant liquid (method of *Longhi*, 1977). This yielded a range of olivine compositions from 76.9 to 68.0 throughout the period that olivine was on the liquidus. During the first 7% of plagioclase fractionation (up to the peritectic) a composition of An92, Ab8 was used. After the peritectic a plagioclase of composition An79, Ab19 was fractionated. Such compositions were based upon the zonation observed in mare basalt plagioclases. The composition of chromite was estimated from analyzed 14321 HA mare basalts (Mg# = 12.6) as was ilmenite (Mg# = 9.4), and these compositions were kept constant throughout. Pyroxene compositions in a crystallizing magma will evolve from Mg-

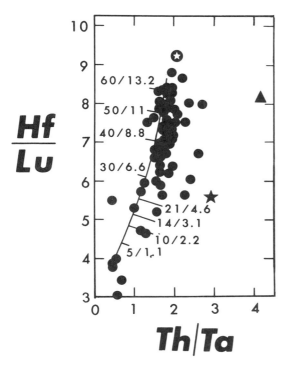

Fig. 9. A plot of Hf/Lu versus Th/Ta which *Shervais et al.* (1985, Fig. 14) used to demonstrate that 15386 was the only feasible contaminant to generate their 14321-type basalts. This plot includes *all* Apollo 14 HA basalt data with an AFC path calculated from 14321,1422 using the phase proportions deduced from the major element modeling. The numbers represent percentage crystallized/percentage assimilated. Note that the ilmenite ferrobasalt cannot be generated by using 15386 as the contaminant. If this basalt has a petrogenesis involving AFC, the KREEP composition of *Warren et al.* (1983) is a more likely composition.

$$C_m/C_m^o = F^{-z} + \left(\frac{r}{r-1}\right)\left(\frac{C_a}{zC_m^o}\right) 1 - F^{-z}$$

where r = mass assimilated/mass crystallized; $z = r + D - 1/r - 1$; D = bulk distribution coefficient; C_m^o = concentration of element in parental magma; C_m = concentration of element in residual magma; C_a = concentration of element in assimilated component; F = proportion of liquid remaining.

We have attempted to model critical trace elements from the REE, compatible elements (Sc), LIL (Ba), and HFS (Hf, Th, and Ta) groups. The parental HA basalt composition is again taken as 14321,1422. The proportions of fractionating phases are those calculated from the major element modeling. Crystal/liquid partition coefficients used and results obtained are presented in Table 4. In the following sequence of diagrams, our new basalt data are plotted with other Apollo 14 HA mare basalt data [*Dickinson et al.* (1985); *Shervais et al.* (1985)] from breccia 14321. The Tridymite Ferrobasalt of *Shervais et al.* (1985) is not included in this study as it is a low-Al mare basalt.

In our modeling, the ratio of mass assimilated to mass crystallized (r value) of 0.22 is used. This provides the best fit with our data, using the fractionating phases in the proportions described above. Figure 8 is a chondrite-normalized plot of the range in REE patterns for the Apollo 14 HA basalts and the calculated REE patterns using the model outlined above. Note the important effect pyroxene fractionation has on the HREE, generating the 14321-type basalts of *Shervais et al.* (1985) and Groups 1 and 2 of *Dickinson et al.* (1985). The

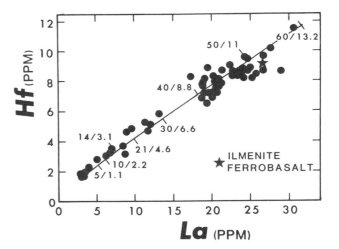

Fig. 10. This plot of La versus Hf is how *Dickinson et al.* (1985) presented their evidence for 5 groups of HA mare basalts at the Apollo 14 site. In this figure all Apollo 14 HA basalt data are plotted, demonstrating that 5 groups do not exist. It may be argued that 2 groups can be defined, but it is evident that as more data are generated, the number of "groups" decreases. As with Fig. 8, we feel that as more basalts are analyzed from the Apollo 14 site, these compositional gaps will disappear. The AFC path has been calculated using the parameters as in previous diagrams. The numbers represent percentage crystallized/percentage assimilated.

rich/Ca, Fe-poor to Ca and Fe-rich (e.g., *Lindsley and Andersen*, 1983). As a result of such crystallization, an average pyroxene composition of Mg# = 48 is used. This is permissible since olivine fractionation has already depleted the magma in Mg. Results of this modeling are presented in Fig. 7 and Table 4. The calculated crystallization path encompasses all the new Apollo 14 HA mare basalt data presented in this study.

Trace Element Modeling

The trace elements have been modeled by combined assimilation and fractional crystallization, using equation (6a) of *DePaolo* (1981):

highest concentration of REE represents *60% crystallization of parent magma 14321,1422 and 13.2% KREEP assimilation.*

However, the HREE abundances of the highly evolved Ilmenite Ferrobasalt of *Shervais et al.* (1985) cannot be generated using this method. This basalt contains the largest negative Eu anomaly of all the Apollo 14 HA mare basalts which argues for a greater proportion of plagioclase fractionation. The flattening of the REE profile in the HREE could reflect a sympathetic decrease in pyroxene fractionation within the general AFC model.

Figure 9 depicts Hf/Lu versus Th/Ta which is similar to Fig. 14 of *Shervais et al.* (1985), who used this plot to demonstrate that 15386 IKFM was assimilated by a primitive magma to form the 14321-type basalts. We have plotted all Apollo 14 HA basalts on this diagram to demonstrate there are no groups depicted, and an AFC path may be calculated between 14321,1422 and 15386 IKFM (KREEP). The most evolved basalts (i.e., those plotting closest to KREEP) are generated by approximately *60% crystallization of parent magma 14321,1422 and 13.2% KREEP assimilation.* The Group 5 HA

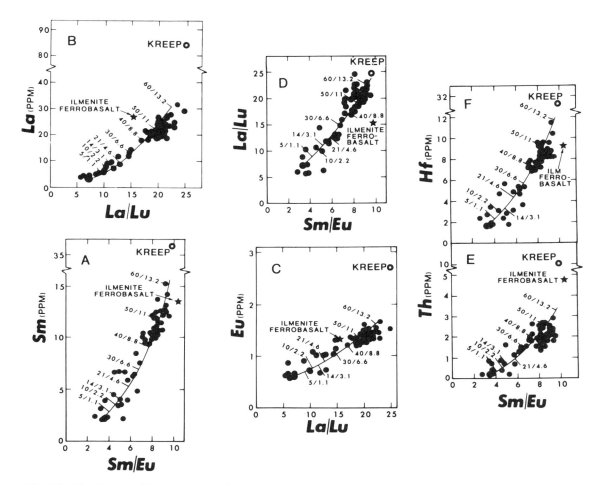

Fig. 11. Six plots involving the rare earth and HFS elements to illustrate that the calculated AFC path is consistent for different elements. In this sequence of diagrams, it is evident that no "groups" of basalts may be defined. It is also evident that each correlation can be consistently represented by our AFC model. The numbers along the AFC path represent percentage crystallized/percentage assimilated. KREEP is shown as an open star symbol. **(a)** Sm versus Sm/Eu is used as a measure of differentiation. Such correlations have been ascribed to short-range unmixing, but with the large range in both Sm and Sm/Eu this process alone is unlikely to be responsible for the Apollo 14 HA mare basalt compositions. These compositions are the result of a more complicated AFC process. **(b)** La versus La/Lu was used by *Shervais et al.* (1985) to demonstrate that their 14321-type basalts could not be generated by fractional crystallization alone. Our figure presents all Apollo 14 HA basalt data and the reasonable fit of an AFC path suggests this process could be responsible for all compositions, not just the "14321-type." **(c)** Eu versus La/Lu which demonstrates the increase of Eu with the steepening of the REE profile (increased La/Lu ratio). **(d)** La/Lu versus Sm/Eu demonstrates that the deepening of the negative Eu anomaly is coupled with the steepening of the REE profile. Such a relationship can be modeled effectively by our AFC scheme. **(e and f)** Th and Hf versus Sm/Eu demonstrate the increase of the HFS elements with the deepening of the negative EU anomaly. Both these relationships may also be effectively represented by our AFC model.

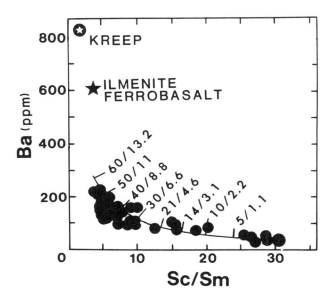

Fig. 12. Ba versus Sc/Sm illustrates the importance of plagioclase fractionation in suppressing Ba enrichment. The correlation between these elements can also be accounted for in our AFC model. The numbers along the AFC path represent percentage crystallized/percentage assimilated. Note that the ilmenite ferrobasalt of *Shervais et al.* (1985) plots between our most evolved basalt and KREEP.

basalts of *Dickinson et al.* (1985) plot below ,1422 on this diagram. These may also be used as parent magmas but would not alter the overall results obtained from our model.

Figure 10 is a plot of La versus Hf abundances, the same as in Fig. 1 of *Dickinson et al.* (1985), who depicted their 5 groups on this basis. All Apollo 15 HA basalt data have been included, and note the general continuum represented. Although one can argue for two groups of basalts (there is a compositional break between La 14-17 ppm and Mg 5.5-7 ppm), five groups cannot be identified. With further data, we expect the compositional gap to be filled and thus argue for a continuum of HA basaltic compositions at the Apollo 14 site. Also, even the Ilmenite Ferrobasalt of *Shervais et al.* (1985) falls on this positive trend, which may argue that the petrogenesis of this rock-type is related to other HA mare basalts. In subsequent diagrams, all previously analyzed Apollo 14 HA basalts are plotted with our new data, to illustrate that 5 groups of basalts do not exist, but are part of a continuum of compositions.

The plot of Sm versus Sm/Eu (Fig. 11a) can be used as an indicator of fractional crystallization. The strong positive correlation would tend to support such a process, but KREEP has both a higher Sm/Eu ratio and Sm abundance which could suggest this correlation represents a mixing path. An AFC path has been calculated between our parent magma and 15386 KREEP. The excellent fit of this path with the data supports our model for HA mare basalt generation at the Apollo 14 site by an AFC process. It is important to note that the Sm/Eu ratio (a measure of the negative Eu anomaly in these basalts) cannot be satisfactorily generated without plagioclase fractionation.

A plot in Fig. 11b of La (ppm) versus La/Lu is similar to Fig. 11 of *Shervais et al.* (1985) who argued that fractionation alone cannot produce this positive correlation. We concur with their conclusion that assimilation of KREEP (represented by 15386 IKFM) by a primitive magma produces their 14321-type basalts. However, we extend this conclusion to include all HA basalt data, except the ilmenite ferrobasalt. This rock-type will be discussed in detail at the end of this discussion. Again the AFC path generated by our model depicts a reasonable fit to the data presented.

The development of the negative Eu anomaly in these basalts is illustrated in Figs. 11c,d, where La/Lu is plotted against Eu and Sm/Eu. These diagrams illustrate that the increase of Eu is primarily controlled by plagioclase fractionation, and the steepening of the REE profile is coupled with the deepening of the negative Eu anomaly. In fact, increases in HFS elements can also be related to the deepening of the negative Eu anomaly (Figs. 11e,f). In all these diagrams, the AFC paths calculated from our model presented above show good agreement with the analyzed HA basalts.

The diagram presented in Fig. 12 is of Ba versus Sc/Sm ratio. This serves to illustrate two points: (1) If the process for basalt evolution is bulk mixing between KREEP and 14321,1422 alone, the Ba abundances would increase too rapidly. Plagioclase fractionation is required to inhibit the Ba increase; (2) Sc is controlled primarily by pyroxene and oxide fractionation, the decreasing Sc/Sm ratio being primarily due to increasing Sm. Our calculated AFC path illustrates the importance of these observations.

DISCUSSION

The above modeling demonstrates how observed HA mare basalt compositions can be generated from a single source (concordant with *Binder*, 1982, 1985). The low r values are in accordance with the previous low estimates of required KREEP contamination (see *Binder*, 1982; *Dickinson et al.*, 1985), and our model requires approximately half the KREEP assimilation of *Dickinson et al.* (1985) (13.2% versus 27%) and considerably lower than the 20% proposed by *Shervais et al.* (1985). The fact that AFC paths (r = 0.22) can be generated for LIL, REE, and HFS elements (Figs. 8-12) strongly indicates assimilation of a KREEP component by a magma fractionating olivine, orthopyroxene, plagioclase, and oxide (calculated from major element modeling) during HA basalt petrogenesis at the Apollo 14 site.

This approach extends the scheme of *Binder* (1982, 1985) by demonstrating the feasibility of crystallizing plagioclase plus oxide, as well as olivine and pyroxene. If only olivine and pyroxene were fractionated, Ba, Eu, and Hf abundances and the Sc/Sm ratio would be much higher than observed, and the larger negative Eu anomaly observed in the more evolved basalts would not be produced. If plagioclase and ilmenite are fractionated, Ba and Hf enrichments are suppressed, while Sc is decreased in the residual liquid after olivine fractionation to produce observed Sc/Sm ratios.

Ilmenite Ferrobasalts

The ilmenite ferrobasalt of *Shervais et al.* (1985) does not show good agreement with our proposed model. However, the petrogenesis of this rock-type can be understood within an AFC model by reference to Figs. 8-12. In Fig. 9, the ilmenite ferrobasalt plots to the right of the main bulk of basalt data. If it had a petrogenesis involving KREEP assimilation, 15386 does not represent this component. Rather, the composition reported by *Warren et al.* (1983) would be a better candidate. In Fig. 11a-f, this rock-type plots either above or below the calculated AFC paths, but its petrogenesis is still consistent with an AFC process with the KREEP composition of *Warren et al.* (1983). However, as noted in Fig. 8, the ilmenite ferrobasalt requires a decrease in the proportion of pyroxene with a subsequent increase in plagioclase fractionation in order to account for the flattening of the HREE and increased Sm/Eu ratio. Figure 10 is the only occurrence of the ilmenite ferrobasalt agreeing with our model. This may be because of similar La/Hf ratios in both 15386 and the KREEP composition of *Warren et al.* (1983). In Fig. 12, the ilmenite ferrobasalt plots above the calculated AFC path. We have argued for increased plagioclase fractionation on account of the REE, but this will restrict Ba in the resulting liquid. Therefore, an increased r value (of approximately 0.5) is required if the ilmenite ferrobasalt is generated by an AFC process. *However, the possibility cannot be ruled out that this rock-type represents a melt from a different source region and is unrelated to the other high-Al mare basalts at the Apollo 14 site.* Our qualitative discussion above is to illustrate the possibility of fitting this rock-type into the general confines of a petrogenesis by an AFC process.

Short-Range Unmixing

A problem encountered in the study of lunar basalts is one of sample size. A study by *Haskin and Korotev* (1977) has demonstrated the effects *short-range unmixing* has on a solidifying lava. This process is similar to fractional crystallization, except there is no large-scale separation of crystallized products from the residual parent liquid (*Lindstrom and Haskin*, 1978). Separations that do occur are far from complete and involve short distances. In effect, short-range unmixing amounts to a sampling problem (*Lindstrom and Haskin*, 1978) because it deals with centimeter-sized inhomogeneities in materials that may be homogeneous on a meter scale. *Lindstrom and Haskin* (1978) noted that correlations between chemical elements among different samples do occur due to short-range unmixing alone. These authors illustrated that in the case of the Apollo 15 basalts, there is a positive correlation between Sm/Eu and Sm (ppm), accountable by short-range unmixing. However, the ranges in Sm/Eu (2.5-4.5) and Sm (3.5-4.5) are much smaller than those in the Apollo 14 HA basalts (2.7-9.3 and 2.1-15.4, respectively). We consider that while short range unmixing may have occurred at the Apollo 14 site, it cannot account for the wide range in trace element abundances observed in these basalts.

CONCLUSIONS

(1) The range of major and trace element compositions in Apollo 14 HA mare basalts can be generated by a combined assimilation and fractional crystallization process involving KREEP. This process produces a continuum of basaltic compositions, rather than strict groupings.

(2) Bulk mixing or assimilation alone between primitive mare basalt and KREEP cannot explain all observed basaltic compositions at the Apollo 14 site.

(3) Major element modeling illustrates that olivine, pyroxene, plagioclase, and oxide are important liquidus phases in the generation of the HA mare basalt suite at the Apollo 14 site.

(4) The proportions of these phases change as the AFC process proceeds. During initial crystallization, only olivine (90%) and chromite (10%) are fractionated. After 15% crystallization of the parental magma, the cotectic is reached and plagioclase (50%) becomes an important liquidus phase with the reduction of olivine (40%), and chromite (10%) remains constant. After 21% crystallization, the peritectic reaction point is reached, and pyroxene (60%) replaces olivine and ilmenite (10%) replaces chromite. Plagioclase is reduced to 30%.

(5) Calculated AFC paths between KREEP (represented by 15386 IKFM) using an r value of 0.22 and the crystallizing phases calculated by major element modeling, show good agreement with observed correlations within the HA mare basalt suite. These indicate that the most evolved HA mare basalts were generated by 60% crystallization of a parental magma (14321,1422) with 13.2% KREEP assimilation.

(6) Short-range unmixing cannot generate the wide range of trace element compositions observed in these basalts.

(7) Ilmenite ferrobasalt cannot be generated within the confines of our model. However, a modified AFC process can account for this composition. The possibility that this rock-type has a petrogenesis unrelated to other Apollo 14 HA basalts cannot be ruled out.

(8) It is envisaged that AFC occurred in the lower lunar crust beneath the Apollo 14 site. Periodic eruptions of mare basalts then occurred that produced the variety of compositions seen in the Apollo 14 breccias.

Acknowledgments. This study would not have been possible without the capable assistance of the Planetary Materials Curatorial Staff at Johnson Space Center, especially Ms. Kim Willis. The sawing of the breccias, the elaborate documentation, and the actual plucking were performed by Kim. We thank Randy Korotev at Washington University for assistance in INA analyses, and Tom See at JSC for help in the fusion process of the basalts. Acknowledgments are due to John W. Shervais, Randy Korotev, and John Jones for constructive reviews of this paper. The research reported in this paper was supported by NASA Grants NGR 9-62 to L. A. Taylor and NAG 9-56 to L. A. Haskin.

REFERENCES

Arth J. G. and Hanson G. N. (1975) Geochemistry and origin of the early Pre-Cambrian crust of northeastern Minnesota. *Geochim. Cosmochim. Acta*, 39, 325-362.

Beaty D. W., Hill S. M. R., Albee A. L., Ma M.-S., and Schmitt R. A. (1979) The petrology and geochemistry of basaltic fragments from the Apollo 11 soil, Part I. *Proc. Lunar Planet. Sci. Conf. 10th*, pp. 41-75.

Bence A. E. and Papike J. J. (1972) Pyroxenes as recorders of lunar basalt petrogenesis: Chemical trends due to crystal-liquid interactions. *Proc. Lunar Sci. Conf. 3rd*, pp. 431-469.

Binder A. B. (1982) The Mare Basalt magma source region and Mare Basalt magma genesis. *Proc. Lunar Planet. Sci. Conf. 13th*, in *J. Geophys. Res.*, 87, A37-A53.

Binder A. B. (1985) Mare Basalt genesis: Modeling trace elements and isotopic ratios. *Proc. Lunar Planet. Sci. Conf. 16th*, in *J. Geophys. Res.*, 90, D19-D30.

DePaolo D. J. (1981) Trace element and isotopic effects of combined wallrock assimilation and fractional crystallization. *Earth Planet. Sci. Lett.*, 53, 189-202.

Dickinson T., Taylor G. J., Keil K., Schmitt R. A., Hughes S. S., and Smith M. R. (1985) Apollo 14 aluminous mare basalts and their possible relationship to KREEP. *Proc. Lunar Planet. Sci. Conf. 15th*, in *J. Geophys. Res.*, 90, C365-C374.

Drake M. J. and Weill D. F. (1975) Partition of Sr, Ba, Ca, Y, Eu^{2+} and Eu^{3+}, and other REE between plagioclase feldspar and magmatic liquid: An experimental study. *Geochim. Cosmochim. Acta*, 39, 689-712.

El Goresy A., Taylor L. A., and Ramdohr P. (1972) Fra Mauro crystalline rocks: Mineralogy, geochemistry, and subsolidus reduction of opaque minerals. *Proc. Lunar Sci. Conf. 3rd*, pp. 333-348.

Green D. H., Ware N. G., Hibberson W. O., and Major A. (1971) Experimental petrology of Apollo 12 basalts, I. Sample 12009. *Earth Planet. Sci. Lett.*, 13, 85-96.

Haskin L. A. and Korotev R. L. (1977) Test of a model for trace element partition during closed-system solidification of a silicate liquid. *Geochim. Cosmochim. Acta*, 41, 921-939.

Irving A. J. and Frey F. A. (1984) Trace element abundances in megacrysts and their host basalts: Constraints on partition coefficients and megacryst genesis. *Geochim. Cosmochim. Acta*, 48, 1201-1221.

Kesson S. E. (1975) Mare basalts: Melting experiments and petrogenetic interpretations. *Proc. Lunar Sci. Conf. 6th*, pp. 921-944.

Lindsley D. H. and Anderson D. J. (1983) A two pyroxene thermometer. *Proc. Lunar Planet. Sci. Conf. 13th*, in *J. Geophys. Res.*, 88, A887-A906.

Lindstrom M. M. (1984) Alkali gabbronorite, ultra-KREEPy melt rock, and the diverse suite of clasts in North Ray Crater feldspathic fragmental breccia 67975. *Proc. Lunar Planet. Sci. Conf. 15th*, in *J. Geophys. Res.*, 89, C50-C62.

Lindstrom M. M. and Haskin L. A. (1978) Causes of compositional variations within mare basalt suites. *Proc. Lunar Planet. Sci. Conf. 9th*, pp. 465-486.

Lindstrom D. J. and Korotev R. L. (1982) TEABAGS: Computer programs for instrumental neutron activation analysis. *J. Radioanal. Chem.*, 70, 439-458.

Longhi J. (1977) Magma oceanography 2: Chemical evolution and crustal formation. *Proc. Lunar Sci. Conf. 8th*, pp. 601-621.

McKay G. A., Wagstaff J., and Yang S.-R. (1986) Zr, Hf, and REE partition co-efficients for ilmenite and other minerals in high-Ti lunar mare basalts: An experimental study. *Proc. Lunar Planet. Sci. Conf. 16th*, in *J. Geophys. Res.*, 91, D229-D237.

Nyquist L. E., Wooden J. L., Shih C.-Y., Weismann H., and Bansal B. M. (1981) Isotopic and REE studies of lunar basalt 12038: Implications for petrogenesis of aluminous basalts. *Earth Planet. Sci. Lett.*, 55, 335-355.

Ringwood A. E. and Kesson S. E. (1976) A dynamic model for mare basalt petrogenesis. *Proc. Lunar Sci. Conf. 7th*, pp. 1697-1722.

Rhodes J. M. and Hubbard N. J. (1973) Chemistry, classification, and petrogenesis of Apollo 15 mare basalts. *Proc. Lunar Sci. Conf. 4th*, pp. 1127-1148.

Ryder G., Norman M. D., and Score R. A. (1980) The distinction of pristine from meteorite-contaminated highland rocks using metal compositions. *Proc. Lunar Planet. Sci. Conf. 11th*, pp. 471-480.

Schnetzler C. C. and Philpotts J. A. (1970) Partition coefficients of REE between igneous matrix material and rock-forming mineral phenocrysts - II. *Geochim. Cosmochim. Acta*, 34, 331-340.

Shervais J. W., Taylor L. A., and Laul J. C. (1983) Ancient crustal components in the Fra Mauro breccias. *Proc. Lunar Planet. Sci. Conf. 14th*, in *J. Geophys. Res.*, 88, B177-B192.

Shervais J. W., Taylor L. A., and Laul J. C. (1984a) Very high potassium (VHK) basalt: A new type of aluminous mare basalt from Apollo 14 (abstract). In *Lunar and Planetary Science XV*, pp. 768-769. Lunar and Planetary Institute, Houston.

Shervais J. W., Taylor L. A., Laul J. C., and Smith M. R. (1984b) Pristine highland clasts in consortium breccia 14305: Petrology and geochemistry. *Proc. Lunar Planet. Sci. Conf. 15th*, in *J. Geophys. Res.*, 89, C25-C40.

Shervais J. W., Taylor L. A., and Lindstrom M. M. (1985) Apollo 14 mare basalts: Petrology and geochemistry of clasts from consortium breccia 14321. *Proc. Lunar Planet. Sci. Conf. 15th*, in *J. Geophys. Res.*, 90, C375-C395.

Smith J. V., Anderson A. T., Newton R. C., Olson E. J., Wyllie P. J., Crewe A. V., Isaacson M. S., and Johnson D. (1970) Petrologic history of the moon inferred from petrography, mineralogy, and petrogenesis of Apollo 11 rocks. *Proc. Apollo 11 Lunar Sci. Conf.*, 897-925.

Taylor L. A., Shervais J. W., Hunter R. H., Shih C.-Y., Nyquist L., Bansal B., Wooden J., and Laul J. C. (1983) Pre-4.2 AE mare basalt volcanism in the lunar highlands. *Earth Planet. Sci. Lett.*, 66, 33-47.

Taylor S. R. and Jakes P. (1974) The geochemical evolution of the Moon. *Proc. Lunar Sci. Conf. 5th*, pp. 1287-1305.

Usselman T. M., Lofgren G. E., Donaldson C. H., and Williams R. J. (1975) Experimentally reproduced textures and mineral chemistries of high-titanium mare basalts. *Proc. Lunar Sci. Conf. 6th*, pp. 997-1020.

Vaniman D. T. and Papike J. J. (1980) Lunar highland melt rocks: Chemistry, petrology, and silicate mineralogy. *Proc. Conf. Lunar Highlands Crust*, 271-337.

Villement B., Jaffrezic H., Joron J. L., and Treuil M. (1981) Distribution coefficients of major and trace elements: Fractional crystallization in the alkai basalt series of Chaine des Puys (Massif Central, France). *Geochim. Cosmochim. Acta*, 45, 1997-2016.

Walker D., Longhi J., and Hays J. R. (1972) Experimental petrology and origin of Fra Mauro rocks and soil. *Proc. Lunar Sci. Conf. 3rd*, pp. 797-817.

Walker D., Longhi J., Grove T. L., Stolper E., and Hays J. F. (1973) Experimental petrology and origin of rocks from the Descartes Highlands. *Proc. Lunar Sci. Conf. 4th*, pp. 1013-1032.

Warren P. H. (1985) The magma ocean concept and lunar evolution. *Ann. Rev. Earth Planet. Sci.*, 13, 201-240.

Warren P. H. and Warren J. T. (1979a) The origin of KREEP. *Rev. Geophys. Space Phys.*, 17, 2051-2083.

Warren P. H. and Wasson J. T. (1979b) Effects of pressure on the crystallization of a "chondritic" magma ocean and implications for the bulk composition of the Moon. *Proc. Lunar Planet. Sci. Conf. 10th*, 2051-2083.

Warren P. H., Taylor G. J., Keil K., Kalleyman G. W., Shirley D. N., and Wasson J. T. (1983) Seventh foray: Whitlockite-rich lithologies, a diopside-bearing troctolitic anorthosite, ferroan anorthosites, and KREEP. *Proc. Lunar Planet. Sci. Conf. 14th*, in *J. Geophys. Res.*, 88, B151-B164.

Materials and Formation of the Imbrium Basin

Paul D. Spudis
Branch of Astrogeology, U.S. Geological Survey, Flagstaff, AZ 86001

B. Ray Hawke and Paul G. Lucey
Planetary Geosciences Division, Hawaii Institute of Geophysics, University of Hawaii, Honolulu, HI 96822

The lunar Imbrium basin has been an important target of geologic investigation for many years. To understand better the nature and effects of the Imbrium impact event, we have made a multidisciplinary study of Imbrium basin deposits, involving geologic mapping, analysis of orbital geochemical data, and Earth-based near-infrared spectroscopy, and analytical modeling. Imbrium basin deposits are widely distributed over the Moon and are subdivided on the basis of morphology. The morphologic facies of Imbrium basin continuous deposits display an unusual (and currently not understood) bilateral symmetry with respect to the main rim of the basin, which we interpret to be composed of the Apennine-Carpathian-Alpes mountain ranges and 1160 km in diamter. The Imbrium ring structure consists of six concentric rings ranging from 550-3200 km in diameter; the large-scale, concentric structural pattern on the lunar nearside is probably related to Imbrium, casting doubt on the reality of the proposed "Procellarum basin." The chemical and petrological composition of Imbrium basin deposits is heterogeneous; the deposits are composed of nearly the entire spectrum of known lunar rock types, from ferroan anorthosites to Mg-suite and KREEP-rich lithologies. Apollo missions 14, 15, and 16 landed within or near mappable Imbrium basin deposits; the best candidates of primary Imbrium ejecta are from the Apollo 15 site, in the form of mafic, LKFM melt breccias that were probably derived from the Imbrium basin impact melt sheet. The Apollo 14 and 16 sample collections probably contain Imbrium primary ejecta that remain unrecognized. The Imbrium basin can be modeled as a proportional-growth crater, resulting in a reconstructed excavation cavity 685 km in diameter, a maximum excavation depth of about 80 km, and a total ejected volume of 12×10^6 km^3. Over 90% of the total ejecta volume is derived from depths of 45 km or less.

INTRODUCTION

The Imbrium basin is a conspicuous feature of the lunar nearside (Fig. 1) and has been the focus of intensive investigation by lunar geologists ever since *G. K. Gilbert* (1893) recognized its impact origin. The deposits around the basin form a widespread stratigraphic datum relative to which other lunar geological units may be dated (*Shoemaker and Hackman*, 1962; *Wilhelms*, 1984). The Imbrium impact is considered such a key event in lunar history that Apollo missions 14 and 15 were sent to landing sites chosen specifically to address problems of Imbrium basin geology.

More significantly, as a relatively well preserved multiring basin, Imbrium provides an important target for geologic investigation of the processes involved in lunar basin formation and evolution. One purpose of this paper is to review current knowledge of the stratigraphy and morphology of the Imbrium basin, the composition of its deposits, and the relations among materials collected at the Apollo highland sites adjacent to Imbrium deposits and other basins. Our second purpose is to summarize what we can infer about the formational mechanics of Imbrium and other lunar basins in general.

STRATIGRAPHY AND MORPHOLOGY OF THE IMBRIUM BASIN

Although Imbrium is significantly modified by mare flooding, parts of the basin interior have not been flooded, and its associated deposits outside the basin are well preserved in many regions. Thus, a reasonably complete reconstruction of Imbrium basin stratigraphy and morphology is possible (Fig. 2) if we keep in mind that several key relations among basin units may be obscured by postbasin geologic units.

Imbrium basin deposits are subdivided into several formal and informal rock-stratigraphic units (*Wilhelms*, 1970) that compose the *Imbrium Group* (provisionally named by *Spudis*, 1986; see also *McCauley et al.*, 1981). The informally named *massif material* is exposed as kipukas protruding through the mare basalts that cover the basin interior (Fig. 2) and as segments forming the main basin rim (Figs. 1, 2, 3a). Additionally, isolated massifs occur outside the main rim (Fig. 2) that are probably related to large, external rings (discussed below). Massifs consist of large mountains that are, in some cases, transitional with large hummocks included within other rock units. At Imbrium no unit of the basin interior is recognized that corresponds to the Orientale basin's Maunder Formation (basin impact-melt sheet; *Head*, 1974a; *McCauley*, 1977). Although it has been proposed that the planar Apennine Bench Formation might be this basin impact-melt unit (*Wilhelms*, 1980a; *McCauley et al.*, 1981), geologic and remote-sensing studies (summarized in *Spudis and Hawke*, 1986) have shown that this material probably consists of postbasin volcanic KREEP basalts. However, isolated pools of cracked material along the base of the Apennine scarp may be Imbrium basin impact-melt deposits (*Wilhelms et al.*, 1977; *Spudis*, 1978; *Wilhelms*, 1980a).

Continuous basin deposits are subdivided on the basis of morphology. The informally named *Apenninus material* (*Wilhelms and McCauley*, 1971) consists of coarsely textured deposits that occur only in the southern Apennines (Fig. 3b).

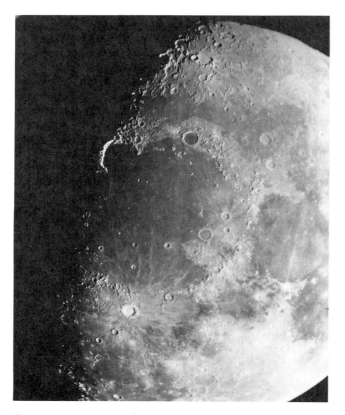

Fig. 1. Basin-centered, telescopic view of the Imbrium basin. Earth-based rectified photograph courtesy of the Lunar and Planetary Laboratory, University of Arizona.

This material appears to be about 1-2 km thick near the Apennine crest (*Carr and El-Baz*, 1971; *Spudis and Head*, 1977); its texture in this area, trending concentrically with the Imbrium rim, suggests that postbasin slumping may be responsible for its morphology (*Spudis and Head*, 1977). The Apenninus material appears to be gradational with the more widely distributed *Fra Mauro Formation* (Fig. 3b,c; *Eggleton*, 1970; *Wilhelms and McCauley*, 1971). The morphology of this unit ranges from strongly lineated textures (radial to Imbrium) east of Mare Vaporum to a more random, hummocky texture elsewhere (e.g., near the Apollo 14 landing site; Fig. 3c; *Eggleton*, 1970). The Fra Mauro Formation appears to be about 1 km thick in the Apennine backslope and thins to a featheredge nears its distal margins, where it is intercalated with smooth plains material. There is little evidence for flow lobes or melt ponds associated with the Fra Mauro Formation, but it commonly appears to have overridden previously formed secondary impact craters and chains (*Eggleton*, 1970; *Head and Hawke*, 1975), suggesting that some type of ground flow was important in its emplacement.

Another extensive Imbrium stratigraphic unit (Fig. 2) is the knobby-textured *Alpes Formation* (Fig. 3d; *Page*, 1970). The knobs average about 10 km across and occur within a matrix of undulatory terra plains. The Alpes Formation extends locally as far as 800 km from the basin rim (discussed below); such a distribution is consistent with an origin by basin-ejecta deposition. *Head* (1974a,b; 1977) proposed that knobby, Alpes-like basin deposits were produced by movement of ejecta on regional slopes after deposition. On the basis of morphologic similarity, other workers have equated the Imbrium Alpes Formation with the Orientale basin's Montes Rook Formation (*McCauley et al.*, 1981). However, the distribution of the Alpes is much more extensive than that of the Montes Rook which is almost completely confined within the topographic basin; occurrence of the latter beyond the Cordillera rim rarely exceeds 50 km (*McCauley*, 1977). Because of the Alpes' great extent and apparent emplacement at the same time as other Imbrium deposits, we interpret the Alpes Formation to be a facies of Imbrium basin *continuous deposits* (ejecta and locally reworked material). The cause of its distinctive morphology remains unknown.

Figure 2 shows a bilateral symmetry of basin deposit morphologic facies that may be of some importance, although its significance is unknown at present. Fra Mauro Formation/Apenninus material occurs in the northwest and southeast quadrants of the basin deposits; the Alpes Formation is confined primarily to the northeast and southwest quadrants (*Spudis*, 1986). Moreover, sparse geochemical data for Imbrium basin deposits (discussed below) suggest that a real compositional difference exists between the Fra Mauro and Alpes Formations. This distribution of deposit facies is discussed later in relation to the problem of the Imbrium basin ring system.

The outermost of the Imbrium basin deposits consist of the light plains *Cayley Formation* (Fig. 3e; *Morris and Wilhelms*, 1967) and numerous secondary craters (*Wilhelms*, 1980a; *Eggleton*, 1981). The Cayley Formation is gradational with the distal margins of both Fra Mauro and Alpes materials. Results of the Apollo 16 mission and subsequent cratering studies (e.g., *Oberbeck*, 1975) suggest that the Cayley plains were emplaced contemporaneously with deposition of basin ejecta by a "debris surge" that closely followed the edge of the ballistic ejecta curtain. The Cayley Formation is widely distributed as fill deposits in primary craters and in large-basin secondary craters (only regional exposures are shown in Fig. 2). The fraction of primary Imbrium ejecta present in these plains is still under debate (cf. *Oberbeck*, 1975, and *Schultz and Gault*, 1985).

RING SYSTEM OF THE IMBRIUM BASIN

The Imbrium basin displays one of the most complex ring systems of any lunar multiring basin. Controversy has arisen over even the location of the main basin ring, a parameter critical to modeling basin excavation cavities and ejecta volumes. It was first proposed (*Hartmann and Kuiper*, 1962) that the Apennine and Carpathian Mountains represent the main rim of the Imbrium basin in the south-southeast, but that the rim is represented by a trough (Mare Frigoris) north of the basin. *Wilhelms and McCauley* (1971) suggested that the north shore of Mare Frigoris is the basin-rim remnant in the north. Both *Dence* (1976) and *Wilhelms et al.* (1977) decreased the main rim diameter of Imbrium by suggesting that the Alpes-Iridum terra block might represent the north portion of the basin rim. *Whitford-Stark* (1981) agreed with the assignment

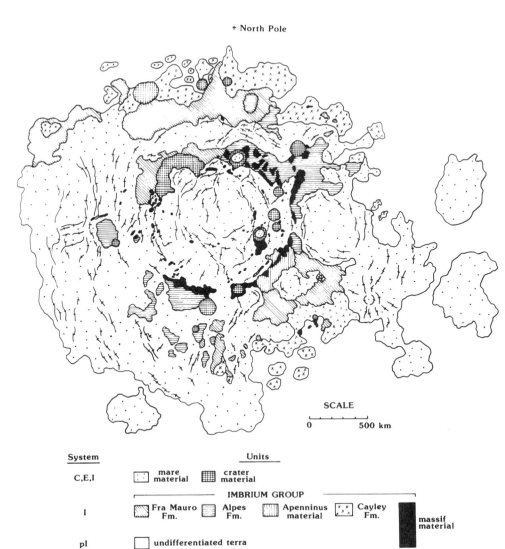

Fig. 2. Geologic sketch map of Imbrium basin deposits, adapted from *Wilhelms and McCauley* (1971), *Scott et al.* (1977), *Lucchitta* (1978), *Eggleton* (1981), and new mapping in progress. Secondary crater field omitted for clarity. Base is a Lambert Equal-area projection centered on 35°N, 17°W (Imbrium basin center; see text).

of the Alpes mountains as the main rim; however, he suggested that it is not in its original position, but is displaced laterally inward by horizontal crustal movements postdating the basin impact. *Wilhelms* (1980b) proposed that the Imbrium rim is in fact a "split" ring; in this model, both the Alpes-Iridum block and the north shore of Frigoris represent the main basin rim.

The problem of the location and size of the main rim of the Imbrium basin is closely related to the question of the complete ring configuration of the basin. It has long been recognized (e.g., *Wilhelms and McCauley*, 1971; *Mason et al.*, 1976) that Imbrium is situated within a large, regional, concentric pattern of scarps, massif lines, and mare shorelines that extends over nearly the entire lunar nearside. The initial interpretation of this relation was that this structural pattern represents Imbrium basin rings external to the main rim, as observed and mapped around many other lunar basins (*Hartmann and Kuiper*, 1962; *Wilhelms and McCauley*, 1971).

An unexpected result of the Apollo missions was the discovery of the concentration of KREEP in the Oceanus Procellarum region of the lunar nearside. To explain this unusual concentration, *Cadogan* (1974) hypothesized that the western shore of Oceanus Procellarum is a remnant of a gigantic, ancient basin ("Gargantuan" basin) that underlies the Imbrium basin; this ancient basin stripped off the original anorthositic crust in this region and was subsequently resurfaced by volcanic KREEP basalts. This model was developed in detail by *Whitaker* (1981), who concluded that a large regional concentric pattern is present on the lunar nearside and that the Imbrium basin is eccentrically displaced from the center of this regional concentric pattern. Whitaker named this postulated ancient structure the "Procellarum" basin. Apparently, he did not advocate a specific configuration for the Imbrium main rim, but his postulated center for Imbrium (near 37°N, 17°W) closely corresponds with the Imbrium configuration proposed by *Wilhelms and McCauley* (1971; basin center 36°N, 18°W; main rim diameter 1500 km).

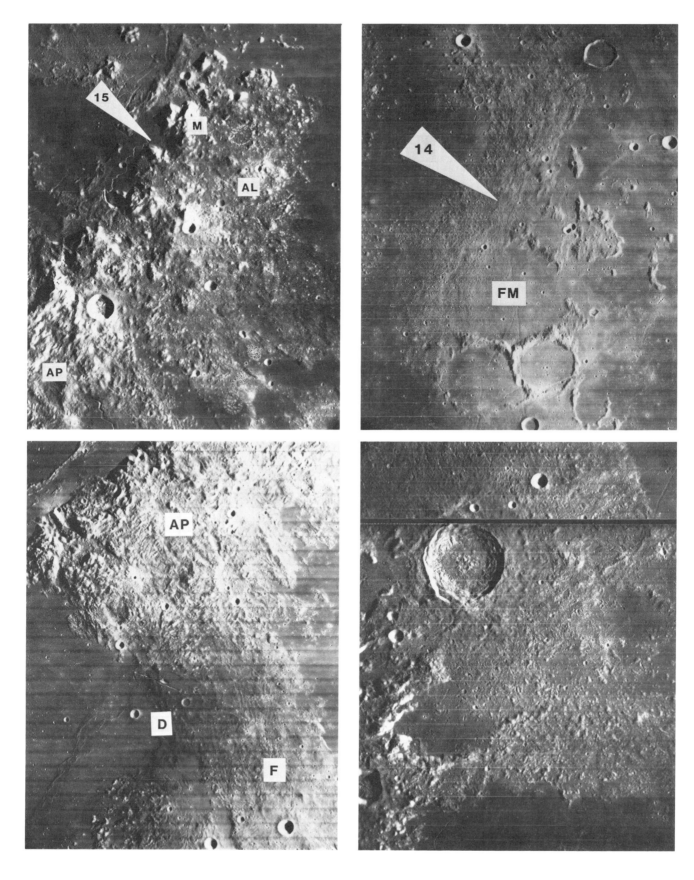

In our opinion, none of these configurations for the Imbrium ring system are completely satisfactory. Although major departures in circularity are associated with many multiring basins, any ring model for Imbrium must account for the major concentric structural patterns and landforms. Such an interpretation of the Imbrium ring structure is shown in Fig. 4; the main topographic ring, 1160 km in diameter, consists of the Carpathian and Apennine mountains in the south and southeast, and the southern Caucasus, Alpes and Iridum rim to the north and west. This arrangement, which we favor, was first proposed by *Dence* (1976) and later by *Wilhelms et al.* (1977); it avoids eccentric, egg-shaped Imbrium main rims (*Wilhelms and McCauley,* 1971) and forms a nearly perfect

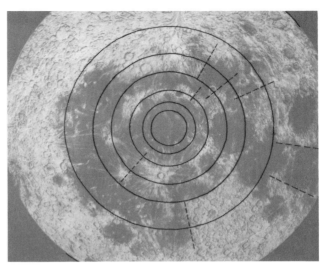

Fig. 4. Ring map of the Imbrium basin, showing six concentric rings of diameters 550, 790, 1160, 1700, 2250, and 3200 km. Major radial megastructures are indicated by dashed lines. Base is a Lambert Equal-area projection centered on 35°N, 17°W (Imbrium basin center).

Fig. 3. Geologic units of the Imbrium Group. North at top and framelet width 12 km in each. **(a)** Northern Apennines, showing massifs (M), the knobby Alpes Formation (AL) and coarsely textured Apenninus material (AP). Apollo 15 landing site shown by arrow. LO IV-102H$_1$. **(b)** Southern Apennines. Apenninus material (AP) gradational with hummocky Fra Mauro Formation (F). Dark material (D) is postbasin, mare pyroclastic mantling deposits. LO IV-109H$_2$. **(c)** The hummocky Fra Mauro Formation near its type area, the pre-Imbrian crater Fra Mauro (FM; 95 km diameter). Apollo 14 landing site shown by arrow. LO IV-120H$_3$. **(d)** The knobby Alpes Formation northeast of the Imbrium basin rim. Large crater is Eudoxus (67 km dia.) LO IV-98H$_2$. **(e)** The light-plains Cayley Formation (C) in the central lunar highlands. The large filled crater is Flammarion (75 km diameter) of pre-Imbrian age. It is cut by gouges (arrows) that make up the "Imbrium sculpture" (basin secondary craters). LO IV-108H$_3$.

circle centered at 35°N, 17°W. If this interpretation is correct, then mare ridge and massif patterns (Fig. 2) would define five additional Imbrium basin rings, two inside the topographic rim and three outside (Fig. 4). Although the outer rings have been suggested by some workers to represent an older megabasin ("Procellarum"; *Whitaker,* 1981; *Wilhelms,* 1983), at least part of the reason for the postulation of this basin was the eccentric position of Imbrium within a large regional concentric pattern (*Whitaker,* 1981). By redrawing the main Imbrium ring as shown (Fig. 4), this pattern is easily explained as Imbrium related (*Schultz and Spudis,* 1985). In this interpretation, Imbrium has six rings ranging from 550 to 3200 km in diameter. As such, it would be the largest multiring basin on the Moon.

Another implication can be derived from drawing the main Imbrium basin ring as shown in Fig. 4. *Wilhelms* (1980b, 1984) suggested that Imbrium displays an asymmetrical ejecta pattern, having ejecta of greater radial extent southeast of the basin. This supposed asymmetry of deposit radial extent has been used to infer that Imbrium was formed by a very low angle (oblique) impact (*Wilhelms,* 1980b, 1984). In its extreme form, this model suggests the orbital decay and impact of an ancient lunar satellite (*Runcorn,* 1982). However, comparison of the geologic map (Fig. 2) and the ring map (Fig. 4) indicates that this supposed ejecta radial-extent asymmetry does not exist; in fact, Imbrium basin deposits are of roughly equal radial extent around the basin rim (about a basin radius, 500-600 km) and Imbrium basin deposits display a bilateral symmetry of deposit morphological facies (Fig. 2). Imbrium basin deposits are symmetrical with respect to a main basin rim 1160 km in diameter centered at 35°N, 17°W. We suggest that Imbrium is a "normal" impact basin, displaying ejecta radial-extent symmetry and moderately coherent ring organization.

REGIONAL COMPOSITION OF IMBRIUM DEPOSITS

Orbital Geochemical Data

Orbital geochemical data for Imbrium basin deposits are few and are confined to the Apennines and Central Highlands/Fra Mauro region. The large field of view of the chemical detectors results in the inclusion of many postbasin materials, such as mare basalts and pyroclastic deposits, in regional compositions. Even so, these data are our primary source of information about variations in the regional chemical composition of Imbrium basin ejecta (Table 1).

TABLE 1. Regional chemical composition of the Apennines*.

	North (Alpes Formation)	South (Apenninus Material)
TiO_2	2.5%	3.5%
Al_2O_3	24.5%	27.5%
FeO	11.0%	11.0%
MgO	9.0%	8.0%
Th	3.0 ppm	7.6 ppm

*Fe and Ti data from *Davis* (1980); Al and Mg from *Clark and Hawke* (1981); Th from *Metzger et al.* (1979).

Geochemical mixing-model results for Imbrium basin deposits are presented in Table 2; models for the Apennines were made for this study and results for the Fra Mauro and Ptolemaeus regions are from *Hawke and Head* (1978). Tables 1 and 2 show that the composition of Imbrium deposits is regionally variable. Within the continuous deposits, the northern Apennines are dominated by Alpes Formation material (Fig. 3a) and the southern Apennines consist of Apenninus (Fra Mauro) material (Fig. 3b). These two units appear to be compositionally different (Table 2). The Alpes Formation is dominated by norite with minor quantities of KREEP. The Apenninus material is more KREEP-rich at the expense of norite. Both units show substantial quantities of anorthosite and mare

TABLE 2. Geochemical mixing-model results for Imbrium basin deposits by region.

Component*	Continuous deposits		Discontinuous deposits	
	N.Apennines	S.Apennines	Fra Mauro†	Ptolemaeus†
Anorthosite	21%	30%	-	9%
Anorthositic gabbro	-	-	5%	41%
Norite	42%	19%	-	-
A14 KREEP	9%	30%	82%	50%
Mare basalt‡	28%	21%	13%	-

*End members defined in *Taylor* (1975) and *Spudis and Hawke* (1981).
†Results of *Hawke and Head* (1978).
‡Apollo 11 high-Ti basalt used in continuous deposits; Apollo 12 low-Ti basalt used by *Hawke and Head* (1978) for Fra Mauro region.

basalt; most of the mare basalt component is probably caused by discontinuous dark-mantle deposits (postbasin basaltic pyroclastics) in the Apennines (*Hawke et al.*, 1979). The Imbrium basin discontinuous deposits (Table 2) are even more diverse, probably as a result of local mixing dominating the distal ends of the Imbrium ejecta blanket (*Oberbeck*, 1975). Even so, the dominance of KREEP in the Fra Mauro and Ptolemaeus regions and in its presence in the southern Apennines (Table 2) suggests that at least some of this component may be related to Imbrium primary ejecta.

These geochemical results for Imbrium deposits suggest that they are composed primarily of norite, with subequal amounts of KREEP and anorthosite. At least some mare basalt is also present in the basin deposits (*Davis and Spudis*, 1985), but the amount is difficult to quantify from the orbital data. The composition of the Imbrium basin deposits contrasts with those of most lunar basin ejecta blankets, where anorthositic material dominates over other rock types (*Spudis et al.*, 1984b). However, Imbrium deposits are not as noritic as Serenitatis basin ejecta (*Spudis and Hawke*, 1981), where anorthosite is virtually absent.

Spectral Reflectance Studies of the Hadley-Apennine Region

Detailed information concerning the mineralogical composition of the surface material in the Hadley-Apennine region is necessary in order to understand the nature and origin of Imbrium basin-related units. The purpose of this section is to present the results of a detailed analysis of near-infrared spectra obtained for a variety of surface units in this area.

Observational methods. Near-infrared reflectance spectra were obtained with the Planetary Geosciences Division indium antimonide spectrometer at the Mauna Kea Observatory 2.24

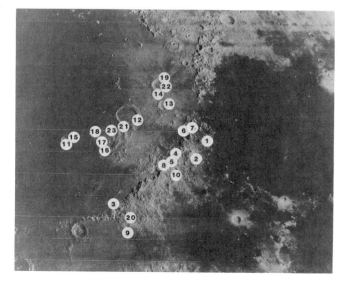

Fig. 5. Regional view of the Apennines showing the location of near-infrared spectra analyzed for this study. Numbers are keyed to locations listed in Table 3.

TABLE 3. Spectral characteristics of the Apennine region, Imbrium basin.

Number*	Location	Class†	Band Minimum	Depth	Width	Cont.
1	Joy	1	0.938	13.20	0.234	0.604
2	Aratus		0.930	9.95	0.226	0.564
3	Apennine Front 10		0.933	9.51	0.196	0.576
4	Apennine Front 11		0.928	10.96	0.234	0.533
5	Apennine Front 2		0.920	11.17	0.218	0.525
6	Hadley Mountain	2	0.934	8.03	0.259	0.566
7	Hadley East		0.941	7.25	0.239	0.596
8	Apennine Front 1		0.922	6.02	0.264	0.578
9	Marco Polo C		0.924	5.17	0.278	0.593
10	Conon South Wall		0.933	5.67	0.242	0.618
16	Apennine Bench 1	3	0.953	5.87	0.331	0.486
17	Apennine Bench 2		0.950	5.39	0.314	0.512
18	Beer		0.961	6.08	0.328	0.509
11	Timocharis East Wall	4	0.953	5.15	0.293	0.586
12	Archimedes East Wall		0.949	7.62	0.295	0.664
13	Autolycus East Wall		0.941	6.69	0.287	0.636
14	Aristillus South Rim		0.968	5.50	0.302	0.565
15	Timocharis East Ejecta		0.966	8.36	0.296	0.611
19	Aristillus Wall dark streak	5	0.962	5.89	0.369	0.620
20	Marco Polo F		0.957	4.24	0.369	0.583
21	Archimedes South Rim		0.960	4.14	0.379	0.764
22	Aristillus East Wall		0.944	5.60	0.281	0.517
23	Bancroft		0.950	7.63	0.280	0.532

*Number refers to spot number on index map (Fig. 5).
†See Fig. 6 for representative appearance of spectral classes.

m telescope. This instrument successively measures intensity in each of 120 wavelength channels covering the 0.6-to 2.5-m region by rotating a filter with a continuously variable band pass. In the mode used for these observations, each scan lasted 90 seconds, and two successive scans were added before the spectrum was recorded on tape. The lunar spots were observed two or three times. Each final spectrum represents the average of four to six individual measurements. The observations were made with a 2.3 arc sec aperture that outlined a 4.5 km diameter area on the lunar surface.

Frequent measurements were made of the Apollo 16 reflectance standard area during the course of each evening's observations. These standard area observations were used to monitor the atmospheric extinction throughout each night. Extinction corrections were made using the methods described by *McCord and Clark* (1979) and the interactive computer program described by *Clark* (1979). This procedure produces spectra representing the reflectance ratio between the observed area and the Apollo 16 site. These relative spectra were converted to absolute reflectance by means of the reflectance curve of an Apollo 16 regolith sample. Analyses of the spectra band parameters and continuum slopes were made using the Gaussian band-fitting technique and other methods described by *McCord et al.* (1981). Mineralogical and petrological interpretations were made on the basis of previous results of many workers (e.g., *Adams*, 1975; *Pieters*, 1983; *Singer*, 1981; *Singer and Blake*, 1983).

Twenty-three near-infrared spectra were obtained for various features in the Hadley-Apennine region (Fig. 5). Six spectra were collected for relatively fresh surfaces in the Apennine Mountains (Apennine front) and five spectra were obtained for fresh impact crater deposits on the Imbrium backslope. Twelve spectra were obtained for highland units within the Apennine ring (Imbrium interior) at the following locations: two fresh, small (diameter <5 km) craters on the Apennine Bench, Archimedes, Aristillus, Autolycus, and Timocharis craters (eastern wall deposits), the south rims of Archimedes and Aristillus, the dark streak on the wall of Aristillus, Timocharis (eastern ejecta deposits), Bancroft crater, and Beer crater.

Results. The spectral band parameters and continuum slopes were determined for all spectra (Table 3). When plotted on variation diagrams (discussed below), the various spectral parameters form clusters and trends of data points that indicate the presence of distinct spectral classes. There are four distinct

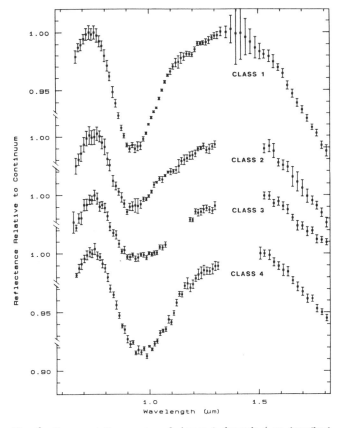

Fig. 6. Representative spectra of classes 1 through 4 as described in text, with a straight line continuum removed. Locations are: class 1-Aratus; class 2-Apennine Front 1; class 3-Apennine Bench 2; class 4-eastern ejecta of Timocharis crater.

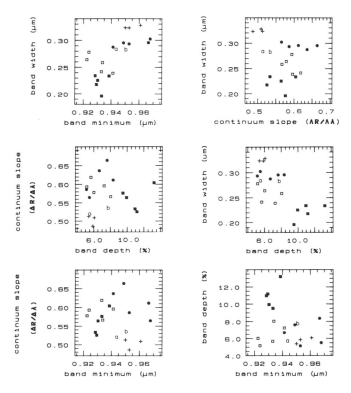

Fig. 7. Variation diagrams of the six combinations of the four spectral parameters listed in Table 1. Filled squares are class 1, open squares are class 2, crosses are class 3, filled circles are class 4. **(a)** and **(b)** represent the east wall of the crater Aristillus and the crater Bancroft respectively. These locations appear to be mixtures of Apennine Bench material and either class 2 or class 4 material.

spectral types or classes in the area studied; we believe that these spectral types represent compositions of regional extent because of the correlation of spectral classes with specific geologic units in and around the Imbrium basin. Representative spectra (after continuum removal) for each class are shown in Fig. 6. In addition, there are five spectra that do not clearly fall into any of the major classes. In some cases, this may be due to the presence of mixing between the classes. In others, the presence of pyroclastic glass, impact glass, or olivine may be indicated.

As a group, the spectra for features in the Apennine Mountains and on the backslope (Imbrium exterior) exhibit many similar characteristics (nos. 1-10, Fig. 5). All exhibit "one micron" band centers ranging between 0.92 and 0.94 microns, typical highlands continuum slopes, and a variety of "one micron" band strengths. However, two distinct classes can be identified. Class 1 spectra (filled squares in Fig. 7) exhibit very deep (9.5-13%) bands, narrow band widths (0.20-0.23 μm), and band centers ranging between 0.92 and 0.94 μm. These characteristics indicate a feldspar-bearing mineral assemblage dominated by orthopyroxene. While class 2 spectra (open squares in Fig. 7) exhibit a similar range of band centers, they have shallower band depths (5-8%) and moderate band widths (0.24-0.28 μm). *Hawke and Lucey* (1984) presented evidence that the differences between class 1 and class 2 spectra are largely due to maturation effects (i.e., a decrease in grain size and an increase in agglutinatic glass content). Additional analysis has indicated that class 2 spectra represent assemblages with slightly less modal orthopyroxene and higher modal feldspar than class 1. Still, in some instances, maturation effects or the presence of impact glass may be the sole difference between the two classes. In terms of lunar rock types, class 1 mineral assemblages correspond to norites and many class 2 assemblages are anorthositic norites.

Class 1 spectra (nos. 1-5 on Fig. 5) include those obtained for Joy and Aratus craters as well as for three separate areas on steep slopes of the Apennine front. Class 2 (nos. 6-10 on Fig. 5) comprises the spectra for the south wall of Conon crater, Marco Polo C crater, and three separate areas in the Apennine Mountains including Hadley Mountain. The spectrum of one small crater (Marco Polo F; 4 km dia.) does not fit into any of the four major classes. The characteristics of this spectrum are consistent with the presence of a relatively large amount of olivine, but an Fe-bearing glass component cannot be ruled out. This spectrum is not of high quality and the observation will be repeated in the near future.

Spectra for features inside the Apennine ring (Imbrium interior) exhibit both a longer "one micron" absorption band center as well as a greater band width than spectra for areas on the Apennine front and backslope. This suggests that nonmare units on the Imbrium interior have more calcic pyroxene assemblages than do the Apennine Mountain and backslope units. Two major spectral classes were identified within the Imbrium interior (classes 3 and 4; Fig. 5; Table 3). Class 3 spectra (crosses in Fig. 7) exhibit moderate (5-6%) band depths, moderate band widths (~0.32 μm), and band minima that range between 0.95 and 0.96 μm. The widths, centers, and overall shapes of the "one micron" bands of members of this class indicate a mixture of orthopyroxene with one or more components that have a longer band center. Candidates include Fe-bearing glass and clinopyroxene. All three of the spectra in the class are for craters that have excavated Apennine Bench material. The Apennine Bench has been shown to have a surface chemical composition nearly identical to the Apollo 15 KREEP basalts (*Spudis and Hawke*, 1986). These KREEP basalts contain clinopyroxene (pigeonite) as well as intersertal glass. Hence, we suggest that class 3 spectra are dominated by KREEP basalt.

The spectrum of the eastern ejecta of Timocharis crater is shown in Fig. 6 as representative of class 4 (nos. 11-15 in Fig. 5). Spectra in this class (filled circles in Fig. 7) exhibit moderate (5-8.3%) band depths, moderate band widths (0.29-0.30 μm) and band minima between 0.94 and 0.97 μm. We interpret these spectra to represent feldspar-bearing assemblages with a mixture of subequal amounts of orthopyroxene and high-Ca clinopyroxene. These assemblages correspond to noritic gabbros or gabbroic norites. The contamination of a highland composition with a mare basalt component cannot be ruled out in every case.

Two minor classes were defined for the purposes of this study. Class 5 contains three spectra characterized by broad, relatively shallow "one micron" bands. However, we do not think that these spectra are genetically related. It was suggested above that Marco Polo F (no. 20 in Fig. 5) may expose olivine-rich material. On the basis of spectral and radar data, *Lucey et al.* (1984) interpreted the Archimedes south rim deposits to contain a component of pyroclastic glass. *Smrekar and Pieters* (1985) interpreted the Aristillus dark streak to consist of pyroxene and Fe-bearing feldspar, primarily in the form of lithic clasts and recrystallized melt, as well as significant amounts of Fe-bearing impact melt glass.

Class 6 contains the spectra for the east wall of Aristillus crater and Bancroft crater ("a" and "b" in Fig. 7). These areas appear to be mixtures of Apennine Bench material with either class 4 or class 2 material.

Summary of remotely sensed compositional data. Results of studies of both orbital geochemical information and Earth-based spectral data suggest that Imbrium basin deposits are compositionally heterogeneous. The observed diversity of compositions around Imbrium is in marked contrast to our previous results for the deposits of the Orientale basin (*Spudis et al.*, 1984b); that study found a dominance of anorthositic rocks with a restricted range of compositions (anorthositic norite to noritic anorthosites). Imbrium, however, displays almost the entire spectrum of known lunar highland rock types, including anorthositic norites, norites and gabbronorites, KREEP basalts and low-K Fra Mauro (LKFM) basalts, and possibly ultramafic, olivine-bearing rocks of the Mg-suite and exotic, KREEP-rich granitic rocks. These results are even more striking in that only part of all the currently exposed Imbrium deposits were studied for this effort (the Apennine region), suggesting that the basin deposits of Imbrium are probably complex and heterogeneous throughout their extent. The observations of the compositions of Imbrium basin deposits presented here suggest that the Imbrium target and surrounding regions were petrologically heterogeneous and probably reflect a complex and protracted pre-Imbrian geologic evolution for this part of the Moon. More complete data for a larger fraction of the exposed Imbrium deposits will be acquired and analyzed in future work.

IMBRIUM BASIN DEPOSITS AND APOLLO HIGHLAND SAMPLES

Apennine Massifs—Apollo 15

The Apollo 15 site lies close to the main (Apennine) ring of the Imbrium basin. The highlands were sampled at Hadley Delta, a massif that is part of the main basin ring. Mixing-model results for Imbrium deposits (Table 2; N. Apennines) suggest that noritic rocks dominate this region. This noritic component may be reflected by the LKFM composition of most Apennine Front regolith material (*Spudis and Ryder*, 1985). Anorthosite is present in minor amounts at the Apollo 15 site (*Spudis and Ryder*, 1985), and orbital data suggest that here it may indeed be part of the Imbrium ejecta (Table 2). KREEP is a minor component in Imbrium deposits in this region (Table 2), but geologic evidence suggests that the abundant KREEP at the Apollo 15 site is related mostly to post-Imbrium basin KREEP volcanism (*Spudis and Hawke*, 1986) and not to the Imbrium ejecta.

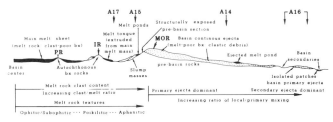

Fig. 8. Inferred geologic relations for lunar basin ejecta deposits. PR is the peak-ring, IR is the intermediate ring, and MOR is the main topographic rim of the basin (see *Croft*, 1981). Relative positions of Apollo landing sites shown at top; "bx" refers to "breccias." Apollo 17 position relative to Serenitatis; all others are shown relative to Imbrium. For details, see *Spudis* (1981).

Because the total expected thickness of Imbrium ejecta at the crest of the Apennines (~1 km; *Carr and El-Baz*, 1971) is less than the 4-km relief of the Apennine scarp, it is likely that pre-Imbrian material was collected at Hadley Delta ("MOR" on Fig. 8). At the Apollo 15 site, this material would consist mostly of Serenitatis ejecta, but because Serenitatis material is dominantly noritic (*Spudis and Hawke*, 1981; *Spudis et al.*, 1984a) it may be difficult to identify purely clastic material. Impact melts may exist that could be identified (*Ryder and Spudis*, 1987), although less total impact melt is expected at this position in the basin than was sampled in the Serenitatis basin by Apollo 17 (Fig. 8). There is probably no coherent "melt sheet" mantling the massifs at the Apollo 15 site, but discontinuous patches and pods of ejected impact melt are probably present and may have been sampled (*Ryder and Bower*, 1977). Pre-Serenitatis material is probably present and may be represented in the samples by small, feldspathic granulitic breccias and rocks derived from them (*Spudis and Ryder*, 1985).

Two of the most important Apollo 15 samples are the so-called "black and white" rocks, 15445 and 15455 (*Ryder and Bower*, 1977; *Ryder and Wood*, 1977). These rocks are aphanitic impact melt breccias that are relatively clast rich. The chemistry of the melt phase of the black and white rocks is mafic, low-K Fra Mauro basalt, considerably more magnesian (MgO = 16%) than the comparable Apollo 17 poikilitic melts (MgO = 11%). The Apollo 15 melts also contain a clast assemblege of deep crustal or upper mantle origin (*Ryder and Bower*, 1977; *Herzberg*, 1978) and include pristine norites, troctolites, and spinel troctolites. Based in part on analogy to the Apollo 17 melt sheet, the Apollo 15 black and white rocks have been interpreted as fragments of the Imbrium basin impact melt sheet (*Ryder and Wood*, 1977; *Ryder and Spudis*, 1987). The Apollo 15 melts are more mafic than the Apollo 17 melts and thus may reflect derivation from deeper within the Moon (*Ryder and Wood*, 1977), probably due to the larger size of the Imbrium basin impact. This interpretation fits the observed chemistry and petrology of the Apollo 15 melts as well as the possible presence of Imbrium basin melt ponds along the margins of the Apennine front.

The Apollo 15 mission may have been successful in attaining the premission goals of sampling both Imbrium basin ejecta and pre-Imbrium rocks. The samples of Imbrium impact melt (interpreted here as represented by 15445 and 15455) extend the petrologic database for lunar basin impact melts and provide direct information on the lower crustal composition in the Imbrium basin target site. The exposed section of pre-Imbrium rocks within the Apennines (*Spudis*, 1980) may be composed mostly of Serenitatis basin ejecta, although some pre-Serenitatis material may be represented by extremely old rocks from the site (e.g., 15415). Thus, Apollo 15 may be the key site in the unraveling of Imbrium basin geology, and the results of this mission provide important constraints on the model for basin development.

Fra Mauro Formation—Apollo 14

The Apollo 14 landing site is located within the continuous deposits of the Imbrium basin (Fig. 3d), near the type area of the Fra Mauro Formation (*Swann et al.*, 1977). The landing site is near a fresh crater (Cone) that presumably was excavated in a ridge of Fra Mauro material below the local regolith. This ridge has been interpreted as an accumulation of Imbrium basin ejecta (*Swann et al.*, 1977) and alternatively, as the rim of a large, local pre-Imbrian crater covered by thin Imbrium deposits (*Hawke and Head*, 1977). The debate over the percentage of locally derived material versus primary basin ejecta in the Fra Mauro Formation hinges on this interpretation. Numerous Imbrium basin secondaries appear to be buried by the continuous Fra Mauro Formation in this region (*Eggleton*, 1970; *Head and Hawke*, 1975). Thus, some local mixing probably occurred (*Morrison and Oberbeck*, 1975; *Oberbeck*, 1975) and the problem involves determining which samples are locally derived.

The most important group of samples from the Apollo 14 site is that collected at the rim of Cone crater, which presumably represents the local Fra Mauro "bedrock." The rocks here are dominated by breccias having a crystalline matrix and an extremely high clast content; these samples have been collectively named "Fra Mauro breccias" (*Wilshire and Jackson*, 1972). They have a bulk composition of KREEP basalt and contain clasts of brecciated feldspathic rocks (Mg-suite), high-alumina basalt, and mare basalt (*Wilshire and Jackson*, 1972; *Duncan et al.*, 1975; *Simonds et al.*, 1977). Petrographically, the breccias are impact melts (*Simonds et al.*, 1977) and appear to have undergone multiple episodes of brecciation (*Wilshire and Jackson*, 1972; *Duncan et al.*, 1975). The shock levels of the Fra Mauro breccias are consistent with their derivation from the "ground-surge" continuous deposits (*Oberbeck*, 1975) emplaced during the Imbrium impact (shock level less than 100 kb; *Austin and Hawke*, 1981). The clast compositions of these breccias, dominated by basalts and reworked breccia clasts, suggest a relatively shallow depth of derivation.

Other rocks collected at the Apollo 14 site have petrography and chemistry consistent with local derivation. A large collection was made of vitric-matrix breccias (*Wilshire and Jackson*, 1972), which are essentially soil breccias and contain Fra Mauro breccia fragments as clasts; these rocks probably were formed by post-Imbrium regolith production on the Fra Mauro Formation. A group of poorly-lithified breccias, informally called the "White Rocks" (*Swann et al.*, 1977), are highly shocked fragmental breccias that may be analogous to the feldspathic fragmental breccias collected at the Apollo 16 site at North Ray crater. If this analogy is correct, the White Rocks are the lithified products of the impact that produced Cone crater.

It is not certain whether the goal of directly sampling Imbrium basin ejecta was accomplished at the Apollo 14 site. The Fra Mauro breccias indicate that the Fra Mauro region was KREEP-rich and may have been extensively resurfaced by KREEP and mare basalt lavas. These data confirm the inferences based on the interpretation of the remote-sensing data that Imbrium basin continuous deposits are petrologically heterogeneous. Although local materials at the Apollo 14 site may dominate the total volume of the deposit, primary basin ejecta within the Fra Mauro continuous deposits probably exist that may have been directly sampled by Apollo 14; such material has not yet been positively identified.

Cayley Formation—Apollo 16

The notion persists that the samples returned from the Apollo 16 landing site may be mostly Imbrium ejecta (*Wilhelms et al.*, 1980; *Hodges and Muehlberger*, 1981). However, there are compelling reasons to believe that most Apollo 16 samples are derived ultimately from Nectaris basin deposits (*James*, 1981; *Spudis*, 1984; *Stöffler et al.*, 1985). If Imbrium basin material exists in the Apollo 16 sample collection, it has not been recognized to date. Mainly on the assumption that KREEP is found on the Moon only in the region of the Imbrium basin, *Wetherill* (1981) postulated that Apollo 16 KREEP, represented mostly by LKFM poikilitic impact melts, is ejected Imbrium basin melt. However, KREEP deposits are found in many highland regions, including the lunar farside (*Spudis*, 1979). Moreover, orbital chemical data show that an LKFM-like chemical component is present within Nectaris basin deposits (*Spudis and Hawke*, 1981), strongly suggesting that some type of LKFM existed in the Nectaris basin target (*Maurer et al.*, 1978; *Spudis and Hawke*, 1981; *Spudis*, 1984). Thus, an Imbrium basin origin for Apollo 16 LKFM is not required, although it is not excluded.

There is little to constrain possible exotic Imbrium basin contributions to the Apollo 16 samples. Anorthosite, found in abundance at the Apollo 16 site, is sparse to absent at the unambiguously Imbrium-related Apollo 14 and 15 sites; moreover, the Apollo 14 anorthosites are more alkalic than the Apollo 16 ferroan anorthosites (*Warren and Taylor*, 1981). The samples at Apollo 16 are probably locally derived.

THE IMBRIUM EXCAVATION CAVITY

Because the Imbrium basin was formed by impact, the problem of reconstructing the impact event is tied to the more general problem of scaling complex craters (*Croft*, 1985). There is wide disagreement among workers in this field as to the original dimensions of basin cavities. Some workers (*Baldwin*, 1974; *Murray*, 1980; *Wilhelms*, 1984) have suggested that the main basin rim (the Apennine ring at

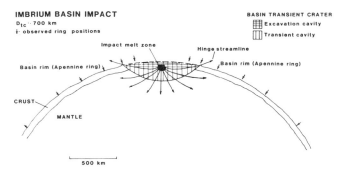

Fig. 9. Scale diagram of the excavation phase of Imbrium basin formation according to the proportional-growth model. Note almost completely crustal provenance of excavated material. Small arrows show location of present Imbrium rings, which did not exist at this stage of basin formation.

Imbrium; 1160 km dia.) corresponds to the rim of the original crater of excavation. In contrast, evidence from terrestrial impact craters (*Grieve et al.,* 1981; *Grieve and Garvin,* 1984), analytical modeling (*Croft,* 1981, 1985), and lunar basins (*Spudis et al.,* 1984b; *Spudis and Davis,* 1985), suggests that the fundamental shape of an excavated cavity is size invariant. The resulting model, called the *proportional-growth model,* has been shown to be valid for crater size ranges from laboratory-scale experimental craters to terrestrial complex craters, such as the Ries (*Stöffler et al.,* 1975; *Hörz et al.,* 1983; *Croft,* 1985). In the absence of any compelling evidence to the contrary, it is assumed to hold for the Imbrium basin in this discussion (Fig. 9).

For a lunar basin with a crater-rim diameter of 1160 km, such as Imbrium, the proportional-growth model suggests excavation-cavity diameters (D_e) ranging from 604 ± 200 km (*Croft,* 1985) to 685 ± 88 km (*Spudis and Davis,* 1985). The relation of *Grieve et al.* (1981), derived from study of terrestrial impact structures, predicts a cavity 667 ± 87 km in diameter. The maximum depth of excavation (d_e) of material from this cavity is related to De as $d_e = 0.09$ to 0.12 D_e (*Grieve et al.,* 1981). Applied to the Imbrium basin ($D_e \sim 685$ km), the maximum depth of crater excavation is on the order of 62 to 82 km. The geometry of the Imbrium excavation cavity is not known, but the "Z-model" (*Maxwell,* 1977; *Croft,* 1981) suggests that a spherical cap segment excavating a spherical moon is a reasonable approximation (*Spudis et al.,* 1984b). Analysis of this geometric figure indicates that the total excavated volume of Imbrium is on the order of 12×10^6 km^3. Moreover, although a crater with such excavation depths would have excavated mantle material (assuming an average crustal thickness of 55 km; see *Bills and Ferrari,* 1976), the mantle ejecta would constitute less than 4% of the total ejected volume. Additionally, 90% of the total ejecta volume would be derived from depths of less than 45 km (Fig. 10).

Thus, the proportional-growth model predicts that the Imbrium basin impact excavated most of the crustal column at its target site; it further predicts that most ejecta were derived from the upper two-thirds of the basin crustal target. This is consistent with what we know from lunar samples (*Taylor,* 1982; *Wilhelms,* 1984). It should be noted that basin-forming models that equate the main rim of the basin with the excavation cavity (*Baldwin,* 1974; *Murray,* 1980; *Wilhelms,* 1984) must invoke mechanisms that produce excavation-cavities that become progressively shallower with increasing crater size to explain the paucity of deep-crustal or mantle material in the lunar samples (e.g., *Head et al.,* 1975). The search for evidence of such mechanisms at basin scales has been inconclusive.

SUMMARY AND CONCLUSIONS

The results of geologic mapping of Imbrium basin deposits, study of the ring configuration and morphology of basin structural elements, analysis of remotely-sensed compositional data, and considerations of cratering models have enabled us to conclude the following about the materials and formation of the Imbrium basin:

1. The target for the Imbrium basin impact had a long and protracted geologic evolution, which included formation of the original ferroan anorthosite crust, intrusion by a complex series of Mg-suite magmas, ancient mare and KREEP volcanism, and a long period of bombardment, resulting in a series of petrologically complex breccias that make up the basin deposits.

2. The Imbrium impact, occurring about 3.85 b.y. ago, formed a basin predominantly by a proportional-growth crater-forming mechanism. This model predicts a transient cavity about 685 km in diameter, material excavated from as deep as 80 km within the Moon, and a total volume of ejecta of about 12×10^6 km^3. The deposits of the Imbrium basin appear to be radially symmetric with respect to extent from the main

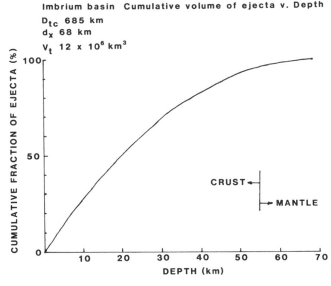

Fig. 10. Cumulative fraction of Imbrium basin ejecta as a function of depth. Over 90 percent of the ejecta is derived from depths of less than 45 km. Crustal thickness estimate from *Bills and Ferrari* (1976).

basin rim; moreover, these deposits display a "bilateral symmetry" of morphological facies distribution that is currently not understood.

3. The Imbrium basin displays six rings, ranging in diameter from 550 km to 3200 km; the main basin rim is 1160 km in diameter. This configuration makes the Imbrium basin the largest multi-ring basin on the Moon and casts doubt on the reality of the proposed "Procellarum (Gargantuan) basin."

4. Three Apollo highland sites are within or near mappable Imbrium basin deposits. Of the samples from these sites, Apollo 15 samples are the most likely to contain *recognizable* Imbrium primary ejecta; this includes the mafic, LKFM melt rocks 15445 and 15455 that are probably fragments of Imbrium basin impact melt. The Apollo 14 and 16 samples are of less certain provenance, but they probably contain primary Imbrium basin ejecta that remain unrecognized.

Acknowledgments. We thank M. Cintala, F. Hörz, and O.B. James, who do not necessarily agree with all of our interpretations, for review of the manuscript. This work is supported by the NASA Planetary Geology and Geophysics Program.

REFERENCES

Adams J. B. (1975) Interpretation of visible and near-infrared diffuse reflectance spectra of pyroxenes and other rock-forming minerals. In *Infrared and Raman Spectroscopy of Lunar and Terrestrial Minerals*, (C. Karr, ed.), pp. 91-116, Academic Press, New York.

Austin M. G. and Hawke B. R. (1981) Tentative speculations on lunar sample transport (abstract). In *Lunar and Planetary Science XII*, pp. 34-36, Lunar and Planetary Institute, Houston.

Baldwin R. B. (1974) On the origin of the mare basins. *Proc. Lunar Sci. Conf. 5th*, 1-10.

Bills B. G. and Ferrari A. J. (1976) Lunar crustal thickness. *Proc. Lunar Sci. Conf. 7th*, Frontispiece.

Cadogan P. H. (1974) Oldest and largest lunar basin? *Nature, 250*, 315-316.

Carr M. H. and El-Baz F. (1971) Geologic map of the Apennine-Hadley region of the Moon. *U.S. Geol. Survey Map I-723*, sheet 1.

Clark P. E. and Hawke B. R. (1981) Compositional variation in the Hadley Apennine region. *Proc. Lunar Planet. Sci. 12B*, 727-749.

Clark R. N. (1979) A large scale interactive one-dimensional array processing system. *Publ. Astron. Soc. Pac., 92*, 221-224.

Croft S. K. (1981) The excavation stage of basin formation: A qualitative model. *Proc. Lunar Planet. Sci. 12A*, 207-225.

Croft S. K. (1985) The scaling of complex craters. *Proc. Lunar Planet. Sci. Conf. 15th*, in *J. Geophys. Res., 90*, C828-C842.

Davis P. A. (1980) Iron and titanium distribution on the moon from orbital gamma-ray spectrometry with implications for crustal evolutionary models. *J. Geophys. Res., 85*, 3209-3224.

Davis P. A. and Spudis P. D. (1985) Petrologic province maps of the lunar highlands derived from orbital geochemical data. *Proc. Lunar Planet. Sci. Conf. 16th*, in *J. Geophys. Res., 90*, D61-D74.

Dence M. R. (1976) Notes toward an impact model for the Imbrium basin. *Interdisciplinary Studies by the Imbrium Consortium, 1*, 147-155.

Duncan A. R., Grieve R. A. F., and Weill D. V. (1975) The life and times of Big Bertha: lunar breccia 14321. *Geochim. Cosmochim. Acta, 39*, 265-273.

Eggleton R. E. (1970) Geologic map of the Fra Mauro region of the Moon. *U.S. Geol. Surv. Misc. Geol. Inv. Map I-708*, sheet 1.

Eggleton R. E. (1981) Map of the impact geology of the Imbrium basin of the Moon. In Geology of the Apollo 16 Area - Central Lunar Highlands. *U.S. Geol. Survey Prof. Paper, 1048*, plate 12.

Gilbert G. K. (1893) The Moon's face. Philos. Soc. Wash. Bull., 12, 241-292.

Grieve R. A. F. and Garvin J. B. (1984) A geometric model for excavation and modification at terrestrial simple impact craters. *J. Geophys. Res., 89*, 11561-11572.

Grieve R. A. F., Robertson P. B., and Dence M. R. (1981) Constraints on the formation of ring impact structures, based on terrestrial data. In *Multi-ring Basins, Proc. Lunar Planet. Sci. 12A*, 37-57.

Hartmann W. K. and Kuiper G. P. (1962) Concentric structures surrounding lunar basins. *Commun. Lunar Planetary Lab. 1*, 55-66.

Hawke B. R. and Head J. W. (1977) Pre-Imbrian history of the Fra Mauro region and Apollo 14 sample provenance. *Proc. Lunar Sci. Conf. 8th*, 2741-2761.

Hawke B. R. and Head J. W. (1978) Lunar KREEP volcanism: Geologic evidence for history and mode of emplacement. *Proc. Lun. Planet. Sci. Conf. 9*, 3285-3309.

Hawke B. R. and Lucey P.G. (1984) Spectral reflectance studies of the Hadley-Apennine (Apollo 15) region: Preliminary results (abstract). In *Lunar and Planetary Science XV*, pp. 352-353.

Hawke B. R., MacLaskey D., McCord T. B., Adams J. B., Head J. W., Pieters C.M., and Zisk S. H. (1979) Multispectral mapping of the Apollo 15-Apennine region: The identification and distribution of regional pyroclastic deposits. *Proc. Lunar Planet. Sci. Conf. 10th*, 2995-3015.

Head J. W. (1974a) Orientale multi-ringed basin interior and implications for the petrogenesis of lunar highland samples. *The Moon, 11*, 327-356.

Head J. W. (1974b) Morphology and structure of the Taurus-Littrow Highlands (Apollo 17): Evidence for their origin and evolution. *The Moon, 9*, 355-395.

Head J. W. (1977) Regional distribution of Imbrium basin deposits: Relationship to pre-Imbrium topography and mode of emplacement. In *Interdisciplinary Studies by the Imbrium Consortium*, LPI Contrib. No. 268D, 120-125.

Head J. W. and Hawke B. R. (1975) Geology of the Apollo 14 region (Fra Mauro): Stratigraphic history and sample provenance. *Proc. Lunar Sci. Conf. 6th*, 2483-2501.

Head J. W., Settle M., and Stein R. S. (1975) Volume of material ejected from major basins and implications for the depth of excavation of lunar samples. *Proc. Lunar Sci. Conf. 6th*, 2805-2829.

Herzberg C. T. (1978) The bearing of spinel cataclasites on the crust-mantle structure of the Moon. *Proc. Lunar Planet. Sci. Conf. 9th*, 319-336.

Hodges C. A. and Muehlberger W. (1981) Summary and critique of geologic hypotheses. In Geology of the Apollo 16 Area, Central Lunar Highlands. *U.S. Geol. Survey Prof. Paper 1048*, 215-230.

Hörz F., Ostertag R., and Rainey D. (1983) Bunte breccia of the Ries: Continuous deposits of large craters. *Rev. Geophys. Space Phys. 21*, 1667-1725.

James O. B. (1981) Petrologic and age relations of the Apollo 16 rocks: Implications for subsurface geology and the age of the Nectaris basin. *Proc. Lunar Planet. Sci. 12B*, 209-233.

Lucchitta B. K. (1978) Geologic map of the north side of the Moon. *U.S. Geol. Survey Map I-1062*.

Lucey P. G., Gaddis L., Bell J., and Hawke B. R. (1984) Near-infrared spectral reflectance studies of localized lunar dark mantle deposits (abstract). In *Lunar and Planetary Science XV*, pp. 495-496. Lunar and Planetary Institute, Houston.

Mason R., Guest J. E., and Cooke G. N. (1976) An Imbrium pattern of graben on the Moon. *Proc. Geol. Assoc. 87*, 161-168.

Maurer P., Eberhardt P., Geiss J., Grogler N., Stettler A., Brown G., Peckett A., and Krähenbühl U. (1978) Pre-Imbrian craters and basins: ages, compositions and excavation depths of Apollo 16 breccias. *Geochim. Cosmochim. Acta, 42,* pp. 1687-1720.

Maxwell D. E. (1977) Simple Z model of cratering, ejection, and overturned flap. In *Impact and Explosion Cratering* (D. J. Roddy, R. O. Pepin, and R. B. Merrill, eds.), pp. 1003-1008, Pergamon, New York.

McCauley J. F. (1977) Orientale and Caloris. *Phys. Earth and Planet. Interiors, 15,* 220-250.

McCauley J. F., Guest J. E., Schaber G. G., Trask N. J., and Greeley R. (1981) Stratigraphy of the Caloris basin, Mercury. *Icarus, 47,* 184-202.

McCord T. B. and Clark R. N. (1979) Atmospheric extinction 0.65-2.50 μm above Mauna Kea. *Publ. Astron. Soc. Pac., 91,* 571-576.

McCord T. B., Clark R. N., Hawke B. R., McFadden L. A., Owensby P. D., Pieters C. M., and Adams J. B. (1981) Near-infrared spectral reflectance, a first good look. *J. Geophys. Res. 86,* 10883-10892.

Metzger A. E., Haines E. L., Etchegaray-Ramierz M. I., and Hawke B. R. (1979) Thorium concentrations in the lunar surface: III. Deconvolution of the Apenninus region. *Proc. Lunar Planet. Sci. Conf. 10th,* 1701-1718.

Morris E. C. and Wilhelms D. E. (1967) Geologic map of the Julius Caesar quadrangle of the Moon. *U.S. Geol. Survey Map I-510.*

Morrison R. H. and Oberbeck V. R. (1975) Geomorphology of crater and basin deposits - emplacement of the Fra Mauro Formation. *Proc. Lunar Sci. Conf. 6th,* 2503-2530.

Murray J. B. (1980) Oscillating peak model of basin and crater formation. *Moon and Planets, 22,* 269-291.

Oberbeck V. R. (1975) The role of ballistic erosion and sedimentation in lunar stratigraphy. *Rev. Geophys. Space Phys., 13,* 337-362.

Page N. J. (1970) Geologic map of the Cassini quadrangle of the Moon. *U.S. Geol. Survey Map I-666.*

Pieters C. M. (1983) Strength of mineral absorption features in the transmitted component of near-infrared reflected light: First results from RELAB. *J. Geophys. Res., 88,* 9534-9544.

Runcorn S. K. (1982) Primeval displacements of the lunar pole. *Phys. Earth Planet. Interiors, 29,* 135-147.

Ryder G. and Bower J. F. (1977) Petrology of Apollo 15 black-and-white rocks 15445 and 15455 - Fragments of the Imbrium impact melt sheet? *Proc. Lunar Sci. Conf. 8th,* 1895-1923.

Ryder G. and Spudis P. D. (1987) Chemical composition and origin of Apollo 15 impact melts. *Proc. Lunar Planet. Sci. Conf. 17th,* in *J. Geophys. Res., 92,* E432-E446.

Ryder G. and Wood J. A. (1977) Serenitatis and Imbrium impact melts: Implications for large-scale layering in the lunar crust. *Proc. Lunar Sci. Conf. 8th,* 655-668.

Schultz P. H. and Gault D. E. (1985) Clustered impacts: Experiments and implications. *J. Geophys. Res., 90,* 3701-3732.

Schultz P. H. and Spudis P. D. (1985) Procellarum basin: A major impact or the effect of Imbrium? (abstract). In *Lunar and Planetary Science XVI,* 746-747. Lunar and Planetary Institute, Houston.

Scott D. H., McCauley J. F., and West M. N. (1977) Geologic map of the west side of the Moon. *U.S. Geol. Survey Map I-1034.*

Shoemaker E. M. and Hackmann R. J. (1962) Stratigraphic basis for a lunar timescale. In *The Moon,* pp. 289-300. Academic Press.

Simonds C. H., Phinney W. C., Warner J. L., McGee P. E., Geeslin J., Brown R. W., and Rhodes J. M. (1977) Apollo 14 revisited, or breccias aren't so bad after all. *Proc. Lunar Sci. Conf. 8th,* 1869-1893.

Singer R. B. (1981) Near-infrared spectral reflectance of mineral mixtures: Systematic combinations of pyroxenes, olivine and iron oxides. *J. Geophys. Res., 86,* 7967-7982.

Singer R. B. and Blake P. L. (1983) Effects of mineral grain size and physical particle size on spectral reflectance of basalts (abstract). In *Lunar and Planetary Science XIV,* pp. 706-707. Lunar and Planetary Institute, Houston.

Smrekar S. and Pieters C. M. (1985) Near-infrared spectroscopy of probable impact melt from three large lunar highland craters. *Icarus, 63,* 442-452.

Spudis P. D. (1978) Composition and origin of the Apennine Bench Formation. *Proc. Lunar Planet. Sci. Conf. 9th,* 3379-3394.

Spudis P. D. (1979) The extent and duration of lunar KREEP volcanism. *Conf. Lunar Highlands Crust,* Lunar and Planetary Institute, Houston, 157-159.

Spudis P. D. (1980) Petrology of the Apennine Front, Apollo 15: Implications for the geology of the Imbrium impact basin (abstract). *Papers Presented to the Conference on Multi-ring Basins,* pp. 83-85, The Lunar and Planetary Institute, Houston.

Spudis P. D. (1981) The nature of lunar basin ejecta deposits inferred from Apollo highland landing site geology (abstract). In *Reports of Planetary Geology Program—1981,* NASA TM-84211, 120-122.

Spudis P. D. (1984) Apollo 16 site geology and impact melts: Implications for the geologic history of the lunar highlands. *Proc. Lunar Planet. Sci. Conf. 15th,* in *J. Geophys. Res., 89,* C95-C107.

Spudis P. D. (1986) Materials and formation of the Imbrium basin. In *Workshop on the Geology and Petrology of the Apollo 15 Landing Site* (P. D. Spudis and G. Ryder, eds.), pp. 100-104. LPI Tech. Rpt. 86-03, Lunar and Planetary Institute, Houston.

Spudis P. D. and Davis P. A. (1985) How much anorthosite in the lunar crust?: Implications for lunar crustal origin (abstract). In *Lunar and Planetary Science XVI,* pp. 807-808, Lunar and Planetary Institute, Houston.

Spudis P. D. and Hawke B. R. (1981) Chemical mixing model studies of lunar orbital geochemical data: Apollo 16 and 17 highlands compositions. *Proc. Lunar Planet. Sci. 12B,* 781-789.

Spudis P. D. and Hawke B. R. (1986) The Apennine Bench Formation revisited. In *Workshop on the Geology and Petrology of the Apollo 15 Landing Site* pp. 105-107. (P. D. Spudis and G. Ryder, eds.) LPI Tech. Rpt. 86-03, Lunar and Planetary Institute, Houston.

Spudis P. D. and Head J. W. (1977) Geology of the Imbrium basin Apennine mountains and relation to the Apollo 15 landing site. *Proc. Lunar Sci. Conf. 8th,* 2785-2797.

Spudis P. D. and Ryder G. (1985) Geology and petrology of the Apollo 15 landing site: Past, present and future understanding. *Eos Trans. AGU, 66,* 721-726.

Spudis P. D., Hawke B. R., and Jackowski T. (1984a) Geochemical mixing-model studies of ejecta from Lunar farside basins: Implications for crustal models (abstract). In *Lunar and Planetary Science XV,* pp. 812-813. Lunar and Planetary Institute, Houston.

Spudis P. D., Hawke B. R., and Lucey P. (1984b) Composition of Orientale basin deposits and implications for the lunar basin-forming process. *Proc. Lunar Planet. Sci. Conf. 15th,* in *J. Geophys. Res., 89,* C197-C210.

Stöffler D., Gault D. E., Wedekind J. A., and Polkowski G. (1975) Experimental hypervelocity impact into quartz sand: Distribution and shock metamorphism of ejecta. *J. Geophys. Res., 80,* 4062-4077.

Stöffler D., Bischoff A., Borchadt A., Deutsch A., Jessberger E. K., Ostertag R., Palme H., Spettel B., Reinold W., Waker K., and Wänke H. (1985) Composition and evolution of the lunar crust in the Descartes highlands, Apollo 16. *Proc. Lunar Planet. Sci. Conf. 15,* in *J. Geophys. Res., 90,* C449-C506.

Swann G. A., Bailey N. G., Batson R. M., Eggleton R. E., Hait M. H., Holt H. E., Larson K. B., Reed V. S., Schaber G. G., Sutton R. L., Trask N. J., Ulrich G. E., and Wilshire H. G. (1977) Geology of the Apollo 14 landing site in the Fra Mauro highlands. *U.S. Geol. Survey Prof. Paper 880,* 103 pp.

Taylor S. R. (1975) *Lunar Science: A Post-Apollo View*, Pergamon, New York. 372 pp.

Taylor S. R. (1982) *Planetary Science: A Lunar Perspective*, Lunar and Planetary Institute, Houston. 481 pp.

Warren P. H. and Taylor G. J. (1981) Petrochemical constraints on lateral transport during lunar basin formation. In *Multi-ring Basins, Proc. Lunar Planet. Sci. 12A*, 149-154.

Wetherill G. W. (1981) Nature and origin of basin-forming projectiles. In *Multi-ring Basins, Proc. Lunar Planet. Sci. 12A*, 1-18.

Whitaker E. A. (1981) The lunar Procellarum basin. In *Multi-ring Basins, Proc. Lunar Planet. Sci. 12A*, 105-111.

Whitford-Stark J. L. (1981) Modification of multi-ring basins—the Imbrium model. In *Multi-ring Basins, Proc. Lunar Planet. Sci. 12A*, 113-114.

Wilhelms D. E. (1970) Summary of lunar stratigraphy—Telescopic Observations. *U.S. Geol. Survey Prof. Paper 599-F,* 47 pp.

Wilhelms D. E. (1980a) Stratigraphy of part of the lunar nearside. *U.S. Geol. Survey Prof. Paper 1046-A,* 71 pp.

Wilhelms D. E. (1980b) Irregularities of lunar basin structure (abstract). *Reports Planetary Geology Program*, NASA TM-81776, 25-27.

Wilhelms D. E. (1983) Effects of the Procellarum basin on lunar geology, petrology, and tectonism (abstract). In *Lunar and Planetary Science XIV,* 845-846.

Wilhelms D. E. (1984) Moon. In *The Geology of the Terrestrial Planets* (M. H. Carr, ed.), NASA SP-469, 106-205.

Wilhelms D. E. and McCauley J. F. (1971) Geologic map of the nearside of the Moon. *U.S. Geol. Survey Map I-703.*

Wilhelms D. E., Hodges C. A., and Pike R. J. (1977) Nested crater model of lunar ringed basins. In *Impact Explosion Cratering*, Pergamon, 539-562.

Wilhelms D. E., Ulrich G. E., Moore H. J., and Hodges C. A. (1980) Emplacement of Apollo 14 and 16 breccias as primary basin ejecta (abstract). In *Lunar and Planetary Science XI*, 1251-1253. Lunar and Planetary Institute, Houston.

Wilshire H. G. and Jackson E. D. (1972) Petrology and stratigraphy of the Fra Mauro Formation at the Apollo 14 site. *U.S. Geol. Survey Prof. Paper 785,* 26 pp.

Apennine Front Revisited: Diversity of Apollo 15 Highland Rock Types

Marilyn M. Lindstrom*
Code SN2, NASA Johnson Space Center, Houston, TX 77058

Ursula B. Marvin
Harvard-Smithsonian Center for Astrophysics, Cambridge, MA 02138

Scott K. Vetter and John W. Shervais
Department of Geology, University of South Carolina, Columbia, SC 29208

*Also at Department of Earth and Planetary Sciences, Washington University, St. Louis, MO 63130

The Apollo 15 landing site is geologically the most complex of the Apollo sites, situated at a mare-highland interface within the rings of two of the last major basin-forming impacts. Few of the Apollo 15 samples are ancient highland rocks derived from the early differentiation of the Moon, or impact melts from major basin impacts. Most of the samples are regolith breccias containing abundant clasts of younger volcanic mare and KREEP basalts. The early geologic evolution of the region can be understood only by examining the small fragments of highland rocks found in regolith breccias and soils. Geochemical and petrologic studies of clasts and matrices of three impact melt breccias and four regolith breccias are presented. Twelve igneous and metamorphic rocks show extreme diversity and include a new type of ferroan norite. Twenty-five samples of highland impact melt are divided into groups based on composition. These impact melts form nearly a continuum over more than an order of magnitude in REE concentrations. This continuum may result from both major basin impacts and younger local events. Highland rocks from the Apennine Front include most of the highland rock types found at all of the other sites. An extreme diversity of highland rocks is a fundamental characteristic of the Apennine Front and is a natural result of its complex geologic evolution.

INTRODUCTION

The Apollo 15 landing site is at the interface between the mare basalts of Palus Putredinus, which are exposed at Hadley Rille, and the highlands of the Apennine Front, exposed at Hadley Delta (*ALGIT*, 1972). The Apennine Front represents the rim of the Imbrium basin, but also falls within the Serenitatis basin (*Spudis and Ryder*, 1985). Highlands materials at the Front are therefore expected to consist of Imbrium ejecta overlying Serenitatis ejecta and possibly pre-Serenitatis materials (*Spudis*, 1980). Although some of the returned samples are rocks from the ancient highland crust or melt breccias representing major basin impacts, most of the samples are younger mare and KREEP basalts or regolith breccias, which are complex mixtures of lithic fragments, glasses, and soils (*A15PET*, 1972; *Chamberlain and Watkins*, 1972). The Apollo 15 Workshop (*Spudis and Ryder*, 1986) focused attention on the unsolved problems of the provenance and petrogenesis of Apollo 15 samples. Numerous research projects, including this one, were undertaken to resolve the unanswered questions at the Hadley-Apennine site. Results of some of those studies have already been published (*Korotev*, 1987a; *Ryder*, 1987; *Ryder and Spudis*, 1987; *Simon et al.*, 1986). This volume contains reports of several other studies. This paper presents petrologic and compositional studies of matrices and clasts from seven Apollo 15 polymict breccias, which demonstrate the diversity of rock types at the site in comparison with rocks found at other sites.

CHARACTERIZATION OF APENNINE FRONT HIGHLAND ROCKS

Sampling and Analytical Procedures

This paper presents the results of a survey of Apennine Front rocks (*Lindstrom*, 1986) and detailed consortium studies of slabs from two regolith breccias: 15459, from the Apennine Front (*Lindstrom and Marvin*, 1987), and 15498, from the mare plain (*Vetter et al.*, 1987a). Sampling techniques and experimental procedures of the three studies differ.

Six polymict breccias collected at Station 7 on the Apennine Front were chosen to survey the materials at the Front. These included three regolith breccias (15426, 15435, 15459) and three impact melt breccias (15405, 15445, 15455). The impact melt breccias were studied previously by *Ridley* (1977; *Ridley et al.*, 1973) and by the Imbrium Consortium (*Wood*, 1976). *Ryder and Bower* (1977) described the black and white breccias (15445, 15455) as LKFM impact melts containing clasts of Mg-suite plutonic rocks. Sufficient major and trace element analyses of these clasts were available (*Ryder*, 1985), so no new clast studies were done. Further characterization of the bulk compositions of the impact melt matrices was done on three matrix splits from each breccia. Breccia 15405 was described by *Ryder* (1976) as an impact melt rock consisting of KREEP basalt and its differentiates. He identified its most conspicuous clasts as quartz monzodiorite, and they have since been studied extensively (*Nyquist et al.*, 1977; *Taylor et al.*,

TABLE 1. Compositional and petrographic characterization of samples.

Generic	INAA*	Pet†	Type‡	Composition/Texture
15405	170	170	C	Plutonic alkali norite
	171	9010	C	LKFM/ungrouped impact melt
	173		M	KREEP/A impact melt
	174		M	KREEP/A impact melt
	175		M	KREEP/A impact melt
	181	9009	C	Cataclastic alkali anorthosite
15426	137	9010	C	Plutonic Mg-troctolite
15435	54-1		C	LKFM/B impact melt
	54-2		C	LKFM/B impact melt
15445	244		M	LKFM/D impact melt
	245		M	LKFM/D impact melt
	246		M	LKFM/D impact melt
15455	258		M	LKFM/D impact melt
	259		M	LKFM/D impact melt
	260		M	LKFM/D impact melt
15459	231	231-2	C	LKFM/E impact melt
	231W	9008	C	Cataclastic Mg-anorthosite
	237	237	C	MKFM green glass
	238	9012	C	Granular Fe-anorthositic norite
	239A	9009	C	Contaminated Fe-anorthosite
	239W	9010	C	LKFM/D impact melt
	241		C	KREEP/A impact melt
	242	9011	C	KREEP/A impact melt
	274	337	C	Cataclastic Fe-anorthosite
	279	339	C	Granular Fe-norite
	286	340	C	LKFM/D impact melt
	288	289	C	Fe-granulitic breccia
	290	342	C	LKFM/A impact melt
	292	343	C	Granular Fe-norite
	305	345	C	LKFM/B impact melt
	309	346	C	Mg-poikilitic granulite
	313	347	C	LKFM/B impact melt
	315	316	C	Quartz monzodiorite
	317	318	C	LKFM/ungrouped impact melt
	320	348	C	LKFM/E impact melt
	329	349	C	LKFM/B impact melt
15498	158	236	C	MKFM green glass
	192	193	C	LKFM/C impact melt
	209	210	C	LKFM/C impact melt

*INAA subsample number.
†Thin section subsample number.
‡Clast (C) or matrix (M) sample.

1980). We analyzed three splits of matrix and three white clasts, none of which had been studied previously.

The regolith breccias included two friable clods (15426 and 15435) and one brown glass matrix breccia (15459). 15426 consists mostly of primitive mare green glass (*Delano*, 1979; *Ma et al.*, 1981; *Ryder*, 1985). We analyzed a bulk sample of the clod and two clasts. Only one of the clasts is a highland rock. Results for the mare materials will be presented elsewhere. 15435 was the pedestal for anorthosite 15415. We analyzed a bulk sample and two coherent clasts. Brown glass matrix breccia 15459 was the largest highland rock returned from the Apollo 15 site. It contains two large clasts, a 5-cm

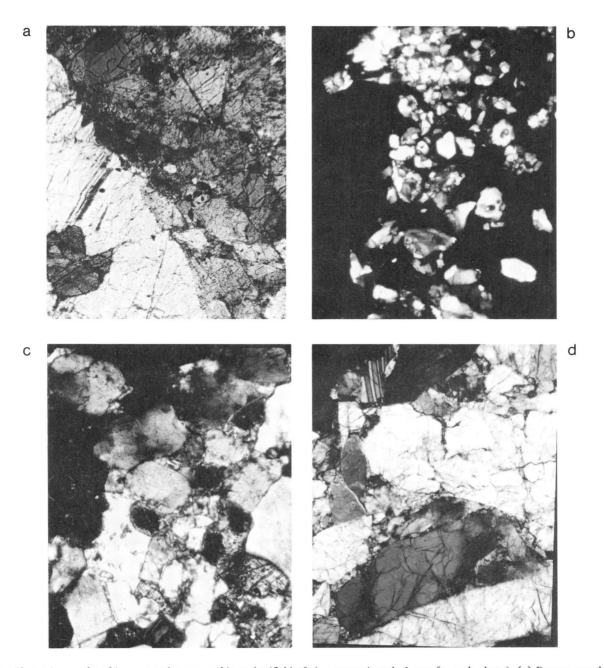

Fig. 1. Photomicrographs of igneous and metamorphic rocks (field of view approximately 1 mm for each photo). (a) Ferroan anorthosite 15459,337 (partially cross-polarized light) showing cataclastic texture, crushed mosaic grains without granulitic recrystallization. Note offset of twin lamellae. (b) Ferroan noritic anorthosite 15459,239a (cross-polarized light) showing crushed granular texture and matrix contamination. (c) Ferroan anorthositic norite 15459,238 (cross-polarized light) showing blocky plagioclase grains with intergranular pyroxene. (d) Magnesian troctolite 15426,137 (cross-polarized light) showing large blocky olivine grains with scattered twinned plagioclase. (e) Magnesian anorthosite 15459,231w (crossed polarized light) showing strained coarse plagioclase and exsolved pyroxene in granular groundmass. (f) Alkali anorthosite 15405,181 (cross-polarized light) showing large blocky plagioclase grains in granular groundmass. Note large exsolved pyroxene grain at left. (g) Alkali gabbronorite 15405,170 (cross-polarized light) showing coarse grains of plagioclase and pyroxene. (h) Quartz monzodiorite 15459,316 (backscattered electron image, field of view 4 microns) showing exsolved pyroxene and interstitial minerals. (i) Ferroan norite 15459,339 (partially cross-polarized light) showing cataclastic texture and lack of recrystalization of plagioclase (gray to white) and pyroxene (bright with dark rims). (j) Ferroan norite 15459,343 (transmitted light) showing coarse igneous texture offset along a fault plane indicated by arrows. The plagioclase (white) is granulitic, the darker material is predominantly orthopyroxene. (k) Granulite 15459,289 (partially polarized light) showing fine-grained granular recrystallized texture. (l) Poikilitic granulite 15459,346 (Partially polarized light) showing coarse poikilitic texture.

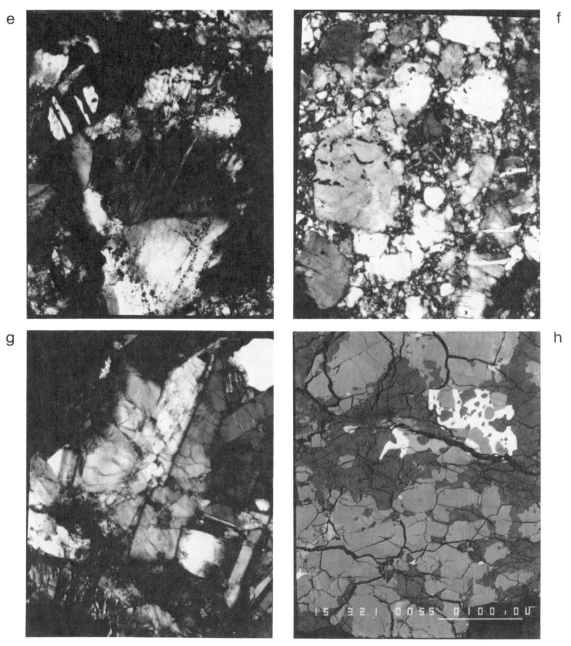

Fig. 1. (continued)

greenish mare basalt and a 3-cm white highland clast. The brief petrographic description (*Ridley,* 1977), and comparison of its bulk composition to those of other Apollo 15 regolith breccias (*Korotev,* 1985 and unpublished data), suggested that clasts of highland plutonic and impact melt lithologies might be more abundant than in other breccias. We selected 14 samples of clasts and matrix of 15459 for analysis.

The Apennine Front survey samples were analyzed by INAA at Washington University using the procedures of *Korotev* (1987c). Selected samples were later analyzed at the University of Missouri, Columbia for short-lived isotopes. Most of these samples were then thin-sectioned for petrographic studies. Petrographic characterization and microprobe analyses of 15405 clasts were done by Shervais and Vetter. Similar studies of the 15426 clast were done by D. Mittlefehldt at JSC (personal communication, 1987), while studies of 15459 clasts were done by Marvin.

The initial study of breccia 15459 was sufficiently productive that a consortium study of a slab was initiated by Lindstrom. The slab was mapped by Marvin and samples selected for extraction. Larger clasts were split for thin sections and INAA. Petrographic characterization was done by Marvin using a

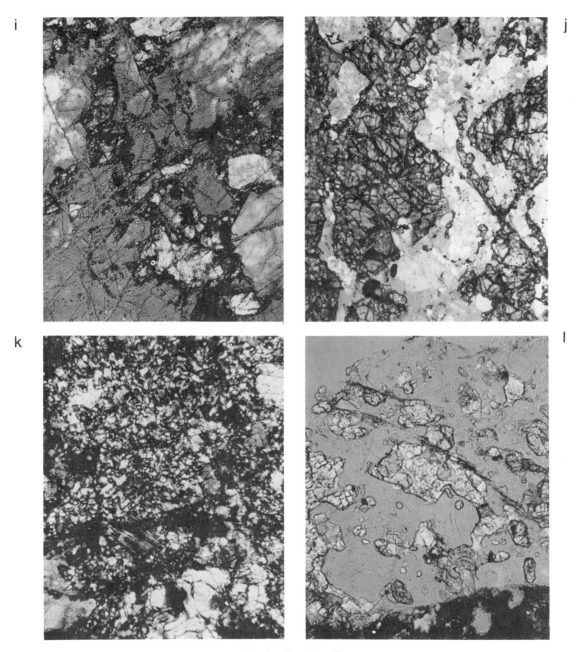

Fig. 1. (continued)

JEOL733 microprobe. Trace element analyses were done by INAA at Washington University. Major element analyses of two clasts (15459,279 and ,292) were determined by Shervais and Vetter using microprobe analyses of fused beads.

Consortium study of breccia 15498 from Station 4 on the mare plain was organized by Shervais, who mapped the slab and selected clasts. Studies of the mare clasts are reported in *Vetter et al.* (1987b). Analyses of three highland clasts from this mare breccia are reported here. The clasts were split for thin sectioning prior to analysis. Petrographic studies were done by Shervais and Vetter. Splits for bulk analyses were ground and split again for major element analysis of fused beads and INAA, which was done at JSC using the procedures of *Jacobs et al.* (1977).

Table 1 presents a brief petrologic and compositional characterization of each of the samples. Samples are listed in Table 1 by generic sample number. When a clast was split for duplicate analyses it is listed only once, but both analyses are given later. When a polymict sample was separated into several lithologies, they are listed separately in Table 1. Detailed discussions in the following sections are organized by rock classification based on a combination of compositional and petrographic characteristics. This paper describes clasts of highland origin and includes 12 igneous and metamorphic rocks, 25 melt breccias, and 2 glasses. Studies of seven clasts of mare origin will be reported elsewhere.

Igneous and Metamorphic Rocks

Twelve clasts of highland crustal rocks are characterized in the following section. Ten of the rocks have at least relict plutonic textures, while two are more recrystallized metamorphic rocks. Photomicrographs of these clasts are shown in Fig. 1. Major mineral compositions are compared to previously defined fields for pristine rocks in Fig. 2, a plot of Mg' in mafic minerals versus An in plagioclase (*Warren et al.*, 1983; *Lindstrom et al.*, 1984). Bulk compositions of the clasts are presented in Table 2. Many of the clasts are examples of familiar rock types. There are plutonic samples belonging to ferroan, magnesian, and alkali suites of highlands rocks. One rock is a highly differentiated quartz monzodiorite. Metamorphic rocks are lunar granulites similar to those found at all highland sites and in the lunar meteorites (*Warner et al.*, 1977; *Lindstrom and Lindstrom*, 1986). Two clasts are plutonic-textured ferroan norites with mineral compositions falling between the fields of the familiar pristine rock suites. These are discussed in detail here, while petrographic and compositional characteristics of rocks belonging to the familiar suites are only briefly summarized and compared to other examples of similar rock types.

Ferroan anorthositic rocks. Three of the clasts are ferroan anorthosites. Cataclastic anorthosite 15459,274/337 has a coarse-grained (up to 1.5 mm) mosaic texture (Fig. 1a), consisting almost entirely of plagioclase of uniform composition (An_{97}). Minute (1 micron) inclusions of pyroxene ($En_{66}Fs_{32}Wo_2$; $En_{44}Fs_{11}Wo_{45}$) occur in the plagioclases, and tiny interstitial grains of ilmenite are also present. The bulk composition of this anorthosite is very nearly that of pure plagioclase. Concentrations of transition metals, incompatible elements, and meteoritic siderophiles are all very low. The REE pattern (Fig. 3a) shows the low concentrations and positive Eu anomaly characteristic of plagioclase. The bulk and mineral compositions of this anorthosite are similar to those of 15415 (*James*, 1972; *Haskin et al.*, 1981), but its texture shows less recrystallization and annealing.

Clast 15459,239a is a cataclastic anorthosite with a small amount of matrix contamination. The thin section contains numerous mineral and lithic fragments, most of which are plagioclase (Fig. 1b). Some plagioclase fragments are crushed, others have a fine-grained granulitic texture. Scattered mafic minerals are found in small patches of matrix contamination and not associated with the bulk of the plagioclase. The bulk composition reflects the matrix contamination in its lower Al_2O_3 and CaO concentrations and higher transition metal and incompatible element concentrations. The REE pattern (Fig. 3a) is flat, with a positive Eu anomaly and concentrations higher than those of most mafic anorthosites. However, the plagioclase-dominated REE pattern limits matrix contamination to a few percent.

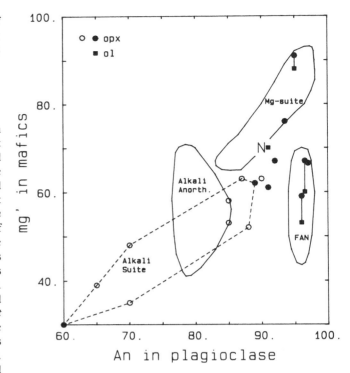

Fig. 2. Compositions of major minerals in igneous and metamorphic rocks compared to those of pristine rocks (*Warren et al.*, 1983). Plotted are Mg' = [Molar Mg/(Mg + Fe) × 100] versus An = [Molar Ca/(Ca + Na) × 100] for Apennine Front samples. Shaded symbols are from this study, open symbols represent other related rocks: ferroan norites (*Ridley*, 1977); alkali gabbronorites (*James et al.*, 1987); and quartz monzodiorite (*Ryder*, 1976). Alkali norite, gabbronorite, and quartz monzodiorite define a new alkali-suite field. Ferroan norites plot at the extension of this field. Ferroan norites plot at the extension of this field. N is Apollo 17 norites.

Clast 15459,238 is a cataclastic anorthositic norite consisting of 61% plagioclase, 29% pyroxene, and 10% olivine with accessory Al-Ti-Mg-chromite and Mg-ilmenite. Plagioclase (An_{95-98}) occurs predominantly in grains up to 1.4 mm long that have been crushed and partially randomized optically. Other patches of plagioclase approach granulitic texture (Fig. 1c). Irregular grains of pyroxene, up to 600 microns across, show fine exsolution lamellae with compositions of $En_{67}Fs_{31}Wo_2$ and $En_{41}Fs_{13}Wo_{46}$. Olivine (dominantly Fo_{59-61}) occurs in individual grains, up to 100 microns long, that are brownish-orange in color. Although crushed and partially recrystallized, the relict texture of this rock indicates a plutonic origin. The bulk composition is that of a mafic anorthosite. Concentrations of transition metals are three to four times higher than 15459,239a mafic anorthosite, but LREE concentrations are lower and reflect little or no contamination (Fig. 3a). The composition and texture of this plutonic anorthositic norite are similar to those of 15243,17 (*Ryder et al.*, 1987). The composition is almost identical to that of 15418, the large severely shocked and recrystallized

TABLE 2. Compositions of Apennine Front igneous and metamorphic rocks.

	Ferroan Anorthosites			Mg-suite		Alkali Suite			Ferroan Norites		Metamorphic	
	FA 15459 ,274	FAN 15459 ,239a	FAN 15459 ,238	Mgt 15426 ,137	MgA 15459 ,231w	AA 15405 ,181	AN 15405 ,170	QMD 15459 ,315	FN 15459 ,279	FN 15459 ,292	GRAN 15459 ,288	POIK 15459 ,309
TiO_2 (%)			0.17	0.31	0.48		0.36					
Al_2O_3		27.8	24.2	12.7	30.7	30.4	24.4		24.4	30.4		
FeO	0.226	1.71	6.27	7.47	1.51	0.302	4.30	12.9	6.88	10.9	4.22	5.82
MgO		1.5	6.1	31.3	1.8		1.0	4.4				
CaO	19.0	17.0	15.1	5.8	17.4	17.6	16.4	11.3	13.1	10.9	16.3	12.6
Na_2O	0.334	0.452	0.297	0.177	1.02	1.81	0.91	0.86	0.607	0.525	0.384	0.578
K_2O			<.12	<.14	<.7							
Sc (ppm)	0.676	3.46	11.92	5.41	3.53	0.591	9.25	29.6	9.84	18.6	6.60	7.17
Cr	26.0	520	740	630	170	5.6	570	1180	2730	2360	390	660
Co	0.235	6.60	14.4	38.9	5.05	0.36	18.0	10.0	26.4	30.1	6.09	18.0
Ni	<6	145	65	110	35		<50	<40	28	51	<40	88
Rb	0.36	<8		<10	<3	<2	5	46	<9	9	<4	<7
Sr	170	150	150	80	340	580	230	190	160	120	170	170
Cs	0.097	0.52	0.11	0.05	0.14	0.19	0.26	1.25	0.12	0.13	0.034	0.073
Ba	7.2	80	28	180	140	180	890	1100	140	110	30	160
La	0.183	2.16	1.43	17.1	8.03	15.1	470	108	5.77	4.46	2.36	8.35
Ce	0.47	6.10	3.80	38.3	19.7	39.2	1254	280	16.6	14.5	6.26	20.7
Nd		2.37	2.00	15.7	9.4	24.0	780	160	7.7	5.5	4.1	13
Sm	0.105	0.84	0.71	4.10	2.72	7.08	213	47.2	2.29	2.98	1.12	3.41
Eu	0.80	0.94	0.74	1.23	2.37	4.85	4.00	2.26	1.40	1.78	0.820	1.34
Tb	0.143	0.151	0.166	0.80	0.53	1.48	42.0	10.7	0.59	1.03	0.224	0.74
Yb	0.052	0.63	0.61	5.69	1.24	2.08	94.0	36.9	2.33	5.15	0.750	2.80
Lu	0.0069	0.112	0.104	1.00	0.187	0.320	11.9	5.07	0.361	0.775	0.119	0.448
Zr	<6	110	<80	1030	50	80	220	1510	120	50	40	100
Hf	0.029	1.65	0.55	22.9	1.42	2.29	11.0	36.5	2.63	1.05	0.74	2.65
Ta	0.0045	0.12	0.37	2.73	0.216	0.08	0.96	4.77	0.48	0.058	0.110	0.395
Th	0.0048	0.23	0.14	2.4	0.76	0.53	39.4	22.3	1.23	0.15	0.258	0.83
U	<0.01	0.06	<.07	3.64	0.21	0.07	1.6	6.4	0.60	0.05	0.064	0.17
Ir (ppb)	<1	<2	1.8	<2	<2	<1	<3	<4	<2	<2	<2	<3
Au	<1	<2	<1	<1	<1	<1	<6	<7	<2	<2	<2	<2

Analyses by INAA, one sigma uncertainties based on counting statistics are: 1-2% for Al_2O_3, FeO, Na_2O, Sc, Cr, Co, La, Sm, Eu; 3-10% for CaO, Sr, Ba, Ce, Tb, Yb, Lu, Hf, Ta, Th; 10-40% for TiO_2, MgO, K_2O, Ni, Cs, Nd, U, (Ir, Au).

anorthositic gabbro (*A15PET*, 1972). This similarity in composition supports suggestions by *Nord et al.* (1977) and *Lindstrom and Lindstrom* (1986) that 15418 is a metamorphosed plutonic rock.

Magnesian suite rocks. Clast 15426,137 is a magnesian troctolite. It is a coarse-grained plutonic rock (Fig. 1d) consisting of 65% olivine (Fo_{88}), 35% plagioclase (An_{95}), minor pyroxene and chromite, and accessory Fe-metal and apatite. The olivine is orange, with the color intensifying toward the margins of crystals. However, the cores and rims show no difference in composition that might reflect alteration. Rare pyroxene occurs as interstitial grains of orthopyroxene (En_{88}) and augite ($En_{49}Fs_4Wo_{47}$). The bulk composition of the clast is highly magnesian with high concentrations of Co and Ni, which are incorporated into olivine, but not Sc and Cr, which enter pyroxene or chromite. The REE pattern is unusual for such a magnesian rock (Fig. 3a). Concentrations are high for a Mg-troctolite and the pattern has an unusual U-shape. The presence of accessory apatite, which was observed in thin section, and zircon, not observed but undoubtedly the cause of high Zr and Hf concentrations, account for the unusual REE pattern. Other clasts of Mg-suite troctolite were found in impact melt breccias 15445 and 15455 (*Ryder and Bower*, 1977; *Ryder*, 1985). These show large variations in major and REE abundances due to variations in proportions of both major and accessory phases.

15459,231w is a magnesian anorthosite. The thin section of the sample analyzed by INAA is essentially a grain mount (Fig. 1e). It consists mainly of plagioclase (An_{90-93}) with scattered olivines (Fo_{69-72}) and clinopyroxenes ($En_{47}Fs_{11}Wo_{42}$). Mineral compositions are more evolved than those of Apollo 14 magnesian anorthosites (*Lindstrom et al.*, 1984) and more similar to those of Mg-norites than Mg-troctolites. The bulk composition is similar to those of slightly mafic magnesian anorthosites. The REE pattern is like those of REE-poor Apollo 14 magnesian anorthosites (*Lindstrom et al.*, 1984), but falls between those of Mg-troctolite 15426,137 and Mg-norites and troctolites from 15445-15455 (Fig. 3a). *Simon et al.* (1987) report mineral compositions, but no bulk composition, for a Mg-anorthosite fragment.

Clasts of Mg-suite norites are also found at the Apennine Front, notably as clasts in the black and white breccias 15445-

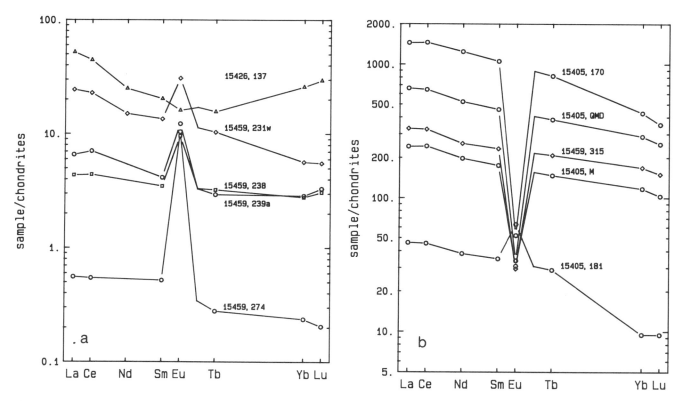

Fig. 3. Chondrite-normalized REE patterns for Apennine Front samples. (**a**) Ferroan anorthosites and Mg-suite rocks. Ferroan anorthositic rocks have low REE concentrations with large positive Eu anomalies, while magnesian anorthosite has a plagioclase pattern with much higher concentrations and a smaller Eu anomaly. Mg-troctolite has a REE-enriched pattern with unusual U-shape. (**b**) Alkali-suite rocks. Samples include clasts and matrix for 15405 and QMD 15459,315 compared with literature data for 15405 QMD clast (*Nyquist et al.,* 1977). Alkali anorthosite 15405,181 is similar to Apollo 14 alkali anorthosites in composition. Alkali norite 15405,170 has twice the REE concentrations as 15405 QMD clast. (**c**) Ferroan norites. Data for ferroan norites 15459,279 and ,292 are compared to alkali norite 15405,170 and Apollo 17 Mg-norites 77075, 77077 and 77215 (*Chao et al.,* 1976; *Warren and Wasson,* 1978). (**d**) Impact melt rocks. Means of individual analyses of representative melt rocks from this study are plotted as points. Ranges of analyses for each group are plotted and show the continuum of melt compositions in B-C-D region.

15455. Analyses of these clasts are reported by *Ridley* (1977; *Ridley et al.,* 1973), *Blanchard et al.* (1976), and *Warren and Wasson* (1978, 1979, 1980). Other clasts have been found in Apennine Front regolith breccias 15306, 15465, and 15565. These clasts tend to be fairly feldspathic norites with 20-25% Al_2O_3 and REE concentrations lower than those of the troctolite presented here.

Alkali suite rocks. Three clasts are members of the alkali suite. These include an alkali anorthosite, an alkali norite, and a quartz monzodiorite. 15405,181 is a cataclastic alkali anorthosite. The sample consists almost entirely of plagioclase (An_{84}) with large relict twinned grains (up to 0.6 mm) in a matrix of granulated plagioclase. Minor ilmenite and rare phosphate are the only accessory minerals, no mafic silicates were observed in thin section (Fig. 1f). The plagioclase composition falls at the Ca-rich end of the range for alkali anorthosites, but the absence of mafic silicates precludes plotting the sample in Fig. 2. The bulk composition is similar to that of other alkali anorthosites with very high concentrations of Al and Ca and very low concentrations of transition metals and siderophiles. REE concentrations are moderate, with LREE enrichment and a positive Eu anomaly (Fig. 3b), as is observed for other alkali anorthosites (*Lindstrom et al.,* 1984).

15405,170 is a feldspathic alkali norite with plutonic texture. Cumulus plagioclase (An_{89}, up to 0.6 × 0.25 mm) and pigeonite ($En_{61}Fs_{37}Wo_2$ with minor exsolution of $En_{42}Fs_{15}Wo_{43}$, up to 1.0 × 0.4 mm) are enclosed by postcumulate plagioclase and pyroxene (Fig. 1g). Interstices between the major minerals contain an abundance of accessory minerals, including large ilmenites and small grains of Fe-metal and troilite, an intergrowth of K-feldspar and silica, zircon, ZrO_2, and a large (300 × 500 micron) whitlockite. The bulk composition of the clast is that of a moderately feldspathic alkalic rock. Al_2O_3 is moderately high (24%) but transition metal concentrations are distinctly higher than in alkali anorthosites. Trace element characteristics of this clast are dependent on the amounts of accessory minerals. REE concentrations are extremely high (Fig. 3b) due to the abundance of whitlockite. Concentrations are a factor of five higher than KREEP basaslt, a factor of two to five higher than quartz monzodiorite and alkali anorthosites *Warren et al.,* 1983), and nearly as high as unique alkali anorthosite 14313c (*Haskin et al.,* 1973). Concentrations of

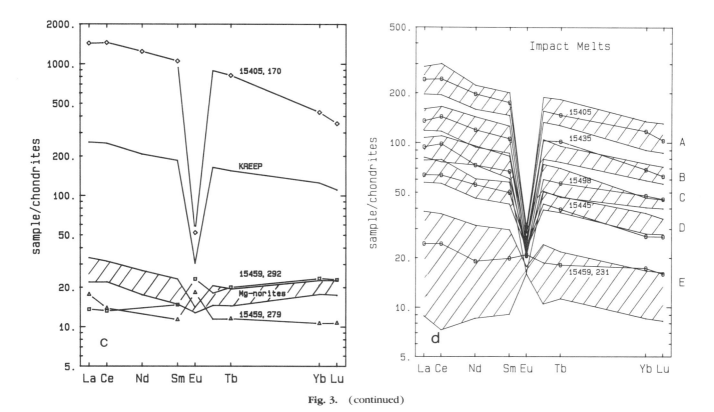

Fig. 3. (continued)

other incompatible elements are variable: Th is a factor of three enriched over KREEP basalt, while Sr and Ba are at the same level, and Rb, Cs, U, Zr, Hf, and Ta are depleted relative to KREEP basalts. This clast bears much similarity to alkali gabbronorite clasts in 67975 (*James et al.*, 1987), but it has slightly more primitive plagioclase and pyroxene compositions than the magnesian alkali gabbronorites and contains very little clinopyroxene. The high and variable trace element characteristics resemble some members of that suite.

15459,315/316 is a somewhat recrystallized quartz monzodiorite consisting of 59% feldspar, 39% pyroxene, 2% silica, and accessory ilmenite, whitlockite, and zircon. Crushing and deformation have obscured the original texture, reduced the grain size, and produced wavy extinction in both the feldspars and pyroxenes, but have not homogenized the mineral compositions (Fig. 1h). The most abundant feldspar is sodic plagioclase (An_{59-88}), but ternary feldspar ($An_{72}Ab_{20}Or_7$ to $An_{44}Ab_{10}Or_{46}$) and orthoclase ($An_1Ab_2Or_{97}$) are also present. Pyroxene grains have closely-spaced exsolution lamellae with end-member compositions of $En_{28}Fs_{68}Wo_4$ and $En_{23}Fs_{36}Wo_{41}$. These mineral compositions are very similar to those of quartz monzodiorite clasts in 15405 (*Ryder*, 1976). The bulk composition of our clast is somewhat more mafic than that of the 15405 clasts (*Ryder*, 1985), and a factor of two lower in REE concentrations (Fig. 3b).

These three clasts are the first true alkali suite clasts reported from Apollo 15. Two of the clasts were found in KREEP impact melt breccia 15405. *Ryder* (1976) describes KREEP basalts, quartz monzodiorite, and granite clasts in that breccia. He also mentions seeing granular anorthositic clasts in hand specimen, but not finding any in thin section. He comments on a high-K white clast analyzed by *Ganapathy et al.* (1973) and speculates on whether to assign it to the ANT suite or the KREEP basalt suite. The wide range of clast types in this breccia (KREEP basalt, alkali anorthosite, alkali norite, quartz monzodiorite, and granite) represent extrusive KREEP basalt and its plutonic equivalents and differentiates. *Ryder* (1976) recognized that the differentiates were related to the basalts, but did not identify the alkali anorthosite or norite and relate the alkali suite to KREEP basalts. The alkali suite forms an extensive differentiation trend on Fig. 2 from alkali norite through gabbronorite and anorthosite to quartz monzodiorite. Breccia 15405 provides concrete evidence for a genetic link between KREEP basalt and alkali suite plutonic rocks.

While providing important information on KREEP and alkali suite rocks, breccia 15405 is so unusual that *Ryder* [1976] considered it to be exotic to the site. *Ryder and Spudis* (1987) suggest that it is a young (1 b.y.) impact melt from the Apennine Bench. It is by no means typical of Apennine Front material. The third alkali suite rock in this study was not a clast in this unique breccia, but was instead a clast in a typical Apennine Front regolith breccia. This clast is also the first QMD clast not found in 15405, suggesting that alkali suite rocks are not as uncommon at Apollo 15 as previously assumed, and that 15405 may not be exotic. Certainly KREEP basalts are common at the Apollo 15 site (*Ryder*, 1987) and the link between alkali suite rocks and KREEP basalts would suggest that alkali suite rocks may also be present near the site.

Ferroan norites. Two ferroan norites having unusual mineral compositions were found in 15459. 15459,279/339 is a shocked and granulated norite consisting of 65% plagioclase (An$_{93}$), 33% pyroxene (dominantly orthopyroxene En$_{66}$Fs$_{32}$Wo$_2$, with rare augite En$_{42}$Fs$_{13}$Wo$_{45}$) with accessory ilmenite, chromite, silica, K-feldspar, whitlockite, zircon, and an unidentified Fe-Ca-Zr-Ti oxide. Chromite compositions cluster around (Mg$_{0.10}$Fe$_{0.90}$) (Al$_{0.18}$Ti$_{0.07}$Cr$_{0.75}$)$_2$O$_4$, and Mg-ilmenites around (Mg$_{0.10}$Fe$_{0.90}$)TiO$_3$. Original 1-2 mm grains of plagioclase and pyroxene have been shocked and granulated to subsets of angular fragments 500 to 700 microns long, with wavy extinction (Fig. 1i).

15459,292/343 is a coarse-grained norite consisting of 50% pyroxene (dominantly orthopyroxene En$_{60}$Fs$_{37}$Wo$_3$, with minor augite En$_{41}$Fs$_{15}$Wo$_{44}$), 48% plagioclase (An$_{91}$), and accessory ilmenite, chromite, zircon, silica, K-feldspar, troilite, and metal. Individual pyroxene grains originally 1-2 mm have been granulated to blocky mosaics, and plagioclase crushed and recrystallized to granulitic texture. At a later stage the fabric was offset by 0.5 mm along a smoothly curving fault trace (Fig. 1j). Patches rich in silica and K-feldspar occur in the crushed zones, but most of the accessory minerals over 10 microns in size are embedded in the pyroxene. These include zircon in euhedral or subhedral crystals, chromite (Mg$_{0.12}$Fe$_{0.88}$)(Al$_{0.18}$Ti$_{0.11}$Cr$_{0.71}$)$_2$O$_4$, and ilmenite (Mg$_{0.14}$Fe$_{0.86}$)TiO$_3$. Sparse grains of Fe metal and troilite, about 10 microns across, occur mainly in plagioclase.

These two norites are petrographically very similar, both in the compositions of their major minerals and in their suites of accessory minerals. Another ferroan norite clast was found in 15459 by *Ridley* (1977). The compositions of the major minerals in the norites are plotted in Fig. 2 compared to familiar plutonic rocks. The ferroan norites plot in a region intermediate to the ferroan, magnesian, and alkali suites. A few other unusual rocks also plot in this region: alkali norite 15405,170 and Mg-anorthosite 15459,231w (this study), Apollo 17 ferroan anorthosite 76504,18 (*Warren et al.,* 1986), and mare gabbros in 15459 (*Ridley,* 1977) and mare peridotite rake samples 15385 and 15387 (*Dowty et al.,* 1973; *Ryder,* 1987). The significance of these similarities will be discussed after the compositional characterization of the norites.

The modes of the two norite clasts differ, and that difference is reflected in their bulk compositions and normative mineralogies (Table 3). The norms correspond quite well to the modes of the norites. Norite 279/339 is a felsic norite with 22.9% Al$_2$O$_3$, while clast 292/343 is more mafic with 16.9% Al$_2$O$_3$. Concentrations of transition metals are moderate, while those of meteoritic siderophiles are low. Concentrations of Co and Ni are reasonable for pristine norites and those of Ir and Au are below detection limits for INAA. Concentrations of REE are also moderate. The REE patterns (Fig. 3c) reveal a flat pattern with a positive Eu anomaly for the felsic norite and a HREE-enriched pattern with a positive Eu anomaly for the mafic sample. This HREE-enrichment reflects the moderately high pyroxene distribution coefficients for HREE. The REE pattern of the felsic norite is similar to that of Mg-norites such as those in 15445-15455, but mafic Mg-norites do not display HREE-enrichment as observed in the mafic

TABLE 3. Bulk analyses and cation % norms of Fe norites.

Wt %	15459, 279/339	15459, 292/343
SiO$_2$	48.50	49.50
TiO$_2$	0.25	0.31
Al$_2$O$_3$	22.90	16.90
FeO	6.97	10.99
MgO	7.35	10.03
CaO	13.38	11.28
Na$_2$O	0.613	0.506
K$_2$O	0.095	0.118
Cr$_2$O$_3$	0.365	0.286
MnO	0.142	0.201
P$_2$O$_3$	0.065	0.039
Total	100.63	100.16
Mg' atomic	0.65	0.62
Norm		
Olivine	0	0
Pyroxene	33%	50%
En	62	57
Fs	30	33
Wo	8	10
Plagioclase	65%	49%
An	91	89
Ab	8	9
Or	0.8	1.4
Quartz	0.8	0.6
Ilmenite	0.3	0.4
Chromite	0.4	0.3
Apatite	0.1	0.08

ferroan norite. Some mafic ferroan anorthositic rocks do show HREE-enrichments at an order of magnitude lower concentrations.

Relationships between the ferroan norites and other lunar endogenous rocks can be evaluated based on petrographic and compositional characteristics. Among the samples noted as having mineral compositions intermediate to major ferroan, magnesian, and alkali suites, relationships with ferroan anorthosite 76504,18, Mg-anorthosite 15459,231w, and mare gabbros and peridotites are unlikely because these samples all have olivine and augite as their dominant mafic minerals, and not low-Ca pyroxene which is dominant in the norites.

Some relationship may exist between the ferroan norites and either Mg-norites or alkali norite 15405,170. The alkali norite has mineral compositions quite similar to those of the feldspathic norite, while several Apollo 17 Mg-norites [77075, 77077 (*Warren and Wasson,* 1978); 77215 (*Chao et al.,* 1976)] have mineral compositions only slightly more magnesian than the mafic norite. All of these norites have low-Ca pyroxenes as the dominant mafic mineral, with only rare augite as lamallae in orthopyroxene or in small separate grains. The rocks all display a similar assemblage of accessory minerals usually found in mesostasis. These include ilmenite, troilite, Fe-metal, K-feldspar, silica, zircon, and whitlockite. Bulk compositions of the ferroan norites are plotted on the pseudoternary diagram

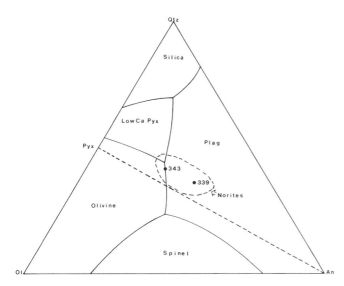

Fig. 4. Pseudoternary diagram of *Walker et al.* (1973) showing that bulk compositions of ferroan norites both plot in the norite field of *Taylor* (1982). The igneous texture and cotectic composition of 15459,292/343 suggest that it is a simple plutonic rock. Clast 15459,279/339 is a plagioclase-rich norite.

of *Walker et al.* (1973) in Fig. 4. Both lie in the norite field as depicted by *Taylor* (1982). Mafic norite 15459,343 plots on the cotectic line, while feldspathic norite 15459,339 falls in the plagioclase field. Alkali norite 15405,170 also lies in the plagioclase-rich part of the norite field, while KREEP basalts and norites plot near the peritectic point. This suggests that all these norites may be plutonic norites, with samples 15459,279/339 and 15405,170 being somewhat enriched in plagioclase.

Comparison of the norite REE patterns in Fig. 3c reveals some complications. The REE concentrations of the alkali norite are extremely high and, as noted above, are due to an unusually high abundance of whitlockite (calculated as 5% from the bulk and whitlockite concentrations of La). REE patterns of the Mg-norites overlap with those of the Fe-norites, but their pattern shapes are clearly distinct. The Mg-norites and 15459,279 have LREE-enriched patterns, while norite 15459,292 has a HREE-enriched pattern. The Mg-norites have negative Eu anomalies, while both Fe-norites have positive Eu anomalies. Preliminary model calculations were done to evaluate the possibility that Mg and Fe norites might be derived from similar parent norite but contain different proportions of minerals and residual liquid. The Mg-norites would contain a higher proportion of residual liquid to account for the LREE-enriched patterns and the Fe-norites would contain a higher proportion of plagioclase to account for the positive Eu anomalies. Although it is possible to generate both LREE- and HREE-enriched patterns and positive and negative Eu anomalies by varying the proportions of minerals and residual liquid, we have not been able to match the patterns in detail. Another problem is that the calculated variations in mineral proportions are larger than allowed by observed modal proportions or bulk compositions. We cannot

at this point resolve the question of possible relationships between ferroan norites and Mg- and alkali-suite norites. The ferroan norites may be unusually ferroan Mg-norites, or they may represent the Mg-rich end of the alkali suite, which now forms an extensive differentiation trend subparallel to the Mg-suite (Fig. 2).

Metamorphic rocks. The two metamorphic clasts are granulitic breccias as discussed by *Warner et al.* (1977) and *Lindstrom and Lindstrom* (1986). Clast 15459,288/289 is a highly feldspathic granulite consisting of 90% plagioclase (An_{94-96}), 10% olivine and pyroxene, and accessory Mg-ilmenite and Fe metal. Rare large feldspar grains are visible and some 200-300 micron zones within the plagioclase show well-developed granulitic texture, but most of it displays a fine intergrowth of irregular grains suggestive of an early stage in the development of granulitic texture (Fig 1k). Grains of olivine (Fo_{52-54}), 10-20 microns across, and a few large pyroxene grains ($En_{59}Fs_{37}Wo_4$ and $En_{44}Fs_{19}Wo_{37}$) are scattered through the plagioclase. This clast is a feldspathic ferroan granulite with low REE concentrations, similar to more mafic samples of the ferroan anorthosite suite as well as the ferroan granulites of *Lindstrom and Lindstrom* (1986).

The second metamorphic clast, 15459,309/346, is a poikilitic rock consisting of 60% plagioclase (An_{93}), 30% pyroxene ($En_{76}Fs_{20}Wo_4$; $En_{47}Fs_9Wo_{44}$), 9% olivine(Fo_{77}), and accessory ilmenite, chromite, zircon, and whitlockite. Fine-grained plagioclase is interspersed among long branching chains of poikilitic pyroxene (Fig. 1l). Olivine occurs linked to pyroxene chains or as minute inclusions in pyroxene. The rock was shocked after formation of the poikilitic texture. This clast is a mafic magnesian granulite with REE concentrations a factor of two higher than in typical granulitic breccias (*Lindstrom and Lindstrom,* 1986).

Granulitic anorthositic norites are fairly common at the Apennine Front. Anorthositic gabbro 15418 (*Nord et al.*, 1977; *Lindstrom and Lindstrom,* 1986) is a severely shocked example. Analyses of such clasts reported by *Warren and Wasson* (1978) include a REE-poor ferroan granulite and a REE-rich magnesian clast. Other clasts of metamorphic-textured anorthositic norites were found in 15459 (*Ridley,* 1977) and 15465 (*Cameron and Delano,* 1973), but only petrographic studies were reported.

Highland Impact Melt Rocks and Glasses

Twenty-five new analyses of highland impact melt rocks are presented in Table 4. The samples include the matrices of the three large impact melt breccias (15405, 15445, 15455) and clasts from the regolith breccias. Most of the clasts have been examined in thin section and found to be fine- to medium-grained impact melts. Most of the impact melts are noritic in bulk composition, having 15-20% Al_2O_3 and 8-11% FeO. MgO has not been measured in all of the samples, but existing analyses reveal variations in Mg' from 55 to 78. There are also major variations in REE and other incompatible element concentrations. REE patterns of some of the more representative melt rocks (Fig. 3d) show variations from ~20 to 300 times chondrites, with all except the lowest patterns having

TABLE 4. Compositions of Apennine Front impact melts and glasses.

	A* 15459 ,242	A 15405 ,173	A 15405 ,174	A 15405 ,175	A 15459 ,290	A 15459 ,241	B 15459 ,305	B 15459 ,329	B 15435 ,54-1	B 15435 ,54-2	B 15459 ,313	C 15498 ,192	C 15498 ,209	D 15459 ,286
TiO_2 (%)	1.74	1.99		2.03		1.70				1.58		0.78	0.72	
Al_2O_3	15.3	14.8		15.5		15.1				16.1		22.0	21.4	
FeO	10.9	10.9	10.8	10.1	9.51	10.9	10.4	10.1	8.98	9.09	6.69	6.21	6.48	9.58
MgO	10.5	7.3		8.4		13.7				11.6		8.49	8.50	
CaO	9.8	10.3	9.4	10.5	11.8	11.2	10.0	10.0	9.7	10.6	13.0	12.94	12.87	10.7
Na_2O	0.620	0.856	0.861	0.829	0.811	0.563	0.51	0.595	0.69	0.66	0.561	0.656	0.636	0.513
K_2O	<3		1.2	0.19					0.49	0.53		0.29	0.29	
Sc (ppm)	20.8	22.3	22.1	20.7	19.1	23.4	20.5	19.9	18.0	17.3	12.4	11.4	12.6	18.0
Cr	1410	1630	1580	2000	1450	1710	2420	1830	1780	1760	1230	910	900	1660
Co	20.5	18.4	17.7	20.8	23.5	38.7	37.4	30.9	22.1	30.4	22.5	18.5	16.7	30.4
Ni	120	60	45	50	170	370	350	240	100	95	85			170
Rb	22	24	29	19	17	16	12	14	14	17		13	9	9
Sr	140	190	180	150	190	200	160	120	140	170	180	200	170	180
Cs	1.27	1.05	1.05	0.75	0.55	0.60	0.38	0.53	0.62	0.57	0.16	0.23	0.22	0.30
Ba	840	880	880	680	660	600	430	460	550	480	320	330	310	270
La	95.0	83.4	87.3	67.6	67.7	64.6	49.1	44.8	44.3	45.2	39.1	34.1	27.8	24.5
Ce	260	222	229	180	181	168	132	120	129	119	103	94	75.6	64.0
Nd	140	130	137	105	110	100	79	67	84	66	62	50	42	37
Sm	40.7	37.5	38.9	29.9	31.3	29.6	22.6	20.5	22.3	20.3	18.0	15.1	12.1	11.3
Eu	2.18	2.34	2.39	2.39	2.34	2.26	1.93	1.69	1.86	1.84	2.22	1.62	1.54	1.41
Tb	9.4	7.95	8.24	6.45	6.42	6.52	4.70	4.59	5.87	4.46	3.94	3.13	2.72	2.30
Yb	29.5	27.5	28.0	21.7	20.1	19.8	15.2	15.1	15.2	15.4	11.23	10.7	10.3	7.91
Lu	4.42	3.75	3.80	2.90	2.76	2.94	2.09	2.05	2.18	2.06	1.51	1.55	1.53	1.19
Zr	1170	1100	1100	800	990	770	680	640	630	630	420	400	350	350
Hf	33.4	31.2	29.5	23.9	24.0	22.7	15.3	15.2	17.3	16.6	10.38	10.8	10.3	8.62
Ta	4.35	3.50	3.60	2.82	2.50	2.19	2.99	1.95	2.01	2.00	1.34	1.17	1.11	1.08
Th	22.0	16.0	16.2	11.6	10.5	9.55	7.32	7.71	8.75	8.16	4.81	6.70	6.10	4.94
U	5.90	4.29	4.43	3.14	2.60	2.90	2.00	2.25	2.60	2.14	1.26	1.70	1.70	1.14
Ir (ppb)	2.4	<2	<2	1.1	4.6	5.3	5.9	2.9	<3	1.2	1.6	<4		3
Au	2.2	1.5	<2	<2	2.2		7.4	6.1	<1.4	<1.6	<5	<7		<4

characteristic KREEP slopes and negative Eu anomalies. Many of the samples have siderophile element (Ir, Au) concentrations detectable by INAA (>2 ppb), but much lower than those of Apollo 16 POIK and VHA impact melts (*Korotev*, 1987b).

Ryder and Spudis (1987) divided their limited set of 14 Apennine Front impact melts into five groups (A to E) largely according to REE concentrations. The studies reported in this volume add 48 new analyses of Apollo 15 impact melts (12 in *Laul et al.*, 1987; 11 in *Ryder et al.*, 1987; and 25 here) for a total of 62 recent analyses of Apollo 15 melt rocks. Several of these samples have only partial major element analyses, but there is generally sufficient information to compare them with other impact melts. Using this expanded dataset, the populations of the five groups change considerably and the divisions become less distinct. Group A has increased from 3 to 11 analyses, group B from 5 to 21, group C from 5 to 8, group D from 2 to 12, and group E from 1 to 10. The numbers are somewhat misleading because some of the data represent multiple analyses of the same impact melt. Eliminating multiple analyses reduces the numbers to 8, 21, 6, 5, and 8, respectively. The group B impact melts are by far the most abundant melts at the Apennine Front. Figure 3d shows ranges of REE concentrations for each of the melt rock groups and examples of each type from the analyses presented here. The REE concentrations are almost continuous throughout the entire suite. Whether these melt groups actually represent distinct impact melts or a continuum of melt compositions can be evaluated by comparisons of their major element compositions. The first four melt groups generally have noritic bulk compositions with 15-20% Al_2O_3 and Mg' ranging from ferroan (55-65) in the group A melts to moderately magnesian (69-78) in the other three groups. The group E melts have anorthositic norite bulk compositions (20-24% Al_2O_3), with Mg' (59-72) spanning most of the range of the group A-D melts. The REE patterns of the group E melts are distinct from the other melts, varying from LREE-enriched patterns with negative Eu anomalies to HREE-enriched patterns with positive Eu anomalies. The group A and E melts are therefore distinct from the group B-C-D melts, but a continuum appears to exist in the noritic magnesian melt rock suite.

The significance of the continuum of melt compositions encompassing the groups B-C-D melts is not well understood. These three types may be described as LKFM (low-K Fra Mauro) melts, having noritic bulk compositions similar to that of samples from the Fra Mauro formation (Apollo 14), with REE and other incompatible element ratios similar to that of KREEP, but at lower concentration levels. LKFM melt rocks of similar compositions are found at other highland sites, and are generally

TABLE 4. (continued)

	†	D	D	D	D	D	D	†	D	E	E	E	Green glass	Green glass
	15459 ,317	15445 ,244	15445 ,245	15445 ,246	15455 ,258	15455 ,259	15455 ,260	15405 ,171	15459 ,239w	15459 ,231-1	15459 ,231-2	15459 ,320	15498 ,237	,158
TiO_2 (%)		1.24		1.62		1.71		0.38	1.76		2.14		1.11	1.64
Al_2O_3		20.6		17.1		16.3		22.8	19.8		20.1	19.2	16.9	
FeO	8.83	7.98	8.28	9.17	9.53	9.79	9.66	8.61	8.38	8.22	8.07	7.58	10.2	9.05
MgO		12.9		16.0		16.2		7.0	11.2		11.6		8.2	9.15
CaO	14.4	11.4	11.3	9.8	10.1	9.4	9.3	13.3	12.6	11.7	11.7	12.3	12.9	10.81
Na_2O	0.772	0.551	0.550	0.523	0.552	0.548	0.527	0.537	0.666	0.661	0.693	0.679	0.621	0.72
K_2O		<0.2	0.15	<0.7	0.14	<.6		<.8			<.3		0.44	0.44
Sc (ppm)	21.7	13.4	14.9	16.9	17.3	17.9	17.8	11.2	16.6	17.1	17.4	15.4	19.2	18.0
Cr	640	1380	1500	1940	1660	1700	1700	1130	1130	1100	1090	1000	1570	1670
Co	15.9	35.4	29.0	30.5	32.1	37.7	38.9	11.6	21.0	20.0	20.5	19.5	55.6	20.9
Ni	55	320	270	270	260	330	350	<40	110	66	82	76	830	
Rb	3.6	<9	6	5	<10	<7	<9	<12	<9		<7	<9	18	17
Sr	210	190	160	170	160	180	170	150	170	220	200	190	190	150
Cs	0.063	0.27	0.18	0.21	0.13	0.18	0.14	0.10	0.21	0.12	0.13	0.10	0.95	0.49
Ba	220	200	200	230	250	250	230	160	220	120	130	120	400	530
La	21.9	16.5	19.4	20.7	21.8	22.0	22.3	20.0	18.9	8.10	8.49	7.49	38.0	51.1
Ce	56.5	43.0	50.0	51.9	56.6	57.0	56.6	52.4	49.0	21.0	22.0	20.6	101	141
Nd	38	25	32	34	33	36	36	27	29	13	14	9.9	54	90
Sm	11.1	7.65	8.80	9.30	10.3	10.3	10.3	8.65	8.61	4.02	4.35	3.74	17.3	23.1
Eu	2.08	1.63	1.72	1.72	1.87	1.85	1.81	1.16	1.76	1.62	1.64	1.58	1.36	2.07
Tb	2.38	1.64	1.86	1.97	2.12	2.07	2.15	1.85	1.94	0.94	1.00	0.85	3.96	4.97
Yb	7.57	5.03	5.59	6.01	6.90	6.45	6.45	6.66	6.16	3.60	3.87	3.40	13.0	16.2
Lu	1.15	0.74	0.84	0.88	0.95	0.93	0.95	0.99	0.94	0.538	0.578	0.489	1.83	2.38
Zr	350	190	250	270	270	230	260	250	270	110	100	100	460	650
Hf	8.23	5.72	6.64	7.70	8.12	7.84	7.76	6.67	6.46	3.25	3.55	2.78	13.8	18.1
Ta	1.10	0.85	0.94	0.96	1.01	1.04	1.03	0.86	0.90	0.57	0.61	0.53	1.69	2.02
Th	2.82	2.85	3.26	3.00	2.82	3.32	3.45	4.73	3.05	1.38	1.46	1.60	7.78	9.0
U	0.70	0.60	0.68	0.65	0.75	0.62	0.65	1.25	0.75	0.50	0.44	0.35	2.03	2.50
Ir (ppb)	<2	3.7	2.5	2.6	2.0	3.6	2.8		1.6			1.5	2.3	
Au	<4	<2	2.1	2.5	2.2	3.2	2.4	<1		<1.4		<1.4	6.7	

Analyses by INAA, one sigma uncertainties based on counting statistics are: 1-2% for Al_2O_3, FeO, Na_2O, Sc, Cr, Co, La, Sm, Eu; 3-10% for CaO, Sr, Ba, Ce, Tb, Yb, Lu, Hf, Ta, Th; 10-25% for TiO_2, MgO, K_2O, Ni, Cs, Nd, U, Ir, Au.
*Impact melt groups indicated by letters A-E (*Ryder and Spudis*, 1987).
†Ungrouped impact melts.

interpreted to represent major basin impacts because they are most common at the Apollo 15 and 17 sites, which are located on the ejecta from Imbrium and Serenitatis basins. The average composition of the highlands in those regions, as measured by the orbital geochemistry experiments, is very similar to the LKFM composition. At the Apollo 16 and 17 sites, impact melt compositions cluster into distinct groups rather than form a continuum of melt compositions (*McKinley et al.*, 1984; *Ryder and Wood*, 1977). The clustering of a large number of samples in the Apollo 15 group B (high REE) melts suggests that group B may represent the dominant impact melt at the site. At the other end of the continuum are the low-REE group D melts, whose dominant members are 15445 and 15455, the only two large samples of LKFM impact melt returned from the Apennine Front. Their clast assemblage is restricted to mafic magnesian pristine rocks one of which has been dated at 4.5 b.y. (*Ryder*, 1985). *Ryder and Bower* (1977) interpret these clasts as deep crustal rocks that could only be brought to the surface by basin-sized impacts. They suggest that these impact melts are ejecta from the Imbrium impact. The group B and D impact melts may represent the major basin impact melts at the site, and the group C melts with intermediate compositions result from smaller local impacts that mix the ejecta from the basin-forming impacts.

The group A impact melt is a ferroan melt (Mg' 55-65) that has the bulk composition of KREEP basalt. There is considerable variation in both Mg' and REE concentrations in samples of the melt, but the same variation is seen in splits of individual KREEP basalts (15382,15386; *Ryder*, 1985). Group A melts appear to be fairly common at the Apollo 15 site, and include breccia 15405 as its most prominent member. The Ar-Ar age of 1.3 b.y. for the 15405 matrix (*Bernatowictz et al.*, 1978) is very young. This age, coupled with the variations in melt composition, suggest that samples are small local impact melts and not major basin melts, which should be more homogeneous in composition. KREEP basalt flows underlying mare basalts at the site, and at the Apennine Bench, are probably the sources of group A impact melts.

Group E melts do not represent a single impact melt, but a suite of melts that have more feldspathic bulk compositions than do the other melt groups. The more REE-rich samples have slopes parallel to that of KREEP and may represent the mixture of feldspathic highland rocks and a small amount of KREEP. The REE-poor group E melts have patterns distinct from KREEP, more similar to those of some plutonic and metamorphic rocks. There is not presently sufficient information on these melt rocks to determine whether there are distinct subgroupings of these feldspathic highland melt rocks, but an origin as impact melts of pre-Serenitatis plutonic and metamorphic highland rocks is likely for these group E melt rocks.

Two analyses of highland glasses are also presented in Table 4. The glasses are homogeneous green glasses that have noritic bulk compositions with K_2O (0.4%) and Mg' (60-65) characteristic of the moderate K Fra Mauro basalt glass in Apennine Front soils (*Reid et al.*, 1973). These glasses and a third Apollo 15 highland green glass (*Ryder et al.*, 1987) have REE concentrations in the range of group B impact melts, but their low Mg' (63) distinguishes them from group B melts (Mg' 72). These glasses may represent the average composition of Apollo 15 highland rocks, which are a mixture of magnesian basin-derived impact melts with ferroan KREEP basalts and a variety of plutonic rocks.

IMPLICATIONS OF DIVERSITY OF APENNINE FRONT ROCKS

Provenances of Apollo 15 Igneous Rocks

The Apollo 15 highland rock suite is the most diverse of the highland suites sampled by the Apollo program. The igneous rocks include members of the three major types of highland plutonic rocks, previously unidentified ferroan norites, and extrusive KREEP basalts. Metamorphic rocks are similar to the granulitic breccias found at all highland sites. A very wide variety of impact melt compositions occur at the site; some of them (groups A and E) undoubtedly are local in origin, but others (groups B and D) are probably basin-related. Evaluation of the provenances of the Apennine Front igneous rocks can be based on comparison to rocks of known provenance.

The Apollo 15 ferroan anorthosite suite is similar to that from Apollo 16 in both bulk rock and mineral compositions. Although fewer in number, the Apollo 15 ferroan samples span the range from pure anorthosite to anorthositic norite, as do the Apollo 16 ferroan samples. Recent isotopic studies have succeeded in dating these important highland rocks at 4.5 b.y. (*Hanan and Tilton*, 1987; *Lugmair*, 1987), confirming their suspected ancient age. In the generally-accepted magma ocean model (*Warren*, 1985), these ferroan anorthositic rocks make up much of the earliest outer shell of the Moon's crust and would be an important pre-Serenitatis component at Apollo 15.

Many of the Apollo 15 magnesian suite samples resemble noritic and troctolitic rocks from Apollo 17 in both mineral and bulk composition. Members of each suite have been dated as ancient (older than 4.2 b.y.). At both the Apollo 15 and 17 sites the Mg-suite rocks occur as clasts in LKFM impact melts. *Ryder and Wood* (1977) discussed these impact melts and their magnesian clasts and concluded that the Mg-suite plutonic rocks are deep crustal materials that could be brought to the surface only by basin-forming impacts. Based on a small but significant difference in melt composition they concluded that the two suites could not be derived from the same impact. They assigned the Apollo 17 highland rocks to Serenitatis and the Apollo 15 suite to Imbrium. Despite the fact that these group D impact melts are not the most abundant melt group at the site, the arguments are sound and led *Ryder and Spudis* (1987) to reiterate that conclusion. The Mg-troctolite and anorthosite presented here and one Apollo 17 troctolite (*Winzer et al.*, 1974) are enriched in incompatible elements and more closely resemble the Apollo 14 Mg-suite troctolites and anorthosites (*Lindstrom et al.*, 1984). These probably result from assimilation of KREEP components in shallower plutons.

The Apollo 15 alkali suite bears strong resemblance to the Apollo 14 alkali suite *Goodrich et al.*, 1986) and to Apollo 16 alkali gabbronorites (*Lindstrom*, 1984; *James et al.*, 1987). Members of the suite have evolved mineral compositions and abundant late-stage accessory minerals. Alkali suite rocks are an abundant highland component at the Apollo 14 site, but only minor components at the Apollo 15 and 16 sites. They occur in varied associations, in pre-Imbrium fragmental breccias at Apollo 16, and with post-Imbrium local KREEP basalts at Apollo 15. It is probable that alkali suite rocks are scattered throughout the lunar crust, perhaps in near-surface plutons that can be brought to the surface by smaller impacts.

The ferroan norites are not sufficiently similar to other known highland rocks to assign them to a specific geologic location. We might infer from the fact that they are ferroan plutonic rocks that appear to be unrelated to ferroan anorthosites that they are derived from moderately shallow plutons, but whether they are local or exotic can only be determined when we know how common they are at the Apollo 15 site. We have begun a search for noritic rocks among Apollo 15 samples to evaluate this question.

Relative Abundances of Components in the Apennine Front

The relative abundances of rock types at the Apennine Front can be approximated in several ways. The lithologic approach uses the distribution of rock types among a large number of rock fragments that have been subjected to petrologic and chemical studies. This distribution may vary with different sampling strategies. The compositional approach attempts to model the average composition of the material as a mixture of various components using least-squares mixing calculations. The results are as good as the choice of components. Ideally the two approaches should be combined, with the results of the lithologic studies being used to select components for the mixing calculations. Complications inevitably arise because intermediate rock types that may be mixtures of other rock types, such as impact melts and granulitic breccias, are often the most common components. This is indeed the case for Apennine Front samples. Three lithologic studies of Apollo 15 highland rocks (this study; *Ryder et al.*, 1987; *Simon et al.*, 1987 and *Laul et al.*, 1987) all demonstrate that impact melts are the most abundant rock type, making up about half of the nonregolith lithic fragments. As previously discussed, compositions of impact melts vary widely, as do their inferred components and origins.

Among igneous fragments, KREEP basalt is the dominant rock type in both the studies of *Ryder et al.* (1987) and *Simon et al.* (1987). KREEP basalts are post-Imbrium volcanic rocks whose proportion we intended to limit by selecting KREEP-poor regolith breccia 15459 for our detailed study of the ancient Apennine Front components. The number of well-characterized Apollo 15 plutonic rocks remains small: These new studies add about 20 to the number available at the time of the Apollo 15 workshop. Most of the fragments are small and may not be representative of their rock units, but mineral compositions and clusters in bulk composition can be used to define typical compositions. Ferroan and magnesian plutonic rocks are approximately equally abundant, with alkali suite rocks a minor component except in breccia 15405. The ferroan rocks include ferroan anorthosites and anorthositic norites, but also the unusual ferroan norites, which are apparently unrelated to ferroan anorthosites. The proportions of mafic minerals in the ferroan anorthositic rocks tend to be higher than in Apollo 16 rocks. The Mg-suite rocks, especially those in impact melts 15445 and 15455, are highly magnesian members of the suite. They include both troctolitic and noritic varieties and tend to be more feldspathic than their Apollo 17 counterparts. The Apollo 15 alkali suite samples extend the differentiation trend from more primitive norite to highly evolved quartz monzodiorite. It is difficult to select appropriate plutonic components for use in compositional modeling of the Apennine Front.

Korotev (1987a) reviewed the compositional approach to the Apennine Front and showed that the selection of input components influences the conclusions. He found that compositional variations in Apollo 15 soils could be modeled as mixtures of five components: KREEP basalt, mare basalt, green glass, meteorite and an Apennine Front Soil Component (AFSC). AFSC was defined as the Apennine Front soil containing the least mare basalt. The composition of AFSC corresponds more closely to that of LKFM glass (*Reid et al.*, 1973) than either do to the compositions of individual rocks at the Apennine Front. AFSC is obviously a mixture of preexisting rock types, including the various types discussed above. The fact that a single component is sufficient to describe variations in soil composition implies that the Apennine Front rocks are well mixed and their proportions do not vary. All variations in soil composition seem to be accounted for by variations in the amounts of young volcanic products: KREEP, mare basalts and green glass.

Korotev (1987a) did not present mixing models giving the proportions of components in the AFSC, but used compositional arguments to suggest which rock types might be important components. The rocks most similar in composition to AFSC are the group D impact melts (15445,15455), which are themselves mixtures dominated by Mg-suite and ancient KREEP components. Although generally similar in bulk composition, these impact melts are distinctly more mafic and more magnesian than AFSC (Al_2O_3 16 versus 20%; Mg' 75 versus 65). Korotev showed that if the group D impact melts are an important component of the Apennine Front, there must be a corresponding ferroan component that raises the Al_2O_3 and lowers the Mg'. He concluded that ferroan anorthosite alone cannot contribute enough Fe or Mg to change the Mg' of the bulk composition and that a more mafic ferroan rock is required. He suggested anorthositic norite 15418, but added that another mafic ferroan rock could also provide the necessary constituents. The ferroan norites found in 15459 represent another such mafic ferroan rock whose bulk composition is similar to that of AFSC. The proportions of other ferroan rocks (mare basalt, KREEP, and alkali norite) are strictly limited by their high concentrations of either transition metals and REE.

Some of the relationships between regolith samples and their components can be seen in Fig. 5, a plot of Sm versus Sc for Apennine Front samples. The major types of igneous rocks form a triangle outlining the compositions of polymict rocks. Pristine plutonic rocks of the three major suites spread outward from the origin. Mare basalts and green glass plot at the high Sc, low Sm corner of the triangular array, while KREEP basalts are at the top, having high Sm and moderate Sc. Apennine Front impact melts (triangles) form arrays between the plutonic rocks and KREEP. Regolith samples from Apennine Front Stations 2, 6, and 7 are plotted as fields for the soils and individual points (zeros) for the regolith breccias (*Korotev*, 1987a, and personal communication, 1986). The arrays of regolith breccia and soil compositions extend outward from

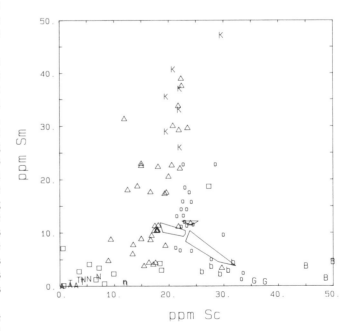

Fig. 5. Sm versus Sc for Apennine Front samples. Igneous rocks from literature data plotted by letters representing their composition: ANT-g-m=highland plutonic rocks; K=KREEP basalt; BbG=mare basalt and glass (*Ryder*, 1985). Regolith components: o = breccias, soil fields for Stations 2, 6, 7 outlined (*Korotev*, 1987a). New data: boxes=igneous rocks; triangles=impact melt rocks.

the AFSC central point toward the younger volcanic rocks: KREEP, green glass, and mare basalts. Group D melt rocks lie just to the left of the AFSC point, with group E melts below and other groups above. Various proportions of impact melts contribute most of the Sm in AFSC because the concentrations are low in the plutonic rocks. As stated above, the major element variations require significant proportions of mafic ferroan plutonic rocks. Ferroan anorthosites and Mg-suite rocks have very low Sc concentrations and need to be balanced by high Sc plutonic rocks. Ferroan norite, which plots near the group E impact melts, is the only such component. It appears likely that AFSC is a mixture of many components, among them KREEP-rich impact melts (groups B and D), Mg-suite plutonic rocks, and ferroan anorthosite, anorthositic norites, and norites. Such a complex mixture is difficult to model quantitatively, but is unlikely to be dominated by any individual igneous component.

In summary, the suite of highland rocks at the Apennine Front is the most diverse of any yet studied. Igneous rocks include members of the three major suites of highland plutonic rocks, a newly identified ferroan norite, and younger extrusive KREEP basalts. Impact melt rocks have a similarly wide range in composition, extending from those that resemble some plutonic rocks to melts of KREEP basalt composition. None of these rock types appears to dominate the average composition of the front. Diversity is a fundamental characteristic of the Apennine Front that results from its complex geologic history.

Acknowledgments. The efforts of numerous people at our four institutions were instrumental in the study. The curator's staff, Kim Willis and Charlie Galindo, extracted the clasts used in this study. Assistance with analyses was provided by R. Korotev and D. Lindstrom (WU), A. V. Murali (LPI), D. Mittlefehldt (Lockheed), and J. Wittke (USC). Discussions of Apollo 15 rocks with G. Ryder, R. Korotev, and D. Lindstrom were invaluable. This work was supported by NASA through grants NAG 9-56 (Haskin), NAG 9-29 (Marvin) and NAG 9-169 (Shervais).

REFERENCES

A15PET (Apollo 15 Preliminary Examination Team) (1972) The Apollo 15 lunar samples: a preliminary description. *Science, 175,* 363-375.

ALGIT (Apollo Lunar Geology Investigation Team) (1972) Geologic setting of the Apollo 15 samples. *Science, 175,* 407-415.

Bernatowicz T., Hohenberg C. M., Hudson B., Kennedy B. M., and Podosek F. A. (1978) Argon ages for lunar breccias 14064 and 15405. *Proc. Lunar Planet. Sci. Conf. 9th,* 905-919.

Blanchard D. P., Brannon J. C., Haskin L. A., and Jacobs J. W. (1976) Sample 15445 Chemistry. In *Interdisciplinary Studies by the Imbrium Consortium,* vol. 2, pp. 60-62. Smithsonian Astrophysical Observatory, Cambridge.

Cameron K. L. and Delano J. W. (1973) Petrology of Apollo 15 consortium breccia 15465. *Proc. Lunar Sci. Conf. 4th,* 461-466.

Chamberlain J. W. and Watkins C., eds. (1972) *The Apollo 15 Lunar Samples,* The Lunar Science Institute, Houston. 525 pp.

Chao E. C. T., Minkin J. A., and Thompson C. L. (1976) The petrology of 77215, a noritic impact ejecta breccia, *Proc. Lunar Sci. Conf. 7th,* 2287-2308.

Delano J. W. (1979) Apollo 15 green glass: Chemistry and possible origin. *Proc. Lunar Planet. Sci. Conf. 10th,* 275-300.

Dowty E., Conrad G. H., Green J. A., Hlava P. F., Keil K., Moore R. B., Nehru C. E., and Prinz M. (1973) Catalogue of Apollo 15 rake samples from stations 2 (St. George), 7 (Spur Crater, and 9a (Hadley Rille). *Institute of Meteoritics Special Publication No. 8,* University of New Mexico, Albuquerque. 75 pp.

Ganapathy R., Morgan J. W., Krahenbuhl U., and Anders E. (1973) Ancient meteoritic components in lunar highland rocks: clues from trace elements in Apollo 15 and 16 samples. *Proc. Lunar Planet. Sci. Conf. 4th,* 1239-1261.

Goodrich C. A., Taylor G. J., Keil K., Kallemeyn G. W., and Warren P. H. (1986) Alkali norite, troctolites and VHK mare basalts from breccia 14304. *Proc. Lunar Planet. Sci. Conf. 16th,* in *J. Geophys. Res., 91,* D305-D318.

Hanan B. B. and Tilton G. R. (1987) 60025: relict of primitive lunar crust? *Earth Planet. Sci. Lett., 84,* 15-21.

Haskin L. A., Blanchard D. P., Jacobs J. W., and Telander K. (1973) Major and trace element abundances in samples from the lunar highlands. *Proc. Lunar Sci. Conf. 4th,* 1275-1296.

Haskin L. A., Lindstrom M. M., Salpas P. A., and Lindstrom D. L. (1981) On compositional variations among lunar anorthosites. *Proc. Lunar Planet. Sci. 12B,* 41-66.

Jacobs J. W., Korotev R. L., Blanchard D. P., and Haskin L. A. (1977) A well-tested procedure for instrumental activation analyses of silicate rocks and melts. *J. Radioanal. Chem., 40,* 93-114.

James O. B. (1972) Lunar anorthosite 15415: Texture, mineralogy, and metamorphic history. *Science, 175,* 432-436.

James O. B., Lindstrom M. M., and Flohr M. K. (1987) Petrology and geochemistry of alkali gabbronorites from lunar breccia 67975. *Proc. Lunar Planet. Sci. Conf. 17th,* in *J. Geophys. Res., 92,* E314-E330.

Korotev R. L. (1985) Geochemical studies of Apollo 15 regolith breccias (abstract). In *Lunar Planetary Science XVI,* pp. 459-460, Lunar and Planetary Institute, Houston.

Korotev R. L. (1987a) Mixing levels, the Apennine Front soil component, and compositional trends in the Apollo 15 soils. *Proc. Lunar Planet. Sci. Conf. 17th,* in *J. Geophys. Res., 92,* E411-E431.

Korotev R. L. (1987b) The meteorite component of Apollo 16 noritic impact melt breccias. *Proc. Lunar Planet. Sci. Conf. 17th,* in *J. Geophys. Res., 92,* E491-E512.

Korotev R. L. (1987c) National Bureau of Standards coal flyash (SRM 1633a) as a multielement standard for instrumental neutron activation analysis. *Proc. Internat. Conf. Nucl. Anal. Chem.,* in *J. Radioanal. Chem.,* in press.

Laul J. C., Simon S. B., and Papike J. J. (1987) Chemistry and petrology of the Apennine Front, Apollo 15, Part II: Impact melt rocks. *Proc. Lunar Planet. Sci. Conf. 18th,* this volume.

Lindstrom M. M. (1984) Alkali gabbronorite, ultra-KREEPy melt rock and the diverse suite of clasts in North Ray crater feldpathic fragmental breccia 67975. *Proc. Lunar Planet. Sci. Conf. 15th,* in *J. Geophys. Res., 89,* C50-C62.

Lindstrom M. M. (1986) Diversity of rock types in Apennine Front Breccias (abstract). In *Lunar and Planetary Science XVIII,* pp. 486-487. Lunar and Planetary Institute, Houston.

Lindstrom M. M. and Lindstrom D. J. (1986) Lunar granulites and their precursor anorthositic norites of the early lunar crust. *Proc. Lunar Planet. Sci. Conf. 16th,* in *J. Geophys. Res., 91,* D263-D276.

Lindstrom M. M. and Marvin U. B. (1987) Geochemical and petrologic studies of clasts in Apennine Front breccia 15459 (abstract). In *Lunar and Planetary Science XVIII,* pp. 554-555. Lunar and Planetary Institute, Houston.

Lindstrom M. M., Knapp S. A., Shervais J. W., and Taylor L. A. (1984) Magnesian anorthosites and associated troctolites and dunite in Apollo 14 breccias. *Proc. Lunar Planet. Sci. Conf. 15th,* in *J. Geophys. Res., 89,* C41-C49.

Lugmair G. W. (1987) The age of the lunar crust: 60025-Methuselah's legacy (abstract). In *Lunar Planetary Science XVIII*, pp. 584-585. Lunar and Planetary Institute, Houston.

Ma M.-S., Liu Y.-G., and Schmitt R. A. (1981) A chemical study of individual green glasses and brown glasses from 15426: Implications for their petrogenesis. *Proc. Lunar Planet. Sci. 12B*, 523-534.

McKinley J. P., Taylor G. J., Keil K., Ma M.-S., and Schmitt (1984) Apollo 16: Impact melt sheets, Contrasting nature of the Cayley Plains and Descartes Mountains, and geologic history. *Proc. Lunar Planet. Sci. Conf. 14th*, in *J. Geophys. Res., 89*, B513-B524.

Nord G. L., Christie J. M., Lally J. S., and Heuer A. H. (1977) The thermal and deformational history of Apollo 15418, A partly shock-melted lunar breccia. *The Moon, 17*, 217-231.

Nyquist L. E., Weismann H., Shih C.-Y., and Bansal B. M. (1977) REE and Rb-Sr analysis of 15405 quartz-monzodiorite (Super-KREEP) (abstract). In *Lunar Science VIII*, pp. 738-740. The Lunar Science Institute, Houston.

Reid A. M., Warner J., Ridley W. I., and Brown R. W. (1973) Major element composition of glasses in three Apollo 15 soils. *Meteoritics, 7*, 395-415.

Ridley W. I. (1977) Some petrological aspects of Imbrium stratigraphy. *Phil. Trans. Roy. Soc. Lond., A285*, 105-114.

Ridley W. I., Hubbard N. J., Rhodes J. M., Weismann H., and Bansal B. (1973) The petrology of lunar breccia 15445 and petrogenetic implications. *J. Geol., 81*, 621-631.

Ryder G. (1976) Lunar sample 15405: remnant of a KREEP basalt-granite differentiated pluton. *Earth Planet. Sci. Lett., 29*, 255-268.

Ryder G. (1985) *Catalog of Apollo 15 Rocks, Parts 1, 2, and 3.* Curatorial Branch Publication 72, JSC 20787. 1296 pp.

Ryder G. (1987) Petrographic evidence for nonlinear cooling rates and a volcanic origin for Apollo 15 KREEP basalts. *Proc. Lunar Planet. Sci. Conf. 17th*, in *J. Geophys. Res., 92*, E331-E339.

Ryder G. and Bower J. F. (1977) Petrology of Apollo 15 black-and-white rocks 15445 and 15455-fragments of the Imbrium melt sheet? *Proc. Lunar Sci. Conf. 8th*, 1895-1923.

Ryder G. and Spudis P. (1987) Chemical composition and origin of Apollo 15 impact melts. *Proc. Lunar Planet. Sci. Conf. 17th*, in *J. Geophys. Res., 92*, E432-E446.

Ryder G. and Wood J. A.(1977) Serenitatis and Imbrium impact melts: implications for large-scale layering in the lunar crust. *Proc. Lunar Sci. Conf. 8th*, 655-668.

Ryder G., Lindstrom M. and Willis K. (1987) The reliability of macroscopic identifications of lunar coarse-fines particles and the petrogenesis of 2-4 mm particles in Apennine Front sample 15243. *Proc. Lunar Planet. Sci. Conf. 18th*, this volume.

Simon S. B., Papike J. J., Gosselin D. C., and Laul J. C. (1986) Petrology, chemistry, and origin of Apollo 15 regolith breccias. *Geochim. Cosmochim. Acta, 50*, 2675-1591.

Simon S. B., Papike J. J., and Laul J. C. (1987) Chemistry and petrology of the Apollo 15 Apennine Front I: KREEP basalts and plutonic rocks. *Proc. Lunar Planet. Sci. Conf. 18th*, this volume.

Spudis P. D. (1980) Petrology of the Apennine Front, Apollo 15: implications for the geology of the Imbrium impact basin (abstract). In *Papers Presented to the Conference on Multiring Basins*, pp. 83-85, Lunar and Planetary Institute, Houston.

Spudis P. D. and Ryder G. (1985) Geology and petrology of the Apollo 15 landing site: past, present, and future understanding. *Eos Trans. AGU, 66*, 721-726.

Spudis P. D. and Ryder G., eds. (1986) *Workshop on the Geology and Petrology of the Apollo 15 Landing Site*, LPI Tech Rpt. 86-03, Lunar and Planetary Institute, Houston, 126 pp.

Taylor S. R. (1982) *Planetary Science: A Lunar Perspective.* Lunar and Planetary Institute, Houston, 481 pp.

Taylor G. J., Warner R. D., Keil K., Ma M.-S., and Schmitt R. A. (1980) Silicate liquid immiscibility, evolved lunar rocks and the formation of KREEP. In *Proc. Conf. Lunar Highlands Crust*, (R. B. Merrill and J. J. Papike, eds.) pp. 339-352. Pergamon, New York.

Vetter S., Shervais J., and Lindstrom M. M. (1987a) Petrology of mare basalt and highland clasts from breccia 15498 (abstract). In *Lunar and Planetary Science XVIII*, pp. 1040-1041. Lunar and Planetary Institute, Houston.

Vetter S. K., Shervais J. W., and Lindstrom M. M. (1987b) Petrology and geochemistry of olivine-normative and quartz-normative basalts from regolith breccia 15498: new diversity in Apollo 15 mare basalts. *Proc. Lunar Planet. Sci. Conf. 18th*, this volume.

Walker D., Grove T. L., Longhi J., Stolper E. M., and Hays J. F. (1973) Origin of lunar feldspathic rocks. *Earth Planet. Sci. Lett., 20*, 325-326.

Warner J. L., Phinney W. C., Bickel C. E., and Simonds C. H. (1977) Feldspathic granulitic impactites and pre-final bombardment lunar evolution. *Proc. Lunar Sci. Conf. 8th*, 2051-2066.

Warren P. H. (1985) The magma ocean concept and lunar evolution, *Ann. Rev. Earth Planet. Sci. 13*, 201-240.

Warren P. H. and Wasson J. T. (1978) Compositional-petrographic search for pristine nonmare rocks. *Proc. Lunar Planet. Sci. Conf. 9th*, 185-217.

Warren P. H. and Wasson J. T. (1979) The compositional-petrographic search for pristine nonmare rocks: Third foray. *Proc. Lunar Planet. Sci. Conf. 10th*, 583-610.

Warren P. H. and Wasson J. T. (1980) Further foraging for pristine nonmare rocks. *Proc. Lunar Planet. Sci. Conf. 11th*, 431-470.

Warren P. H., Taylor G. J., Keil K., Kallemeyn G. W., Rosener P. S., and Wasson J. T. (1983) Sixth foray for pristine non-mare rocks and an assessment of the diversity of lunar anorhtosites. *Proc. Lunar Planet. Sci. Conf. 13th*, in *J. Geophys. Res., 88*, A615-A630.

Warren P. H., Shirley D. N., and Kallemeyn G. W. (1986) A potpourri of pristine moon rocks, including a VHK mare basalt and a unique, augite-rich Apollo 17 anorthosite. *Proc. Lunar Planet. Sci. Conf. 16th*, in *J. Geophys. Res., 91*, D319-D330.

Winzer S. R., Nava D. F., Schuhmann S., Kouns C. W., R. K. L. Lum, and Philpotts J. A. (1974) Major, minor and trace element abundances in samples from the Apollo 17 Station 7 Boulder: Implications for the origin of early lunar crustal rocks. *Earth Planet. Sci. Lett., 23*, 439-444.

Wood J. A., ed. (1976) *Interdisciplinary studies by the Imbrium Consortium, Vols. 1 & 2*, Smithsonian Astrophysical Observatory, Cambridge, 280 pp.

Chemistry and Petrology of the Apennine Front, Apollo 15, Part I: KREEP Basalts and Plutonic Rocks

S. B. Simon and J. J. Papike

Institute for the Study of Mineral Deposits, South Dakota School of Mines and Technology, Rapid City, SD 57701

J. C. Laul

Battelle, Pacific Northwest Laboratories, Richland, WA 99352

A major objective of the Apollo 15 mission was determination of the petrology and chemistry of the highland component of the Apennine Front. Toward this end we have determined the mineralogy, petrology, and chemistry of rock fragments from the Apennine Front coarse fines (10-4 and 4-2 mm). KREEP basalts have medium-K KREEP compositions and have mineral and bulk compositions similar to those of previously described samples. Data are consistent with a single eruptive event that produced several flows, giving rise to the variety of textures observed. Plutonic rocks fall into previously recognized compositional groups and most are ferroan in nature; of ten samples only three appear to belong to the Mg-suite, and no alkali anorthosites were observed. It is unlikely that any abundant rock types remain undiscovered among the returned Apennine Front samples. Unlike regolith breccia clast populations, ferroan anorthosite is clearly dominant among our samples, perhaps supplied to the regolith by the formation of Spur Crater, after formation of the regolith breccias. The abundance among these samples of rocks with ferroan affinities may be indicative of a prebasin ferroan anorthosite crust, as suggested in a hypothetical geologic cross-section by *Spudis* (1980) and *Spudis and Ryder* (1985).

INTRODUCTION

The Apollo 15 mission marked several firsts in manned exploration of the Moon. It was the first lunar mission mostly devoted to science, the first to use a lunar rover, and the first to land at a highland/mare interface. Therefore a larger area could be visited, and more ambitious scientific goals could be set. These goals included collection of basalt samples and the first deep drill core, investigation of Hadley Rille, and adequate sampling of the Apennine Front to allow determination of the petrology and chemistry of its highland component. The major features of the site are illustrated in Fig. 1. The lunar module (LM) landed on a mare (basalt) plain that is bordered to the west by Hadley Rille, a collapsed lava tube or channel that at the site is 1500 m wide and 300 m deep (*Greeley and Spudis,* 1986), and is bordered to the south by Hadley Delta mountain and the lithologic unit known as the Apennine Front. The Front was sampled at the base of the mountain, and the samples collected there probably represent material from higher on the mountain.

A major objective of the Apollo 15 mission was investigation of the highland component of the Apennine Front. Unfortunately, the highland samples returned from the Front are generally small, and due to their complexity and the pressure of samples from closely spaced later missions, have not been studied in detail. As *Ryder and Spudis* (1985) point out, this major objective of the Apollo 15 mission will remain unfulfilled until this is done; we still do not have a good understanding of the highland petrology of the Apennine Front. Specifically, new petrologic and chemical data for new rock types as well as for previously recognized plutonic rocks, KREEP basalts, and impact melt rocks will greatly improve the data base and bring

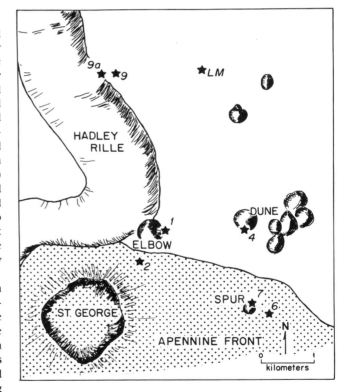

Fig. 1. Schematic map of the Apollo 15 landing site showing the major geologic features, sampling stations (numbered), and location of the Lunar Module (LM). Samples used for this study are from Stations 2, 6, and 7.

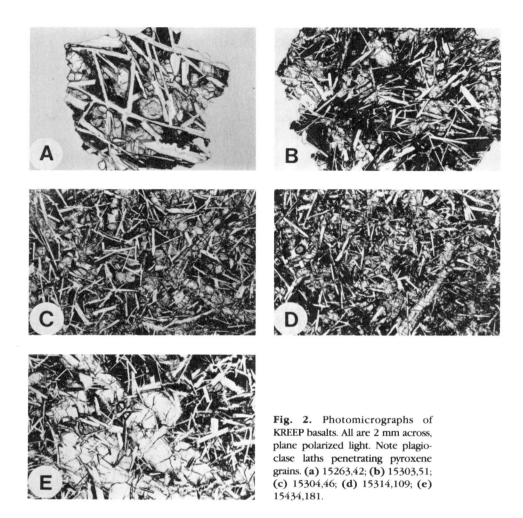

Fig. 2. Photomicrographs of KREEP basalts. All are 2 mm across, plane polarized light. Note plagioclase laths penetrating pyroxene grains. (a) 15263,42; (b) 15303,51; (c) 15304,46; (d) 15314,109; (e) 15434,181.

us much closer to a determination of the petrologic and chemical variability of the highland component of the Apennine Front. This can best be accomplished by investigation of coarse fines (10-4 and 4-2 mm fragments) collected at the Apennine Front sampling Stations 2, 6, and 7 (ref. Fig. 1), as well as studies of clasts in breccias (*Lindstrom*, 1986). The present study examines approximately 40 fragments that were hand-picked (under a binocular microscope) from Apennine Front coarse fines. The sample suite includes rocks of plutonic origin, KREEP basalts, and impact melt rocks. The impact melt data will be reported in a companion paper (*Laul et al.,* 1987), and the other rock types are discussed in this paper.

ANALYTICAL METHODS

Approximately 2 g of 4-2 mm fragments from each of the Apennine Front Stations 2, 6, and 7 were allocated for picking in addition to 10-4 mm fragments that were specifically requested based on published binocular descriptions (*Powell,* 1972). Nonmare 4-2 mm samples were hand-picked under a binocular microscope in a nitrogen-filled glove box at the Curatorial Facility, NASA Johnson Space Center. Enough samples were allocated to us so that all of the best potential highland samples could be selected from the picking pots. It is therefore unlikely that even a few plutonic samples, which are plagioclase-rich, were overlooked. Rock types more likely to be underrepresented by the selection process are KREEP basalts and impact melt rocks, both of which are generally mafic-rich, gray in color, and fine-grained. Some of the 10-4 mm fragments had previously been analyzed; the others, and all the 4-2 mm fragments, were chipped in the Curatorial Facility, one split for a thin section and the other for chemical analysis (INAA). Genealogy of the samples discussed in this paper is summarized in the Appendix. The thin sections were examined in transmitted and reflected light with a Zeiss photomicroscope. Modal data were collected with a Swift automated point counter. Mineral analyses were obtained with an automated ETEC (MAC-5) electron microprobe operated at 15 KV. Probe data were reduced according to the method of *Bence and Albee* (1968).

Bulk compositions were not determined for all samples discussed in this paper. A total of 25 samples (including impact melts) were selected for analysis by INAA using a high-efficiency Ge(Li) detector (25% FWHM 1.8 KeV for 1332 KeV of ^{60}Co),

TABLE 1. Modal mineralogy of KREEP basalt fragments from A-15 coarse fines.

	15263,42	15303,51	15304,46	15314,109	15434,181
Pyroxene	34.0	28.4	39.8	42.0	41.7
Plagioclase	41.4	49.0	46.1	43.3	41.1
Mesostasis	15.8	20.8	8.6	11.3	13.7
Cristobalite	4.0	-	3.0	-	1.8
Ilmenite	4.6	0.8	2.1	3.3	1.5
Sulfide	tr	0.4	tr	tr	0.2
Metal	0.2	-	0.1	tr	tr
Phosphate	-	-	0.3	-	-
Other	-	0.6	-	0.1	-
Total	100.0	100.0	100.0	100.0	100.0

a 4096 channel analyzer, and coincidence-noncoincidence Ge(Li)-NaI(Tl) counting systems. The coincidence-noncoincidence approach enables us to improve our detection limits for the normally difficult to measure noncoincidence γ-rays produced by the isotopes of Nd, Ni, Tm, Zr, and Sr. The details of our INAA procedure and the systematics of coincidence-noncoincidence counting are described by *Laul et al.* (1979) and *Laul* (1979). The samples were irradiated at the Oregon State University TRIGA reactor with USGS standards BCR-1, BHVO-1, GSP-1, and PCC-1 used as controls.

KREEP BASALT

Small fragments of KREEP (K-, REE-, and P-rich) basalt are fairly abundant at the Apollo 15 site. Also known as feldspathic basalts (*Dymek*, 1986), 60 samples have been analyzed, the largest being rake sample 15386 with a weight of 7.5 g (*Irving*, 1977). Most are much smaller. The KREEP basalts are distinguished from mare basalts by their relatively high orthopyroxene and feldspar contents, low ilmenite and TiO_2, and enrichments in incompatible elements, especially large-ion lithophiles (LIL). KREEP basalts are discussed in review papers by *Meyer* (1977), *Irving* (1977), and *Dymek* (1986). Our present sample suite includes five KREEP basalt fragments.

TABLE 2. Representative analyses of pyroxenes in Apollo 15 coarse fines.

	KREEP Basalts					Ferroan Anorthosites								Mg-Suite				
	15263,42		15434,181			15223,50		15223,51		15303,53		15303,103		15314,125	15303,104	15223,48	15303,55	15264,19
	1.	2.	1.	2.	3.	1.	2.	1.	2.	1.	2.	1.	2.	1.	1.	1.	1.	1.
SiO_2	54.86	49.16	55.68	51.91	49.63	53.16	52.54	52.52	52.74	53.93	52.09	54.30	51.97	54.22	51.68	53.77	55.62	54.62
Al_2O_3	2.08	1.66	1.55	1.77	1.60	0.77	1.20	0.98	0.76	0.69	1.38	0.72	1.47	0.80	0.41	1.11	0.70	0.75
TiO_2	0.0	0.29	0.30	0.90	1.26	0.46	0.65	0.27	0.38	0.34	0.74	0.42	0.84	0.22	0.45	0.44	0.39	0.48
FeO	9.41	20.86	9.38	18.00	17.50	17.59	7.20	23.35	11.55	18.54	8.34	17.16	7.67	16.25	24.74	12.41	13.12	14.64
MnO	0.28	0.27	0.17	0.29	0.36	0.43	0.22	0.56	0.25	0.36	0.31	0.44	0.23	0.27	0.51	0.29	0.14	0.19
MgO	30.48	10.41	31.81	23.27	13.33	24.81	15.59	20.68	14.33	22.80	14.71	24.79	16.07	26.66	18.18	29.02	28.58	27.23
CaO	1.52	15.09	1.25	2.14	14.60	1.39	21.54	1.36	20.84	2.80	21.07	0.85	20.64	1.11	2.42	1.51	0.95	1.99
Na_2O	0.16	0.02	0.06	0.09	0.04	0.14	0.14	0.06	0.0	0.0	0.20	0.11	0.0	0.02	0.10	0.07	0.0	0.16
Cr_2O_3	0.72	0.33	0.60	0.64	0.34	0.37	0.39	0.08	0.19	0.24	0.43	0.37	0.43	0.39	0.16	0.70	0.39	0.57
	99.51	98.09	100.80	99.01	98.66	99.12	99.47	99.86	101.05	99.70	99.27	99.16	99.32	99.94	98.65	99.32	99.89	100.63
Si	1.939	1.947	1.940	1.927	1.921	1.962	1.955	1.975	1.962	1.990	1.950	1.989	1.935	1.966	1.988	1.937	1.983	1.958
Al	0.061	0.053	0.060	0.073	0.072	0.033	0.045	0.025	0.033	0.010	0.050	0.011	0.064	0.034	0.012	0.046	0.017	0.030
Total Tet.	2.000	2.000	2.000	2.000	1.993	1.995	2.000	2.000	1.995	2.000	2.000	2.000	1.999	2.000	2.000	1.983	2.000	1.988
Al	0.025	0.024	0.003	0.004	0.000	0.000	0.007	0.018	0.000	0.019	0.010	0.020	0.000	0.000	0.006	0.000	0.011	0.000
Ti	0.000	0.008	0.007	0.025	0.036	0.012	0.017	0.006	0.010	0.008	0.020	0.011	0.023	0.005	0.012	0.011	0.009	0.012
Fe	0.278	0.690	0.273	0.559	0.565	0.542	0.224	0.733	0.359	0.571	0.260	0.525	0.238	0.493	0.796	0.373	0.390	0.438
Mn	0.008	0.008	0.004	0.008	0.011	0.013	0.006	0.018	0.007	0.011	0.010	0.013	0.006	0.007	0.016	0.008	0.004	0.005
Mg	1.606	0.614	1.653	1.287	0.769	1.364	0.865	1.158	0.793	1.254	0.820	1.353	0.892	1.441	1.042	1.558	1.520	1.455
Ca	0.057	0.639	0.046	0.085	0.605	0.054	0.858	0.054	0.830	0.110	0.845	0.033	0.823	0.042	0.099	0.058	0.035	0.076
Na	0.010	0.001	0.003	0.005	0.002	0.009	0.010	0.004	0.000	0.000	0.013	0.007	0.000	0.001	0.006	0.004	0.000	0.010
Cr	0.020	0.009	0.016	0.018	0.010	0.009	0.011	0.002	0.005	0.006	0.012	0.009	0.012	0.010	0.004	0.019	0.010	0.016
Total Oct.	2.004	1.993	2.005	1.991	1.998	2.003	1.998	1.993	2.004	1.979	1.990	1.971	1.994	1.999	1.981	2.031	1.979	2.012
Total Cations	4.004	3.993	4.005	3.991	3.991	3.998	3.998	3.993	3.999	3.979	3.990	3.971	3.993	3.999	3.981	4.014	3.979	4.000
Wo	2.9	32.9	2.3	4.4	31.2	2.8	44.1	2.8	41.9	5.7	43.9	1.7	42.1	2.1	5.1	2.9	1.8	3.9
En	82.7	31.6	83.8	66.6	39.7	69.6	44.4	59.5	40.0	64.8	42.6	70.8	45.7	72.9	53.8	78.3	78.1	73.9
Fs	14.3	35.5	13.8	28.9	29.1	27.6	11.5	37.7	18.1	29.5	13.5	27.5	12.2	24.9	41.1	18.8	20.1	22.2

Mineralogy and Petrography

The KREEP basalts generally consist of blocky pyroxenes, plagioclase laths, and Si-, K-rich mesostasis in a mostly subophitic texture. A variety of grain sizes is observed, as illustrated in Fig. 2. Cristobalite, phosphate, ilmenite, troilite, and Ni-Fe metal are also present in minor amounts, and modal data are given in Table 1. As Fig. 2 shows, plagioclase occurs as tabular grains and as narrow to acicular laths. Some small laths are completely enclosed by orthopyroxene. An explanation for the coexistence of coarse Mg-rich orthopyroxene and fine plagioclase laths and mesostasis in many A-15 KREEP basalt samples is a two-stage cooling history, consisting of slow cooling and orthopyroxene accumulation followed by rapid cooling (*Ryder*, 1987). This virtually requires a volcanic origin for the phenocryst-bearing basalts and is not consistent with an impact origin. Uniform cooling rates can produce porphyritic textures, but only if there is a fairly large interval over which the phenocryst phase is the only mineral crystallizing and the nucleation density is low (*Lofgren et al.*, 1974). Apollo 15 KREEP basalt compositions are approximately multiply saturated (pyroxene-plagioclase) at the liquidus (*Irving*, 1977) and would not tend to form a porphyritic texture with a linear cooling rate. A volcanic origin is consistent with the lack of clasts in the samples and, as discussed below and by *Irving* (1977), the chemical systematics.

Mineral Chemistry

Representative pyroxene analyses are presented in Table 2. All analyses are summarized in Fig. 3, and they display a trend from Mg-rich orthopyroxene to pigeonite. In some of the samples the trend continues through augite, and in three of the samples small crystals of Fe-rich pyroxene were found in the mesostasis. Our analyses extend the observed compositional range of orthopyroxene to slightly more Mg-rich compositions (~En_{87}) than previously observed (*Dymek*, 1986). These pyroxenes probably began crystallizing prior to eruption and grew slowly.

Plagioclase is also zoned, from ~An_{90} to ~An_{75} (Fig. 4), with only slight increases in the Or component; K_2O becomes concentrated in the mesostasis (5-6 wt % K_2O). Representative plagioclase analyses are given in Table 3.

The other phases, including ilmenite, phosphate, and cristobalite, are found in the mesostasis and apparently

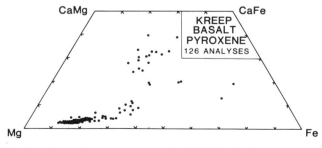

Fig. 3. Compositions of pyroxenes in KREEP basalt fragments.

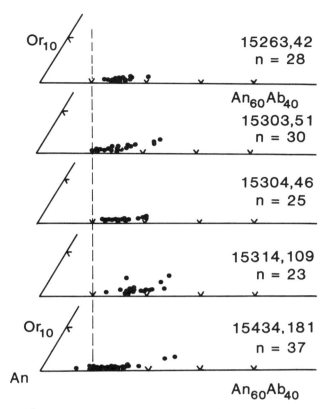

Fig. 4. Feldspar compositions in KREEP basalts.

crystallized late. Analyses of whitlockite from 15434,181 are given in Table 4, along with analyses from other KREEP basalts. They all have similar La_2O_3 contents and therefore presumably similar REE contents, but the A-15 samples have higher Fe/Mg ratios than the KREEPy Apollo 14 impact melt rock 14310.

Bulk Chemistry

The bulk compositions of the KREEP basalts are tabulated in Table 5. Four of the five are compositionally similar despite their textural variations and have K_2O contents of ~0.5 wt %, typical of A-15 KREEP basalts (*Irving*, 1977). Sample 15314,109 may be an unrepresentative chip in that not only is it compositionally different from our other KREEP basalt analyses, it is also different from a previous analysis of a different chip (15314,81) from the same fragment (15314,34) by *Hubbard et al.* (1973), which gave more typical values.

Irving (1977) reported that bulk atomic 100 × Mg/(Mg+Fe) values for A-15 KREEP basalts range from 34.6 to 73.8. The values for the present samples range from 59 to 66.4, but overall there is not much variation in the major elements.

REE contents, normalized to chondrites, are summarized in Fig. 5. Four of the five appear to be medium-K KREEP, with La at 200× chondrites.

TABLE 3. Representative analyses of plagioclase in Apollo 15 coarse fines.

	KREEP Basalts				Ferroan Anorthosites						Mg-Suite		
	15303,51		15434,181		15223,50		15223,51		15303,104		15264,19		15223,48
	1.	2.	1.	2.	1.	2.	1.	2.	1.	2.	1.	2.	1.
SiO_2	47.45	50.32	47.55	47.96	45.24	45.78	44.45	45.20	44.78	44.75	45.98	46.67	46.82
Al_2O_3	33.71	30.85	32.37	32.22	35.73	35.43	35.71	35.42	34.73	34.46	33.73	33.75	33.03
FeO	0.38	0.71	0.30	0.32	0.0	0.0	0.0	0.09	0.03	0.14	0.08	0.08	0.03
CaO	17.77	15.02	17.89	16.95	19.43	19.14	19.41	18.98	19.23	19.17	18.73	18.34	19.07
Na_2O	1.40	2.23	1.13	1.51	0.14	0.31	0.29	0.58	0.35	0.61	0.57	0.78	0.68
K_2O	0.12	0.43	0.09	0.13	0.02	0.0	0.03	0.0	0.03	0.05	0.12	0.13	0.09
	100.83	99.56	99.33	99.09	100.56	100.66	99.89	100.27	99.15	99.18	99.21	99.75	99.72
Si	2.165	2.308	2.200	2.218	2.071	2.092	2.053	2.078	2.083	2.085	2.134	2.151	2.162
Al	1.813	1.667	1.765	1.757	1.928	1.908	1.943	1.919	1.905	1.893	1.846	1.834	1.798
Total	3.978	3.975	3.965	3.975	3.999	4.000	3.996	3.997	3.988	3.978	3.980	3.985	3.960
Fe	0.013	0.027	0.011	0.012	0.000	0.000	0.000	0.002	0.001	0.005	0.003	0.003	0.001
Ca	0.868	0.737	0.887	0.840	0.953	0.937	0.959	0.935	0.959	0.957	0.931	0.906	0.944
Na	0.123	0.198	0.100	0.134	0.012	0.026	0.025	0.051	0.032	0.055	0.051	0.070	0.061
K	0.006	0.024	0.005	0.006	0.000	0.000	0.001	0.000	0.002	0.003	0.007	0.008	0.005
Total	1.010	0.986	1.003	0.992	0.965	0.963	0.985	0.988	0.994	1.020	0.992	0.987	1.011
Total Cations	4.988	4.961	4.968	4.967	4.964	4.963	4.981	4.985	4.982	4.998	4.972	4.982	4.971
Or	0.6	2.4	0.5	0.6	0.0	0.0	0.1	0.0	0.2	0.3	0.7	0.8	0.5
Ab	12.3	20.7	10.1	13.7	1.2	2.7	2.5	5.2	3.2	5.4	5.2	7.1	6.0
An	87.1	76.8	89.4	85.7	98.8	97.3	97.4	94.8	96.6	94.3	94.1	92.1	93.5

Apollo 15 KREEP basalts typically contain ~20 ppm Ni. Because Ni is relatively abundant in meteorites, it is a good indicator of meteoritic contamination. Table 5 shows that four of the KREEP basalts we analyzed have low Ni contents, except for 15314,109. We conclude that our chip of this sample contains a slight amount of non-KREEPy, Ni-rich material, probably regolith adhering to the side. This would account for the sample's relative depletion in KREEP-related components and enrichment in Ni. Otherwise, the low Ni contents of A-15 KREEP basalts further support a volcanic origin, as opposed to impact melting.

Discussion

Petrographic observations, mineral chemical data, and bulk chemical data are all consistent with a volcanic origin for Apollo 15 KREEP basalts. The virtually identical fractionation trends combined with textural and grain-size variations are consistent with derivation from a single large flow or from different cooling units within a series of cogenetic flows.

Multiple saturation of the melt is indicated by: (1) textures in which plagioclase laths penetrate and are enclosed by opx crystals; (2) pyroxene Ti-Al systematics that do not indicate an incoming of plagioclase (*Bence and Papike*, 1972); and (3) the bulk compositions, which generally plot on or near the pyroxene-plagioclase cotectic on an En-An-SiO$_2$ plot (*Irving*, 1977). However, KREEP basalt petrogenesis has not been successfully modeled. One problem is the combination of relatively unfractionated pyroxene and plagioclase compositions in a fairly fractionated, REE-rich melt, which require contradictory amounts of partial melting of the source (*Dymek*, 1986).

Three A-15 KREEP basalts have been dated, and they give $^{87}Rb/^{86}Sr$ isochron ages of approximately 3.9 b.y. (*Nyquist et al.*, 1973; *Papanastassiou and Wasserburg*, 1976). It is possible that the formation of the Imbrium basin triggered the generation and eruption of the A-15 KREEP basalts through pressure-release melting (e.g., *Spudis and Ryder*, 1985; *Dymek*, 1986).

TABLE 4. KREEP basalt whitlockite compositions.

	15434,181		15386*	14310†
	1.	2.		
MgO	1.67	1.58	0.9	1.88
Al_2O_3	0.06	0.05	-	-
SiO_2	0.45	0.29	-	0.39
P_2O_5	42.21	41.14	42	43.22
CaO	40.26	40.91	42	38.68
FeO	3.83	4.29	4	3.35
MnO	0.13	0.07	-	-
Y_2O_3	2.60	2.50	8	3.01
La_2O_3	0.95	0.84	0.6	1.02
Ce_2O_3	-	-	1.2	

*Takeda et al., (1984).
†Brown et al. (1972).

TABLE 5. Bulk compositions of KREEP basalts.

	15263,42	15303,51	15304,46	15314,109	15434,181
Sample Wt (mg)	8.0	20.9	28.4	11.1	30.9
TiO_2 (wt %)	1.7	1.6	1.7	1.3	1.9
Al_2O_3	15.3	15.7	15.1	15.0	14.6
FeO	9.90	9.2	10.0	12.1	10.4
MgO	9.6	10.2	9.2	9.9	8.8
MnO	0.140	0.126	0.136	0.155	0.150
Cr_2O_3	0.355	0.340	0.340	0.330	0.350
CaO	9.4	9.3	9.3	10.2	9.0
Na_2O	0.74	0.70	0.74	0.49	0.81
K_2O	0.54	0.48	0.53	0.25	0.56
Sc (ppm)	20.2	18.5	20.2	24.0	21.2
V	65	60	63	70	60
Co	20.4	19.4	20.0	35.5	19.0
Ba	750	650	710	370	780
Sr	230	180	210	310	180
La	68.0	61.0	71.0	33.0	74.0
Ce	170	160	180	85	190
Nd	110	95	105	61	110
Sm	31.6	26.0	29.5	15.4	33.0
Eu	2.60	2.50	2.75	1.65	2.70
Gd	36	33	39	18	40
Tb	6.10	5.50	6.27	3.20	6.50
Dy	39	33	37	20	42
Ho	8.6	6.9	8.5	4.4	9.1
Tm	3.1	2.7	3.1	1.7	3.4
Yb	20.4	18.5	20.5	11.0	21.8
Lu	3.05	2.65	3.00	1.55	3.20
Zr	880	860	800	400	980
Hf	22.7	20.0	22.4	10.3	23.9
Ta	2.8	2.4	2.8	1.4	2.9
Th	12.0	10.9	12.2	6.0	12.8
U	3.3	2.9	3.3	1.6	3.5
Ni	<20	<20	<20	120	<20

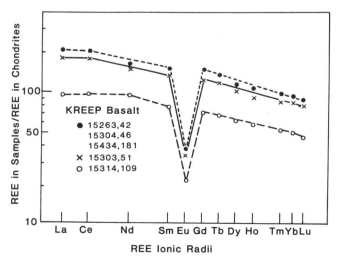

Fig. 5. Chondrite-normalized rare earth element patterns of KREEP basalts. Chondritic values used for normalization are: La = 0.34, Ce = 0.87, Ne = 0.64, Sm = 0.195, Eu = 0.073, Gd = 0.26, Tb = 0.047, Dy = 0.33, Ho = 0.078, Tm = 0.032, Yb = 0.22, and Lu = 0.034.

PLUTONIC ROCKS

One of the goals of the Apollo 15 mission was the sampling of ancient crustal rocks excavated by the Imbrium event and presumably present in the Apennine Mountains. Rocks of plutonic origin are present in the Apennine Front, but they are not abundant. We found 10 fragments, including some breccias, of probable plutonic origin. Representative pyroxene, plagioclase, and olivine analyses are given in Tables 2, 3, and 6, respectively.

The modal mineralogies of plutonic samples fine-grained enough to give a reasonably representative mode are given in Table 7. At the top of each column is the rock name according to the scheme of *Prinz and Keil* (1977), in which they are named based on their plagioclase:mafic ratio and their olivine:opx:cpx ratio. For example, anorthosites are defined as having ≥ 90 vol % plagioclase. Rocks with 77.5-90% plagioclase are troctolitic, noritic, or gabbroic anorthosites if the dominant mafic mineral is olivine, orthopyroxene, or clinopyroxene, respectively. However, modes and therefore bulk chemistry can vary within a rock. Also these rocks are very old, have

TABLE 6. Representative analyses of olivine in Apollo 15 coarse fines.

	15223,50		15303,103	15314,125		15303,104	
	1.	2.	1.	1.	2.	1.	2.
SiO_2	36.96	36.21	37.45	38.55	38.50	33.97	34.86
Al_2O_3	0.06	0.11	0.08	0.12	0.18	0.11	0.00
TiO_2	0.04	0.13	0.05	0.03	0.03	0.06	0.11
FeO	30.02	31.70	28.46	27.48	27.15	41.63	41.97
MnO	0.30	0.37	0.26	0.30	0.26	0.34	0.44
MgO	32.32	30.29	33.62	35.57	33.88	22.65	21.82
CaO	0.09	0.09	0.09	0.09	0.12	0.10	0.12
Cr_2O_3	0.0	0.0	0.06	0.00	0.06	0.16	0.23
	99.79	98.90	100.07	102.14	100.18	99.02	99.55
Si	1.000	1.000	1.001	1.001	1.019	0.985	1.009
Al	0.001	0.003	0.002	0.003	0.005	0.003	0.000
Ti	0.000	0.002	0.001	0.000	0.000	0.001	0.001
Fe	0.679	0.731	0.637	0.597	0.600	1.010	1.015
Mn	0.006	0.008	0.005	0.006	0.005	0.007	0.010
Mg	1.303	1.246	1.339	1.379	1.337	0.979	0.941
Ca	0.002	0.002	0.002	0.002	0.003	0.002	0.003
Cr	0.000	0.000	0.001	0.000	0.001	0.003	0.005
Total Cations	2.991	2.992	2.998	2.988	2.970	2.988	2.984
Fo	65.7	63.0	67.8	69.8	69.0	49.2	48.1
Fa	34.3	37.0	32.2	30.2	31.0	50.8	51.9

experienced one or more impact events, and generally have been brecciated and/or recrystallized and rarely retain their original (possibly cumulate) textures. Therefore a somewhat more dependable method of rock identification is with mineral chemistry.

It has been known for some time (e.g., *Warner et al.*, 1976; *Warren and Wasson*, 1977) that lunar plutonic rocks define two major trends on a plot of average Mg′ [atomic Mg/(Mg+Fe)] in mafic silicates versus average anorthite content in plagioclase (Fig. 6). One trend, the "Mg-suite," has a positive slope and mainly includes non-anorthosites (e.g., norites, troctolites, gabbroic anorthosites, etc.) with magnesian mafics. The members of the other group, the ferroan anorthosites, have extremely calcic plagioclase but have ferroan mafics relative to the Mg-suite. They exhibit a wide range of mafic Mg′ values with little corresponding variation in plagioclase composition. As described by *Raedeke and McCallum* (1980), the large amount of plagioclase relative to the mafics may have buffered the plagioclase composition. Alternatively, the anorthosites may simply be reflecting very low Na_2O contents of the parent liquid (*Ryder*, 1982). The abundance of plagioclase of uniform composition represented by the lunar ferroan anorthosites is one line of evidence in support of a virtually Moonwide magma ocean. Concentration of plagioclase by flotation on a large magma body after some fractional crystallization has enriched the melt in FeO (increasing its density) is one way to account for the presumed volume of plagioclase and its coexistence with ferroan mafic silicates.

Unlike the ferroan anorthosites, the Mg-suite members exhibit a wide range of plagioclase composition and a trend

TABLE 7. Modal mineralogy of plutonic rocks (vol %).

	Anorthosites			Noritic Anorthosites		Troctolitic Anorthosite	Anorthositic Gabbro	Anorthositic Norite	Granulite
	15223,50	15223,51	15303,53	15264,19	15303,103	15314,125	15303,104	15223,48	15303,59
Plagioclase	90.1	91.9	91.4	85.4	78.0	81.9	75.3	69.8	86.8
Pyroxene	5.4	8.1	5.5	13.9	10.8	5.5	19.1	30.1	8.7
Olivine	4.5	-	2.9	-	9.8	12.4	4.3	-	4.3
Ilmenite	-	-	-	tr	1.0	0.2	0.2	tr	tr
Sulfide	tr	-	tr	0.1	-	-	tr	-	tr
metal	-	-	0.2	-	0.4	-	-	-	0.2
Other	-	-	-	0.6	-	-	1.1	-	-
Total	100.0	100.0	100.0	100.0	100.0	100.0	100.0	99.9	100.0

tr = <0.1

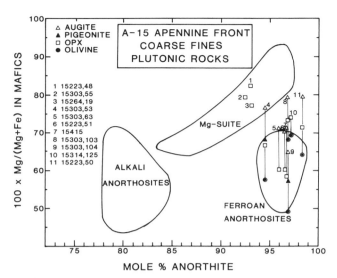

Fig. 6. Average Mg' in mafic silicates versus average anorthite content in plagioclase in plutonic rocks. Most samples have ferroan anorthosite affinities. Data for "Genesis Rock," anorthosite 15415 (*Smith and Steele*, 1974), are plotted for comparison.

of decreasing Mg' with decreasing An. Based on the differences in slopes, the gap between the two suites (*Warren and Wasson*, 1980a) and the diversity of lithologies in the Mg-suite (*James*, 1980), it is unlikely that the suites are comagmatic. Although the sources of the Mg-suite magmas have not been identified, they are thought to be partial melts of the lower crust, postdating the ferroan anorthosites. Various degrees of assimilation of the anorthositic crust by the Mg-suite magmas as they were intruded gave rise to the observed variety of lithologies, norites and troctolites being the most abundant (*Warren*, 1985).

Over the last few years a field for alkali anorthosite, a KREEP-related type with relatively sodic (An_{80-85}) plagioclase, has been added. This relatively rare rock type was first described by *Hubbard et al.* (1971).

Figure 6, a plot of Mg' versus An for our plutonic rock mineral compositions has been used to group our samples. All the analyses plot either with the Mg-suite or the ferroan anorthosites, and show no evidence of mixing of rock types or KREEP addition. The ferroan anorthosites and Mg-suite rocks that we analyzed are discussed below.

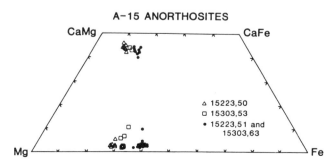

Fig. 7. Compositions of pyroxene in ferroan anorthosites from the A-15 coarse fines.

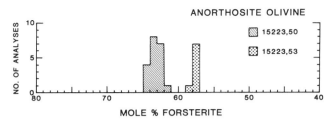

Fig. 8. Compositions of olivine in ferroan anorthosites from the A-15 coarse fines.

Ferroan Anorthosites

Anorthosite has been known for a long time to be a minor but, as samples of the ancient lunar crust, important component of the Apennine Front. The first "large" sample of lunar anorthosite to be returned, rock 15415, was collected at Station 7. It yielded an Ar-Ar age of ~4.1 b.y. (*Husain et al.*, 1972). This anorthosite is of a type termed "ferroan" because of the relatively Fe-rich nature of the mafic silicates, reflecting their crystallization from a fractionated trapped liquid during the formation of the early lunar crust. Other than 15415, ferroan anorthosites are found as small fragments in the Apennine Front regolith.

Figure 7 summarizes the compositions of pyroxenes in the ferroan anorthosite (based on modes and mineral compositions) samples. Note the absence of Mg-rich orthopyroxenes that are typical of most other highland rock types. The wide miscibility gap between the opx and augite is consistent with slow cooling. Olivine compositions are illustrated in Fig. 8, and like the pyroxenes show no compositional overlap among samples. These samples seem to define a crystallization trend

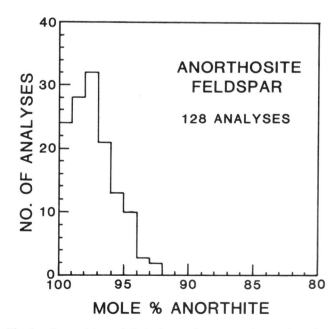

Fig. 9. Compositions of plagioclase in ferroan anorthosites from the A-15 coarse fines.

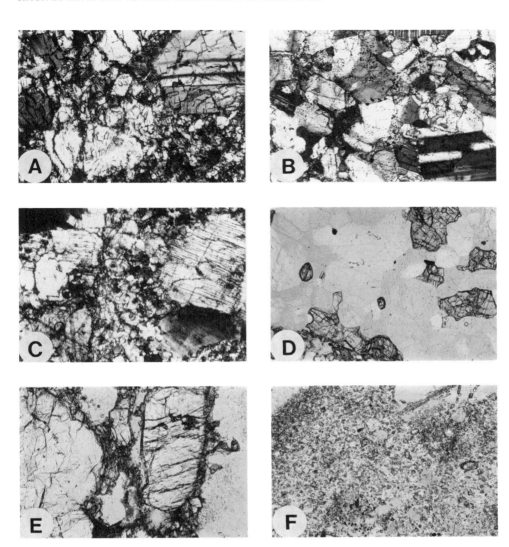

Fig. 10. Photomicrographs of plutonic rock fragments. All are 1 mm across. (**a**) 15264,19, x-nic; (**b**) 15303,103, x-nic; (**c**) 15314,125, x-nic; (**d**) 15303,104; (**e**) 15223,48. Grains with cracks are orthopyroxene; white grains without cracks are maskelynite. (**f**) Granulite 15303,59.

on the basis of their mafic silicate compositions, although the corresponding feldspars do not co-vary in a similar way. This is a characteristic of the ferroan anorthosites, which exhibit a wide range of Fe/Mg ratios in mafic silicates with no corresponding Na increase in plagioclase.

It is clear from Fig. 9 how extremely Ca-rich the feldspar is in these rocks. Most compositions are more calcic than An_{94} despite coexistence with relatively iron-rich mafic silicates.

On the basis of mineral compositions, several other samples are probably members of the ferroan suite: 15303,103, 15303,104, and 15314,125. On Fig. 6 they plot in the ferroan anorthosite. Photomicrographs are shown in Fig. 10. Although the modes (Table 7) indicate <90 vol % plagioclase, they may be anorthosites, ferroan norites, or gabbronorites.

Sample 15303,103 consists of several large lithic clasts separated by a fine-grained granulated matrix. The clasts have a cumulate texture, with some recrystallization of plagioclase. The pyroxene in this sample is more Mg-rich than that in the ferroan anorthosites, and is similar to that in 15314,125 (Fig. 11). They also have similar olivine (Fo_{68-70}) and plagioclase

(Fig. 12) compositions. Although the mafic silicates are similar to those in Mg-norites (*James and Flohr,* 1983) the plagioclase is too calcic, and the samples plot closer to the ferroan anorthosites than to the Mg-suite (Fig. 6). These samples (,103 and ,125) may be members of the magnesian subgroup of ferroan anorthosites (*James,* 1987).

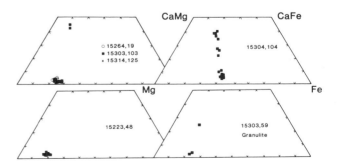

Fig. 11. Compositions of pyroxenes in plutonic rocks.

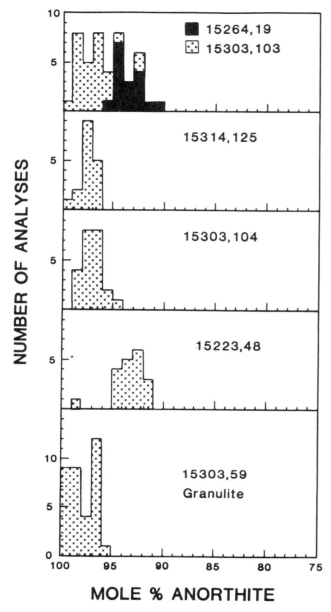

Fig. 12. Compositions of plagioclase in plutonic rocks.

oxene and plagioclase that is slightly less calcic than that in the ferroan anorthosites. The mineral compositions in these samples indicate noritic affinities.

All three samples show evidence of shock deformation. Sample 15264,19 (Fig. 10a) has a brecciated, cataclastic texture in which the largest plagioclase grain is 0.5 mm across. Plagioclase in 15303,55 has fractures and wavy extinction. Sample 15223,48 (Fig. 10e) has been strongly shocked; the plagioclase has been converted to maskelynite and the pyroxene exhibits mosaic-like extinction. The sample was probably originally fairly coarse-grained, as some of the surviving pyroxene grains are 1 mm in length. Pyroxene compositions (Fig. 11) show a very limited range, and plagioclase (maskelynite) compositions (Fig. 12) are consistent with classification of this sample as a norite.

Sample 15303,55 is a small chip of a fairly coarse (~1.5 mm) shocked fragment. It looks like an anorthosite, but its plagioclase and pyroxene plot with the Mg-suite (Fig. 6), and it may be a norite. However, also present in the sample is a phosphate grain that is very similar in composition to a phosphate in a Mg-anorthosite reported by *Lindstrom et al.* (1984), with ~2 wt % Y_2O_3 and 1 wt % La_2O_3. Based on the coarse texture and the mineral compositions, 15303,55 may be the first Mg-anorthosite found at the Apollo 15 site.

Granulite

Granulites are considered to be true metamorphic rocks, in the sense that they have experienced thermally-induced solid state recrystallization. They have been found at all lunar landing sites (*Ostertag et al.*, 1987). Sample 15303,59 is a typical lunar

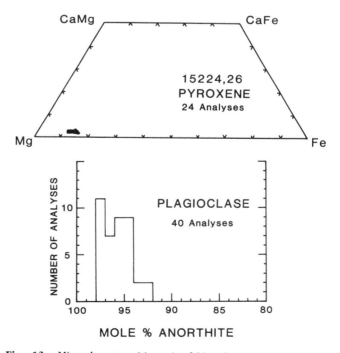

Fig. 13. Mineral compositions in feldspathic cataclastic rock 15224,26.

Sample 15303,104 has an annealed texture. The plagioclase is recrystallized with many polygonal grains and triple junctions. Exsolution lamallae can be observed in the pyroxene, and the compositions (Fig. 11) show the effect of Fe/Mg reequilibration. The olivine is also highly equilibrated and ferroan, Fo_{48-9}. The plagioclase compositions (Fig. 12) range from An_{94} to An_{98}. This sample plots well within the ferroan anorthosite field (Fig. 6).

Mg-Suite

Three samples (15223,48, 15303,55, and 15264,19) clearly plot with the Mg-suite (Fig. 6), having magnesian orthopyr-

TABLE 8. Bulk compositions of selected plutonic rocks.

	15233,50	15223,51	15314,125	15303,104	15303,59
Sample Wt (mg)	11.3	22.0	22.4	20.3	10.8
TiO_2 (wt %)	<0.1	<0.1	<0.1	0.30	0.15
Al_2O_3	32.1	31.4	29.7	27.2	29.0
FeO	2.30	1.43	2.50	5.40	3.50
MgO	1.50	1.0	3.3	3.5	4.7
MnO	0.030	0.023	0.032	0.078	0.053
Cr_2O_3	0.023	0.013	0.025	0.108	0.114
CaO	17.7	17.1	16.3	16.5	16.8
Na_2O	0.28	0.23	0.25	0.30	0.35
K_2O	0.012	0.007	0.014	0.015	0.034
Sc (ppm)	1.7	1.1	2.8	11.0	5.2
V	<10	<10	<10	20	25
Co	3.6	1.5	5.8	10.0	22.5
Ba	<10	<10	<20	30	50
Sr	150	150	150	140	160
La	0.25	0.19	0.21	1.15	2.20
Ce	0.56	0.44	<1	3.0	5.4
Nd	<1	<1	<1	2.2	<4
Sm	0.084	0.074	0.10	0.64	0.98
Eu	0.76	0.75	0.73	0.76	0.94
Gd	<1	<1	<1	<1	1.3
Tb	0.020	0.017	0.025	0.14	0.23
Dy	<0.3	<0.3	<0.3	1.0	1.6
Ho	<0.3	<0.3	<0.3	0.24	<0.5
Tm	<0.1	<0.1	<0.1	<0.2	<0.2
Yb	0.058	0.055	0.11	0.71	0.95
Lu	0.0090	0.0085	0.017	0.10	0.14
Zr	<10	<10	<10	<30	<40
Hf	<0.1	0.050	0.082	0.40	0.63
Ta	<0.02	<0.02	<0.05	0.050	0.10
Th	<0.05	<0.05	<0.05	0.13	0.45
U	<0.1	<0.1	<0.01	<0.1	0.11
Ni	<10	<10	<10	30	480

granulite in that it has fine-grained, recrystallized mafics and feldspar (Fig. 10f), and is rich in extremely calcic plagioclase (>An_{95}). The few mafics large enough to analyze, along with the calcic plagioclase, suggest that it is a ferroan granulite (*Lindstrom and Lindstrom*, 1986).

Feldspathic Cataclastic Rock

As the name implies, this sample (15224,26) is a breccia with abundant angular feldspar clasts in a glassy matrix. This type of breccia is common at the Apollo 16 site and is generally derived from extremely feldspathic rocks. The mineral compositions in 15224,26 (Fig. 13) indicate that it probably is a monomict breccia of a noritic rock. The pyroxene is Mg-rich opx with a very limited compositional range, and the plagioclase is slightly less calcic than that of ferroan anorthosites.

Bulk Chemistry of Plutonic Rocks

Several of the plutonic samples were analyzed for 32 major, minor, and trace elements by INAA and the analyses are given in Table 8. Low Ni contents, except for the granulite, indicate a lack of meteoritic contamination. The results are consistent with the petrographic observations, and are typical of the various rock types. Chondrite-normalized REE abundances are illustrated in Fig. 14. Lunar plutonic rocks have low REE contents and positive Eu anomalies. Anorthosites are the most extreme, with the lowest REE contents and the largest Eu anomalies. The other plutonic rocks, because they have less plagioclase than anorthosites, have smaller Eu anomalies and somewhat higher REE contents. Mg-norites range up to 30× chondrites and Mg-gabbronorites up to ~60× chondrites, with negative Eu anomalies (*James and Flohr*, 1983), whereas granulites exhibit positive Eu anomalies (*Lindstrom and Lindstrom*, 1986), as does our sample 15303,59.

DISCUSSION

For reasons explained earlier, our discussion of the plutonic samples has been organized based on their mineral compositions. In most cases there is good agreement between the rock type indicated by the mode and that indicated by mineral compositions, although there are three exceptions.

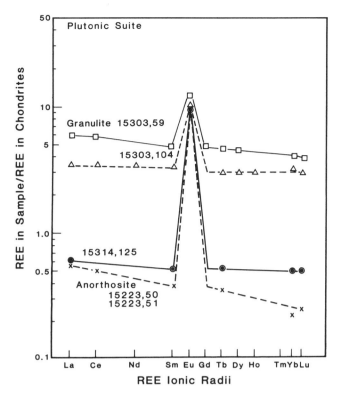

Fig. 14. Chondrite-normalized rare earth element patterns for plutonic rock fragments.

to our analysis of another chip of that sample (29.7 wt % Al_2O_3, 0.21 ppm La; Table 8). Recently Ryder (personal communication, 1985) called it a "poikilitic feldspathic norite...not an anorthosite." The sample does have poikilitic pyroxene, but is probably too olivine-rich to be considered noritic. James (personal communication, 1987) suggested that ,103 and ,125 may be coarse-grained granulites, but the low Ni content of ,125 (Table 8) indicates that it does not have the meteoritic contamination typical of lunar granulites (*Lindstrom and Lindstrom*, 1986; *Ostertag et al.*, 1987). As Fig. 6 shows, both ,103 and ,125 plot in the upper part of the ferroan anorthosite field. They may, as suggested by James (personal communication, 1987), belong to the magnesian subgroup of ferroan anorthosites (*James*, 1987).

Thus, in spite of the various modes, most of the plutonic samples have ferroan anorthosite affinities, with no indication of a KREEP component or alkali anorthosite. This is similar to our findings for Apollo 11 highland coarse fines (*Simon et al.*, 1983), and in sharp contrast to our results for the A-12 coarse fines (*Simon and Papike*, 1985), in which alkali anorthosites are present and no ferroan anorthosites were found. This shows the regional, or local, nature of the highland component at each site.

One of our objectives in this work is the comparison of the samples we studied to rocks from other sites. A convenient way to do this is to plot Sm versus Eu anomaly, or [(Sm/Eu) in sample]/[(Sm/Eu) in chondrites]. *Wakita et al.* (1975) noted a linear trend on this plot for highland samples, with a change in slope as KREEPy (high Sm) compositions are approached. *Warren and Wasson* (1980b) claimed that this plot showed that A-12 and A-14 samples were enriched in

Samples 15303,103, 15303,104, and 15314,125 have plagioclase and mafic silicates like those of ferroan anorthosites, but have less than 90 vol % plagioclase. The parent rock(s) may have been a ferroan anorthosite with a heterogeneous pyroxene and plagioclase distribution, or it may have been unrepresentatively chipped. Another possibility is that these fragments represent rare rock types, as discussed below.

Sample 15303,104 may be a ferroan anorthositic norite, which can be described as a ferroan anorthosite that had more trapped liquid and formed more mafics (*Lindstrom*, 1983). Mineral compositions reported for one such sample, 67215 (*Taylor*, 1982) from the Apollo 16 site are identical to those in 15303,104: $An_{96.5}$, Fo_{50}, and exsolved pyroxenes with compositions of $Wo_{42}En_{39}Fs_{19}$ and $Wo_2En_{57}Fs_{41}$. Data for this rock type are too scarce to make a definite identification, but this would account for the ferroan mineralogy and relatively plagioclase-poor mode. Ferroan rocks with as little as 83% plagioclase, which is fairly close to the mode of 15303,104, are known (*Warren and Wasson*, 1980b).

Samples 15303,103 and 15314,125 have similar mineral compositions and may be related, although they prove difficult to classify, especially 15314,125, which was first described by *Simonds et al.* (1975) as an annealed feldspathic norite with poikilitic pyroxene. However, *Hubbard et al.* (1973) had previously published an analysis of chip 15314,61, which they described as "anorthositic," and the analysis is indeed anorthositic (29.6 wt % Al_2O_3, 0.15 ppm La) and very similar

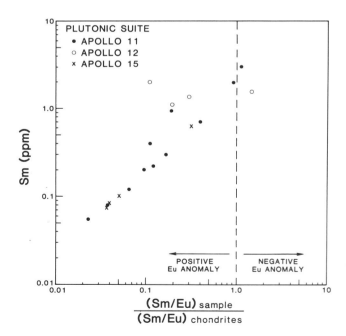

Fig. 15. Plot of Sm concentration versus magnitude of Eu anomaly for plutonic samples from A-11, A-12, and A-15 coarse fines.

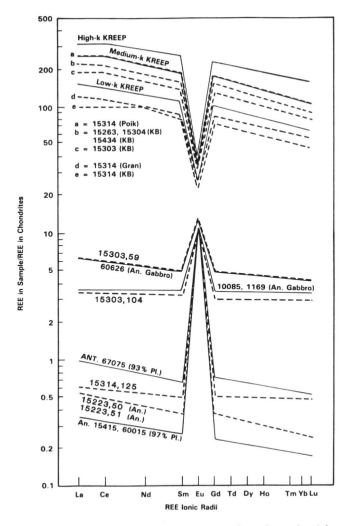

Fig. 16. Chondrite-normalized REE patterns of samples analyzed for this study (dashed lines) compared to previously analyzed samples (solid lines). Reasonable matches are found for each sample; no new rock types are indicated.

Eu relative to "eastern" samples with the same Sm content. *Laul* (1987) and *Lindstrom et al.* (1984) argued against a systematic, Moonwide petrologic variation. In Fig. 15 we have plotted data for A-11 samples (*Laul et al.*, 1983), A-12 samples (*Laul*, 1986), and several new analyses of A-15 coarse fines; all analyses were done at Battelle under the same conditions. With a little scatter the data fall on a single linear trend, especially the A-11 and A-15 samples. Three of the A-12 samples are displaced toward higher Eu values, but one is slightly displaced toward lower Eu. The A-12 plutonic samples are somewhat more REE-rich (*Laul*, 1986) and alkalic (*Simon and Papike*, 1985) than the A-11 and A-15 plutonics, which are non-KREEPy and have extremely Ca-rich plagioclase. We agree with *Lindstrom et al.* (1984) that regional pristine rock associations are observed, whereas a systematic east-west petrologic variation remains doubtful.

In Fig. 16 we compare the REE contents of our samples to previously analyzed samples. The KREEP basalts are compared to high-K, medium-K, and low-K KREEP. Four of the five plot close together, between low-K and medium-K KREEP. The plutonic samples we analyzed are also similar to previously described rock types. Anorthosites 15223,50 and 15223,51 have REE abundances and patterns very similar to the large anorthosites 15415 and 60015. Our analysis of 15314,125 plots above the anorthosites and has a flatter slope. Sample 15303,104 is very similar to a sample from the A-11 coarse fines (*Laul et al.*, 1983). Granulite 15303,59 is almost identical to an anorthositic gabbro from Apollo 16. Detailed studies such as these commonly reveal new rock types, but, based on bulk chemistry, such is not the case for the samples analyzed by INAA. The KREEP basalts are typical A-15 KREEP basalts, and the plutonic rocks correspond to previously recognized rock types.

SUMMARY

Among the returned Apennine Front samples, rake samples and coarse fines have been studied, and clasts in breccias are also being investigated (e.g., *Lindstrom and Marvin*, 1987). Although it is difficult to unambiguously interpret the data from such a wide variety of small rock fragments, a clearer picture is beginning to emerge. In addition to Mg-suite rocks, two major pristine rock types in the Front are ferroan anorthosite and KREEP basalt. Both predate the eruption of mare basalt at the site.

The mineralogy, chemistry, and textures of A-15 KREEP basalts are consistent with a volcanic origin (*Ryder*, 1987). Analysis of numerous small fragments reveals some textural variations but minor compositional differences. They probably formed in a volcanic eruption ~3.9 b.y. ago that produced either a single large flow or several cooling units. These basalts may crop out west of the landing site as the lighter-colored Apennine Bench Formation (*Hawke and Head*, 1978).

Ferroan anorthosites in the coarse fines are similar to larger anorthosites found at the site. The Mg-suite rocks in the coarse fines we studied are generally noritic (opx-rich) in nature. It is probable that the plutonic fragments found at the Apennine Front represent the prebasin crust that is now covered by Serenitatis and Imbrium ejecta (*Spudis and Ryder*, 1985).

CONCLUSIONS

It appears that Mg-suite rocks, such as gabbroic anorthosite and anorthositic norites, are more abundant as clasts in regolith breccias than are anorthosites (*Lindstrom*, 1986; *Lindstrom and Marvin*, 1987), whereas in the coarse fines the reverse is true, seven of ten plutonic fragments studied are ferroan. This difference suggests that two different source areas are being sampled. A survey of A-15 sampling localities shows that all the individual anorthosites were collected at Station 7 (*Ryder*, 1986) near Spur Crater, "a fresh crater on a fresh slope" (*Phinney et al.*, 1972). The crater is 100 m in diameter and probably penetrated the talus and excavated bedrock

(*Swann*, 1986). Thus Spur Crater is likely a major source of the ferroan anorthosite in the Apennine Front. It is a young crater, and therefore most ferroan anorthosites are relatively recent additions to the regolith at the site, after most of the regolith breccia formation, explaining their relative abundance in the coarse fines compared to the breccia clast population. The close similarity of several of the anorthosites analyzed for this study to larger anorthosites (note comparison with 15415 in Fig. 6) suggests that they come from the same source. Chemical mixing models (*Walker and Papike*, 1981) for A-15 regolith (<1 mm) indicate that the soils, like the regolith breccias, have smaller anorthosite components than melt rock (LKFM) or KREEP basalt components.

Impact melt rocks are another important highland component in the Apennine Front (*Spudis and Ryder*, 1985; *Ryder and Spudis*, 1987). Results of our work on impact melts form the coarse fines are reported in a companion paper (*Laul et al.*, 1987) in this volume.

Acknowledgments. Samples were hand-picked and chipped in the Curatorial Facility with the valuable assistance of J. Dietrich and especially K. Willis. This work was inspired by a suggestion by G. Ryder, who also supplied allocation histories of the 10-4 mm fragments. Comments and suggestions from D. Gosselin, O. James, M. Lindstrom, and J. Shervais improved the clarity and content of the text. We would also like to thank L. Thomas for efficient typing and G. Duke for drafting. This research was supported by NASA grants NAG 9-22 (JJP) and NAS 9-15357 (JCL) and funding is gratefully acknowledged.

APPENDIX

TABLE A1. Cross-referencing and genealogy of coarse fines particles.

Thin section (this paper)	Chemical split (this paper)	Original particle	Other thin sections or potted butts	Other splits	Collection location (station)
4-2 mm					
15223,48	,26	,25	—	—	2
15223,50	,30	,29	—	—	2
15223,51	,32	,31	—	—	2
15263,42	,41	,22	—	—	6
15303,51	,50	,32	—	—	7
15303,53	,52	,33	—	—	7
15303,55	,54	,34	—	—	7
15303,59	,58	,36	—	—	7
15303,63	,62	,38	—	—	7
15303,103	,60	,37	,61	—	7
15303,104	,64	,39	,65	—	7
10-4 mm					
15224,26	,24	,6	,25	,6	2
15264,19	,32	,5	—	,18 ,20	6
15304,46	,67	,6	,45	—	7
15314,109	,155	,34	,108	,81	7
15314,125	,154	,15	,126	,61	7
15434,181	,156	,8	,188	,155	7

REFERENCES

Bence A. E. and Albee A. L. (1968) Empirical correction factors for the electron microanalysis of silicates and oxides. *J. Geol.*, 76, 382-403.

Bence A. E. and Papike J. J. (1972) Pyroxenes as recorders of lunar basalt petrogenesis: Chemical trends due to crystal-liquid interaction. *Proc. Lunar Sci. Conf. 3rd*, 431-469.

Brown G. M., Emeleus C. H., Holland J. G., Peckett A., and Phillips R. (1972) Mineral-chemical variations in Apollo 14 and Apollo 15 basalts and granitic fractions. *Proc. Lunar Sci. Conf. 3rd*, 141-157.

Dymek R. F. (1986) Characterization of the Apollo 15 feldspathic basalt suite. In *Workshop on the Geology and Petrology of the Apollo 15 Landing Site* (P. D. Spudis and G. Ryder, eds.), pp. 52-57. LPI Tech. Rpt. 86-03, Lunar and Planetary Institute, Houston.

Greeley R. and Spudis P. D. (1986) Hadley Rille, lava tubes and mare volcanism at the Apollo 15 site. In *Workshop on the Geology and Petrology of the Apollo 15 Landing Site* (P. D. Spudis and G. Ryder, eds.), pp. 58-61. LPI Tech. Rpt. 86-03, Lunar and Planetary Institute, Houston.

Hawke B. R. and Head J. W. (1978) Lunar KREEP volcanism: geologic evidence for history and mode of emplacement. *Proc. Lunar Planet. Sci. Conf. 9th*, 3285-3309.

Hubbard N. J., Gast P. W., Meyer C., Jr., Nyquist L. E., Shih C., and Wiesmann H. (1971) Chemical composition of lunar anorthosites and their parent liquids. *Earth Planet. Sci. Lett.*, 13, 71-75.

Hubbard N. J., Rhodes J. M., Gast P. W., Bansal B. M., Shih C., Wiesmann H., and Nyquist L. E. (1973) Lunar rock types: The role of plagioclase in non-mare and highland rock types. *Proc. Lunar Sci. Conf. 4th*, 1297-1312.

Husain L., Schaeffer O. A., and Sutter J. F. (1972) Age of a lunar anorthosite. *Science*, 175, 428-430.

Irving A. J. (1977) Chemical variation and fractionation of KREEP basalt magmas. *Proc. Lunar Sci. Conf. 8th*, 2433-2448.

James O. B. (1980) Rocks of the early lunar crust. *Proc. Lunar Sci. Conf. 11th*, 365-393.

James O. B. (1987) Magnesian members of the lunar ferroan anorthosite suite (abstract). In *Lunar and Planetary Science XVIII*, pp. 456-457. Lunar and Planetary Institute, Houston.

James O. B. and Flohr M. K. (1983) Subdivision of the Mg-suite noritic rocks into Mg-gabbronorites and Mg-norites. *Proc. Lunar Planet. Sci. Conf. 13th*, in *J. Geophys. Res.*, 88, A603-A614.

Laul J. C. (1979) Neutron activation of geological materials. *Atomic Energy Review IAEA*, 17, 603-695.

Laul J. C. (1986) Chemistry of the Apollo 12 highland component. *Proc. Lunar Planet. Sci. Conf. 16th*, in *J. Geophys. Res.*, 91, D251-D261.

Laul J. C., Lepel E. A., Vaniman D. T., and Papike J. J. (1979) The Apollo 17 drill core: Chemical systematics of grain size fractions. *Proc. Lunar Planet. Sci. Conf. 10th*, 1269-1298.

Laul J. C., Papike J. J., Simon S. B., and Shearer C. K. (1983) Chemistry of the Apollo 11 highland component. *Proc. Lunar Planet. Sci. Conf. 14th*, in *J. Geophys. Res.*, 88, B139-B149.

Laul J. C., Simon S. B., and Papike J. J. (1987) Chemistry and petrology of the Apennine Front, Apollo 15, Part II: Impact melt rocks. *Proc. Lunar Planet. Sci. Conf. 18th*, this volume.

Lindstrom M. M. (1983) Ferroan anorthositic norite: An important highland cumulate. In *Workshop on Pristine Highlands Rocks and the Early History of the Moon* (J. Longhi and G. Ryder, eds.), pp. 56-57. LPI Tech. Rpt. 83-02, Lunar and Planetary Institute, Houston.

Lindstrom M. M. (1986) Petrology and geochemistry of highlands samples from the Apennine Front. In *Workshop on the Geology and Petrology of the Apollo 15 Landing Site* (P. D. Spudis and G. Ryder, eds.), pp. 70-74. LPI Tech. Rpt. 86-03, Lunar and Planetary Institute, Houston.

Lindstrom M. M. and Lindstrom D. J. (1986) Lunar granulites and their precursor anorthositic norites of the early lunar crust. *Proc. Lunar Planet. Sci. Conf. 16th*, in *J. Geophys. Res., 91*, D263-D276.

Lindstrom M. M. and Marvin U. B. (1987) Geochemical and petrologic studies of clasts in Apennine Front breccia 15459 (abstract). In *Lunar and Planetary Science XVIII*, pp. 554-555. Lunar and Planetary Institute, Houston.

Lindstrom M. M., Knapp S. A., Shervais J. W., and Taylor L. A. (1984) Magnesian anorthosites and associated troctolites and dunite in Apollo 14 breccias. *Proc. Lunar Planet. Sci. Conf. 15th*, in *J. Geophys. Res., 89*, C41-C49.

Lofgren G. E., Donaldson C. H., Williams R. J., Mullins O., Jr., and Usselman T. M. (1974) Experimentally reproduced textures and mineral chemistry of Apollo 15 quartz-normative basalts. *Proc. Lunar Sci. Conf. 5th*, 549-567.

Meyer C., Jr., (1977) Petrology, mineralogy, and chemistry of KREEP basalt. *Phys. Chem. Earth, 10*, 239-260.

Nyquist L. E., Hubbard N. J., Gast P. W., Bansal B. M., Wiesmann H., and Jahn B. (1973) Rb-Sr systematics for chemically defined Apollo 15 and 16 materials. *Proc. Lunar Sci. Conf. 4th*, 1823-1846.

Ostertag R., Stöffler D., Borchardt R., Palme H., Spettel B., and Wänke H. (1987) Precursor lithologies and metamorphic history of granulitic breccias from North Ray Crater, Station 11, Apollo 16. *Geochim. Cosmochim. Acta, 51*, 131-142.

Papanastassiou D. A. and Wasserburg G. J. (1976) Rb-Sr age of troctolite 76535. *Proc. Lunar Sci. Conf. 7th*, 2035-2054.

Phinney W. C., Warner J. L., Simonds C. H., and Lofgren G. E. (1972) Classification and description of rock types at Spur Crater. In *The Apollo 15 Lunar Samples* (J. W. Chamberlain and C. Watkins, eds.), pp. 149-153. The Lunar Science Institute, Houston.

Powell B. N. (1972) *Apollo 15 Coarse Fines (4-10 mm): Sample Classification, Description, and Inventory*. MSC 03228, NASA Manned Spacecraft Center, Houston. 91 pp.

Prinz M. and Keil K. (1977) Mineralogy, petrology and chemistry of ANT-suite rocks from the lunar highlands. *Phys. Chem. Earth, 10*, 215-237.

Raedeke L. D. and McCallum I. S. (1980) Lunar fractionation trends are not anomalous (abstract). In *Lunar and Planetary Science XI*, pp. 908-910. Lunar and Planetary Institute, Houston.

Ryder G. (1982) Lunar anorthosite 60025, the petrogenesis of lunar anorthosites, and the composition of the Moon. *Geochim. Cosmochim. Acta, 46*, 1591-1601.

Ryder G. (1986) Samples at the Apollo 15 landing site: Types and distribution. In *Workshop on the Geology and Petrology of the Apollo 15 Landing Site* (P. D. Spudis and G. Ryder, eds.), pp. 86-90. LPI Tech. Rpt. 86-03, Lunar and Planetary Institute, Houston.

Ryder G. (1987) Petrographic evidence for nonlinear cooling rates and a volcanic origin for Apollo 15 KREEP basalts. *Proc. Lunar Planet. Sci. Conf. 17th*, in *J. Geophys. Res., 92*, E331-E339.

Ryder G. and Spudis P. D. (1985) Materials in the Apennine Front, Apollo 15 (abstract). In *Lunar and Planetary Science XVI*, pp. 722-723. Lunar and Planetary Institute, Houston.

Ryder G. and Spudis P. D. (1987) Chemical composition and origin of Apollo 15 impact melts. *Proc. Lunar Planet. Sci. Conf. 17th*, in *J. Geophys. Res., 92*, E432-E446.

Simon S. B. and Papike J. J. (1985) Petrology of the Apollo 12 highland component. *Proc. Lunar Planet. Sci. Conf. 16th*, in *J. Geophys. Res., 90*, D47-D60.

Simon S. B., Papike J. J., Shearer C. K., and Laul J. C. (1983) Petrology of the Apollo 11 highland component. *Proc. Lunar Planet. Sci. Conf. 14th*, in *J. Geophys. Res., 88*, B103-B138.

Simonds C. H., Warner J. L., and Phinney W. C. (1975) The petrology of the Apennine Front revisited (abstract). In *Lunar Science VI*, pp. 744-746. The Lunar Science Institute, Houston.

Smith J. V. and Steele I. M. (1974) Intergrowths in lunar and terrestrial anorthosites with implications for lunar differentiation. *Am. Min., 59*, 673-680.

Spudis P. D. (1980) Petrology of the Apennine Front, Apollo 15: Implications for the geology of the Imbrium impact basin (abstract). In *Papers Presented to the Conference on Multi-Ring Basins*, pp. 83-85. Lunar and Planetary Institute, Houston.

Spudis P. D. and Ryder G. (1985) Geology and petrology of the Apollo 15 landing site: Past, present, and future understanding. *Eos Trans. AGU, 66*, 721, 724-726.

Swann G. A. (1986) Some observations on the geology of the Apollo 15 landing site. In *Workshop on the Geology and Petrology of the Apollo 15 Landing Site* (P. D. Spudis and G. Ryder, eds.) pp. 108-112. LPI Tech. Rpt. 86-03, Lunar and Planetary Institute, Houston.

Takeda H., Mori H., Miyamoto M., and Ishii T. (1984) Mesostasis-rich lunar and eucritic basalts with reference to REE-rich minerals (abstract). In *Lunar and Planetary Science XV*, pp. 842-843. Lunar and Planetary Institute, Houston.

Taylor G. J. (1982) A possibly pristine, ferroan anorthosite (abstract). In *Lunar and Planetary Science XIII*, p. 798. Lunar and Planetary Institute, Houston.

Wakita H., Laul J. C., and Schmitt R. A. (1975) Some thoughts on lunar ANT-KREEP and mare basalts. *Geochem. J., 9*, 25-41.

Walker R. J. and Papike J. J. (1981) The Apollo 15 regolith: Chemical modeling and mare/highland mixing. *Proc. Lunar Planet. Sci. 12B*, 509-517.

Warner J. L., Simonds C. H., and Phinney W. C. (1976) Genetic distinction between anorthosites and Mg-rich plutonic rocks: New data from 76255 (abstract). In *Lunar Science VII*, pp. 915-917. The Lunar Science Institute, Houston.

Warren P. H. (1985) The magma ocean concept and lunar evolution. *Ann. Rev. Earth Planet. Sci., 13*, 201-240.

Warren P. H. and Wasson J. T. (1977) Pristine nonmare rocks and the nature of the lunar crust. *Proc. Lunar Sci. Conf. 8th*, 2215-2235.

Warren P. H. and Wasson J. T. (1980a) Early lunar petrogenesis, oceanic and extraoceanic. *Proc. Conf. Lunar Highlands Crust* (J. J. Papike and R. B. Merrill, eds.), pp. 81-99. Pergamon, New York.

Warren P. H. and Wasson J. T. (1980b) Further foraging for pristine nonmare rocks: Correlations between geochemistry and longitude. *Proc. Lunar Planet. Sci. Conf. 11th*, 431-470.

Chemistry and Petrology of the Apennine Front, Apollo 15, Part II: Impact Melt Rocks

J. C. Laul

Battelle, Pacific Northwest Laboratories, Richland, WA 99352

S. B. Simon and J. J. Papike

Institute for the Study of Mineral Deposits, South Dakota School of Mines and Technology, Rapid City, SD 57701

The sources of impact melt rocks found in the Apennine Front have remained elusive since the return of the samples. Petrographic study of 21 impact melts hand-picked from Apennine Front coarse fines reveals a variety of textures: some are poikilitic, some are more fine-grained than others, two have plagioclase laths, and five have no olivine. Clasts are mostly monomineralic and unshocked. With a few exceptions, pyroxene, olivine, and plagioclase compositions are quite similar to each other. Bulk compositions are somewhat LKFM-like but most are lower in Al_2O_3. Eight of the samples fall into groups B, C, and D of *Ryder and Spudis* (1987), but we cannot associate them with specific events or sources. With additional analyses a compositional continuum may be observed. The bulk major and trace element compositions indicate that the A-15 impact melts are mixtures of LKFM, KREEP, and plutonic components. The Ni/Ir ratios of the melt rocks are greater than chondritic values, indicating "ancient" and/or iron meteorite components. Eight samples have ratios between 30,000 and 40,000; this may be the signature of the Imbrium event. These ratios are similar to values for A-17 melt rocks (24,000-30,000), suggesting that the Serenitatis and Imbrium projectiles were similar in nature, although the former may have had a slightly more chondritic (lower) Ni/Ir ratio.

INTRODUCTION

The Apollo 15 mission had several ambitious scientific goals, one of which was to sample impact melts and ejecta related to the formation of the Imbrium and Serenitatis basins. The site is strategically located between these basins (Fig. 1) in the Apennine Mountains, which form one of the rings around the Imbrium basin. The site is well within the range of ejecta from the Serenitatis basin as well as from the Imbrium basin. It is likely that materials related to the formation of both of these basins are present in the return samples of A-15 regolith. In addition to the excavated plutonic rocks, detailed work on impact melt rocks from this site will give us a better understanding of the geology of the site and the surrounding area. Study of the impact melts can potentially define the number of major impacts that have affected the site. Ideally, we would like to identify impact melts related to the Imbrium and Serenitatis events to determine the relative contributions of these materials to the site and to learn about the composition of the target materials. This in turn will provide information bearing on regional geochemistry and geologic history and, in general, the ancient lunar crust. Identification of Serenitatis materials among A-15 samples would allow comparison with Serenitatis ejecta among A-17 materials. This would improve our understanding of large-scale melt sheets, especially with regard to their textural and compositional homogeneity.

In a recent study of 14 A-15 impact melt rocks, *Ryder and Spudis* (1987) found five compositional groups, each possibly representing an impact event. Although possible candidates were suggested, none could unambiguously be related to a basin-forming event. To complement that study and to further define the petrologic and chemical variability of the Apennine Front and of A-15 impact melt rocks, we have studied 21 melt rock fragments hand-picked from Apennine Front coarse fines (10-4 and 4-2 mm).

Impact melt rocks are but one component of the highland lithologic unit known as the Apennine Front, which was sampled at the base of Hadley Delta mountain. In addition to mare basalts, KREEP basalts and plutonic rocks are also present. These latter two highland rock types are discussed in a companion paper (*Simon et al.*, 1987). For additional background information on the A-15 landing site and mission objectives, see that paper or *Spudis and Ryder* (1985).

ANALYTICAL METHODS

Approximately 2 g of 4-2 mm fragments from each of the Apennine Front Stations 2, 6, and 7 were allocated for picking in addition to 10-4 mm fragments that were specifically requested based on binocular descriptions by *Powell* (1972). The 4-2 mm samples were hand-picked under a binocular microscope in a nitrogen-filled glove box at the Curatorial Facility of the NASA Johnson Space Center. Some of the 10-4 mm fragments had previously been analyzed; the others, and all the 4-2 mm fragments, were chipped in the Curatorial Facility, one split for a thin section and the other for chemical analysis (INAA). The thin sections were examined in transmitted and reflected light with a Zeiss photomicroscope. Modal data were collected with a Swift automated point counter. Mineral analyses were obtained with an automated ETEC (MAC-5) electron microprobe operated at 15 KV. Probe data were reduced according to the method of *Bence and Albee* (1968).

Thirteen impact melt samples were analyzed by INAA using a high-efficiency Ge(Li) detector (25% FWHM 1.8 KeV for 1332 KeV γ of ^{60}Co), a 4096 channel analyzer, and coincidence-

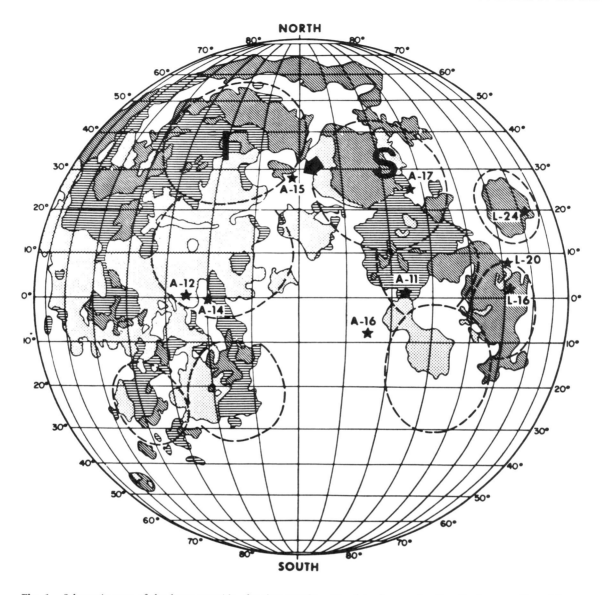

Fig. 1. Schematic map of the lunar nearside, showing sampling sites (stars), mare basalt units (patterned), and large basins (dashed lines). The proximity of the A-15 site (arrow) to the Imbrium (I) and Serenitatis basins (S) is clear. After *Papike et al.* (1982).

noncoincidence Ge(Li)-NaI(Tl) counting systems. The coincidence-noncoincidence approach enables us to improve our detection limits for the normally difficult to measure noncoincidence γ-rays produced by isotopes of Nd, Ni, Tm, Zr, and Sr. The details of our INAA procedure and the systematics of coincidence-noncoincidence counting are described by *Laul et al.* (1979) and *Laul* (1979). The samples were irradiated at the Oregon State University TRIGA reactor with an Allende standard and USGS standards BCR-1, BHVO-1, GSP-1, and PCC-1 used as controls.

The chemistry splits were assigned sample numbers different from the thin section numbers. For clarity only the latter are used in the discussion. Table 1a gives the genealogy of the coarse fines fragments, including the chemical splits analyzed for this paper.

PETROGRAPHY AND MODAL MINERALOGY

Impact melts were identified petrographically based on texture, grain size, and the presence of mineral and lithic clasts. Photomicrographs are shown in Fig. 2. Most samples contain only monomineralic clasts, some of which are rounded, in a fine (~10-30 μm) matrix of plagioclase, pyroxene, and in most

TABLE 1a. Cross-referencing and genealogy of coarse fines particles.

Thin Section (this paper)	Chemical Split (this paper)*	Original Particle	Other Thin Sections or Potted Butts	Other Splits
4–2 mm fragments				
15223,21	,22	,7	-	-
15223,44	,14	,13	-	-
15223,45	,16	,15	-	-
15223,47	,20	,19	-	-
15303,69	,68	,41	-	-
15303,102	,48	,31	,49	-
15303,105	,66	,40	,67	-
10–4 mm fragments				
15294,20	,18	,6	,19	,6
15304,48	-	,7	,16,47	,29,66
15314,100	-	,30	,101	,77,146
15314,106	,152	,32	,105	-
15314,135	,147	,27	-	-
15314,150	,149	,29	-	,29
15434,140	-	,13	,141	,13
15434,145	,177	,15	,144	,69
15434,182	,164	,157	,163	,157
15434,183	,166	,158	,165	,158
15434,184	,168	,159	,167	,159
15434,185	,170	,160	,169	,160
15434,186	,172	,161	,171	,161
15434,187	,174	,162	,173	,162

*Note: Not all chemical splits were analyzed.

samples, olivine. Plagioclase clasts commonly have ragged boundaries that in places are continuous with matrix plagioclase. This makes it difficult to unambiguously distinguish small clasts from larger matrix grains. No vesicles were observed in any of the samples. A few samples have plagioclase laths but most have fine, tabular plagioclase.

The modal mineralogies of the samples are listed in Table 1b. Except for a quenched feldspathic melt (Fig. 2a) all the samples contain pyroxene, have no glass, and all but four have olivine. The olivine:pyroxene ratio varies widely from 0 to 6:1. All samples have ilmenite, and most have minor amounts of sulfide and metal.

MINERAL CHEMISTRY

Plagioclase

Plagioclase compositions are summarized in Fig. 3 (~50 analyses per sample) and representative analyses are given in Table 2. With a few exceptions most analyses for each sample are between An_{89} and An_{96}. Feldspathic melt 15223,44 and samples 15294,20 and 15303,69 have plagioclase that is somewhat more An-rich, and samples 15314,100 and 15434,184 have more plagioclase that is more sodic than An_{85}. The relatively sodic plagioclase, as compared to that in plutonic rocks, probably indicates the influence of a KREEP component in these rocks. Overall, the compositional ranges are similar to those of comparable A-17 rocks (*Simonds et al.,* 1974), whereas feldspar in A-16 melt rocks is restricted to the more calcic ($>An_{90}$) compositions (*Bence et al.,* 1973; *Vaniman and Papike,* 1980). *Simonds* (1975), in a study of A-17 impact melts, noted that in several samples the feldspar clasts tended to be more calcic than the matrix feldspar. In the A-15 samples we observe fairly complete overlap between the two. Sodic clasts are present as well as calcic ones, and are probably from KREEPy rocks in the source area.

Pyroxene

Representative pyroxene analyses are given in Table 3, and all analyses are summarized in Fig. 4. Olivine compositions are indicated by bars on the En-Fs join. Many of the pyroxene analyses fall in the range $Wo_{2-10}En_{70-80}$ with varying degrees of Fe-Mg equilibration. Several samples also have magnesian Ca-rich pyroxene. The pyroxene compositions of most samples are similar to those determined by *Simonds* (1975) for A-17 impact melts.

Two samples are different enough from the others to merit discussion. Sample 15294,20 has Mg-rich low-Ca pyroxene,

TABLE 1b. Modal mineralogy of impact melts in A-15 coarse fines.

	15223,21	15223,44	15223,45	15223,47	15294,20	15303,69	15303,102	15303,105	15304,48	15314,100	15314,106
Pyroxene	10.4	-	6.5	15.3	10.8	18.6	15.2	38.2	37.2	22.3	51.9
Olivine	27.8	-	39.7	18.0	43.1	32.6	25.4	24.3	17.2	32.3	-
Plagioclase	59.0	76.2	51.2	47.4	45.1	46.2	58.0	36.6	43.6	44.8	46.6
Ilmenite	2.8	0.4	2.6	2.7	0.9	2.6	1.0	0.8	1.7	0.2	1.0
Sulfide	-	-	-	tr	tr	tr	0.2	tr	0.1	-	0.3
Metal	-	-	-	0.1	tr	tr	0.1	tr	0.2	0.4	0.2
Spinel	-	-	-	-	0.1	-	-	-	-	-	-
Mesostasis	-	20.8	-	-	-	-	-	-	-	-	-
Glass	-	2.6	-	-	-	-	-	-	-	-	-
Lithic clasts	-	-	-	16.5	-	-	-	-	-	-	-
Other	-	-	-	-	-	-	-	-	-	-	-
	100.0	100.0	100.0	100.0	100.0	100.0	99.9	99.9	100.0	100.0	100.0

TABLE 1b. (continued)

	15314,135	15314,150	15434,140	15434,145	15434,182	15434,183	15434,184	15434,185	15434,186	15434,187
Pyroxene	24.9	35.8	47.3	53.5	30.5	46.8	36.3	28.2	26.3	35.1
Olivine	9.7	15.3	-	3.8	12.6	-	-	10.9	21.6	21.0
Plagioclase	61.6	46.5	50.1	40.7	52.3	51.8	57.6	58.5	49.7	42.3
Ilmenite	3.2	1.7	2.2	1.8	4.4	0.4	1.3	2.4	2.2	1.3
Sulfide	0.1	0.1	0.1	tr	0.2	0.1	0.3	tr	-	-
Metal	-	0.1	0.2	tr	tr	0.4	tr	-	-	0.3
Spinel	-	-	0.1	0.2	tr	tr	0.3	-	-	tr
Mesostasis	0.4	-	-	-	-	-	3.5	-	-	-
Glass	-	-	-	-	-	-	-	-	-	-
Lithic clasts	-	-	-	-	-	-	-	-	-	-
Other	0.1	0.5	-	-	-	0.5	0.7	-	0.2	-
	100.0	100.0	100.0	100.0	100.0	100.0	100.0	100.0	100.0	100.0

olivine in the range Fo_{80-90}, and plagioclase that is generally more calcic than that of the other samples (Fig. 3). These observations are consistent with formation of this sample from a source different from those of the samples, one that was more magnesian and generally less fractionated.

On the other hand, 15434,184 has relatively Fe-rich pigeonite ($Wo_{10}En_{55-60}$) and more sodic plagioclase, with only three analyses more calcic than An_{90}. This sample probably represents a more fractionated target than most of the other samples. The majority of the samples formed from similar sources, although the pyroxene compositions indicate differing equilibration and exsolution histories.

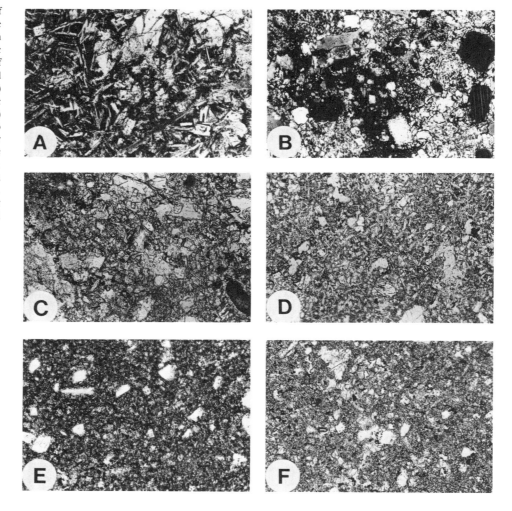

Fig. 2. Photomicrographs of impact melt rocks illustrating the textural variety. All views are 1 mm across. (**a**) 15223,44. Feldspathic impact melt. Note presence of skeletal plagioclase crystals and coarse plagioclase clasts. (**b**) 15223,45, x-nic. Note poikilitic texture and unshocked clasts. (**c**) 15294,20. This sample is also poikilitic and contains spinel (lower right). (**d**) 15314,100. Note the abundance of fine plagioclase laths. (**e**) 15434,186. This sample has a fine-grained, relatively mafic matrix. (**f**) 15434,187. Note the irregular clast boundaries, and the rounded clast (lower center).

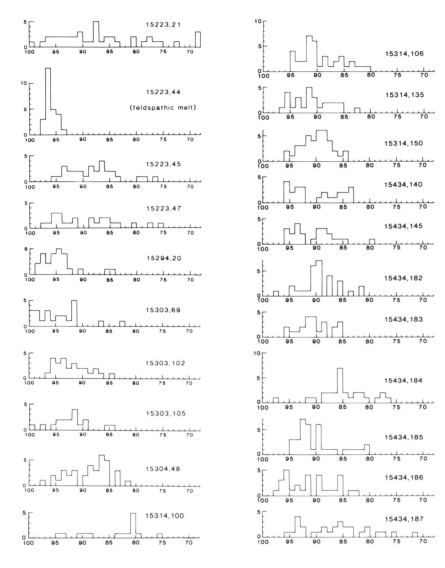

Fig. 3. Compositions of plagioclase in A-15 impact melt rocks. The analyses indicate a variable KREEPy component and lack of Ca-Na equilibration.

Olivine

Olivine compositions (Fig. 5) are even more equilibrated than the pyroxenes and make a sharp contrast with the variability of plagioclase compositions. Most samples have olivine in the range Fo_{72-76}. Only 15294,20 falls outside this range. Representative analyses are given in Table 4.

To compare the degree of olivine/pyroxene equilibration we have plotted mg' [defined as $Mg/(Mg + Fe)$] of olivine versus mg' of pyroxene in Fig. 6. There have been many studies of partitioning of Mg and Fe^{2+} between olivine and pyroxene (e.g., *Medaris*, 1969; *Matsui and Nishizawa*, 1974; *Wood*, 1975) and they have shown that the partitioning is not strongly temperature- or pressure-dependent, but it is affected by the Mg/Fe ratio of the bulk melt (*Grover and Orville*, 1969). For Fe-rich bulk compositions, Mg is partitioned into pyroxene; with increasing bulk Mg the preference weakens (*Brown*, 1980). Our analyses are consistent with the experimental data and show that in the impact melt rocks the pyroxenes have slightly higher mg' values than coexisting olivines. Of the 16 samples with both pyroxene and olivine, 12 form a tight cluster indicating crystallization from melts with similar compositions, probably under similar conditions.

Data for 15294,20 plot at higher mg' values and closer to a K_D of one than the other samples. This does not necessarily indicate crystallization at a higher temperature. Although the

TABLE 2. Representative analyses of plagioclase in A-15 impact melt rocks.

	15223,21		15223,44		15294,20		15303,102		15314,135		15314,150		15434,182		15434,184	
	1.	2.	1.	2.	1.	2.	1.	2.	1.	2.	1.	2.	1.	2.	1.	2.
SiO_2	46.44	46.73	44.83	44.93	44.52	45.44	45.55	45.91	45.91	48.07	46.38	47.99	45.73	46.08	47.77	48.13
Al_2O_3	34.16	34.04	35.20	34.71	35.09	34.05	34.71	33.54	34.45	33.48	34.12	33.32	34.52	34.39	33.27	32.09
FeO	0.12	0.14	0.18	0.15	0.15	0.28	0.20	0.34	0.14	0.18	0.15	0.18	0.09	0.22	0.22	0.22
MgO	n.a.	n.a.	n.a.	n.a.	n.a.	n.a.	n.a.	n.a.	0.08	0.06	0.09	0.12	0.13	0.07	0.08	0.10
CaO	18.92	17.57	19.50	19.15	19.12	18.85	18.85	18.01	18.86	18.24	18.85	17.31	18.52	18.15	17.10	16.11
Na_2O	0.77	1.48	0.33	0.52	0.32	0.71	0.59	1.00	0.60	1.04	0.72	1.38	0.77	1.01	1.45	1.66
K_2O	0.03	0.22	0.06	0.05	0.09	0.13	0.11	0.21	0.05	0.12	0.11	0.24	0.08	0.10	0.22	0.32
	100.44	100.18	100.10	99.51	99.29	99.46	100.01	99.01	100.09	101.19	100.42	100.54	99.84	100.02	100.11	98.63
Si	2.130	2.146	2.069	2.085	2.070	2.110	2.100	2.138	2.113	2.182	2.128	2.190	2.110	2.122	2.189	2.233
Al	1.847	1.843	1.916	1.899	1.923	1.864	1.887	1.841	1.869	1.792	1.846	1.793	1.877	1.867	1.798	1.755
Total	3.977	3.989	3.985	3.984	3.993	3.974	3.987	3.979	3.982	3.974	3.974	3.983	3.987	3.989	3.987	3.988
Fe	0.005	0.005	0.007	0.006	0.006	0.011	0.008	0.013	0.005	0.007	0.006	0.007	0.003	0.008	0.008	0.009
Mg	-	-	-	-	-	-	-	-	0.005	0.004	0.006	0.008	0.009	0.005	0.005	0.007
Ca	0.930	0.865	0.964	0.952	0.952	0.938	0.931	0.899	0.930	0.887	0.927	0.847	0.915	0.895	0.840	0.801
Na	0.068	0.132	0.030	0.047	0.029	0.064	0.053	0.090	0.054	0.092	0.064	0.122	0.070	0.090	0.129	0.149
K	0.002	0.013	0.004	0.003	0.005	0.008	0.006	0.012	0.003	0.007	0.006	0.014	0.005	0.006	0.013	0.019
Total	1.005	1.015	1.005	1.008	0.992	1.021	0.998	1.014	0.997	0.997	1.009	0.998	1.002	1.004	0.995	0.985
Total cations/ 8 ox.	4.982	5.004	4.990	4.991	4.985	4.995	4.986	4.993	4.979	4.971	4.983	4.981	4.989	4.993	4.982	4.973
Or	0.2	1.3	0.4	0.3	0.5	0.8	0.6	1.2	0.3	0.7	0.6	1.4	0.5	0.6	1.3	2.0
Ab	6.8	13.1	3.0	4.7	2.9	6.3	5.4	9.0	5.5	9.3	6.4	12.4	7.1	9.1	13.1	15.4
An	93.0	85.6	96.6	95.0	96.5	92.9	94.0	89.8	94.2	90.0	93.0	86.2	92.4	90.3	85.5	82.7

n.a. = not analyzed.

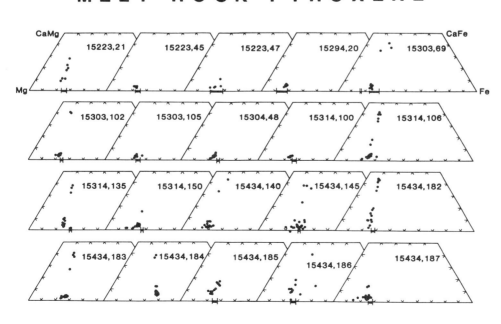

APOLLO 15 MELT ROCK PYROXENE

Fig. 4. Compositions of pyroxene in A-15 impact melts. Most show evidence of Fe-Mg equilibration Coexisting olivine compositional ranges are indicated on the En-Fs joins.

TABLE 3. Pyroxene compositions in A-15 impact melt rocks.

	15223,21	15294,20	15303,102	15304,48		15314,100		15434,140			15434,145		15434,182		15434,184	
	1.	1.	1.	1.	2.	1.	2.	1.	2.	3.	1.	2.	1.	2.	1.	2.
SiO$_2$	51.43	54.91	55.11	55.38	53.67	54.68	55.23	55.30	53.39	52.84	54.14	52.66	52.39	51.61	51.28
Al$_2$O$_3$	1.88	2.05	1.29	0.60	2.43	0.63	0.51	0.81	3.04	1.94	1.41	1.38	2.04	0.72	0.82
TiO$_2$	1.62	1.07	0.87	0.38	0.75	0.33	0.35	0.49	0.69	0.70	0.56	0.68	0.95	0.74	0.68
FeO	11.59	10.36	12.91	14.81	14.25	14.19	11.14	12.14	11.20	16.54	16.36	12.80	6.79	21.87	20.77
MnO	0.20	0.18	0.22	0.24	0.17	0.28	0.21	0.28	0.20	0.25	0.29	0.14	0.04	0.43	0.34
MgO	22.11	28.82	27.43	27.06	26.69	27.55	30.24	28.77	28.56	24.71	24.05	24.70	17.48	20.26	20.11
CaO	9.55	2.54	1.91	1.39	1.84	1.38	1.33	1.79	2.40	1.92	3.47	6.68	20.39	2.89	4.14
Na$_2$O	0.38	0.0	0.57	0.0	0.20	0.0	0.13	0.0	0.0	0.13	0.04	0.0	0.0	0.13	0.05
Cr$_2$O$_3$	0.32	0.71	0.66	0.23	0.31	0.30	0.25	0.30	0.37	0.46	0.30	0.37	0.41	0.37	0.45
	99.08	100.64	100.97	100.09	100.31	99.34	99.39	99.88	99.85	99.49	100.62	99.41	100.49	99.02	98.64
Si	1.898	1.932	1.955	1.988	1.925	1.976	1.970	1.971	1.901	1.935	1.963	1.925	1.916	1.957	1.950
Al	0.081	0.068	0.045	0.012	0.075	0.024	0.021	0.029	0.099	0.065	0.037	0.059	0.084	0.031	0.036
Total tet.	1.979	2.000	2.000	2.000	2.000	2.000	1.991	2.000	2.000	2.000	2.000	1.984	2.000	1.988	1.986
Al	0.000	0.016	0.008	0.013	0.027	0.002	0.000	0.004	0.028	0.018	0.023	0.000	0.004	0.000	0.000
Ti	0.044	0.029	0.022	0.009	0.020	0.008	0.008	0.012	0.018	0.018	0.015	0.018	0.025	0.021	0.019
Fe	0.357	0.305	0.382	0.444	0.427	0.428	0.332	0.362	0.333	0.506	0.496	0.390	0.207	0.693	0.661
Mn	0.005	0.005	0.006	0.006	0.004	0.008	0.006	0.008	0.005	0.007	0.008	0.004	0.001	0.013	0.010
Mg	1.216	1.511	1.450	1.448	1.426	1.483	1.608	1.528	1.516	1.349	1.300	1.346	0.952	1.145	1.139
Ca	0.377	0.098	0.072	0.052	0.070	0.053	0.050	0.068	0.091	0.075	0.134	0.260	0.798	0.117	0.168
Na	0.027	0.000	0.038	0.000	0.012	0.000	0.008	0.000	0.000	0.008	0.002	0.000	0.0	0.009	0.003
Cr	0.008	0.020	0.018	0.006	0.008	0.008	0.006	0.008	0.009	0.013	0.008	0.010	0.012	0.011	0.013
Total	2.034	1.984	1.996	1.978	1.994	1.990	2.018	1.990	2.000	1.994	1.986	2.028	1.999	2.009	2.013
Total cations/6 ox.	4.013	3.984	3.996	3.978	3.994	3.990	4.009	3.990	4.000	3.994	3.986	4.012	3.999	3.997	3.999
Wo	19.3	5.1	3.8	2.7	3.6	2.7	2.5	3.5	4.7	3.9	6.9	13.0	40.8	6.0	8.5
En	62.4	78.9	76.2	74.5	74.2	75.5	80.8	78.0	78.1	69.9	67.4	67.4	48.6	58.6	57.9
Fs	18.3	15.9	20.1	22.8	22.2	21.8	16.7	18.5	17.2	26.2	25.7	19.5	10.6	35.4	33.6

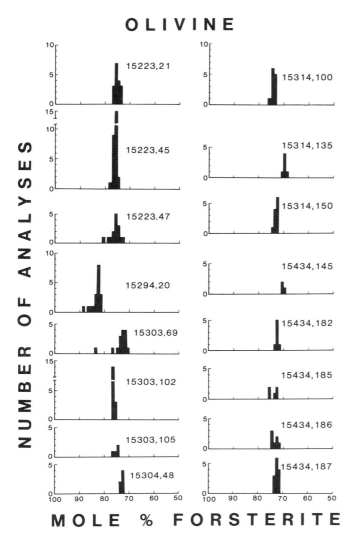

Fig. 5. Summary of olivine compositions in A-15 impact melt rocks. Most are highly equilibrated.

K_D might be expected to approach one with increasing temperature as crystallographic sites become less selective, K_D also approaches one with increasing mg' of the bulk composition (e.g., *Grover and Orville*, 1969).

Oxide Minerals

Ilmenite occurs in all the melt rock samples. Table 5 shows it is rather magnesian, especially compared to mare basalt ilmenite, which commonly has <2 wt % MgO (*Papike et al.*, 1976). Sample 15294,20 has the most magnesian ilmenite (~8-9 wt % MgO).

Chromian spinel is also present in several of the samples (Table 5). The Ti-Cr-Al variations in these spinels are summarized in Fig. 7, which shows their Cr-rich nature, except for the spinel in 15294,20. On this diagram an igneous crystallization trend would be from Al-rich compositions toward Cr and to Ti enrichment. Thus, in addition to its less fractionated silicate mineral compositions, 15294,20 also has unfractionated ilmenite and spinel compositions.

The spinels in the other samples are dominantly chromites, with some Mg and Al. Unfortunately, they are not by themselves diagnostic of a source rock type, because chromite occurs in many different types of lunar rocks. For example, chromites similar to those in 15434,187 have been found in anorthositic norite 15308 and KREEP basalt 15359 (*Nehru et al.*, 1974). Spinel in 15434,145 encloses euhedral ilmenite, but most occur as single grains.

Other Phases

Ryder and Spudis (1987) noted that all but one of their samples had KREEP-like incompatible element patterns, similar to the present samples (see below). The KREEP component is manifested petrologically not only by relatively sodic plagioclase, but also by trace amounts of K-rich and P-rich phases. Several samples contain ternary feldspars that are dominantly calcic (~15 wt % CaO) but have ~1-5 wt % K_2O. Fine grains (~5-10 μm) of Ca-phosphate are also present. Sample 15434,183 contains a phosphate grain that is unlike those in A-15 KREEP basalts (*Simon et al.*, 1987) in that it is essentially a pure chlorapatite with no La and 0.2-0.3 wt % each of MgO, FeO, and Y_2O_3.

CHEMICAL SYSTEMATICS

The data for 34 major, minor, and trace elements for 13 melt rocks along with the USGS standards BCR-1 and BHVO-1 are shown in Tables 6 and 7. Table 6 lists the major and minor element abundances (oxide wt %) and Table 7 gives

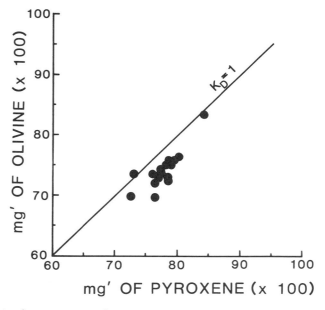

Fig. 6. Comparison of coexisting olivines and pyroxenes. The tight cluster indicates crystallization from melts with similar Fe/Mg ratios.

TABLE 4. Representative olivine compositions in A-15 impact melt rocks.

	15223,21	15223,45		15223,47	15294,20		15303,69	15303,102	15434,145	15434,185		15434,187	
	1.	1.	2.	1.	1.	2.	1.	1.	1.	1.	2.	1.	2.
SiO_2	38.74	38.35	38.20	38.32	40.58	40.39	38.91	39.12	37.77	37.80	38.13	38.46	38.52
Al_2O_3	0.13	0.11	0.11	0.18	0.25	0.42	0.10	0.04	0.05	0.10	0.07	0.10	0.05
TiO_2	0.05	0.04	0.15	0.09	0.09	0.13	0.09	0.05	0.02	0.17	0.11	0.12	0.07
FeO	22.18	22.47	22.83	21.23	15.25	16.20	24.65	22.12	27.41	24.25	25.09	25.25	26.05
MnO	0.18	0.21	0.27	0.27	0.17	0.07	0.33	0.19	0.40	0.23	0.32	0.32	0.25
MgO	39.27	38.85	38.49	39.00	43.35	42.57	36.39	39.20	35.56	38.30	37.10	36.68	37.16
CaO	0.04	0.06	0.06	0.15	0.18	0.30	0.24	0.06	0.06	0.12	0.12	0.17	0.05
Cr_2O_3	0.11	0.02	0.06	0.09	0.03	0.14	0.18	0.03	0.13	0.0	0.02	0.07	0.00
	100.70	100.11	100.17	99.33	99.90	100.22	100.89	100.81	101.40	100.97	100.96	101.17	102.15
Si	0.997	0.996	0.994	0.998	1.018	1.014	1.013	1.005	0.991	0.983	0.995	1.000	0.994
Al	0.003	0.003	0.003	0.005	0.007	0.012	0.003	0.001	0.001	0.003	0.001	0.003	0.001
Ti	0.001	0.001	0.002	0.001	0.001	0.002	0.001	0.001	0.000	0.003	0.001	0.002	0.001
Fe	0.477	0.487	0.497	0.462	0.319	0.339	0.536	0.475	0.601	0.528	0.547	0.549	0.562
Mn	0.003	0.005	0.005	0.005	0.003	0.001	0.007	0.003	0.009	0.005	0.007	0.007	0.005
Mg	1.507	1.503	1.493	1.514	1.621	1.593	1.411	1.501	1.391	1.485	1.443	1.421	1.429
Ca	0.001	0.001	0.001	0.003	0.004	0.007	0.006	0.001	0.001	0.003	0.003	0.005	0.001
Cr	0.001	0.000	0.001	0.001	0.000	0.002	0.003	0.000	0.002	0.000	0.000	0.001	0.000
Total cations/4 ox.	2.990	2.996	2.996	2.989	2.973	2.970	2.980	2.987	2.996	3.010	2.997	2.988	2.993
Fo	76.0	75.5	75.0	76.6	83.6	82.4	72.5	76.0	69.8	73.8	72.5	72.1	71.8
Fa	24.0	24.5	25.0	23.4	16.4	18.6	27.5	24.0	30.2	26.2	27.5	27.9	28.2

the trace element abundances (ppm). Typical errors of analysis for various elements are also shown. For TiO_2, Al_2O_3, FeO, Na_2O, K_2O, MnO, Sc, Co, Hf, and Ta, BCR-1 values were used as internal standards. GSP-1 = 105 ppm for Th, PCC-1 = 45.7% for MgO, 0.410% for Cr_2O_3, and 0.25% for Ni, and Allende standard for Au = 140 ppb and Ir = 780 ppb were used as internal standards. These are the values recommended by *Flanagan* (1973).

TABLE 5. Ilmenite and spinel compositions in A-15 impact melt rocks.

	Ilmenite					Spinel						
	15434,145	15434,182		15434,183	15434,185	15294,20		15434,183		15434,187	
	1.	1.	2.	1.	1.	1.	2.	1.	2.	1.	2.
TiO_2	52.15	52.69	52.67	51.99	51.80	0.62	0.55	5.61	6.36	8.00	7.85
Al_2O_3	0.11	0.11	0.09	0.09	0.37	56.42	55.62	10.44	10.73	13.66	13.40
Cr_2O_3	1.19	0.49	0.10	0.08	0.09	11.73	12.87	48.07	45.20	39.20	39.10
FeO	37.74	39.97	40.44	39.80	39.25	10.98	11.75	30.14	30.88	32.79	32.05
MgO	5.67	4.91	4.50	6.08	4.68	18.52	18.24	5.21	5.39	5.64	5.62
MnO	0.43	0.39	0.30	0.41	0.44	0.18	0.09	0.35	0.36	0.40	0.46
CaO	0.08	0.26	0.09	0.04	0.38	0.21	0.11	0.06	0.06	0.07	0.10
SiO_2	0.19	0.10	0.07	0.07	0.58	1.07	0.34	0.11	0.07	0.15	0.09
	97.56	98.92	98.26	98.56	97.59	99.73	99.57	99.99	99.05	99.91	98.67
Ti	0.973	0.976	0.986	0.966	0.970	0.011	0.011	0.143	0.164	0.202	0.200
Al	0.002	0.002	0.002	0.002	0.010	1.727	1.722	0.417	0.433	0.539	0.536
Cr	0.023	0.009	0.001	0.001	0.001	0.241	0.267	1.292	1.227	1.039	1.049
Fe	0.783	0.823	0.841	0.823	0.817	0.238	0.257	0.857	0.887	0.919	0.910
Mg	0.209	0.180	0.167	0.223	0.173	0.717	0.714	0.263	0.275	0.281	0.284
Mn	0.008	0.008	0.005	0.008	0.008	0.003	0.001	0.009	0.009	0.011	0.013
Ca	0.001	0.006	0.002	0.000	0.009	0.005	0.002	0.001	0.001	0.002	0.003
Si	0.004	0.002	0.001	0.001	0.014	0.027	0.008	0.003	0.001	0.005	0.002
Total cations	2.003	2.006	2.005	2.024	2.002	2.969	2.982	2.985	2.997	2.998	2.997
No of oxygens	3	3	3	3	3	4	4	4	4	4	4

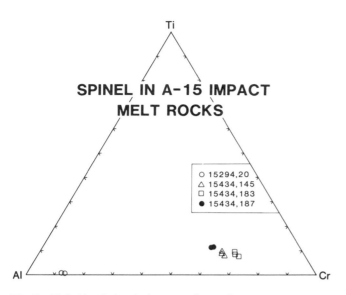

Fig. 7. Ti-Cr-Al variations in impact melt spinels.

Major Elements

The abundances of the major elements and selected trace elements are summarized in Fig. 8, which also includes data from *Ryder and Spudis* (1987). Figure 8 shows a high degree of overlap between our analyses and those of *Ryder and Spudis* (1987). The lack of tight groupings is also evident.

It has been known for some time that glasses with compositions similar to that of Fra Mauro basalt but with lower K_2O contents are present in the Apollo 15 regolith (*Reid et al.*, 1972). Modeling of regolith compositions has shown that this low-K Fra Mauro (LKFM) component increases in the regolith with decreasing distance from the Apennine Front (*Duncan et al.*, 1975; *Walker and Papike*, 1981). Therefore LKFM is an important component in the Apennine Front, although an indigenous rock with that composition has not been identified; it has been found only as impact melt rocks and glasses. Figure 9 compares A-15 melt rocks with LKFM glass compositions from different sites (*Reid et al.*, 1977). An average of 82 Apollo 15 glasses (*Reid et al.*, 1972) gives an average Al_2O_3 content of 18.8 wt % (one standard deviation = 1.85). Therefore, as Fig. 9 shows, many, but not all, of the melt rock analyses are within one standard deviation of LKFM with respect to Al_2O_3. Note that the A-15 LKFM composition is most similar to the melt rock compositions, but it is higher in Al_2O_3 than most of the melt rocks, and the other LKFM compositions (A-16, A-17, Luna 16, and Luna 20) are even higher in Al_2O_3. The melt rocks with lower Al_2O_3 tend to have higher K_2O and Na_2O (Table 6) than LKFM, suggesting a K-rich component, probably medium-K KREEP.

Rare Earth Elements (REE)

Chondrite-normalized REE abundances of the 13 samples analyzed for this study are shown in Fig. 10. Their concentrations vary by a factor of about 8. As is typical of lunar rocks, with decreasing REE contents the Eu anomaly becomes less negative. Sample 15434,145 has the highest REE contents (La at 150×) and has a pattern typical of low-K KREEP. Other low-K KREEP samples are 15434,187, 15434,183, 15314,150, 15314,106, 15303,69, (La at ~120×), and 15303,105 (La at 95×). Samples 15294,20 (La at 55×), 15223,45, and 15434,182 (La at ~35×) are approximately of LKFM composition. Samples that do not fit into these groups are 15303,102 (La at ~28×) and samples 15223,21 and 15434,185 (La at 20×, positive Eu anomaly).

Ryder and Spudis (1987) observed variations by a factor of 27 in the REE contents of 14 melt rocks. Based on the REE and elemental correlations (e.g., TiO_2 versus Sm versus Sc), Ryder and Spudis recognized five chemical groups (A to

TABLE 6. Major and minor abundances (%) via INAA in Apollo 15 melt rocks.

Sample	Wt. (mg)	TiO_2	Al_2O_3	FeO	MgO	CaO	Na_2O	K_2O	MnO	Cr_2O_3
15223,45	25.0	0.86	17.5	6.65	10.6	11.3	0.61	0.16	0.088	0.150
15223,21	9.94	1.1	19.2	6.70	9.0	12.0	0.60	0.088	0.099	0.136
15294,20	29.9	1.0	19.3	7.80	21.5	9.4	0.36	0.14	0.100	0.250
15303,102	15.1	0.96	19.7	7.00	11.8	11.7	0.52	0.11	0.090	0.142
15303,105	8.62	0.83	15.0	9.60	14.8	9.3	0.48	0.23	0.120	0.270
15303,69	7.97	1.4	16.0	10.1	14.6	10.3	0.60	0.255	0.135	0.254
15314,150	27.6	1.2	15.2	10.2	16.8	9.4	0.40	0.18	0.125	0.440
15314,106	30.7	0.97	16.2	8.70	13.2	9.7	0.55	0.44	0.122	0.250
15434,182	33.3	2.0	19.3	7.90	11.1	11.6	0.64	0.12	0.104	0.141
15434,183	17.0	0.91	16.8	9.30	11.5	10.9	0.56	0.35	0.125	0.260
15343,185	16.9	1.4	20.8	7.10	10.0	12.4	0.69	0.14	0.090	0.152
15434,187	16.8	1.1	12.9	11.2	15.7	8.2	0.58	0.094	0.150	0.275
15434,145	21.3	1.0	16.0	9.60	11.8	10.4	0.57	0.35	0.127	0.271
Controls										
BCR-1		≡2.20	13.6	≡12.2	3.0	6.90	≡3.20	1.70	≡0.180	0.0020
BHVO-1		2.8	13.8	11.3	7.6	11.5	2.20	0.55	0.170	0.042

Estimated errors based on counting statistics are: ±0.5-3% for TiO_2, Al_2O_3, FeO, MnO, Na_2O, and Cr_2O_3; ± 5% for MgO, CaO, and K_2O.

TABLE 7. Trace abundances (ppm) via INAA in Apollo 15 melt rocks.

Sample	Sc	V	Co	Ba	Sr	La	Ce	Nd	Sm	Eu	Gd	Tb	Dy	Ho	Tm	Yb	Lu	Zr	Hf	Ta	Th	U	Ni	Ir ppb	Au ppb
15223,45	11.0	30	20.1	160	160	11.0	26	17	4.50	1.50	5.9	1.00	6.6	1.5	0.60	4.20	0.61	110	3.40	0.55	2.12	0.57	100	3.0	<1
15223,21	14.3	30	16.0	110	170	6.60	17	12	3.40	1.50	4.2	0.74	4.8	1.2	0.50	3.20	0.48	70	2.34	0.47	1.20	0.31	80	<2	<1
15294,20	13.0	50	32.4	210	120	18.5	48	32	8.90	1.25	10.3	1.80	11	2.5	0.90	5.90	0.85	230	6.00	0.80	3.00	0.73	240	3.2	1.8
15303,102	11.0	25	21.0	130	150	9.20	23	14	4.10	1.30	5.0	0.92	6.3	1.4	0.53	3.70	0.53	140	3.30	0.56	2.20	0.62	170	4.8	1.1
15303,105	16.0	45	37.5	300	130	32.6	82	52	14.5	1.90	18	2.90	20	4.3	1.6	10.2	1.50	340	9.45	1.2	5.10	1.4	320	5.0	6.2
15303,69	19.0	40	29.5	450	170	41.6	110	65	19.4	2.20	23	3.90	26	5.2	2.1	13.5	1.95	680	19.0	1.7	7.00	1.7	230	4.5	4.1
15314,150	16.0	60	61.0	350	150	41.0	105	70	18.3	1.90	23	3.70	22	4.9	1.8	11.0	1.70	400	11.0	1.5	6.00	1.5	620	15	12.7
15314,106	18.0	60	30.5	450	180	40.5	100	64	17.2	1.95	22	3.50	23	5.0	2.0	13.0	1.90	420	12.1	1.5	6.70	1.6	210	2.9	4.0
15434,182	18.0	30	18.0	140	180	11.0	27	18	5.10	1.70	6.8	1.20	7.6	1.9	0.73	4.7	0.71	140	4.00	0.69	2.20	0.55	60	2.0	<2
15434,183	18.0	50	47.0	460	160	42.0	105	70	18.5	2.10	23	3.60	23	5.2	2.0	13.5	1.91	410	11.0	1.4	7.00	1.6	420	12	5.7
15434,185	15.0	30	18.0	160	210	10.0	25	17	4.70	1.90	6.4	1.00	6.8	1.5	0.64	4.00	0.60	150	3.50	0.63	1.90	0.45	70	<2	<2
15434,187	19.0	50	35.0	400	120	43.0	110	70	19.5	1.75	24	3.90	25	5.9	2.0	13.1	1.95	420	12.5	1.7	7.20	1.8	360	4.7	6.9
15434,147	20.0	50	25.7	480	160	49.0	125	80	22.0	2.30	28	4.70	30	6.4	2.2	15.0	2.05	350	9.50	1.3	6.05	1.6	160	5.0	2.0
Controls																									
BCR-1	≡32.0	≡400	≡36.0	670	320	26.0	55	30	6.70	2.00	7.3	1.00	6.3	1.3	0.60	3.40	0.53	180	≡4.70	≡0.80	6.3	1.7	-	-	-
BHVO-1	31.0	320	44	140	-	15.2	36	25	6.20	1.95	8.0	1.1	5.4			2.00	0.30	-	4.4	1.20	1.1	-	120	-	-

Estimated errors based on counting statistics are: ±5-5% for Sc, Co, La, Sm, Eu, Yb, and Lu; ±5-10% for V, Ba, Ce, Nd, Tb, Dy, Ho, Hf, Ta, Th, and Ni; ±10-20% for Sr, Gd, Tm, Zr, and U.

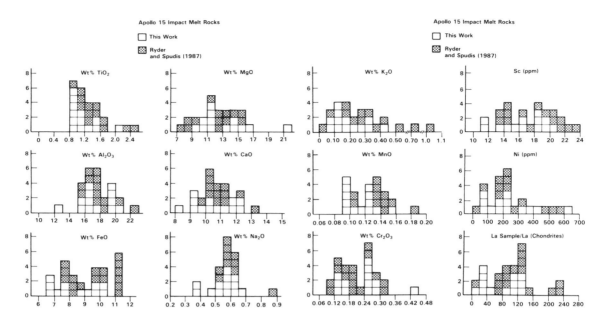

Fig. 8. Summary of major and selected trace element abundances in A-15 impact melts, including data from *Ryder and Spudis* (1987).

E), A being the highest in REE (La at 240×) with a strong negative Eu anomaly, and E (one sample) being the lowest in REE (La at 8.5×) with a positive Eu anomaly. Figure 11 shows a plot of chondrite-normalized La versus chondrite-normalized Sm/Eu ratios of melt rocks. Eight of our samples fall into their B, C, and D groups, which are closely spaced. Groups D and E are far apart, and five melt rocks plot between them: 15223,45, 15434,182, and 15303,102 at 30× La with a negative Eu anomaly and 15223,21 and 15434,185 20×-30× La with a slight positive Eu anomaly. These data points and groupings are shown in another plot of Sm versus Sc/Sm ratio (Fig. 12). Again, groups B, C, and D form a continuum. Groups A and E are well separated with our two new groups in between groups D and E. However, we believe that with additional chemical data on melt rocks, the discrete groupings will merge to form a continuum.

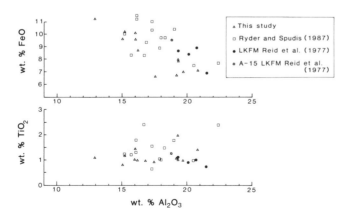

Fig. 9. Comparison of A-15 impact melts with various LKFM glass compositions. The A-15 LKFM composition (star) is more similar to the impact melts than the A-16, A-17, L-16, and L-20 compositions (solid circles), which tend to be more Al_2O_3-rich than the melt rocks.

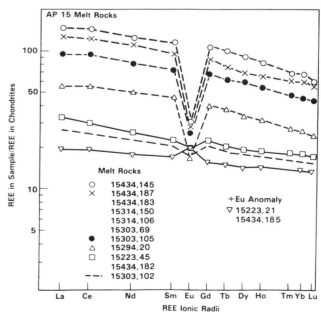

Fig. 10. Chondrite-normalized REE abundances for 13 impact melt rocks.

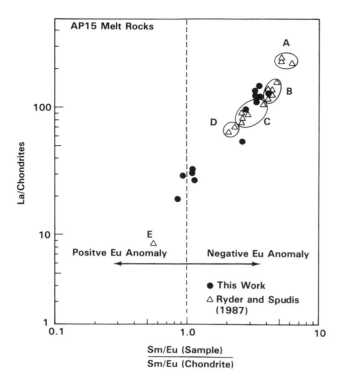

Fig. 11. Chondrite-normalized La versus magnitude of Eu anomaly. The analyses and groups (A-E) of Ryder and Spudis are indicated.

IMPACT MELT SOURCES

As illustrated in Fig. 9, even with a loose definition of LKFM, most of the melt rocks are not simply melts of LKFM composition. But because LKFM is an important component and probably is itself a mixture, the melt rock compositions are difficult to model petrologically. Mineral compositions, major element abundances, and REE abundances show the influence of a KREEP component. Furthermore, the plot of La versus Eu anomaly (Fig. 11) is suggestive of plutonic-KREEP mixing. *Ryder and Spudis* (1987) interpret their Group A as melts of A-15 KREEP basalt formed by impacts into the local Apennine Bench Formation. The other analyses (theirs and ours) trend away from Group A toward their Group E, with lower REE by a factor of 10 and a positive Eu anomaly. Of course, these plutonic and KREEPy "endmembers" could themselves be combinations of more than one plutonic rock, and more than one KREEPy lithology, respectively.

MgO contents in the melt rocks range from 7 to 21.5 wt %, and most are in the 11-15% range. LKFM averages ~9-11 wt % (*Reid et al.*, 1977), and A-15 KREEP basalts have 7-8 wt % MgO (*Meyer*, 1977; *Dymek*, 1986). Thus a magnesian component is indicated, possibly troctolite. A large-scale impact would be likely to excavate plutonic rocks, and an olivine-rich source would be consistent with the abundances of olivine found in the melt rocks.

Sample 15294,20 has either the highest troctolite component or a different source than the other impact melts. It has primitive mineral compositions, is olivine-rich (43 vol %), and is MgO-rich (21.5 wt %).

Also, the major elements and REE in 15434,185 indicate a large plutonic component. This sample has low K_2O and REE abundances and a positive Eu anomaly—no KREEP signature.

PROJECTILES AND THE NI/IR RATIO

Ryder and Spudis (1987) suggest that their groups B through E are compositionally tight enough to each represent a single impact. This is largely based on definition of the inferred target materials from the rock bulk compositions. Another way to compare the impact melts and obtain information bearing on the number of events is by estimating the composition of the projectile. This can be done by analysis of siderophile element abundance in the melt rocks. *Korotev* (1987) used this approach in a study of A-16 noritic impact melts. Except for a small indigenous component of ~10 to 20 ppm Ni (e.g., *Simon et al.*, 1987) and <0.3 ppb Ir (*Warren and Wasson*, 1980), the Ni and Ir in impact melt rocks can be attributed to the meteoritic projectile. Furthermore, these are relatively refractory elements and should be less mobile than Au (*Wänke et al.*, 1978). Figure 13 shows that the samples we analyzed

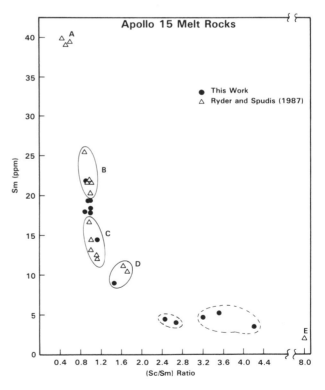

Fig. 12. Plot of Sm versus Sc/Sm showing the separation of Ryder and Spudis groups A and E from B, C, and D.

Apollo 15 Melt Rocks

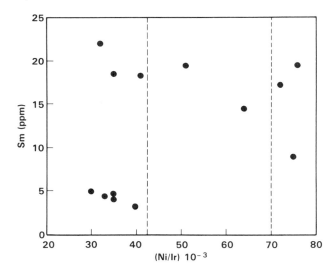

Fig. 13. Plot of Sm versus Ni/Ir showing the distribution of Ni/Ir values.

have a wide range of Ni/Ir ratios, with eight samples in the 30,000 to 40,000 range, and three in the 70,000 to 80,000 range, uncorrected for indigenous Ni. Most chondrites, however, have ratios in the 20,000 to 30,000 range (*Palme*, 1980). Thus, we are not seeing a present-day chondritic component, but rather an "ancient" meteorite component (e.g., *Morgan et al.*, 1974). Another possibility is an iron meteorite component, which can have Ni/Ir ratios as high as 2.5×10^7, though most are probably around 1×10^5 (*Scott and Wasson*, 1975). *Korotev* (1987) concluded that iron meteorites formed the A-16 noritic impact melt breccias, and pointed out the hazards of assuming a chondritic meteorite component.

Ryder and Spudis (1987) selected 15445 and 15455 as their best candidates for samples of Imbrium melt. Based on data in *Ryder* (1985), the melt matrix of 15455 has a Ni/Ir ratio of ~31,000; the abundance of A-15 impact melts with Ni/Ir ratios in the range 30,000–35,000 (as shown in Fig. 13) makes this a likely value for the impactor that formed Imbrium.

It is also interesting to note that many of our samples have Ni/Ir ratios slightly higher than those of A-17 boulders, which are in the 24,000 to 30,000 range (*Palme*, 1980). This similarity may suggest that projectiles that formed Imbrium and Serenitatis were similar, with the projectile that formed the latter probably having a more chondritic (lower) Ni/Ir ratio. Further analysis of these and additional data for A-15 and A-17 melt samples will aid in the interpretation of impact melts from the Imbrium and Serenitatis basins.

CONCLUSIONS

Our analyses show that A-15 impact melt rocks have similar mineral chemistries but bulk compositions are more variable. Most of the samples are approximately of LKFM composition, highlighting the importance of LKFM in the Apennine Front. Other important components in the melt rocks are KREEP and a plutonic component, probably magnesian in nature. While at this time we cannot associate the melt rocks with specific events, seven of our chemically analyzed samples are very similar to the Group B and C samples of *Ryder and Spudis* (1987), who could not rule out these samples as possible Imbrium melts. One of our samples fits into Group D of *Ryder and Spudis* (1987), and this group is their best candidate for Imbrium melt. None of the groups are identical to the Serenitatis impact melts from Apollo 17. The Apennine Front impact melts appear to be complex mixtures, and we agree with Ryder and Spudis that dating of these samples would greatly aid in the understanding of the origins of the A-15 impact melt rocks.

Acknowledgments. Samples were hand-picked and chipped in the Lunar Sample Curatorial Facility at the Johnson Space Center with the valuable assistance of J. Dietrich and especially K. Willis. Samples were irradiated at the Oregon State University TRIGA reactor and we thank R. A. Schmitt and A. G. Johnson for their cooperation. D. Gosselin, M. Lindstrom, and R. A. Schmitt provided helpful comments. L. Thomas and G. Duke are thanked for their efficient typing and drafting, respectively. This work was supported by NASA grants NAG 9-22 (JJP) and NAS 9-15357 (JCL) and funding is gratefully acknowledged.

REFERENCES

Bence A. E. and Albee A. L. (1968) Empirical correction factors for the electron microanalysis of silicates and oxides. *J. Geol., 76*, 382–403.

Bence A. E., Papike J. J., Sueno S., and Delano J. W. (1973) Pyroxene poikiloblastic rocks from the lunar highlands. *Proc. Lunar Sci. Conf. 4th*, 597–611.

Brown G. E., Jr. (1980) Olivines and silicate spinels. In *Reviews in Mineralogy Vol. 5, Orthosilicates* (P. H. Ribbe, ed.), pp. 275–381. Mineralogical Society of America, Washington, D.C.

Duncan A. R., Sher M. K., Abraham Y. C., Erland A. J., Willis A. P., and Ahrens L. H. (1975) Interpretation of the compositional variability of Apollo 15 soils. *Proc. Lunar Sci. Conf. 6th*, 2309–2320.

Dymek R. F. (1986) Characterization of the Apollo 15 feldspathic basalt suite. In *Workshop on the Geology and Petrology of the Apollo 15 Landing Site* (P. D. Spudis and G. Ryder, eds.) pp. 52–57. LPI Tech. Rept. 86-03, Lunar and Planetary Institute, Houston.

Flanagan F. J. (1973) 1972 values for international geochemical reference samples. *Geochim. Cosmochim. Acta, 37*, 1189–1200.

Grover J. E. and Orville P. M. (1969) The partitioning of cations between coexisting single- and multi-site phases with application to the assemblages: orthopyroxene-clinopyroxene and orthopyroxene-olivine. *Geochim. Cosmochim. Acta, 33*, 205–226.

Korotev R. L. (1987) The meteorite component of Apollo 16 noritic impact melt breccias. *Proc. Lunar Planet. Sci. Conf. 17th*, in *J. Geophys. Res., 92*, E491–E512.

Laul J. C. (1979) Neutron activation of geological materials. *Atomic Energy Review IAEA, 17*, 603–695.

Laul J. C., Lepel E. A., Vaniman D. T., and Papike J. J. (1979) The Apollo 17 drill core: Chemical systematics of grain size fractions. *Proc. Lunar Planet. Sci. Conf. 10th*, 1269–1298.

Matsui Y. and Nishizawa O. (1974) Iron(II)-magnesium exchange equilibrium between olivine and calcium-free pyroxene over a temperature range 800°C to 1300°C. *Bull. Soc. Fr. Mineral. Cristallogr., 97*, 122–130.

Medaris L. G. (1969) Partitioning of Fe^{++} and Mg^{++} between coexisting synthetic olivine and orthopyroxene. *Am. J. Sci., 267,* 945-968.

Meyer C., Jr. (1977) Petrology, mineralogy and chemistry of KREEP basalt. *Phys. Chem. Earth, 10,* 239-260.

Morgan J. W., Ganapathy R., Higuchi H., Krahenbuhl U., and Anders E. (1974) Lunar basins: Tentative characterization of projectiles, from meteoritic elements in Apollo 17 boulders. *Proc. Lunar Sci. Conf. 5th,* 1703-1736.

Nehru C. E., Prinz M., Dowty E., and Keil K. (1974) Spinel-group minerals and ilmenite in Apollo 15 rake samples. *Am. Mineral, 59,* 1220-1235.

Palme H. (1980) The meteoritic contamination of terrestrial and lunar impact melts and the problem of indigenous siderophiles in the lunar highland. *Proc. Lunar Planet. Sci. Conf. 11th,* 481-506.

Papike J. J., Hodges F. N., Bence A. E., Cameron M., and Rhodes J. M. (1976) Mare basalts: Crystal chemistry, mineralogy, and petrology. *Rev. Geophys. Space Phys., 14,* 475-540.

Papike J. J., Simon S. B., and Laul J. C. (1982) The lunar regolith: Chemistry, mineralogy, and petrology. *Rev. Geophys. Space Phys., 20,* 761-826.

Powell B. N. (1972) *Apollo 15 Coarse Fines (4-10 mm): Sample Classification, Description, and Inventory.* Publ. No. MSC 03228, NASA Manned Spacecraft Center, Houston.

Reid A. M., Warner J. L., Ridley W. I., and Brown R. W. (1972) Major element composition of glasses in three Apollo 15 soils. *Meteoritics, 7,* 395-415.

Reid A. M., Duncan A. R., and Richardson S. H. (1977). In search of LKFM. *Proc. Lunar Sci. Conf. 8th,* 2321-2338.

Ryder G. (1985) *Catalog of Apollo 15 Rocks.* Curatorial Branch Publ. No. 72, JSC 20787. NASA, Johnson Space Center, Houston.

Ryder G. and Spudis P. (1987) Chemical composition and origin of Apollo 15 impact melts. *Proc. Lunar Planet. Sci. Conf. 17th,* in *J. Geophys. Res., 92,* E432-E446.

Scott E. R. D. and Wasson J. T. (1975) Classification and properties of iron meteorites. *Rev. Geophys. Space Phys., 13,* 527-546.

Simon S. B., Papike J. J., and Laul J. C. (1987) Chemistry and petrology of the Apollo 15 Apennine Front, Part I: KREEP basalts and Plutonic rocks. *Proc. Lunar Planet. Sci. Conf. 18th,* this volume.

Simonds C. H. (1975) Thermal regimes in impact melts and the petrology of the Apollo 17 Station 6 boulder. *Proc. Lunar Sci. Conf. 6th,* 641-672.

Simonds C. H., Phinney W. C., and Warner J. L. (1974) Petrography and classification of Apollo 17 non-mare rocks with emphasis on samples from the Station 6 boulder. *Proc. Lunar Sci. Conf. 5th,* 337-353.

Spudis P. D. and Ryder G. (1985) Geology and petrology of the Apollo 15 landing site: Past, present, and future understanding. *Eos Trans. AGU, 66,* 721, 724-726.

Vaniman D. T. and Papike J. J. (1980) Lunar highland melt rocks: Chemistry, petrology and silicate mineralogy. *Proc. Conf. Lunar Highlands Crust* (J. J. Papike and R. B. Merrill, eds.), pp. 271-337. Pergamon, New York.

Walker R. J. and Papike J. J. (1981) The Apollo 15 regolith: Chemical modeling and mare/highland mixing. *Proc. Lunar Planet. Sci. 12B,* 509-517.

Wänke H., Dreibus G., and Palme H. (1978) Primary matter in the lunar highlands: The case of the siderophile elements. *Proc. Lunar Planet. Sci. Conf. 9th,* 83-110.

Warren P. H. and Wasson J. T. (1980) Further foraging for pristine nonmare rocks: Correlations between geochemistry and longitude. *Proc. Lunar Planet. Sci. Conf. 11th,* 431-470.

Wood B. J. (1975) The application of thermodynamics to some subsolidus equilibria involving solid solution. *Fortschr. Mineral., 52,* 21-45.

The Reliability of Macroscopic Identification of Lunar Coarse Fines Particles and the Petrogenesis of 2-4 mm Particles in Apennine Front Sample 15243

Graham Ryder
Lunar and Planetary Institute, 3303 NASA Road One, Houston, TX 77058

Marilyn Lindstrom[1]
Department of Earth and Planetary Science and McDonnell Center for the Space Sciences, Washington University, St. Louis, MO 63130

Kim Willis[2]
Northrop Services Inc., P. O. Box 34416, Houston, TX 77234

[1] Now at NASA Johnson Space Center, Code SN2, Houston, TX 77058
[2] Now at Lockheed EMSCO, 2400 NASA Road One, Houston, TX 77058

Coarse-fines particles are useful for understanding the nature of the Moon and will probably be returned by future unmanned sample missions to other planetary bodies. The preliminary description of these particles is an essential part of the efficient study of such particles, ensuring that material is not inappropriately allocated or consumed. To check the present reliability of initial macroscopic identifications of lunar coarse-fines (2-4 mm) particles, we studied particles from 15243 sequentially using macroscopic, neutron activation, and petrographic methods. 15243 is from Station 6 on the Apennine Front and was expected to contain a variety of rock types. The results of the petrographic and chemical analyses demonstrate that the macroscopic identifications distinguish rock types very well, and generally allow good identification of the particles, at least for such a sample where some prior information is available. The main difficulty was a propensity to identify Apollo 15 KREEP basalts as mare basalts (which were actually absent from the sample). In addition, a few coherent regolith breccias were identified as polymict breccias distinct from regolith breccias, and two fine-grained KREEP basalts were identified as impact melts. Some of these misidentifications would probably have been avoided if scanning electron microscope examination, including energy dispersive analysis, had been included in the preliminary examination. The KREEP basalts include both extremely fine-grained and porphyritic varieties, yet the chemistry indicates little fractionation among them. The impact melts have a wide range of composition, showing that the Apennine Front is a petrologically complex mix of materials that is as yet little understood. Most of the fragments in 15243 are regolith breccias and glasses that were produced in the small event that produced the 1-m crater from which they were sampled.

INTRODUCTION

The coarse-fines from the lunar regoliths (1-2 mm, 2-4 mm, 4-10 mm fractions) are an important and as yet underused source of information about rock types and petrogenesis on the Moon. They potentially include rare or exotic rock types, which are otherwise not represented among the larger rocks. The 2-4 mm coarse fines in particular provide large sample populations whose individual particles are large enough for the rock type to be petrographically characterized, even for samples that are fairly coarse-grained. However, they are not large enough for a complete petrographic, chemical, and isotopic characterization; a 2-mm rock cube will have a mass less than 25 mg, and a 4-mm rock cube less than 200 mg. Thus the analysis of 2-4 mm particles requires careful management if crucial material or rare rock types are to be identified as such and most efficiently consumed.

The preliminary binocular microscope-assisted characterization of particles is a necessary prelude to selection for any subsequent analysis. At the instigation of the Lunar and Planetary Sample Team (LAPST) we described, processed, and analyzed selected coarse-fines. We wished to assess how good the preliminary characterization of 2-4 mm particles might be at the present time, after several years of experience (by many investigators) with lunar samples. Our main object was to see if known rock types could be correctly identified macroscopically, and whether new varieties could be similarly recognized. We assume that all observers would classify and identify particles in a fairly similar manner, i.e., that the process be objective. Such an assumption is consistent with the misclassifications made in this study being somewhat similar to those made by *Powell* (1972). Without confidence in reasonably observer-independent identifications, the whole concept of preliminary examination would be pointless.

Our intended method was to characterize a large random collection macroscopically (performed by GR), and then make petrographic (thin-section) and chemical (INAA) analyses for a selected subset to determine the actual lithologies. It is clear that good preliminary identifications of any coarse fines sample returned from future unmanned missions (from which the amount of sample is likely to be much less than the Apollo samples) is necessary; the results of our study, therefore, have implications reaching beyond the existing lunar sample collection.

Our choice of coarse fines sample was directed toward regoliths from the Apennine Front collected on the Apollo

Fig. 1. Presampling, cross-sun photograph of the fresh crater at Station 6 from which sample 15420 and other regolith samples were collected. Area above top dashed line is disturbed by astronaut activity. Large closed dashed line is the rim of the 1-m crater. 15240 was collected as two scoops from the central portion of the crater. 15243 is the 2-4 mm fraction obtained by sieving 15240 at the Space Center. (AS-15-86-11610).

15 mission. This was because we have an interest in furthering our understanding of that massif, and because a variety of particles including mare basalts, KREEP basalts, glasses, and varied highlands rocks are potentially present in any 2-4mm sample suite from the Front. The landing site is surrounded by terrains of varied geology, and the massifs have been shown to be petrologically complex (e.g., *Ryder and Spudis*, 1987; *Lindstrom*, 1986; *Lindstrom and Marvin*, 1987). Furthermore, several rays cross the area of the landing site, and therefore we were optimistic of finding new rock types.

Various estimates of the number of 2-4 mm particles per gram suggested that a sample of 20-25 g would be required to obtain the 1000 particles we thought desirable for the study (in actuality we obtained only half that number). Only two samples from the Apennine Front had 2-4 mm fractions large enough without further sieving of the bulk (unsieved) reserve samples. We chose sample 15243 (the 2-4 mm fraction from bulk 15240) from Station 6, although in retrospect the high abundance of regolith breccia and glass in this sample, and the corresponding small proportion of crystalline or highland rocks, was a predictable problem. 15243 had a mass of 31.8 g.

This paper is split into two distinct parts. In Part 1 we describe the sample environment, the methods of analysis, and the assessment of the successes, failures, and limitations of the preliminary identifications made. In Part 2 we describe in more detail the petrology and chemistry of the individual particles, and their implications for petrogenesis at the Apollo 15 landing site.

PART 1. RELIABILITY OF MACROSCOPIC OBSERVATIONS

Sample Collection

Sample 15240 was collected from a 1-m-diameter crater approximately 20 m southeast and upslope from the parked Rover at Station 6. The crater was marked by a concentration of fragments, primarily clods up to 10 cm across on an otherwise smooth surface. Sample 15240 was two scoops of material including obvious splash glass from the floor of the crater. A second regolith sample, 15250, was "very fine light gray" material (*Bailey and Ulrich*, 1975) from the east rim of the crater (Fig. 1). Several other soil samples were taken from Station 6; all are very similar to each other in composition but different from regoliths from other stations on the Front (e.g., *Korotev*, 1987a; *Cuttitta et al.*, 1973). Regolith breccias from Station 6 also have compositions similar to the Station 6 regolith samples (*Korotev*, 1987a).

Macroscopic Inspection

About 20 g of 15243 were taken in two subequal batches, and resieved under a nitrogen atmosphere. About 5 g were of friable material (probably almost all barely lithified regolith breccias) that had broken down since the original sieving and passed into the less than 2 mm fraction. The remaining 15 g consisted of 545 particles. These were transferred to a flow bench and washed in freon to remove most of the adhering regolith dust, which obscures binocular identification. The 545 particles were examined individually in air under a binocular microscope and characterized and classified. The results of this inspection and sorting are shown in Table 1, simplified from the notes taken during inspection.

We identified 373 particles as regolith breccias and another 144 as impact splash glass or regolith breccia with glass splashes. Thus 95% of the particles are regolith products, which is a high proportion consistent with the observations of the astronauts during collection and with most of the glass production and regolith lithification taking place during the small cratering event itself. Only 28 particles were not obvious regolith products.

TABLE 1. Descriptions of sample groups summarized from processing notes.

Particle Split Number	Rock Type	# Particles	Description
,13	Glass and regolith breccias with abundant adhering glass.	144	Glass is exterior, vesicular, black, and cindery-looking mostly. Several have very delicate-appearing forms, although they have survived sieving. Nearly all are irregular or angular.
,14	Regolith breccias.	358	Gray-brown, generally rounded and moderately friable. Glass balls and shards are present, but need high binocular power to see them. Clasts as large as 1 mm are rare. As a group they seem quite homogeneous and similar in color, clast content, and clast size distribution.
,15; ,25; ,43; ,60	*Mare* basalts.	4	Apparently ultramafic and with granular textures. Grain sizes less than 1 mm. White, dark brown, and greenish crystals can be distinguished in some, apparent olivine phenocrysts and appearance similar to olivine-normative mare basalts. Two largest have adhering regolith.
,18; ,26; ,44; ,45; ,61	*Polymict* breccias.	5	Coherent, with clasts a little larger than those in regolith breccias. Matrix may be glassy, but clasts do not appear to be glassy. Some adhering regolith.
,12; ,30; ,38-,42; ,62-,65;	Fine-grained, crystalline impact melts.	11	Homogeneous, dark-colored, clasts not prominent and are small. Two vesicular. A range in color from fairly dark to grayish.
,19 A15	KREEP basalt.	1	Fine-grained, plagioclase laths not distinct. Mafics about 50% and interstitial or granular. Could be a mare basalt.
,27	*Olivine* (or glass).	1	Yellow-green, glassy-looking without clasts or vesicles.
,16	Glass.	1	Orange-brown glass, angular, no clasts or vesicles obvious.
,28	*Granulite?* or *fine-grained impact melt.*	1	Similar to fine-grained impact melts, but a little paler in color and plagioclase clasts are more visible. If granulite, then fine-grained.
,21	Impact melt? or *granulitic.*	1	Angular, coherent, fine-grained, pale-colored. Clast-free or clast-poor poor, could be a granulite. Appears to be about half plagioclase, half mafic minerals.
,17	Cataclastic anorthosite?	1	White, anorthosite, crystalline, rare mafic grains (possibly as much as 20%) are pale-colored. Appears shocked, coherent. Cat-An? Granulite? 15418-like?
,29	Regolith breccia (2-lithologies).	1	Obvious regolith breccia, with white friable clast which has yellowish and dark mafics, and about 80% plagioclase. White clast is possibly shocked granulite.
,20	? 2-lithologies.	1	Two lithologies in sharp contact. (1) aphanitic, very pale-green, clast-free material. (2) Coarser, red-brown, with vug(?). Dark, angular, coherent.
,22; ,46- ,59	Regolith breccias.	15	Picked from ,24 grouping, to be as varied as possible. One is a little different in appearance, more like a welded tuff.

Identifications later shown to be erroneous are surrounded with asterisks.

TABLE 2a. Chemistry of Nonregolith particles from 15243.

	A15 KREEP Basalts							K+R*	Glass[†]			Impact Melts								Glass[‡]		S[§]	CA[¶]
Wt %	,15	,25	,43	,60	,19	,12	,30	,55	,16	,20	,38	,40	,41	,39	,42	,63	,64	,65	,21	,62	,27	,28	,17
Na₂O	.921	.863	.850	.877	.820	.849	.840	.741	.820	1.04	.660	.619	.638	.676	.598	.654	.472	.603	0.702	0.600	0.626	0.511	0.306
K₂O	1.03	-	.57	.68	-	.49	-	-	.69	2.2	0.8	0.43	.38	-	-	.47	.22	-	0.098	-	-	0.28	<0.03
CaO	10.7	11.1	10.0	9.5	10.4	9.3	9.7	10.0	9.3	8.7	9.1	10.6	10.5	11.9	12.1	11.9	10.3	14.0	12.6	11.6	10.2	10.2	16.7
FeO	10.4	10.4	10.5	10.4	10.3	10.1	10.2	10.5	10.5	11.7	10.1	10.1	10.1	8.0	8.3	7.3	11.3	6.0	8.6	10.1	9.8	11.1	3.48
ppm																							
Sc	22.8	21.6	22.3	22.5	21.9	21.7	21.8	21.4	23.3	23.9	17.6	14.9	14.9	19.5	16.6	14.2	23.9	13.3	18.3	29.6	19.9	27.3	8.15
Cr	1905	2190	2140	2045	2240	2180	2060	2210	2235	1270	1340	1370	1370	1200	1250	899	2075	953	1100	967	1940	2037	576
Co	18.6	21.2	19.4	19.7	21.0	20.7	19.5	25.3	21.3	19.1	31.4	80.9	81.3	18.1	25.6	17.3	42.2	16.9	21.4	21.2	24.7	26.3	7.45
Ni	-	-	-	-	-	-	-	53	-	-	215	1230	1260	37	190	77	160	130	58	44	82	119	-
Rb	27	19	14	16	17	14	17	12.9	23	42	-	13	9	8	11	13	-	8	-	-	17	8	-
Sr	-	200	-	225	225	-	230	175	240	-	200	220	-	171	177	181	-	174	-	158	197	221	172
Cs	1.04	0.62	.57	.83	0.60	0.68	0.80	0.55	0.68	0.80	-	0.67	0.62	0.36	0.30	0.48	0.28	0.22	-	-	0.48	-	-
Ba	850	710	800	815	740	720	780	640	780	1000	540	498	500	473	380	515	277	151	140	67	500	334	20
La	83.0	72.8	78.4	80.3	73.3	73.0	75.0	63.4	77.4	87.7	67.4	49.5	49.6	38.9	45.0	42.0	25.8	12.6	8.4	4.15	50.9	41.7	0.78
Ce	220	189	242	221	192	210	197	166.4	204	229	190	128	-	101	109	110	68.4	32.0	23.1	11.7	125	110	2.34
Nd	145	117	130	132	116	116	118	97	128	132	108	80	86	59	61	65	38	19.8	10.5	7.6	71	73	1.9
Sm	36.9	32.3	35.5	35.5	33.1	33.2	33.7	28.5	34.6	38.2	29.2	22.5	22.9	17.5	17.6	18.7	11.7	6.0	4.25	3.37	22.8	18.7	0.425
Eu	2.58	2.68	2.93	2.74	2.69	2.74	2.59	2.44	2.66	2.09	2.04	1.99	2.00	1.90	1.67	1.80	1.24	1.36	1.58	1.36	1.89	1.89	0.789
Tb	9.30	6.83	7.45	7.30	6.64	6.75	6.94	5.15	7.34	8.04	5.76	4.80	4.80	3.92	3.40	3.87	2.43	1.12	0.96	0.86	4.51	3.83	0.080
Yb	26.5	22.2	24.2	25.3	22.3	22.5	23.9	21.8	22.1	31.7	19.7	14.2	14.2	13.0	10.9	12.2	8.22	3.67	4.10	3.43	14.0	11.0	0.353
Lu	3.56	2.97	3.31	3.42	2.99	3.06	3.19	2.62	3.21	4.34	2.66	2.06	2.06	1.91	1.68	1.76	1.17	0.554	0.570	0.561	2.14	1.63	0.058
Zr	1000	840	880	870	840	860	1000	840	1010	1100	600	670	550	540	680	610	300	240	(48)	-	940	450	-
Hf	32.5	26.4	29.5	29.1	26.7	26.1	28.3	23.5	27.6	35.2	20.0	16.6	17.0	13.6	13.2	16.1	8.85	4.18	3.68	2.11	18.1	12.9	0.24
Ta	2.90	2.93	3.19	3.27	2.81	2.89	3.20	2.44	3.01	4.08	2.40	1.93	1.86	1.89	1.51	1.66	1.15	0.56	0.55	0.35	1.86	1.35	-
Th	14.2	11.5	12.5	14.4	11.40	11.2	12.8	9.96	13.2	21.2	12.0	8.29	8.23	6.42	6.61	7.94	5.16	1.88	1.37	0.60	8.23	4.15	-
U	4.1	3.2	3.7	4.2	3.0	3.3	3.6	2.73	3.7	5.8	3.2	2.5	2.2	1.8	1.7	2.3	1.5	0.4	0.42	0.10	2.1	1.1	0.03
ppb																							
Ir	-	-	-	-	-	-	-	-	-	-	-	27	25	-	4.4	-	-	1.7	-	-	-	-	-
Au	<3	<6	<2	<3	4<	3<	<6	-	<4	<6	<6	30	32	<4	2.2	<3	<6	<6	<6	<2	<6	5.2	-
Mass(mg)	29	84	15	17	44	17	35	36	12	26	22	21	14	22	13	22	20	22	60	20	12	17	21

*K+R = KREEP basalt + regolith breccia.
[†]Orange-brown.
[‡]Green.
[§]S = Shock melted impact melt or granulite.
[¶]CA = cataclastic anorthosite.

TABLE 2b. Chemistry of regolith breccia particles from 15243 and of regolith 15271.

Wt %	,29*	,18†	,26†	,44†	,45†	,61†	,22	,46	,47	,48	,49	,50	,51	,52	,53	,54	,56	,57	,58	,59	15271
Na₂O	.521	.472	.452	.488	.455	.485	.477	.492	.454	.449	.500	.551	.523	.449	.469	.480	.463	.482	.461	.476	.449
K₂O	.33	-	-	.19	-	-	-	-	-	-	-	-	-	-	.18	.20	.27	.18	-	-	-
CaO	10.4	10.5	11.3	12.1	11.5	10.1	10.6	10.5	10.2	10.5	11.6	10.4	10.7	10.3	10.2	10.0	10.3	12.3	10.1	11.2	10.4
FeO	11.5	12.42	9.85	9.10	12.0	11.73	12.14	11.80	11.9	11.7	12.7	11.6	11.8	11.8	11.9	12.3	11.7	10.1	12.1	12.2	12.1
ppm																					
Sc	25.4	24.6	17.54	16.61	23.5	22.5	23.6	23.4	23.8	22.4	25.3	23.4	23.9	23.6	23.2	23.5	22.6	19.9	23.6	23.7	23.3
Cr	2100	2170	1660	1670	2260	2120	2580	2240	2220	2200	2390	2180	2140	2300	2240	2380	2190	1780	2240	2280	2230
Co	38.2	39.3	37.8	24.4	38.3	39.4	40.6	37.9	41.7	37.5	40.2	34.7	36.9	40.1	37.9	51.2	38.6	32.8	39.6	39.7	39.5
Ni	98	165	350	75	210	240	260	216	260	210	225	128	181	187	255	220	219	190	199	220	230
Rb	13	-	-	7	-	9	11	12	-	-	14	11	10	11	6.4	9.2	-	5.6	-	-	-
Sr	186	182	197	204	141	167	129	236	141	90	165	170	165	170	181	270	188	-	188	148	184
Cs	0.28	0.29	0.26	0.26	0.31	0.29	0.30	0.40	0.24	0.30	0.14	0.39	0.38	0.30	0.26	0.36	0.23	0.26	0.26	0.27	0.29
Ba	348	283	278	280	290	305	300	310	280	280	340	430	345	287	330	310	300	260	300	300	290
La	27.6	25.6	24.6	26.8	28.7	29.2	27.4	30.2	25.8	27.4	31.6	40.9	32.9	27.5	28.4	28.5	26.5	22.5	27.1	27.1	26.9
Ce	74.2	72.2	65.8	70.7	70.4	81.4	75.0	77.9	66.4	69.7	78.7	108.1	86.9	74.2	80.7	81.5	84.8	61.0	75.5	72.7	74.4
Nd	41	39	39	46	43	45	43	41	38	39	43	61	50	44	40	44	37	37	43	45	46
Sm	12.9	11.42	10.93	10.2	13.4	13.2	12.8	14.3	12.2	13.2	14.8	18.8	15.4	12.9	13.2	12.6	12.1	10.5	12.4	12.4	12.3
Eu	1.66	1.49	1.47	1.50	1.49	1.54	1.53	1.49	1.37	1.48	1.61	1.80	1.60	1.48	1.56	1.49	1.50	1.44	1.46	1.54	1.50
Tb	2.73	2.16	2.12	2.60	2.42	2.41	2.31	2.43	2.12	2.25	2.62	3.53	2.96	2.35	2.42	2.20	2.30	2.04	2.31	2.38	2.36
Yb	9.05	9.56	9.08	8.46	9.08	10.6	10.2	9.63	8.28	8.78	10.04	13.47	10.93	9.22	10.55	10.88	9.78	8.64	10.18	10.00	9.95
Lu	1.33	1.16	1.10	1.16	1.22	1.30	1.23	1.31	1.13	1.21	1.37	1.75	1.41	1.20	1.28	1.27	1.20	1.04	1.23	1.20	1.21
Zr	390	260	280	500	460	310	300	530	450	470	570	530	440	360	320	330	300	270	290	340	320
Hf	9.67	9.45	8.65	8.79	9.83	10.4	10.2	10.4	9.00	9.65	11.3	14.4	11.3	9.67	10.5	10.6	9.47	8.37	9.75	10.0	9.63
Ta	1.28	1.12	1.03	1.06	1.17	1.22	1.22	1.20	1.08	1.20	1.33	1.72	1.40	1.16	1.29	1.29	1.14	1.04	1.17	1.14	1.16
Th	5.04	4.96	4.18	4.20	4.54	5.60	5.54	5.27	4.45	4.67	5.22	7.33	5.17	5.05	5.48	5.77	5.32	3.65	5.41	5.12	4.43
U	1.25	0.98	1.11	1.30	1.05	1.27	1.27	1.25	1.05	1.25	1.25	1.70	1.25	1.10	1.34	1.22	1.18	0.96	1.25	1.16	1.20
ppb																					
Ir	-	6.2	7.7	1.8	4.0	4.8	9.3	4.4	5.6	3.4	6.1	3.4	2.4	6.1	4.0	5.6	10	1.8	4.6	7.0	5.8
Au	<6	2.3	4.5	2.1	3.0	3.0	2.8	2.4	1.9	1.5	2.3	-	7.2	2.9	2.9	1.8	3.9	0.3	1.2	3.3	4.9
Mass(mg)	18	38	57	13	18	37	60	9	10	13	13	22	26	25	29	28	29	58	37	63	33

*Two-lithology particle.
†Originally described as "polymict" breccias.

TABLE 3. Comparison of macroscopic, chemical, and thin section identifications of 2-4 mm particles in 15243.

Particle Split Number	Macroscopic Identification	Chemical Identification	Petrographic Identification (no probe data)
,15	Mare basalt	KREEP basalt	KREEP basalt
,25	Mare basalt	KREEP basalt	KREEP basalt
,43	Mare basalt	KREEP basalt	KREEP basalt
,60	Mare basalt	KREEP basalt	KREEP basalt
,19	KREEP basalt	KREEP basalt	KREEP basalt
,17	Cat-Anorth or Granulite	"Gabbroic Anorthosite"	Cat-Anorthosite
,12	Impact melt	Impact melt? A or KREEP basalt	KREEP basalt
,30	Impact melt	Impact melt? A or KREEP basalt	KREEP basalt
,38	Impact melt	Impact melt? A	Impact melt
,40	Impact melt	Impact melt? B1	Impact melt
,41	Impact melt	Impact melt? B1	Impact melt
,39	Impact melt	Impact melt? B2	Impact melt
,42	Impact melt	Impact melt? B2	Impact melt
,63	Impact melt	Impact melt? B2	Impact melt
,64	Impact melt	Regolith (or impact melt?)	(glassy)
,65	Impact melt	Impact melt? X	Impact melt
,62	Impact melt	Impact melt? Y	Impact melt
,28	Granulite or impact melt		Granulitic norite, or impact melt (shocked melted)
,21	Impact melt (?) or granulitic?	Impact melt or granulite?	Impact melt
,20	2 lithologies: (a) aphanitic (b) coarse red/br	High-K KREEP	Impact melt
,29	2 lithologies: (a) reg. bx (b) white clast	Regolith breccia	Regolith breccia
,16	Orange/brown glass	KREEP	Clast-rich glass
,18	Polymict breccia	Regolith breccia?	Regolith breccia
,26	Polymict breccia	Polymict breccia?	Regolith breccia
,44	Polymict breccia	Polymict breccia?	Regolith breccia
,45	Polymict breccia	Polymict breccia?	Regolith breccia
,61	Polymict breccia	Regolith breccia?	Regolith breccia
,22	Regolith breccia	Regolith breccia	*
,46	Regolith breccia	Regolith breccia	*
,47	Regolith breccia	Regolith breccia	*
,48	Regolith breccia	Regolith breccia	*
,49	Regolith breccia	Regolith breccia	*
,50	Regolith breccia	Regolith breccia	*
,51	Regolith breccia	Regolith breccia	*
,52	Regolith breccia	Regolith breccia	*
,53	Regolith breccia	Regolith breccia	*
,54	Regolith breccia	Regolith breccia	*
,55	Regolith breccia	KREEP basalt (1)	KREEP basalt with regolith breccia
,56	Regolith breccia	Regolith breccia	*
,57	Regolith breccia	Regolith breccia	*
,58	Regolith breccia	Regolith breccia	*
,59	Regolith breccia	Regolith breccia	*

Letter designations following impact melt identifications correspond with the groups of *Ryder and Spudis* (1987), with added designations and subdivisions.
(1) Reinspection shows a basalt particle covered with regolith dust.
*No thin sections cut.

Instrumental Neutron Activation Analysis (INAA) and Petrographic Methods

Chemical analysis of a particle helps to identify the lithology of a particle, and also provides important petrogenetic information. The 28 particles that were not obvious regolith particles constituted a small enough population that all were chosen for analysis by INAA and for subsequent thin sections of the same particles. Fifteen apparent regolith particles, chosen to cover as wide a spectrum of appearance as possible, were also taken for INAA, but were not thin-sectioned. A sample of nearby regolith 15271 was simultaneously analyzed for reference. The INA analyses were carried out at Washington University, St. Louis. The procedures used were those described by *Korotev* (1987c). The analyses are listed in Table 2, along with the mass of each particle analyzed. They range from 9 to 84 mg; most were less than 30 mg.

Following INAA counting the particles were returned to the Johnson Space Center where they were made into thin sections, except for the regolith breccia particles. In some cases some irradiated material remains free of thin-sectioning contamination, but in most any remaining sample is in an epoxy potted butt. The thin sections were inspected with the microscope for identification of lithologies; only a few microprobe analyses have yet been performed on them, as they are generally not necessary for the primary identification purposes of this study. The microprobe analyses were made on the Cameca microprobe at the Johnson Space Center using normal conditions and standards. The thin sections have retained the same split number as the original particle, which is the same as the INAA sample. This split number thus includes the thin section, the remaining potted butt, and any portion of the irradiated sample that was not used for the thin-sectioning.

Results

A comparison of the macroscopic, chemical, and petrographic identifications are shown in Table 3. A more detailed description of the individual samples is given in Part 2 of the paper. The chemical analyses provide good information on what any particle is, despite the small sizes of the analyzed materials. For instance, they show the lack of mare basalts and the presence of KREEP basalts, and regolith compositions are clearly delimited. The chemistry in fact appears to be adequate to discriminate KREEP basalts from impact melts, e.g., the two KREEP basalts that were macroscopically classified as fine-grained impact melts differ from the impact melts in having the higher Cr and Sc, low Co/Fe, and the precise incompatible element concentrations of the KREEP basalts. The thin sections, in many cases quite small (areas of 1 to about 5 mm^2) are large enough in all cases for precise characterization.

The chemical analyses and petrographic observations indicate that the macroscopic observations are useful in grouping like samples, and to a large extent in determining what these groups are lithologically. However, there are some miscues, and some of the discriminations may seem to be inadequate; they are nonetheless probably unavoidable with the technique.

Macroscopic "mare basalts." Arguably the most serious miscue is that four samples identified as mare basalts were in fact KREEP basalts. This is serious in that the petrogenesis of these rock types in the lunar context is quite different, and because we have prior knowledge in this case that these two rock types exist. Nonetheless, the samples were correctly identified as basalts. The characteristics that suggested they were mare basalts were the presence of a partly phenocrystic green mineral, incorrectly presumed to be olivine and in reality orthopyroxene, and an overestimate of the abundance of mafic minerals. This error has also been made previously, e.g., in the 4-10 mm coarse-fines catalog of *Powell* (1972); future inspection of basalts should more carefully consider the possibility of conspicuous green orthopyroxene. It should also more carefully consider the amount of dark groundmass; in the present case at least some of the problem lies in the fact that some of the KREEP basalts are vitrophyric, hence the proportion of dark/light material is higher than normally expected for KREEP. Further, although the samples were washed in freon, thin sections show small amounts of regolith still attached to some particles, again obscuring identification, but apparently negligibly affecting the chemistry. Some of these problems could be solved in future studies by an examination of the particles with SEM techniques.

Macroscopic "polymict breccias." Five samples were macroscopically characterized as polymict breccias, as distinct from regolith breccias. The distinction was made largely on their more coherent nature, and their apparent lack of glass spheres or glassy-looking clasts. Some were paler-colored than the obvious regolith breccias. However, the thin sections show that these "polymict breccias" are indeed regolith breccias, with glass spheres and glassy clasts. Their chemistry is very similar to that of the presumed regolith breccias analyzed in this study, except that two of them have FeO and Sc a little lower. This chemistry is also very similar to that of Station 6 soil, including 15271 analyzed in this study. One of the samples characterized macroscopically as a regolith breccia had a chemistry that had much higher incompatible element abundances than the Station 6 regolith, e.g., Sm 28.5 ppm *cf.* about 12-13 ppm. Reinspection of this particle prior to thin sectioning showed it more likely to be a fine-grained crystalline

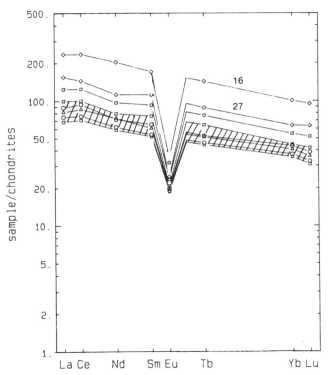

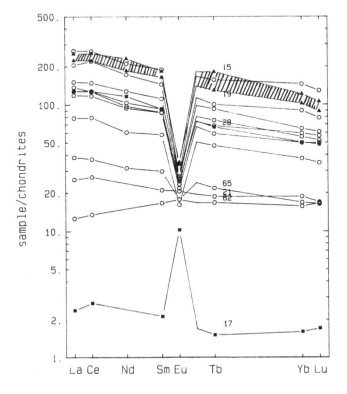

Fig. 2. Rare earth elements in individual particles from 15243. **(a)** KREEP basalts, impact melts, and gabbroic anorthosite. Most individual particle numbers are omitted. KREEP basalts are shown by the shaded range from the highest in rare earths (,15) to the lowest (,19) (filled triangles). ,55, whose analysis includes a significant regolith contamination, is not plotted. Impact melts are open circles, and the gabbroic anorthosite ,17 is filled squares. ,28 (filled squares) is the shock-melted norite, impact melt, or granulite. **(b)** Range for regolith breccias (open squares, with those macroscopically described as "polymict breccias" as open circles) and regolith 15271 (open triangle). All but one of the regolith breccias fall in the shaded zone. The two glass particles (one orange-brown, one green) are also plotted, as open diamonds.

particle largely covered in regolith, and the thin section confirmed this identification. The fine-grained crystalline material is a fine-grained KREEP basalt. The adhering regolith is rather firmly held on, such that the particle appears to have been part of a larger, lithified regolith breccia at some time.

Macroscopic impact melts. Of eleven samples identified macroscopically as fine-grained impact melts, all but two in fact were impact melts, with a range of textures from glassy or very fine-grained to poikilitic. The chemistry shows a wide variation too, although all are distinct from regolith compositions, and are unadulterated highlands lithologies. The two exceptions are fine-grained KREEP basalts that are probably of volcanic, not impact, origin. It is apparent that the macroscopic observation of particles is not going to allow the distinction among different chemical groups of impact melts, and that the distinction of such melts from fine-grained volcanics will be difficult, especially when small phenocrysts or even small amounts of strongly-adhering regolith may mimic clasts.

Imprecise discriminations. Two samples (,16; ,27) were macroscopically characterized as clast-free: one green and probably consisting of olivine or glass, the other orange and probably glass. They contained no vesicles and appeared clear. Their chemistry (Table 2a) shows that neither has the composition of a single mineral, and the thin sections show that both are glass, but surprisingly, clast-rich.

Several other samples gave less clear macroscopic identifications. Two double-lithology fragments appear to have been adequately identified. One fragment (,28), which was identified as a "granulite, or fine-grained impact melt," is a shock-melted, granulitic norite. Another (,21), identified as a probable clast-poor impact melt but possibly a granulite, is a poikilitic impact melt; it is probably more plagioclase-rich than most impact melts and it has low incompatible element abundances. A white particle (,17) had a shocked appearance and was characterized as a cataclastic anorthosite or possibly a granulite; its chemistry and petrography show that it is a shocked anorthosite that is a little more mafic than most.

Conclusions

Preliminary macroscopic characterizations of 2-4 mm particles in this particular regolith, 15240 (for which prior knowledge of the types of lithologies to be expected exists) is quite good, with few errors in general identification. The two main miscues were the failure to distinguish between two known basalt types, and the failure to recognize some polymict breccias more precisely as (polymict) regolith breccias. All distinct samples, e.g., cataclastic anorthosite (?) and granulite (?), were recognized as distinct, and their identification was reasonably accurate. Shock effects in some particles obscured the detailed nature of the sample. It is clear that clast-poor impact melts and fine-grained volcanic rocks will continue to be very difficult to distinguish by binocular examination, but if an absence of any certainly recognizable clast is taken as a characteristic of a volcanic fragment, then the proportion of correctly classified particles should be quite high. No new or exotic rock types were observed in this study; the binocular

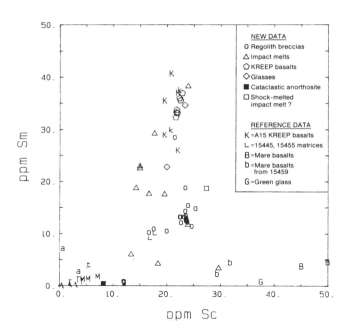

Fig. 3. Sm versus Sc for individual particles from 15243 and reference materials. Letters not in legend are varied anorthositic, noritic, and troctolitic rocks (for which see *Lindstrom et al.*, 1987, for details). Our KREEP basalts cluster quite tightly; the impact melts show a very wide range of compositions. Nearly all particles are more enriched in incompatible elements than is the regolith.

examination did not falsely give the impression that there were any. We are confident that substantially new crystalline lithologies would have been identified if they had existed; it is not clear that distinct basalt types would have been recognized as such.

Examination of the particles with scanning electron microscope techniques, including energy-dispersive analysis, could have produced an even better preliminary classification. The textures of the particles could have been seen, probably allowing the discrimination of fine-grained basalts and impact melts. The distinction of orthopyroxene and olivine phenocrysts could also have been fairly easily made. For such an inspection, particles of this large size would need to be coated; if the coating were carbon, it should not interfere with any except some very specialized potential studies of the particles.

We cannot predict how reliable a preliminary examination of particles from a completely new sample collection (e.g., from a previously unsampled lunar site, an asteroid, or even Mars) would be. The indications, given the variety of material we had in our study, are that a careful inspection of such samples would provide useful preliminary classification of particles for more detailed analysis.

PART 2. REPRESENTIVITY, PETROCHEMISTRY, AND PETROGENESIS OF THE 2-4 MM PARTICLES

The combined chemistry and petrographic data for the particles allows inferences to be made about the petrogenesis of the lithologies present, and their significance in the evolution

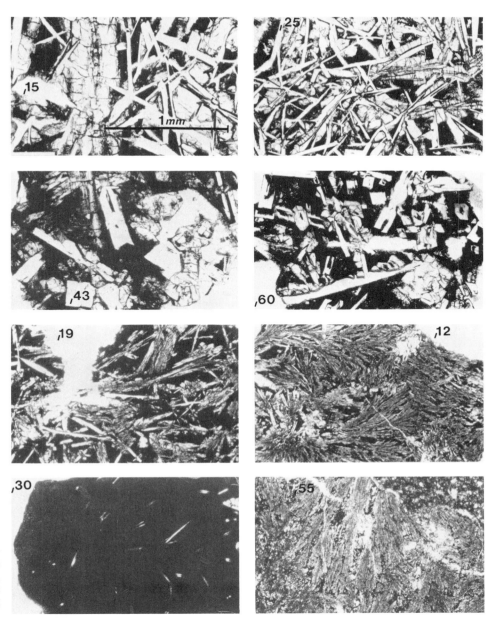

Fig. 4. Plane light photomicrographs of Apollo 15 KREEP basalt particles in 15243. All to same scale, shown for ,15. Numbers correspond with both thin section number and split for chemical analysis.

of the local terrain. The chemical analyses are listed in Table 2 and some of the salient chemical features are summarized in Figs. 2a, 2b, and 3. Photomicrographs of all nonregolith particles are shown in Figs. 4-6.

Representivity

In understanding the petrogenesis of a particle, it is necessary to know how representative a 2-4 mm fragment is of some rock unit of petrological significance. The samples analyzed here were small, 10 to 84 mg (Table 2). From assumptions about distribution of elements among phases, the distribution of phases, and the grain sizes of rock units, one can calculate theoretical values for sample masses required for a chemical analysis that is representative at some specified degree of confidence. (This will actually vary among elements.) Obviously, the coarser the grain-size and the more heterogeneous the rock unit, the larger the sample mass required. In the present study, a more empirical approach is taken.

The regolith breccia particles (Table 2b), ranging from 9 to 63 mg, all have very similar chemical compositions for most elements, and these agree with the bulk composition of the local soil as established from larger masses and averages (e.g., *Korotev*, 1987a). In some cases these regolith breccia particles may not be samples of a larger unit, but as constructional

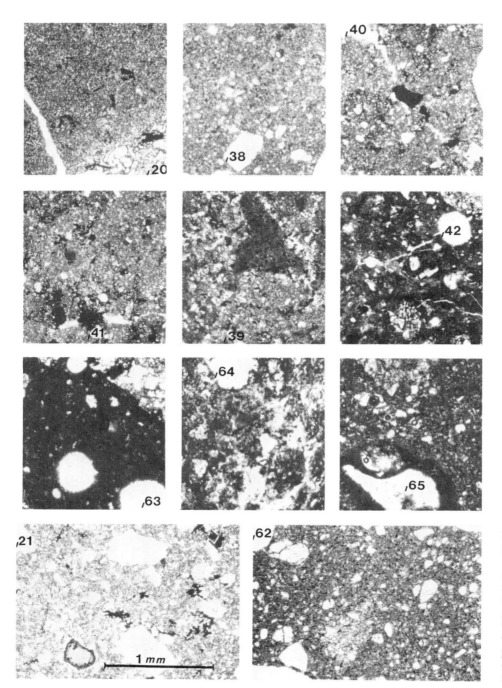

Fig. 5. Plane light photomicrographs of impact melt particles in 15243. All to same scale, shown for ,21. Numbers correspond with both thin section number and split for chemical analysis. ,20 has a coarse grained area in lower right. Round white patches in ,42 and ,63 are vesicles, whereas the larger white patch in ,64 is a glass fragment. Most other white areas are mineral clasts, mainly plagioclases.

fragments they are the unit itself; thus representivity is not a problem for them. All these regolith breccia particles can thus be assumed to be representative.

Ryder and Spudis (1987) found good agreement among replicates of Apollo 15 impact melts for 15 to 40 mg splits, and even between new analyses and published analyses for entirely different subsamples in a similar mass range. *Lindstrom et al.* (1987) found agreement among subsplits of about 100 mg for several Apollo 15 impact melts. Our somewhat smaller sample size (average about 20 mg) means we have somewhat lower confidence, but all our samples are fine-grained (few grains larger than 200 microns). They are also homogeneous both macroscopically and microscopically at a scale similar to the samples analyzed by *Ryder and Spudis* (1987) and *Lindstrom et al.* (1987). They lack large clasts that might distort the analysis. Thus we believe that the analyses are representative. For the glassy particles ,16 and ,27 the analyses can certainly be accepted as representative.

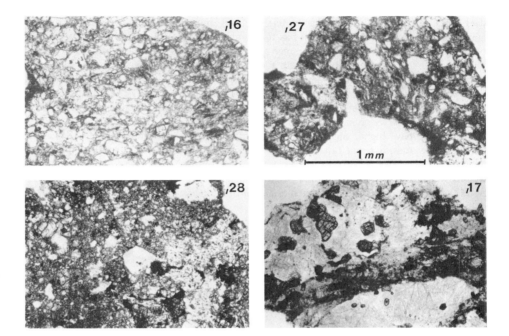

Fig. 6. Plane light photomicrographs of lithic fragments in 15243. All to same scale, shown for ,27. ,16 is clast-rich glass (macroscopically orange-brown) with the composition of Apollo 15 KREEP basalt. ,27 is also a clast-rich glass (macroscopically green) with an impact melt composition. ,28 is a shock-melted fragment which may have been a poikilitic impact melt, or possibly a norite or granulite. ,17 is a cataclastic, "gabbroic" anorthosite, with a shock vein across it.

The KREEP basalt particles include materials coarser-grained than the impact melts. However, their individual homogeneity, their uniformity of composition, and their similarity to previous analyses of Apollo 15 KREEP basalts (including even tinier fragments analyzed by *Korotev*, 1987a) requires their being representative of larger units, unless some coincidence is invoked. For the cataclastic anorthosites, containing coarse plagioclase crystals and scattered mafic and accessory crystals, representivity is much less assured and is not claimed. The composition is merely a guide to the character of the parent unit.

For most of the 2-4 mm particles analyzed here, their homogeneity and fine grain-size, combined with evaluations of previous replicate analyses of similar rock types, allow us to conclude that their chemical analyses and their petrographic characteristics are quite representative of larger rock units that have geological and petrological significance.

Petrochemistry and Petrogenesis

KREEP basalts. Eight of the nonregolith particles are Apollo 15 KREEP basalts, similar in chemistry to those long known (e.g., *Irving*, 1975, 1977; *Dymek*, 1986; *Ryder*, 1976). Seven of the KREEP basalts have remarkably similar chemistry (Table 2, and Figs. 2a and 3), despite their having remarkably different textures ranging from glomeroporphyritic through intersertal through fine-grained, feathery spherulitic (Fig. 4). Comparative grain sizes are easily seen in Fig. 4, as well as absolute grain sizes for the major phases. One porphyritic sample (,43) contains euhedral orthopyroxene crystals embedded in a euhedral plagioclase phenocryst; the other (,60) contains small euhedral pigeonite phenocrysts and very little orthopyroxene along with its lathy and blocky plagioclase phenocrysts. In both cases the groundmass is quenched and much finer grained than most Apollo 15 KREEP basalts. The finest, nonporphyritic sample (,30) was rapidly quenched, containing scattered needles of plagioclase in a glassy and microcrystalline groundmass.

The incompatible element abundances vary by only 10%, even though the sample sizes are small and small amounts of adhering regolith are present (an addition of 10% regolith to a KREEP basalt would lower its trace element contents by only about 6 or 7%, so the effect is not great). In these fragments, Ni, Ir, and Au are below the detection limits of our INAA methods, unlike the impact melts; hence the data are consistent with a volcanic origin for these basalts. The analysis with the highest rare earth abundances also has much higher K_2O (1.0% cf. 0.6%) and Rb (27 ppm cf. 17 ppm) than the other analyses; this analyzed sample (,15) is also the coarsest grained and was only 29 mg. The analysis of the eighth sample (,55) includes a greater but unknown amount of regolith breccia; its incompatible element abundances are 10% lower, and its Ni and Co higher than the others as a result of this regolith contamination.

The chemistry of these KREEP basalt fragments is similar to that of the larger samples of Apollo 15 KREEP basalts 15382 and 15386, or at least the larger masses reported by *Hubbard et al.* (1973, 1974). Other analyses of Apollo 15 KREEP basalts show the same consistency of chemistry indicative of very limited fractionation—for instance, the three "high-Sm" samples of *Korotev* (1987a) (all less than 10 mg), and the seven particles of *Lindstrom et al.* (1977) (21-78 mg). Two samples analyzed by *Warren et al.* (1983) (unstated masses) however have 25% lower incompatible trace element abundances and lower FeO. An orange-brown glass (,16), rich in mineral fragments, analyzed in this study also has the same chemistry as the crystalline KREEP fragments we analyzed.

The small range of incompatible element abundances restricts the amount of fractionation by separation of major mineral phases affecting these basalts to less than 10%. Fractionation was perhaps less, considering the potential for sampling error induced by the coarser-grained and especially the porphyritic samples. These have higher incompatible element abundances than the finer ones, a result not to be expected in a simple fractionation sequence. Rather it appears that all the textural varieties were produced by differential cooling, nucleation, and crystallization histories of a magma of almost constant composition.

Impact melts. Eleven of the particles are fine-grained impact melts (few matrix grains larger than 300 microns), nine of which were identified as such macroscopically. There are also two fragment-containing glasses that are impact melts. The criteria for identifying a particle as an impact melt are those of *Ryder and Spudis* (1987). The impact melt particles in 15243 show a wide variety of both chemistry (Figs. 2a and 3) and texture (Fig. 5), although almost all are fine-grained or glassy. Incompatible element abundances range from La 12x chondrites to La 270x chondrites. All but the one (,62) with the lowest incompatible element abundances have the REE slope like that of KREEP. According to the CaO, FeO, and Sc abundances, none are particularly feldspathic except for glassy sample ,65. We estimate that most contain less than 60% normative feldspar. According to their major and trace element abundances, most of the impact melts thus seem to fall in the low-K to medium-K Fra Mauro basalt compositional range. Several of the impact melts have detectable Ir and Au, and most have Ni greater than 100 ppm, unlike the KREEP basalts.

The most incompatible-element-rich sample (,20) has very high K_2O (2.2%). It was described macroscopically as a two-lithology particle, but the thin section shows mainly one aphanitic lithology that is a clast-free melt. A small patch at one edge is a coarser, basaltic area (plagioclases 200 microns long). Another fragment (,38) with high La (200x chondrites) has a rather granular texture. Two particles (,40 and ,41) have almost identical chemistries (La 150x chondrites), including very high siderophile abundances (Ir 25 and 27 ppb; Au 30 and 32 ppb). They also have very similar textures, with poikilitic pyroxenes, and abundant but very small mineral clasts, and may well be broken pieces of one original coarser fragment. Residual glass is abundant, and ilmenite and phosphate needles are common. These two particles are more similar to some Apollo 16 samples, especially poikilitic impact melt 60315, in both their trace incompatible element abundances and their very high siderophile abundances (a distinctive feature of Apollo 16 melt rocks; *Korotev,* 1987b). A fragment of green, clast-rich glass (,27) has a chemistry very similar to ,40 and ,41, except that it lacks such high siderophiles, and its Sc and Zr abundances are a little higher.

Several of the impact melt samples are glassy, and some of these are vesicular. Some contain undevitrified glass, but in most the glass is devitrified and brown. One (,64) is heterogeneous and contains glass fragments. Its chemistry is identical with that of the local regolith, from which it was undoubtedly derived in a small-scale cratering event. The other glassy samples, however, are not like local regolith. Of the crystalline samples, the coarsest is ,21, which we macroscopically described as either an impact melt or a granulite. This sample is poikilitic, but the oikocrysts are only a few hundred microns across. It does contain clasts larger than those in the other melts; these clasts are mainly plagioclases, but pyroxene, olivine, and rare pink spinel are present. Only a minor amount of interstitial glass is present. On balance we think it is an impact melt. It is chemically distinct in that the incompatible element abundances are low (La 25x chondrites) and a Eu anomaly is lacking. The trace element abundances are not the lowest of the melt samples, however. The fine crystalline sample ,62 has La only 12x chondrites, and has a completely different REE slope, enriched in the heavier elements. It also has the highest Sc of any of the samples analyzed in this study and is probably pyroxene-rich.

Ryder and Spudis (1987) analyzed 14 impact melts from the Apennine Front, finding that they had a wide range of compositions that appeared to cluster into five groups (A to E). The data in the present study suggest that there are more groups, or even that there is a continuum of compositions, as also suggested by recent analyses by *Laul* (1987), *Laul et al.* (1987) and *Lindstrom et al.* (1987). The present data for eleven impact melts and two glasses are more limited in the number of elements analyzed than are those of *Ryder and Spudis* (1987). The sample sizes are also smaller, but the fine grain size of the samples suggests that the chemical data should be quite representative. Group A is represented by orange glass ,50. On the basis of incompatible elements, five of the impact melts fall into the range (La 120 to 150x chondrites) that defined Group B, and the green, fragment-bearing glass sample ,27 also falls within this range. However, Sc is slightly lower in most of them than in the *Ryder and Spudis* (1987) Group B samples. All of the other six samples are different from any of those analyzed by *Ryder and Spudis* (1987). None are very close to the average composition of the Apennine Front, as represented by Station 2 soils (*Korotev,* 1987a), and most are more KREEP-rich, although one (,64) is very similar in bulk chemistry to the local Station 6 regolith. None are at all like the Serenitatis melt or the 15445-group that may be Imbrium melt (*Ryder and Bower,* 1977; *Ryder and Spudis,* 1987). Sample ,21 is roughly like those Apollo 16 samples that are represented by 68415, which is in turn similar to Descartes regolith (*Ryder and Norman,* 1980).

These results demonstrate conclusively the complex nature of the Apennine Front at the Apollo 15 site, as represented by impact melts. The Front appears to contain much more diverse melt compositions than do the Apollo 14, 16, or 17 collections. Most are quite distinct from samples from these other landing sites, indicating a distinct provenance for the Apennine Front. Some, particularly the higher KREEP melts, may be post-Imbrium; most are probably from a wide variety of pre-Imbrium events and were dumped in the Front as part of the Imbrium ejecta. A more elaborate study of these impact melt particles, particularly radiogenic isotope data, would potentially be very informative about the pre-Imbrium crust in the basin area.

Other lithologies. Sample ,17 is a cataclastic, feldspathic sample (Fig. 6); the mode and the chemistry both suggest that the rocklet consists of about 90% plagioclase. The plagioclase is shocked with erratic extinctions, and forms contiguous areas more than a millimeter across; the original grain size cannot be reliably determined. The mafic minerals form small (less than 300 micron), partly clustered grains; their textural relationships are quite unlike those in granulites. Most of the orthopyroxenes and olivines form separate grains; in a few cases the olivines form cores or edges to orthopyroxenes. The opaque mineral content of the fragment is negligible. A finer-grained, apparently shock-produced zone crosses the center of the fragment. The sample is probably a chemically pristine, texturally deformed igneous fragment, and is probably part of the ferroan anorthosite suite: Microprobe analyses show uniform compositions for olivine (Fo65), orthopyroxene (mg′ = 72, CaO = 1.1 wt %), and plagioclase (An 96-97). One discrete grain was augite (mg′ = 81, CaO = 22.0 wt %), and a chromite grain was aluminous (15.2 wt % Al_2O_3). These analyses are consistent with well-equilibrated ferroan anorthosites, and indeed are quite similar to some of the compositions in pristine anorthosite 60025 (*Ryder*, 1982). However, the dominance of orthopyroxene and olivine distinguish 15243,17 from local anorthosite 15415. Insofar as the chemistry is known 15243,17 is similar to the slightly mafic ferroan anorthosites 62236 and 62237 (*Warren and Wasson*, 1978) except that it has slightly higher Na_2O and incompatible elements 3× as high. Indeed its incompatible elements are as much as 3-5 times as high as most ferroan anorthosite samples, although the pattern is very similar to that of most ferroan anorthosites. The chemistry is distinct from that of granulites, except for two of the ferroan varieties 15418 and 67215 (*Lindstrom and Lindstrom*, 1986).

The shock-melted sample ,28 is less distinct in all respects. Its chemistry is not pristine, as Au is present at about 5 ppb, and Ni is 119 ppm. It is very similar to the Group B impact melts in its incompatible element and major oxide abundances, although it differs in having high Sc (27 ppm). The mineralogy of the fairly coarse-grained relics is noritic, with prominent opaques including Fe-metal. Texturally the shock effects obscure the details which might distinguish this fragment as either a granulite or an impact melt. The chemistry and noritic mineralogy suggest that it was a coarse-grained poikilitic impact melt that was subsequently re-shock melted *in situ*.

Regolith breccias. Most of the regolith breccia particles were not thin-sectioned. All the samples identified macroscopically as "polymict breccias" as distinct from regolith breccias are actually regolith breccias, as shown by their chemistry (Table 2) and their petrography (Fig. 6). In addition, one two-lithology sample that was identified as mainly regolith breccia was correctly diagnosed. A white clast observed macroscopically may have been a large crushed Apollo 15 KREEP basalt clast observed in the section or another smaller clast of coarse breccia. The chemistry of these regolith breccia samples is quite uniform and very similar to that of the local regolith and the bulk <1 mm size fraction of 15240 (Table 2; *Korotev*, 1987a). There can be no doubt that these regolith breccias were formed very close to their collection site, as the regolith at Station 7 and elsewhere along the Front is different. 15240 was collected to sample glass and other material in the 1-m crater, and the sample contains a very high proportion of regolith breccia particles. Most of these are poorly lithified, except where bonded by glass. It is evident that most of these regolith breccias must have been produced by the impact that produced the small glass-lined crater itself. Thus even very small impacts suffice to partly lithify regolith, as shown experimentally by *Schaal and Hörz* (1980) and *Simon et al.* (1986). Larger fragments, such as those in the >1 cm fraction 15245 were also produced, and these also have the same composition (summary in *Ryder*, 1985). Some of the regolith breccias must be older, and these possibly include the more lithified fragments that were erroneously identified merely as "polymict breccias." Alternatively these latter particles may have formed in the recent 1-m cratering event but were more strongly lithified. Regolith breccias much larger than a few millimeters presumably formed in much larger events.

Population of the 2-4 mm Particles and Conclusions

Apart from the dominant regolith and regolith glass particles, the 2-4 mm fines 15243 consist predominantly of impact melt fragments and KREEP basalt fragments. Only one probable cataclastic anorthosite was found. Conspicuous by their absence are mare basalt fragments of any kind, crystalline polymict breccias, and granulites. Fragmental breccias such as those in the Apollo 16 collection are either not present or are rare; although such breccias would not be directly observable in 2-4 mm fragments, their constituents would be. At Apollo 16 these constituents are dominated by granulites, cataclastic anorthosites, and aluminous melt fragments, clearly rare in our Apollo 15 sample.

It is apparent that the Station 6 regolith is not close to a mass-weighted average of the 2-4 mm crystalline particles, as such a concoction would be much richer in incompatible elements than is the regolith. As *Korotev* (1987a) found for the 1-2 mm fines 15271, the coarser materials are enriched in KREEP. It is clear that the Station 6 regolith must consist of a finer-grained, comparatively KREEP-poor material and a coarser component consisting of Apollo 15 KREEP basalts and KREEPy impact melt fragments. Apparently a well-mixed Apennine Front had KREEP basalts deposited on it, but not in it, at a time when or from a place from which mare basalt was not available. Some mare basalts and green glasses have been added, but not simultaneously with these KREEP basalts. The rock types that make up the Front as it was produced by the Imbrium impact remain elusive, apart from the impact melts. The relative importance of varied pristine rocks and varied impact melts cannot yet be deciphered.

Acknowledgments. This work is Lunar and Planetary Institute Contribution No. 641. Steve Simon and Jim McGee provided constructive reviews, including the suggestion of future use of SEM techniques.

REFERENCES

Bailey N. G. and Ulrich G. E. (1975) Apollo 15 voice transcript pertaining to the geology of the landing site. *U. S. Geol. Surv. Rept. No. USGS-GD-74-029*, 232 pp.

Cuttitta F., Rose H. J., Jr., Annell C. S., Carron M. K., Christian R. P., Ligon D. T., Jr., Dwornik E. J., Wright T. L., and Greenland L. P. (1973) Chemistry of twenty-one igneous rocks and soils returned by the Apollo 15 mission. *Proc. Lunar Sci. Conf. 4th*, 1081-1096.

Dymek R. F. (1986) Characterization of the Apollo 15 feldspathic basalt suite. In *Workshop on the geology and petrology of the Apollo 15 landing site* (P. D. Spudis and G. Ryder, eds.), pp. 52-57. LPI Tech. Rept. 86-03. Lunar and Planetary Institute, Houston.

Hubbard N. J., Rhodes J. M., Gast P. W., Bansal B. M., Shih C.-Y., Wiesmann H., and Nyquist L. E. (1973) Lunar rock types: the role of plagioclase in non-mare and highland rock types. *Proc. Lunar Sci. Conf. 4th*, 1297-1312.

Hubbard N. J., Rhodes J. M., Wiesmann H., Shih C.-Y., and Bansal B. M. (1974) The chemical definition and interpretation of rock types returned from the non-mare regions of the Moon. *Proc. Lunar Sci. Conf. 5th*, 1227-1246.

Irving A. J. (1975) Chemical, mineralogical, and textural systematics on non-mare melt rocks: implications for lunar impact and volcanic processes. *Proc. Lunar Sci. Conf. 6th*, 363-394.

Irving A. J. (1977) Chemical variation and fractionation of KREEP basalt magmas. *Proc. Lunar Sci. Conf. 8th*, 2433-2448.

Korotev R. L. (1987a) Mixing levels, the Apennine Front soil component, and compositional trends in the Apollo 15 soils. *Proc. Lunar Planet. Sci. Conf. 17th*, in *J. Geophys. Res., 92*, E411-431.

Korotev R. L. (1987b) The meteoritic component of Apollo 16 noritic impact melt breccias. *Proc. Lunar Planet. Sci. Conf. 17th*, in *J. Geophys. Res., 92*, E491-512.

Korotev R. L. (1987c) National Bureau of Standards coal flyash (SRM 1633a) as a multielement standard for instrumental neutron activation analysis. *Proc. Int. Conf. Nucl. Anal. Chem.*, in *J. Radioanal. Nucl. Chem.*, in press.

Laul J. C. (1987) Chemistry of the Apollo 15 Apennine Front: highland lithologies (abstract). In *Lunar and Planetary Science XVIII*, pp. 535-536. Lunar and Planetary Institute, Houston.

Laul J. C., Simon S. B., and Papike J. J. (1987) Chemistry and petrology of the Apennine Front, Apollo 14, Part II: Impact melt rocks. *Proc. Lunar Planet. Sci. Conf. 18th*, this volume.

Lindstrom M. M. (1986) Diversity of rock types in Apennine Front breccias (abstract). In *Lunar and Planetary Science XVII*, pp. 486-487. Lunar and Planetary Institute, Houston.

Lindstrom M. M. and Lindstrom D. J. (1986) Lunar granulites and their precursor anorthositic norites of the early lunar crust. *Proc. Lunar Planet. Sci. Conf. 16th*, in *J. Geophys. Res., 91*, D263-D276.

Lindstrom M. M. and Marvin U. B. (1987) Geochemical and petrologic studies of clasts in Apennine Front breccia 15459 (abstract). In *Lunar and Planetary Science XVIII*, pp. 554-555. Lunar and Planetary Institute, Houston.

Lindstrom M. M., Nielsen R. L., and Drake M. J. (1977) Petrology and geochemistry of lithic fragments from the Apollo 15 deep-drill core. *Proc. Lunar Sci. Conf. 8th*, 2869-2888.

Lindstrom M. M., Marvin U. B., Vetter S. K., and Shervais J. W. (1988) Apennine Front revisited: Diversity of Apollo 15 highland rock types. *Proc. Lunar Planet. Sci. Conf. 18th*, this volume.

Powell B. N. (1972) Apollo 15 Coarse Fines (4-10 mm): Sample classification, description, and inventory. NASA MSC 03228. Manned Spacecraft Center, Houston. 91 pp.

Ryder G. (1976) Lunar sample 15405: remnant of a KREEP basalt-granite differentiated pluton. *Earth Planet. Sci. Lett., 29*, 255-268.

Ryder G. (1982) Lunar anorthosite 60025, the petrogenesis of lunar anorthosites, and the composition of the Moon. *Geochim. Cosmochim. Acta, 46*, 1591-1601.

Ryder G. (1985) *Catalog of Apollo 15 rocks*. Curatorial Branch Publ. 72, JSC 20787. NASA-Johnson Space Center, Houston. 1296 pp.

Ryder G. and Bower J. F. (1977) Petrology of Apollo 15 black-and-white rocks 15445 and 15455-fragments of the Imbrium melt sheet? *Proc. Lunar Sci. Conf. 8th*, 1895-1923.

Ryder G. and Norman M. (1980) *Catalog of Apollo 16 rocks*. Curatorial Branch Publ. 52, JSC 16904. NASA-Johnson Space Center, Houston. 1144 pp.

Ryder G. and Spudis P. D. (1987) Chemical composition and origin of Apollo 15 impact melts. *Proc. Lunar Planet. Sci. Conf. 17th*, in *J. Geophys. Res., 92*, E432-446.

Schaal R. B. and Hörz F. (1980) Experimental shock metamorphism of lunar soil, *Proc. Lunar Planet. Sci. Conf. 11th*, 1679-1695.

Simon S. B., Papike J. J., Hörz F, and See T. H. (1986) An experimental investigation of agglutinate meltling mechanisms: shocked mixtures of Apollo 11 and 16 soils. *Proc. Lunar Planet. Sci. Conf. 17th*, in *J. Geophys. Res., 91*, E64-E74.

Warren P. H. and Wasson J. T. (1978) Compositional-petrographic investigation of pristine non-mare rocks: third foray. *Proc. Lunar Planet Sci. Conf. 9th*, 185-217.

Warren P. H., Taylor G. J., Keil K., Kallemeyn G. W., Shirley D. N., and Wasson J. T. (1983) Seventh foray: whitlockite-rich lithologies, a diopside-bearing troctolitic anorthosite, ferroan anorthosites, and KREEP. *Proc. Lunar Planet. Sci. Conf. 14th*, in *J. Geophys. Res., 88*, B151-164.

The Origin of Pristine KREEP: Effects of Mixing Between UrKREEP and the Magmas Parental to the Mg-Rich Cumulates

Paul H. Warren

Institute of Geophysics and Planetary Physics, University of California, Los Angeles, CA 90024

This paper examines implications of the following paradox: Despite their extraordinarily high incompatible element contents, the pristine KREEP basalts generally have bulk-rock *mg* ratios scarcely lower than the mean *mg* ratio of the lunar crust. Their Ni contents are also "primitive," compared to mare basalts with comparable REE contents. Explaining this paradox appears to require systematic mixing between ancient KREEP precursor materials ("urKREEP" magmasphere residuum?) and primitive Mg-rich melts. The magmasphere hypothesis, as modified to have large portions of the crust form as intrusions of Mg-rich melts into the earliest (ferroan anorthosite) plagioclase flotation crust, provides a ready mechanism for this systematic mixing. The Mg-rich melts had to pass through the crust/mantle boundary, where they probably tended to assimilate or mix with the magmasphere residuum. Satisfactory fits to the compositions of pristine KREEP basalts are obtained from finite-difference models of this mixing, followed by anorthosite assimilation plus fractional crystallization. Its ability to account for the bizarrely magnesian nature of the pristine KREEP basalts represents a further confirmation of the magmasphere hypothesis.

INTRODUCTION

The two main varieties of lunar basalt, mare basalt and KREEP basalt, are seldom difficult to distinguish. Besides having far higher contents of incompatible elements, KREEP basalts generally have higher Al_2O_3 and lower Ca/Al (reflected in lower modal abundances of high-Ca pyroxene). Both types of basalt are widely assumed to be derived by remelting of late products of the Moon's "magma ocean" (a.k.a. magmasphere). But modeling both basalt types simply as products of remelting of a series of cumulates from a single former magma runs into difficulties with the sharpness of the distinctions between the two basalt types. Perhaps the most important of these distinctions is a paradox that has long been at least dimly understood, and is considered in detail below: KREEP basalts have far higher contents of incompatible elements than mare basalts, even though both basalt types have similar *mg* [= molar MgO/(MgO+FeO)] ratios.

This *mg* ratio-incompatible elements paradox is clearly one of the key constraints on the origin of KREEP, and therefore on the gross igneous evolution of the Moon. Such a paradox is difficult to explain with any straightforward "serial magmatism" model, in which the Moon's crust forms in the absence of a magmasphere (*Walker*, 1983). Under such a scenario, complexities such as assimilation and magma mixing would occasionally lead to geochemical anomalies, including basalts with high incompatible element contents and moderate *mg* ratios. But such cases would presumably be exceptional. Most basalts with high incompatible element contents would have low *mg*. Assuming that KREEP is derived from magmasphere residual liquid (urKREEP), one possible means of raising its *mg* ratio is by assimilation reactions between urKREEP and lower crustal material (*Warren and Wasson*, 1979). But this model, originally suggested by *Hubbard and Minear* (1976) and *Dowty et al.* (1976), has not, until now, been studied at all quantitatively.

THE PARADOXICAL NATURE OF TYPICAL PRISTINE KREEP COMPOSITIONS

Most estimates of the bulk composition of the Moon hold that its *mg* is less than 0.84 (for a review, and arguments that the true bulk-Moon *mg* is probably closer to 0.88, see *Warren*, 1986a). Most estimates of the bulk *mg* of the crust are of course much lower still. *Taylor's* (1982) estimated bulk highlands crust composition has *mg* = 0.65. One way to estimate the mean *mg* of the upper nonmare crust is to consider the compositions of essentially mare-free soils, which represent products of perfectly random, impact-induced mixing of the upper crust, with the source matter focused around the site where the soil is produced. Such soils generally have *mg* in the range 0.66-0.69, although including the several regolith breccia meteorites extends this range to 0.62-0.73 (*Warren and Kallemeyn*, 1987). Thus, it seems unlikely that the upper nonmare crust's *mg* is much greater than 0.70. In the simple magmasphere-plagioclase flotation model of crustal genesis, the crust grows from the top down, so the lower magmasphere-generated crust presumably acquires an *mg* ratio even lower than that of the upper crust.

As the final residuum of the magmasphere, enriched over bulk-Moon material in U, Th, and similar incompatible elements by a factor of the order 300 (*Warren and Wasson*, 1979), urKREEP should have had an *mg* close to 0, assuming simple closed-system fractional crystallization until f (the fraction of the initial melt remaining molten) becomes <0.01. The urKREEP residuum must have initially collected as a "sandwich horizon" at the top of the mantle. Barring vigorous convective mixing of the mantle, assimilation reactions between urKREEP and the mantle would be unlikely to raise the *mg* ratio of the urKREEP, however, because the cumulates in the uppermost mantle (the presumed sources of the mare basalts) formed late in the differentiation of the magmasphere, and therefore had modest *mg* ratios; moreover, the low density of urKREEP

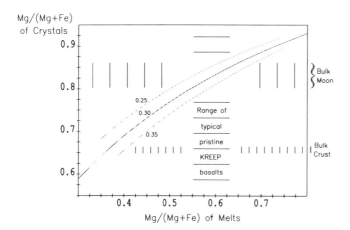

Fig. 1. Relationship between mg of melt and mg of crystals at equilibrium for K_D ranging from 0.25 to 0.35. Included for reference is a range comprising the middle 6 out of 10 recent estimates of the bulk-Moon MgO/(MgO+FeO) ratio (*Warren*, 1986a), and *Taylor's* (1982) estimation of the bulk mg ratio of the nonmare crust.

probably caused it to slowly rise into the crust (*Shirley and Wasson*, 1981).

Within the context of such a simple model of pre-KREEP crustal genesis (i.e., entirely by accumulation over a closed-system magmasphere), it also seems unlikely that any type of assimilation reaction with solid anorthositic norite (or anorthositic gabbro) of the lower crust, where this model implies that mg is <0.70, could suffice to transform the urKREEP residuum into liquids with the mg ratios of the pristine KREEP basalts. *Ryder* (1987) adduces textural observations that appear to confirm earlier indications (mainly from siderophile contents) that the common variety of KREEP basalts from Apollo 15 are pristine. As reviewed by *Irving* (1977), among pristine Apollo 15 KREEP basalts bulk-rock mg ratios appear to range from 0.35 to 0.73, although most are in the range 0.50-0.66. *Irving* (1977) suggests that these basalts in general were derived by fractional crystallization of a more magnesian parent melt with $mg \cong 0.72$. The pristine KREEP basalts found as clasts in Apollo 17 breccia 72275 also have moderate mg ratios, 0.47-0.55 (*Blanchard et al.*, 1975; *Ryder et al.*, 1977; *Salpas et al.*, 1987). Numerous experimental studies, cited by *Warren* (1986a), indicate that the Fe/Mg exchange reaction distribution coefficient (K_D) for both olivine and low-Ca pyroxenes at low pressure and oxygen fugacity is ~ 0.30, and almost certainly within the range 0.25-0.35. Relationships between mg of melt and mg of crystals for this range of K_D are illustrated in Fig. 1. At equilibrium, assuming K_D = 0.30, a solid with mg = 0.70 will coexist with a liquid with mg = 0.41. The eight or so Apollo 15 KREEP basalts with mg >0.65 (*Irving*, 1977) would only be in equilibrium, as melts, with solids with mg >0.86; the Apollo 15 KREEP basalt with mg = 0.73 would be in equilibrium with solids with mg = 0.90. Even assuming K_D = 0.35, the basalts with mg >0.65 would be in equilibrium with solids with mg >0.84, and the basalt with mg = 0.73 would be in equilibrium with solids with mg = 0.89—higher than any reasonable estimate for the bulk-Moon mg ratio, let alone the mg ratio of the Moon's lower crust.

As illustrated by Fig. 2, their primitive mg ratios notwithstanding, the pristine KREEP basalts (unfilled squares) have relatively high contents of incompatible elements such as Sm. Although none of the three highest-mg Apollo 15 KREEP basalts reviewed by *Irving* (1977) (two with mg = 0.69 and one with mg = 0.73) have been analyzed for REE, *Irving* (1977) does cite their K contents. The highest-mg Apollo 15 point shown in Fig. 2 is based on *Irving's* (1977) two mg-0.69 samples with average Sm content inferred by assuming a Sm/K ratio of 0.0071 (*Warren and Wasson*, 1979). Among the many (17) Apollo 15 KREEP basalts for which the sources cited in the caption of Fig. 2 report Sm/K ratios, Sm/K ranges only from 0.0055 to 0.0090. Thus the error bars shown for Sm on the mg-0.69 point are probably conservative.

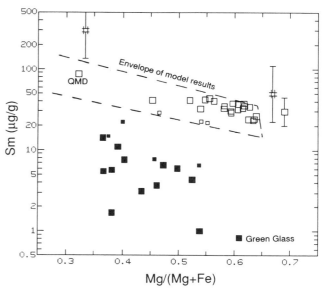

Fig. 2. Bulk-rock mg versus Sm for lunar basalts. Filled squares represent individual varieties of mare basalt (low-Ti varieties with larger symbols). Unfilled large squares represent Apollo 15 KREEP basalts (and QMD), unfilled small squares represent clasts of KREEP basalt from Apollo 17 breccia 72275, and the two unfilled squares with pointed corners (and large error bars for Sm) represent rough estimates inferred for KREEPy (?) melts parental to several Apollo 14 cumulates. Results from modeling are shown as an envelope, which corresponds closely to the actual range among Apollo 15 (and Apollo 17) KREEP basalts. The data for mare basalts come from *Taylor's* (1982) compilation (his Tables 6.3 and 6.4), and from the following additional sources: *Murali et al.* (1976) (for Apollo 16 mare basalt 60639c); *Delano et al.* (1986) (for Apollo 15 yellow-brown glass); *Laul et al.* (1978) and *Ma et al.* (1978) (both for Luna 24 VLT basalt); *Ryder* (1985), *Morgan et al.* (1974), and *Maxwell et al.* (1970) (for Ni contents of Apollo 15 pigeonite basalt, Apollo 17 orange glass, and Apollo 11 high-K basalt, respectively); and *Dickinson et al.* (1985) (for average Al content of Apollo 14 high-Al mare basalt). The data for Apollo 15 KREEP basalts come from *Ryder's* (1985) review (for 15382 and 15386); from *Lindstrom et al.* (1977) (for seven 15003 rocklets); from *Warren et al.* (1983) (for five 15007 rocklets); and from *Irving's* (1977) Table 1 (for eight additional rocklets from various 4-10 mm fines samples). The data for the quartzmonzodiorite ("QMD") found as clasts in Apollo 15 breccia 15405 are also from *Ryder* (1985). The data for three pristine-textured, matrix-free clasts of KREEP basalt from Apollo 17 breccia 72275 come from *Blanchard et al.* (1975) and *Salpas et al.* (1987).

Figure 2 also includes two points for "inferred" Apollo 14 KREEPy melts. The *mg*-0.33 composition is inferred for the parent melt of whitlockite anorthosite 14321c (*Warren et al.,* 1983), and the *mg*-0.67 composition is inferred for the parent melts of several troctolite clasts from breccias 14179, 14303, and 14321 (see Fig. 13 of *Warren et al.,* 1983). These inferred compositions are of course highly model-dependent. The main assumptions are that these rocks are cumulates with reasonable contents of "trapped liquid," that $K_D = 0.30$ for low-Ca pyroxene and olivine, and that the parent melts were KREEPy (which is far from certain in the case of the *mg*-67 composition). The uncertainties suggested by the error bars for these points are only crude estimates. Nevertheless, it seems reasonable to tentatively conclude that the typical KREEP components of the Apollo 14 breccias have even higher Sm, per given *mg*, than the Apollo 15 KREEP basalts. Possibly the Sm contents of KREEPy melts, per given *mg*, are correlated with the KREEP concentrations of the regions where the KREEP basalts were generated. Regional regolith Th (i.e., KREEP) concentrations are 1.9 μg/g for Apollo 17, 4.2 μg/g for Apollo 15, and 8.2 μg/g for Apollo 14 (*Metzger et al.,* 1977).

The paradoxical "primitive" character of the KREEP basalts is also manifested by their Ni contents. Like the *mg* ratio, the Ni content of a melt should generally vary in inverse proportion to its Sm content, because the crystal/melt distribution coefficient for Ni, D_{Ni}, is roughly 15 (dependent upon the melt Mg content) for olivine, and also $\gg 1$ (roughly 3) for low-Ca pyroxene (*Irving,* 1978). Indeed, among mare basalts Ni appears inversely correlated with Sm (Fig. 3). Pristine KREEP basalts, which like mare basalts have extremely low contents of more "noble" siderophile elements such as Re and Ir (e.g., *Morgan et al.,* 1974; *Warren et al.,* 1983), tend to have Ni contents comparable to those of low-Ti mare basalts with Sm contents 5-10 times lower (Fig. 3).

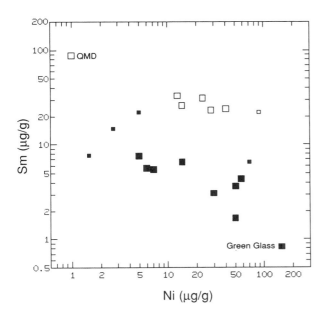

Fig. 3. Bulk-rock Ni versus Sm for lunar basalts. Symbols, data sources analogous to Fig. 2.

PREVIOUS ASSIMILATION—MIXING MODELS

Dowty et al. (1976), who assumed that all of the ancient nonmare (a.k.a. "ANT") cumulates formed prior to the formation of the magma ocean residuum, appealed to assimilation as a process in which "equilibrium was strictly maintained only between the liquid and the outermost surface layer of the ANT crystals," as a result of which, these authors suggested, "the Fe/Fe+Mg ratio would tend to approach that of the melted ANT crystals." In this model, the assimilation process resembles continual ultra-low-degree partial melting of the crustal rocks through which the urKREEP is supposed to percolate. However, in any partial melt, the melt *mg* ratio increases as the degree of melting increases. Thus, the transient sort of equilibrium invoked in this model would probably tend to yield liquids with modest *mg* ratios, even lower than in the case of a slower, more thorough type of assimilation, and in any case far below the mean *mg* ratio of the crust with which the liquid undergoes the assimilative reactions. As discussed above, most of the pristine KREEP basalts have *mg* ratios comparable to, and in a few cases apparently actually greater than, the bulk *mg* ratio of the lunar crust.

About 10 years ago a consensus emerged (mainly from studies of pristine nonmare rocks) that early lunar evolution was more complex than most models of the 1970's had assumed. Consequently, many workers (e.g., *Longhi and Boudreau,* 1979; *Longhi,* 1980; *James and Flohr,* 1983; *Warren and Wasson,* 1979, 1980; *Warren,* 1985, 1986b) have suggested that mixing and assimilation may have had important effects during and/or immediately following the crystallization of a primordial "magma ocean." Most of these authors did not discuss the effects of these processes on the *mg* ratio, per se, but *Longhi* (1980), for example, suggested that during his "Stage IV" of early lunar evolution, shortly after the waning of the magma ocean, primitive melts "with compositions near the olivine-plagioclase low-pressure cotectic... rise [from the upper mantle] and mix with the magma ocean residuum," which later leads to KREEP volcanism.

The *mg* ratio—incompatible elements paradox is readily explained by a model that considers not only the magmasphere, but also its immediate aftermath. A widely accepted model for the origin of the Mg-rich cumulates (i.e., the nonmare cumulates other than ferroan anorthosites; the latter are presumed to have formed atop the magmasphere) holds that they formed in more or less conventional layered intrusions, emplaced into older, ferroan-anorthositic crust within a few hundred Ma of the origin of the Moon (e.g., *Warren and Wasson,* 1980; *James,* 1980). As discussed by *Warren and Wasson* (1980), the source regions of these melts were probably concentrated in the lower-middle mantle, where the *mg* ratio was at least nearly as high as the bulk-Moon *mg* ratio. In any event, these melts yielded cumulates with olivine *mg* ratios frequently >0.90, and occasionally as high as 0.92 (*Warren,* 1986a). Assuming $K_D = 0.30$, olivine with *mg* = 0.90 implies the parent melt's *mg* was 0.73; olivine with *mg* = 0.92 implies the parent melt's *mg* was 0.78 (Fig. 1).

Pristine Mg-rich rocks range in age from 4100 to 4500 Ma, and thermal models suggest that the magmasphere completed 99% of its crystallization, yielding the ferroan anorthosites plus a thin layer of urKREEP immediately below, within 200 Ma of the origin of the Moon (*Warren*, 1985). Thus, the onset of Mg-rich plutonism apparently came shortly after, or even overlapped, the final stages of magmasphere crystallization. The magmasphere's final urKREEP residual liquid, representing only a few tenths of a percent of the original magmasphere, had such high contents of U, Th, and K that it could have remained molten at the base of the crust for many hundreds of Ma. Left undisturbed, its low density would have caused it to slowly migrate toward the surface (*Shirley and Wasson*, 1981). But soon after urKREEP formed, and perhaps even during its final stages of formation, Mg-rich magmas apparently plowed through the urKREEP collection layer (the crust/mantle boundary) on their way to the crust. These rising parcels of Mg-rich melt presumably mixed with, or at least partly assimilated, whatever urKREEP that they passed through. Implications of this mixing for the composition of the melts parental to Mg-rich rocks have been appreciated for some time (e.g., *Warren and Wasson*, 1980). But if mixing between urKREEP and Mg-rich melts was pervasive enough, the composition of most of the basalts derived from urKREEP would also have been affected.

QUANTITATIVE MODELING OF THE EFFECTS OF MIXING AND ASSIMILATION ON URKREEP

The hypothesis that systematic mixing of primitive Mg-rich material with magmasphere residual liquid preceded the genesis of the pristine KREEP basalts, and thus gave the KREEP basalts their paradoxically moderate-high mg ratios, can be tested using finite difference computer modeling of fractional crystallization. The model employed here is adapted from that of *Warren* (1986b), which was in turn a descendant of a model by *Longhi* (1977). The relatively simple phase equilibria are modeled primarily after *Walker et al.* (1973), with embellishments from *Longhi* (1980). Following *Warren* (1986b), the models assume that in addition to mixing with urKREEP, the Mg-rich parent

TABLE 1. Compositions of materials used in, or predicted to form by, models discussed in the text.

Material	Na$_2$O	MgO	Al$_2$O$_3$	SiO$_2$	K$_2$O	CaO	TiO$_2$	Cr$_2$O$_3$	FeO	mg	Sm	Eu	Ce/Yb
Inputs:													
"SI" melt	0.6	27.6	7	46	0.06	5.5	0.3	0.5	12.4	0.80	0.80	0.30	
A16 "komatiite"	0.10	26.92	6.61	45.16	-	5.31	0.54	0.40	14.77	0.76	-	-	
Fer. An'site	0.28	1.03	33.45	44.4	0.01	19.2	0.03	0.01	1.06	0.63	0.08	0.80	
urKREEP	0.89	3.60	12.6	56.4	2.06	8.10	1.85	0.20	13.5	0.32	87.5	2.65	1.000
Results:													
PS1 melt	0.73	10.41	17.77	48.28	0.27	11.55	0.47	0.36	9.94	0.65	13.7	0.72	0.915
PS1(-60) melt	1.32	6.29	13.87	48.39	0.60	12.45	1.18	0.22	15.10	0.43	30.8	1.50	0.962
PS2 melt	0.73	9.97	17.34	49.12	0.47	11.22	0.61	0.34	9.99	0.64	24.2	0.92	0.945
PS2(-30) melt	0.91	8.16	15.87	49.70	0.64	11.25	0.86	0.30	12.0	0.55	33.5	1.22	0.958
KPS2 melt	0.34	9.73	17.82	47.85	0.29	11.52	0.79	0.28	11.17	0.61	23.8	0.88	0.950
KPS2(-30) melt	0.43	7.86	15.74	49.52	0.39	11.14	1.10	0.32	13.21	0.51	32.9	1.17	0.957
PS3 melt	0.74	9.53	16.93	49.90	0.65	10.90	0.74	0.33	10.07	0.63	34.8	1.12	0.975
Actual rocks:													
15382/6 avg.	0.82	8.59	16.0	51.65	0.57	9.52	1.98	0.30	9.71	0.61	32.2	2.65	0.998
72275c avg.	0.42	8.42	14.3	49.18	0.33	10.41	1.10	0.43	13.96	0.52	22.8	1.61	1.087

Units: wt % oxides, molar MgO/(MgO+FeO), μg/g Sm and Eu, and Ce/Yb normalized to urKREEP (9.24).

A16 "komatiite": Composition proposed by *Ringwood et al.* (1987) as a major component of Apollo 16 polymict breccias (see text).

UrKREEP: QMD 15405c, representing composition of assimilated urKREEP (see text).

PS1 melt: First plagioclase-saturated melt predicted by model of AFC in an Mg-rich intrusion, assuming initial melt of SI composition was mixed with urKREEP in a 9:1 mass ratio.

PS1(-60) melt: Melt derived by 60 wt % fractional crystallization of PS1, during which fresh "SI" melt is being added at a rate of 1 g of SI for each 10 g of crystals crystallized.

PS2 melt: First plagioclase-saturated melt predicted by model of AFC in an Mg-rich intrusion, assuming initial melt of SI composition was mixed with urKREEP in a 8:2 mass ratio.

PS2(-30) melt: Melt derived by 30 wt % fractional crystallization of PS2, during which fresh "SI" melt is being added at a rate of 1 g of SI for each 10 g of crystals crystallized.

KPS2 melt: First plagioclase-saturated melt predicted by model of AFC in an Mg-rich intrusion, assuming initial melt of A16 "komatiite" composition was mixed with urKREEP in a 8:2 mass ratio.

KPS2(-30) melt: Melt derived by 30 wt % fractional crystallization of KPS2, during which fresh A16 "komatiite" melt is being added at a rate of 1 g of fresh melt for each 10 g of crystals crystallized.

PS3 melt: First plagioclase-saturated melt predicted by model of AFC in an Mg-rich intrusion, assuming initial melt of SI composition was mixed with urKREEP in a 7:3 mass ratio.

15382/6: Average composition of "large" Apollo 15 KREEP basalts 15382 and 15386 (*Ryder*, 1985).

72275c: Average composition of KREEP basalt clasts from 72275 (see text for references).

melts assimilate ferroan anorthosite country rock while simultaneously crystallizing olivine (thus tending to "make over" the country rock into magnesian troctolites), until this combined "AFC" process eventually brings the melt to the olivine-plagioclase cotectic. Thereafter the melts crystallize first olivine + plagioclase, then orthopyroxene + plagioclase, and later pigeonite + plagioclase. In all models for this work, the AFC was assumed to take place under the "thermally nominal" conditions of *Warren* (1986b).

Among the greatest uncertainties in such a model are the compositions of (1) the primary Mg-rich melt and (2) the urKREEP with which the Mg-rich melt mixes/assimilates. Of course, in reality both of these materials probably varied in composition, depending upon time and place in the Moon. For the primary Mg-rich melt composition, a reasonable choice is the "Standard Initial" (SI) composition used by *Warren* (1986b), who derived it to represent a very high degree partial melt (olivine ± low-Ca pyroxene = sole residual phases) of primitive lunar mantle assumed to be compositionally similar to the Earth's mantle. The extraordinarily low Al_2O_3 of the SI composition (7.0 wt %) is compensated by a high *mg* ratio (0.80). As discussed by *Warren* (1986b), this low Al content causes the melt to assimilate abundant ferroan anorthosite and crystallize abundant olivine—both of which processes drive down its *mg* ratio, and thus more than offset the choice of a high initial *mg* ratio—before the melt reaches the olivine-plagioclase cotectic. The SI melt is assumed to originate several hundred kilometers deep in the lunar mantle, and thus its Si content was constrained to place the melt on the estimated position of the olivine-pyroxene cotectic at 400 km, and its Ca/Al ratio was made slightly subchondritic, to allow for low-Ca pyroxene in the mantle residue (*Warren*, 1986b). More recently, *Ringwood et al.* (1987) have inferred, based on mixing-deconvolution models for the compositions of Apollo 16 polymict breccias, an abundant "komatiite" component which has a composition (Table 1) remarkably similar to *Warren's* (1986b) SI composition for all elements except Na and Ti.

The 15405c quartzmonzodiorite (QMD) was chosen to represent the composition of urKREEP, for several reasons. First of all, the QMD is the most evolved of all the KREEPlike rocks, with twice the REE contents of any other pristine KREEPy rock. Unlike any of the KREEP basalts, the QMD has an *mg* ratio that is nearly as low as expected based on the trend of the mare basalts extrapolated to its Sm content (Fig. 2). Its position on the Ni-Sm diagram (Fig. 3) is completely analogous. Most of the pristine KREEP basalts come from Apollo 15, like the QMD. Last, but not least, *Compston et al.* (1984) report an age of 4.365 ± 0.030 Ga for zircon from the QMD. This age is in excellent agreement with typical Rb-Sr and Sm-Nd model ages for KREEP basalts. Crystallization ages for KREEP basalts are invariably far younger: The range of eight Rb-Sr dates (based on a decay constant of 0.0142 Ga^{-1}) is 3.83-3.93 Ga (*Warren*, 1985). *Ryder* (1976) interpreted the QMD as a late differentiate of a KREEPy-basaltic pluton. This model may well be correct. The QMD has a REE pattern significantly fractionated (La = 1.17 × Lu) relative to "nominal" high-K KREEP (*Warren and Wasson*, 1979). Nevertheless, in the absence of any other KREEPlike rock nearly as evolved as the QMD, it seems like the logical choice to approximate the composition of urKREEP.

Trace element fractionation was modeled using distribution coefficients listed in Table 2. The primary SI melt was assumed to have REE in chondritic proportion to its Al content. The composition of the urKREEP was assumed to have the same Ce, Sm, Eu, and Yb contents as the 15405c QMD (*Ryder*, 1985). As an added complexity, in most models the melt continually receives fresh pulses of the primary SI melt (but only after the completion of the assimilation process, the melt having arrived at plagioclase saturation—no attempt has been made to model crystallization, assimilation, and mixing all happening simultaneously, although in reality such complexities might often have developed). The masses of the pulses are governed by an input variable, but in all models they are assumed to diminish in mass in direct proportion to the mass fraction of the original melt remaining molten *(f)*.

Typical results are illustrated in Fig. 2 and Table 1. Table 1 shows only results for compositions derived in models with the rate of replenishment of "SI" melt equal to 0.1× the rate of crystallization. However, in other models the replenishment factor varied from 0.0 to 0.3, without much impact on the "fit" of the model melts to the actual compositions of the KREEP basalts, albeit the best fits come at different *f* in models with different replenishment factors. Fig. 2 shows an envelope surrounding all SI-based models with the ratio of SI to urKREEP between 0.1 and 0.3, and the replenishment factor between 0.1 and 0.3. The "fit" to the actual KREEP basalts speaks for itself. Note also in Table 1 that the Ce/Yb ratio does not fractionate greatly relative to urKREEP, as the minor loss of heavy REE to low-Ca pyroxene crystallization is approximately offset by the similar loss of light REE to plagioclase crystallization.

TABLE 2. Crystal/liquid distribution coefficients used for four rare earth elements in the AFC models.

	Ce	Sm	Eu	Yb	References
Plagioclase	0.048	0.03	1.1	0.009	*McKay and Weill* (1977); *McKay* (1982a)
Orthopyroxene	0.0075	0.018	0.018	0.113	*McKay and Weill* (1976)
Pigeonite	0.0027	0.021	0.029	0.127	*McKay* (1982b)
Olivine	<0.001	0.0008	0.01	0.027	*McKay* (1986)

Although in the model results the Na contents tend to be too high, and the Ti contents too low, the contents of Na and Ti are poorly constrained for both the SI melt and the urKREEP. The fits for both elements can be improved by choosing an initial melt composition closer to the "lunar komatiite" proposed by *Ringwood et al.* (1987) (Table 1), and/or by exclusion of a modal recombination pseudoanalysis (*Taylor et al.*, 1980) from the average for the 15405c QMD (the assumed composition of urKREEP). A model with Apollo 15 green glass (*Taylor*, 1982) as the initial Mg-rich melt composition gives roughly similar results, but the hydridized KREEPy/Mg-rich melt does not fractionate to the point of plagioclase saturation until its mg ratio has fallen to a value too low (e.g., 0.44, assuming the urKREEP/green glass mixing ratio was 0.2) to fit the pristine KREEP basalts.

Perhaps the most serious discrepancy between the model results and the observed basalt compositions involves Eu. The Eu contents of the model results tend to be considerably lower than the Eu contents of the actual basalts. For example, the "PS3" melt (Table 1) has Eu content only about 0.42× the Eu contents of typical Apollo 15 KREEP basalts (Fig. 4). The significance of this discrepancy is unclear, however, because the sample used to represent urKREEP in the models (the 15405c QMD) may not be truly representative of urKREEP for this particular element. There is a strong correlation between Eu and Sm among lunar basalts (Fig. 4). Yet the QMD has a lower Eu content than numerous Apollo 15 KREEP basalts that have consistently lower Sm contents by a factor of two. In other words, extrapolation from the pristine basalts, and the KREEP basalts in particular, to the Sm content of the QMD would suggest a Eu content roughly twice that of the QMD. Using for urKREEP a Eu content of 4 μg/g instead of the QMD value (2.65 μg/g) would lead to Eu contents of 1.91, 1.66, and 1.58 μg/g for the PS1(-60), PS2(-30), and PS3 melts, respectively, as opposed to the nominal values of 1.50, 1.22, and 1.12 μg/g, respectively (Table 1). On the other hand, assuming the QMD composition is representative of urKREEP for Eu, the model results tend to favor relatively low urKREEP/Mg-rich melt mixing ratios for the intrusions that ultimately (according to this model) were sources of the pristine KREEP basalts.

DISCUSSION

These results show that in terms of combining moderately high mg with high Sm, a wide range of assumed initial urKREEP/SI mixing ratios and SI replenishment factors all tend to lead, at some stage during the fractional crystallization process, to melt compositions approximating those of the pristine KREEP basalts. Note that the f at the stage of optimal "fit" is directly proportional to the model urKREEP/SI mixing ratio. The actual mixing ratio of urKREEP to Mg-rich melts probably was diverse, because urKREEP presumably collected into numerous small pockets as it was trapped between the upward-growing mantle and the downward-growing crust. Some of the pristine KREEP basalts may have been derived by remelting of the upper portions of intrusions produced by mixing with a low ratio of urKREEP to Mg-rich, SI-like melt. Elsewhere KREEP basalts of very similar composition may have been derived by near-total remelting of an intrusion with a high mixing ratio of urKREEP to Mg-rich melt. However, a great diversity of major-element compositions (and especially mg ratios) would probably have resulted if the degrees of remelting of the intrusions were independent of the urKREEP—Mg-rich melt mixing ratio. Thus it appears that the heating involved in the remelting of the KREEPy differentiates tended to occur in proportion to the concentrations of radioactive U, Th, and K in the hybridized urKREEP/Mg-rich materials; i.e., it appears that the heat was derived mainly from U, Th, and K, and not from some more exotic heat source such as impacts or tidal friction.

Isolated patches of urKREEP may have virtually escaped dilution by Mg-rich melt. Intrusions with exceptionally high ratios of urKREEP to Mg-rich melt (and hence high U, Th, and K) would have been slow to crystallize, and might even have remained molten throughout the first several hundred Ma of lunar history. The longevity of the molten urKREEP, as well as the eventual remelting of thoroughly assimilated (diluted) urKREEP, would have been enhanced by the thermal insulation of the Moon's megaregolith (*Warren and Rasmussen*, 1987), an effect that warrants further study.

As extrusive rocks, mare basalts were relatively immune to assimilation because they presumably tended to pass rapidly from mantle source regions to rapid-cooling lava flows, with less opportunity for interaction with materials like urKREEP deep within the crust. Indeed, experimental petrology studies indicate that most mare basalts are not in equilibrium as melts with plagioclase at any lunar pressure (*Green et al.*, 1975). Moreover, most of the original urKREEP may have been removed to the upper crust (and/or disseminated by impacts) by the time that typical mare basalts formed. Even so, *Binder* (e.g., 1982) claims considerable success at modeling the

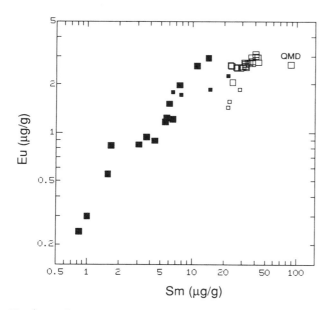

Fig. 4. Bulk-rock Sm versus Eu for lunar basalts. Symbols, data sources analogous to Fig. 2.

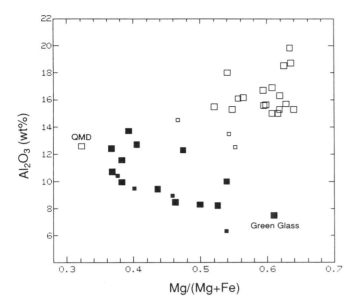

Fig. 5. Bulk-rock mg versus Al_2O_3 for lunar basalts. Symbols, data sources analogous to Fig. 2.

compositions of mare basalts by assuming that they assimilated diverse proportions of "urKREEP residuals" (depleted in light REE by prior episodes of partial melting).

Assimilation and mixing processes of the type outlined above could also help to account for the higher Al_2O_3 contents of the KREEP basalts, in comparison to mare basalts (Fig. 5). The Al_2O_3 content of melts cosaturated with a low-Ca mafic silicate plus plagioclase tends to correlate with the melt mg ratio (e.g., Longhi, 1977). Thus, the Al_2O_3 content of a KREEP basalt derived from a mixture of urKREEP and primitive, high-mg melt will tend to be higher than the Al_2O_3 contents of basalts produced by direct remelting of late-stage, low-mg magmasphere cumulates. In addition, the KREEP/Mg-rich melts probably tended to assimilate plagioclase from ferroan anorthosite country rock as they solidified (Warren, 1986b). Besides increasing the melt Al_2O_3 content, plagioclase assimilation would have a moderating effect on the Ca/Al ratio of the residual liquid.

The proposed solution to the KREEP—mg paradox may also help explain another lunar paradox. Many of the pristine Mg-rich cumulates have surprisingly high contents of REE and other incompatible elements. The most extreme cases, including the two illustrated by the "inferred" points of Fig. 2, tend to be anorthosites and troctolites from Apollo 14 (Warren et al., 1983). The Apollo 14 site is near the center of a large region of extraordinarily high regolith Th (i.e., KREEP) concentration, averaging about 6.3× the global mean (Metzger et al., 1977). The well-known pristine troctolite 76535 (from Apollo 17, where regional regolith Th contents average about 1.5× the global mean) probably formed from a melt with a Sm content of roughly 5 μg/g (Haskin et al., 1974) (this estimate assumes a conservatively high "trapped" liquid content of 12 wt %). Yet the mg-0.88 olivines of 76535 imply that the parent melt had a mg ratio of at least 0.69 (Fig. 1), and this limit can be met only if the trapped liquid content is negligible. For comparison, the Apollo 15 green glass, with roughly 6 times lower Sm, has an mg ratio of 0.61 (Fig. 2). High light-REE/heavy-REE ratios inferred for the parent melts of many Mg-rich cumulates are also incongruous in relation to their high mg ratios (Haskin et al., 1974; Laul and Schmitt, 1975; McKay et al., 1979). This type of paradox is readily explained, however, under the assumption that the melts parental to the Mg-rich cumulates were often severely contaminated by assimilated magmasphere residual liquid (urKREEP), as previously suggested by various authors, most explicitly Warren and Wasson (1980) and James and Flohr (1983). I further suggest that residual melts of these early postmagmasphere intrusions (essentially recycled, mg-enhanced urKREEP) were often later the sources of pristine KREEP basalts.

The paradoxical contrast between the KREEP basalts' uniformly high incompatible element contents and their uniformly moderate-high mg ratios (and Ni contents) clearly requires special pleading in the framework of any model that would account for the genesis of the Moon's crust by purely serial magmatism. Exceptional circumstances might easily lead to local anomalies in a "serial" crust, but the highest-REE lunar rocks should by and large have extremely low mg ratios. In contrast, the paradox is readily explained if we assume (1) that the Moon produced a major portion of its crust by plagioclase flotation over a primordial magmasphere, (2) that the residual liquid from this global differentiation collected mainly as a "sandwich horizon" between the top of the mantle and the bottom of the flotation crust, and (3) that between the magmasphere melting event and the era of KREEP basalt volcanism additional melts loosely akin to the melts that formed the magmasphere continued to rise toward the surface (or at least the crust). In fact, all three of these assumptions have been fixtures of many lunar models for nearly a decade (and the first two assumptions for nearly two decades). The magmasphere hypothesis has long been accepted as a working model, but generally only with reservations (see review by Warren, 1985). Warren (1986b) argued that a clear corollary to the modern magmasphere hypothesis is inevitable anorthosite assimilation by the Mg-rich intrusions; this corollary appears uniquely capable of accounting for the marked geochemical distinctions between the ferroan anorthosites and the other ancient cumulates (the Mg-rich rocks). The suitability of the modern magmasphere hypothesis to account for the bizarrely magnesian nature of the pristine KREEP basalts represents yet another confirmation of that hypothesis.

Acknowledgments. This work was supported by NASA grant NAG 9-87, and benefited considerably from reviews by R. L. Korotev and J. H. Jones.

REFERENCES

Binder A. B. (1982) The mare basalt magma source region and mare basalt magma genesis. *Proc. Lunar Planet. Sci. Conf. 13th*, in *J. Geophys. Res., 87*, A37-A53.

Blanchard D. P., Haskin L. A., Jacobs J. W., Brannon J. C., and Korotev R. L. (1975) Major and trace element chemistry of Boulder 1 at Station 2, Apollo 17. *The Moon, 14*, 359-371.

Compston W., Williams I. S., and Meyer C., Jr. (1984) Age and chemistry of zircon from late-stage lunar differentiates (abstract). In *Lunar and Planetary Science XV*, pp. 182-183. Lunar and Planetary Institute, Houston.

Delano J. W., Hughes S. S., and Schmitt R. A. (1986) Collaborative study of Apollo 15 pristine yellow/brown glass: an interim report (abstract). In *Lunar and Planetary Science XV*, pp. 172-173. Lunar and Planetary Institute, Houston.

Dickinson T., Taylor G. J., Keil K., Schmitt R. A., Hughes S. S., and Smith M. R. (1985) Apollo 14 aluminous mare basalts and their possible relationship to KREEP. *Proc. Lunar Planet. Sci. Conf. 15th*, in *J. Geophys. Res.*, 90, C365-C374.

Dowty E., Keil K., Prinz M., Gros J., and Takahashi H. (1976) Meteorite-free Apollo 15 crystalline KREEP. *Proc. Lunar Sci. Conf. 7th*, 1833-1844.

Green D. H., Ringwood A. E., Hibberson W. O., and Ware N. G. (1975) Experimental petrology of Apollo 17 mare basalts. *Proc. Lunar Sci. Conf. 7th*, 871-893.

Haskin L. A., Shih C.-Y., Bansal B. M., Rhodes J. M., Wiesmann H., and Nyquist L. E. (1974) Chemical evidence for the origin of 76535 as a cumulate. *Proc. Lunar Sci. Conf. 5th*, 1213-1225.

Hubbard N. J. and Minear J. W. (1976) Petrogenesis in a modestly endowed moon. *Proc. Lunar Sci. Conf. 7th*, 3421-3435.

Irving A. J. (1977) Chemical variation and fractionation of KREEP basalt magmas. *Proc. Lunar Sci. Conf. 8th*, 2433-2448.

Irving A. J. (1978) A review of experimental studies of crystal/liquid trace element partitioning. *Geochim. Cosmochim. Acta*, 42, 743-770.

James O. B. (1980) Rocks of the early lunar crust. *Proc. Lunar Planet. Sci. Conf. 11th*, 365-393.

James O. B. and Flohr M. K. (1983) Subdivision of the Mg-suite noritic rocks into Mg-gabbronorites and Mg-norites. *Proc. Lunar Planet. Sci. Conf. 13th*, in *J. Geophys. Res.*, 88, A603-A614.

Laul J. C. and Schmitt R. A. (1975) Dunite 72415: A chemical study and interpretation. *Proc. Lunar Sci. Conf. 6th*, 1231-1254.

Laul J. C., Vaniman D. T., and Papike J. J. (1978) Chemistry, mineralogy and petrology of seven >1 mm fragments from Mare Crisium. In *Mare Crisium: The View from Luna 24* (R. B. Merrill and J. J. Papike, eds.), pp. 537-568. Pergamon, New York.

Lindstrom M. M., Nielsen R. L., and Drake M. J. (1977) Petrology and geochemistry of lithic fragments from the Apollo 15 deep-drill core. *Proc. Lunar Sci. Conf. 8th*, 2869-2888.

Longhi J. (1977) Magma oceanography 2: Chemical evolution and crustal formation. *Proc. Lunar Sci. Conf. 8th*, 601-621.

Longhi J. (1980) A model of early lunar differentiation. *Proc. Lunar Planet. Sci. Conf. 11th*, 289-315.

Longhi J. and Boudreau A. E. (1979) Complex igneous processes and the formation of the primitive lunar crustal rocks. *Proc. Lunar Planet. Sci. Conf. 10th*, 2085-2105.

Ma M.-S., Schmitt R. A., Taylor G. J., Warner R. D., Lange D. E., and Keil K. (1978) Chemistry and petrology of Luna 24 lithic fragments and <250 μm soils: Constraints on the origin of VLT mare basalts. In *Mare Crisium: The View from Luna 24* (R. B. Merrill and J. J. Papike, eds.), pp. 569-592. Pergamon, New York.

Maxwell J. A., Peck L. C., and Wiik H. B. (1970) Chemical composition of Apollo 11 lunar samples 10017, 10020, 10072 and 10084. *Proc. Apollo 11 Lunar Sci. Conf.*, 1369-1374.

McKay G. A. (1982a) Partitioning of REE between olivine, plagioclase, and synthetic basaltic melts: implications for the origin of lunar anorthosites (abstract). In *Lunar and Planetary Science XIII*, pp. 493-494. Lunar and Planetary Institute, Houston.

McKay G. A. (1982b) Experimental REE partitioning between pigeonite and Apollo 12 olivine basaltic liquids. *Eos Trans. AGU*, 63, 1142.

McKay G. A. (1986) Crystal/liquid partitioning of REE in basaltic systems: Extreme fractionation of REE in olivine. *Geochim. Cosmochim. Acta*, 50, 69-79.

McKay G. A. and Weill D. F. (1976) Petrogenesis of KREEP. *Proc. Lunar Sci. Conf. 7th*, 2427-2447.

McKay G. A. and Weill D. F. (1977) KREEP petrogenesis revisited. *Proc. Lunar Sci. Conf. 8th*, 2339-2355.

McKay G. A., Wiesmann H. and Bansal B. (1979) The KREEP-magma ocean connection (abstract). In *Lunar and Planetary Science X*, pp. 804-806. Lunar and Planetary Institute, Houston.

Metzger A. E., Haines E. L., Parker R. E., and Radocinski R. G. (1977) Thorium concentrations in the lunar surface. I: Regional values and crustal content. *Proc. Lunar Sci. Conf. 8th*, 949-999.

Morgan J. W., Ganapathy R., Higuchi H., Krähenbühl U., and Anders E. (1974) Lunar basins: Tentative characterization of projectiles, from meteoritic elements in Apollo 17 boulders. *Proc. Lunar Sci. Conf. 5th*, 1703-1736.

Murali A. V., Ma M.-S., and Schmitt R. A. (1976) Mare basalt 60639, another eastern lunar basalt (abstract). In *Lunar Science VII*, pp. 583-584. Lunar and Planetary Institute, Houston.

Ringwood A. E., Seifert S., and Wänke H. (1987) A komatiitic component in Apollo 16 highland breccias: implications for the nickel-cobalt systematics and bulk composition of the Moon. *Earth Planet. Sci. Lett.*, 81, 105-117.

Ryder G. (1976) Lunar sample 15405: Remnant of a KREEP basalt-granite differentiated pluton. *Earth Planet. Sci. Lett.*, 29, 255-268.

Ryder G. (1985) *Catalog of Apollo 15 Rocks*. NASA Johnson Space Center Curatorial Branch Publication No. 20787, Houston. 1296 pp.

Ryder G. (1987) Petrographic evidence for nonlinear cooling rates and a volcanic origin for Apollo 15 KREEP basalts. *Proc. Lunar Planet. Sci. Conf. 17th*, in *J. Geophys. Res.*, 92, E331-E339.

Ryder G., Stoeser D. B., and Wood J. A. (1977) Apollo 17 KREEPy basalt: a rock type intermediate between mare and KREEP basalts. *Earth Planet. Sci. Lett.*, 35, 1-13.

Salpas P. A., Taylor L. A., and Lindstrom M. M. (1987) Apollo 17 KREEPy basalts: evidence for nonuniformity of KREEP. *Proc. Lunar Planet. Sci. Conf. 17th*, in *J. Geophys. Res.*, 92, E340-E348.

Shirley D. N and Wasson J. T. (1981) Mechanism for the extrusion of KREEP. *Proc. Lunar Planet. Sci. 12B*, 965-978.

Taylor G. J., Warner R. D., Keil K., Ma M.-S., and Schmitt R. A. (1980) Silicate liquid immiscibility, evolved lunar rocks and the formation of KREEP. In *Proceedings of the Conference on the Lunar Highlands Crust* (R. B. Merrill and J. J. Papike, eds.), pp. 339-352. Pergamon, New York.

Taylor S. R. (1982) *Planetary Science: A Lunar Perspective*. Lunar and Planetary Institute, Houston. 481 pp.

Walker D. (1983) Lunar and terrestrial crust formation. *Proc. Lunar Planet. Sci. Conf. 14th*, in *J. Geophys. Res.*, 88, B17-B25.

Walker D., Grove T. L., Longhi J., Stolper E. M., and Hays J. F. (1973) Origin of lunar feldspathic rocks. *Earth Planet. Sci. Lett.*, 20, 325-336.

Warren P. H. (1985) The magma ocean concept and lunar evolution. *Ann. Rev. Earth Planet. Sci.*, 13, 201-240.

Warren P. H. (1986a) The bulk-Moon MgO/FeO ratio: a highlands perspective. In *Origin of the Moon* (W. K. Hartmann, R. J. Phillips, and G. J. Taylor, eds.), pp. 279-310. Lunar and Planetary Institute, Houston.

Warren P. H. (1986b) Anorthosite assimilation and the origin of the Mg/Fe-related bimodality of pristine Moon rocks: Support for the magmasphere hypothesis. *Proc. Lunar Planet. Sci. Conf. 16th*, in *J. Geophys. Res., 91*, D331-D343.

Warren P. H. and Kallemeyn G. W. (1987) Geochemistry of lunar meteorite Yamato-82192: Comparison with Yamato-791197, ALHA81005, and other lunar samples. *Proc. Symposium Antarct. Meteorites (Tokyo) 11th*, pp. 3-20.

Warren P. H. and Rasmussen K. L. (1987) Megaregolith insulation, internal temperatures, and bulk uranium content of the Moon. *J. Geophys. Res., 92*, 3453-3465.

Warren P. H. and Wasson J. T. (1979) The origin of KREEP. *Rev. Geophys. Space Phys., 17*, 73-88.

Warren P. H. and Wasson J. T. (1980) Early lunar petrogenesis, oceanic and extraoceanic. In *Proceedings of the Conference on the Lunar Highlands Crust* (J. J. Papike and R. B. Merrill, eds.), pp. 81-99. Pergamon, New York.

Warren P. H., Taylor G. J., Keil K., Kallemeyn G. W., Shirley D. S., and Wasson J. T. (1983) Seventh Foray: Whitlockite-rich lithologies, a diopside-bearing troctolitic anorthosite, ferroan anorthosites, and KREEP. *Proc. Lunar Planet. Sci. Conf. 14th*, in *J. Geophys. Res., 88*, B151-B164.

The Formation of Hadley Rille and Implications for the Geology of the Apollo 15 Region

Paul D. Spudis and Gordon A. Swann*

Branch of Astrogeology, U. S. Geological Survey, Flagstaff, AZ 86001

Ronald Greeley

Department of Geology and Center for Meteorite Studies, Arizona State University, Tempe, AZ 85287

*Now at Dept. of Geology, Northern Arizona University, Flagstaff, AZ 86001

The origin of lunar sinuous rilles via the lava channel/tube mechanism is accepted by most investigators, although the precise mode of formation is a matter of contention. We have combined the results of studies of terrestrial lava tube systems and the regional and detailed site geology of the Apollo 15 area to develop a model for the formation of Hadley Rille. The earliest post-Imbrium basin unit in the region is the pre-mare Apennine Bench Formation; remote-sensing and lunar sample studies have shown that this unit most likely consists of volcanic KREEP basalt flows. Numerous structural depressions in the Apennine Bench Formation were probably produced by regional deformation associated with basin adjustment. These structural and topographic flows provided a network of depressions that were later followed by basalt flows that established two major sinuous rilles in the region, the Mozart and Hadley Rilles. In this model, the rilles are primarily constructional features that developed along preexisting structural depressions. As such, substantial thermal or mechanical erosion by lava during the formation of these rilles did not occur. A consequence of this model for Hadley Rille formation is that the observed outcrop of mare basalt in the rille walls at the Apollo 15 site may represent the total thickness of basalt in that area (less than 50 m). The formation of lunar sinuous rilles by lava erosion, along with associated assimilation of highland debris into mare basalt magmas, may be a much less widespread phenomenon than has been proposed.

INTRODUCTION

Sinuous rilles have long attracted attention from the lunar science community. Although numerous hypotheses for the origin of rilles flourished in the pre-Apollo years (see review in *Howard et al.*, 1972a), the post-Apollo period saw a general consensus reached by the community, viz., that lunar sinuous rilles are lava tubes and/or channels formed during the eruption of mare basaltic lavas on the Moon (*Kuiper et al.*, 1966; *Oberbeck et al.*, 1969; *Greeley*, 1971a).

Although there is general agreement on the lava tube/channel mode of origin for sinuous rilles, there is considerable disagreement among workers regarding which of several genetic processes is dominant in rille formation. A notable case in point is the Hadley Rille (Fig. 1), one of the largest sinuous rilles on the Moon and a specific target of the Apollo 15 landing site selection process (*Wilhelms*, 1986). The first published model on the origin of Hadley Rille was by *Greeley* (1971a). This model advocates a purely constructional lava channel origin: (1) a massive basalt eruption floods the Hadley-Apennine valley, (2) cooling of this flow proceeds from the edges inward; (3) the cooling margins confine the flow to a channel, portions of which are roofed over to form a tube (*Greeley*, 1971a). After the Apollo 15 mission, *Howard et al.* (1972a,b) slightly modified the *Greeley* (1971a) model by suggesting that a large, standing lava lake had been drained by Hadley Rille. Thus, in the model of Howard et al., the rille is not purely constructional, but at least near the Apollo 15 site it served mostly as a drainage channel for the standing lava inferred to have been present (see evidence listed in *Howard et al.*, 1972b).

An alternative hypothesis for the formation of rilles emerged after the Apollo 15 mission. Several investigators have suggested that the process of thermal erosion is important in sinuous rille formation. *Carr* (1974) presented computer models of the thermal erosion of lava channels by mostly laminar flow of lava. He concluded that Hadley Rille could have formed under such conditions at eruption rates comparable to terrestrial basaltic eruptions (5×10^{-3} km^3/hr). Other studies of thermal erosion by lava, emphasizing the low viscosities expected for lunar basaltic magmas, suggested that turbulent flow of lunar lava was important and greatly facilitated thermal erosion (*Hulme*, 1973; *Head and Wilson*, 1980; *Wilson and Head*, 1980). In particular, *Head and Wilson* (1980, 1981) argued for very large mass eruption rates (greater than 10^7 m^3/sec) during mare flooding, which would produce turbulent flow and result in substantial thermal erosion. Applied to Hadley Rille, such an analysis suggests that the total eroded volume of the rille represents only 1-2% of the volume of basalt emplaced in Palus Putredinis (see discussion by J. W. Head in *Spudis and Ryder*, 1986, pp. 18-21). Such a low value implies that the chemical effects of assimilation of highland debris into the Apollo 15 mare basalts may not be detectable.

We believe that certain aspects of each of these models have merit, but that some of the observed geologic relations of Hadley Rille are not totally explained by any of the published models. In this paper, we first discuss terrestrial lava tube formation and basaltic flow emplacement. We next address several aspects of the regional geology of Hadley Rille that have heretofore not been considered and, on the basis of geologic analysis of the Apollo 15 region and evidence from study of terrestrial lava tube systems, we present a new model

Fig. 1. Regional view of the Palus Putredinis/Hadley Rille area showing major features discussed in the text. Apollo 15 metric frames 0415 and 0418.

for the formation of Hadley Rille that addresses these and other problems. Finally, we use this new formational model of rille development to draw some specific conclusions regarding the geology of the Apollo 15 landing site, in addition to some general implications for the formation of sinuous rilles elsewhere on the Moon.

TERRESTRIAL LAVA TUBE SYSTEMS

The study of terrestrial lava tubes and channels can provide insight into the general geology of similar features on the Moon. Such insight is particularly important in regard to the emplacement of associated lava flows, configuration with regard to vent systems, and their relationship to the terrain in which tubes and channels occur (*Greeley*, 1987).

Lava tube and channel activity can result in both erosion and construction. Mechanical and thermal erosion by flow through lava tubes is well documented, but as terrestrial tubes are extensions of the eruptive vent for the emplacement of flows, they are *predominantly* constructional features. For the most part, they occur within their associated flows. Thus, their locations are governed by the same factors which control flow position, with preflow topography being the principal consideration. Because lava tubes and channels generally occupy the thickest part of the flow, their axes tend to coincide with the lowest preflow topography, be it eroded as in a stream channel, or structurally controlled as within a graben or along a fracture zone.

Relatively few exposures are available to assess the detailed relations among preflow topography, flow configuration, and lava tube position. Field studies by *Greeley* (1971b,c), *Wood* (1981), and others that involve surveying lava tube interiors and the general geology of the flows in which they occur provide some information. The Giant Crater lava tube system in northern California (*Greeley*, 1969; *Greeley and Baer*, 1971) is south of the Medicine Lake highlands. The principal tube originates in a pit crater 300 m in diameter and can be traced more than 28 km south. Much of the flow occupies a graben, although some parts of the flow spill over the bounding fault scarps (Fig. 2). The walls and floors of some sections of the uncollapsed lava tube have broken away to expose preflow rocks. In some such areas, rounded cobbles and gravel deposits suggest that the tube occupied a former stream channel, which followed the graben structure. In other places, surveying of the tube (Fig. 3) strongly suggests that it has entrenched itself tens of meters below the lower flow contact. The location of these sections suggests that entrenchment occurred along segments of the fault planes bounding the graben.

Topographic control of flows and associated lava tubes and channels is commonly observed in many areas. Mapping of the Cave Basalt flow associated with Mount St. Helens (*Greeley and Hyde*, 1972) shows entrenchment of the lava tube downward into the valley floor and laterally into the preflow hillside (Fig. 4). The southwest rift zone of Mauna Loa displays prominent fissures and fractures. The paths of many of the flows erupted in the rift zone are controlled by the location of these structural depressions.

From these and other examples, it is clear that the floors of lava tubes may form at an elevation lower than that of the base of the flows in which they occur. In some cases, tubes appear to entrench along preexisting structures to depths several times greater than the thickness of the flow.

As extensions of the vent system, the areas where flows emerge from lava tubes commonly will have characteristics

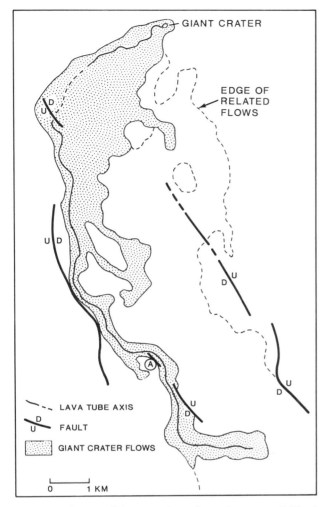

Fig. 2. Sketch map of the Giant Crater lava tube system, California, showing flow associated with the lava tube; the flow system occupies a topographic depression bounded partly by faults. Flow source is Giant Crater and flow direction is south, down the valley. Other flows erupted in the general area of Giant crater shown with dashed line; "A" shows the location of the cross-section of Figure 3.

of "true" vent locations. Low domes, local flows, spatter ramparts, and minor pyroclastic deposits may occur around sections of ruptured lava tube roofs (*Greeley and Hyde,* 1972). Moreover, existing lava tubes that formed in early flows may be reactivated by subsequent eruptions; also, younger flows may intersect collapsed roof sections of older tubes, flow along the older tube, and reemerge downslope (*Macdonald,* 1943). In either case, the composition of the younger flow may be distinctive from the flow in which the tube formed. Thus, if the younger flows were emplaced via older lava tubes, local lava outcrops may appear as petrologically distinctive "islands."

Some terrestrial basalt flows are extremely fluid during emplacement; low viscosity combined with high rates of effusion often produce features more typical of flowing water than of volcanic processes. These include high lava "strand lines" (Figs. 5 and 6). For example, the 1823 flow on the southwest rift of Kilauea was a fissure eruption that involved extremely fluid lavas. *Stearns* (1926) described a Hawaiian village on the coast that was destroyed by the flow; the flow advanced so rapidly that villagers were unable to launch their canoes to escape. As the flow sped down the slope, two cinder cones in the pathway were overridden by the flow. The flow left high-lava marks on the sides of the cones 10 m above the present level of the flow (Fig. 5) and a veneer of lava as thin as 1 cm. Similar "high-lava" marks are visible on a cinder cone along the southwest rift of Mauna Loa which resulted from the 1950(?) eruption. Figure 6 shows a stream channel that confined the 1974 eruption on the rim of Kilauea caldera. High-lava marks are clearly visible on the channel banks.

While it is tempting to treat very fluid lavas in a manner comparable to water, i.e., that flows achieve hydrostatic equilibrium, even very fluid lavas on Earth and the Moon are orders of magnitude greater in viscosity than water. Only in relatively quiet, stagnant ponds or lakes of lava are level surfaces achieved (*Holcomb,* 1971). Active flows associated with tubes and channels or those moving down steep slopes will form local strand lines governed by conditions of flow and topography.

REGIONAL GEOLOGY OF THE APENNINE BENCH FORMATION AND ITS RELATION TO MOZART AND HADLEY RILLES

The Apennine Bench Formation, first described by *Hackman* (1966), is a plains-forming unit characterized by intermediate albedo, numerous secondary craters from the crater Archimedes, and a variety of linear structural features that appear to be formed by collapse (Fig. 1). The structural features consist of: (1) elongate graben-like troughs; (2) linear grooves, some of which, although very narrow, have flat floors suggestive of grabens; (3) scarps; and (4) somewhat elongate depressions bounded by scarps that grade into "hinges." These structural features trend northeast and northwest (Fig. 7), and are radial and concentric to the Imbrium basin.

The subsurface material on which the Apennine Bench Formation was deposited is probably broken-up rock and fallback material from the Imbrium basin impact; the bulk porosity of this debris must have been high for some time after the event. Later regional adjustment of the Imbrium basin and filling of the basin by mare flooding could produce a regional tensile stress field; this basin adjustment could explain the development of collapse features in the Apennine Bench Formation.

The proximity of widespread exposures of Apennine Bench Formation to the north, west, and south of the site, and even closer structures that typify the unit, suggest that the Apennine Bench Formation may underlie the surface material at the Apollo 15 site. *Carr and El-Baz* (1971) mapped extensive exposures of Apennine Bench Formation 40 km northwest of the Apollo 15 landing site that were not recognized by *Hackman* (1966). Two additional exposures of Apennine Bench Formation 50 and 100 km southwest of the site are apparent in Apollo 15 metric and panoramic camera

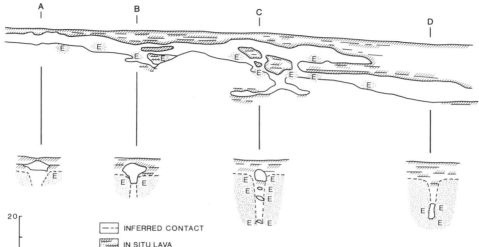

Fig. 3. Diagrams of the Giant Crater tube showing longitudinal profile and selected cross sections showing exposures of "preflow" materials; tube system has apparently entrenched by erosion along a fault plane.

photographs. In addition, the large exposures of Apennine Bench Formation south of Archimedes extends to within 75 km west of the site, not 125 km as mapped by *Hackman* (1966). These determinations by *Carr and El-Baz* (1971) were made from Lunar Orbiter IV photographs and by *Swann* (1986a) from Apollo 15 metric and panoramic camera photographs, none of which were available to Hackman at the time of his mapping. Furthermore, scarps and depressions that typify the Apennine Bench Formation occur in two areas, one 25 km north (Fig. 8) and the other 6 km northeast of the site; these areas appear to be occurrences of Apennine Bench Formation covered by a thin mantle of basalt (*Swann*, 1986a).

The probability that the Apennine Bench Formation underlies the mare basalts at the landing site is also supported by orbital geochemical data and returned-sample information. Several early studies (*Schonfeld and Meyer*, 1973; *Spudis*, 1978a; *Hawke and Head*, 1978) noted a rough chemical correspondence between the Apennine Bench Formation and Apollo 15 KREEP basalt, on the basis of early reduction of the orbital geochemical data. In recent years, these data have been refined (e.g., *Metzger et al.*, 1979; *Clark and Hawke*, 1981) and the resemblance of the Apennine Bench material to Apollo 15 KREEP basalt is remarkable (Table 1). Although we still have not identified Imbrium basin impact melt positively, the matrices of Apollo 15 impact melt rocks 15445 and 15455 remain the best candidates for this composition (*Ryder and Spudis*, 1987). The Apennine Bench Formation does not correspond to this composition; this difference, and other geological arguments (*Spudis and Hawke*, 1986), suggest that the original interpretation of the Apennine Bench Formation as post-Imbrium basin, volcanic KREEP lava flows is correct. Such an interpretation has important implications for understanding Apollo 15 site geology and is discussed in the next section.

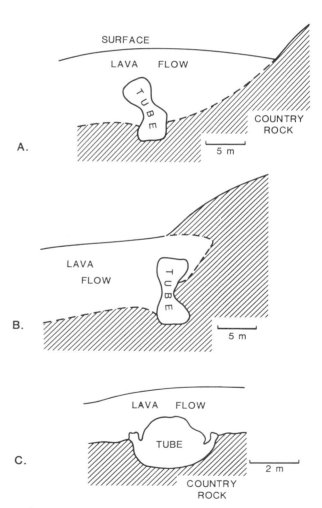

Fig. 4. Cross section of lava tube in the Cave Basalt, Washington (a and b) and a lava tube in Hawaii (c) in which the tube has eroded into preflow terrain (after *Greeley*, 1987).

Fig. 5. Oblique aerial view of the 1823 lava flow (dark area, flow from right to left) from Kilauea volcano which overrode the cinder cones in the center of the view (arrow marks areas where flow overrode the cone in excess of 10 m).

Fig. 6. View of the 1974 Kilauea flow down a preflow stream channel, leaving a veneer of lava (figure in lower left indicates scale).

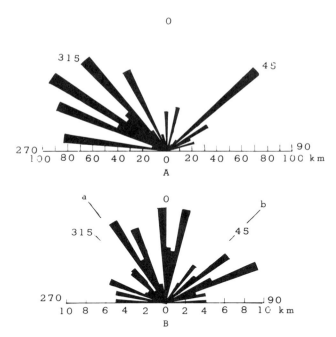

Fig. 7. Rose diagrams of structural trends in the Apennine Bench Formation and Hadley Rille. (a) Structural trends of depressions in Apennine Bench Formation. (b) Structural trends of segments of Hadley Rille. Direction labeled "a" is the general trend of the northern half of Hadley Rille; "b" is the general trend of the southern half of the rille. From *Swann* (1986b).

Mozart Rille, 100 km west of the Apollo 15 landing site, is a 40-km-long sinuous rille confined mostly to the Apennine Bench Formation but extending about 10 km into a small patch of mare, which, other than for a possible connection through Bradley Rille, is isolated from Palus Putredinus by highlands materials and Apennine Bench Formation (Fig. 1). The rille originates in an elongate crater in the high ground of the Apennine Bench, out of which lava probably flowed down the rille into another elongate crater in the mare, a fairly common occurrence during terrestrial basaltic eruptions. The crater in the mare may simply be a "skylight," or an entrance into a lava tube. Near the source of the rille, it branches into two channels to form a diamond shape. Near its terminus in the mare, a small, elongate crater connects to the rille; this crater is probably a "feeder," or additional source crater (Fig. 9).

The general trend of Mozart Rille, and its three associated craters, is approximately aligned with what appears to be a collapse feature extending into the patch of mare from, and at an approximate right angle to, Bradley Rille (Fig. 9). This may be a structural feature along which the two source and one terminal craters are aligned; alternatively, it could be a collapsed roof segment of a lava tube extending beyond the terminal crater. The trends of Mozart Rille appear to be controlled by structural scarps in the Apennine Bench Formation; *Coombs et al.* (1987) reach similar conclusions regarding the structural control of Mozart Rille.

The presence of a faint, dark mantling around the source crater and upper reaches of Mozart Rille suggests some pyroclastic activity emanating from the rille and/or its source crater, as has been suggested for other parts of the area (e.g., *Carr and El-Baz*, 1971). This dark material also could arise from thin, small flows from the rille which have since been mixed with Apennine Bench material by impact gardening. Whatever the cause of the darkening, any significant lava flow appears to have been contained within the rille, at least until it reached the present limits of the mare; hence the form and course of the rille are structurally controlled.

Hadley Rille, although one of the longest and widest sinuous rilles on the Moon, is in some respects typical of many lunar sinuous rilles. It originates at a 15-km-long arcuate cleft at its southern end; the southern part of the cleft is in highlands massif material, and the northern part is in the mare. A smaller arcuate cleft is adjacent to the west side of the larger one (Fig. 1). The rille trends northward from the cleft and merges with two straight rilles of the Fresnel system in Apennine Bench Formation 90 km away. The rille itself, measured along its meanders and not including the cleft, is about 135 km long. It is 2 km across at its widest point, near the cleft, and about 0.5 km across where it is narrowest, near its northern terminus. West of the Apollo 15 landing site (Station 9A), it is about 1.5 km across and 300 m deep. Along its southern part, the rille has a V-shaped profile to about 10 km northwest of the landing site; from there to its northern terminus the profile is U-shaped (*Howard et al.*, 1972b). *Greeley* (1971a) estimates the volume of the rille to be 2.8×10^{10} m^3.

Two reaches of the rille are clearly controlled by preexisting topography. One area exhibiting topographic control is where the rille bends around Hadley Delta below St. George crater; the other is where the rille turns northwest along the base of Fresnel Ridge (Fig. 8). The northwest-trending segment west of the landing site may also be controlled by pre-rille topography. The North Complex is a topographic high that appears to be composed of Apennine material (*Swann*, 1986a), as does the unnamed high area between the North Complex and Fresnel Ridge. Furthermore, the Apollo 15 lunar module landed on a slight northwest trending ridge that may be a pre-mare "high" covered by a thin veneer of mare basalt, all of which suggest that the course of the rille near the landing site was deflected along a pre-mare, topographically high area. In detail, Hadley Rille consists of straight segments connected by rounded corners (Fig. 1). The straight segments have led several authors (e.g., *Howard et al.*, 1972b; *Swann*, 1986a, 1986b) to suggest that the rounded corners connect straight segments that are structurally controlled.

As previously discussed, the Apennine Bench Formation probably is present in the subsurface in the vicinity of the landing site, and this formation is characterized by northeast- and northwest-trending collapse features. We propose that Hadley Rille began its development on Apennine Bench Formation which is now covered by mare basalt. The regional topographic gradient was downhill toward the north and thus the average course of the rille is north. Excursions from this northern direction were caused by northeast- and northwest-trending and structurally controlled topographic features (Fig. 7). The rille maintained its structurally controlled pattern during the period of mare flooding and development of the

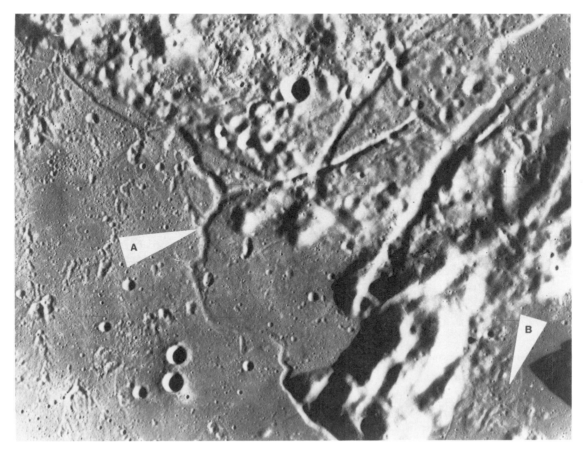

Fig. 8. The northern portion of the Hadley region. "A" shows the convergence of the northern terminus of Hadley Rille with structural features of the Apennine Bench Formation (Fresnel Rilles); at "B" are Apennine Bench-like structures occurring within the maria. Area shown is about 55 by 40 km; north at top. Portion of Apollo 15 metric frame 0414.

rille. This contention is supported by the observation that Hadley Rille at its northern terminus merges with two straight rilles (part of the Fresnel Rille system) in the Apennine Bench Formation (Fig. 8). This confluence of Hadley Rille and the preexisting Fresnel structures is not explained by any previously published model for the formation of Hadley Rille. Had mare flooding continued so that Mozart Rille was also totally engulfed in mare basalts, and if Mozart had remained active for a longer period, it too probably would have continued to develop along its original trends.

TABLE 1. Comparison of compositions of the Apennine Bench Formation, Apollo 15 volcanic KREEP, and probable Imbrium basin impact melt.*

	Apennine Bench Fm†	Apollo 15 KREEP‡	Probable Imbrium Impact Melt§
TiO_2	1.7–2.3%	1.8–2.3%	1.35–1.70%
Al_2O_3	16.0%	14.8–17.8%	16.2–17.5%
FeO	9.5–12.4%	8.6–11.1%	7.9–10.2%
MgO	5.7%	6.3–8.2%	13.4–16.0%
Th	10.7–12.0 ppm	10.5–12.0 ppm	2.4–2.9 ppm

*After *Spudis and Hawke*, 1986.
†Al and Mg values from *Clark and Hawke* (1981), Fe and Ti values from *Davis* (1980); Th values from *Metzger et al.* (1979).
‡Values from *Dowty et al.* (1976) and *Meyer* (1977).
§Values for black matrix of 15445 and 15455, compiled in *Ryder* (1985).

Fig. 9. Mozart Rille. "A" is the pit crater "origin" and "B" is the "side feeder." "C" is a collapse structure. Area shown is about 50 by 50 km; north at top. Portion of Apollo 15 metric frame 0416.

Several lines of evidence indicate that Hadley Rille formed by flow of lava primarily within and along the preexisting structural depressions of the Apennine Bench Formation. First, the oldest postbasin unit in the region (the Apennine Bench material) contains numerous scarps, cracks, and grabens that tend to be oriented in directions radial and concentric to Imbrium basin structural trends. Second, the trends of individual segments of Hadley Rille are coincident with these regional structural patterns (Fig. 7). Third, Mozart Rille provides direct evidence for sinuous rille development along, and control by, preexisting structural patterns within Apennine Bench material. We suggest that Hadley Rille is simply a more evolved version of Mozart Rille; massive mare basalt eruptions onto the surface of Palus Putredinis tended with time to become confined within the preexisting depressions on the Apennine Bench. The Hadley terminus/Fresnel Rille relations (Fig. 8) provide direct evidence for the intimate association of Hadley Rille with these preexisting structures. Thus, although some erosion by lava probably occurred (to connect depression segments), the total eroded volume is much less than would be deduced by interpreting the entire rille to be formed by lava erosion.

If Hadley Rille formed by this mechanism, then the size of rille segments should be comparable to the sizes of structural features found on unflooded portions of the Apennine Bench Formation. Figure 10 shows depth and width measurements both for segments of Hadley Rille (*Howard et al.*, 1972a) and for cracks on the Apennine Bench Formation. It is apparent that the two features of of comparable size. Many Apennine Bench structures appear to be somewhat wider than rille segments (Fig. 10), suggesting that lateral accretion of lava on the walls of preexisting structures may be a more important process during Hadley Rille development than is downward incision by lava erosion. The relatively restricted range in width observed for segments of Hadley Rille (Fig. 10) is very similar to the narrow range of widths observed in terrestrial constructional lava tubes (e.g., data of *Greeley and Hyde*, 1972), again suggesting that Hadley Rille is a lava tube/channel of predominantly constructional origin.

HADLEY RILLE AND THE APOLLO 15 LANDING SITE

The model proposed here for the formation of Hadley Rille has several implications for the geology of the Apollo 15 landing site. In this section, we discuss some geologic relations observed at the site in light of the new model for Hadley Rille formation and we also draw some general inferences regarding sinuous rilles and mare emplacement elsewhere on the Moon.

If Hadley Rille formed by the mechanism proposed here, significant amounts of thermal erosion of highland debris by the mare lavas did not occur. The width/depth relations for Apennine Bench Formation cracks and Hadley Rille (Fig. 10) show that some Apennine Bench cracks are as deep as the deepest portions of the rille, suggesting that downcutting by fluid lava need not have operated on a large scale (cf. *Carr*, 1974). The data of Fig. 10 show that, indeed, a more likely process is the accretion of lava along the walls of preexisting Apennine Bench cracks, generally narrowing the width of the depressions that later became the rille. Undoubtedly, the

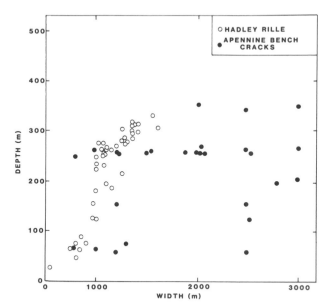

Fig. 10. Morphometry of cracks, grabens, and collapse features on the Apennine Bench Formation (data from Lunar Topographic Orthophotomaps 41A4, 41B4, and 41A3) and segments of Hadley Rille (data from *Howard et al.*, 1972a). These data show that the size of Apennine Bench depressions and Hadley Rille segments are of comparable depths, with Apennine Bench depressions tending to be wider. This observation suggests that the rille-forming mechanism proposed here is feasible and that significant downcutting during lava erosion did not occur.

Fig. 11. Surface telephoto view of the western wall of Hadley Rille at Station 9A, Apollo 15 landing site. At least three basalt units are observed within the upper 60 m of the rille wall (*ALGIT*, 1972). AS15-89-12157.

formation of Hadley Rille by the mechanism proposed would have been a complex, evolving process; different portions of the rille walls could easily expose different geologic sections, from laterally accreted overplating lava to submare Apennine Bench Formation.

The formation of Hadley Rille probably involved much less erosion by lava than predicted by the models of *Hulme* (1973), *Carr* (1974), and *Head and Wilson* (1980, 1981). Such erosion that may have occurred would have been mostly to connect segments of Apennine Bench cracks that were physically separated prior to mare flooding. As such, the quantities of pre-mare debris eroded and assimilated into the lavas of Palus Putredinis would be a minuscule fraction of the total volume of emplaced mare basalt.

A current problem of Apollo 15 site geology is determining the total thickness of the mare basalts within the explored region. The west wall of Hadley Rille observed at the site displays several probable basalt units (Fig. 11), totaling about 60 m thick; at least two of these units (the upper dark unit and middle massive unit) are probably represented in the returned Apollo 15 samples by the olivine-normative and quartz-normative basalts, respectively [Fig. 12; *Apollo Lunar Geology Investigation Team (ALGIT)*, 1972]. The lower portions of the rille wall are covered by talus. One implication of forming Hadley Rille by the manner advocated here is that the exposures of basalt in the rille walls could represent the entire section of mare basalt at that spot. In this case, the upper 2 m consists of olivine basalt and at least the next lower 10-20 m consists of quartz-normative basalt. Dynamic crystallization experiments (*Grove and Walker*, 1977) suggest that the Apollo 15 quartz-normative mare basalts are derived from basalt flows at least 20 m thick, a value similar to the thickness of the middle massive unit observed in the rille wall (*ALGIT*, 1972). Thus, the section less than 25 m thick observed in the upper part of the rille wall could represent the total mare basalt thickness at the site (Fig. 13).

More indirect evidence for the thickness of basalts at the site comes from studies of the modal petrography of Apollo 15 regolith samples. All mare regolith samples from the site contain a significant nonmare component. Basalt fractions in the mare regolith are greatest at the rille edge (Station 9A, about 75%) and lowest near the LM-ALSEP site (Station 8, less than 50%; *Walker and Papike*, 1981). Previous studies have documented the importance of deriving much nonmare debris in mare regoliths via vertical, rather than lateral, mixing (*Rhodes*, 1977; *Hörz*, 1978). The high proportion of nonmare debris observed in the mare soils at the Apollo 15 site is consistent with, and may even require, a relatively thin total basalt thickness at the site (on the order of a few tens of meters or less).

A consequence of the model for rille formation proposed here relates directly to the issue of mare basalt thickness discussed above. An important observation of regolith petrographic studies is that Apollo 15 KREEP basalts make up a substantial fraction of the observed nonmare component in a given mare regolith sample (*Walker and Papike*, 1981); for example, the LM-Station 8 soils are composed of about 30% KREEP basalts (*Basu and Bower*, 1977). We believe that these KREEP basalts are derived mostly by vertical mixing from

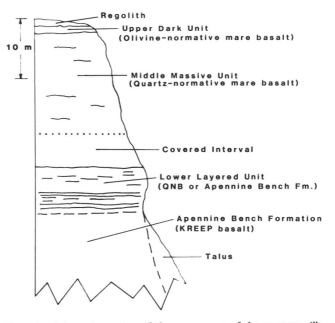

Fig. 12. Schematic section of the upper part of the western rille wall from Station 9A. The upper dark unit consists of olivine-normative basalt, the middle massive unit is quartz-normative basalt (*ALGIT*, 1972). These mare basalts probably directly overlie the Apennine Bench Formation (Apollo 15 KREEP basalt flows). Modified after *Howard et al.* (1972a).

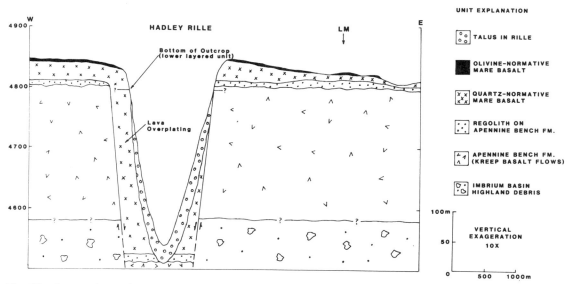

Fig. 13. Interpretive geologic cross section from west to east through Hadley Rille and the Apollo 15 landing site. Because Hadley Rille formed in a preexisting graben or topographic low within the Apennine Bench Formation, the exposures of mare basalt in the west rille wall at Station 9A may reflect total thickness of mare basalt at the site (less than 50 m). Topographic data from ACIC Lunar Topographic Photomap of Apollo 15 site, scale 1:25,000.

beneath the relatively thin layer of mare basalts at the site (Figs. 12, 13). Because Hadley Rille formed in preexisting structures of the Apennine Bench Formation, these observations are consistent with the hypothesis that Apollo 15 KREEP basalt is derived from the Apennine Bench Formation (*Spudis*, 1978a,b; *Hawke and Head*, 1978; *Spudis and Hawke*, 1986). Supporting evidence for this assertion comes from the petrology of sample 15205, a regolith breccia collected from an exotic block at Station 2 on the Apennine Front. This breccia was formed from an immature regolith; lithic clasts consist only of quartz-normative mare basalts and Apollo 15 KREEP basalts to the exclusion of almost all other recognized Apollo 15 rock types (*Dymek et al.*, 1974). The most likely source for 15205 is a small, very recent crater on the western rim of the Elbow crater ejecta blanket (*Swann and Reed*, 1974). This breccia probably formed by mixing of Apennine Bench Formation regolith (KREEP basalt) and quartz-normative mare basalt that directly overlies the Apennine Bench material (G. Ryder, quoted in *Spudis and Ryder*, 1986, p. 19).

The origin of the approximately 80 m "high-lava mark" observed by the Apollo 15 crew at the base of Mt. Hadley is unresolved by the model of rille formation proposed here. However, evidence for surges in the lava effusion rate producing lava marks in terrestrial tube and flow systems suggests that this mechanism is an alternative to the previous model of standing lava lakes discussed by *Howard et al.* (1972a,b). In a lava surge model, temporary increases in the lava effusion rate would overflow the active lava channel of Hadley Rille and could leave a residual scum of lava both on top of inundated terra islands (e.g., North Complex) and along the margins of the flooded lava depositional basin (Mt. Hadley). Either model for the development of the high-lava mark may be accommodated by our proposed model for the formation of Hadley Rille.

The approximately 50 m thickness of basalt observed at the Apollo 15 site may be representative of typical mare thicknesses throughout Palus Putredinis. This conjecture is supported by detailed study of high resolution orbital images of Hadley Rille by *Howard et al.* (1972a,b), who found outcrop patterns along the entire length of the rille similar to those observed at the site. Thus, the average thickness of mare basalt in Palus Putredinis is probably on the order of 50 m or less; ponding of lava in preexisting low spots may have produced isolated thicknesses on the order of a few hundred meters. The total volume of basalt in Palus Putredinis (total area 9200 km^2; variable thicknesses, mostly less than 50 m, but perhaps locally as thick as 100 m) is probably between 460 and 900 km^3; almost all of this basalt was probably emplaced through Hadley Rille.

The formation of Hadley Rille by the mechanism proposed in this paper has some implications for the formation of sinuous rilles elsewhere on the Moon. First, the currently popular invocation of substantial thermal and subsidiary mechanical erosion by lava during the formation of lunar sinuous rilles (*Hulme*, 1973; *Carr*, 1974; *Head and Wilson*, 1981) may not be widely applicable to all lunar rilles. Undoubtedly, mechanical and/or thermal erosion does occur, particularly in association with sinuous rilles incised into highlands terrain (e.g., the rille near Herodotus E; *Carr*, 1974, Fig. 5). However, a number of other sinuous rilles have many features in common with Hadley, including broad-scale preexisting structures and relatively large size (e.g., Schröter's Valley). The regional geologic context of a given rille should be evaluated before it is determined to what degree mechanical and/or thermal erosion has been responsible for its development.

A second implication of this study is that rille depths cannot be used to infer mare thicknesses. In the case of Hadley Rille, such an inference would grossly overestimate the quantity of mare basalt present. The results of this study of the Palus Putredinis mare suggest that isolated maria on the Moon need not be substantially thicker than the thin (on the order of a few tens of meters) lava plains of the irregular maria (e.g., Mare Tranquillitatis) and support previous suggestions that, in general, most lunar mare deposits are relatively thin, typically less than 100 m thick (e.g., *Hörz*, 1978).

Finally, our model for the formation of Hadley Rille suggests that assimilation of highland debris into erupted mare magmas probably is not an important lunar surface process. A significant exception to this generalization is the assimilation of KREEP into some lunar basaltic magmas, as documented for the Apollo 14 high-alumina mare basalts by *Dickinson et al.* (1985). Additionally, the relatively young, high-titanium basalts of the late Imbrium flows (*Schaber*, 1973) may also have assimilated KREEP, as these flows appear to possess substantial Th enrichment (about 8 ppm; *Etchegaray-Ramirez et al.*, 1983). However, most magma genesis models suggest that KREEP is incorporated into mare basaltic magmas deep within the crust (e.g., *Shervais et al.*, 1985), where the primordial "urKREEP" layer is presumed to reside (*Warren and Wasson*, 1979). KREEP has a relatively low melting point and would be the easiest highland material to assimilate into a mare basalt magma. The assimilation of the more refractory highland lithologies (e.g. anorthosite, norite) would be much more difficult and, based on our results that suggest little erosion of upper surface debris during rille formation, much more unlikely.

SUMMARY AND CONCLUSIONS

We have developed a model for the formation of Hadley Rille that is consistent with available information on regional geologic relations and details of the geology of the Apollo 15 site. In essence, we propose the following:

1. Palus Putredinis and the Apollo 15 site was originally flooded by volcanic KREEP basalts (the Apennine Bench Formation) shortly after the Imbrium basin formed (about 3.85 b.y. ago).
2. After the Apennine Bench lavas were emplaced, regional tectonic deformation, caused by Imbrium basin adjustment, produced a complex series of normal faults, cracks, grabens, and collapse features. The trends of these tectonic features were dominantly parallel with radial and concentric tectonic trends of the Imbrium basin.
3. About 3.3 b.y. ago, an eruption of mare basalt began in the vicinity of the Hadley cleft. As the basalts flooded the Hadley-Apennine valley, flow was preferentially confined to the preexisting structural and topographic lows on the underlying Apennine Bench Formation. Rille development proceeded within these lows, with occasional overflow caused by surges in the basalt effusion rate, until the basalts of Palus Putredinis were emplaced. The eruptions associated with Hadley Rille were probably episodic, including the use of the Hadley vent by at least two separate and distinct bodies of magma (represented by the quartz-normative and olivine-normative Apollo 15 mare basalts).
4. After the emplacement of the Palus Putredinis basalts, impact cratering has been the dominant process in the region. Such cratering has slightly modified rille morphology, including collapse of some segments, and also generated regolith, mixing both mare and the underlying KREEP basalts into the surface deposits.

On the basis of this proposed model for the origin of Hadley Rille, we conclude the following about the geology of the Apollo 15 landing site and the formation of lunar sinuous rilles in general:

1. The total thickness of mare basalt at the Apollo 15 landing site is on the order of a few tens of meters, mostly less than 50 m.
2. The ubiquitous KREEP basalt fragments found in the regolith at the Apollo 15 site are mostly of local derivation; specifically, they are derived from the Apennine Bench Formation plains that directly underlie the Apollo 15 mare basalts at the site. The inferred relative thinness of the mare basalts suggests that the Apennine Bench material occurs in the shallow subsurface at the site, i.e., at depths of less than 50 m.
3. The role of thermal erosion in the development of sinuous rilles on the Moon may be of less general importance than is widely assumed.
4. The assimilation of refractory highland rock types (e.g., anorthosite) into mare basaltic magma appears to be a minor lunar process.

Acknowledgments. We thank Charles Barnes, Dave Gosselin, Grant Heiken, and Baerbel Lucchitta for constructive reviews of this manuscript, and Jim Papike for editorial handling. This work is supported by the NASA Planetary Geology and Geophysics Program.

REFERENCES

Apollo Lunar Geology Investigation Team (ALGIT) (1972) Geologic setting of the Apollo 15 samples. *Science, 175*, 407–415.

Basu A. and Bower J. F. (1977) Provenance of Apollo 15 deep drill core sediments. *Proc. Lunar Sci. Conf. 8th*, pp. 2841–2867.

Carr M. H. (1974) The role of lava erosion in the formation of lunar rilles and martian channels. *Icarus, 22*, 1–23.

Carr M. H. and El-Baz F. (1971) Geologic map of the Apennine-Hadley region of the Moon (Apollo 15 pre-mission map). *U. S. Geological Survey Misc. Geol. Inv. Map I-723* (sheet 1).

Clark P. E. and Hawke B. R. (1981) Compositional variation in the Hadley-Apennine region. *Proc. Lunar Planet. Sci. 12B*, pp. 727–749.

Coombs C. R., Hawke B. R., and Wilson L. (1987) Geologic and remote sensing studies of Rima Mozart. *Proc. Lunar Planet. Sci. Conf. 18th*, this volume.

Davis P. A. (1980) Iron and titanium distribution on the Moon from orbital gamma-ray spectroscopy with implications for crustal evolutionary models. *J. Geophys. Res., 85*, 3209–3224.

Dickinson T., Taylor G. J., Keil K., Schmitt R. A., Hughes S. S., and Smith M. R. (1985) Apollo 14 aluminous mare basalts and their possible relationship to KREEP. *Proc. Lunar Planet. Sci. Conf. 15th*, in *J. Geophys. Res., 90*, C365–C374.

Dowty E., Keil K., Prinz M., Gros J., and Takahashi H. (1976) Meteorite-free Apollo 15 crystalline KREEP. *Proc. Lunar Sci. Conf. 7th*, pp. 1833–1844.

Dymek R. F., Albee A. L., and Chodos A. A. (1974) Glass-coated soil breccia 15205: Selenologic history and petrologic constraints on the nature of its source region. *Proc. Lunar Sci. Conf. 5th*, pp. 235-260.

Etchegaray-Ramirez M. I., Metzger A. E., Haines E. L., and Hawke B. R. (1983) Thorium concentrations in the lunar surface: IV. Deconvolution of the Mare Imbrium, Aristarchus and adjacent regions. *Proc. Lunar Planet. Sci. Conf. 13th*, in *J. Geophys. Res.*, 88, A529-A543.

Greeley R. (1969) Geology and morphology of a small lava tube. *Eos Trans. AGU*, 50, 678.

Greeley R. (1971a) Lunar Hadley Rille: Considerations of its origin. *Science*, 172, 722-725.

Greeley R. (1971b) Geology and morphology of selected lava tubes in the vicinity of Bend, Oregon. *Oreg. Dept. Geol. Mineral. Ind. Bull.*, 71. 47 pp.

Greeley R. (1971c) Observations of actively forming lava tubes and associated structures, Hawaii. *Mod. Geol.*, 2, 207-223.

Greeley R. (1987) The role of lava tubes in Hawaiian volcanoes. In *Volcanism of Hawaii* (R. W. Decker, T. L. Wright, and P. H. Stauffer, eds.), pp. 1589-1602, U. S. Geol. Survey. Prof. Paper 1350.

Greeley R. and Baer R. (1971) Hambone, California, and its magnificent lava tubes. In *Geol. Soc. Am., Abstr. with Program*, 3, 128.

Greeley R. and Hyde J. H. (1972) Lava tubes of the Cave Basalt, Mount St. Helens, Washington. *Geol. Soc. Am. Bull.*, 83, 2397-2418.

Grove T. L. and Walker D. (1977) Cooling histories of Apollo 15 quartz-normative basalts. *Proc. Lunar Sci. Conf. 8th*, pp. 1501-1520.

Hackman R. J. (1966) Geologic map of the Montes Apenninus region of the Moon. *U. S. Geol. Survey Misc. Geol. Inv. Map I-463*.

Hawke B. R. and Head J. W. (1978) Lunar KREEP volcanism: Geologic evidence for history and mode of emplacement. *Proc. Lunar Planet. Sci. Conf. 9th*, pp. 3285-3309.

Head J. W. and Wilson L. (1980) The formation of eroded depressions around the sources of lunar sinuous rilles: Observations (abstract). *Lunar and Planetary Science XI*, pp. 426-428. Lunar and Planetary Institute, Houston.

Head J. W. and Wilson L. (1981) Lunar sinuous rille formation by thermal erosion: Eruption conditions, rates, and durations (abstract). *Lunar and Planetary Science XII*, pp. 427-429. Lunar and Planetary Institute, Houston.

Holcomb R. (1971) Terraced depressions in lunar maria. *J. Geophys. Res.*, 76, 5703-5711.

Hörz F. (1978) How thick are lunar mare basalts? *Proc. Lunar Planet. Sci. Conf. 9th*, pp. 3311-3331.

Howard K. A., Head J. W., and Swann G. A. (1972a) Geology of Hadley Rille: Preliminary report. U. S. Geol. Survey Interagency Report, *Astrogeol.*, 41, 57 pp.

Howard K. A., Head J. W., and Swann G. A. (1972b) Geology of Hadley Rille. *Proc. Lunar Sci. Conf. 3rd*, pp. 1-14.

Hulme G. (1973) Turbulent lava flow and the formation of lunar sinuous rilles. *Mod. Geol.*, 4, 107-117.

Kuiper G. P., Strom R. G., and LePoole R. S. (1966) Interpretation of the Ranger records. In *Ranger VIII and IX Experiments' Analyses and Interpretations*, pp. 35-248. JPL Tech. Rpt. 32-800.

Macdonald G. A. (1943) The 1942 eruption of Mauna Loa, Hawaii. *Amer. J. Sci.*, 241, 241-256.

Metzger A. E., Haines E. L., Etchegaray-Ramirez M. I., and Hawke B. R. (1979) Thorium concentrations in the lunar surface: III. Deconvolution of the Apenninus region. *Proc. Lunar Planet. Sci. Conf. 10th*, pp. 1701-1718.

Meyer C. (1977) Petrology, mineralogy and chemistry of KREEP basalt. *Phys. Chem. Earth*, 10, 239-260.

Oberbeck V. R., Quaide W. L., and Greeley R. (1969) On the origin of lunar sinuous rilles. *Mod. Geol.*, 1, 75-80.

Rhodes J. M. (1977) Some compositional aspects of lunar regolith evolution. *Phil. Trans. Roy. Soc. London, A285*, 293-301.

Ryder G. (1985) *Catalog of Apollo 15 Rocks*. Curatorial Branch Publ. 72, JSC Publ. No. 10787, NASA Johnson Space Center, Houston. 1295 pp.

Ryder G. and Spudis P. D. (1987) Chemical composition and origin of Apollo 15 impact melts. *Proc. Lunar Planet. Sci. Conf. 17th*, in *J. Geophys. Res.*, 92, E432-E446.

Schaber G. G. (1973) Lava flows in Mare Imbrium: Geologic evaluation from Apollo orbital photography. *Proc. Lunar Sci. Conf. 4th*, pp. 73-92.

Schonfeld E. and Meyer C. (1973) The old Imbrium hypothesis. *Proc. Lunar Sci. Conf. 4th*, pp. 125-138.

Shervais J. W., Taylor L. A., and Lindstrom M. M. (1985) Apollo 14 mare basalts: Petrology and geochemistry of clasts from consortium breccia 14321. *Proc. Lunar Planet. Sci. Conf. 15th*, in *J. Geophys. Res.*, 90, C375-C395.

Spudis P. D. (1978a) Composition and origin of the Apennine Bench Formation. *Proc. Lunar Planet. Sci. Conf. 9th*, pp. 3379-3394.

Spudis P. D. (1978b) Origin and distribution of KREEP in Apollo 15 soils (abstract). In *Lunar and Planetary Science IX*, pp. 1089-1091. Lunar and Planetary Institute, Houston.

Spudis P. D. and Hawke B. R. (1986) The Apennine Bench Formation revisited. In *Workshop on the Geology and Petrology of the Apollo 15 Landing Site* (P. D. Spudis and G. Ryder, eds.), pp. 105-107. LPI Tech. Rpt. 86-03, Lunar and Planetary Institute, Houston.

Spudis P. D. and Ryder G., eds. (1986) *Workshop on the Geology and Petrology of the Apollo 15 Landing Site*. LPI Tech. Rpt. 86-03, Lunar and Planetary Institute, Houston. 126 pp.

Stearns H. T. (1926) The Keaiwa or 1823 lava flow from Kilauea volcano, Hawaii. *J. Geol.*, 34, 336-351.

Swann G. A. (1986a) Some observations on the geology of the Apollo 15 landing site (abstract). In *Workshop on the Geology and Petrology of the Apollo 15 Landing Site* (P. D. Spudis and G. Ryder, eds.), pp. 108-112. LPI Tech. Rpt. 86-03, LPI, Houston.

Swann G. A. (1986b) Collapse structures in the Apennine Bench Formation and their influence on the formation of Mozart and Hadley Rilles (abstract). In *Lunar and Planetary Science XVII*, pp. 855-856. Lunar and Planetary Institute, Houston.

Swann G. A. and Reed V. S. (1974) A method for estimating the absolute ages for small Copernican craters and its application to the determination of Copernican meteorite flux. *Proc. Lunar Sci. Conf. 5th*, pp. 151-158.

Walker R. J. and Papike J. J. (1981) The Apollo 15 regolith: Comparative petrology of drive tube 15010/15011 and drill core section 15003. *Proc. Lunar Planet. Sci. 12B*, pp. 485-508.

Warren P. H. and Wasson J. T. (1979) The origin of KREEP. *Rev. Geophys. Space Phys.*, 17, 73-88.

Wilhelms D. E. (1986) Selection of the Apollo 15 landing site (abstract). In *Workshop on the Geology and Petrology of the Apollo 15 Landing Site* (P. D. Spudis and G. Ryder, eds.), pp. 116-118. LPI Tech. Rpt. 86-03, Lunar and Planetary Institute, Houston.

Wilson L. and Head J. W. (1980) The formation of eroded depressions around the sources of lunar sinuous rilles: Theory (abstract). In *Lunar and Planetary Science XI*, pp. 1260-1262. Lunar and Planetary Institute, Houston.

Wood C. (1981) Exploration and geology of some lava tube caves on the Hawaiian volcanoes. *Trans. British Cave Res. Assoc.*, 8, 111-129.

Petrology and Geochemistry of Olivine-Normative and Quartz-Normative Basalts from Regolith Breccia 15498: New Diversity in Apollo 15 Mare Basalts

Scott K. Vetter and John W. Shervais
Department of Geology, University of South Carolina, Columbia SC 29208

Marilyn M. Lindstrom
Planetary Materials Branch, Code SN2, NASA Johnson Space Center, Houston, TX 77058

Mare basalt clasts from regolith breccia 15498 are divided into olivine normative basalts (ONB) and quartz normative basalts (QNB), based on major element chemistry. The QNB are characterized by high SiO_2 (46.5-48%), FeO< 21%, TiO_2 of 1.6%-2.6%, and low compatible element concentrations. The QNB suite in 15498 extends to more evolved compositions than previously reported at Apollo 15, and can be divided into four groups: primitive, intermediate/1, intermediate/2, and evolved. Variations within each group can be explained by short-range unmixing of the modal components; variations between groups result from fractional crystallization of olivine (early), pigeonite, and augite (late). The ONB suite in 15498 is divided into three subgroups, based on variations in SiO_2, MgO, and TiO_2: (1) low-SiO_2 ONB, (2) high-SiO_2 ONB, and (3) olivine-pyroxene cumulates. The low-SiO_2 ONB are geochemically identical to the ONB suite described by earlier studies. They are characterized by low SiO_2 (44-46%), high FeO (20-23%), and TiO_2 of 2.2-2.5%. The high-SiO_2 ONB are characterized by high SiO_2 (47-48%), high MgO (9.7-12.7%), low TiO_2 (1.65-1.9%), and low FeO (19.7-20%). The olivine-pyroxene cumulates have the highest MgO and lowest incompatible element concentrations, and one sample contains cumulus augite. Least-squares mixing models show that the more evolved QNBs can be derived from parent magmas similar to the primitive QNBs by crystal fractionation of pigeonite, olivine, and spinel. Variations within the low-SiO_2 ONBs can be explained by olivine fractionation. However, the low-SiO_2 ONBs cannot be parental to any of the QNB lavas. Similar modeling calculations show that the high-SiO_2 ONBs may be parental to the QNB suite, documenting the first link between ONBs and QNBs at the Apollo 15 site. The composition of the high-SiO_2 ONB parent magma is the identical to the hypothetical QNB suite parent postulated by *Chappell and Green* (1973).

INTRODUCTION

A primary goal of the Apollo 15 mission was to sample mare basalts from Palus Putredinis, where mare stratigraphy is exposed to depths of over 500 meters in Hadley Rille, a major lava channel. Over 76 kg of basalt were returned from several widely spaced stations. Although the walls of Hadley Rille were not sampled directly, sampling of crater ejecta near the rille provides reasonable hope that some of the stratigraphic sequence visible in the walls of the rille was sampled.

Early studies of the Apollo 15 mare basalt suite (*Rhodes*, 1972; *Rhodes and Hubbard*, 1973; *Chappell and Green*, 1973; *Helmke et al.*, 1973; *Dowty et al.*, 1973) established that two distinct groups are represented: the olivine-normative basalts (ONB) and the quartz-normative basalts (QNB). Distinctions between these groups are based on both petrography and major element geochemistry. The ONB group is characterized by low SiO_2, higher TiO_2, higher FeO, and lower Mg#s than the QNB group. The ONB group is also characterized by both normative and modal olivine. Modal olivine commonly occurs as large phenocrysts that are apparently in equilibrium with their groundmass (*Papike et al.*, 1976). The QNB lavas have higher SiO_2 and Mg#s and lower TiO_2 and FeO than the ONB, and are characterized by quartz-normative compositions. Pigeonite is the most common phenocryst phase, but relicts of partly resorbed olivine phenocrysts may be found in the more primitive samples (always accompanied by pigeonite).

Chemical variations observed within each of these suites are attributed to near surface crystal fractionation of the liquidus phases—olivine in the ONB, pigeonite in the QNB (*Rhodes and Hubbard*, 1973; *Mason et al.*, 1972; *Chappell and Green*, 1973). This conclusion is supported both by major element mixing calculations and by the small range in trace element concentrations.

The ONB and QNB groups as defined by these early studies cannot be related to one another by low pressure crystal fractionation (*Rhodes*, 1972; *Rhodes and Hubbard*, 1973; *Chappell and Green*, 1973; *Helmke et al.*, 1973; *Dowty et al.*, 1973). The combination of high Mg#, high SiO_2, low TiO_2, and silica saturation in the QNB precludes a relationship to the ONB suite by simple removal of the liquidus minerals (olivine and pigeonite). These characteristics also preclude a similar relationship to the low-Ti mare basalts of Apollo 12.

The average isotopic age data show that the ONB suite is slightly younger than the QNB (3.2 aeons versus 3.3 aeons; *BVSP*, 1981, p. 951). *Duncan et al.* (1975) and *LSPET* (1972) determined that the dominant mare component of the Apollo 15 soils is ONB. The younger isotopic ages and regolith compositions lead to the conclusion that the ONB represents a thin flow or series of flows that stratigraphically overlie the

QNB flow. The number of QNB flows and thus possible number of parent magmas, however, remains ambiguous. Comparison of textural variations in the QNB suite with experimental crystallization studies (*Lofgren et al.*, 1975) suggests that all of these samples may be derived from a single flow two to three meters thick. This conclusion is supported by the small range in major element compositions. However, *Helmke et al.* (1973) cite variations in REE concentrations and Sm/Eu ratio to suggest two or three parent magmas. Further progress on the origin of this important basalt group has been thwarted, however, by the small number of samples available for study.

Several of the regolith breccias returned by the Apollo 15 mission contain abundant clasts of mare basalt that can be used to expand our knowledge of the range of basalt types present and their petrogenesis. The clasts in these breccias were essentially overlooked during the first selection and analysis of Apollo 15 samples in the early 1970's. Recent studies have shown that these breccia clasts can dramatically increase our database at a particular site and yield significant new insights into the origin of mare basalts (e.g., *Taylor et al.*, 1983; *Shervais et al.*, 1985a,b; *Neal et al.*, 1987).

This report presents results of a consortium investigation of 25 mare basalt clasts from a new slab (,141) and existing chips of lunar breccia 15498. Our data show that the olivine-normative basalts at Apollo 15 can be subdivided into three groups: (1) a low-silica group identical to the ONB suite characterized by earlier studies; (2) a high-silica group that is parental to the quartz-normative basalts, and (3) olivine-pyroxene cumulates that may be related to the evolved QNBs. Our data also show that the QNB suite extends to more evolved compositions than were previously recognized. Major element fractionation models for the combined QNB/high-SiO$_2$ ONB groups show that they can be related by early olivine fractionation, followed by pigeonite and then augite fractionation. The low-SiO$_2$ ONB group cannot be related to either of the other groups by low pressure fractionation, as shown by major element models and trace element data.

GEOLOGIC SETTING

The Apollo 15 lunar module landed in Palus Putredinis, a small outlier of mare basalt in the lunar highlands located on the SE margin of the Imbrium Basin. This site offered several important mission objectives, among them the opportunity to observe *in situ* mare basalt stratigraphy in the walls of Hadley Rille. Panoramic photographs of the western wall of the Rille, coupled with astronaut descriptions, show that at least three flows are present in the upper 60 m of the mare across from Stations 9 and 10.

The uppermost flow forms dark hackly outcrops ~6 m thick immediately below the regolith. Based on the high proportion of olivine-normative basalt in the regolith, this flow is commonly interpreted to represent the ONB flow sampled at several locations (*ALGIT*, 1972). Below this flow, and separated from it by a short covered interval, is a massive outcrop, 15-20 m thick, of light-colored basalt. This may represent a single flow since no internal layering is apparent, and because talus blocks thought to derive from it are more than 10 m across.

The lowest exposed "flow" is a layered outcrop 8 to 16 m thick that contains at least 12 distinct layers 1-3 meters thick. These may represent separate flows or flow units, or merely fracture surfaces within a single flow. The lower and middle units may comprise the QNB suite samples that dominate the mare basalt suite at this site (*ALGIT*, 1972).

Fig. 1. View of the new 15498 slab (,141) with representative clasts.

Regolith breccia 15498 was collected on the south rim of Dune Crater (Station 4), approximately 1 km north of the Apennine Front, near boulder samples 15485, 15486, and 15499. Weighing 2.34 kg, 15498 is one of the largest regolith breccias returned by the Apollo 15 mission. Sample 15498 is a coherent glassy-matrix regolith breccia consisting of approximately 16% clasts in a matrix of glass and finely comminuted minerals (Fig. 1). The majority of clasts are mare basalts, although some highland clasts can be observed. Little previous data exist on this breccia, aside from random analyses of matrix material and one clast (*Wänke et al.*, 1977; *Laul and Schmitt*, 1973).

Dune Crater may sample bedrock to a depth of 90 m (*ALGIT*, 1972), making Station 4 an ideal setting to sample and analyze deeper flows. Stratigraphic layering was observed in the walls of Hadley Rille across from Station 9, but could not be sampled there. The hope is that Dune crater penetrated far enough into the mare to sample several of these elusive flows.

METHODS

A new slab of regolith breccia 15498 (,141) was mapped during a visit to the pristine sample lab at JSC to determine the distribution and textural variations of the clasts. Clasts were selected from this slab and existing chips to include the complete textural range observed. The genealogy of each clast is listed in Table 1, which documents the corresponding whole rock and probe mount numbers with the clast # used in this report.

The clasts range from 0.5-3 cm in diameter. Approximately 100-150 mg of matrix-free material was extracted from each clast for whole rock analysis. This sample was ground and

TABLE 1. Genealogy of 15498 basaltic clasts.

Type Clast #	Whole rock #	Probe #	Parent #
QNB			
Primitive			
B-1	,178	,240	,141
B-2	,180	,241	,141
B-6	,190	,245	,141
B-20	,206	,249	,141
B-40	,173	,239	,24
Intermediate- 1			
B-3	,183	,242	,141
B-5	,188	,244	,141
B-11	,198	,246	,141
B-14	,204	,248*	,141
B-24	,219	,252	,141
Intermediate- 2			
B-23	,217	,218*	,141
B-44	,155	,235	,14
Evolved			
B-22	,212	,251	,141
B-45	,151	,234	,14
ONB			
Low-SiO$_2$			
B-4	,186	,243	,141
B-9	,194	,195	,141
B-26	,224	,225*	,141
B-29	,228	,229*	,141
B-43	,166	,238	,19
High-SiO$_2$			
B-10	,196	,197	,141
B-25	,222	,253	,141
B-41	,148	,233	,13
Cumulates			
B-12	,201	,247	,141
B-27	,226	,227*	,141
B-24	,163	,237	,19

*Indicates major elements only.

split into two aliquots for major element (10-40 mg) and trace element (100-120 mg) analyses. Additional material (± matrix) was extracted to make a corresponding probe mount.

All whole rock major element analyses were made on fused glass beads with a Cameca SX-50 electron microprobe at the University of South Carolina. Samples weighing 10-40 mg were crushed in an agate mortar to ensure homogeneity during melting and then fused in Mo boats on an electric strip furnace at NASA/JSC. High purity Ar at 40 psi was used as the surrounding atmosphere. The samples melted within 10-20 sec ensuring little volatilization. Operating conditions for the EMP analyses were 15 KV accelerating potential and beam current of 10 nA. A natural glass standard (VG-2, Smithsonian Institution) was used for all elements except for Cr, Mn, and Mo, for which pure metals were employed. The electron microprobe data was reduced using Cameca's online *PAP* routine. Twenty points were taken on each bead and averaged for the final analysis. A minor Mo correction (usually less than 0.02%) was made on all of the samples.

Mineral analyses were made at either USC or NASA/JSC. The analyses at NASA/JSC were made with a Cameca MBX EMP system using natural and synthetic mineral standards at 15 KV, 30 A. At USC the operating conditions were 15 KV accelerating potential and beam current of 15 nA. Natural mineral standards were used except for Cr and Mn, for which pure metals were used.

Instrumental neutron activation analyses (INAA) were done on 19 selected clasts at NASA/JSC (6 clasts were too small or had sufficient matrix contamination). Uncertainties based on counting statistics at the 1 sigma level are: 1%-2%—Co, Cr, Fe, Sc, Sm, Eu, La, Na; 3%-5%—Ce, Hf, Lu, Yb, Tb, Ta; 20%—Ba, Th; and 30%—Nd and Zr. Both Ir and Au occur below the detection limit of 2 ppb, supporting the textural evidence that these clasts were not formed by impact melting events.

CLASSIFICATION

All of the clasts studied are low-Ti mare basalts similar to those described from the Apollo 15 site by earlier studies (*Rhodes and Hubbard*, 1973; *Chappell and Green*, 1973; *Dowty et al.*, 1973). As recognized by these earlier studies, two groups may be distinguished—one group is olivine normative, the other quartz normative. Petrographically, these samples show a wide range in textures from vitrophyric to coarse grained granular-subophitic.

Geochemically, the 15498 mare basalt suite (Table 2) has a range in FeO content (18.8-23.4 wt %) and Mg# [100Mg/(Mg+Fe) = 29.5-68.6] which is larger than reported previously for Apollo 15 mare basalt (*Rhodes and Hubbard*, 1973; *Dowty et al.*, 1973; *Chappell and Green*, 1973). Based on the geochemical variations, we recognize three subdivisions within the ONB suite, and four subdivisions within the QNB suite:

A. OLIVINE-NORMATIVE BASALTS (ONB)
 (1) Low-SiO$_2$ ONB
 (2) High-SiO$_2$ ONB
 (3) Olivine-pyroxene cumulates

B. QUARTZ-NORMATIVE BASALTS (QNB)
 (1) Primitive QNB
 (2) Intermediate Group 1 QNB
 (3) Intermediate Group 2 QNB
 (4) Evolved QNB

Table 1 summarizes the genealogies of these clasts, and Table 3 presents thin section modes based on approximately 500 points per section.

PETROGRAPHY AND MINERAL CHEMISTRY

A. Olivine-Normative Basalts

1. Low-SiO$_2$ ONB: (B-4, B-9, B-26, B-29). The low-SiO$_2$ ONBs form two distinct textural types. Fine-grained olivine-phyric basalt is represented solely by clast B-4. This clast is characterized by large subhedral olivine phenocrysts (up to

TABLE 2. Whole rock geochemistry and norm calculations of 15498; major elements by EMP, trace elements by INAA.

	Quartz-Normative Basalts (QNB)													
	Primitive					Intermediate- 1					Intermediate- 2		Evolved	
	B-1	B-2	B-40	B-20	B-6	B-24	B-43	B-5	B-3	B-14	B-11	B-23	B-44	B-22	B-45
SiO_2	48.80	47.89	47.62	48.03	47.52	48.06	47.86	47.74	48.09	47.67	46.82	46.59	47.18	47.49	46.88
TiO_2	1.82	2.35	1.95	1.95	2.14	2.17	2.00	2.14	1.93	1.59	2.06	2.17	2.22	2.48	2.63
Al_2O_3	9.64	9.14	9.66	9.68	9.82	10.67	10.07	10.27	10.37	10.20	10.35	11.62	11.03	13.00	12.71
FeO	18.75	20.19	19.76	19.06	19.98	18.96	19.09	19.43	18.96	18.99	20.27	20.55	20.00	19.67	21.00
MnO	0.22	0.24	0.30	0.25	0.30	0.23	0.26	0.26	0.24	0.28	0.26	0.22	0.26	0.22	0.26
MgO	9.27	9.25	9.48	9.09	9.42	8.12	8.43	8.04	8.56	8.56	7.92	5.95	6.98	4.61	4.92
CaO	10.58	10.34	10.27	10.44	10.26	10.78	10.82	11.01	10.68	11.32	10.88	11.80	11.24	11.52	11.56
Na_2O	0.24	0.23	0.26	0.23	0.25	0.37	0.26	0.24	0.25	0.27	0.26	0.37	0.26	0.31	0.27
K_2O	0.04	0.04	0.04	0.04	0.05	0.04	0.04	0.02	0.04	0.05	0.04	0.03	0.05	0.04	0.05
P_2O_5	0.06	0.05	0.06	0.06	0.06	0.08	0.06	0.09	0.05	0.05	0.06	0.03	0.06	0.10	0.06
Cr_2O_3	0.51	0.48	0.55	0.60	0.56	0.54	0.59	0.47	0.58	0.36	0.39	0.25	0.33	0.16	0.16
Total	99.93	100.19	99.94	99.54	100.37	99.79	99.24	99.70	99.76	99.34	99.29	99.35	99.63	99.60	100.50
Mg#	46.9	45.0	46.1	46.0	45.7	43.3	44.1	42.5	44.6	44.5	41.0	34.0	38.4	29.5	29.5
Norms															
Q	2.46	1.33	0.39	2.07	0.27	2.23	2.14	2.13	2.12	0.66	0.45	0.63	1.75	4.41	2.54
Or	0.24	0.24	0.24	0.24	0.30	0.24	0.24	0.12	0.24	0.30	0.24	0.18	0.30	0.24	0.30
Ab	2.03	1.95	2.20	1.95	2.12	3.13	2.20	2.03	2.12	2.28	2.20	3.13	2.20	2.62	2.28
An	25.11	23.79	25.07	25.26	25.53	26.71	25.59	26.89	27.06	26.47	26.96	29.96	28.78	33.96	33.32
Di	22.56	22.77	21.33	21.87	20.92	22.04	23.24	22.82	21.47	24.64	22.44	24.15	22.48	19.28	20.23
Hy	43.19	44.85	46.08	43.22	46.21	40.34	41.05	40.76	42.12	41.24	42.41	36.65	39.27	33.91	36.46
Ol	0.00	0.00	0.00	0.00	0.00	0.00	0.00	0.00	0.00	0.00	0.00	0.00	0.00	0.00	0.00
Il	3.46	4.46	3.70	3.93	4.06	4.12	3.80	4.06	3.67	3.21	3.91	4.39	4.22	4.71	5.00
Ap	0.13	0.11	0.13	0.13	0.13	0.17	0.13	0.20	0.11	0.11	0.13	0.07	0.13	0.22	0.13
	Selected Major and Trace Elements (INAA)														
Na_2O	0.223	0.222	0.244	0.236	0.243	0.237	0.246	0.255	0.254		0.254		0.256	0.309	0.311
FeO	19.12	20.04	19.55	18.58	19.20	19.15	19.06	18.63	18.84		18.73		19.29	19.89	19.79
Sc	48.3	46.7	42.7	41.3	41.3	43.8	43.3	44.2	43.3		41.6		43.8	39.3	37.1
Cr	3100	3100	3500	3500	3400	3500	3700	3200	3600		2700		2300	1100	1100
Co	37	47.5	42	39.2	40.6	40.8	39.6	39.3	40		37.9		36.7	36.9	37.6
Rb	20		20					9							
Sr	180	150	120				130	150	110		150		140	150	
Cs					0.24									0.21	
Ba	89	90	90	90	90	90	130	100	90		60		100	87	100
La	4.93	4.41	5.68	5.63	5.85	5.70	5.48	5.43	5.02		5.74		6.15	7.39	7.67
Ce	14.1	12.1	17.1	16.4	18.0	17.2	16.0	15.5	14.0		16.6		18.5	21.1	22.5
Nd	16	16	13	11	16	16	19	19	20		14		13	17	22
Sm	3.15	3.14	3.54	3.39	3.67	3.57	3.51	3.43	3.27		3.54		3.55	4.55	4.77
Eu	0.68	0.723	0.784	0.739	0.77	0.777	0.764	0.770	0.715		0.803		0.839	0.98	1.001
Tb	0.78	0.76	0.82	0.84	0.83	0.83	0.83	0.82	0.77		0.82		0.85	0.99	1.11
Yb	2.32	2.17	2.61	2.53	2.59	2.62	2.56	2.54	2.38		2.59		2.86	3.08	3.29
Lu	0.34	0.317	0.360	0.352	0.382	0.376	0.37	0.37	0.344		0.384		0.410	0.47	0.48
Zr	130	160	150	170	165	120	100	110	140		110		100	100	200
Hf	2.20	2.33	2.54	2.37	2.67	2.41	2.54	2.45	2.33		2.50		2.71	3.09	3.36
Ta	0.36	0.36	0.41	0.34	0.39	0.35	0.36	0.38	0.32		0.32		0.40	0.45	0.47
Th	0.58	0.43	0.67	0.60	0.67	0.65	0.64	0.65	0.65		0.66		0.74	0.87	1.01

0.9 mm across) set in an aphanitic to fine-grained granular matrix of pigeonite, subcalcic augite, plagioclase, and ilmenite (Fig. 2a). Olivine phenocrysts are sparse (< 5% modally) and range in composition from Fo_{66} to Fo_{68} with no apparent zoning.

The other three low-SiO_2 ONB clasts have coarse ophitic textures with small plagioclase laths (up to 0.5 mm long) totally enclosed by large (up to 1.2 mm long) blocky pyroxene crystals. Ilmenite and Cr-spinel occur both interstitially and intergrown with pyroxene. Olivine comprises up to 10% of the mode in some rocks, but is not found in all of our sections. This may be due in part to sampling problems (our probe mounts of the coarse ophitic low-SiO_2 ONBs all contain less

TABLE 2. (continued)

	Olivine–Normative Basalts (ONB)									
	Cumulates			Low-SiO$_2$ ONB				High-SiO$_2$ ONB		
	B-12	B-42	B-27	B-9	B-4	B-26	B-29	B-41	B-25	B-10
SiO$_2$	43.07	44.44	44.71	45.14	44.54	45.83	45.19	46.98	47.70	47.39
TiO$_2$	0.53	1.50	1.49	2.25	2.46	2.28	2.47	1.65	1.90	1.73
Al$_2$O$_3$	4.42	7.76	8.13	9.37	7.75	8.70	7.96	8.25	8.35	9.01
FeO	19.67	20.77	20.83	20.13	23.44	21.57	23.26	19.71	19.75	20.19
MnO	0.26	0.22	0.24	0.24	0.26	0.28	0.30	0.26	0.22	0.28
MgO	24.10	16.41	13.63	9.31	11.65	7.87	9.49	12.74	11.83	9.73
CaO	6.84	7.84	8.92	10.67	9.20	11.47	9.93	8.91	9.15	10.30
Na$_2$O	0.10	0.18	0.20	0.25	0.20	0.24	0.26	0.21	0.19	0.23
K$_2$O	0.01	0.03	0.03	0.03	0.03	0.04	0.03	0.03	0.03	0.03
P$_2$O$_5$	0.06	0.07	0.04	0.08	0.03	0.05	0.05	0.04	0.06	0.05
Cr$_2$O$_3$	0.98	0.95	0.91	0.56	0.71	0.32	0.54	0.90	0.77	0.54
Total	100.08	100.04	99.13	99.35	100.25	98.65	99.48	99.69	10 0.08	99.48
Mg#	68.6	58.5	53.8	45.2	47.0	39.4	42.1	53.5	51.6	46.2
Norms										
Q	0.00	0.00	0.00	0.00	0.00	0.00	0.00	0.00	0.00	0.00
Or	0.09	0.18	0.18	0.18	0.18	0.24	0.18	0.18	0.18	0.18
Ab	0.85	1.52	1.69	2.12	1.69	2.03	2.20	1.78	1.61	1.95
An	11.57	19.92	21.20	24.60	20.11	22.54	20.44	21.48	21.84	23.46
Di	17.87	15.24	18.88	23.29	21.17	28.74	21.59	18.61	19.23	22.86
Hy	16.19	31.61	33.11	41.46	34.20	37.50	41.62	45.82	50.76	46.37
Ol	50.87	27.16	19.74	2.43	17.11	2.55	7.12	7.27	1.35	0.37
Il	1.08	2.85	3.00	4.27	4.67	4.62	5.00	3.13	3.84	3.49
Ap	0.13	0.15	0.09	0.17	0.07	0.11	0.11	0.09	0.13	0.11
Selected Major and Trace Elements (INAA)										
Na$_2$O	0.077	0.178		0.239	0.238			0.184	0.201	
FeO	18.19	20.5		18.3	22.84			19.19	20.05	
Sc	27.2	31.4		42.3	37.8			38.6	41.7	
Cr	6500	4600		3600	4200			6200	5000	
Co	85.2	65.4		40.3	58.9			55	53.3	
Rb	27								35	
Sr				130				120		
Cs										
Ba	75	160		110	80			78	90	
La	2.11	3.99		5.30	4.63			4.34	4.62	
Ce	7.9	12.2		15.5	13.7			11.5	14.4	
Nd	11	11		13	12			15	7	
Sm	1.25	2.49		3.29	3.43			2.67	2.94	
Eu	0.211	0.544		0.705	0.828			0.54	0.609	
Tb	0.28	0.60		0.76	0.78			0.63	0.69	
Yb	0.91	1.75		2.41	2.04			1.98	2.24	
Lu	0.15	0.266		0.358	0.292			0.286	0.332	
Zr		120		120	220			110	160	
Hf	0.86	1.73		2.36	2.49			1.96	2.04	
Ta	0.20	0.26		0.37	0.37			0.30	0.36	
Th		0.35		0.59	0.42			0.47	0.47	

than 2.5 sq. mm of basalt). The olivine in the coarse ophitic low-SiO$_2$ ONBs is Fo$_{55}$ to Fo$_{62}$—more Fe-rich than in B-4. Pyroxenes span a wide range of compositions, from magnesian pigeonite to subcalcic ferroaugite, but are generally more Fe-rich than pyroxene in the high-SiO$_2$ ONBs (Table 4, Fig. 3a). A few pyroxene grains contain thin exsolution lamellae of another pyroxene, but these are too small to analyze. Plagioclase in all low-SiO$_2$ ONBs ranges from An$_{97}$ to An$_{85}$ in composition (Fig. 4a).

2. High-SiO$_2$ ONB: (B-10, B-25, B-41). The high-SiO$_2$ ONBs are all medium to coarse-grained, with subophitic to ophitic textures. Small plagioclase laths (up to 0.6 mm long)

TABLE 3. Modal compositions of representative mare basalt clasts from Apollo 15 breccia 15498.

	Pyroxene	Olivine	Plagioclase	Opaque	Silica	Groundmass	Other
QNB							
Primitive							
B-1	56	-	39	6	<1	-	-
B-2	55	-	41	2	<1	-	-
B-6	1	7	-	-	-	92	-
B-20	-	7	-	-	-	93	-
B-40*	-	-	-	-	-	100	-
Intermediate-1							
B-3	56	10	28	5	<1	-	<1
B-5	53	-	35	10	<1	-	-
B-11	60	-	36	3	-	-	<1
B-14	54	-	37	8	<1	-	-
B-24	64	-	26	9	1	-	-
B-43	59	-	34	7	<1	-	-
Intermediate-2							
B-23	52	-	46	3	-	-	-
B-44	56	-	35	6	3	-	-
Evolved							
B-22	54	-	35	7	4	-	5
B-45	61	-	30	6	3	-	<1
ONB							
Low-SiO$_2$ B-4	52	9		31	9	-	-
B-9†	50	6	37	7	<1	-	-
B-26†	62	-	33	5	-	-	-
B-29†	62	-	33	5	-	-	-
High-SiO$_2$ B-10†	50	-		32	13	-	-
B-25†	65	-	31	4	-	-	-
B-41	68	10	18	4	-	-	2
Cumulates							
B-12	55	38	1	5	-	-	-
B-27†	75	25	-	-	-	-	-
B-42	45	31	23	5	-	-	4

Groundmass = devitrified glass/fine-grained quenched textured.
Others = late-stage residual areas.
*No phenocryst in probe mount; however, pyroxene (?) phenocryst seen in hand sample.
†Small probe mount with matrix attached. Less than 200 points.

are partly to totally enclosed in pyroxene grains that average about 0.5 mm across. Olivine forms rounded grains up to 0.7 mm across and may include small euhedral Cr-spinels (Fig. 2b). Olivine cores are Fo$_{57}$ to Fo$_{67.5}$, rims Fo$_{45}$ to Fo$_{55}$. Pyroxene ranges from Wo$_7$En$_{66}$ to Wo$_{25}$En$_{34}$ and is generally more magnesian than pyroxene in the low-SiO$_2$ ONBs (Table 4, Fig. 3b). Plagioclase ranges from An$_{94}$ to An$_{87}$ and is lower in potassium than plagioclase with the same An content in the low-SiO$_2$ ONB suite (Fig. 4a).

A major difference in the mineralogy of the high-SiO$_2$ ONBs that distinguishes them from the low-SiO$_2$ ONBs is the presence in two clasts (B-10, B-25) of minor cristobalite as a late, interstitial phase. This indicates that the residual liquids in the high-SiO$_2$ ONBs evolved to silica-oversaturated compositions prior to final crystallization.

3. Olivine-pyroxene cumulates: (B-12, B-27, B-42). These rocks are characterized by cumulus textures with primocrysts of olivine, magnesian pigeonite, Cr-spinel, and (in clast B-12) augite, surrounded by intercumulus plagioclase, ilmenite, and pyroxene (Fig. 5c,d). Postcumulate pyroxene most commonly forms by adcumulate enlargement of cumulate pyroxene grains or as mantles on cumulate olivine. Olivine and pyroxene primocrysts are generally about 1.0 mm in diameter, with blocky euhedral to subheral outlines. Modal plagioclase is less abundant in these rocks than in the basalts (Table 3), reflecting cumulate enrichment in mafic phases.

In contrast to the basalts, the cores of the cumulus phases in the cumulate rocks exhibit restricted compositional ranges. Olivine cores range from Fo$_{71}$ to Fo$_{64}$ (Fo$_{71}$ to Fo$_{69}$ in B-12); rims are somewhat more Fe-rich in B-27 (Fo$_{58}$ to Fo$_{45}$).

TABLE 4. Representative pyroxene compositions in 15498 basalts.

	Quartz-Normative Basalts (QNB)							Olivine-Normative Basalts (ONB)					
	Evolved (B-45)		Intermediate (B-44)		Primitive (B-1)			Cumulate (B-12)		High-SiO$_2$ (B-4)		Low-SiO$_2$ (B-41)	
	Pheno-cryst	Ground-mass	Pheno-cryst	Ground-mass	Pheno-cryst	Pheno-cryst	Ground-mass	Pigeonite	Augite	Pheno-cryst	Ground-mass	Pheno-cryst	Groun-dmass
					(core)	(rim)							
SiO$_2$	52.51	47.52	49.23	48.59	51.93	49.95	46.96	53.45	51.00	53.55	49.40	53.27	48.90
TiO$_2$	0.42	1.05	1.25	0.89	0.49	1.05	1.11	0.31	0.54	0.36	1.06	0.33	1.40
Al$_2$O$_3$	1.66	1.14	4.55	1.39	2.08	3.65	1.17	1.32	3.49	1.34	1.78	1.45	1.75
Cr$_2$O$_3$	0.82	0.11	1.04	0.33	0.99	1.04	0.03	0.74	1.25	0.91	0.49	0.83	0.19
FeO	17.80	35.35	14.17	27.77	17.48	13.83	37.05	16.88	9.34	16.61	25.48	17.03	26.29
MnO	0.34	0.47	0.23	0.41	0.28	0.36	0.45	0.28	0.20	0.32	0.41	0.31	0.34
MgO	21.57	3.99	13.41	11.84	21.09	14.60	3.11	23.02	16.67	23.67	14.15	22.41	10.99
CaO	4.08	10.51	16.93	7.56	4.76	15.39	10.05	3.83	17.87	3.43	6.88	4.53	9.75
Na$_2$O	0.03	0.00	0.03	0.01	0.01	0.94	0.05	0.01	0.06	0.02	0.03	0.02	0.03
Total	99.23	100.14	100.84	98.79	99.11	99.91	99.98	99.84	100.42	100.21	99.68	100.18	99.64
	Formula Based on 24 Oxygens												
Si	7.810	7.815	7.393	7.739	7.747	7.530	7.798	7.853	7.532	7.825	7.685	7.825	7.703
Al	0.290	0.221	0.806	0.261	0.366	0.647	0.229	0.228	0.608	0.231	0.327	0.252	0.324
Ti	0.047	0.129	0.141	0.106	0.055	0.118	0.139	0.034	0.060	0.040	0.124	0.037	0.166
Cr	0.096	0.014	0.124	0.041	0.117	0.124	0.004	0.086	0.146	0.106	0.060	0.096	0.023
Fe	2.215	4.862	1.779	3.699	2.181	1.743	5.145	2.074	1.154	2.030	3.314	2.092	3.463
Mn	0.043	0.065	0.029	0.055	0.035	0.046	0.063	0.035	0.025	0.039	0.055	0.039	0.046
Mg	4.783	0.979	3.002	2.811	4.690	3.278	0.769	5.040	3.670	5.155	3.280	4.907	2.581
Ca	0.650	1.851	2.724	1.290	0.762	2.483	1.788	0.603	2.827	0.537	1.146	0.712	1.645
Na	0.009	0.000	0.008	0.003	0.002	0.010	0.016	0.002	0.016	0.005	0.010	0.005	0.009

Pigeonite primocrysts cluster around Wo$_7$En$_{67}$ with postcumulus rims of Wo$_{26}$En$_{45}$ (Table 4, Fig. 3c). Clast B12 is unusual in that it contains cumulus augite (Wo$_{37}$En$_{46}$) that zones outward towards magnesian pigeonite (Wo$_{17}$En$_{62}$). The apparent cosaturation with both pigeonite and augite is uncommon in mare cumulates (e.g., *Taylor et al.*, 1983).

Plagioclase compositions are generally similar to those in the other ONBs, except in clast B-12, where the plagioclase is more potassic than in the other two ONB groups (Fig. 4a).

These clasts are similar to the feldspathic peridotites and olivine microgabbros described by *Dowty et al.* (1973) from the Apollo 15 rake samples, but their constituent minerals have

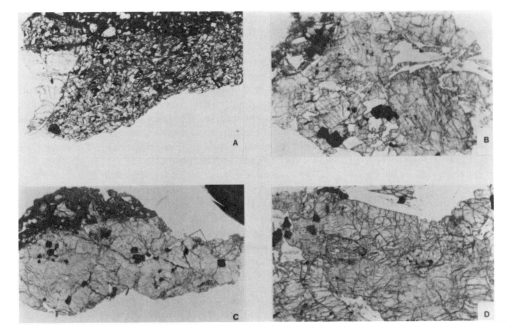

Fig. 2. Plane-polarized light photomicrographs of the ONB suite from 15498. Field of view is 3 mm across. (**a**) = B-4, low-SiO$_2$ ONB; (**b**) = B-41, high-SiO$_2$ ONB; (**c**) = B12, cumulate ONB; (**d**) = B-42, cumulate ONB.

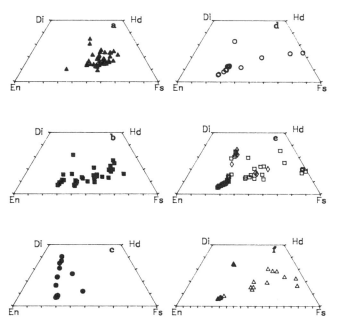

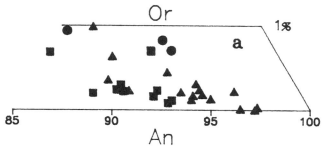

Fig. 3. Pyroxene compositions in the 15498 basalts. (**a**) = low-SiO$_2$ ONBs; (**b**) = high-SiO$_2$ ONBs; (**c**) = cumulate ONBs; (**d**) = primitive QNBs; (**e**) = intermediate-1 and intermediate-2 QNBs; (**f**) = evolved QNBs.

more restricted compositional ranges than the rocks described by Dowty. The origin of these rocks as cumulates from mare basalt parent magmas is suggested by their magnesian mineral compositions, restricted compositional ranges, and low modal feldspar contents.

B. Quartz-Normative Basalts

1. Primitive QNB: (B1, B2, B6, B20, B40). The primitive QNBs exhibit a wide range of textures, including vitrophyric, ophitic, and radial-ophitic. Clasts B-6 and B-20 exhibit textures characteristic of quenched liquids, although they are not true vitrophyres because the groundmass is only partly glass (Fig. 5a,b). These rocks consist of sparse olivine phenocrysts (0.25-0.75 mm across) in an aphanitic to fine-grained variolitic groundmass of pyroxene, plagioclase, opaques, and glass. The olivine phenocrysts, which comprise less than 2% of the mode, have rounded to subhedral shapes and Mg-rich core compositions (Fo$_{66}$ to Fo$_{70}$ in B-6; Fo$_{71}$ to Fo$_{73}$ in B-20). Olivine rim compositions are more Fe-rich in B-6 (Fo$_{48}$ to Fo$_{58}$) but are unzoned in B-20. B-6 also contains a small (0.25 mm), subhedral pigeonite microphenocryst (Wo$_4$ En$_{70}$) that appears to be in equilibrium with the olivine microphenocrysts.

Clast B-40 differs somewhat from the other vitrophyric clasts. The groundmass of B-40 consists of randomly arranged, euhedral to subhedral pigeonite laths surrounded by quench pyroxene (radiating pyroxene grains intergrown with glass and opaques), plagioclase, opaque oxides, and glass. The glass is opaque, with abundant tiny inclusions. The slender pyroxene laths decrease in abundance towards one end of the clast, where glass and quench pyroxene become dominant. This may represent a cooling-rate gradient within the parent lava flow, suggesting proximity to a cooling surface (either top or bottom). Clast B-40 was observed to contain large (< 2.0 mm) phenocrysts of either olivine or pyroxene in hand specimen, but no phenocrysts were included in our probe mount.

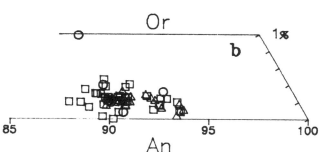

Fig. 4. Plagioclase compositions of the 15498 basalts. (**a**) = ONBs; (**b**) = QNBs. Symbols same as Fig. 6.

The other two primitive QNB clasts (B-1, B-2) have medium to coarse-grained ophitic textures. In B-1, the ophitic texture consists of radiating, fan-like arrays of pyroxene and plagioclase that surround phenocrysts of magnesian pigeonite. Groundmass pyroxene and plagioclase in these clasts and in the vitrophyric clasts have the same compositional ranges as in the other QNB suite clasts (figure 3c and figure 4b).

The textural evidence indicates that B-6, B-20, and B-40 probably represent quenched liquid compositions with little or no olivine accumulation. This conclusion is supported by the composition of the olivine phenocrysts and microphenocrysts, which are Mg-rich and appear to be in equilibrium with the bulk rock Mg/Fe ratios (based on an equilibrium Kd = 0.33). The presence of olivine in these quartz-normative basalts suggests that their compositions are relatively primitive and may be parental to more evolved pyroxene-phyric basalts.

2. Intermediate group 1 QNB: (B-3, B-5, B-11, B-14, B-24, B-43). Four of these clasts (B-3, B-11, B-24, B-43) are characterized by the common occurrence of large pigeonite phenocrysts (and more rarely olivine) set in a finer-grained groundmass of pyroxene, plagioclase, opaques, cristobalite, and glass (Fig. 5c,d). Two samples are aphyric (B-5, B-14), but have textures that are identical to the groundmass in the other intermediate/1 QNBs. The aphyric samples are most likely phenocryst-free portions of of normal pyroxene-phyric flows.

Pigeonite phenocrysts have subhedral, blocky shapes and range in size from 0.5-5.0 mm, with aspect ratios of about 2/1. Olivine phenocrysts range from 0.5-0.7 mm across and have irregular, partly resorbed shapes. Groundmass textures vary from ophitic (Fig. 5c) to radial-ophitic, with radiating fans of slender plagioclase laths enclosed by similar, radiating pyroxene crystals (Fig. 5d). The plagioclase laths may be as long as 1.0 mm, but are rarely wider than 50 microns. Groundmass pyroxene is generally coarser than the plagioclase, but also forms granular aggregates between the feldspar laths. Ilmenite, Ti-rich Cr-spinel, and cristobalite are intergrown with groundmass plagioclase and pyroxene, but Cr-rich spinel also forms inclusions in pigeonite phenocrysts.

Olivine phenocrysts range in composition from Fo_{73} to Fo_{67}. Pyroxene phenocrysts show small compositional ranges, with cores of magnesian pigeonite (Wo_6En_{68} to $Wo_{11}En_{59}$) that commonly display epitactic rims of augite ($Wo_{34}En_{42}$ to $Wo_{31}En_{41}$). Groundmass pyroxenes are more Fe-rich and range in composition from subcalcic ferroaugite to ferrohedenbergite (Fig. 3e). Plagioclase compositions range from An_{94} to An_{89} (Fig. 4b).

3. Intermediate group 2 QNB: (B-23, B-44). The intermediate/2 QNBs display the same range in textures and mineral chemistry as the intermediate/1 QNBs. Major differences in the phenocryst assemblages distinguish the two groups: The intermediate/2 QNBs lack olivine phenocrysts, and one clast (B-44) contains large euhedral augite phenocrysts in addition to phenocrysts of magnesian pigeonite. The augite phenocrysts are generally similar to epitactic augite in composition, but exhibit a wider range, from $Wo_{35}En_{40}$ to $Wo_{31}En_{41}$ (Fig. 3e). Plagioclase ranges in composition from An_{94} to An_{89} (Fig. 4b).

4. Evolved QNB: (B-22, B-45). The evolved QNB clasts studied here are more evolved chemically than any reported previously. One of these clasts (B-45) contains sparse large pyroxene phenocrysts in a radial-ophitic to radial-subophitic groundmass. The other clast (B-22) lacks phenocrysts, but its texture is nearly identical to the groundmass in clast B-45, except that it is coarser-grained. In addition to pyroxene and plagioclase, both of the aphyric QNB contain ilmenite, Ti-rich Cr-spinel, and cristobalite.

The coarser-grained aphyric clast, B-22, is a textbook example of radial subophitic texture. This clast consists of fan-like arrays of slender plagioclase laths up to 2.0 mm long and 50 microns wide intergrown with splays of prismatic pyroxene needles that mold around the ends of the laths, or are molded by adjacent plagioclase (Fig. 5f). The common occurrence of plagioclase laths that mold around pyroxene prisms, despite the modal predominance and larger size of the pyroxene, suggests that pyroxene nucleated before plagioclase or simultaneously with it. The latter hypothesis is supported by

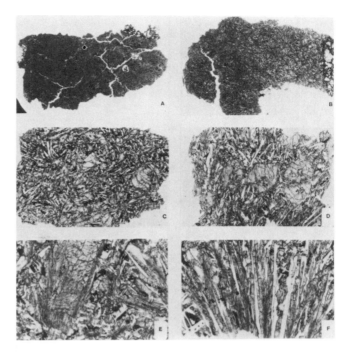

Fig. 5. Plane-polarized light photomicrographs of the QNB suite in the 15498 basalts. (a) = B-6, primitive QNB; (b) = B-20, primitive QNB. Both primitive QNB contain no significant phenocryst phases. (c) = B-3, intermediate-1 QNB; (d) = B-11, intermediate-1 QNB. Both intermediate-1 QNBs contain large pigeonite phenocrysts set in a medium grained groundmass of pigeonite, plagioclase, and residual liquid. (e) = B-44, intermediate-2 QNB (the large phenocryst on the left side is an augite phenocryst); (f) = B-22, evolved QNB; note the radiating intergrowths of pyroxene and plagioclases.

analyses of the cores of pyroxene splays, which show that the cores to consist of finely intergrown pyroxene and plagioclase that cannot be resolved optically.

Pyroxenes in the evolved QNB show the same compositional ranges displayed by pyroxene in the intermediate QNBs, from subcalcic augite to ferrohedenbergite (Fig. 3f). The only major difference between pyroxenes of the evolved QNB group and the other QNBs is a tendency towards more Fe-rich compositions, and fewer magnesian pigeonite phenocrysts. Plagioclase (An_{90} to An_{94}), ilmenite, and Ti-rich Cr-spinel likewise exhibit compositional ranges similar to their counterparts in the intermediate QNB groups.

WHOLE ROCK GEOCHEMISTRY

Major and trace element analyses of the twenty-five mare basalt clasts studied here are presented in Table 2. Major element variations as a function of MgO are shown in Fig. 6, trace element variations in Fig. 7, and chondrite-normalized REE concentrations in Fig. 8.

A. Olivine-Normative Basalts

Major element variations in the ONB suite are generally consistent with olivine control (either subtraction or addition):

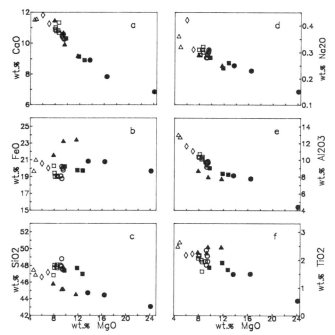
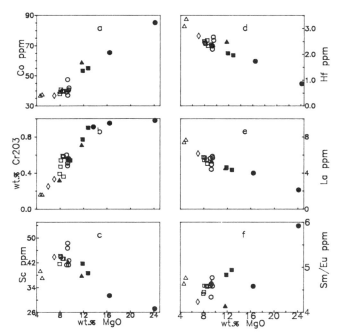

Fig. 6. Major element variations in 15498 basalts, using MgO as a fractionation index. Symbols: Open circles = primitive QNBs; open boxes = intermediate-1 QNBs; open diamonds = intermediate-2 QNBs; open triangles = evolved QNBs; filled circles = cumulate ONBs; filled boxes = high-SiO_2 ONBs; filled triangles = low SiO_2 ONBs.

Fig. 7. Trace element variations in 15498 basalts, using MgO as a fractionation index. (a,b,c) = compatible trace elements (Co, Cr, Sc); (d,e,f) = incompatible trace elements (Hf, La, Sm/Eu). See Fig. 6 for symbols.

SiO_2, Al_2O_3, CaO, Na_2O, and TiO_2 all increase with decreasing MgO, whereas FeO remains constant or decreases slightly (Fig. 6). Trace element variations show similar trends—the REE, Hf, and Sc all increase with decreasing MgO, whereas Cr and Co decrease sharply (Fig. 7). REE concentrations are the same in both the low and high-SiO_2 ONBs (La = 13–16 × chondrite) but are lower in the olivine-pyroxene cumulates (La = 6–12 × chondrite; Fig. 8). Despite this overall similarity, significant differences exist between the three ONB groups described earlier.

1. Low-SiO_2 ONB. The low-SiO_2 ONBs recognized here are identical to the ONB suite described by earlier studies (*Rhodes and Hubbard*, 1973; *Chappell and Green*, 1973; *Dowty et al.*, 1973). The low-SiO_2 ONBs have low SiO_2 contents (44–46 wt %), high TiO_2 (2.2–2.5 wt %), and high FeO (20.1–23.4 wt %) compared to the high-SiO_2 ONB suite and to the QNB suite. Alumina is also low in the low-SiO_2 ONBs relative to these other groups (Fig. 6). MgO concentrations are relatively low (7.9–11.7 wt %), so that their Mg#s range from 47.0–39.4.

2. High-SiO_2 ONB. The high-SiO_2 suite is characterized by high SiO_2 contents (47–48 wt %), low TiO_2 (1.65–1.9 wt %), low FeO (19.7–20.2 wt %), and MgO of 9.7–12.7 % (Fig. 6). Because the low FeO concentrations are coupled with relatively high MgO, Mg#s for these rocks (46.2–53.5) are higher than the low-SiO_2 ONB suite. The amount of normative olivine in the high-SiO_2 group (0.4 to 7.3 wt%) is less than in the low-SiO_2 group (2.6–17 wt %), and is due to the high Mg content, despite their high silica.

3. Olivine-pyroxene cumulates. These cumulates are characterized by low SiO_2 (43–44 wt %), low TiO_2 (0.5–1.5 wt %), and high MgO (16–24 wt %). Their high normative olivine concentrations (up to 50 wt%) are matched by high modal olivine and pyroxene (Table 3). These rocks have high concentrations of Cr and Co (which are compatible elements in spinel and olivine) but are low in Sc (which concentrates in pyroxene; Fig. 7). Incompatible element concentrations are the lowest observed in any ONB or QNB analyzed here, which is consistent with dilution by cumulate mafic phases. REE concentrations range from 6–12 × chondrite (Fig. 8). The data show that these clasts accumulated olivine and thus do not represent liquid compositions.

B. Quartz-Normative Basalts

The QNBs are characterized by high SiO_2 concentrations of 46.5–48 wt %, FeO abundances of < 21 wt %, and TiO_2 of 1.6–2.6 wt %. Mg#s range from 29.5–47, with MgO concentrations between 4.6–9.5 wt %. Two of the QNB clasts studied here have more evolved compositions than any QNBs described previously (*Rhodes and Hubbard*, 1973; *Helmke et al.*, 1973).

Major and trace element variation trends in the QNB suite are not consistent with simple olivine control: CaO, Al_2O_3, FeO, Na_2O, and TiO_2 all increase with decreasing MgO, but SiO_2 decreases (Fig. 6). The REE, Hf, Ta, and Th increase with decreasing MgO, and Cr continues to decrease, as in the ONB suite. In contrast to the ONBs, however, Co concentrations remain constant and Sc concentrations decrease with decreasing MgO (Fig. 7). These trends are consistent with

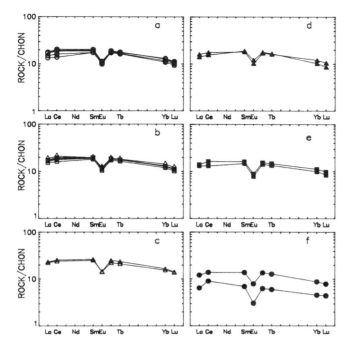

Fig. 8. Chondrite-normalized REE concentrations for 15498 basalts. (a) = primitive QNBs; (b) = intermediate-1 (boxes) and intermediate-2 (diamonds) QNBs; (c) = evolved QNBs; (d) = low-SiO$_2$ ONBs; (e) = high-SiO$_2$ ONBs; (f) = cumulate ONBs.

pyroxene fractionation.

REE concentrations are generally lowest in the primitive QNBs (La = 13-17 × chondrite), and highest in the evolved QNBs (La = 22 × chond-rite). Sm/Eu ratios vary from 4.2-4.8 (Fig. 7), and the slopes of the REE patterns show little variation.

1. Primitive QNB. The primitive QNBs are characterized by Mg#s between 45.0 and 47.0 (Table 2). They have SiO$_2$ contents of 47.5-48.8 wt %, FeO of 18.8-20.3 wt %, and CaO of 10.2-10.6 wt %. These samples have the highest compatible element concentrations (Co, Cr, Sc) in the QNB suite and the lowest incompatible element concentrations.

2. Intermediate group 1 QNB. The intermediate/1 QNBs are slightly more evolved than the primitive QNBs, with Mg#s between 44.5 and 41 (Table 2). They have SiO$_2$ contents of 46.6-48.1 wt %, FeO of 19.0-20.6 wt %, and CaO of 10.8-11.3 wt %. Compatible element concentrations are similar to the primitive QNBs or slightly lower; incompatible elements concentrations are slightly higher than the primitive QNBs.

3. Intermediate group 2 QNB. The intermediate/2 QNBs are slightly more evolved than the intermediate/1 QNBs, with Mg#s of 34 and 38 (Table 2). The SiO$_2$ content is 46.2-47.2 wt %, FeO of 20.0-20.6 wt %, and CaO of 11.2-11.8 wt %. Compatible element concentrations are slightly lower than in the intermediate/1 QNBs, and incompatible element concentrations slightly higher (Fig. 7). REE concentrations are available for only one sample from this group (B-44); these concentrations are slightly higher than the intermediate/1 QNBs and lower than the evolved QNBs (Fig. 8).

4. Evolved QNB. These basalts are more evolved than any reported previously, with Mg#s of 29.5, and 2.5-4.4% normative quartz (Table 2). The SiO$_2$ contents are 46.8-47.5 wt %, FeO of 19.7-21.0 wt %, and CaO of 11.5 to 11.6 wt %. Compatible trace element abundances are the lowest and the incompatible element abundances the highest (e.g., La = 22 × chondrite) among the QNBs.

The drop in CaO relative to the most Ca-rich intermediate/2 QNB terminates the Ca-enrichment trend seen in the less evolved samples (Fig. 6). This drop in CaO is not accompanied by a coincident drop in alumina, suggesting that augite fractionation may be involved.

DISCUSSION

The quartz-normative basalt (QNB) clasts studied here include samples more evolved than recognized before (with up to 4.4 wt % normative quartz), and olivine-phyric quenched liquids that are more primitive than any found in previous studies. The olivine-normative basalt (ONB) suite is considerably more complex. Basalts in the low-SiO$_2$ ONB group are identical to Apollo 15 basalts classified as "ONB" by earlier studies (e.g., *Rhodes and Hubbard,* 1973; *Chappell and Green,* 1973; *Dowty et al.,* 1973). Basalts in the high-SiO$_2$ ONB group do not correspond to any previously described basalt suite at Apollo 15. Their chemical characteristics are similar to those of the QNB suite, suggesting that the high-SiO$_2$ ONBs may be related to that group. The olivine-pyroxene cumulates are enriched in mafic phases and cannot represent liquid compositions.

The relationships between these basalt groups are more complex than has been observed in previous studies of Apollo 15 basalts, and enable us to study the evolution of the Apollo 15 basalt suites over a wider compositional range. These relationships also raise some important questions:

(1) Are the observed variations real, i.e., are they due to fractionation processes or do they merely reflect inadequate sample size (short-range unmixing)?

(2) Are the primitive, olivine-phyric QNBs parental to the more evolved QNBs by normal processes such as crystal fractionation?

(3) Are the QNBs related to the ONB by fractional crystallization? If so, are they related to the low-SiO$_2$ ONB or the high-SiO$_2$ ONB?

(4) Are the high-SiO$_2$ ONB and low-SiO$_2$ ONB related by crystal fractionation?

(5) Are the olivine-pyroxene cumulates related to the high-SiO$_2$ ONBs, the low-SiO$_2$ ONBs, or the QNBs?

(6) How do these basalts relate to other Apollo 15 basalts?

Short Range Unmixing

A major source of uncertainty in the analysis of lunar mare basalts derives from the small size of the samples available for analysis. *Lindstrom and Haskin* (1981) have shown that small samples from a single flow of Icelandic basalt exhibit significant compositional differences that can not be related to fractional crystallization. They propose that these variations are caused by a process they termed "short-range unmixing," in which each sample contains different relative proportions of mafic and felsic phenocrysts, groundmass, and residual liquids.

Lindstrom and Haskin (1978) used the short-range unmixing model to show that among the Apollo 15 ONBs studied previously, variations in chemical modes were ~-10% for pyroxene and plagioclase, 20% for spinel and residual liquid, and 20-30% for olivine, small enough to be accounted for by short-range unmixing of a single, undifferentiated flow. *Lindstrom and Haskin* (1978) also show that the QNB suite samples studied previously are too variable to be accounted for by short-range unmixing of a single, undifferentiated flow, or by fractional crystallization. They suggest that at least two separate flows may have been sampled.

We applied the short range unmixing model of Lindstrom and Haskin to Apollo 15 basalts studied here to test whether or not the intragroup variations observed in the QNB and ONB samples could be explained by short range unmixing, or if these variations require other explanations (e.g., fractional crystallization). Chemical modes were calculated for each rock using a least-squares mixing program and the average phase compositions used by *Lindstrom and Haskin* (1978); these modes were compared to the "average mode" of each group to evaluate the relative modal variations.

Variations observed within both the low-SiO_2 and high-SiO_2 ONB groups are too large to be accounted for solely by short-range unmixing. Olivine shows a variation of greater than 100%, spinel greater than 40%, pyroxene between 2 and 75%, plagioclase up to 50%, and residual liquid of up to 250%. Thus, although each of these groups may represent a single lava, these flows have differentiated by crystal fractionation to create the range in compositions observed. It is also important to note that these groups cannot be related to one another by short-range unmixing either. Two distinct groups are required.

The short-range unmixing model accounts for the variations observed within each of the four QNB groups described above, but cannot account for variations observed between these groups. Thus, the four QNB groups may represent four distinct, undifferentiated flows. The composition of the primitive flow is represented by the average of the primitive QNBs, which include three vitrophyric samples, enriched in residual liquid and two crystalline samples that are depleted in residual liquid relative to the average.

The intermediate/1 QNBs show considerable scatter on major and trace element variation plots (e.g., Fig. 6, 7), but variations in the calculated mode are less than 10% from the average for pyroxene, 10-20% for plagioclase and residual liquid (except for B-14 residual liquid). Thus, these samples may be related to a single, undifferentiated flow whose composition is best represented by the average of all six analyses.

The intermediate/2 and evolved QNB groups have modal compositions that fall far from the average of all QNBs, but modal variations between the two groups are more equivocal. The largest difference is seen in the residual liquid component: when compared to the average of both groups, the intermediate/2 sample B-44 deviates by 4% and the evolved samples deviate by 39%. This suggests that the differences between these samples are real, and result from processes other than short-range unmixing. However, the small number of evolved samples for which we have data (two in each group) means that the averages calculated for each group may vary somewhat from the parent lava flows.

Fractional Crystallization Relationships

Major and trace element variations in both ONB and QNB suites exhibit trends that are consistent with crystal fractionation of the observed liquidus minerals—olivine in ONB suite and pyroxene in QNB suite (Fig. 6 and 7). Our evaluation of short-range unmixing effects, however, shows that much of the scatter evident on these diagrams is the result of modal variations in small samples that may not represent the average composition of their parent lava flow. This effect is most pronounced on the QNB suite samples. In order to avoid the problems associated with limited sample sizes, we have averaged the compositions of all the samples within each QNB group to derive a "best" estimate for the composition of the parent lava flow. These averages are presented in Table 5, and will be used in all of the calculations that follow. Because the ONB suite clasts exhibit differences that are too large to be explained by short-range unmixing, we will continue to consider these compositions individually. It is important to keep in mind, however, that the compositions of these clasts may also deviate somewhat from that of their parent lava flow.

Figure 9 is an MgO variation diagram for SiO_2 which illustrates fractionation trends in the ONB and QNB samples studied here. The large open symbols represent the average QNB data from Table 5, the stars are QNB analyses of *Rhodes and Hubbard* (1973), the closed symbols are ONBs studied here, and the crosses are ONB analyses from *Rhodes and Hubbard* (1973). Representative control lines for olivine (Fo_{70}) and pigeonite (Wo_7En_{67}) fractionation are drawn from the observed liquidus minerals through the most primitive basalt of each group. This

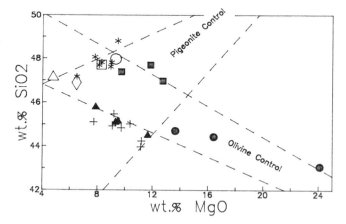

Fig. 9. SiO_2 wt % versus MgO wt % of 15498 basalts. Dashed lines represents olivine (Fo_{70}) and pigeonite (Wo_7,En_{67},Fs_{26}) fractionation control lines. Symbols same as Fig. 6. Larger symbols: Circle = average of the primitive QNBs; box = average of the intermediate-1 QNBs; diamond = average of the intermediate-2 QNBs; triangle = average of evolved QNBs. Previous data of *Rhodes and Hubbard* (1973) are represented by the stars (QNBs) and crosses (ONBs).

diagram is useful because the olivine and pigeonite control lines intersect at high angles, resulting in residual liquid trends that diverge from the presumed parent magma composition.

This diagram illustrates several important points. First, the QNB clasts studied here, and the QNB data of Rhodes and Hubbard, plot on or near a pigeonite control line, suggesting that these samples may be related by pigeonite fractionation. Second, the ONB clasts studied here plot on or near two distinct olivine control lines (Fig. 9). The high-SiO_2 ONBs plot near an olivine control line that intersects the QNB array at its primitive (high MgO) end; the low-SiO_2 ONBs and the ONB data of Rhodes and Hubbard plot near an olivine control line that intersects the QNB array at its evolved (low MgO) end. Thus, the high-SiO_2 ONBs may be parental to QNB suite lavas, but only the most evolved QNBs could be related to the low-SiO_2 ONBs. Finally, the olivine-pyroxene cumulates plot between the high-SiO_2 ONB and low-SiO_2 ONB olivine control lines and may be related to either suite. These trends are evaluated quantitatively in the next section.

Least-Squares Modeling

Least-squares calculations were made for the major elements using the most primitive basalts as parental magmas for each group and the observed liquidus phases (olivine, pigeonite, augite, and Cr-spinel). The calculations take the form parent magma = daughter magma + liquidus phases, with the parent magma as the dependent variable.

Calculations involving the QNB suite use the average group compositions from Table 5. Clast B-4 is chosen to represent the low-SiO_2 parent magma because it lies at the high-Mg end of the low-SiO_2 ONB trend, and because its texture (very fine-grained, olivine-phyric; Fig. 2a) is likely to represent that of a quenched liquid. Clast B-41 is chosen to represent the high-SiO_2 parent magma for the same reasons, although its texture (medium-grained ophitic) makes this interpretation more uncertain. The results of these models are summarized in Table 6 (lavas) and Table 7 (cumulates).

The major element fractionation models calculated here were cross-checked by calculating the relative enrichment of La between the parent and daughter compositions. This fractionation factor, termed F^* in the results, calculates the fraction of liquid remaining for Rayleigh fractionation if the bulk partition coefficient for La = 0 (i.e., $F^* = La_{parent}/La_{daughter}$).

QNB parent magmas. Several possible combinations were tested in which QNB parent magmas fractionate to QNB daughters (Table 6). A parent magma similar to the average primitive QNB requires about 5% fractionation of either olivine or olivine + pigeonite to generate the intermediate/1 QNB group. The results for both cases are equally good (the "olivine-only" solution has a smaller error in the sum, but a larger residual than the "olivine + pigeonite" solution); it is not possible to choose between these two options. Although the primitive QNBs contain phenocrystic olivine, several also contain pigeonite phenocrysts. We favor the two phase fractionation result.

Primitive QNB parents may also fractionate to intermediate/2 QNB (17% crystallization) and evolved QNB (27% crystalli-

TABLE 5. Average of the QNB groups used in least-squares modeling.

	Primitive	Intermediate-1	Intermediate-2	Evolved
SiO_2	47.97	47.71	46.89	47.18
TiO_2	2.04	1.98	2.20	2.55
Al_2O_3	9.59	10.32	11.33	12.85
FeO	19.55	19.28	20.27	20.33
MnO	0.26	0.25	0.24	0.24
MgO	9.30	8.27	6.47	4.77
CaO	10.38	10.91	11.52	11.54
Na_2O	0.24	0.27	0.32	0.29
K_2O	0.04	0.04	0.038	0.05
P_2O_5	0.06	0.06	0.05	0.08
Cr_2O_3	0.54	0.49	0.29	0.16
Total	99.97	99.52	99.49	100.05
Mg#	45.9	43.3	36.2	29.5
Norms				
Q	1.30	1.62	1.19	3.48
Or	0.25	0.23	0.24	0.27
Ab	2.05	2.33	2.67	2.45
An	24.95	26.61	29.37	33.64
Di	21.89	22.78	23.32	19.755
Hy	44.71	41.32	37.96	35.19
Ol	0.0	0.0	0.0	0.0
Il	3.92	3.80	4.31	4.86
Ap	0.13	0.14	0.1	0.18
Trace Elements				
Sc	44.06	43.24	43.80	38.20
Cr	3320	3340	2300	1100
Co	41.26	39.52	36.70	37.25
Rb	8.0	1.80		0.0
Sr	90	108		145
Cs	0.05	0.0		0.11
Ba	89.8	94.0	100.0	93.5
La	5.30	5.47	6.15	7.53
Ce	15.54	15.86	18.50	21.8
Nd	14.40	18	13	19
Sm	3.38	3.46	3.55	4.66
Eu	0.74	0.77	0.84	0.99
Tb	0.81	0.85	1.05	
Yb	2.44	2.54	2.86	3.18
Lu	0.35	0.37	0.41	0.475
Zr	155	116	100	150
Hf	2.42	2.45	2.71	3.23
Ta	0.37	0.35	0.40	0.46
Th	0.59	0.65	0.74	0.94

zation) magma compositions (Table 6). These results are corroborated by the La fractionation factors, which indicate 14% and 30% crystallization, respectively. When more evolved parent magmas are chosen (e.g., intermediate/1 QNB), similar results are obtained (Table 6). Note that these results are additive, so that the result for primitive QNB → evolved QNB is approximately the same as the sum of primitive QNB → intermediate/1 QNB plus intermediate/1 QNB → evolved QNB. Note also that all solutions for evolved QNB and intermediate/2 QNB require augite fractionation. These solutions are consistent

TABLE 6. Representative major element least-squares calculations for 15498 basalts.

Parent	Daughter	% Daughter	% Phases			Cr-Spinel	Sum	Error	Residuals	F*
			Olivine	Pigeonite	Augite					
QNB	QNB									
Prim	Inter-1	93.7	0.9	5.4	-	.06	100.0	3.1	.134	96.9
Prim	Inter-1	95.8	4.1	-	-	.12	100.0	1.0	.700	96.9
Prim	Inter-2	82.5	-	15.5	1.9	.26	100.0	1.7	.033	86.2
Prim	Evolved	73.0	2.4	16.5	7.3	.32	99.50	5.0	.540	70.4
Inter-1	Inter-2	87.9	-	9.2	2.6	.22	99.90	1.2	.188	88.9
Inter-1	Evolved	77.6	1.6	11.8	8.0	.27	99.22	4.1	.230	72.6
Inter-1	Evolved	77.0	-	14.1	7.9	.25	99.25	2.0	.430	72.6
Inter-2	Evolved	88.9	3.0	0.8	6.3	.08	99.17	6.2	.730	82.0
Inter-2	Evolved	89.1	3.5	-	6.4	.09	99.14	2.6	.710	82.0
Low-SiO$_2$ ONB	Low-SiO$_2$ ONB									
B-4	B-29	93.4	7.3	-	-	.04	101.1	1.7	.460	-
B-4	B-26	85.1	14.1	-	-	.09	100.0	3.1	1.77	-
Low-SiO$_2$ ONB	QNB									
B-4	Evolved	66.6	14.7	15.4	-	.85	97.6	26.6	13.5	61.5
High-SiO$_2$ ONB	Low-SiO$_2$ ONB									
B-41	B-4	87.0	-11.1	24.6	-	.02	100.6	16.9	6.00	93.7
B-25	B-4	90.4	-14.3	24.8	-	.01	100.0	13.1	4.39	99.8
B-25	B-29	84.1	-7.1	24.0	-	.26	101.3	15.0	6.30	-
High-SiO$_2$ ONB	QNB									
B-41	Prim	82.6	8.4	8.1	-	.73	99.80	3.8	.090	81.9
B-41	Inter-1	77.5	9.5	11.9	0.08	.80	99.90	1.6	.003	79.3
B-41	Inter-2	68.0	8.5	20.6	1.8	.96	99.80	2.0	.050	70.5
B-41	Inter-2	69.9	7.8	21.7	-	.95	100.0	3.1	.330	70.5
B-41	Evolved	59.8	11.2	20.7	-	1.0	99.10	5.7	.160	57.6
B-41	Evolved	59.8	10.9	21.0	6.6	.99	99.20	3.5	.140	57.6

Residual* = sum of the squares of the residuals.
F* = fraction of liquid remaining based on concentration of La in parent/concentration in daughter with bulk La D ≡ 0.
*Indicates mineral not used in fit.

with the occurrence of augite phenocrysts in one of the intermediate/2 basalts, and with the termination of CaO enrichment seen at low MgO-contents in Fig. 6. We will return to this important point later when we discuss the cumulates.

ONB parent magmas. Four separate problems must be considered here: (1) Can the assumed low-SiO$_2$ ONB parent (clast B-4) fractionate to form more evolved low-SiO$_2$ ONBs? (2) Can a low-SiO$_2$ parent fractionate to form the evolved QNBs? (3) Can a high-SiO$_2$ parent magma fractionate to form the low-SiO$_2$ basalts? and (4) Can the assumed high-SiO$_2$ parent magma (clast B-41) fractionate to form various members of the QNB suite?

The modeling calculations show that the more evolved low-SiO$_2$ ONBs (B-26, B-29) can be derived from a low-SiO$_2$ parent magma similar to B-4 by 7-15% olivine fractionation, with minor Cr-spinel (Table 6). No pigeonite is required, consistent with its absence as an early liquidus phase. However, a low-SiO$_2$ parent magma similar to B-4 cannot be parental to the evolved QNB suite samples. While a parent-daughter relationship appears possible in Fig. 9, the mixing calculations show extremely poor fits, with totals ≅ 98% and the sum of the square of the residuals > 10 (Table 6).

The calculations also show that high-SiO$_2$ ONB parent magmas cannot be parental to the low-SiO$_2$ ONBs (e.g., B-4, B-29). Not only are the fits for these mixes exceptionally bad (sum of the square of the residuals = 4-6), but the solutions all require negative olivine fractionation and substantial pigeonite fractionation (~25%) despite the fact that pigeonite is not a liquidus phase in either suite.

Fractionation of the assumed high-SiO$_2$ ONB parent magma (B-41) results in excellent fits for all QNB suite samples. The primitive QNBs require about 8% each olivine and pigeonite fractionation, a result supported by the La fractionation factor, which indicates 16% crystallization (Table 6). The more evolved QNB samples (intermediate/1, intermediate/2, and evolved QNB) require progressively more pigeonite fractionation of

TABLE 7. Representative major element least-squares calculations for 15498 cumulates.

Cumulate	Trapped Liquid	Trapped Liquid %	% Phases				Sum	Error	Residual*	T.L.*
			Olivine	Pigeonite	Augite	Cr-Spinel				
Cumulates	QNB									
B-12	Prim	33.3	42.5	1.3	19.5	.91	97.5	19.7	.544	40.0
B-12	Prim	33.5	43.5	-*	19.6	.91	97.5	7.3	.707	40.0
B-27	Prim	81.4	20.0	-6.7	3.2	.92	98.8	5.5	.096	-
B-42	Prim	74.9	27.4	-3.9	.56	1.1	100.0	6.2	.994	75.3
B-12	Inter-1	32.7	45.0	-	19.2	.92	97.8	5.9	.798	38.6
B-27	Inter-1	76.1	19.3	-	2.7	.96	99.0	1.5	.304	-
B-42	Inter-1	70.4	28.2	.15	.39	1.1	100.2	4.8	.641	72.9
B-12	Inter-2	28.3	42.3	6.4	19.6	.99	97.7	16.4	.385	34.3
B-27	Inter-2	67.4	19.8	6.1	4.5	1.1	99.0	2.8	.030	-
B-42	Inter-2	61.8	27.1	8.1	1.9	1.3	100.2	5.9	.977	64.9
B-12	Evolved	24.5	43.9	6.3	21.9	.99	97.5	16.8	.476	28.0
B-27	Evolved	59.0	22.1	6.8	9.3	1.2	98.4	5.1	.550	-
B-42	Evolved	54.2	29.6	8.3	6.3	1.3	99.7	2.8	.207	53.0
Cumulates	Low-SiO$_2$ ONB									
B-12	B-4	30.6	41.9	-	23.7	.97	97.1	12.6	4.89	45.6
B-27	B-4	81.6	2.4	8.3	6.1	.41	98.8	22.4	3.34	-
B-42	B-4	76.1	10.4	10.1	2.7	.64	100.0	22.9	6.71	86.2
Cumulates	High-SiO$_2$ ONB									
B-12	B-41	40.7	37.6	-	18.9	.59	97.7	7.3	.333	48.6
B-27	B-41	98.9	11.4	-14.1	2.8	.19	99.2	3.7	.143	-
B-42	B-41	91.3	19.3	-10.6	-	.40	100.4	4.8	.621	91.9

Residual* = sum of squares of the residuals.
T.L.* = amount of trapped liquid based on concentration of La in whole rock/concentration in T.L. with bulk La D ≡ O.
*Indicates mineral not used in fit.

a high-SiO$_2$ parent magma, along with 8%-11% olivine and minor Cr-spinel (Table 6). Although B-41 can fractionate to Inter-2 and Evolved with or without augite, the presence of augite phenocrysts supports the model with augite. La fractionation factors are within 2% of the major element mixing results, which confirms the solutions.

Cumulates. The olivine-pyroxene cumulates are best modeled as cumulate rocks that reflect the mixing of cumulus phases (olivine, pigeonite, augite, Cr-spinel) with trapped intercumulus magma. These calculations take the form: cumulate rock = cumulus phases + trapped liquid. The amount of trapped liquid calculated from the major element modeling can be cross-checked using La to calculate the fraction of trapped liquid by mass balance, assuming the bulk partition coefficient for La is zero (i.e., [La$_{WR}$/La$_{TL}$] = mass fraction trapped liquid). This is shown in Table 7 as T.L.*.

Mixing calculations which assume a trapped liquid composition similar to the low-SiO$_2$ parent magma B-4 all result in poor fits to the actual cumulate rock compositions (Table 7). Errors about the sum are large, and the sums of the squares of the residuals range from 3.3-6.7 on the best solutions. Models which assume a high-SiO$_2$ parent magma (B-41) have negative pigeonite, except for cumulate clast B-12. This clast can be modeled reasonably well if pigeonite is excluded from the fit, but this is not consistent with petrographic evidence for cumulus pigeonite.

The best fits to these cumulates are achieved using QNB suite magmas for the trapped liquid component. Even here, fits which use the primitive and intermediate/1 QNBs as the trapped liquid component yield poor fits (Table 7). The best solutions result when the trapped liquid component is assumed to have an evolved composition, similar to intermediate/2 QNB or evolved QNB (Table 7). These results are confirmed by the La mass balance calculations, which show the closest correspondences for evolved trapped liquids.

Derivation of at least one of these cumulates (B-12) from an evolved parent magma is supported petrographically by presence of cumulus augite. Augite occurs as a phenocryst phase in the intermediate/2 QNBs, but not in less evolved QNBs. The onset of augite saturation at this point is further supported by the termination of the Ca-enrichment trend in the evolved QNBs (Fig. 6).

Relationships to Other Apollo 15 Mare Basalts

Two of the seven mare basalt groups described above have not been reported by previous studies of Apollo 15 mare basalts—the evolved QNB group, and the high-SiO$_2$ ONB group. There are probably two important reasons for this: (1) the geographic location of the sample site, and (2) analytical limitations of the previous data set.

Mare basalts were collected at three main locations: Station 1 (Elbow crater), Station 4 (Dune crater), and Station 9a (Hadley Rille rim). Pyroxene basalts (QNB) were common at all three sites, but olivine basalts were found only at Station 9a. In addition, a few mare cumulates were collected at Station 7. This distribution suggests that pyroxene basalts are laterally contiguous and underlie the entire mare plateau (*ALGIT*, 1972). The olivine basalts are interpreted as a thin flow unit that overlies the pyroxene basalt layer locally, but does not occur near the Apennine Front. Because 15498 was collected from the rim of Dune Crater (Station 4), the basalts present in this sample may come from as deep as 90 m—below the pyroxene basalt flows that underlie most of the site (*ALGIT*, 1972). Thus, the ONB and QNB clasts in 15498 may represent flows that predate the more common mare basalt samples studied previously. This interpretation is strengthened by the fact that 15498 is a breccia produced by impact processes containing random fragments of mare basalt. Thus, a wider range of compositions, from deeper stratigraphic levels, is more likely to be found.

Another possibility is that chemical differences between subgroups within each mare basalt suite (ONB versus QNB) have been masked by limited analytical data, and by chemical data of poor quality. For example, 15065 is a slightly olivine-normative basalt that *Rhodes and Hubbard* (1973) group with the QNBs, based on its chemical similarity to that suite. This rock is very similar chemically to the high-SiO_2 ONBs described here, and may represent the largest single sample of this group (but not necessarily from the same flow).

Ryder and Steele (1987) have recently re-analyzed several Apollo 15 ONBs, and calculated mass-weighted averages of the new and existing analyses. They used the same technique which we employed (fused bead EMP analysis), so our data sets should be comparable. Several of the analyses presented by Ryder and Steele have chemical characteristics similar to the high-SiO_2 ONBs described here: 15274, 15387, 15641, 15651, and possibly 15672. These samples have high SiO_2 for their MgO content, higher Al_2O_3, and lower TiO_2 than the "normal," low-SiO_2 ONBs with which they are correlated. Curiously, three of these samples (15641, 15651, 15672) were classified as "Olivine microgabbro B" by *Binder et al.* (1980), who noted their deviation from the other olivine-phyric samples. The other two samples (15274, 15387) are picrite basalts ("feldspathic peridotite" of *Dowty et al.*, 1973) that are similar to the olivine-pyroxene cumulate B-12. Although these rocks may not be directly related to the high-SiO_2 ONB group studied here, they may represent primitive, olivine-normative parent magmas to other QNB lava flows (or cumulates derived from these QNB flows). In any case, the data imply that more than one suite of olivine-normative basalt is present at the Apollo 15 site. Resolving these distinct ONB groups will be difficult, however, because basalts of the surface-outcropping units dominate the sample suite.

CONCLUSIONS

The data presented here document a greater diversity in Apollo 15 mare basalts than was recognized in earlier studies and provide important new constraints on mare basalt petrogenesis at this site. The wider range of basalt types found here allow us to identify potential parent magmas, test interrelationships between sample groups, and define probable lines of descent. Our main conclusions are:

1. The Apollo 15 quartz-normative basalts (QNB) may be subdivided into four groups based on chemical variations: (a) primitive QNB, (b) intermediate/1 QNB, (c) intermediate/2 QNB, and (4) evolved QNB. The primitive QNB are commonly olivine-phyric and may be parental to the more evolved QNBs. The intermediate/1, intermediate/2, and evolved QNBs represent progressively more evolved compositions that may be related to the primitive QNB group by crystal fractionation of olivine (early), pigeonite, and augite (late). Chemical variations within each of these four groups can be explained by the short-range unmixing model of *Lindstrom and Haskin* (1978, 1981).

2. The Apollo 15 olivine-normative basalts (ONB) comprise three distinct groups: (a) low-SiO_2 ONB, (b) high-SiO_2 ONB, and (c) olivine-pyroxene cumulates. The low-SiO_2 ONBs are the same as the ONB suite described by earlier studies. The high-SiO_2 ONBs are a new type of mare basalt that were not recognized by previous studies. The olivine-pyroxene cumulates correspond to the feldspathic peridotites and olivine gabbros described by *Dowty et al.* (1973).

3. Least-squares mixing calculations show that the high-SiO_2 ONBs may be parental to the QNB suite, documenting the first link between ONBs and QNBs at the Apollo 15 site. The composition of the high-SiO_2 ONB clast B-41 is the identical to the hypothetical QNB suite parent postulated by *Chappell and Green* (1973).

4. Least-squares mixing models show that variations within the low-SiO_2 ONBs can be explained be olivine fractionation. However, the low-SiO_2 ONBs cannot be parental to any of the QNB lavas, nor can the low-SiO_2 ONBs and high-SiO_2 ONBs be related by either short-range unmixing or fractional crystallization.

5. The discovery of new basalt types in breccia 15498 may result from its geographic position near the edge of the mare plain, where normal ONBs are scarce, and from its presumed origin as ejecta from Dune Crater. The ejecta from this crater may come from as deep as 90 m, well below the ONB and QNB flows exposed at the surface. In addition, new data on mare basalts from other stations suggests that the new basalt groups may be more widespread than previously recognized.

6. The cumulates are similar to Apollo 15 picritic basalts (*Ryder and Steele*, 1987) and feldspathic peridotites (*Dowty et al.*, 1973). They also have characteristics similar to ultramafic rocks from the Apennine Front (M. M. Lindstrom, personal communication, 1987). The presence of *cumulus* augite in clast B-12 suggests a relationship to the more evolved QNBs, which is supported by our fractionation models. However, the relationship of the cumulates to the basalts is still imprecise due to the large uncertainties in the results.

Acknowledgments. We wish to thank Charlie Galindo for curatorial assistance in the Pristine Sample Lab, Gordon McKay for access to the JSC EMP and Jerry Wagstaff for help in running it, and Jim Wittke for assistance with the EMP analyses at USC. Discussions with Graham Ryder, John Delano, and Larry Taylor on mare basalt petrogenesis were both educational and appreciated. Gary Lofgren provided access to the strip furnace at JSC, without which this paper would not be possible. We would also like to thank Odette James and Jeff Taylor for their careful reviews which improved the manuscript significantly. This work was supported by NASA grant NAG 9-169 to J.W. Shervais.

REFERENCES

ALGIT (Apollo Lunar Geology Investigation Team) (1972) Geological setting of the Apollo 15 samples. *Science, 175*, 407-415.

Binder A. B. (1976) On the compositions and characteristics of the mare basalt magmas and their source regions. *The Moon, 16*, 115-150.

Binder A. B., Lange M. A., Brant H. J., and Kahler S. (1980) Mare basalt units and the compositions of their magmas. *Moon and Planets, 23*, 445-481.

BVSP (Basaltic Volcanism Study Project) (1981) *Basaltic Volcanism on the Terrestrial Planets.* Pergamon, New York. 1286 pp.

Chappell B. W. and Green D. H. (1973) Chemical composition and petrogenetic relationships in the Apollo 15 mare basalts. *Earth Planet. Sci. Lett., 18*, 237-246.

Dowty E., Prinz M., and Keil K. (1973) Composition, mineralogy, and petrology of 28 mare basalts from Apollo 15 rake samples. *Proc. Lunar Sci. Conf. 4th*, 423-444.

Duncan A. R., Sher M. K., Abraham Y. C., Erlank A. J., Willis J. P., and Ahrens L. H. (1975) Compositional variability of the Apollo 15 regolith (abstract). In *Lunar Science VI*, pp. 220-222. The Lunar Science Institute, Houston.

Helmke P. A., Blanchard D. P., Haskin L. A., Telander K., Weiss C., and Jacobs J. W. (1973) Major and trace elements in igneous rocks from Apollo 15. *The Moon, 8*, 129-148.

Laul J. C. and Schmitt R. A. (1973) Chemical composition of Apollo 15, 16, and 17 samples. *Proc. Lunar Sci. Conf. 4th*, 1349-1367.

Lindstrom M. M. and Haskin L. A. (1978) Causes of compositional variations within mare basalt suites. *Proc. Lunar Planet. Sci. Conf. 9th*, 465-486.

Lindstrom M. M. and Haskin L. A. (1981) Compositional inhomogeneities in a single Icelandic tholeiite flow. *Geochim. Cosmochim. Acta, 45*, 15-31.

Lofgren G. E., Donaldson C. H., and Usselman T. M. (1975) Geology, petrology, and crystallization of Apollo 15 quartz-normative basalts. *Proc. Lunar Sci. Conf. 6th*, 79-99.

LSPET (Lunar Sample Preliminary Examination Team) (1972) The Apollo 15 lunar samples: A preliminary description. *Science, 175*, 363-375.

Mason B., Jarosewich E., Melson W. G., and Thompson G. (1972) Mineralogy, petrology, and chemical composition of lunar samples 15085, 15256, 15271, 15471, 15475, 15476, 15535, and 15556. *Proc. Lunar Sci. Conf. 3rd*, 785-796.

Neal C. R., Taylor L. A., and Lindstrom M. M. (1987) Mare basalt evolution: The influence of KREEP-like components (abstract). In *Lunar and Planetary Science XVIII*, pp. 706-707. Lunar and Planetary Institute, Houston.

Papike J. J., Hodges F. N., Bence A. E., Cameron M., and Rhodes J. M. (1976) Mare basalts: Crystal chemistry, mineralogy, and petrology. *Rev. Geophys. Space Phys., 14*, 475-540.

Rhodes J. M. (1972) Major element chemistry of Apollo 15 mare basalts (abstract). In *The Apollo 15 Lunar Samples*, pp. 250-252. The Lunar Science Institute, Houston.

Rhodes J. M. and Hubbard N. J. (1973) Chemistry, classification, and petrogenesis of Apollo 15 mare basalts. *Proc. Lunar Sci. Conf. 4th*, 1127-1148.

Ryder G. and Steele A. (1987) Chemical dispersion among Apollo 15 olivine-normative basalts. *Proc. Lunar Planet. Sci. Conf. 18th*, this volume.

Shervais J. W., Taylor L. A., and Lindstrom M. M. (1985a) Apollo 14 mare basalts: Petrology and geochemistry of clasts from consortium breccia 14321. *Proc. Lunar Planet. Sci. Conf. 15th*, in *J. Geophys. Res., 90*, C375-C395.

Shervais J. W., Taylor L. A., Laul J. C., Shih C.-Y., and Nyquist L. E. (1985b) Very high potassium (VHK) basalt: Complications in mare basalt petrogenesis. *Proc. Lunar Planet. Sci. Conf. 16th*, in *J. Geophys. Res., 90*, D3-D18.

Taylor L. A., Shervais J. W., Hunter R. H., Shih C.-Y., Nyquist L., Bansal B., Wooden J., and Laul J. C. (1983) Pre-4.2 AE mare basalt volcanism in the lunar highlands. *Earth Planet. Sci. Lett., 66*, 33-47.

Wanke H., Baddenhausen H., Blum K., Cendales M., Dreibus G., Hofmeister H., Kruse H., Jagoutz E., Palme C., Spettal B., Thacker R., and Vilcsek E. (1977) On the chemistry of lunar samples and achondrites Primary matter in the lunar highlands: A re-evaluation. *Proc. Lunar Sci. Conf., 8th*, 2191-2213.

Note added in proof:

It has been pointed out to us that our modeling of the primitive QNB group by fractionation of olivine in pigeonite from a high-Si ONB parent is in conflict with experimental data showing the primitive QNB group to have olivine as the sole liquidous phase. Fractionation models with olivine and CR-spinel as the liquidous phases in a high-Si ONB give results similar to the olivine pigeonite models, although the fits are not quite as good (13% olivine, 0.3% spinel, 86% daughter melt, with 1.3% error and 1.6% residual). The poorer fit probably reflects minor deviations of B-41 from a true liquid composition as a result of short range unmixing. This does not, however, invalidate our basic conclusion that high-Si ONB magmas similar to B-41 are parental to the QNB suite. More samples are needed to better define this liquid composition.

Chemical Dispersion Among Apollo 15 Olivine-Normative Mare Basalts

Graham Ryder and Alison Steele*

Lunar and Planetary Institute, 3303 NASA Rd. 1, Houston, TX 77058

*Also at Department of Geology, Acadia University, Nova Scotia, Canada, BOP IXO

New analyses of Apollo 15 olivine-normative mare basalts for major and minor elements combined with a synthesis of all published data for them provide no support for the hypothesis that the coarser-grained varieties (olivine microgabbros) consist of two chemical groups. Instead, analyses of the coarse and fine basalts are consistent with their being a single group. The dominant control is that of olivine. The magnitude of olivine fractionation required is great enough to suggest that substantial olivine separation rather than merely short-range unmixing is required. The few coarse picritic basalt samples (feldspathic peridotites) appear to be olivine cumulates from the same parent magma. However, the dispersion of the chemical analyses of the olivine-normative mare basalts from a simple olivine-control trend (which is for some elements severe) requires that some combination of analytical sampling problems, short-range unmixing, and perhaps other as yet unrecognized petrogenetic processes are superimposed on the olivine separation. Sample 15388, which contains no olivine, has lower rare earth abundances and a positive Eu anomaly; it is undoubtedly a cumulate, possibly from the olivine-normative basalt magma. Both coarse-and fine-grained olivine-normative basalts span almost the entire compositional range, as do vesicular basalts. The abundance of olivine phenocrysts is inadequate to explain the fractionation, although the picritic basalts demonstrate accumulation in surface flows. Thus extruded magmas are represented by the more magnesian samples, but the magnesian samples probably also include cumulates. More evolved samples include representatives of magmas fractionated during both preextrusion and postextrusion crystallization, and some are apparently cumulates from such magmas. The textures, mineralogical characteristics, and chemical variation of the olivine-normative basalts are consistent with a sequence of thin fractionating flows, all from a common parent. However, the present chemical and petrographic data are inadequate to allow more than a superficial understanding of the near-surface petrogenesis of the olivine-normative basalts, and hence their origin and relationships to other basalts.

INTRODUCTION

Most of the crystalline mare basalts at the Apollo 15 landing site are from one of two distinct series, one olivine-normative, the other quartz-normative (e.g., *Rhodes and Hubbard*, 1973). A large number of chemical analyses for these samples, using varied techniques and varied sample sizes, has been published (compiled in *Ryder*, 1985).The olivine-normative mare basalts have a much greater range in composition than do the quartz-normative basalts. The two groups have similar trace incompatible element abundance patterns, although the olivine-normative basalts tend to have slightly lower abundances; they also have identical ages and isotopic characteristics (summary in *Nyquist*, 1977). However, the two groups cannot be related by simple fractionation from some common parental magma or partial melting of a common source (e.g., *Rhodes and Hubbard*, 1973; *Chappell and Green*, 1973).

The olivine-normative basalts were sampled predominantly at the edge of Hadley Rille (Station 9a), where they include samples chipped from small boulders. Nonetheless, examples were found to be widely distributed over the mare sites. Small fragments are also present on the Apennine Front, whereas the quartz-normative group appears to be absent from the Front (*Ryder*, 1986). Furthermore, the regolith chemical data seems to trend toward olivine-normative mare basalts rather than quartz-normative basalts, suggesting that it is the predominant mare type in the upper regolith over the entire landing site (e.g., *Rhodes and Hubbard*, 1973; *Korotev*, 1986). However, *Korotev* (1987, p. E426) suggested that the data for a few critical elements (e.g., Si) was inadequate to substantiate such a claim, except perhaps for Station 9a, at which the interfering highlands component is small.

The olivine-normative mare basalts have textures ranging from fine-grained (<500 microns) with small olivine phenocrysts through medium-grained gabbroic (>500 microns) to coarser-grained gabbroic (>2 mm). The gabbroic-textured samples contain large crystals of pigeonite, but also include plagioclase that poikilitically encloses small rounded olivine and pyroxene grains. No vitrophyric examples are known.

Opinions about the relationships among the olivine-normative mare basalts have varied, even as to whether they represent derivatives of a single magma or of several magmas that differ slightly in composition. The O'Hara group (e.g., *Humphries et al.*, 1972) considered all the samples to be olivine-plus-pigeonite cumulates from a cotectic liquid. However, the hypothesis that quickly gained credence was that the within-group variation was a product of moderate amounts (<15%) of near-surface fractionation of olivine from one singly-saturated (olivine) magma (*Rhodes and Hubbard*, 1973; *Chappell and Green*, 1973; *Mason et al.*, 1972; *Helmke et al.*, 1973). On the basis of their own analysis, *Rhodes and Hubbard* (1973) concluded that the coarse olivine-rich sample 15385 was not comagmatic with the main olivine-normative group.

Binder (1976) and *Binder et al.* (1980) used a synthesized, hence more comprehensive but less precise, data base. It was less precise because it included microprobe defocussed beam analyses (henceforth referred to as DBAs). These authors separated the olivine-normative basalts into three distinct flow units on the basis of small, but apparently consistent, compositional differences, particularly evident when the data were cast into normative form. Within each flow unit variation was controlled by olivine fractionation. They also concluded that the 15385/15387 and 15388 samples represented yet two more distinct magmas. One stimulus for the *Binder* (1976) distinction was the petrographic distinction that had been made by *Dowty et al.* (1973) between two groups of gabbroic-textured basalts; however, the distinctions made by *Binder* (1976) do not classify specific samples in the same way as *Dowty et al.* (1973), nor indeed in quite the same way as *Binder et al.* (1980).

Haskin et al. (1977) found, for their own suite of analyses of 13 olivine-normative basalts (*Helmke et al.*, 1973), that the chemical dispersion was no more pronounced than that found in horizontal sections of single terrestrial basaltic flows, and that much of the dispersion could be explained by the sampling of the parent rocks for analysis. *Lindstrom and Haskin* (1978), who quantitatively modeled the same data set, concluded that the partial separation of major consitituent phases on a moderate scale (short-range unmixing) could account for the dispersion among analyses; hence the samples could be from a single, undifferentiated lava flow. Further analyses led *Ma et al.* (1978) to conclude that neither shallow fractional crystallization nor varied partial melting of a homogeneous source explained the observed trace element fractionation trends. They suggested that short-range unmixing (filter pressing of residual liquid in particular) could be the dominant process, but they did not rule out the possibility of several distinct flows.

In none of the studies described above were the petrographic characteristics given more than fleeting relevant attention. For instance, it was rarely noted whether the sample was fine-grained or coarse-grained, or whether the sample might be a cumulate or liquid composition.

In several cases there have been chemical analyses that are anomalous; the authors have concluded that they represent separate flows (e.g., 15634 and 15643) despite being petrographically similar to the olivine-normative basalts. Most of such analyses are on rather small samples.

In the present study, we made chemical analyses of 12 mare basalts, all but one of them known or believed to be olivine-bearing, using INAA and microprobe fused bead techniques. Our main objective was to provide data adequate to test the *Binder* (1976) hypothesis that the olivine microgabbro groups represent two magmas distinct from each other and from the finer-grained olivine phenocrystic basalts. We also analyzed one of the basalts with apparently anomalous compositions (15643), two coarse-grained rocks from the Apennine Front (15387 and 15388), and a 4-10-mm particle from the Apennine Front (15274,3). We used masses large enough to assure a reasonable representivity. Using the compilation of data in *Ryder* (1985), a few more recently published analyses, and the new analyses reported here, we made mass-weighted averages for all analyzed olivine-normative mare basalts. With these we

TABLE 1. New analyses of Apollo 15 mare basalts by microprobe fused bead (MFB) and INAA.

MFB wt %	INAA wt %	15274 ,21	15387 ,9	15387 ,10	15388 ,15	15620 ,21	15623 ,1	15633 ,19	15641 ,21	15643 ,17	15651 ,10	15663 ,16	15663 ,16S	15668 ,17	15672 ,20
SiO_2		46.9	45.3	47.6	44.2	45.8	45.4	44.6	46.5	45.8	46.8	45.8	-	46.7	45.7
TiO_2		1.39	1.67	1.44	5.91	1.95	2.27	2.70	1.90	1.98	2.04	2.05	-	2.37	2.0
Al_2O_3		7.3	5.8	5.5	11.1	9.2	8.2	8.5	9.3	8.8	9.9	8.3	-	9.1	8.9
FeO		21.1	22.7	21.0	19.1	21.5	22.5	23.0	21.2	22.2	20.3	21.8	-	21.9	21.6
	FeO	21.3	21.5	21.3	19.8	21.8	22.5	23.1	21.5	22.2	19.8	22.1	22.3	21.7	21.5
MnO		.32	.29	.25	.35	.35	.37	.35	.38	.35	.37	.34	-	.38	.28
MgO		15.0	17.4	16.9	8.0	11.3	11.6	10.5	11.1	11.5	10.3	10.8	-	8.9	11.7
CaO		6.7	6.4	6.2	10.2	9.2	9.6	9.1	9.9	9.4	10.3	9.1	-	9.9	8.9
Na_2O		.22	.16	.12	.32	.23	.28	.23	.34	.29	.33	.23	-	.25	.26
	Na_2O	.20	.15	.16	.32	.26	.24	.26	.25	.27	.28	.26	.25	.26	.27
P_2O_5		.09	.11	.08	.06	.12	.13	.11	.09	.11	.11	.10	-	.12	.12
	ppm														
	Sc	27.5	28.5	30.4	48.6	39.1	39.0	41.1	37.4	39.2	39.8	39.9	44.6	44.6	39.3
	Cr	5280	3660	4609	2323	4526	5481	3963	3749	3718	3422	3865	4333	3462	3963
	Co	66.3	66.5	66.4	41.9	52.8	55.0	54.0	53.0	54.1	46.2	53.7	54.7	46.8	54.3
	Hf	1.69	1.70	1.60	1.82	2.11	2.27	2.39	1.91	2.10	2.20	2.57	2.62	2.63	2.37
	Th	.420	.390	.390	.430	.629	.667	.450	.470	.612	.445	.529	.600	.420	.637
	La	2.95	3.02	2.80	1.75	4.92	5.23	4.93	4.03	4.60	4.56	4.81	5.23	4.95	4.82
	Ce	7.7	7.9	7.6	5.0	13.2	14.3	12.8	11.8	12.1	11.8	11.2	15.4	11.7	13.5
	Sm	2.14	2.22	2.07	1.89	3.30	3.50	3.38	2.80	2.92	3.10	3.20	3.52	3.50	3.23
	Eu	.630	.524	.520	.841	.809	.826	.844	.769	.810	.821	.813	.867	.907	.805
	Tb	.530	.540	.480	.580	.810	.821	.800	.650	.790	.770	.814	.888	.850	.790
	Yb	1.55	1.54	1.47	1.98	2.04	2.22	2.01	1.82	1.98	2.11	2.06	2.40	2.26	2.27
	Lu	.228	.210	.190	.301	.324	.322	.323	.264	.293	.293	.313	.334	.319	.313
Mass (mg)		81	222	81	628	503	254	498	518	495	303	300	196	297	508

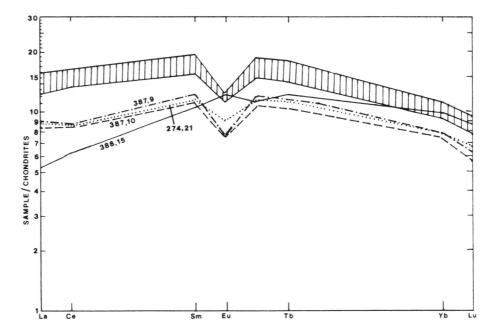

Fig. 1. Chondrite-normalized rare earth abundances in our new analyses of Apollo 15 mare basalts (data from Table 1). The shaded region is that for the entire range of the Station 9a samples. The Apennine Front samples are individually numbered; the prefix 15 is understood.

inspect petrogenetic relationships among the olivine-normative basalts, taking into account some relevant petrographic characteristics.

SAMPLES ANALYZED

We analyzed: (1) nine rake samples from Station 9a—15620, 15623, 15633, 15641, 15643, 15651, 15663, 15668, and 15672; (2) two splits of 15387, one of 15388, and a coarse-fine particle 15274,3 from the Apennine Front.

Most of the rake samples are medium- to coarse-grained and gabbroic. They were analyzed to evaluate the suggestion of *Binder* (1976) and *Binder et al.* (1980) that olivine microgabbros represent two chemical groups distinct from each other and from olivine-phyric basalts; our samples include representatives of both their groups A and B. Their thesis depended substantially on DBAs, which we can show as being insufficiently accurate for distinctions as fine as they made (see Results section below). Except for 15651 these samples had all been analyzed at least partially. Some analyses, such as that of *Ma et al.* (1978), are more recent than the *Binder* (1976) data compilation. However, most splits were of small mass or the analysis lacked critical data such as MgO or useful trace elements. 15668 is distinct in that it is a fine-grained sample with few and tiny phenocrysts. Petrographically it appears to represent a liquid composition. It was reanalyzed because there was little reliable trace element data for it. 15672 was reanalyzed to provide a comparison of our data with that from several other workers. 15643 was selected because *Laul and Schmitt* (1973) and *Ma et al.* (1978) considered it to be anomalous (with a positive Eu anomaly) and necessarily from a distinct magma. These authors did not consider the constraints of the sample size of only 49 mg. We attempted to use sample sizes that were more representative; for six of our analyses we homogenized about 500 mg of sample, but for three analyses we only had 200-300 mg.

15387 and 15388 are coarse basalts texturally distinct from most of the olivine-normative basalts, and their relationship to them is obscure. Petrographically, 15387 is indistinguishable from 15385; both are olivine-rich and were called feldspathic peridotites by *Dowty et al.* (1973) and in most subsequent references. Given the zoning and high Ca abundances of their olivines, hence clearly extrusive origin, picritic basalts—as they were called by *Binder* (1976)—is a more appropriate name, and will be used here. The age and isotopic characteristics of 15385, for which there is also some chemical data, are indistinguishable from olivine-normative mare basalts, as is the mineral chemistry. There was no previous chemical analysis of 15387. We analyzed a total of about 300 mg as two splits that turned out to be very similar. 15388 is texturally distinctive and lacks olivine; it was called feldspathic microgabbro by *Dowty et al.* (1973). Existing analyses showed a positive Eu anomaly and low overall incompatible element abundances, but the sample sizes used were very small considering the coarse grain size of the sample. We analyzed a split from 600 mg of homogenized powder. Coarse-fine (2-4 mm) particle 15274,3 (,21) was identified macroscopically as an olivine basalt by *Powell* (1972); no thin section of it has been made. Because of its tiny size, we used only 80 mg of it.

ANALYTICAL METHODS

We inspected the allocated chips in air under a binocular microscope, checking for possible extraneous material (such as adhering regolith), and describing them. Using agate tools, we crushed each sample by hand into a fine powder to homogenize it. A small (<15 mg) portion was fused into a glass bead, in an argon atmosphere (40 psi) and on a Mo strip, at the NASA Johnson Space Center. The equipment and techniques used were described by *Brown* (1977). The major elements in these glass beads were analyzed on an JEOL

Superprobe 733 at the Dalhousie University, Nova Scotia. The analyses were made at 15 Kv accelerating potential, a beam current of 5 na, and a beam diameter of 10 microns. Data reduction was made using Tracor Northern programs for ZAF corrections. Uncertainties are about 2% for all elements. The glasses were all very homogeneous and analyses of four points per glass were adequate.

Forty to 120 mg of the remaining homogenized powder for each sample was used for the Instrumental Neutron Activation analyses. Samples were irradiated at the University of Missouri Research Reactor; counting and reduction was done at the NASA Johnson Space Center using standard techniques. BCR was used as the standard for all elements except Cr, for which the standard was BHVO.

RESULTS

Our analyses by both microprobe fused bead and INAA are listed in Table 1. Our rare earth data are plotted in Fig. 1. Our analyses by both techniques for FeO and Na_2O in any sample are in very good agreement, hence we consider both sets of data to be good analyses of the powders. It is evident that fusing of the powder did not result in significant sodium loss. Table 2 is a comparison of previous analyses for 15672, a so-called olivine microgabbro B (*Binder,* 1976; *Binder et al.,* 1980), for critical elements. In this sample, pyroxenes reach 2 mm long. There is fairly good agreement for the major elements among all the analyses (including the present one) except the DBA, although the incompatible trace elements vary by over 20%. The DBA differs significantly in Ti, Al, and Mg from the similarity of the major elements of the other analyses. For this reason we consider the faith of *Binder* (1976) in the data base on which he made his conclusions to be unjustified. [All the group B olivine microgabbros of *Binder et al.* (1980) had only DBAs]. Some other DBAs do not differ from other analyses so much, but many do; the point is that they cannot be considered reliable for the fine distinctions made between possible groups.

Two analyses were made on two of the samples. Sample 15663,16 had a saw-cut surface, and after mild crushing it was separated into two portions: one with sawn surfaces (146 mg; 15663S), the other without sawn surfaces (350 mg;

TABLE 2. Comparative analyses of 15672.

	(1)	(2)	(3)	(4)	(5)	(6)
TiO_2	2.75	2.16	2.1	2.2	1.9	2.02
FeO	21.6	21.8	22.5	21.2	21.1	21.5
MgO	9.3	12.0	(12)	11.9	12.2	11.7
CaO	9.9	9.4	8.6	9.1	9.2	8.9
Al_2O_3	10.4	8.6	8.7	9.0	8.4	8.9
Sc	-	-	31	40	39	40	39
Sm	-	-	3.0	2.9	2.6	3.2
Mass (mg)	-	252	120	325	336	508

References: (1) DBA—*Dowty et al.* (1973); (2) *Christian et al.* (1972); (3) *Laul and Schmitt* (1973); (4) *Ma et al.* (1976)—Analysis a; (5) *Ma et al.* (1976)—Analysis b; and (6) present study.

15663). A fused bead was not made of the sawcut-containing sample, but the FeO and Na_2O abundances measured by INAA of the two portions are virtually identical. Both compatible and incompatible trace elements are about 10% higher in 15663S, but there is no evidence for any real contamination from the sawn surfaces; e.g., there is no enrichment in Co, and Ni was not detected. Thus both analyses are used in the mass-weighted averages (below). Two splits of 15387, the coarse-grained picritic basalt sample, had masses of 222 and 81 mg. Despite these small sizes and the coarse grain size of the rock, the two analyses are virtually identical for all elements except Cr. Chromium is concentrated in the small and sparse chromite grains, which can be expected to vary among such small splits. Both analyses are used in the mass-weighted averages (below).

NEW DATA AND COMPARISON WITH EXISTING DATA

The present analyses confirm that all the Station 9a samples are typical olivine-normative mare basalts (Table 1, Fig. 1). Most have very similar compositions with MgO ranging from 10.3 to 11.7%, except for the very fine-grained sample 15668, which has only 8.9% MgO (in agreement with earlier analyses). However, 15651 appears to have anomalously low FeO, which in the absence of other chemical data apart from the DBA is enigmatic, but might suggest a plagioclase excess in the sample. The incompatible trace elements vary among the samples by only 15%, except for 15641, which is almost 15% lower than any of the others. Chromium is the most varied compatible trace element, consistent with the uneven distribution of chromite. The only major element analysis that is significantly different from those previously published is that of 15651, for which the FeO abundance is lower than the others, about 20% instead of about 22%; both INAA and fused bead give lower FeO, hence it is a characteristic of the split, not an analytical peculiarity. The (Al_2O_3+CaO) is correspondingly higher, and it may be a plagioclase-enriched sample. Sample 15643, proposed by *Laul and Schmitt* (1973) to be anomalous, does not in fact have a positive Eu anomaly and is instead quite ordinary, as suggested by an earlier major element analysis by *Christian et al.* (1972). These data emphasize the limitations of making inferences based on partial analyses of small samples; the sample used by *Laul and Schmitt* (1973) was only 49 mg.

The coarse-fine sample 15274,3 (our analysis ,21), for which no thin section yet exists, was described macroscopically by *Powell* (1972) as an olivine basalt with about 20% olivine. The analysis of 81 mg shows it to be very similar to the picritic basalt 15387 for all elements, including the pattern and abundance of incompatible trace elements. Its Al_2O_3 is slightly higher and its MgO slightly lower, but it surely represents the same parent rock type. The data also agree reasonably well with the major element analysis for 15385 by *Ma et al.* (1976), but not with that by *Rhodes and Hubbard* (1973); the incompatible trace elements of 15385 by *Ma et al.* (1976) are 20-25% higher than in our 15387 analysis, and those of *Rhodes and Hubbard* (1973) are twice as high. A clast from breccia 15299 (,201) analyzed by *Warren et al.* (1987) has a composition very similar (remarkably so given the fact that its mass was only 49 mg). We suggest that all these samples represent essentially the same cumulate rock, but the coarse grain size has resulted in some unrepresentative analyses. Unlike the other two samples (15385 and 15387), coarse-fine particle 15274,3 (,21) and breccia 15299 were collected at Station 6, not Station 7. Thus these samples expand the known geographical range of this rock type and make it more likely that it is of local rather than exotic genesis.

15388 is petrographically unique among the Apollo 15 samples (*Dowty et al.*, 1973). Our new analysis confirms the positive Eu anomaly and low incompatible element abundances found in the two previous analyses (*Laul and Schmitt*, 1973; *Ma et al.* 1976). Our larger mass confirms that the sample is also high in TiO_2 (about 5%) and that the sample is aluminous. Its petrogenesis will be discussed below.

Mass-Weighted Averages

Numerous chemical analyses of Apollo 15 olivine-normative mare basalts have been made, including many on the same rock sample. They include neutron activation, XRF, and wet chemical analyses, and sample sizes ranging from a few milligrams to a few grams. Each technique covers a different spectrum of elements with varied precision and accuracy. There are few analyses for both major and trace elements on reasonably representative sample sizes. In most of the discussions of the petrogenesis of the basalts (summarized in the introduction), each analytical group has tended to use only its own data, thus restricting both the number of samples and elements used. The synthesis of *Binder* (1976) and *Binder et al.* (1980) is an exception.

In an attempt to overcome the analytical deficiencies, we have made mass-weighted averages of all of the Apollo 15 olivine-normative samples as a basis for assessing the relationships among and the petrogenesis of these basalts. In principal this method is similar to that used by Binder (1976) and *Binder et al.* (1980). Our data base, which comprises more than 50 separate rock samples, differs in that it includes more recent analyses, including our own, which are on reasonably representatively-sized samples for the grain sizes, and does not include DBAs, for the reasons outlined above. For our own data we use the FeO and Na_2O derived by INAA rather than by fused bead, although the difference is not significant. Apart from unreliable data, which is totally excluded, we have assumed that all analyses are equally valid estimates of the particular split analyzed, although this certainly may not be true; for instance, most major element analyses by INAA will not be as precise as those derived from XRF. However, making a totally objective calculation of how to weight analyses according to technique is difficult and would probably not create any major differences in our results, except as noted in the discussions. The unreliable data includes the DBAs, the rare earth data of *Christian et al.* (1972) (which is consistently much higher than that from other laboratories), some of the magnesia values derived from INAA, and a few other odd analyses. For the mass used, an attempt was made to find out how much sample the analyzed split represented; e.g., if 500 mg of sample was ground up and 60 mg was analyzed by INAA, then the sample mass used for this analysis in the weighted average was 500 mg. In most cases use of the published information and the Curatorial data packs allowed retrieval of this information, but in a few cases this was not possible and a conservative estimate of the probable mass used was made.

Our mass-weighted averages do suffer for petrogenetic purposes in that variation within a sample is suppressed (such that the scale of variation of this suite remains unknown) and different elements rely on different splits and different sample sizes, e.g., the rare earth data are rarely based on as much mass as the major elements. The averages also suffer from the interlaboratory biases. The present data set would not compare in value with a set of analyses for many major and trace elements on splits of homogenized samples of many basalts, using several grams from each. Nonetheless, we consider our averages to be the best available estimates of the bulk composition of each of the Apollo 15 olivine-normative mare basalts.

Chemical Dispersion and Relationships Among Olivine-Normative Mare Basalts

Samples classified as microgabbro groups A and B by *Binder et al.* (1980) for which analyses other than DBAs exist are plotted in Fig. 2, which summarizes most of the major element data for them. A similar plot was used by *Binder et al.* (1980, Fig. 8e) to show the distinction of Group A from Group B. Our data base shows no discrimination of the groups, either when our own data (small diamonds) or average data (large diamonds) are inspected. This overlapping of the two "groups" is seen in other plots (not shown) as well. There is also no distinction of these microgabbros from the finer-grained,

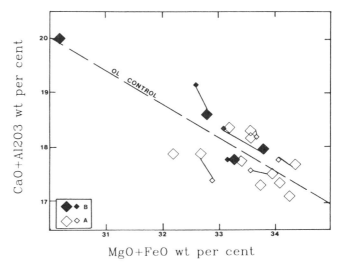

Fig. 2. $CaO + Al_2O_3$ versus $MgO + FeO$ for samples designated "Olivine-microgabbros A and B" by *Binder et al.* (1980) for which data is available other than DBAs; groups are distinguished by symbol. Large diamonds are the mass-averaged analyses, small diamonds are our new analyses. The olivine control line is only one among an infinite number that could have been drawn almost parallel to each other.

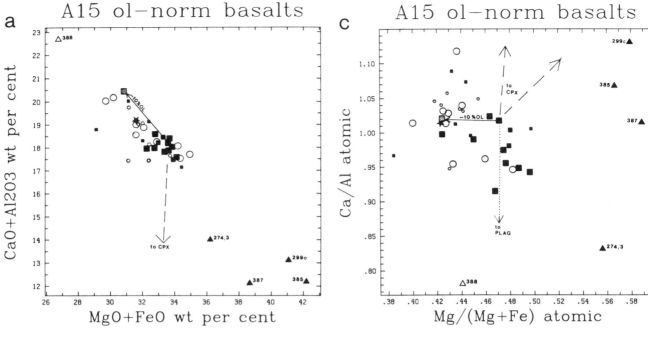

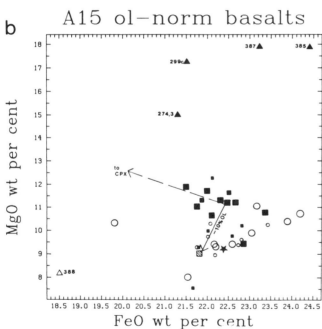

Fig. 3. Major element plots for the mass-weighted averages of the Apollo 15 olivine-normative mare basalts and 15388. Open circles are fine-grained basalts; solid squares are medium- and coarse-grained basalts. Largest symbols are for analyses comprising over 500 mg, intermediate symbols are 250-500 mg, and smallest symbols are less than 250 mg. Solid triangles are picritic basalts (numbered; the prefix 15 is understood). Open triangle is 15388. Star is the impact melted sample 15256. Fractionation lines are drawn from 15555 as an arbitrary starting point; for olivine, ten per cent subtraction is indicated by the shaded square. The olivine assumed is about Fo 65; the clinopyroxene assumed is a calcic pigeonite with Mg/(Mg+Fe) about 55; the composition of the assumed plagioclase is irrelevant.

phenocryst-bearing basalts on this plot (see Fig. 3a). Recalculation of the data into normative form, and plotting on the several ternary diagrams used by *Binder et al.* (1980), shows the same lack of discrimination (Binder, personal communication, 1987). Hence the data base as it presently stands allows no distinction among possible separate groups of olivine-normative mare basalts on the basis of major element data.

The data in Fig. 2 appear to be dispersed along an olivine control line, but such a line on this plot is similar to either a plagioclase control line or simply a closure-sum line. (By closure-sum line we mean that analyses, summing near to 100%, will plot along a line in element plots whose data dominates the analysis, even if they are multiple analyses of the same object. As "FeO + MgO" goes up by an absolute x%, then "Al_2O_3 + CaO," which compose the bulk of the remainder except SiO_2, will go down by about x%. It so happens that olivine-control and plagioclase-control lines on Figs. 2 and 3a have slopes similar to closure-sum lines, i.e., about 1.) The spread of data is probably too wide for simple closure, and indeed the Binder (personal communication, 1987) normative plots show that the effect is that of olivine, not plagioclase. However, there are complications shown by the data, as discussed below.

At present we have chosen to further assess the synthesis, mass-weighted data mainly from element-element plots; we are not certain of the validity of using more quantitative, mulitivariate methods (including least-squares techniques) on such a synthesized data base. On our plots (Figs. 3, 4) we use different symbols to make some assessment of sample size (hence representivity). Where different sample sizes were available for different elements on a plot, the symbol represents the size for the smaller mass. We also use different symbols to distinguish samples with fine grain sizes with small olivine

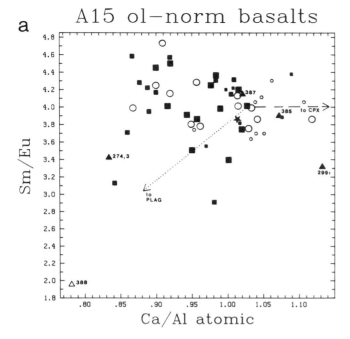
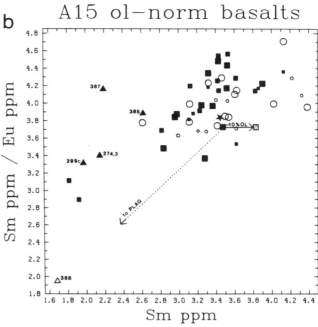

phenocrysts from those with coarser grain sizes, on the assumption that the fine-grained samples probably have compositions similar to that of their parent magmas, whereas the coarser ones may or may not.

The major elements, despite the dispersion (which is less when only larger sample sizes are considered), suggest that the samples form a single compositional group that is related mainly by olivine fractionation (Fig. 3a). This sequence trends in the olivine-richer direction to the coarse picritic basalt samples, and in the opposite direction to the 15388 sample (Fig. 3a). However, some of the data is apparently not precise enough to confirm simple olivine fractionation; there is a large degree of scatter introduced by either petrogenetic processes other than fractionation, or sampling problems induced by small or merely different subsamples, or both. For instance, on Fig. 3b (MgO versus FeO) the picritic basalts (especially the larger mass-analyzed ones 15385 and 15387) are consistent with being olivine cumulates from the main series. (Further discussion of these coarser samples will be deferred to the following section.) However, within the main series there is no consistent olivine trend. Furthermore, the coarser samples form a group that is distinct from the finer ones, in the direction of pyroxene (+ olivine?) addition, on Fig. 3b. The finer ones themselves form a line roughly consistent with plagioclase variation (Fig. 3b). The petrographic characteristics of these basalts and the experimental data (e.g. *Humphries et al.,* 1972) for these compositions (which show plagioclase far removed from the liquidus of even the most evolved of these compositions) preclude the separation of plagioclase crystals.

Olivine fractionation should not result in variation of either Ca/Al or Sm/Eu ratios. Figure 3c shows that there is a variation of the Ca/Al ratio; however, this variation is not correlated with Sm/Eu variation (Fig. 4a) as it would be from plagioclase variation. Pyroxene has a small effect on the Ca/Al ratio, but

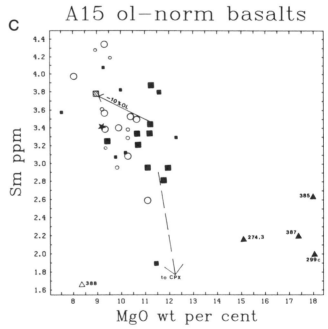

Fig. 4. Variation of some rare earth element parameters for mass-weighted averages of Apollo 15 olivine-normative mare basalts and 15388. Symbols as in Fig. 3.

not enough to cause anywhere near the amount required. We conclude that the data for Ca and Al are simply not good enough to assess the petrogenetic relations among these basalts, or that some process we do not understand is the cause of the variation. Figure 4b also shows the dispersion in the Sm/Eu ratio, and the trend appears to be a plagioclase trend. But it is also the trend expected of random variation of Sm around a constant Eu abundance. Indeed all the analyses do have Eu just under 1 ppm. The total variation in Sm is much greater than can be expected from olivine fractionation, and suggests some other cause—analytical, sampling, or alternate petrogenetic processes.

Simple olivine fractionation should result in an anticorrelation of precisely measured incompatible elements with MgO. Figure 4c (Sm v. MgO), like Fig. 4b, shows the much wider dispersion of Sm than can be explained by reasonable olivine fractionation, although there is a rough increase of Sm with decreasing MgO. The coarser-grained samples are on average richer in MgO than are the finer-grained samples. The scatter in the main (nonpicritic) group is more aligned with pyroxene than olivine, although olivine could have contributed to this effect.

The major elements thus seem to be paradoxical; in sum they are compatible with olivine fractionation (e.g., Fig. 3a), but in detail olivine fractionation cannot explain the data (e.g., Fig. 3b,c). The incompatible trace elements, represented by the precisely-measured Sm, show a much greater variation than can be accommodated by reasonable amounts of olivine fractionation. The element correlations we have considered cannot readily distinguish the causes of this variation. Assuming analytical problems are minimal, then the possibilities are sample size problems, the hodge-podge nature of the data base (synthesising data from different splits of a rock), or more complex petrogenetic processes.

One of the potential processes that may influence the chemistry of a sample is short-range unmixing. *Lindstrom and Haskin* (1978) invoked this process to explain their own data as being consistent with samples from a single undifferentiated flow. In their model each analysis is a mixture of crystal phases and a common residual liquid; least-squares modeling showed that the analyses were consistent with varying mixes of olivine, spinel, pyroxene, plagioclase, and residual liquid. They found olivine to be the most variable phase, with samples up to 30 relative % more or less than the average abundance (actually, 5.5 absolute % minimum, 9.5 absolute % maximum), and that spinel and residual liquid also had large relative variations (spinel being a minor phase and residual liquid able to move relatively easily, e.g., by filter pressing). The amount of variation they found appears to be at the upper limit of that expected from a single flow, and the data set used comprises only part of the known range of these basalts. For instance, the large, well analyzed, and finer-grained sample 15556 was not included in their analyses, yet is the most evolved of the well-characterized samples, e.g., Mg' about 0.40 (see *Ryder,* 1985). Its inclusion, as well as inclusion of other more evolved samples such as 15668, would probably cause the model to exceed the criteria for a single flow.

There is a difficulty in distinguishing the effects of short-range unmixing and sampling problems; indeed they are conceptually the same thing except that short-range unmixing includes relative movement of phases (especially early-crystallized ones and residual liquids). Both cases are a question of scale: the sample size relative to the unmixing package size. The present data base, especially where either multiple analyses are combined or only one analysis on a small sample is available, does not allow the estimation of the scale of short-range unmixing, hence whether the dispersion is merely sampling error or real within-sample-size unmixing. However, the reasonable within-sample consistency of at least some comparative analyses of single rocks, such as 15672 and 15556, shows that there are differences between samples (unless all splits were taken from a very restricted portion), which means either different flow compositions, traditional fractionation processes (e.g., crystal settling), or unmixing at scales larger than the samples (i.e., tens of centimeters). However, we suggest that a much better, unified data base for these basalts is required before their relationships can be understood adequately in terms of the petrogenetic causes of chemical variation. This improved data base would comprise major and trace element analyses on large splits, including duplicate or even replicate analyses, of many of these samples.

Picritic basalts

On all of the elemental plots (Figs. 3 and 4), 15385 and 15387 fall in a position consistent with their origin by addition of about 30% olivine to almost any member of the olivine-normative basalt suite. The smaller samples 15274,3 and 15299c show more dispersion but in most plots are similarly roughly consistent with only olivine accumulation. Considering the small sample sizes (80 and 49 mg, respectively), their analyses are likely to be grossly unrepresentative and subject to modal phase variation. The major element data, the similarity of the rare earth patterns, the mineral chemistry, and the similarity of the radiogenic age of 15385 (Ar-Ar 3.39 Ga; *Husain,* 1974) to the olivine-normative basalts suggest that these basalts are in fact olivine cumulates from an olivine-normative basalt magma. The whole-rock Rb-Sr isotopic data also plot within analytical error on the same isochron as the olivine-normative basalts. It is apparent that the subsample of 15385 from which *Rhodes and Hubbard* (1973) concluded that this basalt was unrelated to the olivine-normative basalts was unrepresentative.

15388

15388 is an unusual basalt in that it has a positive Eu anomaly. The only reasonable hypothesis for a positive Eu anomaly in lunar samples is that of plagioclase accumulation, and on all of the element plots (Figs. 3 and 4) 15388 falls in the direction of plagioclase addition from the olivine-normative mare basalts. However, simple plagioclase accumulation is not the explanation; the sample is also higher in TiO_2 and it lacks olivine completely. The lower trace element contents and lower Sm/Eu ratio suggest about 50% plagioclase accumulation, but the high FeO and MgO contents are not compatible with such simple dilution effects. Because olivine-normative mare basalt compositions have only olivine among silicate minerals on their liquidi (e.g., *Humphries et al.,* 1972), a relationship between 15388 (assuming that the analyses represent a rock rather than coarse-scale short-range unmixing) and these basalts first requires separation of olivine to produce a more fractionated liquid. Then accumulation of pyroxene, plagioclase, and Ti-oxides with a comparative scarcity of actual liquid would be required to account for the Mg/(Mg+Fe), the high Al and low Sm/Eu, and the high TiO_2 of 15388. Such an origin would be consistent with the apparent flow alignment of mineral phases in 15388, and the mineral chemistry; testing this hypothesis awaits more quantitative modeling and the acquisition of isotopic data. Such an origin would be unusual; it requires that a liquid more fractionated than any known sample produced the most obviously cumulate sample in the suite. Such an origin would directly influence our concepts about the physical volcanology of the Apollo 15 olivine-normative basalts.

OLIVINE FRACTIONATION AND SURFACE FLOWS

The data analysis above appears to be generally consistent with the concept that the compositions of the olivine-normative mare basalts are related to each other largely by olivine fractionation, as proposed by *Rhodes and Hubbard* (1973), *Chappell and Green* (1973), and others. This hypothesis is also consistent with the experimental phase equilibria data (*Humphries et al.,* 1972), which show that even for evolved compositions olivine (+ spinel) is at the liquidus at temperatures much higher than that of pyroxene. Thus pyroxene separation (although not pyroxene effects in sampling problems or short-range unmixing) cannot be expected to have influenced element abundances in the analyses. However, the petrography of the samples brings into question the nature and location of such a fractionation process. Fine-grained basalts, which are likely to reasonably represent actual liquid compositions, occur from almost the most magnesian samples to the least magnesian samples, so that the most magnesian samples do not consist solely of cumulate rocks. Furthermore, vesicular samples (of which this suite contains some spectacular examples) are also represented by a wide compositional range, and it is less likely that a very vesicular rock is a cumulate. Thus it appears that compositions among the most magnesian were extruded as liquids, not that the magnesian samples all represent cumulates from some average composition.

The range of compositions of the main olivine-normative mare basalts corresponds with about 15% olivine fractionation. However, few rocks contain anywhere near that much olivine. In phenocryst form the amount of olivine is nearly always only a few percent; much of the olivine occurs as small anhedral crystals grown with pyroxenes and poikilitically enclosed in plagioclases. This texture in itself is quite unusual. The more fractionated basalts do not seem to have a consistent pattern of less modal olivine, although the data at present is poor for quantitative comparison. These characteristics suggest to us that most of the olivine fractionation deduced to have taken place did so prior to the lavas reaching the Apollo 15 landing area, perhaps prior to extrusion. Thus the basalts represent a range of lava compositions flooding as multiple thin flows into the area, perhaps from a common parent that was fractionating in a subsurface magma chamber. Nonetheless, the picritic basalts provide evidence that substantial (30%) olivine accumulation did occur at the surface in some lavas. The total picture is thus quite complicated; the Apollo 15 olivine-normative mare basalts cannot be imagined easily as a single erupted flow that crystallized olivine after extrusion and changed its composition uniformly as it cooled.

This complicated picture is enhanced by a consideration of the distribution of the basalts around the landing site. In particular, as detailed by *Ryder* (1986), olivine-normative basalts are found on the Apennine Front, whereas the quartz-normative basalts have not so far been identified among Front samples (Stations 6, 6a, and 7). These Front samples include the clearly cumulate samples. Olivine-basalts might be present at 100 m up the Front because they originally flowed there, or ponded up to that height. Alternatively they may have been thrown there by impact, in which case they would need to be the locally dominant flow(s) at the very surface, or quartz-normative basalts should be equally abundant on the Front. The understanding of this distribution promises evaluation of the surface flow history of the basalts filling Palus Putredinis.

CONCLUSIONS

New analyses and a synthesis of previous ones for olivine-normative mare basalts from the Apollo 15 landing site provide no evidence that they comprise more than one chemical group, such as the four (olivine-phyric, microgabbro A, microgabbro B, and picritic basalts) suggested by *Binder* (1976) and *Binder et al.* (1980). The single group includes vesicular and both coarse-grained and fine-grained basalts. The general trend is that of olivine separation. For the entire suite, rather than the more limited sets investigated by most workers, the dispersion of compositions along the olivine trend is probably too great to be explained by short-range unmixing of an unfractionated flow. Thus the olivine separation is a genuine effect, presumably by crystal settling. The petrographic characteristics of the basalts suggests that much of the olivine that caused the dispersion is not present in the samples, and therefore much of it may be in subsurface magma chambers. Multiple thin flows, each of which may have had its own surface fractionation, are consistent with many of the petrographic features of the samples.

There is considerable dispersion among analyses in our synthesised data base over and above that produced by olivine separation. The reasons could be analytical, from sampling problems, from the combining of analyses from different splits subject to different sampling problems, or from petrogenetic processes other than simple fractionation. One such process is short-range unmixing. The present data base, which includes few analyses for both major and trace elements on large subsamples of any of the basalts, is inadequate (despite covering more than 50 samples) to distinguish among these possibilities or to assess the scale at which short-range unmixing took place, if it did at all. The understanding of the nature and causes of dispersion among this suite of basalts requires a thorough reanalysis of the basalts for both major and trace elements on representative subsamples of many members of the Apollo 15 olivine-normative mare basalt suite.

Acknowledgments. Drs. C. Goodrich and G.J. Taylor provided useful reviews of the manuscript that considerably improved the clarity of the presentation. We appreciate the support of personnel in the JSC INAA laboratory, the JSC Experimental Petrology laboratory, and the Dalhousie University microprobe laboratory. This work was partly conducted under a NASA grant to the senior author and is Lunar and Planetary Institute Contribution #637. The Lunar and Planetary Insitute is operated by the Universities Space Research Association under Contract No. NASW-4066 with the National Aeronautics and Space Administration.

REFERENCES

Binder A. B. (1976) On the compositions and characteristics of the mare basalt magmas and their source regions. *The Moon, 16,* 115-150.

Binder A. B., Lange M. A., Brandt H.-J., and Kahler S. (1980) Mare basalt units and the compositions of their magmas. *Moon and Planets, 23,* 445-481.

Brown R. W. (1977) A sample fusion technique for whole rock analysis with the electron microprobe. *Geochim. Cosmochim. Acta, 41,* 435-438.

Chappell B. W. and Green D. H. (1973) Chemical compositions and petrogenetic relationships in Apollo 15 mare basalts. *Earth Planet. Sci. Lett., 18,* 237-246.

Christian R. P., Annell C. S., Carron M. K., Cuttitta F., Dwornik E. J., Ligon D. T., Jr., and Rose H. J., Jr. (1972) Chemical composition of some Apollo 15 igneous rocks (abstract). In *The Apollo 15 Lunar Samples,* pp. 206-209. The Lunar Science Institute, Houston.

Dowty E., Prinz M., and Keil K. (1973) Composition, mineralogy, and petrology of 28 mare basalts from Apollo 15 rake samples. *Proc. Lunar Sci. Conf. 4th,* pp. 423-444.

Haskin L. A., Jacobs J. W., Brannon J. C., and Haskin M. A. (1977) Compositional dispersions in lunar and terrestrial basalts. *Proc. Lunar Sci. Conf. 8th,* pp. 1731-1750.

Helmke P. A., Blanchard D. P., Haskin L. A., Telander K., Weiss C., and Jacobs J. W. (1973) Major and trace elements in igneous rocks from Apollo 15. *The Moon, 8,* 129-148.

Humphries D. J., Biggar G. M., and O'Hara M. J. (1972) Phase equilibria and origin of Apollo 15 basalts (abstract). In *The Apollo 15 Lunar Samples,* 103-107. The Lunar Science Institute, Houston.

Husain L. (1974) ^{40}Ar-^{39}Ar chronology and cosmic ray exposure ages of the Apollo 15 samples. *J. Geophys. Res., 79,* 2588-2606.

Korotev R. L. (1986) Chemical components of the Apollo 15 regolith. In *Workshop on the Geology and Petrology of the Apollo 15 Landing Site* (P. D. Spudis and G. Ryder, eds.), pp. 75-79. LPI Tech. Rept. 86-03, Lunar and Planetary Institute, Houston. Korotev R. L. (1987) Mixing levels, the Apennine Front soil component, and compositional trends in thfe Apollo 15 soils. *Proc. Lunar Planet. Sci. Conf. 17th,* in *J. Geophys. Res., 92,* E411-E431.

Laul J. C. and Schmitt R. A. (1973) Chemical composition of Apollo 15, 16, and 17 samples. *Proc. Lunar Sci. Conf. 4th,* pp. 1349-1367.

Lindstrom M. M. and Haskin L. A. (1978) Causes of compositional variations within mare basalt suites. *Proc. Lunar Planet. Sci. Conf. 9th,* pp. 465-486.

Ma M.-S., Murali A. V., and Schmitt R. A. (1976) Chemical constraints for mare basalt petrogenesis. *Proc. Lunar Sci. Conf. 7th,* pp. 1673-1695.

Ma M.-S., Schmitt R. A., Warner R. D., Taylor G. J., and Keil K. (1978) Genesis of Apollo 15 olivine-normative mare basalts: trace element correlations. *Proc. Lunar Planet. Sci. Conf. 9th,* pp. 523-533.

Mason B., Jarosewich E., Melson W. G., and Thompson G. (1972) Mineralogy, petrology, and chemical composition of lunar samples 15085, 15256, 15271, 15471, 15474, 15476, 15535, 15555, and 15556. *Proc. Lunar Sci. Conf. 3rd,* pp. 785-796.

Nyquist L. E. (1977) Lunar Rb-Sr chronology. *Phys. Chem. Earth, 10,* pp. 103-142.

Powell B. N. (1972) Apollo 15 coarse fines (4-10mm): Sample classification, description, and inventory. *NASA Manned Spacecraft Center MSC 03228,* NASA Johnson Space Center, Houston, 91 pp.

Rhodes M. J. and Hubbard N. J. (1973) Chemistry, classification, and petrogenesis of Apollo 15 mare basalts. *Proc. Lunar Sci. Conf. 4th,* pp. 1127-1148.

Ryder G. (1985) Catalog of Apollo 15 rocks, Parts 1, 2, and 3. *Curatorial Branch Publication 72, JSC 20787,* NASA Johnson Space Center, Houston, 1296 pp.

Ryder G. (1986) Apollo 15 olivine-normative mare basalts on the Apennine Front: Observations and possible interpretations (abstract). In *Lunar and Planetary Science XVII,* pp. 740-741.Lunar and Planetary Institute, Houston.

Warren P. H., Jerde E. A., and Kallemeyn G. W. (1987) Pristine moon rocks: a "large" felsite and a metal-rich ferroan anorthosite. *Proc. Lunar Planet. Sci. Conf. 17th,* in *J. Geophys. Res., 92,* E303-E313.

Petrology and Provenance of Apollo 15 Drive Tube 15007/8

Abhijit Basu
Department of Geology, Indiana University, Bloomington, IN 47405

David S. McKay
NASA Johnson Space Center, Houston, TX 77058

Tammie Gerke
Department of Geology, Indiana University, Bloomington, IN 47405

The petrology of submillimeter fractions of soils from 13 levels of Apollo 15 drive tube 15007/8, which was selected for consortium studies, was investigated in detail. We have performed grain size analysis, modal analysis of the 90–150 μm sieve fraction, and electron probe microanalysis of monomineralic fragments of olivine, pyroxene, and plagioclase from these soils. Our data show that there is no petrologic layering in this drive tube. Variations in petrologic components with depth do not correspond to units defined on the basis of color, ferromagnetic resonance (FMR), and other maturity indicators. The origin of the soils of drive tube 15007/8 is dominated by intricate mixing processes that overwhelmed any *in situ* reworking/gardening process; mixing was further complicated by downslope grain flow of soil components. We therefore do not expect any stratigraphic correlation of this core with others taken from the Apollo 15 site. Positive correlation between the abundances of mare basalt fragments and KREEP basalt fragments, and the lack of correlation between KREEP basalt and ANT fragments from the Apennine Front, indicate that the Front is not the source of KREEP basalts; rather, mare basalt and KREEP basalt sources are closely related and may even be interbedded with each other. Major and minor element chemistry of monomineralic olivine, pyroxene, and plagioclase indicate contributions from KREEP basalts, mare basalts, and highland rocks to the core soils. Some olivine and pyroxene fragments with mare basalt affinity are optically and chemically unzoned and may have been derived from unusual source rocks.

INTRODUCTION

Three cores from the Apollo 15 site have been opened to date: two (15001–15006; 15010/11) are from the mare embayment and one (15007/8) is from the Front (Fig. 1). A fourth, unopened core (15009) is also from the Front. The core sites are on or near different geomorphic features and the thickness of the regolith varies from site to site: Station 9a has a very thin regolith (~ 1 m), whereas the Front may have much thicker regolith. A complete and extensive study of these cores has the potential of finding correlations between soil horizons across the Apollo 15 site and finding if the lateral petrologic variability of the local regolith is also reflected vertically. While the complete site study must await the opening and examination of core 15009, much information on lateral and vertical variation already exists from the three opened cores.

Core 15007/8 is the only core taken close to the margin between a lava rille and a highland front on the moon; thus the petrology of this core material is likely to be a product of mixing two materials of different compositions and ages. This study provides an opportunity of comparing and contrasting the petrology of a nearly pure mare regolith (15010/11) and that of a ray-like hump (15001–15006) with the petrology of this core. *Bogard et al.* (1982) measured the ferromagnetic resonance (FMR) and abundances and isotopic ratios of noble gases in samples from various depths of this core. Their data show that this ~57-cm-long core is made up of four depositional units with boundaries at

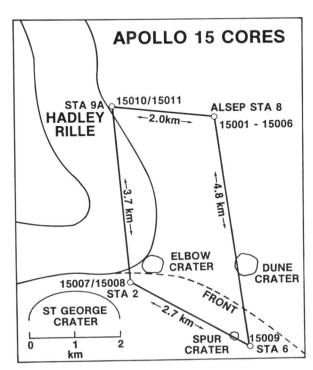

Fig. 1. Sketch map of the Apollo 15 landing site showing locations of drill core sites (from *McKay et al.*, 1980, Fig. 1).

TABLE 1. Submillimeter grain size analysis results in grams (gm) and percent sieve fraction; mean size (M_z) is for subcentimeter fraction.

DSM # (Depth cm)		500-1000	250-500	150-250	90-150	75-90	45-75	20-45	<20 µm	% yield	M_z µm
260	gm	.02476	.03720	.04136	.05391	.01945	.06169	.07500	.11600	95.95	94
(0.7)	%	5.77	8.66	9.63	12.56	4.53	14.37	17.47	27.02		
267	gm	.01468	.03373	.03624	.05100	.01951	.05890	.08383	.11771	94.15	88
(5.3)	%	3.53	8.12	8.72	12.27	4.69	14.17	20.17	28.32		
274	gm	.02292	.03100	.03719	.05193	.02293	.06189	.08765	.12082	96.60	60
(9.3)	%	5.25	7.10	8.52	11.90	5.26	14.18	20.09	27.69		
281	gm	.02005	.02895	.03700	.04953	.02017	.06095	.08060	.11282	95.90	122
(13.7)	%	4.89	7.06	9.02	12.08	4.92	14.86	19.66	27.51		
288	gm	.01413	.03007	.03672	.05439	.02052	.05959	.08647	.11050	94.12	54
(18.3)	%	3.43	7.29	8.90	13.19	4.98	14.45	20.97	26.80		
295	gm	.01712	.02778	.03559	.05118	.02334	.06688	.08600	.11287	94.61	54
(22.3)	%	4.07	6.60	8.46	12.16	5.55	15.90	20.44	26.83		
406	gm	.02508	.03469	.03798	.04964	.02033	.06042	.08108	.11649	95.01	58
(25.3)	%	5.89	8.15	8.92	11.66	4.78	14.19	19.05	27.36		
407	gm	.01900	.03036	.03517	.04961	.01991	.06492	.08445	.12494	95.42	53
(31.9)	%	4.44	7.09	8.21	11.58	4.65	15.16	19.71	29.17		
408	gm	.01793	.03423	.03894	.05505	.02105	.06018	.08919	.11398	95.62	54
(41.9)	%	4.16	7.95	9.04	12.79	4.89	13.98	20.72	26.47		
409	gm	.02266	.03523	.03976	.05282	.01996	.06084	.08882	.11760	97.35	113
(48.3)	%	5.18	8.05	9.08	12.07	4.56	13.90	20.29	26.87		
410	gm	.03133	.04251	.04278	.05444	.03864	.06034	.07685	.10231	99.85	104
(50.3)	%	6.97	9.46	9.52	12.12	8.60	13.43	17.11	22.78		
411	gm	.04200	.04184	.04051	.05091	.01893	.05843	.06968	.11087	96.72	102
(54.3)	%	9.70	9.66	9.35	11.75	4.37	13.49	16.90	25.60		
412	gm	.02876	.03983	.03845	.05117	.01959	.05971	.06885	.12619	97.25	79
(55.9)	%	6.65	9.21	8.89	11.83	4.53	13.80	15.92	29.17		

approximate depths of 18, 49, and 55 cm. These contacts correlate fairly well with those observed by *Nagle* (1981) during initial dissection of the core. There is a significant change in the cosmogenic ^{131}Xe/^{126}Xe ratio across the 18-cm contact; other contacts, however, do not exhibit any significant shift in the abundances of noble gases or their isotopes. The other contacts are based on substantial changes in ferromagnetic resonance properties (I_s/FeO), especially at 49 cm and at 55 cm. This interval of the core is immature. Other intervals of the core are submature to mature.

We have characterized the core petrologically and have measured the grain-size distribution and the modal abundances of mineral, rock, and glassy fragments, and have also determined the mineral chemistries of monomineralic fragments throughout the core. The ultimate purpose of our investigation is to better understand the petrology and the evolution of Apollo 15 regolith, especially at Station 2. This involved a search for layers that can be identified on the basis of mineralogy and petrology. We also related petrologic variations to layers defined by maturity and irradiation properties. A detailed provenance study of the core was carried out to test the interpretation of *Bogard et al.* (1982) that "such cores are less likely to contain old or exotic soil layers." Regolith breccia 15205 was collected from Station 2 but is not similar to the surface soils (*Dymek et al.*, 1974; *Simon et al.*, 1986). Does the petrology of regolith breccia 15205 match that of the core soils better, or is 15205 simply atypical and/or really exotic to both surface soils and core soils? Finally, we would like to relate the petrology of this core to that of the regolith at the Apollo 15 site as a whole, which has been well characterized (*Heiken et al.*, 1976; *Basu and Bower*, 1977; *Basu and McKay*, 1979; *Grüffiths et al.*, 1981; *McKay et al.*, 1980; *Simon et al.*, 1986; etc.).

METHODS AND RESULTS

Samples of core soils taken at various depths were sieved in the Lunar Sample Curatorial Facility and about 0.5 g of the submillimeter fraction of 13 soils earmarked for consortium studies were allocated to us. Each sample was wet sieved in freon into eight size fractions (1000-500 µm; 500-250 µm; 250-150 µm; 150-90 µm; 90-75 µm; 75-45 µm; 45-20 µm; and <20 µm) at the Johnson Space Center following a standard procedure. Yield was mostly about 95.5% but varied from 94.12% to 99.85%. Data from the Curator's initial sieving were used for the 1 mm to 1 cm fraction. Results and grain size distribution (*Folk and Ward*, 1957) are given in Table 1 and

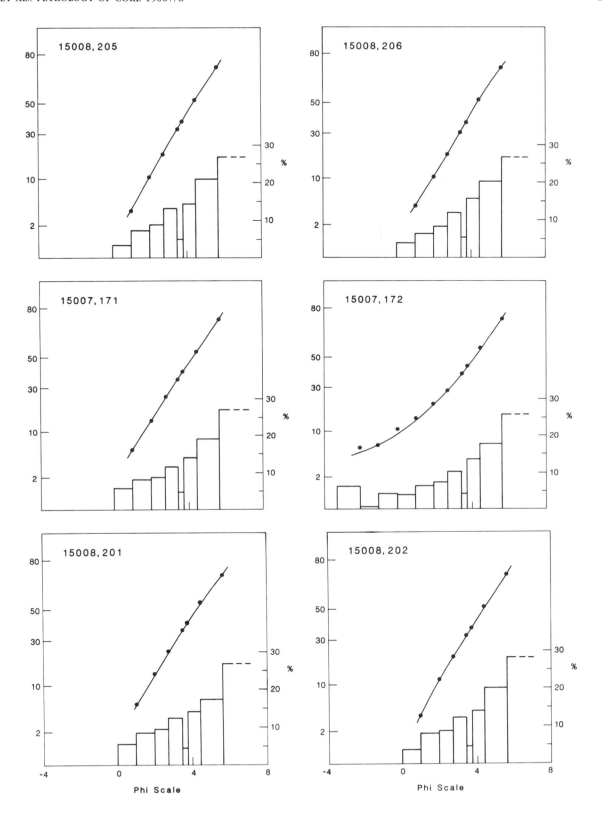

Fig. 2. Histograms and cumulative curves showing the grain size distribution of 13 drill core soils.

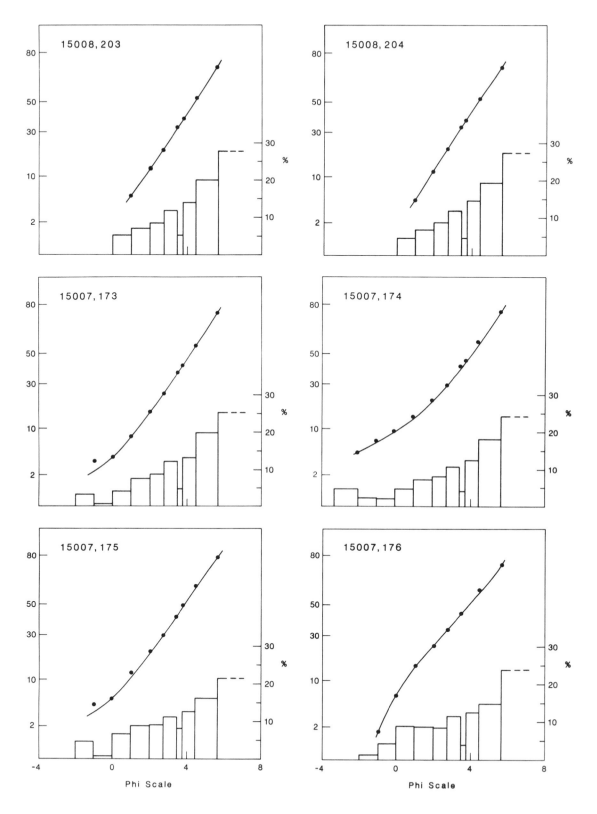

Fig. 2. (continued)

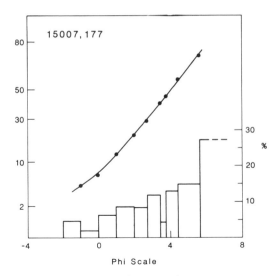

Fig. 2. (continued)

Fig. 2. Note that the sieve size intervals are not identical either on the mm or the φ scale. Consequently, these histograms should not be used for genetic interpretation. Derived histograms having equal size intervals are more suitable for interpretation and comparisons, and these histograms can be generated from the cumulative curves. The cumulative curves (Fig. 2) are generally smooth and linear on probability plots; a few show very minor inflections that indicate departure from unimodal gaussian size distribution.

An aliquot of each size fraction of each of the 13 soils was used to prepare polished grain mounts for petrographic study. Detailed modal analysis of 4179 particles in terms of 33 particle types in the 90–150 μm fraction was performed using both transmitted and reflected light and at a microscope magnification of 500×. Whereas no single grain size fraction may be representative of all lunar soils, the 90–150 μm grain size fraction is probably optimum for detailed regolith petrology. Grain size fractions around 60 μm, the approximate average mean grain size of lunar soils (*Heiken*, 1975), are too small to retain any identifiable rock fragment. Grain size fractions that are much coarser, e.g. >150 μm, are depleted in agglutinate and monomineralic fragments and are not suitable for soil maturity interpretations; and historically, most soil petrology and other studies have concentrated on the 90–150 μm fraction (*Morris et al.*, 1983). Any particle with any single mineral occupying >90% of its surface area was considered to be monomineralic. Anorthosites, which can be monomineralic, were counted as rock fragments. Other rock types include KREEP basalts and mare basalts with or without olivine. The primary breccia categories are regolith matrix (= vitric) and crystalline matrix breccias. These two groups were further subdivided on the basis of mineralogy and texture. Examples of these particle types are given in Fig. 3 and may be compared with those of *Papike et al.* (1982), *Simon et al.* (1986), *Ryder* (1985), *Basu and Bower* (1977), etc. Agglutinates were recognized solely on the basis of the criteria set up by *McKay et al.* (1972), *Heiken* (1975), and *McKay and Basu* (1983).

Glass, if free of clasts and crystals, were classified on the basis of color. Results are given in Table 2, which also provides a list of all particle categories.

Electron probe microanalysis was performed with an ETEC Autoprobe at 15 kV with a regulated beam current of approximately 20na on brass. Appropriate mineral or glass standards were used to perform six-element (Na, Al, Si, K, Ca, and Fe) analysis of 425 plagioclase grains, and eight-element (Mg, Al, Si, Ca, Ti, Cr, Mn, and Fe) analysis of 460 pyroxene and 124 olivine grains in the 90–150 μm size fraction. Two background measurements on either side of the peak position of each element were read for each analysis, but no detailed modeling of the continuum was undertaken. Representative compositions of different minerals are given in Table 3 and all of the data are presented in Figs. 4–6.

DISCUSSION

Depositional Units and Petrologic Layering

Petrologic versus irradiation properties. Different depositional units of the lunar regolith may show petrologic layering in a core. In fact, depositional units are defined initially on the basis of macroscopic observation during dissection in the curator's laboratory. *Nagle* (1981) defined four depositional units for the core 15007/8. These units were also confirmed with FMR and noble gas studies (*Bogard et al.*, 1982). Because micrometeoritic bombardment with its ensuing comminution and agglutination accompanies surface exposure and maturity of the regolith, the mean grain size, agglutinate abundance, and the I_s/FeO ratios of all the samples are plotted for comparison (Fig. 7). There is a good correlation between the agglutinate content and the I_s/FeO ratio in samples below ~18 cm. This is expected because agglutinates are enriched in single domain iron that contributes to FMR. Correlation between mean grain size, which should decrease with maturity, and either agglutinate content or I_s/FeO (cf. *Morris*, 1978; *McKay et al.*, 1977) is poor in this section of the core. However, there appears to be a good negative correlation between mean grain size and agglutinate content in the upper ~18 cm of the core. On the other hand, no systematic relationship exists between I_s/FeO ratio and mean grain size in samples from this part of the core. Extremely puzzling is the soil sample 15008,204 (DSM # 281) at a depth of ~13.5 cm, which has a mean grain size of 122 μm (coarsest of all core soils) and 30.4% agglutinate, both indicating that the soil is rather immature, and yet has an I_s/FeO ratio of 69 (highest value in this core), indicating that it is very mature. This sample also has a very high abundance of monomineralic plagioclase, which indicates that it is an immature soil, consistent with grain size and agglutinate data. Possibly a sampling problem contributed to this lack of correlation; the FMR data were not determined on the same aliquot used for our detailed studies, although they were from the same core depths. Another possibility is that this lack of correlation may represent mixing of two or more soils of very disparate maturity. One possible scenario would be where intermittent or episodic soil creep down the Apennine Front is interspersed between relatively

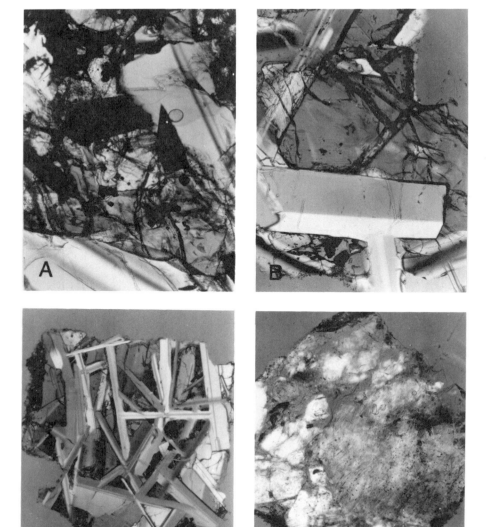

Fig. 3. Photomicrographs of representative rock types from soils in drill core 15007/8; all are taken with obliquely crossed polarized light. Each photomicrograph is 0.66 mm × 0.44 mm. (**a**) olivine basalt; (**b**) olivine-free basalt; (**c**) KREEP basalt; (**d**) anorthosite; (**e**) crystalline matrix breccia; (**f**) regolith breccia; (**g**) agglutinate; (**h**) green glass clod.

quiescent soil maturation periods. Thus we envisage an inhomogeneous mixing of soil components of various maturity levels down the Front. This scenario will likely cause destruction of petrologic layering at a soil component scale. Why then do we see four units of different color and texture at a macroscopic scale? We suggest that this reflects the relative rates at which soil creep and soil maturation operated on the Front. The soils are darker if maturation and irradiation processes outpaced creep. Indirect support for this scenario is provided by *Bogard et al.* (1982), who find "no compelling evidence for a gardening zone" in this core, and find it plausible to have "fast downslope movement of soil away from the core site." We would like to emphasize here that downslope movement of soil does not necessarily imply *en masse* movement without any internal shear. It may be that grain flow and not debris flow is the dominant process.

Variation of petrologic components with depth. Vertical profiles of other petrologic entities show some interesting covariance (Fig. 8). The most obvious negative correlation is between the abundances of highland crystalline rocks (ANT: anorthosite, norite, troctolite) and mare basalts. The source of ANT material is clearly the Front, and the local sources of mare material are the areas near the rille. When downslope deposition predominates, ANT material should increase and mare material should decrease, and *vice versa* when impact deposition from nearby mare areas predominate. More significantly, the abundance of mare basalt and KREEP basalt show a fairly good correlation throughout the core. This

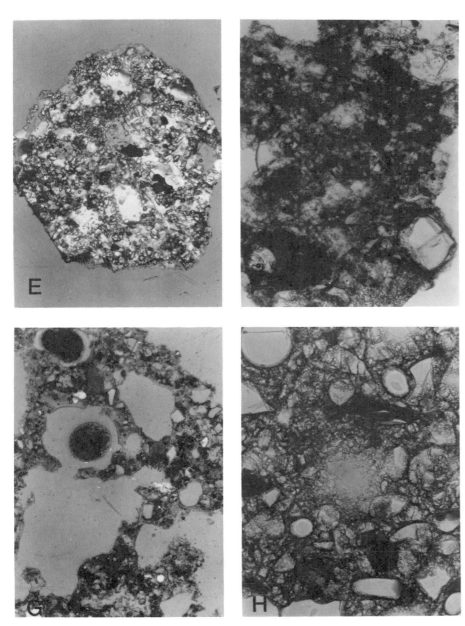

Fig. 3. (continued)

suggests that mare basalts and KREEP basalts at this Station had the same provenance. We believe that the Apennine Bench Formation below the mare basalt flows in this region is relatively rich in KREEP basalts and supplied the KREEP basalts found in the core. Mare basalts above and intercalated with the Apennine Bench formation were the source of the mare basalt fragments in the core (*Spudis*, 1978; *Basu*, 1986). The fact that KREEP basalts are not correlated with ANT is evidence that these KREEP material did not come from the Front. In the lower 40 cm of the core ANT and crystalline breccias show a good positive correlation; this is also expected because both are derived from highlands. However, in the top 20 cm ANT and crystalline breccias show a definite negative correlation; in this upper part of the core crystalline breccias and mare basalts show a good correlation. It is likely that an additional source of crystalline breccias from the mare side of Station 2 supplied a large fraction of these fragments. A regolith breccia rich in brecciated highland material ejected from below the thin basaltic cover of Apollo 15 site may be such a source. Note, however, that abundances of regolith breccias correlate with those of ANT and crystalline breccias only in the lower part of the core. The abundance of glass increases about threefold at a depth between 50 and 55 cm; most of this increase is due to an increase in the abundance of crystal- and clast-free green glass spherules (Table 2). This increase in the green glass population is probably due to the breakdown of a green glass clod such as those found in other parts of the Front. *Nagle* (1981) noted the presence of green

TABLE 2. Modal analysis data of 13 soils from Apollo 15 core 15008/7 (90-150 μm)

	15008						15007							Avg
Split #	201	202	203	204	205	206	171	172	173	174	175	176	177	
DSM #	260	267	274	281	288	295	406	407	408	409	410	411	412	
Depth (cm)	0.7	5.3	9.3	13.7	18.3	22.3	25.3	31.9	41.9	48.3	50.3	54.3	55.9	
Monomineralic	*33.2*	*34.2*	*31.3*	*40.7*	*27.9*	*26.5*	*38.5*	*36.3*	*36.8*	*39.4*	*30.5*	*29.5*	*41.7*	*34.6*
Plagioclase	16.9	16.2	12.5	21.1	12.1	10.3	18.5	15.4	18.4	18.8	18.1	15.4	23.2	16.7
Pyroxene	13.9	17.4	17.2	17.4	14.6	12.8	18.8	19.4	16.5	17.5	11.5	13.5	16.4	15.9
Olivine	2.4	0.6	1.3	1.6	1.2	2.2	0.9	0.9	1.9	2.5	0.3	0.6	1.5	1.4
Opaques			(0.3)	(0.6)		(0.9)		(0.6)		(0.6)	(0.6)		(0.6)	(0.3)
Oxides			0.3	0.6		0.9		0.3		0.6	0.6		0.3	0.3
Metal, sulphide								0.3					0.3	tr
Silica phase					0.3	0.3								tr
Crystalline lithics	*7.4*	*6.1*	*8.0*	*7.8*	*8.0*	*10.2*	*6.3*	*8.0*	*6.6*	*8.5*	*7.0*	*8.7*	*9.4*	*7.8*
ANT suite	(0.9)	(0.9)	(1.2)	(1.8)	(2.5)	(0.6)	(1.9)	(0.9)	(1.3)	(1.2)	(0.9)	(0.6)	(1.5)	(1.2)
Anorthosite	0.6	0.6	0.3	0.9	1.9	0.6	1.3	0.3	0.6	0.3	0.3	0.6	0.6	
Noritic	0.3	0.3	0.9	0.9	0.6		0.6	0.6	1.3	0.6	0.6	0.3	0.9	0.6
Mare basalt	(5.0)	(4.0)	(4.6)	(2.8)	(2.5)	(5.9)	(1.6)	(3.4)	(1.8)	(2.7)	(1.8)	(2.1)	(1.5)	(3.1)
Olivine bearing	2.3	1.2	0.3			0.6		0.6	0.3	1.2	0.6	0.9	0.6	0.7
Olivine-free	(1.2)	(1.2)	(1.5)	(0.9)	(0.6)	(1.9)	(1.3)	(2.2)	(1.2)	(0.6)	(0.3)	(1.2)	(0.9)	(1.2)
Microgabbroic	0.3	0.3	0.3							0.3				0.1
Porphyritic, etc.	0.6	0.6	0.6	0.9	0.3	1.6	1.3	1.6	0.6	0.3	0.3	0.3	0.6	0.7
Ophitic, etc.	0.3	0.3	0.6		0.3	0.3		0.3	0.3	0.3		0.6	0.3	0.3
Intersertal/granular									0.3					tr
Other	1.5	1.6	2.8	1.9	1.9	3.4	0.3	0.6	0.3	0.9	0.9		1.2	1.3
Non-mare basalt	(0.6)	(0.9)	(1.9)	(1.6)	(1.5)	(2.5)	(1.9)	(3.4)	(2.2)	(3.4)	(3.7)	(3.8)	(3.7)	(2.4)
Feldspathic										—				
KREEP	0.6	0.9	1.9	1.6	1.5	2.5	1.9	3.4	2.2	3.4	3.4	3.8	3.7	2.4
Plagioclase-phyric					0.3									tr
Indeterminate	0.6	0.3	0.3	1.6	1.5	1.2	0.9	0.3	1.3	1.2	0.6	2.2	1.5	1.0
Breccias	*16.9*	*13.3*	*11.3*	*9.9*	*12.1*	*9.6*	*13.8*	*12.2*	*11.1*	*12.9*	*13.8*	*9.1*	*15.8*	*12.4*
Regolith	(12.7)	(9.9)	(5.6)	(6.5)	(9.0)	(9.0)	(8.2)	(7.2)	(6.3)	(6.8)	(7.2)	(5.0)	(10.3)	(8.0)
Dark matrix	11.8	9.0	4.7	5.6	8.4	7.8	6.9	6.9	4.4	6.2	5.3	4.4	8.4	6.9
Light matrix	0.9	0.9	0.9	0.9	0.6	1.2	1.3	0.3	1.9	0.6	1.9	0.6	1.9	1.1
Crystalline	(4.2)	(3.4)	(5.7)	(3.4)	(3.1)	(0.6)	(5.6)	(5.0)	(4.8)	(6.1)	(6.6)	(4.1)	(5.5)	(4.5)
Poikilitic	1.2	1.2	1.3		0.3		0.3	1.6	2.5	0.6	1.9	0.6	0.9	1.0
Melt matrix	0.9	0.6	0.6	0.3	0.6		0.9	0.3	1.0	1.8	2.8	1.6	0.9	
Other	2.1	1.6	3.8	3.1	2.2	0.6	4.4	3.1	1.3	3.7	1.9	1.9	4.6	2.6
Agglutinate	*30.5*	*32.4*	*36.9*	*30.4*	*42.1*	*43.9*	*27.0*	*27.9*	*32.7*	*27.7*	*19.3*	*18.5*	*19.5*	*29.9*
Glass	12.0	13.8	12.5	10.9	9.9	9.5	14.2	15.1	12.7	11.4	28.6	33.8	12.7	15.2
Ropy/Clast-laden	2.7	4.0	3.8	1.6	2.8	3.1	2.2	2.5	3.2	3.4	2.2	1.3	3.4	2.8
Quench-crystal (vitrophyres)	(1.5)	(0.6)		(0.3)	(0.3)	(0.6)	(2.9)	(0.9)	(0.9)	(0.6)	(1.5)	(1.9)	(0.0)	(0.9)
Green glass					0.6	0.3	1.6		0.6	0.3	0.9	1.3		0.4
Other	0.9	0.6		0.3	0.3	0.3	1.3	0.9	0.3	0.3	0.6	0.6		0.5
Devitrified (cryptocrystalline)	1.2	1.2	2.5	0.6	1.2	0.6	0.9	0.3	1.0	1.6	0.9	2.5	1.1	
Crystal or clast free	(6.6)	(7.1)	(6.2)	(8.4)	(5.6)	(5.2)	(8.2)	(11.4)	(7.6)	(7.4)	(23.3)	(29.7)	(6.8)	(10.3)
Green	3.6	5.0	3.1	5.3	3.4	3.7	5.0	6.6	5.4	3.7	20.9	26.0	4.6	7.4
Yellow	2.1	1.2	1.9	1.9	1.9	0.3	1.9	1.6	0.3	2.2	1.2	0.6	0.3	1.3
Gray, colorless	0.9	0.6	0.3	0.9	0.3	0.3	1.3	1.9	1.3	1.5	0.6	3.1	1.9	1.1
Brown, black, etc.	0.3	0.9	0.3		0.9		1.3	0.6		0.6			0.4	
Miscellaneous		*0.6*		*0.3*			*0.3*	*0.3*			*0.6*	*0.6*	*0.6*	*0.3*
No. of particles	331	321	320	322	323	321	319	319	315	325	321	319	323	4179

glass clods at this level of the core during initial dissection. Green glass clods are also seen in the 500-1000 μm fraction of core soils at this depth (Fig. 3h). Therefore, the sudden and remarkable increase in green glass population at this depth does not signify any layer; rather the presence and disintegration of a random clod is the cause of the observed enrichment. It is this large increase in glass population that brings about an apparent decrease in the abundances of many other soil components at a depth of about 55 cm (Fig. 8). Plagioclase and pyroxene are the most abundant monomineralic particles

TABLE 3. Selected microprobe analysis of monomineralic pyroxene, olivine, and plagioclase in core 15007/8 (blank entry denotes not analyzed).

DSM #	Grain #	Na$_2$O	MgO	Al$_2$O$_3$	SiO$_2$	K$_2$O	CaO	TiO$_2$	Cr$_2$O$_3$	MnO	FeO	Total
					Pyroxene							
406	118		17.0	2.0	51.4		8.6	0.76	0.60	0.28	19.34	99.95
406	71		28.9	1.0	53.8		1.3	0.42	0.64	0.34	13.5	99.99
406	14		6.7	1.6	47.8		0.44	1.0	0.18	0.45	29.9	100.08
407	173		18.0	2.3	50.0		11.04	0.80	0.78	0.25	16.5	99.73
407	71		13.7	2.0	49.1		14.5	1.8	0.61	0.34	17.9	99.91
407	49		29.7	2.7	53.5		1.7	0.62	0.84	0.23	10.7	99.95
408	201		21.7	1.7	51.0		5.4	0.61	0.69	0.30	18.8	100.04
408	43		12.3	5.2	46.4		12.9	1.5	0.74	0.34	20.5	99.73
409	54		28.9	1.4	53.9		1.3	0.34	0.76	0.16	13.15	99.88
410	1		12.6	2.3	49.3		13.6	1.4	0.15	0.41	20.0	99.73
274	272		15.9	1.8	49.4		10.2	1.0	0.60	0.41	20.5	99.75
260	263		30.3	2.9	54.3		1.1	0.43	0.92	0.20	9.3	99.88
					Olivine							
406	17		49.4	0.09	41.0		0.07	0.00	0.00	0.16	9.5	100.21
407	10		46.7	0.05	38.9		0.02	0.04	0.08	0.11	13.5	99.41
408	251		46.0	0.04	38.4		0.03	0.07	0.00	0.20	16.0	100.68
409	143		28.8	0.05	34.9		0.30	0.04	0.09	0.30	37.3	101.75
409	89		29.5	0.00	35.5		0.34	0.05	0.08	0.36	33.9	99.69
409	219		49.6	0.02	39.8		0.05	0.05	0.00	0.13	9.6	99.27
412	34		43.7	0.00	38.3		0.09	0.04	0.00	0.12	17.0	99.30
412	199		30.0	0.02	35.7		0.32	0.00	0.15	0.30	33.1	99.55
412	69		44.8	0.07	39.6		0.09	0.07	0.00	0.15	14.9	99.73
267	100		45.8	0.10	38.9		0.06	0.11	0.00	0.23	15.2	100.46
274	174		31.3	0.03	36.2		0.26	0.00	0.12	0.35	32.1	100.23
274	226		30.8	0.07	36.3		0.20	0.10	0.07	0.28	32.1	99.91
					Plagioclase							
409	17	0.48		35.7	44.6	0.7	19.0				0.00	99.86
409	38	0.39		35.3	44.7	0.12	19.2				0.02	99.69
409	66	0.20		35.7	43.3	0.06	20.2				0.23	99.70
408	293	1.41		33.0	47.9	0.15	16.9				0.23	99.59
408	198	0.77		33.4	47.0	0.04	18.2				0.68	100.15
407	501	0.98		35.1	45.4	0.00	18.0				0.38	99.84
407	276	2.6		31.4	50.0	0.35	14.8				0.69	99.94
411	291	1.70		31.9	48.8	0.20	17.3				0.27	100.08
267	111	0.34		35.9	43.8	0.05	19.7				0.04	99.78
274	506	0.30		36.1	43.5	0.07	19.8				0.02	99.86
281	271	0.30		36.1	44.0	0.00	19.3				0.14	99.92
295	160	0.51		35.5	44.3	0.13	19.6				0.04	100.07

in this core. In the top 50 cm plagioclase and pyroxene show a somewhat poor negative correlation, but in the lowermost 8 cm there is a strong covariance between plagioclase and pyroxene. Abundances of both plagioclase and pyroxene decrease dramatically between 50 to 55 cm. If the correlation at the bottom of the core is assigned to the sudden increase in glass abundance, then the relatively poor negative correlation may reflect major contributions of plagioclase from ANT and those of pyroxene from mare basalts.

It is clear that several soil components show some covariance. However, the correlations between components do not always match up in similar depth zones in the core; i.e., the correlations do not correspond to any depositional unit recognized on the basis of macroscopic observation, or on the basis of FMR and rare gas studies. Because correlations between particle types depend on discrete units having discrete petrologic provenances, there is no *a priori* reason for depositional units to necessarily coincide with breaks and changes in the modal abundance of particles. If there happens to be such a correlation, it would probably indicate a slab by slab regolith development at that locality. If there is no such correspondence, as in this core, it is likely that discrete slabs were not deposited to build up the local regolith. This is compatible with our preferred depositional scenario described above. This is also in agreement with the expectation of *Bogard et al.* (1982) that mixing has been the dominant process of regolith development at this site and the core is "less likely to contain mineralogically and texturally distinct layers."

We also plotted the chemical variations (major and minor elements, and ratios thereof) of monomineralic fragments of pyroxene, olivine, and plagioclase separately for all 13 soils. None of our attempts provided any clue to any systematic

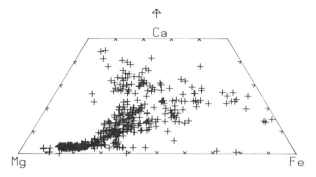

Fig. 4. Compositions of pyroxene from monomineralic fragments from the core soils plotted in the pyroxene quadrilateral. Each point represents one analysis per grain. Note the strong influence of a KREEP basalt trend.

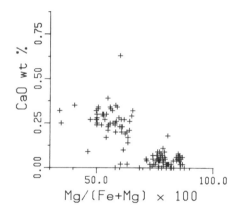

Fig. 5. Plot of CaO content versus forsterite content [Mg/(Mg + Fe) × 100] of monomineralic olivine from the core soils. Each point represents one analysis per grain. Note two distinct clusters. The relatively iron rich cluster is also slightly enriched in CaO.

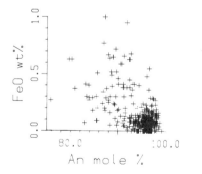

Fig. 6. Plots of FeO content versus An content of monomineralic plagioclase from the core soils. Each point represents one analysis per grain. Plagioclases with high concentrations of Fe indicate that they are derived from mare basalts. Note that all plagioclases have compositions >An_{82}.

change with core depth in the chemistry of these three common minerals. This conforms to our observation above that this core does not show any petrologic layering.

The covariance of certain soil components does suggest that different parent material played a role in the production and supply of the core regolith under examination. This material was likely supplied from the regolith that makes up the Apennine Front and from impacts into nearby mare basalts and possibly underlying KREEP basalt rich material. Downslope movement would likely contribute both Front regolith and non-Front material that had been introduced to the Front as ejecta or breccias. This mixed material would be further mixed with local ejecta at the core site. Downslope movement of the regolith accompanied by further disintegration and mixing of soil components is the likely origin of these soils. If this interpretation is correct, one cannot expect any lateral continuity of soil layers across Apollo 15 site. There would also be no reason to expect any correlation of soil units across the site.

Provenance of Soil Components

Contributions from local and exotic sources. An approximate quantitative estimate of proportions of pristine rocks that contribute to the mineralogic abundance of a regolith may be obtained from mass balance calculations of modal data. Briefly, our method involves converting the abundances of rock fragments in a regolith to their mineralogic constituents to obtain a theoretical distribution and then comparing the actual modal abundances of these minerals with the theoretically expected values. The data of this core show a similarity between actual modal abundances of monomineralic fragments of pyroxene and plagioclase and those calculated from the abundances of rock fragments. No major exotic component has to be invoked to account for the mineralogy of this core.

Mineral chemistry and source rock types. It has been shown by many workers that mineral chemistry is a good guide to decipher the petrogenesis of the principal lunar rocks; some rocks have characteristic patterns of elemental distribution in major rock-forming minerals such as pyroxene, olivine, and plagioclase (e.g., *Bence and Papike,* 1972; *Steele and Smith,* 1975; *Smith,* 1974; *Smith and Steele,* 1976; *Vaniman et al.,* 1979; etc.). Many provenance studies of lunar material are based on this premise (e.g., *Papp et al.,* 1978; *Vaniman and Papike,* 1977; *Basu and McKay,* 1979; *Simon et al.,* 1985, 1986; etc.). Proportions of Ca, Mg, and Fe in monomineralic fragments of pyroxene in the core are shown in Fig. 4. Considering that we estimate that about half of the monomineralic fragments are from KREEP basalts, it is not surprising to find a strong imprint of pyroxene zoning pattern of Apollo 15 KREEP basalts (*Basu and Bower,* 1976; *Dymek et al.,* 1974; *Ryder,* 1976) in the pyroxenes of the core. The presence of several pyroxenes with their compositions in the forbidden zone indicate contribution from mare basalts. There is a small population of magnesian bronzite and diopsidic augite, which is likely to represent pyroxenes from highland rocks that show subsolvus exsolution. In short, contributions from all three major rock

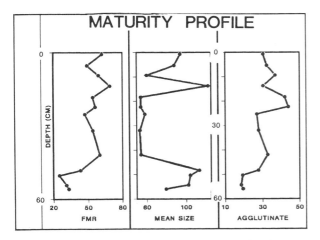

Fig. 7. A vertical profile of the core comparing the three maturity indicators, FMR (I_s/FeO), mean grain size (M_z μm), and agglutinate percent. Note that the expected covariance is not strong; at certain depths there is a negative correlation between agglutinates and both mean grain size and FMR. FMR and grain size are, by and large, positively correlated.

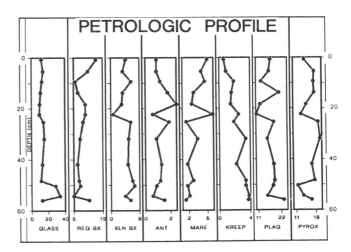

Fig. 8. A vertical profile of the core comparing the abundances of principal petrographic components. Note the general negative correlation between mare basalts and ANT, and the positive correlation between KREEP and mare basalt. See text for a detailed discussion.

types can be inferred from the pyroxene quadrilateral (Fig. 4). Whereas compositional ranges of pyroxenes from different rock types overlap in the standard quadrilateral, some clustering in the compositions of olivines is seen in a plot of CaO versus Fo content (Fig. 5). Two clusters, the first enriched in Mg and very poor in CaO and a second that is more iron- and calcium-rich, are obvious. Most of the high-Mg olivines are likely derived from highland rocks; the rest are from mare basalts because olivine is extremely rare in KREEP basalts (*Dymek*, 1986; *Basu and Bower*, 1976). It is seen that compared to the full range of compositions of olivines from Apollo 15 mare basalts (*Dowty et al.*, 1973, Fig. 6; *Papike et al.*, 1976, Fig. 36), those from the core 15007/8 are relatively restricted and are not as iron-rich. In addition, CaO contents in the olivines from this core are lower than in those from lunar mare basalts (*Papike et al.*, 1976; Fig. 37). Compared to the compositions of monomineralic olivine from Luna 24 regolith, the most pure mare regolith studied so far (>95% mare basalt detritus; Fo varies from about 80 to 0, and CaO from about 0.1 to about 1.2% with increasing Fe; *Papp et al.*, 1978; *Basu et al.*, 1978), both the range of Fo content and the abundance of CaO in our olivine samples are much lower. It has been shown that the Ca content of olivine varies with Fe in olivines, and that the Ca content itself is likely to be a function of cooling rate (*Butler*, 1972; *Smith*, 1974; *Donaldson*, 1975). Therefore, the compositions of the second cluster of our olivines suggest that relative to Luna 24 olivines, and possibly also relative to those in average Apollo 15 mare basalts, some of the olivines in this core with mare basalt affinity cooled *relatively* slowly. Plots of FeO versus An content of monomineralic plagioclase particles from this core (Fig. 6) show a cluster around An_{97} (~60% of all plagioclase); most of these plagioclases are likely to have been derived from the anorthositic rocks of the Apennine Front. It has been shown that Fe content of plagioclases is probably a function of the bulk composition of parent rocks/magmas and therefore high-Fe plagioclases in the lunar regolith indicate a mare basalt source rock, at least chemically. FeO content of plagioclase particles from this core range up to 1.0% and many of the plagioclases are likely from mare basalt. On the other hand, the most sodic plagioclase in the core is only An_{82}; this is much more calcic than An_{75}, which is the most sodic plagioclase in mare basalts in general, including those from Apollo 15 (*Basaltic Volcanism Study Project*, 1981). We believe that this suggests that one major source of plagioclases in this core had a mare basalt chemical affinity. This interpretation is compatible with our olivine data, and does not contradict pyroxene chemistry in any way. Several monomineralic olivine and pink pyroxene grains, the compositions of which fall within the mare basalt range, are optically unzoned. This is also compatible with the possible existence of a relatively slowly cooled source rock, which may have a mare affinity or may even be a ferroan variety of highland plutonic lithology.

Local regolith breccia 15205. Discussions at the Workshop on the Geology and Petrology of the Apollo 15 Site (*Spudis and Ryder*, 1986a) clearly revealed that the regolith breccia 15205 collected from Station 2 and local surface soils are not similar either petrographically or chemically. The principal difference is in the large KREEP basalt content and the corresponding enrichment of incompatible elements in the breccia 15205 relative to local soils. Here we briefly compare the petrology of the regolith breccia 15205 (*Dymek et al.*, 1974; *Simon et al.*, 1986) with that of the core 15007/8 to see if the subsurface soils are any closer to the regolith breccia. Although we have not characterized this breccia in detail, a comparison is in order because we have inferred a very high

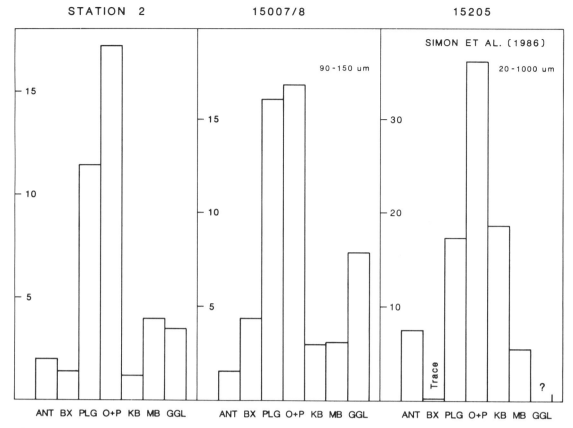

Fig. 9. Bar graphs comparing the relative abundances of selected petrographic components of the averaged modal data of the 15007/8 core soils (this study) with those of the surface soils at Station 2 (*Basu et al.*, 1981) and the clast population of regolith breccia 15205 (*Simon et al.*, 1986). Note that although actual modal percent is used as the ordinate, each bar graph has a different scale for reasons explained in the text. Abbreviations are: ANT = anorthositic, noritic, troctolitic rocks; BX = crystalline matrix breccias; PLG = monomineralic plagioclase; O+P = monomineralic olivine and pyroxene; KB = KREEP basalt; MB = mare basalt; GGL = green glass.

contribution from KREEP basalts (~40%) to the drive tube soils. The principal petrologic components of Station 2 surface soils (*Basu et al.*, 1981), the drive tube soils (this study), and the breccia (*Simon et al.*, 1986) are compared in Fig. 9 where the ordinates represent actual modal abundance of the components.

Note that the scales of the bar graphs for the materials are adjusted so that olivine + pyroxene abundances in the samples *appear* to be similar. There are two important reasons for such a somewhat unorthodox data display. First, this method provides a way to eliminate the effects of different degrees of regolith maturity in different samples; recalculation on a fused-soil-free basis would have altered the actual abundances, which we want to avoid. Second, because olivine + pyroxene abundance is an approximate measure of basaltic contribution to Apollo 15 regolith, a graphical normalization to olivine + pyroxene depicts the distribution patterns of soil components, again *without* altering the figures of actual abundances. In other words, this method retains elements of a table as well as those of a normalized graph.

Because *Simon et al.* (1986) studied the 20–1000 μm fraction and we studied the 90–150 μm fraction, some difference in the abundances of rock fragments and monomineralic fragments are expected (*Basu et al.*, 1981). However, there is a large difference between the relative proportions of monomineralic pyroxene and plagioclase in these two samples, plagioclase being enriched in the soil. KREEP basalts and ANT fragments appear to maintain approximately similar proportions in both the breccia and the soils; mare basalts are, however, relatively enriched in the soils. Clearly, the drive tube soils are not petrographically similar to regolith breccia 15205.

A discrepancy is seen in the abundance of green glasses that constitute a significant fraction of the soils and breccia 15205 appears to have none (*Simon et al.*, 1986). In our cursory examination of one thin section of breccia 15205 we have found a noticeable amount of brownish green glass fragments, which may turn out to have "green glass" composition on chemical analysis. *Dymek et al.* (1974) report a number of chemical analysis of green glass fragments, with and without crystallites, in 15205. It appears that 15205 contains a small amount of green glass that may be identified only on the basis of chemical composition.

Our preliminary optical examination of one thin section of 15205 shows an apparent absence of olivine basalt fragments. In this regolith breccia, *Dymek et al.* (1974) report about 1% olivine basalts, identified mostly on the basis of textural criteria and presence of strongly zoned olivine phenocrysts, and the presence of about 20% olivine-free basalts. In the core soils, olivine basalts are about half as abundant as olivine-free basalts. Thus, the core soils differ significantly from breccia 15205 in the relative proportions of mare basalt types. Considering that soils and rake samples from the mare region of the Apollo 15 site are enriched in olivine basalts (*Dowty et al.,* 1973; *Basu and McKay,* 1979) presumably because olivine basalt forms the top most flow at this site (*Spudis and Ryder,* 1986b), it is interesting to speculate on the origin of 15205. If 15205 were to have formed on the local mare plain, it should have contained far more olivine basalt fragments. On the other hand, 15205 does contain the local types of KREEP basalt, pyroxene phyric basalt, and green glass fragments. Perhaps 15205 formed on the regolith developed locally but *before* any major flooding by olivine basalt lava, or perhaps 15205 formed in a region that was *not* flooded by olivine basalt lava, which is less likely. Further investigation of 15205 with the purpose of understanding the stratigraphy of the Apollo 15 site should be very interesting.

Comparison with Other Soils

Detailed modal analyses of many surface soils, and at least one from each station of the Apollo 15 site, are available in the literature (*Engelhardt et al.,* 1973; *Heiken et al.,* 1976; *Basu and Bower,* 1977; *Basu and McKay,* 1979; *Griffiths et al.,* 1981; *Basu et al.,* 1980, 1981; *McKay et al.,* 1980; etc.). For any properly petrologic and purely provenance related comparison of modal data, all effects of surface maturation should be subtracted. One way of achieving this is to recalculate the data on a "fused soil free basis" (cf. *Papike et al.,* 1982). Here we omit the fused soil components but do not recalculate the rest to 100% (Figs. 9, 10).

Comparison of the modal data of the core with those of surface soils at Station 2 shows that the core is enriched in plagioclase, crystalline breccia, green glass, and KREEP basalts (Fig. 9). Interestingly, this is approximately the opposite of the relationship between core soils and regolith breccia 15205, which is depleted in plagioclase, crystalline breccia, and green glass, but is enriched in KREEP basalt. It is possible that the surface soil at Station 2 is a mixture of Apennine Front soils creeping downslope and some disintegration products of regolith breccias like 15205. If so, KREEP basalt enrichment in both 15205 and in the core soils remains an anomaly.

In comparison to soils from other parts of the Front (Stations 6, 6A, and 7), it is obvious that the eastern Front at the Apollo 15 site is generally enriched in green glass (Fig. 10a). Otherwise, the core soils are more similar to these other Front soils in the distribution patterns of rock types and monomineralic fragments. Apparently, Station 2 surface soils deviate more from the distribution patterns of other Front soils, including core 15007/8. This again could be a result of partial disintegration of regolith breccias such as 15205 at Station 2. The distribution patterns of major petrologic components of all soils from the six stations on the lower mare plain are very similar (Figs. 10b, 10c). There is an enrichment of mafic minerals and mare basalts and a depletion of plagioclase, KREEP basalt, and green glass, signifying relatively high contribution of local mare basalts. The similarity between the soils at the six stations is rather interesting because the edge of the Hadley Rille is quite different from the geomorphic hump at Station 8 and LM. It is likely that local thin flows of mare basalts and the Apennine Bench Formation dominate the detrital modes of these soils, which therefore maintain similar proportions of particle types. Both surface and subsurface soils of Station 2 show patterns that are different from the above.

Petrologic Observation versus Chemical Mixing Models

In our discussion we have considered petrologic entities only. A comparison of chemical components of all soils *vis a vis* petrology is necessary to understand the site geology of any lunar landing site. *Korotev* (1986, 1987) has recently reviewed many of the chemical data pertaining to Apollo 15 soils (e.g., *Chou et al.,* 1975; *Duncan et al.,* 1975; *Fruchter et al.,* 1973; *Laul and Schmitt,* 1973; *Laul and Papike,* 1980; *Walker and Papike,* 1981; etc.). Korotev's study is based primarily on mixing calculations and is, at least in part, model dependent. There are some serious discrepancies between the abundances of observed petrologic entities and those obtained from chemical mixing model calculations. For example, we find approximately 6% green glass in core 15007/8 at a depth of 42 cm, whereas the mixing model shows −1%; we have observed about 2% green glass in soils from the edge of Hadley Rille, whereas the mixing model shows 17% as noted by *Korotev* (1987). Both of our groups realize that we have to reconcile these discrepancies, but it is beyond the scope of this paper to address the issue.

CONCLUSIONS

1. Core 15007/8 does not contain distinct lithologic units. Units based on FMR, rare gas abundances, grain size, and agglutinate maturity do not correlate with variations in petrologic components.
2. Core 15007/8 likely contains soils of mixed maturity in which mixing has greatly dominated in situ reworking.
3. No petrologic layering is present in this core. We do not expect any stratigraphic correlation with petrologic units or layers of other Apollo 15 cores.
4. Abundances of KREEP basalts correlate with mare basalts in this core and not with ANT components from the Apennine Front; the source of KREEP basalt fragments is therefore more closely related to mare basalts and may underlie mare basalts or may be interbedded with them.
5. Distribution patterns of Ca in some olivines, of Fe in some plagioclases, and the presence of some unzoned pink pyroxenes and olivines suggest that a search for contribu-

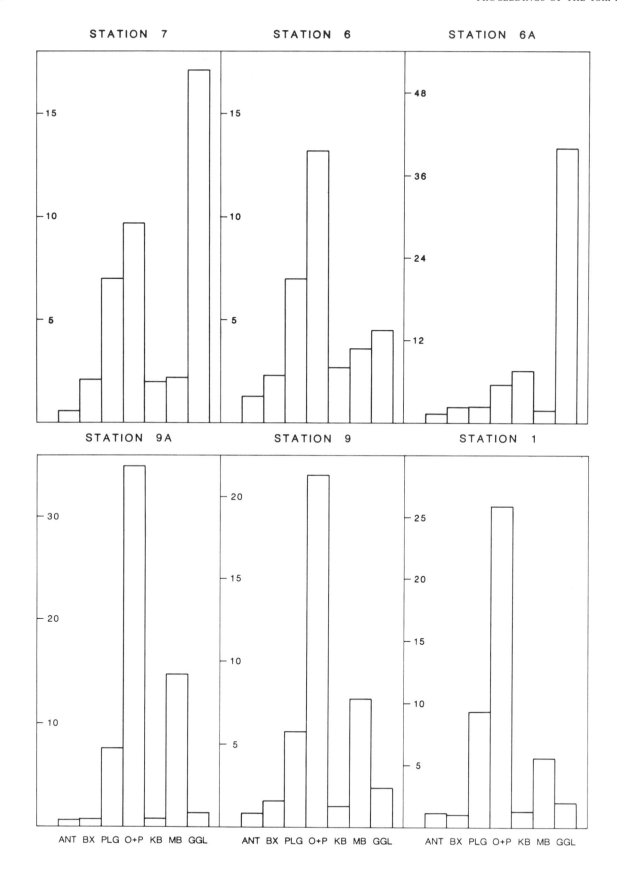

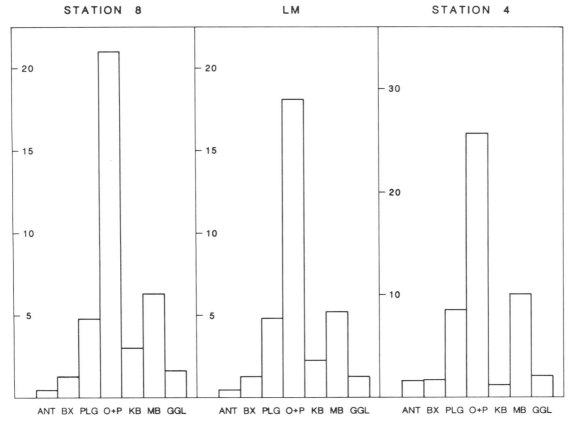

Fig. 10. Bar graphs comparing the relative abundances of selected petrographic components of average Apollo 15 surface soils from each sampling station. (a) Comparison of soils from stations (6, 6A, 7) near the Spur crater on the Apennine Front. (b) Comparison of soils from stations (9A, 9, 1) that are close to the rille. (c) Comparison of soils from stations (8, LM, 4) that are farther from the rille but are on the mare plain. Note that the Front soils (Stations 6, 6A, and 7) are similar to each other except for green glass enrichment in 6A and 7 soils. All other soils show similar patterns of particle type distribution that are different from those of the Front soils. Although actual modal percent is used as the ordinate, each bar graph has a different scale for reasons explained in the text. Abbreviations are identical to those used in Fig. 9.

tions from a previously unidentified relatively slowly cooled rock of mare basalt composition may be in order.

Acknowledgments. This research was supported in part by NASA grant NSG-9077, the Department of Geology, Indiana University, and the I.U. Foundation. Electron microprobe facilities at Indiana University were installed with NSF grants to Dr. C. Klein. K. Walker, B. Hill, and Z. Antolovic assisted in manuscript preparation. Excellent and helpful reviews and editorial comments were provided by Marta Kempa-Flohr, Steve Simon, and Jim Papike. We are grateful to all.

REFERENCES

Basaltic Volcanism Study Project (1981) *Basaltic Volcanism on the Terrestrial Planets*. Pergamon, New York. 1286 pp.

Basu A. (1986) Regolith erosion and regolith mixing at the Apollo 15 site on the moon (abstract). In *Workshop on the Geology and Petrology of the Apollo 15 Landing Site* (P. D. Spudis and G. Ryder, eds.), pp. 29-31. LPI Tech. Rpt. 86-03, Lunar and Planetary Institute, Houston.

Basu A. and Bower J. F. (1976) Petrography of KREEP basalt fragments from Apollo 15 soils. *Proc. Lunar Sci. Conf. 7th*, pp. 659-678.

Basu A. and Bower J. F. (1977) Provenance of Apollo 15 deep drill core sediments. *Proc. Lunar Sci. Conf. 8th*, pp. 2841-2867.

Basu A. and McKay D. S. (1979) Petrography and provenance of Apollo 15 soils. *Proc. Lunar Planet. Sci. Conf. 10th*, pp. 1413-1424.

Basu A., McKay D. S., and Fruland R. M. (1978) Origin and modal petrography of Luna 24 soils. In *Mare Crisium: The View from Luna 24*, pp. 321-337. Pergamon, New York.

Basu A., McKay D. S., Nace G., and Griffiths S. A. (1980) Petrography of lunar soil 15601. *Proc. Lunar Planet. Sci. Conf. 11th*, pp. 1727-1741.

Basu A., McKay D. S., Griffiths S., and Nace G. (1981) Regolith maturation on the earth and the moon with an example from Apollo 15. *Proc. Lunar Planet. Sci. 12B*, pp. 433-449.

Bence A. E. and Papike J. J. (1972) Pyroxenes as recorders of lunar basalt petrogenesis: Chemical trends due to crystal-liquid interaction. *Proc. Lunar Sci. Conf. 3rd*, pp. 431-469.

Bogard D. D., Morris R. V., Johnson P., and Lauer H. V., Jr. (1982) The Apennine Front core 15007/8: Irradiational and depositional history. *Proc. Lunar Planet. Sci. Conf. 13th*, in *J. Geophys. Res., 87*, A221-A231.

Butler P., Jr. (1972) Compositional characteristics of olivines from

Apollo 12 samples. *Geochim. Cosmochim. Acta, 36*, 773–785.

Chou C.-L., Boynton W. V., Sundberg L. L., and Wasson J. T. (1975) Volatiles on the surface of Apollo 15 green glass and trace-element distribution among Apollo 15 soils. *Proc. Lunar Sci. Conf. 6th*, pp. 1701–1728.

Donaldson C. H. (1975) Calculated diffusion coefficients and the growth rate of olivine in a basaltic magma. *Lithos, 8*, 163–174.

Dowty E., Prinz M., and Keil K. (1973) Composition, mineralogy, and petrology of 28 mare basalts from Apollo 15 rake samples. *Proc. Lunar Sci. Conf. 4th*, pp. 423–444.

Duncan A. R., Sher M. K., Abraham Y. C., Erlank A. J., Wills J. P., and Ahrens L. H. (1975) Interpretation of the compositional variability of Apollo 15 soils. *Proc. Lunar Sci. Conf. 6th*, pp. 2309–2320.

Dymek R. F. (1986) Characterization of the Apollo 15 feldspathic basalt suite (abstract). In *Workshop on the Geology and Petrology of the Apollo 15 Landing Site* (P. D. Spudis and G. Ryder, eds.), pp. 52–57. LPI Tech. Rpt. 86-03, Lunar and Planetary Institute, Houston.

Dymek R. F., Albee A. L., and Chodos A. A. (1974) Glass-coated soil breccia 15205: Selenologic history and petrologic constraints on the nature of its source region. *Proc. Lunar Sci. Conf. 5th*, pp. 235–260.

Engelhardt W. von, Arndt J., and Schneider H. (1973) Apollo 15: Evolution of the regolith and origin of glasses. *Proc. Lunar Sci. Conf. 4th*, pp. 239–249.

Folk R. L. and Ward W. C. (1957) Brazos river bar: A study in the significance of grain size parameters. *J. Sed. Petrol., 27*, 3–26.

Fruchter J. S., Stoeser J. W., Lindstrom M. M., and Goles G. G. (1973) Apollo 15 clastic material and their relationship to local geologic materials. *Proc. Lunar Sci. Conf. 4th*, pp. 1227–1238.

Griffiths S. A., Basu A., McKay D. S., and Nace G. (1981) Petrology of Apollo 15 Station 9A surface and drive tube soils. *Proc. Lunar Planet. Sci. 12B*, pp. 475–484.

Heiken G. (1975) Petrology of lunar soils. *Rev. Geophys. Space Phys., 13*, 567–587.

Heiken G. H., Morris R. V., McKay D. S., and Fruland R. M. (1976) Petrographic and ferromagnetic resonance studies of the Apollo 15 deep drill core. *Proc. Lunar Sci. Conf. 7th*, pp. 93–111.

Korotev R. L. (1986) Chemical components of the Apollo 15 regolith (abstract). In *Workshop on the Geology and Petrology of the Apollo 15 Landing Site* (P. D. Spudis and G. Ryder, eds.), pp. 75–79. LPI Tech. Rpt. 86-03, Lunar and Planetary Institute, Houston.

Korotev R. L. (1987) Mixing levels, the Apennine Front soil component, and compositional trends in the Apollo 15 soils. *Proc. Lunar Planet. Sci. Conf. 17th*, in *J. Geophys. Res., 92*, E411–E431.

Laul J. C. and Papike J. J. (1980) The lunar regolith: Comparative chemistry of the Apollo sites. *Proc. Lunar Planet. Sci. Conf. 11th*, pp. 1307–1340.

Laul J. C. and Schmitt R. A. (1973) Chemical composition of Apollo 15, 16, and 17 samples. *Proc. Lunar Sci. Conf. 4th*, pp. 1349–1368.

McKay D. S. and Basu A. (1983) The production curve for agglutinates in planetary regoliths. *Proc. Lunar Planet. Sci. Conf. 14th*, in *J. Geophys. Res., 88*, B193–B199.

McKay D. S., Heiken G. H., Taylor R. M., Clanton U. S., Morrison D. A., and Ladle G. H. (1972) Apollo 14 soils: Size distribution and particle types. *Proc. Lunar Sci. Conf. 3rd*, pp. 983–995.

McKay D. S., Dungan M. A., Morris R. V., and Fruland R. M. (1977) Grain size, petrographic, and FMR studies of the double core 60009/10: A study of soil evolution. *Proc. Lunar Sci. Conf. 8th*, pp. 2929–2952.

McKay D. S., Basu A., and Nace G. (1980) Lunar core 15010/11: Grain size petrography and implications for regolith dynamics. *Proc. Lunar Planet. Sci. Conf. 11th*, pp. 1531–1550.

Morris R. V. (1978) The surface exposure (maturity) of lunar soils: Some concepts and I_s/FeO compilation. *Proc. Lunar Planet. Sci. Conf. 9th*, pp. 2287–2297.

Morris R. V., Score R., Dardano C., and Heiken G. (1983) *Handbook of Lunar Soils*. Planetary Material Branch Publ. 67, JSC Publ. No. 19069. NASA Johnson Space Center, Houston. 914 pp.

Nagle J. S. (1981) Depositional history of core 15008/7: Some implications regarding slope processes. *Proc. Lunar Planet. Sci. 12B*, pp. 463–473.

Papike J. J., Hodges F. N., Bence A. E., Cameron M., and Rhodes J. M. (1976) Mare basalts: Crystal chemistry, mineralogy, and petrology. *Rev. Geophys. Space Phys., 14*, 475–540.

Papike J. J., Simon S. B., and Laul J. C. (1982) The lunar regolith: Chemistry, mineralogy, and petrology. *Rev. Geophys. Space Phys., 20*, 761–826.

Papp H. A., Steele I. M., and Smith J. V. (1978) Luna 24: 90–150 micrometer fraction: Implications for remote sampling of regolith. In *Mare Crisium: The View from Luna 24* (R. B. Merrill and J. J. Papike, eds.), pp. 245–264. Pergamon, New York.

Ryder G. (1976) Lunar sample 15405: Remnant of a KREEP basalt-granite differentiated pluton. *Earth Planet. Sci. Lett., 29*, 255–268.

Ryder G. (1985) *Catalog of Apollo 15 Rocks*. Curatorial Branch Publ. 72. JSC Publ. No. 20787. NASA Johnson Space Center, Houston.

Simon S. B., Papike J. J., Gosselin D. C., and Laul J. C. (1985) Petrology and chemistry of Apollo 12 regolith breccias. *Proc. Lunar Planet. Sci. Conf. 16th*, in *J. Geophys. Res., 90*, D75–D86.

Simon S. B., Papike J. J., Gosselin D. C., and Laul J. C. (1986) Petrology, chemistry, and origin of Apollo 15 regolith breccias. *Geochim. Cosmochim. Acta, 50*, 2675–2691.

Smith J. V. (1974) Lunar mineralogy: A heavenly detective story: Presidential Address, Part I. *Amer. Mineral., 59*, 231–243.

Smith J. V. and Steele I. M. (1976) Lunar mineralogy: A heavenly detective story: Presidential Address, Part II. *Amer. Mineral., 61*, 1059–1116.

Spudis P. D. (1978) Composition and origin of the Apennine Bench Formation. *Proc. Lunar Planet Sci. Conf. 9th*, pp. 3379–3394.

Spudis P. D. and Ryder G., eds. (1986a) *Workshop on the Geology and Petrology of the Apollo 15 Landing Site*. LPI Tech. Rpt. 86-03, Lunar and Planetary Institute, Houston. 126 pp.

Spudis P. D. and Ryder G. (1986b) Geology and petrology of the Apollo 15 landing site: Past, present, and future understanding. *Eos Trans. AGU, 66*, 721, 724–726.

Steele I. M. and Smith J. V. (1975) Minor elements in lunar olivine as a petrologic indicator. *Proc. Lunar Sci. Conf. 6th*, pp. 451–467.

Vaniman D. T. and Papike J. J. (1977) The Apollo 17 drill core: Characterization of the mineral and lithic component (sections 70007, 70008, 70009). *Proc. Lunar Sci. Conf. 8th*, pp. 3123–3159.

Vaniman D. T., Labotka T. C., Papike J. J., Simon S. B., and Laul J. C. (1979) The Apollo 17 drill core: Petrologic systematics and the identification of a possible Tycho component. *Proc. Lunar Planet. Sci. Conf. 10th*, pp. 1185–1227.

Walker R. J. and Papike J. J. (1981) The Apollo 15 regolith: Chemical modelling and mare/highland mixing. *Proc. Lunar Planet. Sci. 12B*, pp. 509–517.

Laser Probe ^{40}Ar-^{39}Ar Dating of Impact Melt Glasses in Lunar Breccia 15466

M. A. Laurenzi*, G. Turner, and P. McConville

Department of Physics, University of Sheffield, Sheffield, England

*Permanent address: Istituto di Geocronologia e Geochimica Isotopica, CNR, Pisa, Italy

A laser probe has been used to measure ^{40}Ar-^{39}Ar and cosmic-ray exposure ages of individual microgram regions of melt glass and clasts in lunar breccia 15466. The ^{40}Ar-^{39}Ar age, 130 ± 90 Ma, is imprecise due to a variable trapped (^{40}Ar/^{36}Ar) ratio, 3.1 ± 0.4, but is indistinguishable from the cosmic-ray exposure age of 60 ± 12 Ma. This indicates that the glass, which constitutes 90% of the breccia, was formed very recently, possibly as recently as the time indicated by the exposure age. A high trapped (^{38}Ar/^{36}Ar) ratio (0.21 compared to the solar wind value of 0.19), together with the high trapped (^{40}Ar/^{36}Ar) ratio, indicates that the glass and the breccia formed from preirradiated near surface soil. The age of the breccia is a candidate age for Spur Crater. Age measurements on clasts occluded by the glass are variable and much older, indicating that they were relatively unaffected by the brief heating by the melt glass. The low ages obtained for the glass by the laser technique contrast with much older ages obtained previously from stepped heating measurements. The high stepped heating ages of the glass probably arise from the presence of small occluded clasts and are thus an artifact. This observation seems to invalidate a number of similar published stepped heating ages on lunar impact glasses.

INTRODUCTION

There have been a number of previous attempts to date lunar soil breccias and impact melt glasses by the ^{40}Ar-^{39}Ar technique. These have used both the stepped heating approach (*Husain*, 1972, 1974; *Huneke et al.*, 1973; *Turner et al.*, 1973; *Eglinton et al.*, 1974) and the laser probe (*Plieninger and Schaeffer*, 1976; *Muller et al.*, 1977; *Eichhorn et al.*, 1978). Stepped heating has typically produced ages based on rather unconvincing (^{40}Ar/^{36}Ar)-(^{39}Ar/^{36}Ar) "isochrons" (see for example *Eglinton et al.*, 1974). The unreliability of these ages arises from the great complexity of soil breccias, which are mixtures of genetically unrelated rock fragments, minerals, glasses, previous breccias, and regolith fragments, held together by impact melt glass, and complicated by the presence of solar wind implanted gases and cosmogenic nuclides. Even large fragments of seemingly uniform melt glass are seen in thin section to be inhomogeneous due to the presence of occluded mineral and rock fragments.

Since the pioneering work of *Megrue* (1967, 1972, 1973), the laser probe has been much improved as a tool for noble gas analysis. The use of more powerful lasers, the availability of more sensitive mass spectrometers with improved background levels, and the use of low volume systems has made it possible to analyse samples about two orders of magnitude smaller than those originally analysed by Megrue. However, virtually all of the laser probe work on breccias published so far has concentrated on the wide range of lithic fragments of which they are composed, rather than on the glass matrix that binds them together and that was molten at the time of brecciation. The aim of the present experiment therefore was to carry out a more detailed analysis of the glassy component of a soil breccia than has previously been carried out. The intention was to determine from the apparent age of the glass the compaction age of the breccia. A further important aim was to make a comparison of stepped heating and laser probe results with a view to establishing the reliability or otherwise of the stepped heating approach to lunar glasses. The laser probe used in this work was capable of analysing argon released from microgram regions of the sample and was thus able to resolve differences in argon composition on a 100-μm scale.

In attempting to date lunar impact melts by the ^{40}Ar-^{39}Ar method we were aware of possible problems associated with incomplete degassing of the molten glass of its preexisting radiogenic argon. Measurements on melt glass vein (*McConville and Turner*, 1984; *McConville*, 1985; *McConville et al.*, 1985, 1987) and on the fusion crust (Laurenzi and Turner, unpublished data, 1987) of the Peace River meteorite indicate that radiogenic argon can be almost quantitatively retained in a melt glass that is rapidly quenched. This turned out not to be a major problem with the lunar glass. Although some solar wind argon, cosmogenic argon, and (presumably) radiogenic argon were retained by the melt, the isotopes were apparently homogenised sufficiently (though not completely) to allow the approximate time of melting to be inferred.

The measurements reported in this paper were carried out on fragments of melt glass from Apollo 15 soil breccia 15466, a dark vesicular glassy breccia collected near the summit of the 100 m Spur Crater (Station 7, *Swann et al.*, 1972). It is remarkable for its high content of glass, which is 90% according to the Apollo 15 sample catalogue. Electron probe analysis indicates that it has a similar composition to the dark vesicular glass matrix of 15465 (*Cameron and Delano*, 1974) and the Appennine Front soil (*LSPET*, 1972). The glass of both 15465 and 15466 was therefore interpreted as shock-melted Appenine Front soil, although the FeO, MgO, and Al$_2$O$_3$ contents are slightly different from the soil at Spur Crater.

Apart from the measurements of *McConville* (1983, 1985), discussed below, there is no published chronological information on 15466. There is information concerning samples collected at the same sampling location, viz 15455 and 15465.

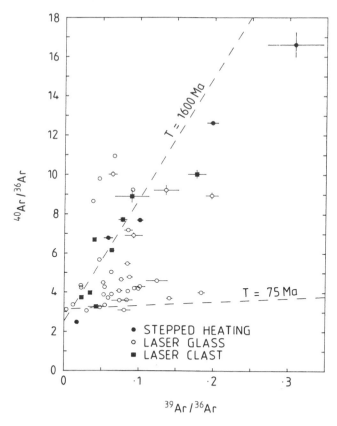

Fig. 1. Correlation diagram of (^{40}Ar/^{36}Ar) against (^{39}Ar/^{36}Ar) obtained from a stepped heating of 15466 glass and multishot laser degassing (previously published in *McConville*, 1983).

^{40}Ar-^{39}Ar stepped heating studies of several fragments of 15455 ("black and white breccia") gave complex release patterns with evidence of argon loss. Minimum ages were estimamted to be of 3770 Ma and 3870 Ma (*Alexander and Kahl*, 1974), indicating the influence of the "lunar cataclysm" on these two samples. Their cosmic ray exposure ages were around 200 Ma and neither appears to be related to 15466. On the other hand, as indicated above, soil breccia 15465, which was collected close to 15466, resembles it closely in appearance and chemistry. *Husain* (1972) reported a complex argon release pattern from which he inferred a "minimum age" of 1190 ± 140 Ma for 15465. Later, *Plieninger and Schaeffer* (1976) carried out extensive laser probe studies on 15465, obtaining ages of 3910 ± 40 Ma and 1910 ± 100 Ma on two heterogeneous groups of different clasts. From the work reported in this paper we can argue that none of these measurements relate to the time of the impact that probably lithified both 15465 and 1546.

McConville (1983, 1985) has carried out both stepped heating and laser probe analyses of 15466. The stepped heating analysis on a fragment of glass produced a very rough (^{40}Ar/^{36}Ar)-(^{39}Ar/^{36}Ar) correlation (Fig. 1), which translated to an apparent age of around 1400 Ma (*McConville*, 1983). His laser probe measurements, when plotted on the same diagram, showed a very large scatter with apparent ages ranging from the cosmic ray exposure age of around 75 Ma up to 2700 Ma. He concluded that the data were consistent with an age of melting for the glass similar to the exposure age. Furthermore, he concluded that the high stepped heating age was in all probability an artifact representing some kind of average of the (^{40}Ar/K) ratio of the glass and occluded old lithic fragments. If this conclusion is correct, it casts serious doubt on virtually all of the published stepped heating data on lunar soil breccia melt glasses.

McConville's measurements were made at a time when the Sheffield laser probe was under development. Due to the lower sensitivity of the spectrometer at that time and higher background levels it was necessary to perform multishot laser extractions involving 30-90 μg of sample. Consequently, McConville's individual data points represent averages over regions that are relatively large compared to the microstructure of 15466. Subsequent improvements in both sensitivity and background levels have made it possible to perform "single shot" age determinations that relate to microgram quantities of sample. The increased spatial resolution made possible has clarified the situation considerably and, as will be seen below, amply confirmed McConville's tentative conclusions.

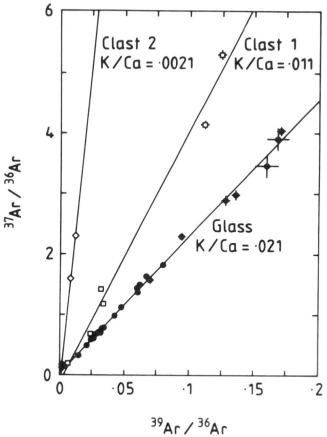

Fig. 2. Correlation diagram of (^{37}Ar/^{36}Ar) against (^{39}Ar/^{36}Ar) showing the different chemistry of glass (solid circles) and clasts (open symbols).

EXPERIMENTAL

The specimen we have analysed, 15466,7, was a 0.49-g sample of glass containing many small lithic inclusions. It was broken with a stainless steel chisel into several pieces and cleaned ultrasonically in methanol to remove, as far as possible, solar-wind-rich soil from the surface of the glass. Three pieces were analysed in the previous study by McConville. Two further pieces were selected for the present study. They were characterised by large areas of freshly broken glass surface and a number of white lithic fragments in contact with the glass.

The glass pieces were wrapped in pure Al foil and irradiated in the Herald reactor at AWRE, Aldermaston. Monitoring was carried out using hornblende standard hb3gr. As usual the package was rotated through 180° midway through the irradiation to minimise the fluence gradient across the sample. Nickle wires monitored the relative fluences of the different sample packages. The J value inferred for samples analysed in this study was 0.0287 ± 0.0003.

After irradiation both samples were mounted under the glass window of the laser port so that the faces to be analysed were horizontal. The laser system (*McConville*, 1985; P. McConville et al., unpublished data, 1987) comprises a non-Q-switched Nd-glass laser capable of delivering 180 $v$sec pulses of up to 4 Joules at a wavelength of 1060 nm. Its position can be controlled by two precision micrometers. Prior to analysis samples were baked using a heating lamp to a temperature slightly over 100°C. Sample cleanup was carried out by an SAES getter integral with the laser port.

For most of the extractions the laser was operated at an energy of 0.35 J and the argon released by the laser admitted directly to the mass spectrometer. Instrumental blanks were determined several times each day during the 10-day period of analysis and were very reproducible, equivalent to 6.7 ± 1.3, 14 ± 2.4, 4.0 ± 1.0, 2.8 ± 0.7, and 460 ± 65 ($\times 10^{-14}$ cc STP) of argon at masses 36, 37, 38, 39, and 40, respectively. During each run masses 35 (Cl) and 41 (hydrocarbon) were monitored to check that the level of instrumental background remained constant. Corrections were applied to the data for blanks/background, discrimination, decay of ^{37}Ar, and neutron interference reactions on Ca and K.

RESULTS

The experimental data are too extensive to publish in full and can be obtained by writing to the authors. Data were obtained from 41 individual laser extractions, backed up by 27 blanks. Twenty-four of the measurements were made on glass, and 17 on clasts and breccia fragments included in or adjacent to the glass. For all but two large white clasts, argon isotopes were dominated by solar wind as a result of the laser melting underlying glass or matrix. The measurements were made in the form of a number of traverses in the glass and across clasts. Concentrations of solar wind argon (see below) varied in an irregular fashion by an order of magnitude over distance scales of less than a millimeter, indicating that this component was not homogenised through the glass during melting. No attempt has been made to relate these profiles to diffusion processes and no further reference will be made to this aspect of the experiment.

CHEMISTRY

Chemical information on the regions analysed is obtained in the form of (K/Ca) ratios inferred from the measured ($^{39}Ar/^{37}Ar$) ratios. The contrasting chemistry between glass and the two clasts for which age information was obtained is illustrated in Fig. 2, which is effectively a graph of ($Ca/^{36}Ar$) against ($K/^{36}Ar$). The well-defined linear array for the glass indicates that it is characterized by a very uniform (K/Ca) ratio of 0.021. This observation agrees with published chemical data and electron probe measurements made on our own samples (Table 1). Both of the clasts, referred to for convenience as #1 and #2, have lower (K/Ca) ratios, 0.011 and 0.0021, respectively, and are clearly anorthositic in character.

Direct determination of absolute abundances using the laser probe is not possible in general on account of the difficulty of determining the amount of sample melted. However, the amounts melted can be inferred indirectly for those samples in which either Ca or K abundances are known independently. The amount of glass melted in each of the 24 laser shots has been calculated assuming a Ca concentration of 7.6%, and the results are shown in the form of a histogram in Fig. 3. Most of the extractions involved melting around 2 μg of glass (note that the four extractions of 0.5 μg were the result of four initial shots at lower power).

The dispersion of points in Fig. 2 is the result of large variations in the concentration of solar wind ^{36}Ar. The concentration of this solar wind ^{36}Ar can also be calculated on the basis of an assumed Ca concentration, and a histogram of concentrations in the glass is shown in Fig. 4. Order of magnitude variations are seen over distance scales of less than

TABLE 1. Mean values of two series of electron probe analyses of 15466 impact melt glass.

(Wt %)	1	2
SiO_2	46.5	46.4
TiO_2	1.26	1.35
Al_2O_3	16.1	16.0
Cr_3O_3	0.32	0.33
FeO	11.6	11.7
MnO	0.18	0.19
MgO	10.6	10.7
CaO	10.2	10.7
Na_2O	0.42	0.42
K_2O	0.21	0.21
Sum	97.4	98.0
Ca	7.29	7.63
Na	0.31	0.31
K	0.18	0.17

Analyses by Dr. F. G. F. Gibb.

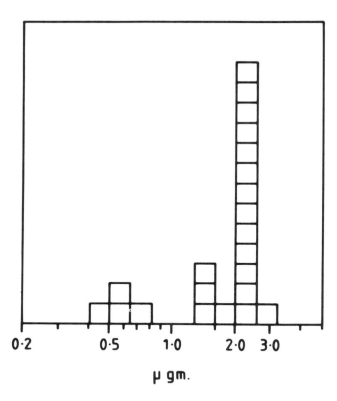

Fig. 3. Histogram of sample weights of glass degassed by 24 laser shots. The four extractions of 0.5 μg were the result of four initial shots at low power.

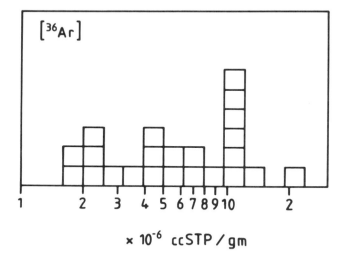

Fig. 4. Histogram of solar wind ^{36}Ar concentrations in the glass, based on (^{36}Ar/^{37}Ar) and an assumed Ca concentration of 7.6%.

a millimeter. In the case of the clasts the solar wind is partly released from the underlying black matrix or glass, which is melted by the passage of the laser beam through the partially transparent anorthite.

COSMIC-RAY EXPOSURE AGES

Cosmogenic argon is produced by spallation of Ca, K, Ti, and Fe. The concentrations of these elements in the glass are 8%, 0.2%, 1%, and 10%, respectively. The relative production rates are energy dependent (especially for Ti and Fe) but are typically in the ratio 1:1:0.1:0.05 (*Hohenberg et al.*, 1978; *Hennessy and Turner*, 1980). Thus Ca accounts for 91% or more of the Ar production. For the clasts, production from elements other than Ca is negligible.

Cosmic-ray exposure ages can be inferred from a plot of (^{38}Ar/^{36}Ar) against (^{37}Ar/^{36}Ar), which is a mixing diagram of solar wind argon, (^{38}Ar/^{36}Ar) = 0.19, and Ca correlated cosmogenic argon, (^{38}Ar/^{36}Ar) = 1.6. The appropriate graphs for clasts 1 and 2 and for the glass are shown in Fig. 5. The exposure age is calculated from (^{38}Ar/^{37}Ar) for the cosmogenic end member, which, based on the (^{38}Ar/^{36}Ar) ratio of the end members, is easily shown to be 1.13 times the slope of the graph.

Clasts 1 and 2 give well-defined correlations with distinct slopes corresponding to exposure ages of 910 Ma and 170 Ma, based on nominal production rates for ^{38}Ar of 1.4×10^{-8} cc STP/g Ca/Ma. The intercepts are, respectively, 0.217 ± 0.009 and 0.203 ± 0.0002. These ratios probably reflect the underlying matrix or glass and are significantly higher than the solar wind value of 0.19.

The correlation for the glass shows considerable scatter, beyond experimental error, and a best-fit slope corresponding to an exposure age of 60 ± 12 Ma (the error figure is based on the observed spread of the data). As for the clast measurements, the intercept, 0.208 ± 0.003, is significantly higher than the solar wind value. Both the scatter and the high intercept can be understood in terms of inherited and incompletely homogenised cosmogenic argon from a period of preirradiation prior to the impact that produced the glass. An additional source of scatter may be the effect of ^{38}Ar produced by (n,γ) reactions on Cl. The exposure age obtained from the present experiment is somewhat lower than that determined by *McConville* (1983), 75 ± 5 Ma, but not significantly so in view of the observed scatter of the data points.

RADIOMETRIC AGES

Radiometric age information is plotted in Fig. 6 in the form of an (^{40}Ar/^{36}Ar) versus (^{39}Ar/^{36}Ar) isochron diagram. As for the cosmogenic isotopes, the picture for the clasts is simpler than that for the glass and is discussed first.

Both clasts give convincing linear correlations, corresponding to K-Ar ages of 1910 ± 70 Ma and 3400 ± 400 Ma. The data points are spread along the mixing lines in Fig. 6 by the release of variable amounts of solar wind from the underlying matrix. The possibility of a corresponding contribution of a (contaminating) radiogenic component is discussed briefly in the next section. In a chronological sense these clast measurements are not particularly significant. The clasts presumably represent early lunar crustal rocks that have

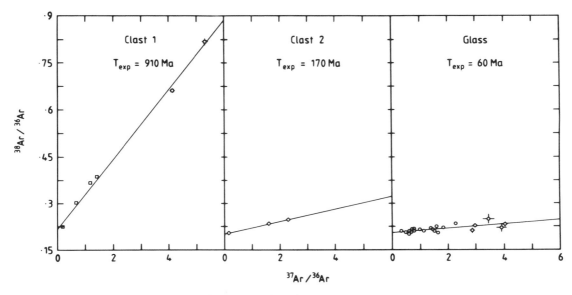

Fig. 5. Correlation diagrams of ($^{38}Ar/^{36}Ar$) against ($^{37}Ar/^{36}Ar$) from which cosmic ray exposure ages are deduced.

been subjected to reheating in one or more minor impact events. The most significant statement that can be made concerns the fact that they have retained some or all of their radiogenic argon despite being heated briefly by the impact melt glass.

The isochron diagram for the glass is less convincing at first glance in that the points show considerable scatter, with ($^{40}Ar/^{36}Ar$) ratios in the range 2.5 to 4.0. The scatter is presumed to arise from incomplete homogenization of solar wind and radiogenic argon in the glass at the time of melting.

There is a marginal correlation with ($^{39}Ar/^{36}Ar$), which corresponds to a radiometric "age" of 130 ± 90 Ma. In spite of the scatter, *one very important first-order observation can be made*; there is absolutely no evidence for the age of 1400 Ma suggested by the stepped heating experiment. It is clear therefore that the stepped heating age was undoubtedly an artifact, resulting probably from the release of radiogenic argon from old clasts incorporated in the glass. It is likely that this important conclusion can be generalised to most if not all of the published stepped heating data on lunar glasses. The

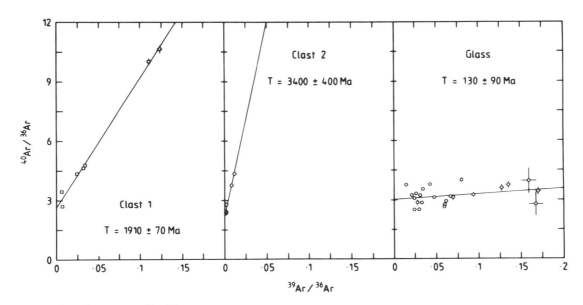

Fig. 6. ($^{40}Ar/^{36}Ar$) versus ($^{39}Ar/^{36}Ar$) isochron diagrams for clasts and glass. The scatter of points for the glass results from incomplete homogeneisation during melting 130 ± 90 Ma ago. The inferred age is very imprecise but clearly indicates that the glass formed very recently.

ill-defined radiometric age of the glass is indistinguishable from the cosmic-ray exposure age, suggesting that the breccia may have been produced only 60 or so million years ago and lain undistrubed on the surface of the Moon at Spur Crater since that time.

MIXED AGES AND THREE-DIMENSIONAL CORRELATION DIAGRAMS

It was argued above that the spread of the data points for the clasts in Figs. 5 and 6 represented a mixing with solar wind gases released from the underlying black matrix, which is a more efficient absorber of the laser pulse than the anorthosite of the clasts. A consequence of this is the possibility that the matrix or glass also contributes cosmogenic (and radiogenic) argon, thereby influencing the slope as well as the intercept of the two graphs.

We have found that a useful method of visualising whether this kind of mixing is important for individual data points is to plot three-dimensional correlation diagrams. In the case of cosmogenic argon the appropriate axes are (^{39}Ar/^{36}Ar), (^{37}Ar/^{36}Ar), and (^{38}Ar/^{36}Ar) and for radiogenic argon are (^{39}Ar/^{36}Ar), (^{37}Ar/^{36}Ar), and (^{40}Ar/^{36}Ar). In each case the two horizontal axes represent the chemistry of the sample and the vertical axis represents the cosmogenic (or radiogenic) contribution. Argon that is a mixture of solar wind with *two* cosmogenic (or radiogenic) components of well-defined chemistry (i.e., K/Ca) will define a mixing plane, whether or not linear correlations are obtained in plots such as those in Figs. 5 and 6. The parameters of the mixing planes define the ages of the end members, provided only that K/Ca for the end members are different and known.

The mixing planes for 15466 glass and clasts are shown in Fig. 7. There is evidence for only one of the clast data points (in clast #1) containing an obvious cosmogenic and radiogenic contribution from the glass. The point in question lies on the correlation plane intermediate between the two extreme clast and glass linear arrays. In light of this evidence it was omitted from Figs. 5 and 6.

Aside from this one point, the existence of two distinct linear arrays in Fig. 7 suggests (but does not prove) that the glass has not contributed to, and therefore lowered, the clast ages inferred from Fig. 5 and 6. We can note in passing that the equation of the correlation plane does in principle establish a link between permitted radiometric and cosmogenic ages and between each of these and independent (e.g., electron probe) measurements of end-member chemistry. However, in view of the lack of detailed significance attached to the clast ages, it is not profitable to pursue this aspect further.

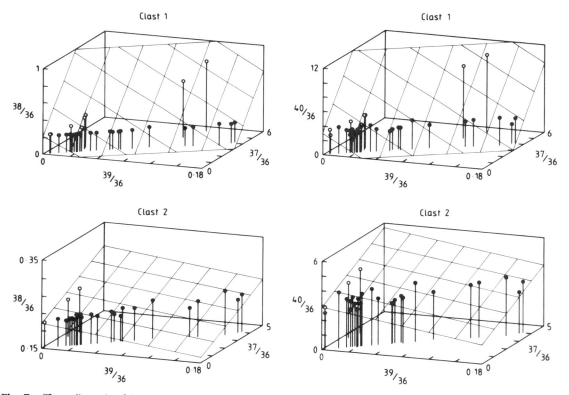

Fig. 7. Three-dimensional isotope correlation diagrams that can be used to investigate the extent of mixing of argon from clasts and glass by the laser outgassing. Contamination of the clast data by radiogenic (or cosmogenic) argon outgassed by the laser from the underlying dark glass would cause the clast points (open circles) to move in the planes indicated toward the array of glass points (solid circles). The age of the clast can still be inferred from the equation of the plane provided the clast chemistry, K/Ca, is known independently.

TABLE 2. Summary of data discussed in this paper.

	K/Ca*	K(%)	Ca(%)	^{39}Ar/^{40}Ar apparent age	CRE age
Glass	0.021	0.18†	7.45†	130 ± 90	60 ± 12
Clast #1	0.011	0.14‡	≡13§	1910 ± 70	910
Clast #2	0.0021	0.03‡	≡13§	3400 ± 400	170

*Inferred from ^{39}Ar/^{37}Ar ratio.
†Electron probe analyses.
‡Inferred from K/Ca ratio.
§Average value for anorthostic rocks.

SUMMARY AND CONCLUSIONS

The chemical and chronological information inferred from this study of lunar breccia 15466 is summarised in Table 2. The work has produced two important general conclusions concerning the data of lunar impact glasses, as well as a number of findings specific to 15466:

1. The most important specific observation is that the impact melt glass that constitutes the bulk of lunar soil breccia 15466 was produced in a recent impact. A previously measured stepped heating age of 1400 Ma for the glass is, as suspected, meaningless. The best estimate of the time of that impact comes from the cosmic-ray exposure age, 60 ± 12 Ma (75 ± 5 according to *McConville*, 1983). The latter is indistinguishable from the K-Ar age of the glass, 130 ± 90 Ma.

2. The high trapped (^{38}Ar/^{36}Ar) ratio (0.208 ± 0.003 compared to the solar wind value of 0.19) and the high trapped (^{40}Ar/^{36}Ar) ratio (3.1 ± 0.4) indicate that the glass formed from a preirradiated near-surface soil.

3. Two anorthositic clasts were analysed, giving radiometric ages of 1910 ± 70 and 3400 ± 400 Ma, and cosmic-ray exposure ages of 910 and 170 Ma. The ages have no intrinsic significance beyond indicating a history of preexposure for the constituents of 15466, together with evidence of only minimal outgassing by the hot melt glass that engulfed them.

4. The exposure age of 15466 is a candidate age for Spur Crater.

Our more general findings relate to the credibility of published ages of lunar glasses and soil breccias:

1. Comparison of our laser probe measurements on 15466 with previous stepped heating measurements made in our laboratory indicates that the stepped heating approach is unreliable when applied to lunar impact melt glasses and soil breccias. The presence of very small inclusions of older material leads to an overestimation of the age of melting of the glass when macroscopic samples of glass are step heated. Published ages of impact melt glasses, which are frequently of the order of 1000 to 2000 Ma, should be viewed with extreme scepticism.

2. Laser probe studies can, when interpreted with caution, provide a more useful approach than stepped heating to dating melt glasses. Even here, however, it is necessary to plan experiments carefully and to avoid overinterpretation of the observations. Both the present studies and our work on the Peace River meteorite indicate that argon is not lost from a melt when melting is followed by rapid quenching. Direct measurement of the time of melting (or as in the present experiment constraining it to a recent event) relies, as for Rb-Sr dating, on a degree of isotopic homogenization in the molten silicate. The Peace River observations indicate that this does not necessarily occur during the short time available.

Acknowledgments. We wish to thank our colleagues David Blagburn and Simon Kelley for help with many aspects of the experimental work. Dr. Fergus Gibb kindly performed the electron probe analyses of 15466. The work was supported by a grant from the Natural Environment Research Council (GR3/5004). M. Laurenzi acknowledges support from a CNR-NATO Fellowship for her visit to a part of Britain that was overlooked by her compatriots 1.9 Ka ago.

REFERENCES

Alexander E. C., Jr. and Kahl S. B. (1974) ^{40}Ar-^{39}Ar studies of lunar breccias. *Proc. Lunar Sci. Conf. 5th*, 1353-1373.

Cameron K. L. and Delano J. W. (1974) Petrology of Apollo 15 consortium breccia 15465. *Proc. Lunar Sci. Conf. 4th*, 461-466.

Eglinton G. et al. (The "European Consortium") (1974) The history of lunar breccia 15015. In *Lunar Sample Studies*, pp. 1-33. NASA SP-418.

Eichhorn G., James O. B., Schaeffer O. A., and Muller H. W. (1978) Laser ^{39}Ar-^{40}Ar dating of two clasts from consortium breccia 73215. *Proc. Lunar Planet. Sci. Conf. 9th*, 855-876.

Hennessy J. and Turner G. (1980) ^{40}Ar-^{39}Ar ages and irradiation history of Luna 24 basalt. *Phil. Trans. Roy. Soc. Lond.*, A297, 27-39.

Hohenberg C. M., Marti K., Podosek F. A., Reedy R. C., and Shirck J. R. (1978) Comparison between observed and predicted cosmogenic noble gases in lunar samples. *Proc. Lunar Planet. Sci. Conf. 9th*, 2311-2344.

Huneke J. C., Jessberger E. K., Podosek F. A., and Wasserburg G. J. (1973) ^{40}Ar/^{39}Ar measurements in Apollo 16 and 17 samples and the chronology of metomorphism and volcanic activity in the Taurus-Littrow region. *Proc. Lunar Sci. Conf. 4th*, 1725-1756.

Husain L. (1972) The ^{40}Ar-^{39}Ar and cosmic-ray exposure ages of Apollo 15 crystalline rocks, breccias and glasses. In *The Apollo 15 Lunar Samples* (J. W. Chamberlain and C. Watkins, ed.), pp. 374-377. The Lunar Science Institute, Houston.

Husain L. (1974) ^{40}Ar-^{39}Ar chronology and cosmic-ray exposure ages of the Apollo 15 samples. *J. Geophys. Res.*, 79, 2588-2606.

Laurenzi M. A. and Turner G. (1987) Laser probe ^{39}Ar-^{40}Ar dating of impact melt glasses in lunar breccia 15466. *Lunar Planet. Sci. Conf. 18th*, 537-538.

LSPET (Lunar Sample Preliminary Examination Team) (1972) The Apollo 15 lunar samples: a preliminary description. *Science*, 175, 363-375.

McConville P. (1983) A laser probe study of argon isotopes in lunar impact melt glasses. *Meteoritics*, 18, 350-352.

McConville P. (1985) Development of a laser probe for argon isotope studies, Ph.D. thesis, University of Sheffield.

McConville P. and Turner G. (1984) The thermal history of the Peace River L6 chondrite based on ^{40}Ar-^{39}Ar measurements. *Meteoritics*, 19, 267-268.

McConville P., Kelley S., and Turner G. (1985) Laser probe ^{40}Ar-^{39}Ar studies of the Peace River L6 chondrite. *Meteoritics*, 20, 707.

Megrue G. H. (1967) Isotopic analyses of rare gases with a laser microprobe. *Science*, 157, 1555-1556.

Megrue G. H. (1972) In situ ^{40}Ar/^{39}Ar ages of breccia 14301, and concentration gradients of helium, neon and argon isotopes in Apollo 15 samples. In *The Apollo 15 Lunar samples* (J. W. Chamberlain and C. Watkins, eds.), pp. 378-379. The Lunar Science Institute, Houston.

Megrue G. H. (1973) Spatial distribution of ^{40}Ar/^{39}Ar ages in lunar breccia 14301. *J. Geophys. Res.*, 78, 3216-3221.

Muller H. W., Plieninger T., James O. B., and Schaeffer O. A. (1977) Laser probe ^{39}Ar/^{40}Ar dating of materials from consortium breccia 73215. *Proc. Lunar Sci. Conf. 8th*, 2551-2565.

Plieninger T. and Schaeffer O. A. (1976) Laser probe ^{39}Ar-^{40}Ar ages of individual mineral grains in lunar basalt 15607 and lunar breccia 15465. *Proc. Lunar Sci. Conf. 7th*, 2055-2066.

Swann G. A. et al. (1972) Preliminary geologic investigation of the Apollo 15 landing site. *Apollo 15 Preliminary Science Report*, pp. 6-1, 5-112. NASA SP-289.

Turner G., Cadogan P. H., and Yonge C. J. (1973) Argon selecochronology. *Proc. Lunar Sci. Conf. 4th*, 1889-1914.

Lunar Mare Ridges: Analysis of Ridge-Crater Intersections and Implications for the Tectonic Origin of Mare Ridges

Virgil L. Sharpton
Lunar and Planetary Institute, 3303 NASA Road One, Houston, TX 77058

James W. Head III
Department of Geological Sciences, Brown University, Providence, RI 02912

A variety of volcanic and tectonic models have been proposed to explain the diverse morphological characteristics of lunar mare ridges. Volcanic models stress the "flow-like" appearance of mare ridges and particularly the manner in which these features modify the interiors of small preexisting impact craters. We have investigated the nature of ridge-crater intersections to gain a better understanding of mare ridge formation. Distinct evidence of ridge-related deformation was observed in 65 small craters within the lunar maria. Our results indicate that ridge deformation within these craters is inconsistent with any type of lateral transport such as volcanism or mass wasting. Instead, these ridge segments appear to consist of crater floor materials uplifted along nearly vertical splay faults. Ridge topography (generated by uplift along splay faults) accommodates most of the strain across mare ridges; additional lateral offsets are usually minor. Such small lateral offsets accompanied by up to 250 m of vertical offset across ridge axes disfavor recent suggestions that all mare ridges are simple thrust faults (*Plescia and Golombek*, 1986). Fault attitudes, measured from the walls of small fresh craters, range from ~15° to near-vertical and also indicate that a broad range of tectonic mechanisms are involved in ridge formation. In light of these findings, the evidence cited to support a volcanic origin for mare ridges is reassessed. Two tectonic models are in accordance with available surface and subsurface data and may represent end members in a broad suite of formative mechanisms: one calls upon vertical faulting to explain offsets across mare ridges, and the other relies upon buckling of the surface to produce ridges with no vertical offsets. The morphology and possibly the occurrence of mare-type ridges appear to be influenced not only by the near-surface stress field, but also by such material properties as competency, thickness, and the presence or absence of layering.

INTRODUCTION

Lunar mare ridges are among the most common landforms observed on the Moon and occur almost exclusively within the mare basins. Adhering to *Strom's* (1972) nomenclature, mare ridges typically consist of two components (Fig. 1): a gently sloping, broad arch, and a sharper but more irregular and discontinuous crenulation or "wrinkle" ridge. The dimensions of these components are variable, but arches usually range up to 20 km in width, 300 km in length, and 0.5 km in relief, whereas the ridge is typically less than 400 m wide and 200 m high. Although the ridge is commonly superposed on the arch, generally in the vicinity of the arch crest, the relationship exhibited by these two components is highly variable. The ridge, for instance, may be situated adjacent to and oriented subparallel to the arch (*Hodges*, 1973), or may extend beyond the arch onto the adjacent mare surface (*Lucchitta*, 1977). Additionally, arches are sometimes present without their ridge counterparts (*Colton et al.*, 1972; *Hodges*, 1973), and numerous examples of ridges without conspicuous arch development have been documented (*Hodges*, 1973; *Maxwell et al.*, 1975). In cross section the arch may appear symmetrical or asymmetrical, and there is sometimes a measurable vertical offset (50 to 250 m) of the mare surface across asymmetrical arches (*Lucchitta*, 1977; *Young et al.*, 1973). Mare ridges, while sometimes occurring as single features, more commonly occur in groups forming a variety of braided (*Young et al.*, 1973; *Scott*, 1973), sinuous (*Bryan*, 1973), or en echelon (*Tjia*, 1970; *Fagin et al.*, 1978) patterns. Regional trends and global patterns have been discussed by *Tjia* (1970), *Fagin et al.* (1978), and *Raitala* (1978). The diversity in small-scale morphology, texture, and associated structures characteristic of these features is thoroughly documented by *Schultz* (1976).

Proposed models for the origin of mare ridges are divisible into three main categories: (1) volcanic; (2) tectonic; and (3) a combination of volcanic and tectonic. Evidence presented in support of categories (1) and (2) is summarized in Table 1. It is interesting to note that terrestrial analogs have been advanced in favor of both the volcanic [lava lake features (*Hodges*, 1973)] and the tectonic [thrust faults (*Plescia and Golombek*, 1986) and folded lavas (*Greeley and Spudis*, 1978; *Lucchitta and Klockenbrink*, 1981; *Watters and Maxwell*, 1985)] hypotheses. It is perhaps the difficulty in refuting or substantiating either of the first two categories, as well as the wide variability in mare ridge surface expression, that has led to a wide acceptance of the third category.

While a tectonic origin is supported by a variety of data, including stratigraphic relationships, relative ages, and cross-sectional analyses (Table 1), it is the flow-like appearance of mare ridges and particularly the style by which ridges modify preexisting impact craters (Fig. 2) that is the keystone to the hypothesis that mare ridges are volcanic landforms (Table 1, V-1). Previous to this report, however, the intersections of mare ridges and impact craters have been described only casually. Here we present the results of our analysis of ridge-crater

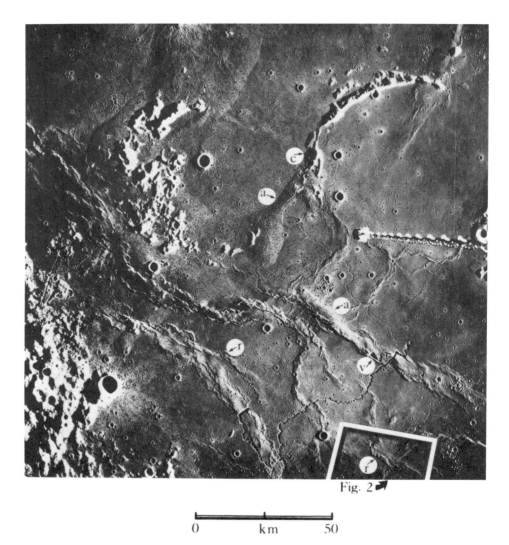

Fig. 1. A complex arrangement of mare ridge systems composed of broad arches (a) and steep-crested ridges (r). The area within the box in the southeastern portion of the image is shown in Fig. 2 and includes a small crater whose interior morphology has been modified by the adjacent mare ridge. This region in southern Oceanus Procellarum, west of Mare Cognitum, also illustrates the common association of mare ridges with buried topography, such as crater rims (c), as well as volcanic features such as sinuous rilles and vents (v). Apollo 16-2838 metric.

intersections in Oceanus Procellarum and Maria Cognitum, Imbrium, and Serenitatis, and document the range of fault attitudes and displacements associated with mare ridge development. As our findings indicate tectonic rather than volcanic involvement, the evidence cited to support a volcanic origin for mare ridges is reassessed and plausible end-member tectonic models are advanced.

RESULTS

In the majority of instances where ridges intersect craters, inferred by their preservation state to predate ridge formation, the craters show no evidence of deformation. Only 65 small craters (< 2% total number of preexisting craters) were found to show unambiguous evidence of mare ridge deformation. None exhibit offsets in the crater rim suggestive of strike-slip faulting and only five show measurable lateral offsets in the crater rim or distortion of the presumed original crater circularity consistent with displacement along low-angle reverse (thrust) faults. An example of a small crater that has apparently been shortened in the direction normal to the mare ridge axis is shown in Fig. 3. If this 700-m-diameter crater were originally circular, it has been shortened by about 150 m, the largest amount of overthrusting we have observed. The other examples of overthrusting associated with mare ridge formation indicate lateral displacements of less than 50 m.

The more typical style of crater modification does not result in significant impairment of the crater circularity, although small offsets in the rim are commonly observed (Fig. 4) and these may have a minor lateral component. Instead, lobate or welt-like mounds of material are present within the crater interior. The 60 ridge-crater intersections expressing this type of modification were observed to have the following characteristics:

1. The morphology, distribution, and surface appearance of ridge segments within crater interiors appear to correlate with the degradation state the crater: Older, more degraded craters consistently contain flat lobate features, whereas

younger, less degraded craters contain narrow welts of material, typically better developed on the rim than on the floor of the crater (Fig. 5). Degraded craters contain lobes with smooth, dark, and fine-grained surfaces, similar to the regolith that mantles other parts of the crater (Fig. 4). In contrast, fresher craters are modified by light-colored material with the bright rubbly texture characteristic of fresh ejecta and impact breccia facies (Figs. 4 and 5). In some cases several craters of various degradation states are modified by a single mare ridge as shown in Fig. 4. These observations indicate that the ridge segments represent material uplifted from the crater floor along splay or tear faults associated with ridge deformation and not material that has been extruded into the crater by volcanic processes.

2. In many craters, ridge segments appear on the steep slopes of the crater walls, rather than the floors where fluid extrusions might be expected to collect. These segments may also occur on the opposite side of the crater from the main ridge.

3. Noticeable vertical displacements of the crater rim associated with ridge deformation are common (Fig. 4).

4. The bright zones often observed on mare ridges and previously interpreted as viscous lavas of an unusual composition (*Strom*, 1972) were determined instead to be concentrations of boulders (Fig. 6) protruding through areas where the finer-grained regolith has either wasted downslope or seeped between the larger blocks during deformation. Numerous occurrences of such zones, containing much fresher looking materials than are associated with the majority of small impact craters on the lunar maria, suggest that ridge deformation continued long after the termination of mare volcanism.

Mare Ridge Deformation Within Craters

The characteristics of all mare ridge-crater intersections examined in this study imply that mare ridge segments extending into adjacent impact craters consist of materials that have been structurally uplifted from the crater rim and floor along high-angle splay faults. These segments can appear as asymmetrical structures involving displacement along a single fault, or symmetrical ridges indicative of uplift along bounding faults that converge with depth as schematically depicted in Fig. 7. We attribute the morphological differences between features in fresh and degraded craters to variations in the crater depth profile as well as the relative competency of the materials inside the craters. Because the bounding faults are converging, the uplifted zone narrows rapidly down the steep slopes of fresh craters, resulting in thin zones of uplift that appear better developed on the higher rim areas than on the floor. In contrast, the depth profile of degraded craters is significantly reduced by accumulations of unconsolidated crater fill and loss of rim topography. This results in a broad relatively uniform width to the uplifted zone that may appear somewhat lobate in plan view.

Fault Displacements and Attitudes

The paucity of overridden or truncated craters indicates that overthrusting is not characteristic of mare ridge formation. There is no indication of a significant strike-slip component to the displacements associated with the mare ridges we examined. While these observations do not preclude local thrusting or shearing along individual splay faults associated with ridge development, they do imply that ridge topography is generated primarily through uplift along steeply-dipping reverse faults, resulting in very small lateral offsets (typically < 20 m). In most cases, the total amount of shortening across a mare ridge does not significantly exceed that contained in the ridge topography.

TABLE 1. Evidence cited for the origin of lunar mare ridges.

Origin	Evidence
Volcanic (V)	(1) Ridges modify adjacent craters by a flow-like mechanism [1-5].
	(2) Braided, discontinuous, overlapping, or ropy patterns suggest an extrusive origin [1,3,5].
	(3) Some ridges are associated with sinuous rilles and other volcanic landforms [1,4].
	(4) Ridges are sometimes associated with positive gravity anomalies suggesting dyke intrusion.
	(5) The broad arch suggests a laccolith [1].
	(6) Ridges express some similarities to "squeeze-up" features in lava lake environments on Earth [3].
Tectonic (T)	(1) Mare surface topography implies tectonic deformation [7].
	(2) Ridges cross volcanic unit boundaries [4,8-10].
	(3) Several ridges extend into highlands as fault scarps [4,9,11,12].
	(4) Vertical offsets in the mare surface occur across some ridges [13].
	(5) Ridges modify fresh craters, thus the latest ridge formation postdates last regional volcanic episode [9,11,14,15].
	(6) Ridges often occur over buried premare topography [3,16,18,19].
	(7) Ridges sometimes deform postmare craters by shortening [2,9].
	(8) Ridge locations correlate with changes in basalt thickness where stresses might be localized [13,16,18-20].
	(9) Regional trends of mare ridges are consistent with global fracture patterns [16,21-23].
	(10) Circular basin trends are consistent with deformation due to loading by mare fill [18,19,22].
	(11) Local ridge trends may parallel regional topographic contours [24].

References: [1] *Strom* (1972); [2] *Howard and Muehlberger* (1973); [3] *Hodges* (1973); [4] *Young et al.* (1973); [5] *Scott* (1973); [6] *Scott* (1974); [7] *Scott et al.* (1978); [8] *Howard et al.* (1973); [9] *Lucchitta* (1976); [10] *El-Baz* (1978); [11] *Wolfe et al.* (1972); [12] *Colton et al.* (1975); [13] *Lucchitta* (1977); [14] *Muehlberger* (1974); [15] *Howard* (1978); [16] *Maxwell et al.* (1978); [17] *Peeples et al.* (1978); [18] *Maxwell* (1978); [19] *Sharpton and Head* (1982); [20] *De Hon and Waskom* (1978); [21] *Raitala* (1978); [22] *Solomon and Head* (1979); [23] *Fagin et al.* (1978); and [24] *Bryan* (1973).

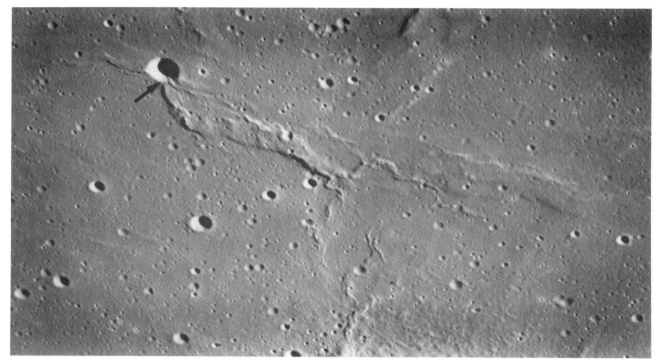

Fig. 2. Oblique photograph of area marked in Fig. 1 shows a small (aproximately 1 km diameter) crater that has been modified by a narrow, welt-like mare ridge lobe extending into the crater interior. Apollo 16-119-19158.

Because lateral offsets are typically very small, displacements along high-angle master fault systems seems to be required in order to account for the distinct vertical offsets (up to 250 m) and topography characteristic of many ridge systems. We

Fig. 3. The small crater (black arrow) appears to be substantially shortened by the mare ridge. White arrow points north and is 1 km in length. Black arrow indicates direction of solar illumination. Apollo 16-5439 pan.

have estimated the dip angle of ridge-related faults by measuring lateral excursions of detectable fault traces with depth inside 21 fresh simple craters, such as those in Fig. 8. Where available, depth control was provided by LTO maps, otherwise a depth-to-diameter ratio of 0.2 (*Pike*, 1977) was assumed. The frequency distribution of these fault attitudes (Fig. 9) also clearly indicates that ridge formation is associated with a broad range of fault attitudes, with high-angle faulting as the predominant mode of deformation. These results do not support recent suggestions by *Plescia and Golombek* (1986) that all mare ridge systems are simple low-angle reverse (thrust) faults.

DISCUSSION AND IMPLICATIONS

The results of our study suggest that the manner in which the "flow-like" ridge segments modify small adjacent impact craters is inconsistent with any type of lateral transport mechanism such as volcanism or mass wasting. Instead, the mare ridge extensions appear to be material that has been structurally uplifted along nearly vertical splay faults that are probably activated impact fracture systems. Thus a major argument for the volcanic model (Table 1, V-1) appears to be more consistent with a tectonic origin.

Assessment of Volcanic Models

The ropy, braided, or overlapping patterns mare ridges exhibit in plan view are suggestive of the influence of volcanic processes (Table 1, V-2). However, our analysis indicates that this appearance is due to a combined occurrence of the lobate features discussed above, and the complex patterns of mare ridge segments within individual systems. Several terrestrial examples of tectonic features with similar morphological attributes such as overlapping and braided arrangements have been discussed by *Plescia and Golombek* (1986). In addition, the lunar maria have sustained a long and intense history of impact bombardment. The frequent association of mare ridge systems with flooded premare topography such as basin rings (*Sharpton and Head*, 1982; *Maxwell*, 1978; *Maxwell et al.*, 1975) and crater rims (*Maxwell*, 1978; *Maxwell et al.*, 1975; *Hodges*, 1973) indicates the strong control large impact events have had on the location of mare ridges. Smaller craters can also fracture the mare and provide depressions in which ponding of subsequent lavas can occur. This suggests to us

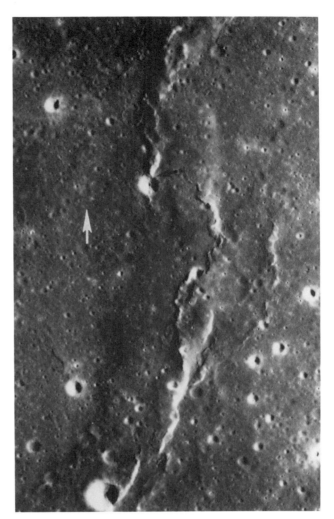

Fig. 5. Fresh crater modified by a narrow mare ridge lobe (black arrow). White arrow points north and is 1 km in length. Black arrow indicates direction of solar illumination. Apollo 16-5428 pan.

Fig. 4. Several craters (black arrows) of various degradation states illustrating the typical style of modification by mare ridges. Also note the apparent vertical offset in the rim of the southern-most affected crater (a). White arrow points north and is 1 km in length. Black arrows indicate direction of solar illumination. Apollo 16-5428 pan.

the possibility that small-scale variations in ridge trends may be controlled by local variations in competency induced by smaller (< 10 km diameter) but more abundant impact craters, thus providing an additional component of sinuosity to lunar ridge trends. An example of this effect is shown in Fig. 10, where a portion of the small crater's rim has been uplifted and integrated into the mare ridge, forming a distinct arc in the ridge trend. The remaining portion of the crater rim is almost totally inundated by subsequent mare deposits. Many similar examples indicate that local anisotropy induced by small craters may play an important role in determining ridge location and morphology.

The spatial association of mare ridges and the source depressions of volcanic sinuous rilles (Table 1, V-3) does not require that they be caused by the same process. The ascent of magma occurs preferentially along zones of weakness or fracture in planetary crusts. Tectonic deformation is also

Fig. 6. Close-up of bright material on a mare ridge slope illustrating the abundance of boulders (black arrows). White arrow points north and is 1 km in length. Black arrows indicate direction of solar illumination. Apollo 15-9358 pan.

associated with such zones. There is abundant evidence for mare ridge formation postdating most sinuous rilles (*Young et al.,* 1973; *Scott et al.,* 1978). Thus a more plausible explanation for this occasional association is simply the well-established correlation in space of tectonic and volcanic features.

The occurrence of positive gravity anomalies over some mare ridges (Table 1, V-4) has been reported by *Scott* (1974). He identified extended linear gravity anomalies, both positive and negative, in Oceanus Procellarum, trending northwest-southeast, roughly parallel to major systems of mare ridges in the area. The interpretation that the mare ridges correlate with positive gravity anomalies and "are the surface expressions of deep-seated systems of basaltic dikes" (*Scott,* 1974) is based on shifting the observed anomalies westward as much as 70 km "to produce a better correspondence, in places, between the positive gravity trends and some of the larger or more dense clusters of mare ridges" (*Scott,* 1974). The observation that mare ridges in this area (and elsewhere) are confined to the mare basalt units and do not modify the frequent "islands" of highland-type material disfavors the hypothesis that mare ridges are related to "basaltic dikes which have intruded fractures and faults in less dense crustal rocks" (Fig. 11a). Instead, we suggest that the elongate gravity anomalies, which parallel the mare-highlands boundary in western Oceanus Procellarum, express regional variations in the thickness of the mare fill due to subsurface topography. These thickness variations determine the local strength of the mare layer, influence the mare stress field, and therefore assert a control over the location of mare ridge systems. This results in a general parallelism between the basalt fill, the trends of major mare ridge systems, and observed gravity anomalies in Oceanus Procellarum.

The broad arch commonly associated with mare ridges has suggested to some investigators the presence of a shallow

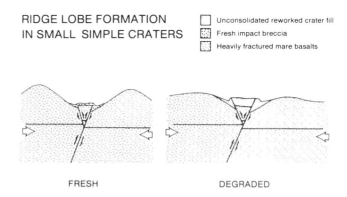

Fig. 7. Sketch of a cross section through a fresh (left) and a degraded (right) simple impact crater, illustrating uplift along high-angle reverse faults and the manner in which the degradation state of the crater affects the morphology of the ridge segments. (See text for details.)

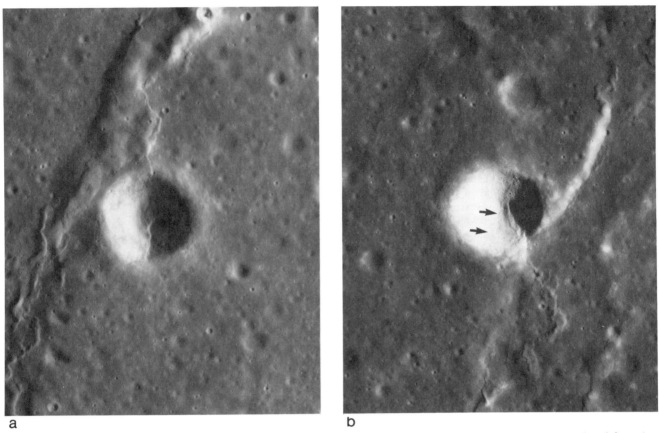

Fig. 8. Two fresh simple craters similar to those used in this study to estimate fault attitudes associated with mare ridge deformation. Both craters are approximately 1 km in diameter; north is up. (**a**) This crater shows a conspicuous near-vertical fault trace along the inner crater wall. Apollo 16-5433 pan. (**b**) The rim of this crater is offset slightly and the prominent fault trace dips approximately 50° W. An antithetic splay fault (located by arrows) appears to dip at approximately 50° E. Apollo 16-5428 pan.

laccolith-type intrusion (Table 1, V-5). The ridges would then be viscous extrusions of lava from extensional fractures along the crest of the arch (Fig. 11b). Several observations, however, pose difficulty for this model: (1) Mare arches typically follow regional topographic trends (Table 1, T-12), implying that the arches, like the variations in regional topography (Table 1, T-1), result from regional deformation of the mare surface (*Lucchitta*, 1977; *Scott et al.*, 1978); (2) ridges and arches frequently occur separately, and when observed together, the ridge segments may not trend parallel to the arch, sometimes crossing the arch or trending off the arch all together (*Hodges*, 1973; *Lucchitta*, 1977), suggesting only an indirect relationship between the two features; and (3) there is commonly a significant topographic offset on either side of the arch implying a tectonic rather than volcanic origin for these features (Table 1, T-5).

Fig. 9. Histogram of fault attitudes estimated from fault traces in 21 small fresh craters. These results indicate that a broad range of fault attitudes is characteristic of mare ridge deformation and that, in this sample, nearly vertical displacements are most common. We estimate the uncertainties associated with these measurements to be approximately ± 10°.

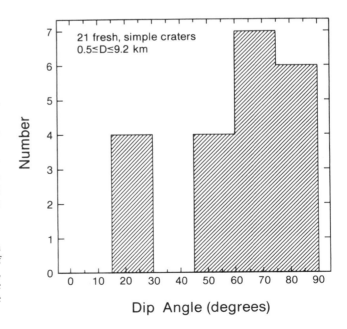

The auto-intrusive or "squeeze-up" hypothesis (Table 1, V-6) is somewhat similar to the laccolith model above. Here the morphology of mare ridges (disregarding monumental differences in scale) has been compared to small-scale features associated with the solidification of terrestrial lava lakes (*Hodges,* 1973). As the upper portion of the cooling lava solidifies, it tends to fracture and sink, squeezing partially solidified material up from below and creating small linear welts (Fig. 11c). However, in addition to the obvious difficulties with the scale differences, two observations seem to lessen the applicability of this analogy: (1) The presence of boulder patches on mare ridges and the modification of very fresh-appearing impact craters indicate that ridge formation continued well after solidification of the mare regions was complete. If mare ridges were squeeze-up features they should all be covered with the same dark, fine-grained regolith that mantles the mare surface and younger craters should show no modification by ridge formation; and (2) ridges are observed crossing distinct volcanic unit boundaries (Table 1, T-2).

Analysis of the lines of evidence supporting a tectonic origin for mare ridges reveals two independent observations that apparently preclude a volcanic origin for mare ridges. The first is that the lobate or ropy extensions ubiquitous to mare ridges modify adjacent features in a manner that is inconsistent with volcanic emplacement. The second is the observation that mare ridges cross known volcanic unit boundaries. Figure 12 shows two ridge segments that cross a distinct volcanic unit boundary in southeastern Mare Serenitatis. The portions of these ridges that modify the light mare unit to the north are also light and distinct; where the ridges cross into the dark southern unit, however, they assume the same dark, subdued mature appearance as that unit. Thus the surface texture, morphology, albedo, and color of mare ridges imply that they are a

Volcanic Models

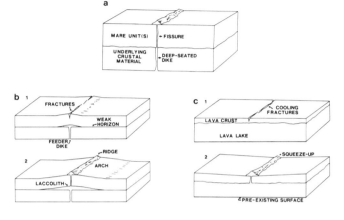

Tectonic Models

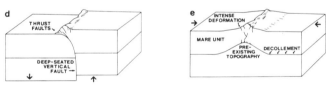

Fig. 11. Models of mare ridge formation; models **(a)**-**(c)** invoke volcanic processes, whereas **(d)** and **(e)** call upon tectonic processes. **(a)** Dike Model (*Scott,* 1972): Intrusions form deep within the crust beneath the mare basalts and continue upward through the volcanic units until lava is extruded onto the surface to form the mare ridge. **(b)** Laccolith Model (*Strom,* 1972): Intrusion of magma along a weak horizon between mare units domes the upper unit (1). As doming continues, bending stresses produce fractures through which lavas are ultimately extruded, forming the mare ridge (2). The bulbous and steep-sided nature of the ridge is explained by attributing a high viscosity to the lava. **(c)** Lava Lake Model (*Hodges,* 1973): The mare unit is emplaced as a deep lava lake. As the lava cools, a crust forms on the surface and fractures develop due to thermal contraction, convection or magma withdrawal (1). The denser crust begins to sink, squeezing molten or partially molten material up from below, creating the mare ridge. **(d)** Vertical Faulting Model (*Lucchitta,* 1976): Deep-seated vertical displacement is translated into numerous reverse splay faults at the surface, forming offsets in the mare surface across the ridge axis. **(e)** Buckling Model: Horizontal compressive stresses associated with volcanic loading result in buckling of the mare surface. Intense deformation, characterized by uplift and local thrusting along numerous splay faults, is concentrated in regions where the mare layer is weak, such as over buried topography or preexisting fracture systems. Buried regolith layers could decouple mare layers and act as *décollements* during buckling.

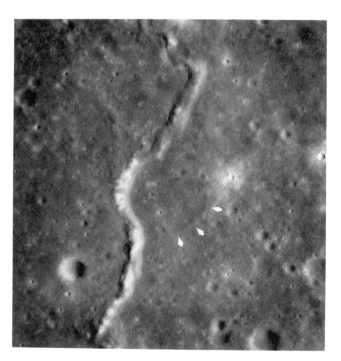

Fig. 10. Arcuate excursion in mare ridge trend apparently induced by a flooded impact crater (~1.2 km diameter). The western third of the crater rim has been uplifted and integrated into the narrow ridge system. Remnants of the eastern portions of the crater rim remain as a subtle arcuate ridge marked by arrows. Apollo 16-5427 pan.

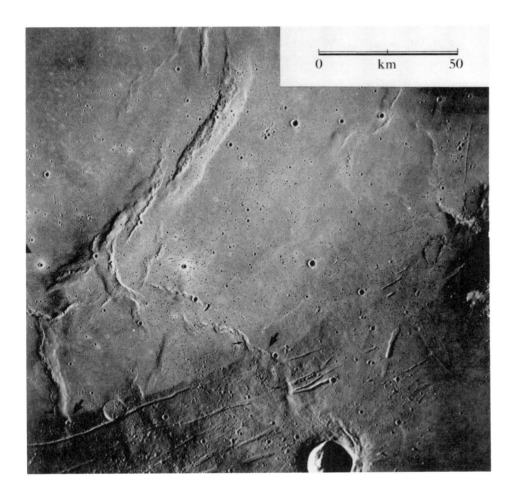

Fig. 12. Mare ridge systems crossing a prominent volcanic unit boundary in southeastern Mare Serenitatis. The intersection of the ridges and stratigraphic boundaries are marked by the larger black arrows; small arrows denote the trace of the ridges into the older, darker mare unit to the south. Apollo 17-0450 metric.

fundamental part of the individual mare units on which they occur. Because mare ridges cross stratigraphic boundaries, however, they cannot be related to the emplacement of these mare units. The volcanic model appears unable to resolve this dilemma; however, these observations are consistent with the tectonic origin of mare ridges.

Tectonic Models

The lunar maria contain a diverse suite of tectonic expressions, including concentric graben systems along the flanks of mare basins (*Lucchitta and Watkins*, 1978), postemplacement tilting of mare surfaces (*Muehlberger*, 1974, 1977; *Scott et al.*, 1978), and downwarping, faulting, and tilting of buried stratigraphic boundaries (*Sharpton and Head*, 1982). Figures 11d and 11e depict two generalized tectonic models for mare ridge formation that call upon either high-angle faulting or buckling associated with regional flexure. *Lucchitta* (1976) proposed that those mare ridges associated with asymmetrical arches and vertical offsets in the mare surface across the arch result from displacements along deep-seated vertical faults that may become low-angle reverse faults toward the surface, presumably because of compressive bending stresses at the surface (*Solomon and Head*, 1979) as shown in Fig. 11d. The second model (Fig. 11e) applies to those ridges not expressing appreciable vertical offsets or asymmetrical arches. Here buckling of the mare surface associated with downwarping and regional flexure could be localized over buried topography (Fig. 11e) or along other zones of weakness within the mare units. Thus the broad arch, when present, could result from regional flexure of the mare surface, and the ridge is the site of more intense (perhaps later) strain that results in deformation along discrete fault planes. Subsurface information across Mare Serenitatis collected by the Apollo 17 Lunar Sounder Experiment (ALSE) supports both tectonic models (*Sharpton and Head*, 1981). Of the four mare ridge systems transected by the ALSE ground track, the three systems that are asymmetrical in profile and are associated with discrete offsets in the mare surface are also associated with displacements in subsurface reflecting horizons consistent with high-angle faulting along preexisting zones of weakness (*Sharpton and Head*, 1981, 1982). The fourth mare ridge system exhibits a broad, well-defined arch, symmetrical in profile with no apparent vertical offsets. The subsurface characteristics of this feature suggest it occurs over the buried peak ring of the Serenitatis impact basin, that the volcanic units are significantly thinner over this topography, and that this region underwent continual deformation during the mare infilling stages of basin evolution (*Sharpton and Head*, 1982). Thus the two tectonic models presented in Figs. 11d and 11e appear to be consistent with all surface and subsurface observations to date, and may represent end-members in a continuous suite of tectonic mechanisms ranging from shallow buckling in response to regional flexure to near-vertical offsets along basement faults.

CONCLUSIONS

Although lunar mare ridges possess various morphological characteristics that might suggest an origin through volcanic processes, compilation and analysis of the observations in Table 1 reveal serious inconsistencies with the volcanic model. Results of this analysis clearly support the alternative model that mare ridges and their characteristic morphologies are the expressions on the mare surface of tectonic deformation within the lithosphere; associated volcanic features are of minor importance and unrelated to the basic genesis of mare ridges. The modification of fresh-appearing impact craters and the blocky, bright regions associated with mare ridges indicate that deformation of the mare surface continued well beyond the stage of mare infilling. Fault attitudes associated with mare ridge formation range from ~15° to nearly vertical. Topography accounts for almost all of the shortening across these ridges; large lateral displacements do not appear to be characteristic of these features. While local thrust faulting clearly accompanies the formation of some mare ridges studied, the negligible amounts of lateral offset typically associated with ridge deformation, coupled with vertical offsets of up to 250 m across ridge axes, disfavor simple thrust fault models as the predominant process in ridge formation. The general near-surface picture of mare ridges that emerges from our analysis is one of ridges as complex, diffuse zones within which deformation occurs along numerous small fault planes rather than a single discrete fault scarp.

The vertical faulting model and the buckling model depicted in Figs. 11d and 11e appear to account for the major characteristics of mare ridge morphology in a manner that is consistent with surface observations and subsurface data. If these models can be taken as end-member cases, then the morphological diversity expressed by lunar mare ridges would be influenced by a number of factors, including (1) variations in the magnitude and direction of the local stress field; (2) variations in the thickness of the volcanic layers; (3) the presence of preexisting zones of weakness such as buried craters, faults or buried topography; (4) the occurrence of buried regolith or pyroclastic layers that might act as *décollements* and decouple various mare units during buckling (Fig. 11e); and (5) variations in the density of impact craters that fracture and weaken the surface of the mare.

The strong control these material contrasts have over ridge morphology is demonstrated in the distinct morphological changes ridges undergo at certain volcanic unit boundaries (Fig. 12) as well as at the mare-highlands boundaries, where conspicuous fault scarps, not mare-type ridges, mark the continuations of faulting into the highlands. This indicates that the absence of mare-type ridges in highlands regions may not be regarded as evidence against compressional deformation in these regions. Furthermore, the occurrence of morphologically analogous features on other planets suggests that basic similarities in surface properties and tectonic environment exist between these ridged units and the lunar maria. Such similarities might include variables such as relative competency, thickness, or the presence of layering and detachment surfaces. In any case, mare ridges are surely complex features reflecting a complex tectonic setting, and no one variable or model is likely to account for all their characteristics.

Acknowledgments. Partial support for this research was provided through a grant from the Planetary Geology Program of the National Aeronautics and Space Administration (NGR-40-002-088) to J.W.H., which the authors gratefully acknowledge. Portions of this research were done while V.L.S. was a Visiting Scientist at the Lunar and Planetary Institute, which is operated by the Universities Space Research Association under Contract No. NASW-4066 with the National Aeronautics and Space Administration. This paper is Lunar and Planetary Institute Contribution No. 634.

REFERENCES

Bryan W. B. (1973) Wrinkle ridges and deformed surface crust on ponded mare lava. *Proc. Lunar Sci. Conf. 4th*, 93-106.

Colton G. W. (1978) The Maria (Figure 85). In *Apollo Over the Moon*, (H. Masursky, G. W. Colton, and F. El-Baz, eds.) pp. 67-103. NASA SP-362, NASA, Washington, D.C.

Colton G. W., Howard K. A., and Moore H.J. (1972) Mare ridges and arches in southern Oceanus Procellarum. In *Apollo 17 Preliminary Science Report*. pp. 29-90 to 29-93. NASA SP-315, NASA, Washington, D.C.

De Hon R. A. and Waskom J. D. (1978) Geologic structure of the eastern mare basins. *Proc. Lunar Sci. Conf. 7th*, 2729-2746.

El-Baz F. (1978) The Maria (Figure 79). In *Apollo Over the Moon* (H. Masursky, G. W. Colton, and F. El-Baz, eds.) pp. 67-103. NASA SP-362, NASA, Washington, D.C.

Fagin S. W., Worrall D. M., and Muehlberger W. R. (1978) Lunar mare ridge orientations: implications for lunar tectonic models. *Proc. Lunar Planet. Sci. Conf. 9th*, 3473-3479.

Greeley R. and Spudis P. D. (1978) Ridges in western Columbia Plateau, Washington—Analogs to mare-type ridges? (abstract) In *Lunar and Planetary Science IX*, pp. 411-412. Lunar and Planetary Institute, Houston.

Hodges C. A. (1973) Mare ridges and lava lakes. In *Apollo 17 Preliminary Science Report*, pp. 31-12 to 31-21. NASA SP-330, NASA, Washington, D.C.

Howard K. A. (1978) The Maria (Figure 78). In *Apollo Over the Moon* (H. Masursky, G. W. Colton, and F. El-Baz, eds.) pp. 67-103. NASA SP-362, NASA, Washington, D.C.

Howard K. A. and Muehlberger W. R. (1973) Lunar thrust faults in the Taurus-Littrow region. In *Apollo 17 Preliminary Science Report*, pp. 31-22 to 31-25. NASA SP-330, NASA, Washington, D.C.

Howard K. A., Carr M. H., and Muehlberger W. R. (1973) Basalt stratigraphy of southern Mare Serenitatis. In *Apollo 17 Preliminary Science Report*, pp. 29-1 to 29-12. NASA SP-330, NASA, Washington, D.C.

Lucchitta B. K. (1976) Mare ridges and related highland scarps—Results of vertical tectonism? *Proc. Lunar Sci. Conf. 7th*, 2761-2782.

Lucchitta B. K. (1977) Topography, structure, and mare ridges in southern Mare Imbrium and northern Oceanus Procellarum. *Proc. Lunar Sci. Conf. 8th*, 2691-2703.

Lucchitta B. K. and Klockenbrink J. L. (1981) Ridges and scarps in the equatorial belt of Mars. *Moon and Planets, 24*, 415-429.

Lucchitta B. K. and Watkins J. A. (1978) Age of graben systems on the moon. *Proc. Lunar Planet. Sci. Conf. 9th*, 3459-3472.

Maxwell T. A. (1978) Origin of multi-ring basin ridge systems: An upper limit to elastic deformation based on finite element analysis. *Proc. Lunar Planet. Sci. Conf. 9th*, 3541-3559.

Maxwell T. A., El-Baz F., and Ward S. H. (1975) Distribution, morphology, and origin of ridges and arches in Mare Serenitatis, *Geol. Soc. Am. Bull.*, 86, 1273-1278.

Muehlberger W. R. (1974) Structural history of southeastern Mare Serenitatis and adjacent highlands. *Proc. Lunar Sci. Conf. 5th*, 101-110.

Peeples W. J., Sill W. R., May T. W., Ward S. H., Phillips R. J., Jordan R. L., Abbott E. A., and Killpack T. J. (1978) Orbital radar evidence for lunar subsurface layering in Maria Serenitatis and Crisium, *J. Geophys. Res.*, 83, 3459-3470.

Pike R. J. (1977) Apparent depth/diameter relation for lunar craters *Proc. Lunar Sci. Conf. 8th*, 3427-3436.

Plescia J. B. and Golombek M. P. (1986) Origin of planetary wrinkle ridges based on the study of terrestrial analogs. *Geol. Soc. Am. Bull.*, 97, 1289-1299.

Raitala J. (1978) Tectonic patterns of mare ridges of the Letronne-Montes Riphaeus region of the moon. *Moon and Planets*, 19, 457-477.

Schultz P. H. (1976) Moon Morphology. Univ. of Texas, Austin, 626 pp.

Scott D. H. (1973) Small structures of the Taurus-Littrow region. *Apollo 17 Preliminary Science Report*, pp. 31-25 to 31-29. NASA SP-330, NASA, Washington, D.C.

Scott D. H. (1974) The geological significance of some lunar gravity anomalies. *Proc. Lunar Sci. Conf. 5th*, 3025-3036.

Scott D. H., Watkins J. A. and Diaz J. M. (1978) Regional deformation of mare surfaces. *Proc. Lunar Planet. Sci. Conf. 9th*, 3527-3539.

Sharpton V. L. and Head J. W. (1981) The origin of mare ridges: evidence from basalt stratigraphy and substructure in Mare Serenitatis. In *Lunar and Planetary Science XII*, pp. 961-963. Lunar and Planetary Institute, Houston.

Sharpton V. L. and Head J. W.(1982) Stratigraphy and structural evolution of southern Mare Serenitatis: A reinterpretation based on Apollo Lunar Sounder Experiment data. *J. Geophys. Res.*, 87, 10,983-10,998.

Solomon S. C. and J. W. Head (1979) Vertical movement in mare basins: Relation to mare emplacement, basin tectonics, and lunar thermal history. *J. Geophys. Res.*, 84, 1667-1682.

Strom R. G. (1972) Lunar mare ridges, rings and volcanic ring complexes. In *The Moon*, (S. K. Runcorn and H. C. Urey, eds.), pp. 187-215.

Tjia H. D. (1970) Lunar wrinkle ridges indicative of strike slip faulting, *Geol. Soc. Am. Bull.*, 81, 3095-3100.

Watters T.R. and Maxwell T. A. (1985) Mechanisms of basalt-plains ridge formation. *Reports of Planetary Geology and Geophysics Program, 1984*, pp. 479-481. NASA TM-87463, NASA, Washington, D.C.

Wolfe E. W., Freeman V. L., and Muehlberger W. R. (1972) Apollo 17 Exploration at Taurus-Littrow. *Geotimes*, 17, 14-17.

Highland Contamination and Minimum Basalt Thickness in Northern Mare Fecunditatis

William H. Farrand

Lunar and Planetary Laboratory and Department of Geosciences, University of Arizona, Tucson, AZ 85721

Orbital X-ray fluorescence data, multispectral images, and Luna 16 and 20 sample chemistry were used to investigate the extent and origin of the contamination of mare basalts by highland materials in northern Mare Fecunditatis. Various combinations of multispectral and geochemical data were used to construct classification maps of the region using an unsupervised classifier. The classifier distinguished different mare, highland, and mixed units, but no systematic trend of highland contamination was evident. Using Al/Si as the sole discriminant of highland contamination produced better results. The Al/Si ratios that would result from various admixtures of highland materials in selected mare basalts were calculated and these numbers were used as thresholds on the Al/Si image to obtain maps that display the level of highland contamination in the mare. The agreement between these maps and the photogeologically derived isopach map of Mare Fecunditatis suggests that *vertical* mixing is the dominant process responsible for the highland component in mare soils. Using a theoretical model of regolith evolution that relates the percentage of highland materials in the mare regolith to the thickness of the basalt fill, lower limits of mare basalt thickness were established. For different types of basalt that might fill the northern portion of the Fecunditatis basin, minimum basalt volumes ranging from 2,140 km^3 to 20,540 km^3 were calculated.

INTRODUCTION

The Soviet Luna 16 spacecraft landed in the northeast corner of Mare Fecunditatis at 0°41′S and 56°18′E and returned a 101 g mare regolith sample, about 25% of which was made up of lunar highlands material (*Rhodes*, 1977). Mare regolith samples returned by Apollo 11 and 12 also contained 25% to 70% of nonbasaltic materials (*Rhodes*, 1977), while the regolith sample returned from Mare Crisium by Luna 24 contained a 2% to 10% highland component (*Bence and Groves*, 1978; *Hörz*, 1978). Early speculation presumed that the highland component in the Luna 16 sample was ejecta from the crater Langrenus, which is located 280 km to the southeast of the Luna 16 landing site (*Reid et al.*, 1971). Such an explanation is plausible, but it presumes that Luna 16 landed on a ray from Langrenus. Such a special case seems unlikely in view of the nonmare component at the Apollo 11, 12, and Luna 24 landing sites, which are even further from major highland craters than is the Luna 16 site. There is also evidence from remote sensing data that a significant highland component can be found throughout the lunar maria.

Orbital X-ray fluoresence measurements of surface Al/Si and Mg/Si intensity ratios and gamma ray measurements of Fe and Th abundances indicate that a significant highland component is common in lunar maria. *Hubbard and Wolozyn* (1977) noted that no lunar mare overflown by the Apollo spacecraft had Fe values as high or Th values as low as are found in mare basalt hand samples. Similarly, examination of the Al/Si database shows that Al/Si ratios are uniformly higher in the maria than would be expected if these areas were covered by pristine mare basalt. Therefore, a key question is raised as to the origin of this nonmare component. Has it been spread laterally over the mare as ejecta from impacts into the surrounding highlands or has it been vertically excavated from beneath the mare basalts by impacts within the mare?

Hörz (1978) and *Rhodes* (1977) have argued in favor of a vertical mixing origin for the highland component. Among the evidence cited by the above authors against extensive lateral mixing is the observation of sharp contacts between mare and highland terrains. Orbital measurements of surface chemistry show relatively sharp edges between maria and highlands (*Adler et al.*, 1974) as do Earth-based albedo measurements (*Pohn and Wildey*, 1970). These observations are supported by samples collected at the Apollo 15 and 17 landing sites, both of which border mare and highland terrains. Within 4–6 km of the mare/highland boundary, the highland component in the regolith at both sites drops to levels comparable to those observed at intramare landing sites.

Petrologic and geochemical studies on the regolith sampled at the Apollo landing sites have also indicated that lateral transport is an inefficient process (*Labotka et al.*, 1980; *Laul and Papike*, 1980). These studies concluded that the highland component sampled at mare landing sites was derived by vertical mixing.

A vertical mixing origin that would explain the highland component within the maria would require mare basalts to be thin enough to allow craters a few hundred meters to a few kilometers in size to excavate highland materials from the basin floor. However, the thickness of lunar mare basalts is still uncertain. Circular mascon basins such as Imbrium, Serenitatis, and Crisium must have basalt thicknesses in excess of 1 km in order to account for their associated positive gravity anomalies. Alternatively, irregular basins such as Tranquilitatis and Fecunditatis, which have no gravity anomalies, can have much thinner basalt fills. Where mare basalt thicknesses are greater than 1 km, extensive vertical mixing should not occur (*Hörz*, 1978). To account for the presence of highland material in mare regolith samples, either great expanses of mare basalt are less than 1 km thick or lateral mixing is a more efficient process than has been purported. One goal of this study was

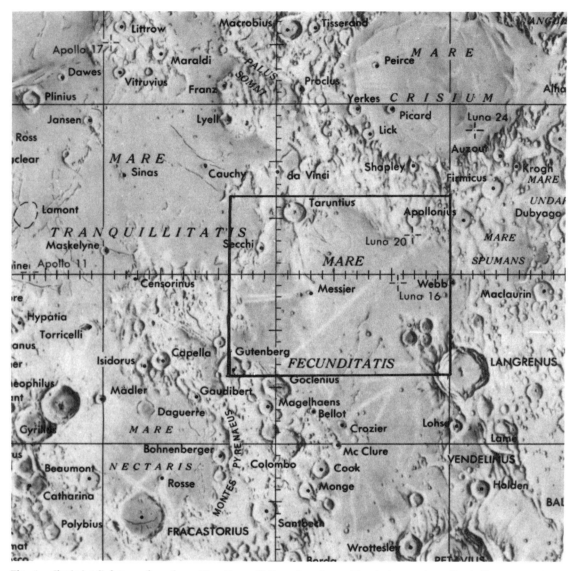

Fig. 1. Shaded relief map of northern Mare Fecunditatis and environs. The study area is outlined. (base map - NASA LPC-1)

to map the distribution of highland materials in the regolith of northern Mare Fecunditatis and see if it could be explained most reasonably by a lateral or a vertical mixing origin.

The northern Mare Fecunditatis region shown in Fig. 1 is particularly well suited for an examination of the extent and cause of highland contamination of mare soils. There is good orbital X-ray and gamma ray coverage of the region and there are two sites within the area that have been sampled. In addition to the Luna 16 mission, Luna 20 landed in the highlands 120 km north of Luna 16 at 3°32′N and 56°33′E. Therefore, there is ground-truth data for both the mare and the highland material that may have contaminated it.

GEOLOGICAL SETTING

Mare Fecunditatis is an irregularly shaped basin, probably of pre-Nectarian age, that is filled with young Imbrian-age mare basalts (*Wilhelms*, 1984). *Pieters* (1978) has identified two spectrally distinct basalts in northern Mare Fecunditatis. Luna 16 landed in an embayment of basalts given the label hDW by *Pieters* (1978). The h refers to a medium-high UV/visible ratio, D indicates a low albedo, and W denotes a weak 1 μm band. This embayment also shows up in the color-difference photography of *Whitaker* (1972). The basalts sampled by Luna 16 were of two types: an iron-rich, low-Ti basalt and a high alumina basalt (*Cimbalnikova et al.,* 1976). The crystallization ages of Luna 16 basalts have been dated at about 3.4 b.y. (*Basaltic Volcanism Study Project,* 1981). This age is in agreement with the 3.4 ± 0.1 b.y. age determined for most of Mare Fecunditatis by the crater-morphology dating technique of *Boyce* (1976).

The basalts covering the rest of northern Mare Fecunditatis are designated mIG for their medium UV/visible ratio, their intermediate albedo and the average strength of their 1 μm band (*Pieters,* 1978). The mIG basalts share the spectral characteristics of low-Ti basalts sampled by Apollo 12. Consequently, in some of the modelling discussed below, Apollo 12 low-Ti basalts are used as possible analogues of uncontaminated Mare Fecunditatis basalts.

Mare ridges within northern Mare Fecunditatis trend approximately northwest and northeast (*Whitford-Stark,* 1986). A notable exception is the arcuate Dorsa Geike ridge in the north-central portion of the mare that is generally north trending.

Prominent in the surrounding highlands are the Copernican-aged craters Langrenus and Taruntius, which have radii of 66 and 28 km, respectively. Both craters have continuous ejecta blankets and ray patterns that overlap parts of the mare.

REMOTE SENSING DATABASES

To map the distribution and amount of highland materials over the length and breadth of northern Mare Fedunditatis requires the regional view offered by remote sensing data. The images of surface chemistry derived from the orbital geochemical data of the Apollo 15 and 16 missions are particularly well suited for studying regional trends in regolith composition. Of the elements mapped from orbit, aluminum and iron are the two that are best suited for discriminating between highland and mare materials, because iron is generally low in the highlands and high in the maria with the reverse being true for aluminum. The spatial resolution of the Fe data is on the order of 300 km, while that of the Al/Si data is more on the order of 30 km. Since the study area shown in Fig. 1 is only about 570 by 360 km in size, the Fe data is not of much use in detecting the local changes in surface chemistry sought in this study. Moreover, there is a very strong inverse linear correlation between iron and aluminum in the lunar sample collection (*Miller et al.,* 1974), so similar results should have been achieved using either iron or aluminum. Since the Al/Si data does have the better spatial resolution, the image of Al/Si ratios produced by *Clark and Hawke* (1981) was the primary data set used in this study.

The Mg/Si X-ray database is not as efficient in discriminating between highland and mare materials as is Al/Si because of the very small range of Mg values and their overlap in mare and highland rocks. Also, the Mg/Si data is generally less reliable than the Al/Si data. Uncertainties in the Mg/Si data are on the order of 20%, while those of the Al/Si data are about 15%. Even so, Mg/Si proved useful as a second parameter to complement Al/Si, particularly in the mixing models described below.

Conversions from Al/Si and Mg/Si intensity ratios to the corresponding concentrations ratios and hence to Al_2O_3 and MgO were achieved through the use of a ground truth normalization scheme similar to that used by *Bielefeld* (1977) but adapted for use with the more recent *Clark and Hawke* (1981) X-ray database. Table 1 shows the Apollo and Luna sampling areas used along with the Al/Si and Mg/Si concentration ratios of soil samples from those areas. Conversions from oxide to concentration ratio were made assuming Si is constant at 21 wt % (*Bielefeld,* 1977). Also listed in Table 1 are the locations of regions considered by the author to be representative of the areas sampled along with the average Al/Si and Mg/Si intensity ratios for those regions. In some cases, these regions are slightly removed from the landing sites so that the mixing of the X-ray signal from mare and highland terrains might be avoided. Figure 2 shows the regression lines for the Al/Si and Mg/Si normalizations. In a preliminary investigation into the problem of highland contamination in northern Mare Fecunditatis (*Farrand,* 1987),

TABLE 1. Al/Si and Mg/Si concentration and intensity ratios used in ground truth normalization.

Ground truth area	$(Al/Si)_c$*	$(Mg/Si)_c$*	Central pixel location†	$(Al/Si)_i$	$(Mg/Si)_i$
Apollo 11	0.34	0.23	30°E, 10°N	0.88 ± .12	1.24 ± .19
Luna 16	0.39	0.25	56°E, 0°	1.00 ± .11	1.23 ± .19
Luna 20	0.59	0.26	56°E, 3°N	1.54 ± .11	1.25 ± .07
Apollo 15 mare	0.32	0.31	1.5°E, 26.5°N	0.85 ± .12	1.25 ± .24
Apollo 15 highlands	0.41	0.32	2.5°E, 23°N	1.04 ± .08	1.30 ± .13
Apollo 17 mare	0.28	0.27	29.5°E, 21.5°N	0.83 ± .13	1.23 ± .13
Apollo 17 highlands	0.41	0.28	32°E, 19°N	1.16 ± .12	1.16 ± .11

*Soil compositions from *Laul and Papike,* 1980; *Basaltic Volcanism Study Project,* 1981; *Duncan et al.,* 1975; *Rhodes et al.,* 1974.

†Where possible, intensity ratios were averaged within a 7 × 7 pixel (1.75° × 1.75°) box over the area sampled.

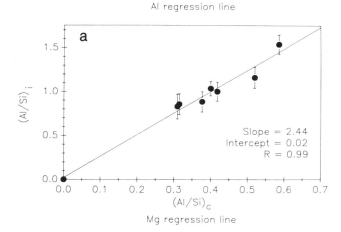

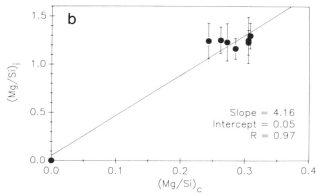

Fig. 2. (a) Ground truth normalization of Al/Si concentration (($Al/Si)_c$) to Al/Si intensity (($Al/Si)_i$) ratios. (b) Ground truth normalization of Mg/Si concentration (($Mg/Si)_c$) to Mg/Si intensity (($Mg/Si)_i$) ratios.

the $Al/Si-Al_2O_3$ conversion formulas of *Clark and Hawke* (1987) were used. However, the Al_2O_3 values calculated from those formulas were finally deemed too high; hence, the switch to the ground truth normalization.

Also used in this study were three images of the lunar nearside taken from Earth through 0.38 μm, 0.56 μm, and 0.89 μm filters (L. A. Soderblom, unpublished data, 1978). The spatial resolution of these images is essentially that of the orbital X-ray data (P. A. Davis, personal communication, 1986), and they were used to complement the Al/Si data in discriminating geologically distinct surface units. The utility of the 0.56 μm image is that it can be used as a measure of surface albedo and as such it can serve as an additional tool for discriminating between highland and mare terrains. The ratio of 0.89/0.38 μm can be used to distinguish between young and old terrains because it is sensitive to the agglutinate content of the regolith (*BVSP*, 1981). The 0.38/0.56 μm ratio is most strongly related to TiO_2 content (*Johnson et al.*, 1977), and thus is useful for distinguishing between high, low, and very low Ti-basalts.

CLASSIFICATION OF MARE UNITS

A first attempt at mapping highland contamination in Mare Fecunditatis was made using the unsupervised cluster analysis program devised by *Jayroe et al.* (1976), which was first used on lunar remote sensing data by *Bielefeld et al.* (1978). It was hoped that the cluster analysis would subdivide the mare into units that could each be characterized by a different level of highland contamination. Various combinations of multispectral and orbital geochemical images of northern Mare Fecunditatis and its environs were analysed by the classifier. The assignment of image pixels into feature space clusters (which ultimately translate into map units) was made on the basis of a majority decision rule among linear discriminant function, nearest neighbor and maximum likelihood classification algorithms.

The two most promising maps of the northern Mare Fecunditatis area that resulted from the classification process are presented as Figs. 3a and 3b. Parameters for the units in both maps are given in Table 2. The resulting units in both maps bear a close resemblance to the geologic units in U.S.G.S. quadrangle maps of the region (*Elston*, 1972; *Wilhelms*, 1972; *Hodges*, 1973; *Olsen and Wilhelms*, 1974), indicating that the parameters used in the classification are representative of the region's geology.

Four different mare surface units are evident on Fig. 3a: Unit a1 blankets most of the mare and corresponds to the mIG basalt. Unit a2 occurs most notably in the northeastern portion of the mare and seems to correspond to the hDW basalt sampled by Luna 16. In the southeast, the high albedo unit a3 corresponds to the continuous ejecta blanket of Langrenus. Unit a4 is low in albedo but high in the 0.38/0.56 μm ratio, which could indicate the presence of high Ti basalt.

Adding Al/Si data to the classification process has some interesting consequences. While units b1, b2, and b3 are analogous to a1, a2, and a3, the areal extent of b2 and b3 have grown at the expense of b1. Unit b4, which seems totally unrelated to unit a4, occurs near the borders of the mare, is high in Al/Si and albedo and thus seems to be contaminated by highland material. While units b5, b6, and b7 appear within the boundaries of the mare, their Al/Si and albedo values are more reminiscent of the highlands. The questionable provenance of these units is best illustrated in the scattergram plot of Mg/Si versus Al/Si ratios shown in Fig. 4. The first four units from Fig. 3b plot solidly within the mare cluster while units b5, b6, and b7 plot between the mare highland clusters. The occurrence of units b6 and b7 at the southeastern corner of the map can be interpreted as the recognition of ejecta from, and the northwestern rim of, the crater Langrenus. Other occurrences of these units within the mare could be attributed either to very high levels of highland contamination or to misclassification.

In order to gain a qualitative understanding of the petrologic character of the units, a least squares mixing model (*Bryan et al.*, 1968) with two oxides, Al_2O_3 and MgO, and two end members was used. One end member, representing the

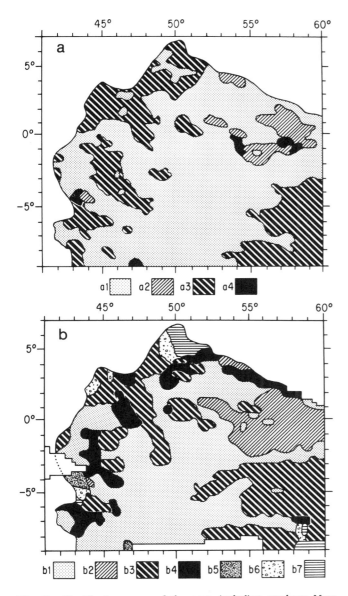

Fig. 3. Classification maps of the area, including northern Mare Fecunditatis and the northwestern rim of Langrenus, which were constructed using an unsupervised classification program. Map units represent areal distributions of pixel clusters resulting from classification of (a) 0.56-μm albedo, 0.38/0.56-μm ratio, and 0.89/0.38-μm ratio, and (b) the preceding three databases in addition to the Al/Si database.

TABLE 2. Parameters used to construct classification maps of Fig. 3.

Class	$(Al/Si)_i$	Albedo	.38/.56 μm	.85/.38 μm
a1	-	0.095	1.11	0.99
a2	-	0.087	1.17	0.95
a3	-	0.106	1.08	1.03
a4	-	0.100	1.12	0.94
b1	1.00	0.096	1.11	0.99
b2	1.07	0.086	1.15	0.95
b3	1.04	0.101	1.08	1.01
b4	1.08	0.110	1.11	1.03
b5	1.23	0.113	1.13	0.97
b6	1.19	0.115	1.09	1.02
b7	1.26	0.115	1.07	1.05

highland contaminant, had the composition of the Luna 20 soil sample (*BVSP,* 1981) while each of the four basalts listed in Table 3 were tested as the second end member. The best-fit basalt for all the map units turned out to be an average of Apollo 12 low-Ti basalts (*BVSP,* 1981) although the average high-alumina basalt of *Cimbalnikova et al.* (1976) fit the model equally well for some units.

The mixing model results showed that all the mare units have substantial amounts of highland materials. However, it was difficult to identify any spatial relation among the most highly contaminated units that would indicate whether vertical or lateral mixing was more responsible for the contamination. In order to get a less ambiguous picture of highland contamination in northern Mare Fecunditatis, in the following analysis only the Al/Si data utilized.

In this next step, an "inverse mixing model" approach was tried. In the mixing model calculation just discussed, the concentrations of Al_2O_3 and MgO in the end members and in the mare units were used to calculate the relative proportions of the end members in each unit. In the inverse mixing model, some uniform basalt composition is assumed to cover all of northern Mare Fecunditatis. Al/Si concentration ratios for the region are then calculated for 0%, 15%, 30%, and 45% admixtures of the highland component used in the mixing models into this basalt. These Al/Si ratios are then converted to image grey level and used as thresholds on the image of Al/Si intensity ratios to segment the image into a series of units, each with a different level of highland contamination. The boundaries between these units can be thought of as contours of highland contamination and the maps presented in Fig. 5 are referred to below as highland contamination maps.

The work of *Pieters* (1978) indicates that the assumption of a single basalt type covering northern Mare Fecunditatis could be a false one. Nevertheless, the spread in aluminum values between the mare units, a1-a4 and b1-b4, was minimal, and all the map units could be modelled as having the composition of single basalt. Therefore, the assumption of a single basalt type seems reasonable. Even if false, it should not significantly affect the overall contamination trends since the hDW basalt (unit a2 and/or b2 from Fig. 3) only covers the northeastern corner of the mare; therefore, contamination trends over most of the mare will be unaffected.

With the above assumption in mind, Fig. 5a shows the effect on the Al/Si image of thresholds calculated for a mare covered with a basalt having the Al content of the high-Al basalt sample B1 (*BVSP,* 1981). If Fecunditatis were covered with a less aluminous basalt, such as the average Apollo 12 low-Ti basalt used in the mixing models, higher levels of highland contamination would be required to match the observed Al/Si signal. Notice in Fig. 5b that the 0% contamination unit has disappeared for this case. Alternatively, if the Fecunditatis

TABLE 3. End members used in mixing model calculations.

End members	Al_2O_3	MgO
Basalts		
Apollo 12 low-Ti*	9.00	10.38
Luna 24 VLT†	13.19	6.50
Luna 16 sample B1†	13.75	7.05
Luna 16 high-Al‡	16.25	8.86
Highland Component		
Luna 20 soil†	23.44	9.19

*Average of ilmenite, olivine, and pigeonite basalts from *BVSP* (1981).
†*Basaltic Volcanism Study Project* (1981); where more than one sample was listed, an average was used.
‡*Cimbalnikova et al.* (1976).

considers Fig. 5a as coming closest to representing the true state of highland contamination in Mare Fecunditatis. Since the Apollo 12 low-Ti basalt fit in well with the mixing models and was noted by *Pieters* (1978) as being spectrally similar to the mIG basalt, a case could be made for Fig. 5b being the best representation. However, the levels of highland contamination required by a mare covered by the Apollo 12 low-Ti basalt seem excessively high. Likewise, the high-Al basalt of *Cimbalnikova et al.* (1976) is so aluminous that a mare covered by it would need almost no highland contamination, and this, too, is unlikely.

HIGHLAND CONTAMINATION AND RELATIVE BASALT THICKNESS

If highland materials are excavated from beneath the mare fill, then the lowest levels of highland contamination should occur where the basalt flows are thickest. Remotely sensed geochemical data could then be used as a tool to estimate the relative thickness of mare basalts.

The most comprehensive mapping of the thickness of the basalt fill in Mare Fecunditatis was conducted using photogeological techniques. Craters partially buried by mare basalts were analyzed by *DeHon* (1975), *DeHon and Waskom* (1976) and *Hörz* (1978). *DeHon* (1975) made two major assumptions in his study. First, he assumed that all the craters used in his analysis had the depth, diameter, and rim height relations of fresh craters. Second, he assumed that the craters he examined were formed on basin floor materials. Basalt thickness was then taken to be equal to the theoretical rim height minus the observed rim height over the mare surface. *Hörz* (1978) assumed that the flooded craters were in a variety of degradational states at the time of basalt flooding. Thus using a population of craters in random states of degradation, he decreased *DeHon's* (1975) thickness estimates by a factor of two. The isopach map for northern Mare Fecunditatis with the thickness values of *Hörz* (1978) is reproduced here as Fig. 6.

A natural step in this investigation was to compare the isopach map to the highland contamination maps. By comparing Fig. 6 with Figs. 5a and 5b, one can see that the lowest Al/Si ratios in the mare correspond very well with the photogeo-

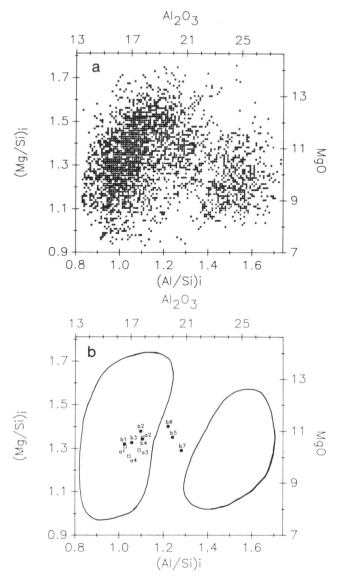

Fig. 4. (a) Scattergram plot showing the relation between Mg/Si and Al/Si intensity ratios for northern Mare Fecunditatis and the immediately surrounding highlands. Each dot represents a single pixel. The left cluster represents mare areas while the right cluster represents highland areas. (b) Positions of classification map units from Fig. 3 are shown relative to outlines of mare and highland clusters.

basin were covered with an extremely aluminous basalt, such as the high-Al basalt of *Cimbalnikova et al.,* (1976), Fig. 5c shows that large portions of the mare would be uncontaminated. Only the very edges would require contamination levels as high as 45%.

It is unclear which one of the highland contamination maps most accurately depicts the true situation. Since the high-Al basalt B1 was collected from northern Mare Fecunditatis and is listed in references such as the *BVSP* (1981) as the sample most representative of the Luna 16 high-Al basalts, this author

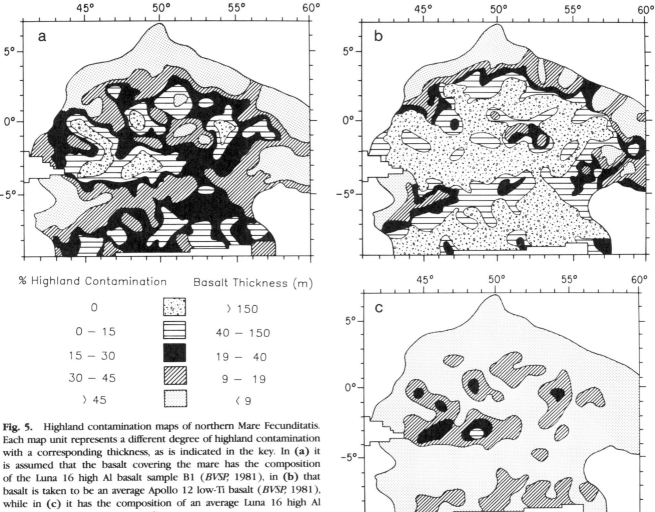

Fig. 5. Highland contamination maps of northern Mare Fecunditatis. Each map unit represents a different degree of highland contamination with a corresponding thickness, as is indicated in the key. In **(a)** it is assumed that the basalt covering the mare has the composition of the Luna 16 high Al basalt sample B1 (*BVSP,* 1981), in **(b)** that basalt is taken to be an average Apollo 12 low-Ti basalt (*BVSP,* 1981), while in **(c)** it has the composition of an average Luna 16 high Al basalt (*Cimbalnikova et al.,* 1976).

logically determined deepest portion of the mare. Attention should be drawn to the fact that *DeHon* (1975) placed the deepest part of the basin somewhat west of the mare's center (on the basis of the supposition that the Dorsa Geike mare ridge represents the buried rim of a large crater). While not precisely matching DeHon's placement, the highland contamination maps indicate that the area least contaminated by highland materials (corresponding to the area with the thickest basalt cover in a vertical mixing case) also lies west of center. The isopach map also shows that there is a roughly east-west trending trough running west from the deepest portion of the mare. Again, on the highland contamination maps an east-west trending swath with low levels of highland contamination occurs in the vicinity of the trough. If lateral mixing were responsible for the highland contamination, the lowest contamination levels would occur at a point equidistant from the surrounding highlands, i.e., at the exact center of the mare. Also, if there were an elongate swath of low contamination in the lateral mixing case, it would trend north-south (paralleling the north-south elongation of Mare Fecunditatis) rather than in the roughly east-west direction that is evident in Fig. 5.

A more analytical comparison between the relation of basalt thickness to Al/Si was made by digitizing the isopach map and using the bands between isopachs as masks on the Al/Si image. The results are given in Table 4. With one exception, Al/Si ratios decreased from shallow to deep. The exception was a band that included the southern portion of the study area. The highland contamination maps show this as an area with less highland materials and, hence, thicker basalts than might be expected from examination of the isopach map.

The correspondence between the geochemically derived highland contamination maps and the photogeologically derived isopach map lends some credence to both methods. Most importantly, it shows that *vertical* mixing is responsible for the highland component in mare soils. Consequently, remotely-sensed geochemical information can be used to map relative basalt thicknesses in the maria.

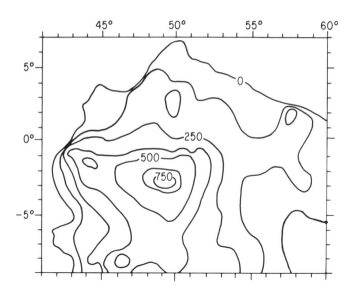

Fig. 6. Isopach map of northern Mare Fecunditatis with basalt thicknesses in meters from *Hörz* (1978).

TABLE 4. Aluminum averages for isopach map units.

Isopach unit	$(Al/Si)_i$	Al_2O_3
0-125 m	1.15	18.3
125-250 m	1.02	16.2
250-375 m	0.96	15.2
375-500 m	0.98	15.6
500-625 m	0.97	15.4
625-750 m	0.84	13.3
> 750 m	0.78	12.3

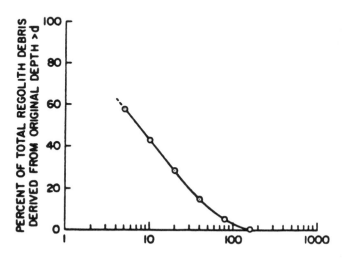

Fig. 7. Relation between the percentage of regolith volume and the depth from which any such percentage might originate for a regolith with a median depth of 4.7 m and an age of approximately 3.65 b.y. (after *Quaide and Oberbeck*, 1975).

Given certain caveats, a semiquantitative statement can be made about the thickness of mare basalts based on the amount of highland material in the mare regolith. *Quaide and Oberbeck* (1975) developed a model of mare regolith evolution using a Monte Carlo computer program that simulated regolith evolution from pristine basaltic bedrock to maturity. They noted that regolith thickness can be related to the total number of craters in a given region and that most regolith debris is locally derived. They were also able to establish a relation between the percentage of regolith material and the depth below which it could be derived. Therefore, given some percentage of highland debris in a mare regolith, its depth of origin (i.e., the thickness of overlying basalt) can be estimated. Quaide and Oberbeck's relation is reproduced here as Fig. 7. The percentages of highland contaminant that were used in the inverse mixing model can be translated into basalt thickness via Fig. 7; these thicknesses are included in the key to Fig. 5.

It should be noted that the crater distribution and median regolith thickness implicit in Fig. 7 are for an area including the Apollo 11 landing site. On the relative age scale of *Boyce* (1976), the Apollo 11 site is older than the Luna 16 site and northern Mare Fecunditatis as a whole. According to *Quaide and Oberbeck*, (1975) "the proportion of deeply derived debris increases as the regolith evolves." Hence, highland materials in the younger mare Fecunditatis regolith should come from shallower depths than are indicated by Fig. 7. However, factors discussed below more than compensate for these "overestimates" of basalt thickness.

DISCUSSION

The thicknesses of the continuous basalt fill in Mare Fecunditatis as derived from Fig. 7 are far *thinner* than any previous such estimates. Indeed, they are almost certainly *too* thin, and there are a number of reasons why they should be viewed as lower limits on the thickness of the basalt fill. One reason is that the basalt fill in the mare is not vertically continuous, but more probably consists of a number of basalt flows each topped by an ancient regolith containing some highland component. Thus an impact would not have to penetrate the entire basalt fill in order to excavate some highland materials. Another reason is that the mare basalts do not cover a perfectly flat basin floor. This is especially true in an old, irregular basin such as Fecunditatis. Impacts that strike buried topographic highs or crater rims would be able to excavate highland materials without going through too much basalt.

An example of excavated basin floor materials on the mare surface can be found directly to the northwest of the area of lowest highland contamination. This is the location of the post-mare craters Messier and Messier A, which both lie at 2°S and at 47°40' and 46°50'E, respectively. *Andre et al.* (1979) have already shown that Messier A excavates a subsurface magnesium-rich basalt. As shown in Fig. 8, there is an appearance of relatively high Al/Si ratios in the midst of a region of otherwise low Al/Si. This indicates that one or both

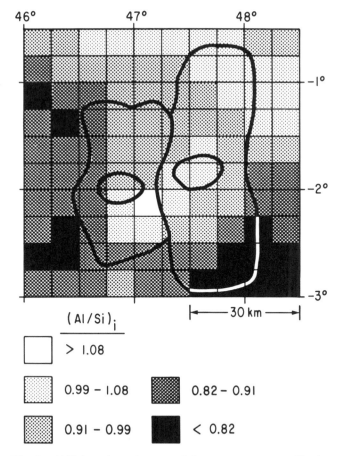

Fig. 8. Al/Si intensity ratios around the post-mare craters Messier and Messier A. Each box represents a single pixel. The increase in Al/Si at the top of the figure indicates a regional decrease in mare basalt thickness while the increase in the vicinity of the craters is due to recent excavation of basin floor material.

of these craters penetrated through the basalt fill and excavated basin floor materials as well as the subsurface high-Mg basalt.

The example of the Messier craters highlights a weakness in the geochemical method for mapping mare basalt thickness. In those regions where the highland component has been assimilated into the regolith over time by limited lateral migration and "gardening," the fit with *Quaide and Oberbeck's* (1975) model and the thicknesses predicted by Fig. 7 is good. However, immediately around those craters that are excavating materials from the basin floor, there will be higher percentages of highland materials than predicted by the model and the resulting thicknesses will be spuriously low. This effect could also be contributing to the very jagged contours of the highland contamination maps.

Whether the actual thickness contours in Mare Fecunditatis are better represented by DeHon's map or by the highland contamination maps is open to debate. Of the two methods, the geochemical mapping has a more consistent spatial representation than the crater distributions within Fecunditatis,

but is subject to spurious local effects. The craters used by DeHon to estimate basalt thickness are irregularly spaced and very often much more than 30 km apart. However, there are also errors inherent in the geochemical mapping of basalt thickness.

In addition to the aforementioned uncertainties inherent in the use of Fig. 7, there is the ever-present uncertainty in the X-ray data. The uncertainties in the X-ray data are such that they more than outweigh small scale surface effects that might alter a more precise data set. One such surface effect is the enrichment of "highland materials" in the < 10-μm fraction of mare regolith. Some of the feldspars that enrich this fraction are derived from the basin floor and some are the comminuted remains of mare basalts. This size fraction only takes up 10-15% of the regolith (*Laul and Papike*, 1980) and not all of that percentage is "highland material." Hence, less than 10% of the regolith will be < 10-μm-sized particles of highland material so that the contribution of these particles to the overall level of highland contamination will be relatively minor.

Using the contamination/thickness units from the highland contamination maps, it is possible to calculate volumes for the mare basalts covering the northern Mare Fecunditatis area. The intermediate thickness units of Fig. 5a yield a minimum volume of 8,900 km^3. The highly contaminated, thin basalts of Fig. 5b would have a volume of 2,140 km^3, while the largely uncontaminated basalts of Fig. 5c would have a minimum volume of 20,540 km^3. All three of these volumes should be considered minimums for their respective cases for the same reasons as stated above. In addition, while the 0% contamination unit in Figs. 5a and 5c could have a basalt thickness anywhere in excess of 150 m, for the purposes of the volume calculations the lower limiting thickness of 150 m was used. Hence, even using the limiting thicknesses of Fig. 7, the volumes for the cases indicated by Fig. 5a and 5c could easily be much larger.

If the basalt volumes quoted above are minimum values, then what is the maximum value? A good candidate would be the volume of 99,400 km^3 derived from the isopach map of Fig. 6 using the thicknesses of *DeHon* (1975). If the craters examined in the studies of DeHon and *Hörz* (1978) do not have the depth-diameter relations of fresh craters, then DeHon's estimates are too thick. However, if some or all of the craters examined were formed on some early basalt flow rather than on the basin floor, then Hörz's estimates are too thin. A good compromise should allow for the fact that some of Hörz's now degraded craters might have formed on a relatively thin ancient basalt flow. If that is the case, then the total basalt thicknesses should be closer to DeHon's estimates. Thicknesses greater than those postulated by DeHon are still possible; however, if true, such a thick basalt fill would hamper the vertical mixing process which this paper indicates is in operation.

SUMMARY

Two main techniques were used in an effort to map the distribution of highland materials in northern Mare Fecunditatis. The use of images of 0.56-μm albedo, 0.38/0.56-μm and 0.89/0.38-μm ratios, and Al/Si in an unsupervised cluster analysis produced maps of surface units of distinct geologic character. The amount of highland material in the mare units were estimated through the use of a least squares mixing model, and all the units were found to have a substantial highland component.

An inverse mixing model was used to map highland contamination in the mare. In this technique, a basalt type was assumed to cover the entire mare and various fractions of a highland component were admixed to model the observed Al/Si ratios. The simulated Al/Si ratios were converted to image grey level and used as thresholds on the Al/Si image to produce maps of highland contamination. For the high-Al basalt sample B1 (*BVSP*, 1981), an average Apollo 12 low-Ti basalt (*BVSP*, 1981), and an average high-Al basalt (*Cimbalnikova et al.*, 1976), three different patterns of highland contamination were produced. Of the three, the map representing a mare covered by a basalt with the aluminum content of the Luna 16 sample B1 was deemed most likely to reflect the actual pattern of highland contamination in the mare. In broad outline, these highland contamination maps resemble the published isopach map of the region (*DeHon and Waskom*, 1976; *Hörz*, 1978). Moreover, an inverse relation is observed between Al/Si and thickness of the basalt fill. This suggests that *vertical* mixing is the main process responsible for the highland component in mare soils.

By using *Quaide and Oberbeck's* (1975) model of regolith evolution and a relation established therein (Fig. 7), it proved feasible to assign to the contamination map units thickness values ranging from 9 m to more than 150 m. Basalt volumes for northern Mare Fecunditatis ranging from 2,140 km^3 to 20,540 km^3 were calculated using the area and calculated thickness of each contamination map unit for each of the three candidate basalts.

While the basalt thicknesses and volumes calculated in this study are minimum values, they are significant in that they were derived using geochemical measurements rather than through photogeological means. These basalt thicknesses are best viewed as the lower limits on basalt thickness in Mare Fecunditatis with the upper limits being those values quoted by *DeHon* (1975).

Acknowledgments. The author wishes to thank a lot of people without whom this project could not have been accomplished. Special thanks go to Paul Spudis and Phil Davis for getting me started on this project, for help in using the HINDU unsupervised classifier at the USGS offices in Flagstaff, for our discussions about lunar geology, and for their reviews of the manuscript. Thanks also go out to Pamela Clark for her assistance in working with the Apollo X-ray data. For help in image processing work done at the University of Arizona's Digital Image Analysis Laboratory, I wish to thank Charles Glass, Robert Schowengerdt, and Thomas Bortone. For his assistance and reviews, I want to thank Michael J. Drake. Most especially, I want to thank John S. Lewis for help, encouragement, and his willingness to support my work on such a "useless body" as the Moon.

REFERENCES

Alder I., Podwysocki M., Andre C., Trombka J., Eller E., Schmadebeck R., and Yin L. (1974) The role of horizontal transport as evaluated from the Apollo 15 and 16 orbital experiments. *Proc. Lunar Sci. Conf. 5th*, pp. 975-979.

Andre C. G., Wolfe R. W., and Adler I. (1979) Are early magnesium-rich basalts widespread on the Moon? *Proc. Lunar Planet. Sci. Conf. 10th*, pp. 1739-1751.

Basaltic Volcanism Study Project (1981) *Basaltic Volcanism on the Terrestrial Planets*. Lunar and Planetary Institute, Houston. 1286 pp.

Bence A. E. and Groves T. L. (1978) The Luna 24 highland component. In *Mare Crisium: The View from Luna 24* (R. B. Merrill and J. J. Papike, eds.), pp. 429-444. Pergamon Press, New York.

Bielefeld M. J. (1977) Lunar surface chemistry of regions common to the orbital X-ray and gamma-ray experiments. *Proc. Lunar Sci. Conf. 8th*, pp. 1131-1147.

Bielefeld M. J., Andre C. G., Eliason E. M., Clark P. E., Adler I., and Trombka J. (1977) Imaging of lunar surface chemistry from orbital X-ray data. *Proc. Lunar Sci. Conf. 8th*, pp. 901-908.

Bielefeld M. J., Wildey R. L., and Trombka J. I. (1978) Correlation of chemistry with normal albedo in the Crisium region. In *Mare Crisium: The View from Luna 24* (R. B. Merrill and J. J. Papike, eds.), pp. 33-42. Pergamon Press, New York.

Boyce J. M. (1976) Ages of flow units in the lunar nearside maria based on Lunar Orbiter IV photographs. *Proc. Lunar Sci. Conf. 7th*, pp. 2717-2728.

Bryan W. B., Finger L. W., and Chayes F. (1968) Estimating proportions in petrographic mixing equations by least-squares approximation. *Science, 163*, 926-927.

Clark P. E. and Hawke B. R. (1981) Compositional variations in the Hadley Apennine region. *Proc. Lunar Planet. Sci. Conf. 12B*, pp. 727-750.

Clark P. E. and Hawke B. R. (1987) The relationship between geology and geochemistry in the Undarum/Spumans/Balmer region of the Moon. *Earth, Moon & Planets* (in press).

Cimbalnikova A., Palivcova M., Frana J., and Mastalka A. (1976) Chemical composition of crystalline rock fragments from Luna 16 and Luna 20 fines. In *The Soviet-American Conference on Cosmochemistry of the Moon and Planets*, NASA SP-370, 263-275.

DeHon R. A. (1975) Mare Fecunditatis: Basin configuration (abstract). In *Conference on Origins of Mare Basalts and their Implications for Lunar Evolution*, pp. 29-31. Lunar Science Institute, Houston.

DeHon R. A. and Waskom J. D. (1976) Geologic structure of the eastern mare basalts. *Proc. Lunar Sci. Conf. 7th*, pp. 2729-2746.

Duncan A. R., Sher M. K., Abraham Y. C., Erlank A. J., Willis J. P., and Ahrens L. H. (1975) Interpretation of the compositional variability of Apollo 15 soils. *Proc. Lunar Sci. Conf. 6th*, pp. 2309-2320.

Elston D. P. (1972) Geologic map of the Colombo quadrangle of the Moon. *U.S. Geol. Survey Map I-714*.

Farrand W. H. (1987) Vertical versus lateral mixing of highland materials and minimum basalt thickness in northern Mare Fecunditatis (abstract). In *Lunar and Planetary Science XVIII*, pp. 282-283. Lunar and Planetary Institute, Houston.

Hodges C. A. (1973) Geologic map of the Langrenus quadrangle of the Moon. *U.S. Geol. Survey Map I-739*.

Hörz F. (1978) How thick are lunar mare basalts? *Proc. Lunar Planet. Sci. Conf. 9th*, pp. 3311-3331.

Hubbard N. J. and Wolozyn D. (1977) Orbital gamma-ray data and large scale lunar problems. *Phys. Earth Planet. Inter., 15*, 287-302.

Jayroe R. R., Atkinson R., Dasarathy B. V., Lybanon M., and Ramapryian H. K. (1976) *Classification Software Technique Assessment*. NASA TN D-8240.

Johnson T. V., Saunders R. S., Matson D. L., and Mosher J. A. (1977) A TiO$_2$ abundance map for the northern maria. *Proc. Lunar Sci. Conf. 8th*, pp. 1029-1036.

Labotka T. C., Kempa M. J., White C., Papike J. J., and Laul J. C. (1980) The lunar regolith: comparative petrology of the Apollo sites. *Proc. Lunar Planet. Sci. Conf. 11th*, pp. 1285-1305.

Laul J. C. and Papike J. J. (1980) The lunar regolith: Comparative chemistry of the Apollo sites. *Proc. Lunar Planet. Sci. Conf. 11th*, pp. 1307-1340.

Miller M. D., Pacer R. A., Ma M. S., Hawke B. R., Lockhart G. L., and Ehmann W. D. (1974) Compositional studies of the lunar regolith at the Apollo 17 site. *Proc. Lunar Sci. Conf. 5th*, pp. 1079-1086.

Olsen A. B. and Wilhelms D. E. (1974) Geologic map of the Mare Undarum quadrangle of the Moon. *U.S. Geol. Survey Map I-837.*

Pieters C. M. (1978) Mare basalt types on the front side of the Moon: A summary of spectral reflectance data. *Proc. Lunar Planet. Sci. Conf. 9th*, pp. 2825-2849.

Pohn H. A. and Wildey R. L. (1970) A photoelectric-photographic study of the normal albedo of the Moon. *U.S. Geol. Surv. Prof. Pap. 559E*, 1-20.

Quaide W. and Oberbeck V. (1975) Development of the mare regolith: Some model considerations. *The Moon, 13*, 27-55.

Reid J. B., Taylor G. J., Marvin U. B., and Wood J. A. (1971) Luna 16: Relative proportions and petrologic significance of particles in the soil from Mare Fecunditatis. *Earth Planet. Sci. Lett., 13*, 286-298.

Rhodes J. M. (1977) Some compositional aspects of lunar regolith evolution. *Philos. Trans. R. Soc. London, A285*, 293-301.

Rhodes J. M., Rodgers K. V., Shih C., Bansal B. M., Nyquist L. E., Wiesmann H., and Hubbard N. J. (1974) The relationship between geology and soil chemistry at the Apollo 17 landing site. *Proc. Lunar Sci. Conf. 5th*, pp. 1097-1117.

Whitaker E. A. (1972) Lunar color boundaries and their relationship to topographic features: A preliminary survey. *The Moon, 4*, 348-355.

Whitford-Stark J. L. (1986) The geology of the lunar Mare Fecunditatis (abstract). In *Lunar and Planetary Science XVII*, pp. 940-941. Lunar and Planetary Institute, Houston.

Wilhelms D. E. (1972) Geologic map of the Taruntius quadrangle of the Moon. *U.S. Geol. Survey Map I-703.*

Wilhelms D. E. (1984) *The Geology of the Terrestrial Planets*. NASA SP-469, 107-205.

Thickness and Volume of Mare Deposits in Tsiolkovsky, Lunar Farside

Robert A. Craddock and Ronald Greeley

Department of Geology and Center for Meteorite Studies, Arizona State University, Tempe, AZ 85287

In the 193 km crater Tsiolkovsky on the lunar farside, dark mare deposits overlie a brighter unit presumed to be impact melt. Craters impacting through the mare materials commonly excavate the underlying impact melt and form a high-albedo ejecta deposit. Apollo metric photographs of Tsiolkovsky were divided into 25 km^2 bins, and within each bin a crater was statistically determined to be the minimum size necessary to excavate underlying impact-melt materials. This "threshold" crater diameter was measured, and the corresponding depth of excavation was determined based on *Maxwell's* (1973) Z-model calculations. Crater excavation depths were related to mare thicknesses and an isopach map was derived that shows an average mare thickness of 22.3 m and a maximum thickness of >70 m. The maps also shows asymmetry in thickness located in the southeast portion of the mare deposit as expected from an oblique impact. The mean thickness between isopach lines was multiplied by the related surface area between lines, and a volume of 212.6 km^3 was calculated. Assessment of topographic data shows that a range of mare thicknesses and volumes are possible. Based on these data, the most probable average thickness is 116 m with a corresponding mare volume of 1182 km^3. Comparing topographic data to those more accurately determined through impact crater excavation indicates that a possible floor resurgence of up to 95 m may have occurred prior to and during mare emplacement.

INTRODUCTION

Determining the thicknesses and volumes of lunar maria is important in understanding the Moon's thermal history, the loading on the lithosphere, the depth of associated impact basins, and interpreting lunar seismic and gravity data. Mare deposits cover 17% of the surface area of the Moon (*Head,* 1975) and have been subdivided by composition (e.g., *Pieters,* 1978) and age (e.g., *Boyce,* 1976). Quantifying the volumes of mare deposits will aid in understanding the history of mare emplacement and magma evolution. Thickness and volume estimates for the mare deposits in Tsiolkovsky are presented by using a technique based on impact crater excavation. These values are then compared to values determined through topographic data.

Previous Work

The most familiar method of estimating mare thickness and volume is the use of flooded or "ghost" impact craters. *De Hon* (1974, 1975, 1979) and *DeHon and Waskom* (1976) measured the remnant rim height and diameter of "ghost" craters and compared the values to *Pike's* (1967) depth-to-diameter ratios for fresh lunar craters. The differences in rim height and Pike's depth measurements were assumed to be the local mare thickness. *Hörz* (1978) reduced DeHon's thickness values by a factor of 2 based on calculated depth-to-diameter ratios for degraded craters. Both authors assumed that the craters have not rebounded significantly and that the craters impacted on the basin floor, not an underlying mare deposit. Similarly, *Whitford-Stark* (1979) calculated thicknesses and volumes for the isolated crater-fill mare deposits composing Mare Australe. His method estimates the thickness of an interior mare deposit based on depth-to-diameter ratios for fresh lunar craters calculated by *Wood and Andersson* (1978). Spacecraft gravity data were used by *Thurber and Solomon* (1978) to estimate mare thicknesses on the nearside. The Apollo 17 lunar sounder experiment measured the thickness of mare deposits through the use of radar signals transmitted into the lunar subsurface and recorded as reflected signals (*Phillips et al.,* 1973). By this means, *Peeples et al.* (1978) determined mare/basin interfaces and several possible mare/mare horizons for Mare Serenitatis and Mare Crisium. By combining this information with Apollo 15 X-ray fluorescence and other remote sensing data, *Andre et al.* (1978) inferred the composition and areal distribution of Mare Crisium and underlying units. Recently *Farrand* (1987) has used similar remote sensing data to determine the extent of vertical mixing of highland materials with those of Mare Fecunditatis. He assumes that less mixing relates to thicker mare deposits and has generated an isopach map showing the distribution of possible mare thicknesses. Together, these techniques increase the understanding of the evolution and emplacement history of nearside mare volcanism. However, mare volcanism as a lunar process has been quantified only partly. Accurate estimates for the thickness and volume of mare deposits on the farside are necessary to understand fully this important geologic process.

Tsiolkovsky

Tsiolkovsky (Fig. 1) is a 193 km diameter crater formed by an oblique impact (*Guest and Murray,* 1969; *Howard and Wilshire,* 1975; *Hawke and Head,* 1977) on the lunar farside at lat. 21°S, long. 129°E. Tsiolkovsky has terraced walls, a large central peak, and mare deposits that partly cover the floor. Age estimates for the mare deposit range from Erastosthenian (*El-Baz and Worden,* 1972) to late Imbrium (*Wilhelms and El-Baz,* 1977). Geologic mapping (*Guest and Murray,* 1969) shows eight units, including "dark floor material" (informally referred to as Mare Tsiolkovsky in this paper) and the "edge of floor material," interpreted as being impact melt of country

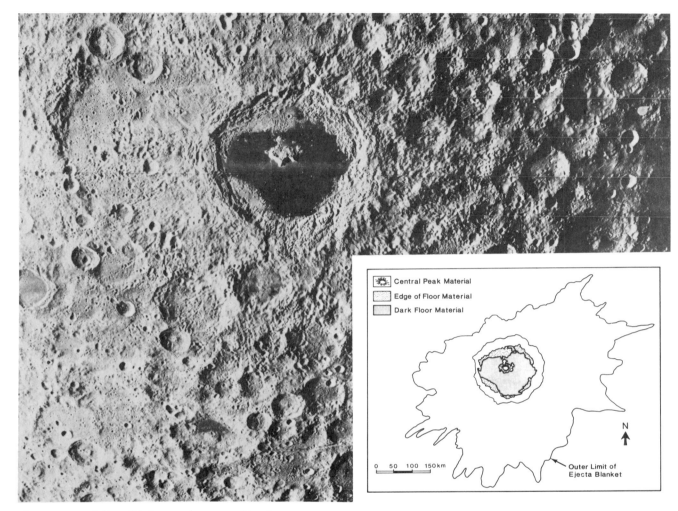

Fig. 1. Lunar Orbiter III photograph M-121 of Tsiolkovsky; north is at the top. Inset shows the distribution of crater ejecta, impact-melt deposit, and mare deposit. (Inset reprinted with permission from *Planet. Space Sci., 17,* J. E. Guest and J. B. Murray, Nature and origin of Tsiolkovsky Crater, lunar farside, copyright 1969, Pergamon Journals, Ltd.)

rock and the oldest of the floor materials within the crater. Mare Tsiolkovsky is typical of the mare deposits occurring as isolated crater-fill common on the farside.

Using *Whitford-Stark's* (1979) technique, *Whitford-Stark and Hawke* (1982) calculated a 250 m average thickness and an approximate volume of 2750 km^3 for Mare Tsiolkovsky. Using a very similar technique, *Walker and El-Baz* (1982) calculated a maximum thickness of 1.78 km and a volume of 36,000 km^3. These values are based on crater dimension. However, several problems exist with these estimates: A single value for the thickness is impossible to determine because the rim-crest varies in height by ~2.6 km (*Defense Mapping Agency,* 1973a,b,c,d). Also, these estimates cannot be corrected for resurgance of the crater floor that occurred prior to and during the mare emplacement (*Wilbur,* 1978), and volume estimates have been obtained that include the surface area of the central peak (*Hawke and Walker,* personal communication, 1987). A more accurate thickness and volume estimate can be made based on impact crater excavation, as presented here.

TECHNIQUE

Schultz and Spudis (1979, 1983), *Bell and Hawke* (1984), and *Hawke et al.* (1985) determined that lunar dark-haloed craters are the result of impacts excavating through ejecta from large impacts that have buried ancient mare materials. The subsequent smaller impacts excavated through the overlying high-albedo ejecta deposits, incorporated underlying mare materials into their ejecta, and exposed dark materials on the bright surface (*Schultz and Spudis,* 1979). Geologic sequences at Tsiolkovsky deal with an inversion of similar stratigraphic units (i.e., dark material burying older lighter materials). Lighter materials are easily visible against a dark background, and laboratory evidence indicates that even very small amounts of bright material deposited on a dark surface could raise the overall albedo significantly (*Wells et al.,* 1984). Thus, we assume that bright-ejecta resulting from the excavation of impact-melt should be visible on the surface of Mare Tsiolkovsky (Fig. 2). Admixing of ejecta deposits from secondary craters as described

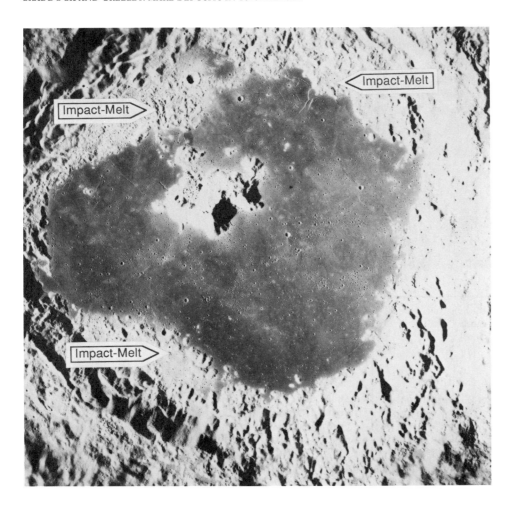

Fig. 2. Apollo 15 metric photograph 1030 of Tsiolkovsky. Note the undulating topography of the impact-melt deposit and the distribution of the mare deposit. Craters on the mare surface excavating into the underlying impact melt show as bright ejecta patterns.

by *Oberbeck* (1975) and *Oberbeck and Morrison* (1976) would be negligible due to the small average size (< 1.0 km) of most craters on Mare Tsiolkovsky.

Apollo 15 metric photographs 0478 to 0484 (sun angle 28° to 35°) and 1026 to 1032 (sun angle 15° to 23°) were used in the analysis. A 5 × 5-km square grid was applied to the photographs, and a total of 403 of these 25 km² bins was used to cover the surface (9550 km²) of Mare Tsiolkovsky. The "threshold" crater (Fig. 3) is defined as the smallest crater that has excavated enough of the underlying impact-melt material to form bright ejecta. Craters larger than this "threshold" size will excavate relatively bright impact melt and will also have a bright ejecta pattern, while smaller craters have dark ejecta. The "threshold" crater diameter was determined for 359 of the total 403 bins by plotting crater type (i.e., bright ejecta or dark ejecta) and diameter versus

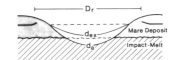

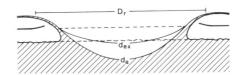

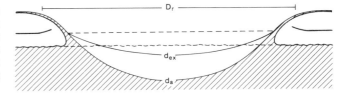

Fig. 3. Small craters excavate into the mare deposit; large craters are capable of excavating into the underlying impact-melt deposit. The minimum size crater capable of excavating into the impact-melt deposit is termed the "threshold" crater. Because the apparent crater depth (d_a) does not correspond to the excavation depth (d_{ex}), a crater must be deeper than the mare deposit before it is capable of excavating impact melt (*Stöffler et al.*, 1975).

A

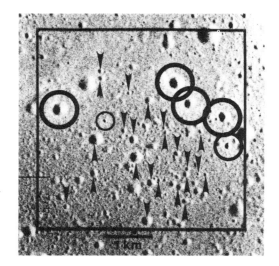

B

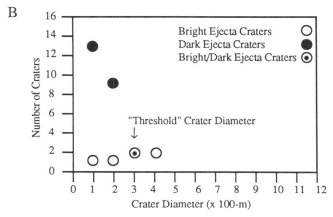

Fig. 4. (a) Enlargement of typical 25 km² bin. Craters with dark ejecta are shown by an arrow and bright ejecta craters are circled. Note that only the sharp-rim craters are used. (b) Craters from this bin are plotted by type and diameter versus their corresponding number. The "threshold" crater diameter is determined by the occurrence of the largest dark ejecta crater, which avoids using the smaller bright ejecta craters that are probably "fresh" impacts.

number (Fig. 4). Bright ejecta from fresh impact craters were separated from the "threshold" crater and subsequent larger bright ejecta craters by defining the "threshold" crater as the last (or largest) occurrence of a dark eject crater (Fig. 4). This makes use of the fact that the youngest and most common impact craters are generally the smallest. A total of 44 bins were not used due to an insufficient number of total craters or because of "contamination" by fresh impacts.

Mare thicknesses for the remaining bins were determined through properties of crater morphology. First, the corresponding apparent depth for the "threshold" crater was determined through *Pike's* (1974) method. For craters <15 km diameter, Pike showed

$$d_a = 0.196 \, D_r^{1.010} \quad (1)$$

where d_a is the apparent crater depth and D_r is the rim-crest diameter. This translates into

$$d_a \cong 0.2 \, D_r \quad (2)$$

However, the apparent crater depth does not represent the excavation depth but merely the maximum depth of the crater because some materials remain in the crater and undergo displacement inward during crater formation (*Dence et al.,* 1977). *Stöffler et al.* (1975) showed that for simple craters, the excavation depth is commonly less than the apparent crater depth. *Croft* (1980) found that by using *Maxwell's* (1973) Z-model flow field and values for Z between 2.5 and 3 (typical of expected lunar cratering flows), the excavation depth of a given crater is approximately half the transient crater depth or

$$d_{ex} \cong 0.1 D_r \quad (3)$$

where d_{ex} is the depth of excavation. *Grieve et al.* (1981) has shown that for terrestrial craters

$$d_{ex} = 0.1 D_r \pm 0.02 \quad (4)$$

The "threshold" crater depth determined through equation (3) was used as the mare thickness for any given bin. Although equation (4) shows that these values have a slight error, with typical "threshold" crater diameters of 100 to 1000 m the error is insignificant for the purposes of this study.

RESULTS AND DISCUSSION

An isopach map of mare thickness was generated for Mare Tsiolkovsky (Fig. 5). Based on the distribution of ejecta mapped by *Guest and Murray* (1969) evident in Fig. 1, the object that formed Tsiolkovsky travelled from the northwest and impacted obliquely toward the southeast. In such a crater, excavation should be greatest in the southeast portion allowing a greater accumulation of mare material. Maximum mare thickness (>70 m) occurring in this area conforms with this interpretation. Other shallower concentrations of mare thicknesses probably reflect the uneven crater floor evidenced by the unflooded impact-melt deposit (Fig. 2). The average thickness for Mare Tsiolkovsky based on Fig. 5 is 22.3 m, which is much thinner than the 250 m calculated by *Whitford-Stark and Hawke* (1982); the maximum depth of >70 m is also far less than the 1.78 km estimate of *Walker and El-Baz* (1982).

These values are realistic, however, when compared to possible values determined from existing topographic maps of Tsiolkovsky. For craters >15 km, *Pike* (1974) showed that the apparent depth of a crater can be determined from its rim-crest diameter by the equation

$$d_a = 1.044 \, D_r^{0.301} \quad (5)$$

At $D_r = 193$ km for Tsiolkovsky, the corresponding apparent depth should be 5089 km. Based on topography (*Defense Mapping Agency,* 1973a,b,c,d; *U.S. Geological Survey,* 1981),

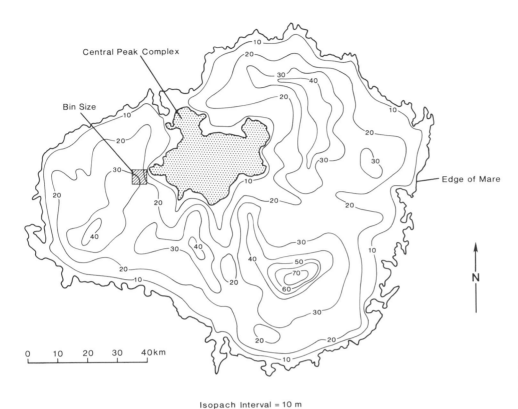

Fig. 5. Isopach map showing the distribution of mare thickness within Tsiolkovsky. Location of the sample bin used in Fig. 4 is shown.

the maximum rim-crest height is 12,003 m; the minimum rim-crest height is 9408 m. By subtracting the minimum crater elevation (5732 m) from these values, the depth of the crater is determined to be between 3676 m to a maximum of 6271 m. Possible mare thicknesses can be calculated by subtracting the apparent depth determined through equation (5) from the topographically measured crater depth; this results in mare thickness values ranging from 0 m to 1182 m (Fig. 6). The mean rim-crest height is 10,706 m with a corresponding crater depth of 4974 m and a predictable mare thickness of 116

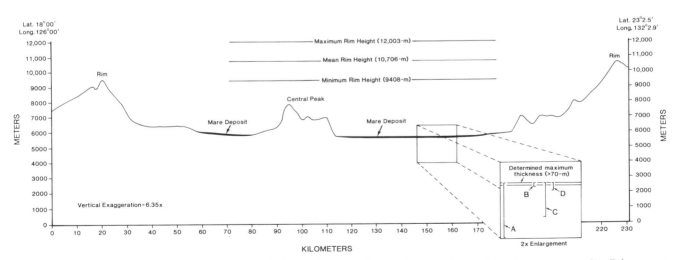

Fig. 6. Topographic profile of Tsiolkovsky showing the thickness of the mare deposits determined through impact crater excavation. Enlargement shows: **(a)** the maximum thickness of Mare Tsiolkovsky as determined by *Walker and El-Baz* (1982; 1780 m); **(b)** the thickness associated with the mean rim height (116 m); **(c)** the thickness associated with the minimum rim height (1128 m); and **(d)** the average thickness determined by *Whitford-Stark and Hawke* (1982; 250 m). (Profile based on Defense Mapping Agency, 1973a,b,d.)

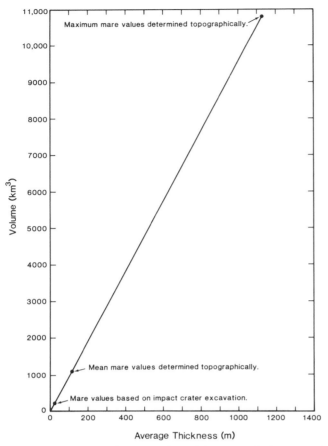

Fig. 7. Plot showing the range of possible average thicknesses and associated volumes for Tsiolkovsky based on topographic data. Volume is based on a surface area of 9550 km². Values determined through impact crater excavations are also shown.

m (Fig. 6). Assuming that this should be the true mean thickness of Mare Tsiolkovsky, the 22.3-m average thickness calculated from Fig. 5 indicates a floor resurgence of up to approximately 95 m.

The volume of Mare Tsiolkovsky was determined from the isopach map. Mean values between isopach lines were multiplied by the associated surface area. The total surface area for Mare Tsiolkovsky was determined as being 9550 km², and the total volume was determined as being 212.6 km³. Based on mare thicknesses derived from the topography, the volume may be as large as 11288 km³ (Fig. 7). However, a more likely upper limit to the mare volume is associated with the mean thickness of 116 m; this would yield a volume of 1108 km³. Comparing these estimates with those determined through Fig. 5 indicates that approximately 900 km³ of floor material may have been uplifted during resurgence and mare emplacement.

SUMMARY

Local changes in the minimum size of craters with bright ejecta indicate variations in the thickness of Mare Tsiolkovsky. Excavation depths of these "threshold" craters were determined through the equation $R_e = 0.1 D_r$ (equation (3)). An isopach map (Fig. 5) showing the distribution of mare thicknesses was generated, and the volume of Mare Tsiolkovsky determined from this isopach map was 212.6 km³. These estimates are much more accurate than the range of thicknesses and volumes possible based on topographic data (Fig. 7). However, comparison of the two data sets allows estimates of floor resurgence. Future work will include determination of the thicknesses and volumes of other mare deposits on the lunar farside.

Acknowledgments. We wish to thank Paul Spudis for his useful suggestions, Don Wilhelms and B. Ray Hawke for their thorough reviews, John Guest for use of the Tsiolkovsky geologic map (Fig. 1 inset), Dan Ball for his skilled photographic work, and Sue Selkirk for her patient drafting. This work was supported by the Office of Planetary Geosciences, National Aeronautics and Space Administration.

REFERENCES

Andre C. G., Wolfe R. W., and Adler I. (1978) Evidence for a high magnetism substrate basalt in Mare Crisium from orbital X-ray fluorescene data. *Mare Crisium: The View from Luna 24* (J. J. Papike and R. B. Merrill, eds.), pp. 1-12. Pergamon, New York.

Bell J. F. and Hawke B. R. (1984) Lunar dark-haloed impact craters: Origin and implications for early mare volcanism. *J. Geophys. Res.*, 89, 6899-6910.

Boyce J. M. (1976) Ages of flow units in the lunar nearside maria based on Lunar Orbiter IV photographs. *Proc. Lunar Sci. Conf. 7th*, 2717-2728.

Croft S. K. (1980) Cratering flow fields: Implications for the excavation and transient expansion stages of crater formation. *Proc. Lunar Planet. Sci. Conf. 11th*, 2347-2378.

Defense Mapping Agency (1973a) Lunar topographic orthophotomap of Tsiolkovskij Borealis. *Sheet LTO101B2*, Edition 2, Washington, D.C.

Defense Mapping Agency (1973b) Lunar topographic orthophotomap of Tsiolkovskij Australis. *Sheet LTO101B3*, Edition 1, Washington, D.C.

Defense Mapping Agency (1973c) Lunar topographic orthophotomap of Patsaev. *Sheet LTO102A1*, Edition 3, Washington, D.C.

Defense Mapping Agency (1973d) Lunar topographic orthophotomap of Fesenkov. *Sheet LTO102A4*, Edition 1, Washington, D.C.

DeHon R. A. (1974) Thickness of mare material in the Tranquillitatis and Nectaris basins. *Proc. Lunar Sci. Conf. 5th*, 53-59.

DeHon R. A. (1975) Mare Spumans and Mare Undarum: Structural depositional history (abstract). In *Lunar Science VI*, pp. 181-183. The Lunar Science Institute, Houston.

DeHon R. A. (1979) Thickness of the western mare basalts. *Proc. Lunar Planet. Sci. Conf. 10th*, 2935-2955.

DeHon R. A. and Waskom J. D. (1976) Geologic structure of the eastern mare basins. *Proc. Lunar Sci. Conf. 7th*, 2729-2746.

Dence M. R., Grieve R. A. F., and Robertson P. B. (1977) Terrestrial impact structures: Principal characteristics and energy considerations. In *Impact and Explosion Cratering* (D. J. Roddy, R. O. Pepin, and R. B. Merrill, eds.), pp. 247-275. Pergamon, New York.

El-Baz F. and Worden A. M. (1972) Visual observations from orbit. In *Apollo 15 Preliminary Science Report*, pp. 25-1 to 25-27. NASA SP-289, NASA, Washington, D.C.

Farrand W. H. (1987) Highland contamination and minimum mare basalt thicknesses in northern Mare Fecunditatis. *Proc. Lunar Planet. Sci. Conf. 18th*, this volume.

Grieve R. A. F., Robertson P. B., and Dence M. R. (1981) Constraints on the formation of ring impact structures, based on terrestrial data.

In *Multi-Ringed Basins* (P. H. Schultz and R. B. Merrill, eds.), *Proc. Lunar Planet. Sci. 12A*, pp. 37-57. Pergamon, New York.

Guest J. E. and Murray J. B. (1969) Nature and origin of Tsiolkovsky crater, lunar farside. *Planet. Space Sci., 17,* 121-141.

Hawke B. R. and Head J. W. (1977) Impact melt on lunar crater rims. In *Impact and Explosion Cratering* (D. J. Roddy, R. O. Pepin, and R. B. Merrill, eds.), pp. 815-841. Pergamon, New York.

Hawke B. R., Spudis P. D., and Clark P. E. (1985) The origin of selected lunar geochemical anomalies: Implications for early volcanism and the formation of light plains. *Earth, Moon and Planets, 32,* 257-273.

Head J. W. (1975) Lunar mare deposits: Areas, volumes, sequence, and implications for melting in source areas (abstract). In *Papers Presented to the Conference on Origins of Mare Basalts and Their Implications for Lunar Evolution,* pp. 61-65. The Lunar Science Institute, Houston.

Hörz F. (1978) How thick are lunar mare basalts? *Proc. Lunar Planet. Sci. Conf. 9th,* 3311-3331.

Howard K. A. and Wilshire H. G. (1975) Flows of impact melt at lunar craters. *J. Res. U.S. Geol. Survey, 3,* no. 2, 237-251.

Maxwell D. E. (1973) *Cratering Flow and Crater Prediction Methods.* Tech. Memo TCAM 73-17, Physics International, Calif., 50 pp.

Oberbeck V. R. (1975) The role of ballistic erosion and sedimentation on lunar stratigraphy. *Rev. Geophys. Space Phys., 13,* 337-362.

Oberbeck V. R. and Morrison R. H. (1976) Candidate areas for *in situ* ancient lunar materials. *Proc. Lunar Planet. Sci. Conf. 7th,* 2983-3005.

Peeples W. J., Sill W. R., May T. W., Ward S. H., Phillips R. J., Jordan R. L., Abbott E. A., and Killpack T. J. (1978) Orbital evidence for lunar subsurface layering in Maria Serenitatis and Crisium. *J. Geophys. Res., 83,* 3459-3468.

Phillips R. J., Adams G. F., Brown W. E., Jr., Eggleton R. E., Jackson P., Jordan R., Linlor W. I., Peeples W. J., Porcello L. J., Ryu J., Schaber G., Sill W. R., Thompson T. W., Ward S. H., and Zelenka J. S. (1973) Apollo lunar sounder experiment. In *Apollo 17 Preliminary Science Report,* pp. 22-1 to 22-26. NASA SP-330, NASA, Washington, D.C.

Pieters C. M. (1978) Mare basalt types on the front side of the moon: A summary of spectral reflectance data. *Proc. Lunar Planet. Sci. Conf. 9th,* 2825-2849.

Pike R. J. (1967) Schroeter's rule and the modification of lunar impact morphology. *J. Geophys. Res., 72,* 2099-2106.

Pike R. J. (1974) Depth/diameter relations of fresh lunar craters: Revision from spacecraft data. *Geophys. Res. Lett., 1,* 291-294.

Schultz P. H. and Spudis P. D. (1979) Evidence for ancient mare volcanism. *Proc. Lunar Planet. Sci. Conf. 10th,* 2899-2918.

Schultz P. H. and Spudis P. D. (1983) Beginning and end of lunar mare volcanism. *Nature, 302,* 233-236.

Stöffler D., Gault D. E., Wedekind J., and Polkowski G. (1975) Experimental hypervelocity impact into quartz sand: Distribution and shock metamorphism of ejecta. *J. Geophys. Res., 80,* 4062-4077.

Thurber C. H. and Solomon S. C. (1978) An assessment of crustal thickness variations on the lunar nearside: Models, uncertainties, and implications for crustal differentiation. *Proc. Lunar Planet. Sci. Conf. 9th,* 3481-3497.

U. S. Geological Survey (1981) Special topographic map of the moon. *Topographic Map of the Moon (Eastern Region-LOC-3),* preliminary version.

Walker A. S. and El-Baz F. (1982) Analysis of crater distribution in mare units on the lunar far side. *Moon and Planets, 27,* 91-106.

Wells E. N., Veverka J., and Thomas P. (1984) Mars: Experimental study of albedo changes caused by dust fallout. *Icarus, 58,* 331-338.

Whitford-Stark J. L. (1979) Charting the Southern Seas: The evolution of the lunar Mare Australe. *Proc. Lunar Planet. Sci. Conf. 10th,* 2974-2994.

Whitford-Stark J. L. and Hawke B. R. (1982) Geologic studies of the lunar far side crater Tsiolkovsky (abstract). In *Lunar and Planetary Science XIII,* pp. 861-862. Lunar and Planetary Institute, Houston.

Wilbur C. L. (1978) Volcano-tectonic history of Tsiolkovskyj (abstract). In *Lunar and Planetary Science IX,* pp. 1253-1255. Lunar and Planetary Institute, Houston.

Wilhelms D. E. and El-Baz F. (1977) Geologic Map of the East Side of the Moon. *U.S. Geological Survey Map I-948,* Arlington, Virginia.

Wood C. A. and Andersson L. (1978) New morphometric data for fresh lunar craters. *Proc. Lunar Planet. Sci. Conf. 9th,* 3669-3689.

Geologic and Remote Sensing Studies of Rima Mozart

Cassandra R. Coombs, B. Ray Hawke, and Lionel Wilson*

Planetary Geosciences Division, Hawaii Institute of Geophysics, University of Hawaii, Honolulu, HI 96822

*Also at Department of Environmental Science, University of Lancaster, Lancaster, LA1 4YQ, England

Many questions concerning the nature and origin of lunar sinuous rilles remain unanswered. In order to understand better the processes responsible for the formation of lunar sinuous rilles, we have conducted a detailed study of Rima Mozart using a variety of geologic, photographic, and remote sensing data. Rima Mozart is a 40-km-long sinuous rille located near the southeast rim of the Imbrium basin, approximately 100 km southwest of the Apollo 15 landing site (25°21'N, 359°03'W). Several lines of evidence suggest that sinuous rilles in general, and Rima Mozart in particular, are lava channels, or collapsed lava tubes. Geologically, the Rima Mozart region is very diverse (mountains, benches, graben, sinuous rilles, and collapse features), and tends to be dominated by a strong northwest/southeast trend in many of the features. The stratigraphy of the area is very similar to that of the Apollo 15 landing site with mare basalts and pyroclastics superposed on units of the Apennine Bench Formation, Imbrium ejecta, and Serenitatis ejecta (respectively). Construction of Rima Mozart began with an explosive eruption at Kathleen, an elongate crater (3 × 5 km) located at the head of the rille. Some pyroclastic material was deposited around the vent while turbulent erupting lava began carving the rille to the southeast. Lava from this and later eruptions thermally eroded two channels to form the present bifurcated rille. Explosive eruption(s) at Ann, the second elongate (1 × 3 km) source vent, also deposited pyroclastic material and resulted in the formation of a shallow, secondary rille that is joined to the main channel at a point 10 km northwest of Michael, a sink crater. Michael is also elongate (1.5 × 3 km) and may be connected to Patricia, a large elongate depression (1.5 × 10 km) to the southeast through an underground plumbing system or network of fractures. The distal end of Rima Mozart is embayed by "Lacus Mozart," a mare pond. Photographic and topographic data also indicate the presence of numerous smaller pyroclastic vents and several smaller rilles aligned with Ann along a northwest/southeast trending fissure zone. It was once speculated that Rima Mozart initially formed as a lava tube; however, our calculations indicate that it is unlikely that a complete lava tube could have existed along the rille. Such a tube would have had dimensions on the order of 100 m deep × 500 m wide, implying an eruption rate of 6×10^5 m^3/s. More reasonably, we calculated a total eruptive volume of 6372 km^3 for an open channel or tube, with an eruption rate of about 8×10^4 m^3/s and duration of 947 days. Radar (3.8- and 70-cm) and near-infrared spectral reflectance data for Rima Mozart also indicate that volcanic activity was responsible for the formation of the rille. There is a strong indication in both the radar and spectral data that pyroclastic deposits are present around Kathleen and Ann as well as at the base of the Apennines.

INTRODUCTION

Since 1684, when Christian Huygens first discovered Hyginus Rille, linear negative relief features have been known to exist on the lunar surface (*Cruikshank and Wood*, 1972). Since then, numerous other rilles have been discovered on the Moon and were classified according to their morphology. *Schultz* (1976) developed two main classes based on a rille's plan: curvilinear and rectilinear. While the nature and origin of lunar sinuous rilles remain the subject of major controversy, the linear, arcuate, and rectilinear rilles have been interpreted to be of structural origin (e.g., *Golombek*, 1979; *McGill*, 1971).

Lunar sinuous rilles typically occur on plains-forming units including mare deposits, "pooled" regions within the highlands, and the floors of large flooded craters (*Schultz*, 1976). Proposed origins for lunar sinuous rilles have varied widely from erosion by nuées ardentes (*Cameron*, 1964), to tectonic features (*Quaide*, 1965), to a fluvial origin (*Urey*, 1967; *Gilvarry*, 1968). More recently, the idea that lunar sinuous rilles were formed as a result of, or in conjunction with, volcanic eruptions has become the most popular (*Kuiper et al.*, 1966; *Oberbeck et al.*, 1969; *Greeley*, 1971; *Howard et al.*, 1972; *Carr*, 1974; *Hulme*, 1973, 1982; *Head and Wilson*, 1981; *Coombs and Hawke*, 1986, 1987; *Spudis et al.*, 1987). Of major concern today is whether the rilles are constructional features formed during an eruption (*Kuiper et al.*, 1966; *Greeley*, 1971), constructional features formed along preexisting structural weaknesses (*Spudis et al.*, 1987), or formed as a result of the incision of the channels by thermal erosion and/or turbulent flow through the channels (*Hulme*, 1973, 1982; *Head and Wilson*, 1981; *Coombs and Hawke*, 1986, 1987).

Thus, in order to understand better the processes responsible for the formation of lunar sinuous rilles, we have conducted a detailed study of Rima Mozart using a variety of geologic, photographic, and remote sensing data. In this paper we present the results of our analyses of these data and propose a geologic history for Rima Mozart and its immediate vicinity.

GEOLOGIC SETTING AND REGIONAL STRATIGRAPHY

Rima Mozart is located near the southeast rim of the Imbrium basin, approximately 100 km southwest of the Apollo 15 landing site (25° 21' N, 359°03'W; Fig. 1). It lies between the second and third (Apennine) rings of the Imbrium basin, just south of Palus Putredinus. Rima Mozart is partly surrounded by and incised into the Apennine Bench Formation (*Hackman*, 1966).

The age of Rima Mozart is estimated to lie in the range of 3.3-3.8 Ga, younger than the Apennine Bench Formation,

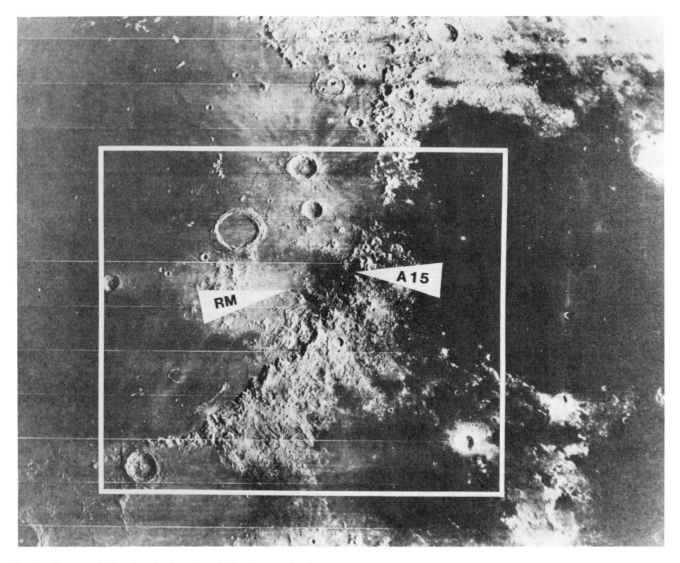

Fig. 1. Photograph showing the location of Rima Mozart (RM) in relation to the Apollo 15 landing site (A15) and the Apennine Bench Formation. Outline designates the area shown in Fig. 2.

the oldest plains forming unit in the area (3.8 Ga; *Spudis and Hawke,* 1986), and older than the nearby Hadley Rille (~3.3 Ga). This age estimate for Rima Mozart is based on a comparison of the crater densities, geologic sequence, and the relative amount of degradation along Rimae Hadley and Mozart.

Geologically, the Rima Mozart region is very diverse. Features present include: mountains (Apennine Mountains), benches (Apennine Bench Formation), volcanic plains (Lacus Mozart, Palus Putredinus), volcanic channels (Rima Hadley, Rima Mozart), graben (Rima Bradley), numerous secondary craters, mare wrinkle ridges, and collapse features. A northwest/southeast structural trend (radial to the Imbrium basin) is dominant in the region with a minor secondary trend running northeast/southwest (Figs. 1 and 2; *Swann,* 1986a,b; *Spudis et al.,* 1987).

Stratigraphically, the Rima Mozart area is very similar to that of the Apollo 15 site. Underlying the mare basalt and pyroclastic units associated with the rille is the Apennine Bench Formation. *Wilhelms* (1980) and *McCauley et al.* (1981) suggested an impact melt origin for the Apennine Bench Formation based upon the surface morphology and comparisons with similar impact melt deposits in the Orientale basin. Others (*Hackman,* 1966; *Hawke and Head,* 1978; *Spudis,* 1978a; *Spudis and Hawke,* 1986; *Ryder,* 1987) have suggested that this unit is of volcanic origin. Recently refined orbital geochemical data have shown a remarkable resemblance between the composition of the Appenine Bench Formation and the Apollo 15 KREEP basalts. This suggests that a very strong KREEP component exists in the Apennine Bench Formation (*Spudis,* 1978b; *Hawke and Head,* 1978; *Metzger et al.,* 1979; *Davis,* 1980; *Clark and Hawke,* 1981; *Spudis and Hawke,* 1986). This KREEP component is thought to be represented in the lunar sample collection by the Apollo 15 KREEP basalts. *Irving* (1977), *Dowty et al.* (1976), and *Ryder* (1987) have presented

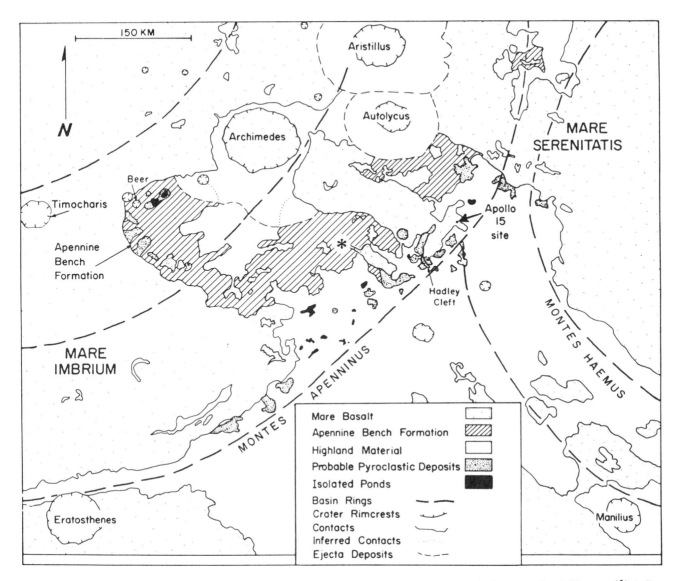

Fig. 2. Geologic map of the Rima Mozart region. The area shown is outlined in the photograph shown in Fig. 1. The star (*) indicates the head of Rima Mozart. (U.S.G.S Map I-703, Map 41A351.)

strong evidence in support of a volcanic origin for the Apennine Bench Formation by showing that the Apollo 15 KREEP basalt samples are pristine, endogenically-generated volcanic rocks. Hence it is highly likely that this unit consists of post-Imbrium volcanic KREEP basalt lava flows (*Spudis and Hawke*, 1986). These flows of variable thickness are now thought to be present beneath the Apollo 15 landing site and the Hadley Rille area (*Spudis*, 1978b; *Swann*, 1986a; *Spudis et al.*, 1987).

On the surface the Apennine Bench Formation is an undulatory or rolling smooth plains deposit marred by numerous secondary craters and crater chains as well as a variety of collapsed linear structural features (i.e., graben, grooves, scarps, and irregular depressions) that trend predominanately northwest/southeast. This irregular surface was most likely the result of regional deformation caused by basin adjustment (*Swann*, 1986b).

Underlying the undulatory plains-forming Apennine Bench Formation is an unknown thickness of Imbrium ejecta (*Hawke and Head*, 1978; Figs. 3, 4, and 5). Material slumped from the Apennine scarp (*Carr and El-Baz*, 1971) is also included in this unit for the purposes of this study. This unit of broken-up ejecta and slump blocks is thought to have been riddled by an enormous fracture system that resulted in a very porous deposit. Adjustment and resettling of this unit may have caused the collapse features seen on the surface, as well as having filled in some of the fractures and reduced the overall porosity of this unit. However, it is not thought that the unit would have been completely restructured and the cracks and fissures completely filled in by basin settling and seismic shaking related to nearby impacts. These cracks and fissures as well as the related high porosity play an important role in our model for the formation of Rima Mozart.

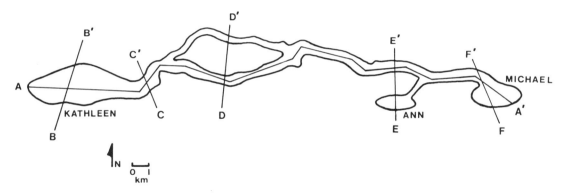

Fig. 3. A reference sketch map indicating the location and orientation of the rille cross-sections (A-A', B-B', C-C', D-D', E-E', and F-F') shown in Figs. 4 and 5.

Beneath the fractured Imbrium ejecta is another unit of impact-generated ejecta material from the Serenitatis basin (Figs. 4 and 5). This unit is also of undetermined depth in our study area.

Just north of the rille, but not stratigraphically related to it, is Palus Putredinus, a mare flood basalt (*Wilhelms,* 1980, 1985) that may well have been emplaced, at least in part, by the eruptions that formed Hadley Rille. This basalt unit may reach a thickness as great as 50 m in some areas (*Spudis et al.,* 1987).

RESULTS OF GEOLOGIC AND REMOTE SENSING STUDIES

A variety of data sets were used to help investigate the geologic history and composition of Rima Mozart and its immediate area. Photographic and topographic data were utilized to determine the morphology, morphometry, and overall character of the rille (i.e., its size, direction of flow, and whether or not it was a channel or tube). Remote sensing data such as Earth-based radar images and telescopic spectra further helped to determine the composition of units associated with the rille and surrounding areas, as well as to investigate the presence of pyroclastic deposits in the vicinity of the rille.

Geology of Rima Mozart

Rima Mozart extends for 40 km between the craters Kathleen and Michael. Topographically, the rille floor is very flat, though several "humps" occur along the floor. The gradient of the rille is a minute 0.1 degree (west to east) with an overall regional slope of 0.3 degrees (west to east). Rima Mozart's main course generally follows the dominant structural trend in the region: northwest/southeast, with some deviations along the minor northeast/southwest trend (*Swann,* 1986a,b). Beginning at Kathleen, an elongate, irregularly shaped depression interpreted to be a source vent, Rima Mozart extends northeast for 3 km before bifurcating, with one channel going southeast (A) and the second, smaller, shallower one (B) going northeast (Figs. 6 and 7). The two channels, A and B, remain separated and outline a rough diamond shape for 10.5 km where they rejoin to form a single channel for the next 20 km.

A careful examination of the Lunar Orbiter IV, Apollo 15 Metric, Panoramic, and Hasselblad photographs has revealed that two hanging valleys mark the proximal and distal junctions of segment B to segment A. At these points the hanging valley floors (of segment B) are roughly 20 m above the bottom of segment A. Several possibilities are suggested by the presence of these hanging valleys: (1) Two (or more) distinctly different

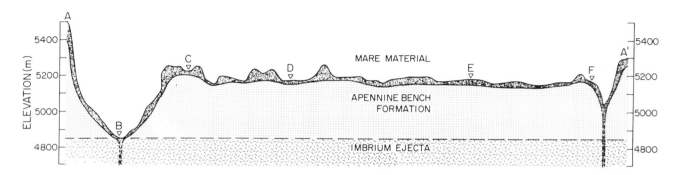

Fig. 4. A geologic cross-section of Rima Mozart (A-A'), showing the relative thickness of the underlying units. Vertical exaggeration = 1.69. Location and orientation of A-A' shown in Fig. 3.

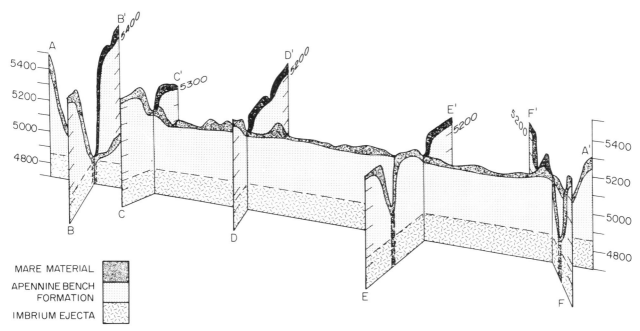

Fig. 5. A geologic panel diagram of Rima Mozart and the area immediately adjacent to it. Locations and orientations of sections B-B', C-C', D-D', E-E', and F-F' are shown in Fig. 3. Units shown are of relative thickness. Vertical exaggeration = 1.69.

eruption episodes may have formed this rille. That is, the first eruption formed either one or both of the channel segments before its supply of lava was cut off. The second eruptive phase may have produced a more voluminous and/or turbulent flow that would then have undercut the previously formed segment B. (2) A blockage at the mouth of segment B may have simply prevented the lava from the second eruption from entering the channel, in which case all the lava from the second eruption would have flowed down segment A. (3) Regional tectonic activity or resettling may have shifted the crustal block containing rille B upward about 20 m, creating the hanging valleys at both ends. Much tectonic activity has occurred in this region (as is noted by the numerous graben in the area). Depending upon the timing of the tectonic reshuffling in relation to the formation of the lava channels, segment B could have been formed before future eruptive episodes at Kathleen created segment A. (4) Some thermal erosion occurred in channel A to undercut channel B. Throughout our arguments in this paper we elude to the role of thermal erosion in the formation of Rima Mozart. If the volume flux was high enough in the flow in channel A, the flow could easily have undercut segment B. The remaining possibility is that (5) some combination of the above occurred.

Though all of the above are viable solutions, the available evidence suggests that the "hanging rille" was formed by two or more different eruptive episodes emanating from Kathleen, the second of which was much hotter and more voluminous. Hence, this second flow was able to thermally erode a deeper channel as it flowed along its path. It is unlikely, however, that the hanging valleys along Rima Mozart are due to structural readjustment in the area. Had the structural reshuffling occurred it would have displaced more than just a small stretch of land between the proximal and distal ends of segment B. No other such features can be discerned in the Apollo or Orbiter IV photographs for this area.

Photographic data indicate the presence of two major source vents for Rima Mozart, Kathleen and Ann (Fig. 6). Now girdled by dark material, these vents appear to have originally formed in the Apennine Bench Formation. Several dark (pyroclastic, spatter) deposits can be seen around these vents, though it appears that much of these deposits has been altered by local impact gardening. Most prominent is the dark spatter around the western limb of Ann (Fig. 8). The existence of these spatter deposits is strong evidence for a volcanic origin for this crater and its associated rille segment.

Kathleen, the largest and primary source vent, is an elongate crater (3 × 5 km) located at the head of the rille (Figs. 6 and 7). From this crater the rille follows a southeasterly, downslope (0.1 degree from west to east) course until reaching its terminus at Michael. Pyroclastic deposits and some minor spatter present around Kathleen and at the mouth of the rille support a volcanic origin for this feature (Fig. 8). Topographic data indicate that this vent crater has a depth of 550 m and is the deepest crater associated with the rille.

Ann, the second source vent, connects to the main channel of Rima Mozart via a shallow, secondary rille at a point 10 km northwest of Michael, the sink crater (Figs. 6 and 7). This secondary rille is 2 km long and has an average depth of 30 m. It is parallel to several other northwest/southwest trending segments of Rima Mozart, suggesting some influence by preexisting structures. This segment is also left hanging, as

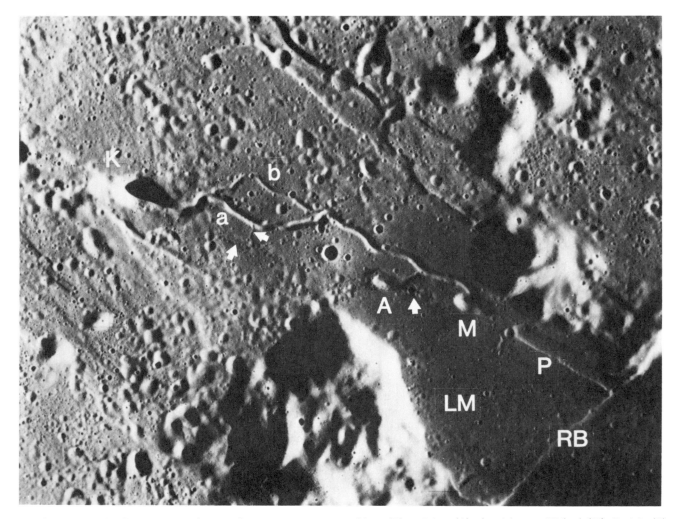

Fig. 6. Photograph of Rima Mozart showing the two source vents, Kathleen (K) and Ann (A), the sink vent, Michael (M), Patricia (P), Lacus Mozart (LM), Rima Bradley (RB), and segments A (a) and B (b). Arrows indicate the location of the third rille, fissure vents, and smaller volcanic vents associated with Ann. Rille length is 40 km.

is the bifurcated segment up-channel, indicating that Ann and its associated secondary channel formed prior to the postulated late-stage eruptions of Rima Mozart.

Michael, the sink structure, is also an elongate, irregularly shaped crater (3 × 1.5 km; Figs. 6 and 7) with a depth of ~400 m. The main segment of Rima Mozart joins Michael on the north side of the crater. The junction position and the nature of the structure suggest that Michael may have initially begun as a small collapse feature and was later eroded to its present state by the hot turbulent lava that flowed through it.

Michael may also be connected to Patricia, a large elongate depression to the southeast, through an underground plumbing system or network of fractures (Figs. 6 and 9). Patricia is 10 km in length, about 2 km wide, and has an average depth of 90 m. The long axis of this feature is parallel to the dominant northwest/southeast regional trend. Based on the calculations of *Wilson and Head* (1981), these fractures would need only to be of the order of 0.5 m wide to allow free passage of the molten lava for distances of tens of kilometers into the lunar lithosphere without excessive cooling.

Embaying the lower portion of Rima Mozart, including Ann and Michael, and continuing off to the southeast, is "Lacus Mozart" (unofficial name, Figs. 6 and 7). This mare "lake" has a very low albedo, is smooth textured, covers an area of 330 km^2, and appears to drape over a segment of Rima Bradley. There are numerous indications that Lacus Mozart was once flooded to a higher level and that drainage occurred. Presently, the lava in Lacus Mozart is estimated to be about 50 m thick, making the total volume of lava present in Lacus Mozart 16,500 km^3. Several different origins for this lake have been postulated: (1) Rima Mozart flooded the entire lower-lying area. Some magma may have drained into the subsurface through Michael and other possible sinks including the fractures associated with Patricia and Rima Bradley. (2) The area was flooded by mare basalt from Palus Putredinus to the north. This could have

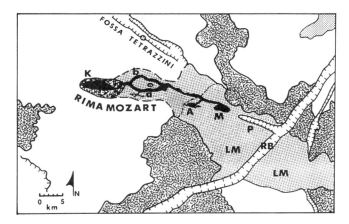

Fig. 7. A geologic sketch map of Rima Mozart and vicinity showing the location of Kathleen (K), Ann (A), Michael (M), Patricia (P), Lacus Mozart (LM), and Rima Bradley (RB). Segments A and B are indicated by (a) and (b).

covered up any evidence of previous episodes of lava flooding and/or draining from the area. (3) A now-buried vent (or vents) somewhere underneath Lacus Mozart erupted to flood the entire region after the formation of Rima Mozart. (4) Some combination of the above may have happened.

Although all are viable solutions as to why Lacus Mozart exists, the second and/or third solution best explains the absence of drainage fractures along the drapes and the present shallow depth of Patricia and Lacus Mozart. However, it is entirely possible that eruptions from Rima Mozart formed Lacus Mozart without the aid of outside sources.

Photographic and topographic data also indicate the presence of numerous smaller eruptive vents aligned with Ann, along the northwest/southeast trend (Figs. 6 and 7). At least five vent-like craters can be seen immediately east of Ann, and at least three extend westward from the western rim. These vents may be evidence of late-stage minor eruptive activity associated with Ann. Each is thought to have been a separate event as they all form individual craters with distinct irregular rims that appear to have been formed by spatter.

Also aligned along this northwest/southeast trend is what may be a small fissure zone with at least three vents visible on the Apollo 15 and Orbiter IV photographs (two left arrows on Fig. 6). Emanating from one of these fissure vents are two small rille-like features, one of which joins the main rille of Rima Mozart. The other channel runs south/southeast and terminates in the volcanic unit mapped in this area (Fig. 7). These channels are less than 20 m deep and approximately 200 m wide. Several other small dark-rimmed craters are also present along the fissure vent, but no rilles are associated with them. These features appear to have been produced by a fissure eruption associated with the formation of Rima Mozart.

Further, in constructing the cross-section and 3-D model for Rima Mozart we discovered irregularities on an otherwise relatively smooth rille floor. Several humps, noticeably higher than adjacent spots, rise from the rille floor (Figs. 4 and 5). This discovery led us to question their origin. Possible explanations are: (1) rille-wall subsidence/collapse occurring since the rille formed, (2) crater ejecta from nearby impacts partially infilled the rille, and (3) remains of partially collapsed tube roof with the possibility of some uncollapsed sections still remaining.

To address these issues, we have constructed 13 cross-sections normal to segment A, the longest of Rima Mozart's channels. From these we found that the average slope for the rille walls is 21°, with the greatest slope along the wall being 26°. This wall material is clearly fragmental debris that has spread completely or almost completely across the floor of the rille in all places, making it unlikely that we are seeing the true rille floor. Thus we are unable to determine unambiguously the original width and depth of the channel (though we can, of course, estimate its cross-sectional area).

Secondly, we noticed that some nearby impacts have emplaced ejecta along the rille floor and/or caused the subsidence and collapse of the rille walls in these areas. At two particular areas on segment B (Fig. 6) crater impacts on the rille wall have obliterated the channel walls and filled in adjacent sections close to the impact. At two other locations, however, one on segment A and one on segment B, the rille floor is obscured by what may be an intact lava tube roof. No impact craters are present in these two areas.

Though the presence of a lava tube is possible, some questions remain: (1) How can a tube roof sustain itself over a distance of 830 m (the present average width of channel) if the area underneath has been drained of all support material, particularly over billions of years of resettling and bombardment of the lunar surface? (2) If the tube did exist how large would it have had to be in order to accommodate the large volume flux of lava that flowed through the channel? (3) If the tube did/does exist, are the higher areas (humps) the remains of an excessively thick part of the tube-roof that did not collapse, or simply a very large talus slope that formed when the adjacent walls collapsed?

So we are left with the question: Did a tube once exist below the channel? And, if so, what would its aspect ratio have been? To determine this we looked at the 13 cross-sections normal to the rille where spot elevations are known on the rille floor. Analysis of these sections revealed that if a lava tube were to have existed in the rille it would have been enormous: on the order of 100 m deep × 500 m wide; a very large tube, even under lunar conditions. Problems exist with a tube this large, namely the support necessary to keep the roof of a lava tube of this size intact. But, if a tube were to have formed, would it have to have been just one tube? There is also the possibility that more than one tube may have existed at one time, as is currently happening in the 1987 flows from Kilauea's C-vent on Hawaii. If more than one tube existed to accommodate the large volume flux of lava, each tube would have been fairly large too, but not as unreasonable in size as the single tube would have been.

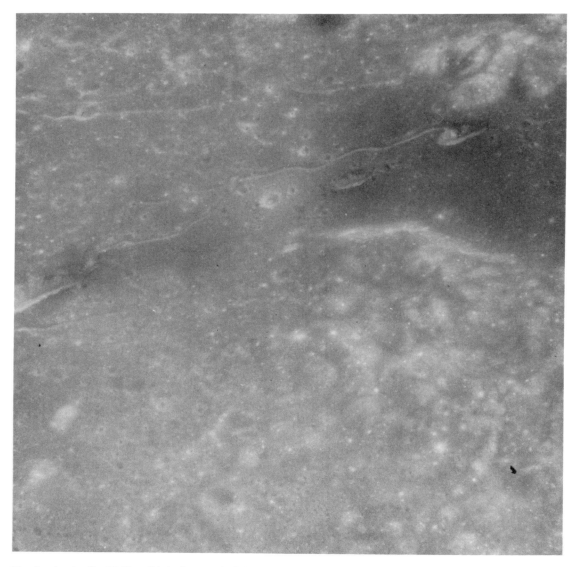

Fig. 8. An Apollo 15 Hasselblad photograph (AS15-97 13238) of Rima Mozart showing Kathleen and Ann and the spatter and pyroclastic material present around their rims.

However, after careful examination of the slopes of the rille walls and the width variations along segment A (largest channel), we feel that it is unlikely that (a) tube(s) existed (at least along the entire length of the present channel). Rima Mozart apparently was formed for the most part as an open channel. The raised portions of the rille floor that we now see are most likely the result of subsidence along the rille walls or crater ejecta. However, the possibility that a tube formed along a portion of this channel cannot be ruled out.

Thermal Erosion

Although many features of lunar flows are similar to terrestrial flows, there are differences (i.e., size, shape). These may be the result of several differences between these two bodies:

(1) The lunar gravity field is one-sixth that of the Earth, allowing the lava to travel more slowly on similar slopes and to form flows almost twice as deep for similar eruption rates (*Hulme*, 1973; *Hulme and Fielder*, 1977; *Wilson and Head*, 1983); (2) lunar flows occur in a vacuum making the lava outgas much more rapidly than on the Earth (*Hulme*, 1973); (3) lunar basaltic compositions are quite different from those of terrestrial basalts, the lunar lavas being generally less viscous (*Hulme*, 1973); and (4) lunar lava effusion rates are generally higher than terrestrial rates (*Wilson and Head*, 1983), resulting in longer flows (i.e., *Walker*, 1973).

A number of studies (*Hulme*, 1973; *Carr*, 1974; *Head and Wilson*, 1980, 1981; *Wilson and Head*, 1980; *Hulme*, 1982) have shown that thermal erosion may have occurred in conjunction with the formation of lunar sinuous rilles. *Carr*

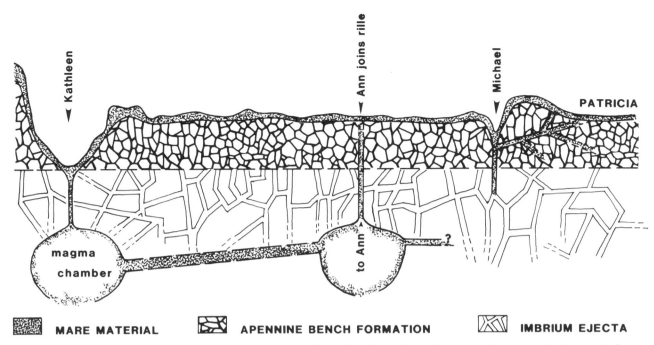

Fig. 9. Two-dimensional diagram of Rima Mozart showing the relationship of the rille to the surrounding and underlying terrain-fractures. This figure illustrates one possible escape route for the large volume of lava erupted.

(1974) demonstrated that such erosion could occur if the lava moved in a laminar manner provided the eruption lasted long enough. *Hulme* (1973, 1982), however, has shown that turbulent erosion would lead to more efficient thermal erosion.

In a laminar lava flow each part of the lava flows along a path roughly parallel to the ground. Heat is lost from the lava by radiation from the surface and by conduction into the underlying rock. Within the flow, heat is transferred by conduction perpendicular to the flow direction since there is no movement of lava in that direction to allow the convection of heat. The surface temperature need only be slightly higher than the temperature of the surroundings for the heat radiated from the surface to equal the heat conducted upwards from the center of the flow. If both the lava and underlying substrate are of similar conductivity, the downward heat flow is equal in the lava and the ground, the temperature gradient in each is about the same, and the interface temperature would be about halfway between the central lava temperature and the original ground temperature (*Jaeger*, 1959; *Hulme*, 1973). Thus the two surfaces of the flow are at temperatures well below the solidus of the lava and a layer of solid lava forms at each one. A solid skin then forms over the flow while lava is deposited beneath it (*Hulme*, 1973).

In a turbulent lava flow, material from different depths is constantly being mixed, so that heat is transferred by convection. The surface loses heat much quicker in a turbulent lava flow since the surface temperature is somewhat higher than in a laminar flow. The solid skin that forms on the surface is very thin and easily disrupted by the turbulence. At the base of turbulent flows is a boundary layer where the flow velocity falls to zero at the ground. The laminar flow zone at the base of this boundary layer is termed the laminar sublayer. Here, heat is transferred by conduction alone. Since the thickness of this layer is small compared with the distance over which the temperature falls to the original temperature of the ground, the interface temperature is only a few degrees less than the mean temperature of the lava. Therefore the lava at the base of the flow is likely to be above its solidus temperature and may also be above the solidus temperature of the underlying rock (*Hulme*, 1973).

In order for thermal erosion to form a rille, an initial lava flow is required in which a high effusion rate causes turbulent motions. This maximizes the efficiency of heat transfer to the basal layer and leads to the heating of the substrate rocks above their solidus temperature (*Head and Wilson*, 1981). The flow sinks progressively into a deepening channel as partial melting and mechanical erosion excavate the underlying terrain. The vertical erosion rate is always greatest near the vent where lava is hottest and is always zero at some finite distance

TABLE 1. Table of values.

Symbol	Quantity	Value
k	thermal conductivity of lava	1 Wm^{-1}K^{-1}
c	specific heat of lava and of ground	730 Jkg^{-1}K^{-1}
ρ	density of lava and of ground	3 × 10^3 kg/m^3
A	constant in viscosity relationship	1635 K
n	constant in viscosity relationship	15
T_{mg}	melting temperature of ground	1360 K
L	latent heat of fusion of ground	5 × 10^5 j/kg
σ	Stefan constant	5.6703 ×10^{-8} Wm^{-2} K^{-4}

*From *Hulme* (1973).

downstream (*Hulme*, 1973). Beyond this point, the rille ceases to deepen by thermal erosion and the lava continues to flow over the surface between levees as normal. In this study, we have used available topographic data to apply *Hulme's* (1973) model to Rima Mozart. Hulme's complete model for the formation of lunar sinuous rilles is not given here; rather, a synopsis of the steps used in our analysis is given (see Table 1 and the Notation for the definition of variables).

The volume rate of flow per unit width of channel, Q, can be determined by

$$Q = \frac{3 \sigma \epsilon x_m}{\rho c \left(\frac{1}{T_{mg}^3} - \frac{1}{T_{lo}^3} \right)} \quad (1)$$

where ρ_l is the density of the lava, c the specific heat, ϵ the emissivity (for the lava surface), (T_{lo}) the temperature of the lava at the source, (T_{mg}) the solidus temperature of the ground, X_m the length of the rille, and σ the Stefan constant. The depth, d, of the flowing lava (which in general does not fill the rille once sufficient erosion has taken place) is given by

$$d = Q^{2/3} \left(\frac{f}{2 g \alpha} \right)^{1/3} \quad (2)$$

where α is the slope of the ground, f the friction factor, and g the acceleration due to gravity. d can then be used to calculate the mean flow velocity in the channel, u,

$$u = \left(\frac{2 g \alpha d}{f} \right)^{1/2} \quad (3a)$$

or, more simply,

$$u = \frac{Q}{d} \quad (3b)$$

If the lava density and viscosity are ρ and η, respectively, the Reynolds number based on depth for the flow will be

$$Re_d = \rho \frac{Q}{\eta} \quad (4)$$

If Re_d is of the order of 10^3 or greater, the flow will be turbulent, a necessary condition for thermal erosion to occur.

Using the standard engineering treatment of heat transfer from a fluid in a pipe, and converting to an open channel geometry

$$h_o = 0.017 \, k^{0.6} \, c^{0.4} \, \rho_l^{0.8} \, Q^{2/15} \left(\frac{2 g \alpha}{f} \right)^{1/3} \left(\frac{T_{lo}}{A} \right)^{2n/5} \quad (5)$$

where h_0 is the heat transfer coefficient at the head of the stream, k the thermal conductivity, and A a constant equal to 1635 K (Table 1), which determines the temperature variation of lava viscosity via

$$\eta = \left(\frac{A}{T_l} \right)^n \quad (6)$$

The actual thermal erosion rate, E, is given by

$$E = h_o \frac{(T_{lo} - T_{mg})}{\rho_g L} \cdot \frac{N}{(1+N)} \quad (7)$$

where L is the latent heat of fusion of the ground and N is defined by

$$N = \frac{L}{c_g (T_{mg} - T_0)} \quad (8)$$

Assuming values of T_{mg} and T_{lo} of 1360 K and 1400 K, respectively, flow parameters for Rima Mozart were calculated as follows. Rima Mozart is 40 km in length so Q is about 93.8 m^2/s according to equation (1). With an average width of 830 m the total eruption rate is then 7.78 × 10^4 m^3/s. The viscosity of the lava is 10 kgm^{-1}s^{-1} at the head of the rille from equation (6). With a lava density of 3 × 10^3 kg/m^3, Re_d from equation (4) would be 2.74 × 10^4 (fully turbulent). To calculate f, the friction factor, it was assumed that the floor was relatively smooth. f was found to be 0.005 for Re_d = 2.74 × 10^4 (*McAdams*, 1954). The overall gradient of the channel, measured from the Lunar Topophotomap 41A3S1, is minute at 0.1 degree.

Using equation (2) the mean depth of the flowing lava was estimated at 18.9 m. The mean velocity of the flow, determined by equation (3a) or (3b), is then 4.96 m/s. Before calculating the total amount of material erupted it is necessary to know how long the eruption lasted. If the rate at which the rille floor was lowered is known, as well as the depth of the rille floor, then an estimate of how long the eruption lasted may be made.

Equations (5) and (7) can be combined to calculate the erosion rate at the head of the rille. L, the latent heat of fusion of the ground, was taken from *Hulme* (1973) as 5 × 10^5 J/kg; following *Hulme* (1973), the fraction of the ground material that would have melted before the remaining material became mechanically unstable and was swept away in the flow is assumed to be approximately 40%, reducing L to 2 × 10^5 J/kg.

h_0 was then calculated to be about 113.4 Wm^{-2}K^{-1} and E to be about 1.55 × 10^{-6} m/s. The period of eruption is then taken to be the time required to erode 127 m (mean depth to flat floor) at a rate of 1.55 × 10^{-6} m/s, or 8.18 × 10^7 s (946.9 days). The total amount of material erupted is then estimated at 6372 km^3, which is comparable to values obtained by *Head and Wilson* (1981) in their study of thermal erosion in lunar sinuous rilles. A discussion as to the possible whereabouts of this large volume of erupted material is given in the next section.

We would like to note again that it is very unlikely that a tube existed along Rima Mozart. Not only are there problems in creating and supporting a roof this size, there are problems in sustaining it. As mentioned previously, had the tube existed it would have had minimum dimensions on the order of 100 m × 500 m. For the tube roof to have remained stable, the lava flowing through the tube must have remained at a constant depth of nearly 100 m. As such, the volume flux of material erupted would have been phenomenal. To illustrate this we calculated the eruption rate for a lava tube with the dimension 100 m × 500 m. With a depth of lava in the tube of 100 m, the mean velocity of the flow would be just over 11 m/s (nearly three times that of our previous calculation), and the volume flux nearly 6×10^5 m/s. These conditions would produce a thermal erosion rate of 2.8×10^{-6} m/s. As such, the support for the tube roof would be withdrawn, due to subsidence of the flowing material, at a rate of about a quarter of a meter per day (during which time about 50 km^3 of magma would have been erupted!).

Remote Sensing

Radar. The 3.8-cm and 70-cm radar collected by *Zisk et al.* (1974) and *Thompson* (1987), respectively, were used in this study of the Rima Mozart region. In particular, we were interested in the distribution of any pyroclastic material in the region. Of the two radar data sets, the 3.8-cm wavelength radar, with its 2-km resolution, is most sensitive to small-scale roughness (~1-50 cm; *Zisk et al.*, 1974). Local slope and the relative amount of diffuse components have been interpreted as the two major factors controlling lunar radar backscatter (*Thompson and Zisk*, 1972). Thus topographic features facing away or toward the radar beam will produce shadows or highlights on the radar map. For areas where topography is not responsible for echo strength differences, diffuse scattering [controlled in part by the presence of surface scatterers 1-50 cm in size (i.e., rocks and boulders) and bulk electromagnetic properties] is considered to be dominant (*Thompson*, 1979).

All currently recognized regional pyroclastic mantling deposits on the lunar surface are distinguished by very weak to nonexistent echoes on the depolarized 3.8-cm radar maps of *Zisk et al.* (1974). These low depolarized returns are thought to be due to the lack of scatterers on the smooth surfaces of pyroclastic mantling deposits (*Gaddis et al.*, 1985; *Zisk et al.*, 1977; *Pieters et al.*, 1973). Depolarized returns on 3.8-cm radar maps have been used by a number of workers to map the areal extent of pyroclastic units (*Gaddis et al.*, 1984, 1985; *Hawke et al.*, 1979; *Zisk et al.*, 1977; *Pieters et al.*, 1975), and may be used to map localized pyroclastic deposits.

Figures 10a,b show the polarized and depolarized 3.8-cm radar images for that portion of the lunar surface that includes Rima Mozart (*Zisk et al.*, 1974). The source vent, Kathleen, is clearly visible as a radar enhancement in both of these images. In the depolarized radar image, Kathleen appears to be surrounded by material with a much lower radar return. This material may be pyroclastic in origin. The rille itself is visible as a weak enhancement surrounded by material with lower radar returns (darker in Fig. 10b). This darker material is part of a much more extensive zone or relatively radar-dark material that trends west/northwest from the Apollo 15/Hadley rille area to the vicinity of Bancroft crater on the southwest rim of Archimedes. The nature and origin of this radar unit remains uncertain. It crosses a variety of geologic units of diverse ages and origins (*Hackman*, 1966; *Carr and El-Baz*, 1971). The darkest portions of this zone correlate well with the low-albedo pyroclastic deposits mapped by *Carr and El-Baz* (1971) and *Hawke et al.* (1979). The westernmost portion of this zone includes an area on the south rim of Archimedes that *Lucey*

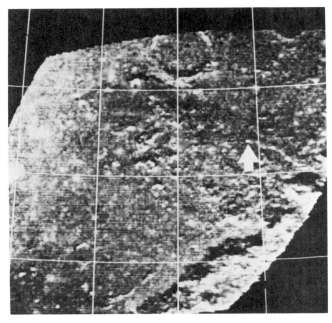

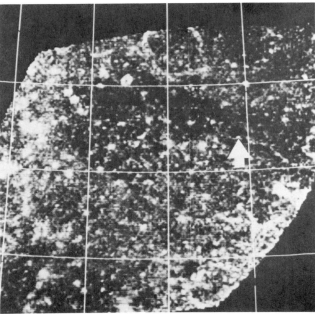

Fig. 10. 3.8-cm radar photographs of Rima Mozart. **(a)** Polarized; **(b)** depolarized. Arrows point to the primary source vent for Rima Mozart, Kathleen.

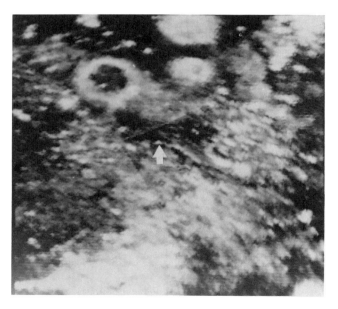

Fig. 11. 70-cm radar image of the Rima Mozart region. Arrow points to the primary source vent for Rima Mozart, Kathleen.

et al. (1984) determined as having a major component of pyroclastic glass. Also included in this zone is a major pyroclastic vent in the Apennine Bench Formation and the associated dark mantling deposits mapped by *Swann* (1986a). While the existence of local dark mantling deposits of pyroclastic origin appear to play a major role in the formation of this zone, they cannot account for all of it.

The new high-resolution 70-cm radar data obtained by Thompson between 1981 and 1984 (*Thompson*, 1987) were also investigated (Fig. 11). Kathleen and Rima Mozart are visible as areas of relatively high 70-cm radar return. An excess of meter-sized blocks is indicated. While Kathleen is largely surrounded by areas of relatively low 70-cm radar returns, the rille is only partly surrounded by these dark areas. A rather complex pattern is seen in Lacus Mozart. As with the 3.8-cm data, there is a zone exhibiting relatively low 70-cm radar values that extends from the Apollo 15/Hadley rille area to the south rim of Archimedes crater. Although the nature and origin of this zone is unclear, we suspect that pyroclastic activity has played some role in its formation.

Spectral data. In order to investigate the composition of the units associated with Rima Mozart, six near-infrared spectra (0.6-2.5 μm) were collected in October, 1986, of features associated with this rille. Unfortunately, adverse atmospheric conditions during data collection resulted in poor quality spectra. Still, some general, qualitative observations can be made:

1. The steep continuum slopes exhibited by spectra obtained for the dark deposits immediately adjacent to Kathleen and at the foot of the Apennines are consistent with a pyroclastic composition.

2. The relatively shallow continuum slope of the Kathleen spectrum suggests that relatively immature material is exposed on the walls of this feature.

3. Though "noisy," several of the Rima Mozart spectra appear to exhibit a very broad "one micron" band. A large Fe-bearing glass or olivene component may be present.

We will attempt to obtain high quality spectra during future observing runs in order to more fully investigate the composition of the Rima Mozart region.

ORIGIN OF RIMA MOZART

Based upon our analysis of the available topographic, photographic, geologic, and remote sensing data it is clear that the Rima Mozart region is the product of a complex series of geologic processes. After the formation of Imbrium basin, the Apennine Bench Formation was emplaced as a series of volcanic flows of KREEP composition (*Ryder*, 1987; *Spudis and Hawke*, 1986; *Spudis*, 1978b; *Hawke and Head*, 1978). Stratigraphically, this unit overlies an extremely fractured Imbrium ejecta unit. This area underwent tectonic settling and readjustment in conjunction with or just after the flooding of the region by the Apennine Bench material (*Swann*, 1986b). Structures resulting from this adjustment and resettling can be seen in the area. These include the graben and other irregular collapse features.

Construction of Rima Mozart began with an explosive eruption at Kathleen crater. A limited amount of pyroclastic debris was emplaced around this source vent. Lava erupted at Kathleen, carved a channel to the southeast, and spread out in the low-lying region now occupied by Lacus Mozart. Photographic evidence indicates that this first channel trended to the northeast before turning southeast to form segment B of the rille. Lava from this first eruption may have formed a tube along the channel. If so, almost all of this tube is now collapsed. The volume of lava erupted during this first eruption is hard to determine because of the uncertainty as to how much lava from the second eruption flowed into this channel. The lava from this first eruption was most likely deposited in the Lacus Mozart area.

Kathleen erupted at least one more time to form the second channel, or segment A. This eruption may have also had an explosive phase which contributed to the blanket of pyroclastic material surrounding the crater. Lava from this eruption appears to have been more turbulent. Photographic and topographic evidence shows the amount of downcutting along this path is much greater than that accomplished during the first eruption. The high temperature and low viscosity, as well as the turbulent nature of this flow, were all conditions that promoted thermal erosion along this channel. Some tube formation may have also occurred along this segment, allowing the lava to flow at higher temperatures and with greater turbulence over a longer distance. Eruption(s) at Ann also appear to have been pyroclastic in origin and voluminous enough to have formed a connecting channel to the main rille. Spatter and some limited pyroclastic debris around the fringes of the vent support its explosive origin.

According to our calculations, a large volume of material was erupted (6372 km^3) and flowed through Rima Mozart.

This led us to question where such an enormous quantity of lava could have gone. A number of possibilities exist for its dispersal:

1. The molten lava reached and flowed into Michael, the sink crater, and was absorbed by the numerous fissures underneath in the fractured Imbrium and Serenitatis ejecta.

2. The flowing, turbulent lava overflowed the confines of the channel and flooded the surrounding area to form Lacus Mozart.

3. A linear, elongate feeder dike(s) underneath the rille and Patricia drained, allowing Patricia and Michael to collapse. The magma emanating from Kathleen and Ann was then able to drain into this newly opened plumbing system. Similar conditions were seen in the 1959 eruption of Kilauea Iki (*McDonald et al.,* 1983), where magma erupted explosively, formed a channel, then flowed back into the ground a short distance later. This lava presumably entered a fissure system known to exist under Kilauea volcanic complex.

4. Patricia is tectonic in origin and related to the other structures present in the region. When Patricia subsided the lava flow drained into the fractures created along its margins and continued down into the underlying fracture system in the Imbrium and Serenitatis ejecta (Fig. 9).

As we have suggested previously, much of the lava probably reentered the ground through preexisting fissures (Fig. 9); some of it also helped to flood the Lacus Mozart region and may have draped the rims of both Patricia and Rima Bradley as lava drainage occurred. However, it is unlikely that these three means of dispersal were enough to account for 6372 km^3 of molten lava. It is possible that some of the lava may have flowed downslope (north/northeast) through the channel that the preexisting Rima Bradley provided through the Imbrium highlands to Palus Putredinus. The albedos of the Rima Mozart basalts and the basalts at the mouth of Rima Bradley are very similar, as are the textures of their surfaces and their crater densities. There is one obstacle in this theory, however, and that is the presence of a small area of hummocky ground just past the mouth of Rima Bradley where it opens up into Palus Putredinus. This area of hummocky ground stands just over 100 m above the surrounding terrain and is aligned along the northwest trend of other wrinkle ridges and collapse features present within Palus Putredinus. Thus we may explain the presence of this high ground as being due to a later upheaval and/or resettling of the area, in which case the path to Palus Putredinus from Rima Bradley would have been unobstructed during the period of Rima Mozart's eruptive history and subsequent flooding of Palus Putredinus.

CONCLUSIONS

The major conclusions of our study of the Rima Mozart region are as follows.

1. Rima Mozart is largely of volcanic origin.

2. The Rima Mozart region was an active volcanic zone at one time as evidenced by the presence of: (a) two major source vents, Kathleen and Ann; (b) a small volcanic fissure zone 5 km south of the main rille, with at least three vents; (c) numerous small volcanic vents adjacent to Ann; and (d) pyroclastic and spatter deposits flanking Kathleen and Ann.

3. Rima Mozart was constructed through a complex series of events. Beginning with the pyroclastic eruption at Kathleen, two channels were formed. Segment B was the first channel formed, and is shallower and narrower than the later formed Segment A. Segment A also truncates Segment B to form two hanging valleys. Later eruptions at Ann formed a smaller, secondary rille to connect Ann to the main channel. Earlier downfaulting along preexisting weaknesses or faults may have set the initial pattern for the rille shape; however, it was not responsible for the total formation of the rille.

4. One or more buried source vents may be present under Lacus Mozart. If so, these vents may have been responsible for some of the flooding of the Lacus Mozart area. Few fractures, cracks, or fissures are evident along Patricia and Rima Bradley, suggesting that something occurred at a later time to flood them or to mask any original features; i.e., once flooded to a higher level, the mare material draped the sides of Patricia and Rima Bradley as it drained from Lacus Mozart.

5. Lava tubes may have been present along some sections of the rille, although calculations suggest that this is unlikely.

6. Rounded corners, wider outside curves, the depression at the mouth of Kathleen, the presence of the hanging valleys at the distal and proximal ends of segment B, as well as the overall rille shape suggest that some thermal erosion occurred along this rille. Our calculations indicate that the duration of the eruption was 947 days (though if the channel were at least partly roofed over into a tube, heat losses would have been less, thermal erosion would have been more efficient, and the duration would have been shorter by a factor of 2).

NOTATION

The following are the variables used in the equations to determine thermal erosion (after *Hulme,* 1973):

A	constant defined in equation (6)
c	specific heat of lava (c_g of ground)
d	depth of flow
E	erosion rate
f	friction factor
g	acceleration due to gravity
h	heat transfer coefficient at interface
k	thermal conductivity—defined in Table 1
L	latent heat of fusion—defined in Table 1
N	constant defined in equation (8)
n	constant defined in equation (6)
Q	volume rate of flow per unit width of channel
Re	Reynolds number
T_l	lava temperature
T_{l_0}	lava temperature at source
T_{mg}	solidus temperature of ground
T_0	initial temperature of ground
u	mean flow velocity
W	width of flow
x_m	length of rille
α	slope of ground
ε	emissivity of lava surface—defined in Table 1
η	viscosity of lava
ρ	density, $ρ_l$-of lava, $ρ_g$-of ground
σ	Stefan constant—defined in Table 1

Acknowledgments. This research was supported by NASA grant NAGW-237. The authors wish to thank Pam Owensby for her help in the collection and reduction of the spectral reflectance data, T. Thompson and S. Zisk for providing the 70- and 3.8-cm data, and the U.H. 88″ Telescope Scheduling Committee for the telescope time allotted to us. Helpful discussions and reviews provided by Lisa Gaddis, Paul Spudis, Gordy Swann, Jim Whitford-Stark, and Nancy Hulbirt aided in the preparation of the manuscript and figures and are gratefully acknowledged.

REFERENCES

Cameron W. S. (1964) An interpretation of Schroter's Valley and other lunar sinuous rilles?. *J. Geophys. Res., 69,* 2423-2430.

Carr M. H. (1974) The role of lava erosion in the formation of lunar rilles and martian channels. *Icarus, 22,* 1-23.

Carr M. H. and El-Baz F. (1971) Geologic maps of the Apennine-Hadley region of the Moon. *U.S.G.S. Map I-723,* Washington, D.C.

Clark P. and Hawke B. R. (1981) Compositional variation in the Hadley-Apennine region. *Proc. Lunar Planet. Sci. 12B,* pp. 727-749.

Coombs C. R. and Hawke B. R. (1986) Preliminary results of geologic and remote sensing studies of Rima Mozart. In *Reports of Planetary Geology and Geophysics Program—1986,* pp. 214-216. NASA TM-89810.

Coombs C. R. and Hawke B. R (1987) Geologic and remote sensing studies of Rima Mozart: Early results (abstract). In *Lunar and Planetary Science XVIII,* pp. 195-196. Lunar and Planetary Institute, Houston.

Cruikshank D. P. and Wood C. A. (1972) Lunar rilles and Hawaiian volcanic features: possible analogs. *The Moon,3,* 412-447.

Davis P. (1980) Iron and titanium distribution on the moon from orbital gamma ray spectrometry with implications for crustal evolutionary models. *J. Geophys. Res., 85,* 3209-3224.

Dowty E., Keil K., Prinz M., Gros J., and Takahashi H. (1976) Meteorite-free Apollo 15 crystalline KREEP. *Proc. Lunar Sci. Conf. 7th,* pp. 1833-1844.

Gaddis L. R., Lucey P. G., Bell J. F., and Hawke B. R. (1984) Remote sensing analyses of localized lunar dark mantle deposits. In *Reports of Planetary Geology and Geophysics Program, 1984, Planetary Geology Principal Investigators,* pp. 399-401. NASA TM-87563.

Gaddis L. R., Pieters C. M., and Hawke B. R. (1985) Remote sensing of lunar pyroclastic mantling deposits. *Icarus, 61,* 461-489.

Gilvarry J. J. (1968) Observational evidence for sedimentary rocks on the Moon. *Nature, 218,* 336-341.

Golombek M. P. (1979) Structural analysis of lunar graben and the shallow crustal structure of the Moon. *J. Geophys. Res., 84,* 4657-4666.

Greeley R. (1971) Lunar Hadley Rille: considerations of its origin. *Science, 172,* 722-725.

Hackman R. J. (1966) Geologic map of the Montes Apenninus region of the Moon. *U.S. Geol. Survey Misc. Geol. Inv. Map I-463.*

Hawke B. R. and Head J. W. (1978) Lunar KREEP volcanism: geologic evidence for history and mode of emplacement. *Proc. Lunar Planet. Sci. Conf. 9th,* pp. 3285-3309.

Hawke B. R., MacLaskey D., McCord T. B., Adams J. B., Head J. W., Pieters C. M., and Zisk S. H. (1979) Multispectral mapping of the Apollo 15-Apennine region: The identification and distribution of regional pyroclastics. *Proc. Lunar Planet. Sci. Conf. 10th,* pp. 2995-3015.

Head J. W. and Wilson L. (1980) The formation of eroded depressions around the sources of lunar sinuous rilles: observations (abstract). In *Lunar and Planetary Science XI,* pp. 426-428. Lunar and Planetary Institute, Houston.

Head J. W. and Wilson L. (1981) Lunar sinuous rille formation by thermal erosion: eruption conditions, rates and durations (abstract). In *Lunar and Planetary Science XII,* pp. 427-429. Lunar and Planetary Institute, Houston.

Howard K. A., Head J. W., and Swann G. A. (1972) Geology of Hadley Rille. *Proc. Lunar Sci. Conf. 3rd,* pp. 1-14.

Hulme G. (1973) Turbulent lava flow and the formation of lunar sinuous rilles. *Mod. Geol., 4,* 107-117.

Hulme G. (1982) A review of lava flow processes related to the formation of lunar sinuous rilles. *Geophys. Surveys, 5,* 245-279.

Hulme G. and Fielder G. (1977) Effusion rates and rheology of lunar lavas. *Phil. Trans. Roy. Soc. Lond., A285,* 227-234.

Irving A. J. (1977) Chemical variation and fractionation of KREEP basalt magmas. *Proc. Lunar Planet. Sci. Conf. 8th,* pp. 2433-2448.

Jaeger J. C. (1959) Temperature outside a cooling intrusive sheet. *Amer. J. Sci., 257,* 44-52.

Kuiper G. P., Strom R. G., and LePoole R. S. (1966) Interpretation of the Ranger Records. *JPL Tech. Rept. 32-800,* 35-248.

Lucey P. G., Gaddis L. R., Bell J. F., and Hawke B. R. (1984) Near-infrared spectral reflectance studies of localized dark mantle deposits (abstract). In *Lunar and Planetary Science XV,* pp. 495-496. Lunar and Planetary Institute, Houston.

Lunar Topophotomap, *Rima Mozart,* 41A3S1 (50), 1974, Defense Mapping Agency, Topographic Center, Washington, D.C.

McAdams W. H. (1954) *Heat Transmission.* McGraw Hill, New York.

McCauley J. F., Guest J. E., Schaber G. G., Trask N. J., and Greeley R. (1981) Stratigraphy of the Caloris Basin, Mercury. *Icarus, 47,* 184-202.

McDonald G. A., Abbot T. A., and Peterson F. L. (1983) *Volcanoes in the Sea.* 2nd edition, University of Hawaii Press, Honolulu. 517 pp.

McGill G. E. (1971) Attitude of fractures bounding straight and arcuate lunar rilles. *Icarus, 14,* 53-58.

Metzger A. E., Haines E. L., Etchegaray-Ramirez M. I., and Hawke B. R. (1979) Thorium concentrations in the lunar surface: III. Deconvolution of the Apenninus region. *Proc. Lunar Planet. Sci. Conf. 10th,* pp. 1701-1718.

Oberbeck V. R., Quaide W. L., and Greeley R. (1969) On the origin of lunar sinuous rilles. *Mod. Geol., 1,* 75-80.

Pieters C. M., McCord T. B., Zisk S. H., and Adams J. B. (1973) Lunar black spots and the nature of the Apollo 17 landing area. *J. Geophys. Res., 78,* 5867-5875.

Pieters C. M., Head J. W., McCord T. B., Adams J. B., and Zisk S. H. (1975) Geochemical and geological units of Mare Humorum: definition using remote sensing and lunar sample information. *Proc. Lunar Sci. Conf. 6th,* pp. 2689-2710.

Quaide W. L. (1965) Rilles, ridges, and domes—clues to maria history. *Icarus, 4,* 374-389.

Ryder G. (1987) Petrographic evidence for nonlinear cooling rates and a volcanic origin for Apollo 15 KREEP basalts. *Proc. Lunar Planet. Sci. Conf. 17th,* in *J. Geophys. Res., 92,* E331-E339.

Schultz P. H. (1976) *Moon Morphology, Interpretations Based on Lunar Orbiter Photography.* Univ. of Texas Press, Austin. 626 pp.

Spudis P. D. (1978a) Composition and origin of the Apennine Bench Formation. *Proc. Lunar Planet. Sci. Conf. 9th,* pp. 3379-3394.

Spudis P. D. (1987b) Origin and distribution of KREEP in Apollo 15 soils (abstract). In *Lunar and Planetary Science IX,* pp. 1089-1091. Lunar and Planetary Institute, Houston.

Spudis P. D. and Hawke B. R. (1986) The Apennine Bench Formation revisited (abstract). In *Workshop on the Geology and Petrology of the Apollo 15 Landing Site* (P. D. Spudis and G. Ryder, eds.), pp. 105-107. LPI Tech. Rpt. 86-03, Lunar and Planetary Institute, Houston.

Spudis P. D., Swann G. A., and Greeley R. (1987) The formation of Hadley Rille and implications for the geology of the Apollo 15 region. *Proc. Lunar Planet. Sci. Conf. 18th*, this volume.

Swann G. A. (1986a) Some observations on the geology of the Apollo 15 landing site (abstract). In *Workshop on the Geology and Petrology of the Apollo 15 Landing Site* (P. D. Spudis and G. Ryder, eds.), pp. 108-112. LPI Tech. Rpt. 86-03, Lunar and Planetary Institute, Houston.

Swann G. A. (1986b) Collapse structures in the Apennine Bench Formation and their influence on the formation of Mozart and Hadley Rilles (abstract). In *Lunar and Planetary Science XVII*, pp. 855-856. Lunar and Planetary Institute, Houston.

Thompson T. W. (1979) A review of Earth-based radar mapping of the Moon. *Moon and Planets, 20*, 179-198.

Thompson T. W. (1987) High-resolution lunar radar map at 70-cm wavelength. *Earth, Moon, and Planets, 37*, 59-70.

Thompson T. W. and Zisk S. H. (1972) Radar mapping of lunar surface roughness (Chapter 1c). In *Thermal Characteristics of the Moon* (J. Lucas, ed.), from *Progress in Aeronautics and Astronautics*, pp. 83-117. MIT Press, Cambridge, Mass.

Urey H. C. (1967) Water on the Moon. *Nature, 216*, 1094.

Walker G. P. L. (1973) Lengths of lava flows. *Phil. Trans. Roy. Soc. Lond., A274*, 107-118.

Wilhelms D. E. (1980) Stratigraphy of part of the lunar near side. *U.S.G.S. Prof. Paper 1046A*, 71 pp.

Wilhelms D. E. (1985) Geologic setting of Apollo 15 (abstract). In *Workshop on the Geology and Petrology of the Apollo 15 Landing Site* (P. D. Spudis and G. Ryder, eds), pp. 119-123. LPI Tech. Rpt. 86-03, Lunar and Planetary Institute, Houston.

Wilhelms D. E. and McCauley J. F. (1971) Geologic map of the near side of the Moon. *U.S.G.S. Map I-703*, Washington, D.C.

Wilson L. and J. W. Head (1980) The formation of eroded depressions around the sources of lunar sinuous rilles: theory (abstract). In *Lunar and Planetary Science XI*, pp. 1260-1262. Lunar and Planetary Institute, Houston.

Wilson L. and Head J. W. (1983) A comparison of volcanic eruption processes on Earth, Moon, Mars, Io and Venus. *Nature, 302*, 663-669.

Zisk S. H., Pettengill G. H., and Catuna G. W. (1974) High-resolution radar map of the lunar surface at 3.8-cm wavelength. *The Moon, 10*, 17-50.

Zisk S. H., Hodges C. A., Moore H. J., Shorthill R. W., Thompson T. W., Whitaker E. A., and Wilhelms D. E. (1977) The Aristarchus-Harbinger region of the Moon: surface geology and history from recent remote-sensing observations. *The Moon, 17*, 59-99.

A Remote Mineralogic Perspective on Gabbroic Units in the Lunar Highlands

Paul G. Lucey and B. Ray Hawke

Hawaii Institute of Geophysics, Planetary Geosciences Division, University of Hawaii, Honolulu, HI 96822

Three regions in the highlands that have been shown to contain abundant high-Ca pyroxene are studied in order to constrain the chemistry of the pyroxene assemblage of gabbroic regions, identify the rock type present in the returned sample collection responsible for gabbroic regions in the highlands, and assess the state of gabbroic materials in the highlands crust. The derived pyroxene chemistries are shown to be consistent with those of sample highland pyroxenes. A compositional gap between ortho- and clinopyroxenes is inferred from the spectral data that corresponds to that observed in lunar rocks. Tycho has a clinopyroxene/total pyroxene ratio probably in excess of 0.8. Assuming typical highland clinopyroxene chemistry for the Aristarchus region and for five locations in the eastern interior of Imbrium basin, the clinopyroxene/total pyroxene ratio of these areas are 0.5-0.6 and 0.4-0.7, respectively. Assuming extreme Wo or Fs contents the latter areas will still have clinopyroxene/total pyroxene contents in excess of 0.2. The mineralogical characteristics of the few rock types in the sample collection that contain significant high-Ca pyroxene are compared to the remotely acquired data and it is shown that Mg-gabbronorite or KREEP basalt with abundant high-Ca pyroxene are the most likely rock types of those thus far identified that may be responsible for gabbroic areas in the highlands. It is shown that the western lunar highlands are distinct from the east in mineralogy. The east is primarily noritic with occasional exposures of anorthosite in locations inferred to have been derived from 5-15 km in depth. The mineralogy of small craters in the west is similarly noritic; however, the deposits of some large craters in the west contain abundant gabbroic material. Anorthosite is thus far unidentified in the west but exposures of olivine-rich material have been identified at Copernicus and Aristarchus. The compositional dichotomy between deposits of large and small craters in the western lunar highlands is speculated to have been caused by the introduction of gabbroic intrusions at moderate depths after the bulk of megaregolith formation. The change with time of chemistry of crystallizing magma from noritic to gabbroic in the west is speculated to have been due to the cessation of anorthosite assimilation by Mg-suite liquids due to exhaustion of anorthosite. Anorthosite may have originally been less abundant in the west due to magma ocean processes or crustal thinning by single or multiple giant impacts. Anorthosite was sufficiently abundant in the east to support assimilation and production of norite until the end of Mg-suite magmatism.

I. INTRODUCTION

The pyroxene assemblage of the rocks returned from the lunar highlands is dominated by low-Ca pyroxene. For this reason, *Warren et al.* (1983) stated that "One of the fundamental characteristics of the nonmare crust is that it contains more low-Ca than high-Ca pyroxene." The scarcity of gabbroic highland rocks has puzzled petrologists. *Longhi* (1978, 1981) pointed out that parent melts of the nonmare crust should have crystallized abundant high-Ca pyroxene assuming chondritic Ca/Al ratios. *Longhi* (1981) hypothesized that either the parent melts had subchondritic Ca/Al ratios or that the melts formed by large degrees of partial melting. *Warren* (1986), however, showed that even with low initial Ca/Al ratios, crystallization of troctolites would inevitably increase the Ca/Al ratio of a cooling magma, again eventually producing rocks with abundant high-Ca pyroxene.

Despite their rarity in the sample collection, highland rocks with significant amounts of high-Ca pyroxene do occur, including Mg-gabronorites (*James and Flohr*, 1983; *Taylor et al.*, 1983; *Warren*, 1986), alkali anorthosites (*Hubbard et al.*, 1971; *Warren et al.*, 1983), and the quartz monzodiorite clasts in sample 15405 (*Irving*, 1977; *Ryder*, 1976).

Remote sensing data seem inconsistent with the scarcity of gabbroic materials observed in the highland samples. Orbital geochemistry studies show the need for abundant materials with gabbroic affinities to account for the chemical character of portions of the highland crust (e.g., *Taylor*, 1975; *Davis and Spudis*, 1985). *Davis and Spudis* (1985) invoked mare basalt to account for this chemical component that is abundant on the eastern limb because of photogeologic evidence for buried mare basalts in the vicinity. However, they did acknowledge the possibility (considered by them unlikely) that highland gabbros could account for this mafic component in portions of the lunar crust.

The dominance of low- over high-Ca pyroxene in the highlands was recognized in telescopic remote mineralogic measurements of the lunar surface (*McCord et al*, 1981). However, in contrast to the scarcity of highland rocks with a calcic pyroxene assemblage in the returned sample collection, Earth-based telescopic near-infrared reflectance spectroscopy has shown that there are several highland locations in the lunar nearside with abundant high-Ca pyroxene. *McCord et al.* (1981) presented spectroscopic measurements for Aristarchus crater that showed a pyroxene absorption feature with a wavelength minimum position indicative of the presence of high-Ca clinopyroxene. Since this initial discovery, several other highland areas have been found to have pyroxene assemblages dominated by high-Ca pyroxene (*Pieters*, 1986; *Lucey et al.*, 1986; *Hawke et al.*, 1983). *Pieters* (1986) classified as

TABLE 1. Spectral parameters.

Location name	Wavelength of Relative Minimum (μm)	Relative absorption depth (%)	Continuum slope ($\Delta R/\Delta \lambda$)	Absorption width at half height(μm)
Aristarchus A	.964	16.3	.290	.280
Aristarchus southwest wall	.964	11.4	.288	.283
Aristarchus northwest wall	.954	12.6	.208	.280
Aristarchus south floor	.988	7.31	.240	.354
Aristarchus east wall	.977	7.94	.248	.351
Aristarchus central peak	.995	6.87	.202	.365
Timocharis east wall	.953	5.15	.586	.293
Archimedes east wall	.949	7.62	.664	.295
Autolycus east wall	.941	6.69	.636	.287
Aristillus south rim	.968	5.503	.565	.302
Timocharis east ejecta	.966	8.36	.611	.296
Tycho: Peak, high resolution	.982	16.70	.292	.256
Floor	.970	7.75	.427	.244
Wall	.971	9.92	.366	.252
Floor	.994	9.73	.432	.296
East rim	.980	8.65	.413	.243
Peak, low resolution	.978	12.16	.408	.270
Southwest wall	.968	7.44	.457	.250
Southwest floor	.986	8.53	.461	.245
East floor	.982	8.89	.401	.271
North rim	.963	9.84	.460	.260

"gabbroic" 18 locations out of 75 studied and remarked that though gabbros "...have not been abundantly recognized in the collection as potential pristine crustal samples, they may represent a significant pristine rock type for the Moon."

This discrepancy between the remote sensing and sample data bases in the apparent relative abundance of materials with significant amounts of high-Ca pyroxene (high-Ca pyroxene/total pyroxene > 0.1) and materials with little or no high-Ca pyroxene has encouraged us to study further the remote mineralogic data in an effort to quantify better the nature and distribution of units with unusually large high-Ca pyroxene/total pyroxene as detected remotely. If these materials are indeed underrepresented in the sample collection, then the remote sensing data may better characterize the global nature of the crust than is possible with the existing sample collection. Preliminary results of this effort were presented in *Lucey and Hawke* (1987a,b,c). (A note on terminology: Throughout this paper the term "gabbroic" will denote a pyroxene assemblage with high-Ca pyroxene/total pyroxene > 0.5, while "noritic" will denote a pyroxene assemblage with high-Ca pyroxene/total pyroxene < 0.5. This terminology is not meant to imply the presence of norite or gabbro *sensu stricto*.)

This paper consists of four major sections. The first will integrate remote sensing studies of selected individual gabbroic regions. Spectral data will be used to better characterize the pyroxene chemistry of gabbroic regions. The second part will use the compositional interpretations derived in this work and other studies to attempt to determine which rock type or types in the sample collection dominate gabbroic areas. The third section will investigate the dependence of lunar surface compositions on longitude and depth of origin by reexamining the data in the global survey of *Pieters* (1986). Finally, hypotheses will be presented to account for the composition and distribution of gabbros.

II. PYROXENE COMPOSITIONS OF GABBROIC REGIONS

Several recent remote sensing studies of regions that exhibit variable amounts of gabbroic material have placed the interpreted compositions into geologic contexts. They are the study of the Aristarchus region by *Lucey et al.* (1986), the study of Tycho crater and its deposits by *Hawke et al.* (1986), and the study of the Imbrium basin by *Hawke and Lucey* (1984) and *Spudis et al.* (1987). Each study presents a large number of spectra obtained for a fairly small region so that conclusions can be based on more than a single observation. The spectral parameterization technique of *Lucey et al.* (1986) was applied to the data of each study, allowing direct comparison of data sets.

The compositions exposed on the Aristarchus Plateau were interpreted from reflectance spectra by *Lucey et al.* (1986). Three highland lithologies were detected by those workers: a gabbro or anorthositic gabbro (Aristarchus Spectral Class 1); an olivine gabbro (Aristarchus Spectral Class 2); and a troctolite, possibly dunite (Aristarchus Spectral Class 3). The presence of these three lithologies and the high-KREEP component detected by the orbital geochemistry experiment prompted the authors to hypothesize that the Aristarchus Plateau was the site of an exposed differentiated gabbroic pluton.

Spectra in the eastern portion of Imbrium basin were presented and interpreted by *Hawke and Lucey* (1984) and *Spudis et al.* (1987). It was concluded that spectra for certain locations within the Apennine ring represented gabbroic norites or noritic gabbros due to band centers near 0.95 μm, implying the presence of subequal amounts of orthopyroxene and clinopyroxene.

Hawke et al. (1986) presented preliminary analyses of the spectra of the interior and exterior deposits of the crater Tycho. They concluded that the spectra represented gabbros, impact melt, and mixtures of the two.

Lucey et al. (1986) presented a detailed method for deriving four important spectral parameters from reflectance spectra of the Moon and presented the results of this analysis for spectra of the Aristarchus region. This technique was used by *Spudis et al.* (1987) to quantify the eastern Imbrium spectra and we have applied the same analysis to the Tycho data. Table 1 lists the values of these parameters for gabbroic classes of *Lucey et al.* (1986), *Spudis et al.* (1987), and those derived from the Tycho spectra, excluding spectra obviously dominated by impact melt.

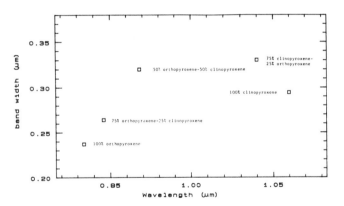

Fig. 2. Plot of band width in micrometers versus band center in micrometers for orthopyroxene-clinopyroxene mixtures of *Singer* (1981).

Figure 1 shows a series of variation diagrams in which six combinations of these parameters are plotted against each other. It can be seen that the classes and locations plot as independent clusters. Figure 1d plots the two parameters that are least affected by soil maturity: band minimum and band width. In this diagram, the Aristarchus olivine gabbro points are distinctly separated from the other spectral types. The other three classes plot much closer to each other relative to the olivine gabbro. The nonolivine gabbro classes exhibit a trend of decreasing band width with increasing wavelength of band minima. Eastern Imbrium interior spectra display the widest bands with the shortest band minima, Aristarchus gabbroic spectra are intermediate, and Tycho spectra display the narrowest bands with the longest minima.

Most pyroxene-rich lunar samples contain two pyroxenes. It is very likely that the locations measured remotely also have this characteristic. In order to place quantitative limits on the high-Ca pyroxene/total pyroxene ratio and the chemistry of the pyroxenes at these locations the spectral effects of pyroxene composition and relative abundance of two pyroxenes must be separated. While the wavelength of the minimum of the absorption band center relative to the continuum slope is a crucial compositional parameter, in a mixture of two pyroxenes that parameter is a function of the composition of each pyroxene and the relative abundance of the two. It is not possible to use this parameter to distinguish between the presence of a single pyroxene and a mixture of two pyroxenes. Thus the compositional calibration of *Adams* (1974) and *Hazen et al.* (1978), derived for single pyroxenes, cannot be used directly to determine Ca or Fe content if a mixture is suspected.

Fortunately, the width of the absorption band is sensitive to the presence of mixtures. Figure 2 is derived from analysis of the orthopyroxene-clinopyroxene mixture series of *Singer* (1981). The widths of the mixtures are not linear averages of the width of the endmembers. Note that the pure endmember bands are are narrow while the mixtures show wider bands. This is due to the nonlinear nature of reflectance in intimate mixtures experiencing multiple scattering. In such mixtures, darker particles have a disproportionate effect upon

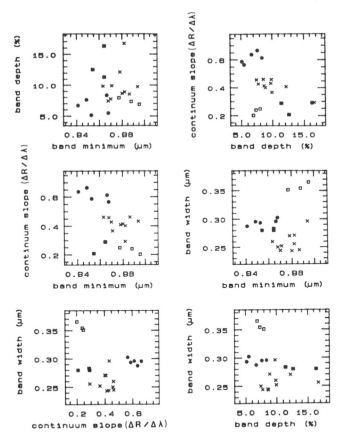

Fig. 1. Variation diagrams derived from values tabulated in Table 1. Symbols represent classes listed in Table 1 as follows: Filled squares are Aristarchus noritic gabbro; open squares are Aristarchus olivine gabbro; crosses represent Tycho spectra; and filled circles are Imbrium interior gabbroic location.

the reflectance because multiple scattering due to bright particles increases the probability of encountering a dark, photon-absorbing grain. In the spectra of mxitures of two materials with absorption band center separations on the order of the width of the absorption band, the low reflectance of the two components near their band centers suppresses the wings of the composite band, which results in widening of the composite band. The nonlinearity of the process is further demonstrated by the observation that the wavelength of the minima of the mixtures is not proportional to the relative amounts of the two components. The change in mininmum is small near the endmembers but relatively rapid at intermediate values.

Qualitatively, the trend of decreasing band width with increasing band center observed in Fig. 1d resembles a portion of the trend observed in Fig. 2 and suggests that this relationship is due to variation in high-Ca pyroxene/total pyroxene between the three areas. Figure 2 cannot be quantitatively compared to the remotely obtained lunar data because the clinopyroxene endmember of Singer (1981) is much more calcic than most lunar pyroxenes (*Papike*, 1980) and thus is centered at a much longer wavelength than is likely to be encountered. However, Fig. 2 is illustrative of the important effect of mixing on band width. Figure 3 plots the gabbroic classes shown in Fig. 1d (excluding the Aristarchus olivine gabbro) as well as the noritic classes reported in the eastern Imbrium region by *Spudis et al.* (1987). The scatter is relatively large in both the widths and minima. This reflects both the error in the measurement and probable effects of other minerals or glass present at these locations. However, the general trend is similar to that shown in Fig. 2. From comparison of Figs. 2 and 3 we suggest that the trend shown in Fig. 3 is due to the mixing of varying amounts of clino- and orthopyroxene. There is an apparent "gap" between 0.945 μm and 0.960 μm where narrow bands are absent. This suggests that spectra exhibiting this range of band center are composed of mixtures of clino- and

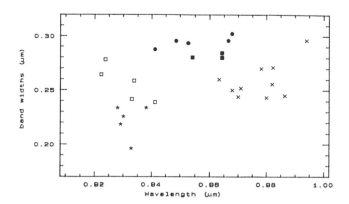

Fig. 3. Plot of band width versus band minimum for gabbroic areas mentioned in text (excluding Aristarchus olivine gabbro), and noritic areas in the east Imbrium exterior. Filled squares are Aristarchus noritic gabbro, crosses represent Tycho spectra, filled circles are Imbrium interior gabbroic locations, open squares and stars are noritic classes of *Spudis et al.* (1987).

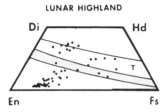

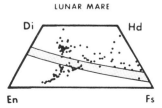

Fig. 4. Pyroxene compositions of lunar samples with remotely determined compositional ranges as explained in the text. The area labeled "T" represents the range of compositions for the Tycho region. The stippled area represents the "gap" where only wide bands implying the presence of mixtures of clino- and orthopyroxene are present. Sample data are after BVSP (1981).

orthopyroxene and that single pyroxenes with compositions corresponding to these band centers are not present in the portions of the crust thus far measured.

Figure 4 shows data on highland and mare pyroxenes plotted on the pyroxene quadrilateral [*Basaltic Volcanism Study Project (BVSP)*, 1981]. The compositional region corresponding to the observed gap in spectral data is plotted as derived from the calibration of *Hazen et al.* (1978). The gap corresponds well to the compositional gap between clino- and orthopyroxenes observed in both mare and highland rocks.

The narrowness of the bands in the Tycho spectra suggest the presence of a mineral assemblage that has a mafic component of almost pure high-Ca pyroxene. If this suggestion is correct, then the band centers measured for Tycho correspond to the compositional range shown in Fig. 4. The wide range of possible compositions is due to the fact that the 1-μm band center is a function not only of Wo composition, but also, to a lesser extent, of Fs. If the pyroxene assemblage at Tycho is not composed entirely of the clinopyroxene endmember (high-Ca pyroxene/total pyroxene < 1) then the composition of the clinopyroxene endmember will be somewhat farther from the En apex than the low Wo portion of the range depicted in Fig. 4. Thus the line nearest the En corner of the pyroxene quadrilateral shown in Fig. 4 for Tycho compositions constitutes a lower limit on Wo and Fs content.

To estimate clinopyroxene/total pyroxene for the Aristarchus and east Imbrium interior regions we must assume a calcic endmember chemistry (orthopyroxenes show quite restricted ranges in wavelength minima over relatively large changes in chemistry). We assume a composition equal to the mean of the highland pyroxenes shown in Fig. 4 from BVSP (1981). The highland pyroxene data shown in Fig. 4 can be used to estimate a likely band center of ~1.0 μm for the clinopyroxene endmember of a highland rock. With this assumption, Aristarchus gabbroic material has an approximate clinopyroxene/total pyroxene ratio of 0.5-0.6 and the Imbrium interior locations, ~0.4-0.7. If the clinopyroxene endmembers of Aristarchus or eastern Imbrium are more calcic than assumed, then the clinopyroxene/total pyroxene values for Aristarchus and Imbrium will be lower than the above estimates. They could perhaps be as low as 0.2, but certainly not lower.

III. CORRELATION OF REMOTELY DETECTED GABBROIC REGIONS WITH KNOWN PRISTINE ROCK TYPES

This section is intended to determine which rock type identified in the sample collection, if any, may be responsible for the gabbroic compositions in the lunar highlands. This determination has important implications for the abundance and distribution of these rock types and the role they have played in the evolution of the lunar crust.

Our approach is as follows: (1) Identify pristine rock types or samples in the sample collection characterized by a high value of high-Ca pyroxene/total pyroxene and classify these as candidates; (2) examine each candidate rock type or sample for consistency with the mineralogical and compositional data derived for gabbroic regions in this and in other works; (3) eliminate candidates inconsistent with the remote sensing data; and (4) report the remaining candidates, if any.

Each of the following samples or rock types are characterized by a high value of high-Ca pyroxene/total pyroxene: Mg-gabbronorite (*James and Flohr,* 1983), rare KREEP basalts (*BVSP,* 1981), the quartz monzodiorite clasts in 15405 (*Ryder,* 1976), alkali anorthosites (*Warren et al.,* 1983), one ferroan anorthosite (*Warren et al.,* 1986), and mare basalts (*BVSP,* 1981).

Anorthosites

The depth of the pyroxene feature of the most immature of the locations in the Aristarchus and Tycho regions strongly indicate that pyroxene is very abundant. It is very unlikely that the abundance of pyroxene at these locations is less than 5 wt %. *Lucey et al.* (1986) gave 10% pyroxene by weight as the lower limit to pyroxene abundance for the Aristarchus gabbroic locations. This pyroxene abundance eliminates anorthosites as candidate materials.

KREEP Basalt

The presence of the Th anomaly at Aristarchus (*Etchegaray-Ramirez et al.,* 1982) made KREEP basalt or its derivatives an attractive candidate for the identity of this remotely detected highland gabbro. The pyroxene assemblage of KREEP basalt is generally characterized by low-Ca pyroxene only. KREEP basalts with abundant high-Ca pyroxene are rare. However, relative abundance of a rock type in the sample collection cannot be used to exclude a particular rock type because the remote sensing data suggest that highland gabbroic materials are underrepresented in the sample collection. The only spectra of an area that is likely to be composed of KREEP basalt, the Appenine Bench Formation (*Spudis,* 1978), are unlike those considered here and have band shapes suggesting the presence of abundant glass (*Spudis et al.,* 1987). However, KREEP basalt with abundant clinopyroxene that is spectrally unlike the Appenine Bench Formation cannot be eliminated as a candidate rock type.

The abundant high-Ca pyroxene and very high level of incompatibles of the quartz monzodiorite clasts in 15405 suggested that the QMD is an especially good candidate. *Ryder* (1976) and *Irving* (1977) interpreted this clast to be the result of differentiation of a magma of KREEP basalt composition. These workers showed that to derive the quartz monzodiorite from a KREEP basaltic magma, fractional crystallization removing large quantities of low-Ca pyroxene was required. At no time does olivine appear in the crystallization sequence. If the evidence presented by *Lucey et al.* (1986) is correct and the Aristarchus Plateau is the site of an excavated, differentiated pluton, the complete absence of orthopyroxene-dominated units and the presence of olivine-rich lithologies in the Aristarchus region argue against the interpretation of the Aristarchus compositions as the result of differentiation of an intrusion of KREEP basalt composition.

Mare Basalt

Mare basalt is by far the most common sample type that contains abundant high-Ca pyroxene. Spectrally, however, mare basalts are distinct from the highland gabbros reported here. The spectra of mare basalts are characterized by IR continuum slopes that are much steeper than those exhibited by the Tycho central peak and Aristarchus locations (*Pieters,* 1980). The continuum slopes exhibited by the spectra of large craters in the interior of Imbrium are closer to mare values but have bands that are weaker than those exhibited by fresh mare craters. Excepting albedo, the measured parameters of the Imbrium interior locations more closely resemble mature mare than fresh mare. However, the morphology and albedo of the locations measured show clearly that they are much fresher surfaces than mature mare. The IR data suggest that none of the measured gabbroic areas are composed of mare basalt. Visible spectra of Aristarchus and Tycho presented by *Pieters* (1977) are quite different than spectra of fresh mare craters. A visible spectrum of Archimedes, one of the Imbrium interior locations, was classified by *Pieters* (1977) as a highlands fresh crater. The current understanding of the spectral characteristics of mare basalts and highland gabbros suggests that the gabbroic regions are not composed of mare basalt.

Mg-Gabbronorite

All of the compositional characteristics that can be determined by remote sensing techniques are consistent with Mg-gabbronorite. The high-Ca pyroxenes contained in the six Mg-gabbronorites of *James and Flohr* (1983) have values consistent with the chemistries estimated for Tycho (Fig. 4). The range of clinopyroxene/total pyroxene values given for all the gabbroic regions (from 0.2 to 1.0) are also consistent with the values reported by *James and Flohr* (1983). The hypothesis of *Lucey et al.* (1986) that the Aristarchus compositions may be the result of excavation of a differentiated gabbroic pluton is consistent with the presence of Mg-gabbronorite, if the undifferentiated plutonic magma was KREEPy. Mg-gabbronorites show highly variable KREEP signatures (*James and Flohr*, 1983).

Summary

The only two candidates that have not been eliminated on the basis of their spectral characteristics are Mg-gabbronorite and KREEP basalt with high relative abundances of high-Ca pyroxene. Both of these rock types are rare in the sample collection, though because of the apparent undersampling of gabbroic material in the sample collection, there is no *a priori* reason to choose one of these types over the other. In future, a more careful look at the geochemical data may yield some clues that may allow resolution of the ambiguity.

IV. CHARACTERIZATION OF LONGITUDINAL DEPENDENCE OF MINERALOGY

It has been suggested that the western portion of the lunar nearside is compositionally distinct from the eastern nearside based on sample studies (*Warren and Wasson*, 1980), orbital geochemistry data (e.g., *Metzger et al.*, 1973), and spectral reflectance measurements (*Pieters*, 1986). The samples show differences both in isotopic chemistries and in relative abundance of the various rock types between the western landing sites and those in the east. The Apollo orbital geochemistry experiment revealed a concentration of radioactive elements in the west. Spectral reflectance measurements show that gabbroic spectral classes are concentrated in the west. This section is intended to further investigate changes in mineralogy with longitude.

Mineralogic changes as a function of longitude were studied using data of *Pieters* (1986). Three mineralogic parameters were considered: pyroxene chemistry, distribution of olivine, and distribution of anorthosite. East-west changes in relative abundance of low- and high-calcium pyroxene can be investigated by plotting band minimum versus longitude.

Figure 5 presents the results of the investigation. Including only locations with spectra dominated by pyroxene absorptions, it is clear that the spectra obtained for the western portion of the nearside show more scatter in band centers than those of areas on the eastern portion. On Fig. 5, the spectra of small craters (5-15 km in diameter) and spectra of deposits of large craters (15-100 km in diameter) are plotted separately. The spectra of the craters in the two different size ranges show different trends of band minimum versus longitude. The spectra of small craters show relatively constant band minima with longitude. These band positions are indicative of the dominance of low-Ca pyroxene in the mafic mineral assemblages of small craters across the lunar nearside. In contrast, central peaks, rims, and walls of large craters all showed similar behavior: a general increase in wavelength minimum, and thus increase in proportion of high-Ca pyroxene, toward the west. It is not clear whether this increase is due to an abrupt change to longer wavelengths as the central meridian is crossed or whether the change is more gradual or even linear with longitude.

A correlating asymmetry is also found in the distribution of olivine and anorthosite. Pure anorthosites have been detected in the eastern portion of the lunar nearside but are undetected in the western portion between the western limb at Orientale (Inner Rook Ring) and near the central meridian at Alphonsus (*Pieters*, 1986; *Spudis et al.*, 1984). Olivines are detected only in the western portion of the nearside thus far.

The change in composition from large to small craters implies a change in composition with depth. A similar suggestion has been made for the Moon as a whole by *Pieters* (1986). We differ in that our analysis suggests that a change in composition with depth holds only for the lunar west, and that this change is inferred not because of a difference between central peaks and other geologic settings, but between small craters and deposits related to large craters.

On the basis of this analysis we conclude that the pyroxene chemistry of the eastern nearside highlands is generally homogeneous and noritic to depths of about 15 km with the exception of occasional occurrences of anorthosite in the upper crust below the megaregolith (e.g., central peak of Petavius). The western nearside highlands, in contrast, has a generally noritic near-surface zone (probably representing the megaregolith) but the evidence suggests that the upper crust below the surface zone is gabbroic in composition. In addition, the west exhibits olivine-rich material derived from the upper crust (e.g., Copernicus central peaks) and present in the near-surface environment (e.g., the mountain Herodotus X on the Aristarchus Plateau).

V. SPECULATIONS UPON THE ORIGIN OF VERTICAL AND HORIZONTAL ASYMMETRIES IN THE DISTRIBUTION OF GABBROIC MATERIAL ON THE NEARSIDE

Three important observations have been made from the remote mineralogy data. First, gabbroic regions are more abundant on the nearside than would be inferred from the abundance of this material relative to noritic rock types in the lunar highlands sample suite. Second, gabbroic regions are more abundant in the west than in the east. Third, in the west, gabbroic material seems concentrated at depth as inferred from the apparent concentration of gabbroic material in the deposits of large craters. In this section we will present hypotheses to explain these observations.

Vertical Compositional Distribution

The change in composition with depth observed in the west has an obvious implication as pointed out by *Pieters* (1986): The megaregolith as presently observed cannot be derived from the material that presently exists at depths of 5-15 km. Pieters listed a number of caveats that caution us against accepting this observation too readily. However, these are the best data now available and the possible significance of the observation deserves comment. There are two alternatives that may explain

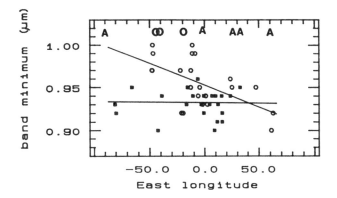

Fig. 5. Distribution of wavelength of band minimum versus longitude. Filled squares represent spectra of craters 5-15 km in diameter. Open circles represent spectra of deposits of large craters 15-100 km in diameter. Lines are linear least squares fits to the two size ranges. Letters "A" and "O" show the longitude of anorthosite and olivine-rich locations respectively (after *Lucey and Hawke*, 1987a).

the observation. Either the noritic upper 1-2 km was emplaced after formation of a gabbroic crust in the west, or the entire upper 15 km was originally noritic but the lower portion of this zone evolved compositionally with time.

The first alternative is a possibility. Pieters (personal communication, 1987) suggests that the noritic upper crust could be due to blanketing of the nearside by the ejecta of the last few basins that presumably formed from impacts into noritic targets. The volume of material necessary to cover one-third of the Moon 1.5 km deep (equivalent to the most of the area accessible to Earth-based spectroscopy) is over four times the volume excavated by Imbrium Basin as estimated by *Spudis et al.* (1987). This suggestion is viable if multiple large basins were formed in noritic targets after most of the gabbroic highland intrusive activity ceased.

We present the second alternative. We propose that Mg-suite noritic intrusive (and probably extrusive) activity into and on the ancient ferroan anorthosite crust in the vicinity of 4.4 Gy and earlier gave the ancient megaregolith its noritic character. After the rapid decline in megaregolith-forming impact flux, this noritic activity was followed by gabbroic intrusive activity, which has been suggested here to account for certain gabbroic anomalies and the Aristarchus troctolitic anomaly. Thus the vertical compositional dichotomy may have been due to a compositional evolution of the upper crust below the megaregolith subsequent to the formation of the megaregolith. This evolution was accomplished through emplacement of gabbroic plutons into the upper curst.

If this hypothesis is correct, it explains both the megaregolith/submegaregolith compositional dichotomy and the dearth of gabbroic material in the highland sample collection. An insufficient number of impacts have occurred to mix significant amounts of gabbroic material into the upper 1-2 km. Exposures of gabbroic material at the surface would be expected to be highly localized (which is observed) and gabbroic material would not be expected to be collected at a landing site unless the site was very near one of these isolated exposures. This situation is directly analogous to that of the sampling of mare basalts. *Pieters* (1977) showed that unsampled types of mare basalts are common on the nearside of the Moon. These types are simply too recent for the impact process to have transported much of this material to any of the landing sites.

For this hypothesis to be supported evidence for the occurrence of gabbroic intrusive activity should be found, and this activity must be generally younger than noritic activity. Remote sensing data has shown that gabbroic plutonic activity has occurred in the western nearside (*Lucey et al.*, 1986; *Lucey and Hawke*, 1987b) and that this activity may have been widespread (*Lucey and Hawke*, 1987a). The presence of plutonic gabbros in the sample collection has been established (*James and Flohr*, 1983).

The best evidence for a difference in relative age is that the Mg-gabbronorites thus far measured are younger than the norites (*James and Flohr*, 1983). If the interpretation of the gabbros detected remotely as being due to Mg-gabbronorite is accepted, then it is possible that the remotely detected gabbros are younger than remotely detected noritic material. Sharp compositional boundaries exist between highland compositions in the Aristarchus region (*Lucey et al.*, 1986), the site of the best documented gabbroic area. If this is indeed the site of an exposed pluton, these sharp boundaries suggest a low degree of mixing and imply that this region escaped homogenization expected to be attendant with megaregolith formation. Thus there is good evidence that gabbroic activity has occurred, and there is some inferential evidence that this activity is younger than noritic plutonism.

This scenario is sensitive to the relative timing between the decline in impact flux and the introduction of gabbroic material into the upper crust. *Hartmann* (1980) pointed out that the formation of the lunar crust was contemporaneous with an intense but rapidly declining impact flux associated with the accretion of the Moon. *Hartmann* (1980) presented an impact flux model assuming a saturated crater population in the highlands and a smooth, exponential decline in cratering rate since the accretion of the Moon. According to this model, at 4.2 Gy, for example, rocks emplaced below a kilometer would survive to the present (not be homogenized into the megaregolith). If correct, this flux model generally supports the hypothesis that gabbroic material emplaced in the vicinity of 4.2 Gy escaped the homogenization attendant with the waning stages of accretion. For this hypothesis to be valid, a "terminal cataclysm" (*Tera et al.*, 1974) in which there is a large increase in the number of impactors at all size ranges could not have occurred in the vicinity of 4.0 Gy. However, cataclysms that include only, or primarily, basin-forming events are permissible and perhaps necessary to excavate (but not homogenize) material emplaced at moderate (10-20 km) depths in the vicinity of 4.2 Gy.

Horizontal Mineralogic Asymmetries

Compositional differences between the eastern and western portions of the lunar nearside were underscored above. The west seems to have experienced considerably more igneous activity resulting in gabbroic lithologies in its history than the lunar east. The crust in the west seems deficient in noritic and anothrositic material.

The observation may well be due to the relative thinness of the nearside western crust (*Bills and Ferrari,* 1976). In considering the closely related problem of olivine at Copernicus, *Pieters and Wilhelms* (1985) suggested that the proposed Procellarum impact may have caused uplift that brought a gabbroic lower crust near the surface. This uplift enabled large craters to excavate heterogeneous lower crustal material, including troctolites and gabbros. However, rather compelling evidence has been presented that the lower crust is noritic, not gabbroic (*Ryder and Wood,* 1977; *Spudis et al.,* 1984). If our suggestion that the compositional dichotomy between megaregolith and upper crust is due to emplacement of gabbroic plutons is correct, then crustal thinning must have predated this process. Impact thinning postdating this process would have stripped off the noritic upper crust. A thin early anorthosite crust in the west may explain the east/west dichotomy within the context of our gabbroic intrusion hypothesis. *Warren* (1986) suggested that the norite/gabbronorite distinction is due to assimilation. In his model, gabbronorite parent liquid assimilated anorthosites, which lowered the Ca/Al ratio of the compositionally mixed magma sufficiently to inhibit crystallization of high-Ca pyroxene and yield norites. If this model is correct the production of norite through assimilation can be limited by availability of anorthosite. If the primordial abundance of anorthosite in the west was low, either because of magma ocean processes (*Warren and Wasson,* 1980) or thinning of ancient crust through single or multiple giant impacts (*Pieters and Wilhelms,* 1985), and, if the volume of intrusion was great enough during an epoch of gabbroic magma intrusion, most of the western anorthosite may have been consumed in the assimilation process. Subsequent gabbroic magma intruding into this anorthosite-depleted western crust would then remain relatively unchanged in major element composition and would crystallize high-Ca pyroxene. In the east, the volume of anorthosite was sufficient to prevent near-total depletion of anorthosite by assimilation. The observation of norite and anorthosite in the east and norite, troctolite, and gabbro in the west is consistent with this suggestion. Detection of anorthosite in the west and gabbro in the east would invalidate this suggestion, at least as a driver of hemispheric mineralogic composition.

CONCLUSION

The remote mineralogic observations have constrained the composition of gabbros in the highlands far from the sample return sites. Of the rocks identified in the returned sample collection only KREEP basalts with unusually high clinopyroxene/total pyroxene ratios of Mg-gabbronorite are viable candidates for the rock type responsible for remotely detected gabbroic regions. Gabbroic material seems concentrated at depth in the lunar west. Though other hypotheses may be correct, we feel that this distribution reflects an ancient deficiency in anorthosite in the west, allowing eventual crystallization of gabbroic magmas unchanged in major element chemistry by anorthosite assimilation.

The geochemical identity of the widespread gabbroic highland rock type and the apparent dependence of mineralogy on longitude and depth have important implications for the evolution of the lunar crust. Many additional measurements should be made to confirm or deny the reality of this observation. While this task will be accomplished by the proposed LGO mission, groundbased imaging spectrometers can also perform the necessary measurements in this case.

Acknowledgments. Very helpful reviews were provided by G. J. Taylor, C. M. Pieters, and P. D. Spudis. This work was supported by NASA grants NAGW 237 and NSG 7312.

REFERENCES

Adams J. B. (1974) Visible and near-infrared diffuse reflectance spectra of pyroxenes as applied to remote sensing of solid objects in the solar system. *J. Geophys. Res., 79,* 4829-4836.

Basaltic Volcanism Study Project (1981) *Basaltic Volcanism on the Terrestrial Planets.* Pergamon, New York, 1286 pp.

Bills B. G. and Ferrari A. J. (1976) Lunar crustal thickness. *Proc. Lunar Sci. Conf. 7th,* frontispiece.

Davis P. A. and Spudis P. D. (1985) Petrologic province maps of the lunar highlands derived from orbital geochemical data. *Proc. Lunar Planet. Sci. Conf. 16th,* in *J. Geophys. Res., 90,* D61-D74.

Etchegaray-Ramirez M. I., Metzger A. E., Haines E. L., and Hawke B. R. (1982) Thorium concentrations in the lunar surface: IV. Deconvolution of the Mare Imbrium, Aristarchus, and adjacent regions. *Proc. Lunar Planet. Sci. Conf. 13th,* in *J. Geophys. Res., 87,* A529-A543.

Hartmann W. K. (1980) Dropping stones in magma oceans: Effects of early lunar cratering. *Proc. Conf. Lunar Highland Crust,* (J. J. Papike and R. B. Merrill, eds.), pp. 155-171. Pergamon, New York.

Hawke B. R. and Lucey P. G. (1984) Spectral reflectance studies of the Hadley Apennine (Apollo 15) region: Preliminary results (abstract). In *Lunar and Planetary Science XV,* pp. 352-353. Lunar and Planetary Institute, Houston.

Hawke B. R., Lucey P. G., McCord T. B., Pieters C. M., and Head J. W. (1983) Spectral studies of the Aristarchus region: Implications for the composition of the lunar crust (abstract). In *Lunar and Planetary Science XIV,* pp. 289-290. Lunar and Planetary Institute, Houston.

Hawke B. R., Lucey P. G., Bell J. F., Jaumann R., and Neukum G. (1986) Spectral reflectance studies of Tycho crater: Preliminary results (abstract). In *Lunar and Planetary Science XVII,* pp. 999-1000. Lunar and Planetary Institute, Houston.

Hazen R. M., Bell P. M., and Mao H. K. (1978) Effects of compositional variation on absorption spectra of lunar pyroxenes. *Proc. Lunar Planet. Sci. Conf. 9th,* 2919-2934.

Hubbard N. J., Gast P. W., Meyer C., Nyquist L. E., Shih C., and Weismann H. (1971) Chemical composition of lunar anorthosites and their parent liquids. *Earth Planet. Sci. Lett., 13,* 71-73.

Irving A. J. (1977) Chemical variation and fractionation of KREEP basalt magmas. *Proc. Lunar Planet. Sci. Conf. 8th,* 2433-2448.

James O. B. and Flohr M. K. (1983) Subdivision of the Mg-suite noritic rocks into Mg-gabbronorites and Mg-norites. *Proc. Lunar Planet. Sci. Conf. 13th,* in *J. Geophys. Res., 88,* A603-A614.

Longhi J. (1978) Pyroxene stability and the composition of the lunar magma ocean. *Proc. Lunar Planet. Sci. Conf. 9th,* 285-306.

Longhi J. (1981) Preliminary modelling of high pressure partial melting: Implications for early lunar differentiation. *Proc. Lunar Planet. Sci. 12B,* 1001-1018.

Lucey P. G. and Hawke B. R. (1987a) Characterization of mineralogical changes with longitude on the lunar nearside based on spectral reflectance measurements (abstract). In *Lunar and Planetary Science XVIII,* pp. 574-575.

Lucey P. G. and Hawke B. R. (1987b) Probable outcrops of Mg-gabbronorite in the lunar highlands detected by near-infrared remote sensing (abstract). In *Lunar and Planetary Science XVIII,* pp. 578-579. Lunar and Planetary Institute, Houston.

Lucey P. G. and Hawke B. R. (1987c) Speculations on the origin of possible compositional layering of the upper ten kilometers of the lunar crust (abstract). In *Lunar and Planetary Science XVIII,* pp. 580-581. Lunar and Planetary Institute, Houston.

Lucey P. G., Hawke B. R., Pieters C. M., Head J. W., and McCord T. B. (1986) A compositional study of the Aristarchus region of the Moon using near-infrared reflectance spectroscopy. *Proc. Lunar Planet. Sci. Conf. 16th,* in *J. Geophys. Res., 91,* D344-D354.

McCord T. B., Clark R. N., Hawke B. R., McFadden L. A., Owensby P. D., Pieters C. M., and Adams J. B. (1981) Moon: Near infrared spectral reflectance, the first good look. *J. Geophys. Res., 86,* 10883-10892.

Papike J. J. (1980) Pyroxene mineralogy of the Moon and meteorites. In *Reviews in Mineralogy, Volume 7: Pyroxenes* (C. T. Prewitte, ed.) pp. 495-525. Mineralogical Society of America, Washington, D. C.

Pieters C. (1977) Characterization of lunar mare basalt types-II: Spectral classification of fresh craters. *Proc. Lunar Planet. Sci. Conf. 8th,* 1037-1048.

Pieters C. M. (1980) Near infrared spectra: patterns in the increasing data set (abstract). In *Lunar and Planetary Science XI,* pp. 879-881. Lunar and Planetary Institute, Houston.

Pieters C. M. (1986) Composition of the lunar crust from near infrared spectroscopy. *Rev. Geophys., 24,* 557-578.

Pieters C. M. and Wilhelms D. E. (1985) Origin of olivine at Copernicus. *Proc. Lunar Planet. Sci. Conf. 14th,* in *J. Geophys. Res., 90,* C415-C420.

Ryder G. (1976) Lunar sample 15405: Remnant of a KREEP basalt-granite differentiated pluton. *Earth Planet. Sci. Lett., 29,* 255-268.

Ryder G. and Wood J. A. (1977) Serenitatis and Imbrium impact melts: Implications for large scale layering in the crust. *Proc. Lunar Sci. Conf. 8th,* 655-688.

Singer R. B. (1981) Near infrared spectral reflectance of mineral mixtures: Systematic combinations of pyroxenes, olivine, and iron oxides. *J. Geophys. Res., 86,* 7967-7982.

Spudis P. D. (1978) Composition and origin of the Apennine Bench Formation. *Proc. Lunar Planet. Sci. Conf. 9th,* 3379-3394.

Spudis P. D., Hawke B. R., and Lucey P. G., Materials and formation of Imbrium basin. *Proc. Lunar Planet. Sci. Conf. 18th,* this volume.

Spudis P. D., Hawke B. R., and Lucey P. (1984) Composition of Orientale Basin deposits and implications for the lunar basin forming process. *Proc. Lunar Planet. Sci. Conf. 15th,* in *J. Geophys. Res., 89,* D197-D210.

Taylor L. A., Shervais J. W., Hunter R. H. (1983) Ancient ($\cong 4.2$ AE) highlands volcanism: The gabbronorite connection? (abstract). In *Lunar and Planetary Science XIV,* pp. 777-778. Lunar and Planetary Institute, Houston.

Taylor S. R. (1975) Lunar Science: A post-Apollo view. Pergamon, New York. 372 pp.

Tera F., Papanastassiou D., and Wasserburg G. (1974) Isotopic evidence for a terminal lunar cataclysm. *Earth Planet. Sci. Lett., 22,* 1-21.

Warren P. H. (1986) Anorthosite assimilation and the origin of the Mg/Fe-related bimodality of pristine moon rocks: Support for the magmasphere hypothesis. *Proc. Lunar Planet. Sci. Conf. 16,* in *J. Geophys. Res., 91,* D331-D343.

Warren P. H. and Wasson J. T. (1980) Further foraging for pristine nonmare rocks: Correlations between geochemistry and longitude. *Proc. Lunar Planet. Sci. Conf. 11th,* 431-470.

Warren P. H., Shirley D. N., and Kallemeyn G. W. (1986) A potpourri of pristine moon rocks, including a VHK mare basalt and a unique, augite rich Apollo 17 anorthosite. *Proc. Lunar Planet. Sci. Conf. 16th,* in *J. Geophys. Res., 91,* D319-D330.

Warren P. H., Taylor G. J., Keil K., Kallemeyn G. W., Rosener P. S., and Wasson T. T. 91983) Sixth foray for pristine nonmare rocks and an assessment of the diversity of lunar anorthosites. *Proc. Lunar Planet. Sci. Conf. 13th,* in *J. Geophys. Res., 88,* A614-A630.

Electromagnetic Energy Applications in Lunar Resource Mining and Construction

David P. Lindroth and Egons R. Podnieks
Twin Cities Research Center, Bureau of Mines, U.S. Department of the Interior, Minneapolis, MN 55417

Past work during the Apollo Program and current efforts to determine extraterrestrial mining technology requirements have led to the exploration of various methods applicable to lunar or planetary resource mining and processing. The use of electromagnetic energy sources is explored and demonstrated using laboratory methods to establish a proof of concept for application to lunar mining, construction, and resource extraction. Experimental results of using laser, microwave, and solar energy to fragment or melt terrestrial basalt under atmospheric and vacuum conditions are presented. Successful thermal stress fragmentation of dense igneous rock was demonstrated by all three electromagnetic energy sources. The results show that a vacuum environment has no adverse effects on fragmentation by induced thermal stresses. The vacuum environment has a positive effect for rock disintegration by melting, cutting, or penetration applications due to release of volatiles that assist in melt ejection. Consolidation and melting of basaltic fines are also demonstrated by these methods.

INTRODUCTION

The real need for using extraterrestrial resources to supply lunar bases supporting the Manned Mars Mission and future space activities requires us to search for rock fragmentation methods for mining operations that are compatible with the lunar environment. Planning for new lunar and planetary missions has created a multitude of concepts for utilizing extraterrestrial resources. In most instances, these involve mining or construction on lunar or planetary surfaces. Significant changes in conventional equipment and methods and even novel mining systems will be required. Electromagnetic energy methods are adapted to extraterrestrial requirements, especially in fragmenting solid rock. Lunar regolith, consisting of glass particles and glass-bonded aggregates (agglutinates) (*Taylor*, 1975), may be difficult to use as a source of lunar minerals. Therefore, the underlying rock formations should also be considered as mineral resources since the thickness of the regolith varies considerably: During Apollo 15, outcrops were observed in the wall of Hadley Rille, and at the Apollo 16 landing site the regolith on the rim of North Ray Crater is only a few centimeters thick.

Using electromagnetic energy to fragment rocks has been found by the Bureau of Mines to be effective in the terrestrial environment. For lunar applications this energy form has the advantage of relatively low input energy and is less affected by environment than conventional mechanical methods requiring equipment with many moving parts. Electromagnetic energy is well suited to vacuum because there is no attenuation or dispersion during propagation and remote generators can beam energy to work locations with minimal loss. Electromagnetic energy methods also have the potential to integrate lunar rock fragmentation and rock material processing for extraction of volatiles on site.

The application of concentrated electromagnetic energy for mining and excavation generates thermal energy in the lunar rock or regolith. A rapid increase in thermal energy induces thermal stresses that cause fracture and fragmentation. Melting and penetration of the rock material occur above a threshold energy value. Preliminary tests were performed on igneous Earth-rock material using three electromagnetic energy sources to establish the feasibility of working various materials on the Moon: (1) CO_2 laser, (2) microwave, and (3) solar. Failure of rocks by electromagnetic energy is grouped into two main classes: (1) fragmentation by thermal stresses and (2) disintegration by melting including melting to form solid blocks from fines.

ELECTROMAGNETIC FRAGMENTATION

Basic studies by the Bureau of Mines during the Apollo program (*Thirumalai and Lindroth*, 1969, pp. 72-80; *Thirumalai and Demou*, 1970; *Thirumalai*, 1970; *Bacon et al.*, 1973; *Khalafalla*, 1973; *Lindroth*, 1974; *Lindroth and Krawza*, 1971; *Atchison and Schultz*, 1968; *Podnieks et al.*, 1972) on fragmentation of rocks by thermal methods indicated that fragmentation depended on thermophysical properties such as thermal diffusivity, thermal expansion, and melting (solidus) temperature. The rock types used are part of a suite of 14 previously chosen by the Bureau of Mines to include the range most likely to be representative of those on the lunar surface (*Fogelson*, 1968). Element analysis and petrography of the basalts used in this study are given in Tables 1 and 2.

Thermal diffusivity above 25°C, up to 110°C, and at pressures from ambient to 10^{-9} torr was measured for six igneous rock types: tholeiitic basalt, obsidian, rhyolite, granodiorite, gabbro, and dunite (*Lindroth*, 1974). Values ranged from a low of 5.3×10^{-3} cm^2/sec for rhyolite at 110°C and 2×10^{-7} torr to a high of 13.4×10^{-3} cm^2/sec for dunite at 25°C and 760 torr. Tholeiitic basalt was in the lower range with a value of 6.2×10^{-3} cm^2/sec at 110°C and 2×10^{-7} torr. Diffusivity values were, as expected, inversely proportional to temperature and directly proportion to pressure, which allows electromagnetic fragmentation to create a more localized high temperature region within dense, low porosity rock.

Thermal diffusivity data at 1 atmosphere obtained by *Horai et al.* (1970) for crystalline igneous rocks (type A) from Apollo

TABLE 1. Chemical analysis of various basalts, %.

	Dresser basalt			Tholeiitic basalt*	Basalt hornfels†	Lunar sample 10084‡
	Original	Melt 1	Melt 2			
SiO_2	47.5	48.1	48.4	51.0	41.31	41.22
TiO_2	1.8	1.8	1.8	2.7	7.04	7.49
Al_2O_3	14.6	15.3	15.8	14.0	12.12	13.82
Fe_2O_3	7.2	8.5	7.4	3.4	3.52	0.00
FeO	7.0	7.0	6.5	8.8	14.57	15.74
MnO	0.19	0.18	0.19	0.25	0.21	0.20
MgO	7.0	6.6	6.7	4.4	6.58	7.90
CaO	6.3	9.1	9.0	8.0	11.07	11.98
Na_2O	3.8	2.8	2.7	3.4	2.06	0.44
K_2O	0.8	0.59	0.59	1.7	0.16	0.14
P_2O_5	ND	ND	ND	1.4	0.63	0.10
H_2O^+	5.2§	1.7	1.7	0.76	0.44	0.00
H_2O^-	ND	ND	ND	0.1	0.06	0.00
CO_2	1.5	1.2	0.9	0.03	0.05	0.00
S	0.02	0.016	0.01	0.004	0.10	0.13
Total	102.9	102.9	101.7	99.9	99.9	100.2

ND = not detected at 0.02 for P and 0.1 for H_2O^-.
*Fogelson (1968).
†Buddington (1939).
‡Goldrich (1971).
§5.4 Fe++ oxidized during loss on ignition (LOI) at 1000°C, LOI adjusted to 5.2.

11 lunar specimens is in close agreement with that of three terrestrial rocks measured: tholeiitic basalt, obsidian, and rhyolite. Three other rock types (dunite, granodiorite, and gabbro) have thermal diffusivities significantly higher (40-100%).

Within the 5% error of measurement, meaningful decreases in diffusivity at reduced pressure and elevated temperature were observed for granodiorite, dunite, and rhyolite. Tholeiitic basalt demonstrated a decrease less than 2%. In general, the decrease in diffusivity from 1×10^{-2} torr to less than 7×10^{-7} torr is insignificant. This pressure effect is concentrated in the range 1×10^{-2} torr and was also noted by *Cremers* (1973) and 10^{-4} torr by *Horai and Fujii* (1972).

Thermal expansion is a significant parameter in the creation of thermal stresses within the rock. Rock fracture caused by internal thermal stresses during heating in atmosphere and vacuum environment was investigated (*Thirumalai and Demou*, 1970). Thermal expansion and response of rock to induced thermal stresses was measured and shown to be independent of reduced pressure to 10^{-5} torr. Thermal stresses induced during thermal expansion cycling from 25°C to 300°C, however, caused structural damage to the rock material both at 1 atmosphere and 10^{-5} torr.

Thermal stresses within a rock can induce rock fragmentation using the various electromagnetic energy sources. However, failure induced by thermal stresses is limited by the temperature at which the rock begins to soften (solidus temperature) (*Mysen*, 1981). At this temperature (and above) the thermal stresses are relieved and the rock has an inelastic behavior. Experimental determination of the effect of reduced pressure to 10^{-5} torr on the solidus temperature was conducted on Dresser basalt, obsidian, and granodiorite (*Thirumalai and Lindroth*, 1969). The results indicated that this temperature was not significantly affected by the vacuum environment.

All measurements at reduced pressure were performed after the vacuum pump down curve (P versus t) became asymptotic (*Redhead et al.*, 1968, p. 386). Although the pumping time was short in terms of days, the rock samples were considered

TABLE 2. Petrography of basalt samples.

Commercial name	Source location	Bulk density, g/cm³	Apparent porosity, %	Mineral constituents	Volume percent	Appearance
Tholeiitic basalt (Fogelson, 1968)	Northeast of Madres, OR	2.84	2.0	Plagioclase:		Color: black
				An_{67} & An_{54}	39	
				microlites	12	Texture: aphanitic
				Olivine	13.5	
				Augite	10.5	Structure: massive columnar
				Magnetite	8	
				Glass	12	
				Chlorite	4	Some chlorite grains are marginal alterations of pyroxenes.
				Quartz	1	
				Apatite	<1	
				Epidote	<1	
Dresser basalt (Krech et al., 1974)	Dresser, WI	3.01	0.19	Plagioclase	41.4	Color: dark grayish green
				Pyroxene	40.4	
				Chlorite	6	Texture: ophitic, massive, fine-grained pyroxene altered to amphibole chlorite
				Magnetite	5.3	
				Amphibole	1.9	
				Sericite	4.7	
				Hematite	0.3	

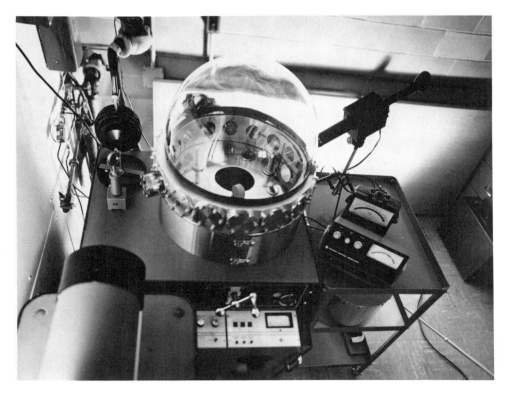

Fig. 1. CO_2 laser-vacuum system.

adequately outgassed. In outgassing, porosity is an important physical parameter since the interstitial gas pressure modifies the thermal conduction. The results of the above studies indicate that a vacuum environment has no adverse effects on fragmentation of rocks by induced thermal stresses.

Laser Beams

Tests using a CO_2 laser (Fig. 1) were conducted to fragment samples of tholeiitic basalt, Dresser basalt, obsidian, and granodiorite in vacuum (10^{-6} torr) and at 1 atmosphere. Major element analysis of the basalts is given in Table 1. The laser beam was reflected into the vacuum chamber through a germanium window and onto a focusing mirror. Samples were exposed to power densities of 100 W/cm^2 with a spot diameter of 1.1 cm and 200 W/cm^2 with a spot diameter of 0.8 cm. Successful fragmentation was achieved in both environments. Figure 2 shows an example of a fractured 3 × 3 × 6-cm obsidian prism.

Microwaves

High-power microwaves can penetrate rock to a depth inversely proportional to the product of frequency, dielectric constant, and loss tangent. A thermal inclusion is formed inside the rock as the electromagnetic energy is converted to heat. The thermal inclusion is a constrained inner volume of specific geometry depending on the rock formation and is subjected to rapid internal heating. The tangential stresses developed at the thermal inclusion boundary are tensile and result in fragmentation of the rock.

Most current mining systems break rock through the use of compressional and shear stresses. However, rocks are far weaker in tension. The ability of the microwaves to create tensile stresses in a rock face results in more economical fragmentation using less energy.

Fig. 2. Fragmented obsidian by CO_2 laser at 10^{-6} torr.

Fig. 3. High-power microwave test chamber.

Current Bureau microwave fragmentation research at 1 atmosphere is being conducted using a 2.45 GHz continuous wave generator with power levels up to 25 kW (Fig. 3). Rock samples (center) are irradiated at various power levels and power densities inside a copper screen room. Power densities up to 670 W/cm^2 are achieved at the rock surface. Microwaves are transmitted from the generator into the screen room by a waveguide (left). Various waveguide apertures and antennas are used to produce different beam shapes. Water wall absorbers (right) are used to minimize reflected radiation back to the generator.

As a sample of microwave fragmentation a fractured 73-kg block of Dresser basalt is shown in Fig. 4a. This approximately 0.3-m cube was irradiated at the 20 kW level for a total of 300 sec. The irradiated area of the most intense part of the beam is circled and the less intense outer part of the beam is elliptically marked indicating the total area of irradiation. The fracturing process started within 120 sec, and a 2-cm-diameter circular red spot (600°C) developed in less than 300 sec.

A second example of fragmentation is shown in Fig. 4b. The 64-kg block of Dresser basalt was irradiated at the 20 kW level for up to 1080 sec with a different beam shape. After the block fractured (<300 sec), the irradiation was continued until the molten thermal inclusion erupted through the front surface. The rock surface around the eruption bulged outward creating a volcanic cone appearance that indicated strong internal expansion forces due to the molten inclusion. The discolored area around the solidified magma shows the extent of the high intensity irradiated area. The solidus temperature for Dresser basalt is 1150°C ± 50°C with the liquidus temperature at 1250°C (*Thirumalai*, 1970).

Solar Energy

The ability of solar energy to fragment rock by thermal stresses was demonstrated at 1 atmosphere using a 390 W solar beam. A Fresnel lens (70 cm × 94 cm) was mounted in an altitude-azimuth adjustable stand and the rock was placed at the focal point (1.0 cm diameter). The lens was sloped at 35°, facing directly south at lat. 45°N during 12 noon–1 pm solar time. Solar incidence data for Minneapolis (*Duffie*

Fig. 4. Microwave fragmentation of Dresser basalt. (**a**) 73-kg block, 300 sec. (**b**) 64-kg block, 1080 sec.

Fig. 5. Solar energy fragmentation of Dresser basalt.

and Beckman, 1974) for early April gives approximately 600 W/m^2. Figure 5 shows the fragmentation results on a 15 × 15 × 20-cm block (2.8 kg) of Dresser basalt. The surface was irradiated with a calculated intensity of 500 W/cm^2 for 260 sec using Earth's rotation rate (15°/hr) to traverse the spot a distance of 2.24 cm. The initial large diagonal crack formed at 120 sec. Vapor phase ejection was observed from the frothing melt pool. The temperature, measured with a micro-optical pyrometer, was 2370°C ± 50°C.

EFFECTS OF VOLATILES

Volatiles like H_2O and HF are usually bound in minerals such as amphibole and mica. If the breakdown temperature (with attendant release of the volatile as a separate fluid phase) of such minerals is lower than the rock solidus temperature in which they are found, the volatile species will be present as a free phase at the rock solidus temperature. Since water is highly soluble in silicate melts, this component should help increase the pressure that the melt exerts on the surrounding rock and aid in stress for fragmentation by microwaves. For laser and solar irradiated material, recent work by *Simons* (1986) shows that during pyrolysis, large pressures must develop within the internal porous structure in order to transport the volatile gases to the free surface. These high internal pressures may cause macroscopic fracture and/or solid-phase mass removal prior to complete pyrolysis or vaporization. He used a transport model to assess the conditions under which an enhanced mass removal mechanism will occur. Critical energy densities ($\cong$kJ/g) are predicted to be relatively insensitive to the physical parameters involved. The threshold energy density (J/cm^2 absorbed by the target) is sensitive to pulse duration, pyrolysis rate, and absorption depth.

The laboratory experiments using electromagnetic energy methods of laser, microwave, and solar beams have shown that igneous rock types can be fragmented by these methods. The energy requirements are quite minimal and can be made available in the lunar mining or construction setting. The use of solar energy has a particular advantage due to its ready availability on the lunar surface.

ELECTROMAGNETIC MELTING

Melting is strongly dependent on the material's chemical composition, total pressure, heat transfer properties, and activities of volatiles such as H_2O, CO_2, CO, HF, etc. Melting (solidus) temperature and thermal diffusivity are directly proportional to pressure. Previous measurements (*Lindroth*, 1974) on the thermal diffusivity of six igneous rock types from 25°C to 110°C and pressures from 1 atmosphere to 10^{-9} torr showed variations in diffusivity of over 100%. For solid dense rock, the decrease in thermal diffusivity due to pressure alone was most significant over the range from 1 atmosphere to 10^{-2} torr.

When incident electromagnetic energy is applied to a rock face, melting of the surface rock forms an insulating layer at 1 atmosphere. This prevents exposure of the energy input to a new free face. Removal of the melt after its formation poses technical problems unless high-power pulsing is used. However, this method appears to be effective only to limited depths.

Since water is highly soluble in silicate melts, the solidus temperature of a water-bearing rock is lowered by several hundred degrees relative to that of a volatile-free rock. The effect of CO_2 is negligible. An important consequence of the great solubility of H_2O in silicate melts is the temperature interval between the solidus and liquidus may be expanded compared to a water-free rock. Because of the much lower solubility of CO_2 than H_2O in basaltic magmas, a CO_2-rich vapor will evolve from the basaltic melt at temperatures near its liquidus (*Mysen*, 1981). This will result in blowoff of molten rock in vacuum.

Rock melting using electromagnetic energy sources serves a different purpose than rock fragmentation. It can be used in molding fine particles into solid blocks of different shapes (*Meek et al.*, 1987; *Wright et al.*, 1986; *Vaniman et al.*, 1986) or to create cavities in rock faces. Rock melting can also be used for rock disintegration purposes.

Laser Beams

The effect of vacuum is well demonstrated in experiments with basaltic rocks (tholeiitic basalt and Dresser basalt) at a vacuum of 2 × 10^{-6} torr. At that vacuum level, a continuous violent outgassing, frothing, and self-ejection of the melt creating a continued exposure of new free rock face to incident electromagnetic energy was observed. Figure 6 is a comparison of rock disintegration by melting for tholeiitic basalt in atmosphere and in vacuum using a CO_2 laser beam. Both disk samples (5.4-cm-diameter × 1.17-cm-thick) were irradiated with 100 W for 10 sec, at a power density of 250 W/cm^2 in the system shown in Fig. 1. Temperatures measured with a micro-optical pyrometer were 1475°C ± 50°C in atmosphere and 1675°C ± 50°C in vacuum. Figure 6a shows the glass melt produced at 1 atmosphere and Fig. 6b the melt ejection with associated production of solid microspheres in 2 × 10^{-6} torr vacuum. Both melt diameters are 0.7 cm. The crater

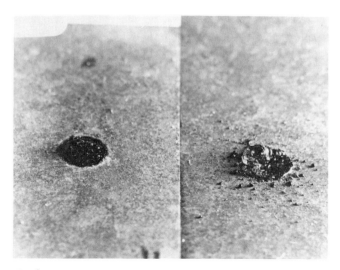

Fig. 6. CO_2 laser melting of tholeiitic basalt. ATM (left); VAC (right).

produced in vacuum has a depth/diameter ratio (0.16) twice that of the one at 1 atmosphere. The ejection of the melt in vacuum is caused by release of volatiles resulting in a violent agitation in the melt cavern. Melt ejection and progression of cavity depth continued as long as the laser beam was applied.

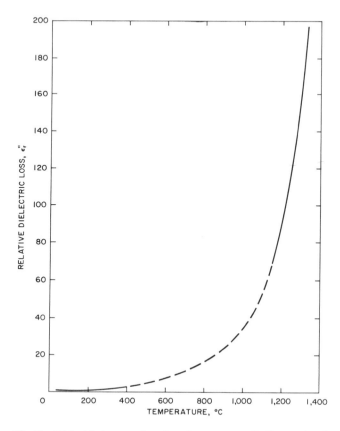

Fig. 7. Dielectric loss as a function of temperature for Dresser basalt.

The crater and the melt microspheres produced by the laser beam in vacuum, as shown in Fig. 6b, resemble the lunar microcraters and spherical ejecta caused by hypervelocity impacts of micrometeorites as shown by D. S. McKay in *Taylor* (1975, pp. 70-71, 90). In both cases the ejected spheres are typical of shapes assumed by splashed liquids.

Microwaves

Basaltic rocks have favorable properties for interaction with microwave energy. The thermal runaway phenomena occurs for Dresser basalt in the 2.45 GHz frequency region. Measured values of the relative dielectric constant (ϵ'_r) and relative loss factor (ϵ''_r) of ilmenite show a very high peak occurring for these values at 2.45 GHz (14). For the entire rock, the ϵ''_r was calculated from values of specific resistivity (*Bacon et al.*, 1973), relative dielectric constant (ϵ'_r), and loss tangent (*Khalafalla*, 1973) as a function of temperature up to 1350°C, as shown in Fig. 7. Interpolation is indicated by the dashed line from 400°C to 1150°C. For a fixed frequency and material thermal properties, the microwave heating rate is a direct function of the available power ($\epsilon_0 E^2$) and the material relative loss factor ($\epsilon'_r \tan \delta$) as given by the heating rate formula

$$\frac{dT}{dt} = \frac{2\pi f \epsilon_0 E^2 \epsilon'_r \tan\delta}{\rho C_p}$$

where

C_p = specific heat
E = electric field strength
ϵ_0 = dielectric constant of vacuum
f = frequency
ρ = density
$\tan\delta$ = loss tangent = ϵ''_r/ϵ'_r
T = temperature
t = time.

Previous studies (*Wright et al.*, 1986; *Tucker et al.*, 1985) using terrestrial alkali basalt indicated the need to use a coupling agent in order to heat this material. Recent studies (*Meek et al.*, 1987), however, have shown the ability to heat poorly-coupling lunar soil simulants with a higher power microwave system during soil sintering experiments. In the current Bureau study, melting of low TiO_2 basalt was easily achieved with a high power (25 kW) and power density system, with either a solid rock block or finely crushed material. High power accelerates the temperature to the knee of the loss versus temperature curve (Fig. 7) where an asymptotic increase in the loss factor occurs (thermal runaway).

Electrical property variation in dry silicates show first that electrical conductivity is a monotonically increasing function of temperature. The dry relative dielectric constant is independent of all parameters except bulk density. Finally, the loss tangent is linearly proportional to density and strongly controlled by chemistry (e.g., ilmenite), temperature, and frequency. Pressure has little or no effect on dielectric properties and electrical conductivity (*Olhoeft*, 1981).

TABLE 3. Evolved vapor species from Dresser basalt.*

Mass Temp.(°C)	2 H_2	18 H_2O	23 Na	24 Mg	28 Si	32 O_2	39 K	44 CO_2/SiO	47 PO	55 Mn	56 Fe	63 Cu
610	14	10,000						125				
800	300	1900						36				
950	198	4200				3		21				
1100	12	18	16			5	6	13				
1200			12			20	3	3				
1300			24			116	11					2
1400			102			124	36	—	6			6
1500			325			195	120	8	20		2	18
1600			880			156	204	16	47		8	30
1700			1800	11	14	110	365	153	82		77	26
1800			2900	51	48	100	594	510	125	5	195	12
1900			3240	148	186	57	1400	2800	85	25	1850	8

*Ion signals (mv.).

Except for metallic or semiconducting minerals such as sulfides and iron oxides, pyroxene is the most significant unhydrated mineral. *Olhoeft* (1981) showed that pyroxene has a strongly polarizable crystal structure that significantly increased the high temperature dielectric constant of pyroxene-containing materials such as basalt.

Chemical analyses of the basalts used in this study are compared to the basalt hornfels and lunar sample 10084 in Table 1. The analysis indicates that the test materials with low TiO_2 and FeO contents represent the lower values for electromagnetic energy applications. These materials would be classified as very-low-Ti basalts (*Taylor*, 1986).

During the melting process using microwave energy, volatiles are being discharged. To determine the vapor phase given off in vacuum from molten Dresser basalt, measurements were made as a function of temperature over the range from 600°C to 2000°C, by a Nuclide mass spectrometer equipped with a Knudsen cell (V. D. Altemose, personal communication, 1987).

The results indicate that in the temperature range from 600°C to 1000°C water, hydrogen, and CO_2 are predominantly evolved (Table 3). However, above 1000°C the primary vapor species are the alkalis, sodium, and potassium. The accompanying oxygen is probably from the disassociation of the alkali oxides. This basalt is composed of 50% plagioclase, which is a Na,Ca aluminosilicate. All of the observed species, except copper, are present in the chemical analysis. Even though these peaks were quite small, the isotopic ratios of mass 63 and 65 agreed very well with those of copper. Dresser basalt contains trace amounts of rare earths and elements of which copper averages about 227 ppm (*Cavaleri*, 1987). In addition to the vapors shown, trace amounts of either CO or N_2 were observed below 1000°C. A very small signal believed to be SO_2 was observed from 90° to 1100°C.

Microwave beam melt penetration in Earth environment is achievable with the low TiO_2 basalt, as shown in Fig. 8. An 82-kg cube of Dresser basalt was irradiated at the 20 kW level with an elliptical beam shape for 900 sec. A dark red spot (700°C) was observed after 420 sec of irradiation. Magma then flowed from the area and ran down the face of the block. A molten cavity was receding into the block, when at 570 sec a vapor phase was ejected from the cavity in the form of brilliant flame. Additional vapor phase ejections were observed before the test was terminated.

Dresser basalt fines, less than 6.5 mm, were melted in alumina crucibles and pans to determine the feasibility of agglomerating, sintering, and fusing this material by microwave. An example of melting fine material to form cast products is shown in Fig. 9 where 335 g of fines were melted in a 10-cm high-alumina crucible with a 5-cm-diameter base for 600 sec at the 6 kW level. Arcing and melting were observed on top of the crucible at 300 sec of irradiation. After solidification, a 25-mm-diameter × 5-cm-long core was drilled from the center

Fig. 8. Microwave melt penetration in Dresser basalt.

Fig. 9. Microwave melt of basalt fines.

Fig. 10. Effect of microwave power density variation on Dresser basalt fines.

of the crucible. The cored melt material closely resembles obsidian with a compressive strength of 6500 psi. A chemical analysis of Melt 1 is shown in Table 1. A second test was performed under the same conditions except that irradiation time was 570 sec. Arcing and melting started at 195 sec during this test. The elemental analysis of this product is given under Melt 2 in Table 1. Notice the voids created (Fig. 9) by volatile release in the viscous melt that were trapped upon freezing of the melt. Some cracking and fragmentation was introduced by drilling the core.

By controlling the power density of the microwave beam either sintering or melting is achieved. For this test, 1800 g of fines were placed in a 13 × 25 × 5-cm-deep alumina pan and irradiated from the top at the 9 kW level for 600 sec. Magma flow was observed after 300 sec of irradiation. The natural beam divergence from a tapered waveguide antenna provided the power density gradiant across the sample. In Fig. 10, the gradual transition from agglomeration of basalt particles to a complete melt is shown from right to left.

Solar Energy

Preliminary tests were performed with Dresser basalt to investigate using a solar energy source to melt, cut, and seal the rock. As in fragmentation tests, during melting vapor ejection was taking place indicating a discharge of volatiles. The rock material was traversed beneath the focal spot of the solar beam concentrated by the Fresnel lens to a calculated power density of 500 W/cm^2. Figure 11 shows the results of a seam-welding test on a 15 × 15 × 5-cm block of Dresser basalt using a beam traverse rate of 0.7 mm/sec. A 3-mm weld seam depth was produced by the 1.0-cm-diameter spot. Sealing of cracks in a rock face may become important in lunar shelter construction.

Since the intensity of solar radiation on the Moon (1353 W/m^2) is more than twice that used in these tests, it will be advantageous to use this natural resource and minimize energy source transportation costs from the Earth. A simple 3-m-diameter parabolic dish solar collector system could produce a respectable 8.6 kW solar beam assuming a 10% loss.

CONCLUSIONS AND RECOMMENDATIONS

The investigations on the use of electromagnetic energy sources in lunar resource utilization has yielded promising results. The ability of three different electromagnetic energy beams to fragment igneous rock by thermal stress was demonstrated. A vacuum environment has no adverse effects on fragmentation of dense igneous rocks by induced thermal stresses. The vacuum environment has a positive effect for rock disintegration by melting, cutting, or penetration applications because the release of volatiles assist in melt ejection.

The following conclusions can be drawn: (1) The CO_2 laser beam is effective in fracturing igneous lunar rocks and melting/penetrating the rock face; (2) the microwave energy beam at higher power density is capable of fracturing and also melting low TiO_2 (or ilmenite) basaltic rocks; (3) the use of a solar energy beam of nominal power density (500 W/cm^2) can fracture, melt, and fuse cracks in basaltic rocks; (4) any one of these energy sources can be directly used to extract volatiles from the basaltic rocks; and (5) in all three cases the input power requirements are minimal and can be generated in the lunar setting.

Fig. 11. Solar seam welding of Dresser basalt.

These preliminary tests indicate that the electromagnetic energy application for lunar resource mining and extraction has a high potential and warrants further research and development prior to the planning and design of prototype systems. The tests demonstrate that lunar vacuum has a very limited effect on electromagnetic energy use in rock fragmentation. However, vacuum conditions have a significant effect on other fragmentation and processing events, especially those associated with rock surface properties.

REFERENCES

Atchison T. C. and Schultz C. W. (1968) Bureau of Mines research on lunar resource utilization. *Proceedings of 6th Annual Meeting, Working Group on Extraterrestrial Resources*, pp. 65-74. NASA SP-177, NASA, Washington, DC.

Bacon J. S., Russell S., and Carstens J. P. (1973) Determination of rock thermal properties. *BuMines Contract Report No. HO220052*, Fig. 23, p. 43.

Buddington A. F. (1939) Adirondack igneous rocks and their metamorphism. *Geol. Soc. Amer. Mem. 7*, 36 pp.

Cavaleri M. (1987) Petrogenesis of Chengwatana volcanics near Taylors Falls, Minnesota. *Geol. Soc. Amer. Abstr. with Program, 19*, 192.

Duffie J. A. and Beckman W. A. (1974) *Solar Energy Thermal Processes*. Wiley, New York.

Cremers C. J. (1973) Thermophysical properties of Apollo 12 fines. *Icarus, 18*, 294-303.

Fogelson D. E. (1968) Simulated lunar rocks. *Proceedings of 6th Annual Meeting, Working Group on Extraterrestrial Resources*, pp. 75-95. NASA SP-177, NASA, Washington, DC.

Goldrich S. S. (1971) Lunar and terrestrial ilmenite basalt. *Science, 171*, 1245-1246.

Horai K. and Fujii N. (1972) Thermophysical properties of lunar material returned by Apollo missions. *The Moon, 4*, 447-475.

Horai K., Simmons G., Kanamori H., and Wones D. (1970) Thermal diffusivity, conductivity, and thermal inertia of Apollo 11 lunar material. *Proc. Apollo 11 Lunar Sci. Conf.*, 2243-2249.

Khalafalla A. S. (1973) Effect of frequency and temperature on rock dielectric properties. *BuMines Contract No. HO220054*, p. 61.

Krech W. W., Henderson F. A., and Hjelmstad K. E. (1974) A standard rock suite for rapid excavation research. *BuMines RI 7865*, pp. 10-12.

Lindroth D. P. (1974) Thermal diffusivity of six igneous rocks at elevated temperatures and reduced pressures. *BuMines RI 7954*, 33 pp.

Lindroth D. P. and Krawza W. G. (1971) Heat content and specific heat of six rock types at temperatures to 1,000°C. *BuMines RI 7503*, 24 pp.

Meek T. T., Vaniman D. T., Blake R. D., and Godbole M. J. (1987) Sintering of lunar soil simulants using 2.45 GHz microwave radiation (abstract). In *Lunar and Planetary Science XVIII*, pp. 635-636. Lunar and Planetary Institute, Houston.

Mysen B. O. (1981) Melting curves of rock and viscosity of rock forming melts. In *Physical Properties of Rocks and Minerals* (Y. S. Touloukian and C. Y. Ho, eds.), Chap. 11. McGraw-Hill/Cindas Data Services on Material Properties, McGraw-Hill, NY.

Olhoeft G. R. (1981) Electrical properties of rocks. In *Physical Properties of Rocks and Minerals* (Y. S. Touloukian and C. Y. Ho, eds.), Chap. 9. McGraw-Hill/Cindas Data Services on Material Properties, McGraw-Hill, New York.

Podnieks E. R., Chamberlain P. G., and Thill R. E. (1972) Environmental effects on rock properties. In *Basic and Applied Rock Mechanics* (K. E. Gray, ed.), pp. 141-215. Proc. 10th Symposium on Rock Mechanics, Soc. of Min. Eng., AIME, Denver, CO.

Redhead P. A., Hobson J. P. and Kornelsen E. V. (1968) *The Physical Basis of Ultra High Vacuum*. Chapman and Hall, Ltd., New York.

Simons G. A. (1986) Pyrolysis-induced fragmentation and blowoff of laser-irradiated surfaces. *AIAA J., 24*, 1861-1866.

Taylor L. A. (1986) Rocks and minerals of the moon: Materials for a lunar base. *Symposium 86—The First Lunar Development Symposium*, p. 1. Atlantic City, NJ.

Taylor S. R. (1975) *Lunar Science: A Post-Apollo View*. Pergamon, New York.

Thirumalai K. (1970) Potential of internal heating method for rock fragmentation. *Twelfth Symposium on Rock Mechanics*, pp. 697-719. Rolla, MO.

Thirumalai K. and Demou S. G. (1970) Effect of reduced pressure on thermal expansion behavior of rocks and its significance to thermal fragmentation. *J. Appl. Phys., 41*, 5147-5151.

Thirumalai K. and Lindroth D. P. (1969) Effect of vacuum on fragmentation of rocks by thermal stresses and rock disintegration by thermal melting. *Annual Report, Twin Cities Research Center*. Bureau of Mines, U.S. Dept. of the Interior, Minneapolis, MN.

Tucker D. S., Vaniman D. T., Anderson J. L., Clinard F. W., Jr., Feber R. C., Frost H. M., Meek T. T., and Wallace T. C. (1985) Hydrogen recovery from extraterrestrial materials using microwave energy. In *Lunar Bases and Space Activities of the 21st Century* (W. W. Mendell, ed.), pp. 583-590. Lunar and Planetary Institute, Houston.

Vaniman D. T., Meek T. T., and Blake R. D. (1986) Fusing lunar materials with microwave energy—Part II: Melting of a simulated glassy Apollo 11 soil (abstract). In *Lunar and Planetary Science XVII*, pp. 911-912. Lunar and Planetary Institute, Houston.

Wright R. A., Cocks F. H., Vaniman D. T., Blake R. D., and Meek T. T. (1986) Fusing lunar material with microwave energy—Part I: Studies of Doping media (abstract). In *Lunar and Planetary Science XVII*, pp. 958-959. Lunar and Planetary Institute, Houston.

Detecting a Periodic Signal in the Terrestrial Cratering Record

Richard A. F. Grieve
Geophysics Division, Geological Survey of Canada, Ottawa, Canada K1A 0Y3

Virgil L. Sharpton
Lunar and Planetary Institute, 3303 NASA Road One, Houston, Texas 77058

James D. Rupert and Alan K. Goodacre
Geophysics Division, Geological Survey of Canada, Ottawa, Canada K1A 0Y3

Time-series analyses of periodic, random, and a mixture of periodic and random ages indicate that it is difficult to consistently detect a period if age uncertainties are present. This problem is present in real data taken from the terrestrial cratering record, which always has attached age uncertainties. In a model of 50:50 periodic and random data, the correct period can only be detected at the 99% confidence level over 50% of the time when the age uncertainties are <10% of the period being sought. Analyses of compilations of observed crater ages; one by Shoemaker and Wolfe (1986), the other presented in this study, fail to detect a consistent ~30-m.y. period, the most common periods found being ~16 m.y. and ~18-20 m.y., respectively. Data truncation according to criteria of age uncertainty and age indicate that these are relatively stable periods but that the ~18-20-m.y. period is largely controlled by events in the last 40 m.y. Given the problems with age uncertainties and the evidence furnished by siderophile trace element data, the significance of these periods, particularly the most common period at ~18-20 m.y., is doubtful and its statistical detection may be fortuitous. Whatever the case, the evidence for a consistent ~30-m.y. period, previously reported as coincident with other geological phenomena, is weak, and the question of periodicities in the cratering record is debatable.

INTRODUCTION

Following *Raup and Sepkoski's* (1984) report of a periodicity in marine faunal extinctions over the past 250 m.y., a number of workers have proposed several imaginative astrophysical causes centering around periodic cometary showers (e.g., *Whitmire and Jackson, 1984; Davis et al., 1984*). Time-series analyses of various subsets of the known terrestrial cratering record have suggested a periodic component, which in turn has been cited in favor of these mechanisms (*Alvarez and Muller,* 1984; *Rampino and Stothers,* 1984a,b). The subsets analyzed were selected from the then current lists of known terrestrial impact structures (*Grieve,* 1982), using a variety of selection criteria (e.g., age, age uncertainty, diameter).

We have argued that the significance of these periods, generally around 30 m.y., is questionable because of the inherent uncertainties and biases in the known cratering record (*Grieve et al.,* 1985a). It was also noted that the period, phase, and statistical significance of these periods was dictated by the selection criteria used to generate the subsets of crater ages and that the cratering rate based on the observed flux of earth-crossing bodies is similar to that calculated from known craters. We have further argued that siderophile evidence for impacting body compositions did not favor a cometary impact origin for a number of the craters contributing to the periodic signal (*Grieve et al.,* 1985b).

Recently, the hypothesis that the terrestrial cratering record has an ~30 m.y. periodic component again figured prominently in a number of papers in *Smoluchowski et al.* (1986). In most cases, these papers draw upon earlier studies or a reanalysis of the data in *Grieve* (1982). *Shoemaker and Wolfe* (1986), however, analyzed their own updated compilation of terrestrial craters with ages <250 m.y. with <20-m.y. uncertainty and diameters >5 km. They concluded that the record is "moderately cyclical" and may consist of a mix of random (asteroidal and cometary) and periodic (cometary) impacts. Most recently, *Trefil and Raup* (1987) compared the results of time-series analysis of model simulations of periodic and random ages with an analysis of the data used by *Grieve et al.* (1985a). They duplicated the occurrence of periods at ~20

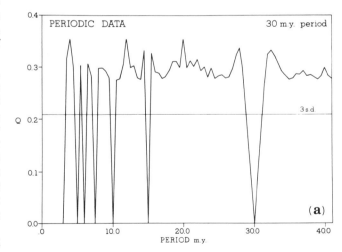

Fig. 1. Plot of variable Q against period, in m.y., for (a) 30 m.y. periodic data, (b) random data, and (c) 50:50 random and 30 m.y. periodic data. A value of Q below the line 3 s.d. (three standard deviations) is considered significant in terms of a period.

MODEL DATA

Periodic and Random Events

For easy comparison with our previous work (*Grieve et al.,* 1985a), we have continued to use an adaptation of the method described by *Broadbent* (1955, 1956) to search for a constant time interval (period) between events. Briefly, the procedure is to examine the hypothesis that a series of crater ages, y_i, may be expressed as follows

$$y_i = A + r_i t \quad (i = 1, 2 \ldots) \quad (1)$$

where A (phase) and t (period) are constants, and r_i is zero or a positive integer. A similar model for periodicity was used by *Rampino and Stothers* (1984a,b). As a measure of the goodness-of-fit of the data to a period, we use the variable Q, which is an r.m.s. measure of q_i, the departure of the individual observed ages divided by the period in question from an integer, such that

$$Q = \left\{ 1/n \sum_{i=1}^{n} (q_i)^2 \right\}^{1/2} \quad (2)$$

For perfectly periodic data, Q is zero. An example is shown in Fig. 1a, for 26 crater ages with a model 30-m.y. period. Zero Q values at 15 m.y., 10 m.y. and smaller periods (Fig. 1a) can be considered as harmonics of the 30-m.y. period. They are an arithmetical result stemming from the fact that ages equal to n × 30 are also divisible by 15, 10, and smaller numbers to yield integers and thus low Q values.

For random data, Q is an approximately normally distributed variable with a mean of 0.29 and a standard deviation of $0.13/n^{1/2}$. An example is shown in Fig. 1b, for 26 random crater ages chosen from a uniform distribution ranging from 0 to

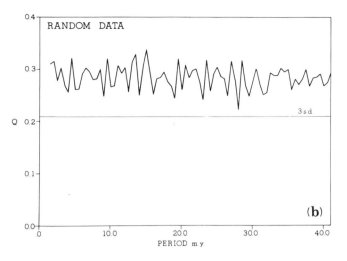

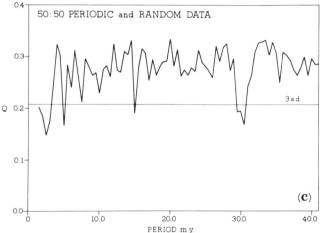

Fig. 1. Plot of variable Q against period, in m.y., for **(a)** 30 m.y. periodic data, **(b)** random data, and **(c)** 50:50 random and 30 m.y. periodic data. A value of Q below the line 3 s.d. (three standard deviations) is considered significant in terms of a period.

m.y and ~30 m.y., noted by *Grieve et al.* (1985a), but ascribed the ~20-m.y. period to random events. They concluded that the cratering record has a periodicity of ~30 m.y., but it is diluted by a substantial number (50-66%) of random events.

We have investigated this hypothesis with respect to whether it is possible to detect consistently a significant period from such a mix of random and periodic impacts. The philosophical approach is similar to that of *Trefil and Raup* (1987). In detail, we undertake time-series analysis of model periodic data, where the period and phase are known, and consider specifically the effect of age uncertainties and random ages in the detection of a periodic signal. An equivalent analysis is then performed with observed data on crater ages and compared with the model data. Finally, the effects of the temporal distribution of crater ages on the results from time-series analysis is investigated. The results of these studies are then used to evaluate the hypothesis that the terrestrial cratering record has a periodic signal.

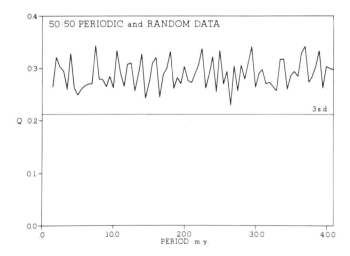

Fig. 2. Typical plot of variable Q against period, in m.y., for 50:50 random and 30 m.y. periodic data but with ± 20 m.y. age uncertainty attached to periodic data. See text for details. Note no 30-m.y. period is detected and plot is similar to that with random data (Fig. 1b).

TABLE 1. Percentage probability of detecting model 30-m.y. period, for 26 ages with uncertainties.

Uncertainty, m.y.	Periodic Data	50:50 Periodic and Random Data
±20	1	1
±15	1	1
±10	1	1
±9	1	1
±8	3	1
±7	20	5
±6	25	9
±5	95	18
±4	100	30
±3	100	44
±2	100	52
±1	100	59
±0	100	65

250 m.y. Perfectly periodic ages combined with random ages yield Q values between 0.29 and zero, depending on the relative proportions of periodic and random data. We consider Q values less than 0.29-3 s.d. as indicating the signal of a statistically significant period. This corresponds to less than 1 chance in 100 that the detection of a specific period is the result of a fortuitous combination of random data. A similar level of significance is quoted (*Alvarez and Muller*, 1984; *Rampino and Stothers*, 1984a) for periods determined from analyses of subsets of the observed cratering record. An example is shown in Fig. 1c, for 13 periodic 30-m.y. ages and 13 uniformly distributed random ages. It is apparent that such a 50:50 mixture of periodic and random ages, which is similar to in proportion to that suggested by *Shoemaker and Wolfe* (1986) and the maximum periodic component suggested by *Trefil and Raup* (1987), still yields a clear periodic signal. It assumes, however, that the periodic ages are known exactly.

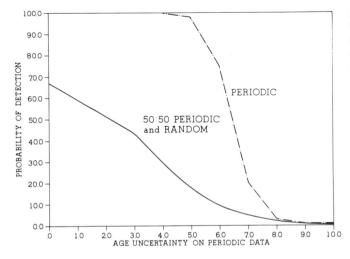

Fig. 3. Plot of probability of detecting a 30-m.y. period against age uncertainty attached to 30 m.y. periodic data and 50:50 random and 30 m.y. periodic data.

The Effect of Age Uncertainties

If the terrestrial cratering record has a periodic signal it will not be perfect. This is due in part to orbital dynamics, which will smear out over several million years the time of impact of an injection of cometary bodies (*Hut*, 1984). More importantly, real age data always have an attached uncertainty, due to experimental error, variability of isotopic ages from different samples and, in some cases, uncertainties in the absolute age of biostratigraphic indicators. To simulate age uncertainties, the algorithm determining Q for each possible phase and period was run 100 time with noise added to each age. The noise was added in such a manner that after 100 cycles it conformed to a normal distribution around each age, with a standard deviation equal to the desired uncertainty.

The effect of age uncertainties was first evaluated on the model data set of 26 ages with a 30-m.y. period but with ± 20-m.y. uncertainty attached to each age. This is equivalent to the maximum uncertainty permissible for "good" crater ages in previous time-series analyses. It was found that the model 30-m.y. period was detected only ~1% of the time, using the 3 s.d. criterion. A typical result (Fig. 2) resembled that from random data. Expanding the simulations to model jiggling of the possible period by orbital forces, increases the chance of detecting a period at, for example, 30 ± 2 m.y., to ~5%. As the attached uncertainty is reduced, the chances of detecting the model 30-m.y. period increases (Table 1; Fig. 3), but it is only detected in more than 50% of the runs when the uncertainty is reduced to ±6 m.y., i.e., <20% of the period in question.

If the simulations are a mixture of random and periodic data with attached age uncertainties, the ability to consistently detect the model period is greatly reduced over that of periodic data with age uncertainties (Table 1; Fig. 3). It is apparent that age uncertainties have to be <10% of the period being sought for the correct periodic signal to be detected in at least 50% of the trials from a 50:50 mixture of random and periodic ages. Age uncertainties were also considered in the model simulations by *Trefil and Raup* (1987). They modeled age uncertainties as ranges of ±2.5 m.y. This range is <10% of the period in question and, therefore, unlikely to affect significantly the detection of the model period (see Fig. 3 in *Trefil and Raup*, 1987). They did not specifically investigate the effect of varying the age uncertainty range.

OBSERVATIONAL DATA

We have applied the linear test for periodicity to the crater ages cited by *Shoemaker and Wolfe* (1986) and to our latest compilation of ages, which is restricted to isotopic ages (Table 2). We have not repeated the analysis on *Trefil and Raup's* (1987) data, as they used the data in *Grieve et al.* (1985a). Both data sets are limited to craters with diameters, D, >5 km with ages <250 m.y., and quoted uncertainties <±20 m.y. The lowest Q for Shoemaker and Wolfe's data was for a phase of +3 m.y. and period of 16 m.y. Marginally higher value of Q was noted for a phase of +4 m.y. and period of 30 m.y. Both just satisfy the 3 s.d. criterion (Fig. 4). Different crater

TABLE 2. Impact structures used in time series analysis.

Name	Diameters, km	Shoemaker and Wolfe Age, m.y.	This study Age, m.y.
Bosumtwi	10.5	1.04± 0.11	1.3 ± 0.2
Zhamanshin	10	1.07 ± 0.05	0.75 ± 0.6
Elgygytgyn	23	4.5 ± 0.1	3.5 ± 0.5
Karla	12	5 ± 3*	—
Bigatch	7	—	6 ± 3
Ries	24	14.7 ± 0.4	14.8 ± 0.7
Haughton	20	20 ± 5	21.5 ± 1.2
Popigai	100	30.5 ± 1.2	39 ± 9
Wanapitei	8.5	32 ± 2	37 ± 2
Mistastin	28	38 ± 4	38 ± 4
Logoisk	17	—	40 ± 5
Logancha	20	—	50 ± 2
Montagnais	45	—	52.8 ± 3
Goat Paddock	5	55 ± 3*	—
Rogozinskaja	8	—	55 ± 5
Kara	60	57 ± 9	57 ± 9
(Ust Kara	25	57 ± 9	57 ± 9)
Kamensk	25	65 ± 3	—
Manson	32	67.5 ± 2.5	61 ± 9
Lappajärvi	14	78 ± 2	77 ± 4
Steen River	25	95 ± 7	95 ± 7
Boltysh	25	100 ± 5	100 ± 5
Dellen	15	100 ± 2	109.6 ±1
Carswell	37	117 ± 8	117 ± 8
Mien	5	119 ± 2	118 ± 3
Gosses Bluff	22	133 ± 3	142.5 ± 0.5
Vyapryai	8	150 ± 16*	—
Rochechouart	23	160 ± 5	165 ± 15
Puchezh-Katunki	80	183 ± 3	183 ± 5
Obolon'	15	183 ± 17*	—
Manicouagan	100	212 ± 3	212 ± 2
St. Martin	23	—	220.5 ± 18

*Stratigraphic age. Most of the differences in ages between data sets result from *Shoemaker and Wolfe's* (1986) preference for fission track ages over K/Ar ages for structures <100 m.y. in age. The ages used in this study are all isotopic and are an updated version of those in *Grieve* (1982). They include some additional ages and modified ages from recent isotopic dating.

ages contribute to these periods and they are not harmonics of each other. Our most recent compilation of ages, however, indicates no significant period at 30 m.y., with the only Q value meeting the 3 s.d. criterion occurring for periods of ~18-20 m.y. (Fig. 5). Considering other "impact" events, such as those hypothesized for the Cretaceous-Tertiary and Eocene-Oligocene boundaries in addition to the crater ages, does not affect this conclusion. It could be argued that the 3 s.d. criterion is too stringent and a lower level of significance would improve detection of a period. While a lower level of significance would result in an apparent improvement in the "strength" of these periods, it would also increase the number of other periods that satisfy the significance criterion (Figs. 4 and 5).

As with the model data, the effect of the quoted age uncertainties in both data sets (Table 2) on detecting these periods was examined. Simulating the quoted age uncertainties as before, the 30-m.y. period in Shoemaker and Wolfe's data was detected only in 4% of the trials. Interestingly, the ~16-m.y. period was detected more often (Table 3). Using our data, the 18-20-m.y. period was detected in 6% of the trials (Table 3). This is not surprising in view of the results from model simulations (Table 1; Fig. 3) and the fact that the average quoted age uncertainties, treated as standard deviations, for the observational data sets are ~7 m.y. (Table 2).

In an effort to reduce the randomising effect of age uncertainties, the observational data sets were culled to remove the crater ages with large uncertainties. As an example, Shoemaker and Wolfe's data set with ages with uncertainties of <±10 m.y. and <± 5 m.y. exhibited the 30-m.y. period at a phase of 4 m.y. in 8% and 6% of the trials, respectively. This is only a marginal improvement in period detection. At <±3 m.y. uncertainty in ages, which is 10% of the period being sought and is the uncertainty limit suggested for ~50% probability of detection (Fig. 3), the 30-m.y. period was detected in only 1% of the trials. This is little better than expected from random data, using the 3 s.d. criterion. Although

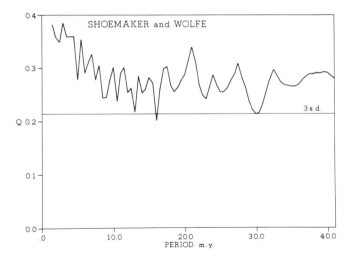

Fig. 4. Plot of variable Q against period, in m.y., for Shoemaker and Wolfe's data (Table 2). Plot is for a phase of 4 m.y. The lowest Q occurs as a period of 16 m.y., although a 30 m.y. also results in a Q value satisfying the 3 s.d. criterion.

culling the data set does not significantly improve the detection of a 30-m.y. period, it does, however, improve slightly the detection of an ~16-m.y. period (Table 3).

Culling our data set to remove crater ages with the large uncertainties does lead to an improvement in the detection of the ~18-m.y. period. It is detected in ~50% of the trials, when only ages with <±3-m.y. age uncertainties are considered (Table 3). *Trefil and Raup* (1987) analyzed a culled data set based on *Grieve et al.* (1985a), removing craters with ages <5 m.y. and with uncertainties of >±10 m.y. and adding random noise to simulate a ±2.5-m.y. age uncertainty on each age. Their results [Fig. 2 in *Trefil and Raup* (1987)] indicated periods at ~19 m.y. and ~29 m.y. that were detected in

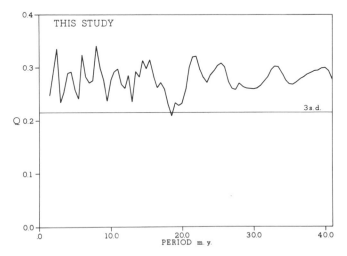

Fig. 5. Plot of variable Q against period, in m.y., for the data in this study (Table 2). Plot is for a phase of 1 m.y. The lowest Q occurs for a period of 18.5 m.y. Note there is no indication of a 30-m.y. period.

approximately equal proportions in >50% of the trials. These results are similar to ours. The data sets, however, are not exactly the same. They also used 2 s.d. as their significance criterion and a fixed age uncertainty. *Trefil and Raup* (1987) chose to ascribe the ~29-m.y. period to a real periodic component and the ~19-m.y. period to the cumulative effects of random impacts, although they stated that the ~19-m.y. period "was not simply a statistical fluctuation."

Lutz (1985) has argued that a circular time-series model is more appropriate when searching for periods. Details of the computational method are given in *Lutz* (1985). He reanalyzed the magnetic reversal record and found no compelling evidence for periodicity, although a previous analysis by *Raup* (1985) had suggested a 30-m.y. period. He also suggested that a circular model be applied to other data sets, such as the impact record, for which periodicity had been suggested. We have analyzed the two compilations of crater ages discussed here using the *Lutz* (1985) circular model. The results are analogous to those discussed earlier, using the *Broadbent* (1955, 1956) method.

Lutz also expressed concern over the long-term variation in the frequency of events over the record length. Due to

TABLE 3. Percentage probability of detecting period taking into account age uncertainties in observational data.

Data, uncertainty	Shoemaker and Wolfe, 1986	This study
All data, <±20 m.y.	11(3, 16.0)[*]	6(2, 17.5)
Culled data, <±10 m.y.	9(0, 16.5)	21(3, 17.5)
Culled data, <±5 m.y.	12(0, 16.5)	30(2, 17.5)
Culled data, <±3 m.y.	24(1, 16.0)	56(2, 17.5)

[*](a,b) refers to phase (a) and period (b) detected most often. Note a ~16-m.y., not ~30-m.y. period is detected most often in Shoemaker and Wolfe's data.

the severe effects of subsequent geological processes on the retention and recognition of impact structures (*Grieve*, 1984), the subset of the impact record with "good" ages represents such a case, with the record strongly skewed to more recent ages. For example, the average number of events in the first four 30-m.y. periods (0-120 m.y.) in *Shoemaker and Wolfe's* data (1986) is 5, compared to 1.25 for the next four 30-m.y. periods (120 to 240 m.y.). A similar variation in sampling frequency occurs in our data (Fig. 6). Such effects can induce apparent periodicities, which are simply harmonics of these long-term variations (*Lutz*, 1985). We have investigated the effects of variation in the sampling frequency of data in the cratering record by sequentially removing craters from each subset in a manner similar to that suggested by *Lutz* (1985) for the magnetic record. For consistency, however, we continue to display the results using the *Broadbent* (1955, 1956) method. *Lutz* (1985) also noted so-called "wrapping" effects, which arise from the fact that the record length is not necessarily an integer multiple of the trial period. In our case, this is not a major concern, as there are relatively few events with old ages (Table 2; Fig. 6).

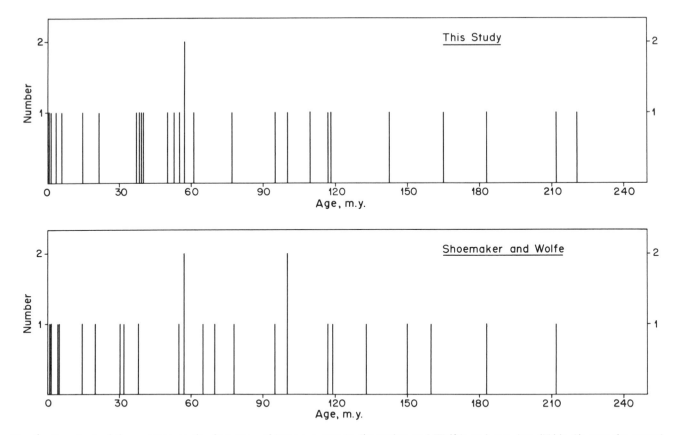

Fig. 6. Bar chart of the variation in the frequency of impact events in Shoemaker and Wolfe's and our data (Table 2) over the record length of 250 m.y. Note the reduction in the frequency of occurrence of events with increasing age.

In our data, backward truncation tends to strengthen the significance of the ~18-20 m.y. period (Fig. 7a). The strength of the ~18-m.y. peak is weakened with increased forward truncation (Fig. 7b) but is stable in period and phase. As the data are truncated forward, a ~21-m.y. peak appears but there is no indication of a peak around 30 m.y. Our results indicate that the ~18-20-m.y. peak is probably not related to long-term variations but is strongly dependent upon events contained in the first 40 m.y. of the record. These results also argue against Trefil and Raup's conclusion that their ~19-m.y. period was due to random events, whereas the 30-m.y. peak was a true periodicity obscured by a random component.

With backward truncation of the *Shoemaker and Wolfe* (1986) data, the 30-m.y. peak undergoes substantial fluctuation in period and shape (Fig. 8a). With forward truncation of the *Shoemaker and Wolfe* (1986) data, the 30-m.y. peak is weakened when the first two craters are removed. When the first four craters are removed from the record, the 30-m.y. peak is strengthened, but the phase has shifted from 4 m.y. to 6 m.y. (Fig. 8b). In contrast, the 16-m.y. peak is stable against forward and reverse truncation and consistently more significant than the 30-m.y. peak. These results indicate that the 30-m.y. peak is vulnerable to moderate changes in the sample record, whereas the 16-m.y. peak is significantly more robust. Neither of the peaks, however, appear to be directly attributable to long-term sampling frequency variations.

CONCLUSIONS AND DISCUSSIONS

From our model simulations of the cratering record, it is apparent that it is difficult to consistently detect a period from a mixture of periodic and random ages, unless the record is either dominated by periodic ages or the ages have small uncertainties (<10%) with respect to the period in question.

When two compilations of known crater ages are examined, the results of trials to find a period are inconsistent with respect to the occurrence of the previously suggested ~30-m.y. period. This is similar to our previous conclusion that slightly different data sets give different results (*Grieve et al.*, 1985). In addition, the Shoemaker and Wolfe data set, which has been used to indicate an ~30-m.y. period, fails to indicate such a period when age uncertainties are considered. The most common period found is at ~16 m.y., which is similar to an ~18-20-m.y. period determined from our compilation of ages. When the data are culled to account for variations in the frequency of events over the record, similar results are obtained. The 30-m.y. period in Shoemaker and Wolfe's data is not particularly stable and, in addition to a relatively stable period at ~16 m.y., there is a weak indication of a period at ~20 m.y. When our compilation is culled for sampling frequency variations, the ~18-20-m.y. period appears to be relatively stable. If such a period is real, it is inconsistent with that of other quoted periods of ~30 m.y. for other supposedly related phenomena in the geologic record. In the case of the marine extinction record, it is in phase at only approximately one out of three mass extinctions.

The craters that contribute most to the ~18-20-m.y. period in our data set: Zhamanshin (0.75 m.y.), Bosumtwi (1.3 m.y.), Wanapitei (37 m.y.), Mistastin (38 m.y.), Popigai (39 m.y.) and Kara and Ust-Kara (57 m.y.)—are few in number. With the exception of Kara and Ust-kara, all have been examined with respect to the siderophile trace element content of their respective melt rocks. Only Wanapitei shows a well-defined chondritic signature (*Wolfe et al.,* 1980). While siderophile data are open to interpretation, variations in their relative

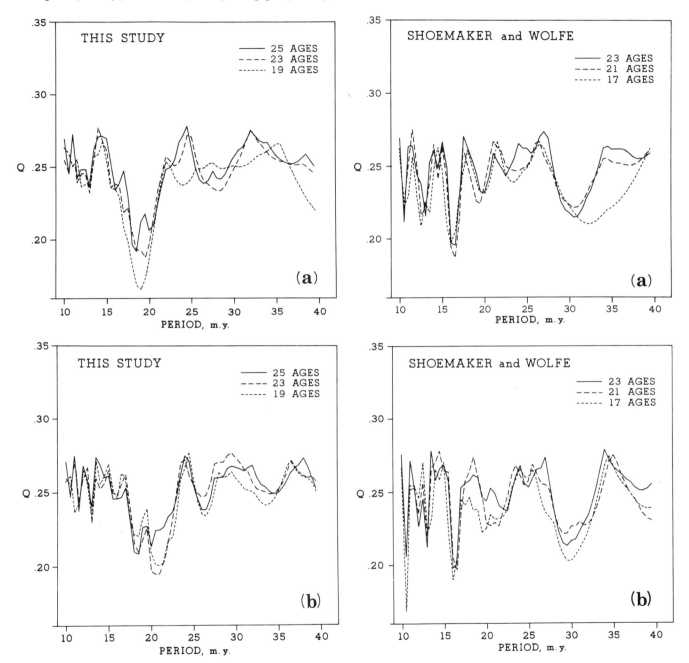

Fig. 7. Plot of the minimum value of the variable Q, against period, in m.y., for any phase, using the data in this study (Table 2). **(a)** Data are progressively truncated by removing the two youngest, four youngest, and then the eight youngest ages. **(b)** Data are progressively truncated by removing the two oldest, four oldest then the eight oldest ages. See text for discussion.

Fig. 8. Plot of the minimum value of the variable Q, against period, m.y., for any phase, using Shoemaker and Wolfe's data (Table 2). **(a)** Data are progressively truncated by removing the two youngest, four youngest and then the eight youngest ages. **(b)** Data are progressively truncated by removing the two oldest, four oldest, and then the eight oldest ages. See text for discussion.

abundances at these craters are not consistent with a common (cometary) source for the projectiles that produced these craters.

Shoemaker and Wolfe prefer fission-track over K-Ar ages for young craters. One could debate the relative merits of the ages and, thus, the best data set to use for analysis. For example, *Shoemaker and Wolfe* (1986) use fission track ages of 4.5 and 30.5 m.y. for Elgygytgyn and Popigai, respectively, compared to our K-Ar ages of 3.5 and 39 m.y. (Table 2). However, *Komarov et al.* (1983) and *Komarov and Raykhlin* (1976) report fission track ages of 3.4 and 38.9 m.y. for Elgygytgyn and Popigai, respectively, which agree with the K-Ar ages. We believe that such debates over which data set is better would resolve little at this stage, considering the effects of age uncertainties in detecting a consistent period.

In closing, we find no compelling evidence for the occurrence of a period in the cratering record due to cometary showers. Statistical indications of a period are not strong and, when age uncertainties are considered, period detection may be fortuitous. We consider the question of periodicity in the cratering record as still highly debatable. It may also be a difficult question to answer definitively. Not only are additional ages required, they must also be precise ages, if age uncertainties are treated in the analysis for periodicity.

Acknowledgments. A portion of this work was conducted while R. A. F. Grieve was the H. C. Urey Visiting Fellow at the Lunar and Planetary Institute, which is operated by the Universities Space and Research Association under contract No. NASW-4066 with the National Aeronautics and Space Administration. Receipt of the fellowship is gratefully acknowledged, as is the assistance of Lunar and Planetary Institute support staff. This paper is Lunar and Planetary Institute Contribution No. 633 and contribution of the Geological Survey of Canada 21587. Reading of an earlier version of this paper by B. Robertson and H. Weber is appreciated, as are formal reviews by A. Hildebrand and A. Woronow.

REFERENCES

Alvarez W. and Muller R. A. (1984) Evidence from crater ages for periodic impact on the Earth. *Nature, 308,* 718-720.

Broadbent S. R. (1955) Quantum hypotheses. *Biometrika, 42,* 45-57.

Broadbent S. R. (1956) Examination of a quantum hypothesis based on a single set of data. *Biometrika, 43,* 32-45.

Davis M. P., Hut P., and Muller R. A. (1984) Extinction of species by periodic comet showers. *Nature, 308,* 715-717.

Grieve R. A. F. (1982) The record of impact on the earth: Implications for a major Cretaceous/Tertiary impact event. *Geol. Soc. Am. Spec. Pap., 190,* 25-37.

Grieve R. A. F. (1984) The impact cratering rate in recent time. *Proc. Lunar Planet. Sci. Conf. 14th,* in *J. Geophys. Res., 89,* B403-B408.

Grieve R. A. F., Sharpton V. L., Goodacre A. K., and Garvin J. B. (1985a) A perspective on the evidence for periodic cometary impact on Earth. *Earth Planet. Sci. Lett., 76,* 1-9.

Grieve R. A. F., Sharpton V. L., Goodacre A. K., and Garvin J. B. (1985b) Periodic cometary impacts and the terrestrial cratering record. *Eos Trans. AGU, 66,* 813.

Hut P. (1984) How stable is an astronomical clock which can trigger mass extinctions on earth? *Nature, 311,* 638-641.

Komarov A. N. and Raykhlin A. I. (1976) Comparison of fission-track and potassium-argon dating of impactites (in Russian). *Dok. Acad. Nauk SSSR, 228,* 673-676.

Komarov A. N., Kolcva T. V., Gurova E. P. and Gurov E. P. (1983) Determination of fission-track ages of impact glasses of the Elgygytgyn crater (in Ukrainian). *Dok. Acad. Nauk Ukran. RSR, Ser. B, 10,* 11-13.

Lutz T. M. (1985) The magnetic reversal record is not periodic. *Nature, 317,* 404-407.

Rampino M. R. and Stothers R. B. (1984a) Terrestrial mass extinctions, cometary impacts and the Sun's motion perpendicular to the galactic plane. *Nature, 308,* 709-712.

Rampino M. R. and Stothers R. B. (1984b) Geological rhythms and cometary impacts. *Science, 226,* 1427-1431.

Raup D. M. (1985) Magnetic reversals and mass extinctions. *Nature, 314,* 341-343.

Raup D. M. and Sepkoski J. J., Jr. (1984) Periodicity of extinctions in the geologic past. *Proc. Natl. Acad. Sci., 81,* 801-805.

Shoemaker E. M. and Wolfe R. F. (1986) Mass extinctions, crater ages, and comet showers. In *The Galaxy and Solar System* (R. Smoluchowski, J. N. Bahcall, and M. Matthews, eds.), pp. 338-386. University of Arizona Press, Tucson.

Smoluchowski R., Bahcall J. N., and Matthews M., eds. (1986) *The Galaxy and the Solar System.* University of Arizona Press, Tucson. 483 pp.

Trefil J. S. and Raup D. M. (1987) Numerical simulations and the problem of periodicity in the cratering record. *Earth Planet. Sci. Lett., 82,* 159-164.

Whitmire D. P. and Jackson A. A. (1984) Are periodic mass extinctions driven by a distant solar companion? *Nature, 308,* 713-715.

Wolfe R., Woodrow A., and Grieve R. A. F. (1980) Meteoritic material at four Canadian craters. *Geochim. Cosmochim. Acta, 44,* 1015-1022.

Crater Identification and Resolution of Lunar Radar Images

H. J. Moore

U.S. Geological Survey, Menlo Park, CA 94025

T. W. Thompson

Jet Propulsion Laboratory, California Institute of Technology, Pasadena, CA 91109

Our study has three principal results: (1) The percentage of craters that can be identified increases with diameter or relief for any given resolution, but craters are not identified at all diameters and relief. (2) Relations between the percentage of identified craters and their dimensions depend on the size-frequency distributions of both diameters and relief and their interrelations. (3) Identification of craters is strongly dependent on the resolutions of the radar images. Our data also indicate that crater identification depends on geologic age, that the effect of background terrain is uncertain, and that angle of incidence of radar illumination has a modest effect, if any, on crater identification. The ability to recover an actual crater size-frequency distribution from radar images diminishes with increasing size of the resolution elements because of the results listed above. Good agreement between the actual distribution and the one derived from radar images is attained if the crater diameters are at least 10 times larger than the resolution of the radar images. Our results are important considerations in geologic interpretations of radar images because conclusions about geologic processes, estimates of ages of planetary surfaces based on crater statistics, and assessments of size-frequency distributions of other landforms depend on resolution. We expect that the high-resolution radar images of Venus to be acquired from orbit during the planned Magellan mission will reveal at least 64 times more craters and volcanic landforms than are known today.

INTRODUCTION

The principal objective of this study is to examine the relations between landform identification and the resolution of radar images. An understanding of these relations is essential to geologic interpretation of radar images. Conclusions about geologic processes, the importance of the processes, and relative ages of the geologic units produced by the processes depend on the resolutions of the radar images. Here we explore the relations between the identification of the dominant lunar landform, craters, and radar images at three resolutions and present some results on lunar mountains. We also speculate on the outcome of the high-resolution radar images of Venus that will be acquired by the planned Magellan orbiting spacecraft.

Resolution, as used here, is the radar cell size. Radar cell size can be thought of as the imaged size of a very small, but efficient, reflector (such as a corner reflector), and it is roughly equivalent to a pixel in a Voyager or Viking photograph. No matter how small the reflector, it appears as large as the radar cell size. Azimuth and ground-range resolutions of synthetic aperture radar (SAR) images are equivalent to radar cell size (*Moore*, 1983). Photographic resolutions are often quoted as "line-pair" resolutions. For the radar images, the equivalent line-pair resolutions are 2.2 times the radar cell sizes (W. Johnson, personal communication, 1987) because two equally bright, very small reflectors would have to be separated by 2.2 times the cell size to be seen as two individual targets.

Radar images of the Moon are an ideal source of data for this study because of the availability of (1) three sets of images that cover the entire Earth-facing hemisphere with a large range of cell sizes, (2) ample topographic and photographic data on the dimensions of craters less than 1 km to hundreds of kilometers in diameter, (3) well-established relative ages of craters, (4) topographic backgrounds that vary from rough to smooth, and (5) angles of incidence of radar illumination from several to 90°.

We used three sets of polarized radar images of the Moon: (1) high-resolution 3.8-cm wavelength images (cell sizes of 1–2 km; *Zisk et al.*, 1974), (2) high-resolution 70-cm wavelength images (cell sizes of 2.4–5.5 km; *Thompson*, 1987), and (3) low-resolution 70-cm wavelength images (cell sizes of 5–10 km; *Thompson*, 1974). These images sample the Moon from longitude 90° E to 70° W between latitudes 32° N and 16° S (Fig. 1); resolution within each set is nearly uniform. Techniques and procedures employed in acquiring and displaying the images have been described (*Thompson*, 1965, 1978, 1979; *Pettengill et al.*, 1974) and thus will not be presented here, but two comments are relevant. First, the images are derived from polarized echoes that are reflected from tilted surfaces much larger than the wavelengths of the radars; these echoes are distinct from depolarized echoes (*Hagfors*, 1967) that are scattered from fine-scale roughnesses comparable to the wavelengths of the radars. Second, the radar echo strengths of the images are scaled logarithmically and the display values are set so that the strongest echoes are whites, the weakest echoes are blacks, and the average echoes present gray tones. Thus, in the radar images of the Moon at small to moderate angles of incidence, slopes tilted toward the radar create highlights and slopes tilted away from the radar appear dark and resemble shadows, although true radar shadows may not be present. These modulations of the polarized echo strengths in the radar images, which are created by the walls and flanks of a crater, allow the crater to be identified.

We use a classification scheme that reflects both our interpretations of craters portrayed (or not portrayed) by the

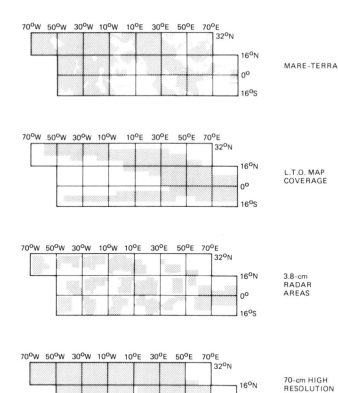

Fig. 1. Generalized lunar maps showing mare and terra distribution and coverage by Lunar Topographic Orthophotomaps (L.T.O.) and radar images. Stippled patterns indicate mare areas, L.T.O. map coverage, and radar areas used in this study.

three sets of radar images and our comparisons of the radar interpretations with actual craters shown on topographic maps and photographs. The classification scheme is as follows:

1. *Clearly identified*—Crater is resolved and its interpretation on radar image is clearly correct.
2. *Inferable*—Crater is resolved and its interpretation on radar image is probably correct.
3. *Resolved*—Crater is resolved but its interpretation on radar image is uncertain.
4. *Detected*—Crater is detected by the radar but its elements are not resolved.
5. *Undetected*—Crater is not detected by radar because no echoes are present, the background is "speckled," or both.
6. *Ambiguous*—A complex array of craters and mountains is resolved but interpretation is ambiguous.
7. *Fictitious*—The radar image portrays a crater that does not exist or differs significantly from the actual crater.

In the following sections, craters of classes 1 and 2 are combined as "identified"; all other craters are considered to be unidentified.

Our data permit tests that show how identifications of the craters vary with size, geologic age, background terrain, and angle of incidence. Diameters and interior relief of the craters were obtained from Apollo topographic maps and Lunar Orbiter IV photographs; diameters of identified and fictitious craters were measured on the radar images. Geologic ages and background terrains were obtained from U.S. Geological Survey 1:1,000,000-scale and 1:5,000,000-scale maps and other sources (*Wilhelms*, 1980). The ages are broadly grouped into Copernican and Copernican-Eratosthenian (C and CE), Eratosthenian and Eratosthenian-Imbrian (E and EI), and Imbrian and pre-Imbrian (I and pI). Background terrains include maria (smooth); terrae (rough); crater ejecta, floors, and walls (rough and bright); and volcanic mantles (dark). Selenographic coordinates supplied the information for computing angles of incidence of radar illumination.

TABLE 1. Total number of craters and identified craters (in parentheses) in selected sample subsets for three sets of radar images.

Data subset	Radar Image Set		
	3.8-cm high resolution	70-cm high resolution	70-cm low resolution
All craters	1841 (322)	1484 (115)	840 (35)
Geologic ages* for all background terrains and angles of incidence			
C+CE	475 (64)	316 (16)	125 (1)
E+EI	542 (68)	404 (14)	202 (1)
I+pI	824 (190)	764 (85)	513 (33)
Background terrains† for all ages and angles of incidence			
Mare	547 (81)	493 (25)	258 (2)
Terra	1089 (197)	870 (82)	518 (33)
Crater	201 (44)	115 (8)	61 (0)
Angles of incidence for all ages and background terrains			
0°–40°	943 (145)	651 (42)	372 (9)
41°–71°	588 (130)	601 (49)	337 (9)
72°–90°	310 (47)	232 (24)	131 (7)
Angles of incidence for E+EI craters and all background terrains			
9°–29°	233 (28)	137 (6)	57 (1)
30°–80°	223 (32)	242 (7)	132 (0)
Special cases for all angles of incidence			
E+EI, mare	231 (33)	226 (8)	122 (1)
E+EI, terra	255 (29)	151 (4)	60 (0)
I+pI, terra	689 (145)	620 (72)	413 (33)
Mountains	284 (206)	314 (93)	225 (5)

*Geological age symbols: C+CE (Copernican and Copernican-Eratosthenian), E+EI (Eratosthenian and Eratosthenian-Imbrian), and I+pI (Imbrian and pre-Imbrian).

†One terrain (volcanic) is excluded because of small numbers.

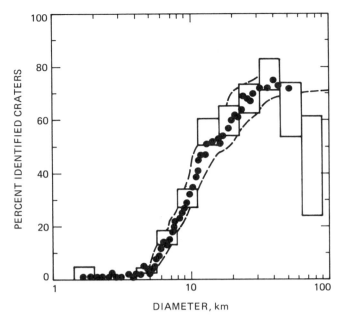

Fig. 2. Percentage of identified craters seen on 3.8-cm radar images as a function of diameter, calculated in two ways. Filled circles represent percentages and mean diameters; short dashes indicate diameter range for the frequency bins; calculations are in frequency bins of 100; one-standard-deviation error bars are ±5% at 50% identified craters; elsewhere they are less. Some points in this figure and subsequent similar figures are not plotted because of overlap. Rectangles represent percentages of identified craters in logarithmic diameter bins (vertical edges) and one-standard-deviation error bars for percentages in the diameter bin (horizontal edges); average percentages are midway between error bars. Both plots exhibit "toes" or small percentages for diameters below 4 km, a steep slope between 4 and 12.5 km, a moderate slope from 12.5 to 40 km, and then a negative slope above 40 km.

In our data analyses, the percentages of craters in each identification class were calculated as a function of geometric mean diameter or relief in specified frequency bins. Maximum and minimum diameters or relief in each frequency bin were governed by the bin size. Calculations were made in frequency steps that are 1/10 the size of the frequency bin. This procedure can be followed for a variety of subsets that include the following options.

Radar image set: 3.8-cm high resolution,
   70-cm high resolution, or 70-cm low resolution
Dimensions: diameter or relief (interior for craters)
Landform type: all, craters only, or mountains only
Geologic age: all, C and CE, E and EI, or I and pI
Background terrain: all, mare, terra, crater, or volcanic
Incidence angle: all or specified ranges of angles

We limit our discussion to selected subsets and require that the total number of craters exceed 200 and the number of identified craters exceed 25 in each subset. Table 1 gives the total number of craters and identified craters for selected subsets.

Plotting of the percentage of craters in each class as a function of a diameter (or relief) results in a curve defined by a series of points (Fig. 2). Larger bin sizes result in smoother curves, lower maximum percentages recorded, smaller statistical errors in percentages, and larger ranges of diameters in each bin than smaller bin sizes, but the major portions of the curves are not displaced. Typically, the curves have relatively steep slopes at the larger diameters and a "toe" that extends to smaller diameters at small percentages. Similar curves are obtained by computing the percentage in each class in logarithmic diameter intervals (Fig. 2).

Because craters either are or are not identified, the binomial distribution applies to our results. Thus, the one-standard-deviation error bars in Fig. 2 are equal to ± the square root of the product of the percentage of identified craters and the percentage of unidentified craters divided by the total number of craters in the diameter bin. For the frequency bin plot, the error bar is ± 5% at 50% of identified craters because the bin size is 100; elsewhere, the error bars are less.

AN ILLUSTRATION OF RESOLUTION EFFECTS

The effects of resolution on crater identification and thus on our portrayal of the lunar surface are illustrated by the Theophilus region of the Moon. A photograph of this region (Fig. 3), at a resolution of about 0.5 km, shows the rugged

TABLE 2. Examples of craters in the Theophilus region of the Moon protrayed by the three sets of radar images and their identification classes.

Identification class	3.8-cm high-resolution image (Fig. 4)	70-cm high-resolution image (Fig. 5)	70-cm low-resolution image (Fig. 6)
1. Clearly Identified	Alfraganus (21) T, C	Ibn-Rushd (31) T, pI	—
2. Probably Identified	Andel F (8) T, E	Madler (28) C, E	Abulfeda (65) T, pI
3. Resolved	Torricelli C (11) M, E	Descartes A (15) T, C	Delambre (51) T, E
4. Detected	Saunders A (7) T, C	Dollond (11) T, E	Tacitus B (12) T, E
5. Undetected	Kant C (20) T, pI	Kant C (20) T, pI	Torricelli (30) M, I
6. Ambiguous	Zollner J (10) T, I	Andel (35) T, pI	Hypatia (41) T, pI
7. Fictitious	Unnamed (19) T, na	Part of Cyrillus (47) T, na	Part of Cyrillus (47) T, na

Numbers in parentheses by crater names indicate crater diameters in kilometers. The names and locations of these examples are given in Fig. 3; their identification class numbers are given in Figs. 4–6. Letter symbols below names indicate background terrain and geologic age.

*Background terrain symbols are C=crater, M=mare, and T=terra. Geologic age symbols are C=Copernican, E=Eratosthenian, I=Imbrian, and pI=pre-Imbrian.

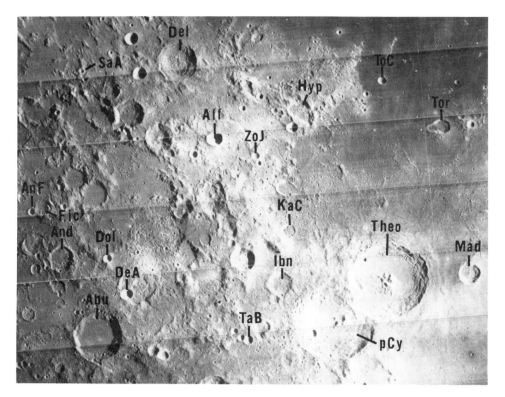

Fig. 3. Photograph of the Theophilus region of the Moon. Symbols indicate craters (see text and Table 2): SaA - Saunders A, AnF - Andel F, Fic - fictitious crater, And - Andel, Dol - Dollond, DeA - Descartes A, Abu - Abulfeda, Del -Delambre, Alf - Alfraganus, ZoJ - Zollner J, Hyp - Hypatia, KaC - Kant C, Ibn -Ibn-Rushd, TaB -Tacitus B, Tor - Torricelli, ToC - Torricelli C, Theo - Theophilus, Mad - Madler, pCy - part of Cyrillus. Note rugged terra and smooth maria in northeast and southeast corners (upper and lower right). Craters as small as 1.5 km can be identified. Scale is about 1:50,000,000. Sun incidence angle is about 68 degrees; sun is to right. Resolution is about 0.5 km. Lunar Orbiter IV frame M84.

terra, relatively smooth maria, craters about 1.5 to 100 km across, and a spectrum of hills and mountains. Comparisons of the photograph with the radar images (Figs. 4–6) show the loss of detail that accompanies progressively lower resolutions. (Examples of our identification classes are indicated in Figs. 4–6 and listed in Table 2; names and locations are given in Fig. 3.) In all three radar images, the rugged terra and smooth maria are apparent. We can identify some craters 4.5 km across in the 3.8-cm radar image (Fig. 4), but many others of about the same size cannot be identified. Craters that are tens of kilometers across and larger are generally well portrayed. We can identify some craters 12 km across in the 70-cm high resolution image (Fig. 5); again, many other craters of about the same size cannot be identified. Portrayals of the largest craters, except Cyrillus, and some craters several tens of kilometers across are good. In the 70-cm low-resolution image (Fig. 6), an experienced radar-image interpreter would probably identify the craters Abulfeda (No. 2 at lower left, 65 km across) and Theophilus (100 km across), but he could not locate or identify smaller craters without previous knowledge.

RESULTS

Below, we briefly show how the identification of craters is related to (1) diameter for all identification classes, (2) size-frequency distributions of diamters and interior relief, (3) resolution of the radar, (4) geologic age, (5) background terrain, and (6) angle of incidence of radar illumination. Also, we briefly describe our results for a small sample of lunar mountains, for the relations between diameters measured on radar images and those on maps and photographs, and for crater size-frequency distributions.

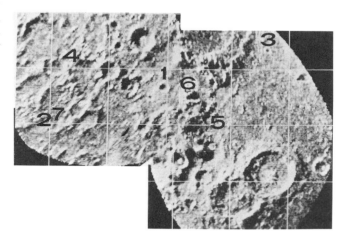

Fig. 4. A 3.8-cm high-resolution radar image of part of the Theophilus region. Large craters at lower right are Theophilus and Cyrillus. Separation of rugged terra and smooth mare in northeast corner (upper right) is not as clear as in Fig. 3. Many of the large craters visible in Fig. 3 and some craters as small as 4.5 km across can be identified. Numbers are centered above examples of identification classes listed in Table 2. Scale is about 1:61,000,000. Incidence angle is about 22°; illumination is from upper left. Resolution is about 1.5 km.

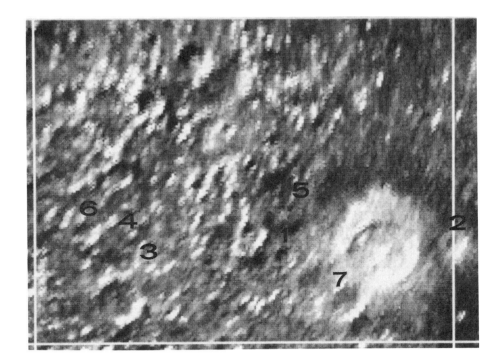

Fig. 5. A 70-cm high-resolution radar image of the Theophilus region. Large, bright crater at lower right is Theophilus. Separation of rugged terra from smooth mare in northeast corner (upper right) is not as clear as in Fig. 4. Some of the large craters visible in Fig. 3 and some craters as small as 12 km across can be identified. Numbers are centered above examples of identification classes listed in Table 2. Scale is about 1:50,000,000. Incidence angle is about 22°; illumination is from upper left. Resolution is about 4 km.

Crater Identification Classes and Diameter

Our data show that the percentage of craters that can be identified increases with diameter (or relief) but that craters cannot be identified at all diameters and relief at any given resolution (Fig. 7a,c and e). In the 3.8-cm radar images, for example, only 6% of the craters are identified whose mean diameter is 6 km, but 60% are identified whose mean diameter is 20 km. The percentages of craters that are simply not detected and those that are detected only as bright spots dominate at the smaller diameters and become smaller at the larger diameters (Fig. 7a, 7c, and 7e). Diameters of the other three identification classes are plotted separately (Fig. 7b,d and f).

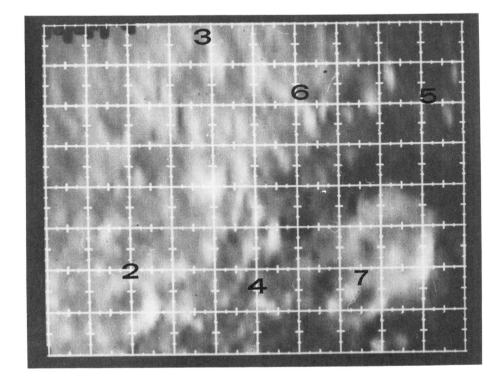

Fig. 6. A 70-cm low-resolution radar image of the Theophilus region. Large, bright crater at lower right is Theophilus. Separation of rugged terra from smooth mare in northeast corner (upper right) is clear. Other areas around Theophilus appear to be as smooth as those in northeast corner. Only two of the craters in Fig. 3 would probably be identified: Abulfeda and Theophilus. Numbers are centered on examples of identification classes listed in Table 2. Scale is about 1:50,000,000. Incidence angle is about 22°; illumination is from upper left. Resolution is about 7.5 km.

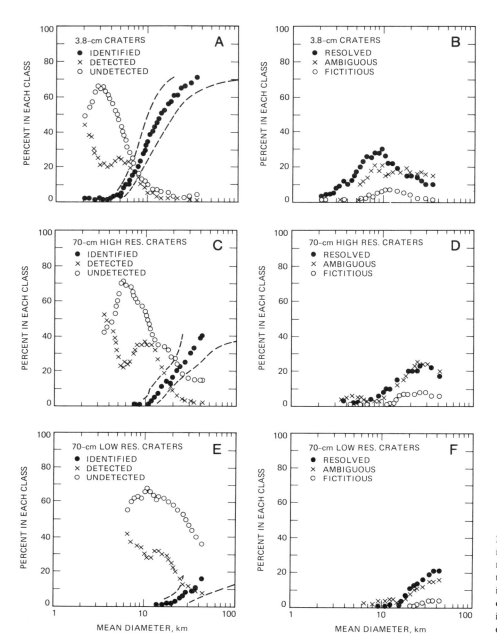

Fig. 7. Variations in percentages of identification classes as a function of mean crater diameter for all resolutions, backgrounds, ages, and angles of incidence. Calculations are in frequency bins of 200. Dashed lines indicate range of diameters in frequency bins.

Size-Frequency Distributions of Diameters and Interior Relief

These distributions and their interrelations are shown in our results in three ways. First, the highest resolution radar "sees" greater maximum percentages of identified craters than the lower resolution radars (Fig. 8a) if the bin sizes are the same. Second, the slopes of the curves for identified craters versus diameter are different for the data sets (Fig. 8a): The slope is steep in the lower part of the 3.8-cm curve and flat in the 70-cm high-resolution curve and the upper parts of the other two curves. In contrast, the curve for identified craters versus relief for the 3.8-cm data exhibits one slope (Fig. 8b) and the 70-cm high-resolution curve is more or less parallel to it. (Little can be said for the 70-cm low-resolution curve.) Third, results for various subsets of data are different. In Fig. 9, the 3.8-cm curve for old, degraded, and shallow I and pI craters with terra backgrounds lies below the curve for all craters, so that the steeper portion is less evident. E and EI craters represent a well-behaved distribution of diameters and relief because they have depth-diameter ratios close to 0.17. The slopes of the curves for both diameter and relief of E and EI craters are the same (Fig. 10a,b), in contrast to those for all craters (Fig. 8a,b).

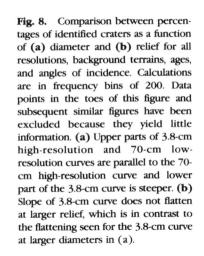

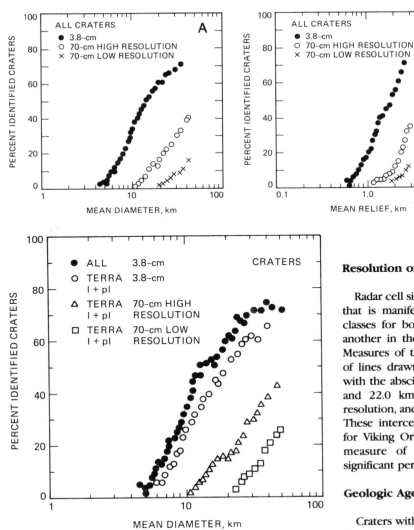

Fig. 8. Comparison between percentages of identified craters as a function of (a) diameter and (b) relief for all resolutions, background terrains, ages, and angles of incidence. Calculations are in frequency bins of 200. Data points in the toes of this figure and subsequent similar figures have been excluded because they yield little information. (a) Upper parts of 3.8-cm high-resolution and 70-cm low-resolution curves are parallel to the 70-cm high-resolution curve and lower part of the 3.8-cm curve is steeper. (b) Slope of 3.8-cm curve does not flatten at larger relief, which is in contrast to the flattening seen for the 3.8-cm curve at larger diameters in (a).

Resolution of the Radar

Radar cell sizes have a profound effect on crater identification that is manifested in one way: curves for the identification classes for both diameter and relief are displaced from one another in the same sense as the cell sizes (Figs. 7 and 8). Measures of these displacements are the diameter intercepts of lines drawn through the lower linear parts of the curves with the abscissas (Fig. 8). The intercepts are about 4.8, 10.2, and 22.0 km for the 3.8-cm high-resolution, 70-cm high-resolution, and 70-cm low-resolution radar images, respectively. These intercepts, which are similar to the detection indices for Viking Orbiter pictures (*Dial and Schaber*, 1981), are a measure of the "identification threshold" above which significant percentages of craters are identified.

Geologic Age

Craters within the various geologic age groups have different distributions of diameter and relief, which affect their identification. This is illustrated by 3.8-cm diameter data (Fig. 11a). The E and EI craters are as readily identified as the relatively young C and CE craters, but craters of both these age groups are more readily identified than older I and pI

Fig. 9. Illustration of the effect of diameter distributions of identified craters. For 3.8-cm radar images, slopes of curve for all craters are different than those of shallow, degraded I and pI craters. Steep slopes of upper parts of 3.8-cm curves are about the same as those of the 70-cm curves. Calculations are in frequency bins of 100.

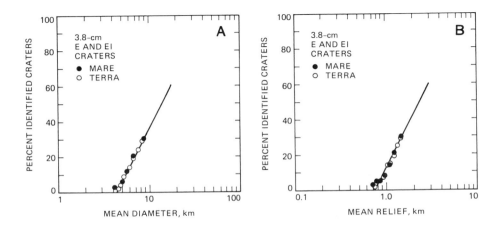

Fig. 10. Results for the well-behaved population of Eratosthenian (E) and Eratosthenian-Imbrian (EI) craters, which have nearly constant depth-to-diameter ratios, as a function of diameter and relief. Straight lines indicate expectations for craters with a depth-diameter ratio of 0.17. Calculations are in frequency bins of 100.

Fig. 11. Effects of geologic age and background terrain on crater identification. **(a)** Percentage of identified craters on all background terrains and at all angles of incidence for three relative age groups (see Table 1 for identification of age symbols). **(b)** Percentage of identified craters of all ages at all angles of incidence for three background terrains. Calculations are in frequency bins of 100.

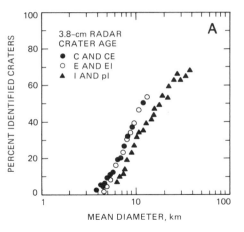
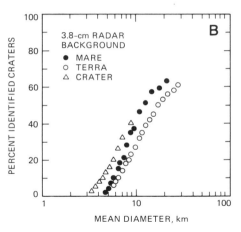

craters. When these age groups are plotted against relief, the positions of the curves are reversed. This reversal should be expected because diameters of the generally degraded I and pI craters are larger than those of the younger craters with the same relief.

Background Terrain

The effect of background terrain may be more apparent than real, as illustrated by the 3.8-cm data (Fig. 11b). Craters on crater backgrounds appear to be more readily identified than those on mare and terra backgrounds, and those on terra backgrounds are least readily identified. On the other hand, the plots of E and EI craters (Fig. 10) indicate that little difference can be attributed to background terrain. Thus, the relations in Fig. 11b may reflect differing distributions of diameters and relief.

Angle of Incidence of Radar Illumination

Angle of incidence has a modest effect, if any, on crater identification. In general, our data suggest that craters are more readily identified when the angles of incidence are near 60°, but this result could be due to different distributions of

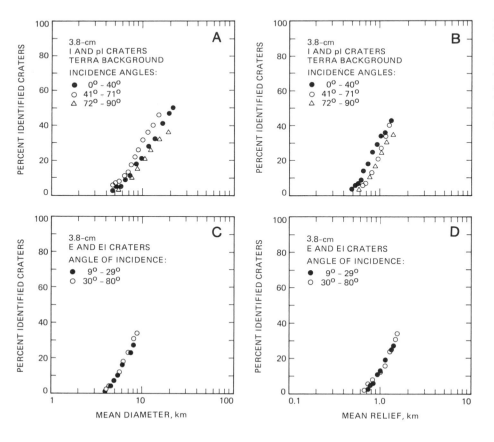

Fig. 12. Unverified effects of angle of incidence on crater identification on 3.8-cm radar images. **(a)** and **(b)** show I and pI craters on terra background terrain; **(c)** and **(d)** show E and EI craters on all background terrains (see Table 1 for identification of age symbols). Calculations are in frequency bins of 100.

diameters and relief in the angle of incidence bins. This problem is illustrated by comparing the percentage of identified I and pI craters on terra backgrounds with diameter and relief for three ranges of incidence angles (Figs. 12a,b). Here, craters appear to be more readily identified by this diameter when the angles of incidence are 41-71 degrees (Fig. 12a), but this relation is not supported when relief is used as a criterion (Fig. 12b). Slopes of the diameter and relief curves for paired subsets are different (Figs. 12a,b). We also examined E and EI craters because they have well-behaved relations between diameter and relief. In this age group, craters appear to be equally well identified on the basis of diameter (Fig. 12c) and relief (Fig. 12d) at the incidence angles shown. Thus, we find no demonstrable effects related to angles of incidence, although there might be modest effects. However, two conditions are not addressed by our data. The first is for very small angles of incidence where image layover and other problems are serious, and the second is for very large angles of incidence, especially in rough terrain, where radar shadows obscure craters.

Mountains

We report our results on mountains because they exhibit the same trends as craters, but we are far from satisfied with the data sets on them. Lunar mountains are difficult to work with because few of their shapes are simple in plan view. Most isolated mountains in the maria are elongate and those in the

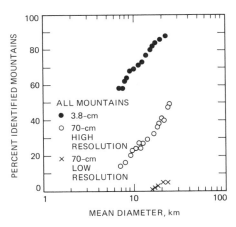

Fig. 13. Percentage of identified mountains seen in the three sets of radar images as a function of diameter. Included are mountains of all ages in all background terrains imaged at all angles of incidence. Calculations are in frequency bins of 100. There are no toes because samples do not extend to cell sizes of radars.

terrae have complex plan views. Fig. 13, despite the small sample (Table 1), shows significant margins of difference among the three data sets for mountains having the same horizontal dimensions. Similar results are obtained when relief is used. In view of our limited sample of mountains, extrapolations of the curves to the abscissas are unjustified.

TABLE 3. Results of linear regressions on crater and mountain diameters measured on radar images and maps and photographs.

Radar Image Set	Number of Data Points N	Y-axis Intercepts a	Slope b
3.8-cm high resolution	528	1.26 ± 0.29	0.978 ± 0.007
70-cm high resolution	208	2.53 ± 0.85	0.983 ± 0.014
70-cm low resolution	40	3.05 ± 2.62	0.997 ± 0.027

Error limits for constants are for 95% confidence. See text for equation.

Measured Diameters

In general, the diameters of smaller craters and mountains measured on radar images are larger than those measured on maps and photographs by about one to one-half the cell-size of the radar, but agreement at diameters near 400-500 km is within a few percent. Table 3 gives the results of regressions to equations of the form

$$D_r = a + b D_o$$

where D_r is the diameter of craters and mountains measured on the radar images, D_o is the diameter of craters and mountains measured on maps and photographs, a is the Y-axis intercept, and b is the slope.

Crater Size-Frequency Distributions

As would be predicted from our previous results (Fig. 8), the ability to recover an actual crater size-frequency distribution diminishes with increasing cell size (Fig. 14). Below some diameter, the numbers of craters are underestimated by progressively larger amounts as their diameters decrease to several times the radar cell size because the percentage of identified craters also decreases with decreasing diameters. Thus, if one chooses to define identification resolution as the diameter where error bars of the observed and radar frequency distributions touch, the identification resolution would be 10-40 times larger than the cell size of the radar.

CONCLUSIONS

Perhaps our most unexpected result is that only some of the craters and mountains in radar images are identified at all sizes. Thus, simple rules of thumb, such as identification resolution being 3 to 4 times the radar cell size, are not valid. Such rules would apply to an "identification threshold" where we begin to identify significant percentages of landforms. As might be expected, the percentage of identified landforms increases as their diameters increase above the cell sizes of the radars. A similar result is obtained when relief is considered separately. The minimum relief where we begin to identify a significant percentage of landforms is probably related to

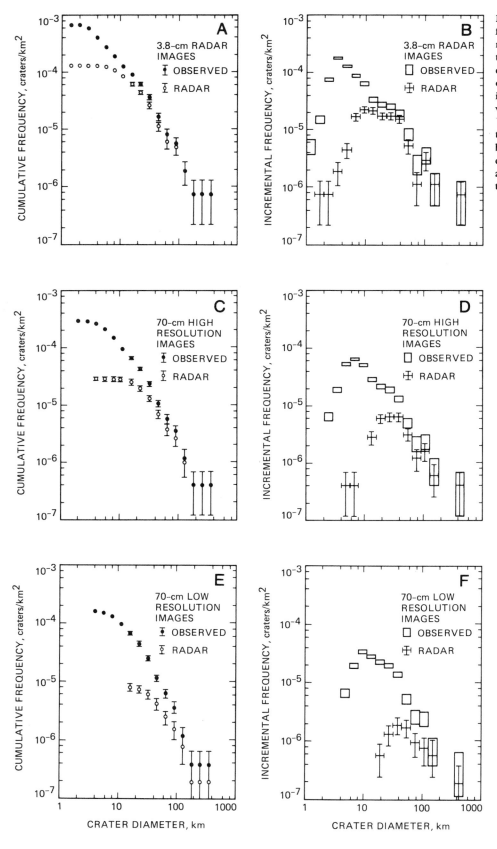

Fig. 14. Contrasting crater size-frequency distributions observed on maps and photographs and seen on the three sets of radar images. Cumulative distributions are shown in **a, c,** and **e**; incremental distributions are shown in **b, d,** and **f.** Standard procedures were used (*Crater Analysis Techniques Working Group,* 1979); radar distributions employ observed diameters because they are not significantly different from those of the radar images at diameters more than 10–40 times the cell sizes.

relations between diameters and relief. For example, the curve for relief of E and EI craters (Fig. 10) intersects the abscissa at about 0.74 km and corresponds to an E and EI crater 4.5 km in diameter because the depth-diameter ratio is 0.17.

Intuitively, we can see that the size-frequency distributions of landform diameters and relief should affect landform identification. The change in slope of the diameter curve for the 3.8-cm radar-image data (Figs. 8 and 9) reflects the change in depth-diameter ratios of fresh lunar craters (*Pike*, 1980) and subsequently degraded ones. Other effects, such as the following, are produced. Variations in the distributions may give false indications of the effects of background terrains and angles of incidence that appear to be negated when a well-behaved population of landforms (E and EI craters) is considered separately. Image-display procedures may be the cause of the lack of differences related to angles of incidence. Identification of young craters is enhanced because they have large depth-to-diameter ratios and they have large interior and exterior slopes that create larger echo strength modulations than those from old, shallow craters.

It is difficult to quantify the effect of cell size on the identification of craters because of the results discussed above. One would prefer well-behaved curves for large samples from the three radar image sets, but they are not evident in our data. Diameter intercepts of the curves with the abscissas are a fair measure of identification thresholds above which significant percentages of craters are identified. Because average cell sizes for the three sets of radar images are near 1.5, 3.4, and 7.5 km, and the corresponding diameter intercepts are near 4.4–5.5, 10–12, and 21–23 km, identification threshold diameters are about 3 to 4 times larger than the cell sizes of the radars.

Crater size-frequency distributions are affected in three ways. First, the distributions are truncated at diameters that are near the radar cell size. Second, reasonable agreement, taken as that diameter above which error bars of the observed and radar distributions overlap, is not achieved until the diameters of the craters are 10 to 40 times the radar cell size. Third, diameters of the craters at the small cell sizes tend to be roughly one-half to one cell size larger than the actual diameters, but errors are small at diameters 10–40 times the cell size.

SPECULATIONS ON LANDFORM IDENTIFICATION ON VENUS

Landforms on Venus can be observed only from orbit or from Earth with radar imaging systems because of its dense cloud cover. The feasibility of obtaining images of Venus from Earth has been demonstrated by the radio telescopes at Arecibo (e.g., *Campbell and Burns*, 1980) and Goldstone (e.g., *Jurgens et al.*, 1980). Radar images of Venus have been acquired from orbit by Pioneer Venus at 17-cm wavelength with very coarse (about 30 km) resolutions (*Pettengill et al.*, 1979) and by the Venera 15 and 16 spacecraft at 8-cm wavelength with 1–2 km resolutions and angles of incidence near 10° (*Barsukov et al.*, 1986). The Magellan mission is planned to obtain iamges at 12.6-cm wavelength with ground-range resolutions of 116 to 326 m, an azimuth resolution of about 120 m, and angles of incidence from 16° to 49° (*Johnson and Edgerton*, 1985).

Our lunar results provide some insights as to how Venusian landforms are imaged by radar because the variations of reflectivity with angle of incidence from Earth-based data for both the Moon and Venus are similar, although the total radar cross-section of Venus is larger than that of the Moon (Fig. 15). In Fig. 15a, the Venus cross-sections are represented by an empirical formula (*Muhleman*, 1964) that fits the observed cross-sections at 12.5 cm (*Campbell and Burns*, 1980) and is currently used by the Magellan project for design purposes (W. Johnson and R. Jurgens, personal communication, 1987). The total cross-section of Venus is near 0.11–0.12 (*Pettengill*, 1968). The lunar cross-sections (Fig. 15a) are obtained from tabular data (*Hagfors*, 1967, 1970) based on measurements by the MIT Lincoln Laboratory on the Millstone Hill and Haystack radio telescopes (*Evans and Pettengill*, 1963); these cross-sections are adjusted so that their total equals 0.07. Rates of change of cross-section (Fig. 15b) are important because modulation of the echoes by local slopes creates the highlights and tones that make landforms visible in radar images. Here, the derivatives for the cross-sections are approximated by fitting the adjusted Hagfors data with the empirical formula (*Muhleman*, 1964) and differentiating the resulting curves. When this is done, the rates of change of cross-sections with

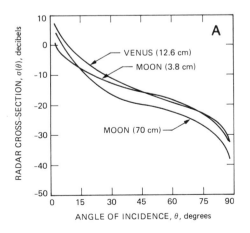
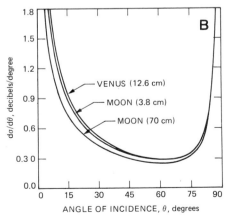

Fig. 15. Similarity of radar reflectivities (**a**) and estimated rates of change of radar cross-sections with angle of incidence (**b**) for Venus and the Moon.

angle of incidence for Venus and the Moon are nearly identical. Thus, the Moon is a good analog for Venus in terms of modulation of radar echoes.

Some wavelength-dependent differences should be expected between radar echoes from Venus and those from the Moon because of different root-mean-square (rms) slopes, variations in dielectric properties of the surface materials, and rock populations. Rms slopes of Venus at 17-cm wavelength average about 4° and range from 1° to 10° (*Pettengill et al.*, 1980); thus they are not significantly different from lunar rms slopes at 13-cm wavelength (*Moore et al.*, 1980). Average rms slopes of the Moon at 3.8-cm wavelength (9°) are much larger than those of Venus, however. Average bulk dielectric constants inferred from the total cross-sections are much higher for Venus (4.2) than for the Moon (2.9). Also, the ubiquitous dry lunar regolith presents a uniform dielectric material to the radar, but such a uniform material does not appear to be present on Venus (*Garvin et al.*, 1985). Surfaces around the Venera landers have large concentrations of rock (*Basilevsky et al.*, 1985) that are higher than those of most lunar surfaces, but comparable to the Surveyor VII site (*Shoemaker et al.*, 1969). The Venera 9 site has an unusually large concentration of rocks. Despite these differences, the terrestrial data (Fig. 15) show that the Moon is a reasonable analog for Venus.

Magellan radar images of Venus will probably reveal more craterlike forms than the Venera images by about two orders of magnitude, because the resolutions will be 4–10 times better. Identification threshold diameters will be near 0.5 to 1 km instead of 3–8 km. The currently known Venusian crater size-frequency distribution, which extends down to 8 km (*Ivanov et al.*, 1986), could be extended down to about 1 km. If the crater distribution on Venus is like that of the average lunar mare with a population index of -2 (*Hartmann*, 1966), the total number in the distribution would be about 64 times greater than that in the distribution presently known (*Ivanov et al.*, 1986). The same conclusion would be reached if the Venusian craters were volcanic, because the population index of terrestrial volcanic craterlike forms is also near -2 (*Pike and Clow*, 1981; *Burns and Campbell*, 1985). We would also expect the Magellan images to reveal a multitude of previously unidentified volcanic edifices and unrecognized types of volcanic edifices, such as low shields, maar craters, and slotted vents (*Pike*, 1978). It is unlikely that detailed morphologies of lava flows will be seen by Magellan because their relief is small, but additional flowlike patterns (*Barsukov et al.*, 1986) will likely be discovered and the known ones will be better portrayed. For the wide variety of deformed terrains of Venus (*Basilevsky et al.*, 1986), new and intriguing details will be portrayed by the Magellan radar images. An appreciation for the improvement that the higher resolution Magellan images will provide for Venus can be gained by comparing Figs. 3 and 4.

Acknowledgments. This study was conducted under contracts between the National Aeronautics and Space Administration, the Jet Propulsion Laboratory, California Institute of Technology, and the U.S. Geological Survey (contract W-15-462). The authors express their gratitude to J. M. Boyce for suggesting this study and to W. T. K. Johnson for his discussions on the resolutions of the Venera and Magellan radar imaging systems. Pazzetta Jones helped us during the initial part of this study. We appreciate the thoughtful reviews of P. J. Mouginis-Mark, G. G. Schaber, R. J. Pike, and an anonymous reviewer.

REFERENCES

Barsukov V. L. et al. (1986) The geology and geomorphology of the Venus surface as revealed by the radar images obtained by Veneras 15 and 16. *J. Geophys. Res.*, *91*, 378–398.

Basilevsky A. T., Kuzmin R. O., Nikolaeva O. V., Pronin A. A., Ronca L. B., Avduevsky V. S., Uspensky G. R., Cheremukhina Z. P., Semenchenko V. V., and Ladygin V. M. (1985) The surface of Venus as revealed by the Venera landings: Part II. *Geol. Soc. Am. Bull.*, *96*, 137–144.

Basilevsky A. T., Pronin A. A., Ronca L. B., Kryuchkov V. P., Sukhanov A. L., and Markov M. S. (1986) Styles of tectonic deformations on Venus: Analysis of Venera 15 and 16 data. *Proc. Lunar Planet. Sci. Conf. 16th*, in *J. Geophys. Res.*, *91*, D399–D411.

Burns B. A. and Campbell D. B. (1985) Radar evidence for cratering of Venus. *J. Geophys. Res.*, *90*, 3037–3047.

Campbell D. B. and Burns B. A. (1980) Earth-based radar imagery of Venus. *J. Geophys. Res.*, *85*, 8271–8281.

Crater Analysis Techniques Working Group (1979) Standard techniques for presentation and analysis of crater size-frequency data. *Icarus*, *37*, 467–474.

Dial A. L. and Schaber G. G. (1981) Viking Orbiter images: their proper selection for the determination of crater-production rates on geologic units of diverse origins. *Third Internatl. Colloq. on Mars*, 62–64.

Evans J. V. and Pettengill G. H. (1963) The scattering of the Moon at 3.6, 68, and 784 centimeters. *J. Geophys. Res.*, *68*, 423–447.

Garvin J. B., Head J. W., Pettengill G. H., and Zisk S. H. (1985) Venus global radar reflectivity and correlations with elevation. *J. Geophys. Res.*, *90*, 6859–6871.

Hagfors T. (1967) A study of the depolarization of lunar radar echoes. *Radio Sci.*, *2*, 445–465.

Hagfors, T. (1970) Remote probing of the Moon by infrared and microwave emissions and by radar. *Radio Sci.*, *5*, 189–227.

Hartmann W. K. (1966) Early lunar cratering. *Icarus*, *5*, 406–418.

Ivanov B. A., Basilevsky A. T., Kryuchkov V. P., and Chernaya I. M. (1986) Impact craters of Venus: Analysis of Venera 15 and 16 data. *Proc. Lunar Planet. Sci. Conf. 16th*, in *J. Geophys. Res.*, *91*, D413–D430.

Johnson, W. T. K. and Edgerton A. T. (1985) Venus radar mapper (VRM): multimode radar system design. *J. Soc. Photo-Opt. Instrum. Eng.*, *589*, 158–164.

Jurgens R. F., Goldstein R. M., Rumsey H. R., and Green R. R. (1980) Images of Venus by three-station interferometry—1977 results. *J. Geophys. Res.*, *85*, 8282–8294.

Moore R. K. (1983) Radar fundamentals and scatterometers. *Am. Soc. Photogram. Man. Remote Sens.*, *1*, 369–427.

Moore H. J., Boyce J. W., Schaber G. G., and Scott D. H. (1980) Lunar remote sensing and measurements. *U.S. Geol. Surv. Prof. Pap. 1046B*, 78 pp.

Muhleman D. O. (1964) Radar scattering from the Moon and Venus. *Astron. J.*, *69*, 34–41.

Pettengill G. H. (1968) Radar studies of the planets. In *Radar Astronomy* (J. V. Evans and I Hagfors, eds.), Chap. 6, McGraw-Hill, New York.

Pettengill G. H., Zisk S. H., and Thompson, T. W. (1974) The mapping of lunar scattering characteristics. *The Moon*, *10*, 3–16.

Pettengill G. H., Ford P. G., Brown W. E., Kaula W. M., Masursky H., Eliason E., and McGill G. E. (1979) Venus: Preliminary topographic and surface-imaging results from Pioneer Orbiter. *Science*, *205*, 91–93.

Pettengill G. H., Eliason E., Ford P. G., Loroit G. B., Masursky H., and McGill G. E. (1980) Pioneer Venus radar results: altimetry and surface properties. *J. Geophys. Res., 85,* 8261–8270.

Pike R. J. (1978) Volcanoes in the inner planets: Some preliminary comparisons of gross topography. *Proc. Lunar Planet. Sci. Conf. 9th,* 3239–3273.

Pike R. J. (1980) Geometric interpretation of lunar craters. *U.S. Geol. Surv. Prof. Pap. 1046-C.* 77 pp.

Pike R. J. and Clow G. D. (1981) Revised classification of terrestrial volcanoes and catalog of topographic dimensions, with new results on edifice volume. *Open-File Rept. 81-1038.* U.S. Geological Survey. 40 pp.

Shoemaker E. M., Morris E. C., Batson R. M., Holt H. E., Larson K. B., Montgomery D. R., Rennilson J. J., and Whitaker E. A. (1969) 3. Television observations from Surveyor. In *Surveyor Program Results.* NASA SP-184, pp. 19–28.

Thompson T. W. (1965) *A study of radar scattering of lunar craters.* Ph.D. thesis, Cornell Univ., Ithaca, New York.

Thompson T. W. (1974) Atlas of lunar radar maps at 70 cm wavelength. *The Moon, 10,* 51–85.

Thompson T. W. (1978) High resolution lunar radar maps at 7.5 meter wavelength. *Icarus, 36,* 174–188.

Thompson T. W. (1979) A review of Earth-based mapping of the Moon. *The Moon and Planets, 20,* 179–198.

Thompson T. W. (1987) High-resolution lunar radar map at 70-cm wavelength, *Earth, Moon and Planets, 37,* 59–70.

Wilhelms D. E. (1980) Stratigraphy of part of the lunar near side. *U.S. Geol. Surv. Prof. Pap. 1046-A.* 71 pp.

Zisk S. H., Pettengill G. W., and Catuna G. W. (1974) High-resolution radar maps of the lunar surface at 3.8-cm wavelength. *The Moon, 10,* 17–50.

A Search for Water on the Moon at the Reiner Gamma Formation, A Possible Site of Cometary Coma Impact

Ted L. Roush* and Paul G. Lucey

Planetary Geosciences Division, Hawaii Institute of Geophysics, University of Hawaii, 2525 Correa Road, Honolulu, HI 96822

*Now at NASA Ames Research Center, MS 245-3, Moffett Field, CA 94035

Earth-based telescopic measurements of the Reiner Gamma Formation in the 3-μm region were made to search for evidence of water at this proposed site of cometary interaction with the lunar surface. Comparison of the spectra of Reiner Gamma and laboratory measurements of a variety of minerals with adsorbed water or structural OH$^-$ show that there is no evidence for water at the locations measured. While these measurements rule out the presence of abundant bound H_2O at these locations, they do not exclude the presence of OH$^-$-bearing minerals that have absorptions not detectable through the wet terrestrial atmosphere. The possible effects of high lunar temperatures on the reflectance spectra were examined and it was concluded that these effects would not obscure a bound water absorption if it were present. Upper limits on the amount of water that could be present at the locations observed and remain undetected are 2.25 average wt % for a mixture consisting of discrete patches of water-bearing and water-free minerals and 0.01 wt % for an intimate mixture of these two materials.

INTRODUCTION

The lunar surface is extremely dry. No unequivocal evidence for indigenous water has been found in the returned sample collection (*Taylor*, 1982). One sample ("rusty rock" 66095) was returned that contained the mineral akaganeite (β-FeO-OH), but the source water was concluded to be terrestrial in origin (*Taylor et al.*, 1973).

Because of the importance of indigenous chemical fuel in economically transporting lunar resources from the surface, the possibility of H_2O being present in special environments or at restricted locations on the lunar surface has received recent attention (*Tucker et al.*, 1985). Though cold trapping of water in permanently shadowed regions near the lunar pole has received the most attention (*Arnold*, 1979, 1987), the introduction of exogenous water at the impact sites of carbonaceous chondrites and comets has also been suggested (*Arnold*, 1979).

Bell (1984) and *Bell and Hawke* (1984, 1987) performed a remote sensing investigation in the 0.6- to 2.5-μm wavelength region to search for the presence of carbonaceous chondritic or cometary debris on the lunar surface. These workers found no evidence for the spectral signature of carbonaceous chondrite-like material at lunar craters with dark ejecta blankets. They concluded that origin by cometary impact was the only proposed mechanism that was consistent with the spectral reflectance of the Reiner Gamma Formation. The Reiner Gamma Formation is the only easily observable example of lunar swirl belts (*Schultz and Srnka*, 1980a; *Bell and Hawke*, 1981), which are characterized by the unusual geometries of their albedo patterns (*Hood*, 1981; *Schultz*, 1976) and their association with major lunar magnetic anomalies (*Hood et al.*, 1979a,b; *Schultz and Srnka*, 1980a,b).

One hypothesis concerning the origin of Reiner Gamma and other lunar swirls relies upon cometary impact (*Schultz and Srnka*, 1980a,b; *Srnka and Schultz*, 1980; *Schultz et al.*, 1980). According to this model, low-density gas or dust streamers in the inner coma struck the lunar surface at high velocity and produced changes in the upper regolith. In this model the swirls are relatively young and their magnetism was induced by the compressed cometary field during impact (*Schultz and Srnka*, 1980a,b; *Srnka and Schultz*, 1980; *Schultz et al.*, 1980). The absence of an identifiable impact crater associated with the Reiner Gamma Formation excludes the possibility that the cometary nucleus impacted in the immediate vicinity of the albedo feature.

Comets are known to contain large amounts of H_2O (*Whipple and Huebner*, 1976; *Delsemme and Rud*, 1977), and *Arnold* (1979) discusses the possible occurrence of cometary impacts as a source of H_2O on the lunar surface. The specific nature of chemical processes occurring during the impact of an icy body into a silicate target is poorly constrained at present. It is undetermined whether the kinetics of reactions that produce minerals with structural OH$^-$ or bound water are appropriate to form such minerals from ice and dry igneous rock during an ice-silicate impact event. If such minerals are produced during such an impact event, then because of the extreme depletion of the lunar surface in native water and the strong spectral signature of water and hydroxyl in certain wavelength regions, the presence of the absorptions of OH$^-$ or H_2O may provide a unique opportunity to identify lunar cometary impact sites using spectral reflectance techniques. The absence of water- or OH$^-$-bearing minerals at Reiner Gamma would not disprove the hypothesis concerning the origin of Reiner Gamma since the formation of such materials is suggested as a possibility, not a requirement, during the impacting event.

Bell (1984) and *Bell and Hawke* (1987) did not identify any absorptions due to OH$^-$ or H_2O in the 0.7- to 2.5-μm spectra of the Reiner Gamma Formation. However, absorptions due to these molecules in that wavelength region are overtone and/or combination modes (*Hunt*, 1977) that are less intense than the fundamental modes, easily masked by opaques (*Clark*, 1983), and eliminated by dehydration before the fundamental modes (*Bruckenthal*, 1987). Recent laboratory investigations of the spectral signatures of OH$^-$ and H_2O-bearing silicates (*Roush and Singer*, 1985; *Roush et al.*, 1986, 1987; *Bruckenthal*, 1987; *Bruckenthal and Singer*, 1986, 1987) indicate that the 2.4- to 3.5-μm wavelength region is especially well suited for identification of OH⁻- and/or H₂O-bearing minerals. The hydroxyl group (OH⁻) and H₂O molecule have fundamental vibrational modes which result in absorptions that occur in this wavelength region (*Hunt*, 1977). Because of the intense nature of these fundamental absorptions, the presence of minor amounts of these groups or molecules is readily apparent in the spectra of minerals that contain them.

Because of this potential for identifying the presence of water on the lunar surface, a telescopic observing program was initiated to measure portions of the Reiner Gamma formation in the 2.5- to 3.5-μm wavelength region, wavelengths not covered by *Bell* (1984) and *Bell and Hawke* (1987). Spectra were also collected of a mare region several hundred kilometers away to provide a standard that we assume is completely free of water or hydroxyl. We note that this method will not detect absorptions if they are present in both the standard and Reiner Gamma locations. Thus our investigation does not address the possibility of OH⁻ being produced on a global scale by the interaction of solar wind protons and lunar minerals.

OBSERVATIONS

Telescopic Observations

Telescopic observations were made at Mauna Kea Observatory using the University of Hawaii 2.2 m telescope. The spectrometer consisted of a circular variable filter, with 1.5% spectral resolution ($\Delta\lambda/\lambda$) and a liquid nitrogen cooled InSb detector. This same instrument was also used in the laboratory to obtain reference sample spectra. Data were collected shortly after lunar dawn at Reiner Gamma in order to reduce thermal background on 18 August, 1986, UT. Individual spots are approximately 26 km in diameter. A region in Oceanus Procellarum approximately on the same isophote as Reiner Gamma (to minimize thermal differences) is used as the standard to which the Reiner Gamma spectra are compared. Figure 1 shows the relative location of Reiner Gamma and the standard spot on the lunar surface. We believe this local standard is representative of a typical lunar mare region and therefore does not contain any OH⁻- or H₂O-bearing materials, hence the ratio of Reiner Gamma spectra to this standard should show residual absorptions if water- or hydroxyl-bearing minerals are present at Reiner Gamma.

The Reiner Gamma Formation is shown in Fig. 2 along with the spot locations observed at Reiner Gamma. The vidicon UV-VIS ratio images presented by *Bell and Hawke* (1981) show a distinct UV dark halo that almost completely surrounds the main albedo feature. In the context of the cometary impact hypothesis, we interpreted this halo as possibly being due to oxidation of some of the Fe^{2+} in the lunar surface to Fe^{3+} by free oxygen produced during the proposed cometary impact. If this is the case, we believe that the darkest portions of this UV anomaly had the greatest likelihood of retaining other evidence of the cometary impact event, i.e., OH⁻- and/or H₂O-bearing minerals. Based on this line of reasoning, the spots observed at Reiner Gamma were chosen specifically to sample this UV halo.

Fig. 1. Northwest quadrant of the moon showing the location of Reiner Gamma (RG) and the standard mare spot (Std).

The reflectance spectra of various Reiner Gamma spots, relative to the lunar mare spot, are shown in Fig. 3. Each spectrum plotted represents the average of 15 individual spectra of the Reiner Gamma spot. We have deleted the data in the 2.53- to 2.82-μm interval because they are derived from measurements with essentially zero signal due to the extreme opacity of the terrestrial atmosphere. This eliminates the potential detection of some OH⁻-bearing species on the lunar surface based on this data set. (An instrument capable of measuring reflectance in this wavelength region and located outside the terrestrial atmosphere would be capable of detecting minerals with sharp OH⁻ absorption features near 2.75 μm). The data in the 3-μm region are of sufficient quality to address the presence of H₂O-bearing species on the lunar surface.

Laboratory Measurements

The Planetary Geosciences spectrogoniometer facility (*Singer*, 1981) was used to collect the laboratory data for comparison with the telescopic measurements. The instrument used at the telescope was also used to collect the laboratory data. Two terrestrial clay minerals (pyrophyllite and montmorillonite) and a basalt were included in the laboratory study for comparison to the telescopic measurements. It is not our intent to imply that clays are formed on the lunar surface by the impact of cometary ice, but rather that these samples conveniently illustrate the spectral properties of

Fig. 2. Lunar Orbiter photograph of the Reiner Gamma Formation. Note the albedo anomaly associated with Reiner Gamma. The spots observed for this study, and their associated nomenclature, are indicated by the ellipses. These locations fall within the ultraviolet halo (*Bell*, 1984) as discussed in the text.

minerals with bound water or hydroxyl. All samples were dry sieved to a 38-45-μm grain size fraction. This grain size was not chosen to simulate the lunar regolith but is adequate for the purposes intended here.

Since the lunar surface environment is extremely dessicated, the clay samples were heated in order to dehydrate the samples and better simulate conditions of the lunar surface. This was done by placing all samples in an environment chamber that was flushed with dry nitrogen gas during data collection. Each sample was equilibrated in the nitrogen environment for at least thirty minutes at room temperature, then the sample was heated in the nitrogen environment at 200°C for thirty minutes and allowed to cool to room temperature before spectral measurements were collected.

The pyrophyllite spectrum shown in Fig. 4 is dominated by the OH^- fundamental absorption band located near 2.75 μm. This is due to the crystal structure of pyrophyllite, which consists of tetrahedral (T) sheets of SiO_4 tetrahedra bonded to both sides of an octahedral (O) layer containing Al and OH^- ions. Since there is little cationic substitution, this T-O-T crystal structure prohibits the presence of much interlayer and/ or adsorbed H_2O and the spectrum is dominated by absorptions due to OH^-. The montmorillonite spectrum shown in Fig. 4 is dominated by H_2O absorptions in the 3-μm region. This is a result of the crystal structure of montmorillonite which consists of a basic T-O-T structure, like the pyrophyllite, but because there is relatively abundant cationic substitution in the montmorillonite, these T-O-T structural sandwiches are linked together by interlayer cations. It is this interlayer location

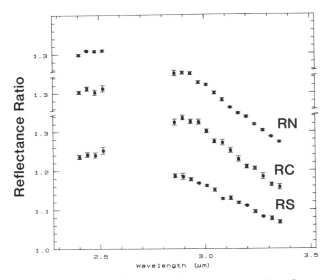

Fig. 3. Ratio of the reflectance of Reiner Gamma to the reflectance of the lunar mare standard. (top) Reiner Gamma North (RN), (middle) Reiner Gamma Center (RC), (bottom) Reiner Gamma South (RS).

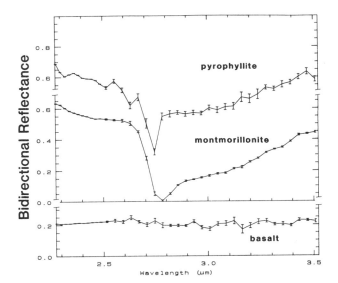

Fig. 4. Reflectance spectra of samples studied in the laboratory. (top) the reflectance spectrum of pyrophyllite after heating to 200°C in a nitrogen atmosphere illustrates the strong relatively narrow absorption near 2.75 μm due solely to OH^-; (middle) the reflectance spectrum of Montmorillonite after heating to 200°C in a nitrogen atmosphere illustrates the strong broad water absorption in the 3-μm wavelength region; (bottom) the reflectance spectrum of basalt illustrates the absence of OH^- and/or H_2O absorptions.

that can readily adsorb water, and as a result the spectrum is dominated by absorptions due to H_2O. In contrast to the clays, the reflectance spectrum of a basalt, also shown in Fig. 4, is relatively flat throughout the 3-μm region, indicating the absence of OH^- and/or H_2O-bearing materials.

DISCUSSION

The Reiner Gamma spectra in Fig. 3 do not exhibit distinct absorptions and instead are characterized by a slight negative slope. For comparison with the laboratory measurements we simulated the telescopic data as shown in Fig. 3 by dividing the clay spectra by the basalt spectrum. The result is shown in Fig. 5, where the obvious OH^- and H_2O absorptions are still readily apparent in the ratio spectra. The lack of any detectable residual absorptions in the telescopic data implies that if H_2O-bearing species are present at Reiner Gamma, then their abundance is too low for detection by our technique. However, there is a possibility that the thermal flux from the lunar surface masks absorptions due to H_2O.

Effects of Temperature on Relative Flux Spectra

Because of high lunar temperatures, it is possible that apparent absorptions can be suppressed by the infilling of such features by the emitted thermal radiation. Although the standard spot and Reiner Gamma are nearly on the same isophote, and these observations were obtained at the lunar dawn to minimize this effect, there is a small difference in the angle of the incoming radiation and the relative albedos of the two areas. Therefore, thermal flux differences are not compensated for in our ratios. It is possible that a given combination of surface temperatures and albedos could result in the total or partial suppression of observable absorption bands.

We assessed the extent of this effect by modeling the measured flux from the lunar surface and assuming that the basalt and clays studied in the laboratory were present at the

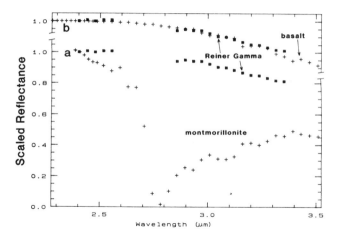

Fig. 6. Calculated reflectance of samples including modeled thermal component located at Reiner Gamma relative to calculated reflectance of basalt at standard location, **(a)** Montmorillonite at Reiner Gamma (+) compared to measured RC (squares), and **(b)** Basalt at Reiner Gamma (+) compared to measured RC (squares). All modeled and measured data has been scaled to 1.0 at 2.40 μm.

locations observed. The telescopic measurements can be expressed as

$$\text{Relative Flux} = \frac{[\pi P(T)(\epsilon) + RF] \text{ at Reiner Gamma}}{[\pi P(T)(\epsilon) + RF] \text{ at Mare Standard}} \quad (1)$$

where $P(T)$ is the Planck blackbody function for a given surface temperature, ϵ is the emissivity of the surface ($\epsilon = 1 - R$), R is the reflectance of the surface, and F is the solar flux incident on the surface. F is derived from data presented in the CRC (*Weast*, 1981), which was convolved to the wavelengths used in the current study. We used the lunar thermal infrared maps of *Saari and Shorthill* (1967) to determine the surface temperatures of Reiner Gamma (~314.4K) and our standard mare spot (326.2-338.0K) for the lunar phase angle (-15°) at which the data were collected. We used the laboratory measured reflectance of OH^- and H_2O-bearing materials and basalt to calculate their emissivity as a function of wavelength. We then calculated a worst case scenario, i.e., the largest reasonable temperature difference between Reiner Gamma and the mare standard, and assumed the OH^- and H_2O materials were located at Reiner Gamma and the basalt located at the mare spot. The results of these calculations are shown in Fig. 6. It is clear that the OH^- and/or H_2O absorption bands would still be identifiable in the telescopic data if these materials are located at Reiner Gamma. The spectrum at the top of Fig. 6 assumes basalt is located both at Reiner Gamma and the mare spot and is most consistent with the telescopic data.

Detection Limit on Abundance of H_2O

In an attempt to place limits on the amount of H_2O that could be present at Reiner Gamma and remain undetected

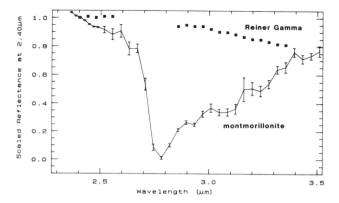

Fig. 5. Ratio of laboratory montmorillonite reflectance to laboratory basalt reflectance (line with error bars) compared to typical Reiner Gamma reflectance relative to lunar mare spectrum (data points). Both data sets have been scaled to 1.0 at 2.40 μm.

by the spectroscopic measurements, we have considered two types of mixing that can occur on a planetary surface. The first is the mixing model of *Singer and McCord* (1979), which assumes the total reflectance of a surface is a linear combination of the relative areal coverage of each surface component multiplied by the individual component's reflectance. In this model the materials occur as discrete patches of centimeter size, or larger, but at a spatial scale finer than the spatial resolution of the telescopic measurements. The second model assumes the homogeneous intimate mixing of materials and we rely on a theoretical study presented by *Pollack et al.* (1978), which relates the amount of water present in a sample to the reflectance of the mixture.

In the areal mixture model we used the montmorillonite as one end-member and a grey reflector, with a reflectance of 0.2 as the other end-member. We chose the 0.2 grey reflector to minimize noise because although the measured reflectance of the basalt was relatively featureless throughout the 2.4- to 3.5-μm wavelength region, it is noisy. The telescopic spectra have an average error of about 2%; therefore, to be conservative, we require that an absorption of the 3-μm region must have a band depth of >4% before it is detected. At 10% areal coverage montmorillonite, the 3-μm band depth barely meets our criterion, but by 15% areal coverage montmorillonite the 3-μm band depth exceeds 4%. This puts an upper limit of 1.5 to 2.25 average wt % H_2O based on the areal mixture case.

In order to investigate the intimate mixture case we rely on a study of *Pollack et al.* (1978) that assumed a homogeneous surface composed of particles with a radius of 35 μm and then computed the reflectance of the surface that contained varying mass fractional amounts of bound H_2O. That work shows that 0.01 wt % bound H_2O yields about a 6% 3-μm band depth. Thus for the intimate mixture case the upper limit of bound H_2O is less than 0.01 wt %.

CONCLUSIONS

Comparison of telescopic data of Reiner Gamma to laboratory spectra constrains the abundance of H_2O proper at the locations measured at Reiner Gamma. If H_2O is present at these locations, then it does not compose more than 2.25 wt %, assuming an areal mixture, or 0.01 wt %, assuming an intimate mixture, of the optically active surface sampled by the telescopic measurements. Several alternatives present themselves to explain the nondetection of H_2O-bearing materials if the Reiner Gamma formation is the site of the collision of a cometary coma: (1) water-bearing minerals were not formed, (2) water-bearing minerals were formed at Reiner Gamma but were not durable enough to survive millions or billions of years of lunar vacuum and temperature conditions, (3) hydroxyl-bearing minerals were formed but were not detectable because of atmospheric opacity, and (4) water- or hydroxyl-bearing minerals are present at locations in the Reiner Gamma Formation other than those measured.

The data presented here cannot eliminate the possibility of the presence of OH^--bearing minerals. Because of the vital importance of hydrogen to lunar resource exploitation, measurements at 2.75 μm from lunar or Earth orbit should have a high priority in coming years. Measurements from Earth using highly efficient infrared array spectrometers of Reiner Gamma and other locations on the Moon near the poles or the east limb swirl belts during favorable librations will be more sensitive to the presence of H_2O or OH^- than the measurements presented here.

Acknowledgments. This research was supported by NASA Planetary Astronomy and Planetary Geology and Geophysics grants NSG 7312 and NSG 7590. We would like to thank Michael Gaffey for useful discussions and suggestions during the inception of this project. We also thank C. Coombs and J. Bell for reviewing an early version of the manuscript and P. Schultz and B. R. Hawke for outside reviews. This is PGD publication number 505.

REFERENCES

Arnold J. R. (1979) Ice in the lunar polar regions. *J. Geophys. Res., 84,* 5659-5668.

Arnold J. R. (1987) Ice in the lunar poles revisited (abstract). In *Lunar and Planetary Science XVIII,* pp. 29-30, Lunar and Planetary Institute, Houston.

Bell J. F. (1984) A search for ultraprimative material in the solar system, *Ph.D. Dissertation,* Univ. Hawaii, Honolulu. 151 pp.

Bell J. F. and Hawke B. R. (1981) The reiner gamma formation: Composition and origin as derived from remote-sensing studies. *Proc. Lunar Planet. Sci. 12B,* pp. 679-694.

Bell J. F. and Hawke B. R. (1984) Lunar dark-haloed craters: Origin and implications for early mare volcanism. *J. Geophys. Res., 89,* 6899-6910.

Bell J. F. and Hawke B. R. (1987) Recent comet impacts on the Moon: The evidence from remote sensing studies. *Publ. Ast. Soc. Pacific,* in press.

Bruckenthal E. A. (1987) Spectral Effects of Dehydration on Phyllosilicates. *Master of Science Thesis,* Univ. Hawaii, Honolulu. 182 pp.

Bruckenthal E. A. and Singer R. B. (1986) Spectral effects of dehydration upon phyllosilicates. *Eos Trans. AGU, 67,* 1271.

Bruckenthal E. A. and Singer R. B. (1987) Spectral effects of dehydration on phyllosilicates (abstract). In *Lunar and Planetary Science XVIII,* pp. 135-136. Lunar and Planetary Institute, Houston.

Clark R. N. (1983) Spectral properties of mixtures of montmorillonite and dark carbon grains: Implications for remote sensing minerals containing chemically and physically adsorbed water. *J. Geophys. Res., 88,* 10,635-10,644.

Delsemme A. H. and Rud D. (1977) Low-temperature condensates in comets. In *Comets, Asteroids, Meteorites* (A. H. Delsemme, ed.) pp. 529-535. Univ. Toledo, Toledo.

Hood L. L. (1981) Sources of lunar magnetic anomalies and their bulk directions of magnetization: Further evidence from Apollo orbital data. *Proc. Lunar Planet. Sci. 12B,* pp. 817-830.

Hood L. L., Coleman P. J., and Wilhelms D. E. (1979a) The Moon: Sources of the crustal magnetic anomalies. *Science, 204,* 53-57.

Hood L. L., Coleman P. J., and Wilhelms D. E. (1979b) Lunar nearside magnetic anomalies. *Proc. Lunar Planet. Sci. Conf. 10th,* pp. 2235-2257.

Hunt G. R. (1977) Spectral signatures of particulate minerals in the visible and near-IR. *Geophysics, 42,* 501-513.

Pollack J. B., Witteborn F. C., Erickson E. F., Strecker D. W., Baldwin B. J., and Reynolds R. T. (1978) Near-infrared spectra of the Galilean satellites: Observations and compositional implications. *Icarus, 36,* 271-303.

Roush T. L. and Singer R. B. (1985) The spectral reflectance of mafic and phyllosilicates from 2.5-5 µm. *Bull. Am. Astron. Soc., 17*, 703.

Roush T. L., Singer R. B., and McCord T. B. (1986) The spectral refelctance of mafic and phyllosilicates from .6-4.6 µm. *Eos Trans. AGU, 44*, 1271.

Roush T. L., Singer R. B., and McCord T. B. (1987) Reflectance spectra of selected phyllosilicates from 0.6 to 4.6 µm (abstract). In *Lunar and Planetary Science XVIII*, pp. 858-859. Lunar and Planetary Institute, Houston.

Saari J. M. and Shorthill R. W. (1967) Isothermal and isophotic atlas of the Moon: Contours through a lunation. *NASA Contractor Report*, NASA CR-855, Washington, D. C. 185 pp.

Schultz P. H. (1976) *Moon Morphology*. Univ. of Texas Press, Austin. 422 pp.

Schultz P. H. and Srnka L. J. (1980a) Cometary collisions on the Moon and Mercury. *Nature, 284*, 22-26.

Schultz P. H. and Srnka L. J. (1980b) Reply to comments concerning cometary collisions on the Moon and Mercury. *Nature, 287*, 86-87.

Schultz P. H., Srnka L. J., Pai S. I., and Menon S. (1980) Cometary collisions on the Moon and Mercury (abstract). In *Lunar and Planetary Science XI*, pp. 1009-1011. Lunar and Planetary Institute, Houston.

Singer R. B. (1981) Near-infrared spectral reflectance of mineral mixtures: Systematic combinations of pyroxenes, olivine, and iron oxides. *J. Geophys. Res., 86*, 7967-7982.

Singer R. B. and McCord T. B. (1979) Mars: Large scale mixing of bright and dark surface materials and implications for analysis of spectral reflectance. *Proc. Lunar Sci. Conf. 10th*, pp. 1835-1848.

Srnka L. J. and Schultz P. H. (1980) A cometary origin of Reiner-γ magnetic anomalies (abstract). In *Lunar and Planetary Science XI*, pp. 1076-1078. Lunar and Planetary Institute, Houston.

Taylor L. A., Mao H. K., and Bell P. M. (1973) "Rust" in the Apollo 16 rocks. *Proc. Lunar Sci. Conf. 4th*, pp. 1835-1848.

Taylor S. R. (1982) *Planetary Science: A Lunar Perspective*. Lunar and Planetary Institute, Houston. 481 pp.

Tucker D. S., Vaniman D. T., Anderson J. L., Clinard F. W., Jr., Feber R. C., Jr., Frost H. M., Meek T. T., and Wallace T. C. (1985) Hydrogen recovery from extraterrestrial materials using microwave energy. In *Lunar Bases and Space Activities of the 21st Century* (W. W. Mendell, ed.), pp. 583-590. Lunar and Planetary Institute, Houston.

Weast R. C. (1981) *CRC Handbook of Chemistry and Physics*. CRC Press Inc., Boca Raton. 2454 pp.

Whipple F. L. and Huebner W. F. (1976) Physical processes in comets. *Ann. Rev. Astron. Astrophys., 14*, 143-172.

Water Content of Tektites and Impact Glasses and Related Chemical Studies

Christian Koeberl
Institute of Geochemistry, University of Vienna, A-1010 Vienna, Austria

Anton Beran
Institute of Mineralogy and Crystallography, University of Vienna, A-1010 Vienna, Austria

---

To study the difference in water content between genetically different types of tektites and impact glasses, we have analysed six Muong Nong type indochinites and five samples from the Zhamanshin impact crater, using the infrared absorption spectrometry method. Muong Nong type tektites are different from normal tektites in several respects, including higher volatile trace element abundances, which is further confirmed by our analyses. We find that Australasian Muong Nong tektites are also enriched in H_2O by a factor of about 1.5 relative to splash-form tektites from the Australasian strewn field. The Zhamanshin glasses show similar results: irghizites, known to be the most homogeneous and to have the lowest volatile element abundances among Zhamanshin impact glasses, also have the lowest water content, but are distinctly higher than even Muong Nong type tektites. Removal of water from sediments (the alleged precursor rocks of tektites and impact glasses) to form the dry tektites has been cited as a problem for the impact model, but published analyses for atomic bomb glass give a water content similar to tektites. Thus, dry glass can be formed in a single high temperature event. We also show that the water content in impact glasses and the intensity of shock melting experienced by the glass are inversely correlated, in accordance with the volatile trace element contents. Lunar rocks are completely devoid of water, thus the measured water content of tektites further supports a terrestrial impact origin.

INTRODUCTION

Impact glasses and tektites are rather water-poor natural glasses: approximate ranges of water contents in tektites are 0.002-0.02 wt %, and 0.02-0.06 wt % H_2O for impact glasses (*Friedman*, 1963; *Gilchrist et al.*, 1969; *v. Engelhardt et al.*, 1987). Most other terrestrial natural glasses, such as obsidian, contain a lot more water than impact glasses, around 0.1-1 wt % H_2O (*Gilchrist et al.*, 1969; *v. Engelhardt et al.*, 1987). Most terrestrial sediments, generally cited as the precursor rocks for tektite and impact glass formation (e.g., *Taylor*, 1973; *Koeberl*, 1986b), contain up to a few percent H_2O. Different sediments from the Ries crater area, for example, which may have been among the precursor rocks for moldavites, show a water content ranging from about 0.2-12 wt % water (*Luft*, 1983). Since it is theoretically difficult to understand how water is removed from the glass in the formation process (*O'Keefe*, 1964), the dryness of tektites has been cited as an argument against a terrestrial origin for tektites (e.g., *O'Keefe*, 1976). The analyses of lunar rocks, however, have revealed that water is practically nonexistent on the Moon, with abundance levels several orders of magnitude below those of tektites. No indigenous lunar water has so far been identified, and all water found in lunar samples at all has been attributed to terrestrial contamination or from reaction with hydrogen from the solar wind (*Epstein and Taylor*, 1974; *Taylor*, 1975, p. 228ff).

The early water analyses of tektites, performed with a variety of different techniques, yielded results that were not always consistent with each other or modern data. Chemical methods (*Clarke and Carron*, 1961) yielded a H_2O content of 0.02 wt % for a Georgia tektite, while *King* (1964) cites an even higher water content for another Georgia tektite, which he himself considered to be doubtful. This was in contrast to the results from the manometric method of *Friedman* (1958), who heated cleaned pieces of tektites and impact glasses (and other natural and artificial glasses) in Pt vessels under vacuum conditions to temperatures of up to 1450°C, and then measured the released H_2O. With a few exceptions, he has measured an average water content of 0.004%, with a spread of 0.0003-0.01%. This would mean that tektites have an average water content of much below 100 ppm, and before Apollo this may have appeared consistent with lunar rocks.

However, the application of a nondestructive method for water analysis, namely, infrared spectrometry, to tektites and impact glasses, yielded somewhat different results. *Gilchrist et al.* (1969), in a careful investigation of tektites and similar materials using the infrared technique, showed that the water content of tektites is in fact higher and clusters around 0.01 wt %. Obviously, the manometric method failed to release all the water bound to the silicates. This water content is still very low, but compared to lunar samples it is higher by many orders of magnitudes. Thus the results of *Gilchrist et al.* (1969) are consistent with a terrestrial origin of tektites. It would seem implausible to find a source for the water enrichment if they are of lunar origin.

The study of the water content of tektites and impact glasses has some very interesting implications for the impact theory. In accordance with conclusions drawn from other chemical studies, the water content of impact materials should vary between different groups. Muong Nong tektites are known to contain higher abundances of volatile elements such as As, Sb, the halogens, Zn, Cu, B, and others (*Chapman and Scheiber*, 1969; *Müller and Gentner*, 1973; *Koeberl et al.*, 1984a,b). Also, they contain a much larger amount of bubbles per volume than normal splash-form tektites. Their Fe(III)/Fe(II) ratios are higher than in normal tektites (*Koeberl et al.*, 1984c).

They are inhomogeneous, and in contrast to splash form tektites, mineral inclusions have been found (*Glass and Barlow*, 1979). All this has been interpreted as indicative of a lower peak pressure and temperature during the formation of the Muong Nong type tektites (*Koeberl*, 1986a). A higher water content in Muong Nong type tektites would be in line with those arguments.

Another interesting question associated with the Muong Nong type indochinites concerns the colored layers. Typically, light layers contain higher abundances of almost all elements except Si, while dark layers are depleted in a complimentary way (*Koeberl*, 1985). This is surprising, because even elements usually thought to be associated with color changes, such as iron, are enriched in the light and not in the dark layers. A difference in water content between the colored layers would be an alternative explanation for the color variation, maybe due to oxidation state changes.

Zhamanshinites, which are impact glasses from the Zhamanshin impact crater (Khazakhstan, Aral Region, USSR), are similar to Muong Nong type indochinites. They are also rather inhomogeneous and have a similar appearance, structure (abundant bubbles), and chemistry. Furthermore, in the case of the Zhamanshin Crater we do have a sequence of different impact materials, including several different impact glass varieties. Irghizites are among the most homogeneous materials and have probably experienced the highest formation temperatures and pressures. Si-rich zhamanshinite and blue zhamanshinites (*Koeberl et al.*, 1986) have experienced less shock melting, and have thus remained less homogeneous, and contain more volatiles. The blue zhamanshinites have a chemistry that puts them (in some respects) between the irghizites and Si-rich zhamanshinites (*Koeberl et al.*, 1986). The least impact-metamorphosed glasses are the Si-poor zhamanshinites, which are the least homogeneous glasses found at the crater. If the current understanding of the formation of impact glasses is correct, then one should also expect a related behaviour of the water content.

To test the aforementioned ideas and look for possible correlations with other chemical features, we have analysed six Muong Nong type indochinites from Ubon Ratchathani, Thailand (samples numbers MN 8302, MN 8304, MN 8308, MN 8310, MN 8317, and MN 8319; for additional information on these samples, see *Koeberl et al.*, 1984a-d), one irghizite (USNM 6200; *Koeberl and Fredriksson*, 1986), one Si-poor zhamanshinite (Zh 57/4b), and three blue zhamanshinites (Zh 31/6b, BZ 8601, BZ 8602; *Koeberl et al.*, 1986) for water and a number of elements that may be of interest in comparison. Some of the samples have already been analysed in more detail for other elements, especially the Muong Nong type indochinites (*Koeberl et al.*, 1984a-d).

ANALYTICAL METHODS

The water content of the samples was determined using infrared spectrometry, similar to the method used by *Gilchrist et al.* (1969). The infrared absorption spectrum of Si-rich glass contains peaks related to the O-H stretching vibration and H-O-H bending vibration modes, at 2.73 μm and 6.2 μm. Usually in glasses similar to tektites only the 2.73 μm peak is found, indicating the predominance of the O-H stretching vibration due to voids in the network structure of silica. The addition of alkali metal causes some modification in the network, leading to hydrogen-bonding of the hydroxyl ions (caused by changes in electronegativity) and to the displacement of the 2.73 μm peak towards higher wavelengths. The measurement of that peak yields information pertaining to the water content.

Tektite and impact glass samples were prepared as thick sections (thickness about 1 mm). The instrument used was a computer controlled Perkin-Elmer PE 580 B, with 8× beam condenser. Accumulated scans were taken to improve the signal/noise ratio. The calculations of the water content followed a standard Lambert-Beer law method, similar to that of *Gilchrist et al.* (1969). We have checked the extinction coefficient given by *Gilchrist et al.* (1969) and found, in agreement with their data, that 75 ± 21 mol^{-1}cm^{-1} is the best value. This is very close to the numbers reported for other silica glasses (*Gilchrist et al.*, 1969). Small variations, depending on the exact structure of the glass, are possible but will not exceed the uncertainty given above. The calculations were performed using an average tektite and impact glass density of 2.40 g cm^{-3}, which is the average of the densities of these glasses (see also *Gilchrist et al.*, 1969; and *v.Engelhardt et al.*, 1987), with deviations of less than a few percent. The density does not enter the calculations in a direct way, and thus small variations in the density affect the end result very little.

Other elements reported here have been determined using a variety of techniques. Lithium, Be, Cu, and Zn have been determined using atomic absorption techniques (see *Koeberl et al.*, 1984b,d for details); B has been determined using a tetrafluoroborate selective electrode (*Kluger and Koeberl*, 1985); F was analysed by F-selective electrode technique (*Koeberl et al.*, 1984a); and As and Br by neutron activation analysis. The data on the Muong Nong type indochinites are part of a larger dataset, including 19 samples at present.

RESULTS

The absorption spectra in the region near 2.8 μm are given in Fig. 1a and Fig. 1b for one of the Muong Nong tektite samples (MN 8302) and one of the blue zhamanshinites (BZ 8601). The transmission characteristics of the two impact glasses are different, as is the location of the peak maximum and the shape of the curve. Table 1 gives the peak location (in wavenumbers, cm^{-1}), and the linewidth at half maximum (also in cm^{-1}) for all the samples. From the data given there it is clear that none of the samples has its peak maximum at the ideal value of 2.73 μm (corresponding to 3660 cm^{-1}) but all are shifted towards higher wavelengths (or smaller wavenumbers). The extent of the shift is different for the different tektite and impact glass groups, giving a peak maximum range of 2.77-2.82 μm. This means that the O-H stretching vibration is affected by other crystal field effects, leading to displacement of the peak maximum. One suggestion, which was also cited by *Gilchrist et al.* (1969), for the explanation of the displacement of the peak maximum in

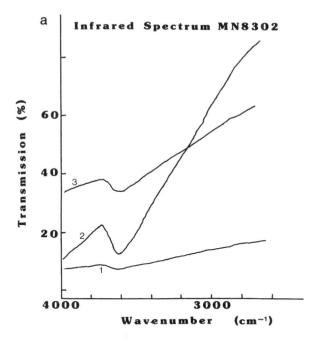

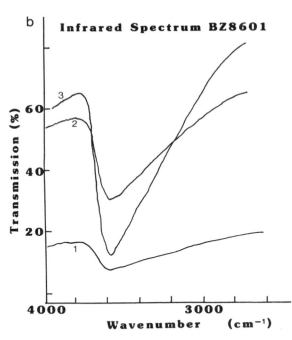

Fig. 1. Infrared absorption spectra of (a) a Muong Nong type indochinite and (b) a blue zhamanshinite. The sample numbers are given in the figure. The three curves from bottom to top are (as they start from the left ordinate): a single scan spectrum (1), an accumulated scan (5 scans) in a stretched version (2), and the original accumulated scan (5 scans) (3). The accumulated scans clarify the shape and height of the peak. The transmission scale is valid in a strict sense only for the original (single scan) spectra. The difference in shape and peak form and location between the two glass varieties, as evident from the figures, is explained in the text.

TABLE 1. Wavenumbers and linewidth for the O-H stretching vibration peak, which was used for H_2O determination.

Sample	Wavenumber (cm^{-1})	Linewidth (cm^{-1}) (half maximum)
MN8302	3610	210
MN8304	3600	220
MN8308	3590	280
MN8310	3610	220
MN8317	3600	230
MN8319	3590	220
USNM6200	3590	310
Zh 57/4b	3550	450
Zh 31/4b	3580	460
BZ8601	3575	410
BZ8602	3580	410

TABLE 2. Water content of six Muong Nong type indochinites and five Zhamanshin impact glasses.

Sample	Water Content (Wt %)
Muong Nong tektites	
MN8302	0.015
MN8304	0.013
MN8308	0.017
MN8310	0.017
MN8317	0.009
MN8319	0.011
Glasses from the Zhamanshin Crater	
USNM6200 (irghizite)	0.026
Zh 57/4b (Si-poor zhamanshinite)	0.034
Zh 31/6b (blue zhamanshinite)	0.063
BZ8601 (blue zhamanshinite)	0.050
BZ8602 (blue zhamanshinite)	0.050

tektites versus pure silicate glasses, involves the addition of alkalis.

The Muong Nong tektites show the usual displacement to about 3600 cm^{-1}, similar to the results given by *Gilchrist et al.* (1969), but the Zhamanshin glasses show greater shifts and also larger linewidths. This cannot be due to the alkali effect alone, and points to a different structural environment for the O-H bonding between the tektite and the impact glass. In addition, slightly different absorption characteristics may add to the effect. The exact cause for the difference between the two groups is not yet known. Unfortunately, *Gilchrist et al.* (1969) do not give wavenumbers and linewidths for the impact glasses and tektites they measured, so we cannot conclude that there is a systematic difference or variation between the O-H bonding in impact glasses and tektites. Within one group, however, the alkali effect seems to be present. Irghizites have lower total alkali abundances than Si-rich zhamanshinites (or blue zhamanshinites), while Si-poor zhamanshinites usually have the highest total alkali abundances (especially Na). Table 1 shows that in fact the irghizite shows the smallest displacement (to 3590 cm^{-1}), while the Si-poor zhamanshinite shows the largest displacement (to 3550 cm^{-1}).

Table 2 gives the results of the water determinations in all 11 samples. Since the extinction coefficient does not vary between different samples by more than a few percent, the analytical precision and the accuracy of the measurement are both within ±5% deviation. The average water content for Muong Nong indochinites, calculated from our six specimens, is 0.014 ± 0.003 wt % H_2O.

The Zhamanshin samples show a higher water content than the Muong Nong type tektites and are more variable. There is a considerable difference between the three different impact glass varieties measured, with the irghizite showing the lowest water content.

DISCUSSION

Gilchrist et al. (1969) have analysed two Muong Nong type tektites, one with 0.017 wt % H_2O, and another one with 0.008 wt %. From that they concluded that Muong Nong tektites have approximately the same water content as other tektites. The average water content of their indochinites (three samples) is 0.008 ± 0.003 wt % H_2O, and of all the Australasian tektites they measured (12 samples, excluding one anomalous philippinite) is 0.011 ± 0.005 wt % H_2O. Tektites from the North American strewn field, as measured by *Gilchrist et al.* (1969), seem to have a slightly higher water content than the average Australasian tektites, and thus change the overall average quoted by *Gilchrist et al.* (1969).

If we compare the water content in Muong Nong type indochinites, as given in the previous section, with the average water content of Australasian tektites (*Gilchrist et al.,* 1969), we see that Muong Nong type tektites show an enrichment compared with splash-form tektites. The comparison with the average Australasian tektite datum gives an enrichment factor of about 1.3, while a comparison with the splash-form indochinites (*Gilchrist et al.,* 1969), which are geographically closely related, would give an enrichment factor of 1.75. We feel that since we have a larger database than *Gilchrist et al.* (1969), who analysed only two Muong Nong tektites (one of which is consistent with the upper limit of our range, and the other one with the lower limit of our range), we can make the conclusion that there is in fact an enrichment of water in Muong Nong type tektites compared with splash-form tektites.

This result is in excellent agreement with our expectations based on the enrichment in other volatile elements. The enrichment of water is smaller than for other volatile elements. Table 3 gives the analyses for eight more or less volatile elements in the six Muong Nong tektites also analysed for water. In addition, the same information is given for an average australite. The enrichment factor varies: it is very high for Zn or Br, and less than 2 for Li or B. The halogens do not behave consistently, although one would expect a different behaviour for F and Br. Fluorine is enriched by a factor of about 2-2.5, and since F in silicate structures shows some similarities to OH$^-$, it is interesting to note that water is enriched by a similar order of magnitude. Water is, of course, one of the most volatile compounds, and thus impact processes are expected to affect it easily. The water removal during a high temperature effect such as an impact may be so thorough that the difference between the water contents in Muong Nong tektites and splash-form tektites is not larger.

We have also tried to address the question of the different colored layers in the Muong Nong tektites, and areas of about 1 mm in diameter have been analysed in some sections, where the layers have been evident. The results are ambiguous. In most cases we have not been able to detect a difference in the water content of light and dark layers. This may well be due to the intergrowths of the layers at this scale. More ideal samples and the use of a smaller beam diameter (which we are planning for future investigations) may give better results. In a few cases, however, we have found indications that the water content in the dark layers seems to be slightly lower than in the light layers (e.g., 0.013 versus 0.014 wt %), but at present we are not able to draw any conclusions from this possible difference.

The Zhamanshin glasses show a range of water contents of 0.026-0.063 wt %. The irghizite, in accordance with the expectations, has the lowest water content of all Zhamanshin samples measured (0.026 wt %), but it is above the tektite average. One irghizite has previously been analysed for water by *King and Arndt* (1977), who give 0.051 wt %. Unfortunately, they have measured only one sample and give no comparison data for other zhamanshinites or chemical data, so the results

TABLE 3. Concentration (in ppm) of various volatile elements in six Muong Nong type indochinites.

Element	MN8302	MN8304	MN8308	MN8310	MN8317	MN8319	Ave. Australite
Li	35.0	37.2	48.0	50.5	47.2	48.9	40
Be	3.2	2.7	4.4	3.4	4.3	5.5	2.2
B	30.8	31.7	30.0	77.5	48.3	37.0	19
Cu	151	20.1	22.1	7.5	9.3	10.8	6.5
Zn	78.3	56.2	63.9	65.6	62.2	79.0	2.0
As	4.6	3.0	3.2	3.6		5.6	
F	90.3	99.2	55.9	72.5	124	121	36
Br	4.5	3.5		7.9			0.18

Data for the average australite from *Koeberl* (1986b).

are probably not directly comparable. The four zhamanshinites we analysed have water contents of 0.034-0.063 wt %, which are higher than that of the one irghizite we analysed. This is consistent with the lower volatile element content of irghizites relative to zhamanshinites. The Si-poor zhamanshinite is intermediate in water content between the blue zhamanshinites and the irghizites. As of now, no Si-rich zhamanshinite has been measured, but related work is in progress. The expectation is that the water content of Si-rich zhamanshinites is in the same range as for blue zhamanshinites. There seems to be a connection between the volatile element content, water content, and a sequence of impact glasses, depending on the temperature and pressure they have experienced during their formation.

The mechanism for driving out water from the sediments during the impact melting process is still not known in detail. There are some obvious theoretical problems (*O'Keefe*, 1964). On the other hand, it is known that impact glasses, associated with impact craters, have low water contents (although not as low as tektites, but clearly they have experienced a similar process on a larger scale). *Gilchrist et al.* (1969) report H_2O contents for Darwin glass of 0.047 wt %, or Aouelloul glass of 0.025 wt %, which is close to that we observe for Zhamanshin glasses. Obsidians have a water content that is even higher, in the 0.X wt % range (*Friedman*, 1963; *Gilchrist et al.*, 1969; *v. Engelhardt et al.*, 1987). But even if there is as yet no explanation for this glass formation, it is definitely possible to produce a dry homogeneous reduced glass in a single high temperature event: glass produced during an atomic bomb explosion in Nevada contains 0.007 wt % H_2O (*Glass et al.*, 1986). *Gilchrist et al.* (1969) report water contents of around 0.04 wt % for synthetic glasses produced from materials including soil by solar furnace melting. *Luft* (1983) performed melting experiments under different conditions (e.g., not under atmospheric pressure, but under near vacuum conditions), using Tertiary sands as precursor material (which, as cited above, contain up to 12 wt % H_2O) and found no water at all in the resulting glasses. In conclusion, we feel that all these results clearly support the impact model, or pose no contradiction to it, and it seems more appropriate to seek to improve the glass-making theory.

SUMMARY

We have measured the water content of 11 tektite and impact glass samples and have been able to show that Muong Nong type tektites have a higher average water content than splash-form tektites from the same strewn field. This is consistent with the analyses of volatile trace elements in Muong Nong tektites that we report. Impact glasses from the Zhamanshin crater also show a clear connection between the water content and volatile element content, and homogeneity, and thus follow a genetic sequence. Impact glasses generally have a higher water content than tektites. This is a further indication that the processes leading to the production of tektites are similar to the processes leading to impact glasses, but have probably been at least an order of magnitude more violent, in agreement with what we know from the chemistry and petrology. The water content of tektites is in full agreement with the terrestrial impact model.

Acknowledgments. C. K. thanks M. A. Nazarov and D. D. Badyukov (Vernadsky Institute for Geochemistry and Analytical Chemistry, USSR, Academy of Sciences, Moscow) for some of the Zhamanshin samples. The investigation was supported by the Austrian "Fonds zur Förderung der wissenschaftlichen Forschung," Project No. 3735.

REFERENCES

Chapman D. R. and Scheiber L. C. (1969) Chemical investigation of Australasian tektites. *J. Geophys. Res.*, 74, 6737-6776.

Clarke R. S., Jr. and Carron M. K. (1961) Comparison of tektite specimens from Empire, Georgia, and Martha's Vineyard, Massachusetts. *Smithson. Misc. Collect.*, 143, 1-18.

Engelhardt W.v., Luft E., Arndt J., Schock H., and Weiskirchner W. (1987) Origin of moldavites. *Geochim. Cosmochim. Acta*, 51, 1425-1443.

Epstein S. and Taylor H. P., Jr. (1974) Oxygen, silicon, carbon, and hydrogen isotope fractionation processes in lunar surface materials (abstract). In *Lunar Science V*, pp. 212-214. Lunar and Planetary Institute, Houston.

Friedman I. (1958) The water, deuterium, gas, and uranium content of tektites. *Geochim. Cosmochim. Acta*, 14, 316-322.

Friedman I. (1963) The physical properties and gas content of tektites. In *Tektites* (J. A. O'Keefe, ed.), pp. 130-136. University of Chicago, Chicago.

Gilchrist J. Thorpe A. N., and Senftle F. E. (1969) Infrared analysis of water in tektites and other glasses. *J. Geophys. Res.*, 74, 1475-1483.

Glass B. P., and Barlow R. A. (1979) Mineral inclusions in Muong Nong type indochinites: implications concerning parent material and process of formation. *Meteoritics*, 14, 55-67.

Glass B. P., Muenow D. W., and Aggrey K. E. (1986) Further evidence for the impact origin of tektites. *Meteoritics*, 21, 369-370.

King E. A. (1964) New data on Georgia tektites. *Geochim. Cosmochim. Acta*, 28, 915-919.

King E. A. and Arndt J. (1977) Water content of Russian tektites. *Nature*, 269, 48-49.

Kluger F. and Koeberl C. (1985) Determination of boron at low abundance levels in geological materials with a tetrafluoroborate-selective electrode. *Anal. Chim. Acta*, 175, 127-134.

Koeberl C. (1985) Geochemistry of Muong Nong type tektites VII: chemistry of light and dark layers - first results (abstract). In *Lunar and Planetary Science XVI*, pp. 449-450. Lunar and Planetary Institute, Houston.

Koeberl C. (1986a) Muong Nong type tektites from the moldavite and North American strewn fields? *Proc. Lunar Planet. Sci. Conf. 17th*, in *J. Geophys. Res.*, 91, E253-E258.

Koeberl C. (1986b) Geochemistry of tektites and impact glasses. *Annu. Rev. Earth Planet. Sci.*, 14, 323-350.

Koeberl C. and Fredriksson K. (1986) Impact glasses from Zhamanshin crater (USSR): chemical composition and discussion of origin. *Earth Planet. Sci. Lett.*, 78, 80-88.

Koeberl C., Kluger F., Kiesl W., and Weinke H. H. (1984a) Geochemistry of Muong Nong type tektites I: fluorine and bromine (abstract). In *Lunar and Planetary Science XV*, pp. 445-446. Lunar and Planetary Institute, Houston.

Koeberl C., Berner R., and Kluger F. (1984b) Geochemistry of Muong Nong type tektites II: lithium, beryllium, and boron (abstract). In *Lunar and Planetary Science XV*, pp. 441-442. Lunar and Planetary Institute, Houston.

Koeberl C., Kluger F., and Kiesl W. (1984c) Geochemistry of Muong Nong type tektites V: unusual ferric/ferrous ratios. *Meteoritics, 19,* 253-254.

Koeberl C., Kluger F., and Kiesl W. (1984d) Geochemistry of Muong Nong type tektites IV: selected trace element correlations. *Proc. Lunar Planet. Sci. Conf. 15th,* in *J. Geophys. Res., 89,* C351-C357.

Koeberl C., Badyukov D. D., and Nazarov M. A. (1986) Blue glass from Zhamanshin impact crater (USSR) (abstract). In *Lunar and Planetary Science XVII,* pp. 430-431. Lunar and Planetary Institute, Houston.

Luft E. (1983) *Zur Bildung der Moldavite beim Ries-Impakt aus tertiären Sedimenten.* F. Enke Verlag, Stuttgart. 202 pp.

Müler O. and Gentner W. (1973) Enrichment of volatile elements in Muong Nong type tektites: clues to their formation history. *Meteoritics, 8,* 414-415.

O'Keefe J. A. (1964) Water in tektite glass. *J. Geophys. Res., 69,* 3701-3707.

O'Keefe J. A. (1976) *Tektites and Their Origin.* Elsevier, Amsterdam. 254 pp.

Taylor S. R. (1973) Tektites: a post-Apollo view. *Earth Sci. Rev., 9,* 101-123.

Taylor S. R. (1975) *Lunar Science: a post-Apollo view.* Pergamon, New York.

The Effects of Impact Velocity on the Evolution of Experimental Regoliths

Mark J. Cintala and Friedrich Hörz

Experimental Planetology Branch, Solar System Exploration Division, NASA Johnson Space Center, Houston, TX 77058

Fragmental targets consisting of a coarse-grained gabbro were subjected to multiple impacts with stainless-steel spheres at 0.7, 1.4, and 1.9 km/s in an attempt to evaluate the effects of impact velocity on the generation and evolution of experimental regoliths. The low-velocity impactors were more efficient in terms of both mass comminution and creating new surfaces. The comminuted material formed by the faster projectiles, however, possessed smaller mean grain sizes and larger proportions of fine-grained debris. In all cases, the 2-4 mm material exhibited a mass "excess" relative to the adjacent size fractions, probably due to an enhancement of comminution around the gabbro's constituent grains, which average ~3 mm in dimension. The varied results of the three series are attributed to (1) differences in projectile penetration, (2) the rate of energy deposition into the target, and (3) the mean grain size presented to the projectile as each target evolved. Lunar secondary impacts are easily capable of comminuting considerable masses of material; high-velocity secondaries will create more fines, thus adding to the regolith's fractionation history while contributing relatively little to the overall agglutinate population. Low-velocity impacts into asteroids might result in significant quantities of fractionated, fine-grained material. The portion that hich is not permanently ejected will cover exposed surfaces and might exert an important effect on visible and near-infrared spectra.

INTRODUCTION

Regoliths on the Moon and, by inference, asteroids have been studied through analysis and interpretation of lunar and meteoritic samples. These approaches, however, have been limited in the sense that quantitative descriptions of many of the parameters involved in the evolution of those impact deposits are exceedingly difficult to extract from the samples themselves. Construction of the detailed history of a regolith in terms of the size distribution, velocity spectrum, and composition of the impacting population, for example, is simply too complicated a proposition for current understanding of regolith processes. So too is the inversion of data from a given regolith sample to the evolutionary path taken by a parameter as straightforward as the grain-size distribution (e.g., *McKay et al.*, 1974; *Morris*, 1978).

Techniques exist that allow the acquisition of some information on selected aspects of regolith development. Computer models, for instance, have been applied to studies of regolith thickness (*Quaide and Oberbeck*, 1975) and gardening rates (*Housen et al.*, 1979a,b). Myriad geochemical, petrographic, and petrologic procedures have been used in efforts to identify source rocks, "exotic" components, and meteoritic contamination in various lunar regoliths (see the review of *Papike et al.*, 1982). Both of these methods unfortunately have limitations imposed upon them by the extreme complexity of the processes involved in regolith formation and development.

Another approach toward the problem can be found in experimentation. Impact experiments have long been employed in investigations of cratering in regoliths (e.g., *Gault et al.*, 1968; *Stöffler et al.*, 1975; *Quaide and Oberbeck*, 1968), while shock experiments have helped to clarify both the physical requirements for the fusion of regolith components (e.g., *Ahrens and Cole*, 1974; *Schaal and Hörz*, 1980) and the mechanisms involved in agglutinate formation (*Simon et al.*, 1986). An avenue of investigation initiated recently employs granular targets that are repeatedly impacted under controlled conditions (*Hörz et al.*, 1984). This experimental approach is reviewed briefly in the next section to establish the background for this contribution, which investigates the effects of impact velocity on the comminution of planetary-surface materials.

BACKGROUND: THE FIRST EXPERIMENTAL REGOLITH SERIES

Hörz et al. (1984) performed a regolith-evolution experiment in which a coarsely fragmental gabbro was impacted 200 times with stainless-steel spheres (6.35 mm in diameter, 1.02 g in mass) at a nominal velocity of 1.4 km/s. The initial target was composed of fragments 2-32 mm in dimension, with a mean size of 15.2 mm; following the final shot, the mean grain size was determined to be 0.4 mm. The comminuted mass increased linearly as the number of shots accumulated until about shot 60, after which the "comminution efficiency" began to decline. (The "comminution efficiency" is defined here as the comminuted mass per unit incident impact-energy.) This change was attributed to a transition from wholesale disruption of large chunks of gabbro to a more familiar cratering environment in a fine-grained medium. This interpretation assumes that more energy would be expended in compressing and possibly ejecting material in the finer medium, because the coarser target had little porosity on a scale smaller than that of the projectile. Thus, energy available for comminuting the coarser fragments would be expended as work in compressing the porous, fine-grained target material, leading to a reduced rate of comminution in the process. An additional factor probably involved in this change was the buffering effect of the finer debris, as has been cited in actual lunar regolith processes (*McKay et al.*, 1974). Introduction of a major fine-grained component to the target frustrates effective interaction between the projectile and the increasingly less abundant, larger fragments, thus limiting the overall efficiency of their disruption.

In addition, work performed in collapsing pore spaces would sap the strength of the shock front, causing more rapid pressure-attenuation with distance from the impact point. At roughly the same time, the mean grain size of the evolving target was moving toward the average size of the mineral grains composing the gabbro itself. As this dimension was being approached, it is likely that intragranular fragmentation began to dominate intergranular crack generation; insofar as grain boundaries are sites of poor lattice fits even in monomineralic rock, it follows that intergranular fracturing along such locations of weakness would be accomplished more easily than disruption of individual crystals.

It is perhaps appropriate from the standpoint of motivation to summarize the results of chemical and petrographic analyses performed on the impact products from the 200-shot series, although comparable data will not be presented in this paper. Mineral-specific comminution was observed even after the first impact, with feldspar fracturing with greater ease than pyroxene. The result was an enrichment of feldspar in the finer fractions of the debris, which was the cause of a substantial chemical fractionation in the resulting "regolith." Relative to their whole-rock abundances, the feldspar fraction increased by 20-30% in the 125-250 μm fraction while the pyroxene portion dropped by 20-25%. These relative fractionations grew as smaller grain sizes were considered. It was concluded by Hörz et al. (1984) that these effects reduced the necessity for lateral and vertical transport of "exotic" components to lunar soil-sampling sites, a process invoked earlier by investigators on the basis of chemical mixing models (e.g., Evensen et al., 1974). Finally, glass-welded aggregates bearing a morphological resemblance to lunar agglutinates were formed during these impact experiments. Upon petrographic and microprobe analysis, however, it was found that the abundance of glass in those objects was far below that typical for lunar agglutinates, and that the experimental glasses were predominantly monomineralic in nature, with most being virtually pure feldspar in composition. A subsequent study of agglutinate-like particles formed in similar experiments conducted at a higher impact velocity (5.4 km/s) has since been performed, the results of which are presented in a companion paper (See and Hörz, 1987).

Since this first series of experiments was performed at essentially a constant impact velocity, it was natural to question whether the observed physical effects were dependent on the projectile's velocity or, equivalently, on its energy or momentum. The present sequence of experiments was therefore initiated in an effort to document the effects of impact velocity on the evolution of experimental regoliths. This contribution describes the physical results of these subsequent investigations, offers some interpretations of the data, and advances a few extrapolations to planetary and asteroidal regoliths.

EXPERIMENTAL CONDITIONS

The principal objective of this study was the evaluation of the effects of impact velocity on the comminution of fragmental targets, which was conducted with an experimental arrangement similar to that of the 200-shot series. Stainless-steel spheres 6.35 mm in diameter and 1.02 g in mass were launched at the 4000-g targets, which were held in stainless-steel buckets equipped with lids and baffle systems to minimize loss of ejecta; the chamber pressure (normal atmospheric gases) was kept at a constant 30-mm Hg to aid sabot separation from the projectile. (A plastic sleeve, or "sabot," is used to protect the projectile and gun barrel from mutual abrasion.) The gabbro used in these experiments (from the Bushveld Complex, South Africa) is relatively coarse-grained, with most crystals ranging from ~1.5 to 5 mm in dimension; the mean crystal-size is on the order of 3 mm. Major phases include plagioclase (54%), orthopyroxene (22%), and clinopyroxene (13%); orthoclase (5%) and quartz (5%) are also present as minor phases, yielding a bulk density for the gabbro of 2.82 g/cm^3. Additional details regarding the experimental arrangement and the gabbro itself can be found in Hörz et al. (1984).

These experiments differed from those of the 200-shot series in two important aspects. First, while the 200-shot series employed a constant nominal impact velocity of 1.4 km/s, three discrete velocities were employed here: 0.7 km/s (half that of the first series), 1.4 km/s, and 1.9 km/s (the highest velocity comfortably attainable with the JSC Vertical Gun and this projectile mass). Since identical projectiles were used throughout, these velocities yielded nominal energies of 1, 4, and 7.4 times that of the 0.7 km/s projectiles, respectively. The second principal difference resided in the initial size distributions of the targets. The 200-shot series employed gabbro chunks between 2 and 32 mm in dimension; the experiments described below utilized initial target fragments from a more constrained size range, namely, 16-32 mm (-4 to -5ϕ). In this manner, the mass decrease of a single initial size fraction could be monitored, thus avoiding the complications of tracking four separate initial size bins (Hörz et al., 1984).

Each target was dry sieved after the first shot in the series, and subsequently after shots 5, 10, 15, 20, and 25. (A procedural miscue resulted in the sieving of the 0.7 km/s charge after shot 6 instead of shot 5; this is noted in the text and figure captions when appropriate.) The sieved fractions were then weighed and recombined in the target bucket, which was inverted 10 times in succession prior to each shot to mix the debris. Each series was terminated at shot 25, as earlier experience demonstrated that trends of interest are generally established by that time.

THE DATA

Grain-Size Distributions

Among the changes induced in the target by these repetitive impacts, perhaps the most obvious is the gradual decrease in the average grain size. While all three cases exhibited basically similar overall behavior in the evolution of their grain-size distributions, some important differences arose. Figure 1 presents the basic data for the 10 size bins investigated in this study; Fig. 1a treats the whole charge, while Fig. 1b addresses only the comminution products by ignoring the

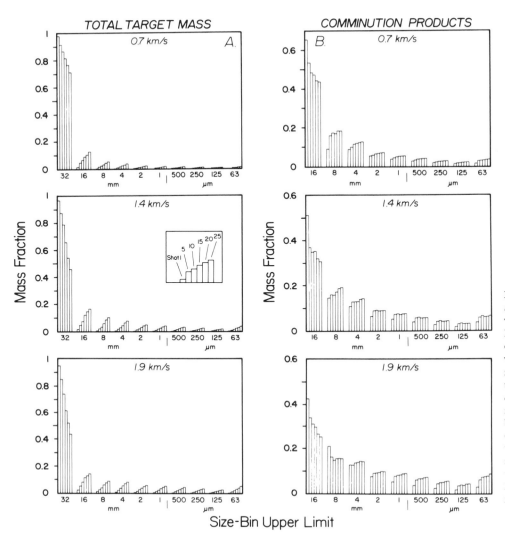

Fig. 1. (a) Fraction of the *total target mass* occupied by material within the indicated size bins. (b) Fraction of the *comminuted mass* occupied by material within the various size bins; the number representing the size bin refers to the upper limit to that bin [e.g., the cluster of histograms labeled "8" represents the material in the size bin between 4 mm and 8 mm in dimension (4-8 mm)]. Note (1) the different scales on the ordinates, and (2) that the second bar from the left in each cluster for the 0.7 km/s experiment represents shot 6 from that series.

material in the 16-32 mm size-range. (The terms "comminution products" and "comminuted fraction" will be used below interchangeably in reference to all debris in size fractions not present before the first shot.)

The changes in the size distributions of all three charges exhibit the same trends, although they differ in magnitude and "rate." Each was subjected to a decrease in the initial size fraction as the masses in the finer bins increased. Because the mass of a fragment is proportional to the cube of its characteristic dimension, only a few coarse pieces are sufficient to equal or exceed the mass of many more, smaller particles. This geometric effect was responsible for the preponderance of the coarser fractions in Figs. 1a and 1b and, due to the statistical "weighting" effect of the larger masses of the coarser debris, the finer fractions grew at a more sedate pace in comparison. (The apparent overabundance of material in the <0.063 mm (<4φ) bin is due to the fact that it represents the cumulative mass of all material smaller than that size, as opposed to the other bins, which refer to fragments in restricted size-ranges.) It is interesting to note the abundance of the 2-4 mm (-2φ to -1φ) fraction, particularly in the 1.9 km/s case. Although it is not readily apparent in this figure, both the 0.7 and 1.4 km/s targets incurred similar enhancements of the 2-4 mm fractions; this will be illustrated below.

Inspection of Fig. 1b sheds additional light on the evolution of the size distributions. For example, while the mass of coarser debris grew at a higher rate than any of the fines when the whole target was considered, it suffered the fastest relative *decrease* of all fractions >16 mm in dimension when only the comminution products are examined. These seemingly contradictory results can be reconciled if it is kept in mind that the differences are relative and not absolute: It is obvious from Fig. 1a that the mass of the 8-16 mm fraction incessantly grew as impact energy accumulated. Insofar as the masses depicted in Fig.1 are constantly normalized to unity and the inevitable effect of the impacts was to create progressively finer material, the contribution of the coarsest component to the comminuted mass must have undergone a relative decline in

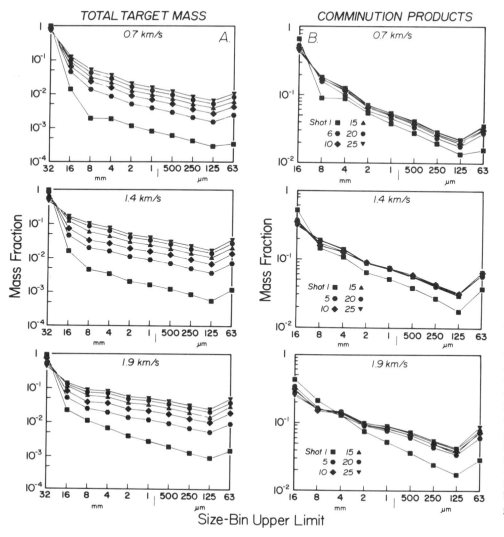

Fig. 2. (a) Fraction of the *total target mass* occupied by material within the indicated size bins. Note the differences in overall slope between shot 1 and the remainder of the 1.9 km/s series. Compare this to the similarity between the corresponding impacts of the 0.7 km/s series. (b) Fraction of the *comminuted mass* occupied by material within the indicated size bins. Compare the first and remaining shots of the series as in Fig. 2a. Note the excess of material in the 2-4 mm fraction, which brackets the mean grain size of the gabbro's constituent minerals.

order to compensate for the growth in the finer fractions. In other words, comminution is a one-way process, and the continuous creation of finer materials dilutes the contribution of the coarsest component to the comminuted mass.

As the rate of decrease in the coarsest fraction in Fig. 1b stabilized in all three cases, so did the rate of increase of the finer portions. These trends indicate that changes in the overall size distributions of the comminuted material were occurring at a reduced pace. The somewhat haphazard variations observed during the early times in each series were due to random factors characteristic of the impact process in such coarse-grained targets. Variables such as the angle of impact, the mass ratio between the projectile and the typical target-fragment, microcracks in the gabbro, and projectile ricochet all conspire to produce quantitatively unpredictable results on a shot-by-shot basis. As the impacts accumulated, however, the quantity of debris grew, and each subsequent shot made a progressively smaller contribution relative to the total mass of this new "regolith." Deviations from the average behavior at these later times were therefore accommodated by the large existing mass of comminuted material without significant excursions from the established trends. These effects manifest themselves in Fig. 1b as "asymptotic" tendencies toward limiting values in the histogram clusters.

The same data are presented in Fig. 2, with a logarithmic ordinate employed to facilitate comparisons between shots. Figure 2a illustrates the evolution of the entire experimental charge as a function of shot number, while Fig. 2b again treats only the comminuted fraction. As in Fig. 1, the displayed trends are fundamentally similar, although differences do exist. The low-velocity shots, for instance, produced less fine-grained ejecta than the higher-velocity impacts, a fact reflected in the steeper slopes for the majority of each of the low-velocity distributions. The 0.7 km/s series also produced less overall debris, which is reflected in the small complementary decrease in the 16-32 mm bin. A drop in comminution efficiency similar to that observed in the 200-shot experiment (*Hörz et al.*, 1984) is also apparent in Fig. 2a as progressively narrower spacing between the distributions, being most dramatic in the higher velocity series. The factors summarized earlier—the change in energy partitioning, development of intragranular fracturing regimes, and accumulation of fine-grained debris as a buffering agent—hindered destruction of the coarser fragments at later times in the series, resulting in smaller increments in the fine-grained portions of the distributions as the experiments progressed. It is also important that, as the coarse fragments decreased in abundance, so did the chances that one or more of them would take part in an impact event.

Figure 2b is particularly suited to displaying the approach toward an "equilibrium" distribution as the impacts accumulated. The difference between the products of each series' first and later shots is readily apparent; although this is still somewhat true of shot 5 in the 0.7 and 1.9 km/s

experiments, the 1.4 km/s target's trend appears to have been well established as early as that. A hint of the unpredictable effects associated with single shots into these targets is apparent in the results for the first shot of each series, particularly in the coarser sizes.

It is difficult to discuss slopes of distributions that are as irregular as those in Fig. 2, especially in quantitative terms. Qualitatively, however, it can be noted that the "slopes" of the distributions decrease as the higher impact-velocities are considered. Figure 3 presents distributions for the comminuted masses as they existed after the first and final shots in each series. The first impact in each series occurred in virtually identical targets, and it is apparent in Fig. 3a that, except for the coarsest and finest fractions, the normalized size distributions of the debris from these shots were very similar in shape. At the end of the experiments, however, the two high velocity series possessed substantially more fine-grained material, causing the net decrease in the distributions' slopes. The similarity of the trends for the 0.7 km/s series' initial and final shots (Fig. 3b) is consistent with the observation that insufficient fines were produced by shot 25 to force a significant change in the overall comminution behavior; that is, the role of the fine-grained material in the comminution process was still minimal at the end of the 0.7 km/s series.

Comminuted Masses

The efficiencies of the different projectiles in comminuting the target can be compared through parameters such as the projectile's kinetic energy or momentum. Since all of the projectiles were identical, allowances for such variables as impactor dimensions and mass are not necessary in interpreting the results of these experiments. The following observations therefore will be made on the basis of kinetic energy.

In a straightforward plot of comminuted mass versus the cumulative energy deposited into the target (Fig. 4), it is immediately apparent that, for a given energy, more material was fragmented by the low-velocity projectiles than by either of the other two impactors. Indeed, there is a well-delimited relationship between the rate of comminution and the projectile velocity, with the higher-velocity impactors having been less efficient overall at fragmenting a given mass of the target. (The more energetic projectiles, of course, created a greater absolute mass of debris, as reflected in the figure.)

These same data are presented in logarithmic form in Fig. 5, minus the point representing the first shot of each series to avoid random effects often associated with the earliest impacts. Included in the figure are curves fit to the data, the equations of which are

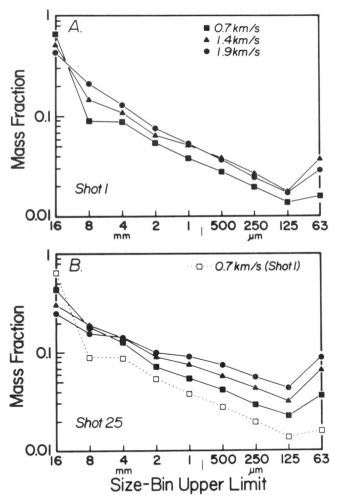

Fig. 3. (a) Size distribution of the comminuted mass after the first shot of each series. (b) Size distribution of the comminuted mass after the final shot of each series. Note the general similarity in the trends after the first shot contrasted to the decrease in "slope" after the final impact as the higher velocity cases are considered. The size distribution after shot 1 of the 0.7 km/s series is included in (b) for comparison.

$$M_c = 1.3^{+0.1}_{-0.1} \times 10^{-6} \, E_c^{0.83 \pm 0.01} \quad (0.7 \text{ km/s}) \quad (1a)$$

$$M_c = 6.7^{+1.2}_{-1.0} \times 10^{-8} \, E_c^{0.92 \pm 0.04} \quad (1.4 \text{ km/s}) \quad (1b)$$

$$M_c = 4.7^{+0.4}_{-0.3} \times 10^{-7} \, E_c^{0.83 \pm 0.02} \quad (1.9 \text{ km/s}) \quad (1c)$$

where M_c is the total comminuted mass and E_c is the cumulative impact energy; the uncertainties are 1-σ values. The random

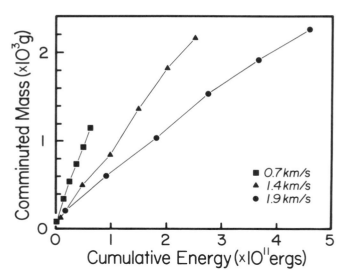

Fig. 4. Disrupted mass as a function of the cumulative energy for the three experiments, illustrating the increasing efficiency of comminution as the lower-velocity impacts are considered.

nature of impacts into such coarse-grained targets is clear from the loose constraints on some of the quantities in equation (1b), which were affected by the anomalous comminution between shots 6 and 10; expansive conclusions made on the basis of these fits would be uncertain at best. Nevertheless, the relative effectiveness of the three projectile groups at rupturing their targets are illustrated clearly by this figure.

Surface Areas

Anyone who has broken a rock into small pieces is (sometimes painfully) aware that the finer the desired end-product, the greater the effort required to obtain it. This is due to the fact that the formation of a unit surface area of a given material requires the input of a specific amount of energy and, since small fragments possess more surface area per unit mass than do large ones, a given quantity of fines demands correspondingly more work to obtain than does a similar mass of bigger pieces. During an impact event, energy is expended in making new surfaces through the breaking of bonds and shearing between fragments. Because of this, a more accurate reflection of the energy expenditure in the target should be given by the quantity of new surface area rather than by the mass of displaced material or the size distribution of fragments. The amount of surface area generated by impact events has been studied in the past (e.g., *Gault and Heitowit,* 1963; *Cintala and Hörz,* 1984), and has been found to be a useful and instructive parameter.

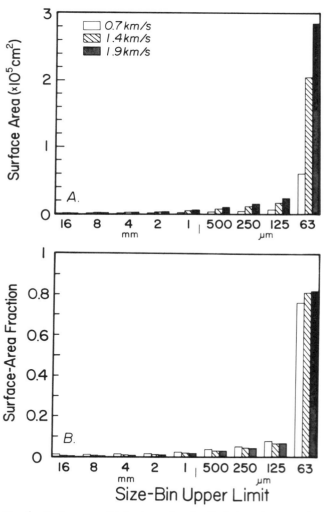

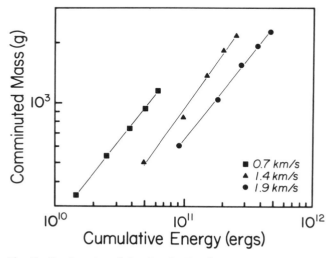

Fig. 5. Log-log plot of the data in Fig. 4, minus that of the first shot in each series to minimize random contributions. The "Comminuted Mass" ordinate increases upward. Coefficients and slopes for the least-squares fits are given in equations (1a)-(1c).

Fig. 6. Surface-area distributions after the final shot of each series in terms of (a) absolute area and (b) fractions of the total area of the comminution products. Moving left to right in each cluster, the bars represent the 0.7, 1.4, and 1.9 km/s series, respectively. Note the strong increase in the surface-area content as the sizes decrease, especially in light of the small masses of the finer fractions (Fig. 1).

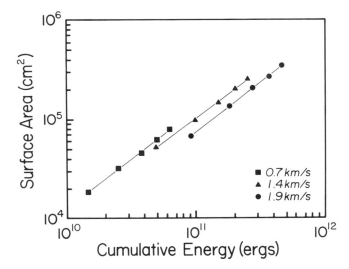

Fig. 7. Surface-area equivalent of Fig. 5; the coefficients and slopes for the least-squares fits are given in equations (2a)-(2c). Note the tighter grouping of these fits when compared to that of Fig. 5.

here again, only the results from shots 5 through 25 are considered to avoid random effects associated with the early events. Logarithmic fits to the three data-sets result in

$$A_s = 1.6^{+0.2}_{-0.2} \times 10^{-6} \, E_c^{0.99 \pm 0.02} \quad (0.7 \text{ km/s}) \quad (2a)$$

$$A_s = 2.1^{+0.3}_{-0.2} \times 10^{-6} \, E_c^{0.97 \pm 0.02} \quad (1.4 \text{ km/s}) \quad (2b)$$

$$A_s = 5.9^{+0.5}_{-0.4} \times 10^{-7} \, E_c^{1.00 \pm 0.01} \quad (1.9 \text{ km/s}) \quad (2c)$$

where A_s is the surface area created by the incidence of cumulative impact energy E_c. As with equations (1a)-(1c), the uncertainties are 1-σ deviations. The slopes are identical within the uncertainty of the fits, implying that there is no significant change in the energy required by the process creating the new surfaces, at least over the velocity and energy ranges studied here.

DISCUSSION

Grain-Size Distributions

Under the conditions prevailing during this investigation—a single projectile composition and size, and a target material common to all three experimental series—the peak shock-stresses were, to a good approximation, proportional to the

The surface areas created in these experiments were calculated by assuming that the individual particles were spherical and applying this approximation to the size distributions discussed above to obtain the area contained in each size bin (*Cintala and Hörz*, 1984). Clearly, the assumption of sphericity is unrealistic. On the other hand, any other approximation of the particle shape would entail further assumptions regarding relative dimensions, and thus would introduce added uncertainties. Insofar as the absolute areas will not be emphasized below, any shape assumption would be valid provided that it is applied equally to all three experiments. In doing so, the constants employed in defining the particle shapes will cancel as comparisons are made between experiments. The size distribution found from the sieve data of *Hörz et al.* (1984) for material in the <0.063-mm bin was utilized in determining the surface areas for that fraction.

The strong relationship between particle size and surface area is evident in Fig. 6, which reflects data from the last shot in each series. As was the case when the masses were considered, the quantities of new surface area created by the impacts were related directly to the impact energy, with the slower projectiles having created fewer new surfaces (Fig. 6a). The relative size-distributions (Fig. 6b) illustrate interesting differences in the production of fines between the low- and high-velocity experiments. While the relative areas contained in each of the size bins were virtually identical for the 1.4 and 1.9 km/s series, more area was contained in the coarser fractions in the 0.7 km/s target. This is essentially another means of displaying the data presented in Fig. 3b, which showed the differences in the abundances of fines in the different targets.

The absolute surface area created by the impacts is presented in Fig. 7 as a function of the cumulative kinetic-energy of the projectile; this plot is the surface-area equivalent of Fig. 5, and

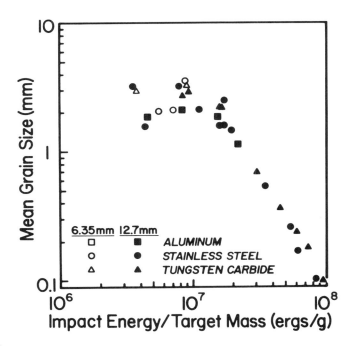

Fig. 8. Mean grain size of fragmentation products as a function of specific impact energy for experiments involving 7.5-cm granodiorite cubes as targets. Two projectile sizes and three compositions were used in these experiments. The cluster of points around 10^7 ergs/g is caused by spallation of large chunks from the targets at these energy densities. The well-defined relation results from near-catastrophic to "overkill" impacts. (After *Cintala and Hörz*, 1984.)

square of the impact velocity (e.g., *Gault and Heitowit*, 1963). Only when fine-grained material began to represent a significant fraction of the target were the characteristics of shock propagation and projectile penetration likely to change. The 1.4 km/s projectiles, for example, generated a peak shock stress of 15.0 GPa (150 kb) when the target was solid gabbro, but only 5.7 GPa when fine-grained debris was impacted. [Maximum shock stresses cited in this report have been calculated under the assumption of one-dimensional, normal impact. Information on the shock behavior of lunar regolith (*Ahrens and Cole*, 1974) was employed in determining the value for the fine-grained target.]

This change in induced shock stress as the grain size of the target decreases will have two principal effects. First, all other things being equal, higher stresses generally imply higher stress gradients. Insofar as the characteristic size of the resulting debris should be proportional to the stress gradient (*Öpik*, 1971; *Oberbeck*, 1975), the higher velocity impactors would be expected to create a greater proportion of fines. Second, the size of the typical target fragment relative to the kinetic energy of the impactor should also play an important role in determining the outcome of a disruptive event. Figure 8, for instance, illustrates the results obtained from impact experiments employing 7.5-cm granodiorite cubes as targets (*Cintala and Hörz*, 1984). At specific impact-energies (i.e., projectile kinetic energy per unit mass of target) less than about 1.5×10^7 ergs/g, spallation of large fragments gives rise to a scattered pattern of mean grain sizes for the displaced material. As the specific energy grows, however, a well-defined and unambiguous trend is established, with higher specific energies yielding correspondingly smaller fragments. Taking a mean (geometric) mass of 17 g for the initial (16-32 mm) target fragments and extrapolating the high-energy trend of Fig. 8, the 0.7, 1.4, and 1.9 km/s impacts would yield mean grain sizes of 0.06, 0.005, and 0.002 mm (60, 5, and 2 μm), respectively, if only a single fragment were involved in each impact. (If a single target fragment were involved in an average 0.7 km/s impact, for example, the specific impact-energy would have been $\sim 1.5 \times 10^8$ ergs/g, which accounts for the small projected mean grain sizes.) Two observations, however, illustrate that this was not the case: (1) Simple energy-mass considerations (Fig. 4) indicate that many fragments were disrupted by each shot, although the average 0.7 km/s impact appears to have involved only a few pieces (at an average of ~ 17 g each). (2) A plot of the mean grain size (using the method of *Folk and Ward*, 1957) as a function of the cumulative impactor energy (Fig. 9) makes it clear that these predicted mean sizes have not even been approached by the comminution products at the end of each series.

Thus, the following intermediate conclusions can be made: (1) Even in such blocky targets, the results of single-target, single-impact experiments can be applied only qualitatively, and then only as a general guide to the disruption process in fragmental targets. (2) At all three impact velocities studied here, the comminution products are more coarse-grained, gram for gram, than those from single-target experiments. (3) Inspection of Figs. 4 and 9 together indicates that, although the lower-velocity impactors are more efficient at comminuting

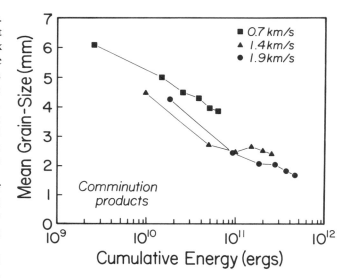

Fig. 9. The mean grain size of each target's comminuted fraction as a function of the logarithm of the cumulative impact energy deposited into the target. These mean dimensions are much larger than would have been the case had the targets behaved as coherent masses.

a given mass of material, they are less effective at producing fine-grained debris. Thus, for a given cumulative impact energy, low-velocity projectiles yield more and coarser comminuted material than do their higher-velocity counterparts.

A third factor can now be considered, and might shed some light on the reasons for these observations. Just as the peak shock stresses affect the target material's comminution behavior, so, too, do they determine the subsequent dynamics of the projectile. In virtually all cases, the impactors in the 0.7 km/s series were removed intact from the target during the sieving operation. Some had evidence of incipient cracking, but none actually ruptured. The 1.4 km/s projectiles were typically broken into a few (2-5) pieces, but late in the series they were removed unscathed with increasing regularity. Only during the very last shots did a few 1.9 km/s spheres emerge unfragmented, and the general rule at this velocity was complete projectile pulverization; most of the remnants of these impactors had to be removed from the sieved fractions magnetically.

The greatest stresses suffered by the projectile occur very early in the impact event, when the relative velocity between the impactor and target is the highest. Thus, it is virtually certain that disruption of the impactor in these experiments took place during its first encounter with a target fragment, provided the local angle of impact was not so small as to result in ricochet. The lower velocity projectiles, on the other hand, obviously survived this phase of the event, retained their integrity during the initial penetration, and even survived through subsequent ricochets and collisions, albeit at reduced velocities and energies.

More than one gabbro fragment was involved in each 0.7 km/s collision, since an average of almost 50 g of target material was disrupted per shot. After projectile contact with the first few chunks of target, however, any subsequent interactions must have occurred at lower velocities, and therefore lower shock stresses and stress gradients. This multiple-collision but reduced-energy environment led to relatively large masses being comminuted, but nevertheless accompanied by only small fractions of fine-grained material. Shattering of the projectiles at the higher velocities inhibited their ability to participate in further interaction with the target as coherent bodies. Thus, a large portion of the high-velocity impactors' energy was deposited in a restricted region around the impact point, resulting in more thorough pulverization of that volume. Projectile fragments undoubtedly participated in further collisions on a smaller scale, although some probably possessed sufficient energy to cause considerable secondary damage to the remainder of the target; the greater absolute target masses comminuted by the higher-velocity impacts attest to the effectiveness of these "secondary" collisions. The reduced energy levels at which they occurred, however, would likely have created coarser debris for reasons described in the discussion of low-velocity impacts. In addition, it is likely that high-velocity target fragments also participated in these secondary collisions, but were probably less effective than the more competent projectile material at disrupting surrounding pieces of gabbro.

Another factor implicitly related to differences in impact velocity also enters into this discussion. Inspection of Fig. 1a shows that, after shot 25, the 0.7 km/s target possessed >60% more of the coarsest material than did either of the other two charges. Since more of these large fragments were still available for disruption even near the end of the low-velocity series, the reduction in comminution efficiency was considerably less drastic than in the other two experiments.

These differences due to impact velocity (or, equivalently in this study, kinetic energy or shock stress) are illustrated in Fig. 10, which presents the final size distributions of the 0.7 and 1.9 km/s series relative to that of the 1.4 km/s experiment through the parameter

$$R = (f_\phi^v/f_\phi^{1.4}) - 1 \qquad (3)$$

In this equation, $f\phi$ is the mass of material in the bin whose upper size limit is ϕ, normalized to the total comminuted mass formed in the experiment at the impact velocity v. The contrasting trends of the high- and low-velocity cases are very well defined, with the abundance of fines in the 1.9 km/s charge as notable as their sparsity in the 0.7 km/s debris. The 1.4 km/s data in this figure, of course, plot as the straight line through zero relative abundance.

It was noted earlier that the 2-4 mm fraction in all three experimental charges exhibits an "excessive" mass relative to the rest of the size fractions (Fig. 2), and that the mean grain size of the gabbro's constituent minerals is near 3 mm. Hörz et al. (1984) concluded that comminution proceeded more efficiently early in the 200-shot series because, among other reasons, intergranular fragmentation required less energy than intragranular crack propagation and disruption. Assuming this to be correct, it would follow that stresses in the target would be relieved most effectively by separation along the planes of least resistance, which in the case of the gabbro would be along grain boundaries. Since these boundaries occur around grains averaging about 3 mm in dimension, the 2-4 mm fraction received more than its share of material than would have been the case if the target were, say, an aphanitic basalt.

In summary, these experiments have demonstrated that, given a single projectile-target combination, higher impact-velocities will produce a size distribution of ejecta that is richer in fine-grained material than the debris produced by slower impactors. This difference is due to higher shock stresses resulting from the more energetic impacts and correspondingly steeper stress gradients, to the changing depth of projectile penetration, and to the rate of energy deposition into the target. In addition, the lower rate of comminution in the low-velocity case resulted in a coarser target persisting farther into that series, with subsequent behavior of the target being different from the more rapidly evolving and finer-grained, high-velocity charges.

Comminuted Masses

The random nature of impacts into these coarse-grained targets was cited earlier during descriptions of the data. Ten equivalent impacts into identical, blocky targets can yield very different results when influenced by any of a variety of factors. Oblique impacts, for instance, result in ricochets, with energy transfer being less efficient than if the target were a single object. A single fragment, in some cases, can absorb most of the projectile's energy while propagating little to surrounding debris, or a layer of fine-grained material can buffer the brunt of an impactor's energy before it encounters a large fragment. The number of variables involved under such conditions alone would virtually guarantee "random" behavior during the early stages of these experiments; as the number of impacts

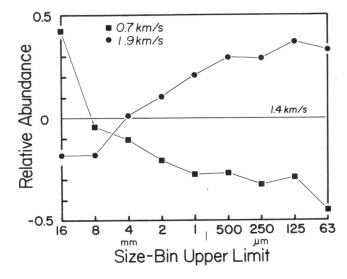

Fig. 10. Size distributions of the 0.7 and 1.9 km/s series after the final impacts, normalized to that of the 1.4 km/s experiment. The differences in the relative abundances of fines is conspicuous.

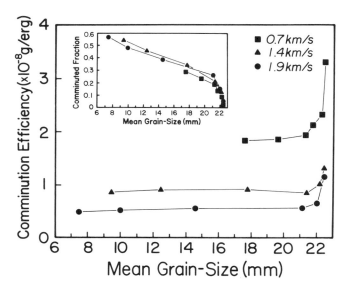

Fig. 11. Comminution efficiency (expressed in units of 10^{-8} g/erg) as a function of the mean grain size of the total target. The coarser sizes correspond to the earlier impacts in the three series, when little debris had been generated. Note the very gradual decline in the comminution efficiency for each case below a mean grain size of ~21 mm. The slight rise in the 1.4 km/s trend after the 21-mm mark is due to the unusually small quantity of material disrupted between shots 6 and 10 in that series. See the text. (Inset) The relationship between the mean grain size and the comminuted fraction of the target's mass for the three series. Note that the mean grain size of 21 mm corresponds to a comminuted fraction of only 20-27% of the bulk target.

accumulated and the target's average grain-size decreased, however, a more familiar environment of cratering in a fairly homogeneous medium would be approached.

In examining the amount of mass comminuted during events such as those studied here, then, it must be kept in mind that equal energy expenditures can result in very different comminuted masses, depending on the size distributions of the fragmentation products. The principal factor governing the energy requirements in these impact experiments is the quantity of energy necessary to produce the new surfaces: Less energy certainly is required to split a 50 g block into three large pieces than to pulverize a 25 g fragment. Even so, a greater comminution efficiency would result in the case of the former. With this understanding, the utility of the comminuted mass (or the comminution efficiency) in a causal sense becomes somewhat diluted. Nevertheless, a brief consideration of the differences in comminuted mass between the three series yields some interesting results.

It is apparent from Fig. 4 that the lower velocity impactors comminuted more material per unit energy than did their higher energy counterparts. The curves fit to the data illustrated in Fig. 5 (equations (1a) and (1b)), however, indicate that the three targets behaved similarly: Excluding the 1.4 km/s case from consideration for the moment, the identical slopes imply that the 0.7 and 1.9 km/s impactors possessed similar relationships between incident energy and comminuted mass.

The different coefficients, on the other hand, confirm that the slower projectiles were more efficient by a substantial factor (~7.6). These results, particularly the similarity of the slopes, point to similar overall processes causing disruption of the targets. This is not surprising, since these impact velocities are readily comparable in the overall context of the material's mechanical behavior during stress-wave passage. Although comparably little energy was partitioned into entropy in all three series, more "waste heat" is generated at the higher velocities, and is a possible factor contributing to the differences in efficiency noted between the three series (see section on "Implications for Planetary Regoliths"). Only at much higher velocities are such factors as plastic deformation and entropy production expected to exert truly major influences on the fundamental energy-partitioning during impact events.

The deviation of the 1.4 km/s case from this trend is taken here to be a manifestation of the randomness of such impacts as discussed earlier. Specifically, less material was disrupted between shots 6 and 10 of that series than would have been expected on the basis of the other experiments; this deficiency in comminution products is evident in Fig. 4, and undoubtedly contributed to both the steepness of the fit's slope and the large uncertainties in the fit's parameters (equation (1b)).

The comminution efficiency as defined above is plotted against the mean grain size of the evolving target in Fig. 11. It is vividly apparent that a rapid decrease in the comminution efficiency takes place until a mean grain size of ~21 mm is attained; the inset in Fig. 11 shows that this value occurs when only 20-27% of the target has been comminuted. The 0.7 and 1.9 km/s series display a constantly declining efficiency, though the rate of decrease becomes very gradual as the mean grain

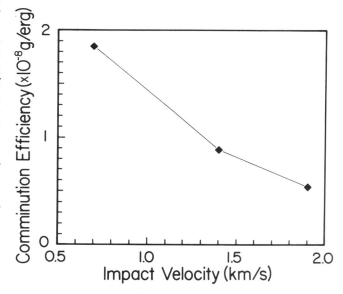

Fig. 12. The average comminution efficiency (determined for the points with mean grain sizes <21 mm in Fig. 11) as a function of impact velocity. The standard deviations of the averages used in determining these comminution efficiencies are smaller than the data points at this scale.

sizes grow smaller. A slight increase occurs in the 1.4 km/s trend after shot 10, but this again appears to have been due to the unusually small comminuted mass resulting from shots 6 to 10. The data points at the smaller grain sizes, particularly those of the 1.4 and 1.9 km/s experiments, become more closely spaced as the target becomes finer-grained, further indicating that less mass was being fragmented with time. It is likely that the fines, occupying the comparatively large gaps between coarse fragments, acted as "energy absorbers" in the sense that high-velocity spallation products expended significant fractions of their energy and momentum in penetrating such accumulations before encountering other large blocks.

Finally, a simple depiction graphically illustrating the differences in comminution efficiency can be obtained by taking the average values of the comminution efficiencies below the 21-mm point and plotting them against their respective impact velocities. The results are presented in Fig. 12 as a single curve; in light of the discussion above, it is uncertain as to whether the "dogleg" at 1.4 km/s is significant. It is also noted that, were the 0.7 km/s series extended to a greater number of impacts, its comminution efficiency would doubtlessly decrease, forcing that point to a lower position in the figure.

Surface Areas

As was done for the comminuted mass in Fig. 11, the efficiency of producing new surfaces is presented in Fig. 13 as a function of the target's mean grain size. Again, the 21-mm mark appears to be the significant value in these experiments, with the efficiency of surface-area production

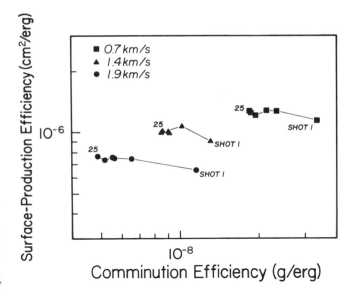

Fig. 14. Log-log plot of the efficiency of surface-area production (increasing upward) versus the mass-comminution efficiency (increasing to the right). The smaller variation in the surface-area parameter when compared to the range of comminution efficiencies for each series is an indication that the energy required to create new surfaces exerted a stronger control on the results of these events than did that necessary to yield these comminuted masses.

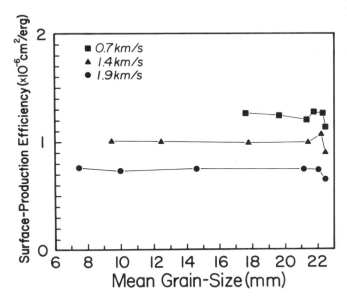

Fig. 13. Surface-area equivalent of Fig. 11. Note the lack of dependence on the mean grain size below the 21-mm mark. Only the 0.7 km/s case shows any tendency toward variation, and only weakly at best. Note that the point representing the results of shots 6 to 10 in the 1.4 km/s experiment appears to be better behaved here than in the cases where the dependent variable is the comminuted mass.

becoming remarkably constant and independent of the target's average grain size. The only variation can be found in the 0.7 km/s series, and that is attributed here to a combination of the coarseness of that target even after the final shot of the experiment and to the penetration characteristics of the slower projectiles as discussed above. It is interesting to note that, even though greater masses of fines were produced by the higher-velocity impacts, the efficiency of surface-area production is still the highest for the slow impactors.

The trend from high to low efficiency is evident in the figure, but the differences are much smaller than those in the mass-comminution plot (Fig. 11). An indication of the relative stabilities of the comminution and surface-area production efficiencies can be obtained from Fig. 14, which displays the efficiency of surface-area formation relative to that of mass comminution. [Note that the point representing the results of shots 6 through 10 of the 1.4 km/s series, which was at some variance with the remainder of the mass comminution trend (Figs. 4 and 5), appears to be well-behaved from the surface-area point of view.] In all cases, the variation in the surface-area production efficiency is substantially less than that of the mass-comminution efficiency, and, except for shot 5 of the 1.4 km/s series, could justifiably be termed "remarkably less." This suggests that the quantity of mass comminuted during an impact into a fragmental target is a parameter with inherently less predictive capability than the measure of new surfaces created by the event. Nevertheless, and especially in view of all of the potential perturbing factors discussed earlier, the mass-comminution efficiency can still be a useful parameter.

This is particularly true when a knowledge of changes in

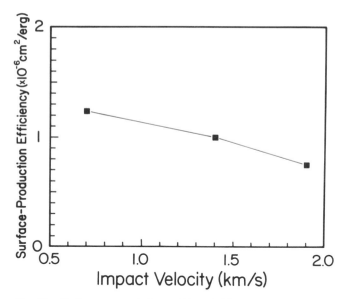

Fig. 15. Surface-area equivalent of Fig. 12. This parameter exhibits a substantially more mild variation than does the comminution efficiency as displayed in Fig. 12, and is more weakly dependent on the impact velocity.

the target's grain-size distribution is necessary to apply the efficiency of surface-area production with meaning. It is likely that the mass-comminution efficiency would be a parameter with increasing dependability in the case of a target whose mean grain size is small relative to the dimensions of the projectile. In such a situation, the random effects of comparatively large blocks on the fate of the impactor and its energy deposition would be minimal, and the target would behave more as a homogeneous medium. Many more target fragments would be involved in the impact event, leading to smaller variances from a statistical point of view. There is no certain indication that any of the targets in this study attained such a sufficiently fine-grained state, but the 200-shot target was so pulverized during the latter stages of the experiment that this condition was approached, if not indeed met (cf. Figs. 3 and 4 of *Hörz et al.*, 1984).

The surface-area equivalent of Fig. 12 can be found as Fig. 15, which, as might be expected after reflection upon Figs. 13 and 14, displays a relationship much less dependent on the impact velocity than does the mass-comminution efficiency. Indeed, the 0.7 and 1.9 km/s mass-comminution efficiencies vary by 39% and 109% from the 1.4 km/s value, while the surface-area cases differ only by 23% and 25%, respectively. Although the difference is less pronounced when the production of new surfaces is considered, the lower-velocity impacts still appear to have been more efficient at rupturing fragmental targets.

Implications for Planetary Regoliths

The results of this study illustrate fundamental trends in the comminution behavior of fragmental deposits as the impact velocity is varied. It would be ill-advised, though, to extrapolate the results directly to planetary impact conditions. While it is safe to assume that the proportion of fine-grained material grows with the impact velocity, other factors begin to enter the picture. Very high impact-velocities will indeed generate greater stresses, which would be expected to yield even more fine-grained debris. Stress levels around 50 GPa, however, will begin to fuse most silicates, while porous targets would start to melt at much lower pressures (e.g., *Stöffler*, 1972; *Kieffer*, 1975; *Schaal et al.*, 1979). Not only will this entropy production consume major fractions of energy that would otherwise be available for comminution, but the accompanying change in material behavior would favor plastic deformation over fragmentation in regions of very high stress. Therefore, depending on the physical and chemical state of the target, valid extrapolation of these empirical trends will be limited to those velocities and stresses below which entropy production is negligible.

Entropic effects become obvious with maskelynite formation in massive feldspar-bearing rocks; this occurs at stresses near 30 GPa (*Ostertag*, 1983), corresponding to a silicate impact into silicate at slightly over 4 km/s. Intergranular melt appears in experimentally shocked lunar soils at stresses below 20 GPa (*Schaal and Hörz*, 1980), which would be attained by the impact of a silicate into regolith at velocities under 3.5 km/s. Indeed, the 200-shot series yielded regolith melts at stresses that could have been as low as 5.5 GPa if the target were powder at the impact point (*Hörz et al.*, 1984). Since the 1.9 km/s projectiles in this study most likely produced shock stresses between 15 and 22 GPa, it is clear that thermal mechanisms were at work even in these experiments. Extrapolation therefore must be done with caution, especially since the degree of entropy production becomes extremely sensitive to the impact velocity at the low end of the velocity spectrum (e.g., *O'Keefe and Ahrens*, 1977; *Grieve and Cintala*, 1981).

Typical impacts of asteroidal debris onto the Moon take place at about 14 km/s (e.g., *Hartmann*, 1977), while the average relative velocities between objects in the asteroid belt are a substantially lower 5 km/s (*Hartmann*, 1977). The latter is much closer to those employed in these experiments, although still somewhat higher. Even so, some ground is gained by the fact that the stainless-steel impactors generated higher stresses than silicate projectiles would at the same velocities. As mentioned above, the peak stress attainable in the 1.9 km/s case would have approached 22 GPa if the impactor hit a solid piece of gabbro at normal incidence. Had the projectile also been gabbro, an impact at almost 4 km/s would have been required to effect an equivalent stress. On this basis, most primary lunar impacts appear to be much out of the range of direct extrapolation, but secondary projectiles are well within the limits of these data. An impact of gabbro into regolith at lunar escape-velocity (2.35 km/s) would generate ~9.7 GPa, comparable to the 9.4 GPa resulting from a 1.4 km/s stainless-steel impact into fine-grained debris.

The results of these three experimental series indicate that, even over their constrained range of impact velocities, obvious differences occur in the comminution of a typical igneous rock. Perhaps most importantly, the production of fines grows directly with increasing impact velocity and shock stress. This observation can be extended to planetary and asteroidal conditions after consideration of two principal factors. First, most impactors in the asteroid belt and all secondary projectiles on the Moon will be silicates, which possess much lower tensile strengths and densities than the stainless steel used in these experiments. This would make the analogy between the high-velocity steel and the natural silicate impactors closer in the sense that fast silicate projectiles would disrupt upon impact and deposit their energy at shallower levels. Comminution products with a significant fine-grained fraction should then result. Second, the combination of weak impactor material and higher velocities will give rise to a transition in the impact process from the brittle sort of behavior characteristic of these experiments to one increasingly dominated by plastic or hydrodynamic flow, accompanied by the severe thermal effects discussed above. Such phenomena cannot be addressed through these experiments, and must remain implicit during application of the empirical results. The net effect would be lower efficiencies of comminution and surface-area production; that is, less debris and surface-area produced per unit impact-energy at planetary impact-velocities.

Fast, nonescaping lunar ejecta will impact the surface at velocities yielding stresses comparable to those prevailing during these experiments, and would easily be capable of comminuting surface materials. Those with higher velocities would disrupt greater masses and create more fines than would slower secondaries, but would be less efficient at doing so because of the heat produced in the process. It was demonstrated by *Hörz et al.* (1984) that secondary impact velocities are sufficient to induce significant fractionation trends in the resulting fines, and that the degree of chemical fractionation grows as finer fractions are considered. Thus, with greater quantities of fine material being produced by the faster secondaries, fractionation of the regolith would be assisted to a greater extent by these projectiles; this fractionation would occur with relatively minor target-heating and, therefore, a minimum of agglutinate formation. Those materials participating in agglutinate formation at these shock levels should be enriched in feldspar due to that mineral's melting behavior during shock passage; this aspect of regolith evolution is discussed in greater detail by *See and Hörz* (1987).

The asteroidal environment will be somewhat different in the sense that fine-grained debris will accumulate at a slower pace than it would on a body with a relatively strong gravitational field (e.g., *Cintala et al.,* 1978, 1979). As detailed by *Oberbeck* (1975), higher stresses create finer ejecta, which will also possess high particle velocities. Since a large fraction of the highly shocked material would thus escape the weak gravitational fields of small bodies, a low rate of regolith accumulation would follow, especially in terms of the finer fractions.

Secondary impacts on the vast majority of asteroids can occur only at very low velocities (meters per second), so little comminution would occur in comparison with the lunar case. At the other extreme, those primary impacts occurring at very high velocities cannot be treated effectively on the basis of a one-to-one correspondence with these experiments. If consideration were limited only to velocities below the 5 km/s average, then conclusions similar to those reached for lunar secondary projectiles could be advanced. The likelihood of coarser regoliths on the asteroids would be a prominent departure from the lunar case; indeed, the analogy might be better between a typical asteroidal regolith and a very young lunar deposit.

In the absence of a working knowledge of ejection velocities in such targets, quantitative speculation as to the rate of retention of fine-grained debris during impact events is unwarranted. Nevertheless, the flow field generated by an impact invariably includes material—some experiencing very high shock-levels—that remains in or near the final crater (e.g., *Maxwell*, 1973, 1977). Thus, it is certain that some fine-grained material would remain on the target body.

Even if the retained fraction of fine-grained material were small, it is probable that the bulk of these fines would reside on exposed surfaces, until ejection or "blanketing" by subsequent impacts. It is more than likely that they would be fractionated, with the enriched phase depending on the petrology of the asteroid (or region of the asteroid) in question. The consequences for remote sensing, especially visible and near-infrared spectroscopy, could be significant (*Hörz et al.,* 1984). [*Basu and Battacharya* (1986) also propose a similar situation for lunar-like regoliths due to ballistic differentiation of ejecta.] A layer of fractionated dust covering the parent lithology (or lithologies), whether ubiquitous or discontinuous, would do little to clarify the spectral signature of an asteroid. It thus might be rewarding to obtain laboratory spectra of fractionated meteorite powders in attempts to obtain matches between meteorite and asteroid spectra.

Acknowledgments. This paper benefitted from reviews by Tom See, Steve Simon, and Herb Zook; a detailed review and discussion by Kevin Housen was particularly illuminating. Contributions in the lab by Frank Cardenas, Will Carswell, and Bill Davidson are very much appreciated. This work was performed under NASA grant 152-412, which is acknowledged with thanks.

REFERENCES

Ahrens T. J. and Cole D. M. (1974) Shock compression and adiabatic release of lunar fines from Apollo 17. *Proc. Lunar Sci. Conf. 5th,* 2333-2345.

Basu A. and Battacharya R. N. (1986) A probabilistic approach to ballistic differentiation of surficial soils on Moon-like planets (abstract). In *Lunar and Planetary Science XVII,* pp. 32-33. Lunar and Planetary Institute, Houston.

Cintala M. J. and Hörz F. (1984) Catastrophic rupture experiments: Fragment-size analysis and energy considerations (abstract). In *Lunar and Planetary Science XV,* pp. 158-159. Lunar and Planetary Institute, Houston.

Cintala M. J., Head J. W., and Veverka J. (1978) Characteristics of the cratering process on small satellites and asteroids. *Proc. Lunar Planet. Conf. 9th,* 3803-3830.

Cintala M. J., Head J. W., and Wilson L. (1979) The nature and effects of impact cratering on small bodies. In *Asteroids* (T. Gehrels, ed.), pp.579-600. University of Arizona, Tucson.

Evensen N. M., Murthy V. R., and Coscio M. R., Jr. (1974) Provenance of KREEP and the exotic component: Elemental and isotopic studies of grain size fractions in lunar soils. *Proc. Lunar Sci. Conf. 5th*, 1401-1417.

Folk R. L. and Ward W. C. (1957) Brazos River Bar: A study of the significance of grain size parameters. *J. Sed. Petrol.*, 27, 3-26.

Gault D. E. and Heitowit E. D. (1963) The partition of energy for hypervelocity impact craters formed in rock. *Proc. 6th Hypervelocity Impact Symposium*, 2, 419-456.

Gault D. E., Quaide W. L., and Oberbeck V. R. (1968) Impact cratering mechanics and structures. In *Shock Metamorphism of Natural Materials*, (B. M. French and N. M. Short, eds.), pp. 87-99, Mono, Baltimore.

Grieve R. A. F. and Cintala M. J. (1981) A method for estimating the initial impact conditions of terrestrial cratering events, exemplified by its application to Brent crater, Ontario. *Proc. Lunar Planet. Sci. 12B*, 1607-1621.

Hartmann W. K. (1977) Relative crater production rates on planets. *Icarus*, 31, 260-276.

Hörz F., Cintala M. J., See T. H., Cardenas F., and Thompson T. D. (1984) Grain-size evolution and fractionation trends in an experimental regolith. *Proc. Lunar Planet. Sci. Conf. 15th*, in *J. Geophys. Res.*, 89, C183-C196.

Housen K. R., Wilkening L. L., Chapman C. R., and Greenberg R. (1979a) Asteroidal regoliths. *Icarus*, 39, 317-351.

Housen K. R., Wilkening L. L., Chapman C. R., and Greenberg R. J. (1979b) Regolith development and evolution on asteroids and the Moon. In *Asteroids* (T. Gehrels, ed.), pp. 601-627, University of Arizona, Tucson.

Kieffer S. W. (1975) From regolith to rock by shock. *The Moon*, 13, 301-320.

Maxwell D. E. (1973) Cratering flow and crater prediction methods. *Tech. Memo TCAM 73-17*, Physics International Co., San Leandro, CA.

Maxwell D. E. (1977) Simple Z model of cratering, ejection, and the overturned flap. In *Impact and Explosion Cratering* (D. J. Roddy, R. O. Pepin, and R. B. Merrill, eds.), pp. 1003-1008, Pergamon, New York.

McKay D. S., Fruland R. M., and Heiken G. (1974) Grain size and the evolution of lunar soils. *Proc. Lunar Sci. Conf. 5th*, 887-906.

Morris R. V. (1978) *In situ* reworking (gardening) of the lunar surface: Evidence from the Apollo cores. *Proc. Lunar Planet. Sci. Conf. 9th*, 1801-1810.

Oberbeck V. R. (1975) The role of ballistic erosion and sedimentation in lunar stratigraphy. *Rev. Geophys. Space Phys.*, 13, 337-362.

O'Keefe J. D. and Ahrens T. J. (1977) Impact-induced energy partitioning, melting, and vaporization on terrestrial planets. *Proc. Lunar Sci. Conf. 8th*, 3357-3374.

Öpik E. J. (1971) Cratering and the Moon's surface. In *Advances in Astronomy and Astrophysics*, (Z. Kopal, ed.), pp. 108-337. Academic, New York.

Ostertag R. (1983) Shock experiments on feldspar crystals. *Proc. Lunar Planet. Sci. Conf. 14th*, in *J. Geophys. Res.*, 88, B364-B376.

Papike J. J., Simon S. B., and Laul J. C. (1982) The lunar regolith: Chemistry, mineralogy, and petrology. *Rev. Geophys. Space Phys.*, 20, 761-826.

Quaide W. L. and Oberbeck V. R. (1968) Thickness determinations of the lunar surface layer from lunar impact craters. *J. Geophys. Res.*, 73, 5247-5270.

Quaide W. L. and Oberbeck V. R. (1975) Development of the mare regolith: Some model considerations. *The Moon*, 13, 27-55.

Schaal R. B. and Hörz F. (1980) Experimental shock metamorphism of lunar soil. *Proc. Lunar Planet. Sci. Conf. 11th*, 1679-1695.

Schaal R. B., Hörz F., Thompson T. D., and Bauer J. F. (1979) Shock metamorphism of granulated lunar basalt. *Proc. Lunar Planet. Sci. Conf. 10th*, 2547-2571.

See T. H. and Hörz F. (1987) Formation of agglutinates in an experimental regolith. *Proc. Lunar Planet. Sci. Conf. 18th*, this volume.

Simon S. B., Papike J. J., Hörz F., and See T. H. (1986) An experimental investigation of agglutinate melting mechanisms: Shocked mixtures of Apollo 11 and 16 soils. *Proc. Lunar Planet. Sci. Conf. 17th*, in *J. Geophys. Res.*, 91, E64-E74.

Stöffler D. (1972) Deformation and transformation of rock-forming minerals by natural and experimental shock processes. I. Behavior of minerals under shock compression. *Fortschr. Mineral*, 49, 50-113.

Stöffler D., Gault D. E., Wedekind J., and Polkowski G. (1975) Experimental hypervelocity impact into quartz sand: Distribution and shock metamorphism of ejecta. *J. Geophys. Res.*, 80, 4062-4077.

Formation of Agglutinate-like Particles in an Experimental Regolith

Thomas H. See
Solar System Exploration Department, Lockheed-EMSCO, C/23, 2400 NASA Road One, Houston, Texas 77058

Friedrich Hörz
Experimental Planetology Branch, SN4, NASA Johnson Space Center, Houston, Texas 77058

Particles composed predominantly of glass (i.e., >50% and commonly >75%) were produced from a fragmental gabbro target that was repetitively impacted by Ni-alloy projectiles (3.18 mm in diameter, 5.4 km/s impact velocity). The morphology and many textural aspects of these particles are identical to lunar agglutinates. These "agglutinate-like particles," some as large as 8 mm, were recovered even after the first impact. After 50 impacts (i.e., after a cumulative kinetic energy of 9.7×10^{11} ergs had been deposited into the target) approximately 1% (weight) of the original 4000 g target had been converted into these constructional particles. The experimental glasses, however, are much more heterogeneous in composition than their lunar counterparts. They are dominated by incomplete mixing of melted component minerals, if not by glasses of monomineralic composition. Plagioclase-rich compositions dominate these impact melts, many being nearly identical to that of the feldspar end member. On average, most particles are variable mixtures of the major component minerals, albeit highly enriched in feldspar and substantially fractionated relative to the initial bulk target, as well as to the specific size fractions of the soil from which they were recovered. The magnitude of this fractionation, commonly a factor of two feldspar enrichment over that of the initial bulk target, clearly mandates preferential melting of feldspar and cannot be attributed solely to the preferential fusion of the fractionated "fine-fines" (i.e., <0.063 mm fraction). Characterized by preferential melting of feldspar over mafic components, the experimentally-produced glasses are products typical of shock stresses in the 40-70 GPa range. It is implied that fractionation trends observed within lunar agglutinitic-glasses may, in part, be due to phase-specific melting, rather than solely to the preferential fusion of the (fractionated) finest fraction. Furthermore, it is suggested that secondary cratering processes may play an important role in generating fractionated fine-fines via impact comminution.

INTRODUCTION

We have initiated an experimental program to study some aspects of regolith evolution in the laboratory. Fragmental targets of well-characterized initial fragment-size distribution are being repetitively impacted to investigate the comminution behavior of planetary surface rocks. Experimental variables include target materials, initial fragment-size distribution, impact velocity, and projectile density and mass.

We previously described the comminution and fractionation of a coarse-grained, fragmental gabbro target that was impacted 200 times at the NASA-JSC Vertical Impact Facility utilizing stainless-steel projectiles with an average velocity of 1.4 km/s (*Hörz et al.*, 1984). To evaluate the comminution behavior of the same gabbro at higher projectile velocities, an identical target was impacted 50 times at 5.4 km/s utilizing the NASA-Ames Vertical Ballistic Gun Range. We report here on the characterization of agglutinate-like, impact-melt particles that were an expected by-product of the 5.4 km/s impact experiments.

Agglutinates are unique lunar soil particles largely composed of variable amounts of clastic detritus bonded by impact-melt glass (*McKay et al.*, 1971). The modal frequency and other properties of agglutinates are widely accepted as a measure of soil "maturity" and as tracers of small-scale lunar surface processes (*McKay et al.*, 1974; *Morris*, 1978; *Papike et al.*, 1982; and others).

Largely stimulated by the work of *Rhodes et al.* (1975), detailed chemical characterization of agglutinitic glasses and their relation to the local soil/bedrock were initiated (e.g., *Gibbons et al.*, 1976; *Hu and Taylor*, 1977). More recent studies have demonstrated that agglutinitic glasses are fractionated relative to their host soils with glass compositions falling on a mixing line between that of the "bulk" soil (generally defined as the <1 mm grain-size fraction) and that of the "finest" grain-size fractions (<0.020 mm), the latter being enriched in feldspar and mesostasis compared to the "bulk" soil (*Evensen et al.*, 1973, 1974; *Korotev*, 1976; *Laul et al.*, 1981; *Papike et al.*, 1982; *Basu and McKay*, 1985).

The occurrence of these fractionation trends led *Papike et al.* (1981) to conclude that the finest soil fractions (i.e., <10 µm) melt preferentially. As a result, they suggested "Fusion of the Finest Fraction" (F^3) to be an important process during the formation of lunar agglutinates. The experimental work of *Simon et al.* (1985, 1986) offers evidence in support of the F^3 model. Furthermore, it has been shown that fractionation of the finer grain-size fractions is a natural consequence of mineral-specific comminution behavior during grinding (*Korotev and Haskin*, 1975) and/or repetitive impact (*Hörz et al.*, 1984). It appears possible, therefore, to generate fractionated fine-fines, as well as fractionated impact melts, entirely by *in situ* processes (*Papike et al.*, 1982). This view contrasts with the first-order interpretations of *Evensen et al.* (1973, 1974), who postulated that "exotic" components from

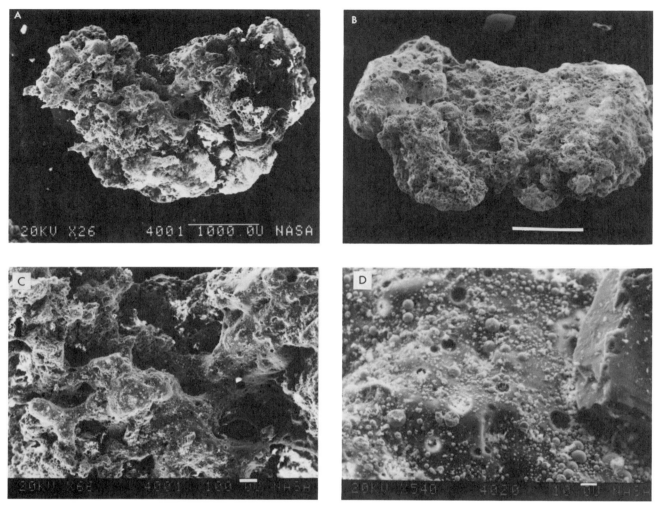

Fig. 1. SEM photographs of agglutinate-like particles. (**a**) and (**b**) Low magnification overviews. Scale bars = 1 mm. (**c**) Detailed particle surface showing ropy, vesicular nature of these melts. Scale bar = 0.1 mm. (**d**) High-magnification view displaying detailed surface characteristics. Note the large clast adhering to the exterior of the particle, which, in turn, has smaller detritus adhering to its surface, and the myriad of droplets covering the exterior of the glass. Scale bar = 0.01 mm.

lithologically different terranes were added to the local soils by ballistic sedimentation. A better understanding of the processes that might produce fractionated fines and associated agglutinitic glasses, therefore, will contribute to the important question concerning the efficiency of lateral mass transport by impact processes (e.g., *Shoemaker,* 1971; *Arvidson et al.,* 1975; *Oberbeck,* 1975; *Pieters et al.,* 1986). What materials represent local bedrock and which components are "exotic"?

The experimental comminution and melt products generated at 5.4 km/s appear suitable to address a number of questions regarding lunar regoliths, such as: (1) Are the fractionation trends exhibited by the fine fractions generated at 5.4 km/s similar to those produced in our initial experiment at 1.4 km/s? (2) How do the compositions of the agglutinitic glasses relate to that of the initial bulk target and to that of the specific grain-size fractions from which they were recovered? (3) Does the amount of fractionation evolve with time and is it related to cumulative bombardment history?

EXPERIMENTAL CONDITIONS

Procedures for the 5.4 km/s experiments were essentially identical to those of the 1.4 km/s series (*Hörz et al.,* 1984). The same gabbro (i.e., from the Bushveld Complex, South Africa) was employed; its dominant minerals are 1.5-5 mm in size and consisted of plagioclase (54%), orthopyroxene (22%), clinopyroxene (13%), orthoclase (<5%), quartz (<5%), and biotite, ilmenite, and magnetite (<2%). The initial size distribution of the fragmental target (2-32 mm) and total mass (4000 g) were also identical to that of the previous 1.4 km/s series.

Use of a substantially smaller Ni-alloy projectile (i.e., 3.18 mm in diameter and weighing 0.1415 g) was necessary to reach the desired 5.4 km/s velocity; these values compare to 6.35 mm and 1.02 g, respectively, for the stainless-steel projectiles of the 1.4 km/s experiments. Impact velocities were reproduced to within 10% of the desired 5.4 km/s and, generally, to within <5%. A series of 50 impacts was undertaken, each of which was nominally twice as energetic as one of the 1.4 km/s impacts. Thus, the cumulative impact energy of the 50 high-velocity impacts duplicated that of the first 100 impacts of the 1.4 km/s experiment. Shock-stress calculations for the 5.4 km/s impacts yield 110 GPa for solid gabbro, the prevailing scenario during the early stages when the target was composed solely of boulders, and some 65 GPa for a particulate gabbro target of 40% porosity (assumed by analogy with lunar regoliths; *Ahrens and Cole*, 1974); corresponding pressures for the 1.4 km/s impacts were 14.4 and 5.5 GPa, respectively. In order to obtain peak stresses of 110 GPa and 14.4 GPa using silicate projectiles, impact velocities of ~8 km/s and ~2 km/s, respectively, would be required. The two experimental series differ substantially, therefore, in peak pressure and associated decay of the shock stress (*Ahrens and O'Keefe*, 1977). On the basis of shock recovery experiments on basaltic targets (*Schaal and Hörz*, 1977, 1980; *Schaal et al.*, 1979) and on equation-of-state studies on lunar fines (*Ahrens and Cole*, 1974), the production of impact melt was expected for the 5.4 km/s impacts.

Sieving of the comminution products was performed following events 1, 5, and subsequently after every fifth impact; representative aliquots (0.5 g) for all grain-size fractions <1 mm were removed following each sieving operation. Sieving, microprobe, SEM, and modal analyses of the various grain-size fractions were performed as described by *Hörz et al.* (1984). Bulk compositions of all materials were determined via the fused-bead techniques of *Brown and Mullins* (1975). The following materials were analyzed.

1. Bulk-compositions of specific grain-size fractions for shots 1, 30, and 50, as well as some >0.25 mm fractions from the earlier 1.4 km/s experiment (previously not analyzed).

2. Bulk compositions of handpicked experimental agglutinates from the 0.125-0.25 mm and 0.5-1.0 mm grain-size fractions after event 50 (sufficient quantities of experimental agglutinates to manufacture glass beads could not be recovered from the 0.5 g aliquots extracted following the earlier shots).

3. Individual experimental agglutinates were selected as follows for microprobe analysis: (a) 0.125-0.25 mm grain-size fraction—9 particles after shot 1, 4 particles after event 30, and 3 particles after event 50; (b) 0.5-1.0 mm grain-size fraction—4 particles after shot 1, 3 particles after event 30, and 3 particles after event 50.

MORPHOLOGY OF AGGLUTINATE-LIKE PARTICLES

The average of some 20 thin-sectioned experimental agglutinates yields >75% glass, the remainder being composed of diverse clastic detritus. The experimental products resemble genuine lunar agglutinates in most major morphologic aspects (Figs. 1 and 2). They are recognized macroscopically by their

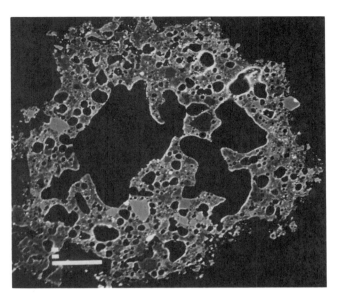

Fig. 2. Backscattered-electron image of a typical experimental particle in cross-section illustrating the percentage of melt (generally >75%), numerous clasts, and vesicles. Scale bar = 0.1 mm.

ropy, bulbous appearance, combined with a distinct glassy luster and abundant detritus clinging to their surfaces; many glassy areas appear transparent to silky. Generally, they are of higher albedo than their host soils, while most lunar agglutinates are darker than their host soils due to finely dispersed, paramagnetic phases (*Morris*, 1978). Dissemination of projectile-melt droplets is pervasive in the experimental melts (Fig. 2), but seems to have little effect on the overall albedo of the particles.

Binocular-microscopic observations reveal that these particles occur in all sieve fractions up to 8 mm, even after the first impact. They remain relatively rare in the coarser materials even after 50 impacts, but become increasingly more abundant in the smaller grain-size fractions. It is possible, if not likely, that some of the experimental agglutinates, especially the larger ones, were destroyed by subsequent impacts, as well as during sieving, owing to their friable nature (discovered during handpicking). The formation of these experimental agglutinates over a wide size range (i.e., <0.063 mm to 8 mm) from a coarse-grained target by lunar standards—if not from "rock" during the very first impact—points out that a "fine" soil fraction is not a mandatory prerequisite for the formation of particles with agglutinate-like morphologies. They can indeed form during impacts into solid rock.

MODAL ANALYSES

Detailed modal analyses of the 0.125-0.25 mm size fraction in the 1.4 and 5.4 km/s experiments are illustrated in Fig. 3. Point counting was aided by qualitative microprobe phase chemistry to distinguish reliably between clinopyroxene (cpx) and orthopyroxene (opx) (see *Hörz et al.*, 1984); some 300-500 grains were counted per aliquot. The melt-particle content is already 6% after the first impact and remains fairly constant following additional impacts. The modal frequency of

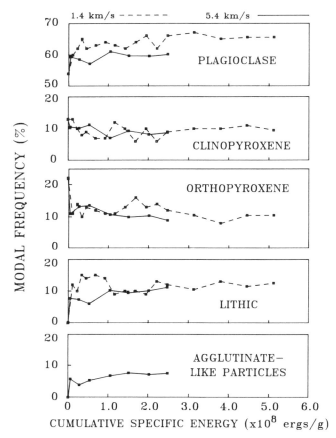

Fig. 3. Modal composition for the 0.125-0.25 mm size fractions from the 1.4 km/s and 5.4 km/s experiments versus specific energy (i.e., model time). The modest decrease in plagioclase enrichment in the 5.4 km/s series is accommodated by the formation of the agglutinate-like particles. Values at 0 ergs/g represent modal composition of the initial target-material.

monomineralic and lithic components indicates similar fractionation trends at 1.4 and 5.4 km/s; plagioclase is enriched in the fine-grained comminution products relative to the initial target while pyroxenes are depleted. In detail, however, the degree of plagioclase enrichment is less pronounced within the high-velocity experiments, a trend that will be explained following chemical analysis. Both experimental series yield similar proportions of lithic detritus. From the modal analyses we conclude that significant, mineral-specific fractionation occurs over a wide range of impact velocities during the comminution of planetary surfaces, with the fine-grained comminution products being enriched in feldspar.

The modal abundance of experimental agglutinates was also determined for the 0.063-0.125 mm grain-size fraction and is plotted together with that of the 0.125-0.25 mm fraction in Fig. 4a. The smaller grain-size fraction exhibits a slightly larger modal content that, like the abundance of 0.125-0.25 mm sized particles, remains fairly constant. Similar values have been reported for immature lunar regoliths (e.g., *Morris*, 1976). The mean grain size of the experimental regolith is plotted also in Fig. 4a in order to convey that a substantial change in the mean grain size (i.e., from ~14.8 to 2.8 cm) may occur without significantly affecting the modal soil composition (Fig. 3) and/or the frequency of these experimental agglutinates (Fig. 4a).

The increase in total comminuted mass, and that of the 0.063-0.125 mm and 0.125-0.25 mm size fractions are illustrated in Fig. 4b. If these masses are multiplied by the modal content of experimental melt particles (Fig. 4a), the increase in total melt mass can be calculated for these two size fractions (Fig. 4c). Thus, after only 50 impacts, the amount of melt mass within these two size fractions alone is some 0.75%. We estimate conservatively that another 0.25% melt mass resides in the unmeasured grain-size fractions. Thus, more than 1% of the total target had been molten.

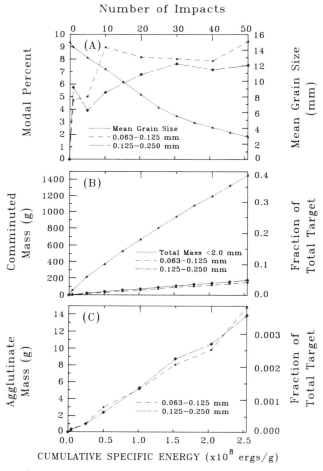

Fig. 4. Modal and absolute contents of melt particles in the experimental regolith. (**a**) Modal occurrence of the melt particles for the 0.063-0.125 mm and 0.125-0.25 mm size fractions and the evolution of the mean grain size as a function of time. Note that the Y-axis for the mean grain size appears on the right. (**b**) Total comminuted mass for all materials <2.0 mm and the 0.063-0.125 mm and 0.125-0.25 mm size fractions. (**c**) Absolute mass of experimental agglutinates for the 0.063-0.125 mm and 0.125-0.25 mm size fractions.

TABLE 1. Bulk and phase compositions for initial target (i.e., rock) and the various grain-size fractions following shots 1, 30, and 50.

	Bulk Target	Major Phases of Target				Shot Number and Size Fraction (mm)											
		Plag	Cpx	Opx	K-Spar	1 <0.063	1 >0.063	1 >0.125	1 >0.500	30 <0.063	30 >0.063	30 >0.125	50 <0.063	50 >0.063	50 >0.125	50 >0.250	50 >0.500
Na₂O	2.62	4.29	0.29	0.02	1.51	2.79	3.04	2.94	2.08	2.86	2.85	2.85	2.78	2.90	2.72	2.53	2.60
K₂O	0.83	0.34	0.01	0.01	14.24	1.11	0.92	0.73	0.61	1.10	0.84	0.71	1.06	0.85	0.70	0.64	0.71
CaO	9.25	12.38	21.09	1.16	0.16	9.15	9.66	9.90	9.33	9.11	9.49	9.84	9.05	9.54	9.61	9.67	9.35
SiO₂	55.16	52.82	50.92	52.23	64.60	54.56	54.73	54.10	53.33	55.44	54.70	54.35	54.68	54.39	53.69	53.11	54.28
Al₂O₃	17.08	29.65	1.11	0.45	18.51	19.01	20.35	19.59	14.11	18.91	19.20	19.16	18.58	19.63	18.38	17.22	17.08
FeO	8.15	0.60	10.89	25.43	0.09	7.12	5.95	6.87	10.78	6.43	6.90	7.48	6.47	6.38	7.59	8.54	8.78
MgO	6.44	0.00	14.98	20.06	0.00	4.64	4.18	4.82	8.42	4.42	4.58	5.15	4.50	4.51	5.29	6.47	6.74
TiO₂	0.29	0.02	0.45	0.10	0.00	0.57	0.42	0.39	0.40	0.47	0.52	0.42	0.48	0.45	0.44	0.30	0.33
MnO	0.15	0.00	0.25	0.53	0.00	0.12	0.11	0.15	0.23	0.12	0.13	0.13	0.13	0.12	0.15	0.16	0.17
Cr₂O₃	0.03	0.04	0.02	0.03	0.01	0.05	0.02	0.02	0.01	0.03	0.02	0.02	0.02	0.02	0.02	0.02	0.01
Ni	0.04	0.01	0.01	0.17	0.01	0.14	0.25	0.22	0.23	1.23	0.68	0.43	0.80	0.58	0.46	0.39	0.12
Total	100.04	100.15	100.02	100.19	99.13	99.27	99.64	99.72	99.53	100.14	99.90	100.52	98.55	99.37	99.03	99.06	100.18

Ni was utilized for tracking projectile contamination; Ni content of the projectile was ~60%.

CHEMICAL ANALYSES

A number of materials were analyzed for comparative purposes: (1) host-rock (i.e., initial bulk-target), (2) specific grain-size fractions including the experimental agglutinates in their modal proportions yielding "bulk-soil" compositions, (3) pure melt-particle separates including clastic detritus for determination of "bulk-particle" compositions, and (4) individual melt particles for the characterization of pure melts. "Glass composition" refers to individual probe analyses; the various glass compositions obtained from a single particle were averaged to yield melt compositions for "individual" particles and the pooling of a number of particles from within the same aliquot resulted in "overall glass" average compositions. The objectives of these chemical studies were to determine changes in soil composition as a function of grain size, to evaluate the homogeneity/heterogeneity of the experimental glasses for individual particles and between particles (i.e., melt variability in a specific soil fraction), and to compare melt particles of various sizes to their respective "host" grain-size fractions. Normalization of these compositions to that of the initial target rock permits identification of *in situ* processes.

Soil Compositions as a Function of Grain Size

The composition of specific grain-size fractions for events 1, 30, and 50 are listed in Table 1; the major findings after shot 50 are illustrated in Fig. 5 by plotting the enrichment/depletion of mafic components relative to the target rock. For comparison, some low-velocity data from *Hörz et al.* (1984) are presented as well, including new data for grain sizes >0.25 mm. All grain sizes <0.125 mm are depleted in pyroxenes; plagioclase enrichment in the fine grain-size fractions is essentially identical for both the low- and high-velocity experiments, amounting to some 20-25%. Detailed analysis of grain sizes <0.02 mm for the 1.4 km/s impacts (see *Hörz et al.*, 1984) revealed progressive fractionation with decreasing grain size, ultimately approaching ~40% plagioclase enrichment over that of the initial target (i.e., whole rock). We did not analyze specific fractionation trends in the <0.063 mm size fractions for the 5.4 km/s experiments. The larger grain sizes (i.e., 0.25-0.5 mm and 0.5-1.0 mm) exhibit pyroxene enrichment with the low-velocity impacts being substantially more enriched than their high-velocity counterparts (see section entitled "Discussion" below).

It follows from Figs. 3 and 5 that comminution of gabbro by repetitive impact leads to "fines" that are enriched in plagioclase and to coarse components that are pyroxene-enriched. These experimental fractionation trends are consistent with lunar observations but the degree of fractionation may differ. Lower impact velocities lead to more fractionation, according to Fig. 5, where mineral-specific effects are particularly pronounced at low-stress levels (i.e., much of the comminution occurs under tensile forces <0.1 GPa; *Cohn and Ahrens*, 1979). The high-stress regimes produced at cosmic

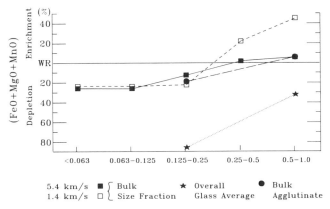

Fig. 5. Enrichment/depletion of mafic components relative to the initial gabbro in the experimental regolith materials resulting from the 1.4 km/s (shot 100) and 5.4 km/s (shot 50) experiments.

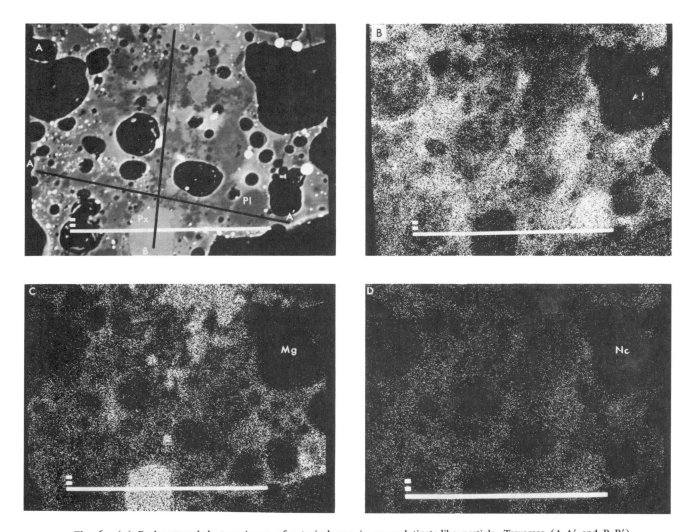

Fig. 6. (**a**) Backscattered-electron image of a typical area in an agglutinate-like particle. Traverses (A-A' and B-B') refer to data depicted in Figs. 7a and 7b. (**b-d**) X-ray images illustrating intensities and concentrations of Al, Mg, and Na, respectively. Note concentration of Al and Na in plagioclase clasts and Mg in pyroxene clasts. Scale bars = 0.1 mm.

velocities will ultimately pass through low-stress regimes upon decay of the shock front, implying that every natural impact is, in principal, capable of causing fractionation. However, the ratios of crater volumes subjected to high and low stresses vary as a function of velocity (*Ahrens and O'Keefe*, 1977). Low-velocity impacts will produce relatively more mass subjected to low stresses than high-velocity impacts; thus, low-velocity impacts will yield more fractionated materials.

Glass Compositions

Some simple qualitative observations are introduced first to illustrate the extreme and unusual compositional heterogeneity of the experimental glasses. Figure 6a depicts a typical melt particle with its highly vesicular texture; various density contrasts in this backscattered-electron image are readily apparent within the glassy materials, as are individual mineral clasts. Unusually bright spots represent disseminated projectile material. Figures 6b-d are X-ray intensity maps for Al, Mg, and Na, respectively, and qualitatively illustrate the variable proportions of plagioclase versus mafic components contributing to the melts on highly localized scales. Note that traverse A terminates in a large feldspar clast and that some glass compositions illustrated in Fig. 7a exhibit similar Al concentrations to that of the feldspar clast; conversely, traverse B encompasses a pyroxene clast and essentially pure pyroxene melts in its vicinity (Fig. 7b). The relative intensities of Al, Mg, and Fe vary sympathetically, indicative of poorly-mixed mineral melts, if not monomineralic glasses. The scale of heterogeneity is <5 μm, as the traverses were run in step-widths approaching the diameter of the focused electron beam (i.e., 2-3 μm).

The extreme chemical heterogeneity associated with the experimental melts is unlike that found within lunar

TABLE 2. Average compositions of experimental agglutinate melts for the >0.125-0.25 mm and >0.5-1.0 mm size fractions following shots 1, 30, and 50 and the "bulk" agglutinate compositions for the same size fractions after shot 50.

	Agglutinate-like Particles						Bulk	
	Glass							
Shot	1	30	50	1	30	50	50	50
Size	>0.125	>0.125	>0.125	>0.500	>0.500	>0.500	>0.125	>0.500
Na_2O	3.02	2.75	3.66	3.38	2.74	3.29	2.44	2.35
	(0.0-6.5)	(0.0-5.1)	(0.2-5.6)	(0.0-5.9)	(0.0-5.8)	(0.0-5.7)	(2.2-2.5)	(2.3-2.5)
	1.26	1.95	1.16	1.37	2.03	1.94	0.08	0.04
K_2O	1.01	0.39	0.67	0.40	0.27	0.29	0.64	0.72
	(0.0-13.8)	(0.0-2.6)	(0.1-10.5)	(0.0-1.9)	(0.0-1.6)	(0.0-0.6)	(0.5-0.7)	(0.7-0.8)
	2.31	0.49	1.80	0.30	0.26	0.17	0.05	0.02
CaO	10.56	11.38	11.67	11.67	9.50	10.16	9.46	9.31
	(0.4-21.4)	(0.1-22.5)	(0.2-19.4)	(1.0-21.9)	(0.8-22.2)	(0.8-22.0)	(9.3-9.8)	(9.1-9.5)
	3.87	5.95	2.96	3.66	5.90	5.06	0.11	0.08
SiO_2	55.26	54.82	54.23	53.00	53.26	53.32	55.30	53.99
	(49.3-83.0)	(49.6-98.3)	(51.1-67.4)	(48.9-60.4)	(51.0-60.9)	(51.0-57.3)	(54.3-56.1)	(53.6-54.4)
	5.67	7.62	3.04	1.94	1.86	1.80	0.41	0.19
Al_2O_3	22.28	18.75	27.20	24.06	18.91	21.97	18.55	16.26
	(0.5-31.0)	(0.1-30.6)	(0.9-30.5)	(0.5-30.0)	(0.4-30.3)	(0.3-30.5)	(18.3-18.7)	(16.1-16.5)
	8.74	13.05	7.10	9.48	13.56	12.12	0.10	0.09
FeO	3.93	5.89	1.24	3.78	8.02	5.49	6.27	8.63
	(0.2-26.2)	(0.2-25.1)	(0.2-12.4)	(0.2-26.2)	(0.3-26.0)	(0.3-25.1)	(6.0-6.9)	(8.5-8.8)
	5.32	8.30	2.88	6.21	10.74	9.11	0.21	0.09
MgO	3.05	5.41	1.00	2.92	6.71	4.54	5.62	6.90
	(0.0-20.2)	(0.0-20.7)	(0.0-15.1)	(0.0-19.9)	(0.0-20.6)	(0.0-20.5)	(5.5-5.9)	(6.8-7.1)
	4.70	7.65	3.46	5.41	8.94	7.80	0.08	0.06
TiO_2	0.18	0.15	0.04	0.14	0.10	0.10	0.42	0.49
	(0.0-1.2)	(0.0-1.7)	(0.0-0.2)	(0.0-1.5)	(0.0-0.7)	(0.0-1.3)	(0.4-0.5)	(0.4-0.5)
	0.21	0.25	0.06	0.26	0.11	0.23	0.03	0.03
MnO	0.08	0.14	0.03	0.08	0.18	0.11	0.20	0.25
	(0.0-0.6)	(0.0-0.7)	(0.0-0.3)	(0.0-0.6)	(0.0-0.6)	(0.0-0.6)	(0.1-0.3)	(0.2-0.3)
	0.12	0.19	0.06	0.15	0.23	0.19	0.03	0.04
Cr_2O_3	0.01	0.01	0.01	0.01	0.02	0.01	0.01	0.02
	(0.0-0.1)	(0.0-0.1)	(0.0-0.1)	(0.0-0.1)	(0.0-0.1)	(0.0-0.0)	(0.0-0.1)	(0.0-0.1)
	0.02	0.01	0.02	0.01	0.02	0.01	0.02	0.02
Ni	0.32	0.96	0.21	0.11	0.52	0.39	1.09	1.08
	(0.0-4.7)	(0.0-4.3)	(0.0-0.9)	(0.0-1.0)	(0.0-3.1)	(0.0-1.2)	(0.6-1.6)	(0.6-1.6)
	0.60	0.97	0.22	0.23	0.56	0.29	0.23	0.24
Total	99.69	100.66	99.98	99.56	100.21	99.66	100.00	100.00
(n)	76	50	53	47	49	54	29	31
(N)	13	4	5	6	3	4	15	9

Numbers in parentheses give compositional ranges based on individual analyses, while the number below the range represents the average standard deviation of the range. n = number of individual analyses, N = number of experimental agglutinate particles analyzed.

agglutinates, the latter being characterized by relatively homogeneous glasses (e.g., *Gibbons et al.*, 1976; *Hu and Taylor*, 1977; *Papike et al.*, 1982; *Basu and McKay*, 1985; and others). While our experimental particles are excellent analogs for lunar agglutinates in a morphologic and textural sense, they are poor analogs in terms of the chemical and physical heterogeneity/homogeneity of the impact melt. The significant difference with lunar agglutinate melts is clearly illustrated by the ranges in glass compositions (Table 2) and by a comparison of averaged standard deviations (Fig. 8). Compared to the variability of lunar agglutinitic glasses, standard deviations for most oxides in the experimental particles are a factor of five larger, and that of Al_2O_3 by a factor of ten. This suggests that dramatically different melting and/or homogenization processes must have been involved in the formation of the experimentally-produced agglutinates compared to the lunar case.

The quantitative microprobe data listed in Table 2 are graphically illustrated in Fig. 9. Most individual glass analyses (Figs. 9a,b) are consistent with, or cluster near, pure monomineralic compositions associated with the major phases of the initial target. More analyses cluster around the plagioclase end member than around the two pyroxenes. The smaller particles tend to exhibit more "mixed" compositions than their large-grained counterparts. Particles resulting from the first

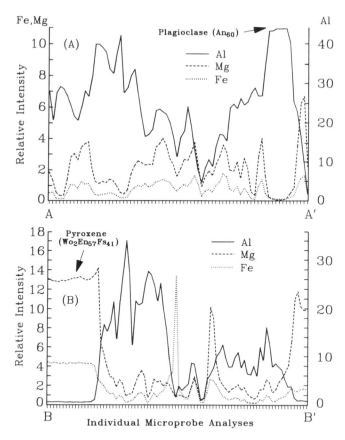

Fig. 7. Relative intensities of Al, Mg, and Fe determined during the microprobe traverses indicated in Fig. 6.

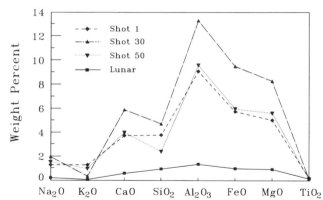

Fig. 8. Plot of average standard deviation for experimental and natural agglutinitic glasses. Standard deviations for lunar glasses taken from *Basu and McKay* (1985) obtained on particles ranging in size from 0.25-1.0 mm, while data from this study include particles from 0.125-1.0 mm. The experimental data represent glass averages at specific times in the evolution of the experimental regolith.

impact exhibit the highest proportion of mixed melts. We cannot offer self-consistent explanations for some of the time-related trends visible within Fig. 9 and consider them of secondary importance, possibly related to small sampling statistics. Nevertheless, 9 of the 27 particles analyzed are essentially feldspathic, while only one exhibits melt of almost pure pyroxene composition (Figs. 9c,d). The remainder are of mixed composition representing variable mixtures of the major phases of the gabbro. Most of these particle averages are biased toward plagioclase, as are the overall melt averages that combine a number of particles from the same aliquot. In addition, the average glass composition of individual particles recovered from a specific grain-size fraction vary widely, encompassing a larger compositional range than has been reported from individual agglutinates from specific lunar soil fractions (e.g., *Taylor et al.*, 1979; *Basu and McKay*, 1985). In summarizing the glass results, the major points are (1) the experimental glasses are characterized by incompletely homogenized mineral melts to a much greater degree than their lunar counterparts, and (2) the experimental glasses are much more enriched in feldspar than are the glasses associated with lunar agglutinates.

As can be seen in Fig. 9c, the "bulk-particle" compositions (including mineral and lithic clasts) cluster much closer to the composition of the "bulk target" (i.e., BT in Fig. 9) than do the individual glass averages for the experimental agglutinates. Similar results can be seen for the 0.5-1.0 mm particles (Fig. 9d) where the modest displacement towards excessive pyroxene is, in part, an artifact caused by disseminated projectile material that was rich in Fe. The "bulk-particle" compositions match the composition of their host soils fairly well (Fig. 5). The addition of clasts that are complementary to the fractionated "melts" can do nothing but bring the bulk-particle composition closer to its progenitor.

In contrast, the "overall glass average" after event 50 (Fig. 5) is so severely biased toward feldspar that it is impossible to derive these compositions by preferential fusion of fractionated fines alone. The glasses within the 0.125-0.25 mm particles have feldspar enrichments approaching a factor of three relative to the initial bulk target and a factor of two relative to their host soil fraction. Even the 0.5-1.0 mm particles are enriched in plagioclase by some 30% relative to the initial bulk target and by a still larger degree relative to the 0.5-1.0 mm soil fraction. Preferential melting of feldspar, rather than melting of the fractionated fines, dominates the melt compositions within our experimental agglutinates.

DISCUSSION

The experimental particles are morphologically and, in many respects, texturally identical to lunar agglutinates; however, the glasses associated with these experimentally-produced particles are more heterogeneous than their lunar counterparts. The experimental glasses are dominated by incompletely mixed mineral melts, totally unlike lunar agglutinitic glasses.

We interpret the preponderance of poorly homogenized, if not predominantly monomineralic, melts to be the result of peak shock stresses barely sufficient to initiate melting of component phases. Such melts are typical for shock stresses in the 50-70 GPa range for experimentally-shocked basaltic targets (*Kieffer et al.*, 1976), and are found in association with

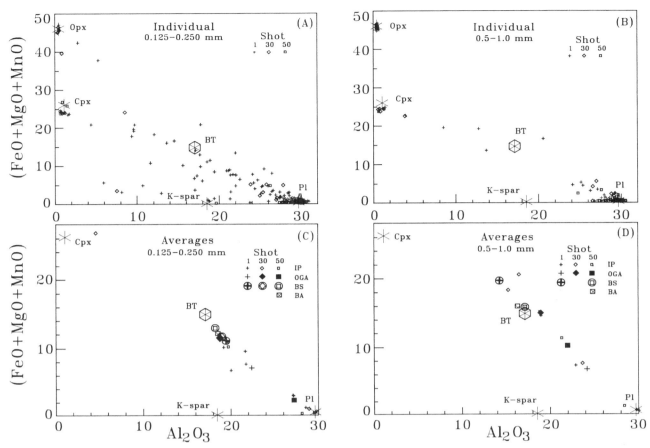

Fig. 9. (FeO + MgO + MnO) versus Al_2O_3 variation diagram of individual analyses (**a and b**) and various average compositions (**c and d**) for the 0.125-0.25 mm and 0.5-1.0 mm grain-size fractions. Data in plots (**a**) and (**b**) represent individual microprobe analyses. In plots (**c**) and (**d**): IP = average glass-composition for individual particles; OGA = overall glass average for particles from a given shot; BS = bulk composition of size fractions after a given shot; BA = bulk composition of experimental agglutinates following shot 50; BT = initial bulk-target composition for all plots.

powdered targets that have been subjected to shock stresses of 20-40 GPa (*Gibbons et al.,* 1976; *Schaal et al.,* 1979). Feldspar, being the more compressible phase, melts at lower shock stresses (i.e., 45-50 GPa; *Ostertag,* 1983) than pyroxene (i.e., >70 GPa; *Stöffler,* 1972; *Schaal and Hörz,* 1977). The preferential melting of tecto- versus inosilicates is characteristic for naturally and experimentally shocked rocks. The dominant stresses generated during our regolith evolution experiments simply did not extend sufficiently into the field of "whole rock" melting (>70 GPa in dense basalts), or "whole soil" melting (>40 GPa in porous targets) to allow all phases to contribute in their original modal proportions to the resulting impact melt.

In contrast, lunar agglutinate glasses are much more homogeneous and more closely resemble the composition of their progenitor soil(s). The formation of homogeneous glasses resembling the soil would be expected with impacts involving cosmic velocities where the peak shock stresses would range well into the whole-rock and whole-soil melt regimes. Additionally, typical lunar soils possess a much smaller mean grain size than our experimental regolith, thereby constituting a more "homogeneous" target in a modal and physical sense and, thus, facilitating the formation of relatively homogeneous melts on very small scales. Furthermore, calcium-rich feldspars (>An_{90}) are exceedingly resistant to shock-induced melting (*Harrison and Hörz,* 1981; *Ostertag,* 1983) requiring peak pressures approaching that of pyroxene to undergo solid-liquid phase transitions. As a consequence, preferential melting of feldspar relative to pyroxene may be less pronounced in lunar soils where such calcium-rich feldspars are common. Thus, the generation of relatively homogeneous impact melts during natural impacts into lunar soils is to be expected.

We are nevertheless surprised that *all* (!) lunar agglutinate melts are generally portrayed to be very homogeneous. During every natural impact the decaying shock wave passes through intermediate stress regimes amenable to mineral-specific melting described from these and many other experiments. Following *Ahrens and O'Keefe* (1977), the target volume engulfed by the 30-60 GPa isobars is approximately equal to the volume subjected to stresses >70 GPa. Where are the "heterogeneous" lunar soil melts that should be generated at these intermediate pressures? The fact that the experimental melt particles resemble lunar agglutinates in morphology—even following the first impact into essentially "rock"—precludes the possibility that lunar agglutinates may be biased (during sample processing) toward homogeneous "high-pressure" melts on the basis of morphology. Very fine-grained progenitor soils yielding physically and chemically "homogeneous" targets on very small scales seem to be the most plausible explanation for the homogeneous lunar melts. Such a conclusion is supported by the F^3 model of *Papike et al.* (1981), which suggests that a substantial portion of the melt is derived from progenitor materials <20 μm in diameter. In addition, many lunar agglutinates, especially in mature soils, may be the products of multiple impacts and melting events (*Basu and McKay*, 1985) facilitating the production of increasingly more homogeneous melts. This explanation, however, surely does not apply to every lunar agglutinate, especially to those extracted from relatively immature soils. *Basu and McKay* (1985) find no difference between lunar agglutinates extracted from immature and mature lunar soils with respect to the melt homogeneity/heterogeneity. Thus, we have no plausible explanation to the apparent absence of natural agglutinates possessing textural and compositional heterogeneities akin to those reported here.

The compositions of our experimental particles are heavily biased toward feldspar, not an unexpected result considering the experiment-specific conditions (i.e., impact velocity, peak stress, feldspar composition). While calcium-rich feldspars tend to melt at shock stresses >50 GPa (An_{86}; *Schaal and Hörz*, 1977) and >70 GPa (An_{95}; *Harrison and Hörz*, 1981), they nevertheless clearly melted before pyroxene in the lunar-basalt experiments of Schaal and Hörz. It is possible that part of the feldspar enrichment relative to progenitor material(s) reported from lunar agglutinates (e.g., *Laul et al.*, 1981; *Papike et al.*, 1982; *Basu and McKay*, 1985) is caused by the preferential melting of feldspar. Phase-specific melting, as well as preferential melting of exceedingly small grain sizes (*Papike et al.*, 1982) may contribute to the observed fractionation trends. While the relative roles of these two processes are difficult to assess, they may both lead to *in situ* fractionation during shock melting of fine-grained powders. Our experimental melts illustrate that not all fractionation may be due solely to the preferential fusion of the finest fraction (i.e., F^3). Phases that may partition preferentially into lunar-soil melts include, next to feldspars, all preexisting glasses, including mesostasis and agglutinates.

Lastly, we address two major inferences derived from the modal analyses illustrated in Figs. 3 and 4 and from the bulk compositions of specific soil fractions illustrated in Figs. 5 and 9:

1. The 5.4 km/s experiments produced fractionation trends similar to those observed in the 1.4 km/s experiments of *Hörz et al.* (1984). Due to the effects of mineral-specific comminution, feldspar is enriched in the fine-grained soil fractions and conversely, pyroxene is enriched in the coarse fractions. The magnitude of fractionation, however, seems to depend on the impact velocity, with the low-velocity regime producing larger degrees of fractionation. Not unexpectedly, mineral-specific behavior is particularly manifested at modest stresses where the differences in the physical properties of the various minerals are amplified relative to high-velocity/high-stress regimes. On the basis of the shock-decay profiles of *Ahrens and O'Keefe* (1977), the normalized crater-volume subjected to stresses at, or modestly above the dynamic tensile strength of typical silicates (i.e., 0.05 to 0.1 GPa) is larger for low-velocity impacts than is the case for high-velocity events. As a consequence, low-velocity impacts seem to produce larger quantities of fractionated material. This leads to the conclusion that secondary impact processes may have contributed substantially to the fractionated nature of lunar soils.

2. Having generated agglutinate analogs under controlled laboratory conditions, it seems natural to apply these findings to the production rate of natural agglutinates. In its most simplistic form, an "agglutinate production curve" relates the absolute modal abundance of melt particles to the absolute surface-residence time of a specific lunar soil (e.g., *Mendell and McKay*, 1975; *Morris*, 1976; *Basu*, 1977; *McKay and Basu*, 1983). Our 5.4 km/s experiments provide evidence that the modal soil composition, and especially the modal agglutinate content, may remain fairly constant for a substantial period of time (see Figs. 3 and 4; a constant modal composition for the 0.125-25 mm size fraction was also observed in the previous 1.4 km/s experiments). Accordingly, the modal composition remains fairly constant during the early evolutionary history of a regolith, closely resembling that following the first impact(s) into boulders and/or bedrock. While we cannot quantify the time period at which the modal composition begins to change, specifically by an increase in agglutinate content, this early stage may last for substantial periods. Changes in the overall mean grain size during this initial period were observed to be a factor of five for the 5.4 km/s series and a factor of 35 for the 1.4 km/s experiments; however, the modal composition remained relatively constant for both series. It follows that a substantial amount of mechanical processing and associated surface residence time must be necessary before the modal content of agglutinates increases measurably over that typical for single impacts into unprocessed materials.

These experiments cannot contribute to the question of whether there is a dominant crater size responsible for lunar agglutinate production. On the other hand, our investigation does refute the suggestions of *Lindsay* (1976) in which he postulated that each agglutinate results from a single cratering event and that the mass of the agglutinate is related to mass of the impactor. Clearly, a large number of agglutinates may be generated during a single impact and their mass-frequency in no way provides a measure of the mass-frequency of micrometeoroids.

In conclusion, our regolith-evolution experiments on fragmental gabbro targets at an impact velocity of 5.4 km/s have strengthened some previously-held views but have also raised new questions regarding lunar agglutinates and the evolution of planetary regoliths in general. Among the arguments strengthened is that impact comminuted fine-fines are indeed fractionated relative to their progenitor rocks and that the composition of soil-derived impact melts is intermediate between the host soil and those fractionated fines. These are the principle features of the F^3 agglutinate formation model (*Papike et al.,* 1981, 1982). New insights and questions relate to the strong possibility that phase-specific melting may contribute to these fractionation trends as well. Furthermore, our experiments suggest that secondary cratering may have played a substantial role in developing the fractionation trends observed within the lunar regolith. Lastly, the question remains as to why lunar-agglutinitic glasses representing intermediate stress regimes akin to our experimental particles are not observed. Such heterogeneous melts are characteristic of many naturally- and experimentally-shocked materials. Is this simply a sampling and/or reporting bias, or does it reflect some fundamental property of the agglutinate-forming process?

Acknowledgments. The able assistance of the NASA-Ames Vertical Ballistic Gun Range crew (Earl Brooks, Ben Langedyke, and John Von Gray) is appreciated, as are the contributions of M. J. Cintala and S. Smrekar. The constructive comments of A. Basu and an anonymous reviewer are appreciated.

REFERENCES

Ahrens T. J. and Cole D. M. (1974) Shock compression and adiabatic release of lunar fines from Apollo 17. *Proc. Lunar Sci. Conf. 5th,* 2333-2345.

Ahrens T. J. and O'Keefe J. D. (1977) Equation of state and impact-induced shock-wave attenuation on the Moon. In *Impact and Explosion Cratering* (D. J. Roddy, R. O. Pepin, and R. B. Merrill, eds.), pp. 639-656. Pergamon, New York.

Arvidson R., Drozd R., Hohenberg C. M., Morgan D. J., and Poupeau G. (1975) Horizontal transport in the regolith, modification of features, and erosion rates of the lunar surface. *The Moon, 13,* 67-79.

Basu A. (1977) Steady state, exposure age, and growth of agglutinates in lunar soils. *Proc. Lunar Sci. Conf. 8th,* 3617-3632.

Basu A. and McKay D. S. (1985) Chemical variability and origin of agglutinitic glass. *Proc. Lunar Planet. Sci. Conf. 16th,* in *J. Geophys. Res., 90,* D87-D94.

Brown R. and Mullins O. (1975) Electron microprobe whole rock analysis of 20mg samples. *Proc. Natl. Conf. Electron Probe Anal. 10th,* Paper 32A-32D.

Cohn S. N. and Ahrens T. J. (1979) Dynamic tensile strength of analogs of lunar rocks (abstract). In *Lunar and Planetary Science X,* pp. 224-226. Lunar and Planetary Institute, Houston.

Evensen N. M., Murthy V. R., and Coscio M. R. (1973) Rb-Sr ages of some mare basalts and the isotopic and trace element systematics in lunar fines. *Proc. Lunar Sci. Conf. 4th,* 1707-1724.

Evensen N. M., Murthy V. R., and Coscio M. R. (1974) Provenance of KREEP and the exotic component: Elemental and isotopic studies of grain-size fractions in lunar soils. *Proc. Lunar Sci. Conf. 5th,* 1401-1417.

Gibbons R. V., Hörz F., and Schaal R. B. (1976) The chemistry of some individual lunar soil agglutinates. *Proc. Lunar Sci. Conf. 7th,* 405-422.

Harrison W. J. and Hörz F. (1981) Experimental shock metamorphism of calcic plagioclase (abstract). In *Lunar and Planetary Science XII,* pp. 395-396. Lunar and Planetary Institute, Houston.

Hörz F., Cintala M. J., See T. H., Cardenas F., and Thompson T. D. (1984) Grain-size evolution and fractionation trends in an experimental regolith. *Proc. Lunar Planet. Sci. Conf. 15th,* in *J. Geophys. Res., 89,* C183-C196.

Hu H. N. and Taylor L. A. (1977) Lack of chemical fractionation in major and minor elements during agglutinate formation. *Proc. Lunar Sci. Conf. 8th,* 3645-3656.

Kieffer S. W., Schaal R. B., Gibbons R. V., Hörz F., Milton D. J., and Dube A. (1976) Shocked basalt from Lonar Crater, India, and experimental analogs. *Proc. Lunar Sci. Conf. 7th,* 1391-1412.

Korotev R. L. and Haskin L. A. (1975) Inhomogeneity of trace element distributions from studies of the rare earths and other elements in size fractions of crushed basalt 70135 (abstract). In *Papers Presented to the Conference on the Origin of Mare Basalts,* pp. 86-90. The Lunar Science Institute, Houston.

Korotev R. L. (1976) Geochemistry of grain-size fractions of soils from the Taurus-Littrow Valley floor. *Proc. Lunar Sci. Conf. 7th,* 695-726.

Laul J. C., Papike J. J., and Simon S. B. (1981) The lunar regolith: Comparative studies of the Apollo and Luna sites. Chemistry of soils from Apollo 17, Luna 16, 20 and 24. *Proc. Lunar Planet. Sci. 12B,* 389-407.

Lindsay J. F. (1976) *Lunar Stratigraphy and Sedimentology.* Elsevier, New York. 302 pp.

McKay D. S. and Basu A. (1983) The production of agglutinates in planetary regoliths. *Proc. Lunar Planet. Sci. Conf. 14th,* in *J. Geophys. Res., 88,* B193-B199.

McKay D. S., Morrison D. A., Clanton U. S., Ladle G. H., and Lindsay J. F. (1971) Apollo 12 soil and breccia. *Proc. Lunar Sci. Conf. 2nd,* 755-773.

McKay D. S., Fruland R. M., and Heiken G. H. (1974) Grain size and evolution of lunar soils. *Proc. Lunar Sci. Conf. 5th,* 887-906.

Mendell W. W. and McKay D. S. (1975) A lunar soil evolution model. *The Moon, 13,* 285-292.

Morris R. V. (1976) Surface exposure indices of lunar soils: A comparative FMR study. *Proc. Lunar Sci. Conf. 7th,* 315-335.

Morris R. V. (1978) In situ reworking (gardening) of the lunar surface: evidence from the Apollo cores. *Proc. Lunar Planet. Sci. Conf. 9th,* 1801-1811.

Oberbeck V. R. (1975) The role of ballistic erosion and sedimentation in lunar stratigraphy. *Rev. Geophys. Space Phys., 13,* 337-362.

Ostertag R. (1983) Shock experiments on feldspar crystals. *Proc. Lunar Planet. Sci. Conf. 14th,* in *J. Geophys. Res., 88,* B364-B376.

Papike J. J., Simon S. B., White C., and Laul J. C. (1981) The relationship of the lunar regolith <10 μm fraction and agglutinates. Part I: A model for agglutinate formation and some indirect supportive evidence. *Proc. Lunar Planet. Sci. 12B,* 409-412.

Papike J. J., Simon S. B., and Laul J. C. (1982) The lunar regolith: Chemistry, mineralogy, and petrology. *Rev. Geophys. Space Phys., 20,* 761-826.

Pieters C. M., Adams J. B., Mouginis-Mark P. J., Zisk S. H. Smith M. O., Head J. W., and McCord T. B. (1986) The nature of crater rays: the Copernicus example. *J. Geophys. Res., 90,* 12393-12413.

Rhodes J. M., Adams J. B., Blanchard D. P., Charette M. P., Rodgers K. V., Jacobs J. W., Brannon J. C., and Haskin L. A. (1975) Chemistry of agglutinate fractions in lunar soils. *Proc. Lunar Sci. Conf. 6th,* 2291-2307.

Schaal R. B. and Hörz F. (1977) Shock metamorphism of lunar and terrestrial basalts. *Proc. Lunar Sci. Conf. 8th,* 1697-1729.

Schaal R. B. and Hörz F. (1980) Experimental shock metamorphism of lunar soil. *Proc. Lunar Planet. Sci. Conf. 11th,* 1679-1695.

Schaal R. B., Hörz F., Bauer J. F., and Thompson T. D. (1979) Shock metamorphism of granulated lunar basalt. *Proc. Lunar Planet. Sci. Conf. 10th,* 2547-2571.

Shoemaker E. M. (1971) Origin of fragmental debris on the lunar surface and the history of bombardment of the Moon. *Instituto de Investigaciones Geologicas de la Diputacion Provincial, Universidad de Barcelona, XXV,* pp. 27-56.

Simon S. B., Papike J. J., Hörz F., and See T. H. (1985) An experimental investigation of agglutinate melting mechanisms: Shocked mixtures of sodium and potassium feldspars. *Proc. Lunar Planet. Sci. Conf. 16th*, in *J. Geophys. Res., 90*, D103-D115.

Simon S. B., Papike J. J., Hörz F., and See T. H. (1986) An experimental investigation of agglutinate melting mechanisms: shocked mixtures of Apollo 11 and 16 soils. *Proc. Lunar Planet. Sci. Conf. 17th*, in *J. Geophys. Res., 91*, E64-E74.

Stöffler D. (1972) Deformation and transformation of rock-forming minerals by natural and experimental shock processes, I: Behavior of minerals under shock compression. *Fortschr. Mineral., 49*, 50-113.

Taylor G. J., Warner R. D., and Keil K. (1979) Agglutinates as recorders of fossil soil compositions. *Proc. Lunar Planet. Sci. Conf. 10th*, 1159-1184.

Dehydration Kinetics of Shocked Serpentine

James A. Tyburczy
Department of Geology, Arizona State University, Tempe, AZ 85287

Thomas J. Ahrens
Seismological Laboratory, California Institute of Technology, Pasadena, CA 91125

Experimental rates of dehydration of shocked and unshocked serpentine were determined using a differential scanning calorimetric technique. Dehydration rates in shocked serpentine are enhanced by orders of magnitude over corresponding rates in unshocked material, even though the impact experiments were carried out under conditions that inhibited direct impact-induced devolatilization. Extrapolation to temperatures of the Martian surface indicates that dehydration of shocked material would occur 20 to 30 orders of magnitude more rapidly than for unshocked serpentine. The results indicate that impacted planetary surfaces and associated atmospheres would reach chemical equilibrium much more quickly than calculations based on unshocked material would indicate, even during the earliest, coldest stages of accretion. Furthermore, it is suggested that chemical weathering of shocked planetary surfaces by solid-gas reactions would be sufficiently rapid that true equilibrium mineral assemblages should form.

INTRODUCTION

Impacts into volatile-bearing rocks and minerals may have strongly influenced planetary accretion and the development of primitive planetary atmospheres. A number of accretion models have been proposed that focus on the importance of the extent of impact-induced devolatilization as a function of impact velocity (or planetary size) in determining the course of planetary atmospheric and thermal evolution (*Arrhenius et al.*, 1974; *Benlow and Meadows*, 1977; *Jakosky and Ahrens*, 1979; *Lange and Ahrens*, 1982a; *Abe and Matsui*, 1985). Impact-induced devolatilization of minerals was first observed by *Boslough et al.* (1980) in hydrous silicates. Subsequent studies have extended the observations in hydrous minerals to higher shock pressures (*Lange et al.*, 1985) and have investigated shock devolatilization in carbonates (*Boslough et al.*, 1982; *Lange and Ahrens*, 1986) and meteorites (*Tyburczy et al.*, 1986).

TABLE 1. Chemical composition of Globe, Arizona lizardite.

Oxide	Wt %
SiO_2	42.28
TiO_2	0.04
Al_2O_3	0.42
CaO	0.03
MgO	41.94
FeO	0.36
MnO	0.08
K_2O	0.02
Na_2O	0.01
H_2O	12.7
Total	97.88

H_2O analysis by TGA. All other oxides analysed by electron microprobe.

Shock-enhanced chemical reactivity is an additional effect of impact that is of importance in planetary evolution. Shock deformation and comminution lead to increased reactivity in oxides, metals, and organic polymers (*Morosin and Graham*, 1982). *Boslough et al.* (1980, 1986) suggested that shocked silicate minerals will be similarly affected. Infrared spectra of shocked serpentine suggest a reduction in the bonding strength of water in the mineral (*Boslough et al.*, 1980; *Lange et al.*, 1985). Enhanced solid-gas reaction rates are of importance in terms of chemical buffering of atmospheric composition through interactions with surface minerals and in terms of surface weathering processes on planets lacking liquid water. We have undertaken a study of the rate of thermally-induced dehydration of shocked lizardite serpentine in order to quantify the degree of enhancement of solid-gas reaction rates in shocked silicate materials.

EXPERIMENTAL

The impact experiments were performed on a nearly pure magnesium end member polycrystalline lizardite serpentine [ideal formula $Mg_3Si_2O_5(OH)_4$] collected near Globe, Arizona (Caltech Mineral Collection, Specimen Number 3221). The bulk chemical analysis is shown in Table 1. The density of the mineral is 2.50 ± 0.02 g/cm^3.

Solid-recovery impact-induced devolatilization experiments were performed on the 40-mm powder gun at the California Institute of Technology using the stainless steel sample chamber assembly sketched in Fig. 1. Using this assembly, sample masses on the order of 0.5 to 1 gram could be impacted and recovered. The sample was inserted into the assembly as a powder, and then pressed in by the containing plug. Sample densities were calculated from the mass of material added and the final volume of the sample. Densities of approximately 85% of the bulk mineral density were attained. Sample dimensions were about 1.0 cm in diameter by 0.5 cm thick. The impacted surface-to-sample distance was approximately 0.5 cm. An attempt was

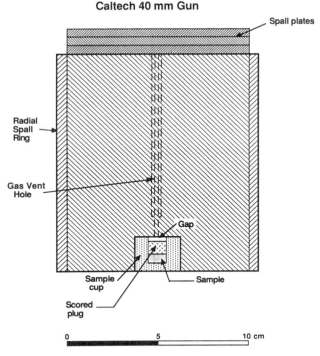

Fig. 1. Schematic diagram of the large-volume, solid recovery sample chamber assembly. Sample cup and plug, 304 stainless steel. Spall plates and ring, cold-rolled steel.

made to provide venting by scoring the sides of the containing plug behind the sample and by providing an escape tube out the back of the assembly. The results indicate that this attempt was not entirely successful and the data from this study (shown in Fig. 2) are indicated as partially vented. For impact velocities up to 1.25 km/s (shots 692 and 693, P_{peak} = 23.0 and 27.5 GPa, respectively) there was no impact-induced devolatilization, while an impact velocity of 1.55 km/s (shot 694, P_{peak} = 35.4 GPa) resulted in 16% devolatilization. Previous results (*Lange et al.*, 1985; *Tyburczy et al.*, 1986) indicate that 10% to 40% devolatilization would be expected had the chamber been adequately vented (Fig. 2). *Lange et al.* (1985) and *Tyburczy and Ahrens* (1985, 1986) discuss the problems of vented versus unvented sample assemblies in impact-induced devolatilization experiments.

Shock pressures were calculated using a one-dimensional impedance match solution (*Rice et al.*, 1958). We report both the initial shock pressure and the peak reverberated shock pressure in Table 2. Problems with nonplanar propagation of shock waves and design of recovery fixtures are discussed by *Graham and Webb* (1984). Equation-of-state parameters for stainless steel were taken from *Marsh* (1980); for serpentine we used initial (nonporous) density ρ_0 = 2.50 g/cm^3, C_0 = 5.30 km/s, and s = 0.78 (Tyburczy and Ahrens, in preparation), where shock velocity U_s and particle velocity u_p are related by $U_s = C_0 + s\, u_p$. The Hugoniot of porous serpentine was calculated using the relation

$$P_{H,p} = P_{H,d} \{1 - \gamma[(\rho_H/\rho_0) - 1]/2\}/\{1 - \gamma[(\rho_H/\rho_{00}) - 1]/2\} \quad (1)$$

(*Ahrens and O'Keefe*, 1972). In equation (1) $P_{H,p}$ is the Hugoniot pressure of the porous material, $P_{H,d}$ is the Hugoniot pressure (at ρ_H) of the nonporous material, γ is the Grüneisen parameter, ρ_H is the Hugoniot density, ρ_0 is the initial density of the nonporous material, and ρ_{00} is the initial porous material density. The Grüneisen parameter is calculated using $\gamma = \gamma_0 (\rho_0/\rho_H)^n$ with γ_0 = 1.0 and n = 1.0 (Tyburczy and Ahrens, in preparation). Initial densities, impact velocities, and calculated initial and peak shock pressures are listed in Table 2. Experimentally determined errors in ρ_0, ρ_{00}, and projectile

TABLE 2. Experimental results.

Shot Number	Initial Density g/cm^3	Projectile Velocity km/s	Initial Shock Pressure GPa	Peak Shock Pressure GPa	DSC Heating Rate K/min	Tmax of Water-Loss Peak deg C	Activation Energy kJ/mole	log (Ao) Ao in 1/sec	Impact-Induced Devolatilization %
Unshocked	-	-	-	-	25	722			
					10	699			
					5	684	340	16.1	
					2	667	±15	±0.8	-
692	2.13	1.07	7.5	22.7	25	670			
	±0.04	±0.02	±1.8	±0.2	10	641			
					5	619	220	10.3	
					2	598	±10	±0.6	0
693	2.17	1.25	9.3	27.2	25	669			
	±0.04	±0.02	±2.0	±0.2	10	641			
					5	620	235	11.2	
					2	600	±11	±0.7	0
694	2.09	1.55	12.5	35.1	25	673			
	±0.04	±0.03	±2.4	±0.4	10	643			
					5	621	220	10.2	
					2	600	±11	±0.6	16

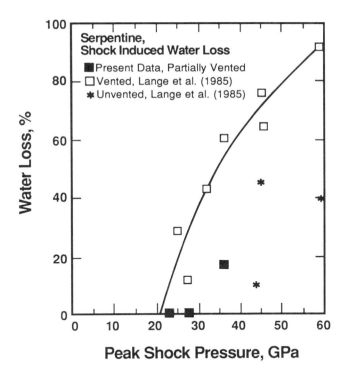

Fig. 2. Impact-induced volatile loss for shocked serpentine. Solid curve is an empirical fit to the vented assembly data of *Lange et al.* (1985).

of the water-loss peak for the shocked material relative to the unshocked material, consistent with the more qualitative results of *Lange et al.* (1985) and *Lange and Ahrens* (1986). Furthermore, there is a systematic increase in the temperature and width of the water loss peak with increasing heating rate.

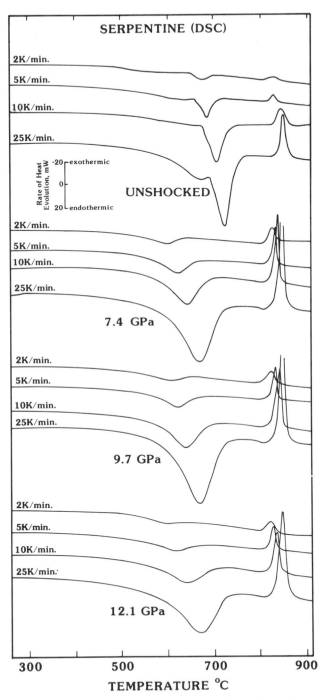

Fig. 3. Differential scanning calorimetry data for different heating rates for shocked and unshocked serpentine. The endothermic (downgoing) peak corresponds to the water loss. Pressure values correspond to initial shock pressures (Table 2).

velocity V_p were assumed to be independent and were propagated through the impedance match relations and equation (1) to yield uncertainties in initial and peak shock pressures (*Bevington*, 1969). Estimates of peak- and postshock temperature, calculated using the serpentine equation of state parameters (Tyburczy and Ahrens, in preparation) and incorporating the effects of sample porosity, are $T_{peak} \sim 640$, 680, and 920 K and $T_{postshock} \sim 510$, 530, and 680 K for shots 692, 693, and 694, respectively.

Powder X-ray diffraction analyses of the shocked serpentine are consistent with previous results for shocked serpentine (*Boslough et al.*, 1980). The shocked material shows a reduction in intensity of the serpentine peaks and an increase in background level, but there is no variation in the position of the peaks; that is, there is no variation in the lattice parameters of the shocked material.

Differential scanning calorimetry (DSC) and thermogravimetric analysis (TGA) of shocked and unshocked samples were performed over the temperature range 300° to 1100°C using a Mettler Thermoanalyzer 2000C. Sample size was approximately 20 mg. Analysis was performed in a nitrogen atmosphere. The extent of impact-induced devolatilization was determined by comparison of TGA analyses of shocked and unshocked material. Figure 3 shows the DSC curves obtained using heating rates of 2, 5, 10, and 25 K/min for the unshocked and shocked materials. In each analysis, the endothermic (downward) peak occurring at temperatures between about 600°C and 700°C is the water-loss peak. The results indicate that for equal heating rates, there is a 50°C to 70°C reduction in the temperature

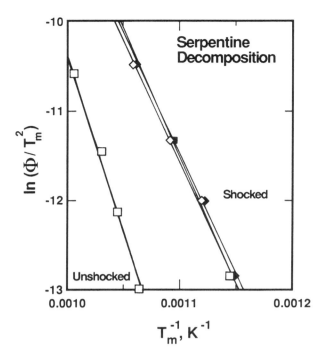

Fig. 4. Plot of ln (ϕ/T_m^2) versus T_m^{-1}, where ϕ is the experimental heating rate and T_m is the temperature of the maximum of the DSC water loss peak. The lines represent linear least squares fits to the data for each shock pressure.

This variation indicates that previous attempts to derive kinetic and thermodynamic information on shocked materials from TGA curves without consideration of the experimental heating rate effects give only qualitative results (*Lange and Ahrens*, 1982b, 1986; *Lange et al.*, 1985). This study is a direct quantification of shocked serpentine water-loss kinetics.

RESULTS AND DISCUSSION

Kinetic information can be obtained from the variation in DSC peak position with heating rate by means of the following analysis, which follows that of *Kissinger* (1957). For a first-order decomposition reaction such as

$$\text{Solid}_1 \rightarrow \text{Solid}_2 + \text{Gas} \qquad (2)$$

the rate of reaction can be expressed as

$$dx/dt = A_o (1 - x) \exp(-E_a/RT) \qquad (3)$$

where x is the fraction of Solid 1 that has decomposed at a given time t, A_o is the preexponential constant, E_a is the activation energy for the decomposition reaction, R is the gas constant, and T is absolute temperature. The time dependence of the reaction rate during DSC is given by

$$d/dt\,(dx/dt) = dx/dt\,\{E_a\phi/RT^2 - A_o \exp(-E_a/RT)\} \qquad (4)$$

where ϕ is the experimental heating rate. The maximum rate of decomposition occurs at the temperature of the maximum in the DSC curve T_m. By setting $d/dt\,(dx/dt)$ to zero and rearranging one obtains

$$\ln(\phi/T_m^2) = -E_a/RT_m + \ln(A_oR/E_a) \qquad (5)$$

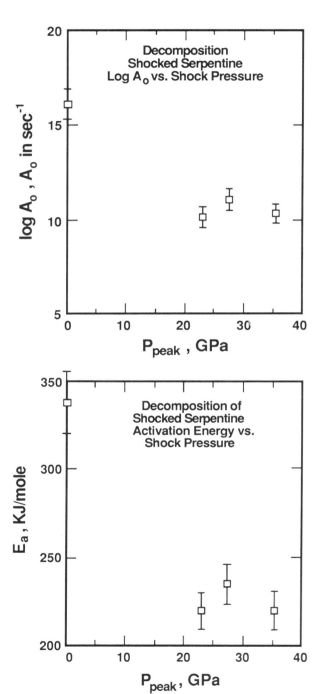

Fig. 5. Variation of preexponential term A_o and activation energy E_a for dehydration of serpentine with peak shock pressure.

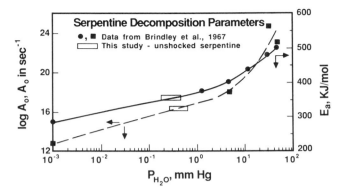

Fig. 6. Serpentine decomposition parameters A_o and E_a as a function of P_{H_2O} from *Brindley et al.* (1967) along with the corresponding parameters for unshocked serpentine from this study. Solid and dashed lines are empirical fits to the data of *Brindley et al.* for E_a and log A_o, respectively.

Thus a plot of ln (ϕ/T_m^2) versus $1/T_m$ will have a slope equal to $-E_a/R$ and an intercept of ln $(A_o R/E_a)$.

Figure 4 is such a plot of the DSC data of this study for the serpentine decomposition reaction

$$Mg_3Si_2O_5(OH)_4 \rightarrow 3MgO \cdot 2SiO_2 + 2H_2O \qquad (6)$$
$$\text{Serpentine} \qquad \text{Solids} \qquad \text{Gas}$$

The solid products are written as $3MgO \cdot 2SiO_2$ instead of forsterite plus enstatite ($Mg_2SiO_4 + MgSiO_3$) because recrystallization may not take place on the time scale of the analysis. The figure shows that for each sample (shocked or unshocked) the data fall on a straight line, and therefore values of A_o and E_a can be obtained using equation (5). These data are listed in Table 2. As shown in Fig. 5, values of A_o are 5 to 6 orders of magnitude lower and values of E_a are nearly a factor of 2 lower for the shocked material relative to the unshocked material. There is no apparent systematic variation in A_o or E_a with shock pressure. Thus, even though the impact experiments were carried out under conditions that inhibit impact-induced devolatilization, the shocked materials exhibit greatly different dehydration kinetics from the unshocked mineral.

Serpentine dehydration kinetics under varying water vapor pressure conditions have been studied by *Brindley et al.* (1967). They showed that the rate of serpentine dehydration is strongly dependent on ambient P_{H_2O}. Surface layers of chemically adsorbed H_2O inhibit the dehydration reaction. The thermal analyses of this study were performed in a nitrogen stream (flow rate ~20 ml/min) with no direct control of P_{H_2O}. Figure 6 is a plot of A_o and E_a versus P_{H_2O} from the results of *Brindley et al.* (1967) and the results for unshocked serpentine of this study. Our results are consistent with those of *Brindley et al.* (1967) and indicate an ambient P_{H_2O} during thermal analysis of approximately 0.2 to 0.3 mm Hg (~2.5 to 4×10^{-4} bar). This experimental value of P_{H_2O} may be caused by released H_2O trapped in the interstices of the reacting powder. Note that estimates of P_{H_2O} may be caused by released H_2O trapped in the interstices of the reacting powder. Note that estimates

of P_{H_2O} on the present-day Martian surface lie between about 6×10^{-3} bars (the vapor pressure of ice at 273 K) and 4×10^{-8} bars (*Gooding*, 1978) with an observed maximum of 3×10^{-6} bars (*Farmer et al.*, 1976). Thus it is reasonable to apply our results directly to calculations concerning the Martian surface.

Figure 7 is a plot of the logarithm of the relative first order rate constant (shocked versus unshocked) versus reciprocal temperature, where the rate constant k is given by

$$k = A_o \exp(-E_a/RT) \qquad (7)$$

Note that a reduction in values of A_o yields a reduction in dehydration rate, whereas a reduction in activation energy E_a results in an increase in dehydration rate. The net effect of the lowered values of A_o and E_a is a rate of dehydration for shocked serpentine greater than for unshocked serpentine. Over the temperature range in which the dehydration occurs during DSC analysis, about 500° to 800° C, dehydration of shocked serpentine proceeds at a rate up to about 10 orders of magnitude greater than that of unshocked serpentine. The rate enhancement is most pronounced at lower temperatures. Extrapolation to the temperature of the surface of Mars (about 170 to 300 K) yields a rate of dehydration of shocked serpentine 20 to 30 orders of magnitude greater than for unshocked serpentine.

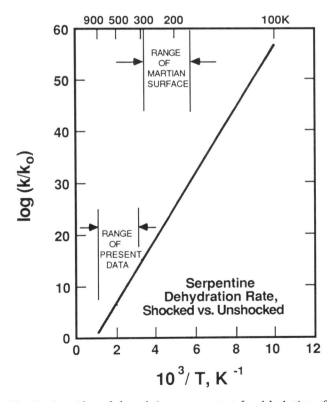

Fig. 7. Logarithm of the relative rate constant for dehydration of shocked compared to unshocked serpentine, log (k/k_o), versus reciprocal temperature.

The large shock enhancement of the bulk serpentine dehydration rate is caused by the large number of lattice defects (and, hence, weakened chemical bonds) introduced during shock deformation and by the increase in surface area caused by shock comminution of the starting material. The relative magnitudes of these effects on the reaction rate are not differentiated in this study. Note that the rate of reaction of a gas with a solid, i.e., the reverse of reaction (2), will also be enhanced by these same shock effects, although to what extent cannot be determined from results of this study. However, these results indicate that the rate of attainment of chemical equilibrium will be greatly enhanced in shocked materials.

The effects of shock enhancement of solid-gas reaction rates are important in planetary development. Atmosphere-lithosphere interactions will be rapid on a planetary surface strongly affected by impacts. Chemical buffering of atmospheric speciation during planetary accretion will be facilitated. Thus, even in the very early, very cold stages of accretion, protoatmosphere-surface reaction rates would very likely have been sufficiently rapid to maintain chemical equilibrium with the shocked near-surface rocks.

Shock enhancement of reaction rates will also have a significant influence on chemical weathering of present-day planetary surfaces, as has been suggested by *Boslough et al.* (1980, 1986). *Boslough and Cygan* (1987) investigated the enhancement of dissolution rates of shocked minerals. On the Martian surface, hydrous magnesian silicates are metastable with respect to magnesium carbonate (magnesite) plus quartz plus H_2O gas (*Gooding*, 1978). Gooding points out that on Mars, reactions between atmospheric gases and solid minerals rather than between aqueous solutions and minerals may control the rates and products of chemical weathering. He postulates that metastable mineral assemblages (i.e., hydrous magnesium silicates) might occur as intermediates during the weathering of igneous rocks and then remain unreacted because of low reaction rates. The results of this study suggest that in areas that have undergone impact, attainment of an equilibrium mineral assemblage is probable.

CONCLUSIONS

Values of both the preexponential term A_o and the activation energy E_a for the rate constant for dehydration of serpentine are greatly reduced by impact. Net dehydration rates are orders of magnitude greater for shocked samples compared to unshocked samples even though the shock experiments were performed in assemblies that inhibited impact-induced devolatilization. At temperatures of the Martian surface, the decomposition rate of shocked serpentine is 20 to 30 orders of magnitude greater than that for unshocked material. Thus planetary surfaces that are strongly affected by impacts will rapidly reach chemical equilibrium with the atmosphere. Buffering of atmosphere composition by surface materials would have occurred during even the earliest, coldest stages of accretion. Long term persistence of metastable mineral assemblages is very unlikely on highly impacted surfaces.

Acknowledgments. We thank Professor George Rossman for the use of his laboratory facilities and Dr. Mickey Atzmon for suggesting the experimental approach. E. Gelle, M. Long, and C. Manning provided expert experimental assistance. Sue Selkirk provided graphic artistry. We appreciate a constructive review by M. B. Boslough. Supported under NASA grant NGL-05-002-105. Caltech Division of Geological and Planetary Sciences contribution #4481.

REFERENCES

Abe Y. and Matsui T. (1985) The formation of an impact-generated H_2O atmosphere and its implications for the early thermal history of the Earth. *Proc. Lunar Planet. Sci. Conf. 15th*, in *J. Geophys. Res.*, 90, C545-C559.

Ahrens T. J. and O'Keefe J. D. (1972) Shock melting and vaporization of lunar rocks and minerals. *The Moon*, 4, 214-249.

Arrhenius G., De B. R., and Alfven H. (1974) Origin of the ocean. In *The Sea* (E. D. Goldberg, ed.), pp. 839-861. Wiley, New York.

Benlow A. and Meadows A. J. (1977) The formation of the atmospheres of the terrestrial planets by impact. *Astrophys. Space Sci.*, 46, 293-300.

Bevington P. R. (1969) *Data Reduction and Error Analysis for the Physical Sciences*, McGraw-Hill, New York. 336 pp.

Boslough M. B. and Cygan R. T. (1987) Shock-enhanced dissolution of silicate minerals and chemical weathering on planetary surfaces. *Proc. Lunar Planet. Sci. Conf. 18th*, this volume.

Boslough M. B., Weldon R. J., and Ahrens T. J. (1980) Impact-induced water loss from serpentine, nontronite, and kernite. *Proc. Lunar Planet. Sci. Conf. 11th*, pp. 2145-2158.

Boslough M. B., Ahrens T. J., Vizgirda J., Becker R. H., and Epstein S. (1982) Shock-induced devolatilization of calcite. *Earth Planet. Sci. Lett.*, 61, 166-170.

Boslough M. B., Venturini E. L., Morosin B., Graham R. A., and Williamson D. L. (1986) Physical properties of shocked and thermally altered nontronite: Implications for the Martian surface. *Proc. Lunar Planet. Sci. Conf. 15th*, in *J. Geophys. Res.*, 91, E207-E214.

Brindley G. W., Narahari Achar B. N., and Sharp J. H. (1967) Kinetics and mechanism of dehydroxylation processes: II. Temperature and vapor pressure dependence of dehydroxylation of serpentine. *Amer. Mineral.*, 52, 1697-1705.

Farmer C. F., Davies D. W., and LaPorte D. D. (1976) Mars: Northern summer ice cap—water vapor observations from Viking 2. *Science*, 194, 1339-1441.

Gooding J. L. (1978) Chemical weathering on Mars. *Icarus*, 33, 483-513.

Graham R. A. and Webb D. M. (1984) Fixtures for controlled explosive loading and preservation of powder samples. In *Shock Waves in Condensed Matter—1983* (J. R. Asay, R. A. Graham, and G. K. Straub, eds.), pp. 211-214. North-Holland, New York.

Jakosky B. M. and Ahrens T. J. (1979) The history of an atmosphere of impact origin. *Proc. Lunar Planet. Sci. Conf. 10th*, pp. 2727-2739.

Kissinger H. E. (1957) Reaction kinetics in differential thermal analysis. *Anal. Chem.*, 29, 1702-1706.

Lange M. A. and Ahrens T. J. (1982a) The evolution of an impact-generated atmosphere. *Icarus*, 51, 96-120.

Lange M. A. and Ahrens T. J. (1982b) Impact induced dehydration of serpentine and the evolution of planetary atmospheres. *Proc. Lunar Planet. Sci. Conf. 13th*, in *J. Geophys. Res.*, 87, A451-A456.

Lange M. A. and Ahrens T. J. (1986) Shock-induced CO_2 loss from $CaCO_3$: Implications for early planetary atmospheres. *Earth Planet. Sci. Lett.*, 77, 409-418.

Lange M. A., Lambert P., and Ahrens T. J. (1985) Shock effects on hydrous minerals and implications for carbonaceous meteorites. *Geochim. Cosmochim. Acta*, 49, 1715-1726.

Marsh S. P., ed. (1980) *LASL Shock Hugoniot Data.* University of California Press, Berkeley. 327 pp.

Morosin B. and Graham R. A. (1982) Shock-induced inorganic chemistry. In *Shock Waves in Condensed Matter—1981* (W. J. Nellis, L. Seaman, and R. A. Graham, eds.), pp. 4-13. American Institute of Physics, New York.

Rice M. H., McQueen R. G., and Walsh J. M. (1958) Compression of solids by strong shock waves. *Solid State Phys., 6,* 1-63.

Tyburczy J. A. and Ahrens T. J. (1985) Partial pressure of CO_2 and impact-induced devolatilization of carbonates: Implications for planetary accretion (abstract). In *Lunar and Planetary Science XVI,* pp. 874-875. Lunar and Planetary Institute, Houston.

Tyburczy J. A. and Ahrens T. J. (1986) Dynamic compression and volatile release of carbonates. *J. Geophys. Res., 91,* 4730-4744.

Tyburczy J. A., Frisch B., and Ahrens T. J. (1986) Shock-induced volatile loss from a carbonaceous chondrite: Implications for planetary accretion. *Earth Planet. Sci. Lett., 80,* 201-207.

Shock-Enhanced Dissolution of Silicate Minerals and Chemical Weathering on Planetary Surfaces

Mark B. Boslough and Randall T. Cygan*

Sandia National Laboratories, Albuquerque, NM 87185

*Present address: Department of Geology, University of Illinois, Urbana, IL 61801

Shock-recovery experiments were performed on samples of bytownite, oligoclase, and hornblende in order to assess the interrelationships between shock-activation and chemical weathering of natural materials. The three silicate minerals were recovered from peak pressures of 7.5, 16, and 22 GPa, and were subjected to analysis by optical microscopy, scanning electron microscopy (SEM), X-ray diffraction, and BET gas adsorption. The dissolution response of three major elements (silicon, aluminum, and calcium) into a pH-buffered aqueous solution was examined for the shocked and unshocked samples. The rate of mass-normalized silicon release increased for all three minerals and, in some cases, was enhanced by a factor of more than 20. Only part of this behavior was due to increased surface area; all three minerals demonstrated an enhancement in surface-area-normalized dissolution of silicon, with two- to seven-fold increases of the rate. The results imply that shock-activation may play a significant role in the chemical weathering of planetary surfaces (e.g., Mars) where impact-cratering and rock-water interactions are both important processes. Estimates of volume amounts of material shocked by impacts onto Mars suggest that the unmelted fraction of ejecta can play an important role in soil formation. Selective weathering of meteoritic material can explain the mafic composition of the Martian regolith. Shock-enhanced authigenesis may also explain the smaller-than-expected quantity of crystalline ejecta associated with the clay layer at the Cretaceous-Tertiary boundary. Shock-enhanced dissolution rates may have an effect on groundwater chemistry near nuclear test sites and natural impact craters on Earth.

INTRODUCTION

There is evidence that when many materials are subjected to intense shock-loading, they undergo changes that affect subsequent chemical reaction processes (see *Graham et al.*, 1986, for recent review). In some cases, chemical reaction rates in the shocked materials are increased by orders of magnitude (e.g., catalysis of CO and methanol by TiO_2 and ZnO; *Golden et al.*, 1982; *Williams et al.*, 1986a). Recent research suggests that this chemical response is due to the shock-induced generation of unusually high densities of defects and dislocations, increasing the number of reactive sites on grain surfaces (e.g., *Adaduro and Gol'danskii*, 1981, for polymerization and thermal decomposition reactions). Other research has also demonstrated that shock-loading of crystalline solids can increase defect densities by orders of magnitude (*Davison and Graham*, 1979; *Graham*, 1981), and defects are known to affect the dissolution process (*Lasaga and Blum*, 1986; *Casey et al.*, 1987). Chemical reactions that are controlled by surface processes will certainly be affected by shock-loading. Both net increases and decreases in specific surface areas have been observed for shocked inorganic powders due to comminution or agglomeration of grains, respectively, with an increase of over an order of magnitude in the case of shocked aluminum oxide (*Williams et al.*, 1986b).

Boslough et al. (1986) suggested that this "shock-activation" may be an important surface process for minerals on planets where impact-induced shocks and atmosphere-regolith and/or groundwater-regolith interactions are significant. For example, a shock-induced enhancement of weathering and hydrothermal alteration rates of the Martian surface material would have affected the evolution of the regolith and determined its present mineralogical composition. Shock-activation of minerals would also have a bearing on the evolving chemistry of groundwaters and other solutions associated with the weathering of shocked phases near terrestrial impact craters and nuclear test sites.

Although no comprehensive study has been performed on shock-modified silicate minerals, research on ore-processing alternatives by *Murr and Hiskey* (1981) examined the enhanced leaching rates of sulfide minerals with shock pressure. They subjected samples of chalcopyrite to well-defined shock states of 1.2 and 18 GPa, and found that leaching rates correlated with dislocation density. However, their observations were complicated by variations in chemisorption mechanisms on the mineral surface and interference of a precipitation product.

We have undertaken the present study to examine the effect of shock-loading on rock-water interactions and, in particular, to determine if there is experimental evidence to support shock-enhancement in the dissolution of silicate phases. We have not yet attempted to determine detailed mechanisms for such shock-enhancement, nor have we attempted to quantify dislocation densities, strains and other material properties modified by shock-loading; our present goal is merely to identify and quantify changes in dissolution rate in a small sample of rock-forming minerals, and to correlate them with shock pressure.

EXPERIMENTAL

We carried out shock-recovery experiments on three silicate minerals using the Sandia "Bear" explosive loading fixtures to achieve well-characterized shock states. The recovery fixtures allow samples to be shocked in a controlled, reproducible manner, and the peak shock states were determined by numerical simulations (*Graham and Webb*, 1984, 1986). Two plagioclase minerals (oligoclase and bytownite) and one mafic

TABLE 1. Schedule of experiments.

Shot	Mineral	Fixture	Explosive	Sample Compact Density (Mg/m^3)	Sample Compact Density (%)	Peak Pressure (GPa)	Estimated Mean Bulk Temp. (K)
1B866	Oligoclase	Momma Bear	Baratol	1.70	65	5-10	90-110
2B866	Oligoclase	Momma Bear A	Baratol	1.70	65	14-20	125-175
3B866	Oligoclase	Momma Bear A	Comp B	1.70	65	19-26	250-500
1B876	Bytownite	Momma Bear	Baratol	1.40	51	5-10	300-325
2B876	Bytownite	Momma Bear A	Baratol	1.40	51	14-20	300-410
3B876	Bytownite	Momma Bear A	Comp B	1.40	51	19-26	450-700
4B866	Hornblende	Momma Bear	Baratol	2.30	77	5-10	25
5B866	Hornblende	Momma Bear A	Baratol	2.30	77	14-20	50-75
6B866	Hornblende	Momma Bear A	Comp B	2.30	77	19-26	125-275

mineral (hornblende) were examined. The oligoclase is from Bancroft, Ontario, and the hornblende is from Mineral County, Nevada. Both of these minerals were obtained from Wards Natural Science Establishment, Inc., and were ground and sieved to a size distribution of 37 to 149 microns. Electron microprobe analyses indicated that the hornblende has an approximate composition of $Ca_2(Mg_{3.3}Fe_{1.7})(Si_{7.5}Al_{0.5})O_{22}(OH)_2$ (hornblende-tremolite solution) and the oligoclase has a plagioclase composition of An_{25}. The bytownite is from Pueblo Park, New Mexico, and has been characterized by *Casey* (1987). The initial grain size range of the bytownite was from 125 to 425 microns, and the composition is An_{60}. All three minerals were recovered from mean peak shock pressures of 7.5, 16, and 22 GPa. Mean-bulk shock temperatures were estimated from the calculations of *Graham and Webb* (1986) and depend on initial sample packing density (see Table 1).

The dissolution experiments were performed after characterizing each sample with BET specific surface area measurements, X-ray diffraction, and optical and electron microscopy. The unshocked control sample and the three shocked samples for each of the three minerals were reacted in batch dissolution cells with a pH-buffered standard solution of 0.01 N potassium hydrogen phthalate (pH = 4). Approximately one gram of sample was reacted with 200 mL of the standard solution. The shocked mineral samples were hand-selected from the recovery fixtures in order to ensure that material came from the most uniformly shocked ("bulk" of *Graham and Webb*, 1984) portion of the compact. Samples were sonically washed in distilled water and acetone prior to dissolution. The dissolution cells were placed in a constant temperature bath at 25°C and were kept agitated at a frequency of 1 Hz. Continuous agitation of the cells disrupts any chemical diffusion layer in the solution at the mineral interface and thereby promotes a surface-controlled dissolution mechanism. The experiments were maintained at these conditions for up to 21 days. One milliliter of the reacted solution was periodically (initially daily) sampled from each of the dissolution cells with a precision micropipette; identical volumes of the standard solution were added to replenish the sampled solution. The extracted solution was diluted and split for analysis. Silicon analyses were completed using the molybdate-blue spectroscopic method and plasma emission spectroscopy was used to determine aluminum and calcium concentrations.

RESULTS

Scanning electron microscopic (SEM) examination of the unshocked and shocked silicate samples indicates substantial brittle disaggregation to have occurred in some of the phases due to shock-loading. This is most apparent for the case of bytownite, in which significant comminution and reduction in mean grain size is obvious from the sequence of photomicrographs (Fig. 1). Still apparent, but less obvious from the photomicrographs, is the disaggregation of the shocked oligoclase (Fig. 2). By contrast, the hornblende does not show any discernable change under SEM examination due to shock (Fig. 3). This result may be due to the differences in elastic/plastic response between the double chain silicate structure of the hornblende and the dense framework structure of the plagioclase phases.

These changes in microstructure were quantified to some extent by the determination of specific surface areas by the BET (multipoint gas adsorption) method (Table 2 and Fig. 4). We assigned a 6.5% uncertainty to the surface area measurements, which corresponds to the overall coefficient of variation in a series of BET area determinations for various powders by *Ace and Parsons* (1979). This uncertainty includes both instrument and powder sampling variation and is a conservative estimate. The laboratory that performed the analyses (Quantachrome Corp.) assigned an uncertainty of 1% to the values determined for our samples. The bytownite demonstrated the largest gain in specific surface area, with an increase of nearly seven-fold for the 7.5 GPa sample. The oligoclase showed increases by factors between about 3 and 4 in all cases, and the hornblende surface areas increased by no more than 36%. The decrease in specific surface area of the hornblende at the high shock pressure may represent the agglomeration of the mineral grains due to localized shock-melting and/or plastic deformation. However, the observed decrease is small (less than the 6.5% measurement uncertainty).

X-ray diffraction analysis of the shocked samples failed to demonstrate the occurrence of any additional crystalline phases.

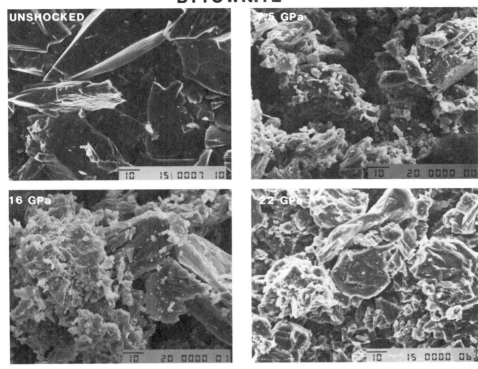

Fig. 1. Scanning electron photomicrographs of bytownite prior to shock and as recovered from three shock experiments. Bar in lower right-hand corner of each photomicrograph indicates scale in microns.

Fig. 2. SEM photomicrographs of unshocked and shocked oligoclase.

HORNBLENDE

Fig. 3. SEM photomicrographs of unshocked and shocked hornblende.

There was significant line-broadening observed in the shocked samples, implying the introduction of residual strain, and/or increase in dislocation densities by the shock loading; this has not yet been quantified. There was no evidence of shock-generated glass from the X-ray data, despite the microscopic observation of small quantities of glass and melt texture in the highest pressure oligoclase sample (Fig. 5). Care was taken to use only a glass-free sample in the dissolution experiment.

Examples of the dissolution behavior for the three analyzed elements (aluminum, silicon, and calcium) are shown in Figs. 6 and 7, respectively, for unshocked and shocked (16 GPa) hornblende. Similar data were obtained for bytownite and oligoclase. In both cases, the calcium is easily leached into solution with its concentration rapidly approaching a constant value, implying that a saturation level has been reached. However, the aluminum and silicon concentrations show continuous increases, and approach an approximately linear dependence on time, in agreement with existing models and observations for various minerals (e.g., *Holdren and Berner*, 1979; *Lasaga*, 1984). It is notable that the ratio of Si and Al molar concentrations is not equal to 15, which would be expected if the dissolution process were congruent, but is about one-fourth of that value, implying that Al enters the solution at a higher rate than Si, relative to their abundances in the mineral. Furthermore, this ratio is greater for the shocked sample than for the unshocked sample by a factor of about two. Thus, the dissolution process is incongruent, and the dissolution response for silicon is more strongly enhanced by the shock than that for aluminum. Because silicon is the major structural component of these minerals, we will emphasize the silicon kinetics in the following discussion.

In Figs. 8-10, the experimental results are presented for silicon release from the three minerals in each of the four states: unshocked and the three shock conditions. A linear least-squares regression was used to determine the rate of concentration increase in each case. The concentration data after 166, 145, and 122 hours were used for hornblende, bytownite, and oligoclase, respectively, since these appear to

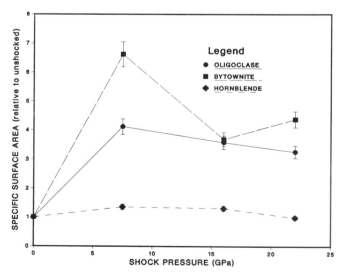

Fig. 4. BET specific surface areas of three shocked minerals relative to the unshocked phases.

TABLE 2. BET specific surface areas and silicon dissolution rates.

Mineral	Mean Shock Pressure (GPa)	Mass-normalized Dissolution Rate (10^{-12} mol/g sec)		BET Specific Surface Area (m^2/g)		Area-normalized Dissolution Rate (10^{-12} mol/m^2 sec)	
		Absolute	Relative	Absolute	Relative	Absolute	Relative
Oligoclase	0.0	0.6 ±3.0	1.0	0.73 ±0.05	1.00	0.8 ±4.2	1.00
Oligoclase	7.5	14.7 ±1.6	25.4 ±2.7	3.01 ±0.20	4.12 ±0.27	4.9 ±0.8	6.18 ±1.06
Oligoclase	16.0	12.3 ±2.6	21.3 ±4.6	2.61 ±0.17	3.58 ±0.23	4.7 ±1.3	5.95 ±1.67
Oligoclase	22.0	14.3 ±1.9	24.8 ±3.3	2.38 ±0.15	3.26 ±0.21	6.0 ±1.2	7.61 ±1.49
Bytownite	0.0	1.0 ±0.6	1.0	0.26 ±0.02	1.00	3.9 ±2.7	1.00
Bytownite	7.5	22.7 ±1.0	22.4 ±1.0	1.72 ±0.11	6.62 ±0.43	13.2 ±1.4	3.39 ±0.37
Bytownite	16.0	22.2 ±0.8	21.9 ±0.8	0.96 ±0.06	3.69 ±0.24	23.1 ±2.4	5.72 ±0.59
Bytownite	22.0	19.4 ±0.8	19.2 ±0.8	1.14 ±0.07	4.38 ±0.28	17.0 ±1.8	4.37 ±0.45
Hornblende	0.0	6.0 ±2.2	1.0	4.93 ±0.32	1.00	1.2 ±0.5	1.00
Hornblende	7.5	17.3 ±1.0	2.9 ±0.2	6.72 ±0.44	1.36 ±0.09	2.6 ±0.3	2.12 ±0.26
Hornblende	16.0	25.8 ±0.7	4.3 ±0.1	6.44 ±0.42	1.31 ±0.09	4.0 ±0.4	3.31 ±0.30
Hornblende	22.0	21.8 ±2.0	3.7 ±0.3	4.86 ±0.32	0.99 ±0.06	4.5 ±0.7	3.71 ±0.58

be the times at which the data approach linearity. The hornblende (7.5 GPa) datum at 238 hours was discarded as a bad analytical point. The slope of the linear fit to Si concentration-versus-time data was used in each case to determine the first-order kinetics of the dissolution process, and the mass-normalized and specific-area-normalized dissolution rates.

The mass-normalized rate, which does *not* correct for surface area effects, is given by

$$r = \frac{\Delta c}{\Delta t} \frac{V}{m}$$

where $\Delta c/\Delta t$ is the slope of the linear section of the dissolution curve in molar concentration increase per second, V is the solution volume in liters, and m is the initial mass of mineral in grams. This rate is divided by the initial specific surface area (A) in m^2/g to obtain the proper specific (surface-area-normalized) rate (k = r/A) in moles/m^2s. In principal the sample mass and specific surface areas are variables during the course of the dissolution experiments; however, the rates are so low and times so short that changes in these values are extremely small and can be ignored. Both of these dissolution rates are listed in Table 2 and plotted in Figs. 11 and 12. The plotted rates are relative to the rates of the unshocked materials. Because the dissolution process associated with unshocked samples is so slow, the concentrations remain low and the fractional errors are large. This is especially true for the oligoclase, as the unshocked sample had particularly noisy data. For this reason, the rate data plotted in Figs. 11 and 12 are plotted with the upper bound of the rate range for the unshocked samples (the best fit rate plus the standard deviation). This emphasized that, even under the most conservative interpretation, the rates are enhanced by the shock.

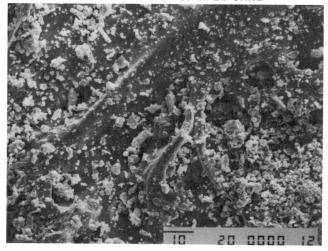

Fig. 5. SEM photomicrograph of oligoclase shocked to 22 GPa, showing morphology indicative of glass formation.

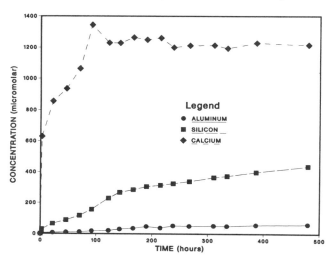

Fig. 7. Concentration history for three elements entering solution from hornblende recovered from a shock pressure of 16 GPa.

Note, however, that the dissolution rates for the shocked samples in these figures are relative to the best fit unshocked sample rates.

Figure 13 shows SEM photomicrographs of unshocked and shocked (22 GPa) oligoclase. The highly deformed surface of the shocked material with a greater number of edge sites is evident in the comparison of reacted material. We believe that the increased defect densities normally associated with such plastic deformation is responsible for the observed increases in dissolution rate.

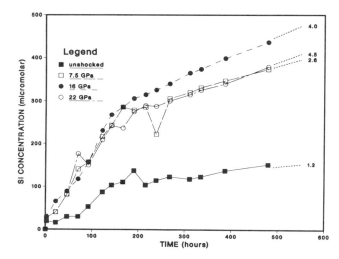

Fig. 8. Silicon concentration history for dissolution from hornblende in each of four conditions: unshocked and recovered from three shock states. Dotted line is least-squares fit to last nine data points. Associated numeral represents surface-area-normalized dissolution rate in 10^{-12} mol/m^{2}sec.

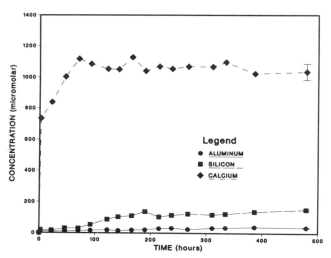

Fig. 6. Concentration history for three elements entering solution from unshocked hornblende.

These results emphasize the *relative* change in dissolution rate of three silicate minerals as a function of shock pressure. Although the absolute rates are of the same order-of-magnitude as those obtained by previous dissolution studies of the plagioclase phases (*Busenberg and Clemency*, 1976; *Holdren and Berner*, 1979), it is difficult to directly compare the kinetic data. Sample characterization, solution chemistry, organic complexing, precipitating phases, and other phenomena associated with dissolution experiments cannot be appropriately addressed here, and will be the subject of a future paper

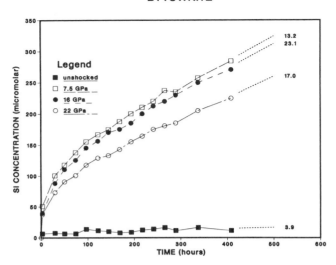

Fig. 9. Silicon concentration history for dissolution from bytownite in each of its four conditions.

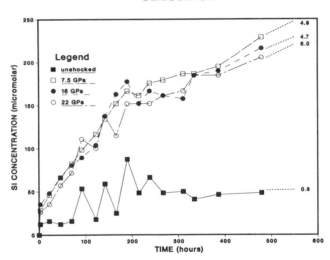

Fig. 10. Silicon concentration history for dissolution from oligoclase in each of its four conditions.

by *Cygan and Boslough* (in preparation, 1987). Nevertheless, by maintaining identical conditions for each of the dissolution experiments, this study provides a means of directly comparing the relative rates for three different silicate minerals.

DISCUSSION

Implications for the Martian Surface

There is a consensus among planetary geologists that the Martian soil consists mainly of mafic silicate alteration products, but the details of the mineralogy are still under debate. The

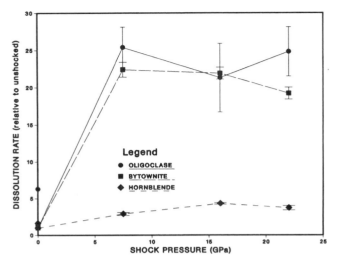

Fig. 11. Mass-normalized silicon dissolution rates for three shocked minerals, relative to the unshocked phases. Symbols on vertical axis represent upper bound of measurement error for associated unshocked phases.

initial interpretation, based on Viking X-ray fluorescence data, was that the fine surface material consists mainly of iron-rich smectite clays and their degradation products, along with ferric oxides, probably in the form of maghemite (*Toulmin et al.*, 1977). These researchers suggested a hydrothermal alteration source for the smectite due to interaction between basaltic magmas and subterranean ice deposits. However, according to *Singer* (1982), Earth-based spectral reflectance data are not consistent with iron-rich clays, but instead strongly suggest the presence of palagonite, an X-ray amorphous weathering product of mafic volcanic glass. Because palagonite is also consistent with the observed bulk chemistry and particle size of the

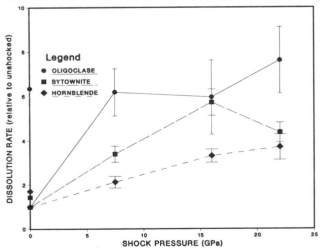

Fig. 12. Surface-area-normalized silicon dissolution rates for three shocked minerals, relative to the unshocked phases.

REACTED OLIGOCLASE
AFTER 480 HOURS OF DISSOLUTION

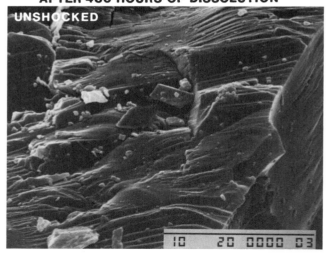

Fig. 13. SEM photomicrographs of partially reacted oligoclase from two dissolution experiments: unshocked and shocked to 22 GPa.

Martian regolith, *Allen et al.* (1981) proposed a mechanism involving subglacial volcanic eruptions resulting in a glassy tephra that is altered at a shallow depth to form palagonite. *Newsom* (1980) also incorporated the idea of hydrothermal alteration of melt, except that in his model the melt is from impact rather than volcanic processes. In a model proposed by *Kieffer and Simonds* (1980), the alteration of impact melt occurs penecontemporaneously with the hot volatiles generated by the impact. The idea of further modification of the alteration product by subsequent impacts was introduced by *Weldon et al.* (1982).

As pointed out by *Gooding and Keil* (1978), models such as those mentioned above assume the existence of a volatile-bearing regolith on Mars, which "presupposes extensive alteration of the regolith to hydrous phases by unspecified weathering processes." One way to achieve such a state is to postulate that the Martian paleoenvironment was more Earth-like, but any such model "invites the burden of proof for the required climatic changes."

A few hypotheses have been advanced that avoid these inherent problems by allowing alteration products to form under present Martian conditions. Calculations by *Gooding and Keil* (1978) show that gas-solid weathering of silicate glass to clay is thermodynamically favored in the present surface environment on Mars, and that the total amount of altered glass from a lunar-like protoregolith would yield a clay layer with a mean-global thickness of 10 to 100 cm. However, according to thermodynamic calculations of *Gooding* (1978), the formation of clays from common rock-forming minerals under the same conditions in the absence of liquid water is unfavorable without some sort of chemical activation process. By extrapolation of experimental data for magnetite, *Huguenin* (1974) proposed a mechanism in which mafic silicates are chemically activated by ultraviolet light, allowing them to react with atmospheric water vapor to form hydrates as well as to oxidize.

Boslough et al. (1986) proposed shock-modification of minerals due to impacts as another possible means of chemical activation to enhance weathering rates. The experimental results reported in the present study provide support for this hypothesis. While these results are very limited, comparing dissolution rates in a liquid aqueous solution at specific temperature and pH, we believe the shock-enhanced chemical weathering demonstrated by the measurements is significant. The enhancement of dissolution rates for all three minerals appears to be a general phenomenon that must be considered when analyzing the weathering of the relevant minerals associated with planetary surfaces. Further support is provided by the data of *Tyburczy and Ahrens* (1987) on the dehydration kinetics of shocked serpentine. Extrapolation of these data to Martian surface temperatures implies that shocked material will reach chemical equilibrium under these conditions much faster than unshocked material; faster by 20 to 30 orders of magnitude for serpentine.

To make a convincing argument that shock-enhancement of weathering rates of crystalline minerals is an important process on Mars, it is important to determine both volumes and rates of production of shock-modified minerals. These calculations are beyond the scope of the present paper, but it is reasonable to assume that the volume of shock-activated minerals would be within an order of magnitude of the volume of impact melt, estimated by *Newsom et al.* (1986) to be equivalent to a global layer 46 m thick. Since most of the cratering of the Martian surface occurred during the period of early bombardment (*Hartmann*, 1977), most of the shocked material was generated at that time, and has had at least 3.5 b.y. to become chemically altered. However, impact-generated material that has been hydrothermally altered was limited to the time over which the impact ejecta remained hot—on the order of years. Since the present results for the dissolution of shocked minerals in liquid water indicate increased alteration rates by factors of around 20 (cf. mass-normalized dissolution rates, with increased surface area effects included), a substantially greater volume of altered minerals can be produced than if there were no such effect. Perhaps more important, it is possible that, as a result of the shock modification, energy barriers are sufficiently decreased that the decomposition of the minerals might be allowed thermodynamically under ambient conditions on Mars.

One argument can be made for the relative importance of shock-enchanced weathering of crystalline material on the basis of impact energy partitioning. The calculations of *O'Keefe and Ahrens* (1977) show that for an impact at 5 km/sec, 50% of the impact energy goes to comminution and plastic work, 40% goes to heating of the target and projectile, and the remaining 10% goes to ejecta kinetic energy. Part of the heat fraction goes toward melting material that is subsequently quenched to glass, but most will be radiated and conducted away as the hot matter comes back to thermodynamic equilibrium with its surroundings. The kinetic energy fraction will eventually be divided between the other two categories, except for any fraction that escapes the planet. Of the 50% that goes toward comminution and plastic work, some is given up as heat when annealing takes place, but the rest is available to increase the specific free energy of the crystalline material, thereby chemically activating it.

It is interesting to note that the most highly shocked material, and the majority of the melt, will come from the meteoroid itself. This implies that the composition of the Martian fines should have a large meteoritic component. In the early paper by the Viking inorganic chemistry team, *Toulmin et al.* (1977) point out that one of the good matches of possible mixtures with Martian regolith samples consists of approximately equal proportions of average tholeiitic basalt and type I carbonaceous chondrites. While they believed that the basalt is likely to be a common rock type on Mars, they found no explanation for a large chondritic component. They dismissed this as a coincidence, stating "the apparent good match of about equal proportions of tholeiitic basalt and type I carbonaceous chondrites to Mars soil is probably fortuitous and does not necessarily imply occurrence of large amounts of primitive, undifferentiated material."

We propose a scenario by which the Martian regolith is enriched in undifferentiated material by selective weathering: A significant fraction of the shock-activated material comes directly from the meteorites, and only shock-activated minerals can weather on Mars. This scenario is also consistent with known production rates of fine-grained material on Mars. According to *Arvidson* (1986), Mars has had a "decidedly non-linear history of debris production," and most of the debris was produced in the first billion years of geologic time. This period is coincident with heavy meteorite bombardment. Similarly, *Guinness et al.* (1987) conclude from spectral reflectance studies at Viking lander sites that "soils are created globally by a number of processes that operated at higher rates earlier in geologic time."

There is some evidence for chondritic enrichment of terrestrial craters, which were formed by much higher impact velocities. The highest meteoritic contamination is found in impact melts, up to 8% for Clearwater East, according to *Palme* (1982). With increasing impact velocity, the amount of meteorite contamination decreases, so one would expect greater enrichment on Mars where impact velocities are lower. Unless the Martian soil is entirely volcanogenic, it should be chondritically enriched. The fact that most of the regolith dates from the first billion years argues against a volcanogenic origin, since significant volcanic activity has taken place since that time. In fact, according to *Plescia and Saunders* (1979), the main shield of Alba Patera, one of the earliest of the huge Martian shield volcanoes, is only 1725 m.y. old.

Implications for the Cretaceous-Tertiary Boundary

The fact that weathering of shocked minerals is enhanced also has a bearing on the Cretaceous-Tertiary boundary on Earth. One of the criticisms of the impact hypothesis of *Alvarez et al.* (1980) comes from mineralogical studies of the boundary clay layer. According to *Rampino* (1982) and *Rampino and Reynolds* (1983), if such an impact occurred, the clay would be expected to contain fine-grained mineral fragments derived from the asteroid and impact area, since *O'Keefe and Ahrens* (1982) calculate that there should be about five times as much submicron ejecta in the form of highly shocked solid mineral and rock fragments as glass from melt and condensed vapor. *Rampino* (1982) allows that the glassy fraction might be altered to smectite, in agreement with *Kastner et al.* (1984). The resulting smectite would be hard to distinguish from the volcanogenic clay present in the layer, but *Rampino* (1982) argues that "diagenetic elimination of fine-grained non-clay minerals from the boundary clay seems unlikely because the degree of diagenesis for the clays is very mild," based on the highly smectitic composition of the clays. However, because the mineral grains are highly shock-activated, the conditions and times required for alteration can be relaxed, and this argument is weakened.

Implications for Nuclear Test Sites and Impact Craters

The observed shock-enhanced dissolution has implications for very long-term effects of underground nuclear testing on groundwater chemistry. According to pressure decay functions determined from nuclear explosion data by R. C. Bass (unpublished data, 1987), a 20-kt explosion in wet tuff or dry alluvium will generate a 9080 m^3 volume of material shocked above 7.5 GPa. However, redistribution of the excess energy from the nuclear device melts significantly more rock than would be melted by the shock alone. According to *Schwartz et al.* (1984), 740 metric tons are melted per kiloton of yield. Thus for tuff, with a density of about 1.8 g/cm^3, a 20-kt explosion will melt 8220 m^3 of rock, nearly all of that shocked above 7.5 GPa. If, however, significant shock-activation occurs at pressures above 1 GPa, the data of R. C. Bass (unpublished data, 1987) indicate that a volume of 5.7×10^4 m^3 of rock is shock-activated for a 20-kt explosion in tuff. The amount of melt is small compared to this volume. It may be useful to carry out further experiments in this pressure range on the minerals found in the tuff at the Nevada test site, to see if shock-enhanced dissolution is significant.

Similar estimates can be made of volumes of shocked material near natural impact craters. For impact craters there is no excess thermal energy beyond that generated by shock heating, so the large volumes of highly shocked material are not melted. However, the material that experiences the highest stress is ejected and widely dispersed, and the local groundwater chemistry will be controlled by interaction with minerals that experience lower peak stresses. A comparison of groundwater from known impact sites to that from undisturbed terrain would be useful.

Other Implications

Our observation of variable incongruent dissolution has possible ramifications for chemical fractionation that is dependent on shock history. It implies that shocked and unshocked terrains may differ not only in rate of chemical weathering, but may also differ in chemical composition of the resulting alteration products.

There are other possible consequences of shock modification of silicates that, although unrelated to weathering, should be mentioned here for completeness. Shocked minerals experience large increases in specific free energy, due to high defect densities and increased specific surface areas. Thus we would expect other physico-chemical properties to be altered in addition to dissolution rate. Among the possible effects might be a depression of the melting point, leading to unusual partial melts. Also affected could be diffusion properties, which might result in different trace element partitioning and isotope chronology than would be expected for unshocked material. These speculations provide suggestions for further experiments, as well as a note of caution for interpretations based on properties of planetary materials for which shock effects cannot be ignored.

CONCLUSIONS

This study provides data that demonstrate that there is direct correlation between shock modification of minerals and enchanced weathering rates. Since shock-enchanced dissolution was observed for three different minerals shocked above 7.5 GPa, it is reasonable to expect other minerals relevant to planetary surfaces to demonstrate the effect in this pressure range.

In order to quantify the effects of shock-enchanced dissolution and determine their relative importance in planetary surface evolution, considerable experimentation is necessary on a variety of relevant minerals. Hydrothermal alteration studies should be performed, as well as an investigation of gas/solid weathering processes on shock-activated silicates. Studies of weathering products from naturally shocked minerals will also provide data that may have a bearing on calculations for very long-term effects of underground nuclear tests.

Shock-weathering experiments on chondritic material would also be particularly useful with regard to the composition of the Martian regolith because, according to the weathering scenario suggested in this paper, the Martian regolith should be highly enriched in alteration products of meteoritic minerals.

Any future sample-return mission to the Martian surface should include the capability of measuring platinum-group elements in the regolith at the parts-per-billion concentration level, in order to determine if it is truly enriched in chondritic material.

Since shocked minerals undergo chemical alteration at a higher rate, any restrictions on the quantity of crystalline ejecta in the clay layer of the Cretaceous-Tertiary boundary can be relaxed, thereby weakening one argument against the impact-extinction hypothesis.

There are a number of other possible implications of this study that should be explored, such as the effects of shock-dependence of incongruent dissolution. Effects of shock history on other physico-chemical properties should also be considered.

Acknowledgments. This work was performed at Sandia National Laboratories, supported by the U. S. Department of Energy under contract number DE-AC04-76DP00789. We wish to thank W. H. Casey, H. E. Newsom, and F. Hörz for their valuable comments. The thorough and constructive reviews by J. F. Bauer and R. V. Gibbons are appreciated. We also acknowledge the technical assistance of M. U. Anderson, C. J. Daniel, K. Elsner, P. F. Hlava, V. S. McConnell, and J. L. Krumhansl, and useful discussions with J. A. Tyburczy, H. R. Westrich, and C. T. C. Busters.

REFERENCES

Ace H. L. and Parsons D. S. (1979) Reproducibility of the surface area of some powders as measured by the Monosorb surface-area analyzer using the Brunauer-Emmett-Teller Equation. *J. Test. Eval.,* 7, 334-337.

Adadurov G. A. and Gol'danskii V. I. (1981) Transformations of condensed substances under shock-wave compression in controlled thermodynamic conditions. *Russian Chem. Rev.,* 50, 948-957.

Allen C. C., Gooding J. L., Jercinovic M., and Keil K. (1981) Altered basaltic glass: A terrestrial analog to the soil of Mars. *Icarus,* 45, 347-369.

Alvarez L. W., Alvarez L., Asaro F., and Michel H. V. (1980) Extraterrestrial cause for the Cretaceous-Tertiary extinction. *Science,* 208, 1095-1108.

Arvidson R. E. (1986) On the rate of formation of sedimentary debris on Mars (abstract). In *Lunar and Planetary Science XVII,* p. 17. Lunar and Planetary Institute, Houston.

Boslough M. B., Venturini E. L., Morosin B., Graham R. A., and Williamson D. L. (1986) Physical properties of shocked and thermally altered nontronite: Implications for the Martian surface. *Proc. Lunar Planet. Sci. Conf. 17th,* in *J. Geophys. Res.,* 91, E207-E214.

Busenberg E. and Clemency C. V. (1976) The dissolution kinetics of feldspars at 25°C and 1 atm CO_2 partial pressure. *Geochim. Cosmochim. Acta,* 40, 41-50.

Casey W. H., Carr M. J., and Graham R. A. (1987) Crystal defects and the dissolution kinetics of rutile. *Geochim. Cosmochim. Acta,* in press.

Cygan R. T. and Boslough M. B. (1987) Shock-activation and enhanced dissolution of silicate minerals. *Geochim. Cosmochim. Acta,* in preparation.

Davison L. and Graham R. A. (1979) Shock compression of solids. *Phys. Rep.,* 55, 255-379.

Golden J., Williams F., Morosin B., Venturini E. L., and Graham R. A. (1982) Catalytic activity of shock-loaded TiO_2 powder. In *Shock*

Waves in Condensed Matter—1981 (W. J. Nellis, L. Seaman, and R. A. Graham, eds.), pp. 72-76. American Institute of Physics, New York.

Gooding J. L. (1978) Chemical weathering on Mars. *Icarus, 33*, 483-513.

Gooding J. L. and Keil K. (1978) Alteration of glass as a possible source of clay minerals on Mars. *Geophys. Res. Lett., 5*, 727-730.

Graham R. A. (1981) Active measurements of defect processes in shock-compressed metals and other solids. In *Shock Waves and High-Strain Rate Phenomena in Metals—Concepts and Application* (M. A. Meyers and L. E. Murr, eds.), pp. 375-386. Plenum, New York.

Graham R. A. and Webb D. M. (1984) Fixtures for controlled explosive loading and preservation of powder samples. In *Shock Waves in Condensed Matter—1983* (J. R. Asay, R. A. Graham, and G. K. Straub, eds.), pp. 211-214. North Holland, New York.

Graham R. A. and Webb D. M. (1986) Shock-induced temperature distributions in powder compact recovery fixtures. In *Shock Waves in Condensed Matter—1985* (Y. M. Gupta, ed.), pp. 831-836. Plenum, New York.

Graham R. A., Morosin B., Venturini E. L., and Carr M. J. (1986) Materials modification and synthesis under high pressure shock compression. *Annu. Rev. Mater. Sci., 16*, 315-341.

Guinness E. A., Arvidson R. E., Dale-Bannister M. A., Singer R. B., and Bruckenthal E. A. (1987) On the spectral reflectance properties of materials exposed at the Viking landing sites. *Proc. Lunar Planet. Sci. Conf. 17th*, in *J. Geophys. Res., 92*, E575-E587.

Hartmann W. K. (1977) Relative crater production rates on planets. *Icarus, 31*, 260-276.

Holdren G. R. and Berner R. A. (1979) Mechanism of feldspar weathering—I. Experimental studies. *Geochim. Cosmochim. Acta, 43*, 1161-1171.

Huguenin R. L. (1974) The formation of goethite and hydrated clay minerals on Mars. *J. Geophys. Res., 79*, 3895-3905.

Kastner M., Asaro F., Michel H. V., Alvarez W., and Alvarez L. W. (1984) Did the clay minerals at the Cretaceous-Tertiary boundary form from glass? Evidence from Denmark and DSDP hole 465A. *J. Non-Cryst. Solids, 67*, 463-464.

Kieffer S. W. and Simonds C. H. (1980) The role of volatiles and lithology in the impact cratering process. *Rev. Geophys. Space Phys., 18*, 143-181.

Lasaga A. C. (1984) Chemical Kinetics of water-rock interactions. *J. Geophys. Res., 89*, 4009-4025.

Lasaga A. C. and Blum A. E. (1986) Surface chemistry, etch pits and mineral-water reactions. *Geochim. Cosmochim. Acta, 50*, 2363-2379.

Murr L. E. and Hiskey J. B. (1981) Kinetic effects of particle-size and crystal dislocation density on the dichromate leaching of chalcopyrite. *Metallurg. Trans. B. 12B*, 255-267.

Newsom H. E. (1980) Hydrothermal alteration of impact melt sheets with implications for Mars. *Icarus, 44*, 207-216.

Newsom H. E., Graup G., Sewards T., and Keil K. (1986) Fluidization and hydrothermal alteration of the suevite deposit at the Ries Crater, West Germany, and implications for Mars. *Proc. Lunar Planet. Sci. Conf. 17th*, in *J. Geophys. Res., 91*, E239-E251.

O'Keefe J. D. and Ahrens T. J. (1977) Impact-induced energy partitioning, melting and vaporization on terrestrial planets. *Proc. Lunar Sci. Conf. 8th*, 3357-3384.

O'Keefe J. D. and Ahrens T. J. (1982) The interaction of the Cretaceous/Tertiary extinction bolide with the atmosphere, ocean, and solid Earth. In *Geological Implications of Impacts of Large Asteroids and Comets on the Earth* (L. T. Silver and P. H. Schultz, eds.), pp. 103-120. Geological Society of America, Boulder.

Palme H. (1982) Identification of projectiles of large terrestrial impact craters and some implications for the interpretation of Ir-rich Cretaceous/Tertiary boundary layers. In *Geological Implications of Impacts of Large Asteroids and Comets on the Earth* (L. T. Silver and P. H. Schultz, eds.), pp. 223-233. Geological Society of America, Boulder.

Plescia J. B. and Saunders R. S. (1979) The chronology of Martian volcanoes. *Proc. Lunar Planet. Sci. Conf. 10th*, 2841-2859.

Rampino M. R. (1982) A non-catastrophist explanation for the iridium anomaly at the Cretaceous/Tertiary boundary. In *Geological Implications of Impacts of Large Asteroids and Comets on the Earth* (L. T. Silver and P. H. Schultz, eds.), pp. 455-460. Geological Society of America, Boulder.

Rampino M. R. and Reynolds R. C. (1983) Clay mineralogy of the Cretaceous-Tertiary boundary clay. *Science, 219*, 495-498.

Schwartz L., Piwinskii A., Ryerson F., Tewes H., and Beiriger W. (1984) Glass produced by underground nuclear explosions. *J. Non-Cryst. Solids, 67*, 559-591.

Singer R. B. (1982) Spectral evidence for the mineralogy of high-albedo soils and dust on Mars. *J. Geophys. Res., 87*, 10159-10168.

Toulmin P., III, Baird A. K., Clark B. C., Keil K., Rose H. J., Evans P. H., and Kelliher W. C. (1977) Geochemical and mineralogic interpretation of the Viking inorganic chemical results. *J. Geophys. Res., 82*, 4625-4634.

Tyburczy J. A. and Ahrens T. J. (1987) Dehydration kinetics of shocked serpentine. *Proc. Lunar Planet. Sci. Conf. 18th*, this volume.

Weldon R. J., Thomas W. M., Boslough M. B., and Ahrens T. J. (1982) Shock-induced color changes in nontronite: Implications for the Martian fines. *J. Geophys. Res., 87*, 10102-10114.

Williams F. L., Lee Y. K., Morosin B., and Graham R. A. (1986a) Catalytic Activity of shock modified ZnO for CO oxidation and methanol synthesis. In *Shock Waves in Condensed Matter—1985* (Y. M. Gupta, ed.), pp. 791-796. Plenum, New York.

Williams F. L., Morosin B., and Graham R. A. (1986b) Influence of shock compression on the specific surface area of inorganic powders. In *Metallurgical Applications of Shock-Wave and High-Strain-Rate Phenomena* (L. E. Murr, K. P. Staudhammer, and M. A. Meyers, eds.), pp. 1013-1022. Marcel Dekker, New York.

Microkrystites: A New Term for Impact-Produced Glassy Spherules Containing Primary Crystallites

B. P. Glass and Christopher A. Burns
Geology Department, University of Delaware, Newark, DE 19716

Glassy spherules containing primary crystallites (clinopyroxene-bearing spherules) and diagenetically altered spherules with relict quench textures have been found in late Eocene and Cretaceous-Tertiary boundary sediments, respectively. These spherules have been referred to as "microtektites" by some authors. Although the late Eocene spherules and K-T boundary spherules, like microtektites, are believed to have been formed by impact, they are not microtektites, which by definition do not contain primary crystallites. Furthermore, the late Eocene and K-T boundary spherules can be distinguished from microtektites by their more basic composition, the absence of lechatelierite, and an association with iridium anomalies. In order to avoid confusion with microtektites, we propose that impact-generated spherules containing primary crystallites be referred to as microkrystites.

INTRODUCTION

A layer of crystal-bearing spherules (Fig. 1) has been found in late Eocene marine deposits in close association with the North American microtektite layer (*Glass et al.*, 1985). The major crystalline phase in these spherules is clinopyroxene; therefore, these spherules have been referred to as clinopyroxene-bearing or cpx spherules. Because of the close association with the North American microtektites at several sites, it was originally assumed that the cpx spherules belong to the North American microtektite layer; however, based on additional data, we now believe that they are slightly older (*Glass et al.*, 1985). Several of the late Eocene "microtektite" layers discussed by *Keller et al.*, (1983) are actually composed of cpx spherules.

Diagenetically altered microcrystalline spherules have been recovered from the Cretaceous-Tertiary (K-T) boundary. These spherules have been described as having relict quench textures (Fig. 2) (*Montanari et al.*, 1983; *Smit*, 1984; *Smit and Kyte*, 1984; *Smit and Romein*, 1985; *Kyte and Smit*, 1986). Several authors have referred to the K-T boundary spherules as altered microtektites, microtektite-like spherules, or microtektites (*Kastner et al.*, 1984; *Smit*, 1984; *Smit and Romein*, 1985; *Anders et al.*, 1986). Both the late Eocene and K-T boundary crystal-bearing spherules are believed to have had an impact origin (*Glass et al.*, 1985; *Montanari et al.*, 1983; *Smit*, 1984).

Although many authors have implied that the late Eocene crystal-bearing spherules and the K-T boundary spherules are microtektites, they are not by definition microtektites. We therefore propose a new term to distinguish these impact-generated spherules containing primary crystallites from microtektites.

JUSTIFICATION FOR A NEW TERM

Microtektites are by definition tektites less than one millimeter in diameter. Tektites are characterized by their relative homogeneity, ubiquitous presence of lechatelierite, absence of primary crystallites, and almost complete absence of relict grains (*Chao*, 1963; *Glass*, 1982). Thus the late Eocene and K-T boundary crystal-bearing spherules by definition cannot be microtektites, since they contain primary crystallites but generally do not contain lechatelierite.

An additional distinction between crystal-bearing spherules and microtektites is that the former have been associated with iridium anomalies due to meteoritic contamination, while the latter have not.

We could continue to call the late Eocene and K-T boundary spherules "crystal-bearing spherules," but that phrase is a description rather than a name, and it is rather lengthy and therefore awkward to use. Some workers have proposed to call these spherules "crystalline microtektites," but this term is self-contradictory and it is also rather long and awkward. We therefore feel that it is necessary to propose a new term to refer to these and other impact-generated spherules containing primary crystallites. We suggest that they be called microkrystites. The "krys" prefix is a derivative of the Greek "kyros" or "krystallos" from which the word crystal was derived. Megascopic forms, if they are even found, would be krystities.

DISCUSSION

If the microkrystites and the microtektites are both of impact origin, then the difference in crystallinity between the two groups of spherules is probably primarily a reflection of the composition of the target material. The late Eocene cpx spherules are basic in composition (*Glass et al.*, 1985) and the K-T boundary spherules are also believed to have been formed from basic material (*Montanari et al.*, 1983). The tektites and microtektites, on the other hand, are generally acidic in composition (Table 1) (*Chao*, 1963; *Glass*, 1984). It appears, therefore, that the more basic melts partly crystallized upon cooling, whereas the acidic melts did not. It should be noted, however, that silica content by itself is not the only factor, since there is a great deal of overlap in the silica content between cpx spherules and microtektites (see for example, *Glass et al.*, 1985).

Some workers have argued that since the crystal-bearing spherules and microtektites have the same origin a new term is not needed to distinguish between them. We do not agree.

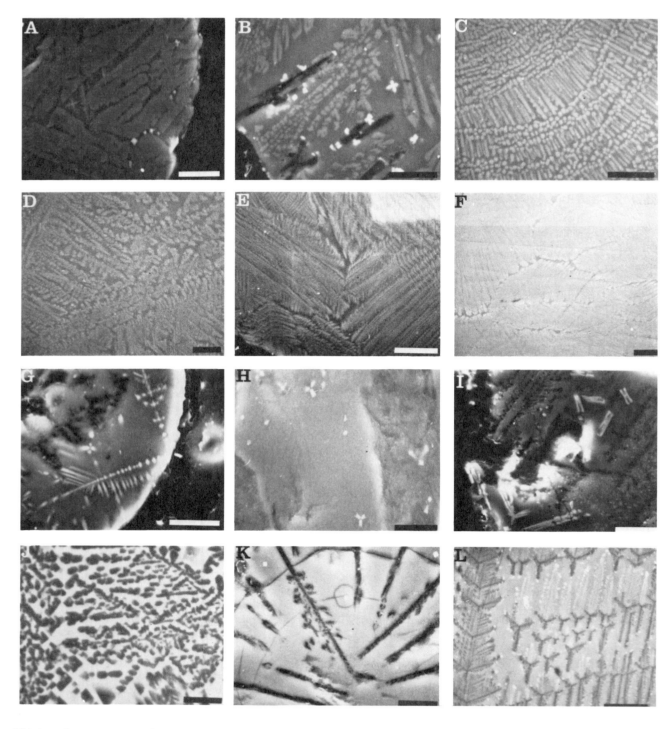

This is analogous to saying that we do not need to have different names for obsidians and basalts, since both are formed by igneous processes.

As mentioned previously, the crystal-bearing spherule layers are associated with iridium anomalies while the microtektite layers are not. This relationship is being obscured in the literature by referring to the crystal-bearing spherules as "microtektites." This has led to a belief that microtektites, and therefore tektites, are associated with iridium anomalies. This is simply not correct. None of the microtektite layers are known to be associated with an iridium anomaly.

Since both the crystal-bearing spherules and the microtektites are thought to be of impact origin, it is not clear to us why both the K-T boundary spherule layer and the late Eocene

Fig. 1. Scanning electron microscope photomicrographs of polished interior sections of clinopyroxene-bearing spherules illustrating some of the many observed textures. (**a**) Sample 210-8, from Deep Sea Drilling Project (DSDP) Site 166 in the central equatorial Pacific Ocean, containing large clinopyroxene crystals (light grey), small equant Fe- and Mg-rich crystals, and fainter Fe-rich skeletal crystals (magnetite?) in a glass matrix (darker grey). (**b**) Sample 74-6, from DSDP Site 166, containing clinopyroxene crystals (light grey) and Fe- and Cr-rich (with some Ni) crystals (white equant or starlike) in a glass matrix (dark grey). The black elongate regions are voids apparently produced when a less stable phase (olivine?) was removed by solution in sea water. (**c**) Sample UD-723-7, from DSDP Site 216 in the eastern equatorial Indian Ocean, containing clinopyroxene crystals (light grey) in a glass matrix (darker grey). (**d**) Sample 206-1, from DSDP 69A in the central equatorial Pacific Ocean, containing feathery or dendritic clinopyroxene crystals (light grey) in a glass matrix (darker grey). (**e**) Sample 208-10, from DSDP Site 292 in the western equatorial Pacific Ocean, containing feathery crystals of clinopyroxene (light grey) in a glass matrix (dark grey). (**f**) Sample 55-17, from DSDP Site 216 in the eastern equatorial Indian Ocean, composed primarily of feathery clinopyroxene crystals (light grey) in a glass matrix (dark grey). (**g**) Sample 56-3, from DSDP Site 70A in the central equatorial Pacific Ocean, containing Fe-rich (with some Cr and Ni) skeletal crystals (white) in a glass matrix (grey). The black regions inside the spherule are voids. (**h**) Sample 164-13, from Core RC9-58 in the Caribbean Sea, containing propeller-shaped, Fe- and Cr-rich (with Ni) crystals (light grey) in a glass matrix (darker grey). The dark grey irregular regions are apparently where a less stable phase has been removed by solution in sea water. (**i**) Sample 183-4, from DSDP Site 94 in the Gulf of Mexico, containing Fe-, Cr- and Ni-rich skeletal crystals (white) and smaller crystals of clinopyroxene (light grey) in a glass matrix (grey). The small black regions, forming geometric patterns, are voids. (**j**) Sample 220-9, from DSDP 94 in the Gulf of Mexico, containing angular voids, forming geometric patterns, in a glass matrix (light grey). (**k**) Sample 207-2, from DSDP Site 69A in the central equatorial Pacific ocean, containing a series of chain-like radiating voids (black) and diamond shaped crystals (white), which may be chromite, in a glass matrix (light grey). (**l**) Sample 220-2, from DSDP Site 94 in the Gulf of Mexico, containing voids producing geometric patterns (dark grey) in a glass matrix (light grey). (**g-l**) The voids were probably produced when an unstable phase was removed by solution in sea water. The scale bars are all 10 µm in length.

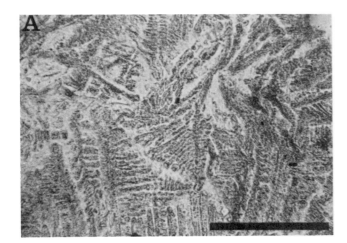

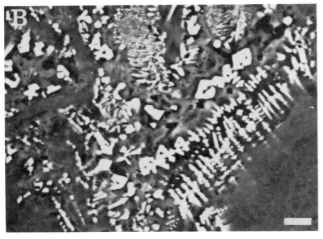

Fig. 2. Photomicrographs of polished sections of Cretaceous-Tertiary boundary spheroids showing quench textures. (**a**) Optical microscope photomicrograph of spheroid from Caravaca, Spain, with dendritic texture (courtesy A. Montanari). (**b**) Backscatter electron image of spheroid of Furlo, Italy, showing skeletal magnetite (courtesy F. T. Kyte).

cpx layers are associated with iridium anomalies, whereas the microtektite layers are not. Since the iridium comes from the impacting body, the above observation suggests that either the impacting bodies that produced the tektites and microtektites were different from those that produced the older crystal-bearing spherule layers or that contamination from the impacting bodies were more widespread during the crystal-bearing spherule events.

Two other groups of spherules that may be examples of microkrystites have been discovered by *Lowe and Byerly* (1986). They described spherules from early Archean deposits in South Africa and western Australia that they believe were formed by impact. The early Archean spherules have undergone recrystallization, replacement, and metasomatism. However, according to *Lowe and Byerly* (1986), many of the spherules show relict quench textures. Thus, some or most of the early Archean spherules described by *Lowe and Byerly* (1986) may be diagenetically altered microkrystites. The early Archean spherule layers contain significant levels of iridium, but it is not presently known if the high iridium is due to meteoritic contamination (D. R. Lowe, personal communication, 1986).

SUMMARY AND CONCLUSION

The late Eocene clinopyroxene-bearing spherules and the K-T boundary spherules have been referred to as "microtektite-like spherules" or "microtektites" by some authors. However,

TABLE. 1. Comparison between microkrystites and microtektites.

Spherule type	Primary crystallites	Lechatelierite	Composition	Associated with iridium anomaly	Age
Microkystites	common	absent	generally basic	yes	late Eocene & older
Microtektites	absent	common	generally acidic	no	late Eocene & younger

microtektites by definition do not contain primary crystallites. Therefore, the late Eocene and K-T boundary crystal-bearing spherules should not be called microtektites. We suggest that they be called microkrystites in order to distinguish them from the microtektites.

The microkrystites not only differ from microtektites in terms of their petrography, but they also differ in terms of their composition and age (Table 1). Furthermore, the microkrystite layers are associated with iridium anomalies and the microtektite layers are not. It is not clear why this difference exists, but it may illustrate our lack of knowledge concerning large impacts and we believe that it is a problem that merits further investigation.

Acknowledgments. We thank the graduate students in the Geology Department at the University of Delaware for their help in choosing an appropriate term for the crystal-bearing spherules. We thank James Gooding and Frank Kyte for critical review of the paper, and James Gooding for suggesting the term krystite. We also thank D. Stöffler, R. A. F. Grieve, D. R. Lowe, A. Montanari, and especially F. Hörz for comments on earlier drafts of the paper. This research supported by NSF grant OCE-8314522.

REFERENCES

Anders E., Wolback W. S., and Lewis R. S. (1986) Cretaceous Extinctions and Wildfires: Response. *Science, 261,* 263-264.

Chao E. C. T. (1963) The petrographic and chemical characteristics of tektites. In *Tektites* (J. A. O'Keefe, ed.), pp. 51-94. University of Chicago Press, Chicago.

Glass B. P. (1982) *Introduction to Planetary Geology.* Cambridge University Press, New York. 469 pp.

Glass B. P. (1984) Tektites. *J. Non-Crystal. Solids, 67,* 333-344.

Glass B. P., Burns C. A., Crosbie J. R., and DuBois D. L. (1985) Late Eocene North American microtektites and clinopyroxene-bearing spherules. *Proc. Lunar Planet. Sci. Conf. 16th,* D175-D196.

Kastner M., Asaro F., Michel H. V., Alvarez W., and Alvarez L. W. (1984) The precursor of the Cretaceos-Tertiary boundary clays at Stevns Klint, Denmark, and DSDP Hole 465A. *Science, 226,* 137-143.

Keller G., D'Hondt S., and Vallier T. L. (1983) Multiple microtektite horizons in upper Eocene marine sediments: No evidence for mass extinctions. *Science, 221,* 150-152.

Kyte F. T. and Smit J. (1986) Regional variations in spinel compositions: An important key to the Cretaceous/Tertiary event. *Geology, 14,* 485-487.

Lowe D. R. and Byerly G. R.(1986) Early Archean silicate spherules of probable impact origin, South Africa and western Australia. *Geology, 14,* 83-86.

Montanari A., Hay R. L., Alvarez W., Asaro F., Michel H. V., and Alvarez L. W. (1983) Spheroids at the Cretaceous-Tertiary boundary are altered impact droplets of basaltic composition. *Geology, 11,* 668-671.

Smit J. (1984) Evidence for worldwide microtektite strewn field at the Cretaceous-Tertiary boundary (abstract). *Geol. Soc. Am. Abstracts with Programs 1984, 16,* p. 659.

Smit J. and Kyte F. T. (1984) Siderophile-rich magnetic spheroids from the Cretaceous-Tertiary boundary in Umbria, Italy. *Nature, 310,* 403-405.

Smit J. and Romein A. J. T. (1985) A sequence of events across the Cretaceous-Tertiary boundary. *Earth Planet. Sci. Lett., 74,* 155-170.

Phase Equilibrium Constraints on the Howardite-Eucrite-Diogenite Association

John Longhi and Vivian Pan

Department of Geology and Geophysics, Yale University, New Haven, CT 06511

Model calculations of fractional crystallization and equilibrium partial melting in the range of 0 to 10 kbar have been carried out for a series of compositions relevant to diogenite and eucrite petrogenesis. The calculated liquidus equilibria are in essential agreement with the experimental work of *Stolper* (1977) and show that olivine reacts with diogenite parent liquids along the plagioclase-absent olivine/low-Ca pyroxene liquidus boundary under conditions of both fractional and equilibrium crystallization up to approximately 2 kbar. Likewise, olivine reacts with eucritic liquids saturated with plagioclase and low-Ca pyroxene to pressures in excess of 2 kbar. Accordingly, low-pressure fractionation of liquids sufficiently magnesian to have crystallized the diogenites produces differentiates with SiO_2 components too high to have quenched to form the eucrites. Pressures on the order of 1 to 2 kbar are required to achieve the proper liquid line of descent. Pressures of at least 2 kbar are required to crystallize the olivine noted in most diogenites. Simple fractional crystallization at 2 kbar can account for many of the mineralogical and chemical features of the diogenite-eucrite association, but its plausibility is strained by the absence of volcanic equivalents of the magmas that crystallized the diogenites, since these liquids would have been less dense than the eucrite liquids and therefore should have erupted if they were present in the same magma chambers. Two scenarios, consistent with phase equilibria constraints are (1) fractional crystallization in a magma ocean developed on a parent body well in excess of 500 km in radius; and (2) partial melting of an olivine-depleted source region to produce eucritic magmas, followed by remelting of the source to produce diogenitic magmas. The radius of the parent body in this latter case might have been as small as 500 km. The scarcity of anorthosites in polymict meteorite breccias lessens the appeal of the magma ocean scenario.

INTRODUCTION

Key parameters, such as similar FeO/MnO ratios (*Wänke et al.*, 1973) and oxygen isotope ratios (*Clayton*, 1977), strongly suggest a common parent body for diogenite (orthopyroxene cumulates) and eucrite (pigeonite-plagioclase cumulates and basalts) meteorites. Mineral and rock fragments of both classes of these meteorites are apparently the major constituents of howardite meteorites, which are generally regarded as regolith breccias developed on the surface of this parent body. Despite the relative chemical and mineralogical simplicity of the howardite-eucrite-diogenite (HED) suite, the relation between diogenites and eucrites remains a subject of active debate focused on two major hypotheses.

One hypothesis is that diogenites are cumulates of primitive magmas that subsequently evolved to produce the less magnesian eucrites (e.g., *Mason*, 1967; *Ikeda and Takeda*, 1985; *Warren*, 1985). The major incentive for this viewpoint is the near continuum in low-Ca pyroxene composition ranging from magnesian orthopyroxene in diogenites to slightly less magnesian pigeonites in cumulate eucrites to ferroan pigeonites in basaltic eucrites. If one includes pyroxenes from howardites (e.g., *Dymek et al.*, 1976; *Ikeda and Takeda*, 1985), there is virtually no gap in composition. *Warren* (1985) has also argued that the proportions of diogenite and eucrite components in howardites are appropriate for the fractional crystallization hypothesis.

The other major hypothesis is that eucrites represent primary, low-pressure partial melts of undifferentiated source regions and that diogenites accumulated from liquids that were produced by remelting the eucrite-depleted source regions (*Stolper*, 1977; *Beckett and Stolper*, 1987). The major incentives for this hypothesis are the clustering of fine-grained eucrite compositions near the low-pressure olivine-pigeonite-plagioclase psuedo-peritectic and the absence of a series of basaltic meteorites with compositions extending back toward the more primitive liquids that would have crystallized the diogenites.

The two hypotheses have important consequences for the parent body composition. As will be shown, the partial melting hypothesis requires a somewhat lower Mg' value [MgO/(MgO + FeO) - molar] for its source, as typified by the eucrite parent body composition proposed by *Hertogen et al.* (1977) with Mg' = 0.68 [EPB2 from *Basaltic Volcanism Study Project (BVSP)*, 1981], whereas the fractional crystallization hypothesis requires more magnesian compositions, such as EPB3 (*Dreibus et al.*, 1977) with Mg' = 0.80.

Since *Stolper's* (1977) original experimental study, new crystal-liquid data have been produced (*Longhi and Pan*, 1987a,b) that span a much wider range of composition. The new data have been incorporated into quantitative models of fractional crystallization and partial melting that permit a wider range of tests of the two competing hypotheses than was possible before. There are two major tests that we wish to apply. The first is to determine if there is in fact a set of conditions under which a fractionating magma can produce diogenites as early cumulates and the eucrites as evolved liquids. *Hewins* (1981) has pointed out that most diogenites contain olivine, so the fractionating magma must crystallize olivine in its early stages but not in its later stages when it approaches eucritic compositions. The second test is to determine if there is a mechanism, consistent with the fractionation hypothesis,

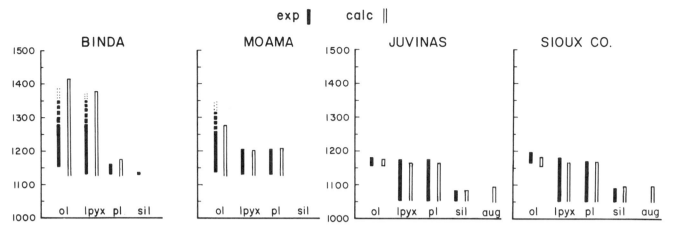

Fig. 1. Comparison of equilibrium crystallization of eucrite compositions determined by melting experiments (solid bars) of *Stolper* (1977) and model calculations from this study. Dashed bars indicate that the high-temperature stability limits of phases were not determined for the Binda and Moama compositions. Input whole rock compositions for Binda and Moama are from *McCarthy et al.* (1974); Juvinas and Sioux County compositions are from *Duke and Silver* (1967).

that can explain the clustering of compositions of the fine-grained eucrites. *Stolper's* (1977) partial melting hypothesis provides a ready explanation for the clustering. One possible mechanism is that of a density filter: Fractionating magmas remain trapped in subsurface chambers until their densities drop below a certain critical value, at which point they intrude upward and erupt onto the surface. *Stolper and Walker* (1980) have proposed such a mechanism to explain the clustering of compositions of midocean ridge basalts. Since calculating the densities of silicate liquids is straightforward, it will be possible to establish whether eucritic liquids are less dense than their putative parental magmas.

MODELING CRYSTALLIZATION

Fractional crystallization cannot be easily investigated with equilibrium crystallization experiments because of the large number of experiments that would be required to approximate the ideal process. Liquid lines of descent derived from controlled-cooling rate experiments can only be applied with confidence to the fractionation that occurs on the scale of a thin section because of unequal nucleation, growth, and diffusion rates among the major minerals. Fractional crystallization can, however, be modeled with a computer once liquidus equilibria spanning a wide range of composition have

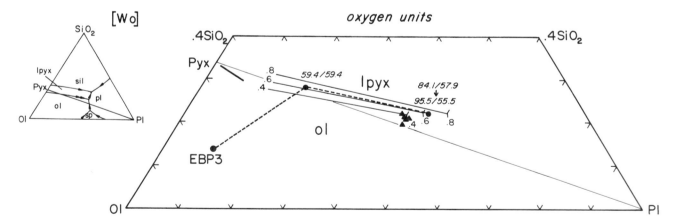

Fig. 2. One bar equilibrium crystallization/partial melting path (dashed line) of EPB3 (*Dreibus et al.*, 1977) projected from the Wollastonite component, Wo, onto a portion of the Ol-Pl-SiO$_2$ diagram in oxygen units. Projection equations are given in the Appendix. Inset triangle shows the entire liquidus diagram for Mg' = 0.6 at values of the Wo (0.05) and alkali feldspar (AFP = 0.1) components appropriate for HED compositions; equations for Wo and AFP are given in the Appendix. Pairs of numbers separated by slashes are total volume percent crystallized and volume percent olivine, respectively. Two sets of numbers are shown at the pseudo-peritectic point and represent crystallization when liquid first reaches the peritectic and at the end of the calculations. Filled circles indicate the appearance of a new phase: ol at the liquidus, lpyx at 59.4% crystallization, and plagioclase at 84.1% crystallization. The heavy line is the range of calculated low-Ca pyroxene (lpyx) compositions. Light solid lines are calculated positions of the olivine/low-Ca pyroxene boundary with Wo = 0.05, AFP = 0.1, and Mg' = 0.4, 0.6, and 0.8. These lines terminate at plagioclase saturation; the Pl component at the terminus decreases as Mg' decreases. Filled triangles are compositions of fine-grained eucrites from *Duke and Silver* (1967); only these are shown to avoid scatter due to interlaboratory bias.

been fitted to analytical expressions. There are several approaches to modeling crystallization of basalts (e.g., *Ghiorso*, 1985; *Nielsen*, 1987). The approach employed in this paper involves direct parameterization of liquidus boundary surfaces in terms of Olivine(Ol) - Plagioclase(Pl) - Wollastonite(Wo) - SiO$_2$ components and has been described by *Longhi* (1987). Modifications to the model are described in the Appendix.

With only minor changes in computer code, the same equations describing liquidus boundaries can be employed to calculate equilibrium crystallization as well. The accuracy of the model calculations can thus be tested by comparing calculated and experimental equilibrium crystallization for a variety of relevant compositions.

Figure 1 illustrates a comparison of phase appearance versus temperature for two cumulate and two noncumulate eucrites previously investigated by *Stolper* (1977). Table 1 lists some representative phase compositions.

The agreement between experimental and calculated phase appearance in Fig. 1 is within 15°C for the most important phases: olivine, low-Ca pyroxene, and plagioclase. An obvious discrepancy is the presence of augite in the calculated crystallization sequences of Juvinas and Sioux County. *Stolper* (1977) also reported Cr-spinel throughout the crystallization interval of Juvinas and Sioux County, yet the calculations produced no spinel at all for these compositions. Neither discrepancy is serious, however, in terms of the model calculations or their petrogenetic implications. For example, an increase of only 0.02 wt % Cr$_2$O$_3$ in the bulk of compositions will produce spinel in the calculations. More importantly, spinel is a minor phase and with the exception of Cr$_2$O$_3$, whose variation is not discussed here, spinel crystallization has only second-order effects on the concentrations of major elements, as demonstrated by the agreement of experimental and calculated Juvinas peritectic phase compositions within experimental-analytical error in Table 1. Likewise, only 6 vol % augite, among a total of 85 vol % crystals, is calculated to be present when silica appears—such a small amount could easily be overlooked in a fine-grained charge, and in any case late-crystallizing augite has little bearing on the important olivine-pigeonite-plagioclase liquidus relations.

More relevant discrepancies exist between the measured and calculated crystallization sequences of Binda and Moama. In both cases *Stolper* (1977) reported complete resorption of olivine and, in the case of Binda, the appearance of silica. However, despite the fact that the calculations predict significant resorption of olivine during the crystallization of low-Ca pyroxene and plagioclase, in neither case is olivine resorbed completely. For example, in the Binda calculations, once low-Ca pyroxene appears, olivine decreases from an initial value of 13 vol % to 5 vol % at 98.5% total crystallization, the point at which the calculation is stopped. Similarly, olivine in the Moama calculations decreases from 15 to 5 vol %. Both of the compositions of Binda and Moama reported by *McCarthy et al.* (1974) are slightly olivine normative, so the results of our calculations are accurate in predicting a small amount of residual olivine.

Differences between the measured and calculated concentrations of TiO$_2$ and Al$_2$O$_3$ at 1277°C in Binda liquids indicate

TABLE 1. Comparison of experimental and calculated phase compositions (wt % oxides, recalculated to 100%).

Run	JV-14*	JV-18*	JV-D&S†	SC-54*	SC-58*	SC-D&S†	BI-2*	BI-MC‡
Temp	1171°C	1171°C	1166°C	1171°C	1171°C	1169°C	1277°C	1277°C
liq								
SiO$_2$	49.98	49.12	49.73	49.90	49.32	49.79	51.20	50.78
TiO$_2$	0.71	0.68	0.72	0.71	0.68	0.65	0.66	0.34
Al$_2$O$_3$	13.23	12.81	12.78	12.98	12.59	12.71	9.09	10.50
Cr$_2$O$_3$	0.33	0.31	0.29	0.31	0.32	0.37	0.67	0.42
FeO	17.56	19.08	18.64	18.07	19.29	18.57	18.84	17.74
MgO	6.49	6.52	6.24	6.51	6.45	6.31	10.21	10.59
MnO	0.58	0.58	0.53	0.58	0.60	0.54	0.65	0.58
CaO	10.60	10.30	10.58	10.50	10.28	10.58	8.36	8.71
K$_2$O	n.a.	n.a.	0.05	n.a.	n.a.	0.04	n.a.	0.01
Na$_2$O	0.54	0.51	0.43	0.44	0.47	0.43	0.32	0.32
ol§	0.654	0.630	0.638	0.654	0.628	0.643	0.76	0.754
lpyx¶	0.045/0.660	0.051/0.636	0.043/0.637	0.049/0.652	0.052/0.640	0.044/0.641	0.03/0.73	0.030/0.756
pl**	0.936	present	0.934	present	present	0.936	absent	absent
cr-sp	present	present	absent	present	present	absent	absent	absent
SiO$_2$[Wo]††	0.214	0.205	0.218	0.219	0.211	0.219	0.265	0.241
Ol[Wo]††	0.336	0.356	0.346	0.343	0.359	0.349	0.427	0.414

n.a. = not analyzed.
*Data from *Stolper* (1977), Tables 1 and 2.
†This study; bulk composition from *Duke and Silver* (1967).
‡This study; bulk composition from *McCarthy et al.* (1974).
§Mg/(Mg + Fe) atomic.
¶Ca/(Ca + Mg + Fe) and Mg/(Mg + Fe) atomic.
**Ca/(Ca + Na + K) atomic.
††Ternary coordinates (Fig. 2) in oxygen units.

that the aliquots melted by *Stolper* (1977) and analyzed by *McCarthy et al.* (1974) had different compositions. Thus sample heterogeneity may have contributed to the discrepancies in phase relations of the cumulate eucrites. However, there may have been other contributions, such as minor iron loss destabilizing olivine in Stolper's experiments or small analytical errors incorporated into our formulations of the liquidus boundaries.

Comparison of the projection coordinates listed in Table 1 indicates a systematic difference in the position of the olivine/low-Ca pyroxene liquidus boundary between this study and *Stolper's* (1977). The plagioclase-undersaturated liquid composition (BI-2) reported by *Stolper* (1977) lies on the high-SiO_2 side of the olivine/low-Ca pyroxene liquidus boundary calculated by our algorithm by approximately 0.02 units of SiO_2[Wo], whereas the plagioclase-saturated peritectic compositions (JV, SC) agree within 0.01 units (Table 1). As a result, the calculated olivine/low-Ca pyroxene liquidus boundary (e.g., Fig. 2) for these compositions is less steep than one that could be inferred from the data of *Stolper* (1977) or *Beckett and Stolper* (1987). We have not as yet determined the relative contributions of experimental and analytical error to this difference, but the consequence to be kept in mind is that our model calculations predict less olivine resorption in low-Al (diogenetic) compositions. Because olivine reaction with liquid was at the heart of *Stolper's* (1977) hypothesis that diogenites and eucrites were not simply related through fractional crystallization, it should be anticipated that, other factors being equal, the calculations presented here are more likely to favor the fractional crystallization hypothesis than are the experimental results.

Equilibrium Crystallization and Partial Melting

In order to appreciate the implications of the fractional crystallization calculations, it is fruitful to examine some aspects of the liquidus boundary parameterizations and their relation to liquidus equilibria. In general the positions of the phase boundaries are treated as functions of Mg', Wo component, and normative alkali feldspar composition (AFP). However, in the case of HED compositions only variations of Mg' have first-order significance and need to be considered. Figure 2 illustrates a calculated liquid line of descent for equilibrium crystallization (or partial melting in reverse) of a eucrite parent body composition, EPB3, proposed by *Dreibus et al.* (1977). For reference, Fig. 2 also illustrates the position of the olivine/low-Ca pyroxene liquidus boundary at three values of Mg', but with constant Wo and normative feldspar components. Initially, only olivine crystallizes and the liquid moves directly away from the Ol component. Once low-Ca pyroxene (lpyx) becomes a saturating phase at 59.4 vol % crystallization, the liquid moves along the olivine/low-Ca pyroxene liquidus boundary, crystallizing orthopyroxene and minor Cr-spinel while resorbing olivine. When the liquid reaches plagioclase saturation at 84.1% crystallization, only 1.5% olivine has been resorbed. In the next 11% crystallization, however, the volume percent of olivine drops sharply by 2.5%. The reason for the difference in the rate of olivine resorption is that there are two independent resorption mechanisms at work: One applies along the olivine/low-Ca pyroxene liquidus boundary and the other applies at the pseudo-invariant point where plagioclase is present.

Consider the latter mechanism first. When olivine coexists with low-Ca pyroxene, plagioclase, and liquid, its reaction relation is controlled by the location of the liquid composition relative to a triangle formed by the compositions of coexisting olivine, low-Ca pyroxene, and plagioclase (one side of the triangle is nearly coincident with the light solid line joining the ideal Pyx and Pl components). When the liquid composition projects within the triangle, the equilibrium is pseudo-eutectic and olivine crystallizaes as heat is withdrawn from the system. When the liquid projects outside of the three-phase triangle, as is the case here, the equilibrium is pseudo-peritectic and olivine dissolves as heat is withdrawn.

Along the olivine/low-Ca pyroxene liquidus boundary, the mechanism by which olivine dissolves or crystallizes is controlled by the liquid trajectory, as described by the Rule of Tangents (e.g., *Morse*, 1980). If a tangent to the liquid path along the boundary curve intersects the line joining the coexisting olivine and low-Ca pyroxene compositions between the two compositions, then olivine will crystallize as temperature falls and the liquidus boundary curve is said to be "even." If the intersection of the tangent with the line joining the two compositions lies on the high-SiO_2 side of the pyroxene composition, as is marginally the case here, then olivine will dissolve as pyroxene crystallizes and the boundary curve is said to be "odd."

Because compositional location and trajectory are independent, the two mechanisms or sets of equilibria are also independent. It also follows that even curves may be associated with peritectics and odd curves with eutectics.

There are several concepts to keep in mind in applying these equilibria to natural systems. First, the nature of a boundary curve may change as a result of changes in composition or pressure. For example, in Fig. 2 it is clear that with decreasing Mg' the olivine/low-Ca pyroxene liquidus boundary eventually changes from odd to even.

Second, the composition of pyroxene changes in response to changing liquid composition and is always displaced from the ideal Pyx component by solid solution of Al_2O_3, mainly in the form of Mg- and Ca-Tschermak components. Accordingly, the Rule of Tangents must not be applied to the projection point of the Pyx component, but rather to the actual pyroxene compositions that lie along the Pyx-Al_2O_3 join. In addition, when low-Ca pyroxene is the only solid phase present, the path of the crystallizing liquid will curve slightly upward toward SiO_2 because of the changing pyroxene composition.

Third, because liquid composition changes during crystallization with respect to parameters not depicted in the projections, such as Mg', the liquid path will not in general follow traces of the boundary curves along which these parameters are held constant (e.g., Fig. 2). In detail the path that a liquid follows during equilibrium crystallization along the olivine/low-Ca pyroxene boundary will have a more negative slope than that of a constant Mg' curve and the true transition from odd to even will take place at lower Mg' than might be inferred from the constant Mg' curves.

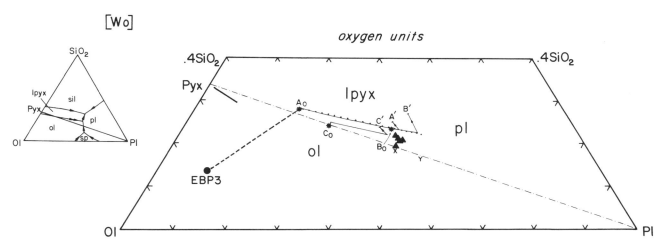

Fig. 3. One bar fractional crystallization paths of three representative liquid compositions indicated by asterisks. Projection scheme as in Fig. 2. A_0 is the liquid composition generated by 41 vol % melting of EPB3: 1436°C, Mg' = 0.68, only olivine (Fo_{86}) on the liquidus. B_0 – 20 vol % melting of EPB3: 1267°C, Mg' = 0.60, olivine (Fo_{82}), opx ($Wo_{2.9}$, En_{81}), Cr-spinel. C_0 – 23 vol % melting of EPB2: 1311°C, Mg' = 0.48, olivine (Fo_{72}). Tic marks at end of fractionation curve (A', B', C') indicate onset of silica crystallization. Dashed and dotted curves are equilibrium melting paths of EPB3 from Fig. 2. "X" and "Y" are mixtures of pyroxene and plagioclase explained in the text. Other symbols as in Fig. 2.

Fourth, the projections of the pseudo-invariant points will also change in response to variation in Mg', Wo, and AFP. Thus the olivine/low-Ca pyroxene/plagioclase peritectic point has three different positions in Fig. 2 reflecting three values of Mg'. Accordingly, Fig. 2 illustrates that low-pressure partial melting of EPB3 produces plagioclase-saturated liquids that are much more magnesian (Mg' = 0.6) and aluminous than inferred eucrite liquids (Mg' = 0.4). However, because Mg' decreases much more rapidly during fractional crystallization than during equilibrium crystallization, it may be anticipated that some liquid along the olivine/low-Ca pyroxene boundary curve, representing a high degree of partial melting of EPB3, might reach plagioclase saturation with Mg' (0.4) appropriate to eucrites. Such behavior is illustrated in Fig. 3.

Fractional Crystallization

In Fig. 3 we present the calculated low-pressure fractional crystallization paths of three liquids, A_0, B_0, and C_0, generated by low-pressure partial melting of proposed eucrite parent bodies, in order to illustrate some of the relationships between bulk composition, Mg', and liquid line of descent relevant to eucrite petrogenesis. Liquids A_0 and B_0 are generated by partial melting of EPB3, whose equilibrium melting/crystallization path is illustrated again as a dotted curve. Liquid A_0 represents approximately 41 vol % partial melting and crystallizes the most magnesian low-Ca pyroxene possible from EPB3. This orthopyroxene ($Wo_{2.5}$ En_{85}) is very similar to the most magnesian orthopyroxene from mineral and rock fragments in the Kapoeta howardite (*Dymek et al.*, 1976). Liquid B_0 represents 20 vol % partial melting of EPB3. Its initial orthopyroxene ($Wo_{2.9}$ En_{80}) is less magnesian than the most magnesian orthopyroxenes observed in Kapoeta (*Dymek et al.*, 1976). Liquid C_0 represents 23 vol % partial melting of EPB2 (*Hertogen et al.*, 1977); it initially crystallizes 3% olivine followed by the most magnesian pyroxene ($Wo_{2.5}$ $En_{71.7}$)

possible from this parent body composition. A_0 and B_0 illustrate the consequences of varying degrees of partial melting of the same source. Contrasting C_0 with A_0 and B_0 provides a means of evaluating the consequences of different source compositions.

In all three cases the fractionating liquids produce very slightly curved paths that trend away from the line of pyroxene compositions in Fig. 3. The curves change direction abruptly when plagioclase begins to crystallize and are shown to terminate when silica begins to crystallize (A', B', C'). Once low-Ca pyroxene precipitates, no olivine crystallizes from the differentiates of A_0 or B_0 until well after silica appears—an indication that the olivine/low-Ca pyroxene boundary is odd for the Mg' values of these liquids. Differentiates of A_0 and B_0 reach plagioclase saturation at SiO_2 coordinates that lie close to the trace of the equilibrium crystallization path and above the group of noncumulate eucrites. Approximately 0.5 % olivine crystallizes from differentiates of C_0, however, so that these liquids traverse an olivine/low-Ca pyroxene cotectic. The line of descent of C_0 trends toward the cluster of eucrite whole rock analyses, but turns away before reaching the cluster because of plagioclase crystallization. This is an indication that compositions with Mg' sufficiently low to have traversed the even portion of the olivine/low-Ca pyroxene liquidus boundary are too low in Mg' to have subsequently crystallized the eucrites at low pressure. In addition, crystallization of C_0 demonstrates that compositions with relatively low Mg', such as EPB2, cannot directly produce pyroxenes as magnesian as those observed in howardites in a single episode of melting and differentiation. Rather, remelting of a basalt-depleted source region is required (e.g., *Stolper*, 1977).

In terms of a diogenite/eucrite linkage, liquid A_0 comes closest to satisfying the requirements of a parental magma: It crystallizes sufficiently magnesian orthopyroxene initially and subsequently reaches plagioclase saturation with a low Mg' (0.36) typical of the eucrites. Because of the olivine reaction

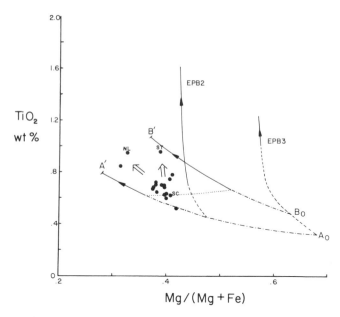

Fig. 4. Plot of wt % TiO_2 versus $Mg' = Mg/(Mg + Fe)$. Filled circles are compositions of noncumulate eucrites from *BVSP* (1981): NL = Nuevo Laredo, ST = Stannern, SC = Sioux County. Curves A_0-A' and B_0-B' are the compositions of fractionally crystallizing liquids from Fig. 3. Curves EPB2 and EPB3 are liquids generated by low-pressure equilibrium partial melting of these two compositions. Solid portions of the various curves indicate plagioclase-saturated liquids. Large arrows indicate Stannern and Nuevo Laredo trends.

relation predicted for high-Mg', low-Al liquids, the line of descent of A_0 at plagioclase saturation has SiO_2 coordinates too high to match the main body of eucrite analyses. Based upon the crystallization paths of A_0, B_0, and C_0, we may generalize that the range of liquids that have sufficiently high Mg' to crystallize the most magnesian pyroxenes observed in howardites and that can also crystallize sufficient amounts of pyroxene to drive Mg' down into the range of the eucrites at plagioclase saturation lie between A_0 and the Ol-SiO_2 join. Such liquids will have lines of descent with slightly higher SiO_2 coordinates than A_0. Accordingly, if one accepts the premise that the whole rock analyses of the fine-grained eucrites are those of quenched liquids, then the calculations indicate that the eucrite compositions cannot have been derived by low-pressure fractional crystallization of liquids that first crystallized the diogenitic orthopyroxenes found in howardites.

Not all authors accept the premise that all the compositions of the fine-grained eucrites are those of quenched liquids, however. *Warren and Jerde* (1987), for example, have suggested that the eucrites that project with the lowest SiO_2 coordinates, such as Sioux County are mixtures of more SiO_2-rich liquids and coaccumulated pigeonite and plagioclase. Although some compositional heterogeneity almost certainly exists in the eucrites, two lines of evidence suggest that the array of compositions evident in Fig. 3 is not caused by variable accumulation of pigeonite plus plagioclase into liquids at the high-SiO_2 end of the array. First, the Sioux County composition projects approximately halfway between the high-SiO_2 end of the array and point X in Fig. 3, which represents the necessary admixture of pigeonite and plagioclase of the partial cumulate hypothesis and implies 20 to 30 vol % accumulation of pigeonite. However, the data of *Stolper* (1977) show that the most magnesian pigeonites produced in melting experiments on Sioux County are virtually identical to the most magnesian pyroxenes observed in Pasamonte (*BVSP*, 1981), which has bulk Mg' similar to Sioux County and is one of the few eucrites in which Fe and Mg in pigeonite are not equilibrated. The similarity of natural and experimental pyroxenes strongly suggests that the composition of Sioux County has not been modified by more than a few percent accumulation of pigeonite, which limits the hypothetical parental liquids to the lower half of the array of compositions in Fig. 3. Second, the calculated line of descent for liquid A_0 along the pigeonite/plagioclase cotectic requires extraction of a pigeonite plus plagioclase mixture that plots near Point Y on Fig. 3. Admixture of such a component is not evident in the array of eucrite compositions.

The partial melting and fractional crystallization calculations also offer some insights into the variation of TiO_2 with Mg', previously examined by *Stolper* (1977), *BVSP* (1981), and *Warren and Jerde* (1987). Figure 4 is a plot of TiO_2 versus Mg' on which are displayed the compositions of the fine-grained eucrites and the calculated compositions of differentiates of A_0 and B_0 discussed above, plus the compositions of liquids produced by partial melting of EPB2 and EPB3. The plagioclase-saturated portions of these curves are shown as solid lines; the plagioclase-undersaturated portions are shown as dashed (equilibrium melting) or dash-dot (fractional crystallization) curves. The dotted line connecting the high-Mg' ends of the A and B curves represents the locus of the most magnesian, plagioclase-saturated compositions that would be produced by fractional crystallization of a series of parental liquids generated by variable degrees of partial melting of EPB3. In comparing the natural and calculated compositions, it is important to keep in mind that the absolute values of calculated TiO_2 and Mg' are strongly dependent upon the uncertain composition of the eucrite parent body, and therefore the calculated patterns, not abundances, provide the most reliable insights.

Two trends among the eucrite compositions were recognized by *Stolper* (1977): Both originate in the vicinity of the Sioux County composition and one extends toward the composition of Nuevo Laredo, whereas the other extends toward Stannern. These arrays of analyses have labeled the "Nuevo Laredo" and "Stannern" trends by *BVSP* (1981). The solid portion of curve A closely parallels the Nuevo Laredo trend in agreement with previous workers who ascribed this variation to fractional crystallization of pyroxene and plagioclase. Although the Stannern trend is usually regarded as representing melts generated by smaller degrees of partial melting of a low Mg' source with residual plagioclase, such as EPB2 (*Stolper*, 1977), the disposition of curves in Fig. 4 permits deriving Stannern by extensive fractional crystallization of a partial melt of EPB3 similar in composition to B_0. Such crystallization would involve sufficient plagioclase to produce the small negative Eu anomaly observed in Stannern (e.g., *BVSP*, 1981). Other members of the Stannern trend are likewise explicable as differentiates of parental liquids intermediate in composition between A_0 and B_0. The Stannern trend could then be explained by limited and highly fortuitous sampling of a series of fractionated magmas. As pointed out in the discussion of Fig. 3, however, if this fractionation were to take place at low pressure, the calculations predict that the proportion of SiO_2 component in the residual liquids would be higher than those observed in the fine-grained eucrites.

In view of the fact that the low-pressure fractional crystallization hypothesis does not provide a satisfactory explanation for the clustering of major element eucrite compositions, but can simply account for the near continuum of pyroxene compositions observed in howardites and possibly for the minor and trace element variations in eucrites, it seems useful at this point to examine fractionation processes at elevated pressures, as has been suggested by *Delaney* (1986) and *Hewins* (1986).

High-Pressure Melting and Crystallization

As a way of introducing the effects of pressure on phase equilibria, Fig. 5 illustrates the traces of liquid compositions along the olivine/low-Ca pyroxene liquidus boundary calculated for equilibrium melting of the EPB3 composition at 2, 5, and 10 kbar (dotted curves). The algorithm (given in the Appendix) that describes the pressure dependence of the olivine/low-Ca pyroxene boundary has been calibrated against the high-pressure melting relations of several low-Ti mare basalt compositions. As with all algorithms describing the pressure dependence of the olivine/low-Ca pyroxene boundary, pressure shifts the boundary away from the SiO_2 component; however, the present formulation differs from previous efforts (e.g., *Warren and Wasson*, 1979) in that there is a rotation of the boundary about a center near the Pyx component as well as a translation. The uncertainty of the algorithm is ±2.9 kbar in the range of 8 to 20 kbar. Much of the uncertainty at these pressures is apparently due to systematic interlaboratory differences in the calibration of the piston-cylinder apparati at Harvard and Stony Brook. This matter is discussed further in the Appendix. Since there is no error associated with pressure measurement at 1 bar, the uncertainty associated with the algorithm at pressures less than 5 kbar can be approximated by determining how well the algorithm predicts 1 bar of pressure for liquid compositions along the 1 bar olivine/low-ca pyroxene liquidus boundary. The four pseudo-peritectic liquids tabulated by *Stolper* (1977) and listed in Table 1 of this study give pressures of 0.1, 1.1, -0.3, and 0.6 kbar for an average deviation of +0.4 kbar. These deviations are derived from small systematic differences between the parameterization of the olivine/low-Ca pyroxene liquidus boundary and the microprobe analyses of *Stolper* (1977). The uncertainty associated with the choice of pressure correction factors adds another ±20% variation to these deviations. Consequently, we adopt an uncertainty of ±1 kbar for compositions near the pseudo-peritectic. The uncertainty is considerably larger for plagioclase-undersaturated compositions (Bi-2 in Table 1 yields a pressure of -2.2 kbar), so we will limit our conclusions to compositions near plagioclase saturation.

As was the case with the low-pressure calculations, EPB3 has too high a Mg' value to produce the eucrites directly by partial melting at any elevated pressure. Figure 5 shows that the 2 kbar melting curve begins well to the right of the array of eucrite compositions—an indication that the initial, plagioclase-saturated melts have both Pl components and Mg' that are too high to match the eucrites. At higher degrees of partial melting the EPB3 melting curve passes through the array of eucrite compositions, but here Mg' of the EPB3 melts is even higher than in the initial melts, so the match with eucrites is poorer. At other pressures the match between EPB3 melts and eucrites worsens. However, the melting calculations do suggest that partial melting of a less magnesian source, possibly EPB2 (not shown), could produce the compositions at the mid- to low-SiO_2 end of eucrite array directly by equilibrium melting in the range of 1 to 2 kbar.

Similarly, calculated high-pressure fractional crystallization of a low-Al magnesian partial melt of EPB3, similar to A_0, at 1

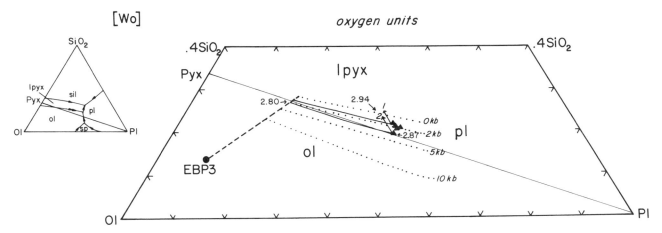

Fig. 5. Calculated partial melting and fractional crystallization at various pressures. Projection scheme and symbols as in Fig. 2. Dotted curves = equilibrium partial melting of EPB3. Solid curves are fractional crystallization paths at 1 and 2 kbar of primitive liquid similar to A_0 in Fig. 3. Melting and fractionation paths do not coincide at the same pressure because of difference in Mg'. Numbers indicate calculated liquid densities along the 2 kbar fractionation curve.

and 2 kbar produces liquid lines of descent that reach plagioclase saturation in the low-SiO$_2$ end of the eucrite array, as shown in Fig. 5. The pyroxenes produced along these lines of descent are virtually identical to those predicted for A$_0$ and its differentiates, so the ability to generate the full range of HED compositions is retained. Likewise, the covariation of TiO$_2$ and Mg' is similar to that of curve A$_0$-A' in Fig. 4. An additional point to note is that the 2 kbar calculations indicate a small amount of olivine cocrystallizing with orthopyroxene—a feature consistent with the petrography of several diogenites (*Hewins*, 1981).

Thus the calculations indicate that it is possible to construct a scenario that relates diogenitic pyroxenes and eucritic liquids through a single episode of partial melting and fractional crystallization. The major proviso of this scenario is that the fractional crystallization take place at a minimum of 2 ± 1 kbar. Having shown that simple fractional crystallization is at least capable of producing the range of HED mineral and liquid compositions, we may now examine the calculated phase equilibria and compositions for explanations for the absence of volcanic equivalents of diogenitic magmas and the apparent clustering of volcanic eucritic compositions.

In Fig. 5, where the calculated fractional crystallization curve reaches plagioclase saturation at 2 kbar, the liquid is marginally saturated with olivine. The olivine is in reaction with the liquid because the liquid still projects outside the olivine/low-Ca pyroxene/plagioclase three-phase triangle. At still higher pressure (4 to 5 kbar) liquids with eucrite-like Mg' and saturated with olivine, pigeonite, and plagioclase would become pseudo-eutectic, but would no longer project in the midst of the eucrite array. Therefore, pseudo-eutectic crystallization cannot be invoked to explain the clustering of eucrite compositions.

At 2 kbar approximately 50 vol % crystallization is required to drive the low-Al parent magma across the diagram to plagioclase saturation. More than an additional 30% crystallization is required, however, to drive the derivative liquids along the short segment of the low-Ca pyroxene/plagioclase cotectic shown in Fig. 5. The displacement per unit of crystallization is relatively small in this portion of the diagram because the solid component extracted from the cotectic liquids is relatively close in composition to these liquids and therefore is not efficient in changing the proportions of the projection components. If we assume random sampling of a fractionating magma chamber, the calculations thus predict an inevitable clustering of compositions on an Ol-Pl-SiO$_2$ projection and an attendant negative correlation of TiO$_2$ and Mg' (e.g., the Nuevo Laredo Trend in Fig. 4), once a liquid similar in composition to Sioux County or Juvinas is produced. If sampling were truly random, however, there should nonetheless be more volcanic compositions spread out along the olivine/low-Ca pyroxene liquidus boundary than are clustered along the low-Ca pyroxene/plagioclase cotectic.

During the crystallization of low-Ca pyroxene and minor olivine at 2 kbar, liquid density calculated by the method of *Bottinga and Weill* (1970) increases from 2.80 to 2.87, as shown in Fig. 5. Cocrystallization of pigeonite and plagioclase causes the liquid density to increase still further, so at no time is a density filter mechanism (e.g., *Stolper and Walker*, 1980) likely to trigger eruption and thereby produce clustered liquid compositions. Therefore, in the absence of some independent physical or tectonic process, diogenetic liquids should have been more likely to erupt to the surface than eucritic liquids.

DISCUSSION

The calculations of low-pressure liquidus equilibria presented here are in substantial agreement with the experiments of *Stolper* (1977) and *Beckett and Stolper* (1987) on two important points: First, olivine is in a reaction relation with liquids in equilibrium with magnesian orthopyroxenes similar to those found in diogenites; second, olivine is in a reaction relation with plagioclase-saturated, eucrite-like liquids with intermediate Mg'. The former equilibrium causes liquids initially saturated with diogenite-like orthopyroxene to reach plagioclase saturation through fractional crystallization with SiO$_2$ components that are too high to match the eucrites. However, the mismatch in SiO$_2$ predicted in the calculations is much less than was predicted by *Stolper* (1977). The latter equilibrium implies that there is no mechanism other than partial melting to produce an isolated cluster of liquid compositions cosaturated with olivine, pigeonite, and plagioclase at low pressure.

Although based on interpolations, calculation of fractional crystallization at 2 kbar indicates that liquids generated by large degrees of partial melting (>40%) of EPB3 are capable of producing the range of low-Ca pyroxenes observed in howardites (magnesian orthopyroxene, magnesian to ferroan pigeonite; *Dymek et al.*, 1976; *Ikeda and Takeda*, 1985), plus evolved liquids that match those at the low-SiO$_2$ end of the eucrite composition array in the Ol-Pl-SiO$_2$ diagram (Fig. 5). Fractional crystallization at approximately 2 kbar thus provides a straightforward explanation for most of the compositional and mineralogical features observed in the diogenite/eucrite association. However, some physical process must be invoked to keep diogenitic magmas trapped in magma chambers while allowing their denser eucritic derivatives to erupt. Two scenarios present themselves. In the first, planet-wide melting produces a series of large magma chambers that crystallize as closed systems to produce diogenitic cumulates and eucritic residual liquids; meteorite impacts then trigger eruption of the dense eucritic liquids. This scenario suffers a loss of plausibility because of the highly fortuitous requirement that the several magma chambers thus tapped and sampled all contained similarly fractionated magmas at the time of impact. The second scenario is that of a magma ocean crystallizing from the bottom up as suggested by *Hewins* (1987). Here fractional crystallization would produce diogenitic cumulates at depth and residual eucritic liquids near the surface. In this case the phase equilibria constraint is that the pressure at the base of the magma ocean be in the range of 1 to 2 kbar when the magma composition reaches that of the eucrites. For a bulk parent body composition such as EPB3, approximately 15 vol % residual liquid is implied. In order to satisfy the pressure/ volume constraint, the radius of the parent body must be well in excess of 500 km and possibly greater than 1000 km. For example, on a parent body with radius of 500 km and density of 3.44, 85 vol % crystallization produces a 26-km-deep residual magma ocean with a pressure at its base of only 430 bars, whereas on a parent body with radius of 1000 km, equivalent crystallization leaves a 53-km-deep column of residual liquid and a basal pressure of 1.7 kbar. Given the depth of the magma column and the high density of eucritic liquids (>2.80), a notable shortcoming of the magma ocean hypothesis is the scarcity of anorthosite in howardites (*Dymek et al.*, 1976; *Ikeda and Takeda*, 1985) and polymict eucrites (*Delaney et al.*, 1984).

By way of contrast, partial melting of a peridotitic source with relatively low Mg' (e.g., EPB2) at pressures of 1 to 2 kbar provides a strong density contrast between melt and country rock as well as a means of clustering eucritic liquid compositions. Furthermore, the eruption of such eucritic liquids would form a relatively low-density basaltic crust that could in turn form a barrier to the eruption of later diogenitic melts, much as proposed by *Stolper* (1977). There is a problem with low Mg' sources, however, pointed out to us by D. Mittlefehldt (personal communication, 1987). The FeO/MnO ratio of EPB2 (52) is much larger than the value of about 35 typical of HED meteorites, and partial melting will not significantly alleviate this discrepancy. Because FeO concentration is generally treated as the most "adjustable" of parameters in calculations of parent body composition, lowering FeO/MnO implies lowering FeO, which in turn increases Mg'. Analogously, *Dreibus et al.* (1977) arrived at the higher Mg' of EPB3 by first deriving Mg/Si/Mn ratios and then adjusting FeO to produce a suitable value of FeO/MnO. If one accepts the apparent coupling of HED FeO/MnO ratio and relatively high Mg' for the parent body, then the source region of the eucrites must itself be differentiated, probably by the prior separation of olivine. If the parent body composition were similar to EPB3, then our calculations show that prior removal of between 45 and 50 vol % olivine would be required to reduce Mg' to its value in EPB2 (0.68). If we assume that the olivine residuum lies at the interior of the planet, then the maximum pressure in the hypothetical eucrite source region on a parent body with radius of 500 km is 1.5 kbar—a marginally acceptable value. This early differentiation might also have contributed to the development of minor heterogeneities in the eucrite source that subsequently led to slightly different trends on En versus An plots (e.g., *Ikeda and Takeda*, 1985).

CONCLUSIONS

In summary, two self-consistent scenarios emerge for the generation of HED meteorites from our modeling of melting and crystallization processes, although petrological considerations lead us to favor the latter. One is that fractional crystallization at the bottom of a magma ocean produces late-stage eucritic liquids overlying diogenitic cumulates (e.g., *Hewins*, 1987). The phase equilibria constraint on this scenario is that the pressure at the base of the magma ocean when it evolves to a eucritic composition must be in the range of 1 to 2 kbar. For a parent body composition such as EPB3 (*Dreibus et al*, 1977), mass balance requires a parent body with radius well in excess of 500 km. The scarcity of anorthosites in polymict HED breccias poses a serious problem for the magma ocean scenario, however, because plagioclase could have easily floated in a dense eucritic magma ocean. The preferred scenario is that partial melting of a source with relatively low Mg' produces the eucrites and remelting of the depleted source produces diogenitic magmas (*Stolper*, 1977). A similar phase equilibrium constraint applies here as well: Partial melting must have taken place at pressures on the order of 1 to 2 kbar in order to produce primary melts with eucritic compositions. Consideration of FeO/MnO ratios (*Dreibus et al.*, 1977), however, suggests that the low Mg' eucrite source region was itself the product of an earlier planet-wide differentiation. In this case mass balance considerations permit a parent body radius of approximately 500 km.

APPENDIX

The models of fractional crystallization and partial melting have been described in detail most recently by *Longhi* (1987). The present version has been modified to handle compositions spanning the range in alkali concentrations from lunar basalts to terrestrial tholeiites and to handle variable pressure.

The modification that enable the model to handle a range of alkali compositions was based upon the experimental study of *Longhi and Pan* (1987b), which also includes extensive data on the compositional dependence of the olivine/low-Ca pyroxene liquidus boundary, the feature of most concern here. As in previous work, the liquidus phase boundaries are parameterized in terms of projection components. In the present study compositions are first converted to mol % oxides and then transformed into four major mineral components: Olivine (Ol), Plagioclase (Pl), Wollastonite (Wo), and SiO_2 with TiO_2, Cr_2O_3, and trace elements being ignored. Of these quaternary coordinates, only Wo is employed in the parameterizations.

$$Wo = (CaO - Al_2O_3 + Na_2O + K_2O)/(SiO_2 - Al_2O_3 - 3 \cdot (Na_2O + K_2O))$$

where the oxide symbols refer to mol %. A subprojection is then made from the Wo component onto the Ol[Wo] - Pl[Wo] - SiO_2[Wo] plane, where [] indicates the component projected from

$$Sum = SiO_2 - CaO - 4 \cdot (Na_2O + K_2O)$$
$$SiO_2[Wo] = (SiO_2 - 0.5 \cdot (FeO + MgO + MnO) - Al_2O_3 - CaO - 5 \cdot (Na_2O + K_2O))/Sum$$
$$Ol[Wo] = 0.5 \cdot (FeO + MgO + MnO)/Sum$$
$$Pl[Wo] = 1 - Ol[Wo] - SiO_2[Wo]$$

These molar projection coordinates are then employed in the parameterizations. The current version differs from previous versions in that Na and K are recast into alkali feldspar components, which in turn are combined with anorthite to form the plagioclase component. Previously, alkalies were ignored—an adaptation to variable alkali loss in experiments in evacuated silica tubes (e.g., *Walker et al.*, 1972). Also included in the parameterizations are Fe' = 1 - Mg' = FeO/(FeO + MgO) and the normative alkali feldspar component, AFP, which in hypersthene (± olivine) normative compositions is given by

$$AFP = 2 \cdot (Na_2O + K_2O)/(Al_2O_3 + Na_2O + K_2O)$$

where the oxides are in mol %.

The various new parameterizations of liquidus boundaries are as follows

Olivine/Al-spinel

$$Ol[Wo] = -0.985 \cdot SiO_2[Wo] + 0.1 \cdot (Fe')^{1/2} + 0.625 - 0.57 \cdot AFP - 0.467 \cdot Wo$$

Olivine/plagioclase

$$Ol[Wo] = -0.528 \cdot (1 - 0.6 \cdot AFP) \cdot (1 + 0.55 \cdot Fe') \cdot SiO_2[Wo] + 0.50 + 0.29 \cdot Fe'' - 0.35 \cdot (1 + 0.18 \cdot Fe') \cdot AFP - 0.05 \cdot AFP^2 + 0.10 \cdot (1 - 1.67 \cdot Fe') \cdot Wo$$

Olivine/low-Ca pyroxene

$$SiO_2[Wo] = (0.40 - 0.32 \cdot (Fe')^{1/2}) \cdot (Ol[Wo] - 0.128) + 0.40 + 0.08 \cdot AFP + 0.25 \cdot (1 - 1.67 \cdot Fe') \cdot Wo - 0.08 \cdot AFP \cdot Ol[Wo] / (1 + 0.08 \cdot AFP)$$

Low-Ca pyroxene/plagioclase

$$Ol[Wo] = -0.68 \cdot (1 + 0.45 \cdot Fe') \cdot (1 - 0.6 \cdot AFP) \cdot SiO_2[Wo] + 0.565 + 0.29 \cdot Fe' - 0.35 \cdot (1 + 0.18 \cdot Fe') \cdot AFP - 0.05 \cdot AFP^2$$

Low-Ca pyroxene/silica

$$SiO_2[Wo] = -(.2787 + 0.15 \cdot Mg') \cdot (Ol[Wo] - 0.32 \cdot Fe' - 0.118) + 0.70 - 0.33 \cdot Fe' + 0.333 \cdot (1 - 0.50 \cdot Fe') \cdot Wo + 0.533 \cdot AFP - 0.533 \cdot AFP \cdot Ol[Wo] / (1 + 0.533 \cdot AFP)$$

Low-Ca pyroxene/augite

$$Wo = 0.945 \cdot Ol[Wo] + 0.3628 \cdot (1 + 0.90 \cdot Fe') \cdot SiO_2[Wo] - 0.204 + 0.15 \cdot AFP - 0.35 \cdot Fe'$$

Olivine/augite

$$Wo = 0.54 \cdot (1 - 0.50 \cdot AFP) \cdot Ol[Wo] - 0.158 \cdot (1 - 0.8 \cdot Fe' - 0.8 \cdot AFP) \cdot SiO_2[Wo] + 0.145 + 0.12 \cdot AFP^2 - 0.20 \cdot Fe'$$

Orthopyroxene/pigeonite

$$Wo = (0.10 - 0.50 \cdot AFP) \cdot Ol[Wo] - 0.161 \cdot SiO_2[Wo] + 0.10 \cdot Fe' - 0.85 \cdot (Fe')^2 + 0.25 \cdot AFP + 0.282$$

Convenience of manipulation of the expressions in the crystallization programs determines whether Ol[Wo] or SiO_2[Wo] is treated as the independent variable in these equations. In order to employ these expressions to generate a phase diagram, one must solve two appropriate equations simultaneously to determine limits of the independent variable. For example, to determine the high SiO_2[Wo] limit of the olivine/Al-spinel boundary, it is necessary to solve the olivine/spinel and olivine/plagioclase equations simultaneously. SiO_2[Wo] at the point where olivine, spinel, and plagioclase coexist will then be a function only of Fe' and AFP.

Although the phase boundary parameterizations are in terms of mole units, oxygen units are preferred for the figures because they closely reproduce volume proportions of minerals and because mole units may appear to vary nonlinearly in projection, as explained by *Longhi and Pan* (1987a,b). Once the molar projection coordinates have been calculated they may easily be converted to oxygen unit coordinates by multiplying each coordinate by the number of oxygen atoms in its respective formula (e.g., Ol[Wo]×4, SiO_2[Wo]×2, Pl[Wo]×8) and then dividing each product by their collective sum.

Pressure is known to cause both rotation and translation of olivine/low-Ca pyroxene liquidus boundary in the CaO-MgO-Al_2O_3-SiO_2 system (e.g., *Longhi*, 1981, Figs. 2 and 3). Accordingly, we have derived an empirical expression for the pressure dependence of the olivine/low-Ca pyroxene liquidus boundary that incorporates both movements

$$SiO_2[Wo] \text{ (high-P)} = SiO_2[Wo] \text{ (low-P)} - 0.0115 \cdot PR \cdot (5.33 \cdot Pl[Wo] + 0.40)$$

where PR is pressure in kilobars. The coefficient of PR, 0.0115, produces the translation, whereas the term in parentheses produces the small rotation illustrated in Fig. 5. The expression contains no term for alkalies and so is appropriate only for lunar and HED compositions. Figure A-1 compares two test versions of the expression against the data set from which it was generated. Pressures are calculated by solving the equation for the olivine/low-Ca pyroxene boundary for PR. Figure A-1 shows that the Harvard data are fit better by a pressure coefficient of 0.0130, whereas the Stony Brook data are fit better by a coefficient of 0.0100. Thus we have adopted a compromise value of 0.0115. These differences suggest an interlaboratory bias with the Stony Brook pressure measurements being approximately 3 kbar higher on similar compositions. As discussed in the main body of text, the uncertainty in estimate of pressure is relevant only at low pressures (<5kbar) and can be estimated by calculating pressure for liquids produced at 1 bar. The uncertainty for four pseudo-peritectic liquids from *Stolper* (1977) that are listed in Table 1 is on the order of ±0.5 kbar. Variation in the choice of the pressure coefficient from 0.0100 to 0.0130 increases the uncertainty by only 20%. Thus we estimate an uncertainty in the pressure calculation at pressure <5 kbar to be ± 1 kbar.

Pressure dependent terms for some of the other boundary surfaces are as follows;

Olivine/plagioclase

$$Ol[Wo] \text{ (high-P)} = Ol[Wo] \text{ (low-P)} - 0.002 \cdot PR$$

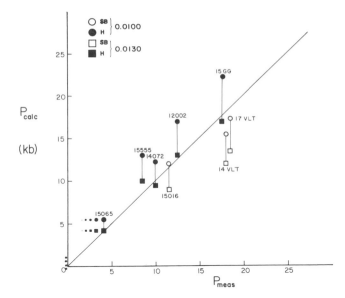

Fig. A1. Comparison of calculated and measured pressures of olivine/low-Ca pyroxene liquidus cosaturation for several low-Ti lunar mare compositions. Experiments conducted in low-purity iron capsules are excluded. Where possible, input compositions are microprobe analyses of starting materials. There are two calculated pressures for each composition: one with a pressure coefficient of 0.010 (circles), the other with 0.013 (squares). Data from the Harvard laboratory are filled symbols: 14072 - *Walker et al.* (1972); Apollo 15 green glass -*Stolper* (1974); 12002 - *Walker et al.* (1976); 15065 and 15555 - *Walker et al.* (1977). Stony Brook data are open symbols: 15016 - *Kesson* (1975); Apollo 14 and 17 VLT glass - *Chen et al.* (1982). Multiple symbols for 15065 indicate that measured pressure is <5 kbar. Dots near the origin along the vertical axis indicate calculated pressures of pseudo-peritectic liquids taken from *Stolper* (1977) and listed in Table 1.

Olivine/augite

$$Wo(high\text{-}P) = Wo(low\text{-}P) - 0.042 \cdot PR \cdot Pl[Wo]$$

Low-Ca pyroxene/plagioclase

$$Ol[Wo] \; (high\text{-}P) = Ol[Wo] \; (low\text{-}P) - 0.002 \cdot PR$$

For the present purposes these are only boundaries for which pressure dependence is relevant.

Acknowledgments. We thank R. Hewins and E. Stolper for sharing and discussing data with us. We also thank J. Beckett and D. Mittlefehldt for very thorough reviews. A Lasaga generously permitted us to use the VAX-750 computer under his supervision. This research was supported by NASA grant NAG-9-93.

REFERENCES

Basaltic Volcanism Study Project (1981) *Basaltic Volcanism on the Terrestrial Planets*. Pergamon, New York, 1286 pp.

Beckett J. R. and Stolper E. M. (1987) Constraints on the origin of eucritic melts (abstract). In *Lunar and Planetary Science XVIII*, pp. 54-55. Lunar and Planetary Institute, Houston.

Bottinga Y. and Weill D. F. (1970) Densities of liquid silicate systems calculated from partial molar volumes of oxide components. *Am. J. Sci., 269,* 169-182.

Chen H.-K., Delano J. W., and Lindsley D. H. (1982) Chemistry and phase relations of VLT volcanic glasses fromApollo 14 and Apollo 15. *Proc. Lunar and Planet. Sci. Conf. 13th*, in *J. Geophys. Res., 87,* A171-A181.

Clayton R. N. (1977) Genetic relationships among meteorites and planets. In *Comets, Asteroids and Meteorites* (A. H. Delsemme, ed.), pp. 545-550. University of Toledo, Ohio.

Delaney J. S. (1986) Phase equilibria for basaltic achondrites (abstract). In *Lunar and Planetary Science XVII*, pp. 164-165. Lunar and Planetary Institute Houston.

Delaney J. S., Prinz M., and Takeda H. (1984) The polymict eucrites. *Proc. Lunar Planet. Sci. Conf. 15th*, in *J. Geophys. Res., 89,,* C251-C288.

Dreibus G., Kruse H., Spettel B., and Wänke H. (1977) The bulk compositions of the moon and the eucrite parent body. *Proc. Lunar Sci. Conf. 8th,* pp. 211-228.

Duke M. B. and Silver L. T. (1967) Petrology of eucrites, howardites, and mesosiderites. *Geochim. Cosmochim. Acta, 31,* 1637-1665.

Dymek R. F., Albee A. L., Chodos A. A., and Wasserburg G. L. (1976) Petrography of isotopically-dated clasts in the Kapoeta howardite and petrologic constraints on the evolution of its parent body. *Geochim. Cosmochim. Acta, 40,* 1115-1130.

Ghiorso M. S. (1985) Chemical mass transfer in magmatic processes—I. Thermodynamic relations and numerical algorithms. *Contrib. Mineral. Petrol., 90,* 107-120.

Hertogen J., Visgarda J., and Anders E. (1977) The composition of the parent body of eucrite meteorites. *Bull. Am. Astron. Soc., 9,* 458-459.

Hewins R. H. (1981) Orthopyroxene-olivine assemblages in diogenites and mesosiderites. *Geochim. Cosmochim. Acta, 45,* 128-126.

Hewins R. H. (1986) Serial melting or magma ocean for the HED achondrites? (abstract). In *Papers Presented to the 49th Annual Meteoritical Society Meeting,* P-8. Lunar and Planetary Institute, Houston.

Hewins R. H. (1987) The howardite parent body: composition and crystallization models (abstract). In *Lunar and Planetary Science XVIII*, pp. 417-418. Lunar and Planetary Institute, Houston.

Ikeda Y. and Takeda H. (1985) A model for the origin of basaltic achondrites based on the Yamato 7308 howardite. *Proc. Lunar Planet. Sci. Conf. 15th*, in *J. Geophys. Res., 90,,* C649-663.

Kesson S. E. (1975) Mare basalt petrogenesis (abstract). In *Papers Presented to the Conference on Origins of Mare Basalts and Their Implications for Lunar Evolution,* pp. 81-84. The Lunar Science Institute, Houston.

Longhi, J. (1981) Preliminary modeling of high pressure partial melting: Implications for early lunar differentiation. *Proc. Lunar Planet. Sci. 12B,* pp. 1001-1018.

Longhi J. (1987) On the connection between mare basalts and picritic volcanic glasses. *Proc. Lunar Planet. Sci. Conf. 17th,,* in *J. Geophys. Res., 92,* E349-E360.

Longhi J. and Pan V. (1987a) Olivine/low-Ca pyroxene liquidus relations and their bearing on eucrite petrogenesis (abstract). In *Lunar and Planetary Science XVIII*, pp. 570-571. Lunar and Planetary Institute, Houston.

Longhi J. and Pan V. (1987b) A reconnaissance study of phase boundaries in low-alkali basaltic liquids. *J. Petrol.,* in press.

Mason B. (1967) Meteorites. *Am. Sci., 51,* 429-455.

McCarthy T. S., Erlank A. J., Willis J. P., and Ahrens L. H. (1974) New chemical analyses of six achondrites and one chondrite. *Meteoritics, 9,* 215-221.

Morse S. A. (1980) *Basalts and Phase Diagrams*. Springer-Verlag, New York.

Nielsen R. L. (1987) The calculation of combined major and trace element liquid lines of descent—I. The temperature dependence of compositionall independent partition coefficients. *Geochim. Cosmochim. Acta*, in press.

Stolper E. M. (1974) Lunar ultramafic glasses. Honor thesis, Harvard University, Cambridge, Mass.

Stolper E. M. (1977) Experimental petrology of eucritic meteorites. *Geochim. Cosmochim. Acta*, 587-611.

Stolper E. M. and Walker D. (1980) Melt density and the average composition of basalt. *Contrib. Mineral. Petrol.*, 74, 7-12.

Walker D., Longhi J., and Hays J. (1972) Experimental petrology and origin of Fra Mauro rocks and soil. *Proc. Lunar Sci. Conf. 3rd*, pp. 797-817.

Walker D., Kirkpatrick R. J., Longhi J., and Hays J. F. (1976) Crystallization history of lunar picritic basalt sample 12002: phase equilibria and cooling rate studies. *Geol. Soc. Am. Bull.*, 87, 646-656.

Walker D., Longhi J., Lasaga A. C., Stolper E. M., Grove T. L., and Hays J. F. (1977) Slowly cooled microgabbros 15555 and 15065. *Proc. Lunar Sci. Conf. 8th*, pp. 1521-1547.

Wänke H., Baddenhausen G., Dreibus G., Jagoutz E., Kruse E., Palme, H., Spettel B., and Teschke F. (1973) Multielement analyses of Apollo 15, 16, and 17 samples and the bulk composition of the Moon. *Proc. Lunar Sci. Conf. 4th*, pp. 1461-1481.

Warren P. H. (1985) Origin of howardites, diogenites and eucrites: a mass balance constraint. *Geochim. Cosmochim. Acta*, 49, 577-586.

Warren P. H. and Jerde E. A. (1987) Comopsition and origin of Nuevo Laredo trend meteorites. *Geochim. Cosmochim. Acta*, 51, 713-725.

Warren P. H. and Wasson J. T. (1979) Effects of pressure on the crystallization of a moon-sized "chondritic" magma ocean. *Proc. Lunar Planet. Sci. Conf. 10th*, pp. 2051-2083.

Partitioning of Siderophile and Chalcophile Elements Between Sulfide, Olivine, and Glass in a Naturally Reduced Basalt From Disko Island, Greenland

W. Klöck
NASA Johnson Space Center, Houston, TX 77058

H. Palme
Max-Planck-Institut für Chemie, Saarstrasse 23, 6500 Mainz, FRG

We have analyzed major and trace elements in coexisting glass, olivines, and metal-sulfide spherules from a chilled margin sample of a strongly reduced basaltic dike from Disko Island, Greenland. Element partitioning between glass (mg = 60) and olivines (mg = 84) indicates that the minerals equilibrated at about 1200°C. The metal-sulfide droplets in the glass show eutectic quench textures and display an excess metal component. Metal-sulfide droplets contain 85% FeS and 15% Fe-Ni metal. They represent metallic liquids that at magmatic temperatures, coexisted with basaltic melt and olivines. Fe-Ni crystals crystallized from the sulfide liquid below 900°C. From trace element analyses of the mineral phases, three sets of partition coefficients can be derived: olivine/silicate liquid and metal-sulfide liquid/silicate liquid partition coefficients established at magmatic temperatures, and FeNi/FeS partition coefficients at lower temperatures. Iron, Mg, Cr, Ni, Co, and Mn olivine/silicate liquid partition coefficients derived from our results agree very well with literature data. In addition we present olivine/silicate liquid partition coefficients for Ca, Ti, Ga, Zn, Cu, As, Sb, and Sc for which only scarce literature data exist. Analyses of sulfides demonstrate high metal-sulfide liquid/silicate liquid partition coefficients for Ni (>5000), Sb (1250), As (1000), Mo (1000), Cu (500), Co (200), and W (25). Metal-sulfide liquid/silicate liquid partition coefficients of Cr (1.1), Ga (0.6), Zn (<0.5), Mn (0.16), and P (<0.16), however, show that these elements do not strongly partition into sulfide liquids. The significance of these data for planetary evolution and the formation of metal- or sulfide-rich metal cores is discussed in terms of P, W, and Mo abundances in the mantles of the Earth, the Moon, the Shergotty parent body, and the Eucrite parent body. The P data may be of particular interest, since P appears to be the best candidate for distinguishing between iron- and sulfur-rich metal cores. The depletions of P in the mantles of the Earth, the Moon, and the Eucrite parent body compared to P abundance in the Shergotty parent body mantle can be explained by removal of P into a metal core. The nearly chondritic P value in SNC meteorites is consistent with the separation of a S-rich metallic core in the Shergotty parent body. It is also shown that Cr is more siderophile and more chalcophile than Mn and Zn. Therefore, core formation cannot change the observed sequence in abundance patterns of these elements (Cr>Mn>Zn) in planetary mantles. Depletions of Mn and Zn must then reflect volatility-dependent processes either in the solar nebula or during accretion of the planet.

INTRODUCTION

The abundances of siderophile and chalcophile elements in the silicate mantles of planets depend on the extent and the nature of formation of an Fe-Ni or FeS core. Formation of an Fe-Ni core leads to depletion of siderophile elements in the residual silicate portion of the planet. A S-rich core, on the other hand, would preferentially take up chalcophile elements, thus effectively removing them from the planetary mantle. The relative abundances of siderophile and chalcophile elements in the silicate portion of a planet may therefore provide evidence for the composition of the core. Highly siderophile and/or chalcophile elements with partition coefficients >10000 are not very useful in determining details of the core formation process. Iridium, for example, will always partition more or less completely into the core. Enhanced Ir contents in the mantle of a planet reflect disequilibrium, inhomogeneous accretion or incomplete segregation of metal (*Wänke*, 1981, *Jones and Drake*, 1986). For the purpose of studying the amount and nature of the segregating phase it is more appropriate to use elements with lower metal/silicate and/or sulfide/silicate partition coefficients, such as Mn, P, Mo, W, As, Ga, Cr, or Cu. Elements that show large differences between sulfide-liquid/silicate and metal/silicate partitioning are particularly useful. If the process that removed siderophile and chalcophile elements from a planetary mantle is understood in sufficient detail, one may try to estimate the abundances of these elements in the bulk planet. The relevant parameters are, of course, partition coefficients among the separating phases. There exist some data on experimentally determined metal/silicate and sulfide-liquid/silicate partition coefficients. However, in many geochemical models only metal/silicate partition coefficients are considered. As sulfur may be a major constituent of the separated metal phase, partition coefficients between sulfide liquid and silicate liquid would be more appropriate.

Here we report partition coefficients of siderophile, chalcophile and lithophile elements between silicate glass, olivines, and metal-sulfide spherules in a basalt sample from Disko Island, West Greenland. In this area, basaltic melt reacted

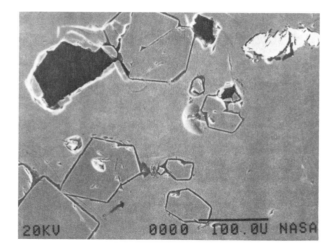

Fig. 1. SEM image showing a sulfide particle and olivines in the glass matrix. Sample was ion etched.

with C-rich sediments, resulting in reduction of the melt and formation of sulfur-rich metallic liquids. This gives a unique opportunity to obtain data on metal-sulfide-liquid/silicate-liquid partitioning in a natural terrestrial system (*Klöck et al.,* 1986). *Pedersen* (1979) has investigated similar samples. He gives a detailed description of the geological setting of the Luciefjeld-dike, which cuts through Precambrian gneisses and Tertiary basalts in South Disko. He also presents petrographic descriptions and chemical analyses of the different mineral phases studied in our paper. Pedersen has concluded that there is equilibrium among the various phases in these rocks and he has extracted a number of partition coefficients between olivine and glass and between metal-sulfide liquid and basaltic melt. In general, the modal abundances and compositions of minerals in our samples are very similar to those described by *Pedersen* (1979). Since the mineral association is unique for a terrestrial rock, we wanted to extend the data base by determining the concentrations of a number of additional elements in the various phases of the rock.

PETROGRAPHIC DESCRIPTION

A few grams of basaltic glass from the Kitdlit-dike at Luciefjeld were available for analysis. The samples were kindly provided by Dr. R. Dietrich, Wiesbaden, Federal Republic of Germany, who collected the samples at the Luciefjeld locality. Detailed geological information about the occurrence of the native iron-bearing basalt, including a map showing the locality on Disko Island, is given by *Pedersen* (1979).

The thickness of the quenched glass at the contact of the basalts and the intruded dike was 0.5 cm-1 cm. The modal abundances of glass, olivines, and sulfides in the quenched glass were quantitatively determined by the following method: polished thin sections were photographed in reflected light and the amount of the different phases were determined by planimetry. Weight percent abundances of the different mineral phases were calculated from the specific weights of the minerals, with the assumption that the area percentages are representative of the volume percent abundances of the minerals.

The sample contains about 90% basaltic glass (Figs. 1 and 2). The color of the glass is slightly greenish or brownish. Some recrystallized parts are dark brown. Olivines make up about 10 wt % of the sample. They are euhedral, mostly in the 50 μm-300 μm size range and are homogeneously distributed within the glass matrix. Most of them contain glass inclusions.

The abundance of metal-sulfide droplets in the glass is about 0.2 wt %, but this number may vary between 0.1 wt % and 0.3 wt % due to heterogeneous distribution of sulfides in the glass on a thin section scale. Metal-sulfide blebs are in the size range from 30 μm to 100 μm, but smaller droplets are occasionally encountered.

Metal-sulfide droplets are composed of troilite and numerous spheroidal grains of nickel-iron smaller than 5 μm (Fig. 3). Some metal-sulfide droplets contain idiomorphic iron-nickel crystals. A chromite crystal adhering to the sulfides was observed in one case only.

Modal analyses of a number of quench-textured sulfide droplets show that they are all nearly identical in composition (85.4% troilite, 14.6% FeNi), suggesting that the initially liquid Fe-Ni-S droplets had the same composition.

ANALYTICAL PROCEDURES

A few grams of basaltic glass were crushed in an agate mortar. One 250-mg piece of glass was used for INAA bulk analyses and about 1.83 grams of powdered glass were used for major element wet chemical analysis. For a number of C and S analyses, several hundred mg of homogenized powder had to be prepared. Finally 240 mg of bulk sample were used for

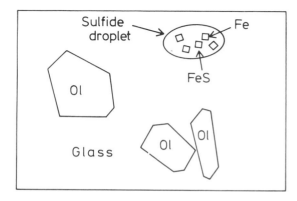

Fig. 2. Schematic drawing of the phase assemblage studied. Idiomorphic olivine crystals and metal-sulfide droplets are imbedded in a basaltic glass matrix. Tiny Fe-Ni crystals grew from the metal-sulfide matrix.

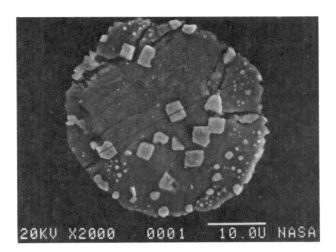

Fig. 3. SEM image of a sulfide droplet in the glass with idiomorphic Fe-Ni crystals smaller than 5 μm.

radiochemical neutron activation analyses of chalcophile elements. Several mg of clean glass fragments, olivines, and sulfides were handpicked from the remaining sample material. Glass and olivines were analyzed by INAA and aliquots were subjected to radiochemical procedures, which are described in more detail below. Neutron activation of samples and standards was done by irradiation in the TRIGA MARK II reactor of the Institute for Nuclear Chemistry of the University of Mainz and in the Geesthacht and Jülich research reactors. INAA procedures have been described by *Wänke et al.* (1970, 1973, 1977) and *Kruse* (1979). In the Geesthacht and Jülich research reactors, the samples were irradiated at a neutron flux of $1 \cdot 10^{14}$ n·cm^{-2} sec^{-1} irradiation times were three days in the Jülich reactor and five days for the samples in the Geesthacht reactor. The neutron flux in the TRIGA reactor was about $7 \cdot 10^{11}$ n·cm^{-2}sec^{-1} for an irradiation time of six hours. The glass fragments were counted both together and individually. Procedures applied to the analysis of olivines and metal-sulfide blebs were similar.

The activated bulk sample, together with a few milliliters of standard solutions of the elements to be determined (Cu, As, Sb, Mo, and Au), was fused with sodium peroxide in a zirconia crucible. The resulting glass was then dissolved in hot 6N HCl over night. Sulfide precipitation was achieved by passing H$_2$S into the solution for about half an hour. The solution and precipitate were transferred into a flask. The flask was then corked to allow a small pressure increase and left overnight in a water bath to complete the sulfide precipitation. The

TABLE 1. Trace elements in glass (in ppm, except where otherwise indicated).

	Bulk 1	Bulk 2	Individual Glass Particles				
	0.53 mg	1.068 mg	20 μg	6.5 μg	10.8 μg	7.6 μg	5.8 μg
Fe (%)	6.53	6.48	7.00	6.53*	6.53*	6.53*	6.53
Ca (%)	6.45	-	6.86	-	-	-	-
Na (%)	1.42	-	1.37	-	-	-	-
K	4434	-	4470	-	-	-	-
Sc	30.0	30.6	32.5	33.1	32.4	32.5	32.8
Cr	780	767	866	860	822	803	820
Co	24.9	24.4	26.2	19.6	19.2	20.4	23.1
Ni	<20	<25	<20	<13	<9	<26	<29
Ga	20.7	-	-	-	-	-	-
As	0.48	-	-	-	-	-	-
Sb	0.05	0.04	-	-	-	-	-
Zn	102	94	89	81	82.4	93	102
Rb	15.2	18.7	-	-	21.9	21.5	-
Sr	175	182	250	-	-	240	-
Ba	162	159	200	-	-	-	-
La	11.2	-	11.2	-	-	-	-
Ce	23.7	22.6	28.7	27.2	27.5	24.8	26.4
Nd	14.3	14.3	15.7	-	18.5	-	-
Sm	3.80	4.60	4.7	-	-	-	-
Eu	1.15	1.13	1.30	1.24	1.25	1.27	1.25
Tb	0.60	0.61	-	0.76	0.67	0.65	0.70
Yb	1.90	1.90	2.30	2.50	2.50	3.0	-
Lu	0.24	0.25	0.25	0.30	0.33	0.29	0.40
Hf	3.20	3.18	3.33	3.59	3.33	3.57	3.38
Ta	0.38	0.36	0.32	-	0.33	0.36	-
Th	2.0	1.8	2.0	2.0	1.7	2.1	2.15
U	0.50	-	0.55	-	-	-	-

Errors based on counting statistics are below 5%, except for Ca, As, Sb, Zn, Rb, Nd, Ta, and U, for which errors are in the range 5% to 15%.

*An iron content of 6.53 wt % was assumed for the small glass particles (<10 μg) and the corresponding sample weights were calculated.

TABLE 2. Trace elements in olivines (uncorrected for contamination, data in ppm).

	1	2	3	4
Na	370	610	-	-
K	135	106	-	-
Sc	7.20	9.31	9.61	8.20
Cr	832	995	940	956
Mn	1364	-	-	-
Fe (%)	12.00	12.23	12.13	11.87
Co	89.3	97.1	96.7	91.0
Ni	140	135	123	116
Zn	86	87	78	-
As	0.1	0.2	-	-
Sb	-	0.01	-	-
Ga	1.2	-	-	-
La	0.22	0.547	0.45	-
Ce	-	1.94	1.43	-
Sm	0.080	0.250	0.31	0.11
Eu	0.024	0.083	0.077	0.038
Yb	-	0.19	0.25	-
Lu	0.033	0.030	0.035	0.031
Au (ppb)	-	1.57	-	-

1 = bulk 1, 3.90 mg (contains 2 wt % of glass).
2 = bulk 2, 0.472 mg.
3 = bulk 3, 0.123 mg, 10 large olivine crystals.
4 = bulk 4, 0.154 mg, 35 small olivine crystals.
2,3,4 contain about 5 wt % of glass.

	Individual Olivine Analyses (Uncorrected Data)			
Weight	10.8 μg	3.0 μg	3.2 μg	2.3 μg
Fe* (%)	12.06	12.06	12.06	12.06
Sc	8.52	9.0	8.56	7.98
Cr	937	947	935	951
Co	103.8	88.6	96.9	90.2
Ni	199	158	93	32
Zn	76	77	72	67

Errors based on counting statistics are below 5%, except for As, Sb, Ga, and Au, for which errors are between 5% and 15%.

*Trace element contents normalized to a Fe content of 12.06 wt % because the samples were too small to weigh accurately.

TABLE 3. Average trace element abundances in glass and olivines (Sc, Ga, Yb and Lu corrected for contribution from glass inclusions, data given in ppm, except where otherwise indicated).

	This Work		Pedersen (1979)	
	Glass	Olivines	Glass	Olivines
Fe (%)	6.53	12.06	-	-
Sc	32.0	7.10	33	<10
Cr	817	937	810	980
Co	22.5	94.2	20	88
Ni	<9	128	9	110
Cu	33.5	18.0	27	14
Zn	92.0	77.6	-	-
Mn	930	1364	-	-
Ga	20.7	0.8	16	<10
As	0.44	0.14	-	-
Sb (ppb)	12.7	4.2	80	-
Au (ppb)	1.34	0.97	-	-
La	11.2	-	13.3	-
Ce	25.8	-	-	-
Sm	4.36	-	4.12	-
Eu	1.23	-	-	-
Yb	2.35	0.10	2.66	-
Lu	0.29	0.018	-	-
Th	1.96	-	2.36	-
U	0.53	-	0.84	-
Hf	3.38	-	3.61	-

Notes: The average iron content of the glass results from the values in column 1 and 2 given in Table 1. The data of all other trace elements in the glass were obtained by averaging all samples regardless of mass. Trace element contents in olivines were calculated treating all samples equally, except for Ni only the average of columns 1-4 in Table 2 is given. Copper, As, Sb, and Au data were obtained by radiochemical analyses. Scandium, Yb, Lu are corrected assuming 5 wt % glass inclusion in olivines. The Ga content was corrected assuming 2 wt % glass contamination (see Table 2).

TABLE 4. Average composition of bulk metal-sulfide spherules (in ppm, except otherwise indicated, normalized to 63% Fe).

Fe (%)	63.00	
Co	5147	(462)
Ni (%)	4.69	(0.54)
Cu (%)	1.60	
Cr	915	(165)
Mn	150	
Zn	<45	
Ga	12.4	
As	403	(53)
Sb	16.4	(3.8)
Mo	86	(12.5)
W	4.7	
Se	46.6	(6.8)
Ir (ppb)	46.8	(7.3)
Au (ppb)	206	(29)

In calculating the average trace element composition of metal-sulfides particles each individual analysis in Table 7 was given the same weight. Values in parentheses are standard deviations of individual particle analyses from the mean sulfide composition, not analytical errors.

precipitate was centrifuged, washed, and finally dissolved in hot aqua regia. The nitric acid was evaporated until dry; then the residue was dissolved in 5N HCl and diluted with 15 ml of distilled water. The solution was finally counted on a Ge(Li) detector. To determine the yields, 1 ml of this solution was transferred into a polyethylene capsule, dried, reirradiated, and counted on a Ge(Li) detector in a precisely defined geometry. The yields were derived by comparison of the measured element contents in the 1 ml of solution with the given element abundances in the standard solutions. The procedure for determination of the chalcophile elements in the glass and in the olivines was similar, except that the mineral samples were irradiated at the Geesthacht research reactor whereas the bulk sample was irradiated in the TRIGA MARK II reactor at the University of Mainz. The As, Sb, and Au data given in Tables 1 and 2 were obtained by INAA; the RNAA results are

TABLE 5. Bulk rock analysis.

Major Elements (in Wt %)	This Work	Pedersen (1979)
SiO_2	51.91	52.09
TiO_2	1.41	1.35
Al_2O_3	14.07	13.81
FeO*	9.60	9.44
MnO	0.18	0.16
MgO	10.13	10.48
CaO	9.17	9.06
Na_2O	1.99	1.82
K_2O	0.55	0.66
P_2O_5	0.18	0.16
H_2O	0.36	0.42
S	0.17	0.10
C†	0.18	0.09
Σ	99.90	99.54
FeO	9.21	
Fe_2O_3	0.44	

Trace Elements (in ppm)	This Work	Pedersen (1979)
Sc	29.7	30
Cr	785	810
Mn	1144	-
Co	41.5	76
Ni	120	135
Cu	73.2	67
Zn	115	-
Ga	17.2	12
As	1.27	-
Sb	0.074	0.03
Mo	0.27	-
Au (ppb)	2.33	-
Rb	14.5	-
Sr	140	142
Ba	162	131
La	10.3	9.1
Ce	22.9	-
Nd	15.2	-
Sm	3.38	3.66
Eu	0.98	-
Tb	0.69	-
Dy	3.37	-
Yb	1.82	1.98
Lu	0.27	-
Hf	3.17	3.24
Ta	0.30	-
Th	2.0	2.18
U	0.5	0.99

Wet chemical major element analysis was done by T. Kost at the University of Mainz. Analytical errors are smaller than 5% relative. Wolfram content of the bulk sample calculated from La/W or W/U correlation of terrestrial basalts amounts to 0.2 ± 0.1 ppm. Errors based on counting statistics are below 5%, except for Ni, Zn, Rb, Nd, Dy, Ta, and U, for which errors are between 5% and 10%.

*Total iron as FeO.
†Carbon due to graphite flakes in the glass.

included in Table 3. The concentrations determined by INAA and by RNAA generally agree within the error limits. However, because of the larger error of the INAA data, due to poorer counting statistics, the RNAA data are more accurate. Molybdenum contents in the glass and in the glass inclusion-bearing olivines could not be determined because of the production of ^{99}Mo by uranium fission. Sulfide particles were analyzed by INAA only.

Carbon and S were analyzed using a Leybold-Heraeus (CSA 2002) automatic C and S analyzer. Mineral analyses were performed using the Cameca microprobe at the Department of Geosciences, University of Mainz. Ca, Ti, Cr, and Ni in olivines were determined by microprobe using the trace element analysis program of the Cameca Corporation. Analytical conditions were 20 keV accelerating voltage, beam currents of 100 nA, and peak counting times of 10 minutes.

PURITY OF MINERAL SEPARATES

As mentioned above, the olivines contain small glass inclusions. The major element geochemistry of glass inclusions in olivines was found through microprobe analyses to be similar to that of the glass matrix. Therefore the contents of a number of trace elements (e.g., Sc, Ga, Yb, Lu) in olivines were corrected by assuming that olivines contain no K or light REE. From the K and light REE contents of olivine analyses, a glass contribution of 2 to 6 wt % was calculated, with an average of 5%. Therefore the Ga, Sc, Yb, and Lu contents of olivines were corrected under the assumption that glass inclusions make up 5 wt % of the olivines. For elements such as As, Sb, Cu, Cr, Mn, and Zn, the contribution from the glass inclusions

TABLE 6. Major elements in glass and olivines determined by microprobe analyses.

	This Work		Pedersen (1979)	
	Glass	Olivines	Glass	Olivines
SiO_2	55.54	38.13	53.11	39.40
TiO_2	1.43	0.027 ± 0.004	1.50	0.06
Al_2O_3	15.03	0.05	15.46	0.07
Cr_2O_3	0.12	0.132 ± 0.009	-	-
FeO	8.04	15.33	8.84	15.69
MnO	0.12	0.21	0.14	0.20
MgO	6.95	45.89	7.16	44.22
CaO	9.56	0.227 ± 0.021	9.64	0.25
Na_2O	2.08	-	2.14	-
K_2O	0.74	-	0.68	-
P_2O_3	0.16	-	0.16	-
H_2O+C+S	-	-	0.36	-
NiO	-	0.0165 ± 0.008	-	-
Σ	99.77	100.21	99.19	99.89
mg'	60.0	84.0	59.1	83.4

TABLE 7. Trace element analyses of metal-sulfide particles (data given in ppm, except where otherwise indicated).

Weight	1 1.5 μg	2 12.0 μg	3 13.0 μg	4 2.7 μη	5 7.0 μg	6 220 μg
Fe* (%)	63.00	63.00	63.00	63.00	63.00	63.00
Co	5077	5308	5360	4995	5354	5338
Ni (%)	4.78	4.92	4.84	4.88	4.98	4.38
Cu (%)	-	-	-	-	-	1.41
Cr	792	707	880	873	856	8300†
Zn	<70	-	-	-	-	-
Mn	-	-	-	-	-	150‡
Ga	-	-	-	-	-	13.9
As	377	388	300	395	-	468
Sb	-	14.4	14.0	-	-	25.4
Mo	76	77	75	76	-	101
W	-	-	-	-	-	6.4
Se	53	53	52.5	56	48	-
Ir (ppb)	59	41	52.5	56	48	
Au (ppb)	213	220	229	212	-	660†
Weight	7 1.0 μg	8 1.5 μg	9 2.5 μg	10 5.0 μg	11 4.0 μg	12 7.5 μg
Fe (%)	63.00	63.00	63.00	63.00	63.00	63.00
Co	4907	4765	4424	6240	4680	5310
Ni (%)	4.68	4.46	3.71	6.0	4.25	4.41
Cu (%)	-	-	-	-	-	1.80
Cr	735	842	1200	1186	979	1020
Zn	<50	<43	<53	<80	<41	<24.5
Ga	-	-	-	-	-	11.0
As	404	453	-	-	-	438
Sb	17.0	15.4	16.1	14.8	14.2	-
Mo	-	-	-	-	85	104
W	-	-	-	-	-	3.1
Se	46.4	38.0	40.7	48.3	42.6	35.0
Ir (ppb)	43.6	44.1	44.3	43	33	50
Au (ppb)	159	166	-	-	240	207

Columns 1-5 represent analysis of individual sulfide particles irradiated at the research reactor at Jülich, FRG.

Column 6 represents the analysis of a number of sulfide particles with a tota l weight of 0.22 mg. Samples were irradiated at the TRIGA reactor, Mainz, FRG.

Columns 7-12 represent analyses of individual sulfide particles irradiated at the Geesthacht research reactor (FRG). Errors of individual analyses based on counting statistics are in general below 3% except for Zn, Ga, Ir, for which errors are in the range of 5%-10%.

*All analyses are normalized to 63.00 wt % Fe because individual sulfide particles were too small to weigh accurately. Microprobe bulk analyses of sulfides showed an average iron content of 63 wt %.

†The bulk sample must contain Cr-and Au-rich phases that could not be identified.

‡The Mn content of sulfides is corrected for glass contamination.

can be neglected. In no case did we observe metal-sulfide droplets associated with olivines. Small metal-sulfide inclusions in separated glass samples, however, cannot be excluded. The Ni content of glass samples, for example, is very sensitive to even the smallest amount of sulfide. We therefore analyzed several tiny glass fragments in order to establish the content and the homogeneity of chalcophile elements. Results are given in Table 1. The weights of samples below 10 μg were determined by assuming a total Fe content of 6.53%, the same Fe content as in the bulk-glass sample. The small variations in the contents of Sc, Cr, Co, Zn, REE, and Hf in all samples demonstrate the homogeneity of the glass. Furthermore the uniform Co content limits the contribution of the sulfides to these samples. The Co content of metal-sulfide blebs is about 5000 ppm (Table 4). The variation in Co concentration among the glass samples is less than 5 ppm, limiting the metal-sulfide contribution to less than 0.1%. This implies that the maximum amount of Ni contributed by the metal-sulfide particles is 40 ppm. The actual measured upper limits in the glass samples are below 20 ppm. This upper limit may be used to estimate the effect of possible sulfide contamination on the olivine/silicate liquid partition coefficients discussed in a later section. The Co partition coefficient would increase from 4.3 to 4.8, Cu from 0.55 to 0.7, As from 0.3 to 0.5, and Sb from 0.3 to 0.7. The actual level of contamination of glass samples with metal-sulfide particles may, however, be considerably lower.

RESULTS

1. Bulk Rock Analysis

The chemical analysis of the bulk rock sample is given in Table 5. The data agree very well with those published by *Pedersen* (1979). Sulfur analyses of several bulk samples showed considerable variations. Sulfur contents range from 0.02 wt % to 0.17 wt %, probably caused by heterogeneous distribution of sulfides. The carbon content of the bulk samples varies from 0.05 wt % to 0.16 wt %.

2. Chemical Analysis of Glass and Olivines

As pointed out above, the glass is very homogeneous in composition. This is confirmed by microprobe analyses. The olivines were also found to be very homogeneous in composition. They are only slightly zoned, ranging from $mg' = 84.0$ in the core to $mg' = 83.5$ at the rim (Table 6, $mg' = $ atomic Mg/Mg + Fe).

In Tables 1 and 2 we present the results of trace element analyses of mg-sized glass and olivine samples, as well as analyses of small glass fragments and individual olivine crystals. Trace element abundances are very similar in the different glass and olivine fractions.

Nickel abundances in mg-sized samples of olivines (corresponding to several hundred olivine crystals) are consistent. Individual olivine grains show considerable scatter in their Ni contents (Table 2). There is a tendency for lower Ni contents in smaller grains. We have mentioned before that olivines are slightly zoned with some FeO enrichment at the rim. This indicates either that the olivines grew by fractional crystallization or that some element exchange occurred at temperatures below the quench temperature. The small olivine crystals may therefore reflect different conditions than the bulk of the olivines.

As discussed previously, Sc, Ga, and REE abundances in olivine separates are probably too high because of glass inclusions in the olivines (Table 2). Table 3 shows the average trace element abundances for the basaltic glass and for the olivines corrected for glass inclusions. In addition this table gives the abundances of chalcophile elements Cu, As, Sb, and Au in glass and olivines as determined by radiochemistry.

3. Composition of the Sulfide Droplets

A number of individual metal-sulfide particles were analyzed by INAA and found to be relatively homogeneous in composition. Table 7 gives the results of analyses of samples irradiated in three different reactors and using a different set of standards each time.

From the composition of the 11 individual particles and the bulk analysis of a large number of sulfides, the average composition is calculated and given in Table 4. In general the standard deviation of individual analysis from the mean is about 15%. Microprobe analyses of the bulk composition of quench-textured metal-sulfide droplets agree well with the INAA data. The average of six microprobe bulk analyses and the composition of Fe-Ni and FeS are given in Table 8. The P content of bulk metal-sulfide droplets is below 0.02 wt % as determined by microprobe. Microprobe analyses show that P in Fe-Ni crystals must also be lower than 0.02 wt %.

From the composition of the metal-sulfide droplets a liquidus temperature between 900°C and 1000°C can be estimated (*Kullerud*, 1963). At lower temperatures Fe-Ni crystals exsolved from the sulfide liquid. The average composition of these Fe-Ni crystals and the FeS matrix are also given in Table 8. The Ni contents of the Fe-Ni crystals vary between 18 wt % and 30 wt %.

The fact that each of the three different mineral phases (olivines, glass, and sulfides) are homogeneous in composition suggests thermodynamic equilibrium between these phases.

DISCUSSION

1. Partition Coefficients Derived from the Data Presented

Assuming that the mineral phases are in equilibrium with each other, three sets of partition coefficients for a number of trace elements can be derived:

(1) K_D olivine/silicate liquid, (2) K_D metal-sulfide liquid/silicate liquid, and (3) K_D Fe-Ni-metal (solid)/Fe-sulfide (solid).

TABLE 8. Average composition of sulfide droplets, Fe-Ni crystals and FeS (microprobe analyses, in wt %).

	Bulk (n = 6)	Fe-Ni (n = 20)	FeS (n = 20)
Fe	63.10 ± 0.85	75 (65-80)	63.5 ± 1.20
Ni	3.95 ± 0.40	21 (18-30)	0.53 ± 0.23
Co	0.48 ± 0.08	2.44 ± 0.06	0.10 ± 0.03
Cu	1.21 ± 0.24	0.90 ± 0.35*	0.37 ± 0.12
Cr	0.06 ± 0.03	0.01	0.1 ± 0.01
As	-	0.17 ± 0.02	0.02
Mo	-	≤0.012	≥0.008
S	30.35 ± 1.10	0.58†	35.70 ± 1.05
Σ	99.15	100.11	100.32

Standard deviations indicate uncertainties due to differences in mineral composition from one analyzed spot to another. Arsenic and Mo in FeS calculated from the bulk composition of the metal-sulfide droplets, the measured As and Mo content in the FeNi-metal and the modal abundance of Fe-Ni crystals and FeS.

*In the rim of FeNi-crystals Cu contents as high as 4 wt % were observed.

†Sulfur in the metal due to excitation of S in adjacent FeS.

TABLE 9. Olivine/silicate liquid partition coefficients.

	This Work	Pedersen (1979)	Literature Data	Ref.
K_D Fe/Mg	0.29	0.28	0.30	(1)
Fe	1.85	1.8	2.1	(2)
Mg	6.60	6.1	6.5	(2)
Mn	1.47	1.4	1.5	(3)
Cr	1.18	1.2	0.84-1.28	(4)
Ca	0.025	0.03	0.033	(5)
			0.020-0.039	(2)
Ti	0.019		0.042	(5)
			0.024	(6)
Ni	>15	12	18.3	(2)
			17.0	(7)
Co	4.3	4.4	4.3	(2)
Cu	0.55	0.5	0.47	(3)
Zn	0.8		0.86	(8)
Ga	0.04	<0.6	0.024	(9)
			0.04	(10)
As	0.3			
Sb	0.3			
Sc	0.23	<0.3	0.14-0.22	(2)
			0.37	(6)
			0.265	(6)
Yb	0.042		0.042	(11)
Lu	0.062			
Au	0.7		≈1	(12)

References: (1) *Roeder and Emslie*, 1970; (2) *Leeman and Scheidegger*, 1977; (3) *Irving*, 1978; (4) *Huebner et al.*, 1976; (5) *Longhi et al.*, 1978; (6) *Lindstrom*, 1976; (7) *Hart and Davis*, 1978; (8) *Bougault and Hekinian*, 1974; (9) *Malvin and Drake*, 1987; (10) *Goodman*, 1972; (11) *McKay and Weill*, 1977; (12) *Jones and Drake*, (1986).

2. Conditions of Mineral Equilibration

From the Fe-Mg distribution between olivines and coexisting glass, an equilibration temperature of 1175°C ± 10°C can be calculated according to the equation of *Roeder and Emslie* (1970). Since the melt was quenched rapidly, we can assume that further elemental partitioning in the silicate phases due to low temperature diffusion of elements is negligible, with the possible exception of the Ni contents of very small olivine crystals, as mentioned before.

From the FeO content of the glass, assuming equilibrium between metal-sulfide liquid and silicate liquid, a formal oxygen fugacity of log P_{O_2} = -13.9 at a temperature of 1175°C can be calculated. This value is obtained from the equation given by *Myers and Eugster* (1983) for the Fe/FeO equilibrium. The activity of FeO in the silicate liquid was assumed to be the same as the mole fraction. A similar assumption was made for the activity of iron in the sulfide liquid. The oxygen fugacity can also be calculated from the olivine/sulfide-liquid equilibrium. Under similar assumptions as those stated above, a value of P_{O_2} of -13.9 is obtained from an equation given by *Nitsan* (1974). The metal-sulfide liquids solidified at about 900°C-1000°C (*Kullerud*, 1963). The equilibration of Fe-Ni crystals and FeS took place probably below 900°C due to elemental diffusion between metals and sulfides. Diffusion, however, must have ceased at temperatures higher than 650°C because Kamacite did not exsolve from Fe-Ni crystals, which have Ni contents of 18%-30% Ni (*Goldstein and Ogilvie*, 1965).

3. Demonstration of Equilibrium Between Olivines and Silicate Glass

Roeder and Emslie (1970) have determined the composition of olivines in equilibrium with a silicate melt. They found that the Fe/Mg distribution coefficient relating the partitioning of iron and magnesium between olivine and silicate liquid in terrestrial samples is about 0.30, independent of temperature.

From the composition of olivines and glass in our sample, a Fe/Mg exchange partition coefficient of 0.29 can be calculated that is close to the value given by *Roeder and Emslie* (1970). We take this to indicate that there is equilibrium between olivines and glass in the basalt sample.

From the data given in Table 3 several olivine/silicate liquid partition coefficients can be calculated. These partition coefficients are listed in Table 9 and are compared with literature data from natural and synthetic systems. There is good agreement in all cases with literature partition coefficients derived from systems similar in composition to ours in terms of major elements.

Since we can demonstrate that the partition coefficients presented for Fe, Mg, Ni, Co, Cr, Mn, and Zn agree with those found by other authors, we are confident that the partition data given for Ca, Ti, Ga, Cu, As, Sb, Yb, Lu, and Sc (for which only a few or no literature data exist), are also reliable. The Ca and Ti contents of the olivines were determined by microprobe and care was taken not to analyze near a glass inclusion. The correction procedure for the Ga, Sc, Yb, and Lu contents in olivines was already described in the analytical section. The As, Sb, Cu, and Au trace element abundances in glass and olivines were determined by a radiochemical procedure (see details in the analytical description).

Calcium and Ti are elements that partition strongly into the basaltic melt. Olivine analyses of these elements are susceptible to contamination. Our measured Ca and Ti partition coefficients are, however, in good agreement with those determined by *Longhi et al.* (1978), *Leeman and Scheidegger* (1977) and *Lindstrom* (1976). Gallium is also incompatible in olivine. Our measured value of the olivine/silicate liquid partition coefficient is 0.04 and is in agreement with D_{Ga} of of 0.024 ± 0.009 given by *Malvin and Drake* (1987) and D_{Ga} of 0.04 by *Goodman* (1972). In contrast to the light REE (La, Sm, Eu), which are effectively excluded from the olivine crystal lattice, Yb and Lu are partitioned into olivine to a small degree. The Yb olivine/liquid partition coefficient presented in this work is similar to that given by *McKay and Weill* (1977). The fact that the Lu olivine/liquid K_D is somewhat higher than that for Yb is in accord with theoretical predictions. The olivine/liquid partition coefficient found for Cu agrees with data listed by *Irving* (1978). The chalcophile elements As and Sb show olivine/liquid partition coefficients similar to Cu, but no literature data exists for comparison.

TABLE 10. Silicate liquid/metal-sulfide liquid partition coefficients.

	This Work	Pedersen (1979)	Jones and Drake (1986)	Newsom and Drake (1987)	
Temp. (°C)*	1175		1260 ± 10	1260	1260
-log f_{O_2}*	13.9		12.2-13.0	12.7	16.1
S (wt %)*	30		25	31	31
Ni	<2·10^{-4}	2.3·10^{-4}	2·10^{-4}	-	-
Co	5·10^{-3}	4.4·10^{-3}	7.5·10^{-3}	-	-
Mo†	1·10^{-3}	-	8·10^{-4}	-	-
Au	6.5·10^{-3}	-	1·10^{-4}	-	-
W‡	0.04	-	1.0	-	-
Ga	1.7	-	0.8	-	-
P	>6	>2.4	0.24	-	-
As	1·10^{-3}	-	-	-	-
Sb	8·10^{-4}	-	-	-	-
Cu	2·10^{-3}	2.6·10^{-3}	-	-	-
Cr	0.9	1.1	-	5	0.01
Zn	>2.0	-	-	-	-
Mn	6.2	3.2	-	25	0.07
V	-	1.7	-	5.5	<0.004

*Temperature and oxygen fugacity of mineral equilibration for the natural sample and for the experiments are given together with the sulfur content of the metal-sulfide liquids.

†Molybdenum in the glass was estimated by subtraction of the Mo content in sulfides from the Mo content of the bulk sample.

‡Tungsten in the silicate glass was estimated from La/W correlation of terrestrial basalts.

Gold behaves as an incompatible element in silicate systems. The olivine/silicate liquid partition coefficient of Au is expected to be distinctly below 1.0 or even lower than 0.1 in MgO-rich systems (*Brügmann*, 1987). We found a surprisingly high value of 0.7, which should be confirmed by further investigations. However, our value agrees with the solid silicate/liquid silicate K_D of Au given by *Jones and Drake* (1986), which was evaluated from element correlations of natural basalts.

4. Metal-Sulfide Liquid/Silicate Liquid Partition Coefficients

In the section above we have shown that partitioning of elements between olivine and silicate liquid reflects a close approach to equilibrium between these phases. Since olivines and glass appear to have equilibrated, one can safely assume equilibrium between silicate liquid and metal-sulfide liquid. Diffusion in the sulfide phase will always be much faster than in the two silicate phases. Therefore, the metal-sulfide droplets, now unmixed into FeS and Fe-Ni crystals, existed at about 1200°C as homogeneous metal-sulfide liquids, in equilibrium with the olivines and the basaltic melt.

In Table 10 we present silicate liquid/metal-sulfide liquid trace element partitioning data derived from the results given in Table 3 and Table 4. Our data are compared to experimental silicate liquid/S-bearing metallic liquid partition coefficients published by *Jones and Drake* (1986) and *Newsom and Drake* (1987). Jones and Drake's metal liquids contained about 25 wt % S, which is somewhat lower than the S content of the natural metal-sulfide liquids analyzed in our work (30 wt % S).

Our measured silicate liquid/metal-sulfide liquid partition coefficients for Ni, Co, Mo, and (to a lesser degree) also Au agree with the data of *Jones and Drake* (1986). The bulk Mo content of our sample is 0.27 ppm (Table 5), while metal-sulfide droplets contain 86 ppm (Table 4). The glass should therefore contain about 0.1 ppm Mo, considering a modal sulfide content of 0.2 wt %.

Major differences from the Jones and Drake partition coefficients are found for W, Ga, and P. Our W content of the glass was estimated from the measured La content of the glass, assuming a La/W ratio of 50, typical for terrestrial and lunar basalts (*Rammensee and Wänke*, 1977). Using the W/Ba ratio of 0.00161 for terrestrial basalts given by *Newsom et al.* (1986), we calculate a silicate/metal-sulfide partition coefficient of W of 0.06 instead of 0.04. It is possible, however, that the actual W content of our glass sample is different from the calculated value. On the other hand, element ratios, such as P/Nd (*Weckwerth*, 1983) and Co/MgO + FeO (*Dreibus and Wänke*, 1984) of the glass sample, are typical for terrestrial basalts and do not support an erratic W/La ratio.

The uncertainty of the Ga content of the sulfides due to varying composition of different particles is below 20%. The

Ga content of the glass is known within 5% to 10%. Therefore the Ga silicate/sulfide partition coefficient should be in the range from 1.5 to 2.0.

For the P content of the metal-sulfide liquids only upper limits were obtained. The exsolved Fe-Ni particles as well as the sulfide matrix contained less than 0.02 wt % of P. The P content of the Fe-Ni particles may, however, be considerably higher than the P content of the sulfide matrix. *Malvin et al.* (1986) have determined P partition coefficients between solid and liquid metal, with variable S contents. From an equation derived by the authors, one can calculate a solid metal/liquid metal P partition coefficient of about 2 for the S content appropriate to the problem here. This would imply a silicate/sulfide-rich metal P partition coefficient of about 6.

This value is much higher than the partition coefficient of 0.24 used by *Jones and Drake* (1986). One possibility to explain this discrepancy is the strong dependence of P activity on the S content of the metallic melt. This is obvious from the partitioning of P between solid and liquid metal at various S contents of the liquid metal, as demonstrated by *Jones and Drake* (1982, 1983). The partition coefficients given by *Jones and Drake* (1986) in Table 1 are calculated for a nominal S content of 25 wt %, whereas our metal-sulfide liquid contained about 30 wt % S. Another possibility is oxygen fugacity. Since P occurs as P_2O_5 in the silicate, its metal/silicate partition coefficient is strongly dependent on oxygen fugacity (*Newsom and Drake*, 1983). The lower silicate/liquid metal partition coefficients for P and Ga in the Jones and Drake experiments would require lower oxygen fugacities. Jones and Drake report a log P_{O_2} of -12.6 at a temperature of 1250°C; our calculated log P_{O_2} is -13.9 at 1175°C. The lower P_{O_2} of our sample would imply lower silicate/metal-liquid partition coefficients, contrary to the observation. However, the lower temperature in our system would produce higher silicate/metal-liquid partition coefficients than those of *Jones and Drake* (1986). The temperature dependence of the P silicate/metal partition coefficient is actually quite strong (*Newsom and Drake*, 1983; *Schmitt*, 1984). According to *Newsom and Drake* (1983), the D_P silicate/metal decreases by more than two orders of magnitude in going from 1190°C to 1300°C. This would correspond to a decrease in D_P from 1175°C to 1260°C. The increase in the P silicate/metal partition coefficient between P_{O_2} -13.9 and -12.6 is about 30 (*Newsom and Drake*, 1983) at a temperature of 1190°C. The lower temperature in our system may therefore partly explain the discrepancy between our P and Ga data and those of *Jones and Drake* (1986).

The difference between our W silicate/metal-sulfide partition coefficient and that of Jones and Drake cannot be explained in this way, but this difference appears to indeed be real. For a W silicate/metal-sulfide partition coefficient of about 1, the W content in our glass sample should be about 4 to 5 ppm. However, W was not observed in the γ-spectra, which means that the W content of our glass sample was less than 0.5 ppm and therefore the observed W silicate/metal-sulfide partition coefficient is smaller 0.1. We see no possibility of reconciling our W partition coefficient with the experimental data. Although the C concentration in the metal-sulfide blebs is probably below 0.3%, the activity of C in the system should

TABLE 11. Fe-Ni (solid)/FeS (solid) partition coefficients.

	Fe-Ni/FeS
Ni	40 ± 10
Co	24.4 ± 5
Cu	2.4 ± 1
Cr	0.1
As	8.7 ± 1
Mo	<1.4

Derived from data in Table 8.

be high (note the graphite flakes mentioned earlier) and could eventually affect silicate/metal-sulfide partition coefficients. More data are necessary to judge the magnitude of the effect of carbon, however.

No comparable data exist for the elements As, Sb, and Cu which behave, as expected, as chalcophile elements. Chromium, Zn, and Mn, however, are not partitioned into metal-sulfide liquids. These data are in good agreement with experimental results published by *Newsom and Drake* (1987). They investigated partitioning of V, Cr, and Mn between solid metal, S-bearing metallic liquid, and liquid silicate at 1260°C and log f_{O_2} = -12.7 and -16.1. They found that Mn is partitioned into sulfide liquids only at very reducing conditions. Vanadium and Cr show a more chalcophile character compared with Mn. Their experimental data from the higher oxygen fugacity runs show an even stronger lithophile behavior of V, Cr, and Mn than reported in this paper. Table 10 includes the V silicate liquid/metal-sulfide liquid partition coefficient reported by *Pedersen* (1979). Chromium and V behave very similarly in our natural system and in the experiments as documented by *Newsom and Drake's* (1987) experimental results.

We cannot compare our results to the Ni, Co, and Cu sulfide/silicate partition coefficients reported by *Rajamani and Naldrett* (1978) and other authors because their experiments were performed at oxygen fugacities above the Fe/FeO buffer. Our metal-sulfide liquid/silicate liquid partition coefficients of Ni, Co, and Cu are distinctly higher than their experimentally determined partition coefficients.

Most of the experimental work is, however, concerned with investigations of partitioning of siderophile elements between stoichiometric sulfides and olivines or silicate glass. The metal-sulfide blebs in our natural sample display an excess metal component, due to the reducing conditions during their formation. *Pedersen* (1979) interpreted the deviation of Ni, Co, and Cu partition coefficients of the Luciefjeld rocks from literature data, to be a result of the excess metal component. Recent experiments on sulfide liquid/silicate liquid partitioning of Ni under controlled temperature, f_{O_2} and f_{S_2} by *Boctor* (1982) and *Boctor and Yoder* (1983) have indeed shown that K_D Ni increases with decreasing oxygen fugacity.

5. Fe-Ni (Solid)/FeS (Solid) Partition Coefficients

As shown in Figs. 2 and 3, tiny Fe-Ni spheroids or Fe-Ni crystals crystallized in the metal-sulfide droplets on cooling. Representative analyses of Fe-Ni crystals and of the iron-sulfide matrix (20 analyses of each) are given in Table 8. The largest variation is found for Ni contents due to zoning of the metal particles. Rims of Fe-Ni crystals have 20-30% (relative %) higher Ni contents than the cores. The average of the Ni contents in the metals is about 21% Ni.

From the bulk composition of the metal-sulfide spherules in the Fe-Ni-S system (*Kullerud*, 1963) a crystallization temperature of the metal-sulfide liquids of about 950°C can be estimated. At or below this temperature the Fe-Ni crystals exsolved from the sulfide.

Analyses of the two phases provide solid metal/solid sulfide partition coefficients for some siderophile elements. In Table 11 the Fe-Ni/FeS partition coefficients of Ni, Co, Cu, Cr, As, and Mo of the sample studied are listed. These partition coefficients are probably representative only for low temperature (<900°C) Fe-Ni (solid)/FeS (solid) partitioning of siderophile and chalcophile elements.

IMPLICATIONS FOR THE FORMATION OF PLANETARY CORES

The general agreement of the partition coefficients determined in this work and those extracted from experiments with synthetic systems is very satisfying and justifies the application of these partition coefficients to natural systems. To demonstrate the usefulness of the partition coefficients (reported above) we shall shortly discuss two aspects of the change in trace element contents of planetary mantles by formation of metal or sulfide-rich metallic cores. The topic has been extensively discussed by *Jones and Drake* (1986) for the Earth and by *Treiman et al.* (1986) for the Earth, the Moon, the Shergotty parent body (SPB), and the eucrite parent body (EPB).

In the first example we want to emphasize the role of P in discriminating between the formation of a pure metal core and a sulfide-rich core. If the P partition coefficient for silicate/sulfur-rich metal liquid is as high as suggested in this study, one should not expect any removal of P from a planetary mantle by the formation of a sulfide-rich core. On the other hand, P is a moderately siderophile element that partitions into metal roughly to the same extent as W (*Schmitt*, 1984, *Newsom and Drake*, 1983). Unfortunately, typical chalcophile elements, such as Mo, are not so well suited in distinguishing between a metal and sulfide-rich core, since these elements are often strongly siderophile.

In Fig. 4 we have compiled data on P, W, and Mo for the Earth, the Moon, the Shergotty parent body (SPB), and the eucrite parent body (EPB). Since all three elements are incompatible with silicates, it is possible to estimate their abundances with some reliability (e.g., *Rammensee and Wänke*, 1977; *Newsom and Palme*, 1984; *Palme and Rammensee*, 1981). Similar depletions of P, W, and Mo in the Earth may be explained within the framework of the inhomogeneous accretion hypothesis (*Wänke et al.*, 1984). The low content of Mo and P in the Moon reflect removal of metal. Models that make the Moon out of terrestrial material require an additional step of metal segregation in the Moon to explain the low contents of P and Mo in lunar rocks (*Newsom*, 1986; *Wänke and Dreibus*, 1986).

Because of the depletion of P and W in eucrites, the separation of a metal core is required. The higher depletion of Mo reflects the higher metal/silicate partition coefficient of Mo.

A completely different picture is found for SNC-meteorites. The high contents of P in these meteorites exclude the segregation of a S-free metal core by a major core formation event. This is consistent with the evidence from other siderophile and chalcophile elements (*Laul et al.*, 1986, *Treiman et al.*, 1986).

A higher degree of oxidation during core formation on the SPB, as suggested by *Treiman et al.* (1986), may produce a similar pattern. This possibility is, however, unlikely since eucrites and shergottites have similar FeO contents, suggesting a similar degree of oxidation for the EPB and SPB mantle. In the second example the influence of core formation on the abundances of Cr, Mn, and Zn will be discussed.

These three elements are listed here in the order of increasing volatility, and it is a matter of considerable interest if the decreasing abundance sequence of Cr, Mn, and Zn in a planetary mantle reflects a primary depletion sequence, or if the position of an element in this sequence can be influenced by core formation (*Dreibus and Wänke*, 1979; *Ringwood and Kesson*, 1977).

The metal/silicate partition coefficient for Cr is considerably higher than that of Mn (*Rammensee et al.*, 1983; *Newsom and Drake*, 1987) and certainly of Zn, according to the free

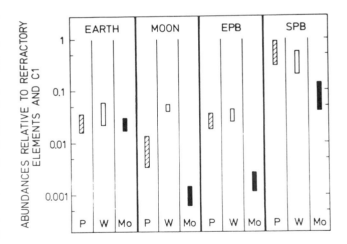

Fig. 4. Estimates of the contents of P, W, and Mo in the mantles of various planets. The abundances are derived from incompatible element ratios, such as P/Nd, W/U, Mo/Nd, as mentioned in the text. Similar estimates for Earth, eucrite parent body (EPB), and Shergotty parent body (SPB) were reported by *Treiman* (1986). Sources of data are as follows: Earth: *Newsom and Palme* (1984), *Weckwerth* (1983); Moon and EPB: *Palme and Rammensee* (1981), *Newsom* (1986); SPB: *Burghele et al.* (1983), *Laul et al.* (1986), and unpublished data from the Mainz Cosmochemistry laboratory.

energy of formation of the oxide. Under extremely reducing conditions, Cr may be partly removed into the metal core during core formation. The effects on Mn and Zn will be much smaller, since their metal/silicate partition coefficients are lower. If, on the other hand, there is no depletion of Cr in the silicate fraction of a planetary body, one can safely assume that Mn and Zn abundances have not been affected by the formation of a metal core.

These arguments are, however, only valid for a metal core. To study the effects of the formation of a sulfur-rich core the knowledge of sulfide/silicate partition coefficients is required. The data presented in this paper indicate that Cr is not only the most siderophile among the three elements, but also the most chalcophile (Table 10). It is therefore clear that a chondritic Cr/Mg-ratio in the silicate fraction of a planet excludes the loss of Mn or Zn by core formation. Lower abundances of these two elements must then reflect volatility-dependent processes in the solar nebula or during accretion.

The C1-normalized abundances (relative to Si) of Mg, Cr, Mn, and Zn in the terrestrial upper mantle are 1.25 (Mg), 0.60 (Cr), 0.28 (Mn), and 0.072 (Zn) (*Jagoutz et al.*, 1979). The Zn-depletion is therefore an intrinsic property of the chemistry of the Earth, either produced by loss of Zn during accretion or as a characteristic signature of the chemistry of the bodies that formed the Earth, i.e., a result of processes in the solar nebula. The same arguments may in part be applied to Mn. There could be some depletion of Mn, simultaneously with some loss of Cr due to formation of a core under extremely reducing conditions, such as envisioned in the two-component model of *Wänke et al.* (1984). It appears unlikely that the depletion of Cr in the Earth's mantle is exclusively produced by volatility related processes. The Mg/Cr ratio in all types of carbonaceous chondrites is constant within a few percent, while Mn and Zn contents, like all other volatile elements decrease in abundance from type 1 to type 3 carbonaceous chondrites. Clearly Cr abundances in meteorites are not volatility controlled (*Kallemeyn and Wasson*, 1981). An additional volatility-dependent depletion is, however, required to explain the lower C1-normalized abundance of Mn, compared to Cr. This possibility was already discussed by *Wänke and Dreibus* (1986).

Acknowledgments. We thank B. Spettel for his valuable analytical advice for the neutron activation analyses, Dr. B. Schulz-Dobrick for his assistance during the microprobe work, and T. Kost for excellent wet chemical analysis. We are grateful to G. Dreibus for the C and S analyses and to Dr. R. Dietrich for generously making available the Disko samples. The samples were activated at the TRIGA reactor of the Institut für Kernchemie, University of Mainz, and at the Geesthacht and Jülich research reactors. This paper was written while W. Klöck was an NRC research associate at the NASA Johnson Space Center in Houston. Reviews by Cyrena A. Goodrich, John H. Jones, and Horton E. Newsom improved the paper considerably.

REFERENCES

Boctor N. Z. (1982) The effect of f_{O_2}, F_{S_2} and temperature on Ni partitioning between olivine and iron sulfide melts. *Carnegie Inst. Washington Year Book, 82*, 366-369.

Boctor N. Z. and Yoder H. S., Jr. (1983) Partitioning of nickel between silicate and iron sulfide melts. *Carnegie Inst. Washington Year Book, 82*, 275-277.

Bougault H. and Hekinian R. (1974) Rift Valley in the Atlantic Ocean near 36°50′: Petrology and geochemistry of basaltic rocks. *Earth Planet. Sci. Lett., 24*, 249-261.

Brügmann G. E., Arndt N. T., Hofmann A. W., and Tobschall H. J. (1987) Noble metal abundances in komatiite suites from Alexo, Ontario and Gorgona Island, Colombia. *Geochim. Cosmochim. Acta, 51*, 2159-2169.

Burghele A., Dreibus G., Palme H, Rammensee W., Spettel B., Weckwerth G., and Wänke H. (1983) Chemistry of shergottites and the shergotty parent body (SPB): Further evidence of the two component model of planet formation (abstract). In *Lunar and Planetary Science XIV*, pp. 80-81. Lunar and Planetary Institute, Houston.

Dreibus G. and Wänke H. (1979) On the chemical composition of the Moon and the Eucrite Parent Body and a comparison with the composition of the Earth, the case of Mn, Cr and V. (abstract). In *Lunar and Planetary Science X*, pp. 315-317. Lunar and Planetary Institute, Houston.

Dreibus G. and Wänke H. (1984) Accretion of the Earth and the inner planets. *Proc. Int. Geol. Congr. 27th, 11*, 1-20.

Goldstein J. I. and Ogilvie R. E. (1965) The growth of the Widmannstätten pattern in metallic meteorites. *Geochim. Cosmochim. Acta, 42*, 1545-1558.

Goodman R. J. (1972) The distribution of Ga and Rb in coexisting groundmass and phenocryst phases of some gasic volcanic rocks. *Geochim. Cosmochim. Acta, 36*, 303-317.

Hart S. R. and Davis K. E. (1978) Nickel partitioning between olivine and silicate melt. *Earth Planet. Sci. Lett., 40*, 203-219.

Huebner S. J., Lipin B. R., and Wiggins L. B. (1976) Partitioning of chromium between silicate crystals and melts. *Proc. Lunar Sci. Conf. 7th*, 1195-1220.

Irving A. J. (1978) A review of experimental studies of crystal/liquid trace element partitioning. *Geochim. Cosmochim. Acta, 42*, 743-770.

Jagoutz E., Palme H., Baddenhausen H., Blum K., Cendales M., Dreibus G., Spettel B., Lorenz V., and Wänke H. (1979) The abundances of major, minor and trace elements in the Earth's mantle as derived from primitive ultramafic nodules. *Proc. Lunar Planet. Sci. Conf. 10th*, 2031-2050.

Jones J. H. and Drake M. J. (1982) An experimental geochemical approach to early planetary differentiation (abstract). In *Lunar and Planetary Science XIII*, pp. 369-370. Lunar and Planetary Institute, Houston.

Jones J. H. and Drake M. J. (1983) Experimental investigations of trace element fractionation in iron meteorites. II: The influence of sulfur. *Geochim. Cosmochim. Acta, 47*, 1199-1209.

Jones J. H. and Drake M. J. (1986) Geochemical constraints on core formation in the Earth. *Nature, 322*, 221-228.

Kallemeyn G. W. and Wasson J. W. (1981) The compositional classification of chondrites -I. The carbonaceous chondrite groups. *Geochim. Cosmochim. Acta, 45*, 1217-1230.

Klöck W., Palme H., and Tobschall H. J. (1986) Trace elements in natural metallic iron from Disko Island, Greenland. *Contrib. Mineral. Petrol., 93*, 273-282.

Kruse H. (1979) Spectra processing with computer graphics. *Mayaguez Conf. Proc.*, United States Department of Energy (DOE) Report 78421, 76-84.

Kullerud G. (1963) The Fe-Ni-S system. *Carnegie Inst. Washington Year Book, 62*, 175-189.

Laul J. C., Smith M. R., Wänke H., Jagoutz E., Dreibus G., Palme H., Spettel B., Burghele A., Lipschutz M. E., and Verkouteren R. M. (1986)

Chemical systematics of the Shergotty meteorite and the composition of its parent body (Mars). *Geochim Cosmochim. Acta, 50,* 909-926.

Leeman W. P. and Scheidegger K. F. (1977) Olivine/liquid distribution coefficients and a test for crystal equilibrium. *Earth Planet. Sci. Lett., 35,* 247-257.

Lindstrom D. J. (1976) Experimental study of the partitioning of the transition metals between clinopyroxene and coexisting silicate liquids. Ph.D. dissertation, University of Oregon. 188 pp.

Longhi J., Walker D., and Hays F. J. (1978) The distribution of Fe and Mg between olivine and lunar basaltic liquids. *Geochim. Cosmochim. Acta, 42,* 1545-1558.

Malvin D. J. and Drake M. J. (1987) Experimental determination of crystal/melt partitioning of Ga and Ge in the system forsterite-anorthite-diopside. *Geochim. Cosmochim. Acta, 51,* 2117-2128.

Malvin D. J., Jones J. H., and Drake M. J. (1986) Experimental investigations of trace element fraction in iron meteorites. III: Elemental partitioning in the system Fe-Ni-S-P. *Geochim. Cosmochim. Acta, 50,* 1221-1231.

McKay G. A. and Weill D. F. (1977) KREEP petrogenesis revisited. *Proc. Lunar Sci. Conf. 8th,* 2339-2355.

Myers J. and Eugster H. P. (1983) The system Fe-Si-O: oxygen buffer calibrations to 1500 K. *Contrib. Mineral. Petrol., 82,* 75-90.

Newsom H. E. (1986) Constraints on the origin of the Moon from the abundance of molybdenum and other siderophile elements. In *Origin of the Moon.* (W. K. Hartmann, R. J. Phillips and G. J. Taylor, eds.), pp. 203-229. Lunar and Planetary Institute, Houston.

Newsom H. E. and Drake M. J. (1983) Experimental investigation of the partitioning of phosphorus between metal and silicate phases: implications for the Earth, Moon and Eucrite Parent Body. *Geochim. Cosmochim. Acta, 47,* 93-100.

Newsom H. E. and Drake M. J. (1987) Formation of the Moon and terrestrial planets: Constraints from V, Cr and Mn abundances in planetary mantles and from new partitioning experiments (abstract). In *Lunar and Planetary Science XVIII,* pp. 716-717. Lunar and Planetary Institute, Houston.

Newsom H. E. and Palme H. (1984) The depletion of siderophile elements in the Earth's mantle: new evidence from molybdenum and tungsten. *Earth Planet. Sci. Lett., 69,* 354-364.

Newsom H. E., White W. M., Jochum K. P., and Hofmann A. W. (1986) Siderophile and chalcophile element abundances in oceanic basalts, Pb isotope evolution and growth of the Earth's core. *Earth Planet. Sci. Lett., 80,* 299-313.

Nitsan U. (1974) Stability field of olivine with respect to oxidation and reduction. *J. Geophys. Res., 79,* 706-711.

Palme H. and Rammensee W. (1981) The significance of W in planetary differentiation processes: Evidence from new data on eucrites. *Proc. Lunar Planet. Sci. 12B,* 949-964.

Pedersen A. K. (1979) Basaltic glass with high-temperature equilibrated immiscible sulfide bodies with native iron from Disko, Central West Greenland. *Contrib. Mineral. Petrol., 69,* 397-407.

Rajamani V. and Naldrett A. J. (1978) Partitioning of Fe, Co, Ni and Cu between sulfide liquid and basaltic melts and the composition of Ni-Cu sulfide deposits. *Econ. Geol., 73,* 82-93.

Rammensee W. and Wänke H. (1977) On the partition coefficient of tungsten between metal and silicate and its bearing on the origin of the Moon. *Proc. Lunar Sci. Conf. 8th,* 399-409.

Rammensee W., Palme H. and Wänke H. (1983) Experimental investigation of metal-silicate partitioning of some lithophile elements (Ta, Mn, V, Cr) (abstract). In *Lunar and Planetary Science XIV,* pp. 628-629. Lunar and Planetary Institute, Houston.

Ringwood A. E. and Kesson S. E. (1977) Basaltic magmatism and the bulk composition of the Moon II. Siderophile and volatile elements in Moon, Earth and Chondrites. Implication for lunar origin. *The Moon, 16,* 425-464.

Roeder P. C. and Emslie R. F. (1970) Olivine-liquid equilibrium. *Contrib. Mineral. Petrol., 29,* 275-289.

Schmitt W. (1984) *Experimentelle Bestimmung von Metall/Sulfid/Silikat Verteilungskoeffizienten geochemisch relevanter Spurenelemente.* Dissertation, University of Mainz. 102 pp.

Treiman A. H. Drake M. J., Janssens M.-J., Wolf R., and Ebihara M. (1986) Core formation in the Earth and Shergotty Parent Body (SPB): Chemical evidence from basalts. *Geochim. Cosmochim. Acta, 50,* 1071-1091.

Wänke H. (1981) Constitution of terrestrial planets. *Philos. Trans. R. Soc. London, A303,* 287-302.

Wänke H. and Dreibus G. (1986) Geochemical evidence for the formation of the Moon by impact-induced fission of the proto-Earth. In *Origin of the Moon* (W. K. Hartmann, R. J. Phillips and G. J. Taylor, eds.), pp. 649-672. Lunar and Planetary Institute, Houston.

Wänke H., Rieder R., Baddenhausen H., Spettel B., Teschke F., Quijano-Rico M., and Balacescu A. (1970) Major and trace elements in lunar material. *Proc. Apollo 11 Lunar Sci. Conf.,* 1719-1727.

Wänke H., Baddenhausen H., Dreibus G., Jagoutz E., Kruse H., Palme H., Spettel B., and Teschke F. (1973) Multielement analyses of Apollo 15, 16, and 17 samples and the bulk composition of the Moon. *Proc. Lunar Sci. Conf. 4th,* 1461-1481.

Wänke H., Kruse H., Palme H., and Spettel B. (1977) Instrumental Neutron activation analysis of lunar samples and the identification of primary matter in the lunar highlands. *J. Radioanal. Chem., 38,* 363-378.

Wänke H., Dreibus G., and Jagoutz E. (1984) Mantle chemistry and accretion history of the Earth. In *Archaean Geochemistry: The Origin and Evolution of the Archaean Continental Crust.* (A. Kröner, G. N. Hanson, and A. M. Goodwin, eds.), pp. 1-24. Springer Verlag, Berlin, New York.

Weckwerth G. (1983) *Anwendung der instrumentellen β-Spectrometrie im Bereich der Kosmochemie, insbesondere zur Messung von Phosphorgehalten.* Diplomarbeit. Max-Planck-Institut für Chemie, Mainz, FRG. 99 pp.

Metal with Anomalously Low Ni and Ge Concentrations in the Allan Hills A77081 Winonaite

Alfred Kracher

Department of Earth Sciences, Iowa State University, Ames, IA 50011

The Ge content of metal in the Allan Hills A77081 winonaite was determined by high-sensitivity electron microprobe analysis. By optimizing analytical conditions for Ge determination, a detection limit of about 75 ppm could be achieved. In A77081 some small kamacite grains contain less Ni and Ge and more Co than coarse-grained metal. These small grains are always associated with sulfide, raising the possibility that anomalous metal is related to eutectic melting. However, when published partition coefficients for Ni and Ge in the Fe-Ni-S system are used to model fractionation of these elements during eutectic melting, one finds that secondary metal should be enriched in Ni and depleted in Ge. Thus the positive Ni-Ge correlation found in this study is the opposite of the expected trend. No explanation for this discrepancy has yet been found. Nonetheless, the existence of anomalous metal is an indication that A77081, and probably other winonaites as well, have undergone some fractionation. This supports the notion that the high-temperature history of winonaites is related to the formation of IAB iron meteorites, whose silicate inclusions are very similar to winonaites.

1. INTRODUCTION

Winonaites are highly metamorphic stony meteorites with primitive bulk compositions. Chemically and (with minor exceptions) texturally they are similar to silicate inclusions in IAB iron meteorites, but their metal and sulfide contents are more similar to chondrites than to metal-rich meteorites.

Various groupings of IAB-like stony meteorites have been suggested. Without considering samples whose status is controversial, there are at least two observed falls (Acapulco and Pontlyfni) and four finds (Winona, Mt. Morris/Wisconsin, Allan Hills A77081, and Tierra Blanca) that are so similar to one another and to IAB silicate inclusion that they are undoubtedly related (*Bild*, 1977; *Prinz et al.*, 1980, 1983).

Perhaps the strongest argument for such a relationship is based on oxygen isotope compositions. Samples that have undergone isotopic exchange of oxygen in a closed system should plot along a mass fractionation line, and thus form an oxygen isotope "family." Each family can be characterized by its Δ value ($\Delta = \delta^{17}O - 0.52\delta^{18}O$), which is a measure of the displacement of this family's mass fractionation line from that of the Earth-Moon system. *Clayton et al.* (1983) reported an average of $\Delta = -0.48 \pm 0.07$ for five winonaites, indistinguishable from IAB silicate inclusions ($\Delta = -0.47 \pm 0.09$) and phosphates from the IIID iron Dayton ($\Delta = -0.49$).

Although the number of stony meteorites related to IAB inclusions is thus greater than five, traditionally the criterion for defining a group (*Wasson*, 1974), there is no consensus on a group name, reflecting in part a lack of consensus about their origin. *Prinz et al.* (1980, 1983) consider them primitive achondrites, while *Wasson* (1974) refers to them as "IAB chondrites." As in previous work (*Kracher*, 1985), I am following *Prinz et al.* (1980) in calling them "winonaites," since this name does not carry genetic implications. *Prinz et al.* (1983) also discuss the taxonomy of winonaites and related meteorites. It is very likely that additional winonaites are among the "anomalous stony meteorites" in our collections, but a more detailed classification has to await a comprehensive study of all potential candidates.

The association between silicate-rich winonaites and IAB irons, which contain widely varying amounts of silicates (from almost zero to more than 20%), raises the question of whether IAB metal and silicate portions are related. Although brecciated meteorites consisting of apparently unrelated components are by no means rare, a number of features of silicate-bearing IAB irons suggest that silicate and metal come from the same parent body and are cogenetic (*Wasson et al.*, 1980; *Kracher*, 1982, 1985).

Two recent models attempt to explain the origin of IAB irons: the impact melt model of *Wasson et al.* (1980) and the partial differentiation model of *Kracher* (1985). Both models allow two possible ways of formation for metal-poor winonaites on the same parent body as IAB irons. Either winonaites represent the primitive parent material from which IAB irons formed by one of the proposed fractionation processes [impact melting according to *Wasson et al.* (1980), partial differentiation according to *Kracher* (1985)], or they have undergone the same fractionation process as IAB irons, but happen to come from a part of the parent body in which the silicate/metal ratio is higher. In the partial differentiation model the most likely source would be the metal-depleted mantle.

Geothermometry of silicates in IAB inclusions (*Bunch et al.*, 1970) and winonaites (*Palme et al.*, 1981; *Schultz et al.*, 1982) indicates metamorphic temperatures above the Fe-S eutectic. Eutectic melting of metal and sulfide should lead to characteristic fractionation patterns between residual metal that remained unmelted and secondary metal that precipitated from the eutectic melt. The composition of metal in winonaites may thus provide clues to both the fractionation process and the relationship of winonaites to IAB silicate inclusions. If winonaites have undergone partial differentiation (*Kracher*, 1982, 1985), we would expect to find chemical fractionations associated with eutectic melting preserved in their Fe,Ni grains, provided these fractionations were not erased by subsequent metamorphism.

Unfortunately, the concentrations of most diagnostic elements are below the detection limits of electron microprobes. The fractionation behavior of Co and Cu is not very characteristic, and most trace elements are only present at the ppm level. Germanium is the only trace element useful and sufficiently abundant to serve as an indicator of partial melting.

Three of the six winonaites mentioned above are too weathered for meaningful metal analysis. Only Allan Hills A77081 (*Schultz et al.,* 1982) and the two observed falls, Acapulco (*Palme et al.,* 1981) and Pontlyfni (*Graham et al.,* 1977), contain relatively unweathered metal. This paper reports the results of a survey of metal in A77081 by high-sensitivity electron microprobe analysis.

2. ANALYTICAL CONDITIONS

Metal in Allan Hills A77081 was analyzed for Fe, Co, Ni, and Ge with the ARL EMX electron microprobe at Iowa State University. Prior to sample analysis, absolute count rates and peak-to-background ratios were determined on pure Ge metal at several acceleration potentials for both the Ge K_α (E = 9.876 keV; $\theta = 18°09'$, LiF crystal) and Ge L_α (E = 1.186 keV; $\theta = 23°48'$, TAP crystal) lines. These measurements showed that the lowest detection limits could be achieved by using the Ge K_α line at 25 kV, the maximum acceleration voltage for our instrument. A sample current of 0.2 µA (on pure Fe) and counting times of 300 sec were used. Experimental problems and analytical conditions are also discussed by *Kracher and Pierson* (1986).

A microprobe study of Ge in iron meteorites was previously undertaken by *Goldstein* (1967), who reached a similar conclusion regarding optimum analytical conditions, with the exceptions of using an acceleration voltage of 40 kV and even longer counting times (up to 1 hour for a detection limit of 25 ppm Ge).

The high level of continuous radiation near the Ge K_α line, which is due to the closeness of the detector to the sample at short X-ray wavelengths, requires that a large number of counts for both the peak and the background are acquired to achieve sufficient statistical precision. If counting times are to be reasonable, high sample currents must be employed. Under these conditions, the use of pure metals as primary standards causes some problems, since count rates for first-order K_α lines are very high. The detector deadtime for K_α lines on pure metal standards (Fe, Co, Ni, Ge) was up to 30%, which exceeds the range over which the deadtime of a gas-filled counter is a linear function of count rate. Although data reduction algorithms routinely correct for detector deadtime, the correction covers only the linear range.

Since looking for correlations of Ge with Ni and Co was an important part of the study, it was not desirable to employ two separate microprobe runs at different sample currents to determine major and trace elements. Therefore, in most runs Fe, Co, and Ni were determined simultaneously with Ge, using third order K_α lines ($K_\alpha \overline{III}$) on an ADP crystal. These lines are sufficiently weaker than the first order lines to eliminate the deadtime problem, and their greater geometric separation allows measurement of Co K_α without the need to correct for Fe K_β overlap.

In order to assess the magnitude of the deadtime correction for Ge, a chip of Landes meteorite was analyzed as a test sample in each run. Landes was chosen because of its relatively homogeneous metal composition and high Ge content. *Wasson* (1974) determined the Ge content of Landes as 414 ppm by NAA. Electron microprobe analyses gave an average Ge content of 340 ppm. As discussed in Appendix 1, the statistical uncertainty in this determination is around ±35 ppm. The electron microprobe values reported here may thus be systematically low, probably because of the high deadtimes on the pure Ge standard. Germanium values are nevertheless reported relative to pure Ge metal. Landes was not used as the primary standard, since the uncertainty in composition and the low concentration of Ge introduces an analytical error

TABLE 1. Total counts on pure Fe, Ni, and Landes secondary standard.

	Counts at		calc. BKG	Net counts	Ge (ppm)
	Ge K_α	Ge K_α ±0.045			
Fe (99.9%)	879769	872240	(Background standards)		
Ni (99.9%)	996124	955320			
Landes	885326	863400	871003	14323	340
uncertainty:					
min.:	$\sqrt{885326 + 863400}$		=	1322	31
max.:	$\sqrt{885326 + 863400 + 879769 + 862240}$		=	1871	44
detection limit:					
min.:	$3 \cdot \sqrt{863400}$		=	2788	66
max.:	$3 \cdot \sqrt{863400 + 879769 + 872240}$		=	4852	115

Typical results represent replicate measurements. Counting time 300 sec. For Fe and Ni, which are free of Ge, the difference between columns 1 and 2 is due to nonlinear background. The calculation of analytical uncertainty and detection limit is explained in Appendix 1.

comparable to that due to the deadtime problem, while much longer standard counting times would have been required.

Although the absolute values may be subject to this systematic error, the element trends reported here are well established, since grain-to-grain comparisons can be made with much better reliability. Table 1 shows the counts obtained at the position of the Ge K_α line and on either side of it for pure Fe and Ni standards and for Landes. As discussed in Appendix 1, estimates of statistical uncertainty and detection limit can be derived from these measurements. These values are expected to be closer to the minimum rater than the maximum bounds quoted in Table 1, and a typical uncertainty of ± 35 ppm and a detection limit of 75 ppm were adopted as reasonable estimates.

3. RESULTS

Most metal in Allan Hills A77081 is kamacite. Taenite only occurs as narrow lamellae within kamacite and rims around some two-phase grains. Large kamacite grains (≥50 μm) are very uniform in composition. Their composition is given in Table 2, together with the bulk metal composition determined by *Schultz et al.* (1982) from NAA of a metal separate.

TABLE 2. Composition of coarse-grained kamacite and bulk metal in Allan Hills A77081.

	Kamacite*	Bulk Metal†
Co (%)	0.45	0.423 ± 0.009
Ni (%)	7.0 ± 0.1	7.71 ± 0.23
Ge (ppm)	230 ± 40	330 ± 70

*Electron microprobe analysis (this work).
†Bulk metal. Neutron activation analysis of metal separate (*Schultz et al.*, 1982).

Relative to the NAA data, microprobe analyses gave lower Ni and lower Ge. The lower Ni is probably due to the inclusion of taenite in the bulk value, although from the small amount of taenite observed the difference is slightly higher than expected. Our lower Ge value may reflect a systematic error due to the deadtime problem already discussed, but the error bars of the two Ge determinations overlap, and the difference may thus be insignificant. A single taenite grain gave a Ge content of 280 ppm, indicating that the Ge content of taenite is probably not significantly higher than that of kamacite. In iron meteorites *Goldstein* (1967) found that kamacite and bulk taenite/plessite have comparable Ge contents, except for the narrow clear taenite (tetrataenite) zone, in which Ge is enriched by up to a factor of 2.

In addition to the homogeneous large metal grains, A77081 contains another metal population that is lower in Ni and Ge and higher in Co. These metal grains are small and without exception associated with much larger troilite grains.

Figure 1 shows kamacite compositions as Ge versus Ni and Co versus Ni plots. Figure 2 is a photomicrograph of an area of A77081 containing both large kamacite with normal composition and some metal-troilite assemblages with small, Ni-poor kamacite.

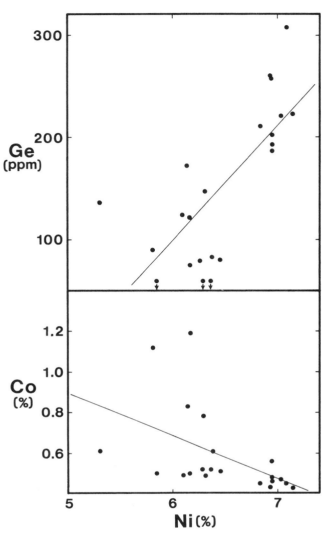

Fig. 1. Diagrams of Ge versus Ni and Co versus Ni for metal grains in Allan Hills A77081.

Table 3 gives the compositions of all grains analyzed in this study together with a size parameter. This is the diameter for equant grains, or the geometric mean of two perpendicular dimensions for irregular shaped ones. All but two grains are associated with troilite, whose dimensions are also given. The two exceptions are large grains, which may have been in contact with troilite outside the plane of the thin section.

The scatter of replicate analyses for Co and Ge in grains no. 8, 11, and 73 is within analytical uncertainty (see Appendix 1), except for Co in no. 8. Omitting the exceptionally high Co values of grain no. 14 and one of the two analyses of no. 8 significantly improves the Co-Ni anticorrelation. However, since no other reason could be found to reject these analyses, they were included in the least squares fit shown in Fig. 1.

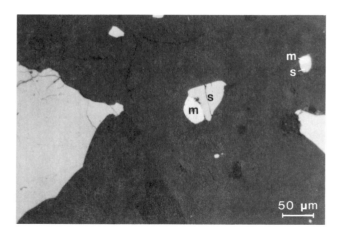

Fig. 2. Textures of metal grains in Allan Hills A77081. The large metal grain on the left is typical of average kamacite (Table 2). Two small grains consisting of metal (m) and sulfide (s) are visible in the center (Ni 6.28%, Co 0.52%, Ge 79 ppm) and in the upper right.

TABLE 3. Contents of Ni, Co, and Ge in kamacite grains from the Allan Hills A77081 winonaite.

Grain No.	Ni (%)	Co (%)	Ge (ppm)	Size (μm) Metal	Size (μm) Sulfide
1	6.14	0.83	172	18	20
6	5.85	0.50	<75	19	29
8	5.31	0.61	136	13	71
	5.81	1.12	90		
9	6.31	0.49	147	8	12
10	6.16	0.50	122	30	37
11*	6.95	0.46	187	212	67
	7.14	0.43	223		
	6.93	0.43	260		
13*	6.95	0.48	192	160	38
14	6.17	1.19	75	14	23
16*	6.83	0.45	211	273	139
17	6.10	0.49	124	23	46
27	6.36	0.52	<75	38	43
30	6.95	0.46	202	41	62
31	6.45	0.51	80	29	135
39	6.29	0.78	<75	44	32
41	6.38	0.61	83	46	74
71	6.28	0.52	79	17	24
72*	6.94	0.56	257	218	absent
73*	7.08	0.45	307	225	absent
	7.03	0.47	221		
Landes (secondary standard)			377		
			303		
			340		
			359		
			340		

*Grains averaged in column one of Table 2.

4. PARTIAL MELTING MODEL

Partition coefficients of Ni and Ge (k_{Ni} and k_{Ge}, as defined in Appendix 2) in the Fe-Ni-S system have been determined experimentally by *Willis and Goldstein* (1982) and *Jones and Drake* (1983). *Narayan and Goldstein* (1982) determined Ge partitioning in the Fe-Ni-P system, and *Sellamuthu and Goldstein* (1984) discussed the combined effects of S and P.

For pure metal, both k_{Ni} and k_{Ge} are somewhat less than unity, but increase with S content of the melt. For a eutectic Fe-S melt, k_{Ni} is around 2. A single determination of k_{Ge} in a Fe-S eutectic (*Jones and Drake*, 1983) gave $k_{Ge} = 300 \pm 100$, although for S contents a few percent lower than the eutectic k_{Ge} is probably only in the teens (*Willis and Goldstein*, 1982; *Sellamuthu and Goldstein*, 1984).

Ideal fractional crystallization presumes perfect mixing of the liquid and no diffusion in the solid. I have argued previously (*Kracher*, 1982) that equilibrium partition coefficients greater than about 20 lead to such strong concentration gradients that these ideal conditions can no longer be fulfilled. Since deviations of real from ideal systems always tend to bring k closer to unity than expected from equilibrium partitioning, the value of k_{Ge} during partial melting of a metal-sulfide assemblage may never be any higher than 20 to 30. Since k_{Ni} is close to unity, it is expected to be close to the experimentally determined value.

If an assemblage such as the primitive precursor of A77081 is heated to about 1250 K, sulfide and some metal will melt to form an S-rich melt. Although strictly speaking melting occurs along the Fe-Ni-S cotectic, the melting interval is very small (probably <5°), and it is justified to treat melting as quasi-eutectic. Hence it is very likely that sulfide will melt completely at a certain temperature. Some of this melt may become mobilized, and migrate in the parent body over distances larger than the size of a meteorite. Therefore individual winonaites may have lost or gained S-rich melt relative to their precursor rocks. Since all unweathered winonaites contain at least some sulfide, melt migration in their parent material was probably restricted.

A rock formed in this way contains two kinds of metal: residual metal that was never molten, and secondary metal, which is the metal complement of the S-rich melt. The concentration C of a minor or trace element in these two kinds of metal is given by the following two equations, whose derivation is given in Appendix 2

$$C_R/C_M = k \cdot q \cdot (1 - \alpha)/[1 - \alpha \cdot k + k \cdot q \cdot (1 - \alpha)] \quad (1)$$

and

$$C_S = C_R/\alpha \cdot k \quad (2)$$

In these equations, q is the metal/sulfide weight ratio in the original assemblage prior to melting and α is the fraction of metal in the melt, i.e., if S-rich melt cools, a weight fraction α will crystallize to metal, and a fraction $(1 - \alpha)$ to sulfide. For the Fe-S eutectic, the value of α is about 0.17.

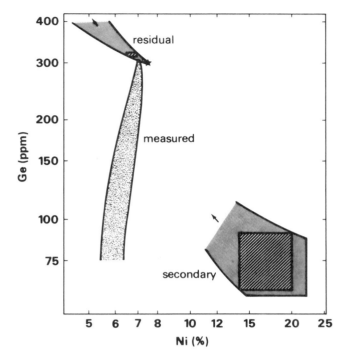

Fig. 3. Correlation of Ge versus Ni expected from the partial melting model. Relative to the metal composition prior to melting (asterisk), eutectic melting leads to residual metal lower in Ni and higher in Ge, and secondary metal higher in Ni and lower in Ge. For A77081's metal-sulfide weight ratio of q = 2, the average compositions are expected to fall inside the shaded boxes. Measured compositions (from Table 3 and Fig. 1) are shown in outline. They define a trend whose slope is opposite to that of a line connecting the predicted residual and secondary metal compositions. The field of measured compositions is truncated at the estimated detection limit of 75 ppm Ge.

It can be seen that if $k<1/\alpha$ (as for Ni), $C_R<C_S$, whereas if $k>1/\alpha$ (as for Ge), $C_R>C_S$. Thus residual metal should have less Ni and more Ge than primitive metal, while the composition of secondary metal must, of course, be complementary.

The fractionation of Ni and Ge observed in A77081 metal grains is not consistent with a simple partial melting model. Figure 3 shows the relationship between residual and secondary metal calculated on the basis of equations (1) and (2) and the range of metal compositions determined in this study.

The assumed composition of original metal, marked by an aterisk in Fig. 3, is $C_M(Ni) = 7.5\%$ and $C_M(Ge) = 300$ ppm. If some S-rich liquid was lost or gained during the high-temperature episode, this composition cannot be uniquely constrained. It can be shown, however, that minor redistribution of S-rich liquid has only a very small effect on bulk metal composition. The above estimate was derived from the composition of coarse kamacite by adjusting the Ni content slightly upward to allow for the presence of taenite. It is also reasonably close to the bulk metal composition determined by *Schultz et al.* (1982).

The envelope of solid lines in Fig. 3 brackets differentiation paths for the range $2 \leq k_{Ni} \leq 3$ and $20 \leq k_{Ge} \leq 30$. For decreasing metal/sulfide ratio q, compositions shift in the direction of the small arrows. The heavily shaded boxes outline the range of possible compositions for q = 2, the approximate ratio now observed in A77081 (*Schultz et al.*, 1982).

The predicted metal compositions in a real system depend on the size scale over which metal and melt equilibrated. Two extreme cases can be distinguished. Either equilibration was on a scale much larger than the size of meteorites. In this case there is a single value of q, and residual and secondary metal compositions should cluster somewhere within the boxes in Fig. 3. At the other extreme, equilibration may only have occurred on the scale of individual grains or small pockets of eutectic melt. In this case there would be a range of q values, and metal grains should form a pair of trends falling within the curved fields in Fig. 3. Neither is the case for A77081 kamacite.

Although this calculation is a gross simplification, the basic features associated with eutectic melting do not change: As long as $\alpha k_{Ni}<1$, which follows necessarily from the Fe-Ni-S phase diagram (*Kullerud et al.*, 1969), and $\alpha k_{Ge}>1$, well established by experiment (*Willis and Goldstein*, 1982; *Jones and Drake*, 1983; *Sellamuthu and Goldstein*, 1984), residual and secondary metal are related by a line of negative slope on a Ge versus Ni diagram.

Even if the measured partition coefficients were drastically in error, or inapplicable in the case of winonaites, this would not eliminate the problem. Since the partial differentiation model tries to explain the compositional trends of Ni and Ge in iron meteorite groups IAB and IIICD by cotectic crystallization of metal and sulfide, these elements were expected to behave similarly on a microscale in winonaites as they do on a macroscale within the whole group IAB. This argument is independent of any experimentally measured partition coefficients. Nickel and Ge are negatively correlated in IAB, but positively correlated among the small metal grains in A77081.

This situation presents a problem to our understanding of the origin of winonaites. Both models relating winonaites and IAB irons, the impact melt model (*Wasson et al.*, 1980) and the partial differentiation model (*Kracher*, 1985), do not predict the observed distribution of Ni and Ge. On the other hand, the observed textures, particularly the troilite/metal ratio associated with the low-Ge metal, and its small size suggest that eutectic melting was involved in the formation of this assemblage.

While it is possible to invent ad hoc explanations that would reconcile either model with the data, such explanations are not likely to be testable and hence are not useful as a guide for further experiments. Investigation of the low-Ni, low-Ge metal with a trace element technique such as PIXE, SIMS, or synchrotron XRF might make it possible to determine additional siderophile elements and shed more light on the fractionation mechanism. However, these techniques have not yet attained the required spatial resolution to study the small, Ni- and Ge-poor grains found in A77081.

5. SUMMARY

The unexpected distribution of Ge in A77081 makes it impossible to use its fractionation behavior as a test for the partial differentiation model. Nevertheless, some significant conclusions can be drawn from this study.

1. A77081 contains small kamacite grains lower in Ni and Ge and higher in Co than coarse-grained metal. They are associated with sulfide, the latter being usually more abundant than metal in an individual assemblage (Table 3). This suggests that anomalous metal is related to eutectic melting.
2. The positive Ni-Ge correlation is the opposite of the fractionation trend expected on the basis of known partition coefficients between S-saturated liquid and metal. No explanation for this discrepancy has yet been found.
3. The distribution of other siderophiles in winonaite metal may provide further clues, but all other trace metals, with the possiblee exception of Cu, are below the detection limit of electron microprobe analysis. More sensitive microanalytical techniques do not yet have the spatial resolution required to analyze the small, Ge-poor metal grains in A77081.
4. The existence of anomalous metal apparently derived from a eutectic melt is an indication that A77081, and probably other winonaites as well, have undergone some fractionation and are not primitive starting material from which IAB irons formed by fractionation. The small size of A77081 (8.56 g, 2 cm) even makes it possible that it is an isolated silicate inclusion.
5. Neither the impact melt model nor the partial differentiation model can account for the composition of the anomalous metal, but neither model is invalidated by it.

APPENDIX 1

Error Analysis of Ge Determination

The uncertainty of each Ge analysis due to counting statistics has two components, peak and background uncertainty. The uncertainty in the peak measurement is simply the square root of the counts in the peak (885326, using the numerical examples from Table 1). The background under the peak cannot be measured directly, but is usually determined by a linear interpolation of measurements taken on either side of the peak. In our runs these measurements were taken at ± 0.045 Å from the peak for a total time equal to the time of the peak measurement. It is usually assumed that the uncertainty in the background determination is the square root of the counts acquired on both sides of the peak ($\sqrt{863400}$), and the resulting total uncertainty is shown in Table 1 under "min."

The underlying assumption that the background is a linear function of wavelength is not strictly correct. In fact, the shape of the background is a complicated function depending on energy, Bragg angle, crystal efficiency, and, to a lesser extent, secondary radiation within the instrument. The composition (mean atomic number) of the target material influences the intensity, but not the shape of the background.

For a high-sensitivity determination the nonlinear shape of the continuum has to be taken into account. This can be done by using a background standard, in this study pure Fe. As the first line of Table 1 shows, counts at the position of the Ge K_α line (879769) are higher by a factor of about 1.009 than the linear interpolation of counts taken on either side (872240). From theoretical considerations, the ratio of these two numbers should be independent of sample composition, and hence for the Landes "sample" (secondary standard) in Table 1 the true background is about $863400 \cdot f = 871003$ counts.

A finite uncertainty is associated with the experimentally determined value of f. If the background were determined only by a single measurement, the uncertainty in f from the laws of error propagation would be $\pm f \cdot \sqrt{1/879769 + 1/872240}$. Approximate formulae for uncertainty and detection limit for this fictitious worst case are given in the lines labeled "max" in Table 1, and are higher by a factor of about $\sqrt{2}$ (uncertainty and $\sqrt{3}$ (detection limit) than the minimal values.

In reality, the uncertainty due to counting statistics lies between the extremes given by the minimal and maximal values in Table 1. It is larger than the low limit, because the shape of the background is not known precisely, but it is smaller than the high limit for two reasons. First, the determination of f depends on several measurements, not just one, reducing the statistical error. Second, on the background standard the measurements at the Ge K_α line and on either side of it, respectively, are not independent of each other, as implicitly assumed in an error propagation calculation, but constrained by the fact that the background is a smooth, nearly linear function of wavelength.

A less tractable source of error in long-time counts is beam current instability. The magnitude of this error is unknown, but a reasonable assessment of the overall reproducibility of our Ge determinations can be made from repeated measurements of the same material. Nonconsecutive replicate measurements of the Landes secondary standards gave a standard deviation of ± 28 ppm ($\sigma = 8\%$, $N = 5$). Analyses of the five largest grains, shown in Table 1, gave 230 ± 40 ppm ($\sigma = 17\%$, $N = 8$). This suggests that the overall analytical uncertainty due to all effects is probably closer to the minimal rather than the maximal value in Table 1. Since the detection limit is directly connected with the analytical uncertainty, we infer that it, too, is close to the minimal value in Table 1. The values of ± 35 ppm for uncertainty and 75 ppm for detection limit adopted in this study are "best estimates" based on this assessment.

APPENDIX 2

Derivation of Fractionation Equations

Definition of Symbols

Phases (subscripts):

M = original metal, prior to any melting event
T = solid sulfide (troilite), both before and after negligible trace element concentrations

L = liquid (metallic, but S-rich)
R = residual metal
S = secondary metal

Ratios:

k = effective partition coefficient, C_{solid}/C_{liquid}
q = weight ratio of metal to sulfide prior to melting
α = weight fraction of S-bearing metallic melt that crystallizes as metal; the complementary fraction $(1 - \alpha)$ crystallizes as troilite

For convenience, we will normalize all masses to troilite, so that prior to melting there is 1 weight unit troilite (T) and q weight units original metal (M). At the eutectic point, troilite and $\alpha/(1 - \alpha)$ weight units of metal melt to form $1/(1 - \alpha)$ weight units liquid (L), and $1 - \alpha/(1 - \alpha)$ weight units residual metal (R). A minor or trace element will partition between solid (residual metal) and liquid so that

$$k = C_R/C_L \quad (3)$$

Also, for mass balance reasons,

$$q \cdot C_M + C_T = [q - \alpha/(1 - \alpha)] \cdot C_R + 1/(1 - \alpha) \cdot C_L \quad (4)$$

For the highly siderophile elements considered in this paper, we can set $C_T = 0$, and obtain

$$C_R/C_M = k \cdot q \cdot (1 - \alpha)/[1 - \alpha \cdot k + k \cdot q \cdot (1 - \alpha 1\,)] \quad (1)$$

The liquid formed in this melting event crystallizes to secondary metal and troilite. The mass balance equation for this reaction is

$$C_L = \alpha \cdot C_S + (1 - \alpha) \cdot C_T \quad (5)$$

Again using $k = C_R/C_L$ and $C_T = 0$, we get

$$C_S = C_R/\alpha \cdot k \quad (2)$$

These equations are given incorrectly in *Kracher* (1985); the single equation on page C693 of that paper should be replaced, in this order, by the two equations (3) and (2) above.

Acknowledgments. I am grateful to Milton L. Pierson for technical help. Thorough reviews by J. H. Jones and an anonymous reviewer led to significant improvements of the manuscript. This study was financially supported by NASA grant NAG9-112.

REFERENCES

Bild R. W. (1977) Silicate inclusions in group IAB irons and a relation to the anomalous stones Winona and Mt. Morris (Wis). *Geochim Cosmochim. Acta, 41,* 1439-1456.

Bunch T. E., Keil K., and Olsen E. J. (1970) Mineralogy and petrology of silicate inclusions in iron meteorites. *Contrib. Mineral. Petrol., 25,* 297-340.

Clayton R. N., Mayeda T. K., Olsen E. J., and Prinz M. (1983) Oxygen isotope groups in iron meteorites (abstract). In *Lunar and Planetary Science XIV,* pp. 124-125. Lunar and Planetary Institute, Houston.

Goldstein J. I. (1967) Distribution of germanium in the metallic phases of some iron meteorites. *J. Geophys. Res., 72,* 4689-4696.

Graham A. L., Easton A. J., and Hutchison R. (1977) Forsterite chondrites: the meteorites Kakangari, Mount Morris (Wisconsin), Ponlyfni, and Winona. *Mineral. Mag., 41,* 201-210.

Jones J. H. and Drake M. J. (1983) Experimental investigation of trace element fractionations in iron meteorites, II: the influence of sulfur. *Geochim. Cosmochim. Acta, 47,* 1199-1209.

Kracher A. (1982) Crystallization of a S-saturated Fe,Ni-melt and the origin of iron meteorite groups IAB and IIICD. *Geophys. Res. Lett., 9,* 412-415.

Kracher A. (1985) The evolution of partially differentiated planetesimals: evidence from iron meteorite groups IAB and IIICD. *Proc. Lunar Planet. Sci. Conf. 15th,* in *J. Geophys. Res., 90,* C689-C698.

Kracher A. and Pierson M. L. (1986) Elements Cu through Ge: cosmochemical significance and microanalysis in meteorites. In *Microbeam Analysis 1986* (A. D. Romig, Jr. and W. F. Chambers, eds.), pp. 151-152. San Francisco Press.

Kullerud G., Yund R. A., and Moh G. H. (1969) Phase relations in the Cu-Fe-S, Cu-Ni-S, and Fe-Ni-S sytems. In *Magmatic Ore Deposits* (H. D. B. Wilson, ed.), pp. 3223-343. *Econ. Geol. Monograph No. 4.* Narayan C. and Goldstein J. I. (1982) A dendritic solidification model to explain Ge-Ni variations in iron meteorite chemical groups. *Geochim. Cosmochim. Acta, 46,* 259-268.

In *Lunar and Planetary Scienc XI,* pp. 902-904. Lunar and Planetary Institute, Houston.

Prinz M., Nehru C. E., Delaney J. S., and Weisberg M. (1983) Silicates in IAB and IIICD irons, winonaites, lodranites, and Brachina: a primitive and a modified-primitive group (abstract). In *Lunar and Planetary Science XIV,* pp. 616-617. Lunar and Planetary Institute, Houston.

Schultz L., Palme H., Spettel B., Weber H. W., Wänke H., Christophe Michel-Levy M., and Lorin J. C. (1982) Allan Hills 77081—an unusual stony meteorite. *Earth Planet. Sci. Lett., 61,* 23-31.

Sellamuthu R. and Goldstein J. I. (1984) Analysis of segregation trends observed in iron meteorites using measured distribution coefficients (abstract). In *Lunar and Planetary Science XV,* pp. 746-747. Lunar and Planetary Institute, Houston.

Wasson J. T. (1974) *Meteorites.* 316 pp. Springer-Verlag, New York.

Wasson J. T., Willis J., Wai C. M., and Kracher A. (1980) Origin of iron meteorite groups IAB and IIICD. *Z. Naturforsch., 35a,* 781-795.

Willis J. and Goldstein J. I. (1982) The effects of C, P, and S on trace element partitioning during solidification of the Fe-Ni system. *Proc. Lunar Planet. Sci. Conf. 13th,* in *J. Geophys. Res., 87,* A435-A445.

Formation of the Lamellar Structure in Group IA and IIID Iron Meteorites

J. A. Kowalik*, D. B. Williams, and J. I. Goldstein

Department of Materials Science and Engineering, Lehigh University, Bethlehem, PA 18015

*Now with Digital Equipment Corporation, Northboro, MA

The Widmanstätten structure occurs in virtually all iron meteorites. The region between the Widmanstätten plates is the decomposed taenite, or plessite. Depending on the cooling cycle and composition of the meteorites, the plessite region decomposes into various morphologies. The lamellar plessite structure typically found in IA and IIID iron meteorites is of interest in this study. The lamellar morphology exists on two quite different scales, <1 μm and >3 μm as found in Group IA and IIID irons, respectively. This lamellar structure was characterized using analytical electron microscopy, light microscopy, and electron microprobe analysis. The α lamellae in Dayton (IIID) contained a compositional gradient from 6.1 ± 0.7 wt % Ni at the center of the α lamellae to 3.6 ± 0.5 wt % Ni at the α/γ interface. At this interface, the γ lamellae has a composition of 50 wt % Ni. The α and γ lamellae in Group IA irons had similar compositions, ~4 wt % Ni in α and ~48 wt % in γ. The orientation relationship at the α/γ interface in the lamellar regions of both Group IA and IIID has been determined by convergent beam electron diffraction and it conforms to the Kurdjumov-Sachs orientation relationship. This orientation is in contrast to the Nishiyama-Wassermann relationship that was found at the Widmanstätten pattern α/γ interface. The phase transformations responsible for the lamellar structure in the IA and IIID chemical groups were examined. The IIID Dayton meteorite underwent a discontinuous transformation $\gamma_{ss} \rightarrow \alpha + \gamma$ between 400°-450°C. The phase transformation that produces the lamellar structure in the IA meteorites is not well established. Both a eutectoid reaction and a discontinuous reaction are feasible. Not until the role of carbon is understood or the existence of a high angle interface is observed will the identification of the phase transformation in IA irons become conclusive.

INTRODUCTION

Iron meteorites are composed primarily of Fe and Ni with small amounts of carbon and phosphorus and have cooled slowly in their parent asteroidal bodies. A Widmanstätten structure arises during cooling within an asteroidal body when kamacite (α) precipitates on the {111} planes of the prior taenite (γ). Upon cooling to low temperatures, the residual taenite regions between the kamacite decompose into various combinations of α and γ, usually termed plessite structures. The structure of the various plessite morphologies depends on the thermal history and composition of the meteorite. A unique lamellar plessite structure is found in carbon-containing Group I irons and one Group IIID iron (Dayton). The purpose of this research is to characterize the lamellar structure and to determine the phase transformations responsible for its formation.

Two forms of lamellar plessite (types A and B) exist in iron meteorites. Type A is a fine lamellar structure <1 μm wide, found in the high-Ni residual taenite of most Group IA meteorites. Representative meteorites from Group IA investigated in this study are: (A) Canyon Diablo, 7.1 wt % Ni, 0.26 wt % P; (B) Odessa, 7.35 wt % Ni, 0.25 wt % P; and (C) Toluca, 8.4 wt % Ni, 0.16 wt % P (*Buchwald*, 1975). Type B is a coarse lamellar structure >3 μm wide, found in the taenite matrix of the IIID Dayton meteorite. Several portions of this meteorite, containing 17.6 wt % Ni and 0.4 wt % P (*Buchwald*, 1975), were investigated. Groups IA and IIID contain very different nickel and phosphorus compositions. The carbon levels of Group IA are generally 0.2 to 2.0 wt % (*Buchwald*, 1975) while the carbon level in the IIID Dayton is 0.05 wt % (*Moore et al.*, 1969).

EXPERIMENTAL PROCEDURE

Meteorite Sample Preparation

Meteorite samples for light microscopy and scanning electron microscopy were prepared using standard metallographic polishing techniques followed by etching in 2% Nital for 30 seconds. Samples for electron microprobe analysis (EMPA) were also prepared by standard metallographic techniques, but the samples were not etched. Scanning electron microscopy (SEM) samples were coated with a thin layer of carbon to reduce charging effects.

Meteorite samples to be examined using the transmission electron microscope (TEM) were first observed in the light microscope in order to choose an area of interest. Samples were then electro-discharge-machined into rods 3 mm in diameter. These rods were then cut with a diamond saw to discs of approximately 15 mils thickness. Final thickness reduction of the discs was obtained by grinding to 3 mils on 600 grit abrasive paper. The sample was then electropolished at +105V using a 2% perchloric/98% ethanol polishing solution at a temperature of -35°C.

Metallographic and Chemical Analysis

To understand the phase transformation responsible for forming the lamellar morphology, the structures of Types A and B were characterized using light, scanning, and transmission electron microscopy. Individual phase compositions were determined by electron probe microanalysis (EPMA) and by X-ray analysis in the analytical electron microscope (AEM) using the Cliff-Lorimer method (*Cliff and Lorimer*, 1975). The chemical analysis of the lamellar structure was interpreted using the Fe-Ni (*Reuter et al.,* 1987a), Fe-Ni-P (*Romig and Goldstein,* 1980), and Fe-Ni-C (*Romig and Goldstein,* 1978) phase diagrams. Convergent beam electron diffraction (CBED) was employed to determine the orientation relationships of kamacite and taenite at the kamacite/taenite interfaces.

RESULTS

Dayton—Group IIID

The structure of Dayton varies extensively over an area of a few centimeters. As seen in Fig. 1, the lamellar morphology (L) is located around large phosphides (P, schreibersite) with residual Widmanstätten plates (W) running through it. The Widmanstätten structure exhibits large variations in size and is located in two regions. One region surrounds the phosphide (schreibersite) and contains a few Widmanstätten plates, with large length to width ratios. These large phosphides, which precipitated at high temperatures (~800°C), contain approximately 15.5 wt % P and 27 wt % Ni and are surrounded by a region of swathing kamacite. No cohenite inclusions were observed. A martensitic plessite structure is also found near the schreibersite (M, Fig. 1). The second region is located approximately 1 cm away from the phosphide and contains the classical Widmanstätten structure.

Figure 2 shows the overlap of the two regions described above at higher magnification. The Widmanstätten pattern (W, Fig. 2) region contains plessite in a lamellar morphology (L, Fig. 2). The lamellar structure in Dayton consists of parallel plates of alternating kamacite and taenite phases. The taenite between the kamacite plates is either one-phase taenite consisting of clear taenite 1 (Fig. 3) or multiphase taenite consisting of a "cloudy zone" sandwiched between two layers of clear taenite 1 (Fig. 4). The detailed structure of clear taenite 1 and the cloudy zone as shown by electron microscopy in Dayton is given by *Reuter et al.* (1987b).

Figure 5 shows the variation of Ni across two taenite lamellae using AEM. The compositional analyses of the kamacite and taenite at the kamacite/taenite, $\alpha+\gamma$, boundary of the lamellar regions are the same as those measured at the kamacite taenite boundary in the Widmanstätten pattern of the Dayton meteorite (*Reuter et al.,* 1987b). Apparently, after nucleation, both structures, lamellar plessite and the Widmanstätten pattern, formed similar microstructures while cooling to the same temperature. *Buchwald* (1975) argues on metallographic evidence that the two structures formed competitively at about the same temperature.

The microscopic orientation relationship between the kamacite and clear taenite 1 phases in the Widmanstätten morphology and in the lamellar morphology was examined. Convergent beam diffraction (CBED) patterns obtained from α and γ regions of the Widmanstätten structure by traversing the sample under the beam show that the $[\bar{1}10]_\alpha$ and the $[\bar{2}\bar{1}1]_\gamma$ are parallel within very small errors (< 1°) (Figs. 6a,b). From Fig. 6 it is evident that the (002) diffraction vector in the α phase, g 002_α, is parallel to g $0\bar{2}2_\gamma$. From the CBED patterns the orientation relation is $[\bar{1}10]_\alpha//[\bar{2}\bar{1}1]_\gamma$, $(002)_\alpha//$

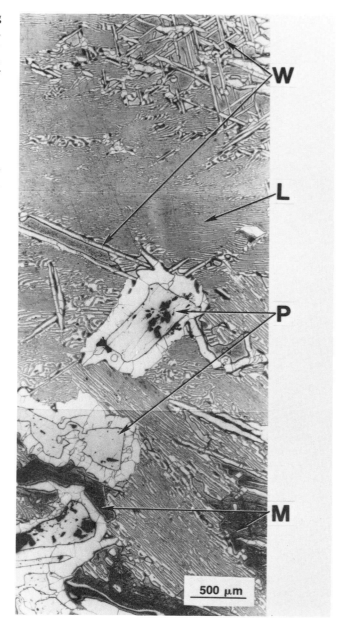

Fig. 1. Light micrograph of the varied structure found in Dayton meteorite showing a large phosphide (P), the lamellar plessite (L), the Widmanstätten pattern (W), and the martensitic plessite (M).

$(0\bar{2}2)_\gamma$, which is a Nishyama-Wassermann orientation relationship. CBED patterns were also obtained for the α/γ interface in the lamellar structure. Figures 7a and 7b show CBED patterns from the $[\bar{2}11]_\alpha$ and the $[\bar{1}2\bar{1}]_\gamma$, indicating that the directions are parallel. From Fig. 7 it is clear that g(11$\bar{1}$)$_\alpha$ is parallel to g(10$\bar{1}$)$_\gamma$. Therefore, from CBED the orientation relationship is $[\bar{2}11]_\alpha // [\bar{1}2\bar{1}]_\gamma$, $(11\bar{1})_\alpha // (10\bar{1})_\gamma$, which is Kurdjumov-Sachs (K-S). Clearly, the orientation relationships of kamaciate and clear taenite 1 phases in the Widmanstätten structure and lamellar plessite are different.

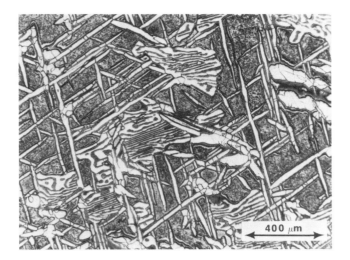

Fig. 2. Light micrograph exhibiting the Widmanstätten and lamellar morphology.

Canyon Diablo, Toluca, and Odessa—Group IA

The Group IA meteorites range from 6.5% to 8.5% wt % Ni, 0.2% to 2 wt % C, and 0.12 to 0.39 wt % P (*Buchwald*, 1975). Because of the lower bulk Ni values, the Widmanstätten α in Group I forms at higher temperatures and has thicker Widmanstätten plates than the higher Ni meteorite Dayton. High Ni content in the residual taenite in Group I irons are caused by the impingement of Ni gradients due to Widmanstätten pattern growth.

The morphology of the plessite region varies extensively in the Group I meteorites. Figure 8 shows several types of lamellar structures that form in the residual taenite of the Group I meteorites. Figure 9 shows a plessite field that is partially cloudy zone and partially lamellar. The lamellar plessite in these micrographs appears to form at the clear taenite 1 boundary.

The lamellar plessite in Group I irons occurs in areas of high Ni content. Electron probe microanalysis indicates that the lamellar plessite forms in residual taenite regions with Ni contents between 25 and 38 wt %. For example, the lamellar structure intertwined with the cloudy zone has a bulk Ni content of ~38 wt % Ni (Fig. 9). Other plessite areas such as acicular plessite fields had lower bulk Ni contents (~19 wt % Ni).

The AEM was used to determine the composition of the $<1\mu m$ lamellar plessite. The lamellae contain a high Ni FCC phase of ~48% Ni and a low Ni BCC phase ~4 wt % Ni. The orientation relationship found at the kamacite/taenite interface of the lamellar structure is Kurdjumov-Sachs. This is the same relationship that was found at the kamacite/clear taenite 1 interface in the lamellar structure of the Dayton meteorite, and also the same relationship observed between the low and high Ni phases in the cloudy zone (*Lin et al.*, 1977).

DISCUSSION

Dayton

Microstructure of Dayton. The phase transformations that occur upon cooling of the Dayton meteorite can be understood by correlating the observed microstructure with the Fe-Ni-P phase diagram (*Doan and Goldstein*, 1970; *Romig and Goldstein*, 1980).

The Dayton meteorite of bulk composition 17.6 wt % Ni and 0.4 wt % P was initially a single crystal of γ above 800°C and transformed to γ + Ph below 800°C. Figure 10 shows a schematic drawing of the P gradient developed as cooling proceeded below 800°C. The length of the P concentration gradient in γ is determined by the basic diffusion equation $x = 2\sqrt{Dt}$, where x is the distance of the P gradient, D is the diffusion coefficient of phosphorous in γ (*Heyward and Goldstein*, 1973), and t is time (~0.5 m.y.). At 705°C, the P concentration gradient is calculated as occurring over ~14

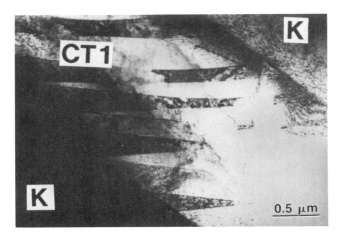

Fig. 3. Transmission electron micrograph of the one-phase taenite lamellae showing kamacite (K) and clear taenite 1 (CT1).

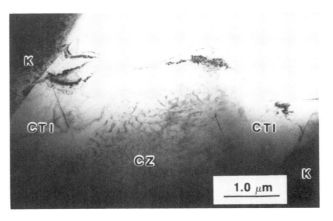

Fig. 4. Transmission electron micrograph of the three-phase taenite lamellae showing kamacite (K), clear taenite (CT1), and the cloudy zone (CZ).

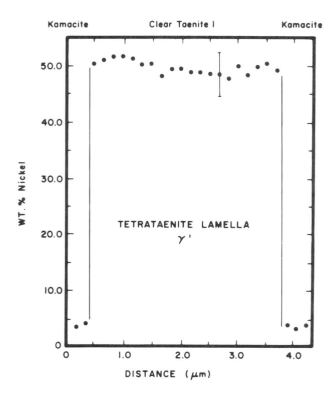

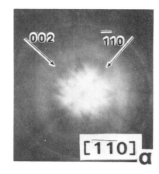 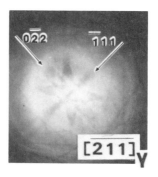

Fig. 6. Widmanstätten α/γ interface analysis in Dayton: **(a)** CBED pattern of low-Ni α $[\bar{1}\bar{1}0]$, **(b)** CBED pattern of high-Ni γ $[\bar{2}\bar{1}\bar{1}]$.

cm. Another large phosphide would most likely nucleate within this P gradient, causing the profiles to overlap and depleting the matrix to the equilibrium phosphorous content between the schreibersite precipitates at 750°C (solid line, Fig. 10).

As the meteorite cooled from 750°C, the schreibersite expanded in size and depleted the matrix of P. The solubility of P and the diffusion distance in γ decreased. The P gradient forming from 750° to 576°C across the γ from the schreibersite is represented in Fig. 10 with a series of hatched lines. Two γ regions appear to form: region A, which is relatively constant in phosphorus at ~0.2 to 0.3 wt %, and region B, which is close to the phosphide and is depleted in phosphorus by the continued growth of this schreibersite. The presence of such a P gradient in region B is also supported through microstructural interpretation (Fig. 1), which shows no α as well as large martensitic plessite fields. These fields form at low temperatures around the schreibersite. The martensite, α_2, forms when γ of low P content crosses the martensite start temperature, M_s, at ~250°C for the Dayton meteorite.

At 575°C, region A, containing 17.6 wt % Ni and ~0.2 wt % P, entered the $\alpha + \gamma + $ Ph phase field (Fig. 11a) while region B, 17.6 wt % Ni and < 0.2 wt % P (Fig. 11a), entered the $\alpha + \gamma$ phase field. The composition in both regions indicates that α is present. Small phosphides precipitated in region A at and below 575°C and helped nucleate Widmanstätten α (*Narayan and Goldstein*, 1984). At 575°C, the large schreibersite crystal would act as a nucleation site in region

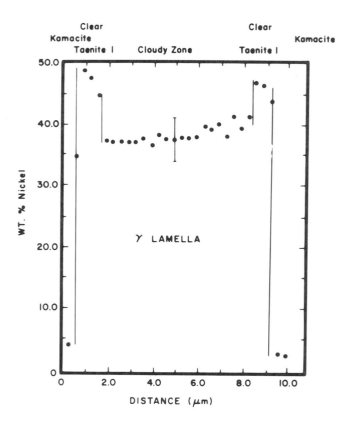

Fig. 5. AEM Ni concentration profile across two taenite lamellae.

 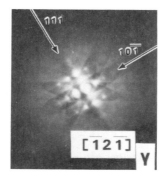

Fig. 7. Lamellar α/γ interface analysis in Dayton: **(a)** CBED pattern of low-Ni α $[\bar{2}\bar{1}\bar{1}]$, **(b)** CBED pattern of high-Ni γ $[\bar{1}\bar{2}\bar{1}]$.

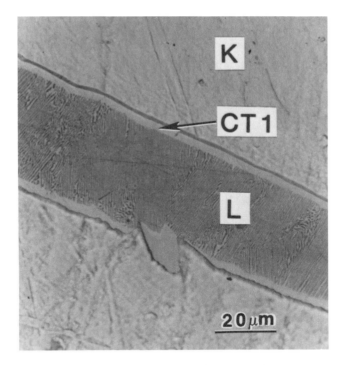

Fig. 8a. Light micrograph of a plessite field in the Toluca meteorite showing kamacite (K), clear taenite 1 (CT1), and a lamellar structure (L).

affect only the precipitation sites and possibly the diffusion kinetics. Further X-ray analysis of the lamellar kamacite in the AEM shows a Ni composition of 6.1 ± 0.7 at the center of the lamella that decreases to 3.6 ± 0.5 wt % Ni at the kamacite/taenite interface. The decreasing Ni content from the center of the lamellar kamacite to the taenite interface indicates that the lamellar structure precipitated at a temperature between 500° and 400°C and then cooled to lower temperatures. The depletion zone at the lamellar α/γ interface would not have formed if precipitation occurred from martensite, α_2, which develops when low P γ cools below the martensite start temperature, M_s, of 250°C. In this case the Ni content in the α would be well below 6.1 wt %. In addition, there is no evidence for the martensite, α_2, structure to decompose into a lamellar morphology. As seen in the work of *Lin et al.* (1977) and *Novotny* (1981), martensite, α_2, normally decomposes into a plessite consisting of small taenite rods in a kamacite matrix.

Two nucleation sites exist for the lamellar structure: (1) small phosphides that formed throughout region A and (2) the interface of the previously precipitated Widmanstätten plates and the γ matrix in regions A and B. Figure 12a shows a residual Widmanstätten plate in the lamellar region B. Figure 12b is a higher magnification photograph of the boxed area in Fig. 12a. The Widmanstätten plate precipitated at 575°C in region B as discussed previously. This Widmanstätten α/γ interface most likely acted as a nucleation site for the lamellar

B, helping to precipitate small amounts of Widmanstätten α and swathing kamacite (SK). The formation of SK, with a high P solubility of 0.28 wt %, further depleted P in the γ adjacent region. Since small phosphides will not form in region B, and since α needs a heterogeneous nucleation site such as phosphide to nucleate (*Narayan and Goldstein*, 1984), most of region B remained as supersaturated γ. Figure 11b shows a schematic microstructure of regions A and B in a meteorite that developed at 575°C according to the precipitation process discussed above.

Between 575° and ~400°C, region A remained in the α + γ + Ph field and region B remained in the α + γ field. At approximately 400°C the low-P region B entered the α + γ + Ph region, and small phosphides precipitated. Parts of region B would undercool further until the martensite start temperature was crossed at ~250°C. These martensite areas are large in relation to the other martensite plessite fields in the meteorite (Fig. 1).

Lamellar morphology in Dayton. We have assumed that carbon has little or no effect on the Ni and P equilibrium compositions described by the Fe-Ni-P phase diagram for the Dayton meteorite. No carbides have been observed, optically or by electron microscopy, in Dayton in our study or by *Buchwald* (1975) either in the phosphides or in the Widmanstätten or lamellar structure. The Fe-Ni-C phase diagram (*Romig and Goldstein*, 1978) suggests C solubilities of ~0.25 wt % in γ to temperatures as low as 500°C. The Dayton iron contains 0.05 wt % C (*Moore et al.*, 1969). Presumably carbon is present in solid solution in the taenite to temperatures below 500°C and perhaps below 400°C. We assume that C would

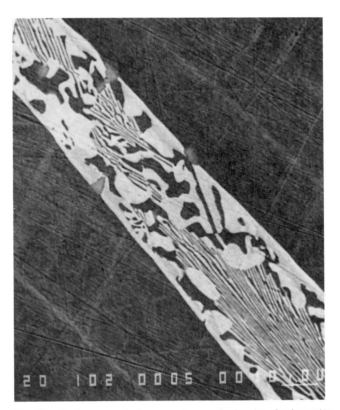

Fig. 8b. Back scattered electron micrograph showing the kamacite (dark), lamellar plessite, and grey phosphide inclusions of the Canyon Diablo meteorite.

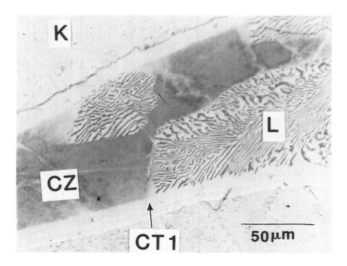

Fig. 9. Light micrograph of the Toluca meteorite showing kamacite (K), clear taenite 1 (CT1), the cloudy zone (CZ), and the lamellar plessite (L).

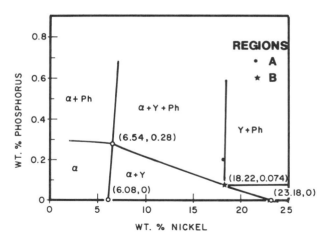

Fig. 11a. Isothermal section of Fe-Ni-P ternary phase diagram at 575°C. Bulk compositions of Regions A and B are noted.

α plates. The analysis of the nucleation and growth mechanism at the interface between the lamellar phase and the Widmanstätten plate is difficult, however. Due to continued diffusion in γ after nucleation of lamellar α, the probable initiation site (α/γ interface) as shown in Fig. 12b cannot be identified.

The kamacite/taenite interface of the lamellae of the Dayton iron forms, as in Group I irons, with a Kurdjumov-Sachs orientation relationship. This relationship is 6° from the Nishiyama-Wassermann orientation of the kamacite/taenite interface in the Widmanstätten pattern. The formation mechanism for the lamellar structure is therefore different from that of the Widmanstätten pattern. We suggest that a discontinuous reaction between 500°C and 400°C controls the formation of the lamellar plessite. An eutectoid-type reaction for the formation of the lamellar plessite is generally ruled out since a high-angle interface (such as that formed by haxonite or cohenite) is not present.

The various types of the discontinuous reactions are discussed by *Williams and Butler* (1981) and are generally defined by a discontinuous change in orientation of the product and a discontinuous change in solute concentration across the advancing interface. In Dayton, the discontinuous precipitation is described as the decomposition of a supersaturated single phase γ_{ss} to a lamellar structure consisting of a precipitate (α) and solute depleted matrix (γ). The γ_{ss} and γ are the same phase (taenite) and crystal structure (fcc), differing only in degree of solute (Ni) supersaturation. The reaction in Dayton is described as $\gamma_{ss} \rightarrow \alpha + \gamma$, where γ_{ss} is the matrix γ-taenite of Dayton containing ~17.6 wt % Ni and $\alpha + \gamma$ are the lamellar α and γ.

Canyon Diablo, Odessa, and Toluca

Composition of lamellar plessite in Group IA. In Group IA meteorites the Widmanstätten pattern began to form around 700°C. As the meteorite cooled, the α grew in width, while

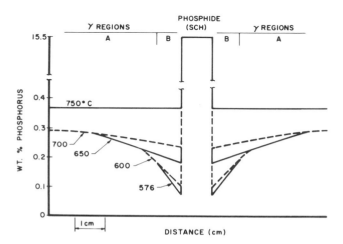

Fig. 10. Schematic P profile across the regions surrounding the large phosphide between 750°C and 576°C.

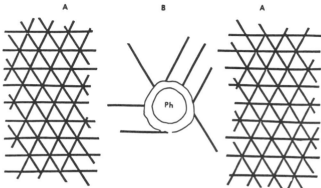

Fig. 11b. Schematic representing the developing morphology in Regions A and B at 575°C.

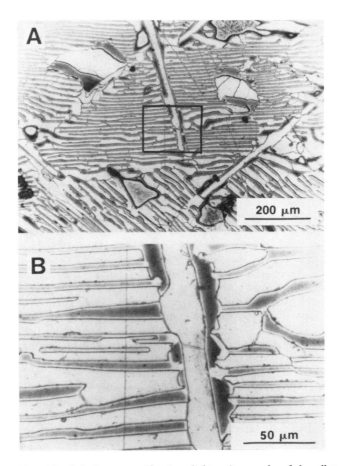

Fig. 12. (a) Low magnification light micrograph of lamellar morphology and residual Widmanstätten plates in Region B of the Dayton meteorite. (b) Higher magnification of the intersection of the lamellae and the Widmanstätten plate.

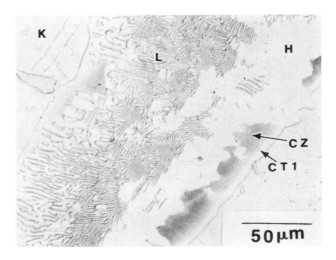

Fig. 13. Light micrograph of a decomposed taenite region in the Toluca meteorite showing kamacite (K), clear taenite 1 (CT1), cloudy zone (CZ), haxonite (H), and the lamellar plessite (L).

rejecting Ni into the γ. The rejected Ni built up at the α/γ interface due to the slow diffusion of Ni in taenite. A typical M-shaped profile developed. The Ni content in the center of the profile will be greater than the bulk Ni content of the meteorite because of impingement. The average Ni content of the lamellar plessite in Group IA was measured as > 25 wt % Ni by area scans in the EPMA. The solubility of P in γ decreases to 0.04 wt % at 500°C, decreasing further with continued cooling (*Romig and Goldstein,* 1980). The carbide cohenite would precipitate at higher temperatures, but an appreciable amount of carbon (~0.25 wt % C) (*Romig and Goldstein,* 1978) can remain in solution in γ at 500°C. Figure 13 shows the lamellar plessite intermixed with haxonite, indicating the presence of C in the plessite region. Therefore the composition of the lamellar plessite in Group IA is approximately > 25 wt % Ni, < 0.04 wt % P, and < 0.25 wt % C.

Lamellar structure in Group IA. The micrograph of Fig. 9 shows that the lamellar structure and the cloudy zone may be competing reactions. Actually, the lamellar structure probably formed at higher temperatures. The size of the γ precipitates in the cloudy zone are ~0.01 μm in comparison to the γ lamellae, which are 0.1 → 1μm in size. *Reuter et al.* (1987b) showed that the cloudy zone forms by a spinoidal reaction at temperatures below approximately 350°C. Since diffusion of Ni is much faster at higher temperatures, the lamellar structure probably precipitated at 400°C to 450°C and continued to grow until 300°C. At ~300°C, the diffusion kinetics of Ni are so slow that the cloudy zone formed in the taenite lamellae through a spinodal decomposition.

As mentioned before, the retained taenite in Group IA irons have an appreciable amount of C in solution at 500°C. Assuming that the carbon precipitated as carbides above 400°C, the carbides could act as a nucleation site for the lamellar structure. Figure 13 shows the carbide mineral haxonite along one side of the plessite field. This low-temperature carbide may assist in nucleating the lamellar structure. Haxonite, however, is not seen in all lamellar fields. A three-dimensional examination of the lamellar plessite is needed to confirm the role of this carbide as the nucleation site of the lamellar plessite.

Two types of nucleation mechanisms may control the lamellar structure formation; an eutectoid or a discontinuous reaction. An eutectoid-type precipitation involves the heterogeneous nucleation of the kamacite phase on a high angle interface (such as haxonite). Upon formation the kamacite will reject Ni into the surrounding matrix; this will enhance the formation of the γ phase or vice versa. The α and γ lamellar structure will then advance into the γ matrix. As indicated previously, the phase transformation that produced the lamellar morphology in Group IA occurred at approximately 400°C. *Chuang et al.* (1986) and *Reuter et al.* (1987a) predict an eutectoid transformation occurring at 390°C (Fig. 14). The temperature of the reaction is in close agreement with our work, but the predicted composition of the eutectoid reaction is at 49.3 wt % Ni. This Ni composition is too high to be directly applicable to the γ phase in the lamellar structure observed in Group IA. An appreciable amount of carbon is present in the lamellar plessite area. The carbon may stabilize

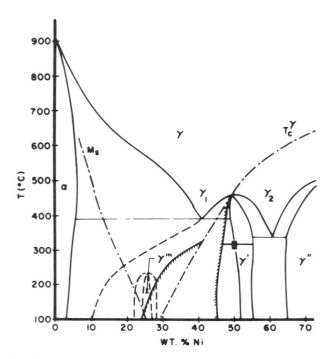

Fig. 14. Low-temperature Fe-Ni diagram (*Reuter et al.,* 1987). The $\alpha/\alpha+\gamma$ and $\alpha+\gamma/\gamma$ boundaries are the same as those given by *Romig and Goldstein* (1980) above ~400°C. M_s represents the martensite start temperature. γ' is FeNi (ordered below 310°C), γ'' is Ni$_3$Fe, γ is Fe$_3$Ni, γ_1 is paramagnetic γ, and γ_2 is ferromagnetic γ. The line T_c^γ is the transformation boundary (Curie temperature) between paramagnetic and ferromagnetic γ. An eutectoid reaction $\gamma \rightarrow \alpha+\gamma$ occurs below 400°C. A miscibility gap between γ and γ' is shown by the dashed lines. Spinodal boundaries are shown by the hatched lines to delineate this transformation boundary.

the γ field, as it does in the Fe-Ni-C phase diagram from 730° to 500°C (*Romig and Goldstein,* 1978), and may possibly shift the eutectoid point from that observed in the Fe-Ni diagram. Carbon may also cause the $\alpha + \gamma$ region to be undercooled to a low enough temperature, whereby a 25% Ni area might undergo the eutectoid reaction. The presence of C may also raise or lower the predicted eutectoid temperature. A ternary eutectoid in the Fe-Ni-C system is also a possibility. Unfortunately, the Fe-Ni-C diagram is not measured below 500°C.

Although the eutectoid reaction mechanism looks favorable, we cannot discount the possibility of a grain boundary discontinuous reaction. More work must be done in the Group I meteorites to determine if there is a high-angle interface between the α in the lamellar structure and the adjacent α in the cloudy zone. If a boundary is present, the reaction is probably a grain boundary discontinuous reaction; likewise, if the boundary is absent, the reaction producing the lamellar structure is probably an eutectoid reaction. The effect of carbide as a nucleation site, γ stabilizer, and/or reaction product is also unknown below a temperature of 500°C. Until we know the role of carbon and whether in fact there is a high-angle interface driving the reaction, the formation of the lamellar morphology in Group IA remains unclear.

SUMMARY

Dayton—Group IIID

1. The lamellae consist of alternating plates of kamacite (α) and taenite (γ) with an interlamellar spacing > 3 μm.

2. The α/γ interface of the lamellar structure conforms to the Kurdjumov-Sachs orientation relationship while the Widmanstätten pattern conforms to a Nishiyama-Wasserman orientation relationship.

3. The central region of the kamacite was a composition of 6.1 ± 0.7% Ni decreasing to 3.6 ± 0.5% Ni at the α/γ interface. At the α/γ interface, the Ni composition in the taenite is 50% Ni decreasing to 38% Ni in the center of the taenite.

4. The formation of the lamellar structure in Dayton, Group IIID is dependent on the presence of large phosphide inclusions. Upon cooling of the meteorite, the schreibersite depleted the surrounding γ matrix of phosphorus. This phosphorus depletion caused the adjacent γ region to undergo a different phase transformation sequence than the rest of the meteorite. In regions away from the schreibersite, the typical sequence consists of the precipitation of small phosphides, in turn nucleating Widmanstätten α as the $\alpha + \gamma + $ Ph phase field was entered at 575°C. The P- depleted areas surrounding the schriebersite did not enter into the $\alpha + \gamma + $ Ph field, but undercooled into the $\alpha + \gamma$ phase field until 500->400°C. In this temperature range, and in the presence of carbon, the meteorite precipitated α in a lamellar morphology. The carbon is present in solid solution in γ, not in carbides. The phase transformation describing the precipitation of the lamellar structure in Dayton is discontinuous of the type $\gamma_{ss} \rightarrow \alpha + \gamma$.

Canyon Diablo, Toluca, and Odessa—Group IA

1. The lamellar morphology of the Group IA iron meteorites consists of alternating plates of kamacite (α) and taenite (γ) with an interlamellar spacing of < 1 μm.

2. The α/γ interface of the lamellar structure conforms to the Kurdjumov-Sachs orientation relationship.

3. The kamacite in the lamellar structure contains ~4 wt % Ni and the high Ni γ phase is ~48 wt % Ni.

4. The composition of the lamellar area is > 25 wt % Ni, < 0.04 wt % P, and < 0.25 wt % C. The Group IA lamellar transformation temperature is between 450° and 400°C.

5. It is unclear whether the lamellar structure in Group IA formed by a grain boundary discontinuous reaction or an eutectoid reaction. Until the identification of a high-angle interface is made, or the effect of C on the Fe-Ni phase diagram below 500°C is determined, the actual phase transformation cannot be conclusively identified.

Acknowledgments. The research was supported through NASA grant NAG 9-45. The authors wish to acknowledge Dr. Roy Clarke of the Smithsonian for his assistance in obtaining the meteorite samples.

REFERENCES

Buchwald V. F. (1975) *Handbook of Iron Meteorites, Their History, Distribution, Composition and Structure.* University of California Press, Berkeley. 1418 pp.

Chuang Y., Chang A., Schmid R., and Lin J. (1986) Magnetic contributions to the thermodynamic functions of alloys and the phase equilibria of the Fe-Ni system below 1200K. *Met. Trans.*, *17A*, 1361-1372.

Cliff G. and Lorimer G. W. (1975) The quantitative analysis of thin specimens. *J. Microsc.*, *103*, 203-207.

Doan A. S. and Goldstein J. I. (1970) The ternary phase diagram, Fe-Ni-P. *Met. Trans.*, *1*, 1759-1767.

Heyward T. R. and Goldstein J. I. (1973) Ternary diffusion in the α and γ phases of the Fe-Ni-P system. *Met. Trans.*, *4*, 2235-2342.

Lin L. S., Goldstein J. I., and Williams D. B. (1977) Analytical electron microscopy study of the plessite structure in the Carlton iron meteorite. *Geochim. Cosmochim. Acta*, *41*, 1861-1874.

Moore C. B., Lewis C. F., and Nava D. (1969) Superior analyses of iron meteorites. D. Reidel, Dordrecht. *Meteorite Research* (P.M. Millman, ed.), pp. 738-748.

Narayan C. and Goldstein J. I. (1984) Nucleation of intragranular ferrite in Fe-Ni-P alloys. *Met. Trans.*, *15A*, 861-865.

Novotny P. M. (1981) An electron microscope investigation of eight ataxite and one mesosiderite meteorites. M. S. thesis, Lehigh University, Bethlehem, PA.

Reuter K. B., Williams D. B., and Goldstein J. I. (1987a) Determination of the Fe-Ni phase diagram below 350°C. *Met. Trans*, in press.

Reuter K. B., Williams D. B., and Goldstein J. I. (1987b) Low temperature phase transformations in the metallic phases of meteorites. *Geochim. Cosmochim. Acta*, in press.

Romig A. D., Jr. and Goldstein J. I. (1978) Determination of the Fe-rich portion of the Fe-Ni-C phase diagram. *Met. Trans.*, *9A*, 1599-1609.

Romig A. D., Jr. and Goldstein J. I. (1980) Determination of the Fe-Ni and Fe-Ni-P phase diagrams at low temperatures (700-300°C). *Met. Trans.*, *11A*, 1151.

Williams D. B. and Butler E. P. (1981) Grain Boundary Discontinuous Precipitation Reactions. *Intl. Metals Rev.*, *3*, 153-183.

Bidirectional Reflectance Properties of Iron-Nickel Meteorites

D. T. Britt and C. M. Pieters

Department of Geological Sciences, Brown University, Providence, RI 02912

Spectral reflectance studies have suggested that elemental iron is a major component of the surface mineralogy of M- and S-type asteroids. Both asteroid types exhibit a red sloped continuum (increasing reflectance with increasing wavelength) that has been interpreted as characteristic of the presence of large amounts of elemental iron. The effects of viewing geometry and small-scale (<1 mm) surface roughness on the bidirectional reflectance spectra of iron-nickel meteorites have been investigated using NASA's RELAB facility. The spectra of metallic surfaces can be divided into three general groups on the basis of surface roughness. Group I consists of surfaces with surface features in the range of 10 μm to 1 mm. This group is characterized by spectra that are similar to diffuse reflectance spectra and show no dependence on viewing geometry. Group II covers the roughness range from 10 μm to 0.7 μm. The surface features in this group are in the same size range as the wavelengths of infrared light and the reflectance spectra are very complex. Group III consists of surfaces smoother than 0.7 μm and is characterized by a two-component reflectance that is strongly dependent on viewing geometry. Group III spectra taken in the specular geometry are very bright and strongly red sloped, while the spectra taken in the nonspecular geometry are dark and flat. The diffuse reflectance spectra of Group I surfaces can be modeled as linear combinations of Group III reflectance components. Published spectra of M-type asteroids exhibit good agreement with Group I spectra, suggesting that the regoliths of these objects act as a diffuse reflector with surface features larger than the wavelength of light.

INTRODUCTION

Using laboratory measurements of metallic meteorites as spectral analogs for asteroidal parent bodies, spectral reflectance studies suggest that elemental iron is a major component of the surface mineralogy of M- and S-type asteroids (*Gaffey and McCord*, 1978; *Zellner*, 1979). Both asteroid types exhibit a red sloped continuum (reflectance increasing with wavelength) characteristic of the iron meteorites and the presence of significant amounts of elemental iron. The actual abundance and nature of the metallic component on asteroidal surfaces is the subject of considerable controversy, particularly for the S-type asteroids (*Feierberg et al.*, 1982; *Gaffey*, 1984, 1986). As asteroids increasingly become targets of remote-sensing observations, both from ground-based telescopes and spacecraft, it is important to understand the bidirectional reflectance properties of naturally occurring metallic surfaces. Knowledge of these reflectance properties is necessary for the interpretation of remotely-sensed spectra and for the understanding of surface processes on metal-rich bodies.

Iron-nickel meteorites are considered to be good spectral analogs for M-type asteroids (*Gaffey and McCord*, 1978) and account for 4.38% of observed meteorite falls (*Graham et al.*, 1985). Geochemical and petrographic studies (*Dodd*, 1981) suggest that many iron-nickel meteorites are samples of the large metallic cores formed by differentiated planetesimals. The M-type asteroids, because of their red sloped but featureless spectra, could be composed largely of iron and nickel (*Gaffey and McCord*, 1978) and may be remnants of these differentiated cores. The main-belt M-type asteroid (16) Psyche has a radar albedo that strongly indicates a surface with an almost completely metallic composition (*Ostro et al.*, 1985). Two M-type asteroids, 1986 DA and 1986 EB, have recently been identified in the Earth-crossing asteroid population (*Gradie and Tedesco*, 1987). Independent radar measurements of 1986 DA strongly suggest a metallic composition (*Ostro et al.*, 1986).

Visible and near-infrared bidirectional reflectance spectroscopy is currently considered to be the best available technique for the remote characterization of the surface mineralogy and petrology of solar system objects. Interpretation of reflectance spectra is based on recognition of electronic and vibrational absorption features that are diagnostic of minerals on the surfaces of remote objects (e.g., *Adams*, 1975). The interpretation of asteroid spectra is greatly enhanced by the study of the spectral properties of meteorites (*Gaffey and McCord*, 1978; *Gaffey*, 1976). Meteorites provide reasonable constraints for the mineralogy of asteroids and the spectra of meteorites provide valuable laboratory analogs for remotely obtained spectra.

However, many laboratory spectra have been obtained under different geometric conditions than those for remotely obtained spectra. Remotely obtained spectral data are bidirectional in nature (*Hapke*, 1981) with the light source (the sun) oriented at a specific incidence angle (i) to the surface in question, and the observer (the telescope or spacecraft) located at a specific emission angle (e). The bidirectional reflectance spectrum of a surface material can vary significantly with changes in viewing geometry (e.g., *Gradie et al.*, 1980). Many laboratory spectra of meteorite samples are directional-hemispherical (diffuse) spectra (*Gaffey*, 1976) obtained with an integrating sphere used to average the sample's reflected radiation over all geometries. By eliminating geometric effects and using all the reflected light, directional-hemispherical reflectance is often able to characterize more precisely the diagnostic absorption features of a sample. However, this method has the disadvantage that it does not mimic the bidirectional nature of remote observations.

Physical properties of a surface that can affect its spectrum are state of compaction, albedo, particle size, and surface roughness (*Adams and Filice*, 1967; *Hapke*, 1981). For cut metallic surfaces (slabs) such as iron-nickel meteorites, the effects of compaction, porosity, and particle size are probably

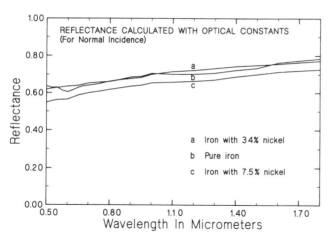

Fig. 1. Reflectance of pure iron and iron-nickel alloys calculated from published optical constants. Constants were measured at normal incidence.

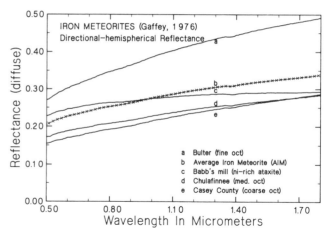

Fig. 2a. Directional-hemispherical (diffuse) reflectance of four iron meteorites from *Gaffey* (1976). The spectrum depicted in Fig. 2b is the average spectrum of the four meteorites.

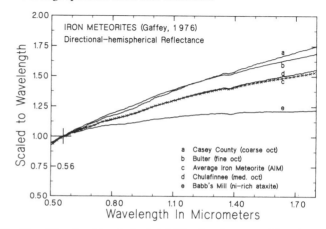

Fig. 2b. Directional-hemispherical (diffuse) reflectance of four iron meteorites and the average iron meteorite scaled to 1.0 at 0.56 μm.

not important parameters. However, changes in surface roughness are known to affect the brightness of metal surfaces (*Gaffey*, 1986). This study examines how viewing geometry and small-scale (<1 mm) surface roughness affect the bidirectional reflectance of iron-nickel meteorites and explores the implications of these effects on our understanding of the surface components of metal-rich (particularly M-type) asteroids.

THEORETICAL MODELS OF METALLIC REFLECTANCE

For the scope of this paper, classical electromagnetic theory based on Maxwell's equations and the Drude model is a good approximation of metallic reflectance properties and will be used in place of the more general but complex quantum theory of absorption and dispersion (*Wooten*, 1972). Briefly, the reflectance of metal in the classical Drude model (*Bohren and Huffman*, 1983) is a result of incident light inducing an electric field in a conducting material. This field penetrates the conductor, but rapidly decays with depth. The depth to which the electric field effectively penetrates, and that the photons of the incident light can sample, is called the skin depth. The skin depth can be considered to be the optically active portion of the metal. For example, a conducting metal such as copper has a skin depth of 0.016 μm for light with a wavelength of 0.19 μm (*Wooten*, 1972). Light that penetrates the metal interacts with the induced field and is subjected to wavelength dependent absorption or reflection. In this way the reflected light acquires information from the metal in the form of a characteristic reflectance spectrum. Metallic reflectance is determined by the relationship between the electric field, the resonance frequency of the free electrons in the metal (the plasma frequency), and the wavelength of incident light. Each metal has its characteristic plasma frequency that, in large measure, determines the reflectance of the metal. For light in the visible and near infrared regions, the Drude model predicts that iron will exhibit a steadily increasing reflectance with increasing wavelength. These results have been verified by the experimental determination of optical constants of iron.

Optical constants for metals are usually not derived from mathematical models, but are generally determined experimentally by measuring normal incidence bidirectional reflectance (i = e = 0) of very smooth, vapor-deposited samples (*Blodgett and Spicer*, 1967). Shown in Fig. 1 are the reflectance spectra for pure iron and several iron-nickel alloys calculated from published optical constants using the Fresnel equations of reflectance for dielectrics (*Born and Wolf*, 1965; *Bolotin et al.*, 1969; *Sasovskaya and Noskov*, 1974). As predicted by the Drude model, iron is characterized by a steadily increasing reflectance with increasing wavelength.

Similar red sloped spectra are seen in laboratory measurements of the spectra of iron-nickel meteorites. Shown in Figs. 2a,b are directional-hemispherical (diffuse) spectra from clean, cut faces of four iron-nickel meteorites (*Gaffey*, 1976). Gaffey found that iron-nickel meteorites exhibit red sloped but featureless spectra. With increasing nickel content the continuum may become less red sloped. Included in Figs. 2a,b is the calculated average spectrum of these four meteorites (shown as an X-ed line). This "average diffuse iron meteorite"

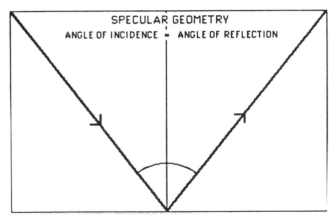

Fig. 3a. Specular reflectance geometry. Angle of incidence equals angle of reflection.

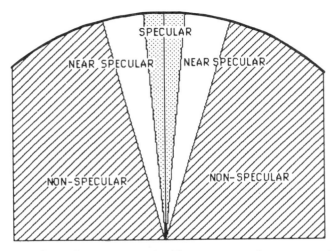

Fig. 3b. Reflectance geometries: The narrow central cone centered around the specular angle of reflection is the specular reflection geometry, the larger cone including the next 10° radius is near-specular reflection geometry, and the remaining orientations of reflected light are the nonspecular reflection geometry.

(AIM) spectrum will be used for comparison with bidirectional laboratory spectra.

SAMPLES AND EXPERIMENTAL PROCEDURES

The reflectance properties of Gaffey's iron-nickel meteorites were measured using a Beckman integrating sphere spectrometer (*Adams and McCord*, 1970). Directional-hemispherical (diffuse) measurements integrate reflected light over all geometries to obtain a single value of reflectance, while bidirectional reflectance is geometry-dependent (*Hapke*, 1981). Bidirectional reflectance measures only one small cone of reflected light and must be described in terms of its geometry with respect to the incident light, the sample surface, and the observer or detector. For this study the following terms will be used to describe the geometric relationships of bidirectional reflectance for source, sample, and observer. The "specular" direction, shown in Fig. 3a, is defined as a reflection geometry for which the angle of incidence equals the angle of reflection. The normal incidence reflection used to experimentally obtain optical constants is a specular geometry. The specular direction can be considered as a narrow cone of light reflecting from a surface in the direction e as illustrated in Fig. 3b. The first one or two degrees in solid angle around the specular geometry (i = e) is included in the specular cone. The angular size of the specular cone is dependent on surface roughness: narrow for smoother surfaces, wider for rougher surfaces. The next bidirectional reflectance zone is the "near-specular" direction. The near-specular includes a cone of light approximately 10 degrees in solid angle around the specular cone. Finally, the geometry describing the remainder of the reflected light is the "nonspecular" direction. This is light for which the angle of incidence is different from the angle of reflection by more than 10 degrees.

In the following discussion, the specular, near-specular, and nonspecular directions will be referred to as components of reflection. In diffuse spectra all the components of reflection are measured simultaneously and integrated into one value. With bidirectional spectra, individual components of reflection can be measured separately. A natural surface reflects some incident light in all directions and therefore into all components, but it is the position of the observer with respect to the incident light and the surface that determines which component of reflection is observed.

Samples of the Gibeon and Canyon Diablo meteorites were used in this study. Gibeon is a fine octahedrite, class IVA, with 7.68% nickel (*Schaudy et al.*, 1972). Canyon Diablo is a coarse octahedrite, class IA, with 6.98% nickel (*Wasson*, 1974). Several cut surfaces on both meteorites were polished with aluminum oxide and diamond abrasive grits to controlled surface roughness. All samples were checked for corrosion and polishing imperfections with a reflected light microscope. After polishing, each sample was immediately washed in ethyl-alcohol and dried to prevent corrosion. A series of bidirectional reflectance spectra were taken at a variety of viewing geometries using NASA's RELAB facility located at Brown University (*Pieters*, 1983).

RESULTS

Overview of Roughness Groups

For this study a wide range of surface textures were examined. Analysis of the resulting measurements suggests that the spectra of metallic surfaces can be divided into three general groups on the basis of surface roughness. These groups are listed in Table 1. Photomicrographs of representative surfaces of each roughness group are shown in Figs. 4a,b,c. Figure 4a is a photomicrograph of the Canyon Diablo meteorite polished with 400 grit aluminum oxide. Surface features average about 40 μm in size. This sample belongs to Group I, which is characterized by macroscopically rough surfaces with features ranging from 10 μm to 1 mm. Because the surface roughness is much greater than the wavelength of light in the visible

TABLE 1. Roughness groups.

Group	Roughness range	Reflectance characteristics
I	10 μm - 1 mm	diffuse
II	0.7 μm - 10 μm	complex scattering
III	<0.7 μm	two-component

and near infrared, these surfaces are shown to be effectively diffuse reflectors. Shown in Fig. 4b is another face of the Canyon Diablo sample polished with 1200 grit aluminum oxide to produce surface relief in the range of 1-5 μm. This is an example of Group II surfaces, which are characterized by microscopically rough surfaces with feature sizes ranging from approximately 0.7 μm to 10 μm, the same size range as the wavelength of infrared light. This group can be considered as a transition zone between rough (Group I) and smooth (Group III) surfaces and tend to be characterized optically by complex scattering behavior. Figure 4c is a Canyon Diablo face polished with 1-μm diamond paste. This level of polish produces features less than 0.4 μm in size that are too small to be resolved with a visible light microscope. Group III consists of surfaces that are smooth relative to the wavelength of infrared light. Feature sizes are generally less than 0.7 μm. The reflectance components of Group III surfaces are distinctly different. The specular component is shown to be very bright and red sloped, while the nonspecular component is dark and flat.

Group I Surfaces

Group I surfaces are the roughest of the three groups with features that range in size from 10 μm to 1 mm. Since these features are much larger than the wavelength of incident light, samples can be considered a collection of randomly oriented reflecting surfaces or facets. Shown in Fig. 5 are scaled spectra measured at a number of viewing geometries of the surface illustrated in Fig. 4a, a face of the Canyon Diablo meteorite polished to 40-μm roughness. This group consists of macroscopically rough surfaces that are, as shown in Fig. 5, characterized by the classic red sloped continuum of iron, but exhibit little or no geometric dependence on reflectance (albedo range 0.20 to 0.29 at a wavelength of 0.6 μm). All spectra, including both specular and nonspecular geometries, show basically the same spectral shape. The spectrum of the average iron meteorite is included in Fig. 5 for comparison. In this roughness range bidirectional spectra agree well with the diffuse spectral reflectance measurements of iron-nickel meteorites.

Similar results are obtained for a number of different surfaces and samples in this roughness range. Shown in Fig. 6 are spectra of fresh filings from the Canyon Diablo meteorite with a wide particle size range of 10 μm to 1 mm. Once again, the spectrum of this metallic surface is characterized by a classic red slope, no geometric dependence on reflection, and close agreement with the average iron meteorite diffuse spectrum (albedo range 0.14 to 0.18 at 0.6 μm).

Fig. 4a. Photomicrograph of a sample of the Canyon Diablo meteorite polished with 400 grit aluminum oxide to produce surface roughness of 40 μm (Group I). Field of view is approximately 600 μm.

Fig. 4b. Photomicrograph of a sample of the Canyon Diablo meteorite polished with 1200 grit aluminum oxide to produce surface roughness of 1-5 μm (Group II). Field of view is approximately 600 μm.

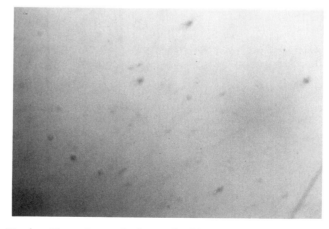

Fig. 4c. Photomicrograph of a sample of the Canyon Diablo meteorite polished with 1-μm diamond paste to produce surface roughness of <0.4 μm (Group III). Field of view is approximately 600 μm.

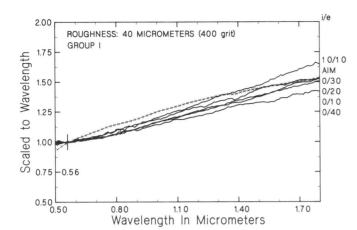

Fig. 5. Bidirectional reflectance spectra of the Canyon Diablo meteorite polished with 400 grit aluminum oxide (Group I surface). Surface roughness is approximately 40 μm. Viewing geometry is indicated as i/e representing "angle of incidence/angle of emission." Also shown is the spectrum of the average iron meteorite (AIM).

An interesting aspect of spectra from the filings sample is the close agreement of all the spectra at different viewing geometries, much tighter than spectra for the 40-μm roughness sample (e.g., Fig. 5). Although both samples belong to the same roughness group and have similar spectral characteristics, the 40-μm sample has a much smaller range of surface roughness than the filings with a 10-μm to 1-mm particle size range. The tight distribution of spectra produced by the filings sample suggests that a wide distribution of particle sizes will tend to homogenize the spectra, perhaps by providing a more random distribution of facets to act as reflecting surfaces. The effect is to reduce further the geometric dependence of the sample's reflectance.

Spectra of additional complex metallic surfaces in Group I are shown in Figs. 7a,b. These surfaces include craters on the Gibeon meteorite created by experimental hypervelocity impacts using the NASA-Ames vertical gun (*Matsui and Schultz,* 1984). Cratering is an important process on planetary bodies and these craters may represent possible surface processes on metal-rich asteroids. Projectile velocity for these experiments was approximately 5 km/sec, which is the average collisional velocity in the asteroid belt (*Davis et al.,* 1979). One crater was created by an aluminum projectile, the other by a basalt projectile. Shown in Fig. 7b are scaled spectra of all Fig. 7a complex surfaces and the average iron meteorite. The different surfaces and samples show a similar red sloped spectrum, little geometric dependence on reflection, and good agreement with the diffuse iron meteorite spectrum (albedo range 0.04 to 0.32 at 0.6 μm).

These data suggest that in this roughness range (10 μm to 1 mm) bidirectional reflectance spectra of metallic surfaces is comparable to diffuse reflectance spectra. The surface textures of macroscopically rough surfaces make them effectively diffuse reflectors.

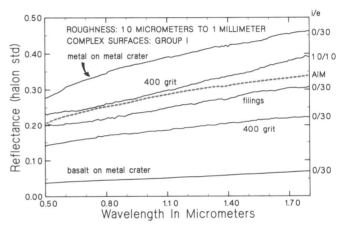

Fig. 7a. Bidirectional reflectance spectra of complex Group I surfaces. Included are spectra of the Canyon Diablo filings and 40-μm prepared surface (400 grit), the Gibeon metal on metal and basalt on metal craters, and the average iron meteorite (AIM).

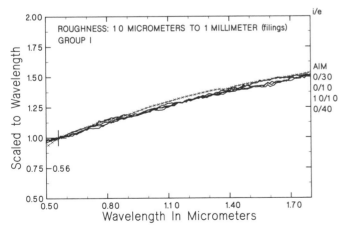

Fig. 6. Bidirectional reflectance spectra of filings of the Canyon Diablo meteorite with roughness range 10 μm to 1 mm (Group I surface). Also shown is the spectrum of the average iron meteorite (AIM).

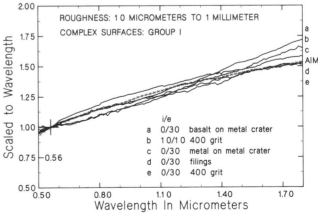

Fig. 7b. Bidirectional reflectance spectra of complex Group I surfaces of Fig. 7a scaled to 1.0 at 0.56 μm.

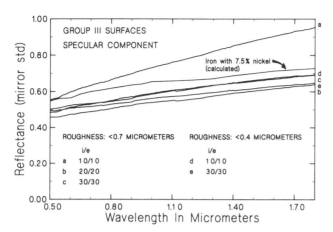

Fig. 8a. Bidirectional reflectance spectra of surfaces of the Canyon Diablo meteorite polished with diamond paste to roughness of <0.7 μm and <0.4 μm (Group III surfaces). All spectra were obtained at a specular geometry (i = e). Reflectance is measured relative to an aluminum mirror standard and is three to four orders of magnitude brighter than the nonspecular component measured relative to halon. Also shown is the spectrum of iron-nickel alloy calculated from optical constants measured at normal incidence.

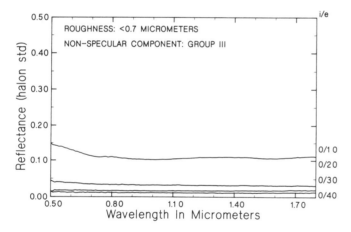

Fig. 8b. Bidirectional reflectance spectra of the Canyon Diablo meteorite polished with diamond paste to roughness of <0.7 μm (Group III surface). All spectra were obtained at a nonspecular geometry (i ≠ e).

Group III Surfaces

Group III surfaces represent the roughness range of less than 0.7 μm, which is as smooth or smoother than the wavelength of visible light. Under high-power optical microscopes these surfaces appear completely featureless. Figure 4c illustrates a smooth Group III surface on a sample of the Canyon Diablo meteorite. Reflectance from these smooth surfaces is strongly dependent on viewing geometry and two components of bidirectional reflectance exhibit very different spectra. The specular component of reflectance exhibits a very bright and red sloped spectrum, while the nonspecular component shows a dark and flat spectrum. The specular component was measured relative to an aluminum mirror standard. The aluminum mirror was calibrated relative to a fresh, vapor-deposited gold mirror and corrected to absolute reflectance using published, experimentally-determined optical constants for gold (*Johnson and Christy*, 1972). Shown in Fig. 8a is the specular component of two Group III surfaces of the Canyon Diablo meteorite measured at a number of specular geometries. All spectra tend to be very bright and exhibit a red sloped continuum. For comparison, the reflectance of an iron-nickel alloy calculated from optical constants is also shown. The specular component of the Canyon Diablo Group III surface measured in RELAB agrees well with the theoretical reflectance of an iron-nickel alloy calculated from optical constants derived from normal incidence reflection.

The specular component of reflection tends to be three to four orders of magnitude brighter than the nonspecular component at all measured wavelengths. The spectral character of the nonspecular component of reflectance is shown in Fig. 8b for the <0.7-μm Group III surface. These spectra were measured relative to a halon standard (halon approximates a diffuse Lambertian reflector) and show a dark, flat spectral shape that exhibits none of iron's characteristic red sloped continuum. These flat reflectance spectra are not predicted by mathematical models of iron reflectance based on classical or quantum mechanics. The source of this flat nonspecular spectrum is currently unknown. The near-specular reflectance component of the <0.7-μm sample exhibits a blue sloped spectrum at short wavelengths. This interesting phenomenon will be discussed further in the next section.

These data for Group III surfaces suggest that the bidirectional reflectance spectra of optically smooth metallic surfaces is strongly dependent on viewing geometry and is characterized by a two-component reflectance model with a bright, red sloped specular component and a dark, flat nonspecular component. Iron's characteristic red sloped continuum is only found in the specular component. In a latter section we will use these components to model the diffuse reflectance of metallic surfaces.

Group II Surfaces

Group II surfaces represent the roughness range between 0.7 and 10 μm and can be considered to be a transition zone between optically smooth and rough surfaces. Surface features are in the same general size range as the wavelength of the incident light. Figure 4b illustrates a face of the Canyon Diablo meteorite polished with 1200 grit aluminum oxide to produce surface features in the 1- to 5-μm range. Bidirectional spectra of this surface are shown in Fig. 9. Group II spectra exhibit similarities and differences with both Group I and Group III spectra. The first and perhaps most striking similarity is that, like Group III, Group II reflectance is strongly dependent on viewing geometry. Group II spectra measured in specular and near-specular geometries (top four spectra) exhibit a more strongly red sloped continuum than those for nonspecular geometries (bottom four spectra). Also, like Group III spectra, these two components of Group II reflectance exhibit different spectral shapes. The specular component is relatively bright and red sloped while the nonspecular component tends to

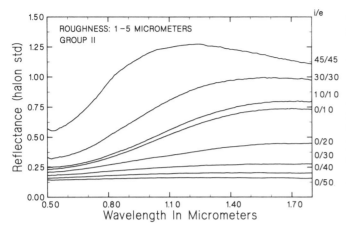

Fig. 9. Bidirectional reflectance spectra of the Canyon Diablo meteorite polished with 1200 grit aluminum oxide (Group II surface).

be darker and flatter than the specular. A significant difference between Group II and Group III is that the nonspecular component, while being darker and flatter than the specular component, still exhibits a red sloped continuum in most viewing geometries. The slope of this continuum, however, shows a strong dependence on viewing geometry. For Group II nonspecular geometries where i = 0, reflectance and slope of the continuum decrease steadily with increasing phase angle. Also, the intensity difference between the specular and nonspecular components is much reduced in Group II spectra from Group III. In Group III the intensity difference between the brightest and darkest spectra was approximately three to four orders of magnitude, while the difference in Group II spectra is about a factor of 10. An important distinction is the complexity of the Group II spectral character. The classic iron red sloped continuum that was ubiquitous in Group I reflectance and the Group III specular component is not readily evident from Group II spectra. The specular component shows a strong red slope only for a portion of the wavelength range; at longer wavelengths the spectra flatten and, in the case of the 45/45 geometry, exhibit a blue slope.

The complexity of spectra for the Group II surfaces is probably related to changes in scattering and reflectance behavior as the wavelength of incident light approaches the size of the surface features. The range of surface roughness for this sample is between 1 to 5 μm while the incident wavelength range is 0.5 to 1.8 μm. To the shorter wavelengths, some of the surface features are sufficiently large and smooth to be specular reflectors and the resulting spectrum exhibits a strong red slope, as the classical model would predict. At longer wavelengths, starting around one micrometer for these samples, the surface begins to appear rough to the incident light and the proportion of the surface that can behave as a specular reflector decreases. As the wavelength increases, a larger proportion of the incident energy is reflected into other nonspecular geometries. The decreasing flux in the specular direction with increasing wavelength causes first a decrease in the red slope of the spectra, and finally, as more energy is lost from the specular geometry, a spectrum with a blue slope.

A related effect that produces a blue sloped continuum is observed in the near-specular reflectance component of some Group II surfaces with smaller surface features. Shown in Fig. 10 are near-specular spectra for three samples: two Group III surfaces and one Group II surface. This Group II sample was first polished with 1200 grit aluminum oxide, then repolished with 0.3-μm aluminum oxide to reduce the roughness range, producing a 1- to 3-μm surface roughness. The most striking feature of this data is the strongly blue sloped spectrum that the Group II surface exhibits over the entire wavelength range. The smoothest surface, a Group III surface with a roughness of <0.4 μm, exhibits a flat spectrum over the wavelength range. The slightly rougher sample (the <0.7-μm Group III sample discussed earlier) exhibits a blue sloped spectrum to 0.7 μm before the spectrum flattens at longer wavelengths. These data suggest that as the incident wavelength increases, this surface becomes increasingly smooth relative to the wavelength of incident light, and that the portion of incident energy that is scattered into nonspecular geometries declines. Scattering continues to decrease with increasing wavelength, producing for nonspecular geometries a blue sloped spectrum that become flat when all surface features are smaller than the incident light.

These results suggest that Group II represents a transition zone between the effectively diffuse reflectance of Group I surfaces and the two-component reflectance of Group III surfaces. Reflectance is strongly dependent on viewing geometry and the size range of surface roughness. Surface features that are in the same size range as the wavelength of incident light produce complex patterns of reflectance that can change significantly as the wavelength changes. The proportions and intensity of specular and nonspecular

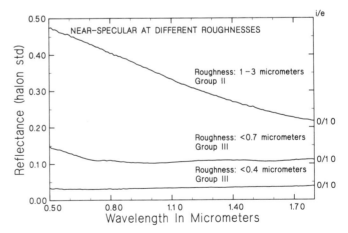

Fig. 10. Bidirectional reflectance spectra of three surfaces of the Canyon Diablo meteorite. The top spectra was obtained from a sample polished with 1200 grit aluminum oxide and then repolished with 0.3 micrometer aluminum oxide (Group II surface). The bottom two spectra were obtained from samples polished with 6-μm and 1-μm diamond paste, respectively (Group III surfaces). All spectra were obtained at a near-specular geometry (i = 0°, e = 10°).

components vary with wavelength and can cause unusual bidirectional reflectance results, such as a blue sloped metallic spectrum.

MODELING THE BIDIRECTIONAL REFLECTANCE OF METALLIC SURFACES

Spectral measurements of Group III surfaces suggest that the reflectance of optically smooth metallic surfaces can be described by two components with very different spectral characteristics: a bright, red sloped specular component and a dark, flat nonspecular component. Any surface with a roughness scale that is significantly larger than the wavelength of incident light, such as a diffuse Group I surface, can be thought of as a random distribution of facets. The facets on this rough surface are large relative to the wavelength of light and can produce a collection of reflective surfaces. At any given angle of incidence to the surface most facets would reflect light in a nonspecular geometry relative to the observer, but a small percentage of the facets would reflect in a specular orientation with respect to the observer. Using hemispherical geometry, it is possible to estimate the probability that a random surface facet is oriented in a specular or nonspecular geometry relative to the observer. A randomly oriented facet will produce a narrow specular cone of reflected light that is, for example, two degrees of solid angle in diameter. This specular cone will subtend an area of 0.06013 steradians, which would be 0.0957% of the area of the hemisphere. In the simplest case, the probability of the observer being in a specular orientation with respect to the facet is a direct function of the area of the hemisphere subtended by the specular cone. In this case the probability would be 0.0957%. Probability estimates for several different diameters of specular cones are detailed in Table 2.

TABLE 2. Probability of specular or nonspecular orientation for a random facet.

Diameter of specular cone (in degrees)	Probability of specular orientation	Probability of nonspecular orientation
1.0°	0.0038%	99.9962%
1.5°	0.0086%	99.9914%
2.0°	0.0957%	99.9043%
3.0°	0.215%	99.7850%

The diffuse reflectance of a complex Group I metallic surface can be described as a linear combination of nonspecular and specular components. Using measurements of the components of reflection from smooth surfaces and the probabilities of specular and nonspecular facet orientations estimated from hemispherical geometry, the reflectance of rough surfaces can be modeled. Spectra of the Group III <0.7-μm-roughness Canyon Diablo meteorite sample were used to calculate a spectrum consisting of a linear mix of 99.99% nonspecular component and 0.01% specular component. For these calculations the 0/20 spectrum shown in Fig. 8b was used for the nonspecular component and the 10/10 spectrum shown in Fig. 8a was used for the specular component. Shown in Fig. 11a is the calculated linear mix spectrum as well as the average iron meteorite spectrum. The linear mix of Group III components and the Group I average diffuse iron meteorite spectra show good agreement. Because the specular component is very bright and strongly red sloped, its contribution to the reflectance dominates the linear mix despite the small proportion of the surface that is in a specular geometry; the resulting spectrum exhibits a classic iron red slope. For comparison, Fig. 11b includes scaled reflectance spectra of the linear mix and several complex surfaces including the metal-on-metal crater and an average of the filings spectra. All spectra of Group I samples show good agreement with the linear mix of the Group III bidirectional laboratory spectra.

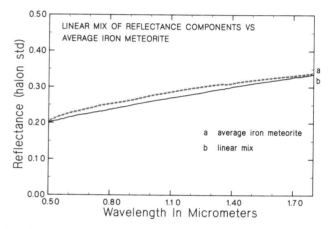

Fig. 11a. Calculated reflectance spectrum produced by using a linear mix of Group III specular and nonspecular reflectance components (solid line) and the spectrum of the average iron meteorite (AIM).

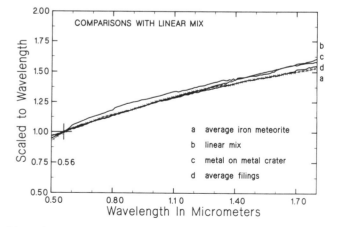

Fig. 11b. Reflectance spectra scaled to unity at 0.56 μm for: Point a = spectrum of average iron meteorite (AIM); point b = calculated reflectance spectrum using a linear mix of Group III reflectance components; point c = bidirectional reflectance spectrum of the metal on metal crater; and point d = the average of bidirectional reflectance spectra of the Group I filings spectra.

DISCUSSION AND CONCLUSIONS

The classical Drude model predicts that iron in the visible and near infrared wavelengths should exhibit an increasing reflectance with increasing wavelength. Bidirectional reflectance measurements of metallic surfaces show that only the specular component of reflectance (i.e., for optically smooth, Group III surfaces) exhibits iron's characteristic red sloped continuum. This result suggests that only the specular component of reflectance interacts optically with the metal and has a chance to acquire spectral information. The nonspecular component of reflectance from optically smooth surfaces, which includes the bulk of viewing geometries, does not appear to interact with the optically active skin depth of the metal. As a result, nonspecular reflectance for optically smooth surfaces displays none of the characteristic and predicted reflectance properties of iron. The dark and flat spectrum of the nonspecular component shows no wavelength-dependence over the studied range.

Surfaces that are much rougher than the wavelength of incident light, such as Group I surfaces, can be considered to be a collection of randomly oriented facets. Each facet contributes to a given viewing geometry a specular or a nonspecular reflectance. Bidirectional reflectance from such a surface will be overwhelmingly nonspecular, but the resulting spectrum will be dominated by the small fraction of bright, red sloped, specularly reflected radiation. Spectra of these Group I diffuse surfaces exhibit iron's characteristic red sloped continuum and can be modeled by linear combinations of facets that are oriented in specular and nonspecular geometries with respect to the observer.

Group II surfaces, with surface roughness in the same size range as the wavelength of incident light, can be considered transitional between optically smooth and rough surfaces. The reflectance characteristics of surfaces in this group vary as the wavelength of light changes and with different viewing geometries. These surfaces exhibit complex reflectance characteristics that can produce flat, red sloped, or even blue sloped metallic spectra.

These laboratory results can be used for the analysis of spectra of asteroidal surfaces that are thought to be metal-rich. Published spectra of M-type asteroids display red sloped, featureless spectra with no apparent strong geometric dependence (*Gaffey and McCord*, 1978). Such characteristics are in good agreement with Group I spectra, suggesting that if the surfaces of M-type asteroids are metal rich, surface particles act as a diffuse reflector, with complex surface features larger than the wavelength of light. This would imply that the surface is either rough metal or covered with a metallic regolith of fragments in the 10-μm to 1-mm size range. These results are supported by optical polarimetry measurements that suggest that the surfaces of M-type asteroids are covered with a regolith of 20-μm- to 50-μm-diameter metallic fragments (*Dollfus et al.*, 1979).

The bidirectional reflectance characteristics of metal-rich asteroids will be determined by the morphology of metal on their surfaces. The morphology of surface particles will control the proportion of facets that are in specular and nonspecular geometries with respect to the observer. For example, a regolith of horizontally oriented flat particles will have much different spectral properties than a regolith of randomly oriented metallic fines. Furthermore, if there are large smooth areas on a metallic object's surface, a sudden surge in reflectance could occur as these areas rotate into a specular geometry with respect to the observer, resulting in a brief several percent increase in the reflectance of an asteroid. However, observation and confirmation of these reflectance surges is very unlikely. The probability of an observer being in the appropriate specular geometry to record a reflectance surge would be even lower than the probabilities listed in Table 2 since an asteroid would be into a spherical rather than hemispherical system. A review of published lightcurves of M-type asteroids shows little evidence of reflectance surges (*Di Martino and Cacciatori*, 1984; *Harris and Young*, 1979; *Scaltriti et al.*, 1978; *Zappala et al.*, 1983; *Zeigler and Florence*, 1985).

The use of laboratory bidirectional spectra of iron-nickel meteorites as spectral analogs for M-type asteroids should be tempered by an appreciation of the effects of surface roughness. Polished or cut meteorite samples will be inappropriate due to the complex reflectance patterns produced by Group II and III surface features. Only rough or fragmental metallic meteorite samples with Group I roughnesses should be used as bidirectional spectral analogs for metallic asteroids.

Acknowledgments. This research would not have been possible without the help, ideas, and suggestions of Steven Pratt and William Patterson. We greatly appreciate their input as well as the funding support from NASA under grant NAGW-28. We would also like to thank M. J. Gaffey, P. Helfenstein, S. L. Murchie, F. Vilas, and M. Zolensky for excellent reviews and many useful suggestions. The RELAB facility is supported under NASA grant NAGW-748.

REFERENCES

Adams J. B. (1975) Interpretation of visible and near-infrared diffuse reflectance spectra of pyroxenes and other rock forming minerals. In *Infrared and Raman Spectroscopy of Lunar and Terrestrial Materials* (C. Karr, ed.), pp. 91-116. Academic, New York.

Adams J. B. and Filice A. T. (1967) Spectral reflectance 0.4 to 2.0 microns of silicate rock powders. *J. Geophys. Res.*, 72, 5705-5717.

Adams J. B. and McCord T. B. (1970) Remote sensing of lunar surface mineralogy: Implications from visible and near-infrared reflectivity of Apollo 11 samples. *Proc. Apollo 11 Lunar Sci. Conf*, pp. 1937-1945.

Blodgett A. J. and Spicer W. E. (1967) Experimental determination of the optical density states of iron. *Phys. Rev.*, 158, 514-523.

Bohren C. F. and Huffman D. R. (1983) *Absorption and Scattering of Light by Small Particles*. Wiley, New York. 530 pp.

Bolotin G. A., Kirillova M. M., and Mayevskiy V. M. (1969) Interband optical absorption in ferromagnetic iron. *Phys. Metal. Metallogr.*, 27, 31-41.

Born M. and Wolf E. (1965) *Principles of Optics*. 3rd ed., Pergamon, Oxford. 803 pp.

Davis D. R., Chapman C. R., Greenberg R., Weidenschilling S. J., and Harris A. W. (1979) Collisional evolution of asteroids: Populations, rotations, and velocities. In *Asteroids* (T. Gehrels, ed.), pp. 528-557. Univ. of Arizona, Tucson.

Di Martino M. and Cacciatori S. (1984) Photoelectric photometry of 14 asteroids. *Icarus*, 60, 75-82.

Dodd R. T. (1981) *Meteorites: A Petrologic-Chemical Synthesis*. Cambridge, New York. 368 pp.

Dollfus A., Mandeville J. C., and Duseaux M. (1979) The nature of the M-type asteroids from optical polarimetry. *Icarus, 37,* 124-132.

Feierberg M. A., Larson H. P., and Chapman C. R. (1982) Spectroscopic evidence for undifferentiated S-type asteroids. *Astrophys. J., 257,* 361-372.

Gaffey M. J. (1976) Spectral reflectance characteristics of the meteorite classes. *J. Geophys. Res., 81,* 905-920.

Gaffey M. J. (1984) Rotational spectral variations of asteroid (8) Flora: Implications for the nature of the S-type asteroids and for the parent bodies for the ordinary chondrites. *Icarus, 60,* 83-114.

Gaffey M. J. (1986) The spectral and physical properties of metal in meteorite assemblages: Implications for asteroid surface materials. *Icarus, 66,* 468-486.

Gaffey M. J. and McCord T. B. (1978) Asteroid surface materials: Mineralogical characterizations from reflectance spectra. *Space Sci. Rev., 21,* 555-628.

Gradie J. C. and Tedesco E. F. (1987) 1986 DA and 1986 EB: Iron objects in near-earth orbits (abstract). In *Lunar and Planetary Science XVIII*, pp. 349-350. Lunar and Planetary Institute, Houston.

Gradie J. C., Veverka J., and Buratti B. (1980) The effects of scattering geometry on the spectrophotometric properties of powdered material. *Proc. Lunar Planet. Sci. Conf. 11th*, pp. 799-815.

Graham A. L., Bevan A. W. R., and Hutchison R. (1985) *Catalogue of Meteorites*. Univ. of Arizona Press, Tucson. 460 pp.

Hapke B. (1981) Bidirectional reflectance spectroscopy, 1, Theory. *J. Geophys. Res., 86,* 3039-3054.

Harris A. W. and Young J. (1979) Photoelectric lightcurves of asteroids 42 Isis, 45 Eugenia, 56 Melete, 103 Hera, 532 Herculina, and 558 Carmen. *Icarus, 38,* 100-105.

Johnson P. B. and Christy R. W. (1972) Optical constants of the noble metals. *Phys. Rev., 6,* 4370-4379.

Matsui T. and Schultz P. H. (1984) On the brittle-ductile behavior of iron meteorites: New experimental constraints. *Proc. Lunar Planet. Sci. Conf. 15th,* in *J. Geophys. Res., 89,* C323-C328.

Ostro S. J., Campbell D. B., and Shapiro I. I. (1985) Mainbelt asteroids: Dual-polarization radar observations. *Science, 229,* 442-446.

Ostro S. J., Campbell D. B., and Shapiro I. I. (1986) Radar detection of 12 asteroids from Arecibo (abstract). *Bull. Amer. Astron. Soc., 18,* 796.

Pieters C. M. (1983) Strength of mineral absorption features in the transmitted component of near-infrared reflected light: First results from RELAB. *J. Geophys. Res., 88,* 9534-9544.

Sasovskaya I. I. and Noskov M. M. (1974) Optical properties of iron-nickel alloys. *Phys. Metal. Metallogr., 37,* 45-52.

Scaltriti F., Zappala V., and Stanzel R. (1978) Lightcurves, phase function and pole of the asteroid 22 Kalliope. *Icarus, 34,* 93-98.

Schaudy R., Wasson J. T., and Buchwald J. F. (1972) The chemical classification of iron meteorites: IV, A reinvestigation of irons with Ge concentration lower than 1 ppm. *Icarus, 17,* 173-192.

Wasson J. T. (1974) *Meteorites*. Springer-Verlag, New York. 316 pp.

Wooten F. (1972) *Optical Properties of Solids*. Academic, New York. 260 pp.

Zappala V., Scalitriti F., and Di Martino M. (1983) Photoelectric photometry of 21 asteroids. *Icarus, 56,* 325-344.

Zeigler K. W. and Florence W. B. (1985) Photoelectric photometry of asteroids 9 Metis, 18 Melpomene, 60 Echo, 116 Sirona, 230 Athamantis, 694 Ekard, and 1984 KD. *Icarus, 62,* 512-517.

Zellner B. (1979) Asteroid taxonomy and the distribution of the compositional types. In *Asteroids* (T. Gehrels, ed.), pp. 783-806. Univ. of Arizona, Tucson.

Nature and Origin of C-rich Ordinary Chondrites and Chondritic Clasts

E. R. D. Scott[1], A. J. Brearley[1], K. Keil[1], M. M. Grady[2], C. T. Pillinger[2], R. N. Clayton[3], T. K. Mayeda[3], R. Wieler[4], and P. Signer[4]

[1]Institute of Meteoritics, Department of Geology, University of New Mexico, Albuquerque, NM 87131
[2]Department of Earth Sciences, Walton Hall, Open University, Milton Keynes, MK7 6AA, England
[3]Enrico Fermi Institute, Department of Chemistry and Department of the Geological Sciences, University of Chicago, Chicago, IL 60637
[4]Swiss Federal Institute of Technology ETH, Sonneggstrasse 5, CH 8092, Zurich, Switzerland

---

We report petrologic, noble gas, and C and O isotopic data for four C-rich ordinary chondrites, Sharps, Allan Hills A77011, Allan Hills A78119, and Allan Hills A81024, which are not paired with each other, and three C-rich chondritic clasts in ordinary chondrites. All of these chondritic samples belong to petrologic type 3 and contain abundant carbon-rich aggregates, which, in at least two of these samples, consist largely of poorly graphitized carbon commonly associated with metallic Fe,Ni. Aggregates in Sharps host and clast range from metal-poor carbon aggregates to large metal grains with minor carbon on grain boundaries. Carbon isotopic analyses were obtained by stepwise combustion of chondrites and chondritic clasts. Bulk $\delta^{13}C$ isotopic values vary considerably from -11‰ for ALH A77011 to -24‰ and -26‰ for Sharps host and clast. Release patterns also vary with major C releases at 200°-600°C in the Sharps clast and at 700°-900°C in ALH A77011 and Sharps host. Oxygen isotopic data suggest that these C-rich chondrites and chondritic clasts belong to, or are closely related to, the H, L, and LL groups. Three of the four C-rich chondrites contain very high concentrations of trapped ^{36}Ar, 70-100 $\times 10^{-8}$ cm^3 STP/g, but ALH A81024 contains only 14×10^{-8} cm^3 STP/g. We conclude that the four C-rich chondrites probably originated on the H, L, and LL parent bodies, but that clasts in Dimmitt (DT1) and Plainview (PV1) and other regolith breccias may come from different bodies. Our data are consistent with the view that graphite was not a common nebular phase.

INTRODUCTION

Most type 3 ordinary chondrites contain 0.1-0.6 wt % C, but two are known to contain significantly higher concentrations (*Grady et al.,* 1982a). Sharps (H3) and Allan Hills A77011 (L/LL3), which both contain 1.0 wt % C, are characterized by abundant carbon-rich aggregates 10-10^3 μm in size (*Scott et al.,* 1981a,b; *McKinley et al.,* 1981). Carbon-rich aggregates are also abundant in five type 3 chondritic clasts found in four H chondrite regolith breccias (*Scott et al.,* 1981b, 1986). In these clasts, which contain up to 13 wt % C, the carbon-rich aggregates form a matrix that encloses the chondrules; the fine-grained, FeO-rich matrix material characteristic of type 3 ordinary chondrites (*Huss et al.,* 1981) is absent. Isolated carbon-rich aggregates can be found in many ordinary chondrite regolith breccias and are probably derived from material similar to the five chondritic clasts.

The carbon-rich aggregates were incorrectly described as "graphite-magnetite" aggregates by *Scott et al.* (1981a,b). The carbon was thought to be graphite because the aggregates contain small, irregularly shaped crystals 1-20 μm in size with the optical anisotropy and color of graphite. However, transmission electron microscopy of aggregates in Sharps and the Plainview clast PV1 shows that the graphitic phase is highly disordered and is more appropriately called "poorly-graphitized carbon" (*Brearley et al.,* 1987a,b). Electron microprobe analyses of areas within the carbon aggregates free of optically visible inclusions of metallic Fe,Ni and other phases commonly show the presence of 5-20 wt % Fe (in addition to minor amounts of Si, Mg, Ca, Ni, and S). X-ray diffraction of an aggregate from Dimmitt also showed the presence of magnetite.

However, this phase was not observed in Sharps or the Plainview clast PV1 by electron microscopy. Instead, *Brearley et al.* (1987a,b) found metallic Fe,Ni grains to be commonly distributed in the Sharps aggregates; in the clast PV1, Fe appears to be dissolved in the poorly graphitized carbon. Both contain minor chromite.

Aside from graphitic carbon, three other C-rich phases have been identified in type 3 ordinary chondrites. Cohenite, Fe_3C, and haxonite, $F_{23}C_6$, are common in Semarkona and two other LL3 chondrites where they occur with magnetite (*Taylor et al.,* 1981). Several other L3 and LL3 chondrites also contain minor amounts of these carbides (*Scott et al.,* 1982). In addition, Semarkona contains calcite in veins (*Hutchison et al.,* 1985). In all these cases, the C-bearing phases probably account for a minor proportion of the total C in the chondrite.

We have discovered abundant carbon-rich aggregates in two additional Antarctic type 3 chondrites, Allan Hills A78119 and Allan Hills A81024. Neither are among the shower of 80-odd specimens paired with Allan Hills A77011 (*Scott,* 1984a, 1986). Here we report petrographic and mineralogical data for ALH A78119 and A81024, Sharps, and a C-rich, type 3 chondritic clast in Sharps. We also studied the abundances and isotopic compositions of C, O, and noble gases in some of these specimens and the two C-rich chondritic clasts from Dimmitt and Plainview, DT1 and PV1. We wish to understand better the origin of these C-rich chondrites and clasts and their genetic relationships with normal type 3 ordinary chondrites. Detailed studies by transmission electron microscopy of the carbon-rich aggregates and a discussion of their origin will be reported elsewhere.

RESULTS

Petrography and Mineralogy

Electron microprobe analyses of between 33 and 66 randomly selected olivine and low-Ca pyroxene grains were analyzed in each of ALH A78119, A81024, and Sharps clast and host. Each mineral was analyzed for Fe, Mg, Ca, and Si using crystal spectrometers at 15kV accelerating voltage on an ARL EMX-SM electron microprobe.

Allan Hills A77011. This chondrite was the first one with large amounts of carbon-rich aggregates (3 vol %) to be identified (*McKinley et al.,* 1981). The number of specimens paired with ALH A77011 has since increased from 34 to 76 (*Scott,* 1984a, 1986). The following criteria were used to pair these specimens during the examination of thin sections. All show major weathering effects: an overall yellow brown color, much metallic Fe,Ni converted to iron oxides and hydroxides, and some or many weathering veins. A wide variety of chondrule types are present and 2-4 chondrules with pink or colorless, clear glass are normally visible within a thin section. Abundances of metallic Fe,Ni are characteristic of L chondrites, after some allowance for weathering. Shock effects are very minor, some or a few grains show undulatory extinction, and troilite has a flaky or coarsely crystalline texture under crossed polars. Fine-grained, FeO-rich matrix material (*Huss et al.,* 1981), either opaque, translucent or recrystallized, is present. Traces of fusion crust are seldom visible in thin section. Finally, the C-rich aggregates described by *McKinley et al.* (1981) are commonly associated with grains of metallic Fe,Ni. In rare cases where there was any uncertainty over a pairing, a second section was obtained. Noble gases were analyzed after these pairings were made, and in all cases the conclusions drawn from both studies were the same.

Allan Hills A78119. This chondrite (Fig. 1), like ALH A77011, contains well-defined chondrules, a few of which have clear glass, some Huss matrix, and shows only very mild shock effects. It differs from A77011 in that the overall weathering color of A78119 is much less yellow, metallic Fe,Ni is less weathered, and weathering veins are absent or rare. Carbon-rich aggregates (Fig. 2) are much less abundant in A78119, and fusion crust in one section (,7) is present at several places. Modal analyses (Table 1) confirm that the carbon-rich aggregates in A78119 are less abundant than in A77011 (0.7 versus 3.2 vol %). Carbon-rich aggregates in ALH A78119 consist of angular to subrounded Fe,Ni metal grains set in a matrix of fine-grained carbon (Fig. 2b) and may occasionally be associated with sulfide grains. Olivines are highly unequilibrated and range in composition from $Fa_{1.0}$ to Fa_{42} (Fig. 3). Low-Ca pyroxenes range in composition from $Fs_{0.6}$ to Fs_{25} with a peak in the histogram (Fig. 3) Fs_{0-4}. Means and standard deviations for these analyses are given in Table 2.

Allan Hills A81024. Like ALH A77011, this chondrite contains abundant carbon-rich aggregates, a few chondrules with clear glass, some Huss matrix, and exhibits extensive weathering effects. However, the overall weathering color is redder than ALH A77011, its chondrules appear to be smaller

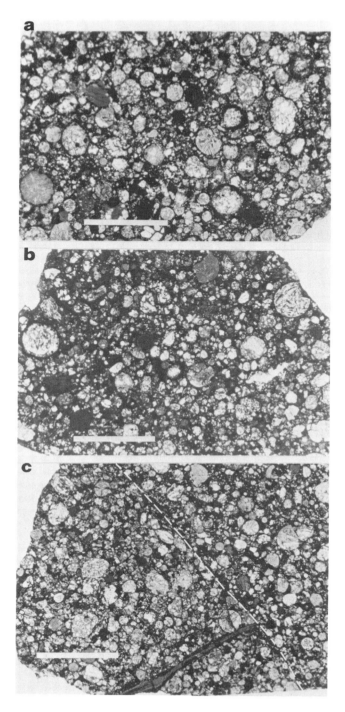

Fig. 1. Photomicrographs (transmitted light) of three ordinary chondrites rich in carbon: **(a)** ALH A78119, **(b)** ALH A81024, and **(c)** Sharps host (right) and a clast (left), which contains a prominent shock vein and is delineated from the host by a white broken line. All three chondrites and the Sharps clast are type 3 chondrites with well-defined chondrules and contain abundant carbon-rich aggregates. Scale bar = 3 mm on all the micrographs.

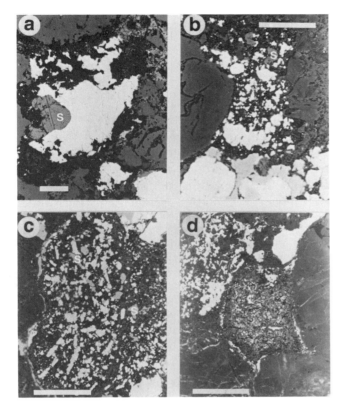

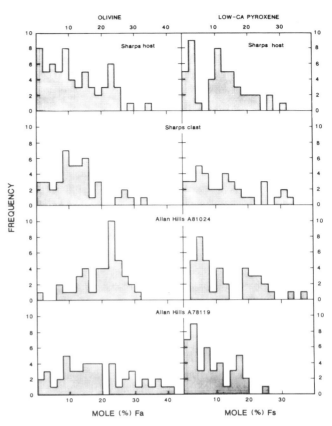

Fig. 2. Photomicrographs showing the range of textures exhibited by carbon-rich aggregates in ALH A78119, ALH A81024 and Sharps (host and clast). (a) Fe,Ni metal grain (white) with associated troilite (S) from ALH A81024 rimmed by dark carbon-rich material, which contains minor metal grains. The surrounding dark gray region consists almost entirely of silicates. (b) Carbon-rich aggregate from ALH A78119 consisting of dispersed Fe, Ni metal grains within a carbon-rich matrix. Minor troilite is also present (S). (c) Carbon-rich aggregate from the Sharps clast consisting of angular and elongate Fe,Ni metal grains and minor troilite dispersed throughout a carbon matrix. (d) A very carbon-rich aggregate, 60 × 80 μm in size, in the Sharps clast; it contains only extremely fine-grained metal, which constitutes <10% of the total volume of the aggregate. Scale bars = 50 μm.

Fig. 3. Histograms showing compositional distribution of olivines in mol % fayalite and low-Ca pyroxenes in mol % ferrosilite from Sharps (host and clast), ALH A81024, and ALH A78119. All four C-rich chondrites contain olivines and low-Ca pyroxenes that display marked compositional heterogeneity, typical of the least equilibrated type 3 ordinary chondrites.

TABLE 1. Modal analyses (wt %) of Allan Hills A78119, A81024, and Sharps host and clast.

Section	ALH A78119 ,15	ALH A81024 ,19	Sharps host USNM 640-1,3	Sharps clast USNM 640-1,3
Silicate	69.9	73.2	66.5	74.7
Metallic Fe,Ni	16.2	14.7	14.3	11.7
Fe-rich matrix material*	6.8	5.2	6.5	2.4
Troilite	6.7	4.6	5.9	3.7
C-rich aggregates[†]	0.4	2.3	1.7	1.7
Metal/C aggregates[‡]	<0.1	<0.1	5.1	5.8

Modal analyses based on 1000 points on each section and calculated using the following densities (g/cm^3): silicate = 3.3; metallic Fe,Ni = 7.7; Huss matrix = 3.6; troilite = 4.75; C-rich aggregates = 2.1; C-metal aggregates = 4.9.
*Opaque and recrystallized matrix (*Huss et al.*, 1981).
[†]Carbon-rich aggregates containing <50 vol % metallic Fe,Ni grains <3 μm in diameter.
[‡]Carbon-rich aggregates containing >50 vol % metallic Fe,Ni grains <3 μm diameter.

TABLE 2. Electron microprobe analyses of olivine and low-Ca pyroxene in three C-rich ordinary chondrites and one C-rich clast.

Sample*	Olivine				Low-Ca pyroxene				
	No. of anal.	Fa (mol %)		CaO (wt %)		No. of anal.	Fe (mol %)		Wo (mol %)	
		mean	σ	mean	σ		mean	σ	mean	σ
ALH A78119	48	18.0	11.0	0.12	0.12	44	8.3	6.4	1.0	0.9
ALH A81024	44	20.0	6.5	0.07	0.08	40	13.2	9.1	0.9	0.8
Sharps	66	13.2	8.5	0.19	0.13	50	11.5	7.4	1.0	0.8
Sharps clast	38	11.5	7.4	0.14	0.07	33	12.5	9.3	1.2	1.2

*Section numbers: 78119,15; 81024,11; USNM 640-1 (Sharps).

and rounder, and the silicates less fractured than those of ALH A77011. In addition, fusion crust was visible on two sides of the section. The carbon-rich aggregates are normally associated with large, irregularly shaped Fe,Ni grains (Fig. 2) and a few contain fine-grained metal. Some aggregates contain up to 30 vol % of metal grains 10-15 μm in diameter. Curiously, ALH A81025 and A81030-81032 are paired with ALH A77011 (*Scott*, 1984a), whereas ALH A81024 is not.

Electron probe analyses of olivines and low-Ca pyroxenes suggest that the silicates are more equilibrated than those of the other chondrites in this survey (Fig. 3, Table 2). The histogram of Fe/(Fe+Mg) ratios in olivine shows a prominent peak at Fa_{22-24}, in the range appropriate to L4-6 chondrites. Since the analyzed section (,11) is rather small and *Sears and Weeks* (1986) claim that ALH A81024 is an H3 chondrite, another section (,19) was obtained from a different part of the specimen. Olivine in the second section shows a similar peak at Fa_{22-24} (B. Epling, personal communication, 1987).

Sharps. Carbon-rich aggregates were originally reported by *Fredriksson et al.* (1969) in Sharps. We have studied two thin sections (USNM 640-1, 640-3) of Sharps and identified a previously unreported type 3 chondritic clast (Fig. 1a), which also contains a significant volume of C-rich aggregates. The clast-host boundary bisects both of the thin sections diagonally. Only after a new section was prepared from a 2 × 5 cm slice of Sharps was it clear which was host material. According to R. S. Clarke (personal communication, 1981), this slice (USNM 640) contains three or four dark carbonaceous clasts of the kind described by *Fredriksson et al.* (1969), and two centimeter-sized, medium dark clasts. One of the latter was studied and proved to be identical to the clast in USNM 640-1 and 640-3; it provided the material for our noble gas and isotopic studies. The clast contains considerable evidence of shock effects, which are not apparent in the host. Occasional local veins of impact melt are present along with veinlets of metal and troilite, which occur throughout the clast. The interface between the host and clast is decorated by veinlets and blebs of metal.

Modal analyses of the clast and host show some small differences as indicated in Table 1. In the clast the silicate content is slightly higher and Huss matrix, metal, and troilite contents lower than in the host. In Sharps the carbon-rich material is almost invariably associated with Fe,Ni metal, but the proportions of the two components vary considerably. In Table 1 we have arbitrarily divided the carbon-rich aggregates into two groups; one containing less than 50 vol % metal grains <3 μm in diameter and the other containing more than 50 vol % metal grains <3 μm in diameter. The proportions of both groups in clast and host are indistinguishable within the errors of the technique and total around 7 wt %.

The C-rich aggregates in Sharps host and clast show a diverse range of textures that cover a whole spectrum from aggregates containing almost no metal to large metal grains that contain small inclusions of C-rich material. The range of textures is summarized in Fig. 2. Figures 2a and 2b are from ALH A78119 and A81024, respectively, but aggregates with similar textures also occur in Sharps. Aggregates with abundant fine-grained, elongate or angular metal grains ranging from 1 μm to 15 μm (Fig. 2c) are also common in Sharps. Such aggregates contain >50% by volume of carbon; the rest consists of Fe,Ni metal and occasional sulfide grains. The aggregates with the highest volume proportion of carbon (Fig. 2d) contain only extremely fine-grained, disseminated metal grains <3 μm in diameter that constitute 10 vol % of the aggregate. This type of occurrence appears to be most abundant in the clast.

We have investigated the mineral chemistry of both clast and host in Sharps to ascertain whether there are any petrological differences between them other than the modal proportions of constituents. Means and standard deviations for clast and host olivines and low-Ca pyroxenes are summarized in Table 2 and histograms of fayalite in olivine and ferrosilite in the low-Ca pyroxenes are shown in Fig. 3. Both clast and host show broad spreads in olivine and pyroxene compositions, typical of petrologic type 3 chondrites. The clast is clearly at least as unequilibrated as the host and is undoubtedly a type 3 chondrite. The modal analysis reveals a significantly lower Fe,Ni metal content in the clast than in the host. However, after allowance for metal associated with carbon aggregates, both clast and host appear to have H group abundances of Fe,Ni metal.

We have also reinvestigated the data of *Huss et al.* (1981) for the FeO-rich silicate matrix in Sharps clast and host that were obtained on USNM section 640-1. Although the data are almost indistinguishable for most of the major elements, the Huss matrix in the clast appears to be depleted in alkalis (K_2O and Na_2O) in comparison with the majority of analyses in the host matrix. This difference may be attributable to shock effects that are apparent in the clast and may have resulted in the loss of volatile alkalis.

Plainview clast PV1 and Dimmitt clast DT1. *Scott et al.* (1981b) described type 3 chondritic clasts in the H chondrite regolith breccias, Plainview, Dimmitt, and Weston, that contain high concentrations of carbon-rich aggregates. They found that the aggregates constitute 14-36 vol % of these clasts, and that there is no Huss matrix present. Consequently these clasts contain very high concentrations of C, 6 and 13 wt % in DT1 and PV1, respectively, considerably higher than C-rich, type 3 ordinary chondrites. The clasts resemble unequilibrated ordinary chondrites in that they contain well-defined chondrules and compositionally heterogeneous olivines and pyroxenes with high Ca contents in the olivines (*Scott et al.*, 1981a,b). The carbon-rich aggregates in the chondritic regolith breccias do differ slightly from those in the type 3 ordinary chondrites in that very little metal is observed associated with the carbon phase in the former. Material from clasts DT1 and PV1 was extracted for isotopic analysis (see below).

Carbon Isotopes

Carbon isotopic analyses were carried out on Sharps host and clast, ALHA 77214 (paired with ALHA 77011), the Plainview clast, PV1, and Dimmitt host by stepwise combustion following the procedures described by *Swart et al.* (1983). Carbon combustion release profiles are presented in Figs. 4a-e and the isotopic data for the samples are reported in Table 3.

Sharps clast and host and ALH A77214. The release profiles for Sharps host and ALH A77214 (Figs. 4a,b) are similar in that most of the C combusts in the temperature range 700°-900°C, and both have smaller release peaks at 200°-400°C. The 700°-900°C peak can be attributed to poorly graphitized carbon, which is abundant in both materials.

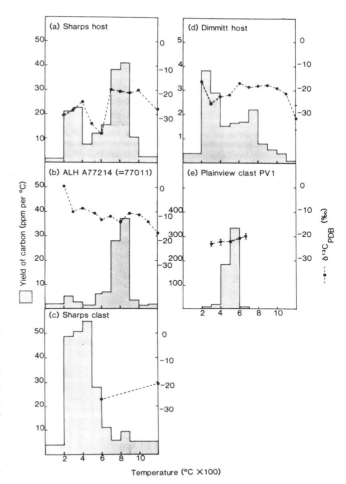

Fig. 4. Yield of C (histogram) and its isotopic composition (broken line) released during stepwise heating of chondrites and chondritic clasts; (a) Sharps host, (b) ALH A77214 (=77011), (c) Sharps clast, (d) Dimmitt host, (e) Plainview clast PV1. Carbon release profiles and isotopic compositions are surprisingly diverse. The most extreme are from the Sharps clast and Allan Hills A77214 with $\delta^{13}C$ values of -26‰ and -12‰, respectively. 2σ errors on isotopic values are less than the size of the spots, unless shown otherwise. Note the changes in scale on the yield axis.

TABLE 3. Carbon abundances and isotopic compositions of some C-rich ordinary chondrites.

Sample	Weight (mg)	Carbon (wt %)	$\delta^{13}C_{PDB}$‰
ALH A77214	22.080	0.99	-11
Sharps host	4.619	1.77	-24
Sharps clast	0.119	2.33	-26
Dimmitt host	47.080	0.16	-21
Plainview clast PV1	1.415	5.65	-21

(*Brearley et al.*, 1987a,b). The small low-temperature peak in ALH A77214 may be due to weathering; that in Sharps is due to organic material, possibly a terrestrial contaminant. The release profile from the Sharps clast (Fig. 4c) is very different in that most of the C is liberated in the temperature range 200°-600°C. Release profiles for synthetic samples of shocked and unshocked poorly graphitized carbons are not available for comparison with those in Fig. 4. However, the profiles are not consistent with graphite, which gives a well-defined peak at 700°-800°C (*Grady et al.*, 1982b).

The measured $\delta^{13}C$ values vary considerably between the three samples: -11‰ for ALH A77214, -24‰ for the Sharps host, and -26‰ for the Sharps clast. It is probable that terrestrial weathering effects have not significantly changed the isotopic composition of ALH A77214, and its heavy isotopic value probably reflects a real difference from the C components in Sharps. In the case of the Sharps clast, only a small amount of sample was available so that insufficient CO_2 was produced at each temperature step for analysis. Consequently the CO_2 from each step up to 600°C was measured together, and the same was done for increments from 700°-1200°C.

Of all the type 3 ordinary chondrites analysed to date, ALH A77214 contains the most isotopically heavy C by around 3‰ (*Grady et al.*, 1982a). The Sharps clast is isotopically lighter than the host with a $\delta^{13}C$ value of -26‰ (cf. -20‰ for the main indigenous C components in the host). This value is isotopically relatively light for type 3 ordinary chondrites and is closer to results from type 4 ordinary chondrites. We note that ALH A77214 and the Sharps clast contain the highest C contents and represent the extreme C-isotopic values of all the type 3 ordinary chondrites.

Dimmitt. Carbon isotopic analyses have been made of the bulk Dimmitt host, but no data are available for the clast DT1. The release profile for the bulk Dimmitt host (Fig. 4d) differs somewhat from other type 3 ordinary chondrite release profiles. Although a significant proportion of the total C is liberated in the graphite temperature range, the rest of the C begins to combust at much lower temperatures, too low to be pure graphite. Well-crystallized graphite has been observed by transmission electron microscopy in acid residues from Dimmitt (*Lumpkin*, 1986), but the residues may not be representative of all the carbon phases present. The isotopic value of the C liberated at 400°-900°C is -18‰, consistent with ordinary chondrite isotopic values.

Plainview clast PV1. The release profile for Plainview (PV1) (Fig. 4e) differs from those of both Dimmitt and the C-rich, type 3 ordinary chondrites. All of the C is liberated by 700°C, most coming off at 500°-600°C. This release temperature is too low for a release pattern from graphite, suggesting that some other phase is responsible. Transmission electron microscopy of clast PV1 (*Brearley et al.*, 1987b) confirms that graphite is not present and that the carbon phase is a poorly-graphitized or turbostratic carbon, which is less well-graphitized than that in Sharps. The $\delta^{13}C$ value for the bulk C in PV1 is -22‰, essentially falling within the range of type 3 ordinary chondrites and comparable with the results from Sharps host and clast.

Oxygen Isotopes

Oxygen isotopic compositions for four specimens paired with Allan Hills A77011, Sharps, the Sharps chondritic clast described above, and the type 3 chondritic clasts, DT1 and PV1, are given in Table 4 and plotted in Fig. 5. The standard techniques outlined by *Clayton et al.* (1976) were used. Data in Table 4 for ALH A77015 are those of *Onuma et al.* (1983).

The four specimens paried with ALH A77011 are all heavily weathered and have weathering grades of B/C or C (*McKinley and Keil*, 1984; Appendix 1 in *Marvin and Mason*, 1980). Their O isotopic compositions define a linear array that is almost parallel to the terrestrial fractionation line. The relative degrees

TABLE 4. Oxygen isotopic composition of some chondrites and chondritic clasts that contain C-rich aggregates.

Sample	$\delta^{18}O$ (‰)	$\delta^{17}O$ (‰)
Allan Hills A77015*,†	4.13	3.22
Allan Hills A77049†	3.23	2.75
Allan Hills A77050†	3.52	2.88
Allan Hills A77241†	2.81	2.25
Dimmitt clast DT1	3.37	2.49
Plainview clast PV1	3.35	2.40
Sharps	3.95	2.73
Sharps clast	3.97	2.81

*Data from *Onuma et al.* (1983).
† Samples paired with Allan Hills A77011 (*Scott*, 1984a).

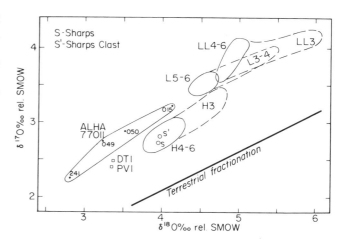

Fig. 5. Three oxygen isotope plot showing isotopic compositions of the C-rich ordinary chondrites, ALH A77015, A77049, A77050, A77241 (all paired with ALH A77011), and Sharps (S), and C-rich chondritic clasts from Sharps (S'), Dimmitt (DT1), and Plainview (PV1). The highly weathered ALH A77011 specimens define a field that is subparallel to the terrestrial fractionation line. Before weathering, their isotopic compositions probably lay close to the L and LL fields. Sharps and the Sharps clast lie within the field of H4-6 chondrites, unlike nearly all other analyzed H3 chondrites. The clasts, DT1 and PV1, which are also somewhat weathered, have O isotopic compositions unlike those of known chondrites.

of weathering of all the samples are not known, but the analyzed sample of 77241 was more weathered than that of 77049. This trend is therefore consistent with measurements on other heavily weathered Antarctic meteorites and the very light O isotopic composition of Antarctic ice, $\delta^{18}O$ of about -40‰ relative to SMOW in the Allan Hills region (*Clayton et al.*, 1984; *Fireman and Norris*, 1982). Thus the preterrestrial composition of the ALH A77011 specimens lies an unknown distance along an extrapolation of this trend toward the L and LL fields.

The Sharps host and clast have O isotopic compositions that are indistinguishable from each other and the mean H4-6 chondrite composition (Fig. 5). However, five out of six previously analyzed H3 chondrites lie on an extension of the H4-6 field toward heavier O isotopic compositions; only Dhajala (H3.8) lies within the H4-6 field (*Gooding et al.*, 1983). Since the $\delta^{18}O$ concentration correlates inversely with the degree of metamorphism (*Clayton et al.*, 1981; *Sears and Weeks*, 1983), it is somewhat surprising that Sharps, which is an H3.4, should lie in the H4-6 field.

Oxygen isotopic compositions of the two clasts PV1 and DT1 are very similar. Terrestrial O contamination in thin sections of PV1 is very minor compared with that of the ALH A77011 samples. The clast of DT1, however, contains 3 vol % weathered metal (*Scott et al.*, 1981b) and its oxygen may have been noticeably affected by terrestrial contamination. The proximity of DT1 and PV1 in Fig. 5 suggests that terrestrial oxygen in DT1 caused only a minor change in its bulk composition. Chondrites with O isotopic compositions like those of the clasts of DT1 and PV1 are not known. A chondritic clast from Mezö-Madaras (L3) with a very similar oxygen isotopic composition is known, but unlike DT1 and PV1 it is rich in fine-grained, FeO-rich matrix and has no observable C-rich aggregates (R.N. Clayton and E.R.D. Scott, unpublished data, 19xx).

Oxygen isotopic compositions of these C-rich type 3 chondrites and type 3 chondritic clasts show a rather broad distribution that in some case is attributable to weathering. However, it seems that they have affinities with both H- and L-group chondrites. It is possible that C-rich ordinary chondrites are isotopically slightly lighter in O than comparable C-poor ordinary chondrites. But in view of the considerable effects of weathering on some samples, we cannot establish unequivocally if this is indeed the case.

Noble Gases

Samples of nine C-rich ordinary chondrites (100-200 mg in size) from Antarctica that were preheated to 100°C for 12 hours were analyzed for He, Ne, and Ar isotopes using the procedures of *Signer et al.* (1977). Errors are ±5% for concentrations and ±1% for isotopic ratios. Table 5 shows that the seven specimens paired with Allan Hills A77011 have very similar concentrations of spallogenic and trapped components (*Wieler et al.*, 1985). These analyses also agree fairly well with those for other chondrites believed to be part of this fall, A77015, A77167, A77214, and A77260 (see *Schultz and Kruse*, 1983). The sample of ALH A81032 appears to have lost much He and small amounts of Ne, but its characteristic concentrations of trapped and spallogenic noble gases confirm the pairing with A77011. Aside from Semarkona (M. G. Zadnik, unpublished data in *Anders and Zadnik*, 1985), Sharps, and ALH A78119 (see below), ALH A77011 is the only ordinary chondrite with >60 × 10^{-8} cm^3 STP/g of trapped ^{36}Ar.

Although there are some similarities between the concentration patterns of ALH A78119 and the A77011 samples, A78119 has almost double the ^{3}He concentration of the specimens paired with A77011. Cosmogenic ^{21}Ne and radiogenic ^{40}Ar are, respectively, about 20% and 40% higher in A78119. By contrast, trapped ^{20}Ne and ^{36}Ar concentrations are significantly lower in A78119. It is highly improbable that one portion of a meteorite lost a considerable part of its cosmogenic and radiogenic gases, while retaining more efficiently its trapped noble gases. Thus noble gas data support the petrographic evidence that ALH A78119 is not part of the A77011 group of paired specimens. Allan Hills A81024 shows vastly higher concentrations of spallogenic noble gases and less than 20% of the trapped ^{36}Ar in A77011. Again, the noble gas concentration data support the petrographic evidence that these are not paired specimens.

Cosmogenic ^{21}Ne concentrations and the ^{22}Ne/^{21}Ne$_{cos}$ ratios were calculated by assuming the following compositions for trapped and cosmogenic Ne: ^{20}Ne/^{22}Ne$_{tr}$ = 8.4, ^{22}Ne/^{21}Ne$_{tr}$ = 32, ^{20}Ne/^{21}Ne$_{cos}$ = 0.91. The correction for trapped ^{21}Ne is less than 1% in all cases; ^{21}Ne$_{cos}$ amounts are therefore not given separately in Table 5. The ^{22}Ne/^{21}Ne$_{cos}$ ratios were used to calculate shielding-corrected ^{21}Ne exposure ages according to *Nishiizumi et al.* (1980).

The specimens of Sharps and the Sharps clast described above were analyzed in Mainz by L. Schultz. The clast has slightly lower trapped ^{36}Ar, but higher ^{40}Ar. *Zadnik* (1987) found higher concentrations of ^{36}Ar and ^{40}Ar in Sharps host, but similar concentrations of the Ne isotopes.

DISCUSSION

Classification of C-rich Chondrites

All of the C-rich chondrites and clasts that we have described show some affinities with ordinary chondrites, and none appear to have significant chemical, mineralogical, or isotopic affinities with CI, CM, or CV chondrites, which are richer in C. Table 6 summarizes the classification of the C-rich chondrites and chondritic clasts by chemical group and petrologic subtype according to various parameters. The problem in classifying type 3 ordinary chondrites by group have been recently discussed by *Sears and Weeks* (1986). Because of their heterogeneous silicates and diverse oxygen isotopic compositions (Fig. 5), type 3 chondrites are best classified using the bulk compositional ranges of type 4-6 chondrites, making allowance for possible systematic differences between type 3 and type 4-6 chondrites of the same group. *Sears and Weeks* (1986) classify ALH A77011 as an LL chondrite with medium confidence, and ALH A81024 and Sharps as H chondrites with high confidence. Of the 38 type 3 chondrites they examined, 24% were classified with high confidence, and 68% with medium confidence.

Modal analyses for metallic Fe,Ni have also been used to classify the C-rich chondrites and clasts using criteria for type 4-6 chondrites (*Gomes and Keil*, 1980). Weathering effects and, for the clast PV1, the presence of shock veins of metal that may not be indigenous to the clast (*Scott et al.*, 1981b) add to the uncertainty of these measurements. Except for ALH A77011, which appears to have too much metallic Fe,Ni for the LL classification of *Sears and Weeks* (1986), the modal analyses for metallic Fe,Ni are generally consistent with classifications from bulk chemistry.

In Table 6, ALH A81024 is classed as an L chondrite on the basis of the prominent peak of Fa 22-24 in its olivine compositional spectrum (Fig. 3). In general such peaks in type 3 chondrites lie in the range appropriate to equilibrated chondrites with the same bulk chemistry (*Scott*, 1984b). However, ALH A81024 appears to be another exception to this rule (*Scott et al.*, 1986).

The petrologic subtypes of the C-rich chondrites and clasts are also given in Table 6, using the criteria of *Sears et al.* (1980, 1982) for four parameters: thermoluminescence sensitivity, standard deviation of fayalite concentrations normalized to mean concentration, bulk C, and trapped ^{36}Ar contents. A fifth estimate of the petrologic subtype is derived using the mean CaO concentration of olivine (*Scott*, 1984b). Except for ALH A81024, all of the subtype estimates for chondrites and clasts are 3.0-3.5. Allan Hills A81024 has olivine compositions and an ^{36}Ar content appropriate to type 3.6-3.7, and a thermoluminescence sensitivity indicative of type 3.5.

TABLE 5. Concentrations of He, Ne, and Ar (10^{-8} cm^3 STP/g) and isotopic ratios in C-rich ordinary chondrites.

Sample	^{3}He	^{4}He	^{20}Ne	^{21}Ne	^{22}Ne/^{21}Ne	^{20}Ne/^{22}Ne	^{22}Ne/^{21}Ne cos	^{36}Ar	^{36}Ar/^{38}Ar	^{40}Ar	^{36}Ar† trap	T$_{21}$ (Ma)
ALH A77047*	3.51	920	2.51	0.76	1.37	2.43	1.09	86.0	5.34	3050	86.0	2.2
	3.48	912	2.57	0.73	1.39	2.52	1.09	90.4	5.35	2690	90.4	2.1
ALH A78015*	3.29	928	2.57	0.73	1.40	2.54	1.09	95.7	5.29	3120	95.6	2.2
ALH A81025*	3.15	873	2.51	0.71	1.39	2.55	1.09	86.8	5.29	2930	86.7	2.0
ALH A81030*	3.51	934	2.63	0.78	1.36	2.45	1.08	90.3	5.28	3080	90.1	2.2
ALH A81031*	3.25	1008	2.37	0.68	1.40	2.48	1.11	85.0	5.28	2600	84.9	2.1
ALH A81032*	1.75	672	2.16	0.61	1.41	2.53	1.10	83.6	5.29	2560	83.5	1.8
ALH A81121*	3.00	882	2.40	0.69	1.41	2.45	1.12	88.9	5.29	2720	88.8	2.3
ALH A78119	6.01	1422	2.19	0.84	1.44	1.81	1.24	68.4	5.27	4050	68.3	4.6‡
ALH A81024	48.3	1497	10.3	9.86	1.15	0.90	1.14	15.2	3.94	6700	14.5	39‡
Sharps host§	30.0	1900	10.4	7.54	1.14	1.21	1.09	99.0	4.97	4180	98.0	23
Sharps clast§	31.1	1850	10.1	7.96	1.13	1.12	1.09	81.1	4.95	5330	80.2	23

*Paired with ALH A77011.
†Calculated by assuming (^{36}Ar/^{38}Ar)$_{tr}$ = 5.35, (^{36}Ar/^{38}Ar)$_{cos}$ = 0.63.
‡Exposure ages of ALH A78119 and A81024 are calculated assuming both are H chondrites (see Table 6).
§Analyzed by L. Schultz (Max-Planck-Institut, Mainz).

TABLE 6. Classification of C-rich chondrites and clasts by chemical group and petrologic subtype according to various parameters.

Specimen	Chemical group				Petrologic subtype				
	Bulk comp.	Modal Fe,Ni	O-isotope comp.	Olivine comp.	TI	σ Fa/mean Fa	Bulk C	^{36}Ar$_p$	CaO in olivine
ALH A77011	LL^a (L^b)	L^b	(L/LL^c)	-	3.5^f	3.5^b	3.0^i	3.0^c	3.6^l
ALH A78119*	-	H^c	-	-	3.4^g	≤3.4^c	-	3.0^c	3.4^c
ALH A81024	H^a	H^c (L^d)	-	L^c	3.5^g	3.6^c	-	3.6^c	3.7^c
Sharps host	H^{a,j}	H^c	H^c	-	3.4^h	≤3.4^c	3.0^j	3.0^c	3.1^c
Sharps clast	-	H^c	H^c	-	-	≤3.4^c	-	3.0^c	3.4^c
DT1 clast	-	(H)	(H^c)	-	-	3.5^e	-	-	3.3^e
PV1 clast	-	(L)	(H^c)	-	-	3.5^e	-	≥3.4^k	3.5^e

Parentheses denote less reliable classification.
Sources of data: (a) *Sears and Weeks* (1986); (b) *McKinley et al.* (1981); (c) this work; (d) *Mason* (1983); (e) *Scott et al.* (1981b); (f) *Sears et al.* (1982); (g) *Sears and Hasan* (1987); (h) *Sears et al.* (1980); (i) see references in *McKinley et al.* (1981); *Fredriksson et al.* (1969); (k) *Caffee et al.* (1987); (l) S. G. McKinley, private communication in *Scott* (1984b).
*Classified as L chondrite by *Moore* (1985).

It is not known whether C and ^{36}Ar were lost during metamorphism (see *Wasson*, 1974). *Anders and Zadnik* (1985) argue that they were not, and that the parameters used to derive subtypes do not correlate perfectly because C and ^{36}Ar, for example, were controlled by nebular and not planetary processes. In general these parameters do correlate very well and C-rich metamorphosed ordinary chondrites are not known. The only type 4-6 chondrites known to contain C-rich aggregates are regolith breccias. The aggregates in these chondrites were probably added to equilibrated materials during regolith mixing processes.

Origin of C-rich Chondrites and Clasts

The above discussion and the presence of small amounts of carbon-rich aggregates in "normal" type 3 ordinary chondrites, such as Allan Hills A77299, Bremervörde, Dhajala, Khohar, Mezö-Madaras, Allan Hills A77278, and Piancaldoli (*Scott et al.*, 1981b) suggest that the C-rich chondrites probably originated in the ordinary chondrite parent bodies. However, there are sufficient petrographic differences between the clasts, DT1, PV1, etc., and the type 3 chondrites to suggest the possibility of an independent origin for the clasts (*Scott et al.*, 1981b). These differences include the small amounts of metallic Fe,Ni associated with carbon-rich aggregates in the clasts, clast abundances of C that are at least three times higher than those of C-rich ordinary chondrites, and the absence of fine-grained, FeO-rich silicate matrix (Huss matrix) in the clasts. We suspect that these clasts represent foreign material introduced into the regolith of the H parent body and possibly also the L parent body.

Electron microscopy of carbon-rich aggregates in the Plainview clast PV1 (*Brearley et al.*, 1987b) shows that the "missing" Fe-rich silicate matrix material is not located inside the carbon. Thus these type 3 chondritic clasts, like EH3 chondrites (*Lusby et al.*, 1986), have no silicate matrix material. Presumably silicate dust was very efficiently converted to chondrules at their nebular formation locations, or efficiently separated after chondrule formation.

Carbon in Ordinary Chondrites

In Sharps, ALH A77011, and ALH A81024, C is concentrated in the carbon-rich aggregates that we have described, but this probably constitutes only a part of the C, albeit a significant fraction. In most type 3 ordinary chondrites, carbon-rich aggregates and carbides are absent or present in very minor amounts, and C is probably concentrated in the FeO-rich matrix material. *Kurat* (1970) found about 2-6 wt % C in matrix rims 10-20 μm wide on Tieschitz chondrules using an electron microprobe. Similarly, *Fredriksson et al.* (1969) report approximately 2-3 wt % C in Sharps matrix material. According to *Allen et al.* (1980), 2 wt % C in 20-μm-wide rims on chondrules that are 250 μm in diameter would account for the typical bulk C content of a type 3 ordinary chondrite. Thus in H-L-LL3 chondrites, C in the FeO-rich matrix probably constitutes a few tenths of a wt % C in the whole rock analysis.

The C-rich phase in the FeO-rich matrix material of type 3 ordinary chondrites has not been characterized by transmission electron microscopy. However, Raman spectroscopy of the matrix rims on Tieschitz chondrules by *Christophe Michel-Levy and Lautie* (1981) shows the existence of a poorly ordered graphitic phase. Christophe Michel-Levy (personal communication, 1982) has also obtained Raman spectra from carbon in the matrix rims of chondrules and the carbon-rich aggregates in ALH A77011. She finds that carbon in the matrix is similar to the poorly ordered graphitic phase in the Tieschitz rims, whereas carbon in the carbon-rich aggregates of ALH A77011 is more ordered. The ratio of the 1605 and 1355 cm^{-1} bands, which is a measure of the size of the graphite layers, is 2.4-3.5 in the aggregates and 1.15 ± 0.05 in the matrix rims of ALH A77011 chondrules. These values give approximate limits of 40 and 100 Å to the respective sizes of the graphite layers in rims and aggregates, using data of *Lespade et al.* (1982).

In our transmission electron microscope studies of the carbon-rich aggregates in Sharps and the Plainview clast PV1 (*Brearley et al.*, 1987a,b), we found poorly graphitized or amorphous carbon; graphite was not observed. Graphite has not been observed in type 4-5 ordinary chondrites, and there are only very limited reports of its occurrence in type 3 ordinary chondrites. *Ramdohr* (1973) identified graphite filling cracks and grain boundaries in the Grady (1937) (H3) chondrite. Since graphite was identified only optically, corroborative studies by electron microscopy or Raman spectroscopy are needed. Graphite of unspecified crystallinity was identified in acid residues of Dimmitt by *Niederer and Eberhardt* (1977) as a carrier of Ne-E. Electron microscopy of a Dimmitt residue by *Lumpkin* (1986) showed that a minor component is micrometer-sized, highly ordered graphite crystals; the major part is partially ordered graphite. Additional electron microscopy of carbon phases in acid residues of carbonaceous chondrites by *Lumpkin* (1986) and *Smith and Buseck* (1981) and in interplanetary dust particles by *Christofferson and Buseck* (1983), *Bradley et al.* (1984), and *Rietmeijer and Mackinnon* (1985) have also failed to find graphite. This lends further support to Nuth's argument (*Nuth*, 1985) that well-crystallized graphite was extremely rare or nonexistent in the solar nebula at the time of accretion.

We originally speculated that some of the carbon-rich aggregates in ordinary chondrites might contain presolar carriers of Ne-E (*Scott et al.*, 1981a), as *Niederer and Eberhardt* (1977) identified a graphitic phase in Dimmitt as a carrier of Ne-E. However, *Caffee et al.* (1987) have analyzed the Ne in carbon-rich aggregates from Dimmitt, Sharps, and the clast DT1, and they find no evidence for the presence of Ne-E in any of these phases. Enhanced concentrations of trapped primordial noble gases and exceptional isotopic effects were also not found.

Acknowledgments. We thank R. S. Clark, Jr. and the Antarctic Meteorite Working Group for providing meteorite samples, and L. Schultz for providing noble gas data for the Sharps samples. This work was supported by NASA grant NAG 9-30 to K. Keil, NSF grant EAR 8316812 to R. N. Clayton, the U. K. Science and Engineering Research Council, and the Swiss National Science Foundation.

REFERENCES

Allen J. S., Nozette S., and Wilkening L. L. (1980) A study of chondrule rims and chondrule irradiation records in unequilibrated ordinary chondrites. *Geochim. Cosmochim. Acta, 44,* 1161-1175.

Anders E. and Zadnik M. G. (1985) Unequilibrated ordinary chondrites: a tentative subclassification based on volatile-element content. *Geochim. Cosmochim. Acta, 49,* 1281-1291.

Bradley J. P., Brownlee D. E., and Fraundorf P. (1984) Carbon compounds in interplanetary dust: evidence for formation by heterogeneous catalysis. *Science, 223,* 56-58.

Brearley A. J., Scott E. R. D., Taylor G. J., and Keil K. (1987a) Transmission electron microscopy of graphite-magnetite aggregates in the Sharps (H3) chondrite (abstract). In *Lunar and Planetary Science XVIII,* pp. 123-124. Lunar and Planetary Institute, Houston.

Brearley A. J., Scott E. R. D., and Keil K. (1987b) Carbon-rich aggregates in ordinary chondrites: transmission electron microscopy observations of Sharps (H3) and Plainview (H regolith breccia) (abstract). *Meteoritics,* in press.

Caffee M. W., Hohenberg C. M., Swindle T. D., and Hudson G. B. (1987) Noble gases from graphite-magnetite inclusions in ordinary chondrites. *Geochim. Cosmochim. Acta,* in press.

Christofferson R. and Buseck P. R. (1983) Epsilon carbide: a low temperature component of interplanetary dust particles. *Science, 222,* 1327-1329.

Christophe Michel-Levy M. and Lautie A. (1981) Microanalysis by Raman spectroscopy of carbon in the Tieschitz chondrite. *Nature*, 292, 321-322.

Clayton R. N., Onuma N., and Mayeda T. K. (1976) A classification of meteorites based on oxygen isotopes. *Earth Planet. Sci. Lett.*, 30, 10-18.

Clayton R. N., Mayeda T. K., Gooding J. L., Keil K., and Olsen E. J. (1981) Redox processes in chondrules and chondrites (abstract). In *Lunar and Planetary Science XII*, pp. 154-156. Lunar and Planetary Institute, Houston.

Clayton R. N., Mayeda T. K., and Yanai K. (1984) Oxygen isotopic compositions of some Yamato meteorites. *Mem. Nat. Inst. Polar Res. (Tokyo)*, 35, 267-271.

Fireman E. L. and Norris T. L. (1982) Ages and compositions of gas trapped in Allan Hills and Byrd core ice. *Earth Planet. Sci. Lett.*, 60, 339-350.

Fredriksson K., Jarosewich E., and Nelen J. (1969) The Sharps chondrite—New evidence on the origin of chondrules and chondrites. In *Meteorite Research* (P. M. Millman, ed.), pp. 155-165. Springer-Verlag, New York.

Gomes C. B. and Keil K. (1980) *Brazilian Stone Meteorites*. University of New Mexico Press. 162 pp.

Gooding J. L., Mayeda T. K., Clayton R. N., and Fukuoka T. (1983) Oxygen isotopic heterogeneities, their petrological correlations, and implications for melt origins of chondrules in unequilibrated ordinary chondrites. *Earth Planet. Sci. Lett.*, 65, 209-224.

Grady M. M., Swart P. K., and Pillinger C. T. (1982a) The variable carbon isotopic composition of type 3 ordinary chondrites. *Proc. Lunar Planet. Sci. Conf. 13th*, in *J. Geophys. Res.*, 87, A289-A296.

Grady M. M., Swart P. K., and Pillinger C. T. (1982b) Carbon isotopes in graphite from graphite-magnetite matrix (abstract). *Meteoritics*, 17, 223-224.

Huss G. R., Keil K., and Taylor G. J. (1981) The matrices of unequilibrated ordinary chondrites: implications for the origin and history of chondrites. *Geochim. Cosmochim. Acta*, 45, 33-51.

Hutchison R., Alexander C., and Barber D. J. (1985) Further evidence against a nebular origin for the ordinary chondrites (abstract). *Meteoritics*, 20, 669-670.

Kurat G. (1970) Zur Genese des kohligen Materials im Meteoriten von Tieschitz. *Earth Planet. Sci. Lett.*, 7, 317-324.

Lespade P., Al-Jishi R., and Dresselhaus M. S. (1982) Model for Raman scattering from incompletely graphitized carbons. *Carbon*, 20, 427-431.

Lumpkin G. R. (1986) High resolution electron microscopy of carbonaceous material from CI, CM, CV, and H chondrites (abstract). In *Lunar and Planetary Science XVII*, pp. 502-503. Lunar and Planetary Institute, Houston.

Marvin U. B. and Mason B. (1980) Catalog of Antarctic Meteorites, 1977-1978. *Smithson. Contrib. Earth Sci.*, 26, 55-71.

Mason B. (1983) In *Antarctic Meteorite Newsletter*, 6, No. 1, p. 14.

McKinley S. G. and Keil K. (1984) Petrology and classification of 145 small meteorites from the 1977 Allan Hills collection. *Smithsonian Contrib. Earth Sci.*, 26, 55-71.

McKinley S. G., Scott E. R. D., Taylor G. J., and Keil K. (1981) A unique type 3 chondrite containing graphite-magnetite aggregates—Allan Hills A77011. *Proc. Lunar Planet. Sci. 12B*, 1039-1048.

Moore C. B. (1985) In *Antarctic Meteorite Newsletter*, 8, No. 2, p. 4.

Niederer F. and Eberhardt P. (1977) A neon-E-rich phase in Dimmitt (abstract). *Meteoritics*, 12, 327-331.

Nishiizumi K., Regnier S., and Marti K. (1980) Cosmic ray exposure ages of chondrites, pre-irradiation and constancy of cosmic ray flux in the past. *Earth Planet. Sci. Lett.*, 50, 156-170.

Nuth J. A. (1985) Meteoritic evidence that graphite is rare in the interstellar medium. *Nature*, 318, 166-168.

Onuma N., Ikeda Y., Mayeda T. K., Clayton R. N., and Yanai K. (1983) Oxygen isotopic compositions of petrographically described inclusions from Antarctic equilibrated ordinary chondrites. *Mem. Nat. Inst. Polar Res. (Tokyo)*, 30, 306-314.

Ramdohr P. (1973) *The Opaque Minerals in Stony Meteorites*. Elsevier, Amsterdam. 245 pp.

Rietmeijer F. J. M. and Mackinnon I. D. R. (1985) Poorly graphitized carbon as a new cosmothermometer for primitive extraterrestrial materials. *Nature*, 316, 733-736.

Schultz L. and Kruse H. (1983) Helium, neon, and argon in meteorites: a data compilation. Max-Planck-Institue für Chemie, Mainz, Germany.

Scott E. R. D. (1984a) Pairing of meteorites found in Victoria Land, Antarctica. *Mem. Nat. Inst. Polar Res. (Tokyo)*, 35, 102-125.

Scott E. R. D. (1984b) Classification, metamorphism, and brecciation of type 3 chondrites from Antarctica. *Smithsonian Contrib. Earth Sci.*, 26, 73-94.

Scott E. R. D. (1986) In *Antarctic Meteorite Newsletter*, 9, No. 2 pp. 20-31.

Scott E. R. D., Taylor G. J., Rubin A. E., Okada A., and Keil K. (1981a) Graphite-magnetite aggregates in ordinary chondritic meteorites. *Nature*, 291, 544-546.

Scott E. R. D., Rubin A. E., Taylor G. J., and Keil K. (1981b) New kind of type 3 chondrite with a graphite-magnetite matrix. *Earth Planet. Sci. Lett.*, 56, 19-31.

Scott E. R. D., Taylor G. J., and Maggiore P. (1982) A new LL3 chondrite, Allan Hills A79003, and observations on matrices in ordinary chondrites. *Meteoritics*, 17, 65-75.

Scott E. R. D., Taylor G. J., and Keil K. (1986) Accretion, metamorphism, and brecciation of ordinary chondrites: evidence from petrologic studies of meteorites from Roosevelt County, New Mexico. *Proc. Lunar Planet. Sci. Conf. 17th*, in *J. Geophys. Res.*, 91, E115-E123.

Sears D. W. G. and Hasan F. A. (1987) The type 3 ordinary chondrites: a review. *Surv. Geophys.*, 9, 43-97.

Sears D. W. G. and Weeks K. S. (1983) Chemical and physical studies of type 3 chondrites: 2. Thermoluminescence of sixteen type 3 ordinary chondrites and relationships with oxygen isotopes. *Proc. Lunar Planet. Sci. Conf. 14th*, in *J. Geophys. Res.*, 88, B301-B311.

Sears D. W. G. and Weeks K. S. (1986) Chemical and physical studies of type 3 chondrites - VI: Siderophile elements in ordinary chondrites. *Geochim. Cosmochim. Acta*, 50, 2815-2832.

Sears D. W., Grossman J. N., Melcher C. L., Ross L. M., and Mills A. A. (1980) Measuring metamorphic history of unequilibrated ordinary chondrites. *Nature*, 287, 791-795.

Sears D. W., Grossman J. N., and Melcher C. L. (1982) Chemical and physical studies of type 3 chondrites - I: Metamorphism related studies of Antarctic and other type 3 ordinary chondrites. *Geochim. Cosmochim. Acta*, 46, 2471-2481.

Signer P., Baur H., Derksen U., Etique Ph., Funk H., Horn P., and Wieler R. (1977) Helium, neon, and argon records of lunar soil evolution. *Proc. Lunar Sci. Conf. 8th*, 3657-3683.

Smith P. P. K. and Buseck P. R. (1981) Graphitic carbon in the Allende meteorite: A microstructural study. *Science*, 212, 322-324.

Swart P. K., Grady M. M., and Pillinger C. T. (1983) A method for the identification and elimination of contamination during carbon isotopic analyses of extraterrestrial samples. *Meteoritics*, 18, 137-154.

Taylor G. J., Okada A., Scott E. R. D., Rubin A. E., Huss G. R., and Keil K. (1981) The occurrence and implications of carbide-magnetite assemblages in unequilibrated ordinary chondrites (abstract). In *Lunar and Planetary Science XII*, pp. 1076-1078. Lunar and Planetary Institute, Houston.

Wasson J. T. (1974) *Meteorites, Classification and Properties*. Springer-Verlag, Berlin. 316 pp.

Wieler R., Baur H., Graf Th., and Signer P. (1985) He, Ne, and Ar in Antarctic meteorites: solar noble gases in an enstatite chondrite (abstract). In *Lunar and Planetary Science XVI*, pp. 902-903. Lunar and Planetary Institute, Houston.

Zadnik M. G. (1985) Noble gases in the Bells (C2) and Sharps (H3) chondrites. *Meteoritics, 20,* 245-257.

Solar, Planetary, and Other Inert Gases in Two Sieve Fractions of a Disaggregated Allende Sample: A Study by Stepwise Heating Extraction

R. L. Palma
Department of Physics, Sam Houston State University, Huntsville, TX 77341

D. Heymann
Department of Geology and Geophysics, Department of Space Physics and Astronomy, Rice University, Houston, TX 77251

Inert gases released by stepwise heating of unaltered, strongly magnetic, and weakly magnetic samples from the 0-64 μm and the 105-250 μm fractions of a disaggregated and sieved sample of the Allende meteorite reveal the occurrence of both solar and planetary neon. The origin of the solar neon is thought to be implantation of solar wind ions. The origin of the planetary neon remains unresolved. Heavy isotope enriched components of krypton and xenon have been detected and there are some indications that a light krypton component may also be present. Other than a larger concentration of ^{129}Xe in the weakly magnetic samples, the signatures of the magnetic separates are isotopically very similar.

INTRODUCTION

At the 16th Lunar and Planetary Science Conference, we reported the discovery of solar wind neon in sieve fractions of a disaggregated bulk sample of 8.24 grams from the Allende meteorite (*Heymann and Palma*, 1985a, 1986). At the 45th Annual Meeting of the Meteoritical Society we reported that five new samples, chips of about 100 mg each from five different spots on the same 150-gram slab, had inert gas contents typical of bulk Allende as reported by *Mazor et al.* (1970), i.e., comparatively small amounts of essentially nonsolar trapped neon (*Heymann and Palma*, 1985b).

Material remained from two of the most gas-rich sieve fractions, 0-64 μm and 105-250 μm, of our earlier experiments on the disaggregated Allende sample. We have continued our study of this material to investigate the following: (1) the puzzling lack of planetary neon seen in our initial data and the suggestion that this component had been lost during the disaggregation process; (2) the siting of the discovered solar neon component; (3) our suggestion that three-dimensional carbon shells may be involved in the trapping of noble gases and the preservation of strange isotopic compositions (*Heymann*, 1986). With these objectives in mind, three samples were prepared from each of the two sieve fractions: (1) unaltered, (2) strongly magnetic, and (3) weakly magnetic. The magnetic separation was done with the aim of obtaining nonmagnetic samples enriched in carbon.

EXPERIMENTAL

The preparation of the original sieve fractions as well as the methodology of the mass spectrometry are reported in detail in *Heymann and Palma* (1986). The following alterations were made. The gas extractions were now done at 500°, 700°, 900°, 1100°, 1300°, and 1500°C. The temperature of the molybdenum crucible was controlled with the power setting of the RF supply and was measured with a tungsten thermocouple.

Aliquots from the 0-64 μm and the 105-250 μm fractions were obtained by using a Franz magnetic separator to generate strongly and weakly magnetic fractions. Aliquots from these and the unaltered samples, a total of six, were weighed, wrapped in aluminum foil, and placed in the sample line of the gas extraction system. Table 1 lists the samples analyzed in this investigation.

The full sample identification number, up to the last letter, refers to the nomenclature and identification system specified in *Heymann and Palma* (1986). The last letter is added here to distinguish among the samples prepared for this study. For conciseness we will identify the samples more briefly throughout the paper by referring to the last number and letter only. For example, (1,U) will be the 0-64 μm unaltered sample, (1,M) the 0-64 μm strongly magnetic sample, etc.

TABLE 1. Samples analyzed.

Full Sample Identification No.	Designation in this Paper	Description	Weight (grams)
HM 196, R4, 1, 1, U	1,U	0-64 μm, unaltered	0.04656
HM 196, R4, 1, 1, M	1,M	0-64 μm, strongly magnetic	0.03988
HM 196, R4, 1, 1, W	1,W	0-64 μm, weakly magnetic	0.06105
HM 196, R4, 1, 3, U	3,U	105-250 μm, unaltered	0.05562
HM 196, R4, 1, 3, M	3,M	105-250 μm, strongly magnetic	0.04457
HM 196, R4, 1, 3, W	3,W	105-250 μm, weakly magnetic	0.04986

RESULTS

Table 2 presents the absolute amounts of ^{3}He and ^{22}Ne and the isotopic compositions of helium and neon for each of the temperature step extractions. Table 3 presents the ^{40}Ar contents and isotopic ratios for argon. Table 4 presents the ^{84}Kr contents and the isotopic composition of krypton. The first sample analyzed (1,W) did not have sufficient gettering prior to the introduction of the krypton gas to the mass spectrometer for the 500°C fraction. This resulted in the krypton peaks being overwhelmed by the hydrocarbon background. Longer Zr,Al gettering for all other temperature releases and samples eliminated this problem. Table 5 presents the ^{132}Xe contents and isotopic ratios for xenon. A dashed line in Tables 2-5 indicates that the amount of gas released at that temperature was essentially equal to the instrumental blank value or less. All gas contents are in units of (cm^3/g) STP.

The customary corrections for interferences at masses 20 and 22 as well as for procedural blanks were made. Corrections for instrumental mass fractionation were also made. The 1 σ errors, which are the numbers in parentheses of Tables 1-5, were compounded from the errors in each correctional step.

TABLE 2. Absolute concentrations and isotopic composition of helium and neon.

T (°C)	^{3}He ($\times 10^{-8}$)	4/3	^{22}Ne ($\times 10^{-8}$)	20/22	21/22
Sample 1,U					
500	7.67 (0.21)	954 (34)	4.29 (0.08)	11.25 (0.08)	0.121 (0.006)
700	2.62 (0.07)	748 (27)	1.53 (0.04)	8.15 (0.46)	0.273 (0.016)
900	0.73 (0.02)	588 (21)	0.73 (0.04)	6.45 (0.47)	0.536 (0.025)
1100	0.19 (0.01)	1110 (40)	1.06 (0.03)	5.39 (0.13)	0.409 (0.017)
1300	-	-	0.52 (0.03)	5.73 (0.26)	0.346 (0.025)
1500	-	-	-	-	-
Sample 1,M					
500	10.62 (0.29)	792 (28)	8.06 (0.10)	10.88 (0.03)	0.0768 (0.0018)
700	2.85 (0.08)	973 (35)	1.23 (0.02)	8.73 (0.18)	0.314 (0.018)
900	1.17 (0.03)	570 (20)	1.33 (0.03)	3.89 (0.36)	0.507 (0.090)
1100	0.19 (0.01)	2600 (90)	2.65 (0.03)	8.29 (0.10)	0.191 (0.008)
1300	0.12 (0.01)	3460 (120)	1.33 (0.02)	8.33 (0.34)	0.361 (0.027)
1500	-	-	-	-	-
Sample 1,W					
500	7.46 (0.20)	975 (35)	5.48 (0.09)	10.93 (0.13)	0.0905 (0.0031)
700	2.87 (0.08)	932 (33)	1.46 (0.02)	9.22 (0.07)	0.268 (0.007)

TABLE 2. (continued)

T (°C)	^{3}He ($\times 10^{-8}$)	4/3	^{22}Ne ($\times 10^{-8}$)	20/22	21/22
900	0.76 (0.02)	628 (22)	1.13 (0.03)	7.81 (0.27)	0.392 (0.018)
1100	0.150 (0.004)	1380 (50)	1.29 (0.02)	6.50 (0.33)	0.359 (0.007)
1300	-	-	-	-	-
1500	-	-	-	-	-
Sample 3,U					
500	5.28 (0.14)	688 (24)	4.65 (0.10)	12.11 (0.09)	0.0647 (0.0028)
700	2.58 (0.07)	606 (21)	1.26 (0.03)	8.85 (0.16)	0.259 (0.009)
900	1.36 (0.04)	370 (13)	1.02 (0.02)	5.01 (0.11)	0.484 (0.009)
1100	0.69 (0.02)	92 (3)	0.69 (0.03)	-	1.01 (0.05)
1300	0.14 (0.007)	116 (4)	0.50 (0.03)	1.43 (0.10)	0.799 (0.025)
1500	0.19 (0.005)	-	-	-	-
Sample 3,M					
500	4.59 (0.12)	294 (11)	4.99 (0.17)	10.18 (0.34)	0.0919 (0.0046)
700	2.06 (0.06)	497 (18)	0.79 (0.02)	6.02 (0.36)	0.361 (0.024)
900	1.53 (0.04)	406 (14)	0.85 (0.02)	3.42 (0.08)	0.619 (0.031)
1100	0.71 (0.04)	388 (14)	0.40 (0.09)	-	1.18 (0.29)
1300	0.26 (0.04)	3100 (250)	3.29 (0.05)	7.03 (0.11)	0.197 (0.005)
1500	0.09 (0.04)	7950 (500)	2.73 (0.04)	9.19 (0.39)	0.070 (0.006)
Sample 3,W					
500	4.83 (0.13)	583 (21)	1.73 (0.03)	11.23 (0.11)	0.103 (0.007)
700	2.68 (0.07)	476 (17)	1.06 (0.03)	8.40 (0.15)	0.364 (0.014)
900	1.45 (0.04)	312 (11)	0.67 (0.04)	-	1.04 (0.09)
1100	0.52 (0.02)	278 (10)	0.51 (0.08)	-	1.19 (0.31)
1300	0.19 (0.01)	174 (6)	0.51 (0.05)	1.38 (0.19)	0.945 (0.068)
1500	0.04 (0.008)	-	0.78 (0.04)	6.21 (0.15)	0.295 (0.011)

Numbers in parentheses are 1σ errors.

DISCUSSION

Comparison with Prior Results

In 1984 we determined the inert gas contents of sieve fractions from HM196, R4, 1 by one-step melting of unaltered samples (*Heymann and Palma*, 1986). These data can be compared to the present results for the two unaltered samples

TABLE 3. Absolute concentrations and isotopic composition of argon.

T (°C)	^{40}Ar (×10^{-6})	36/40	38/40	36/38
Sample 1,U				
500	7.44 (0.16)	0.00400 (0.00001)	0.000774 (0.000005)	5.17
700	1.56 (0.03)	0.01204 (0.00005)	0.00244 (0.00002)	4.93
900	0.82 (0.02)	0.0591 (0.0002)	0.01178 (0.00003)	5.02
1100	0.85 (0.02)	0.06804 (0.00007)	0.01370 (0.00004)	4.97
1300	0.251 (0.006)	0.0319 (0.0001)	0.00651 (0.00006)	4.90
1500	-	-	-	-
Sample 1,M				
500	6.92 (0.06)	0.00365 (0.00001)	0.000705 (0.000004)	5.18
700	1.77 (0.04)	0.01393 (0.00005)	0.00281 (0.00002)	4.96
900	1.26 (0.03)	0.03336 (0.00006)	0.00656 (0.00003)	5.09
1100	1.00 (0.02)	0.05879 (0.00009)	0.01187 (0.00004)	4.95
1300	0.45 (0.01)	0.02181 (0.00007)	0.00448 (0.00004)	4.87
1500	0.125 (0.004)	0.0053 (0.0001)	0.00098 (0.00005)	5.41
Sample 1,W				
500	6.95 (0.15)	0.00400 (0.00001)	0.000773 (0.000007)	5.17
700	1.70 (0.04)	0.0111 (0.0002)	0.00277 (0.00003)	4.89
900	1.22 (0.03)	0.03447 (0.00004)	0.00681 (0.00002)	5.06
1100	1.00 (0.02)	0.0556 (0.0005)	0.01132 (0.00003)	4.91
1300	0.335 (0.008)	0.0263 (0.0001)	0.00538 (0.00004)	4.89
1500	0.123 (0.006)	0.0073 (0.0005)	0.0019 (0.0004)	3.84
Sample 3,U				
500	5.47 (0.05)	0.004382 (0.000008)	0.000822 (0.000002)	5.33
700	2.63 (0.03)	0.0206 (0.0001)	0.00393 (0.00002)	5.24
900	3.71 (0.04)	0.00747 (0.00003)	0.00149 (0.00001)	5.01
1100	3.52 (0.04)	0.0052 (0.0006)	0.00127 (0.00002)	4.09
1300	0.88 (0.01)	0.00458 (0.00002)	0.00112 (0.0002)	4.09
1500	-	-	-	-
Sample 3,M				
500	4.82 (0.05)	0.00199 (0.00001)	0.000408 (0.000006)	4.88
700	1.60 (0.02)	0.00717 (0.00003)	0.00138 (0.00002)	5.20
900	1.85 (0.02)	0.01255 (0.00002)	0.00247 (0.00001)	5.08
1100	2.03 (0.02)	0.00881 (0.00001)	0.00186 (0.00001)	4.74
1300	1.09 (0.01)	0.00662 (0.00003)	0.00155 (0.00002)	4.27
1500	0.50 (0.01)	0.0040 (0.0002)	0.0008 (0.0002)	5.00
Sample 3,W				
500	6.71 (0.06)	0.00130 (0.00001)	0.000233 (0.000006)	5.58
700	3.38 (0.04)	0.00424 (0.00001)	0.000798 (0.000008)	5.31
900	6.83 (0.07)	0.000994 (0.000008)	0.000225 (0.000005)	4.42
1100	4.84 (0.06)	0.00175 (0.00005)	0.000488 (0.000008)	3.59
1300	0.85 (0.01)	-	0.00062 (0.00005)	-
1500	-	-	-	-

Numbers in parentheses are 1σ errors.

(1,U) and (3,U). This comparison is shown in Table 6 along with the total gas concentrations seen in the other samples of the current experiment.

The bulk gas concentrations in the unaltered samples from 1984 and 1985 agree well for ^{3}He and in several other instances. We show later that the ^{3}He is primarily of cosmogenic origin, whereas the other inert gases are primarily from trapped components. Thus the homogeneity of ^{3}He data and the variability seen in neon, argon, krypton, and xenon are not unexpected.

It is striking that for these primarily trapped gases the magnetic separate always contains a greater concentration of gas than the weakly magnetic separate. Most of the trapped gases are probably implanted solar wind ions, which are more strongly held in magnetic than in nonmagnetic phases. In the Orgueil carbonaceous chondrite, the solar gases are contained in magnetite (*Jeffery and Anders,* 1970). Also, while (3,U) can be seen as a composite of (3,M) and (3,W), samples (1,M) and (1,W) both contain greater concentrations of neon, krypton, and xenon than does the unaltered bulk sample (1,U). This is probably the result of the adsorption of atmospheric gas on these very fine-grained samples during the magnetic separation process. This adsorbed gas should be driven off during the 500°C heating. This possibility is supported by the release pattern seen for neon in Fig. 2. The 500°C fractions from samples (1,M) and (1,W) both lie approximately along a mixing line of (1,U) and the atmospheric composition in the three isotope correlation plot. Assuming endpoints of

TABLE 4. Absolute concentrations and isotopic composition of krypton.

T (°C)	^{84}Kr ($\times 10^{-10}$)	80/86	82/86	83/86	84/86
Sample 1,U					
500	2.81 (0.06)	0.49 (0.20)	0.734 (0.021)	0.618 (0.012)	3.21 (0.03)
700	1.38 (0.02)	0.37 (0.15)	0.629 (0.024)	0.593 (0.038)	3.14 (0.09)
900	4.29 (0.06)	0.105 (0.010)	0.634 (0.035)	0.626 (0.029)	3.33 (0.09)
1100	5.06 (0.09)	0.080 (0.010)	0.591 (0.026)	0.612 (0.022)	3.13 (0.06)
1300	0.80 (0.03)	0.29 (0.10)	0.752 (0.075)	0.635 (0.055)	3.20 (0.13)
1500	0.036 (0.004)	-	-	-	-
Sample 1,M					
500	7.88 (0.36)	0.200 (0.058)	0.691 (0.010)	0.657 (0.010)	3.26 (0.02)
700	1.38 (0.03)	0.140 (0.028)	0.539 (0.026)	0.565 (0.024)	3.11 (0.05)
900	4.17 (0.06)	0.211 (0.035)	0.666 (0.013)	0.653 (0.016)	3.21 (0.06)
1100	5.87 (0.08)	0.116 (0.016)	0.628 (0.014)	0.626 (0.011)	3.21 (0.02)
1300	1.03 (0.02)	0.073 (0.021)	0.603 (0.043)	0.583 (0.036)	3.24 (0.05)
1500	0.97 (0.03)	-	0.552 (0.032)	0.557 (0.034)	3.14 (0.12)
Sample 1,W					
500	7.33 (0.22)	peaks lost in hydrocarbon background			
700	1.26 (0.02)	0.089 (0.021)	0.556 (0.050)	0.542 (0.061)	3.24 (0.05)
900	3.91 (0.06)	0.076 (0.020)	0.590 (0.028)	0.641 (0.019)	3.25 (0.04)
1100	5.51 (0.08)	0.094 (0.015)	0.611 (0.018)	0.608 (0.028)	3.16 (0.08)
1300	0.81 (0.02)	0.023 (0.025)	0.568 (0.035)	0.483 (0.083)	3.09 (0.11)
1500	0.17 (0.04)	0.08 (0.14)	0.25 (0.16)	0.40 (0.17)	2.68 (0.71)
Sample 3,U					
500	2.33 (0.04)	0.120 (0.012)	0.612 (0.029)	0.599 (0.033)	3.13 (0.06)
700	0.79 (0.02)	0.144 (0.027)	0.620 (0.049)	0.646 (0.028)	3.20 (0.10)
900	1.86 (0.02)	0.179 (0.031)	0.590 (0.054)	0.572 (0.096)	3.22 (0.33)
1100	2.04 (0.03)	0.133 (0.015)	0.631 (0.029)	0.647 (0.029)	3.27 (0.09)
1300	0.71 (0.02)	0.060 (0.039)	0.651 (0.049)	0.663 (0.073)	3.30 (0.08)
1500	0.07 (0.01)	-	-	-	-
Sample 3,M					
500	2.78 (0.04)	0.148 (0.024)	0.630 (0.033)	0.617 (0.053)	3.17 (0.07)
700	0.68 (0.02)	0.217 (0.065)	0.572 (0.055)	0.581 (0.047)	3.09 (0.08)
900	2.38 (0.03)	0.039 (0.015)	0.605 (0.037)	0.607 (0.023)	3.36 (0.7)
1100	1.87 (0.03)	0.087 (0.017)	0.591 (0.030)	0.602 (0.050)	3.22 (0.10)
1300	1.62 (0.02)	0.178 (0.036)	0.625 (0.044)	0.609 (0.045)	3.23 (0.11)
1500	0.65 (0.02)	0.125 (0.035)	0.699 (0.047)	0.628 (0.087)	3.22 (0.13)
Sample 3,W					
500	1.23 (0.02)	0.166 (0.022)	0.575 (0.038)	0.635 (0.040)	3.19 (0.08)
700	0.21 (0.10)	0.412 (0.082)	0.64 (0.15)	0.45 (0.10)	3.08 (0.28)
900	0.57 (0.02)	0.560 (0.062)	0.663 (0.051)	0.544 (0.083)	3.15 (0.14)
1100	1.43 (0.03)	0.129 (0.012)	0.618 (0.036)	0.659 (0.039)	3.34 (0.07)
1300	0.39 (0.01)	0.173 (0.065)	0.42 (0.13)	0.577 (0.069)	3.25 (0.32)
1500	0.30 (0.01)	0.184 (0.073)	0.44 (0.19)	0.37 (0.18)	3.15 (0.48)

Numbers in parentheses are 1σ errors.

atmospheric neon and the (1,U) composition, a calculation of the average amount of atmospheric contamination in the 500°C fractions of (1,M) and (1,W) was made. This procedure indicated the atmospheric component in (1,M) and (1,W) to be 37% and 28%, respectively. Subtracting this derived atmospheric component, we have recalculated the total gas concentrations of (1,M) and (1,W), and these are indicated in Table 6 by (1,M)* and (1,W)*. There remains a higher concentration of ^{84}Kr and ^{132}Xe in the separates than in the unaltered sample. For ^{84}Kr in (1,W), part of this is undoubtedly the contribution of hydrocarbon background in the 500°C release mentioned earlier, but in general we attribute this to a combination of a different adsorption efficiency for krypton and xenon and the variability of the trapped gas concentrations within the original material.

Gas Release Patterns

Figure 1 shows gas concentration released versus temperature interval of ^{3}He for the six samples of Table 1. The plots for ^{22}Ne, ^{36}Ar, ^{84}Kr, and ^{132}Xe are not shown, because these are very similar to the ^{3}He plots, except that they are "bimodal," i.e., show high releases in the 500°C and 700°C or higher intervals.

TABLE 5. Absolute concentrations and isotopic composition of xenon.

T (°C)	^{132}Xe ($\times 10^{-10}$)	$\frac{124}{136}$	$\frac{126}{136}$	$\frac{128}{136}$	$\frac{129}{136}$	$\frac{130}{136}$	$\frac{131}{136}$	$\frac{132}{136}$	$\frac{134}{136}$
Sample 1,U									
500	2.59	0.063	0.062	0.477	3.02	0.457	1.97	2.43	1.06
	(0.04)	(0.005)	(0.005)	(0.012)	(0.05)	(0.024)	(0.04)	(0.04)	(0.02)
700	1.34	0.051	0.032	0.319	8.56	0.400	2.21	2.69	1.07
	(0.02)	(0.008)	(0.012)	(0.027)	(0.04)	(0.043)	(0.03)	(0.05)	(0.02)
900	4.78	0.016	0.016	0.282	4.79	0.506	2.52	3.07	1.17
	(0.07)	(0.002)	(0.003)	(0.006)	(0.04)	(0.011)	(0.03)	(0.03)	(0.02)
1100	5.68	0.015	0.015	0.263	5.01	0.499	2.51	3.06	1.18
	(0.09)	(0.002)	(0.002)	(0.009)	(0.05)	(0.009)	(0.04)	(0.05)	(0.02)
1300	0.95	0.005	0.009	0.259	6.15	0.479	2.43	2.99	1.19
	(0.02)	(0.008)	(0.007)	(0.029)	(0.11)	(0.017)	(0.07)	(0.05)	(0.04)
1500	0.01	-	-	-	4.71	0.63	2.44	2.82	1.21
	(0.002)				(0.05)	(0.55)	(0.11)	(0.13)	(0.27)
Sample 1,M									
500	17.1	0.360	0.294	2.07	4.13	1.21	2.66	3.04	1.23
	(0.4)	(0.038)	(0.033)	(0.18)	(0.04)	(0.11)	(0.03)	(0.03)	(0.02)
700	1.34	0.071	0.064	0.530	7.45	0.479	2.00	2.40	1.07
	(0.02)	(0.005)	(0.003)	(0.031)	(0.21)	(0.009)	(0.04)	(0.06)	(0.02)
900	4.84	0.026	0.029	0.341	5.35	0.521	2.46	2.97	1.17
	(0.10)	(0.003)	(0.002)	(0.011)	(0.06)	(0.007)	(0.04)	(0.02)	(0.02)
1100	6.74	0.025	0.024	0.309	4.89	0.517	2.48	3.02	1.17
	(0.13)	(0.003)	(0.002)	(0.014)	(0.02)	(0.013)	(0.04)	(0.04)	(0.01)
1300	1.06	0.032	0.036	0.323	6.71	0.542	2.68	3.22	1.25
	(0.03)	(0.012)	(0.020)	(0.016)	(0.21)	(0.022)	(0.11)	(0.15)	(0.11)
1500	0.26	-	-	-	10.8	-	3.57	4.23	2.13
	(0.01)				(1.6)		(0.43)	(0.93)	(0.50)
Sample 1,W									
500	5.98	0.095	0.089	0.650	3.50	0.582	2.29	2.69	1.15
	(0.12)	(0.009)	(0.007)	(0.016)	(0.04)	(0.018)	(0.02)	(0.03)	(0.02)
700	1.36	0.048	0.044	0.437	8.22	0.468	2.12	2.49	1.09
	(0.03)	(0.005)	(0.004)	(0.016)	(0.06)	(0.019)	(0.04)	(0.06)	(0.02)
900	4.11	0.022	0.018	0.292	5.54	0.505	2.50	3.05	1.18
	(0.08)	(0.002)	(0.002)	(0.005)	(0.02)	(0.010)	(0.01)	(0.03)	(0.01)
1100	5.91	0.018	0.017	0.266	4.94	0.496	2.49	3.04	1.17
	(0.11)	(0.002)	(0.002)	(0.004)	(0.04)	(0.010)	(0.02)	(0.02)	(0.02)
1300	0.87	0.008	0.016	0.267	6.35	0.524	2.54	3.11	1.20
	(0.02)	(0.003)	(0.002)	(0.019)	(0.06)	(0.031)	(0.03)	(0.04)	(0.02)
1500	0.10	-	-	-	7.34	0.17	2.09	3.02	1.08
	(0.01)				(0.29)	(0.10)	(0.27)	(0.19)	(0.06)
Sample 3,U									
500	0.76	0.028	0.029	0.332	3.61	0.483	2.33	2.94	1.15
	(0.02)	(0.007)	(0.007)	(0.043)	(0.06)	(0.033)	(0.04)	(0.07)	(0.03)
700	0.51	0.028	0.023	0.239	7.09	0.354	2.14	2.58	1.11
	(0.01)	(0.009)	(0.005)	(0.044)	(0.03)	(0.046)	(0.14)	(0.24)	(0.06)
900	1.80	0.014	0.014	0.251	15.7	0.481	2.41	2.92	1.15
	(0.03)	(0.005)	(0.005)	(0.020)	(0.1)	(0.015)	(0.04)	(0.03)	(0.01)
1100	2.42	0.016	0.016	0.265	6.81	0.492	2.46	3.05	1.16
	(0.04)	(0.002)	(0.005)	(0.007)	(0.04)	(0.012)	(0.02)	(0.05)	(0.01)
1300	0.85	0.012	0.010	0.271	7.69	0.500	2.44	2.96	1.16
	(0.02)	(0.004)	(0.004)	(0.011)	(0.12)	(0.034)	(0.06)	(0.06)	(0.04)
1500	0.11	-	-	0.31	13.6	0.38	2.50	3.06	1.21
	(0.01)			(0.27)	(0.10)	(0.37)	(0.36)	(0.48)	(0.29)

TABLE 5. (continued)

T (°C)	^{132}Xe ($\times 10^{-10}$)	$\frac{124}{136}$	$\frac{126}{136}$	$\frac{128}{136}$	$\frac{129}{136}$	$\frac{130}{136}$	$\frac{131}{136}$	$\frac{132}{136}$	$\frac{134}{136}$
Sample 3,M									
500	0.93	0.064	0.052	0.526	3.65	0.550	2.38	2.86	1.19
	(0.02)	(0.012)	(0.012)	(0.045)	(0.05)	(0.035)	(0.09)	(0.04)	(0.06)
700	0.61	0.037	0.028	0.327	6.54	0.408	2.06	2.56	1.07
	(0.02)	(0.013)	(0.016)	(0.038)	(0.17)	(0.041)	(0.04)	(0.09)	(0.05)
900	2.58	0.012	0.011	0.263	6.07	0.477	2.47	3.02	1.17
	(0.03)	(0.005)	(0.003)	(0.015)	(0.08)	(0.019)	(0.02)	(0.03)	(0.01)
1100	1.80	0.012	0.012	0.228	5.77	0.486	2.41	2.97	1.21
	(0.03)	(0.005)	(0.009)	(0.028)	(0.02)	(0.018)	(0.04)	(0.03)	(0.06)
1300	1.23	0.030	0.042	0.409	7.62	0.566	2.49	2.96	1.21
	(0.02)	(0.012)	(0.010)	(0.023)	(0.05)	(0.032)	(0.06)	(0.06)	(0.03)
1500	0.23	0.149	0.193	1.20	6.34	0.83	2.21	2.71	0.95
	(0.01)	(0.063)	(0.078)	(0.27)	(0.25)	(0.12)	(0.28)	(0.32)	(0.09)
Sample 3,W									
500	0.27	0.053	0.037	0.547	4.56	0.517	2.50	3.23	1.18
	(0.01)	(0.026)	(0.012)	(0.035)	(0.21)	(0.041)	(0.16)	(0.27)	(0.11)
700	0.16	-	-	0.421	19.4	0.470	2.24	2.66	1.05
	(0.01)			(0.075)	(0.6)	(0.073)	(0.15)	(0.14)	(0.06)
900	0.71	0.008	0.015	0.299	60.1	0.483	2.36	2.78	1.13
	(0.01)	(0.006)	(0.007)	(0.031)	(1.1)	(0.034)	(0.06)	(0.08)	(0.03)
1100	1.92	0.015	0.011	0.256	7.64	0.478	2.52	3.10	1.19
	(0.02)	(0.006)	(0.004)	(0.021)	(0.12)	(0.012)	(0.04)	(0.04)	(0.04)
1300	0.59	0.009	0.016	0.268	8.26	0.477	2.42	2.99	1.18
	(0.01)	(0.010)	(0.006)	(0.014)	(0.18)	(0.040)	(0.07)	(0.09)	(0.06)
1500	0.55	0.015	0.024	0.262	5.61	0.450	2.44	3.03	1.12
	(0.01)	(0.012)	(0.017)	(0.023)	(0.07)	(0.021)	(0.03)	(0.08)	(0.05)

Numbers in parentheses are 1σ errors.

In the case of ^{3}He (Fig. 1), we observe a progressive decrease in the amount of gas released with increasing temperature. The amounts of ^{4}He decrease also, but usually not as rapidly as for ^{3}He, so that the ^{4}He/^{3}He ratio in Table 2 tends to increase with temperature. The release pattern seen for helium is expected given its high diffusion rate. The diffusion is more efficient in the smaller grain size sample: two of the three 0-64 μm samples were depleted in ^{3}He after the 1100°C heating, while the other was completely depleted after the 1300°C interval. In contrast, all the temperature steps had measurable amounts of ^{3}He for the 105-250 μm samples.

The ^{22}Ne gas is not depleted as readily as ^{3}He. Although all samples exhibit their greatest gas release at 500°C, much of that gas may be due to atmospheric contamination, as mentioned above. After the initial release temperature, the gas concentrations tend to decrease with temperature, but with only about a factor of two difference between the maximum and minimum values. One notable exception to this can be seen in the two magnetic separates, where spikes due to gas release between 900° and 1300°C can be seen.

Bimodal release patterns similar to that of ^{22}Ne are seen to occur for argon, krypton, and xenon in most samples. That is, there is an initial spike of gas at 500°C with a second release maximum between 700° and 1100°C. The isotopic compositions of the gases released at 500°C tend to be near atmospheric values, hence we conclude again that at least significant fractions of the low temperature releases of these gases are due to terrestrial contamination (*Pepin and Phinney*, unpublished data, 1977; *Liffman*, 1987).

The release patterns for krypton and xenon are very similar

TABLE 6. Total gas concentrations (cm^3/g STP).

Sample	^{3}He ($\times 10^{-8}$)	^{22}Ne ($\times 10^{-8}$)	^{36}Ar ($\times 10^{-8}$)	^{84}Kr ($\times 10^{-10}$)	^{132}Xe ($\times 10^{-10}$)
(1,U)	11.2	8.13	16.2	14.4	15.3
(1985)	(0.3)	(0.35)	(0.7)	(0.8)	(0.7)
(1984)	10.5	8.38	5.51	7.11	7.12
	(0.2)	(0.41)	(0.25)	(0.36)	(0.40)
(1,M)	14.9	14.6	16.1	21.3	31.3
	(0.4)	(0.5)	(0.7)	(1.1)	(1.4)
(1,M)*		11.6	15.2	18.4	25.0
(1,W)	11.2	9.36	15.4	19.0	18.3
	(0.3)	(0.38)	(0.6)	(1.4)	(1.0)
(1,W)*		7.83	14.6	16.9	16.6
(3,U)	10.2	8.12	12.8	7.80	6.45
(1985)	(0.3)	(0.38)	(0.6)	(0.42)	(0.39)
(1984)	9.03	19.5	34.9	5.62	4.73
	(0.12)	(1.0)	(1.2)	(0.28)	(0.31)
(3,M)	9.24	13.0	7.14	10.0	7.38
	(0.23)	(0.5)	(0.31)	(0.6)	(0.33)
(3,W)	9.71	5.26	3.83	4.13	4.20
	(0.28)	(0.21)	(0.19)	(0.20)	(0.24)

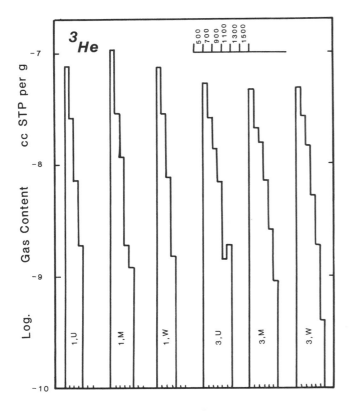

Fig. 1. Amounts of ^{3}He given off by the six samples in the different intervals of heating. The insert in the upper right-hand corner shows, in units of °C, the temperatures at which the gas was given off. These intervals are repeated six times on the abscissa.

in all samples, and the gas concentrations are definitely higher in the 0-64 μm samples than in those of the 105-250 μm samples, even disregarding the 500°C data. Sample (3,W) shows a sharp peak in gas release for both krypton and xenon in the 1100°C heating, whereas in (3,M) the second release peak occurs in the 900°C heating for each. This is in contrast to (1,W) and (1,M), which both have second release peaks for krypton and xenon at 1100°C.

TABLE 7. Cosmogenic helium, neon, and argon concentrations (cm^3/g STP).

Sample	^{3}He(c) (×10^{-8})	^{21}Ne(c)A (×10^{-8})	^{21}Ne(c) B (×10^{-8})	^{38}Ar(c) (×10^{-9})
(1,U)	7.05	1.79	1.72	1.99
(1,M)	9.66	2.36	2.23	1.93
(1,W)	6.78	1.60	1.52	2.08
(3,U)	7.82	2.06	1.99	1.89
(3,M)	7.55	2.31	2.19	1.29
(3,W)	7.69	2.51	2.47	1.31
Average	7.76	2.10	2.02	1.75
Mazor et al. (1970)	7.82	1.88	1.88	

The ^{129}Xe data reveal an indigenous radiogenic component with a bimodal release pattern. All samples show a low-temperature peak in the ^{129}Xe/^{136}Xe ratio at either 700° or 900°C. There is a second peak in this ratio at 1300° or 1500°C. The values for all six samples at both of these release peaks are much greater than the terrestrial atmosphere value, ranging up to 60 in (3,W) at 900°C (see section on isotopic composition of xenon).

Cosmogenic Components

For helium, neon, and argon there are clearly recognizable cosmogenic components in many of the temperature fractions of all samples, especially in the gas fractions extracted at the higher temperatures. Numerical values for these components can only be deduced with selected assumptions. The ^{3}He total consists, in principle, of a cosmogenic and a trapped component. The ^{4}He total consists of radiogenic, trapped, and cosmogenic gas. We have assumed that *all* ^{4}He is trapped and have calculated the cosmogenic ^{3}He from equation (1) below (see *Anders et al.*, 1970)

$$^3\text{He}(c) = {}^3\text{He} - 4.20 \times 10^{-4}({}^4\text{He}) \qquad (1)$$

These are the *minimum* ^{3}He(c) values since they result from making the largest possible correction. We find that these minimum values range from 60% to 81% of the total ^{3}He in the samples. The true ^{3}He content lies somewhere between the minimum ^{3}He(c) and the total ^{3}He contents. We have calculated ^{3}He(c) contents for all temperatures, but present in Table 7 only the sums of the six samples of Table 1, the average value of the six samples, and the bulk value of *Mazor et al.* (1970). We note that there is no evidence whatsoever in the data of Table 7, nor in the results from individual temperature steps, that suggests substantial exposure of matter by galactic and/or solar cosmic rays at the time when the solar trapped gas was acquired.

By assuming that the neon isotopes are simple mixtures of cosmogenic (c) and trapped components, it is possible to derive the following two equations for the percentage of ^{21}Ne that is cosmogenic

$$\left[\frac{^{21}\text{Ne}(c)}{^{21}\text{Ne}}\right]_A = 1.028 - 0.0257 \left[\frac{^{22}\text{Ne}}{^{21}\text{Ne}}\right]_s \qquad (2)$$

$$\left[\frac{^{21}\text{Ne}(c)}{^{21}\text{Ne}}\right]_B = 1.041 - 0.0375 \left[\frac{^{22}\text{Ne}}{^{21}\text{Ne}}\right]_s \qquad (3)$$

In these equations the subscript s refers to the sample ratio. Equation (2) assumes a trapped neon composition of (^{21}Ne/^{22}Ne) = 0.025, the "planetary" or "A" composition. Values for the amount of cosmogenic ^{21}Ne using this assumption are given in the column ^{21}Ne(c)A in Table 7. Equation (3) assumes a trapped neon composition of (^{21}Ne/^{22}Ne) = 0.036, the "solar" or "B" composition. Values for the amount of cosmogenic ^{21}Ne using this assumption are given in the column ^{21}Ne(c)B in Table 7. In both equations the cosmogenic ratio used was

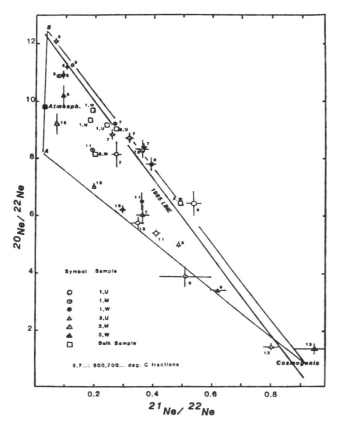

Fig. 2. Three isotope neon plot, $^{20}Ne/^{22}Ne$ versus $^{21}Ne/^{22}Ne$. A and B are the locations of the well-known trapped neon components Ne-A and Ne-B. The Atmosph. point represents the isotopic composition of neon in the Earth's atmosphere. The point labeled "Cosmogenic" represents cosmic-ray produced neon for a meteorite of Allende's composition. The "1985 line" is the linear regression line from work on bulk grain size fractions of Allende reported by *Heymann and Palma* (1985a,b).

$(^{21}Ne/^{22}Ne) = 0.918$, where the trapped and cosmogenic values are from *Anders et al.* (1970). The amounts of cosmogenic ^{21}Ne have been found for all temperature fractions, but only the total contents are shown in Table 7, along with the average value of the six samples and the bulk value of *Mazor et al.* (1970).

As expected, cosmogenic ^{21}Ne is larger assuming a planetary trapped component than a solar trapped component. The cosmogenic component is largest in (3,W), which is also clearly indicated in Fig. 2 by its position on the three isotope neon plot. From the thermal release patterns displayed in Fig. 2, it can be seen that most cosmogenic neon is released at 900°C or higher and that these fractions fall more closely along the Ne-A—cosmogenic Ne mixing line than along the Ne-B—cosmogenic Ne mixing line. Thus, we conclude that the A-based values of Table 7 are closer to the real cosmogenic ^{21}Ne contents than the B-based values.

Much of the cosmogenic ^{21}Ne released and in Table 7 does not appear in the three isotope neon plot. These points, namely the 1100°C fractions of (3,U) and (3,M) and the 900° and 1100°C fractions of (3,W), all have $^{21}Ne/^{22}Ne$ approximately equal to one. For each of these temperature fractions the very low blank amounts of ^{21}Ne permitted the detection of the cosmogenic gas, whereas a like amount of gas at ^{20}Ne was indistinguishable from the blank value variations.

Smith et al. (1977) have determined cosmogenic neon and argon in bulk samples of the Allende meteorite, but also in "matrix," high alkali aggregates, and Ca,Al-rich and other chondrules. Their deduced $^{21}Ne(c)$ contents range from 1.27 × 10^{-8} cm³ STP g⁻¹ in a Ca,Al-rich chondrule to 2.34 × 10^{-8} cm³ STP g⁻¹ in a pyroxene chondrule. A general trend found by *Smith et al.* (1977) is that the "normal" chondrules have the highest $^{21}Ne(c)$ contents, and the Ca,Al-rich chondrules the lowest. Their bulk values are 2.16 × 10^{-8} and 2.22 × 10^{-8} cm⁻³ STP g⁻¹. *Smith et al.* (1977) explain these variations with variations in chemical composition, especially sodium content. We do not see a straightforward method to compare our results with those of *Smith et al.* (1977), except for the bulk samples. In this regard, we note that the value of 1.79 × 10^{-8} cm³ STP g⁻¹ of (1,U) is distinctly below, but the 2.10 × 10^{-8} cm³ g⁻¹ of (3, U) is equal to the *Smith et al.* (1977) values for bulk Allende. We think that the low value of (1, U) is, at least in part, due to grain size effects. The olivine and pyroxene chondrules of *Smith et al.* (1977) are likely to have been "strongly" magnetic in our sense. Hence the relatively high value of 2.36 × 10^{-8} cm³ STP g⁻¹ of (1,M) may well be due to substantial fractions of olivine and pyroxene, but not necessarily in the form of chondrules.

In order to calculate the amount of ^{38}Ar that is of cosmogenic origin, again a simple mixture of trapped and cosmogenic gas was assumed for both ^{36}Ar and ^{38}Ar. The following equation was derived to determine the percentage of ^{38}Ar that is cosmogenic

$$\left[\frac{^{38}Ar(c)}{^{38}Ar}\right] = 1.144 - 0.0216 \left[\frac{^{36}Ar}{^{38}Ar}\right]_s \quad (4)$$

Again the subscript s refers to the sample ratio. Equation (4) assumes $(^{36}Ar/^{38}Ar) = 5.30$ as the trapped component (solar wind value, *Eberhardt et al.,* 1970). The cosmogenic ^{38}Ar contents were found for every temperature fraction, but only the total values for each sample and the average of the six samples are shown in Table 7. Almost all of the cosmogenic argon is released at 900°C or higher temperatures, as in the case of neon.

Isotopic Composition of Neon

Figure 2 is a three-isotope correlation plot of the neon data in Table 2 plus the isotopic ratios of the sums of the neon contents for the six samples (indicated as bulk sample composition in the figure). Also shown are the locations of Neon-A, Neon-B, and cosmogenic neon. The line labeled "1985 line" is the regression line for the 1985 data (*Heymann and Palma,* 1985a, 1986.)

As noted earlier, the data define two different trends. The points nearest the Ne-B composition represent gases released at the lowest temperatures (500°C), and the next gas fraction (700°C) tends to lie along the Ne-B—cosmogenic neon mixing line. Higher temperature releases from some of the samples, notably the magnetic separates, then tend toward the Ne-A—cosmogenic neon mixing line. Thus we see possible chemical differences between the phases holding the Ne-A and Ne-B gases, and also that the Ne-B gas is evidently trapped in sites from which it is more readily released than the Ne-A gas.

In our earlier work we commented that we could not "see" the Ne-A, i.e., that it was masked by the large amounts of Ne-B in our bulk samples from which the gas had been extracted in one step. The strategy employed here has been successful in enabling us to confirm the presence of Ne-A in our gas-rich sample, with trapped ^{20}Ne at roughly the levels found in gas poor Allende samples, i.e., on the order 10^{-8} to 10^{-7} (cm^3/g) STP.

Isotopic Composition of Argon

The ^{36}Ar/^{38}Ar of all gas fractions measured range from a low of 3.59 to a high of 5.58, but there is only one ratio significantly larger than the atmospheric value of 5.35. What we can see is a variable mix of a trapped component with a ratio near 5.3 and a cosmogenic component with a ratio below 1.0. That the smaller values tend to occur for the higher release temperatures is consistent with our earlier assertion that the cosmogenic argon component tends to be released at higher temperatures.

Isotopic Composition of Krypton

There are two known components of trapped krypton with strange isotopic composition in the Allende meteorite. One is "CCF-Kr," a component comparatively rich in ^{86}Kr (see *Lewis et al.*, 1975). The other is a component in which the p-isotopes are present in uncommonly low abundances, or are absent altogether (*Frick*, 1977). We have indications that both strange components may exist in the samples studied.

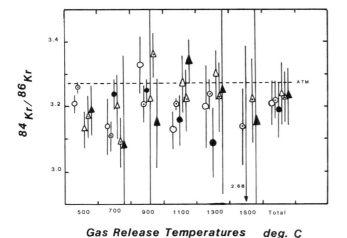

Fig. 3. Isotopic ^{84}Kr/^{86}Kr ratios of gases released. The symbols represent the same samples as in Figure 2. Errors are 1 σ errors.

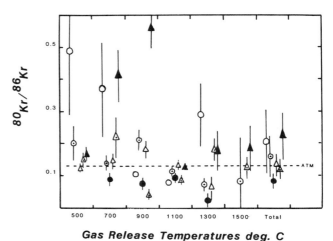

Fig. 4. Isotopic ^{80}Kr/^{86}Kr ratios of gases released. The symbols represent the same samples as in Figure 2. Errors are 1 σ errors.

To look for "CCF-Kr," perhaps krypton directly from stellar nucleosynthesis (*Heymann and Dziczkaniec*, 1980), we have plotted ^{84}Kr/^{86}Kr versus release temperature for all six samples (Fig. 3). The heavy krypton component is most prominently released at approximately 900°C, so if it is present, the ^{84}Kr/^{86}Kr ratio should be below the atmospheric value of 3.273 in either the 900°C or 1100°C fractions. This is observed for all three of the 0-64 μm sieve fraction samples at 1100°C. Although the scatter in the data points precludes a certain identification of this component in these samples, the overall ^{84}Kr/^{86}Kr value in the fine-grained samples is lower than both the atmospheric value and those seen in the 105-250 μm sieve fraction samples.

The search for the isotopically strange "light" krypton component is complicated by the presence of spallation krypton and the hydrocarbon background. In Fig. 4 we have plotted ^{80}Kr/^{86}Kr versus release temperature for all six samples. The terrestrial value of ^{80}Kr/^{86}Kr is 0.130, and we see that in the 900°, 1100°, and 1300°C gas releases several samples fall below the atmospheric ratio. Since a spallation or hydrocarbon component to ^{80}Kr would drive this ratio up, we have an indication that this strange "light" krypton component is also present in these samples.

Isotopic Composition of Xenon

Xenon in the Allende meteorite is known to consist of many components with strange isotopic compositions. The oldest known of these is monoisotopic ^{129}Xe, presumably from in situ decay of now extinct ^{129}I (*Reynolds*, 1960). We find this component to be very evident in each of our samples and at every temperature step. Table 8 displays the overall bulk ratio of ^{129}Xe/^{136}Xe for each of the samples, the standard deviation, and the atmospheric value. In both (3,U) and (3,W) the total xenon content is more than 50% ^{129}Xe. Since it is known that the nonmagnetic minerals melilite, sodalite, and nepheline have especially large ^{129}Xe/^{136}Xe ratios (*Reynolds et al.*, 1980), it is not surprising that (3,W) has a much larger

TABLE 8. ^{129}Xe values in the bulk samples.

	(1,U)	(1,M)	(1,W)	(3,U)	(3,M)	(3,W)	ATM
$\frac{^{129}Xe}{^{136}Xe}$	4.95 (0.08)	4.78 (0.09)	4.92 (0.08)	9.14 (0.13)	5.98 (0.10)	17.3 (0.3)	2.981 (0.008)

^{129}Xe/^{136}Xe value than does (3,M). Although the same trend is marginally evident in (1,W) and (1,M), the adsorption of atmospheric gas on these samples (as discussed in the section on comparison with prior results) has at least smoothed out some of the difference.

The counterpart to the "CCF-Kr" discussed in the previous section is "CCF-Xe." This heavy isotope enriched component of xenon was discovered in bulk samples of carbonaceous chondrites (see *Reynolds and Turner*, 1964; also Pepin and Phinney, unpublished data, 1977), but is most prominently seen in acid resistant residues (see *Lewis et al.*, 1975). Like CCF-Kr, CCF-Xe may result from stellar nucleosynthesis (*Heymann and Dziczkaniec*, 1979). Its hallmark is a ^{134}Xe/^{136}Xe ratio well below the atmospheric value of 1.177. Figure 5 is a plot of ^{134}Xe/^{136}Xe versus release temperature for all six samples. It is clear that all the 700°C fractions fall below the atmospheric value, but with the larger error bars for the 105-250 μm samples, the conclusion that CCF-Xe may be present is most convincing for the 0-64 μm samples.

CONCLUSIONS

In our preceding communications (*Heymann and Palma*, 1985a, 1986), we had concluded that our fragment of the Allende meteorite contains "solar" neon, and that this gas was almost certainly acquired by the irradiation of grains with ions, presumably from the sun. The current work is consistent with this conclusion. By releasing the neon gas in six temperature steps, it became possible to detect the "planetary" neon component. Although the magnetic separation process did not isolate the solar neon component, a greater concentration of neon was found in the strongly magnetic separates than in the weakly magnetic separates. Another distinction between the separates for neon was a significantly greater cosmogenic component in (3,W) than (3,M).

We find indications that the heavy inert gases, krypton and xenon, contain isotopic compositions distinct from those in the Earth's atmosphere or in the solar wind. Some of these components may be the result of direct stellar nucleosynthesis with variable mixing in the interstellar medium, and it is clear that carbon is involved in the preservation of these strange compositions. Unfortunately, with the exception of ^{129}Xe, no significant differences could be detected between the isotopic signatures of the strongly magnetic and weakly magnetic samples for the heavy inert gases. The fine-grained samples did appear to be clearer carriers of the heavy "CCF" krypton and xenon components than did the larger sieve fraction samples.

Acknowledgments. We thank Dr. D. D. Bogard and Mr. P. Johnson of NASA-JSC for their assistance with the measurements. R. L. Palma has received supplemental funding from the Sam Houston State University Faculty Organized Research Fund. Funding has come from NASA grant NAG 9-88.

REFERENCES

Anders E., Heymann D., and Mazor E. (1970) Isotopic composition of primordial helium in carbonaceous chondrites, *Geochim. Cosmochim. Acta, 34,* 127-131.

Eberhardt P., Geiss J., Graf H., Grögler N., Krähenbühl U., Schwaller H., Schwarzmüller J., and Stettler A. (1970) Trapped solar wind noble gases, ^{81}Kr/Kr exposure ages and Kr/Ar ages in Apollo 11 lunar material. *Science, 167,* 558-560.

Frick U. (1977) Anomalous krypton in the Allende meteorite, *Proc. Lunar Sci. Conf. 8th,* 173-292.

Heymann D. (1986) Buckministerfullerene, C60, and siblings: Their deduced properties as traps for inert gas atoms (abstract). In *Lunar and Planetary Science XVII,* pp. 337-338. Lunar and Planetary Institute, Houston.

Heymann D. and Dziczkaniec M. (1979) Xenon from intermediate zones of supernovae, *Proc. Lunar Planet. Sci. Conf. 10th,* 1943-1959.

Heymann D. and Dziczkaniec M. (1980) A first roadmap for kryptology, *Proc. Lunar Planet. Sci. Conf. 11th,* 1179-1213.

Heymann D. and Palma R. L. (1985a) Isotopic systematics of neon in sieve fractions of disaggregated Allende meteorite (abstract). In *Lunar and Planetary Science XVI,* pp. 347-348. Lunar and Planetary Institute, Houston.

Heymann D. and Palma R. L. (1985b) A gas-rich matter in the Allende meteorite. *Meteoritics, 20,* 663.

Heymann D. and Palma R. L. (1986) Discovery of solar wind neon in the Allende meteorite. *Proc. Lunar Planet. Sci. Conf. 16th,* in *J. Geophys. Res., 91,* D460-466.

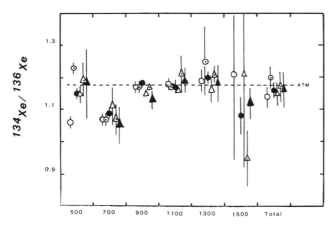

Fig. 5. Isotopic ^{134}Xe/^{136}Xe ratio. The symbols represent the same samples as in Figure 2. Errors are 1 σ errors.

Jeffery P. M. and Anders E. (1970) Primordial noble gases in separated meteoritic minerals - I. *Geochim. Cosmochim. Acta, 34,* 1175-1198.

Lewis R. S., Srinivasan B., and Anders E. (1975) Host phase of a strange xenon component in Allende, *Science, 190,* 1251-1262.

Liffman K. (1986) Atmospheric xenon in the Allende meteorite, *Meteoritics,* in press.

Mazor E., Heymann D., and Anders E. (1970) Noble gases in carbonaceous chondrites, *Geochim. Cosmochim. Acta, 34,* 781-824.

Reynolds J. H. (1960) Determination of the age of the elements, *Phys. Rev. Lett., 4,* 8.

Reynolds J. H. and Turner G. (1964) Rare gases in the chondrite Renazzo, *J. Geophys. Res., 69,* 3263-3281.

Reynolds J. H., Lumpkin G. R., and Jefferey P. M. (1980) Search for ^{129}Xe in mineral grains from Allende inclusions: An exercise in miniaturized rare gas analysis. *Z. Naturforsch., 352,* 257-266.

Smith S. P., Huneke J. C., Rajan R. S., and Wasserburg G. J. (1977) Neon and argon in the Allende meteorite. *Geochim. Cosmochim. Acta, 41,* 627-647.

An Assemblage of a Metallic Particle with a CAI in the Efremovka CV Chondrite

A. V. Fisenko, K. I. Ignatenko, A. Yu. Ljul, and A. K. Lavrukhina

V. I. Vernadsky Institute of Geochemistry and Analytical Chemistry, USSR Academy of Science, Moscow

---

An assemblage of a large (~0.5 mm) metallic particle with a CAI was removed from the Efremovka CV chondrite. The CAI was classified as Type BI based on its structure and the chemical composition of the main mineralogical phases. A polished section of the metallic particle, etched with 2% nital, consisted mainly of taenite, kamacite, phosphide, and some V-rich inclusions, and had an outer rim of V- and Cr-rich minerals. By using instrumental neutron activation analysis it was found that the elements Os, Re, Ir, Mo, Ru, and Pt in the metallic particle were enriched by a mean factor of about 160 relative to CI chondrites, and W by a factor of about 580. The metallic particle is interpreted as a sintered agglomerate of metal and phosphide condensates with V-oxides and V-sulphide. The metal condensate was probably isolated from the gas at ~1450K.

INTRODUCTION

Ca-Al-rich inclusions (CAIs) in carbonaceous chondrites are enriched with refractory siderophile and lithophile elements relative to the CI chondrites (*Grossman et al.*, 1977). Studies of CAIs of the Allende and Leoville CV chondrites have shown that the siderophile elements occur as metallic grains ("nuggets") enriched in the platinum-group elements, and as complex aggregates ("Fremdlinge") (*Palme and Wlotzka*, 1976; *El Goresy et al.*, 1978; *Wark and Lovering*, 1976, 1978; *Armstrong et al.*, 1985; *Wark*, 1986).

Studies of nuggets and Fremdlinge are of great interest since the structure, chemical compositions, and mineralogical compositions of both the objects and the CAI containing them reflect the processes of formation and evolution of early condensates in the solar system or of interstellar grains.

This paper describes an assemblage of a metallic particle enriched in refractory siderophile elements and a CAI of the Efremovka CV carbonaceous chondrite. The chemical and mineralogical characteristics of the meteorite are given in *Kvasha and Charitonova* (1966). Efremovka CAIs are less altered by secondary processes than Allende CAIs (*Nazarov et al.*, 1984). Therefore, constituents of the inclusions, such as metal, are expected to be less altered also.

EXPERIMENTAL PROCEDURE

A polished thick section was made from a fragment removed from a crushed sample of the Efremovka chondrite. Examination of the polished section under the microscope in reflected light showed that the fragments are from an assemblage of a CAI with a metallic particle (Fig. 1). This assemblage is designated as EF1 and the metallic particle as EM1.

Quantitative analyses of different phases in EF1 were carried out with a CAMEBAX electron microprobe analyser at an accelerating voltage of 15 kV and 30 nA current. The K line X-ray intensities were processed with an RDR-11/23 computer applying standard ZAF corrections. The 1δ error in measurements for major elements was 1-1.5% (relative), and 2-3% for the rest.

An ~25-μg sample was cut from EM1 for instrumental neutron activation analysis (INAA). The sample and standards were irradiated by epithermal neutrons (100 hours, $f_n \approx 6 \times 10^{11}$ n/cm^{2}s) to determine As, Au, Ir, Mo, and W, and then by thermal neutrons (20 hours, $f_n = 1.2 \times 10^{13}$ n/cm^2 s) to determine Os, Re, Pt, Ru, Ni, Co, and Cr. The following were used as standards: Dorofeevka and Babb's Mill iron meteorites, standard rocks TB and ANG (*Abbey*, 1980), metallic Mo, and a standard solution of W containing 0.11 mg/ml W. Weights of the standards averaged 5 mg with the exception of metallic Mo (~20 μg) and the standard solution of W (~0.01 ml). The sample and standards were counted with a Ge-Li detector and an LP-4900 (Nokia) multichannel analyser. We have taken into consideration interferences of gamma-ray lines in the treatment of γ-spectra. The error in determining the element content was $\leq$ 10% for Os, W, Re, Ir, Ni, Co, As, and Cr and ~20% for the other elements. These errors are based on replicate measurements.

RESULTS AND DISCUSSION

EF1 is about 2 mm across, but the ragged, chipped surface shows that it has been broken in some places. The metallic particle has an oval shape and its size is 530 × 400 μm. More than half of the volume of EM1 appears to be outside of the silicate inclusion. It is seen from Fig. 1 that only a part of the CAI's outer rim remains. The absence of rim over the CAI's entire perimeter is probably due to breaking of the inclusion before it was incorporated into the meteorite.

In the CAI one can identify a wide melilite-rich mantle and a core that consists of melilite, fassaite, and spinel (Fig. 1b). Grains of anorthite are also found in the mantle and the core, and small Fe-Ni grains ($\leq$10 μm) occur in the melilite and fassaite near EM1. Some Fe-Ni grains form intergrowths with spinel grains.

Melilite. Melilite is the major mineral phase of the CAI, and its chemical composition is given in Table 1. The Ak content of melilite varies from ~60% in the A region (near the core/mantle boundary) up to 10-15% in the mantle and in the melilite in contact with EM1.

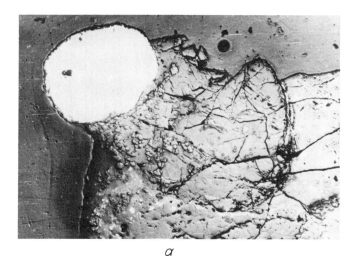

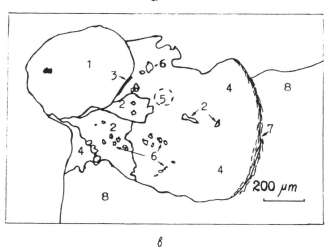

Fig. 1. (a) Polished section of EFl assemblage separated from the Efremovka chondrite; (b) its schematic image. 1 = The EMl metallic particle; 2 = fassaite grains; 3 = fassaite thin rim; 4 = melilite; 5 = the A region (see the text); 6 = spinel grains; 7 = the CAI outer rim; 8 = the matrix.

Fassaite. Two types of fassaite are present. The first occurs in the form of grains in the core and the CAI mantle. The largest grains are in the core. The chemical composition of this fassaite is given in Table 2. There are substantial compositional variations within fassaite grains: The contents of Al_2O_3, TiO_2, and V_2O_3 increase toward the surface of EMl whereas MgO decreases (Fig. 2). It should be noted that the maximum measured contents of Al_2O_3, TiO_2, V_2O_3, and MgO in the large fassaite grains exceed those in the small ones located in the CAI mantle.

The second fassaite type forms a thin (5-10 μm) and short rim near the EMl particle (Fig. 1). The chemical composition of the rim at two points is given in Table 2. A significant difference between this fassaite and the first type is in the second type's high content of V_2O_3 and MgO and low content of TiO_2 and Al_2O_3. A similar tendency is observed in the fassaite rims of Fremdlinge in Allende (*Armstrong et al.,* 1985).

Spinel. Most of the spinel grains are in the fassaite of the CAI core. In the CAI mantle, spinel grains are small (≤ 8 μm) and are concentrated mostly along a boundary between melilite grains (Fig. 1). The largest spinel grains (up to ~30 μm) are in the melilite of the CAI core. The shape of grains varies from idiomorphic to hypidiomorphic. Spinel analyses are given in Table 1 and show the following features: (1) A positive correlation exists between the contents of Cr_2O_3 and V_2O_3 (r = 0.70, with 99.9% confidence level). The content of these oxides in spinel grains increases as the latter approach the surface of EMl (Fig. 3). (2) The Fe oxide content of spinel grains increases as they approach the surface of EMl. An increased Fe oxide content is observed as well in the grains located in the CAI outer rim. The introduction of Fe into spinel most probably occurred in a period of moderate metamorphism as is assumed, for instance, in the case of some CAIs in Allende (*Grossman,* 1975).

We classified the CAI within EFl as Type B1 (*Grossman,* 1980; *Wark and Lovering,* 1982), based on its structure and the chemical composition of its major mineral phases [a wide melilite mantle, large variation of the content of Ak (from ~10% to ~60%)], and a maximum measured value for the TiO_2 content in the fassaite of ~15 wt %. It is commonly accepted that inclusions of this type have been melted to some degree and that the features indicated above are connected with

TABLE 1. Chemical composition of melilite and spinel in the CAI of the Efremovka EFI assemblage (in wt %).

Oxide	Melilite		Spinel	
	Range	Mean (42*)	Range	Mean (31*)
SiO_2	22.80-38.42	27.35	n.d.-0.94	0.17
TiO_2	n.d.-0.08	0.03	0.21-0.82	0.40
Al_2O_3	11.11-35.14	28.50	67.57-72.92	70.28
Cr_2O_3	n.d.-0.07	0.01	0.04-0.64	0.26
V_2O_3	n.d.	-	0.27-0.92	0.52
FeO	n.d.-1.36	0.10	n.d.-2.83	0.90
MnO	n.d.-0.03	0.01	n.d.-0.04	0.01
MgO	1.29-8.86	3.33	24.83-28.71	27.39
CaO	39.00-42.28	40.53	0.02-0.95	0.18
Na_2O	n.d.-0.26	0.06	n.d.-0.02	0.01
Total	96.85-102.82	99.92	96.75-103.93	100.12
Ak,%	10.50-64.85	26.79		
		Cations per 7 oxygens		*Cations per 4 oxygens*
Si		1.246		0.004
Ti		0.001		0.007
Al		1.530		1.973
Cr		-		0.005
V		-		0.010
Fe		0.004		0.018
Mn		-		-
Mg		0.226		0.972
Ca		1.978		0.005
Na		0.006		-
Total		4.991		2.994

n.d. - not detected.
*Number of analyses.

TABLE 2. Chemical compositions of fassaites in the CAI of the Efremovka EF1 assemblage (in wt %).

Oxide	Type I Range	Type I Mean (21*)	Type II 1	Type II 2
SiO$_2$	30.67-45.92	36.74	42.31	48.52
TiO$_2$	1.96-15.45	8.92	5.72	0.79
Al$_2$O$_3$	12.60-24.16	20.93	13.35	10.15
Cr$_2$O$_3$	0.01-0.14	0.05	0.20	0.23
V$_2$O$_3$	0.15-1.32	0.67	2.49	2.46
FeO	n.d.-3.50	0.50	0.69	0.36
MnO	n.d.-0.04	0.02	n.d.	0.05
MgO	5.04-10.36	7.57	10.63	13.11
CaO	23.50-25.53	24.45	24.72	25.26
Na$_2$O	n.d.-0.52	0.04	0.01	n.d.
Total	97.39-102.59	99.89	100.12	100.93
Cations per 6 oxygens				
Si		1.355	1.558	1.751
Ti		0.247	0.158	0.022
Al		0.910	0.579	0.434
Cr		0.001	0.006	0.007
V		0.020	0.074	0.071
Fe		0.016	0.021	0.011
Mn		-	-	-
Mg		0.416	0.583	0.705
Ca		0.966	0.975	0.977
Na		0.003	0.001	-
Total		3.934	3.955	3.978

n.d. - not detected.
*Number of analyses.

fractional crystallization from the surface of the inclusion toward its core (*Wark and Lovering*, 1982; *Kurat et al.*, 1975). However, *Stolper and Paque* (1986) have shown that Type B CAIs were never completely melted.

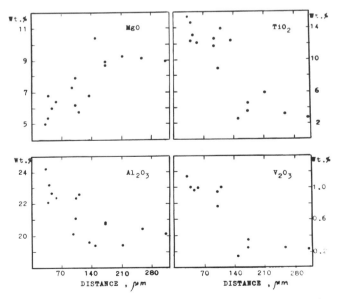

Fig. 2. The variations of the Mg, Al, Ti, and V oxide contents in the fassaite grains, with the distance from the surface of EM1.

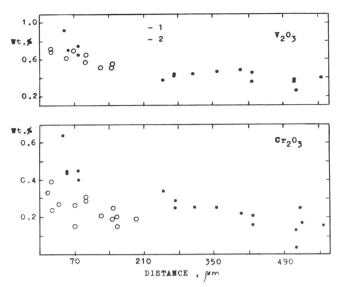

Fig. 3. The variations of the Cr$_2$O$_3$ and V$_2$O$_3$ contents in the spinel grains located in (1) the fassaite and (2) melilite, with their distance from the surface of EM1.

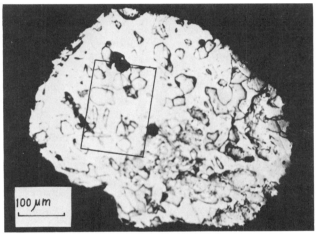

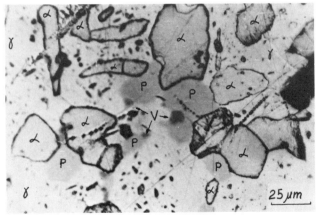

Fig. 4. Structure of EM1. α: Kamacite; γ: taenite; p: phosphide; and V: V-rich phases.

TABLE 3. Chemical composition of kamacite, taenite, and phosphide in EM1 (in wt %).

	Kamacite		Taenite		Phosphide	
	Range	Mean (14*)	Range	Mean (11)	Range	Mean (6)
Fe	90.46-94.81	92.46	54.26-65.46	60.32	32.21-33.41	32.61
Ni	4.04-6.48	6.00	35.37-47.32	40.55	53.32-55.07	54.37
Co	2.14-2.55	2.39	0.42-0.79	0.58	0.18-0.28	0.23
V	n.d.-0.23	0.05	0.02-0.30	0.11	0.02-0.18	0.06
Cr	n.d.-0.05	0.01	0.05-0.16	0.11	n.d.	n.d.
S	0.03-0.06	0.05	n.d.-0.06	0.03	0.03-0.07	0.06
P	n.d.-0.06	0.02	n.d.-0.03	n.d.	14.67-15.38	15.01
Total	99.40-103.11	100.98	99.30-103.00	101.7	101.53-103.06	102.3

n.d. - not detected.
*Number of analyses.

The main focus in our studies was on the EM1 particle, since the metal in CAIs in Efremovka has not been extensively studied. Moreover, the EM1 particle is one of the largest known metallic particles in the CAI of carbonaceous chondrites. The largest (about 1 mm) known metallic particle in a CAI was also found in Efremovka (*Nazarov et al.,* 1987). Possibly, the presence of large metallic particles is typical of Efremovka CAIs only.

Voids of irregular shape that were formed during polishing are present in the EM1 particle. The EM1 structure revealed by etching with 2% nital is complex (Fig. 4). There are numerous inclusions of kamacite, phosphide, and phases rich in vanadium in the taenite. The shapes of the kamacite grains vary from idiomorphic and hypidiomorphic to xenomorphic. In a number of cases the kamacite grains are polycrystalline. As a rule, the phosphide grains are associated with kamacite, and sometimes outlined by the latter (*Fisenko et al.,* 1986).

Chemical compositions of taenite, kamacite, and phosphide are given in Table 3. The phosphorus content in EM1 is, on average, ~2 wt %, based on the EM1 modal composition. We conclude that phosphides precipitated before kamacite in EM1 metal during cooling. This conclusion follows from the Fe-Ni-P-phase diagram (*Doan and Goldstein,* 1969) and from the high contents of P and Ni (20 wt %; see below), and agree with the rimming of the phosphide grains by kamacite.

The size of the V-rich inclusions found in EM1 does not exceed ~10 μm. Two such inclusions are shown in Fig. 5. The chemical compositions of some V-rich inclusions are given in Table 4. Most probably, the inclusions are a mixture of vanadium oxide and sulphide since the qualitative analysis carried out with the microprobe did not reveal significant amounts of any other elements besides those presented in Table 4. Analysis of the distribution of V-rich inclusions in EM1 showed the following: (1) The larger inclusions (up to ~10 μm) are concentrated mostly in the central part of EM1. They appear to be a mixture of vanadium oxide and sulphide, and are similar to the inclusion shown in Fig. 5. Smaller inclusions are distributed in the particle more or less uniformly and consist mostly of vanadium oxide. (2) EM1 has a ≤8-μm-wide outer rim enriched in V and Cr over almost all of its perimeter. A part of this rim is shown in Fig. 6. The rim consists of opaque minerals, some grains of which have an idiomorphic shape. The chemical composition of some of these grains is

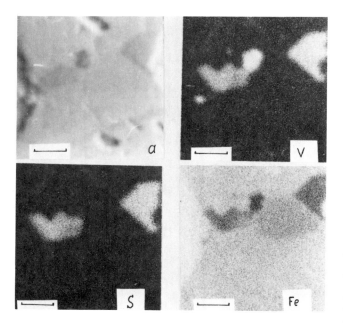

Fig. 5. Secondary electron image of the V-rich inclusions (a) and X-ray scanning pictures (V, S, and Fe). Scale bars measure 5 μm.

TABLE 4. Chemical composition of V-rich inclusions in EM1 (in wt %).

	S-rich		S-poor
	Range	Mean (5*)	(1)
Fe	11.65-15.23	12.85	5.19
Ni	0.80-2.94	1.29	4.81
Co	n.d.-0.11	0.05	0.04
V	27.59-34.08	32.09	54.48
S	36.64-41.52	38.67	0.35
P	0.02-0.08	0.06	0.28
Cr	n.d.-0.94	0.88	4.03
Total	83.39-89.53	85.89	69.18

n.d. - not detected.
*Number of analyses.

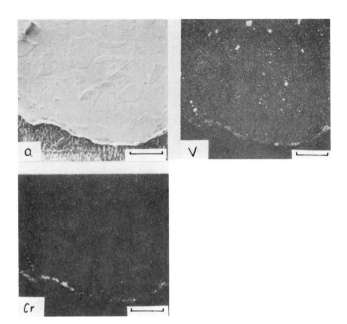

Fig. 6. The secondary electron image of EM1 (a) and X-ray scanning pictures (V and Cr). Scale bars measure 50 μm.

given in Table 5. It should be noted that the V_2O_3 and Cr_2O_3 contents tend to be complementary. By the data obtained (Table 5) one may assume the grains, in a number of cases, are a solid solution of coulsonite with magnetite (grains No. 1, 2; Table 5), or chromite enriched in vanadium (grain No. 6). (3) V-rich inclusions have not been found in the CAI immediately adjacent to EM1.

The neutron activation analyses are presented in Table 6. EM1 is seen to be enriched in the refractory elements W, Os, Re, Mo, Ir, Pt, and Ru, with enrichment factors averaging ~160× CI with the exception of W (~580). These values are considerably smaller than those for nuggets and the Fremdlinge Willy in Allende (*Palme and Wlotzka*, 1976; *Blander et al.*, 1980; *Armstrong et al.*, 1985), but coincide within a factor of ~3 with those for the metal in CAI E38 of Efremovka (*Nazarov et al.*, 1987). All of the indicated elements in EM1, except W, are in approximately cosmic relative abundances. The high content of W may permit the presence in EM1 of a tungsten mineral such as scheelite or W sulfide, which have been found in Willy. Unfortunately, we cannot be sure that the high W is not due to contamination by W-bearing tools. The chromium in EM1 is probably mainly present in chromite or Cr-rich minerals on the surface of the metal particle, since the main minerals in EM1 contain very little chromium (Table 3).

EM1 has virtually no analogues among the nuggets and Fremdlinge in the known CAIs of Allende and Leoville. In contrast with many Allende and Leoville nuggets, EM1 is of a larger size ($\geq$50-fold), but is less enriched in the platinum-group elements. EM1 differs from many Fremdlinge by having cosmic ratios between refractory siderophile elements, an outer rim made of V- and Cr-rich minerals, and by consisting mainly of taenite, kamacite, and phosphide.

EM1 was possibly formed by the agglomeration and subsequent sintering of higher temperature condensates (vanadium oxides, phosphide, FeNi-phase) with lower temperature phases (e.g., vanadium sulphide). The structure, chemical, and mineralogical compositions of EM1 are not inconsistent on the whole with such a suggestion. For instance, the initial FeNi metal equilibrium condensates in gas of solar

TABLE 5. Chemical compositions of opaque minerals on the surface of EM1 (in wt %).

Oxide	1	2	3	4	5	6
SiO_2	0.44	0.39	7.92	11.52	4.81	0.27
TiO_2	n.d.	0.06	0.03	0.04	0.04	n.d.
Al_2O_3	1.58	0.48	1.62	6.26	2.36	0.16
Cr_2O_3	2.99	5.66	26.93	12.59	12.53	67.48
V_2O_3	32.60	39.10	22.39	15.38	26.67	7.83
FeO	56.74	55.25	27.06	28.78	45.63	19.48
MgO	0.49	0.51	2.32	3.86	1.86	1.15
CaO	0.28	0.38	8.81	10.89	1.89	0.22
Na_2O	n.d.	0.01	0.86	0.41	0.16	0.04
Total	95.12	101.84	97.94	89.73	95.95	96.63
Cations per 4 oxygens						
Si	0.018	0.015	0.278	0.421	0.185	0.01
Ti	-	0.002	0.001	0.001	0.001	-
Al	0.078	0.022	0.067	0.270	0.107	0.007
Cr	0.100	0.174	0.747	0.364	0.381	1.966
V	1.101	1.219	0.630	0.450	0.824	0.231
Fe	2.000	1.797	0.794	0.880	1.470	0.600
Mg	0.031	0.029	0.121	0.210	0.104	0.063
Ca	0.013	0.016	0.331	0.426	0.078	0.009
Na	-	0.001	0.059	0.029	0.012	0.003
Total	3.341	3.275	3.028	3.051	3.162	2.889

n.d. - not detected.

TABLE 6. Elements determined by neutron activation in EM1.

	Concentration	EM1/CI*
W	55.6	583
Os	77.0	149
Re	5.5	154
Mo	185.0	202
Ir	60.0	145
Ru	115.0	162
Pt	145.0	143
Ni	20.0	18
Co	0.69	14
Cr	3.7	14
Au	0.14	1
Mn	0.32	1.6
Cu	0.13	10
As	40.1	21

Ni, Co, Cr, Mn, and Cu are in wt %; all others are in g/g.
*Data from *Anders and Ebihara* (1982).

composition at $P = 10^{-5}$ atm may contain up to ~20 wt % of Ni (*Kelly and Larimer*, 1977), and even more Ni at $P = 10^{-3}$ atm (e.g., ~27 wt % at 1470K; *Palme and Wlotzka*, 1976). Therefore, it is not difficult to explain a high content of Ni in EM1. The phosphorus in EM1 could be due to the presence of high-temperature phosphide condensates in the EM1 initial agglomerate. It was shown by *Sears* (1978) that phosphide condensation begins before FeNi metal within a wide range of pressure in solar gas. Note also that the phosphorus content of EM1 does not exceed its content in metallic particles in olivine inclusions of CM carbonaceous chondrites, which are interpreted as primary FeNi condensates (*Olsen et al.*, 1973). from the Os, Re, Ir, and Pt contents in EM1, and calculations of the concentrations of these elements in equilibrium condensates from solar gas at $P = 10^{-3}$ atm (*Palme and Wlotzka*, 1976), we conclude that EM1 metal could have been removed from the condensation sequence at a temperature of ~1450K. The metal in EM1 probably has a similar genesis to nuggets in Allende, which appear to be high-temperature condensates (*Blander et al.*, 1980; *Wark*, 1986). In this case, the differences between EM1 and the nuggets are mainly caused by their differing temperatures of removal from the condensation sequence, and the prior fractionations of each gas.

The EM1 components may have undergone a sintering process, but the final morphology of EM1 may have been acquired as a result of heating during the collision that formed the EF1 assemblage. The EM1 outer rim of V- and Cr-rich minerals could have been formed by a reaction between V and Cr in EM1 with the CAI during heating from the collision or at some subsequent time. The redistribution of V, Al, Ti, and Mg oxides in the large fassaite grains, and the enrichment of V and Cr in the spinel grains, may also have taken place during such heating. We cannot at present exclude the possibility that the V- and Cr-rich rim minerals were formed by a reaction between the metallic particle and the nebular gas.

Alternatively, EM1 might have been formed by intense reduction and secondary processing of a Willy-type proto-Fremdlinge. However, the chemical similarity of the major mineral phases in EF1 to those of other CAIs in Efremovka and Allende make this a less probable scenario for the formation of EM1. Therefore, we prefer the first possibility, that EM1 formed as a result of the agglomeration of high- and low-temperature condensates.

CONCLUSIONS

The following conclusions may be drawn based on our studies of the EF1 assemblage:

1. EM1 differs in chemical and mineralogical compositions, in structure, and in size from the known nuggets and Fremdlinge in Allende and Leoville CAIs.

2. EM1 was possibly formed as a result of the agglomeration and subsequent sintering of higher-temperature condensates (vanadium oxide, phosphides, FeNi-phase) with lower-temperature compounds (vanadium sulphide, for instance). The metal condensates were probably removed from the condensation sequence at a temperature of ~1450K.

3. The EF1 assemblage was formed as a result of the collision of a Type B1 CAI with the EM1 metallic particle, after which the assemblage was incorporated into the Efremovka CV chondrite.

Acknowledgments. We are grateful to Drs. G. B. Baryshnikova and L. F. Semjonova for their helpful comments on this work.

REFERENCES

Abbey S. (1980) Studies in "standard samples" for use in the general analysis of silicate rocks and minerals. Part 6: 1979 edition of "Usable" values. *Geol. Survey of Canada Paper 80-14*, p. 30.

Anders E. and Ebihara M. (1982) Solar-system abundances of the elements. *Geochim. Cosmochim. Acta, 41*, 2263-2280.

Armstrong J. T., El Goresy A., and Wasserburg G. J. (1985) Willy: A prize noble Ur-Fremdlinge—Its history and implications for the formation of Fremdlinge and CAI. *Geochim. Cosmochim. Acta, 49*, 1001-1022.

Blander M., Fuchs L. H., Horowitz C., and Land R. (1980) Primordial refractory metal particles i the Allende meteorite. *Geochim. Cosmochim. Acta, 44*, 217-223.

Doan A. S. and Goldstein J. I. (1969) The formation of phosphides in iron meteorites. In *Meteorite Research* (P. M. Millman, ed.), pp. 763-779.

El Goresy A., Nagel K., and Ramdohr P. (1978) Fremdlinge and their noble relatives. *Proc. Lunar Planet. Sci. Conf. 9th*, 1279-1303.

Fisenko A. V., Ignatenko K. I., Semjonova L. F., and Lavrukhina A. K. (1986) The assemblage metal-Ca-Al-rich inclusion in the Efremovka CV chondrite (abstract). In *Lunar and Planetary Science XVII*, pp. 228-229. Lunar and Planetary Institute, Houston.

Grossman L. (1975) Petrography and mineral chemistry of Ca-rich inclusions in the Allende meteorite. *Geochim. Cosmochim. Acta, 39*, 433-454.

Grossman L. (1980) Refractory inclusions in the Allende meteorite. *Ann. Rev. Earth Planet. Sci., 8*, 559-608.

Grossman L., Ganapathy R., and Davis A. M. (1977) Trace elements in the Allende meteorite, III. Coarse-grained inclusions revisited. *Geochim. Cosmochim. Acta, 41*, 1647-1664.

Kelly W. R. and Larimer J. W. (1977) Chemical fractionations in meteorites, VIII. Iron meteorites and cosmochemical history of the metal phase. *Geochim. Cosmochim. Acta, 41*, 93-111.

Kurate G., Hoinkes G., and Fredriksson K. (1975) Zoned Ca-Al-rich chondrule in Bali: new evidence against the primordial condensation model. *Earth Planet. Sci. Lett., 26,* 140-144.

Kvasha L. G. and Charitonova V. V. (1966) The chemical composition and structure of the stone meteorites. *Meteoritika* (in Russian), *27,* 153-166.

Nazarov M. A., Korina M. I., Ulyanov A. A., Kolesov G. M., and Shcherbovsky E. J. (1984) Mineralogy, petrology and chemical composition of Ca and Al-rich inclusions of Efremovka meteorite. *Meteoritika* (in Russian), *43,* 46-68.

Nazarov M. A., Ulyanov A. A., and Kolesov G. M. (1987) The composition of the metallic phase in Efremovka meteorite white inclusions (abstract). In *Abstracts of XX Meteoritic Conf. USSR,* (10-12 February, 1987, Tallin) *2,* pp. 142-144. Inst. Geochim. and Analyt. Chem. of the USSR, Academy of Science, Moscow.

Palme H. and Wlotzka F. (1976) A metal particle from a Ca-Al-rich inclusion from the meteorite Allende and the condensation of refractory siderophile element. *Earth Planet. Sci. Lett., 33,* 45-60.

Olsen E., Fuchs L. H., and Forbes W. S. (1973) Chromium and phosphorus enrichment in the metal of type II (C2) carbonaceous chondrites. *Geochim. Cosmochim. Acta, 37,* 2037-2042.

Sears D. W. (1978) Condensation and composition of iron meteorites. *Earth Planet. Sci. Lett., 41,* 128-138.

Stolper E. and Paque J. M. (1986) Crystallization sequences of Ca-Al-rich inclusions from Allende: The effects of cooling rate and maximum temperature. *Geochim. Cosmochim. Acta, 50,* 1785-1806.

Wark D. A. (1986) Evidence for successive episodes of condensation at high temperature in a part of the solar nebula. *Earth Planet. Sci. Lett., 77,* 129-148.

Wark D. A. and Lovering J. F. (1976) Refractory/platinum metal grains in Allende calcium-aluminium-rich clasts (CARCs): Possible exotic presolar material (abstract). In *Lunar Science VII,* pp. 912-914. The Lunar Science Institute, Houston.

Wark D. A. and Lovering J. F. (1978) Refractory/platinum metals and other opaque phases in Allende Ca-Al-rich inclusions (CAIs) (abstract). In *Lunar and Planetary Science IX,* pp. 1214-1216. Lunar and Planetary Institute, Houston.

Wark D. A. and Lovering J. F. (1982) The nature and origin of type B1 and B2 Ca-Al-rich inclusions in the Allende meteorite. *Geochim. Cosmochim. Acta, 46,* 2584-2594.

Origin of Fragmental and Regolith Meteorite Breccias—Evidence from the Kendleton L Chondrite Breccia

A. J. Ehlmann*, E. R. D. Scott, and K. Keil
Department of Geology and Institute of Meteoritics, University of New Mexico, Albuquerque, NM 87131

T. K. Mayeda and R. N. Clayton
Enrico Fermi Institute, Department of Chemistry and Department of the Geophysical Sciences, University of Chicago, Chicago, IL 60637

H. W. Weber and L. Schultz
Max-Planck Institut für Chemie, Postfach 3060, D-6500 Mainz, Federal Republic of Germany

*Permanent address: Department of Geology, Texas Christian University, Fort Worth, TX 76129

The Kendleton L chondrite breccia is unusual in that it has most of the characteristics of ordinary chondrite breccias rich in solar-wind gases, but there is no evidence that it contains, or ever lost, these gases. The dark fragmental matrix in Kendleton is largely composed of type 4 material and contains at least four kinds of light- and dark-colored clasts with diverse origins. These clasts are composed of L3, L5, shock-blackened and melt rock material and were probably all derived from normal L chondrite material. The shock-blackened clasts were produced in post-metamorphic impacts, whereas the melt rock formed before the end of metamorphism. We also found a unique tridymite-rich inclusion that has an oxygen isotopic composition consistent with an origin in an H chondrite asteroid. However, the absence of both tridymite-rich inclusions in H chondrites and H chondrite clasts in L chondrites suggests a different origin for the tridymite-rich inclusion. The presence of martensite in some shock-blackened clasts shows that Kendleton, like regolith breccias, was not metamorphosed after compaction. If solar-flare irradiated grains in gas-rich meteorite breccias were irradiated on asteroid surfaces by an early active (T-Tauri) sun, as *Caffee et al.* (1987) argue, then some of these grains were stored inside parent bodies for up to 10^9 years before being mixed with larger volumes of material to form gas-rich breccias. Most of the material in gas-rich chondrite breccias may never have been near the surface of their parent bodies as the mixing could have occurred, for example, during break-up and reassembly of asteroids. Thus meteoritic and lunar regolith breccias may have different origins. Kendleton and other gas-poor, light-dark structured breccias probably formed in an identical fashion to gas-rich breccias, except that the gas-poor breccias did not receive a small admixture of irradiated grains.

INTRODUCTION

Regolith breccias are fragmental rocks that contain implanted solar-wind gases and are widely believed to have formed by lithification of regolith material that once resided at the surface of a body (*Taylor and Wilkening*, 1982). Almost all chondrites that have dark matrices and light- and dark-colored clasts also contain solar-wind gases and are believed to have formed from regolith material on asteroids (*Goswami et al.*, 1984). But a few such light-dark structured chondrites are known that do not contain solar-wind gases (*Keil*, 1982; *Rubin*, 1982). None of these chondrites has been studied in detail. Did such chondrites lose solar-wind gases from heating, or were the gases never acquired? How is it possible for two meteorite breccias to appear identical petrographically, yet differ so much in the abundance of irradiated grains in their hosts?

To elucidate the origin of regolith and fragmental breccias we chose to study the Kendleton L chondrite, which fell in 1939 and contains many prominent clasts. Limited published data (*Taylor and Heymann*, 1969) and our preliminary inspection suggested that Kendleton might be a light-dark structured chondrite that lacks detectable solar-wind gases. Our petrographic and noble gas analyses confirm this and show that Kendleton contains a wide variety of clasts. L3, L5, shock-blackened and melt rock clasts, and a unique tridymite-rich clast were identified. We suggest that Kendleton and some other fragmental breccias lacking solar-wind gases may be cogenetic with gas-rich breccias, except that the former did not receive a small admixture of solar-irradiated grains. We speculate on the origin of fragmental and regolith breccias in the light of the recent studies of *Caffee et al.* (1983, 1986, 1987). We infer from these studies that meteoritic and lunar regolith breccias may have different origins.

METHODS AND MATERIALS

Fifty-two specimens of Kendleton with a total weight of 9.7 kg were inspected at the Monnig Meteorite Collection at Texas Christian University, Fort Worth. Nine were chosen for study because hand-specimen examination revealed the presence of a wide variety of clasts. Polished thin sections, which are now in the collection at the University of New Mexico, were studied microscopically in transmitted and reflected light. Electron microprobe analyses were made using an ARL EMX-SM operated at 15 keV and about 20 nA sample current. We corrected for differential matrix effects using the method of *Bence and Albee* (1968); minerals of well-known compositions were used as standards. Precision was such that 20 analyses of a homogeneous olivine (Fa 17) have a standard deviation of 0.4% Fa.

Fig. 1. Transmitted-light photomicrograph of the Kendleton fragmental breccia (section UNM 711) showing chondritic host (H) and a light-colored, porphyritic, melt rock clast (P); E = epoxy. This meteorite resembles chondritic regolith breccias in having a dark host that contains abundant fragmental material and a few prominent chondrules and encloses light-colored clasts.

Oxygen isotopic compositions of host and clast material from Kendleton were analyzed using standard techniques (*Clayton et al.*, 1976). Noble gas concentrations and isotopic ratios were analyzed using techniques of *Weber et al.* (1982).

RESULTS

Chondritic Host

About 80% of the Kendleton breccia consists of dark chondritic host; discernible clasts greater than 1 mm in size make up the remainder. Some of the chondrules are well defined (Fig. 1) and chondrules and chondrule fragments have mesostases that are microgranoblastic to feathery, as in type 4 chondrites.

Compositions of olivines and low-Ca pyroxenes in the host were measured in nine thin sections (Table 1). All but one have compositional characteristics of type 4 chondrites: percent mean deviations (PMD) of wt % FeO in olivine in eight of the sections are 0.9-3.7. In the other section, UNM 710, which contains a type 3 clast, the host has the compositional heterogeneity of type 3: PMD of FeO in olivine is 10.5. Low-Ca pyroxenes are much more heterogeneous than olivines, implying that most of the host material is low type 4. Histograms

TABLE 1. Mean compositions of olivine and low-Ca pyroxene in nine sections of the chondritic host of Kendleton.

Section No.	Olivine				Low-Ca Pyroxene			
	Fa (mol %)		PMD*	No. Anal.	Fs (mol %)		PMD*	No. Anal.
	Mean	σ			Mean	σ		
649	23.3	1.50	3.7	24	18.7	3.1	12.9	23
650	22.8	0.63	1.5	38	18.3	5.0	19.2	23
709	22.3	0.32	1.1	26	18.3	2.1	8.3	14
710	22.6	4.7	10.5	15	15.1	7.0	41	13
711	22.9	0.34	1.0	28	17.0	3.5	15.4	20
712	21.9	0.27	0.9	53	17.7	3.6	15.1	36
713	22.9	0.32	1.0	36	19.4	3.8	13.2	22
714	22.8	0.79	2.5	19	16.2	3.1	12.8	11
715	22.8	0.50	1.4	24	19.6	5.2	19.9	19
Mean	22.6			263	18.0			181

*PMD, percent mean deviation of FeO analyses in wt %.

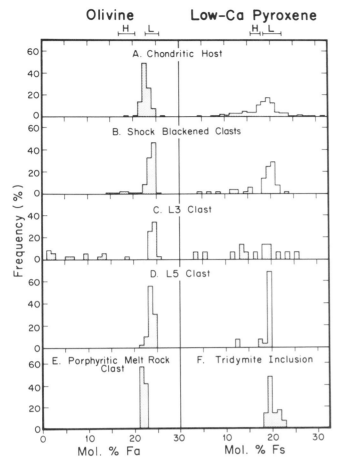

Fig. 2. Histograms of electron probe analyses of randomly chosen olivine and low-Ca pyroxene grains in host (A), and clasts and other objects (B-F) in the Kendleton fragmental breccia. Means and standard deviations are given in Tables 1 and 2. Kendleton host has the compositional characteristics of an L4 chondrite, but it was not metamorphosed to type 4 levels after lithification. Compositional ranges for equilibrated H and L chondrites are from *Gomes and Keil* (1980).

of all the analyses in nine sections are shown in Fig. 2A. Most of the host silicates have iron concentrations that lie in or close to the range of equilibrated L group chondrites (*Gomes and Keil*, 1980); mean concentrations of Fa 22.6 and Fs 18.0 lie at the low end of the L range. Calcium concentrations in host olivines (0.06 ± 0.02 wt % CaO) are marginally higher than those for most equilibrated chondrites (0.01-0.05 wt % CaO; *Busche*, 1975).

Olivines in the host show undulose extinction, implying shock facies b or c and shock pressures of 5-24 GPa (*Dodd and Jarosewich*, 1979). This estimate is consistent with the classification of shock class C by *Taylor and Heymann* (1969), who used X-ray diffraction techniques. Etched sections of host show that kamacite is polycrystalline with occasional taenite blebs; taenite contains partially or completely resorbed cloudy taenite (clear taenite) or plessite. This implies a metallographic shock class of III in the scheme of *Taylor and Heymann* (1969). Pockets of shock melt were not observed in the host, although a few shock veins are visible, especially around some shock-blackened clasts.

Noble gas concentrations were determined on a 0.31 g sample of the Kendleton host. The results are shown in Table 2 with the data of *Taylor and Heymann* (1969) for an uncharacterized sample of Kendleton. The host clearly lacks solar wind gases. The rather high concentrations of trapped planetary gases suggest a type 3 classification for the Kendleton host. The concentrations of trapped ^{36}Ar, for example, are appropriate to type 3.8 ordinary chondrites, according to *Sears et al.* (1980). The cosmic ray exposure age is 2.7 Myr using the ^{21}Ne production rate and shielding correction given by *Nishiizumi et al.* (1980). This value is low for L chondrites; about 95% have higher exposure ages (*Crabb and Schultz*, 1981). The concentration of ^{4}He is somewhat depleted, consistent with the shock level of b or c for Kendleton's host, and the decrease in L chondrites of ^{4}He and ^{40}Ar concentrations with increasing shock level (*Dodd and Jarosewich*, 1979). Data from *Taylor and Heymann* (1969) are broadly similar, although the lower concentration of ^{36}Ar is more consistent with the assignment of petrologic type 4 based on olivine heterogeneity.

An analysis of the oxygen isotopic composition of a host sample yielded values for δ^{18}O of 4.64‰ and δ^{17}O of 3.45‰ relative to standard mean ocean water. These values are entirely consistent with the L group classification derived from FeO concentrations of silicates.

Shock-Blackened Clasts

Five black objects ranging from 3 to 15 mm in diameter were studied in detail; many similar but smaller objects are distributed throughout the host. Although some of these objects appear at first sight to have formed in situ, we argue below that they are clasts of shocked rock. Similarities with black ordinary chondrites (*Heymann*, 1967) indicate that the blackening was produced by shock. Textures vary enormously, as in other heavily shocked chondritic material. A few regions show a faint chondritic texture, partly obscured by troilite veining of silicate fractures. However, much of the black

TABLE 2. Noble gas concentrations (10^{-8} cc STP/g) in Kendleton: (a) host, (b) uncharacterized sample.

Source	^{3}He	^{4}He	^{20}Ne	^{21}Ne	^{22}Ne	^{36}Ar	^{38}Ar	^{40}Ar	^{84}Kr	^{132}Xe
(a) This work	3.25	840	0.84	0.77	0.87	6.8	1.34	5300	0.10	0.14
(b) *Taylor and Heymann* (1969)	3.3	792	1.7	0.92	0.96	3.2	0.68	3380	*	*

*Not determined.

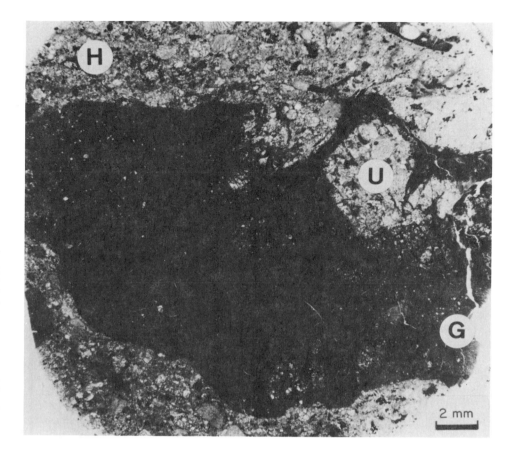

Fig. 3. Transmitted light photomicrograph of the Kendleton fragmental breccia (section UNM 710) showing a large shock-blackened clast, which contains glassy, clast-laden, shock melt (G), light-colored relict chondritic material, and an unequilibrated L3 clast (W), embedded in chondritic host (H). Although its upper horizontal edge is ill defined, this shock-blackened object is a clast of a heavily shocked, preexisting breccia and did not form in situ from the Kendleton host.

material is completely opaque in thin section because of the intrusion of extensive veins of troilite into olivine and low-Ca pyroxene grains (Fig. 3). These veins are generally much less than a micron in width and only a few microns apart. Some regions up to 1 cm across resemble shock melt veins with relict silicates 5-50 μm in size contained in a glassy looking matrix, which contains many submicron metal grains and silicate phenocrysts. Transparent olivine crystals show undulatory to mosaic extinction patterns, indicating diverse shock intensities. Metal and troilite are largely irregular in shape. Composite grains are rare and the dendritic or igneous metal-troilite textures are rare, unlike nearly all other shock-blackened chondritic material that we have observed. However, some regions in Fig. 3 do contain troilite-rimmed globules of metallic

TABLE 3. Mean compositions of olivines and low-Ca pyroxenes in clasts and inclusions in Kendleton.

Object	Section No.	Olivine				Low-Ca Pyroxene			
		Fa (mol %)		PMD*	No. Anal.	Fs (mol %)		PMD*	No. Anal.
		Mean	σ			Mean	σ		
(a) Shock-blackened clasts	649	23.7	1.42	2.6	25	18.7	3.0	12.5	9
	650	23.8	1.35	2.9	23	20.1	-	-	3
	710	22.7	3.0	9.6	38	15.8	5.7	30	18
	712	22.2	0.20	0.6	11	19.4	0.40	1.6	11
	715	24.2	0.4	1.3	20	20.1	0.96	3.9	8
	Mean	23.3	2.0		117	18.1	4.0		49
(b) L3 clast	710	17.9	8.9	43	35	15.9	6.2	30	14
(c) L5 clast	714	23.5	0.66	1.9	59	19.2	2.6	6.7	27
(d) Melt rock clast	711	21.9	0.27	0.9	45	+			
(e) Tridymite-rich inclusion	709	-	-	-	-	20.0	1.03	4.3	29

*PMD, percent mean deviation of FeO analyses in wt %.
+Only high-Ca pyroxenes found with Wo 3.0-12.4.

Fe,Ni that show a martensitic texture on etching. Clast boundaries are well defined in many places, but where the objects contain transparent chondritic material or the host contains shock veins, the boundaries are ill defined (Fig. 3).

Electron microprobe analyses of the larger relict olivine and low-Ca pyroxene crystals are given in Table 3. Olivines in three of the five clasts are distinctly richer in FeO than host olivines (Fig. 2). All but one of the objects have mean fayalite concentrations in the L group range; the exception is only marginally lower. Olivine and pyroxene data suggest that the precursor material was largely type 4 for clasts in two of the sections (649 and 710), but possibly a higher type for two others (712 and 715). The black clast in section 710 also contains an L3 clast (see below), suggesting that the precursor material for the shock-blackened clasts was a breccia or regolith containing type 3 and type 4 or 5 material.

The oxygen isotopic composition of one shock-blackened clast is $\delta^{18}O = 4.88‰$, $\delta^{17}O = 3.70‰$. These values are marginally higher than those of the host and most L chondrites. However, the displacement from the terrestrial fractionation line, $\delta^{17}O - 0.52\ \delta^{18}O$, is 1.16‰, a value which is typical for equilibrated L chondrites (mean is $1.11 \pm 0.08‰$). These data and the silicate compositions (Table 3) show that the shock-blackened clasts were derived from L group chondritic materials.

L3 Chondritic Clast

A light-colored clast or region in the shock-blackened clast shown in Fig. 3 was found to be much more heterogeneous than other host or clast material (Table 3 and Fig. 2C). Olivine heterogeneity suggests a type 3.5 classification (*Sears et al.,* 1982). A single barred olivine chondrule ranges in composition from Fa 1 to Fa 20. The presence of chondrules with mesostases of slightly turbid glass is consistent with a type 3 classification for this clast.

L5 Chondritic Clast

The largest light-colored clast that was observed is 2.5 cm in length and our petrologic studies indicate that it is an L5 chondritic clast. Chondrules are less well defined than those of the host, and olivines and low-Ca pyroxenes (Fig. 2D) are more homogeneous than those in the host. Mean olivine compositions in the clast, Fa 23.5 ± 0.7, are close to those of the neighboring host, Fa 22.8 ± 0.8 (Tables 3 and 1). Metallic Fe,Ni grains on etching show a range of structures like those observed in the host. Residual patches of cloudy taenite show that the L5 material was originally slowly cooled; plessite and polycrystalline kamacite indicate that it was subsequently shock heated. Silicates in the clast indicate a shock facies of b or c according to the scheme of *Dodd and Jarosewich* (1979).

Porphyritic Melt Rock Clast

One well-defined, light-colored clast, 8 mm in length, was found to have a porphyritic texture with little metallic Fe,Ni (Fig. 1). It contains euhedral to subhedral olivine phenocrysts up to 1mm in length, embedded in a matrix of smaller subhedral to anhedral olivine crystals, poorly resolvable pyroxene crystals and turbid brown glass. Microprobe analysis shows that the large and small olivines are homogeneous and have similar compositions of Fa 21.9 ± 0.3 and 0.054 ± 0.026 wt % CaO (Table 3). This fayalite concentration is slightly lower than that of the neighboring host (Fa 22.9 ± 0.3) and is matched by only one host region (section UNM 712; Table 1). Ten pyroxenes were found to have a mean composition of Fs 19 ± 2, Wo 5 ± 3. Shock features in the clast are similar but not identical to those of the host: olivine has undulatory extinction, small grains contain partly resorbed, cloudy taenite

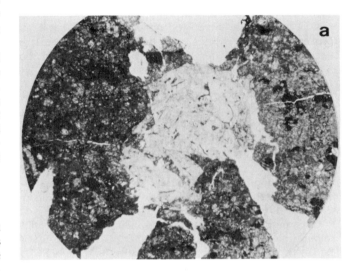

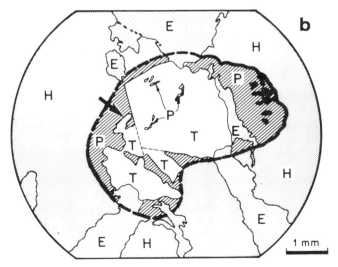

Fig. 4. Tridymite-rich inclusion in the Kendleton breccia (section UNM 709); **(a)** transmitted light photomicrograph, **(b)** sketch showing euhedral and subhedral tridymite (T) enclosed by low-Ca pyroxene (P), host (H) and epoxy (E). Tridymite occurs in four distinctly bounded segments, which are euhedral or subhedral in places; large arrow points to small tridymite crystal with terminating faces. Tridymite is surrounded by an ill defined rim of low-Ca pyroxene, which does not appear to have formed by metamorphic reaction of silica and olivine, unlike other silica-pyroxene occurrences in types 4-6 chondrites.

and tiny taenite grains in kamacite. Lack of plessite and polycrystalline kamacite is more likely to reflect the low abundance of metal rather than a different shock history for host and clast.

This clast is too large and irregular to be a porphyritic chondrule and is very likely to be an impact melt rock clast lacking relict material. In this respect it resembles four melt rock clasts in Dimmitt (*Rubin et al.*, 1983a) and some in other chondrites (*Keil*, 1982).

Tridymite-rich Inclusion

A single tridymite-rich inclusion about 5 mm in diameter was found in the Kendleton host. We describe it as an inclusion, not a clast, because there is no textual evidence that it was broken from a preexisting rock, although due to the large size of the tridymite this is likely. Tridymite occurs in four segments, which are euhedral and subhedral in places, and is largely enclosed by an ill-defined zone of low-Ca clinopyroxene (Fig. 4). Troilite grains are present in the clinopyroxene rim, and traces of troilite, chromite, metallic Fe,Ni and K-feldspar (?) were found within the tridymite. The tridymite is heavily fractured and largely isotropic, due to shock. The optical and electron probe identification of tridymite was confirmed by X-ray powder diffraction (G. Lumpkin, private communication, 1984). The rim of low-Ca pyroxene and the small stringers within the tridymite have identical compositions, Fs 20.0 ± 1.0, Wo 1 (Table 3, Fig. 2F).

Oxygen isotopic analysis of the tridymite (exclusive of visible impurities of low-Ca pyroxene and other phases) gave 5.71‰ for $\delta^{18}O$ and 3.70‰ for $\delta^{17}O$ (relative to SMOW). The displacement from the terrestrial fractionation line, $\delta^{17}O - 0.52 \delta^{18}O$, is 0.73‰. This value is much less than that for L chondrites (1.11 ± 0.08‰) but entirely appropriate for equilibrated H chondrites (0.75 ± 0.08‰).

DISCUSSION

Tridymite-rich Inclusion

The tridymite-rich inclusion in Kendleton is poorly defined (Fig. 4). We do not know whether the outer edge of the low-Ca pyroxene zone marks the edge of the object in which the tridymite crystal formed. This object might include some of the material around the pyroxenes or it is possible that the pyroxene-tridymite inclusion is a fragment of a much larger object.

Silica-pyroxene aggregates greater than a few millimeters in size have been described in several ordinary type 5-6 chondrites: Farmington (*Binns*, 1967), Nadiabondi (*Christophe Michel-Lévy and Curien*, 1965), ALHA76003 (*Olsen et al.*, 1981), Knyahinya and Lissa (*Brandstätter and Kurat*, 1985). In addition silica has been reported in a large silica-pyroxene-plagioclase intergrowth in Vishnupur (*Wlotzka et al.*, 1983) and many chondrules and millimeter and submillimeter sized inclusions (*Fredriksson and Wlotzka*, 1985; *Brigham et al.*, 1986, and references therein). In all cases where the silica polymorph was identified, it was found to be cristobalite, except in

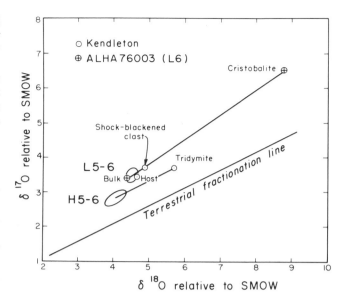

Fig. 5. Oxygen isotopic composition of the host, a shock-blackened clast and tridymite from a tridymite-rich inclusion in the Kendleton breccia; also shown are data from *Olsen et al.* (1981) for Allan Hills A76003 (L6) and its cristobalite inclusion. The host and shock-blackened clast of Kendleton have O isotopic compositions typical of equilibrated L chondrites. The Kendleton tridymite, like the cristobalite in ALHA76003, does not plot on a fractionation line through the L5-6 composition. The composition of the tridymite is consistent with equilibration with H5-6 chondrites at 900°C, however, other evidence suggests that it may have come from an H-like asteroid and not the H chondrite asteroid.

Vishnupur, where it is tridymite (*Wlotzka et al.*, 1983). The occurrence of tridymite in Kendleton is also unusual in that most of the surrounding pyroxene is not a reaction rim between silica and olivine as in Nadiabondi, Farmington, ALHA76003, Knyahinya and Lissa, because the tridymite is euhedral in places. However, the relative homogeneity of the pyroxene associated with the Kendleton tridymite (Fig. 2F, Table 3) may reflect metamorphism in an L4-6 environment.

Conditions favoring silica formation in chondrites have been discussed at length by *Brigham et al.* (1986). They argue that silica-low-Ca-pyroxene objects could not have formed by planetary igneous processes because of the low concentrations of Ca and Al in these objects. Thus, unless the Kendleton inclusion is a very unrepresentative clast of a feldspar-bearing rock, it is unlikely that it formed by planetary igneous fractionation.

Oxygen isotopic compositions of the silica occurrences in Kendleton and ALHA76003 (*Olsen et al.*, 1981) are shown in Fig. 5; other occurrences have not been analyzed. In both cases, the silica could not have formed by any kind of differentiation process that operated on bulk L chondrites. Olsen et al. conclude that their cristobalite must be a foreign grain that accreted into the L chondrite parent body. The Kendleton tridymite has an oxygen isotopic composition that is quite different from that of the ALHA76003 cristobalite, but as noted above, it is compatible with an H chondrite source.

The ^{18}O fractionation between quartz and pyroxene at 900°C has been experimentally determined to be 1.5‰ (*Matthews et al.*, 1983). Hence pyroxene in equilibrium with the Kendleton tridymite at 900°C would have $\delta^{18}O$ = 4.21, $\delta^{17}O$ = 2.92; for associated olivine, values would be a little lower. These values agree well with the means for equilibrated H chondrites: $\delta^{18}O$ = 4.13, $\delta^{17}O$ = 2.90. From the study of *Olsen et al.* (1981), we know that large cristobalite crystals can retain an H group oxygen isotopic composition during metamorphism in an L4-6 environment while the Fe/Mg ratio in associated low-Ca pyroxene changes from an H to an L value.

Four centimeter-sized, igneous-textured clasts with oxygen isotopic compositions close to those of equilibrated H chondrites have also been found in L4-6 chondrites. Two similar olivine-rich clasts in the L6 chondrites, Yamato 75097 and 793241, which are believed to be paired, have mean oxygen isotopic compositions of $\delta^{18}O$ = 4.54, $\delta^{17}O$ = 3.05 and $\delta^{18}O$ = 3.91, $\delta^{17}O$ = 2.46, respectively (*Clayton et al.*, 1984; *Prinz et al.*, 1984; *Mayeda et al.*, 1987). The occurrence of a relict chondrule in the Yamato 793241 clast implies that both clasts were derived from a chondritic source (*Prinz et al.*, 1984). The third clast, which was found in Bovedy (L4), has $\delta^{18}O$ = 4.3 and $\delta^{17}O$ = 2.6, and is thought to have formed from a chondritic source by impact melting (*Rubin et al.*, 1981). The fourth clast was found in the Barwell (L6) chondrite (*Hutchison et al.*, 1986) and has $\delta^{18}O$ = 4.14 and $\delta^{17}O$ = 2.73. All clasts except that in Bovedy, which has heterogeneous silicates, have olivines with compositions appropriate to equilibrated L chondrites. Besides chondrules and inclusions in L3 chondrites (*Gooding et al.*, 1983) and inclusions of carbonaceous chondrites (*Wilkening*, 1977), only one other inclusion in L chondrites has been found with an anomalous oxygen isotopic composition. A 7-mm-long igneous inclusion in Roosevelt County 014 (L5) has an oxygen isotopic composition close to those of EL chondrites (*Scott et al.*, 1986). Chondritic inclusions with H-like oxygen isotopic compositions have not been found in L chondrites.

Some of the four igneous clasts with H-like oxygen isotopic compositions may have been derived from H chondrite or H-like projectiles that melted on impact with the L chondrite parent body, but the Kendleton tridymite inclusion is so different in mineralogy and texture from chondritic impact melts that it is unlikely to have formed in this way. The absence of reports of tridymite inclusions in H chondrites and H chondritic clasts in L chondrites suggests that the Kendleton tridymite inclusion was not derived directly from an H chondrite projectile. L chondritic clasts have not been found in H and LL chondrites, although the presence of an H chondritic clast in an LL chondrite, St. Mesmin, and an LL clast in an H chondrite, Dimmitt (see references in *Clayton et al.*, 1984), suggests that H and LL asteroids may have experienced limited exchange of material. Because there are iron meteorites that formed on asteroids with oxygen isotopic compositions indistinguishable from those of H, L, and LL chondritic asteroids (*Clayton et al.*, 1983), it is possible that the Kendleton tridymite inclusion and some of the four igneous clasts with H-like oxygen are samples of additional asteroids with oxygen isotopic compositions like those of ordinary chondrites. However, it is also possible that the Kendleton inclusion was added to the L body during accretion and was never part of another body.

Shock-Blackened Clasts

We suggest that the shock-blackened objects in Kendleton, like similar objects in Dimmitt (DT7; *Rubin et al.*, 1983a) and Nulles (clast 5; *Williams et al.*, 1985), are clasts from preexisting rocks. However, *Fodor and Keil* (1976) conclude that similar objects in the Plainview regolith breccia, which they call "pseudo-fragments," formed in situ. They note that silicates in these objects are compositionally identical with those of the host, and veins or protrusions of shocked material appear to extend into the host. We have examined three of these objects in Plainview (PV10 in University of New Mexico [UNM] section number 281, PV12 in UNM 274, and one in UNM 282) and conclude that at least two of them are clasts. (Clast PV10 is too close to the edge of the section to make any judgement).

We have compared the Plainview and Kendleton shock-blackened clasts with shocked regions in several heavily shocked chondrites, such as Walters, Orvinio, and Rose City, and conclude that the clasts differ from the regions in two ways. The clasts appear to be largely equant, not vein-like, and textural and mineralogical features in shock-melted parts are not aligned parallel to the edge of the shock-blackened areas. We have observed a large equant pocket of shock melt that contains clasts in the Bluff (L5) chondrite (section L6986 from the Naturhistorisches Museum, Vienna). However, it clearly formed in situ as the edges of olivines adjacent to the melt are shock granulated, and there are extensive protrusions of melt in the surrounding host. For three of the five Kendleton objects, our conclusion that they did not form in situ is strengthened by small but significant compositional differences between clast and host olivines (Fig. 2).

We recognize that the boundaries of shock-blackened objects in Kendleton and Plainview are indistinct in many places, but we attribute this to the host-like chondritic regions within the shock-blackened rock (Fig. 3), which camouflage the clast-host boundary. In addition, in both chondrites there are small shock veins in the host material, which intersect clasts and were formed in situ. These features convey the false impression that the shocked clasts formed in situ.

Thermal and Shock History of Kendleton's Ingredients

To help understand when the host material and clasts were mixed and lithified, we have summarized some of their metamorphic and shock features in Table 4. Host material is largely type 4 with some type 3 material. Since host olivine and low-Ca pyroxene have concentrations of iron that are consistently low for L chondrites, the host material must be derived from only a part of the total L4 reservoir.

The shock-blackened clasts did not experience any planet-wide metamorphism and associated slow cooling because martensite, which is characteristic of rapid cooling (*Heymann*, 1967; *Smith and Goldstein*, 1977), is present. We do not know

TABLE 4. Comparison of metamorphic and shock features of host and clasts in the Kendleton breccia.

Feature	Host Material	Clasts			
		Shock bl.	L5	Melt-rock	SiO$_2$-rich
Metamorphic history					
Petrologic type	4 (3)	4 (3,5)	5	3?	4
Cloudy taenite*	yes	no	yes	yes	-
Shock history					
Silicate shock level	b or c	e or f (b/c)	b or c	b or c	b, c or d
Metallic Fe,Ni†	ct, pk, pl	ct, pk, ma	ct, pk, pl	ct	-

*Cloudy taenite is a submicroscopic mixture of tetrataenite and martensite (*Reuter et al.,* 1986).
†Shock-heating characteristics of metal: ct, clear taenite (i.e., resorbed cloudy taenite); ma, martensite; pk, polycrystalline kamacite; pl, plessite.

whether the L5 clast was mixed with host material after or during late-stage metamorphism, as metallographic cooling rates could not be obtained. The melt rock clast contains cloudy taenite, which is characteristic of slowly cooled irons and chondrites (*Reuter et al.,* 1986), so that the impact melting must have occurred before metamorphism had ended. Turbid glass and heterogeneous pyroxenes in the melt rock clast indicate that this clast was metamorphosed less than type 4 chondrites. Olivine in the clast may have been homogenized during metamorphism, but homogenization during crystallization cannot be excluded. The tridymite-rich inclusion probably experienced type 4 levels of metamorphism in view of the abundance and homogeneity of the clinopyroxene.

Thus the strongest constraint on the time of lithification of the Kendleton breccia is provided by martensite in the shock-blackened clasts. Studies of similar metal in Ramsdorf and Orvinio (*Smith and Goldstein,* 1977) suggest that cooling rates below 700°C were higher than 10^{-4}°C sec^{-1} or 10^{3}°C yr^{-1}. Thus, Kendleton, like chondritic regolith breccias, must have been lithified after planet-wide metamorphism had ended, because metamorphic cooling rates were 10^{-6} to 10^{-3}°C yr^{-1}.

All of the material in Kendleton has been shocked to low or high levels. Thus it is possible that the low shock to b or c levels occurred after lithification of the breccia. The only feature that might require lower levels of shock after lithification is the martensite in parts of the shock-blackened clasts. We do not know the relative rates at which cloudy taenite and martensite are converted into plessite. If martensite decomposes faster, this would suggest that the host was shocked prior to lithification. It is possible that a single event produced the shock-blackened material and shocked the host material, but multiple events cannot be excluded. A much earlier impact formed the porphyritic melt rock.

Comparison with Regolith Breccias

Noble gas analyses of Kendleton (Table 2) show no evidence for the presence of solar-wind gases, nor does it seem likely that solar-wind gases were lost. Shock reheating of Kendleton is responsible for its low ^{4}He and ^{40}Ar retention ages of 2.6 and 3.6 Gyr (*Taylor and Heymann,* 1969). However, heating was relatively mild as resorption of cloudy taenite, which is incomplete in Kendleton host, takes only a few seconds at 700°C in the atmospherically heated rims of iron meteorites (*Buchwald,* 1975, Figs. 45 and 53). The high concentration of trapped ^{36}Ar also suggests that solar wind gases were not removed by shock heating. Finally, because *Binns* (1968) found martensite with 14-18% Ni and kamacite with 7-8% Ni throughout the host and clasts of a gas-rich, regolith breccia, Ghubara, we infer that shock heating much more severe than that experienced by the Kendleton host is insufficient to remove solar-wind gases.

Except for the absence of solar-wind gases and associated irradiation features, Kendleton has nearly all features found in ordinary chondrite regolith breccias (*Keil,* 1982). It has a dark matrix that encloses a wide variety of light- and dark-colored clasts. The chondritic clasts in Kendleton range from type 3 to type 5; some clastic material was formed by impacts during metamorphism, other material was shocked after metamorphism, and the rock contains a unique tridymite-rich inclusion that probably comes from another body. The only petrographic feature that is somewhat unusual for regolith breccias is the high abundance of host, about 80%. In the six regolith breccias for which quantitative estimates have been made, matrix abundances are 40-70% (*Rubin,* 1982; *Rubin et al.,* 1983a; *Williams et al.,* 1985). However, the LL3 regolith breccia, Ngawi, appears to have a matrix abundance that is comparable to that of Kendleton.

Origins of Fragmental and Regolith Breccias

Kendleton has the diversity of clasts expected in regolith breccias but lacks the irradiation features associated with them. Other possible examples of such chondritic breccias include L'Aigle (*Keil,* 1982), Castalia (*Taylor and Heymann,* 1969; *Van Schmus,* 1969), and Waconda (*Weber et al.,* 1983; *Rubin,* 1982). Presumably material in these fragmental breccias, unlike that in regolith breccias, was not close to the asteroid surface for any significant period. However, recent studies of gas-rich meteorites suggest that conventional ideas on the origin of meteoritic regolith breccias may need revision.

Caffee et al. (1983, 1986, 1987) find very large enrichments of cosmogenic Ne in solar-flare irradiated grains in Murchison and Kapoeta, which are absent in nonirradiated grains. They argue that the most likely source is solar cosmic rays from a more active, early sun. Since Kapoeta and other regolith breccias contain some clasts with ages of 3.7 Gyr or less (*Rajan et al.,* 1979; *Schultz and Signer,* 1977; *Keil et al.,* 1980), this requires that irradiated grains were stored inside the parent body for as much as 10^9 years before mixing with other fragmental material.

In order for solar-wind irradiated grains to be introduced into meteorite breccias in this way, these grains must have been loosely consolidated inside the parent body, so they could have been readily mixed with fragmental material over the next 10^9 years. Since many ordinary chondrites are fragmental breccias that were lithified after metamorphism (*Scott et al.,* 1985; *Rubin et al.,* 1983b), there must have been much mixing of weakly lithified or unconsolidated chondritic material after metamorphism as a result of cratering, spallation, and disruption and reassembly of parent bodies. The results of *Caffee et al.* (1983, 1986) do not exclude the possibility that some early irradiated grains were mixed with recently irradiated grains in a regolith as their analyses were made on sets of 6-11 grains. However, it is also possible that most of the material in a meteorite containing solar-flare irradiated grains was never present near the surface of an asteroid. Then the proportion of irradiated grains in a meteorite may have no simple bearing on the proportion in an asteroidal regolith, and the term "regolith breccia" may be inappropriate for some meteorites that contain solar-flare irradiated grains.

One possible test of this scenario would be a search for chemical differences between irradiated and unirradiated grains in the hosts of meteorite regolith breccias. Because most irradiated and unirradiated grains are postulated to have different origins, chemical differences should be detectable in favorable cases. *Wilkening et al.* (1971) analyzed 110 pyroxenes in Kapoeta and did not observe any significant compositional differences between irradiated and unirradiated grains. However, both groups of pyroxenes have a wide compositional range in Kapoeta, so that further searches would be worthwhile.

If gas-rich meteorites did not form in surface regoliths, it is possible that Kendleton and some other gas-poor, fragmental breccias formed in the same way as gas-rich breccias except that Kendleton et al. failed to acquire an admixture of early irradiated grains. If some gas-rich meteorites were lithified in asteroidal regolith, however, all of the constituents of Kendleton et al. may also have resided in a regolith, but for too short a period for surface grains to have acquired detectable solar-wind gases.

Acknowledgments. We thank G. J. Taylor for discussions, G. Lumpkin for technical assistance, and A. E. Rubin for a helpful review. This work was partly supported by NASA Grant NAG-9-30 (Keil) and NSF Grant EAR 8316812 (Clayton).

REFERENCES

Bence A. E. and Albee A. L. (1968) Empirical correction factors for the electron microanalysis of silicates and oxides. *J. Geol., 76,* 382-403.
Binns R. A. (1967) Farmington meteorite: cristobalite xenoliths and blackening. *Science, 156,* 1222-1226.
Binns R. A. (1968) Cognate xenoliths in chondritic meteorites: examples in Mezö-Madaras and Ghubara. *Geochim. Cosmochim. Acta, 32,* 299-317.
Brandstätter F. and Kurat G. (1985) On the occurrence of silica in ordinary chondrites. *Meteoritics, 20,* 615-616.
Brigham C. A., Yabuki H., Ouyang Z., Murrell M. T., El Goresy A., and Burnett D. S. (1986) Silica-bearing chondrules and clasts in ordinary chondrites. *Geochim. Cosmochim. Acta, 50,* 1655-1666.
Buchwald V. F. (1975) *Handbook of Iron Meteorites.* University of California Press. 1418 pp.
Busche F. D. (1975) Major and minor element contents of coexisting olivine, orthopyroxene and clinopyroxene in ordinary chondritic meteorites. Ph.D. dissertation, University of New Mexico, Albuquerque. 75 pp.
Caffee M. W., Goswami J. N., Hohenberg C. M., and Swindle T. M. (1983) Cosmogenic neon from precompaction irradiation of Kapoeta and Murchison. *Proc. Lunar Planet. Sci. Conf. 14th,* in *J. Geophys. Res., 88,* B267-B273.
Caffee M. W., Goswami J. N., Hohenberg C. M., and Swindle T. D. (1986) Pre-compaction irradiation of meteorite grains (abstract). In *Lunar and Planetary Science XVII,* pp. 99-100. Lunar and Planetary Institute, Houston.
Caffee M. W., Hohenberg C. M., and Swindle T. D. (1987) Evidence in meteorites for an active early Sun. *Astrophys. J. Lett., 313,* L31-L35.
Christophe Michel-Lévy M., Curien H., and Goñi J. (1965) Etude à la microsonde électronique d'un chondre d'olivine et d'un fragment riche en cristobalite de la météorite de Nadiabondi. *Bull. Soc. Minér. Cristallogr., 88,* 122-125.
Clayton R. N., Onuma N., and Mayeda T. K. (1976) A classification of meteorites based on oxygen isotopes. *Earth Planet. Sci. Lett., 30,* 10-18.
Clayton R. N., Mayeda T. K., Olsen E. J., and Prinz M. (1983) Oxygen isotope relationships in iron meteorites. *Earth Planet. Sci. Lett., 65,* 229-232.
Clayton R. N., Mayeda T. K., and Yanai K. (1984) Oxygen isotopic composition of some Yamato meteorites. *Mem. Natl. Inst. Polar Res. (Tokyo), 35,* 267-271.
Crabb J. and Schultz L. (1981) Cosmic-ray exposure ages of the ordinary chondrites and their significance for parent body stratigraphy. *Geochim. Cosmochim. Acta, 45,* 2151-2160.
Dodd R. T. and Jarosewich E. (1979) Incipient melting in and shock classification of L-group chondrites. *Earth Planet. Sci. Lett., 44,* 335-340.
Fodor R. V. and Keil K. (1976) Carbonaceous and non-carbonaceous lithic fragments in the Plainview, Texas, chondrite: origin and history. *Geochim. Cosmochim. Acta, 40,* 177-189.
Fredriksson K. and Wlotzka F. (1985) Morro do Rocio: an unequilibrated H5 chondrite. *Meteoritics, 20,* 467-478.
Gomes C. B. and Keil K. (1980) *Brazilian Stone Meteorites.* University of New Mexico Press, Albuquerque. 161 pp.
Gooding J. L., Mayeda T. K., Clayton R. N., and Fukuoka T. (1983) Oxygen isotopic heterogeneities, their petrological correlations, and implications for melt origins of chondrules in unequilibrated ordinary chondrites. *Earth Planet. Sci. Lett., 65,* 209-224.
Goswami J. N., Lal D., and Wilkening L. L. (1984) Gas-rich meteorites: probes for particle environment and dynamical processes in the inner solar system. *Space Sci. Rev., 37,* 111-159.
Heymann D. (1967) On the origin of hypersthene chondrites: ages and shock effects of black chondrites. *Icarus, 6,* 189-221.
Hutchison R., Williams C. T., Din V. K., Paul R. L., and Lipschutz M. E. (1986) An achondritic troctolite clast in the Barwell, L5-6, chondrite (abstract). *Meteoritics, 21,* 402-403.
Keil K. (1982) Composition and origin of chondritic breccias. In

Workshop on Lunar Breccias and Soils and Their Meteoritic Analogs (G. J. Taylor and L. L. Wilkening, eds.) pp. 65-83. LPI Tech. Rpt. 82-02, Lunar and Planetary Institute, Houston.

Keil K., Fodor R. V., Starzyk P. M., Schmitt R. A., Bogard D. D., and Husain L. (1980) A 3.6-b.y.-old impact-melt rock fragment in the Plainview chondrite: implications for the age of the H-group chondrite parent body regolith formation. *Earth Planet. Sci. Lett., 51,* 235-247.

Matthews A., Goldsmith J. R., and Clayton R. N. (1983) Oxygen isotope fractionations involving pyroxenes: the calibration of mineral-pair geothermometers. *Geochim. Cosmochim. Acta, 47,* 631-644.

Mayeda T. K., Clayton R. N., and Yanai K. (1987) Oxygen isotopic compositions of several Antarctic meteorites. *Mem. Natl. Inst. Polar Res. (Tokyo), 46,* 144-150.

Nishiizumi K., Regnier S., and Marti K. (1980) Cosmic ray exposure ages of chondrites, pre-irradiation and constancy of cosmic ray flux in the past. *Earth Planet. Sci. Lett., 50,* 156-170.

Olsen E. J., Mayeda T. K., and Clayton R. N. (1981) Cristobalite-pyroxene in an L6 chondrite: implications for metamorphism. *Earth Planet. Sci. Lett., 56,* 82-88.

Prinz M., Nehru C. E., Weisberg M. K., Delaney J. S., Yanai K., and Kojima H. (1984) H chondritic clasts in a Yamato L6 chondrite: implications for metamorphism. *Meteoritics, 19,* 292-293.

Rajan R. S., Huneke J. C., Smith S. P., and Wasserburg G. J. (1979) Argon 40-argon 39 chronology of lithic clasts from the Kapoeta howardite. *Geochim. Cosmochim. Acta, 43,* 957-971.

Reuter K. B., Kowalik J. A., and Goldstein J. I. (1986) Low temperature studies in the iron meteorite Dayton (abstract). In *Lunar and Planetary Science XVII,* pp. 705-706. Lunar and Planetary Institute, Houston.

Rubin A. E. (1982) Petrology and origin of brecciated chondritic meteorites. Ph.D. thesis, University of New Mexico, Albuquerque. 220 pp.

Rubin A. E., Keil K., Taylor G. J., Ma M.-S., Schmitt R. A., and Bogard D. D. (1981) Derivation of a heterogeneous lithic fragment in the Bovedy L-group chondrite from impact melted porphyritic chondrules. *Geochim. Cosmochim. Acta, 45,* 2213-2228.

Rubin A. E., Scott E. R. D., Taylor G. J., Keil K., Allen J. S. B., Mayeda T. K., Clayton R. N., and Bogard D. D. (1983a) Nature of the H chondrite parent body regolith: evidence from the Dimmitt breccia. *Proc. Lunar Planet. Sci. Conf. 13th,* in *J. Geophys. Res., 88,* A741-A754.

Rubin A. E., Rehfeldt A., Peterson E., Keil K., and Jarosewich E. (1983b) Fragmental breccias and the collisional evolution of ordinary chondrite parent bodies. *Meteoritics, 18,* 179-196.

Schultz L. and Signer P. (1977) Noble gases in the St. Mesmin chondrite: implications to the irradiation history of a brecciated meteorite. *Earth Planet. Sci. Lett., 36,* 363-371.

Scott E. R. D., Lusby D., and Keil K. (1985) Ubiquitous brecciation after metamorphism in equilibrated ordinary chondrites. *Proc. Lunar Planet. Sci. Conf. 16th,* in *J. Geophys. Res., 90,* D137-D148.

Scott E. R. D., Taylor G. J., and Keil K. (1986) Accretion, metamorphism and brecciation of ordinary chondrites: evidence from petrologic studies of meteorites from Roosevelt County, New Mexico. *Proc. Lunar Planet. Sci. Conf. 17th,* in *J. Geophys. Res., 91,* E115-E123.

Sears D. W., Grossman J. N., Melcher C. L., Ross L. M., and Mills A. A. (1980) Measuring metamorphic history of unequilibrated ordinary chondrites. *Nature, 287,* 791-795.

Sears D. W., Grossman J. N., and Melcher C. L. (1982) Chemical and physical studies of type 3 chondrites—I: Metamorphism related studies of Antarctic and other type 3 chondrites. *Geochim. Cosmochim. Acta, 46,* 2471-2481.

Smith B. A. and Goldstein J. I. (1977) The metallic microstructures and thermal histories of severely reheated chondrites. *Geochim. Cosmochim. Acta, 41,* 1061-1072.

Taylor G. J. and Heymann D. (1969) Shock, reheating, and the gas retention ages of chondrites. *Earth Planet. Sci. Lett., 7,* 151-161.

Taylor G. J. and Wilkening L. L. (1982) *Workshop on Lunar Breccias and Soils and Their Meteoritic Analogs.* LPI Tech. Rpt. 82-02. Lunar and Planetary Institute, Houston. 172 pp.

Van Schmus W. R. (1969) The mineralogy and petrology of chondritic meteorites. *Earth Sci. Rev., 5,* 145-184.

Weber H. W., Braun O., Schultz L., and Begemann F. (1983) The noble gas record in Antarctic and other meteorites. *Z. Naturforsch., 38,* 267-272.

Wilkening L. L. (1977) Meteorites in meteorites: evidence for mixing among the asteroids. In *Comets, Asteroids, Meteorites—Interrelations, Evolution and Origins* (A. H. Delsemme, ed.) pp. 389-396. University of Toledo.

Wilkening L. L., Lal D., and Reid A. M. (1971) The evolution of the Kapoeta howardite based on fossil track studies. *Earth Planet. Sci. Lett., 10,* 334-340.

Williams C. V., Rubin A. E., Keil K., and San Miguel A. (1985) Petrology of the Cangas de Onis and Nulles regolith breccias: implications for parent body history. *Meteoritics, 20,* 331-345.

Wlotzka F., Palme H., Spettel B., Wänke H., Fredriksson K., and Noonan A. F. (1983) Alkali differentiation in LL-chondrites. *Geochim. Cosmochim. Acta, 47,* 743-757.

Initial $^{87}Sr/^{86}Sr$ and Sm-Nd Chronology of Chondritic Meteorites

J. C. Brannon and F. A. Podosek

Department of Earth and Planetary Sciences, and McDonnell Center for the Space Sciences, Washington University, St. Louis, MO 63130

G. W. Lugmair

Chemistry Department and Scripps Institution of Oceanography, University of California at San Diego, La Jolla, CA 92093

We present isotopic analyses of the Rb-Sr and Sm-Nd systems in phosphates from ordinary chondritic meteorites (Bjurbole, Mocs, Menow, Elenovka, Ambapur Nagla, and St. Severin). The Sm-Nd data yield an isochron of age 4.55 ± 0.45 Ga and an initial $^{143}Nd/^{144}Nd$ consistent with that of Juvinas. Initial $^{87}Sr/^{86}Sr$ ratios are characteristically (possibly excepting St. Severin) elevated with respect to BABI or ALL. Considering both presently and previously reported data, elevation of initial $^{87}Sr/^{86}Sr$ ratios corresponding to metamorphism intervals of the order of 10^8 years seems quite characteristic of chondrites of metamorphic grade 4 or higher. These formation intervals, however, do not correlate with either degree of metamorphism or with I-Xe ages, hence such data for metamorphic chronologies should be interpreted with caution.

INTRODUCTION

Undifferentiated meteorites (= chondrites) are aggregates of primitive solids from the solar nebula, which have never experienced igneous differentation on a planetary scale. They include the most ancient known macroscopic objects in the solar system, e.g., refractory-element-rich inclusions in carbonaceous chondrites with ~4.56 Ga Pb-Pb ages (cf. *Chen and Wasserburg*, 1981). Chondrites are nevertheless not pristine samples of nebular material, and in general have experienced varying degrees of alteration and/or thermal metamorphism in a parent body environment. They are also not exactly contemporaneous, and there is ample evidence that chondrite "formation," in the usual sense of isotopic closure, occupied a time of the order of the first 10^8 yr of solar system history.

An accurate and reliable chronology of this first 10^8 yr of the history of the solar system is possible within the bounds of current experimental technique, but has so far proved elusive. Methods determining absolute ages on the basis of long-lived radionuclides, e.g., U-Pb, Rb-Sr, K-Ar, are frequently inapplicable or insufficiently precise. There are also several methods that do not yield absolute ages but can give high-resolution relative ages, such as those based on short-lived radionuclides, e.g., ^{129}I, ^{244}Pu, or on the evolution of $^{87}Sr/^{86}Sr$, but these also have so far not provided a fully reliable chronology in the sense of agreement of ages determined by different methods or correlation of age with petrographic or chemical properties.

In this paper we present data for Rb, Sr, Sm, and Nd in phosphates from ordinary chondrites and consider their chronological implications, especially for the relative chronology based on initial $^{87}Sr/^{86}Sr$. Phosphates (merrillite and apatite) are trace or accessory minerals (< 1%) in ordinary chondrites, but they have a substantial importance in geochronology because they are important hosts for several of the trace elements on which chronologies are based (U, Sm and Nd, Pu). While phosphates are not major hosts for Sr, they characteristically have very low Rb/Sr ratios, permitting precise determination of initial $^{87}Sr/^{86}Sr$ ratios. As discussed in more detail below, initial Sr compositions have contributed to the perception that chondrite metemorphism spanned an interval of the order of 10^8 yr (*Wasserburg et al.*, 1969).

SAMPLES

All but one of the phosphate samples described here were originally prepared by P. Pellas (*Kirsten et al.* 1978; *Pellas and Storzer*, 1981), separated from bulk meteorite by a combination of magnetic susceptibility and density (heavy liquid) techniques. The exception is St. Severin phosphate, originally prepared by R. S. Lewis (*Lewis*, 1975) using the magnetic susceptibility technique only.

Grain sizes were generally 20-60 μm. In binocular examination the phosphates were characteristically clear and either colorless, yellow, orange, or amber. They frequently contained small dark opaque inclusions and were occasionally draped by opaque orange stains (neither of which dissolved in dilute HCl). The principal nonphosphate was feldspar; trace amounts of mafic minerals and extraneous nonmeteoritic materials were also present. The mafics and extraneous materials, and substantial amounts of the feldspars, were removed by handpicking under a binocular microscope. The phosphate purity of the analyzed samples can be assessed from the fraction that did not dissolve in dilute HCl (Table 1).

For each sample, 30-60 grains were mounted for EDX analysis. Phosphates were identified by the P signal and identified as either apatite or merrillite by the presence or absence of the Cl signal. Merrillite fractions are given in Table 1. Those grains were not used for any further analysis.

All samples were also examined by cathodoluminescence. Feldspar grains luminesced blue, while phosphates luminesced yellow or orange. Correlation of blue luminescence with optical appearance was used to assist in identifying the feldspars for handpicking. For St. Severin, Menow, Ambapur Nagla, and Mocs, grains examined in the luminoscope were returned to the total

TABLE 1. Analytical results for chondritic phosphates.

Sample	Weight (mg)	Residue (%)*	Merrillite (%)†	Sm‡ (ppm)	Nd‡ (ppm)	$^{147}Sm/^{144}Nd$	ϵ_{143}†
St. Severin L	0.214	3.7		40.74	151.16	0.1871	- 3.7 ± 0.3
Menow	0.359	7.2	93	33.43	106.44	0.1898	- 2.5 ± 0.3
Ambapur Nagla-P	0.256	§	67	24.90	80.62	0.1867	- 3.8 ± 0.3
Ambapur Nagla-W	0.098	<1	87	29.62	95.29	0.1878	- 3.2 ± 0.3
Elenovka	0.429	5.1	63	36.53	118.37	0.1866	- 4.3 ± 0.3
Mocs	0.188	20.7	48	28.35	94.26	0.1818	- 7.0 ± 0.3
Bjurbole	0.262	10.3	89	36.82	120.08	0.1853	- 5.1 ± 0.3

*Fraction of total weight listed that remained undissolved in ~0.5 N HCl for ~30 min; Sm and Nd concentrations are figured on basis of dissolved weight.

†Fraction of phosphates that is merrillite rather than apatite, as determined by absence or presence of Cl signal in EDX examination of ~30-60 grains.

‡Mass spectroscopy performed in La Jolla; Nd analyzed as NdO+, corrected for discrimination by $^{148}Nd/^{144}Nd = 0.241572$, and tabulated as parts in 10^4 deviation from 0.512566 (present-day Juvinas); stated errors are 95% confidence limits for external precision (in-run errors are lower).

§No data for undissolved residue; concentrations calculated on basis of total sample weight.

sample pool. For Elenovka and Bjurbole, grains examined in the luminoscope were not returned to the sample pool.

The relationship between optical appearance, luminescence, and EDX identification was considered as a potential means of separating merrillite and apatite. For Menow and Ambapur Nagla a relationship was found: Merrillite grains were clear and colorless and luminesced orange, while apatite grains were characteristically clear and yellow, orange, or amber (but a small fraction were colorless) and luminesced yellow. For the other samples, it was not possible to make such distinctions, and luminescence was much weaker, making it difficult to distinguish between orange and yellow luminescence. Both Menow and Ambapur Nagla are H chondrites, while the others are L or LL chondrites, which suggests that this distinction reflects chemical class, but with such small numbers of samples it is difficult to make a strong statement.

For Ambapur Nagla we analyzed two fractions. One was "bulk phosphate," the "P" sample in Table 1. The second, the "W" fraction in Table 1, was a collection of clear colorless grains; as seen in Table 1, this sample was significantly enriched in merrillite over apatite, relative to the total phosphates, although still not pure merrillite.

PROCEDURES AND RESULTS

Small fractions (0.1-0.4 mg) of the samples described above, including the "merrillite" fraction of Ambapur Nagla, were taken for Sm-Nd and Rb-Sr analyses. These samples were dissolved in ~1 ml dilute (~.5 N) HCl for ~30 min in teflon beakers. Most (>95%) of the resultant solution was decanted by pipette, leaving undissolved residue and a small fraction of the solution in the beaker; to insure essentially quantitative recovery of the elements in solution, water was added to the residual solution and then mostly decanted, and this procedure repeated at least three times. All the decanted liquids were combined and then spiked for Sm, Nd, Rb and Sr. The undissolved residue was dried and weighed to provide a lower limit on phosphate purity and to allow elemental concentrations to be calculated for the material which dissolved in dilute HCl (taken to be phosphate).

Sample characterization and Sm-Nd data for this suite of analyses are given in Table 1. As seen in Fig. 1, the range of Sm/Nd ratios is small, but nevertheless the data define an isochron corresponding to an age of 4.55 ± .45 Ga with initial $^{143}Nd/^{144}Nd = .5067 ± 5$, which is consistent with the initial ratio in Juvinas (.506705 for a 4.56 Ga age).

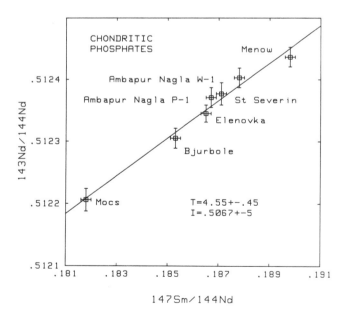

Fig. 1. Sm-Nd isochron diagram for chondritic phosphates (data from Table 1). The line shown is a least squares fit with a slope corresponding to age 4.55 ± 0.45 Ga ($\lambda = 6.54 \times 10^{-12}$ yr^{-1}) and an initial $^{143}Nd/^{144}Nd$ of 0.5067 ± 5 (consistent with initial composition in Juvinas).

TABLE 2. Progressive leaching of 0.061 mg Menow phosphates.

Wash*	ml	min	Recovery (ng)†				ϵ_{143}†	$^{87}Sr/^{86}Sr$†,‡
			Sm	Nd	Rb	Sr		
methanol	50	5	0.01	0.13	§	1.48		0.7087 ± 3
water	10	1			0.013	0.16		
0.25% HNO$_3$	10	1	0.14	0.49	0.007	0.22		
0.5 N HCl	10	1	0.05	0.13	0.015	0.24		
0.5 N HCl	10	1	0.18	0.64	0.028	0.60		
0.5 N HCl	10	15	0.04	0.04	0.019	0.34		
6 N HCl	5	8	1.18	3.74	0.030	1.58	−2.2 ± 0.7	0.7019 ± 3
Total			1.6	5.2	0.11	4.6		
Comparison¶			1.9	6.0	0.08	28.4		

*See text.
†Mass spectrometry performed in St. Louis; Nd analyzed as Nd+ in triple-filament mode; data reduced and cited relative to Juvinas to be comparable to data in Table 1.
‡See notes to Table 4.
§Large hydrocarbon contamination.
¶Quantities corresponding to 0.061 mg of sample and Sm, Nd concentrations from Table 1, and Rb,Sr concentrations from *Brannon et al.* (1987); see text for discussion.

The Rb and Sr data for this suite of analyses are not given here, but are tabulated by *Brannon et al.*, (1987). Strontium concentrations were unexpectedly high (80-766 ppm) and radiogenic ($^{87}Sr/^{86}Sr$ = 0.704-0.710). In light of subsequent analyses described below, these data are clearly not representative of bulk indigenous Sr but rather reflect surficial Sr. The origin of this Sr is not clearly understood, but the most plausible source seems to be contamination during examination in the cathodoluminoscope.

To examine the siting of Sr in these phosphates, a progressive leaching experiment was conducted on another fraction of Menow phosphates. The sample was put into a polycarbonate syringe with a Nuclepore filter at the bottom, and then washed with progressively more aggressive leachants. Each fraction was spiked for Sm, Nd, Rb, and Sr; results are given in Table 2.

As seen in Table 2, in this procedure Sr was released in two principal steps: in the methanol wash, in which it was rather radiogenic, and in the final step of major dissolution, in which it was considerably less radiogenic. This supports association of radiogenic Sr as a surficial component. Also, the total Sr recovered in this procedure is severalfold less than the Sr concentration of *Brannon et al.*, (1987) would suggest, indicating that if contamination is indeed involved, the Menow sample analyzed in Table 2 is an order-of-magnitude less contaminated than the samples analyzed in Brannon et al. Since this sample came from the the same vial, the difference is most likely due to pre-analysis handling, such as the cathodoluminoscope examination. It is noteworthy that of the four elements considered here, the surficial component primarily affects Sr; total recovery of Sm, Nd, and even Rb in this leaching experiment is reasonably comparable to the amounts expected from the Brannon et al. data.

The data in Table 2 are useful only as a guide to siting of a surficial component. It is noted that element recovery corresponding to major dissolution did not occur until well after the phosphates would have been expected to dissolve; this is most plausibly due to sample material having lodged in a gridwork in the syringe and not having been completely exposed to the leachant. Also, blank measurements conducted in parallel with these analyses gave erratic results, but overall suggest that the total blank might be around 1% for Sm and Nd and as high as 10-20% for Rb and Sr.

Guided by these results, we performed a third suite of analyses using a cleaner procedure than that involving the

TABLE 3. Sr results for methanol wash and HNO$_3$ etch of chondritic phosphates.

Sample	Weight (mg)	Methanol		HNO$_3$ etch	
		ng	$^{87}Sr/^{86}Sr$*	ng	$^{87}Sr/^{86}Sr$*
St. Severin	0.238	0.033	0.711 ± 2	2.04	0.70210 ± 28
Bjurbole	0.163	0.017	0.707 ± 2	3.87	0.70258 ± 20
Elenovka	0.066	0.020	0.709 ± 1	1.34	0.70271 ± 24
Ambapur Nagla	0.079	0.066	0.709 ± 1	2.04	0.70189 ± 24

*See notes to Table 4; mass spectrometry performed in St. Louis.

TABLE 4. Rb-Sr results for etched chondritic phosphates.

Sample	Weight* (mg)	Load (ng)[†]		Concentration (ppm)[†,‡]		[†] $^{87}Rb/^{86}Sr$	[†,§] $^{87}Sr/^{86}Sr$	[§,¶] $^{87}Sr/^{86}Sr$	[§,**] $^{87}Sr/^{86}Sr$
		Rb	Sr	Rb	Sr				
St. Severin	0.033	0.013	0.805	0.38	24.3	0.0451±10	0.70182±30	0.70192±28	0.70207±12
Bjurbole	0.042	0.010	0.695	0.24	16.7	0.0418±11	0.70249±28	0.70250±26	0.70299±10
Elenovka	0.022	0.006	0.396	0.24	17.6	0.0423±21	0.70241±30	0.70215±28	
Ambapur Nagla	0.026	0.009	0.666	0.35	25.6	0.0398±12	0.70185±28	0.70188±26	0.70239±30
Blank		0.00028	0.0017					0.709 ±2	

*Equivalent weight of spike aliquot taken from HCl solution after HNO_3 etch (cf. Table 3).
[†]For spike run; mass spectrometry performed in St. Louis. Concentrations are not corrected for undissolved residue or for dissolution during HNO_3 etch (Table 3).
[‡]Corrected for blank, with assigned error equal to 100% of correction.
[§]Adjusted to $^{87}Sr/^{86}Sr = 0.71014$ for NBS-987 (see text).
[¶]Unspiked; mass spectrometry performed in St. Louis.
[**]Unspiked; mass spectrometry performed in La Jolla.

syringe. Phosphate samples in teflon beakers were washed with methanol, then rinsed with water. They were then etched with 0.25% HNO_3 and rinsed again with water, the water then added to the etch solution. The samples were then dissolved (and visually observed to have been dissolved) in 0.5 N HCl. The methanol and HNO_3 (plus rinse water) etch solutions were spiked for Sr and analyzed with the results given in Table 3. From the HCl solutions two aliquots were taken; one was spiked for both Rb and Sr, the second left unspiked; results are given in Table 4. In these experiments the solutions were dried down and directly loaded on filaments for mass spectrometer analysis (i.e., no chemical purification was performed); this procedure was adopted in order to minimize blank. The remainder of the HCl solutions were analyzed for Sr (unspiked) after chemical purification, with results also included in Table 4.

Except for the direct-load analyses noted above, mass spectrometry was preceded by chemical purification by ion exchange chromatography. Some of the mass spectrometric analyses were performed in La Jolla, others in St. Louis; location of analysis is noted in the data tables. La Jolla procedures have been described previously (*Carlson et al.*, 1981). St. Louis analyses were performed on a VG 354 spectrometer. The Rb and Sr data were aquired on a Daly detector operated in pulse-counting mode with a beam of 2×10^5 ions/sec of the largest isotope present. Quoted errors in $^{87}Sr/^{86}Sr$ are two-sigma uncertainties based on in-run statistics (replicate analyses of NBS-987 indicate that reproducibility is consistent with in-run statistics for this mode). In order to facilitate comparison with data from other laboratories, both the La Jolla data and the St. Louis data have been adjusted to a nominal value $^{87}Sr/^{86}Sr$ = 0.71014 for NBS-987 (actual measured values for NBS-987 are 0.71026 for La Jolla Faraday analyses and 0.71010 for St. Louis pulse-counting analyses).

For the direct-load analyses, during Sr measurement there is typically a nontrivial Rb signal, as monitored at mass 85, for which corrections must be made. For the spiked runs in Table 4, Rb and Sr analyses were performed on the same filament load, first Rb at low filament temperature and then Sr at higher temperature; Rb corrections at mass 87 during the Sr analysis were made according to the $^{87}Rb/^{85}Rb$ ratio measured during the Rb analysis. Normal Rb composition was used for corrections in Sr runs which were not spiked for Rb. During Sr analysis, the Rb correction at mass 87 was typically of the order of 1% early in the run, with the Rb signal decreasing to undetectability late in the run. The average Rb correction was typically < 1‰ and we consider that uncertainty in the Rb correction is negligible compared to the statistical errors.

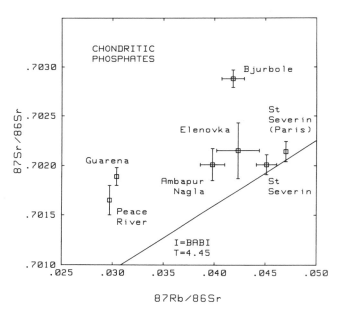

Fig. 2. Rb-Sr isochron diagram for chondritic phosphates (data from Tables 4 and 5), including data from this work and from *Wasserburg et al.* (1969) (Guarena), *Gray et al.* (1973) (Peace River), and *Manhes et al.* (1978) (St. Severin). The line shown is not a fit to the data but a reference isochron of age 4.45 Ga ($\lambda = 1.42 \times 10^{-11}$ yr^{-1}) with initial composition BABI.

TABLE 5. Initial Sr data for chondrites.

Sample		Measured ^{87}Rb/^{86}Sr	Measured ^{87}Sr/^{86}Sr†	Initial ^{87}Sr/^{86}Sr†	Ref.
Phosphates (this work)					
LL6	St. Severin	0.0451	0.70201 ± 10	0.69907 ± 12	*
L4	Bjurbole	0.0418	0.70288 ± 9	0.70015 ± 11	*
L6	Elenovka	0.0423	0.70215 ± 28	0.69939 ± 30	†
H5	Ambapur Nagla	0.0398	0.70201 ± 16	0.69941 ± 18	*
Phosphates (published data)					
H6	Guarena	0.0304	0.70189 ± 9	0.69995 ± 10	1
L6	Peace River	0.0297	0.70165 ± 15	0.69970 ± 10	2
LL6	St. Severin	0.0470	0.70214 ± 10	0.69903 ± 20	3
Other chondrite initial ^{87}Sr/^{86}Sr					
LL4	Soko Banja			0.69959 ± 24	4
LL6	Jelica			0.69959 ± 29	4
H3	Tieschitz			0.69880 ± 20	8
EH5	St. Marks			0.69979 ± 22	5
Reference Compositions					
BABI				0.69899 ± 4	6,7
ALL				0.69877 ± 5	2
H chondrites				0.69877 ± 16	8
LL chondrites				0.69882 ± 8	4
E chondrites				0.69874 ± 22	5

References: (1) *Wasserburg et al.* (1969); (2) *Gray et al.* (1973); (3) *Manhes et al.* (1978); (4) *Minster and Allégre* (1981); (5) *Minster et al.* (1979); (6) *Papastassiou and Wasserburg* (1969); (7) *Birck and Allégre* (1978); (8) *Minster and Allégre* (1979a).

*Measured ^{87}Sr/^{86}Sr is weighted mean of data from Table 4.
†Measured ^{87}Sr/^{86}Sr from unspiked run only (Table 4).
‡^{87}Sr/^{86}Sr ratios from this work are adjusted to 0.71014 for NBS-987. Data from the Pasadena and Paris laboratories are listed as reported in references above, with no adjustments; comparison of interlaboratory standards indicates that the uncertainty thus introduced by interlaboratory biases is small, substantially less than the stated analytical uncertainties.

The only exception is Elenovka: During the spike analysis of Elenovka Sr (Table 4) the Rb signal did not decay to undetectability as the run progressed, and the discrimination-corrected ^{87}Sr/^{86}Sr changed in proportion to Rb signal if spiked Rb composition was used for corrections, but was within errors independent of Rb signal if normal Rb composition was used. For Elenovka (only) then, the Sr composition for the spike run, Table 4, is based on correction using normal rather than spiked Rb composition.

Four blanks were measured following the same procedure except that no sample was present and that all of the HCl solution, rather than just an aliquot, was taken for analysis. Three of the blanks were spiked for Rb and Sr; their average is given in Table 4. The unspiked blank was analyzed for Sr composition, with the result also in Table 4. Sample data in Table 4 are formally corrected for blank, with an added error taken to be 100% of the correction. In practice, the principal effect of the blank correction is in the Rb concentration.

We are confident that the Sr compositions in Table 4 reflect the true indigenous Sr in these phosphates, without any surficial radiogenic component such as found by *Brannon et al.* (1987). This follows from the observation that the surficial component is apparently much less prominent in these specimens than in those of Brannon et al., and that it is removed by methanol (Tables 2 and 3). Also, the HNO$_3$ etch apparently dissolved a significant fraction of the total phosphates, as judged by Sr content (cf. Tables 3 and 4), so that none of the original surface is likely to have survived to the HCl step. It is noteworthy that the Sr compositions in the etch fractions (Table 3) are no more radiogenic than the etch residues (Table 4), indicating that any significant surficial component of radiogenic Sr had already been removed by the methanol + water wash.

In view of the substantial fraction of phosphate dissolved in the HNO$_3$ treatment, the Rb and Sr concentrations in Table 4 are only lower limits to the true phosphate concentrations, and indeed are probably only about half the true concentrations. They are, nevertheless, the appropriate quantities for the etch residue in terms of Rb/Sr ratios corresponding to ^{87}Sr/^{86}Sr ratios, which are the relevant parameters for determination of initial ^{87}Sr/^{86}Sr ratios and the subsequent discussion.

Of potential concern is the possibility of preferential leaching in this chemical treatment, i.e., preferential extraction of Rb relative to Sr, or vice versa, from the phosphates during the HNO$_3$ step or from the nonphosphates (principally feldspar) during the HCl step. If this were a significant effect it would lead to an incorrect Rb/Sr ratio associated with the measured

^{87}Sr/^{86}Sr ratio. We doubt that this is a significant effect, however, since dissolution of the phosphates lattice is so rapid that fractionation of Rb from Sr is unlikely and since the dilute HCl is such a mild treatment for anything but phosphates. In any case, the measured Rb/Sr ratio is sufficiently low that calculation of initial ^{87}Sr/^{86}Sr is not very sensitive to the exact value of Rb/Sr (see below).

Table 4 data are illustrated in Fig. 2, which also includes previously published data for chondritic phosphates. The only meteorite duplicated is St. Severin, for which the present data and that of *Manhes et al.* (1978) are in agreement within errors, particularly in terms of initial ^{87}Sr/^{86}Sr. (The Manhes et al. datum in Fig. 2, like ours, is for St. Severin phosphate originally prepared by R. S. Lewis.) It is evident that, excepting St. Severin, all the phosphate data lie above a reference isochron of age 4.45 Ga with initial composition equal to BABI (Fig. 2).

Initial ^{87}Sr/^{86}Sr compositions are calculated from the Table 4 data by removing the contribution of radiogenic ^{87}Sr/^{86}Sr generated in 4.45 Ga according to the measured ^{87}Rb/^{86}Sr; these are listed in Table 5. The age of 4.45 Ga, rather than, say, 4.56 Ga, is chosen because it more closely approximates published internal isochron ages for chondrites. It is not clear whether this reflects a real age difference between chondritic phosphates and more ancient objects of age ~4.56 Ga or simply an incorrect choice of the decay constant of ^{87}Rb (here taken to be 1.42×10^{-11} yr^{-1}). For present purposes this is not important. For ^{87}Rb/^{86}Sr = 0.045, for example (cf. Fig. 2), the growth of ^{87}Sr/^{86}Sr in 4.45 Ga is only 0.00294. A 2% uncertainty in absolute age, or even a 5% uncertainty in Rb/Sr ratio (Table 4), thus introduces only a small uncertainty in initial ^{87}Sr/^{86}Sr. In all cases, the dominant uncertainty is that in the measurement of present ^{87}Sr/^{86}Sr.

DISCUSSION

In some circumstances, chronological information based on the Rb-Sr system can be obtained with greater resolution from the initial ^{87}Sr/^{86}Sr ratio than from an absolute age derived from isochron analysis. *Papanastassiou and Wasserburg* (1969), for example, reported precise data for whole-rock basaltic achondrites, which have very low Rb/Sr ratios; their data yielded an isochron corresponding to an age 4.30 ± 0.26 Ga and a precise initial ^{87}Sr/^{86}Sr ratio designated BABI (for Basaltic Achondrite Best Initial) (Table 5). The isochron age is sufficient to establish the antiquity of this group, but not to resolve its time of formation from other groups of meteorites of comparable age. Adherence of the data to an isochron relationship also indicates that within errors each meteorite had the same initial ^{87}Sr/^{86}Sr ratio (BABI); since these meteorites evidently formed by differentiation from parent material with a higher Rb/Sr ratio, this condition corresponds to *simultaneous* formation, with a very precise limit on the simultaneity (within 4 Ma if the parent material had chondritic Rb/Sr, or 1.6 Ma for solar Rb/Sr).

Subsequent work has indicated other important marker values (Table 5, Fig. 3) for initial ^{87}Sr/^{86}Sr ratios, including ADOR for the achondrite Angra dos Reis, and ALL, from a

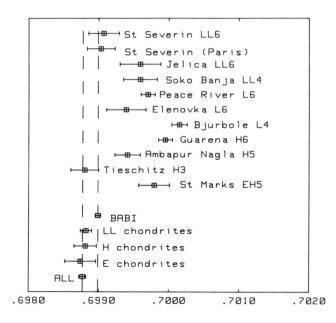

Fig. 3. Initial ^{87}Sr/^{86} ratios for various chondrites, plus reference values for initial compositions (data from Table 5).

refractory inclusion in the Allende carbonaceous chondrite, which is the most primitive (least radiogenic) ^{87}Sr/^{86}Sr ratio yet observed in the solar system. In this same range are initial ^{87}Sr/^{86}Sr ratios inferred from whole-rock isochrons of the E, H, and LL chondrite groups.

Chronological implications of these initial ^{87}Sr/^{86}Sr ratios are best illustrated in a T-I diagram such as Fig. 4. The solar nebula is generally believed to have had a rather high Rb/Sr ratio (Rb/Sr $\approx$ 0.5, ^{87}Rb/^{86}Sr $\approx$ 1.5; *Hauge,* 1972); with this composition the time required to evolve ^{87}Sr/^{86}Sr from ALL to BABI is only 11 Ma. Comparing the chondrite group initial ratios, and their errors, with BABI and ALL (Figs. 3 and 4), the nominal interpretation is that most chondritic meteorites separated from the nebula and achieved their present relatively low Rb-Sr ratios simultaneously, within no more than several Ma, and on the whole-rock scale have been isotopically closed ever since.

Isotopic closure on the spatial scale of a polymineralic whole rock does not necessarily preclude isotopic mobility on the smaller scale of individual mineral phases, and a whole-rock isochron age is only an upper limit to an internal isochron age; correspondingly the whole rock initial ^{87}Sr/^{86}Sr is a lower limit to an internal isochron initial ratio. *Wasserburg et al.* (1969) analyzed separated phases from the chondrite Guarena, obtaining an internal isochron of age 4.46 ± 0.08 Ga and a precise initial composition as well (Table 5). The absolute isochron age is not distinguishable from whole-rock ages determined by Rb-Sr or other techniques, but the initial ^{87}Sr/^{86}Sr ratio is distinctly higher than BABI or the whole-rock chondrite initial ratios (Figs. 2 and 3), indicating that Guarena did not become isotopically closed on a mineral scale until some time later than it became closed on a whole-rock scale.

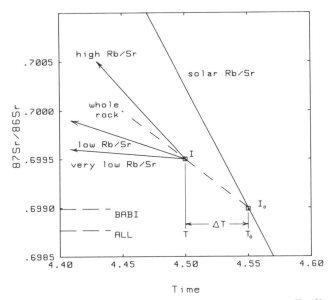

Fig. 4. T-I diagram illustrating model evolution for I = ^{87}Sr/^{86}Sr (ordinate) as a function of time (age) T (abscissa, in Ga, time proceeding from right to left). The illustration is schematic but the numerical values involved are realistic. The solar nebula has high Rb/Sr and I evolves relatively rapidly, passing through the ALL-BABI region around 4.55-4.56 Ga ago. At time T_o some meteorite parent body forms with a lower Rb/Sr ratio and follows a different evolutionary path (such as the dashed line). Because of metamorphism, various mineral phases do not become isotopically closed until time T, later than T_o by ΔT, at which time they show a common initial ratio I that is greater than the value I_o at T_o; subsequently each phase evolves according to its own ratio.

In such cases the time required to achieve isotopic closure on a mineral scale can be evaluated within the context of a model involving parameters as illustrated in Fig. 4. At time T_o parent material becomes isolated from the solar nebula with ^{87}Sr/^{86}Sr ratio I_o, but only at time T, later by ΔT, does it achieve final isotopic closure on a mineral scale, with initial ^{87}Sr/^{86}Sr equal to I. Calculation of the interval ΔT from I requires assumption of I_o and of the average Rb/Sr ratio during this interval, but is not directly dependent on T or To. For Guarena, for example, *Wasserburg et al.* (1969) calculated a formation interval $\Delta T = 74 \pm 12$ Ma, assuming I_o = BABI and Rb/Sr as presently found in Guarena whole rock.

As discussed by *Wasserburg et al.*, this calculation for ΔT is sensitive to the model assumptions. Evolution during ΔT did not necessarily occur with present whole-rock Rb/Sr, or even constant Rb/Sr. Evolution from BABI to Guarena I, for example, might have been more rapid if part of it occurred in a higher Rb/Sr environment, e.g., the solar nebula, and even if all the evolution occurred in a parent body environment at constant (present) Rb/Sr the interval ΔT would be greater or lesser if the assumed I_o were lower or higher.

Evidence from whole rock Rb-Sr isochron analyses (*Minster et al.*, 1979; *Minster and Allégre*, 1979, 1981), however, strengthens the model calculations as presented. If, as the whole rock analyses nominally indicate, chondrites typically have been isotopically closed since a common time of "formation," then the presently observed whole-rock Rb/Sr is indeed the appropriate choice for further evolution during ΔT, and the appropriate choice for I_o is the whole-rock isochron initial ^{87}Sr/^{86}Sr. As *Minster and Allégre* (1981) have stressed, chondritic whole-rock initial ratios I_o are essentially indistinguishable and lower than BABI, more nearly like ALL (Fig. 3), corresponding to longer intervals ΔT than those computed relative to BABI.

There are now several chondrites for which mineral initial ^{87}Sr/^{86}Sr ratios are available with precision useful on the scale of interest here (Table 5). For most of these, including all those reported here, the principal or only control on the initial ratio is analysis of phosphate; the importance of phosphates in this context is their characteristically very low Rb/Sr ratio, which permits a relatively precise determination of *initial* ^{87}Sr/^{86}Sr ratio which is not sensitive to uncertainty in the actual value of Rb/Sr or the absolute mineral closure age T (Fig. 4).

In Table 6 we have listed "formation intervals" ΔT for each meteorite, calculated from the elevation of initial ^{87}Sr/^{86}Sr

TABLE 6. Formation interval calculations.

Meteorite		Initial* ^{87}Sr/^{86}Sr	Whole rock* ^{87}Rb/^{86}Sr	$\Delta T^\dagger$ (Ma)	ΔT(I-Xe)‡ (Ma)
LL4	Soko Banja	0.69959 ± 24	1.563	35 ± 10	3
LL6	St. Severin	0.69907 ± 12	0.134	148 ± 59	8, 30
L4	Bjurbole	0.70015 ± 11	0.792	115 ± 9	≡0
L6	Peace River	0.69970 ± 10	0.749	82 ± 9	
L6	Elenovka	0.69939 ± 30	0.746	55 ± 26	
H3	Tieschitz	0.69880 ± 20	0.361	5 ± 37	
H6	Guarena	0.69995 ± 10	0.869	90 ± 8	
EH5	St. Marks	0.69979 ± 22	0.320	210 ± 45	5

*Data from Table 5 and sources referenced there, plus *Minster and Allégre* (1979b).

†Time required to evolve from ALL (Table 5) to listed initial ^{87}Sr/^{86}Sr in closed system with listed "whole rock" ^{87}Rb/^{86}Sr.

‡Data compiled by *Podosek and Swindle* (1987); these are relative ages arbitrarily referrenced to Bjurbole with positive ages indicating later formation. The two values for St. Severin are for light and dark fractions, respectively.

relative to ALL (cf. Fig. 3) assuming evolution during ΔT at the rate corresponding to observed whole-rock Rb/Sr. We stress that these intervals are subject to the uncertainty involved in these assumptions as well as the experimental errors, but they should nevertheless provide a reasonable evaluation of the time scales involved. The relationship between initial ^{87}Sr/^{86}Sr and whole-rock Rb/Sr is shown in Fig. 5.

A feature evident in Table 6 or Fig. 5 is that Guarena is not unique or even unusual. Essentially *all* the chondrites for which data are available show elevated or evolved initial ^{87}Sr/^{86}Sr ratios and indicate a characteristic time scale of around 10^8 yr, i.e., from a few times 10^7 to a few times 10^8 yr. The only probable exceptions are the unequilibrated chondrite Tieschitz and the highly equilibrated chondrite St. Severin; the initial ratio for St. Severin is indistinguishable from BABI and only slightly elevated above ALL, but its whole rock Rb/Sr ratio is so low that the calculated formation interval is extremely sensitive to the choice of I$_o$.

Wasserburg et al. (1969) suggested that the formation interval ΔT reflected chondrite metamorphism, and subsequent workers have also followed this interpretation. In view of the available whole rock data, this would correspond to temperatures such that Sr could migrate and isotopically equilibrate on the submillimeter-scale of minerals but that chondrites were essentially closed on the whole rock scale, of order centimeters or greater. Qualitatively, such an interpretation is certainly plausible, and no simple alternatives have emerged. In terms of broader models for the early solar system, the ~10^8 yr time-scale which emerges from initial Sr compositions requires either a heat source for metamorphism that persists for that long, or, perhaps more plausibly, a shorter-lived heat source (e.g., ^{26}Al) and metamorphism within parent bodies of the order of at least 200-300 km diameter.

Viewed in isolation, these chronological considerations based on initial Sr composition lead to a reasonably simple and plausible interpretation. In detail, however, and especially in consideration of other constraints, the picture is not simple and is arguably implausible. In the remainder of this discussion we will consider factors that we feel should lead to caution in interpreting results along the simple lines described above.

One such factor is metamorphism: If formation intervals of the order of 10^8 yr are indeed a reflection of Sr isotopic mobility during metamorphism, it is reasonable to expect a correlation between age and the petrographic indications of degrees of metamorphism. In perhaps the simplest model, for example, metamorphism can be imagined to occur in an internally-heated parent body of "onion-shell" structure, the progressively more deeply buried specimens having experienced progressively higher temperatures and thus higher metamorphic grade and having cooled more slowly and thus having progressively younger ages. (In alternative models the heat source is external, and the sense of correlation would be reversed, but a correlation is still expected.)

The database for Sr formation intervals is still relatively meager, especially for comparing members of a given chemical class, but it nevertheless permits some significant evaluations. In some respects the data agree with expectations: Tieschitz, for example, has the lowest metamorphic grade, and also the lowest initial ^{87}Sr/^{86}Sr and the shortest formation interval (if any) of the samples in Table 5, while all the other meteorites are of higher grade and longer ΔT. Also, the next lowest ΔT is for Soko Banja, which is only grade 4. However, the group of meteorites that plot close together in Fig. 5 (Guarena, Peace River, Bjurbole, Elenovka) have a larger ΔT than Soko Banja and include metamorphic grades 4 to 6, and the greatest ΔT is for St. Marks, which is only grade 5. Given that data for any one chemical class are very limited it cannot be concluded that age and metamorphic grade definitely fail to correlate, but certainly no simple relationship has emerged.

Another relevant factor for consideration is comparison with chronologies based on other methods. There are several absolute age techniques applicable to meteorites, e.g., U-Pb, ^{40}Ar-^{39}Ar, Rb-Sr internal isochrons; for samples for which initial Sr formation intervals are available (Table 6), however, the database for unambiguous and precise absolute ages is quite meager. At present, the most relevant comparison seems to be with I-Xe chronology, which, like initial Sr formation intervals, is a relative (rather than absolute) chronology with very high resolution.

There is an extensive database for I-Xe chronology of chondrites (*Podosek and Swindle,* 1987), and there are four meteorites for which both initial Sr and I-Xe formation intervals are available (Table 6). The most obvious feature of the comparison in Table 6 is that the two methods do not agree.

The highest order of disagreement is in the magnitude of

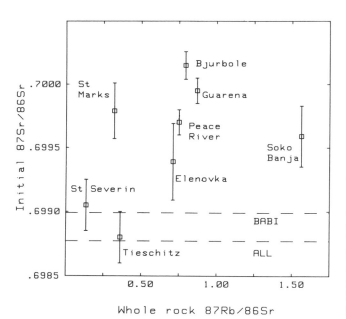

Fig. 5. Relationship between initial ^{87}Sr/^{86}Sr ratio and whole-rock Rb/Sr ratio for chondrites (data from Tables 5 and 6). If formation interval ΔT (cf. Fig. 4) is proportional to elevation of initial ^{87}Sr/^{86}Sr above an original value such as BABI or ALL, and inversely proportional to whole-rock Rb/Sr, then ΔT is proportional to the slope of a line (not shown) from original value on ordinate through the datum in question.

the relative ages: I-Xe ages for chondrites in general, not just those in Table 6, span a characteristic scale of around 10^7 yr (St. Severin is rather an exception in this respect), while the initial Sr ages span a characteristic scale of 10^8 yr. At face value, this implies that isotopic closure for the noble gas Xe occurs earlier (presumably at a higher temperature) than for Sr. A priori, this seems rather implausible, but the subject deserves further investigation. I-Xe correlations persist to remarkably high temperatures, and estimated Xe closure temperatures range from 400°C to 1000°C (*Podosek and Swindle*, 1987), but Xe closure before Sr closure should probably not be accepted in the absence of knowledge of Sr closure temperatures. Alternative scenarios are possible but also seem quite implausible. It could be imagined, for example, that I-Xe ages actually represent a span of ~10^7 yr all occurring around 10^8 yr after original formation in the solar nebula, but this seems unlikely in view of the relatively small spread of I-Xe ages among not only metamorphosed ordinary chondrites but among the unequilibrated ordinary chondrites and various inclusions or phases of carbonaceous chondrites. It might also be imagined that initial Sr formation intervals are actually much shorter than supposed because of evolution in a higher Rb/Sr environment, but this too seems unlikely, since the Rb/Sr ratio would have to be essentially solar, and this seems precluded by the evidence of the whole-rock isochrons for the various chondrite groups.

Initial Sr and I-Xe formation intervals also do not correlate on a more detailed level. Even if one accepts the premise of earlier closure for Xe than for Sr, in any simple monotonic temperature history there should be a one-to-one correspondence of age sequence. There clearly is not.

Again at face value, if the chronological data are correct the inference is a very complicated, nearly chaotic, set of thermal histories. It has been clear for some time that at least for ordinary chondrites there is no correlation between metamorphic grade and I-Xe age, and it seems now that initial Sr intervals correlate with neither metamorphic grade nor I-Xe age. The failure of other chronometers, e.g., metallographic cooling rates, to correlate with metamorphic grade has led to suggestions of just such a chaotic history for ordinary chondrites (e.g., *Scott and Rajan*, 1981), in which various chronometers might indeed become "closed" at times and in environments bearing little coherent relationship to other chronometers or among meteorites of similar class or grade.

An alternative possibility is that there is some significant failure of methodological assumption in either the initial Sr or I-Xe approaches. In some chondrites the Rb-Sr system is "disturbed" in the sense that separate phases do not conform to an isochron relationship, although the nature and timing of the disturbance is not clear. I-Xe systems are also sometimes "disturbed," with the same remarks applicable (although for at least Soko Banja, Bjurbole, and St. Marks there is no sign of disturbance). It has also been suggested that ^{129}I was not homogeneously distributed in the early solar system so that the $^{129}I/^{127}I$ ratio measured in an I-Xe experiment is not directly translatable to an age, but it is difficult to imagine how heterogeneity of ^{129}I could account for a range of I-Xe ages much narrower than initial Sr ages.

Pending clarification of these apparent discrepancies, a literal chronological interpretation of both initial Sr and I-Xe formation intervals, and a simple interpretation that they reflect thermal history in an organized parent body, should be viewed with considerable caution.

Acknowledgments. We are greatly indebted to Paul Pellas and Roy Lewis for providing us with these precious samples. This work was supported in part by NSF grant EAR-8709445 and NASA grant NAG9-49.

REFERENCES

Birck J. L. and Allègre C. J. (1978) Chronology and chemical history of the parent body of basaltic achondrites studied by the ^{87}Rb-^{87}Sr method. *Earth Planet. Sci. Lett., 39*, 37-51.

Brannon J. C., Podosek F. A., and Lugmair G. W. (1987) Strontium, neodymium and plutonium in chondritic phosphates (abstract). In *Lunar and Planetary Science XVII*, pp. 121-122. Lunar and Planetary Institute, Houston.

Carlson R. W., Lugmair G. W., and Macdougall J. D. (1981) Columbia River volcanism: the question of mantle heterogeneity or crustal contamination. *Geochim. Cosmochim. Acta, 47*, 2483-2499.

Chen J. H. and Wasserburg G. J. (1981) The isotopic composition of uranium and lead in Allende inclusions and meteorite phosphates. *Earth Planet. Sci. Lett., 52*, 1-15.

Gray C. M., Papanastassiou D. A., and Wasserburg G. J. (1973) The identification of early condensates from the solar nebula. *Icarus, 20*, 213-239.

Hauge O. (1972) A search for the solar ^{87}Sr content and the solar Rb/Sr ratio. *Solar Phys., 26*, 276.

Kirsten T., Jordan J., Richter H., Pellas P., and Storzer D. (1978) Plutonium and uranium distribution patterns in phosphate from ten ordinary chondrites. In *Fourth Int. Conf. Geochron. Cosmochron. Isotope Geol., Aspen*, pp. 215-219. U.S. Geol. Survey Open-File Report 78-701.

Lewis R. S. (1975) Rare gases in separated whitlockite from the St. Severin chondrite: Xenon and krypton from fission of extinct ^{244}Pu. *Geochim. Cosmochim. Acta, 39*, 417-432.

Manhes G., Minster J.-F., and Allègre C. J. (1978) Comparative uranium-thorium-lead and rubidium-strontium study of the St. Severin amphoterite: consequences for early solar-system chronology. *Earth Planet. Sci. Lett., 39*, 14-24.

Minster J.-F. and Allègre C. J. (1979a) ^{87}Rb-^{87}Sr chronology of H chondrites: constraint and speculations on the early evolution of their parent body. *Earth Planet. Sci. Lett., 42*, 333-347.

Minster J.-F. and Allègre C. J. (1979b) ^{87}Rb-^{87}Sr dating of L chondrites: effects of shock and brecciation. *Meteoritics, 14*, 235-248.

Minster J.-F. and Allègre C. J. (1981) ^{87}Rb-^{87}Sr dating of LL chondrites. *Earth Planet. Sci. Lett., 56*, 89-106.

Minster J.-F., Ricard L.-P., and Allègre C. J. (1979) ^{87}Rb-^{87}Sr chronology of enstatite meteorites. *Earth Planet. Sci. Lett., 44*, 420-440.

Papanastassiou D. A. and Wasserburg G. J. (1969) Initial strontium isotopic abundances and the resolution of small time differences in the formation of planetary objects. *Earth Planet. Sci. Lett., 5*, 361-376.

Pellas P. and Storzer D. (1981) ^{244}Pu fission track thermometry and its application to stony meteorites. *Proc. R. Soc. London, A374*, 253-270.

Podosek F. A. and Swindle T. D. (1987) Extinct radionuclides, cosmochronology and geochronology. In *Meteorites and the Early Solar System* (J. F. Kerridge, ed.), in press.

Scott E. R. D. and Rajan R. S. (1981) Metallic minerals, thermal histories and parent bodies of some xenolithic, ordinary chondrite meteorites. *Geochim. Cosmochim. Acta,* 45, 53-67.

Wasserburg G. J., Papanastassiou D. A., and Sanz H. G. (1969) Initial strontium for a chondrite and the determination of a metamorphism or formation interval. *Earth Planet. Sci. Lett.,* 7, 33-43.

Cosmic-ray-produced Kr in St. Severin Core AIII

B. Lavielle*

Centre d'Etudes Nucleaires de Bordeaux-Gradignan Laboratoire de Chimie Nucleaire, U.A. No. 451 du CNRS, 33170 Gradignan, France

K. Marti

Chemistry Department, B-017, University of California-San Diego, La Jolla, CA 92093

Now at Chemistry Department, B-017, University of California-San Diego, La Jolla, CA 92093

We report Kr isotopic abundances in 10 samples from core AIII of the St. Severin chondrite and discuss the variation with depth of cosmic-ray-produced Kr. We show that the ratio $(^{78}Kr/^{83}Kr)_c$ changes with depth and can be used as an irradiation hardness monitor. We then compare cosmic-ray-produced Kr and Ne in the St. Severin core with ratios observed in bulk chondrites. A linear correlation between $(^{22}Ne/^{21}Ne)_c$ and $(^{78}Kr/^{83}Kr)_c$ exists for bulk samples of chondrites of varying preatmospheric size, but does not exist in the St. Severin core. The calculated ^{83}Kr production rates, P_{83}, are similar in H, L, and LL chondrites for samples of comparable shielding conditions. However, P_{83} rates increase by a factor of about 2 with shielding depth. The maximum observed P_{83} values correspond to $(^{78}Kr/^{83}Kr)_c \lesssim 0.14$.

INTRODUCTION

The production rates of cosmic-ray-produced nuclides in meteorites vary according to the size of the meteoroid and the shielding depth of a sample within the object. Depth dependent production rates for cosmic-ray-produced He and Ne isotopes were reported for cores of the St. Severin (*Schultz and Signer,* 1976) and Keyes (*Wright et al.,* 1973) meteorites. No Kr depth profiles in meteorites are currently available, although it has long been known that cosmic-ray-produced Kr_c, specifically the ratio $(^{78}Kr/^{83}Kr)_c$, varies with shielding depth on the lunar surface (*Marti and Lugmair,* 1971; *Schwaller et al.,* 1971; and *Regnier et al.,* 1979). This study was undertaken in order to rectify this situation.

In chondrites, Kr is produced mainly by high and medium energy particles from Sr, Y, Zr, and Rb. Detailed discussions of the reaction channels were given by *Regnier et al.* (1979, 1982). Several methods were used to calculate production rates in meteorites (e.g., *Nishiizumi et al.,* 1980; *Baros and Regnier,* 1984; *Reedy,* 1985), but the calculated production rate profiles suffer from a lack of reliable data on the secondary neutron fluence and relevant neutron cross-sections. Meteorites generally contain significant trapped gas components that make the determination of cosmic-ray-produced components more difficult. However, in favorable cases such as the St. Severin (LL6) chondrite, the latter is easily recognized, and the relative yields can be determined with reasonable precision. According to *Cantelaube et al.* (1969), the St. Severin meteorite suffered a rather small ablation loss (~25%), compared to losses typically inferred for other meteorites (*Bhandari et al.,* 1980). Samples were taken (courtesy of P. Pellas) from core AIII, which was already studied for light noble gases by *Schultz and Signer* (1976). Their results showed significant variation in the ratio $(^{22}Ne/^{21}Ne)_c$ that may be used as a hardness parameter (e.g., *Nishiizumi et al.,* 1980), as well as in the production rates across the core. It is of interest to see if additional data, specifically the $(^{78}Kr/^{83}Kr)_c$ ratios, might allow estimates of the object size, as well as the depth of a sample within the meteoroid.

EXPERIMENTAL PROCEDURE

The work on the St. Severin core was performed at the CENBG, University of Bordeaux I. Bulk samples from 10 locations across the core AIII, weighing from 0.58 g to 1.98 g were analysed for Kr isotope abundances. The samples were heated in the storage arm of the extraction system at $\sim 10^{-8}$ torr and 100°C for several days to remove atmospheric gases. The samples were then dropped into and heated in a Mo-crucible by electron bombardment. The gas was released in two steps: first, at 450°C for 40 minutes, and then melted at 1700°C for 25 minutes.

Extracted gases were cleaned successively with Ti sponge, Al-Zr (SAES, 450°C) getter, Pd and Ti powders at stepwise decreasing temperature (800°C-200°C) in an all metal system of small volume. Argon, Kr, and Xe were separated cryogenically on charcoal. Two successive Ar-Kr separations were required to eliminate interference from Ar^{2+}/Ar^+ charge exchange and Ar_2^+ ions at mass 80. The mass spectrometer used in this work is a version of Micromass 12, incorporating a small interior volume, a Nier-type source and a (Cu-Be) multiplier detector. It was normally operated at a mass resolution of ~330, as required to partially resolve isobaric hydrocarbon background in the Kr mass region.

Air pipettes were run before and after each sample to determine the mass discrimination and the sensitivity of the instrument. The instrument sensitivity variations (in the range of 7-12%) are included in the uncertainties in Table 1. Two different Air standards, prepared two years apart, agreed to better than 5%.

We also report measurements (carried out at La Jolla) of the Kr isotopic abundances on the Bruderheim (Berkeley) standard, obtained in a one-step 1700°C extraction and in a stepwise release at 750°, 1100°, and 1700°C steps, respectively

TABLE 1. Kr isotopic abundances and calculated cosmic-ray-produced Kr$_c$ yields in samples of St. Severin core AIII.

Depths* (mm)	Weight (g)	^{78}Kr	^{80}Kr	^{82}Kr	^{83}Kr	^{84}Kr	^{86}Kr	^{86}Kr 10^{-12} cm^{3}STP/g	^{83}Kr$_c$	^{78}Kr$_c$	^{80}Kr$_c$	^{82}Kr$_c$	^{83}Kr$_c$	^{84}Kr$_c$
15–17	1.12	7.63 ±.19	31.86 .72	89.0 1.1	92.7 1.1	344.9 4.6	= 100	3.75 ± 0.49	1.05 ± .14	20.3 1.1	68.4 3.7	86.3 5.2	= 100	90. 17.
24–25.3	1.08	6.81 ±.22	29.23 .51	85.7 1.0	89.9 1.0	336.1 3.0	= 100	5.51 ± 0.55	1.38 ± .15	19.5 1.2	65.9 3.3	82.6 5.2	= 100	62. 12.
47–48.5	0.86	7.17 ±.23	39.1 1.0	90.2 1.0	90.20 .90	338.9 3.3	= 100	5.26 ± 0.68	1.34 ± .18	20.6 1.2	104.0 5.4	99.4 5.3	= 100	73. 13.
65–67	1.12	7.45 ±.16	37.76 .69	90.2 1.3	93.4 1.1	333.1 2.3	= 100	4.71 ± 0.61	1.35 ± .18	19.3 9	87.5 4.1	88.3 5.7	= 100	45.6 8.2
117.5–119	0.88	5.89 ±.35	– –	84.0 1.7	91.88 .73	334.8 4.5	= 100	6.64 ± 0.51	1.80 ± .15	14.7 1.4	–	70.3 6.6	= 100	52. 17.
142–146.5	1.25	5.89 ±.12	31.84 .42	85.69 .79	88.2 1.2	336.2 3.4	= 100	7.43 ± 0.97	1.73 ± .24	17.0 1.0	82.1 4.6	88.5 5.7	= 100	66. 15.
166–170.5	1.98	8.16 ±.30	32.6 1.3	95.9 1.8	105.4 1.6	339.9 7.2	= 100	4.67 ± 0.37	1.91 ± .17	15.3 1.0	48.9 3.7	76.3 5.3	= 100	50. 18.
269–274	1.25	6.57 ±.13	29.52 .69	86.1 1.4	90.2 1.6	337.5 6.8	= 100	6.38 ± 0.89	1.61 ± .25	18.3 1.3	66.4 5.0	83.2 7.6	= 100	66. 27.
305–307	0.99	5.64 ±.43	– .83	85.00 .90	83.30 .90	334.9 3.3	= 100	5.87 ± 0.45	1.08 ± .10	20.2 2.5	–	108.4 7.0	= 100	76. 18.
320–322	0.85	–	–	84.6 1.1	85.6 1.9	338.4 5.3	= 100	8.07 ± 0.73	1.67 ± .22	–	–	94. 10.	= 100	83. 27.

The uncertainties in isotopic abundances correspond to 1σ errors.
*Depths are given along the core with zero referring to face II.

TABLE 2. Krypton isotopic abundances in Bruderheim Standard Bru-7-12.

	^{78}Kr	^{80}Kr	^{82}Kr	^{83}Kr	^{84}Kr	^{86}Kr	^{86}Kr×10^{-12}cm^{3}STP/g
Split: 0.474g							
750°	4.45 ± 1.11	21.43 0.38	80.7 1.1	84.36 0.71	343.0 3.0	= 100	5.51 0.55
1100°	10.77 ± 1.12	49.53 1.31	121.5 2.9	139.0 2.9	374.6 2.9	= 100	3.12 0.31
1700°	4.45 ± 0.22	22.62 0.13	78.74 0.70	82.30 0.84	330.7 1.8	= 100	8.62 0.86
Total	5.56 ± 0.42	27.11 0.27	87.13 0.72	93.22 0.72	342.6 1.4	= 100	17.25 1.7
Split: 0.346g	5.28 ± 0.26	24.02 0.84	81.26 1.37	86.44 0.85	336.0 2.1	= 100	20.6 2.0
Kenna (*Wilkening and Marti*, 1976)	1.943 ±0.027	12.795 0.043	65.52 0.16	65.36 0.13	323.21 0.35	= 100	–

Uncertainties in isotopic composition represent 95% confidence limits.

(Table 2). The He and Ne isotopic abundances in this standard have previously been reported (*Nishiizumi et al.,* 1980) incorrectly, because of an error in mass discrimination corrections. We give here the Bru-7-12 He and Ne data again (concentrations in 10^{-8}ccSTP/g):

^{3}He = 48.5 ± 1.5,
^{4}He = 513 ± 15,
^{4}He/^{3}He = 10.58 ± 0.20,
^{22}Ne = 10.90 ± 0.30,
^{20}Ne/^{22}Ne = 0.848 ± 0.088,
^{21}Ne/^{22}Ne = 0.918 ± 0.007.

RESULTS

The Kr fractions released at 400°C carried the isotopic signature of terrestrial Kr and appear to be of atmospheric origin. The Kr isotopic abundances in the 1700°C fractions are given in Table 1, together with the depth of the samples along the core. The data show clearly the sigature of a cosmic-ray-produced component. This Kr_c component was calculated by subtracting trapped and fission Kr components from the data in Table 1. The fission component was estimated from the U and Pu abundances in St. Severin [16 ppb U (*Mason,* 1979) and an adopted ratio ^{244}Pu/U = 0.005] and yields $^{86}Kr_F$ = 2.6 × 10^{-14}cm^{3}STP/g. A fission correction was applied to ^{86}Kr and is negligible for the other Kr isotopes. We adopted the Kr isotopic composition of Kenna (see Table 2) for the trapped component and partitioned the remaining ^{86}Kr into trapped and cosmic-ray-produced components, using (^{86}Kr/^{83}Kr)$_c$ = 0.015 (*Marti and Lugmair,* 1971) to deduce Kr_c abundances. The data for this component are compiled in Table 1 and include all experimental uncertainties (quadratically added). The choice of the trapped component is not critical since, for example, a substitution of AVCC Kr for the Kenna isotopic composition changes the spallation Kr ratios only by ≲20% of the quoted uncertainties. The concentrations of cosmic-ray-produced $^{83}Kr_c$, which amount to between 20% and 40% of total measured ^{83}Kr concentrations, are plotted in Fig. 1. These concentrations (filled circles) show (except for the 305-307 mm data) a smooth dependence with depth of the samples and increases from the surface to the center of the core. The reference curve shown in Fig. 1 represents a parabolic fit to the experimental data. The data at depths >300 mm scatter and additional measurements are planned.

The cosmic-ray exposure age of St. Severin, according to the calibration by *Nishiizumi et al.* (1980), yields a ^{21}Ne age of (16.5 ± 1.8) Ma, if the St. Severin data from the compilation of *Schultz and Kruse* (1983) are used. The uncertainty includes errors in $^{21}Ne_c$ measurements and a 10% uncertainty assumed in P_{21}, quadratically added. The ^{81}Kr-Kr exposure age of St. Severin measured by *Marti et al.* (1969) gives a significantly lower value (13.0 ± 1.4) Ma. A similar disagreement between $^{21}Ne_c$ and ^{81}Kr-Kr ages was already observed by *Eugster et al.* (1987) for another L6 chondrite (Guangrao), which reflects a small degree of shielding. This discrepancy calls for a clarification by additional measurements. For our present purpose, we adopt an average exposure age of (14.8 ± 2.4) Ma for St. Severin. The open squares in Fig. 1 indicate predicted abundances, according to the Reedy-Arnold model adapted to meteorites (*Reedy,* 1985) and our adopted exposure age.

The model predictions are quite good at the surface, but are too low at the center of the core. This discrepancy may have many origins. For example, the model uses a spherical approximation (R = 40 cm), which does not agree very well with the elipsoidal shape of St. Severin, of which the greatest diameter was estimated at 80 cm (*Cantelaube and Pellas,* 1968). Moreover, the neutron cross-sections are not measured. A large diameter, combined with incorrect neutron cross-sections, would result in lowered production rates at the core. However, in light of the uncertainties involved, the general agreement is good.

The ratios (^{80}Kr/^{83}Kr)$_c$ and (^{82}Kr/^{83}Kr)$_c$ of cosmic-ray-produced Kr in our St. Severin core samples are plotted in Fig. 2. The data display a linear correlation that corresponds to the following equation

$$(^{80}\text{Kr}/^{83}\text{Kr})_c = -1.36 + 2.44 \, (^{82}\text{Kr}/^{83}\text{Kr})_c \quad (r = 0.97) \quad (1)$$

The reaction channels for isotopes ^{80}Kr and ^{82}Kr include spallation reactions mainly in the elements Sr, Y, Zr, and Br (n,γ) reactions by low energy neutrons. Since the spallation yields show only small variations [e.g., (^{78}Kr/^{83}Kr)$_c$], the good correlation apparently is chiefly due to Br (n,γ) reactions and shows that Br abundances in our core samples are highly variable.

The minimum values of (^{80}Kr/^{83}Kr)$_c$ and (^{82}Kr/^{83}Kr)$_c$, corresponding to the lowest Br abundance, agree well with (^{80}Kr/^{83}Kr)$_c$ and (^{82}Kr/^{83}Kr)$_c$ data in Table 3 for cosmic-ray-produced Kr components.

The slope (2.44) of the linear correlation agrees well with estimates of 2.25-2.7 deduced for different chondrites (*Eugster et al.,* 1967; *Eugster et al.,* 1969) or estimated from the resonance integrals of ^{79}Br and ^{81}Br in the interval 30 to 300 ev (*Marti et al.,* 1966). The yield of cosmic-ray-produced $^{84}Kr_c$

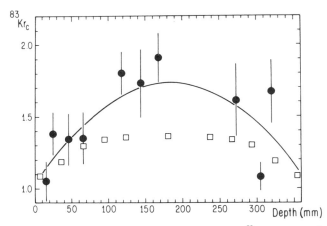

Fig. 1. The concentrations of cosmic-ray-produced $^{83}Kr_c$ in 10 samples of St. Severin core AIII are shown (filled circles) versus depth in the core (O refers to face II of piece A). Open squares represent concentrations predicted according to *Reedy* (1985) and an exposure age 14.8 Ma.

TABLE 3. Calculated isotopic abundances of cosmic-ray-produced Kr_c and $(^{22}Ne/^{21}Ne)_c$ in chondrites.

Meteorite	Class	Source of Kr Data	^{83}Kr $(10^{-12} cm^3 g^{-1})$	^{78}Kr	^{80}Kr	^{81}Kr	^{82}Kr	^{83}Kr	^{84}Kr	$\frac{^{22}Ne}{^{21}Ne_c}$	Source of Ne Data
St. Severin (a)	LL6	(1)	1.30 ±.20	22.3 .8	81.8 3.0	-	84.0 4.5	100	68 13	1.215 12	(1)
(b)		(1)	1.09 ±.17	21.5 1.1	59.4 3.4	1.51 16	73.3 6.6	100	76 19	1.205 12	(1)
Bruderheim (Berkeley Standard.)											
(a) 750°	L6	(2)	1.07 ±.11	12.9 5.7	44.8 2.5	-	80.4 6.4	100	(114)		
1100°			2.34 ±.25	11.8 1.6	49.4 2.6	-	76.4 4.9	100	77.0 4.8		
1700°			1.49 ±.17	14.5 1.5	57.2 2.9	-	78.3 5.5	100	54 11	-	
Total			4.87 ±.50	12.8 1.5	51.0 1.6		77.9 3.2	100	75.2 5.3	-	
(b)		(2)	4.40 ±.46	15.7 1.4	52.8 4.5	-	75.0 7.1	100	67 10	1.089 8	(2)
(c)		(3)	5.0 ± 1.1	16.2 2.2	52.6 8.2	0.71 10	76.9 1.4	100	50 27	1.094 12	(3)
Pulsora	H5	(4)	3.39 ±.86	17.4 2.7	60.2 8.6	-	68. 11.	100	61 39	1.16 3	(8)
Mocs	L6	(4)	2.34 ±.55	15.0 1.4	62.5 4.5	0.97 0.20	77. 12.	100	44 28	1.055 10	(7)
Walters	L6	(4)	0.56 ±.25	13.2 7.0	-	-	-	100	-	1.050 30	(7)
Maziba	L6	(4)	2.54 ±.52	23.7 2.2	57.6 3.2	-	72.6 5.4	100	89 19	1.21 20	(7)
Utzensdorf	H6	(4)	1.20 ±.47	20.2 6.2	56.8 1.7	-	92. 35.	100	- -	1.250 20	(8)
San Juan Capistrano	H6	(5)	2.79 ±.37	21.1 .9	58.3 1.6	0.693 44	76.6 2.2	100	56.9 6.9	1.189 2	(5)
Nan Young Pao	L6	(6)	6.7 ± 1.7	18.0 1.0	56.5 3.0	0.40 11	63.5 7.6	100	-	1.128 6	(6)
Lunan	H6	(6)	2.8 ±.6	15.5 2.1	54.0 6.8	0.65 14	77. 11.	100	-	1.099 5	(6)
Guangrao	L6	(6)	1.83 ±.54	19.3 4.4	60. 7.	1.29 36	79. 26.	100	-	1.180 5	(6)

References: (1) *Marti et al.* (1969); (2) Table 2 and text; (3) *Eugster et al.* (1981); (4) *Eugster et al.* (1969); (5) *Finkel et al.* (1978); (6) *Eugster et al.* (1987); (7) *Eberhardt et al.* (1966); and (8) *Graf* (1967). The ± uncertainties refer to the last digits.

is small compared to the trapped component, and the uncertainties in the ratios $(^{84}Kr/^{83}Kr)_c$ are too large to allow the resolution of a variation with depth.

We would like to add another observation. The Kr data of Bruderheim (Table 2) show that it is possible to significantly enrich the spallation component in a stepwise release of Kr at increasing temperatures. However, our analysis of the cosmic-ray component shows that the relative Kr_c yields are not identical, most probably reflecting different siting of Sr and Zr. Kr_c ratios of temperature fractions should be used carefully

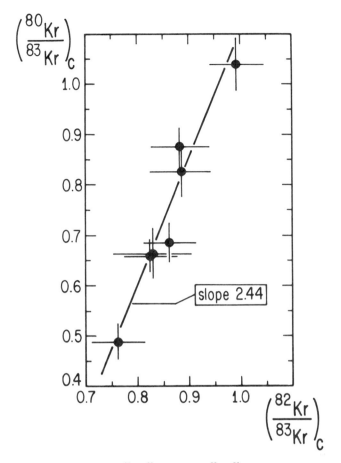

Fig. 2. The ratios $(^{80}Kr/^{83}Kr)_c$ and $(^{82}Kr/^{83}Kr)_c$ of cosmic-ray-produced Kr are correlated and show the effect of varying Br abundances in core AIII samples of St. Severin.

in correlation plots (such as our Figs. 2-6), since bulk chondritic abundances of Sr, Y, and Zr are implied for individual temperature fractions.

DISCUSSION

The production rates of cosmic-ray-produced nuclides in meteorites depend on the size of the meteoroid and the location of the sample within the body. In chondrites the shielding effects can be evaluated by way of the hardness parameter $(^{22}Ne/^{21}Ne)_c$ (*Eberhardt et al.*, 1966; *Schultz and Signer*, 1976; *Nishiizumi et al.*, 1980).

As pointed out earlier, the studies of cosmic-ray-produced nuclides in lunar rocks showed that the ratio $(^{78}Kr/^{83}Kr)_c$ is an indicator of shielding depth. We show that this is also true for chondrites. In Fig. 3 the $^{83}Kr_c$ production rate P_{83} in St. Severin (exposure age 14.8 Ma) is plotted versus $(^{78}Kr/^{83}Kr)_c$ ratios. Filled circles are used for the core AIII samples and filled squares for piece DII (*Marti et al.*, 1969). Figure 3 shows the data for the range $0.14 < (^{78}Kr/^{83}Kr)_c < 0.23$. The production rate P_{83} appears to be linearly correlated with $(^{78}Kr/$

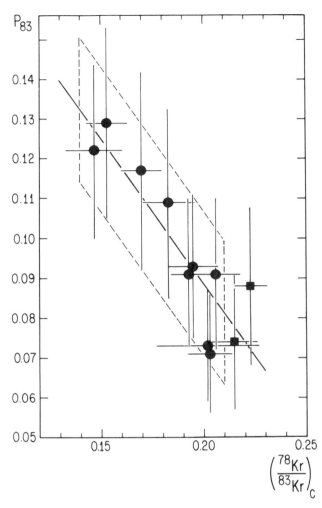

Fig. 3. The production rates P_{83} (in $10^{-12} cm^3 STP\ g^{-1} Ma^{-1}$) are shown versus the shielding parameter $(^{78}Kr/^{83}Kr)_c$. The adopted exposure age is 14.8 Ma. The correlation appears to be linear in the range $0.14 < (^{78}Kr/^{83}Kr)_c < 0.23$. Circles represent core AIII data and squares are data from piece DII (*Marti et al.*, 1969).

$^{83}Kr)_c$ and is fitted by the solid line corresponding to the following equation

$$P_{83}\ (10^{-12}\ cm^3\ STP\ g^{-1}\ Ma^{-1}) = 0.235 - 0.731\ (^{78}Kr/^{83}Kr)_c$$
$$(r = -0.78) \qquad (2)$$

There is a nearly twofold increase in the production rate P_{83} at the centers of meteorites of this size (preatmospheric mass of St. Severin ≈ 272 kg).

We are now interested to see if the Kr spallation systematics obtained from St. Severin core can be extended to other chondrites of varying size and unknown sample locations. In Table 3 we have listed cosmic-ray-produced components in a number of meteorites. In addition, we list the ratios $(^{22}Ne/^{21}Ne)_c$ of cosmic-ray-produced Ne, if measured in the same sample as Kr. According to data reported in *Mason* (1979),

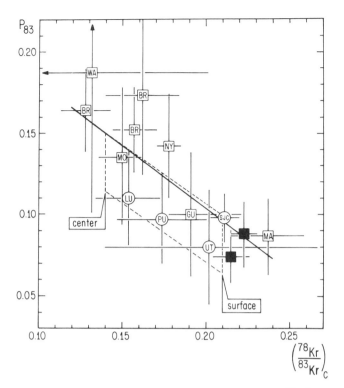

Fig. 4. This figure shows that the production rates P_{83} (in 10^{-12} cm^3 STP g^{-1} Ma^{-1}) for the chondrites listed in Table 3 depend on the ratio (^{78}Kr/^{83}Kr)$_c$. Only total rock data are used; the range of St. Severin core AIII data is indicated by the dashed box. The following abbreviations are used: BR = Bruderheim, GU = Guangrao, LU = Lunan, MA = Maziba, MO = Mocs, NY = Nan Young Pao, PU = Pulsora, SJC = San Juan Capistrano, UT = Utzensdorf, WA = Walters. Filled squares = St. Severin DII, open circles = H chondrites, and open squares = L, LL chondrites. The correlation is approximately linear in the range $0.12 < (^{78}$Kr/^{83}Kr)$_c < 0.24$ (solid line).

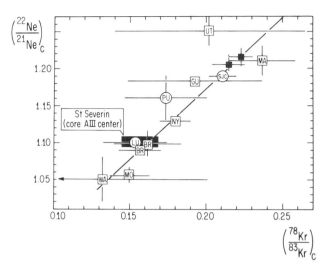

Fig. 5. The shielding parameter (^{22}Ne/^{21}Ne)$_c$ and (^{78}Kr/^{83}Kr)$_c$ for the chondrites listed in Table 3 are shown. The correlation appears to be approximately linear (solid line) in the range $0.13 < (^{78}$Kr/^{83}Kr)$_c <0.24$. For reference, the St. Severin data of core AIII center (filled box; average of three samples) and of piece DII (filled squares) are also shown. For chondrite name labels see Fig. 4.

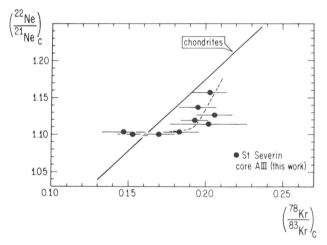

Fig. 6. The Fig. 5 parameters are shown for the St. Severin core AIII samples only. (^{22}Ne/^{21}Ne)$_c$ data for core AIII are interpolated values obtained from the *Schultz and Signer* (1976) data in the same core. The solid reference line is taken from Fig. 5. We note that the core data do not follow the same linear trend.

the H, L, and LL chondrites reveal similar abundances of trace elements Sr, Y, and Zr, which are the chief target elements. All abundances except Rb, show variations $\lesssim 10\%$. These uncertainties are comparable to those in Kr experimental data. Rubidium concentrations are much more variable, and Rb = 0.62 ppm in St. Severin is very low compared to an LL average of 2.3 ppm and averages of 3 ppm in H and L chondrites. Although Rb abundances may not significantly affect the (^{78}Kr/^{83}Kr)$_c$ ratios (*Regnier et al.*, 1979), they will affect the P_{83} production rate, as we discuss later.

We compare the P_{83} rates in different chondrites of varying shielding by using the (^{78}Kr/^{83}Kr)$_c$ ratios. In Table 4 we list the adopted exposure ages and the calculated P_{83} rates (in 10^{-12} cm^2 STP g^{-1}·Ma^{-1}) that are based on the data in Table 3. Whenever possible, the adopted irradiation ages represent averages of ^{21}Ne$_c$ and ^{81}Kr-Kr$_c$ ages. ^{21}Ne$_c$ ages are calculated from Ne data compiled by *Schultz and Kruse* (1983, 1986), using procedures given by *Nishiizumi et al.* (1980). In cases where several measurements are available, the adopted ^{21}Ne age represents a weighted average. Quoted uncertainties reflect experimental errors in ^{21}Ne data and a 10% uncertainty in the ^{21}Ne production rate, quadratically added.

In Fig. 4 we show the P_{83} data versus the ratios (^{78}Kr/^{83}Kr)$_c$. The St. Severin core AIII samples plot in an area outlined in Fig. 3 by the dashed line. Surface and center locations for the core samples are indicated in the figure.

The St. Severin data plot slightly lower than other chondrites. However, following the *Reedy* (1985) model, applied to St. Severin-size meteorites, the low Rb concentration in St. Severin (0.62 ppm) decreases the calculated ^{83}Kr production rates, P_{83}, by about 20% if compared to LL chondrites (average Rb concentration 2.3 ppm). Production rates P_{83} calculated by *Regnier et al.* (1979) (as a function of depth below the lunar surface) show a similar dependence with Rb concentrations.

TABLE 4. Exposure ages and production rates P_{83} in chondrites (see text).

	T_{21} (Ma)	T_{Kr} (Ma)	$T_{adopted}$ (Ma)	P_{83} (10^{-12}cm^{3}STP·g^{-1}Ma^{-1})	
St. Severin (piece DII)	16.5 ± 1.8	13.0 * ± 1.4	14.8 ± 2.4	0.088 ± 20	Sample (a)
				0.074 6	(b)
Bruderheim (Standard)	30.3 ± 3.5	27.0 4.1	28.9 3.5	0.164 25	Sample (a)
				0.152 26	(b)
				0.173 49	(c)
Pulsora	34.8 ± 3.9	-	34.8 3.9	0.097 27	
Mocs	17.9 ± 3.5	16.7 † 3.5	17.3 3.5	0.135 42	
Walters	3.00 ± .33	-	3.00 .33	0.187 86	
Maziba	29.1 ± 3.2	-	29.1 3.2	0.087 23	
Utzensdorf	15.0 ± 3.1	-	15.0 3.1	0.080 35	
San Juan Capistrano	27.7 ± 3.0	28.7 ‡ 2.0	28.4 2.0	0.098 15	
Nan Young Pao	48.5 ± 10	48 § 16	48 10	0.142 32	
Lunan	26.6 ± 5.0	26.8 § 5.0	26.7 5.0	0.105 30	
Guangrao	18.5 ± 3.5	15.1 § 4.0	16.8 3.5	0.110 42	

*Marti et al. (1969).
†Eugster et al. (1969).
‡Finkel et al. (1978).
§Eugster et al. (1987).

As already suggested by *Eugster et al.* (1986), the ^{83}Kr production rate is not constant in chondrites, but depends on the shielding depth and the size of the meteoroid and can be monitored by the ratio $(^{78}$Kr/^{83}Kr$)_c$. Figure 4 suggests the approximate linear relation

$$P_{83} (10^{-12}\text{cm}^3\text{STP g}^{-1}\text{Ma}^{-1}) = 0.26 - 0.76 \, (^{78}\text{Kr}/^{83}\text{Kr})_c$$
$$(r = -0.78) \quad (3)$$

The equation holds for the range $(^{78}$Kr/^{83}Kr$)_c = 0.12$–0.24 and should not be extrapolated, since *Regnier et al.* (1979) showed that P_{83} on the moon steeply decreases at larger shielding depth, documented by smaller $(^{78}$Kr/^{83}Kr$)_c$ ratios. While ^{78}Kr is predominantly produced by high energy particles, ^{83}Kr is augmented by products from secondary neutrons, and the ^{78}Kr production rate is expected to decrease faster than ^{83}Kr production rate for larger shielding depths. Figure 4 indicates that there may be a small difference between P_{83} rates of L, LL, and H chondrites, but the errors overlap. In Fig. 5 we correlate $(^{22}$Ne/^{21}Ne$)_c$ and $(^{78}$Kr/^{83}Kr$)_c$ ratios reported for the same samples of several different chondrites. These data are taken from Table 3. A similar correlation was reported by *Eugster et al.* (1986). For comparison, the shaded box in Fig. 5 represents the center portion of St. Severin core AIII (average of three experimental data points) and St. Severin data of DII (filled squares). Figure 5 reveals a good linear correlation expressed by

$$(^{22}\text{Ne}/^{21}\text{Ne})_c = 0.788 + 1.93 \, (^{78}\text{Kr}/^{83}\text{Kr})_c \quad (r = 0.96) \quad (4)$$

A linear correlation suggests that the parameters $(^{22}$Ne/^{21}Ne$)_c$ and $(^{78}$Kr/^{83}Kr$)_c$ give similar information, namely, the degree of shielding observed in chondritic samples for the range of ratios outlined in Fig. 5. However, if we inspect the St. Severin core data in Fig. 6, a deviation from the linear plot is observed (indicated by the broken line). Why should such a trend be observed only in the St. Severin data? A possible explanation could be ablation. Since most chondrites experienced greater ablation losses than St. Severin, the recovered chondritic samples may represent predominantly interior regions near to the center of meteoroids. If true, the "chondritic" correlation line should fit interior samples from meteoroids of various sizes, as, for example, the center of the core AIII of St. Severin. It is interesting to note that the data of stone DII plot on the "chondritic" line. If they represent samples from a protrusion (the location of stone DII is not known), these data could reveal properties of a small meteorite. This is a model, but if this correlation of the parameters $(^{22}$Ne/^{21}Ne$)_c$

and $(^{78}Kr/^{83}Kr)_c$ within a given meteorite is correct, this would enable us to better separate the parameters of shielding depth of the samples within a meteoroid and of the meteoroid sizes.

Acknowledgments. We dedicate this paper to the late Serge Regnier, who was much interested in this St. Severin core and was instrumental in the design phase of this work. We are indebted to G. N. Simonoff for advice and support, to P. Pellas for supplying the core samples, to O. Eugster and R. C. Reedy for stimulating discussions, to H. Sauvageon and P. Bertin for help with experiments and calculations, and to F. Brout for valuable technical assistance. The research at CENBG was sponsored by the Centre National de la Recherche Scientifique (U.A. No. 451 and ATP Planètologie) and research in La Jolla by NASA Grant NAG 9-41.

REFERENCES

Baros F. and Regnier S. (1984) Measurement of cross-section for ^{22}Na, $^{20-22}$Ne and $^{36-42}$Ar in the proton induced spallation of Mg, Al, Si, Ca and Fe targets. Production ratios of some cosmogenic nuclides in extra-terrestrial materials. *J. Phys.*, 45, 855-886.

Bhandari N., Lal D., Rajan R. S., Arnold J. R., Marti K., and Moore C. B. (1980) Atmospheric ablation in meteorites: A study based on cosmic-ray-tracks and near isotopes. *Nucl. Tracks*, 4, 213-262.

Cantelaube Y. and Pellas P. (1968) Evaluation du taux d'ablation de l'amphoterite de St. Severin par l'etude des traces d' ions lourds du flux primaire du rayonnement cosmique. In *Origin and Distribution of the Elements* (L. H. Arhens, ed.), pp. 479-491. Pergamon Press, New York.

Cantelaube Y., Pellas P., Nordemann D., and Tobailem J. (1969) Reconstitutian de la meteorite de St. Severin daus l'espace. In *Meteorite Research* (P. M. Millman and D. Reidel, eds.), pp. 705-713. Dordrecht, Holland.

Eberhardt P., Eugster O., Geiss J., and Marti K. (1966) Rare gas measurements in 30 stone meteorites. *Z. Naturforschg.*, 21a, 414-426.

Eugster O., Eberhardt P., and Geiss J. (1967) The isotopic composition of krypton in unequilibrated and gas-rich chondrites. *Earth Planet. Sci. Lett.*, 2, 385.

Eugster O., Eberhardt P., and Geiss J. (1969) Isotopic analyses of krypton and xenon in 14 stone meteorites. *J. Geophys. Res.*, 74, 3874-3896.

Eugster O., Grögler N., Eberhardt P., Geiss J., and Kiesl W. (1981) Double drive tube 74001/2: A two-stage model based on noble gases, chemical abundances and predicted production rates. *Proc. Lunar Planet. Sci. 12B*, 541-556.

Eugster O., Shen C., Beer J., Suter M., Wölfli W., Yi W., and Wang D. (1986) Exposure ages of five meteorites from China and ^{81}Kr-Kr based noble gas production rates (abstract). *Meteoritics*, 31, 359.

Eugster O., Shen C., Beer J., Suter M., Wölfli W., Yi W., and Wang D. (1987) Noble gases ^{81}Kr-Kr ages and ^{10}Be of chondrites from China. *Earth Planet. Sci. Lett.*, in press.

Finkel R. C., Kohl C. P., Marti K., Martinek B., and Rancitelli L. (1978) The cosmic ray record in the San Juan Capistrano Meteorite. *Geochim. Cosmochim. Acta*, 42, 241-250.

Graf H. (1967) Cosmic-ray-produced noble gas isotopes. Thesis. University of Bern.

Marti K. and Lugmair G. W. (1971) ^{81}Kr-Kr and ^{40}Ar ages, cosmic ray spallation products, and neutron effects in lunar samples from Oceanus Procellarum. *Proc. Lunar Sci. Conf. 2nd*, 1591-1605.

Marti K., Eberhardt P., and Geiss J. (1966) Spallation fission and neutron capture anomalies in meteoritic krypton and xenon. *Z. Naturforschg.*, 21a, 398.

Marti K., Shedlovsky J. P., Lindstrom R. M., Arnold J. R., and Bhandari N. G. (1969) Cosmic-ray-produced radionuclides and rare-gases near the surface of St. Severin meteorite. In *Meteorite Research* (P. M. Millman and D. Reidel, eds.), pp. 246-266. Dordrecht, Holland.

Mason B. (1979) Data of geochemistry. In *Geol. Surv. Prof. Pap. 440-B-1.* (Michael Fleischer, ed.), p. 118.

Nishiizumi K., Regnier S., and Marti K. (1980) Cosmic ray exposure ages of chondrites, pre-irradiation and constancy of cosmic ray flux in the past. *Earth Planet. Sci. Lett.*, 50, 156-170.

Reedy R. C. (1985) A model for GCR - Particular fluxes in stony meteorites and production rates of cosmogenic nuclides. *Proc. Lunar Planet. Sci. Conf. 15th*, in *J. Geophys. Res.*, 90, C722-C728.

Regnier S., Hohenberg C. M., Marti K., and Reedy R. C. (1979) Predicted versus observed cosmic-ray-produced noble gases in lunar samples: Improved Kr production ratios. *Proc. Lunar Planet. Sci. Conf. 10th*, 1565-1586.

Regnier S., Lavielle B., Simonoff M., and Simonoff G. N. (1982) Nuclear reactions in Rb, Sr, Y and Zr targets. *Phys. Rev.*, C26, 931-943.

Schultz L. and Signer P. (1976) Depth dependence of spallogenic helium, neon and argon in the St. Severin chondrite. *Earth Planet. Sci. Lett.*, 30, 191-199.

Schultz L., and Kruse H. (1983) *Helium, Neon and Argon in Meteorites*. Max-Planck-Institut für Chemie, Mainz.

Schwaller H., Eberhardt P., Geiss J., Graf H., and Grögler N. (1971) The $(^{78}Kr/^{83}Kr)_{sp}$ - $(^{131}Xe/^{126}Xe)_{sp}$ correlation in Apollo 12 rocks. *Earth Planet. Sci. Lett.*, 12, 167-169.

Wilkening L. L. and Marti K. (1976) Rare gases and fossil particle tracks in the Kenna ureilite. *Geochim. Cosmochim. Acta*, 40, 1465-1473.

Wright R. J., Simms L. A., Reynolds M. A., and Bogard D. (1973) Depth variation of cosmogenic noble gases in the 120 kg Keyes chondrite. *J. Geophys. Res.*, 78, 1308-1318.

Spectral Alteration Effects in Chondritic Gas-Rich Breccias: Implications for S-Class and Q-Class Asteroids

Jeffrey F. Bell
Hawaii Institute of Geophysics, University of Hawaii, Honolulu, HI 96822

Klaus Keil
Department of Geology, Institute of Meteoritics, University of New Mexico, Albuquerque, NM 87131

The location of the parent bodies of the ordinary chondrite (OC) meteorites remains controversial. One view is that at least some Class S asteroids are of OC composition, but that some unknown space weathering process alters their uppermost regolith to produce the observed reflection spectra, which are unlike those of powdered ordinary chondrites. We have obtained reflection spectra of gas-rich matrix regions in OC regolith breccias, which are believed to represent lithified portions of asteroid regoliths that were once directly exposed to space. Spectral effects attributable to shock and impact melting were identified; in some portions of particular meteorites they are very strong and would be easily detectable in current asteroid data sets. However, in no case was the curved red continuum observed in Class S asteroid spectra found. These results indicate that while intensely gardened portions of the OC parent body regoliths may possess detectable spectral alteration effects, these effects cannot give such a body the overall appearance of a Class S asteroid. Consequently, it appears very unlikely that any well-observed Class S asteroid can be a source of ordinary chondrites. The newly discovered Class Q asteroids provide a viable alternative source for ordinary chondrites, but the presence of these objects only on Earth-approaching orbits poses some new problems.

INTRODUCTION

The asteroids are of fundamental importance to understanding the origin and evolution of the solar system for several reasons: (1) They probably represent remnants of the population of small bodies that accumulated to form the planets and preserve an otherwise lost intermediate stage in the formation of planetary bodies. (2) They are located between the rocky inner planets and the icy outer solar system and may preserve the compositional transition between these radically different classes of bodies. (3) Some of them have escaped the melting processes that have destroyed evidence of the original geochemistry in planetary rocks. (4) Study of meteorites has provided a vast body of geochemical, mineralogical, and isotopic data that probably refers to some asteroids.

Studies of the spectral distribution ("color") of reflected sunlight provides the best current means of determining asteroidal surface compositions and relating meteorite classes to their parent bodies. Data obtained in the visible spectral region (0.3-1.1 microns wavelength) indicate that a majority of the objects in the inner portion of the asteroid belt belong to spectral class "S" (*Chapman and Gaffey*, 1979; *Zellner et al.*, 1985). Spectra of these objects exhibit reddened continuua in the visible wavelengths that gradually flatten out in the near-IR. Superimposed on these continuua are shallow absorption bands diagnostic of the silicate minerals olivine and pyroxene. Among the 14 asteroid spectral classes identified in the most recent taxonomic analysis (*Tholen*, 1984), the S-types are the only abundant class with such complex spectra. Unfortunately, this class is also the most controversial, with two opposing schools of thought supporting contradictory interpretations of the spectral data.

The conventional method of interpreting asteroid spectra is to pulverize meteorite samples to approximately the texture suggested by polarization studies of asteroid regoliths (*Dollfus and Zellner*, 1979, pp. 177-180), obtain spectra in the laboratory, and compare them with asteroid spectra. This is an imperfect technique, because some meteorite classes are very rare (and therefore are not made available by museum curators for crushing), and because many common types contain large masses or networks of nickel-rich iron alloys (with mechanical properties similar to man-made stainless steel), which makes them difficult to pulverize. Also, this methodology implicity assumes that the regolith-forming process on asteroids changes only the particle size and does not introduce spectral effects due to glass formation, solar wind implantation, and other effects. Nevertheless, an impressive number of very close matches between asteroid and meteorite spectra has been obtained by this method. However, none of the meteorite spectra obtained to date match the S-type asteroid spectra closely. This fact is explained in two ways by different investigators:

(A) (e.g., *Feierberg et al.*, 1982; *Wetherill and Chapman*, 1988, Section 4B) Spectra of the most common meteorites (ordinary chondrites) also contain shallow olivine and pyroxene absorption bands, but lack the steep red slope also present in those of Class S asteroids. During the evolution of the uppermost regoliths on the ordinary chondrite parent bodies the shape of the continuum is altered by some unknown "space weathering" effect that occurs during regolith formation. This spectral alteration effect causes the telescopic spectra (which sample only the very uppermost regolith) to have Class S characteristics. The simulated regoliths prepared from ordinary chondrites found on Earth do not spectrally resemble the natural regoliths on asteroids because their pulverization is an inadequate simulation of the real regolith-altering processes.

(B) (e.g., *Gaffey,* 1984, 1986) The spectral differences between S-type asteroids and OC meteorites represent a real compositional difference. The lack of spectral matches for S-asteroids in the meteorite spectral collection is an artifact of the incomplete nature of that data set. S-asteroids are composed of differentiated stony-iron material similar to the pallasites, lodranites, and silicate-bearing irons, for which no laboratory spectra have been obtained due to the extreme difficulty of pulverizing the samples.

These two opposing schools of thought have radically different implications in many areas. If interpretation A is correct: (1) The most common meteorites (OCs) correspond to the most common inner-belt asteroids (Class S); (2) asteroid spectra refer only to a highly altered regolith and tell us nothing about the bedrock beneath; and (3) most inner-belt asteroids were only slightly heated and metamorphosed. If interpretation B is correct: (1) The ordinary chondrites have no known (i.e., large) parent body in the asteroid belt, and the Class S asteroids are the source of some of the rarest meteorite types; (2) asteroid regoliths are merely pulverized bedrock and asteroid spectra are easily interpretable; and (3) most inner-belt asteroids were strongly heated and melted, but the segregation of silicate and metal components was still incomplete when the heat source decayed and the melt solidified.

The key question is: Do the uppermost few millimeters of an asteroid regolith have the same spectral curve as the bedrock underneath? On the Moon, very strong weathering effects (mostly related to the formation of glassy agglutinates) exist in the uppermost regolith that suppress the absorption bands and introduce a reddened continuum. However, it is generally agreed that lunar-style glass formation is an extremely rare process in asteroidal regoliths. The rarity of impact-produced glasses on asteroids has been amply confirmed by spectral comparisons with the Moon (*Matson et al.,* 1977) and extensive studies of many meteoritic breccias (*Keil,* 1982). This difference is probably due to the very different bombardment environment in the asteroid belt, where the effects of high-velocity glass-forming impacts are swamped by the numerous low-velocity collisions between asteroid fragments on roughly similar orbits. Regolith gardening models for asteroids (*Housen et al.,* 1979) predict that excavation of fresh bedrock predominates over reworking of existing regolith, unlike the lunar case.

The only likely candidate for the S-asteroid reddening agent among known meteoritic materials is metallic nickel-iron. Both iron meteorites and M-class asteroids exhibit a continuum similar to that in a typical S-class asteroid (*Gaffey,* 1976, 1986). Most early advocates of interpretation A proposed that the red slope of S-type spectra could be produced by a gardening mechanism that enhanced the abundance of NiFe metal in the regolith. To test this hypothesis, *Gaffey* (1986) produced simulated ordinary chondrite regoliths in which the metal abundance was enhanced by magnetic separation. Surprisingly, even very metal-rich simulations showed no increase in the red slope thought to be characteristic of nickel-iron metal. Apparently, the spectral signature of metal in undifferentiated meteorites differs from that in differentiated meteorites, for reasons that remain obscure. However, several workers have proposed alternate weathering effects that could cause metal-bearing chondrite bedrock material to develop an upper regolith whose spectrum would exhibit a reddened and curved continuum similar to that of the S-class asteroids (e.g., *King et al.,* 1984; *Pieters,* 1984; *McFadden,* 1983a,b; *Wetherill and Chapman,* 1988).

The question of the metal-like curved red continuum is only one of the problems that face interpretation A. There is abundant evidence that the silicate in most S-type asteroids is also fundamentally different from known chondritic assemblages. *Feierberg et al.* (1982) obtained IR spectra of many S-type asteroids and found that only a few had olivine/pyroxene ratios consistent with those of ordinary chondrites. (To reconcile these data with interpretation A, they postulated the existence of a variety of unknown chondrite types with high olivine/pyroxene ratios comparable to those in carbonaceous chondrites, but without the carbonaceous material that would suppress the silicate absorption bands.) A more recent survey (*Bell et al.,* 1985) found variations in mineralogy even more extreme than those found by *Feierberg et al.* (1982). In particular, some S-types show no evidence for a 2-micron absorption band, indicating a pure-olivine silicate assemblage (*Cruikshank and Hartmann,* 1984). These spectra have been shown to correspond closely to spectra of simulated stony-iron regoliths created by dispersing olivine grains on a metal substrate (*Bell et al.,* 1984). *Gaffey* (1984) made observations of asteroid 8 Flora [which had been nominated by *Feierberg et al.* (1982) as the best match for ordinary chondrites] to search for mineralogic variations across its surface. Such variations were found and exhibit trends not found in chondritic meteorites. Even more extreme mineralogical variations have been found for 15 Eunomia (*Gaffey and Ostro,* 1987). Consequently, most current advocates of interpretation A have adopted a modified version similar to that of *Feierberg et al.* (1982), in which only a few Class S asteroids are composed of known ordinary chondrite types. But it is still necessary for them to postulate some regolith effect to account for the curved red continuum.

A small subset of ordinary chondrites, the solar gas-rich breccias, provide a simple test of all these proposed mechanisms. These meteorites have a pronounced light/dark structure suggesting variations in spectral reflectance (e.g., *Fredriksson and Keil,* 1963), and apparently preserve portions of the uppermost regolith complete with alteration effects characteristic of long exposure to the space environment (e.g., *Keil,* 1982). Typically, the areas of dark fine-grained matrix in these meteorites are rich in implanted solar wind gases, particle tracks left by galactic cosmic rays and solar flares, and small fragments of exotic meteorite classes, apparently projectile material that has survived low-velocity impacts. These effects indicate that this matrix material once resided on the optical surface of an asteroid, was turned over by regolith gardening, lithified by impact, ejected, and transported to Earth. (A similar history is inferred for many lunar samples, including lunar meteorites.) Large light clasts in the same breccias usually are devoid of these effects in their interiors, which were

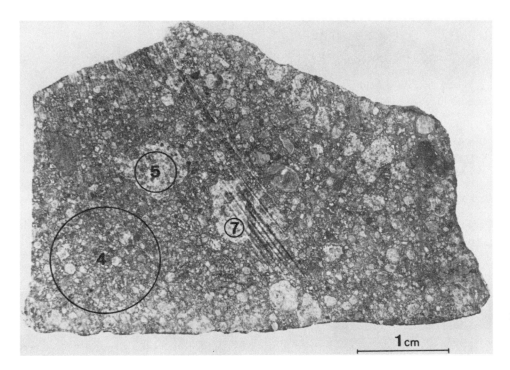

Fig. 1. Sawed surface of slab of Dubrovnik, side 1, showing locations where spectra in Fig. 2 were obtained.

protected from direct exposure to space. If any unknown weathering effects exist in the uppermost regolith of asteroids that significantly affect current interpretations of their composition, it should be evident in a spectral difference between gas-rich matrix and gas-poor clast interiors. We have therefore carried out a systematic search for such effects by measuring reflection spectra of well-characterized matrix and clast areas in gas-rich ordinary chondrite regolith breccias, with emphasis on finding any process that might make ordinary-chondrite material develop the reddened continuum characteristic of the Class S asteroids.

MEASUREMENTS

As described above, meteorite reflection spectra have traditionally been obtained using pulverized samples in an attempt to simulate the actual scattering conditions in an asteroid regolith. This has the disadvantage of destroying the original petrographic texture of the meteorite. In this project, we wished to survey a wide variety of clast and matrix materials in meteorites and conduct later petrographic studies of any material exhibiting S-type spectral properties. Therefore, we measured spectra of sawed or freshly broken surfaces. We emphasize that *these spectra are not directly comparable to those of powdered meteorite samples or asteroids* (they differ in the sense that the continua of our spectra are systematically bluer). However, comparisons within the data set (and particularly between spectra of matrix and clasts on the same slab) should reveal any spectral differences due to space weathering.

All samples studied here are from the meteorite collection of the Institute of Meteoritics, University of New Mexico,

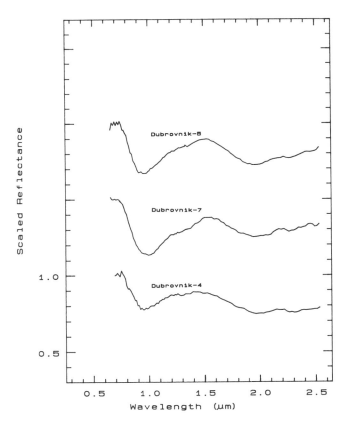

Fig. 2. Spectra of selection regions of Dubrovnik. Locations are shown in Fig. 1.

Fig. 3. Sawed surface of slab of Dubrovnik, side 2, showing locations where spectra in Fig. 4 were obtained.

Albuquerque. Specimen numbers refer to those given in the catalog of the collection (*Scott et al.*, 1986). Microscopic examination of the sawed surfaces on many of the samples revealed that they were covered with a homogenous optically thick coating of fine-grained saw cuttings that could not be identified with any specific petrographic unit. This coating was removed with high-pressure jets of dry nitrogen gas in order to increase the relative spectral contrast in our data set. No other special preparation of the samples was performed.

The reflection spectra obtained are plotted as ratios to the white standard material Halon. All spectra were scaled to a value of 1.0 at 0.7 μm to allow easier comparison of band depths and continuum slopes. In all figures, the vertical scale refers only to the lowermost spectra; others are offset upward for clarity by increments of 0.5.

Dubrovnik (Fall, Feb. 20, 1951). This meteorite is a solar gas-rich regolith breccia whose components range from L3 to L6 (*Williams et al.*, 1986; *Hoinkes et al.*, 1976). Figure 1 shows one face of our slab of Dubrovnik (C207.1). Spectra obtained on this face are plotted in Fig. 2. Spectra 5 and 7 represent the interiors of large light clasts that were never exposed to space, whereas spectrum 4 samples a large area of dark gas-rich matrix material. Besides a lower albedo (not apparent in our scaled reflectance plot), the matrix region exhibits an ~50% decrease in band depth. Spectra 1, 2, and 3 (Figs. 3 and 4) represent dark matrix regions on the opposite face of the same slab. All closely resemble spectrum 4. Finally, in an attempt to detect the spectra effects of terrestrial weathering, we obtained a spectrum of a clearly visible rust stain (spectrum 6 at bottom of Fig. 4). Although this region was clearly reddened to the eye, in the IR it appears similar to the other matrix regions. This is consistent with evidence from powdered sample studies that weathering introduces a

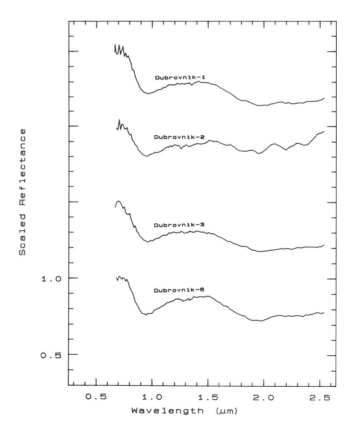

Fig. 4. Spectra of selected regions of Dubrovnik. Locations are shown in Fig. 3.

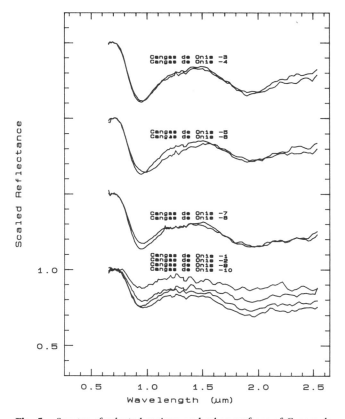

Fig. 5. Spectra of selected regions on broken surfaces of Cangas de Onis.

strong absorption edge in the visible wavelength region without affecting longer wavelengths (*Gaffey*, 1976).

Cangas de Onis (Fall, Dec. 6, 1886). This is a solar gas-rich breccia consisting of H6 clasts in an H5 matrix (*Williams et al.*, 1985). Our specimens were irregular lumps with visible surface weathering. Spectra 1 to 6 (Fig. 5) were obtained from a new surface in the less weathered interior of the 17.5-gram specimen C199.3 (A_1), which was exposed by breaking it in half. Spectra 7 to 10 were obtained from new surfaces similarly exposed in the 32-gram specimen C199.1 (A). At least three petrographic units were identifiable by eye. Spectra 3 and 4 represent the lightest material visible, presumably gas-free bedrock clasts. Spectra 5 and 6 represent a region of this material that was nearer to the original surface and is visibly rusted. No significant differences are apparent. Spectra 7 and 8 represent a darker region, whereas spectra 1, 2, 9, and 10 were obtained on small black veins that are present between the "white" and "grey" clasts. These latter regions show very shallow bands and a slight reddening of the continuum that appears to be correlated with decreasing band strength. We interpret these veins as highly shocked regions. Their spectra appear very similar to those of powdered samples of highly shocked "black chondrites" shown in Fig. 18 of *Gaffey* (1976).

Olivenza (Fall, June 19, 1924). This LL5 chondrite is a fragmental breccia without implanted solar wind gases (*Graham et al.*, 1985). It therefore serves as a useful intermediate case between true regolith breccias and unbrecciated chondrites. Specimen C238.1 consists of a sawed but not polished slab. The spectra in Fig. 7 were taken off the rougher of the two sides (Fig. 6). They are essentially identical to those of the light clasts in Dubrovnik.

Dimmitt (Find, about 1942). This meteorite consists of H3-5 chondrite clasts, impact melt rock clasts, and exotic clasts embedded in a matrix that contains both equilibrated and unequilibrated components (*Rubin et al.*, 1983). Specimen C10.10, a sawed but not polished slab with two nearly parallel faces, exposes a poikilitic melt rock clast (DT4) representing the most extreme form of altered regolith (Fig. 8). A spherical inclusion of metal and schreibersite in the center of the clast has a measured metallographic cooling rate of $30°C/10^3$ years (through $500°C$). This suggests that the clast was originally buried beneath a melt breccia layer about 500 m thick on the floor of a few-kilometers-diameter impact crater. The entire clast is discolored by brownish ferric iron oxide stains of terrestrial origin. We examined this melt clast to test suggestions (e.g., *McFadden*, 1983a,b) that the upper regolith of chondritic asteroids may be dominated by impact melt with S-like spectral properties. Spectra of this clast (Fig. 9) have very deep pyroxene bands similar to those of eucrite spectra. This is consistent with the origin of the clast in impact melting with segregation of the metal component (note the metal nodule in the center of Fig. 8). Apparently a large impact melt component in a chondritic regolith would make its spectra more like the deep-band Class V object Vesta, rather than Class S asteroids.

GENERAL CONCLUSIONS

Our data do not provide any evidence for any unknown weathering effects in asteroidal regoliths that would produce the reddened continuum characteristic of S-type asteroids. Gas-rich matrix material does appear to exhibit a general darkening and lessening of silicate band strength, relative to gas-poor clast material in the same breccia. The only case in which any significant reddening of the continuum occurs is in narrow black veins in Cangas de Onis. This effect appears to be due to high shock levels rather than simple residence at the surface. This effect is well known from previous studies of "black chondrites" in which the entire meteorite exhibits such properties (*Gaffey*, 1976, p. 318). It has never been popular as an explanation of the S-type problem because the reddening is fairly uniform over the entire 0.3-2.5 μm spectral region, and does not produce the characteristic curved S-type spectrum. In addition, the degree of suppression of the silicate absorption bands is too great. However, it is possible that this shock-blackened material does make some minor contribution to the reflectance of the ordinary chondrite parent bodies. The reduction of band depth caused by shock could mimic the similar effect caused by metamorphism (see Fig. 15 of *Gaffey*, 1976) and cause an erroneously high estimate of petrologic grade. Since other breccia studies suggest that different petrologic grades are well mixed in the parent body regoliths, this is unlikely to prove of operational significance.

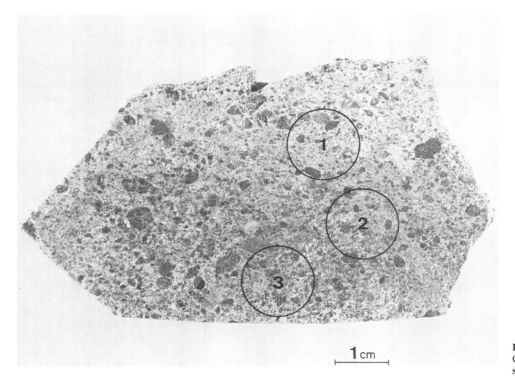

Fig. 6. Sawed surface of slab of Olivenza, showing locations where spectra in Fig. 7 were obtained.

Our spectra of the Dimmitt impact melt clast reveal a previously unsuspected "weathering" effect: Impact melts derived from chondritic parent material may exhibit achondrite-like spectra due to segregation of metal during their formation. With no finely divided metal to reduce silicate band depth, a spectral curve similar to those of eucrites or howardites results. If sizeable pools of impact melt exist on any current OC parent body, they may exhibit achondrite-like spectra until lateral transport of ejecta during later impacts mixes the regolith thoroughly. Both the abundance of melt clasts in regolith breccias and regolith gardening models suggest that this is a very rare and short-lived phenomenon that is very unlikely to affect interpretation of spectra obtained from the Earth, which average the composition of an entire hemisphere.

IMPLICATIONS FOR ASTEROIDS

These results suggest that unknown regolith processes probably do not cover the ordinary chondrite parent bodies with an altered upper regolith exhibiting the spectral properties observed in Class S asteroids. This evidence, combined with the observational data suggesting a large range in olivine/pyroxene ratio among the S-types and large variations across the surfaces of at least some, suggests that none of the well-observed S-types is a plausible parent body of the known ordinary chondrite classes.

Fortunately, a new asteroid class provides a viable alternative. *Tholen* (1984) has defined a new spectral class "Q" of which asteroid 1862 Apollo is the prototype. Although no high-quality IR spectra of a Q-type object are yet available, the data that

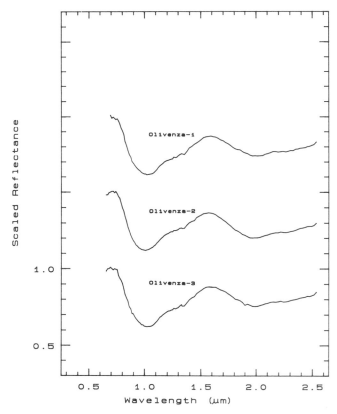

Fig. 7. Spectra of selected regions of Olivenza. Locations are shown in Fig. 6.

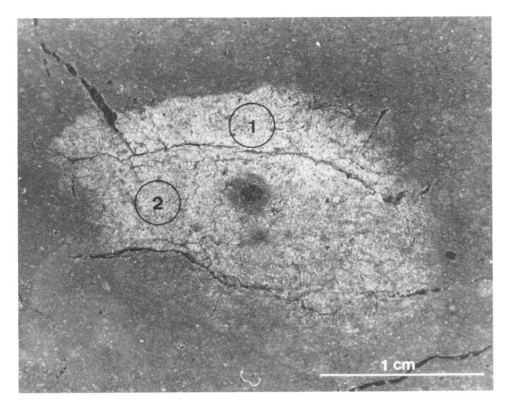

Fig. 8. Sawed surface slab of Dimmitt, showing locations on impact-melt clast where spectra in Fig. 9 were obtained.

do exist for Apollo are more consistent with an ordinary chondrite composition than that for any other object (*McFadden et al.*, 1985). In Fig. 10 we show a portion of the principal component plot derived by *Tholen* (1984) from the Arizona Eight-Color Asteroid Survey spectra, with a possible interpretation in terms of mineralogy. Class A (dunite) asteroids, Class M (iron) asteroids, and the lone Class V (pyroxene-plagioclase) asteroid Vesta map out the corners of a compositional triangle of differentiated objects. The triangle is somewhat distorted because the principal components are purely mathematical descriptions of spectral curves and were not intended to correlate in any simple way with actual mineral abundances. In this scheme Class S asteroids occupy the interior of the triangle (lodranites) with extensions to the left boundary (Steinbach) and the upper right boundary (pallasites). Class Q lies on the olivine-pyroxene line because chondritic metal has the flat spectrum discovered by *Gaffey* (1986) and therefore is "invisible" to Tholen's principal component analysis.

This interpretation raises a new and perplexing problem. About 10-20% of the asteroids on Earth-crossing orbits appear to belong to this class, but it is totally absent among the well-observed asteroids in the main belt. A possible answer lies in the very different size distributions involved. Most observed Earth-crossers are less than 3 km in diameter, while the Eight-Color Survey did not include main-belt S-type objects below about 20 km. So a sizeable population of small Q-class asteroids could exist in the belt and constantly replenish the Earth-crossing population. These objects would be collisional fragments from the small number of ~200 km parent bodies implied by the chemical and cooling-rate data. This idea is

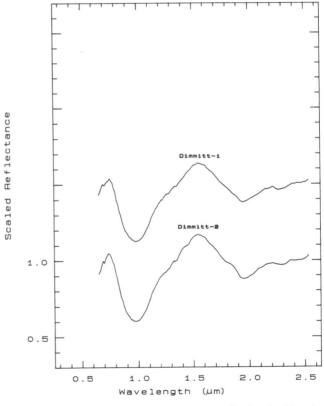

Fig. 9. Spectra of two regions on impact-melt clast in Dimmitt. Locations are shown in Fig. 8.

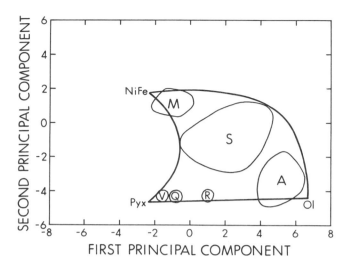

Fig. 10. Domains of the moderate-albedo asteroid classes in the principal component space of *Tholen* (1984), with mineralogical interpretations derived from infrared spectra (superimposed triangle).

consistent with the conclusion of *Wetherill* (1985) that ordinary chondrites are mostly derived from small asteroids (<20 km diameter). The only obvious problem with this explanation is that currently popular collision models require large parent bodies to constantly replenish the small main-belt population.

Another obvious objection to the above interpretation is that it associates very rare classes of stony-iron meteorites with a class of asteroids that is very common, in fact the most common type among the Earth-crossers (*Tedesco and Gradie*, 1987), while the abundant ordinary chondrites are derived from an asteroid class that is very rare, at least in terms of mass. It is possible that the very different mechanical properties of chondrites and stony-irons can account for this. The very large difference in average cosmic-ray exposure ages of iron meteorites and chondrites has traditionally been explained by this effect. There seems to be no reason why differential fragmentation rates should not exist for asteroid-sized objects.

Acknowledgments. This research was supported by grants NAG 9-30 (K. Keil) and NAGW 712 (J. F. Bell) from the National Aeronautics and Space Administration.

REFERENCES

Bell J. F., Gaffey M. J., and Hawke B. R. (1984) Spectroscopic identification of probable pallasite parent bodies. *Meteoritics*, 19, 187-188.

Bell J. F., Hawke B. R., Owensby P. D., and Gaffey M. J. (1985) The 52-color asteroid survey: Results and interpretation. *Bull. Am. Astron. Soc.*, 17, 729.

Chapman C. R. and Gaffey M. J. (1979) Reflectance spectra for 277 asteroids. In *Asteroids* (T. Gehrels, ed.), pp. 665-687. Univ. of Arizona Press, Tucson.

Cruikshank D. P. and Hartmann W. K. (1984) The meteorite-asteroid connection: Two olivine-rich asteroids. *Science*, 223, 281-282.

Dollfus A. and Zellner B. (1979) Optical polarimetry of asteroids and laboratory samples. In *Asteroids* (T. Gehrels, ed.), pp. 170-183. Univ. of Arizona Press, Tucson.

Feierberg M. A., Larson H. P., and Chapman C. R. (1982) Spectroscopic evidence for undifferentiated S-type asteroids. *Astrophys. J.*, 257, 361-372

Fredriksson K. and Keil K. (1963) The light-dark structure in the Pantar and Kapoeta stone meteorites. *Geochim. Cosmochim. Acta*, 27, 717-739.

Gaffey M. J. (1976) Spectral reflectance of the meteorite classes. *J. Geophys. Res.*, 81, 905-920.

Gaffey M. J. (1984) Rotational spectral variations of asteroid (8) Flora: Implications for the nature of the S-type asteroids and for the parent bodies of the ordinary chondrites. *Icarus*, 60, 83-114.

Gaffey M. J. (1986) The spectral and physical properties of metal in meteorite assemblages: Implications for asteroid surface materials. *Icarus*, 66, 468-486.

Gaffey M. J. and Ostro S. J. (1987) Surface lithologic heterogeneity and body shape for asteroid 15 Eunomia (abstract). In *Lunar and Planetary Science XVIII*, pp. 310-311. Lunar and Planetary Institute, Houston.

Graham A. L., Bevan A. W. R., and Hutchinson R. (1985) *Catalogue of Meteorites*. Univ. of Arizona Press, Tucson. 460 pp.

Hoinkes G., Kurat G., and Batic L. (1976) Dubrovnik: Ein L3-6 Chondrit. *Ann. Naturhist. Mus. Wien*, 80, 39-55.

Keil K. (1982) Composition and origin of chondritic breccias (abstract). In *Workshop on Lunar Breccias and Soils and their Meteoritic Analogs* (G. J. Taylor and L. L. Wilkening, eds.), pp. 65-83. LPI Tech. Rpt. 82-02, Lunar and Planetary Institute, Houston.

King T. V. V., Gaffey M. J., and McFadden L. A. (1984) Spectroscopic evidence for regolith maturation. *Meteoritics*, 19, 251-252.

Matson D. L., Johnson T. V., and Veeder G. J. (1977) Soil maturity and planetary regoliths: the Moon, Mercury, and the asteroids. *Proc. Lunar Sci. Conf. 8th*, pp. 1001-1011.

McFadden L. A. (1983a) S-type asteroids and their relation to ordinary chondrites. *Meteoritics*, 18, 352.

McFadden L. A. (1983b) Reflectance characteristics of some Antarctic meteorites and their relation to asteroid surface composition. *Bull. Am. Astron. Soc.*, 15, 827.

McFadden L. A., Gaffey M. J., and McCord T. B. (1985) Near-Earth asteroids: Possible sources from reflectance spectroscopy. *Science*, 229, 160-163.

Rubin A. E., Scott E. R. D., Taylor G. J., Keil K., Allen J. S. B., Mayeda T. K., Clayton R. N., and Bogard D. D. (1983) Nature of the H chondrite parent body regolith: Evidence from the Dimmitt breccia. *Proc. Lunar Planet. Sci. Conf. 13th*, in *J. Geophys. Res.*, 88, A741-A754.

Pieters C. (1984) Asteroid-meteorite connection: Regolith effects implied by lunar reflectance spectra. *Meteoritics*, 19, 290-291.

Scott E. R. D., Keil K., and Nelson L. M. (1986) Catalog of the Meteorite Collection of the Institute of Meteoritics, University of New Mexico. *UNM Institute of Meteoritics Spec. Publ. 23*. 60 pp.

Tedesco E. F. and Gradie J. C. (1987) Discovery of M-class objects among the near-Earth asteroid population. *Astron. J.*, 93, 738-746.

Tholen D. J. (1984) Asteroid taxonomy from cluster analysis of photometry. Ph.D. thesis, University of Arizona, Tucson.

Wetherill G. W. (1985) Asteroidal sources of ordinary chondrites. *Meteoritics*, 20, 1-22.

Wetherill G. W. and Chapman C. R. (1988) Asteroids and meteorites. *Meteorites and the Early Solar System* (J. Kerridge, ed.), Chapter 2. Univ. of Arizona Press Space Science Series, in press.

Williams C. V., Rubin A. E., Keil K., and San Miguel A. C. (1985) Petrology of the Cangas de Onis and Nulles regolith breccias: Implications for parent body history. *Meteoritics*, 20, 331-345.

Williams C. V., Scott E. R. D., Taylor G. J., Keil K., Schultz L., and Wieler R. C. (1986) Histories of ordinary chondrite parent bodies: Clues from regolith breccias. *Meteoritics*, 21, 541.

Zellner B., Tholen D. J., and Tedesco E. F. (1985) The eight-color asteroid survey: Results for 589 minor planets. *Icarus*, 61, 355-416.

Interactions of Light with Rough Dielectric Surfaces: Spectral Reflectance and Polarimetric Properties

S. A. Yon and C. M. Pieters

Department of Geological Science, Brown University, Providence, RI 02912

An understanding of the nature of the interactions of visible and near-infrared radiation (0.6 to 1.8 μm) with the surfaces of rock and mineral samples is necessary for both theoretical modeling and interpretation of remotely sensed data. The expected reflectance and polarization properties of light that has interacted with a rough dielectric surface through both Fresnel and Rayleigh mechanisms are examined. Reflectance and polarization of reflected light were measured in the visible and near-infrared wavelength range for slab samples of obsidian and fine-grained basalt specially prepared to controlled surface roughness. Obsidian was chosen for its optical homogeneity and clarity to allow examination of surface reflected light, while basalt was chosen as a laboratory analog to naturally occurring polycrystalline materials. Comparison of experimentally measured reflectance and polarization from smooth and rough slab materials with the predicted models indicates that single Fresnel reflections are responsible for the largest part of the reflected intensity resulting from interactions with the surfaces of dielectric materials, and that although multiple Fresnel reflections do occur, they are much less important for such surfaces.

INTRODUCTION

Recent years have seen a dramatic growth in the use of visible and near-infrared reflectance spectroscopy for remote mineralogic interpretation of terrestrial and extraterrestrial surfaces and laboratory analysis of rock and mineral samples. Absorption features are observed in reflectance spectra and are produced as radiation is transmitted through the sample. High spectral resolution spectroscopy is capable of distinguishing many rock and mineral types based on the positions and strengths of mineral absorption bands caused by crystal field effects and vibrational resonance (*Burns*, 1970; *Adams*, 1974, 1975; *Hunt and Salisbury*, 1970; *Hunt*, 1977). In addition to providing information that allows specific minerals to be identified, high resolution reflectance spectroscopy can often be used to estimate the composition of constituent minerals (e.g., *Adams*, 1974; *Hazen et al.*, 1978) and has also been shown to provide reasonably accurate estimations of mineral abundance (e.g., *Johnson et al.*, 1983; *Mustard and Pieters*, 1987).

Since soils are the most common surface material found on the terrestrial planets, most laboratory spectroscopic studies have concentrated on particulate samples, laboratory analogs to natural soils. Laboratory measurement of samples consisting of fine particulate material, generally in the size range of 50–250 μm, yields reflectance spectra that exhibit weak to moderate absorption bands superimposed on a background continuum. As examples, reflectance spectra for different particle size samples of obsidian and basalt are shown in Figs. 1a and 1b, respectively. Spectra of the basalt particles exhibit a pyroxene absorption band at ~1.03 μm due to electronic transitions in ferrous iron. Spectra of the obsidian particles exhibit a similar glass absorption band at ~1.14 μm. Spectra of both obsidian and basalt particles exhibit ultraviolet charge transfer absorptions. All of these diagnostic features are caused by preferential absorption of light as it passes through the sample material.

Measurements of the polarization of light reflected from particulate surfaces indicate that there is also a component of reflected light that has not been transmitted but has only interacted with the surface of the constituent particles (e.g., *Dollfus*, 1985). At large phase angles, reflectance is more strongly polarized for the larger, generally darker, particle sizes, and polarization varies inversely with reflectance in an absorption band (*Pieters*, 1974). In the basalt samples shown in Fig. 1b, increasing particle size is accompanied by decreasing reflectance and decreasing absorption band depth, indicating a reduction in the relative amount of the transmitted component in the reflected light (e.g., see *Adams and Filice*, 1967). For these samples, the relative magnitude of first surface or near surface component is greatest when particle size is increased to the point where absorption bands are no longer evident.

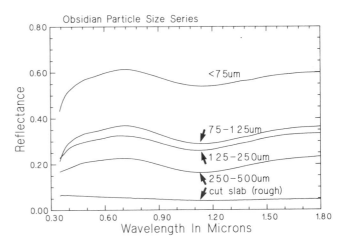

Fig. 1a. Bidirectional reflectance spectra of a prepared slab (rough texture—see text) and for particulate samples of obsidian. L = 0°, D = 30° (Halon Standard).

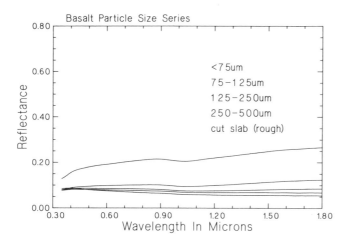

Fig. 1b. Bidirectional reflectance spectra of a prepared slab (rough texture—see text) and for particulate samples of a fine grained basalt. L = 0°, D = 30° (Halon Standard).

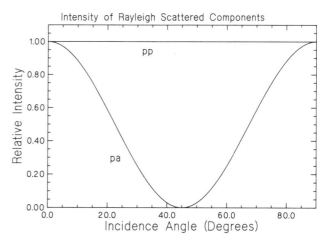

Fig. 2. Relative intensities of the parallel and perpendicular (I_{pa} and I_{pp}) components of Rayleigh scattered light as a function of incidence angle, i.

Rough cut surfaces, or slabs, are laboratory analogs to surfaces composed of large particles. The manner in which light interacts with such rough dielectric surfaces is not fully understood. The background continua for spectra of slab samples (see Figs. 1a,b) are generally low in reflectance and have a relatively blue character (decreasing reflectance with increasing wavelength). Spectra of slab samples typically exhibit only very weak absorption features. When examined as particulate samples, two different materials, such as the obsidian and basalt particles of Figs. 1a and 1b, may be easily identified from observed diagnostic absorption bands. These same two materials, however, are nearly indistinguishable on the basis of spectra from large particles or slabs.

The purpose of the study undertaken here is to isolate some of the properties involved in the complex interaction of light with a rough dielectric surface. There are several reasons to initiate such investigations: (1) Artificial slab surfaces, which represent the extreme in large particle size, are approximate analogs to surface rock outcrops. As such, a more complete understanding of rough surface reflectance properties is important for interpretation of terrestrial and extraterrestrial remotely sensed spectra for regions of exposed bedrock. (2) Theoretical models of reflectance from surfaces, necessary for accurate interpretation of mixtures, are dependent upon understanding the nature of interactions of light with rock and mineral surfaces. The lack of absorption features in spectra from many rough slab surfaces suggests a preponderance of the first surface component in the reflected light. Laboratory examination of artificially prepared rock surfaces should help constrain the type of interactions that occur at the surface of large particles and rocks. (3) A substantial body of material, in particular many lunar rock and meteorite samples, are available for spectroscopic examination only in the form of rough-cut slabs. A more detailed understanding of how light is reflected from such surfaces, and thus the amount and form of petrographic information contained in such samples, is vital to their successful interpretation.

BACKGROUND

For this discussion of reflectance, a distinction is made between the orientation of two types of surfaces that may be encountered. The *bulk surface* is the macroscopic aspect of the sample that is under examination. The bulk surface may be a particulate sample or a single slab. The *optical surface* is the microscopic aspect of the bulk sample with which light actually interacts. The optical surface of a particulate samples is, for the wavelengths for which we are concerned, a small optically smooth area of an individual particle. The optical surface of a slab sample is some small portion of the bulk surface. In the case of a rough slab sample, the optical surface may or may not have the same orientation as the bulk sample.

The angle that the incident rays make with the normal to the optical surface is the incident angle, i, while the angle between the reflected ray and the normal to the optical surface is called the emergent angle, e. The angle between the light source and the normal to the bulk surface is denoted here as L, while the angle between the detector and the normal to the bulk surface is denoted as D. If the optical surface

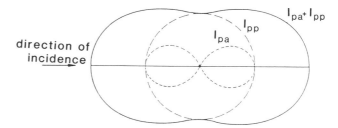

Fig. 3. Polar plot (phase function) of the intensity of Rayleigh scattered light. The incident direction is from the left. Dashed lines indicate the relative intensities of the pa and pp scattered components. The total magnitude of scattered radiation in any direction, indicated by the solid line, is relative to the magnitude at e = 0. (After *Van de Hulst*, 1980).

has the same orientation as the bulk surface, i = L and e = D. When L, D, and the bulk surface normal are coplanar and L and D are opposite sides of the normal, L + D is called the phase angle and is denoted as g.

Any arbitrary light ray can be thought of as composed of two orthogonally polarized rays interfering along their propagation path. When incident upon a bulk surface, it is most convenient to choose these polarization directions as parallel and perpendicular to the plane containing the incident and emergent rays. These two components of the incident ray will be called the parallel and perpendicular rays (pa and pp, respectively). Similarly, emergent light is called parallel or perpendicular depending upon its orientation relative to the plane containing the incident and emergent rays.

Unpolarized light undergoing reflection may become partially polarized (see below) and, inversely, some types of interactions are capable of depolarizing plane polarized incident radiation. Depolarization occurs when a polarized incident ray, such as pa or pp only, results in an emergent ray that is unpolarized (containing equal pa and pp intensities). An emergent component with a polarization orientated perpendicular to the polarization direction of incident light will be referred to as cross-polarized reflected light. The intensity of cross-polarized reflected light is a good measure of depolarization on or within the sample. For fully depolarized reflectance, the cross-polarized fraction is 0.50.

Two types of fundamental interactions of light with optical surfaces will be considered here. These are Rayleigh scattering from edges and corners small in size compared to the wavelength of the incident light, and Fresnel reflection, which is essentially constructive interference of radiation resonantly scattered from an optically flat surface along the specular direction. Laboratory examination of artificially prepared rock surfaces should help to constrain the strength of these interactions.

The intensity of Rayleigh scattered light is a function of both the wavelength of incident radiation and viewing geometry (e.g., *Bohren and Huffman*, 1983). For a specific viewing geometry, the intensity of scattered radiation increases as $(1/\lambda)^4$. For monochromatic unpolarized incident radiation, the intensities of the two components of emergent light (I_{pa} and I_{pp}) are separate functions of the intensity of the incident light (I_o) and the phase angle (g)

$$I_{pp} \propto I_o \qquad (1)$$
$$I_{pa} \propto I_o \cos^2(g)$$

The intensities of these two components of scattered light, as a function of viewing geometry, are shown in Fig. 2. The intensity of scattered light at any phase angle and a specific wavelength is the sum of these two components

$$I(g) \propto I_o(1 + \cos^2(g)) \qquad (2)$$

A phase function for Rayleigh scattered light from a single particle is shown in Fig. 3. As the phase angle approaches 90°, the pa component of scattered light approaches zero, causing the slightly pinched shape of the phase function. The differing intensities of the two components of Rayleigh scattered light indicate that the polarization of scattered radiation is a function of viewing geometry.

The polarization P of Rayleigh scattered light as a function of phase angle, assuming unpolarized incident light, is given by the expression (e.g., *Bohren and Huffman*, 1983)

$$P = \frac{1 - \cos^2(g)}{1 + \cos^2(g)} \qquad (3)$$

Note that the resulting polarization of Rayleigh scattering is independent of the size of the scatterer (although the amount of scattered radiation is clearly size dependent). A graphic representation of this polarization is shown as a dashed line in Fig. 4. The peak polarization of 1.0 results only when each emergent ray has been scattered once. When this occurs, no cross-polarized light is produced. If random multiple interactions have occurred, then cross-polarized light is produced and the peak polarization is reduced to a value less than 1.0 (*Van de Hulst*, 1980). However, the phase angle at which peak polarization is attained should be invariant of wavelength and number of interactions. This peak polarization from Rayleigh scattering always occurs when the phase angle (L + D) equals 90° (Fig. 4).

Fresnel reflection (specular reflection) occurs when the target surface is smooth compared to the wavelength of the incident light. A perfect specularly reflected ray is confined to the direction for which e, the emergent angle, equals i, the incident angle. As in the case of Rayleigh scattering, the intensities of the pa and pp components of the reflected ray are dependent upon the incident angle. The intensities of these components for Fresnel reflection are also dependent upon the complex refractive index of the target material, N = n

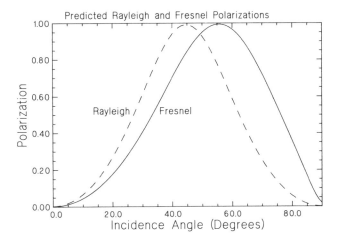

Fig. 4. Predicted polarization of Rayleigh scattered and Fresnel reflected light as a function of incidence angle, i. Fresnel polarization was computed with n (real refractive index) = 1.45 and k (absorption coefficient) = 0.0001. Rayleigh polarization is independent of wavelength.

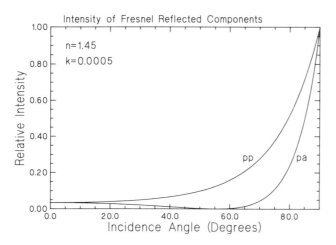

Fig. 5. Relative intensities of the parallel and perpendicular (I_{pa} and I_{pp}) components of Fresnel reflected light as a function of incidence angle, i.

$+ jk$, where n is the real refractive index and k is the absorption coefficient (and j is $\sqrt{-1}$.) In general, n and k are also dependent upon the wavelength of the incident light. For unpolarized incident light, the vector expressions of the two components of Fresnel reflected light from an optically smooth surface are

$$I_{pp} = I_o/2 \left[\frac{\cos i - \sqrt{N^2 - \sin^2 i}}{\cos i + \sqrt{N^2 - \sin^2 i}}\right]^2$$

$$I_{pa} = I_o/2 \left[\frac{N^2 \cos i - \sqrt{N^2 - \sin^2 i}}{N^2 \cos i + \sqrt{N^2 - \sin^2 i}}\right]^2 \qquad (4)$$

Typical Fresnel reflectance as a function of incidence angle is shown in Fig. 5. Again, since both of these components are a function of i, the polarization of a specularly reflected ray is also a function of i. The polarization state of specularly reflected light for unpolarized incident light is

$$P = \frac{I_{pp} - I_{pa}}{I_{pp} + I_{pa}} \qquad (5)$$

A graphic representation of Fresnel polarization from an optically smooth surface, as a function of viewing geometry, is also included in Fig. 4. The geometry at which maximum polarization is achieved for Fresnel reflection is not invariant but depends on n and k and thus on the wavelength of the incident light. Increasing n increases the incident angle at which maximum polarization occurs. For most silicates k is very small and changing k by a factor of 100 has no significant effect on polarization for the range of k's encountered in this study (~0.01-0.0001).

In the case of an optically rough surface, which may be modeled as an area of randomly oriented optically smooth facets (optical surfaces), two types of interactions involving specular Fresnel reflection from the bulk surface are considered: (1) Reflected light has undergone a single specular reflection, but the random orientation of facets produces a geometrically dispersed phase function. For this case, radiation would be reflected in all directions, but a single orientation of the optical surfaces is responsible for all light observed at a specific viewing geometry. Since all light at a specific viewing geometry has thus undergone the same interaction, the polarization state can be described by Fresnel polarization with the effective angle of incidence $i' = g/2$. Cross-polarized light is not produced. (2) Reflected light has undergone multiple specular reflections from randomly oriented surfaces. Again, emergent light will be geometrically dispersed in all directions, but the orientation of the facets responsible for the light observed at a particular viewing geometry is not constant. Various random orientations of facet pairs (or any number greater than two) are responsible for the observed light. This variability of facet orientations allows cross-polarized light to be produced (e.g., *Wolff*, 1975). Light viewed at any geometry is thus largely unpolarized because of the intensity of cross-polarized light.

APPROACH

A series of experiments were designed to determine how light interacts with rock and mineral surfaces and the role of Rayleigh and Fresnel reflections in those interactions. Two different sample materials were chosen. The first, obsidian, was chosen for its optical homogeneity (lack of internal scatterers). By eliminating internally scattered light, the light that has interacted only with the first surface of the sample may be more easily studied. The chemical composition of the obsidian used is given in Table 1. Common geologic materials, however, are not optically homogeneous. Most rock samples likely to be encountered are both polycrystalline and polymineralic. The second sample, a fine-grained basalt, was therefore chosen to allow the effects of internal crystal boundaries on the properties of reflected light to be studied. This basalt is phenocrystic, with two generations of augite phenocrysts, averaging about 0.25 mm in diameter, comprising about 30% of the mode. The groundmass is composed primarily of plagioclase microlites, approximately 0.025 mm in size, and numerous Fe-Ti oxide grains (about 15% of the mode). Chemical analyses

TABLE 1. Electron microprobe analysis of obsidian.

Oxide	Wt %
SiO_2	76.84(38)
TiO_2	0.10(4)
Al_2O_3	12.46(16)
FeO	0.94(16)
MgO	0.02(1)
CaO	0.55(9)
Na_2O	4.30(21)
K_2O	4.71(8)
MnO	0.09(9)

Values are in wt %. Values in parentheses give the measurement error in hundredths of a wt %.

TABLE 2. Electron microprobe analysis of minerals comprising the basalt.

	Fe-Ti Oxides	Clinopyroxene	Plagioclase
SiO_2	—	51.31(1.45)	50.37
TiO_2	20.35(14)	1.32(89)	0.21(3)
Al_2O_3	3.26(5)	4.98(60)	31.31(2)
FeO	68.61(102)	7.41(35)	0.91(11)
MgO	2.74(9)	14.91(212)	0.03(1)
CaO	—	19.79(251)	13.37(22)
Na_2O	—	0.79(17)	3.49(16)
K_2O	—	0.01(2)	0.25(2)
MnO	0.87(6)	0.17(8)	0.03(4)
P_2O_5	—	0.28(9)	0.16(3)
Cr_2O_3	0.25(2)	—	—

Values are in wt %. Values in parentheses give the measurement error in hundredths of a wt %.

of the pyroxene, plagioclase, and magnetite grains are given in Table 2. Reflectance spectra of particulate samples of these two materials are shown in Figs. 1a,b.

The real refractive indices of the two sample materials were determined at 0.6 μm by refractive oils, and at all other wavelengths by computation using a dispersion of -0.0278 μm (*Pollack et al.*, 1973). The absorption coefficients for wavelengths between 0.3 and 1.8 μm were measured for the obsidian using polished plate transmission. The absorption coefficient for the basalt was fixed at approximately ten times that of the obsidian. This is somewhat arbitrary, but fine-grained nature of this sample precludes a straightforward determination of k (since the scattering efficiency is not easily separated from the absorption efficiency).

To prepare the samples for reflectance and polarization measurements, they were first cut into slabs approximately 6 mm thick with a thin section saw, then a controlled surface

Smooth Obsidian

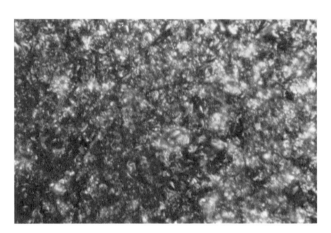

Rough Obsidian

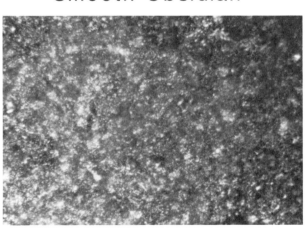

Smooth Basalt

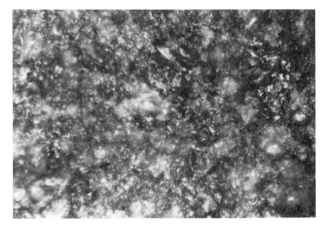

Rough Basalt

Fig. 6a. Microphotographs of saw-cut obsidian surface (upper) and basalt surface (lower) ground with 5 μm Al oxide powder (called the *smooth* obsidian and basalt slabs in the text.) Long dimension of photo ~600 μm.

Fig. 6b. Microphotographs of saw-cut obsidian surface (upper) and basalt surface (lower) ground with 40 μm Al oxide powder (called the *rough* obsidian and basalt slabs in the text.) Long dimension of photo ~600 μm.

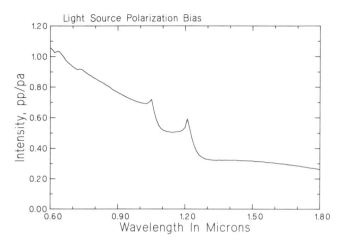

Fig. 7. Polarization bias of RELAB monochrometer between 0.6 and 1.8 μm. This is a measure of the relative intensity of I_{pa} and I_{pp} at the sample surface.

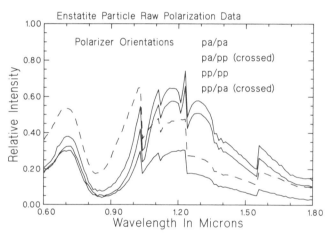

Fig. 8. Relative intensities of the four components of reflected light, required for polarization measurements, from a particulate sample of enstatite (particle size 125-250 μm) measured at g = 90°. Sharp breaks result from instrument gain and lamp changes. Polarizer orientations refer to the sequence (from top to bottom) of intensities shown for wavelengths greater than 1.3 μm.

was produced using Al oxide grit. Two samples were prepared from each material. Slabs produced by grinding with only 40 μm grit will be referred to as the *rough* slabs. Slabs produced by first grinding with 40 μm grit then finishing with 5 μm grit will be referred to as the *smooth* slabs. Microphotographs of the surfaces of the smooth obsidian and basalt slabs are shown in Fig. 6a. Photographs of the rough obsidian and basalt surfaces are shown in Fig. 6b. Microscopic examination of both slab roughnesses reveals that grinding takes place by gouging and chipping on a scale approximately equal to the size of the polishing compound. The scale of surface features appears to be independent of sample material when identically prepared. There is no difference in the microscopic appearance of the two slab materials when grinding with the same grit.

The reflectance and polarization properties of these samples were measured using the RELAB bidirectional reflectance spectrometer (*Pieters*, 1983). Sample reflectance, measured relative to halon and then corrected for halon reflectance (National Bureau of Standards test 232.04/213908), can be measured at phase angles down to 15° with independently variable light source and detector positions. Reflectivities of all four samples were determined at a variety of specular and nonspecular geometries for wavelengths between 0.35 and 1.80 μm.

Polarization determinations were made for the smooth and rough obsidian slabs and for the rough basalt slab at a variety of specular and nonspecular geometries for wavelengths between 0.6 and 1.8 μm. Determination of the polarization

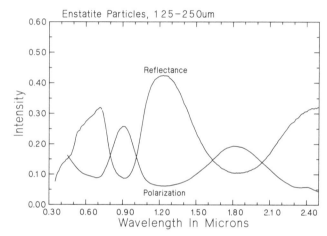

Fig. 9a. Reflectance and polarization of reflected light at g = 90° of particulate enstatite sample. Coincidence of polarization maxima and reflectance minima is due to the higher ratio of surface reflected light relative to the scattered transmitted component in the absorption band.

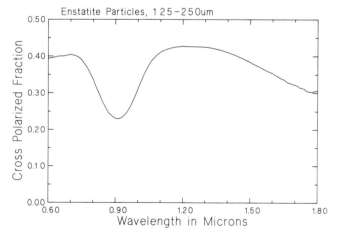

Fig. 9b. Intensity of cross-polarized light relative to total reflected intensity (computed from the four reflective intensities of Fig. 8) for particulate sample of enstatite.

of reflected light requires six separate measurements. Two measurements are required to determine B, the polarization bias of the light source (Fig. 7). The remaining four measurements are sample reflective intensities I, made with polarizing optics in both incident and emergent ray paths (these measurements are not referenced to halon). For each incident polarization direction (pa and pp), the intensities of both emergent components (pa and pp) must be measured. This is done by setting the incident polarizer to either the pa or pp orientation and then measuring the reflected intensities of the sample with the emergent polarizer in both its pa and pp orientations. The measured intensity is a function of the orientation of the polarizer in both the incident and emergent ray paths. For example, $I_{pa,pp}$ was measured with the incident polarizer in the pa orientation and the emergent polarizer in the pp orientation. Examples of these four intensity components for a particulate sample of enstatite at $g = 90°$ are shown in Fig. 8. Most of the structure seen in the component curves is instrumental (e.g., gain and lamp changes). Polarization is calculated by the following equation

$$P = \frac{(I_{pp,pp} + BI_{pa,pp}) - (BI_{pa,pa} + I_{pp,pa})}{(I_{pp,pp} + BI_{pa,pp}) + (BI_{pa,pa} + I_{pp,pa})} \quad (6)$$

The polarization of a sample is computed separately for each wavelength since B and all four I terms are themselves dependent upon wavelength. The polarization and reflectance of light reflected from the enstatite particles at $g = 90°$ is shown in Fig. 9a and the intensity of the cross-polarized fraction of reflected light is shown in Fig. 9b. For this particulate sample, reflected light that is relatively unpolarized also contains a large cross-polarized fraction, implying that the low polarization value is due to the predominance of multiple interactions in the particulate sample.

RESULTS

Spectra from both the basalt and obsidian smooth slabs are very similar. Reflectivities of the smooth obsidian slab at specular geometries (L = D) are shown in Fig. 10a, and the reflectivities of the smooth basalt slab at specular geometries are shown in Fig. 10b. Spectra of both smooth slabs exhibit a marked increase in reflectance toward longer wavelengths. Although off the scale on Fig. 10a, long wavelength reflectance from the smooth obsidian slab increases dramatically, from <50% to >200%, as phase angle increases from 20° to 90°. (Note: since these measurements are relative to halon, >100% reflectance simply means that sample reflectance was greater than that of the halon standard, which approximates a Lambertian surface.) For the smooth obsidian slab, the transition between the low reflectance, short wavelength behavior and

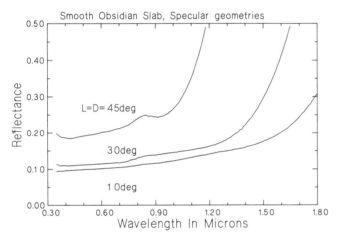

Fig. 10a. Bidirectional reflectance of smooth obsidian slab at three specular geometries (Halon Standard). Peak reflectance at 30° and 45° incidence angles is greater than 100% halon.

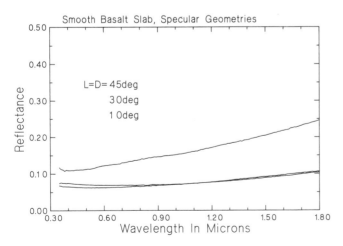

Fig. 10b. Bidirectional reflectance of smooth basalt slab at three specular geometries; angles in degrees (Halon Standard).

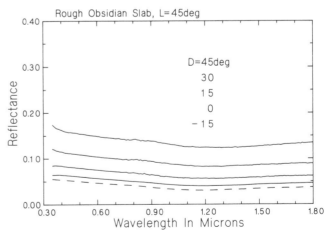

Fig. 11a. Bidirectional reflectance of rough obsidian slab at five geometries, with incident angle (L) = 45°. Emergent angles (D) in degrees (Halon Standard). For clarity, spectra taken at D = -15° are indicated with a dashed line.

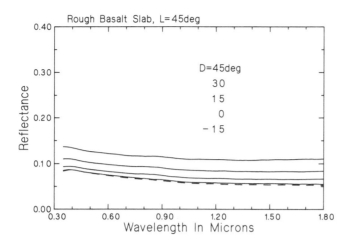

Fig. 11b. Bidirectional reflectance of rough basalt slab at five geometries with incident angle (L) = 45°. Emergent angles (D) in degrees (Halon Standard). For clarity, spectra taken at D = -15° are indicated with a dashed line.

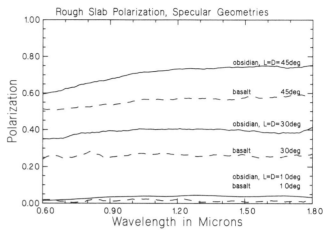

Fig. 14. Polarization spectra at three specular geometries from rough obsidian and basalt slabs between 0.6 and 1.8 μm. For phase angles between 15° and 90°, polarization increases with phase angle and wavelength for both slab materials. For clarity, spectra of basalt slabs are indicated with a dashed line.

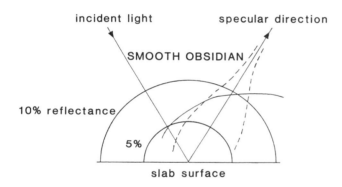

Fig. 12. Generalized phase functions derived from reflectance of smooth obsidian slab with L = 30°. Solid line is for 0.6 μm incident light. Dashed line is for 1.8 μm incident light. Peak reflectance of smooth obsidian slab is 125% halon at D = 30° for 1.8 μm incident light.

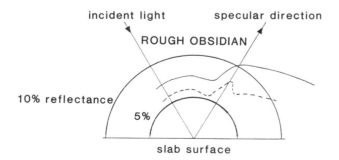

Fig. 13. Generalized phase functions derived from reflectance of rough obsidian slab with L = 30°. Solid line is for 0.6 μm incident light. Dashed line is for 1.8 μm incident light.

the high reflectance, long wavelength behavior migrates to shorter wavelengths as phase angle increases. The behavior of the basalt spectra obtained at specular geometries (Fig. 10b) is similar to those obtained from the smooth obsidian slab, but more subdued. Spectra obtained from these smooth slabs at nonspecular geometries (L≠D) do not exhibit the long wavelength specular behavior seen in the smooth slab spectra.

Spectra of the rough obsidian slab, shown in Fig. 11a, are in general quite similar to spectra of the rough basalt slab, shown in Fig. 11b. These spectra, however, are distinctly different from spectra of the smooth slabs. Rough slab spectra of both materials observed at all geometries are characterized primarily by a slightly blue sloping (decrease in reflectance toward longer wavelengths) background continuum. While the obsidian spectra have a slight indication of an absorption band near ~1.14 μm (similar to that for the spectra of particulate samples shown in Fig. 1a) the basalt slab spectra have no indication of absorption bands even though spectra of the basalt particles have a distinct absorption band near ~1.03 μm (Fig. 1b). The most striking feature of spectra from both rough slabs is the constancy of spectral character for all viewing geometries. As phase angle increases, the entire spectrum increases in brightness relative to the standard halon but the long wavelength specular surge seen in spectra from the smooth slabs is absent.

When the smooth slabs are observed at specular geometries, the reflectance at all wavelengths increases with phase angle. This simple relationship is not found in the reflectance at nonspecular geometries. A graphical representation showing trends of the reflectance phase function obtained from measurements at a single incident angle is shown in Fig. 12. Phase functions with L = 30° for the smooth obsidian slab are shown for both 0.6 and 1.8 μm incident light. The phase function at 0.6 μm is seen to be very dispersed, indicating that incident light is scattered in all directions. The phase function at 1.8 μm is much less uniform with emergent light concentrated along the specular direction. Reflectance measurements were also made at incidence angles of 10° and 45°. The three sets of reflectance measurements (i = 10°, 30°, 45°) show that as the incident angle to the bulk surface, L, increases, the concentration of light along the specular direction becomes more exaggerated for all wavelengths, but increasingly so at longer (1.8 μm) wavelengths. For comparison,

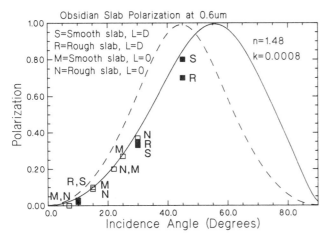

Fig. 15a. Experimentally determined polarization values from both smooth and rough obsidian slabs for incidence angles (L) from 0° to 45° and 0.6 μm incident light. Filled symbols represent data taken at specular geometries. Open symbols represent data taken at nonspecular geometries, for which i' = g/2. Predicted polarizations for Rayleigh scattering and Fresnel reflection are also shown. The Fresnel curve was computed with n = 1.48 and k = 0.0008.

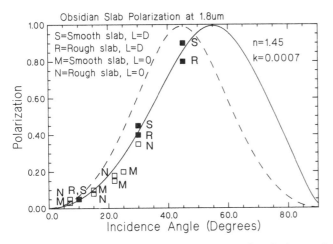

Fig. 15b. Experimentally determined polarizations from both smooth and rough obsidian slabs for incidence angles (L) from 0° to 45° and 1.8 μm incident light. Filled symbols represent data taken at specular geometries. Open symbols represent data taken at nonspecular geometries for which i' = g/2. Also shown are predicted polarizations for Rayleigh scattering and Fresnel reflection. The Fresnel curve was computed with n = 1.45 and k = 0.0007.

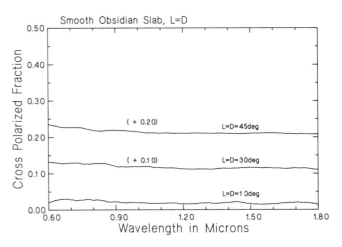

Fig. 16a. Intensity of cross-polarized light ($I_{pa,pp} + I_{pp,pa}$) relative to total reflected intensity from the rough obsidian slab. Data obtained at three specular phase angles are shown. For clarity, data for L = 30° and L = 45° are displaced upward 10% and 20%, respectively.

phase functions with L = 30° for the rough obsidian slab are shown in Fig. 13 for 0.6 and 1.8 μm incident light. The rough obsidian slab exhibits very dispersed reflectance properties, which are nearly identical at long and short wavelengths, and exhibit only a very subdued specular surge at any wavelength.

Polarization spectra obtained from both the obsidian and basalt slabs are shown in Fig. 14. Polarization increases with phase angle and also increases slightly with wavelength. Like the reflectance spectra, these polarization spectra from the obsidian and basalt slabs do not show any indications of the absorption bands seen in the particulate spectra. Polarization values at wavelengths of 0.6 and 1.8 μm were chosen for detailed study because they represent the most extreme values in the range of wavelengths considered.

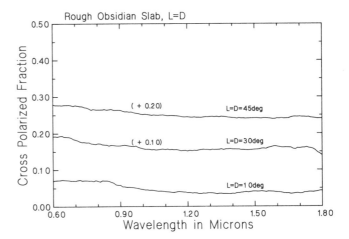

Fig. 16b. Intensity of cross-polarized light ($I_{pa,pp} + I_{pp,pa}$) relative to total reflected intensity from the rough obsidian slab. Data obtained at three specular phase angles are shown. For clarity, data for L = 30° and L = 45° are displaced upward 10% and 20%, respectively.

Shown in Figs. 15a and 15b are polarization values measured from both obsidian slabs for 0.6 μm and 1.8 μm incident light, respectively. Polarization data at incidence angles of 10°, 30°, and 45° were measured at specular geometries (L = D). Data measured with the light source normal to the slab surface (L = O) are also shown in these figures by letting i' = g/2 (e.g., data for D = 15°, 30°, 45°, and 50° are plotted as i' = 7.5°, 15°, 22.5°, and 25°, respectively). Also shown in Figs. 15a and 15b are the polarizations predicted for Fresnel and Rayleigh interactions. The Rayleigh curves are identical at both wavelengths since polarization for this type of interaction is not a function of wavelength. The Fresnel curves, however, were computed separately with equations (4) and (5) using the optical constants computed for obsidian at each wavelength. In general, the measured polarization values follow the theoretical polarization curve for Fresnel reflection and not Rayleigh scattering. Polarization values measured from both smooth and rough obsidian slabs at all geometries approximate the Fresnel polarization curve for each wavelength shown.

There are, however, important small variations of the measured polarization values from the predicted Fresnel curve. At 0.6 μm there is a tendency for the measured polarizations from the obsidian slabs to be less than the predicted values. Also, at both 0.6 and 1.8 μm, the light reflected from the smooth slab often has a notable greater polarization than that reflected from the rough slab at a given phase angle.

The degree to which depolarization of incident radiation occurs on the slab surfaces can be evaluated by examining the magnitude of cross-polarized components. The intensity of cross-polarized light ($I_{pa,pp} + I_{pp,pa}$) relative to total reflected intensity is shown for the smooth obsidian slab in Fig. 16a and for the rough obsidian slab in Fig. 16b. For both smooth and rough slabs, the relative intensity of cross-polarized light is very low but increases with decreasing wavelength for a

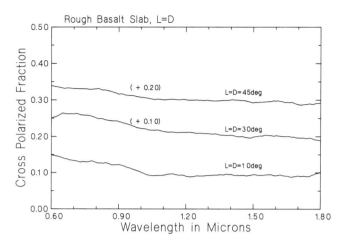

Fig. 18. Intensity of cross-polarized light ($I_{pa,pp} + I_{pp,pa}$) relative to total reflected intensity from the rough basalt slab. Data obtained at three specular geometries are shown. For clarity, data for L = 30° and L = 45° are displaced upward 10% and 20%, respectively. Intensity of cross-polarized light from the rough basalt slab is greater than the cross-polarized light from the rough obsidian slab at any wavelength and phase angle.

given phase angle. Also, for a given phase angle, the relative intensity of cross-polarized light is always greater from the rough obsidian slab than the smooth slab.

The polarization of light measured from the rough basalt slab at 0.6 μm is shown in Fig. 17 as a function of incidence angle. Also shown are predicted Rayleigh and Fresnel polarization curves. The Fresnel curve was computed using equations (4) and (5) and the optical constants derived for basalt. The polarizations of the basalt slab data are generally lower than either of the predicted curves. The intensity of the cross-polarized components of light from the rough basalt slab is shown in Fig. 18. As in the case of the obsidian, the intensity of cross-polarized light from the rough basalt slab increases with decreasing wavelength. The cross-polarized component of reflected light from the basalt, however, is greater than that from either of the obsidian slabs at the same phase angle.

DISCUSSION

Clearly a variety of reflected, transmitted, and scattered components are involved in the interaction of radiation with the boundary of a surface. By measuring the reflectance and polarization properties as a function of wavelength for controlled surface roughness, a few of these interactions have been reasonably well isolated.

At specular geometry (D = L), the amount of cross-polarized light returned from the obsidian slabs is quite low compared to that observed for particulate samples (Figs. 9 and 16). The low intensity of cross-polarized light from the slabs indicates that multiple Fresnel and Rayleigh interactions (which would depolarize radiation) do not contribute significantly to reflected intensity, and single interactions are dominant for such surfaces.

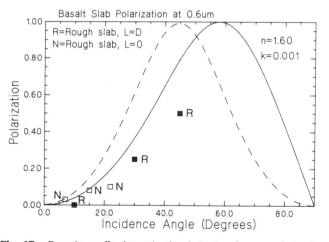

Fig. 17. Experimentally determined polarization from rough basalt slab for 0.6 μm incident light and incidence angles between 0° and 45°. Data were taken at both specular and nonspecular geometries, and are plotted as in Figs. 15a,b. Also shown are predicted polarizations for both Rayleigh scattering and Fresnel reflection. The Fresnel curve was computed with n = 1.6 and k = 0.001.

Two additional lines of evidence indicate that light reflected from the smooth obsidian slab has undergone only a single Fresnel reflection. First, the phase function from the smooth slab at 1.8 μm and large incident angles is almost completely concentrated along the specular direction, as predicted by Fresnel's equations (Fig. 12). Second, the polarization of 1.8 μm light reflected from the smooth obsidian slab is very close to that predicted for a single Fresnel reflection (Fig. 15). Furthermore, since all of the polarization measurements from both the smooth and rough obsidian slabs at a variety of viewing geometries (specular and nonspecular) fall along the same Fresnel polarization curve for a given wavelength, the predominant type of interaction of light with both obsidian slabs is inferred to be a single Fresnel reflection. For the relatively simple obsidian surfaces with well-defined surface roughness studied here, Rayleigh interactions are insignificant.

Fresnel reflection, however, is not simple. The relationship between incident wavelength and slab surface roughness determines the shape of the reflectance phase function. For long wavelengths and very smooth surfaces, slab surface roughness is effectively invisible to the incident light. The bulk surface acts as a mirror and serves to concentrate the reflected light along a specular direction. As surface roughness increases relative to the wavelength of the incident light, incident light begins to interact with individual surface units, or facets rather than the bulk surface. The random orientation of these facets produces a geometrically dispersed phase function (Fig. 13). In both the specular and the dispersed reflection, however, the polarization of reflected light indicates that the dominant interaction type is single Fresnel reflection (Fig. 15).

The intensity of cross-polarized light measured from both obsidian slabs is small (<7%), but is greatest at short wavelengths with large surface roughnesses (Figs. 16a,b). Since multiple Fresnel reflections would contribute cross-polarized light to slab reflectance and reduce the resulting polarization of the light, the observed variation in cross-polarized light from the smooth obsidian slab is likely to be caused by a greater number of interactions of light with surface facets at shorter wavelengths, resulting in a greater number of multiple Fresnel interactions. The phase functions of light reflected from the rough obsidian slab show no indication of specular reflection from the bulk surface. The reflectance from the rough obsidian slab thus appears to be controlled by facet interactions at all wavelengths between 0.6 and 1.8 μm. Compared to that for the smooth obisidan slab, the component of light reflected from the rough slab that has undergone the multiple specular reflections is somewhat larger, causing the observed larger percentage of cross-polarized light.

For both of the obsidian samples, it is interesting to note that although there is a general decrease of polarization with decreasing wavelength, no "negative" polarization (orthogonal to the plane of scattering) was observed at these wavelengths for the measurements g $\leq$ 20° (Figs. 14 and 15). This may indicate that the negative branch of polarization commonly observed in the visible for particulate surfaces (e.g., *Dollfus*, 1985) arises from inter- and/or intraparticle scattering rather than interfacet scattering (Hapke, personal communication, 1987) and merits further investigation.

On a microscopic scale, the facets on both the obsidian and basalt slabs are essentially identical. Interactions of light with the optical surfaces should thus be the same for both materials. Any deviation of the polarization of light reflected from the basalt slab from the predicted Fresnel values must be due to light scattered from the near surface or interior of the slab. If the intensity of internally scattered light from a slab were close to zero, then the measured polarizations should fall along the predicted Fresnel polarization curve. This is observed for both obsidian slabs. What is observed for the rough basalt slab, however, is that the measured polarizations, shown in Fig. 17, are significantly less than the predicted values. Furthermore, the intensity of the cross-polarized light from the rough basalt slab is approximately twice that observed from the rough obsidian slab (Fig. 18). This additional 5-7% depolarization, and the resulting deviation of measured polarization from the predicted values, is inferred to result from light scattered from the interior of the sample, probably originating at near-surface crystal boundaries. This internally scattered light must not penetrate the surface significantly since the rough basalt slab reflectance spectra, which contain up to 14% internally scattered light, do not exhibit any mineral absorption bands.

Interactions of light with the surfaces of these specially prepared samples of obsidian and basalt are representative of most silicates of moderate and low albedo, but this information should be used cautiously when interpreting high albedo materials. The obsidian used in this study was relatively free of internal scatterers, but naturally occurring transparent materials generally have an abundance of internal interfaces. Since an internal (body) component often dominates the interaction for high-albedo materials, the first surface interactions discussed here may be too small to be significant for bright surfaces.

SUMMARY

These initial measurements of the wavelength dependence of polarization and reflectance in the near-infrared provide a few constraints for the complex interaction of radiation with a dielectric surface. The measured reflectivities and polarization of light reflected from the obsidian slabs indicates that light interacts with the first surface primarily through Fresnel single reflections. Multiple Fresnel reflections are less important but increase with the scale of surface roughness relative to the incident wavelength. Unpolarized internally scattered light is shown to be only a minor component of reflected light for the obsidian slab, but increases significantly for the polycrystalline basalt slab.

The amount of mineralogical information contained in slab reflectance spectra depends on the optical and physical properties of the individual samples. The degree to which radiation can penetrate a surface and then scatter back out, an essential criterion for mineralogic determination using reflectance spectra, is dependent not only upon the composition of the material, but also on its physical condition (e.g., sample grain size and surface roughness). The experiments discussed here indicate that first surface interactions of light are likely to be similar for all silicate minerals. To fully understand spectra obtained from slab surfaces and rock outcrops, the internally scattered component of reflectance from polycrystalline surfaces must be better constrained.

Acknowledgments. This work has been supported by NASA grant NAGW-28. The RELAB reflectance laboratory is supported by NASA grant NAGW-748. Many thanks to S. Pratt, without whom this work would not have been possible. Helpful review and comments by B. Hapke and P. Helfenstein are much appreciated.

REFERENCES

Adams J. B. (1974) Visible and near infrared diffuse reflectance spectra of pyroxenes as applied to remote sensing of solid objects in the solar system. *J. Geophys. Res.*, 79, 4829-4836.

Adams J. B. (1975) Interpretation of visible and near-infrared diffuse reflectance spectra of pyroxenes and other rock forming minerals. In *Infrared and Raman Spectroscopy of Lunar and Terrestrial Minerals* (C. Karr, Jr., ed.), pp. 91-116. Academic, New York.

Adams J. B. and Filice A. L. (1967) Spectral reflectance 0.4 to 2.0 microns of silicate rock powders. *J. Geophys. Res.*, 72, 5705-5715.

Bohren C. F. and Huffman D. R. (1983) *Absorption and Scattering of Light by Small Particles*. Wiley and Sons, New York. 530 pp.

Burns R. G. (1970) *Mineralogical Application of Crystal Field Theory*. Cambridge University, London.

Dollfus A. (1985) Photopolarimetric sensing of planetary surfaces. *Adv. Space Res.*, 8, 47-58.

Hazen R. M., Bell P. M., and Mao H. K. (1978) Effects of compositional variation on absorption spectra of lunar pyroxenes. *Proc. Lunar Planet. Sci. Conf. 9th*, 2919-2934.

Hunt G. R. (1977) Spectral signatures of particulate minerals in the visible and near-infrared. *Geophysics*, 42, 501-513.

Hunt G. R. and Salisbury J. W. (1970) Visible and near-infrared spectra of minerals and rocks 1, silicate minerals. *Mod. Geol*, 1, 283-300.

Johnson P. E., Smith M. O., Taylor-George S., and Adams J. B. (1983) A semiempirical method for analysis of the reflectance spectra of binary mineral mixtures. *J. Geophys. Res.*, 88, 3557-3561.

Mustard J. M. and Pieters C. M. (1987) Quantitative abundance estimates from bidirectional reflection measurements. *Proc. Lunar Planet. Sci. Conf. 17th*, in *J. Geophys. Res.*, 92, E617-E626.

Pieters C. M. (1974) Polarization in a mineral absorption band. In *Planets, Stars and Nebulae Studied with Photopolarimetry* (T. Gehrels, ed.), pp. 405-418. Univ. of Arizona Press, Tucson.

Pieters C. M. (1983) Strength of mineal absorption features in the transmitted component of near infrared reflected light: First results from RELAB. *J. Geophys. Res.*, 88, 9534-9544.

Pollack J. B., Toon O. B., and Khare B. N. (1973) Optical properties of some terrestrial rocks and glasses. *Icarus*, 19, 372-389.

Van de Hulst H. C. (1980) *Multiple Light Scattering*. Academic, New York. 739 pp.

Wolff M. (1975) Polarization of light reflected from rough planetary surfaces. *App. Optics*, 14, 1395-1405.

Characteristics and Origin of Greenland Fe/Ni Cosmic Grains

E. Robin
Centre des Faibles Radioactivités, Laboratoire Mixte CEA/CNRS, 91190, Gif sur Yvette, and Laboratoire René Bernas, 91406 Orsay, France

C. Jéhanno
Centre des Faibles Radioactivités, Laboratoire Mixte CEA/CNRS, 91190, Gif sur Yvette, France

M. Maurette
Laboratoire René Bernas, 91406 Orsay, France

Five types of well-preserved Fe/Ni cosmic dust grains have been found in sediment samples collected on the west Greenland ice cap. The chemical analysis of major and trace elements in individual grains was performed using the X-ray spectrometer of a scanning electron microscope and instrumental neutron activation analysis, following a procedure already applied to deep-sea cosmic spherules (*Bonté et al.*, 1987a). In addition to Fe/Ni oxidized grains similar to those previously found in deep-sea sediments (*Brownlee*, 1981), we discovered three new families of grains that have never been reported before, including a new type of oxidized grains (sometimes containing sulfur) and metallic objects appearing either under the form of spherules or unmelted fragments. The characteristics of Greenland Fe/Ni grains are presented and discussed. We conclude that several distinct types of micrometeorites provide the different families: (1) iron sulfide grains as troilite, (2) more or less pure nickel-iron alloys, and (3) stony-iron micrometeorites.

INTRODUCTION

Cosmic grains have been extracted in large numbers from both deep-sea sediments (*Brownlee*, 1981) and Greenland deposits (*Maurette et al.*, 1986). They belong to two families: stony spherules with nearly chondritic composition and spherules of Ni-rich iron oxides (Fe/Ni). The deep-sea collection was previously reported as comprising equal proportions of the two families (*Blanchard et al.*, 1980; *Murrell et al.*, 1980). This is mainly due to the rapid weathering of the chondritic grains after deposition (*Kyte*, 1983). In rather young Greenland sediments, alteration is weak and cosmic grains abundant, so we can expect that all varieties of cosmic material, magnetic or not, melted or unmelted, metallic or oxidized, would be represented without any selection effect. In a complete study of the Greenland cosmic dust population we found that, in the 50-μm size range, the Fe/Ni grains represent only 2-3% of the total infalling cosmic population in this size range, whereas chondritic and glassy spherules represent 70% and unmelted fragments 25% (*Robin et al.*, 1987). This contrasts with the nearly 100% Fe/Ni spherules found by *Kyte* (1983) in old Pacific sediments. In spite of its scarcity, the freshly deposited population of Greenland Fe/Ni grains offers a great diversity. We have identified five families, bearing half of the total iridium brought to the Earth by the cosmic dust grains. Although iridium is the best extraterrestrial indicator, other elements such as chromium show great variations (Cr contents from <40 to 10,000 ppm) and are probably the key for a better understanding of the nature of the incident Fe/Ni micrometeoroid flux entering the atmosphere. In the following, we present the main characteristics and try to explain the origin of the five families of Fe/Ni grains that have been identified so far.

EXPERIMENTAL METHODS

The wet Greenland sediments were disaggregated with a nylon brush on a stainless steel sieve that retains the >50-μm-size mineral fraction. The Fe/Ni cosmic grains were optically sorted from the magnetic fraction. Among them we finally selected 18 grains using a SEM/EDS assembly. These grains, enclosed in an aluminum foil, were irradiated for 24 hours in the 2.3×10^{14} neutrons/cm^2/s flux of OSIRIS reactor at Pierre SUE Laboratory (Saclay) together with a fragment of Negrillos iron meteorite and USGS rock MAG 1 as standards. After a four-day cooling each grain was embedded separately in epoxy resin and we measured the gamma-rays of ^{192}Ir, ^{51}Cr, ^{59}Fe, ^{58}Co, and ^{60}Co with a HP-Ge detector (FWHM: 1.7 keV at 1332 keV to determine the weights of Ir, Cr, Fe, Ni, and Co respectively, with an error less than 5%. Then a polished section was easily obtained to be studied with a JEOL JSM-840 scanning electron microscope equipped with two solid state detectors for Z-contrast image. The major element composition (Fe, Ni, Mg, Al, Si, Ca in wt %) was determined with a LINK energy dispersive X-ray spectrometer. We normalized the INAA data to the EDXA results using Fe from both techniques and then we deduced the trace element concentrations. We used the "polishing step method" as described by *Bonté et al.* (1987a) to determine the depth profile of Ir in order to infer whether this element is homogeneously distributed or concentrated in a nugget (Fig. 1).

RESULTS

The major and trace element contents have been listed in Table 1. In addition, we have reported the Ir/Cr and Ir/Ni

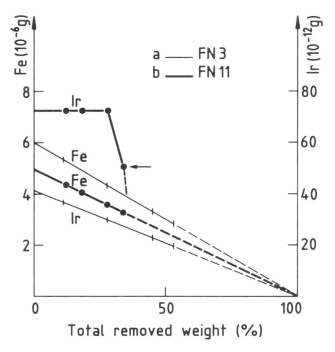

Fig. 1. Step-by-step polishing examples: (a) FN 3: distributed iridium; (b) FN 11: iridium locked in a nugget. Arrow indicates the abrased nugget.

diagrams in Fig. 2. The 18 selected Fe/Ni grains can be divided into 5 distinct species illustrated in Figs. 3 and 4.

Glassy type spheres [labelled G-type by *Brownlee* (1981)] represent 10% of the Fe/Ni particles and are formed with a dominant phase of dendritic magnetite and a minor phase of interstitial chondritic glass (Figs. 3a,b). They contain a Fe/Ni metal bead in which all the Ir is concentrated since polishing steps let Ir weight remain constant as long as the metal bead is not abrased. Chromium is distributed in the magnetite/silicate portion and is particularly abundant in these grains.

Oxidized Fe/Ni spherules (60% of the Fe/Ni grains), labelled as "two-oxide" grains, are composed with two mixed phases of magnetite and wustite (Figs. 3c,d). Their Ir and Ni contents do not show great variations, but from their large variations in Cr we can separate the four low-Cr spheres on the left of the diagram (Fig. 2a) from the six high-Cr spheres on the right. In agreement with earlier work of *Brownlee et al.* (1984) we found that each sphere contains either a tiny nugget or a Ni-core, including all the platinum group metals. Micron-sized nuggets of platinum are sometimes found in the magnetite phase along with the major nugget (*Bonté et al.*, 1987a).

The following types of grains have never been identified in deep-sea sediments:

Oxidized Fe/Ni spherules (15% of the Fe/Ni grains) labelled as "single-oxide" spherules are formed with a single oxide phase determined as Fe_2O_3, probably hematite, from EDS analysis (Figs. 3e,f). They contain low Ni, Co, and Ir and high Cr concentrations (open stars, Fig. 2). One of them includes abundant tiny inclusions of nickel-rich iron sulfides distributed in its mass (Fig. 3f). On the edge of these grains we observed small precipitates indicating that a second oxidation took place. Moreover, we established that, unlike the two-oxide spherules described above, iridium is uniformly dispersed and therefore no nugget was found in these *oxidized* spherules. This means that the oxidation process does not necessarily involve a platinoid nugget formation.

Metallic Fe/Ni spherules (5% of the Fe/Ni grains) can be directly collected in the sediment and contain tiny inclusions of oxidized iron (Figs. 4a,b). These grains have high Ni and very low Cr contents while the Ir and Co contents are the same as in the two-oxide spherules. They can also be extracted from chondritic spherules (Fig. 4a), in which case they contain

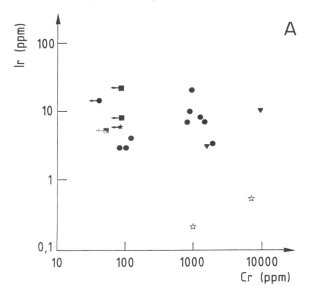

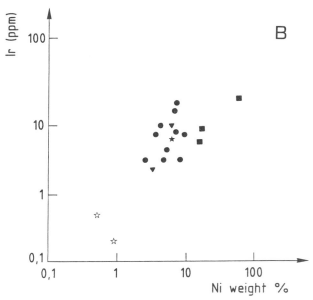

Fig. 2. (a) Ir/Cr diagram; (b) Ir/Ni diagram. Filled triangle = G-type spheres; filled circle = two-oxide spherules; open star = single-oxide spherules; filled box = metallic spherules; filled star = unmelted fragment.

TABLE 1. Greenland Fe/Ni Grain Data.

Type	Grain	X ray analysis			I.N.A.A.				Iridium carrier
		Fe %	Ni %	S %**	Ni %	Co ppm	Ir ppm	Cr ppm	
G-type spheres	FN 25*	53.7	0.2¶		8.3¶	4500	3.1	1600	metal bead
	FN 9†	54.6	2.5		6.2	5100	9.5	9500	metal bead
Two-oxide grains									
A group	FN 6	66.0	6.6		6.6	3200	15.8	<40	nugget
	FN 24	65.0	4.8		4.7	3700	3.0	80	nugget
	FN 2	67.0	2.8		2.5	1800	3.0	100	nugget
	FN 8	65.0	5.6		5.4	3900	4.3	120	nugget
B group	FN 1	67.0	3.5		3.6	1100	6.9	820	nugget
	FN 11	66.0	3.9		4.3	1800	9.8	900	nugget
	FN 16	64.0	6.3		7.3	2600	19.0	980	nugget
	FN 13	68.0	1.6		7.2	3000	8.0	1200	Ni-core
	FN 17	64.0	5.8		9.2	3200	6.8	1350	nugget or Ni-core
	FN 10	64.0	2.7		3.2	2900	3.4	1800	nugget
Single-oxide grains	FN 23	69.0	1.1		0.9	1300	0.2	990	whole grain
	FN 14	68.5	0.8	0.2	0.5	900	0.5	6500	whole grain
	FN 14 inclusions	40.0	55.4	2.0	N.D	N.D	N.D	N.D	
Metallic spherules	FN 3	81	14.7		15.7	3200	5.6	<50	whole grain
	FN 5	82	17.0		16.5	2850	8.3	<100	whole grain
	Fe/Ni-core‡	45	55.0	0.2	55.0	5000	23	<100	whole grain
	Fe/Ni-core inclusions	50	33.0	9.0	N.D	N.D	N.D	N.D	
Unmelted fragment§	Fe/Ni	94	6.0		6.2	4900	6.0	<100	whole grain

N.D. = not determined.
*FN 25: Mg %: 4.3; Al %: 0.6; Si %: 6.4; Ca %: 0.4; Sc ppm: 1.6
†FN 9: Mg %: 3.6; Al %: 0.4; Si %: 5.0; Ca %: 0.4; Sc ppm: 2.1
‡Ga ppm: 10.
§Ga ppm: 14; Ge ppm: 260.
¶Very different Ni contents from both techniques indicate a metal bead.
**No value when < 0.1%.

iron-sulfide inclusions; when these chondritic and metallic components are analysed separately, all the iridium is found in the metal bead (Fig. 4b), leaving the chondritic part completely depleted in this element. The Ir and Ni contents are much more important than in the metal grains directly extracted from the sediment; these grains probably have a different origin. In both types, chromium is strongly depleted compared to the oxidized grains. From the iridium depth profile (FN3, Fig. 1), we deduce that this element is uniformly distributed, indicating the absence of any platinoid nugget.

Unmelted metallic fragments (10% of the Fe/Ni grains) show irregular shapes and contain volatile elements, including Ga and Ge (Figs. 4c,d). The Co, Ni, and Ir contents are comparable to those encountered in the two-oxide grains, while the Cr content is very low (Fig. 2). Iridium is uniformly dispersed in the fragment.

In spite of the very limited statistics, the following general inferences can be made about the distribution of Ir, Cr, Ni, and Co in the grains:

Iridium. Iridium is uniformly distributed in the metallic and single-oxide grains while it is locked in a metallic Fe/Ni core in G-type spheres, and occurs either in a Ni-core or more often in a platinoid nugget in the two-oxide grains (Fig. 1.). The iridium content generally exceeds 3 ppm (with an average value of 7 ppm) except in the single-oxide grains, where values are less than 0.5 ppm. However, Ir concentrations higher than 0.1 ppm can be encountered only in extraterrestrial matter.

Chromium. Compared to iron meteorites, the oxidized grains are often enriched in chromium, ranging from <40 ppm to 10,000 ppm, with the greatest values observed in the G-type and the single-oxide grains. In contrast, Cr is below our detection limit in all metallic grains.

Nickel. Low nickel contents are measured in the single-oxide grains with values less than 1%. The concentration of this element in the two-oxide grains and the metallic unmelted fragments is comparable to that observed in iron meteorites, from 3% to 9%, while it ranges from 15% up to 50% in the metallic *spherules*.

Cobalt. The cobalt content varies from 900 ppm to 5100 ppm, with a mean value of 3000 ppm, the single-oxide grains being the most depleted.

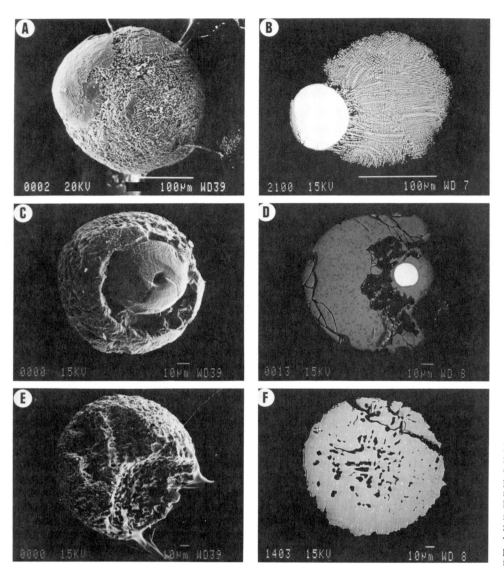

Fig. 3. The oxidized Fe/Ni grain families of Greenland cosmic dust. **(a,c,e)** External pictures; **(b,d,f)** polished sections. **(a and b)** G-type sphere showing an eccentric metal bead. **(c and d)** Two-oxide spherule with a Ni-core. **(e and f)** Single-oxide spherule showing a very porous aspect. The brightest points are tiny inclusions of Ni-rich iron sulfides.

DISCUSSION

We have optically handpicked about 2000 cosmic dust grains from the sediments collected on the Greenland ice-cap; among them, only 2-3% are Fe/Ni grains. This abundance is 25-50% in deep-sea sediments and the difference is probably due to a rapid weathering of chondritic grains in deep-sea conditions (*Kyte*, 1983). In young and frozen Greenland deposits all the species are well-preserved and for this reason the Greenland cosmic dust collection seems to be the best ever found (*Maurette et al.*, 1986). For this same reason, three new families of Fe/Ni particles have been discovered that constitute 30% of the Fe/Ni cosmic dust population larger than 50 μm and include single-oxide spheres, metallic fragments, and metallic spherules. In deep-sea sediments, such grains are probably weathered at a much faster rate than the two-oxide and G-type spheres.

The unmelted metallic fragments are an important fraction of the Fe/Ni grains (10%), consistent with the high abundance (25%) of unmelted fragments that we discovered in the family of chondritic grains (*Bonteet al.*, 1987b). Recently, ^{26}Al and ^{10}Be measurements have shown that unmelted chondritic fragments are micrometeorites (Raisbeck and Yiou, personal communication, 1987). Since spherules are also ablation micrometeoroid products (*Raisbeck et al.*, 1986), there is no doubt that metallic fragments are the parent bodies of most of the iron spherules. This indicates a great probability of survival without melting for micrometeoroids crossing through the atmosphere. Metallic fragments when just melted would produce the metallic spherules collected in the sediments as both have similar Ir, Co, and Cr contents (Table 1). However, these spherules show higher Ni contents somewhat intermediate between the low value of the unmelted fragments and the high value found in the metallic sphere extracted from a chondritic sphere. Presently their origin cannot be specified because the two processes (melting of a fragment or ejection from a cosmic spherule) would explain the measured elemental composition, as well as the iridium dispersion.

When such metallic fragments are melted during ablation, they can generate the two-oxide spherules of the A group (Table 1). The formation of the wustite-magnetite mixture would induce the platinoid nugget or the Ni-core as described by *Brownlee et al.* (1984). *Brownlee et al.* (1983) suggested that metallic spherules ejected from molten silicates can produce the two-oxide grains. This process does not seem important: Fe/Ni two-oxide spherules have much lower Ni contents than ejected metallic spherules; they cannot keep the ^{26}Al lithophile element when they are exsolved from the melted silicate micrometeorites. In agreement with *Czajkowski* (1987) we assume that this process may hardly produce the oxidized Fe/Ni grains.

The two-oxide spherules of the B group (Table 1) show a high Cr concentration. This "anomaly" has already been noted by *Kyte* (1983). They have the same structure and the same platinoid nugget as the A group. Presently, we have no explanation for this abnormal enrichment. We can only propose the following hypothesis: It is possible that a small Cr-rich inclusion (10-μm chromite crystal for example) was contained in the Fe/Ni parent body. This small inclusion would not prevent the oxidation process, but it would disappear during the melting, dispersing its Cr through the remaining spherule.

The single-oxide spheres constitute a specific group, as illustrated in the correlation plots reported in Figs. 2a,b (open stars). As one of them shows a unique homogeneous distribution of Ni-rich iron-sulfide inclusions, we have related their composition to that of the sulfides included in meteorites. Troilite is the most abundant sulfide in meteorites; it contains much less Ir and Ni and much more Cr than metallic phases (*Mason*, 1971), and this trend is observed in the single-oxide spheres. Burning of such sulfides during ablation would generate an unstable iron sequioxide that could be partly transformed into magnetite (*Pascal*, 1967). This would not produce platinoid nuggets. The tiny Ni-rich inclusions observed in FN14 (Table 1) could be relic pentlandite.

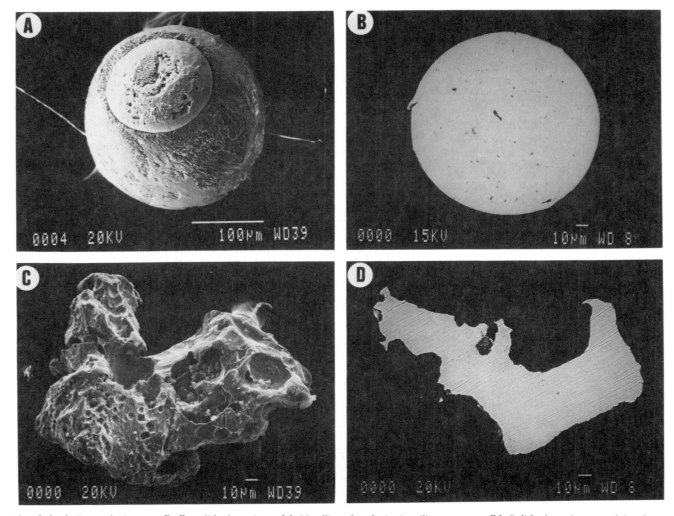

Fig. 4 **(a,c)** External pictures; **(b,d)** polished sections. **(a)** Metallic spherule in its silicate parent. **(b)** Polished section containing iron-sulfide inclusions. **(c and d)** Unmelted metallic fragment showing irregular shape.

The G-type spheres contain a silicate phase distributed between the dendrites of the iron oxide, and consequently high scandium contents (1.6 to 2.1 ppm) are not surprising. Their parent bodies can hardly have a chondritic origin because their Fe, Ni, and Cr contents are very high. The G-type spheres can be derived from parent bodies similar to the stony-iron meteorites.

CONCLUSIONS

The cosmic dust grains from the Greenland sediments have experienced little weathering and give a good representation of the cosmic matter accreted by the Earth in the past. ^{10}Be and ^{26}Al have shown that ablated spherules as well as unmelted fragments are micrometeorites. The study of structures and compositions of new Fe/Ni families that are as yet found only in Greenland lead us to determine three types of Fe/Ni parent micrometeorites: metallic alloys, sulfides, and stony iron grains.

We assume that the Fe/Ni grains are derived from these families except for some metallic spheres that can be ejected from molten chondritic particles.

The study of more grains from each family will be necessary to support the assumptions we set up, to improve the statistics, and to better determine the Fe/Ni grain flux infalling the Earth.

Acknowledgments. We are undebted to Drs. Flynn, McKay, and Sutton for their very helpful review of this manuscript. Financial support from CNRS/INSU and CEA are acknowledged. This is C.F.R. contribution no. 877.

REFERENCES

Blanchard M. B., Brownlee D. E., Hodge P. W., and Kyte F. T. (1980) Meteoroid ablation spheres from deep-sea sediments. *Earth Planet. Sci. Lett., 46,* 178-190.

Bonté Ph., Jéhanno C., Maurette M., and Brownlee D. E. (1987a) Platinum metals and microstructure in magnetic deep-sea cosmic spherules. *Proc. Lunar Planet. Sci. Conf. 17th,* in *J. Geophys. Res., 92,* E641-E648.

Bonté Ph., Jéhanno C., Maurette M., and Robin E. (1987b) A high abundance and great diversity of "unmelted" cosmic dust grains on the west Greenland ice cap (abstract). In *Lunar and Planetary Science XVIII,* pp. 105-106. Lunar and Planetary Institute, Houston.

Brownlee D. E. (1981) Extraterrestrial components in deep-sea sediments. In *The Sea,* vol 7 (C. Emiliani, ed.), pp. 733-762. Wiley, New York.

Brownlee D. E., Bates B. A., and Beauchamp R. H. (1983) Meteor ablation spheres as chondrule analogs. In *Chondrules and Their Origin* (E. A. King, ed.), pp. 10-25. Lunar and Planetary Institute, Houston.

Brownlee D. E., Bates B. A., and Wheelock M. (1984) Extraterrestrial platinum group nuggets in deep-sea sediments. *Nature, 309,* 693-695.

Czajkowski J. (1987) Cosmo and geochemistry of the Jurassic Hardgrounds. Ph.D. thesis, University of California, San Diego.

Kyte F. T. (1983) Analysis of extraterrestrial material in terrestrial sediments. Ph.D. thesis, University of California, Los Angeles.

Mason B. (1971) *Handbook of Elemental Abundances in Meteorites.* Gordon and Breach, New York. 555 pp.

Maurette M., Hammer O., Brownlee D. E., and Reeh N. (1986) Placers of cosmic dust in the blue ice lakes of Greenland. *Science, 233,* 869-872.

Murrel M. T., Davis P. A., Jr., Nishiizumi K., and Millard H. T., Jr. (1980) Deep-sea spherules from Pacific clay: Mass distribution and influx rate. *Geochim. Cosmochim. Acta, 44,* 2067-2074.

Pascal P., ed. (1967) *Nouveau Traité de Chimie Minérale.* Masson, Paris. 762 pp.

Raisbeck G. M., Yiou F., Bourlès D., and Maurette M. (1986) 10Be and 26Al in Greenland cosmic spherules: evidence for a cometary origin (abstract). In *Terra Cognita, 6,* 120.

Robin E., Jéhanno C., Maurette M., and Hammer O. (1987) A micrometeorite "spectrum" for the mass distribution of well preserved Greenland cosmic dust grains (abstract). In *Lunar and Planetary Science XVIII,* pp. 844-845. Lunar and Planetary Institute, Houston.

The Search For Refractory Interplanetary Dust Particles From Preindustrial Aged Antarctic Ice

Michael E. Zolensky
Planetary Materials Branch, SN2/NASA, Johnson Space Center, Houston, TX 77058

Susan J. Webb
Department of Earth Sciences, Memorial University, St. John's, Newfoundland, A1B 3X5 Canada

Kathie Thomas
Lockheed-LEMSCO, 2400 NASA Road One, Houston, TX 77058

In an effort to expand our knowledge of refractory interplanetary dust particles, preindustrial-aged Antarctic ice samples have been collected, melted, and filtered to separate the particle load. Particles containing a significant amount of aluminum, titanium, and/or calcium were singled out for detailed SEM and STEM characterization. The majority of these particles are shown to be volcanic tephra from nearby volcanic centers. Six spherical aggregates were encountered that consist of submicron-sized grains of rutile within polycrystalline cristobalite. These particles are probably of terrestrial volcanic origin, but have not been previously reported from any environment. One aggregate particle containing fassaite and hibonite is described as a probable interplanetary dust particle. The constituent grain sizes of this particle vary from 0.1 to 0.3 μm, making it significantly more fine-grained than meteoritic calcium-aluminum-rich inclusions. This particle is mineralogically and morphologically similar to recently reported refractory interplanetary dust particles collected from the stratosphere, and dissimilar to the products of modern spacecraft debris.

INTRODUCTION

Calcium-aluminum-rich inclusions (CAIs) are considered to be among the most primitive extraterrestrial material available for laboratory examination and contain primitive refractory phases that would be the material most likely to have survived melting or vaporization during the T-Tauri stage of our sun. An important avenue of recent research of CAIs has been the search for special isotopic effects that indicate the presence of presolar materials. However, such effects have been diluted or obscured by subsequent alteration processes that are probably due to the incorporation of CAI grains and clasts into chondrite parent bodies and the subsequent evolution of these bodies.

Since cometary bodies are surely a source of interplanetary dust particles (IDPs), some fraction of the refractory-rich IDP grains collected within the stratosphere should be primary nebula condensate or presolar grains incorporated into cometary bodies that may be essentially unaltered or, at least, far less altered than the material in meteorite parent bodies. Such pristine materials would make the interpretation of the physico-chemical characteristics of the presolar molecular cloud and the early history of the solar system much more straightforward (*Grossman*, 1980; *Dwek and Scalo*, 1980; *Wark and Lovering*, 1982; *Wood*, 1983). Similarly, identification of probable presolar interstellar grains from among these refractory-rich IDPs would add an entirely new dimension to the study of fine-grained extraterrestrial materials.

Unfortunately, limited work has been performed on the intriguing refractory IDPs (*Flynn et al.*, 1982; *Mackinnon et al.*, 1982; *Zolensky*, 1985, 1987; *McKeegan et al.*, 1986; *McKeegan*, 1987) because of their similarities to the products of solid propellant rockets and ablating aerospace debris.

It is well established that aerospace activities have produced a debris belt about the Earth composed largely of metals, refractory oxide and silicate grains, and larger pieces of satellites and rockets (*Kessler and Cour-Palais*, 1978; *Kessler et al.*, 1985). Decay of the orbits of these materials brings them within Earth's atmosphere, where they are collected along with the IDPs by NASA's Stratospheric Cosmic Dust Collection Program (*Zolensky*, 1984). The increasingly frequent launching of solid propellant rockets has resulted in a steadily rising flux of refractory material directly into our atmosphere, and therefore into our stratospheric particle collections (*Brownlee et al.*, 1976; *Zolensky*, 1984; *Zolensky and Kaczor*, 1986; *Zolensky et al.*, 1988). Fortunately, recent examination of the microstructures and chemistry of aerospace materials and the products of solid-rocket propellant engines has indicated that one can differentiate with some confidence between this spacecraft debris material and true extraterrestrial refractory particles (*Thomas-Ver Ploeg and Zolensky*, 1985; *Zolensky*, 1984, 1985, 1987). Briefly, we have found that spacecraft debris material is most commonly composed of metallic aluminum, metastable (γ-, δ- and κ-) and stable (α-) alumina, K- and Zn-silicates, Co-oxides, Mg-silicates, steels, rutile, CaO, plastics, and graphite. More to the point, we have found no evidence that spacecraft debris particles are ever composed of the more interesting refractory oxides and silicates common within CAIs, i.e., hibonite, melilite (akermanite-gehlenite), fassaite, and perovskite. Corundum (α-alumina) is found among both spacecraft debris particles and CAIs, but there appear to be morphological differences between the two occurrences (*Zolensky*, 1984; *Thomas-Ver Ploeg and Zolensky*, 1985).

Using these criteria developed to aid in the identification of refractory IDPs from among a volumetrically larger background of spacecraft debris, Zolensky (1985, 1987) recently identified hibonite, perovskite, and gehlenite within refractory stratospheric particles that he therefore proposed to be IDPs. This conclusion was subsequently verified by McKeegan (McKeegan et al., 1986; McKeegan, 1987), who measured the oxygen isotopic compositions of these same particles and found them to be nonterrestrial.

To further test this discrimination scheme, we decided to examine refractory particles from preindustrial-aged polar ice samples for which there would be no possibility of an artificial origin. Since the abundance of minerals found within refractory IDPs is very low in the terrestrial lithosphere, particles containing these phases should stand out, as compared with the modern polluted stratosphere. Extraterrestrial refractory particles found in preindustrial-aged ice could then be compared with particles presently being collected from the stratosphere, as well as with the known products of aerospace activities.

There have been numerous studies of particles from polar ice, which have been summarized by Thompson (1977). Many of these studies have emphasized particles of probable extraterrestrial origin (notably Hodge et al., 1964; Thompson, 1977; Mosley-Thompson, 1980; Tazawa et al., 1983; Ganapathy, 1983; LaViolette, 1985; Thiel et al., 1986; Maurette et al., 1986; Bonte et al., 1987). Unfortunately, because of the large background of terrestrial material present within ice samples, selection criteria are typically applied that limit the types of extraterrestrial particles that may be found. Thus spherical, metallic, or otherwise magnetic micrometeorites and meteor ablation materials are commonly recovered and studied (Brownlee, 1977), but there have been no systematic attempts to isolate individual nonspherical, nonmagnetic extraterrestrial particles from preindustrial ice samples. A recent exception to this rule is the work of Bonté et al. (1987) on probable meteoric ablation material and IDPs from a Greenland melt catchment basin.

The preliminary results of our search for refractory IDPs from preindustrial-aged Antarctic ice are presented here. Although particles of refractory compositions are the focus of our initial study, other particle types will be sought from among these samples as time and interest permits.

EXPERIMENTAL PROCEDURE

Approximately 50 kg of uncontaminated Antarctic ice was collected by MEZ during the 1985-86 Antarctic field season. These ice samples were collected at two different sites in Victoria Land, these being the Far Western Ice Field (FWI) (76°54'S, 157°01'E) and the Allan Hills Main Ice Field (ALH) (76°41'S, 159°17'E) (see the location map in Fig. 1). Both sites are characterized by a high rate of ice upwelling and ice ablation (~5 cm/yr; Annexstad, 1982), the latter due to the constant katabatic winds. The ice at both sites is preindustrial in age, having probably originated near the Dome C area (approximately 1000 km inland from the sampling localities, see Fig. 1) (Drewry, 1983). The blue ice from these

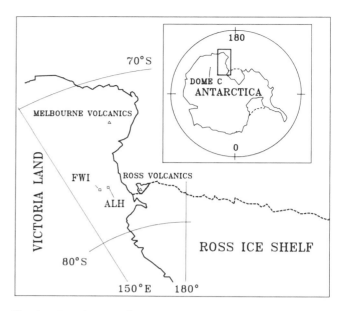

Fig. 1. Location map for Antarctic ice sample sites and volcanic centers mentioned in text.

fields has been dated at <100,000 (FWI) and ~700,000 years old (ALH) (Nishiizumi, 1986). At each sample site, a six inch layer of ice was removed from the surface before sampling began. This precaution, together with the reality of the high annual ice ablation rate (and therefore negligible snow accumulation rate), together eliminated the possibility of contamination from the modern atmosphere. In addition, each sample site lay in the interior of the Antarctic continent, typically upwind from any source of terriginous material. Great care was taken during the collection process to ensure that the ice samples remained uncontaminated, and were never directly touched by tools. All samples were immediately packed into precleaned nylon zip-lock bags for transport out of the field. These samples were transported unthawed to the Johnson Space Center at the close of the field season.

We melted all ice samples on a filtered-air laminar flow bench, and directed the melt water through a sequence of vertically stacked 1-cm-diameter Nucleopore filter membranes with pore sizes measuring 10 and 5 μm, respectively. We initially attempted to catch smaller particles on an additional filter membrane with 1 μm diameter pores, but found that these filters quickly became clogged with aerosol material, which prohibitively complicated the filtration procedure. We changed filter membranes frequently in order to prevent the buildup of particles on any one filter surface, which could interfere with later particle characterization and removal. This filtration procedure resulted in approximately equal particle coverage upon 18 sets of 10- and 5-μm filter membranes, and 5 membranes with 1-μm diameter pores.

We then epoxied the particle laden filters to carbon planchettes and carbon coated them for initial examination in a JEOL-35CF Scanning Electron Microscope (SEM), equipped with standard and windowless X-ray detectors for the energy-dispersive X-ray (EDX) analysis of particle compositions for all elements down (in atomic number) to carbon. Morphological observations of particles were made operating the SEM at 35kV, and accompanying EDX analyses were collected at 10 and 20kV.

Typical 5- and 10- μm filter membranes were found to hold 10^3 to 10^4 particles each. We have not yet examined any of the 1 μm filters because of our greater initial interest in the larger particles. Due to (1) the exceptionally high number of particles on each filter membrane, (2) the probable low relative percentage of extraterrestrial particles among these samples (*Hodge et al.,* 1964), and (3) the finite lifetime of the investigators, we decided to employ an analytical shortcut in the initial particle characterization procedure. Since we were initially interested only in particles of refractory composition, we collected X-ray element "dot maps" of each filter membrane for the elements Al, Ca, and Ti. Only particles with significant concentrations of one or more of these elements were subsequently located on the filter surfaces (not always an easy task) and analyzed.

Following the collection of compositional analyses using the SEM, we picked the most interesting particles off of the filter membranes and deposited them onto holey carbon membranes on Be grids for examination in a JEOL 100CX Scanning Transmission Electron Microscope (STEM). This instrument was also equipped with an EDX detector. We performed all STEM analyses at 100kV. All mineralogical identifications that we will report here are based upon EDX analyses *together* with electron diffraction work, as we believe that accurate identifications can very rarely be made on the basis of EDX analyses alone.

RESULTS

To date we have completely scanned the surfaces of six of the 10-μm filter membranes and one of the 5-μm filters. Thus, only one-fifth of the filter membranes have been completely surveyed. The majority (>95%) of the particles containing significant amounts of Al, Ti, or Ca have been found to be fragments or aggregates of ferromagnesian silicates and glass. When these particles are subtracted, only a few individuals remain as refractory IDP candidates. The major refractory particle types encountered in this study will be described separately.

Ferromagnesian Silicates and Glass Particles

We expected the most abundant terrestrial contaminant in our ice samples to be volcanic tephra of local derivation (within 1000 km), which has been demonstrated to contain abundant basanitic glass, fassaite, and kaersutite (*Zolensky and Paces,* 1986). Unfortunately, these phases all contain significant amounts of the refractory elements Al, Ca, and Ti. Thus, although refractory extraterrestrial material has been shown to contain fassaite or refractory glass (*Zolensky,* 1987), for the purposes of this study particles containing merely fassaite, glass, or kaersutite were viewed with suspicion. Most of the particles (51 out of 63, to date) on the filters that showed predominantly on the Al, Ti, and Ca element maps were found, during initial

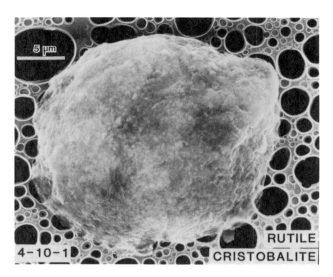

Fig. 2. An example of the spherical particles containing rutile and cristobalite. This is a scanning electron micrograph of particle #4-10-1.

SEM EDX examination, to be predominantly composed of Si. These particular particles were considered to have bulk compositions very near that of the refractory volcanic tephra described by Zolensky and Paces, and consequently were not subjected to further analysis. Two particles picked off of the filters for STEM examination were found to consist solely of fassaite, and were therefore considered to also be volcanic in origin.

Titanium and Silica-rich Particles

An enigmatic new class of particles were encountered in this study, consisting of Ti- and Si-rich spherical aggregates. Six of these particles were found (#'s 1-5-5, 1-5-6, 1-5-7, 1-5-8, 4-10-1, and 4-10-2), two from ALH and four from FWI. These particles range in size from 20 to 40 μm in average diameter, and resemble white snowballs in reflected light. STEM examination revealed that these particles consist of scattered rutile grains, averaging approximately 0.15 μm in diameter, encased within granular cristobalite (with the same approximate grain size). These identifications were verified by electron diffraction, which yielded the diffraction spacings 3.24(16) (110), 2.50(13) (101), 1.60(8) (220), 1.45(7) (310), and 1.31(7)Å (311) for the rutile, and 4.00(20) (101), 3.11(16) (111), 2.48(12) (200), 1.93(10) (113), and 1.56(8)Å (222) for the cristobalite (here and elsewhere the values in parentheses immediately following the measurements are the errors, and the second set of numbers are the Miller indices). SEM and TEM images of one of these particles are shown in Figs. 2 and 3. We are not aware of a report of similar particles. Certainly, similar particles have never been observed from among the stratospheric particle collections. Cristobalite has been commonly reported from volcanic eruptive rocks, but not associated with rutile in this manner. Nevertheless, as such materials have also not been reported

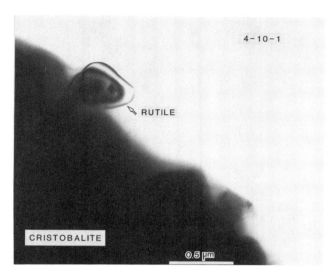

Fig. 3. Transmission electron micrograph of the same particle shown in Fig. 2. Arrow points to a rutile grain within a matrix of cristobalite, all at an electron transparent edge of the particle.

from any extraterrestrial samples, we consider it most probable that these particles are of terrestrial volcanic origin. It is interesting that such obviously rare particles as these would be found within ice from both of our sampling sites. This suggests that the source for these particles is somewhere nearby.

Potential Refractory IDPs

Of the refractory particles from this study that remain when the particles or probable terrestrial origin (the ferromagnesian silicates and cristobalite-rutile spheres) are deleted from consideration, most contain mainly Al, with minor amounts of other elements, most notably Ti and O. One particle (#8-10-5) contains predominently Si, Ca, Ti, Al, and O, with minor Mg. We determined that this particle, a 10-μm-diameter aggregate, contains only fassaite [approximately $Ca(Ti_{0.4}Mg_{0.3}Al_{0.2})(Si_{0.5}Al_{0.5})_2O_6$] and hibonite [approximately $Ca(Al_{0.8}Ti_{0.2})_{12}O_{19}$]. The electron diffraction spacings obtained from these phases were 4.20(21), 3.63(18), 3.30(17), 2.86(14), 2.47(12), 2.30(12), 1.99(10), 1.45(7), and 1.28(6)Å for the hibonite, and 3.70(19) (111), 3.24(16) (220), 2.86(14) (311), 2.50(13) (221), 2.29(11) (31$\bar{1}$), 1.96(10) (13$\bar{2}$), and 1.35(7)Å (043) for the fassaite. The average diameter of the constituent grains within this aggregate range from 0.1 to 0.3 μm (see Figs. 4 and 5). The distribution of each phase within this aggregate has yet to be determined; however, these minerals are intergrown on a scale too fine to permit the gathering of accurate compositional information (the particle has not been crushed or otherwise disaggregated). We will attempt to discover this by thin-sectioning of this particle, and subsequent STEM examination.

The remaining Al-rich particles encountered in this study are presently being characterized. These particles include (1) #7-10-6, a 35-μm aggregate containing major Al and O with minor silicon and trace Fe and Ti; (2) #7-10-7, a 25 × 40-μm fragment containing major Al and O, and trace Cu; (3) #7-10-8, a 30 × 50-μm aggregate containing major Al and O, and minor Si; (4) #7-10-5, a 100-μm fragment containing major Ti and O, and minor Si, Al and Fe; and (5) #3-10-2, a 10 × 5-μm fragment containing only Al, Ti, and O.

DISCUSSION

The most common presumed sources of terrestrial particles in our ice samples should include volcanic tephra, wind-blown sand, and associated mineralogical weathering products that have settled from the atmosphere (*Gaudichet et al.,* 1986), all of which may usually be easily recognized. Since the prevailing direction of the polar katabatic winds in the ice accumulation area from which our samples originated (near Dome C) is toward the periphery of the continent, with only ice lying upwind, the most common terrestrial particles should be volcanic tephra originating from large nearby eruptions. The volcanic fields nearest the ice accumulation region (Dome C) belong to the Ross and Melbourne volcanic provinces (see the location map in Fig. 1), which have been active for 15 m.y. (*Kyle,* 1982), and thus were supplying tephra when our particle samples were initially trapped in accumulating snow.

Zolensky and Paces (1986) examined the mineralogy of particulates from tephra-rich zones in ice adjacent to the sites from which the samples for this study were derived. They report that this tephra consists of fragments of basanitic glass, forsterite (Fo79-85), plagioclase (An variable), fassaite [$Ca(Mg,Fe,Al)(Si,Al)_2O_6$], kaersutite [$NaCa_2(Mg,Fe)_4Ti(Si_6Al_2)O_{22}(OH)_2$], sulfides, spinel, and quartz (in agreement with a similar study by *Katsushima et al.,* 1984). The basanitic glass

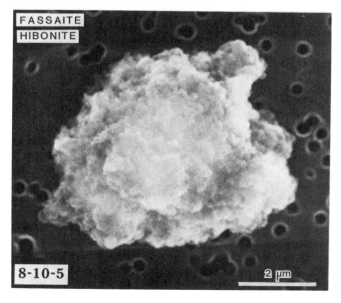

Fig. 4. Scanning electron micrograph of particle #8-10-5, which is probably a refractory IDP. This particle consists of submicron-sized grains of fassaite and hibonite.

Fig. 5. High-magnification detail of the micrograph of particle #8-10-5 shown in Fig. 4, more clearly showing the submicron-sized constituent grains of fassaite and hibonite. Even at this scale the two minerals are intergrown too intimately to be discriminated.

contained an average of 44.24(85) wt.% SiO_2, 4.26(50) wt% TiO_2, 15.53(53) wt % Al_2O_3 and 11.73(65) wt % CaO (Zolensky and Paces, unpublished data, 1986). In their study, Zolensky and Paces (1986) characterized the particle load of ice that was so heavily laden with particles that it was brown in color. The ice selected for our search for IDPs was visibly free of particles. Nevertheless, we found that the vast majority of particles present within these ice samples were identical with those reported by Zolensky and Paces (1986), i.e., were basanitic volcanic particles. The refractory composition of this volcanic tephra made it difficult for us to distinguish between this terriginous material and the refractory IDPs we sought on the basis of bulk particle composition alone.

However, when a promising subset of Al-, Ca-, and Ti-rich particles was subjected to more detailed phase identification through electron diffraction, one particle was found to display clear dissimilarities to the basanitic tephra. Particle 8-10-5 was found to consist of fassaite and hibonite. Although fassaite is found among the basanitic tephra from our ice samples, the presence of hibonite within this particle eliminates the possibility that this sample was derived from volcanic tephra. Hibonite is extremely rare in terrestrial rocks (*Keil and Fuchs,* 1971), where it has not been observed to occur in association with fassaite (*Deer et al.,* 1978, pp. 399-414). Fassaite and hibonite are commonly found to be associated within CAIs in chondrites, but have a grain size in CAIs that is ten times larger (>1 μm) than that in particle 8-10-5 (*Zolensky,* 1987). Extremely fine-grained particles containing hibonite have only been reported from among refractory IDP's (*Zolensky,* 1987), where this phase was found to be associated with perovskite, melilite, and refractory glass. One would expect similarly fine-grained fassaite to be present within refractory IDPs, although this mineral has only previously been found in one chondritic IDP (*Christoffersen and Buseck,* 1986).

Particle 8-10-5 is also completely unlike the characterized or expected products of spacecraft debris. For the purpose of this discussion this debris is taken to include spacecraft paint as well as the products of solid rocket propellant exhaust, rocket engine ablation, and spacecraft hardware ablation. As demonstrated by recent work, spacecraft debris particles do not appear to ever consist of either hibonite or fassaite (or perovskite and melilite, for that matter) (*Zolensky,* 1984, 1985; *Thomas-Ver Ploeg and Zolensky,* 1985). Thus, particle 8-10-5 is mineralogically completely unlike any known natural terrestrial material or spacecraft debris particles. This particle is mineralogically similar to meteoritic CAIs; however, its grain size is at least an order of magnitude smaller than such material. In fact, particle 8-10-5 is both mineralogically and morphologically similar only to refractory IDPs collected from the modern stratosphere. We thus conclude that this particular particle is an IDP, subject to verification by later analysis (such as measurement of the oxygen isotopic composition). If such an origin can be confirmed for this particle this is thus the first report of fassaite within an IDP composed entirely of refractory phases.

The remainder of the aluminum-titanium-rich particles encountered during this study (#'s 7-10-5, 7-10-6, 7-10-7, 7-10-8, and 3-10-2) remain to be properly characterized. However, it is interesting that particles compositionally similar to these have been found among modern stratospheric particle collections, where they have typically been interpreted as spacecraft ablation material (*Zolensky and Mackinnon,* 1985; *Zolensky et al.,* 1988). The demonstrated preindustrial age of the particles located in this study precludes such an origin here, and it is possible that detailed phase characterization of additional refractory stratospheric particles will reveal further examples of previously unknown types of IDPs.

This report also enables us to restate an important, but all too commonly unappreciated fact, namely that most IDPs are not spherical (*Zolensky and Mackinnon,* 1985), but may exist as discrete fragments or aggregates. Thus searches for only spherical IDPs from among polar ice samples will probably miss the bulk of preserved IDPs. In fact, *Bonte et al.* (1987) have recently reported large amounts of nonspherical IDPs from melt-ponds on the Greenland ice cap.

CONCLUSIONS

The most abundant material found among our ice melt particles were found to be volcanic in origin. Volcanic basanitic glass shards and grains of fassaite and kaersutite were found to constitute the bulk of the recovered refractory material as well. We found six spherical aggregates composed of rutile within cristobalite, which we tentatively conclude are of terrestrial volcanic origin. Fortunately, however, we found one particle that was composed of hibonite and fassaite, with a submicrometer grain size. Based upon the similarities between this particle and previously characterized IDPs, as well as its dissimilarities with other natural and artificial material, we propose that this particle is a refractory IDP.

Antarctica presents us with a fourth probable source of IDPs, augmenting our collections from the stratosphere, oceans, and Greenland. Characterization of these particles has only just begun. One ultimate result of these and similar studies will be the development of a data base describing IDPs that will aid in the identification of analogous samples from among the modern stratospheric particle collections. One can forsee a time when our collections of both modern and preindustrial IDPs will be large enough to analyze for possible IDP population and flux variations through time.

Acknowledgments. We thank D. S. McKay for access to the electron microscopy laboratories of the Solar System Exploration Division at the Johnson Space Center. U. Marvin and M. Maurette provided valuable reviews of this paper. MEZ visited Antarctica as a member of the 1985-1986 Antarctic Search For Meteorites at the kind invitation of W. A. Cassidy (NSF grant DPP-8314496 to W. Cassidy). SJW was a Lunar and Planetary Institute Summer Intern.

REFERENCES

Annexstad J. O. (1982) Geography and glaciology of selected blue ice regions in Antarctica. In *Workshop on Antarctic Glaciology and Meteorites* (C. Bull and M. Lipschutz, eds.), pp. 36-37. LPI Tech. Rpt. 82-03, Lunar and Planetary Institute, Houston.

Bonté Ph., Jéhanno C., Maurette M., and Robin E. (1987) A high abundance and great diversity of "unmelted" cosmic dust grains on the West Greenland ice cap. (abstract). In *Lunar and Planetary Science XVIII*, pp. 105-106. Lunar and Planetary Institute, Houston.

Brownlee D. E. (1977) Microparticle studies by sampling techniques. In *Cosmic Dust* (J. A. M. McDonnell, ed.), pp. 295-336. John Wiley and Sons, New York.

Brownlee D. E., Ferry G. V., and Tomandl D. (1976) Stratospheric aluminum oxide. *Science, 191,* 1270-1271.

Christoffersen R. and Buseck P. R. (1986) Refractory minerals in interplanetary dust. *Science, 234,* 590-592.

Deer W. A., Howie R. A., and Zussman J. (1978) *Rock-Forming Minerals, Volume 2A. Single-Chain Silicates.* Second Edition. John Wiley and Sons, New York.

Drewry D. J. (1983) The surface of the Antarctic ice sheet. *Antarctica: Glaciological and Geophysical Folio* (D. J. Drewry, ed.), Sheet 2. Scott Polar Research Institute, Cambridge.

Dwek E. and Scalo J.M. (1980) The evolution of refractory interstellar grains in the solar neighborhood. *Astrophys. J., 239,* 193-211.

Flynn G. J., Fraundorf P., Keefe G., and Swan P. (1982) Aluminum-rich spinel aggregates in the stratosphere: From Earth, or not? (abstract) In *Lunar and Planetary Science XIII,* pp. 223-224. Lunar and Planetary Institute, Houston.

Ganapathy R. (1983) The Tunguska explosion of 1908: Discovery of meteoritic debris near the explosion site and at the South Pole. *Science, 220,* 1158-1161.

Gaudichet A., Petit J. R., LeFevre R., and Lorius C. (1986) An investigation by analytical transmission electron microscopy of individual insoluble microparticles from Antarctic (Dome C) ice core samples. *Tellus, 38B,* 250-261.

Grossman L. (1980) Refractory inclusions in the Allende meteorite. *Ann. Rev. Earth Planet. Sci., 8,* 559-608.

Hodge P. W., Wright F. W., and Langway C. C. (1964) Studies of particles for extraterrestrial origin, 3. Analysis of dust particles from polar ice deposits. *J. Geophys. Res., 69,* 2919-2931.

Katsushima T., Nishio F., Ohmae H., Ishikawa M., and Takahashi S. (1984) Composition of dirt layers in the bare ice areas near the Yamato Mountains in Queen Maud Land and the Allan Hills in Victoria Land, Antarctica. *Proceedings of the Sixth Symposium on Polar Meteorology and Glaciology,* pp. 174-187. National Institute of Polar Research Special Issue No. 34, Tokyo.

Keil K. and Fuchs L.H. (1971) Hibonite [$Ca_2(Al,Ti)_{24}O_{38}$] from the Leoville and Allende chondritic meteorites. *Earth Planet. Sci. Lett., 12,* 184-190.

Kessler D. J. and Cour-Palais B. G. (1978) Collision frequency of artificial satellites: The creation of a debris belt. *J. Geophys. Res., 83,* 2637-2646.

Kessler D. J. et al. (1985) Examination of returned Solar-Max surfaces for impacting orbital debris and meteoroids (abstract). In *Lunar and Planetary Science XVI,* pp. 434-435. Lunar and Planetary Institute, Houston.

Kyle P. R. (1982) Volcanic Geology of the Pleiades, Northern Victoria Land, Antarctica. In *Antarctic Geoscience* (C. Craddock, ed.), pp. 747-754. Univ. of Wisconsin Press, Madison.

LaViolette P. (1985) Evidence of high cosmic dust concentrations in Late Pleistocene polar ice (20,000-14,000 years BP). *Meteoritics, 20,* 545-558.

Mackinnon I. D. R., McKay D. S., Nace G. A., and Issacs A. M. (1982) Al-prime particles in the cosmic dust collection: Debris or not debris? *Meteoritics, 17,* 245.

Maurette M., Hammer C., Brownlee D. E., Reeh N., and Thomsen H. H. (1986) Placers of cosmic dust in the blue ice lakes of Greenland. *Science, 233,* 869-872.

McKeegan K. (1987) Oxygen isotopic abundances in refractory stratospheric dust particles: Proof of extraterrestrial origin. *Science, 237,* 1468-1471.

McKeegan K., Zinner E., and Zolensky M. (1986) Ion probe measurements of O isotopes in refractory stratospheric dust particles: Proof of extraterrestrial origin. *Meteoritics, 21,* 449-451.

Mosley-Thompson, E. (1980) *911 Years of Microparticle Deposition at the South Pole: A Climatic Interpretation.* Institute of Polar Studies Report No. 73, Ohio State University, 134 pp.

Nishiizumi K. (1986) Terrestrial and exposure histories of Antarctic meteorites. In *International Workshop on Antarctic Meteorites* (J. O. Annexstad, L. Schultz, and H. Wanke, eds.), pp. 71-73. LPI Tech. Rpt. 86-01, Lunar and Planetary Institute, Houston.

Tazawa Y., Fujii Y., and Nishio F. (1983) Extraterrestrial dust particles in Antarctic ice. *Eighth Symposium on Antarctic Meteorites,* pp. 23-1 to 23-3. National Institute of Polar Research, Tokyo.

Thiel K., Peters J., Raisbeck G. M., and Yiou F. (1986) Meteoric ablation and fusion spherules in Antarctic ice. In *International Workshop on Antarctic Meteorites* (J.O. Annexstad, L. Schultz and H. Wanke, eds.), pp. 110-112, LPI Tech. Rpt. 86-01, Lunar and Planetary Institute, Houston.

Thomas-Ver Ploeg K. L. and Zolensky M. E. (1985) A comparison of Shuttle solid rocket effluent with aluminum-rich stratospheric particles. *Eos Trans. AGU, 66,* 826.

Thompson L.G. (1977) *Microparticles, Ice Sheets and Climate.* Institute of Polar Studies Report No. 64, Ohio State University, 148 pp.

Wark D. A. and Lovering J. F. (1982) Evolution of Ca-Al-rich bodies in the earliest solar system: Growth by incorporation. *Geochim. Cosmochim. Acta, 46,* 2595-2607.

Wood J. A. (1983) Formation of chondrules and CAI's from interstellar grains accreting to the solar nebula. *Proceedings of the Eighth Symposium on Antarctic Meteorites* (T. Nagata, ed.), pp. 84-92. National Institute of Polar Research Special Issue No. 30, Tokyo.

Zolensky M. E. (1984) The abundance and origin of aluminum-rich particulates in the stratosphere. *Eos Trans. AGU, 65,* 837.

Zolensky M. E. (1985) CAI's among the cosmic dust collection. *Meteoritics, 20,* 792-793.

Zolensky M. E. (1987) Refractory interplanetary dust particles. *Science*, 237, 1466-1468.

Zolensky M. E. and Kaczor L. (1986) Stratospheric particle abundance and variations over the last decade (abstract). In *Lunar and Planetary Science XVII*, pp. 969-970. Lunar and Planetary Institute, Houston.

Zolensky M. E. and Mackinnon I. D. R. (1985) Accurate stratospheric particle size distributions from a flat-plate collection surface. *J. Geophys. Res.*, 90, 5801-5808.

Zolensky M. E. and Paces J. B. (1986) Alteration of tephra in glacial ice by "unfrozen water". *Geol. Soc. Amer. Abstracts with Programs*, 99, 801.

Zolensky M. E., McKay D. S., and Kaczor L. A. (1988) Origin and implications of the ten-fold increase in the abundance of solid particles in the stratosphere, as measured over the period 1976-1984. *J. Geophys. Res.*, in press.

Stratospheric Particles: Synchrotron X-Ray Fluorescence Determination of Trace Element Contents

S. R. Sutton*

Department of the Geophysical Sciences, The University of Chicago, Chicago, IL 60637

G. J. Flynn

Department of Physics, State University of New York—Plattsburgh, Plattsburgh, NY 12901

*Also at Department of Applied Science, Brookhaven National Laboratory, Upton, NY 11973

The first trace element analyses on stratospheric particles using synchrotron X-ray fluorescence (SXRF) are reported. Chromium, Mn, Fe, Ni, Cu, Zn, Ga, Ge, Se, and Br were detected. Concentrations for chondritic particle U2022G1 are within a factor of 1.7 of CI for all elements detected with the exception of Br, which is 37 times CI. Chondritic particle W7029*A27 is also near CI for Cr, Mn, Fe, Ni, Zn, and Ge but enriched in Cu, Ga, Se, and Br by factors of 2.1, 5.8, 3.5, and 8.4, respectively. The third particle of the cosmic dust class U2015G1 also showed high Br enriched relative to CI by a factor of 28. Bromine was detected at a high level in an aluminum-rich particle W7027D2 classified as probable artificial terrestrial contamination but exhibiting a chondritic Fe/Ni ratio. Bromine was not detected in a fifth particle U2022C14 also classified terrestrial and exhibiting a crustal Fe/Ni ratio. If the high Br has a preterrestrial origin, the ubiquity of the effect suggests that a large fraction of the chondritic interplanetary dust particles (IDPs) derive from a parent body (bodies) not sampled in the meteorite collection.

INTRODUCTION

The chemical compositions of interplanetary dust particles (IDPs) collected from the stratosphere by NASA aircraft are valuable in inferring their origins and their relationships to meteorites. Major element compositions are routinely determined in a qualitative sense during curation at the Cosmic Dust Curatorial Facility, NASA Johnson Space Center (JSC) (*Zolensky et al.*, 1984). Particles are then provisionally classified by JSC as cosmic dust, aluminum oxide spheres, artificial terrestrial contamination, or natural terrestrial contamination. Quantitative minor and trace element compositions exist for only five chondritic particles; two analyzed by INAA (*Ganapathy and Brownlee*, 1979) and three by proton-induced X-ray emission (PIXE) (*van der Stap et al.*, 1986). In general, the chondritic IDPs possess compositions near those of CI carbonaceous meteorites. However, two intriguing differences have been reported: (1) a depletion in the refractory Ca (*Fraundorf et al.*, 1982), and (2) enrichments in the volatiles Zn (*Ganapathy and Brownlee*, 1979) and Br (*van der Stap et al.*, 1986) relative to CI.

This paper reports the first use of synchrotron X-ray fluorescence (SXRF) to measure the trace element contents of stratospheric particles. SXRF is well suited for trace element analyses on small specimens offering part-per-million sensitivity for 20-μm beam spots (*Gordon*, 1982). Another advantage is the nondestructive nature of SXRF leaving samples available for complementary analyses such as infrared transmission spectroscopy (*Sandford and Walker*, 1985), ion microprobe (*McKeegan et al.*, 1985), transmission electron microscopy (TEM) (*Bradley and Brownlee*, 1986), and Raman spectroscopy (*Wopenka*, 1987).

EXPERIMENTAL METHOD

Specimens

Five stratospheric particles were analyzed in the present experiment. A summary of the particle characteristics determined at the Cosmic Dust Curatorial Facility at JSC is provided in Table 1. Two of these particles (U2022G1 and W7029*A27, the latter originally numbered W7029K1) exhibited typical chondritic EDS spectra in the JSC characterization, showing major peaks due to Mg, Si, S, Ca, Fe, and Ni as well as detectable Al. W7029*A27 is associated with a cluster (W7027*A) of more than 20 fragments from an original particle estimated to be 60 μm in size. The third particle (U2015G1), also of the cosmic dust class, had a lower S abundance and exhibited detectable amounts of K and Zn in the JSC spectrum. This particle separated into two fragments

TABLE 1. Descriptions of parent stratospheric particles.

Particle	Size (μm)	Class	EDX peaks[‡]
U2022G1	25	C	Mg,Al,**Si**,S,**Fe**
W7029*A27	15	C	Mg,**Si**,S,Ca,**Fe**,Ni
U2015G1	25*	C	Mg,Al,**Si**,S,Fe,Ni,Zn
W7027D2	11	TCA?	**Al**,Si,S,Ca,Ti,Cu
U2022C14	90[†]	TCA?	Na,Mg,Al,**Si**,**S**,Ca,Fe

JSC classifications: C = Cosmic dust, TCA? = probable artificial terrestrial contamination.

*Analyzed as two individual fragments.
[†] A 7-μm fragment of this particle was analyzed.
[‡] X-ray peaks present in JSC EDX spectrum; most intense peak(s) in boldface.

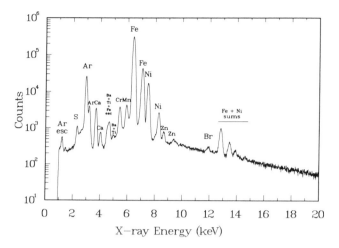
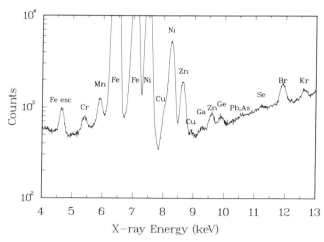

Fig. 1a. Synchrotron X-ray fluorescence spectrum for particle U2022G1 with no detector filter. Argon derives from air along the incident beam path. Three overlapping peaks were detected between 4 and 5 keV: peak 1 = Ba L_α (4.47 keV) + Ti K_α (4.51 keV), peak 2 = Fe escape (4.66 keV), and peak 3 = Ba L_β (4.83 keV) + Ti K_β (4.93 keV). Four peaks labeled "sums" between 12 and 15 keV result from near simultaneous detection of two X-rays from Fe and Ni K transitions. Two peaks near 1 keV labeled "esc" result from Si ionization in the detector by an Ar X-ray and subsequent escape of Si K_α X-ray (1.74 keV).

Fig. 1b. Synchrotron X-ray fluorescence spectrum for particle U2022G1 with 85-μm aluminum detector filter. Krypton derives from the air along the incident beam path. The peak labeled "Pb,As" is either Pb L_α or As K_α, which are indistinguishable at these low levels.

(U2015G1 main and U2015G1 frag 2) during shipping and these were analyzed separately. The remaining two particles were classified by JSC as probable artifically-produced terrestrial contamination. Mounting of particles was performed in a Class 100 clean room. To avoid contamination from handling, each particle was transferred directly from its JSC glass slide shipping container to an individual 7.5-μm-thick Kapton foil upon which the samples were analyzed. No effort was made to remove the silicone oil in which the particles were shipped from JSC since the oil did not interfere with analyses (see section on "Interferences" below).

Apparatus

Synchrotron XRF measurements were performed on beamline X-26C at the National Synchrotron Light Source, Brookhaven National Laboratory, Upton, New York. The synchrotron ring was operated at 2.53 GeV with electron currents between 50 and 150 mA. Specimens were excited with white light (i.e., continuum energy spectrum) filtered by 760 μm Be (vacuum window) and covering the energy range 3 to 30 keV with the maximum flux at 10 keV. The incident X-ray beam, apertured to a 30-μm spot with motor-driven tantalum slits, traversed an air path of 15 cm before striking the specimen. Particle mounts were positioned vertically on a Klinger x-y-z-theta stepper motor stage at 45 degrees to the incident beam. X-ray spectra were obtained in air using a 30-square-millimeter Si(Li) energy dispersive detector. Because the synchrotron radiation is polarized within the electron storage ring plane, scattered background was minimized by placing the detector within this plane and at 90 degrees to the incident beam (*Jones et al.*, 1984). Specimens were positioned at the intersection point of the incident beam and detector axis and were viewed by a horizontally mounted stereozoom binocular microscope equipped with a TV camera.

The X-ray spectrum obtained for U2022G1 (Fig. 1a) showed lines from S, Ar (from air path), Ca, Ba + Ti (overlap), Cr, Mn, Fe, Ni, Cu, Zn, and Br. Pileup peaks were also present due to the high dead time associated with the intense Fe K lines. Because several sum energies lie near fluorescence energies of interest (e.g., Fe K_α + Cr K_α interfering at Br K_α), an 85-μm aluminum filter was placed over the detector for Fe-rich particles, i.e., those from the cosmic dust class. The use of this filter eliminated pileup peaks but restricted the analyses to fluorescence lines above about 4 keV, i.e., K lines above Ti and L lines above Ba. Live acquisition times for the six fragments ranged from 900 to 10^4 seconds (see Table 3).

Data Analysis

Two separate spectral deconvolution programs were used on each spectrum, *Peaksearch* and *Strip*. *Peaksearch*, the Nuclear Data full spectrum peak stripping routine, identifies peaks by Gaussian approximation and fits backgrounds by linear regression between channels on either side of each peak. Subroutines were added to report candidate fluorescence lines within 50 eV of each identified peak, including escape peaks, and to predict the energy and intensity of significant pileup peaks based on observed peak count rates and the time resolution of the detection system. *Strip*, an interactive, graphical peak fitting routine, was used for low intensity and overlapping peaks. Five separate background functions are available in this program, polynomials of degrees 0 to 3 and exponential, and peak areas can be obtained by two approaches: (1) single peak integration using a center of gravity calculation or (2) multiple peak fitting using a Gaussian routine with user-supplied seed values for energy and FWHM.

Among the elements detected from the cosmic dust particles, Fe was in the highest abundance and it was used as the principal standardization element. A standard synthetic glass (AN75, 1480 ppm Fe determined independently by atomic absorption spectrophotometry; A. M. Davis, analyst) was measured under the same conditions as the particles. The mass of Fe in each particle, m(Fe,p), was calculated using the mass of Fe excited in the standard, m(Fe,s), and the intensities of the Fe K_α peaks (net counts per unit incident flux) from the particle and standard spectra, I(Fe,p) and I(Fe,s), respectively

$$m(Fe,p) = m(Fe,s) \frac{I(Fe,p)\, \gamma(s)}{I(Fe,s)\gamma(p)} \quad (1)$$

$\gamma(p)$ and $\gamma(s)$ are self-absorption factors for the particle (about 10 µm thick) and standard (220 µm thick), respectively. The mass of Fe excited in the standard, m(Fe,s), was computed to be 2.9 ± 0.6 ng using

$$m(Fe,s) = c(Fe,s)\, \rho(s)\, A(s)\, t(s)/\sin\theta \quad (2)$$

where t(s) is the thickness of the standard (0.022 cm), $\rho(s)$ is the standard density (2.5 g/cm^3), A(s) is the cross sectional area of the incident beam (2500 µm^2), θ is the angle of incidence (45 degrees), and c(Fe,s) is the Fe concentration (1480 ppm). Self-absorption factors, defined as the intensity from the actual sample divided by that from an ultrathin sample of the same mass, were obtained theoretically using *NRLXRF* (*Criss*, 1977), an X-ray fluorescence program developed at the Naval Research Laboratory and adapted to use the synchrotron energy spectrum as the excitation source. For Fe K_α, $\gamma(p)$ and $\gamma(s)$ results were 0.80 and 0.095, respectively. The uncertainty in $\gamma(p)$ due to morphology is negligible because the particle thickness is small compared to the characteristic absorption depth of K X-rays from Fe and elements of higher atomic number.

Masses for all other elements detected were computed using sensitivities (picograms per count) relative to Fe, S(z), predicted by the *NRLXRF* program. That is

$$m(z,p) = \frac{S(z)m(Fe,s)I(z,p)\gamma(s)}{I(z,s)\, \gamma(p)} \quad (3)$$

Concentrations were computed for cosmic dust particles assuming the major element chemistry was near CI and using the Fe/Si peak height ratio from the JSC EDS spectra to estimate absolute Fe concentrations. Fe/Si for U2022G1 and W7029*A27 relative to that for the Allende standard JSC spectrum are 1.26 and 1.52, respectively, near the 1.3 ratio expected for a CI specimen. We therefore assume an Fe concentration equal to CI (18%) for these two particles. Iron in the third cosmic dust particle, U2015G1, was taken to be 9% based on the observation that the Fe/Si in its JSC spectrum is about 0.5 times that of the other two particles. Approximate concentrations for the suspected terrestrial particles are computed based on optically-estimated volumes and an assumed density of 2 g/cm^3.

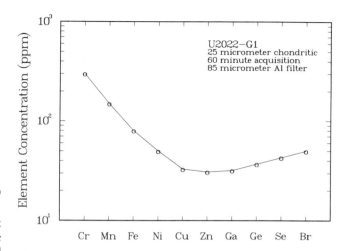

Fig. 2. Minimum detection limits for a one-hour acquisition on chondritic IDP U2022G1 with 85-µm aluminum detector filter. Higher sensitivity can be obtained for longer acquisition times and larger particles.

Detection Limits

Minimum detection limits, defined here as 3 times the square root of the spectral background, correspond to concentrations between 30 and 300 ppm for elements between Cr and Br (Fig. 2) after a one-hour acquisition using the filtering conditions described above on U2022G1. Greatest sensitivity was for Cu, Zn, and Ga. Minimum detectable concentrations become greater at low Z due to absorption by the detector filter and at high Z due to the fall off in incident flux at high energy. These values scale approximately as the square root of acquisition time and particle mass.

Elemental sensitivity for particle analyses is expected to increase more than 10-fold with the addition in 1988 of a focusing mirror, which should increase the flux at the sample by 1000-fold, and a wavelength dispersive detector, which will improve signal to background and eliminate diffraction interference.

Interferences

The incident synchrotron beam excited Kapton and residual silicone oil (the latter used to collect and transport particles) as well as the particle during analysis. The fluorescence contribution from the Kapton and oil was determined in the present work by SXRF analyses under the same experimental conditions employed for particle analyses, i.e., comparable storage ring current, beam size, and acquisition time. A "worst case" measurement on a 100-µm-diameter drop of silicone oil (about 0.3 µg) deposited on Kapton and excited with a 100 × 100-µm beam yielded a spectrum containing small peaks due only to Fe and Zn. The effective mass of these elements within the excitation volume was determined to be 1.6 pg and 10 pg, respectively, corresponding to mean concentrations of 5 and 30 ppm, respectively. Thus, a 30 × 30-µm beam, the size used in particle analyses, would excite 0.1 pg Fe and

0.9 pg Zn. This is a reasonable estimate for the interference from Fe since Kapton-only analyses demonstrate that the Fe resides principally in the Kapton while the oil contains the Zn. Zinc interference will be lower since the oil excited with a typical particle is the amount that adheres to the particle when it is transferred to the Kapton. This volume is much less than that of each particle. A reasonable upper limit for the Zn mass in the adhering oil is 0.1 pg. The Fe interference can be neglected since the lowest particle Fe content was 3 pg. Zinc contents, on the other hand, were 0.12, 0.35, 0.95, 1.7, 1.8, and 35 pg, corresponding to fractional interferences of 0.8, 0.3, 0.11, 0.06, 0.06, and 0.003, respectively. It is therefore likely that the Zn contents given in Table 3 for U2022C14 (terrestrial) and W7029*A27 (cosmic) are overestimates.

Diffraction is a potential interference in the present experimental configuration (i.e., continuum excitation and energy dispersive detection with fixed detector position) (*Sutton et al.*, 1986). Diffraction peaks from single crystals can be readily eliminated by specimen rotation but the fine-grained textures of the chondritic particles more closely approximate powders. For such specimens, a diffraction peak that unfortuitously occurs at a fluorescence energy is identified by the absence of associated companion peaks (e.g., a peak at a K_α energy that lacks a K_β companion). This criterion is, however, inapplicable for small alpha peaks, where the beta peak is below the detection limit. The possibility cannot be ruled out that peaks near the detection limit that occur at K_α energies actually result from diffraction rather than fluorescence. The fact that no unidentified peaks were observed in any of the X-ray spectra strongly suggests that diffraction peaks are rare from these small specimens.

Standard Particles

The theoretical relative sensitivities from *NRLXRF* have yielded concentrations within about 30% of certified values for a wide range of standards over the elemental range of Ca to Pd. An atomic number dependence is observed for the discrepancies, i.e., concentrations for high Z elements such as Sr are overestimated by about 30% while those for elements near Fe are within 10% of certified values. The elemental mass determination technique was further tested by analyzing two fragments of SRM 1874 (K546), a lithium-borate glass containing 0.14% Cr, 0.39% Ni, 0.50% Ge, and 0.52% Zr and reducing the data in the same manner as for the stratospheric particles. K-lines from these four elements encompass the usable energy range in the current analyses and experience grossly different absorption by the detector filter. The ratio of the measured elemental mass and the known concentration gives an estimate of the total particle mass for each of the four elements. The resulting total particle masses, 980 and 770 pg (Table 2), agree within each particle set (scatter of the four estimates are about 5% for each particle) and are at least qualitatively consistent with optically-estimated volumes (assuming a density of 2 g/cm³, optical estimates are 1000 and 400 pg, respectively).

TABLE 2. SRM K546 standard glass particle results.

Particle	Element	Element Mass (pg)	Apparent Total Particle Mass (pg)	Mean Particle Mass (pg)
546-A	Cr	1.43	1020	
	Ni	4.03	1030	980 ±50 (5%)
	Ge	4.67	930	
	Zr	4.98	950	
546-B	Cr	1.07	760	
	Ni	3.00	760	770 ± 50 (6%)
	Ge	3.63	720	
	Zr	4.51	860	

± = standard deviation on the mean.

RESULTS

Elemental contents for the five stratospheric particles are summarized in Table 3. All elements from Cr to Br with the exception of As were detected in the two chondritic particles (U2022G1, 25 µm, and W7029*A27, 15 µm). Elemental concentrations are within a factor of 1.7 of CI abundance for all of these elements for U2022G1 with the exception of Br, which is enriched 37-fold (Fig. 3). W7029*A27 is within a factor of two of CI for Cr through Zn plus Ge but enriched in the volatiles Cu, Ga, Se, and Br (enrichment factors of 2.1, 5.8, 3.5, and 8.4, respectively).

The third particle of the cosmic dust class, U2015G1, has high Zn concentrated in the main fragment (1.2%, also observed in the JSC spectrum). Chromium, Mn, Fe, and Ni are all depleted from CI in this sample by about the same factor (0.3 to 0.5 in the main fragment). Bromine contents of the two pieces of this particle are indistinguishable and enriched relative to

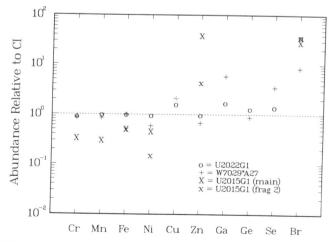

Fig. 3. CI normalized element abundances for particles of the cosmic dust class. Enrichments greater than a factor of two from CI include Br in all particles (8 to 37 times CI), Cu, Ga, and Se in W7029*A27 (2 to 6 times CI), and Zn in U2015G1 (38 times CI). Chromium, Mn, and Ni are depleted in the Fe-poor particle U2015G1.

TABLE 3. Trace element contents of five stratospheric particles.

Particle	Cr	Mn	Fe	Ni	Cu	Zn	Ga	Ge	Se	Br
Cosmic Dust										
U2022G1										
(pg)	12.4	10.2	1090	66	1.1	1.8	0.11	0.24	0.16	0.79
	±2.7	±2.0	±210	±13.2	±0.23	±0.37	±0.03	±0.05	±0.04	±0.16
(ppm)	2060	1700	18%†	1.09%	180	310	18	40	26	130
CI norm	0.92	1.00	1.00	0.96	1.6	0.97	1.7	1.3	1.4	37
W7029*A27										
(pg)	3.3	2.5	300	10.8	0.40	0.35	0.097	0.047	0.105	0.050
	±0.65	±0.48	±57	±2.3	±0.079	±0.068	±0.03	±0.01	±0.025	±0.015
(ppm)	2000	1480	18%†	6500	240	210	58	28	63	30
CI norm	0.88	0.86	1.00	0.59	2.1	0.67	5.8	0.88	3.5	8.4
U2015G1 (main)										
(pg)	4.5	3.1	550	31	0.86	73	<0.09	<0.11	<0.13	0.61
	±1.0	±0.72	±110	±6.2	±0.42	±15				±0.18
(ppm)	730	500	9%†	5000	140	1.2%	<30	<35	<40	100
CI norm	0.33	0.30	0.50	0.45	1.2	38	<3	<1	<2	28
U2015G1 (frag 2)										
(pg)	<1.5	<0.8	230	3.8	<0.15	3.3	<0.15	<0.18	<0.20	0.34
			±40	±0.7		±0.7				±0.10
(ppm)	<600	<300	9%†	1500	<60	1300	<60	<70	<80	130
CI norm	<0.3	<0.2	0.50	0.14	<0.50	4.3	<6	<2	<4	37
Terrestrial?										
W7027D2										
(pg)	0.17	0.11	3.0	0.18	63.5	0.95	<0.02	<0.02	<0.03	0.48
	±0.035	±0.023	±0.59	±0.038	±13.0	±0.20				±0.10
(ppm)*	148	96	2610	157	5.5%	827	<20	<25	<30	420
U2022C14										
(pg)	0.04	0.30	4.5	0.004	<0.004	0.12	<0.005	<0.007	<0.008	<0.01
	±0.011	±0.061	±0.91	±0.003		±0.025				
(ppm)*	112	840	1.3%	11	<10	336	<15	<20	<25	<30

± = 1 standard deviation uncertainty based on known random errors (counting statistics and mass of analyzed standard).

Experimental notes: acquisition times were U2022G1 = 10800 sec, W7029*A27 = 6240 sec, U2015G1 main = 3600 sec, U2015G1 frag 2 = 900 sec, W7027D2 and U2022C14 = 1800 sec; cosmic particles measured with Al detector filter.

*Concentrations based on optically-estimated volumes (W7027D2 = 575 μm^3, U2022C14 = 180 μm^3) and an assumed density of 2 g/cm^3.

†Fe content obtained from JSC EDS spectrum (see text).

CI by 30 times, comparable to the enrichment factor for U2022G1.

The two particles of suspected terrestrial origin, W7027D2 and U2022C14, possess low siderophile abundances (Fe = 0.26% and 1.3%, respectively, Ni <200 ppm in both). Copper accounts for 5.5% of W7027D2 but is undetected in U2022C14. Bromine is high in W7027D2 (420 ppm) but is undetected in the other (<30 ppm).

DISCUSSION

The emerging picture is that chondritic IDPs are near CI composition for a large number of elements of different geochemical character and volatility. Deviations from CI composition center around possible enrichments in volatile elements. Such deviations are important because they may aid in identifying the parent body (bodies).

PIXE results by *van der Stap et al.* (1986) on three chondritic particles showed up to 2-fold enrichments for Cu, Zn, Ge, and Se but 30-fold excesses in Br. They concluded that the volatile elements in general were enriched and suggested formation in a volatile-rich environment. The results reported here confirm that large Br enrichments are widespread in chondritic IDPs. However, a general volatile element enrichment is difficult to demonstrate because of the small and variable magnitude of fractionations in most enriched elements. Moderately volatile elements Cu, Zn, Ga, and Se in chondritic particle U2022G1, our most precise analysis, all deviate from CI abundance by less than a factor of 1.7. Bromine, enriched 37-fold, is the only exception. The data are consistent with Br being a single large anomaly. It is also conceivable that the main effect may instead be a halogen enrichment (although PIXE measurements on two IDPs show CI chlorine contents; *Wallenwein,* 1986). A halogen enrichment may be indigenous to these extraterrestrial specimens or may be a terrestrial contamination.

Measurements on stratospheric aerosols demonstrate the presence of Cl and Br in the present atmosphere (Cl = 2 ppbv; Br = 20 pptv; *Yung et al.,* 1980). Numerous terrestrial halogen sources have been identified, including volcanism, ocean-air interactions, and industrial pollution. Thus high Br in IDPs could reside in a coating acquired during atmospheric entry, stratospheric residence, and/or collection. However, mass balance considerations seem to require unrealistically large Br concentrations for such a coating. Adhering sulphur-rich droplets thought to be stratospheric sulfate aerosols have been observed on the surface of individual grains in chondritic IDPs (*Rietmeijer,* 1986). These sulfate aerosols, because of their small size, have longer stratospheric residence times than the IDPs and might absorb stratospheric halogens. Direct measurement by electron microprobe of the chemical composition of stratospheric sulfate aerosols seldom gives a detectable Cl signal (G. Ferry, personal comunication, 1987), and Br would be expected to be lower than Cl since the stratospheric Br/Cl ratio is 0.01 (*Yung et al.,* 1980). If we use a liberal value for the Br concentration of this coating of 1000 ppm (10 times higher than implied by a Cl detection limit of 1% in the aerosols and a Br/Cl ratio of 0.01) and a surface to volume ratio of 10^4 for a 1-ng particle (the maximum surface/volume ratio for an aggregate of 0.1-μm spheres), a uniform layer about 1000 Å thick is required to account for the observed whole particle content. Coatings of this thickness have not been measured by TEM (*Bradley and Brownlee,* 1986). Direct measurements of the thickness of the contamination on an aluminum oxide (rocket exhaust) sphere collected in the stratosphere were reported by *Mackinnon and Mogk* (1985) using a Scanning Auger Microprobe. The detected sulphur-rich layer contained Cl but was less than 150 Å thick.

Rietmeijer (1986) noted the presence of silica-rich glass and tridymite fragments in a chondritic IDP. Even if these fragments are terrestrial, their miniscule mass relative to the total particle mass requires percent level Br (well above the 100 ppm level reported for some volcanic ash; *Kotra et al.,* 1983) for these to be the source of the bulk Br.

Further evidence suggesting the Br in the cosmic dust particles is not a stratospheric contaminant is that no Br was detected by PIXE in an FSN (Fe, S, Ni) sphere measured by van der Stap (personal communication, 1987). However, since FSN particles have a higher density, and thus a shorter stratospheric residence time, they might accumulate less stratospheric contaminant than chondritic particles.

If the Br enrichment proves to be preterrestrial, the ubiquity of the effect would indicate that samples of the parent body (bodies) of these particles are absent in the meteorite collection. In the carbonaceous chondrites, Br increases by a factor of ten from the CO class (Allende Br = 0.5 ppm; *Clarke et al.,* 1970) to the CI class (Ivuna Br = 5.08 ppm; *Anders and Ebihara,* 1982). Even within the CIs, Br varies by a factor of two from 2.35 ppm in Alais to 5.08 ppm in Ivuna, with the 3.62 ppm concentration in Orgueil near the CI mean (*Anders and Ebihara,* 1982).

High Br contents have been reported for the possibly cometary projectile responsible for the Tunguska event of 1908. *Golenetskii et al.* (1977) attempted to estimate the chemical composition of the projectile by measuring the elemental abundances in layers of a 70-cm-long peat core from the site. While Br was essentially constant in the 27 layers above or below the Tunguska catastrophe layer, the layer itself was enriched by a factor of 15. A Br concentration of 960 ppm in the projectile was estimated, a factor of 7 higher than the 130 ppm in U2022G1 and 300 times CI. It should be pointed out, however, that there are difficulties with their technique. Quantitative determination of the parent body concentration of a trace element depends on an accurate estimate of the total mass of the projectile matter present in the peat layer. Although the accuracy of their total mass estimate is difficult to establish, the sharp positive Br anomaly associated with the Tunguska layer suggests the existence of an extraterrestrial object with Br well above CI concentration.

An assessment of the Br concentration of the meteoritic component in lunar soil suggests a value between 1 and 2 times CI (*Anders et al.,* 1973). Comparisons of these results with those for cosmic dust particles are complicated by the larger particle size, which contributes the majority of the lunar meteoritic component (10^{-2} to 10^{-6} grams; *Hughes,* 1978) and the extent of volatile loss during impact on the Moon.

Substantial Cl was detected by the PUMA-1 mass spectrometer in some particles during the Comet Halley encounter by Vega 1 (*Brownlee et al.,* 1987). This could indicate a cometary enrichment in halogens over the CI abundances, although instrumental contamination has been suggested as a possible source.

The high Br content of the apparently terrestrial aluminum-rich particle W7027D2 may suggest that Br concentration should not be taken to be an indicator of extraterrestrial origin. W7027D2 is unusual, however, in that it exhibits a near chondritic Fe/Ni ratio of 17. The ratio for the other apparently terrestrial particle (U2022C14), on the other hand, is 1100, near the average crustal value of 800. *Flynn et al.* (1982) have noted the existence of a subset of the Al-rich stratospheric particles that exhibit chondritic Fe/Ni and Mg/Si ratios, and suggested this chemical similarity to chondritic meteorites as well as the presence of common meteoritic minerals made these particles possible candidates for extraterrestrial materials. Subsequent work by *Zolensky* (1985) identified Al-rich particles containing minerals typical of refractory inclusions in meteorites. *McKeegan et al.* (1986) established by oxygen isotopic measurements that some of these Al-rich particles are extraterrestrial. Mineralogical and/or isotopic analyses on W7027D2 should determine if this particle is a member of this extraterrestrial class.

Additional analyses on particles of obvious terrestrial origin, such as volcanic ash and aluminum oxide spheres, may help to constrain the magnitude of contamination by terrestrial sources of Br. Experiments designed to reveal the spatial distribution of halogens in these specimens, e.g., etching treatments or ion microprobe analyses, may indicate if Br is a surface-correlated contaminant or is distributed throughout the bulk of the particle.

SUMMARY

Trace element analyses have been made on five stratospheric particles, three of the cosmic dust class and two classified as probable terrestrial contamination, using synchrotron X-ray fluorescence for the first time. Elements between Cr and Br were detected at concentrations as low as 10 ppm after one-hour data acquisition times. Concentrations for chondritic particle U2022G1 are within a factor of 1.7 of CI for all elements detected with the exception of Br, which is 37 times CI. Chondritic particle W7029*A27 is also near CI for Cr, Mn, Fe, Ni, Zn, and Ge but enriched in Cu, Ga, Se, and Br by factors of 2.1, 5.8, 3.5, and 8.4, respectively. The third particle of the cosmic dust class U2015G1 containing 1.2% Zn also showed high Br, enriched relative to CI by a factor of 28. These results for Br confirm previously reported PIXE results on three chondritic IDPs. All five chondritic IDPs analysed by PIXE or SXRF to date possess substantial Br enrichments relative to CI. Bromine was also detected at a high level in an aluminum-rich particle W7027D2 classified as probable artificial terrestrial contamination but exhibiting a chondritic Fe/Ni ratio. The presence of Ca in this Al-rich particle suggests it may be a member of a class of refractory extraterrestrial particles. Bromine was not detected in a fifth particle U2022C14, also classified terrestrial and exhibiting a crustal Fe/Ni ratio. Additional analyses on obvious terrestrial particles are required to evaluate quantitatively the contamination of IDPs by terrestrial halogen sources. Future SXRF analyses will be performed in vacuum or helium atmosphere to eliminate Ar X-rays and allow Cl and Br determinations to be made on the same particle. Further trace elements analyses on cosmic dust are essential to identify particles of common parentage. These results may ultimately aid in distinguishing asteroidal from cometary matter because of suspected differences in their physio-chemical histories.

Acknowledgments. We are grateful to Keith W. Jones, head of the Atomic and Applied Physics Division, Department of Applied Science, Brookhaven National Laboratory, who is responsible for development and operation of the SXRF facility and to the staff of the National Synchrotron Light Source for providing beam time for this research. Mark Rivers (Univ. of Chicago/BNL) provided invaluable technical assistance. Helpful discussions with J. V. Smith, D. Brownlee, M. Zolensky, D. Mogk, and G. Ferry are gratefully acknowledged. We thank the Cosmic Dust Curatorial Facility (JSC) and especially M. Zolensky (JSC) for kindly supplying samples for these measurements. This manuscript benefitted from the reviews by D. Burnett, K. Keil, and S. Sandford. This work was supported by the National Science Foundation (Grant No. EAR-8313683, J. V. Smith) and the National Aeronautics and Space Administration (Grant No. NAG 9-106, J. V. Smith). The Brookhaven facilities are also supported by the U.S. Department of Energy, Division of Materials Sciences and Division of Chemical Sciences (Contract No. DE-ACO2-76CH00016).

REFERENCES

Anders E. and Ebihara M. (1982) Solar system abundances of the elements. *Geochim. Cosmochim. Acta, 46,* 2363-2380.

Anders E., Ganapathy R., Krahenbuhl U., and Morgan J. W. (1973) Meteoritic material on the moon. *The Moon, 8,* 3-24.

Bradley J. P. and Brownlee D. E. (1986) Cometary particles: thin sectioning and electron beam analysis. *Science, 231,* 1542-1544.

Brownlee D. E., Wheelock M. M., Temple S., Bradley J. P., and Kissel J. (1987) A quantitative comparison of Comet Halley and carbonaceous chondrites at the submicron level (abstract). In *Lunar and Planetary Science XVIII,* pp. 133-134. Lunar and Planetary Institute, Houston.

Clarke R. S., Jr., Jarosewich E., Mason B., Nelen H., Gomez M., and Hyde J. R. (1970) The Allende, Mexico, meteorite shower. *Smithson. Contrib. Earth Sci., 5,* 1-53.

Criss J. (1977) *NRLXRF* Naval Research Laboratory Cosmic Program #DOD-00065.

Flynn G. J., Fraundorf P., Keefe G., and Swan P. (1982) Aluminum-rich spinel aggregates in the stratosphere. From Earth or not? (abstract). In *Lunar and Planetary Science XIII,* pp. 223-224. Lunar and Planetary Institute, Houston.

Fraundorf P., Brownlee D. E., and Walker R. M. (1982) Laboratory studies of interplanetary dust. In *Comets,* pp. 383-409. Univ. of Arizona Press, Tucson.

Ganapathy R. and Brownlee D. E. (1979) Interplanetary dust: trace element analyses of individual particles by neutron activation. *Science, 206,* 1075-1077.

Golenetskii S. P., Stepanok V. V., Kolesnikov E. M., and Murashov D. A. (1977) The question about the chemical composition and nature of the Tunguska cosmic body. *Sol. Syst. Res., 11,* 103-113.

Gordon B. M. (1982) Sensitivity calculations for multielemental trace analysis by synchrotron radiation induced x-ray fluorescence. *Nucl. Instrum. Meth., 204,* 223-229.

Hughes D. W. (1978) Meteors. In *Cosmic Dust* (J. A. M. McDonnell, ed.), pp. 123-185. Wiley, New York.

Jones K. W., Gordon B. M., Hanson A. L., Hastings J. B., Howells M. R., Kraner H. W., and Chen J. R. (1984) Application of synchrotron radiation to elemental analysis. *Nucl. Instrum. Meth., 231,* 225-231.

Kotra J. P., Finnegan D. L., Zoller W. H., Hart M. A., and Moyers J. L. (1983) El Chichon: composition of plume gases and particles. *Science, 222,* 1018-1021.

Mackinnon I. D. R. and Mogk D. W. (1985) Surface sulphur measurements on stratospheric particles. *Geophys. Res. Lett., 12,* 93-96.

McKeegan K. D., Walker R. M., and Zinner E. (1985) Ion microprobe isotopic measurements of individual interplanetary dust particles. *Geochim. Cosmochim. Acta, 49,* 1971-1987.

McKeegan K., Zinner E., and Zolensky M. (1986) Ion probe measurements of O isotopes in refractory stratospheric dust particles: proof of extraterrestrial origin. *Meteoritics, 21,* 449-450.

Rietmeijer F. J. (1986) Silica-rich glass and trydimite in a chondritic porous aggregate: Evidence for stratospheric contamination of interplanetary dust particles (abstract). In *Lunar and Planetary Science XVII,* pp. 708-709. Lunar and Planetary Institute, Houston.

Sandford S. and Walker R. M. (1985) Laboratory infrared transmission spectra of individual interplanetary dust particles from 2.5 to 25 microns. *Astrophys. J., 291,* 838.

Sutton S. R., Rivers M. L., and Smith J. V. (1986) Synchrotron x-ray fluorescence: diffraction interference. *Anal. Chem., 58,* 2167-2171.

van der Stap C. C. A. H., Vis R. D., and Verheul H. (1986) Interplanetary dust: arguments in favour of a late stage nebular origin (abstract). In *Lunar and Planetary Science XVII,* pp. 1013-1014. Lunar and Planetary Institute, Houston.

Wallenweim R. (1986) Ph.D. thesis, Ruprecht-Karls-Universität Heidelberg.

Wopenka B. (1987) Raman observations of individual interplanetary dust particles (abstract). In *Lunar and Planetary Science XVIII*, pp. 1102-1103. Lunar and Planetary Institute, Houston.

Yung Y. L., Pinto J. P., Watson R. T., and Sander S. P. (1980) Atmospheric bromine and ozone perturbations in the lower stratosphere. *J. Atmos. Sci., 37*, 339-353.

Zolensky M. (1985) CAI's among the cosmic dust collection. *Meteoritics, 20*, 792-793.

Zolensky M. E., Mackinnon I. D. R., and McKay D. S. (1984) Towards a complete inventory of stratospheric dust with implications for their classification (abstract). In *Lunar and Planetary Science XV*, pp. 963-964. Lunar and Planetary Institute, Houston.

Analytical Electron Microscopy of a Hydrated Interplanetary Dust Particle

David F. Blake[1], A. J. Mardinly[2], C. J. Echer[3], and T. E. Bunch[1]

[1] MS 239-4, NASA/Ames Research Center, Moffett Field, CA 94035
[2] Lockheed Missiles and Space Co., Inc., Sunnyvale, CA 94086
[3] National Center for Electron Microscopy, Lawrence Berkeley Laboratory, Berkeley, CA 94720

A hydrated Interplanetary Dust Particle (IDP), IDP # Ames-Dec86-11, was selected for study from a number of IDPs collected by U-2 aircraft from Ames Research Center. The particle consists primarily of a relatively nonporous aggregate of fine-grained layer silicates, some of which are *in situ* hydrous alteration products of pre-existing grains. The particle shows no apparent alteration due to its deceleration upon atmospheric entry. The layer silicates have a bimodal size distribution, in which "matrix" phyllosilicates have an apparent grain size of 10-50 nm, and phyllosilicates that pseudomorphically replace pre-existing grains have a grain size of 1-10 nm. Despite this order of magnitude difference in crystallite size, both phases are smectites, according to quantitative analytical and electron diffraction data. Euhedral to subhedral pyrrhotites, which have a grain size of 0.1-1.0 μm, have high nickel contents. Pre-existing grains that have been pseudomorphed by clays are commonly surrounded or decorated with fine-grained (10-20 nm) low-nickel pentlandite. Very fine grained (1-10 nm) magnetite occurs in clusters throughout the matrix. Several fragments of a Mg-Fe silicate phase, apparently a glass, are present. The particle is texturally and mineralogically distinct from the matrix of carbonaceous chondrites. It is akin to carbonaceous chondrites only in the sense that it appears to have been subject to alteration processes similar to those described by *Bunch and Chang* (1980) for protoplanetary bodies. If the particle had a cometary origin, its diagenesis/alteration does not appear to fit within the P-T regime typically postulated for such environments. The alteration processes that occurred in this hydrated IDP require far more element mobility than seems possible in the solid state and suggest the presence at some point of a liquid water phase.

INTRODUCTION

With the exception of samples returned during the Apollo program, the only materials available for direct laboratory study that are of confirmed extraterrestrial origin are meteorites and Interplanetary Dust Particles (IDPs). IDPs are a relatively new class of extraterrestrial materials that are collected by high flying aircraft in the stratosphere (*Brownlee*, 1979; *Mackinnon and Rietmeijer*, 1987). These particles, which range from submicron to greater than 50 microns in size, enter the earth's atmosphere at ballistic velocities, but are sufficiently small to be decelerated without burning up.

IDPs often have solar elemental abundances (*Brownlee et al.*, 1974), contain many of the phases and commonly exhibit the phase relationships that have been proposed for presolar system material. One class of IDPs, Chondritic Porous (CP), is composed of loosely aggregated grains of high temperature phases such as olivines and pyroxenes, which appear in at least some cases to have been deposited directly from the gaseous state. However, other IDPs such as the one described in this paper are composed in part of layer silicates which may be products of dissolution-reprecipitation processes in a parent body of some kind.

It is commonly thought that IDPs of the chondritic porous variety are cometary in origin (*Mackinnon and Rietmeijer*, 1987, and references therein). The origin of hydrous IDPs, however, is less certain. High Ni contents and/or elemental abundances close to those for chondritic material (*Anders and Ebihara*, 1982) are unmistakable evidence for their extraterrestrial genesis; beyond this, the exact origin of the material is unknown. The phyllosilicates contained within hydrous IDPs may be sensitive indicators of secondary processes that have occurred in or on their parent bodies.

Early work by *Brownlee* (1978) using X-ray diffraction, and by *Fraundorf et al.* (1981) using infrared spectroscopy established the presence of layer silicates in IDPs. However, these bulk techniques could not be used to elucidate the phase relationships, chemistry, or detailed structure of the phyllosilicates involved. *Mackinnon and Rietmeijer* (1983), *Rietmeijer and Mackinnon*, (1984, 1985) and *Tomeoka and Buseck* (1984a,b, 1985, 1986) used Analytical Electron Microscopy (AEM) to characterize a number of hydrated IDPs. *Mackinnon and Rietmeijer* (1983), and *Rietmeijer and Mackinnon* (1984, 1985) describe layer silicates from CP Aggregate W7029*A that may have grown in a fluid or gaseous medium unhampered by neighboring phases. The chemical and diffraction data obtained from the phyllosilicate phases are consistent with kaolinite, serpentine, and mica. In addition, there are weakly crystalline layer silicates that vary in chemistry and structure and are similar to terrestrial amesite, berthierine, and pyrophyllite. These authors suggest that some of the layer silicates may have formed as alteration products of previously existing silicate phases at low temperatures after aggregate formation. The differences in chemistry between phyllosilicate replacement phases are ascribed to differences in chemistry in the original phases. *Tomeoka and Buseck* (1984a) describe a hydrated particle ("LOW-CA") that consists of a smectite or mica, various Fe-Ni sulfides, magnetite, and minor olivine. These authors conclude that hydrated IDPs are mineralogically and texturally quite distinct from the matrices of CI and CM meteorites, and suggest that the assemblage of minerals present in hydrated IDPs results from a low-temperature alteration process. *Tomeoka and Buseck* (1984b, 1985, 1986) describe a second hydrated particle that, like "LOW-CA," contains smectite, but in addition contains an Fe-rich phyllosilicate and is relatively poor in Fe-Ni sulfides. The authors conclude that the difference in mineralogy between phyllosilicates in CI and CM meteorites and those of hydrous IDPs may be the result of differing alteration in their principal parent bodies, asteroids and comets respectively.

In all of these previous studies, observations were made of crushed grain mounts in whic phase relationships are destroyed or obliterated and sample thickness is largely a function of particle size or resistance to crushing. In the present study, ultramicrotomy is used for sample preparation. This technique results in uniformly thin sections that retain intact the spatial distribution of phases. Interpretations can therefore be made of phase equiilibria, alteration sequences, and the like. However, because the sample preparation technique includes embedding with a carbon containing resin, the ability to analyze for carbon containing phases is compromised.

METHODS

For preparing both the IDP and the Murchison carbonaceous chondrite material, an adaptation of the method described by *Bradley and Brownlee* (1986) was used. First, a four-sided pyramid is fashioned on the tip of a 5/16" plastic rod, and a very small trapezoidal block face is ground on the tip as detailed in *Blake et al.* (1987). Next, a droplet of LR White™ "hard" grade resin is placed on the tip using a glass fiber. The resin is heat-cured at 90°C for 10 minutes until tacky. Individual IDPs or small grains of meteorite material are then transferred to the tacky surface of the block face using a glass fiber. A second thick coat of resin is applied over the sample, and heat cured until hard. Sections of the material can be obtained by microtoming the cured block face. Microtomy was performed using a Porter-Blum MT-2 ultramicrotome equipped with a diamond knife. Sections were cut fairly rapidly until portions of the particle could be seen in the sections floating in the knife trough. Then, the knife speed was slowed considerably and adjustments were made to cut the thinnest possible sections. When the sections appeared silver to gold in reflected light (<90-150 nm thickness), they were picked up on holey carbon support films on 200 mesh 3 mm copper grids. Grids were surveyed for analyzable areas after evaporating a thin coat of carbon on them to eliminate sample charging.

Bright field imaging was performed at 100 kV accelerating voltage, using an Hitachi H-500 Transmission Electron Microscope (TEM) with double tilt specimen stage. High resolution imaging and conventional Scanning Transmission Electron Microscopy/Energy-dispersive X-ray Spectroscopy (STEM/EDS) microanalysis was perfomed at 200 kV on a JEOL 2000FX AEM equipped with a Kevex 8000 microanalysis system. Light element analysis (Na and below) was performed at 200 kV on a JEOL 200CX equipped with a Kevex UTW (Ultra-Thin Window) detector and 8000 analyzer. The Kevex UTW detector is capable of detecting all elements above beryllium in the periodic table. Unless otherwise stated, analyses were recorded using a speciment holder kept at liquid nitrogen temperature to minimize sampling contamination and element volatilization. Quantitative analyses were performed using K-factors generated from standards, and applying light element absorption corrections (*Williams*, 1984). For the absorption corrections, specimen thickness was determined using the contamination spot technique (*Williams*, 1984). Absorption corrections were determined using a computer program provided by Kevex. In general, 1,000 second counting times were used to detect secondary X-rays, which resulted in an analytical precision based on counting statistics alone of less than 0.5% of amount present for the major elements. For quantitative analyses, results are considered accurate to within ±5% of the amount of element present, primarily as a result of uncertainties in K-factor determination and element mobility and electron beam damage during analysis. Semiquantitative analyses were performed without standards, using theoretical K-factors, and have an estimated accuracy of ±20% of the amount of element present.

Meteorite matrix bulk analyses were performed on a Cameca-MBX electron microprobe operated at 15 kV and 5 nA beam current with on-line ZAF data correction. A 20 μm × 20 μm area on polished thin sections was rastered for each sample analysis. Because of the size of the analyzed area, analyses of individual grains of layer silicates were not achieved. The size of the rastered area is roughly equivalent to the average size of a typical IDP.

RESULTS

General Observations

Figure 1 shows a bright field, low magnification TEM image of a microtomed section of IDP Ames-Dec86-11. The particle is approximately 25 μm in its longest dimension, and triangular in shape. It is a Chondritic Filled (CF) particle, according to the terminology of *Brownlee et al.* (1982), and consists of a variety of mineral grains, from 1 μm to less than 1 nm, in a matrix of phyllosilicate minerals. Major phases identified in the particle include Fe-Ni sulfides such as pyrrhotite and pentlandite (10 nm-1.0 μm), clusters of magnetite (1-10 nm), and matrix and grain replacement smectite clays. Additionally, several grains of an unknown phase were found that contain Mg, Si, and Fe, and appear to be silicate glass.

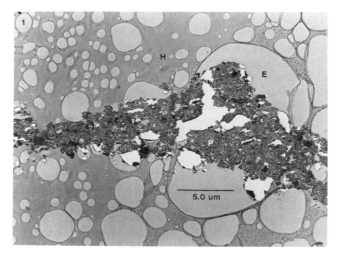

Fig. 1. Low magnification view of Hydrous IDP. P = microtomed IDP, E = embedding medium, H = Holey carbon support film. Scale Bar = 5 μm.

TABLE 1. Cation/silicon ratios for chondritic abundance (from *Anders and Ebihara,* 1982), CM matrix, and Ames-Dec86-011.

Material	Mg/Si	Fe/Si	Ca/Si	Al/Si	S/Si	Ni/Si	Na/Si	K/Si	O/Si
Chondrite	1.03	0.87	0.06	0.08	0.431	0.049	0.055	0.004	4.3
Murray CM*	0.92	0.81	0.03	0.11	0.25	0.12	0.05	0.07	4.27†
Murchison CM*	0.93	0.92	0.03	0.11	0.33	0.12	0.03	0.01	4.31†
AmesDec86-011‡	0.59	0.78	0.04	0.11	0.61	0.05	0.14	0.03	3.80

*Bulk analysis, 20 microprobe analyses, 400 μm^2 each analysis.
†Oxygen by difference.
‡Raster scan of large area of microtomed particle, STEM mode.

An overall analysis of the particle was made in order to compare the elemental abundances of the IDP to chondritic abundances (*Anders and Ebihara,* 1982) and elemental abundance data from two representative carbonaceous chondrites. This analysis was recorded with the sample at liquid nitrogen temperature using a UTW detector. The results are shown in Table 1.

TABLE 2. Cation/silicon ratios for microanalyses of matrix and grain smectite.

Element	Grain Smectite			Matrix Smectite		
	(1G)	(2G)	(3G)	(1M)	(2M)	(3M)
O	2.53	*	*	2.47	*	*
Mg	0.38	0.19	0.22	0.45	0.27	0.16
Al	0.17	0.15	0.09	0.16	0.13	0.06
Si	1.0	1.0	1.0	1.0	1.0	1.0
Fe	0.38	0.36	0.30	0.30	0.32	0.51
Ni	n.d.	n.d.	0.02	n.d.	n.d.	0.06
Zn	0.02	n.d.	n.d.	0.02	n.d.	n.d.

n.d. = not detected.
*Not measured.

Analysis of Phases and Phase Relationships

Phyllosilicates. Phyllosilicates are present both as "matrix" material, and as apparent pseudomdorphic replacements of pre-existing grains. The matrix phyllosilicate crystallite (or mosaic domain) size is on the order of 10-50 nm, while that replacing grains are 1-10 nm in size. Selected area electron diffraction patterns (SAED) of matrix phyllosilicates have characteristic spacings of 0.52, 0.442, 0.253, 0.194, 0.165, and 0.151 nm. A basal spacing of 1.02 nm was measured from lattice fringe images of the matrix material. These spacings are in reasonable accord with those listed for smectite from IDPs in *Bradley et al.* (1987) and *Tomeoka and Buseck* (1984a, 1986).

Quantitative and semiquantitative analyses of both types of clays were performed, the results of which are shown in Table 2. Analyses 1G and 1M are fully quantitative analyses of the grain and matrix material respectively, using a UTW detector. The trace amounts of Zn and Ni in some analyses may be due to the presence of these elements in matrix material intimately intermixed with the clay. Analyses #s 2 and 3 are semiquantitative analyses, performed at ambient temperature using a detector sensitive to all elements above fluorine. These

TABLE 3. Formulas, octahedral/tetrahedral cation ratios, and basal plane spacings for IDP clay and a number of candidate phyllosilicate phases.

Mineral	Composition*	Ratio†	Basal Spacing
Grain Replacement Clay	$[Mg_{1.53}Fe_{1.52}Al_{.68}Zn_{.08}](Si_4)O_{10.12}$‡	0.95	1.02 nm
	$[Mg_{1.31}Fe_{1.30}Zn_{.06}](Si_{3.42}Al_{.58})O_{10.12}$§	0.67	
Matrix Clay	$[Mg_{1.81}Fe_{1.18}Al_{.64}Zn_{.07}](Si_4)O_{9.89}$‡	0.93	1.02
	$[Mg_{1.56}Fe_{1.02}Zn_{.06}](Si_{3.45}Al_{.55})O_{9.89}$§	0.66	
Serpentine	$[M]_6(Si)_4O_{10}(OH)_8$	1.5	0.7-0.73
Septechlorite	$[M,Al]_6(SiAl)_4O_{10}(OH)_8$	1.5	0.7-0.71
Talc	$[M]_3(Si)_4O_{10}(OH)_2$	0.75	0.93
Chlorite	$[M,Al]_6(Si,Al)_4O_{10}(OH)_8$	0.75	1.4
Mica	$\{K,Na\}[M,Al]_3(Si_3Al)O_{10}(OH)_2$	0.75	1.0
Smectite (nontronite)	$[Fe,Al]_2(SiAl)_4O_{10}(OH)_2$	0.5	1.0
Smectite (saponite)	$[Mg,Al]_3(SiAl)_4O_{10}(OH)_2$	0.75	1.0

*{ } = 12-fold coordination, [ ] = octahedral coordination, and ( ) = tetrahedral coordination.
†Ratio of octahedral to tetrahedral cations.
‡All aluminum placed in octahedral sites.
§All aluminum placed in tetrahedral sites.

Fig. 2. Bright field TEM micrograph of replaced grain. Po = pyrrhotite, SG = smectite replacing grain, SM = smectite replacing matrix, Pn = pentlandite grains, M = magnetite in matrix. Scale Bar = 0.5 μm.

results are given to provide some notion of heterogeneity between analyses, since they were recorded from different areas within the IDP. All analyses were of necessity recorded from ~0.1 μm² areas of the clay that possibly contain other phases, since analyzing smaller areas results in increased potential for element loss and beam damage (*Mackinnon et al.*, 1986).

In smectite, neglecting water of hydration, the (Si+Al)/O ratio is approximately 3. If one assumes as a limit that there is no Al in the tetrahedral sites, the Si/O ratio would be about 3. If electron beam damage during analysis resulted in the loss of hydroxyl groups, the Si/O ratio expected would be ~2.5, which is very close to what is observed. Table 3 lists formulae, octahedral/tetrahedral cation ratios, and basal plane spacings for the observed clay minerals, and a number of candidate phyllosilicates for matching. If one compares the basal spacing of the IDP phyllosilicate with that of other minerals, the remaining candidate structures are mica and smectite. Utilizing qualitative and quantitative elemental data, only smectite appears to be a feasable candidate.

Figure 2 is a bright field image showing a pseudomorphically replaced grain (SG) replaced by smectite, in a matrix of other minerals including smectite (SM), pentlandite (Pn), pyrrhotite (Po), and magnetite (M). The difference in grain size (or "mosaic crystallite size") between the matrix and grain replacement smectite is readily apparent, as is the very poorly crystalline character of the material. The upper portion of the replaced grain abuts a grain of pyrrhotite. The juxtaposition of these two grains may be serendipitous, or the pyrrhotite may be a remnant of the original grain that was replaced by the smectite. Arguments for and against this interpretation are presented in the discussion session. Figures 3a and 3b are lattice fringe images of the two types of smectite. Matrix smectite consists of ~5 nm × ~50 nm, crystallites that appear to be oriented relatively randomly. The material surrounding the diffracting crystallites we believe to be a combination of phyllosilicate crystallites that are not in the diffracting condition,

and the embedding medium. Grain smectite consists of packets of 1-3 layers of phyllosilicate, randomly oriented. The material is easily beam damaged, and areas become nearly amorphous in a few seconds observation time. Figure 4a shows a higher magnification image of matrix smectite crystallites. In some areas, octahedral and tetrahedral layers can be discriminated (see arrow). Smaller spacings can be seen (arrowed) which are consistent with an identification of the clay as smectite.

One large, well crystallized grain of phyllosilicate was observed. Figure 4b shows a lattice fringe image of the basal spacing, which was found to be 0.93 nm. No elemental information was obtained from this grain, but a tentative identification of talc is made based on lattice fringe measurements and diffraction patterns.

Fe-Ni sulfides. Iron-nickel sulfides occur throughout the matrix. They range in size from ~1.0 μm to 10 nm, and grains that could be identified by electron diffraction or quantitative

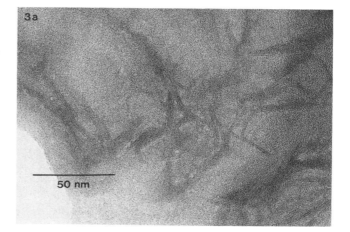

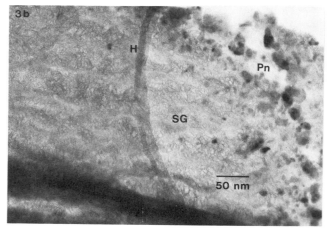

Fig. 3. (a) High resolution, lattice fringe image of matrix smectite. The weakly crystalline character of the matrix material is apparent. The material surrounding individual crystallites of the clay is believed to be clay material that is out of the diffraction condition. Scale bar = 50 nm. (b) High resolution, lattice fringe image of grain replacement smectite. SG = grain replacement smectite, Pn = Pentlandite grains, H = edge of hole in holey C film. Scale bar = 50 nm.

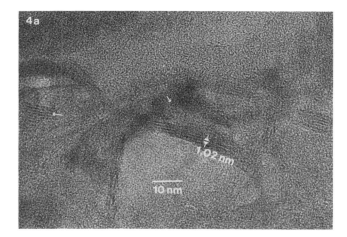

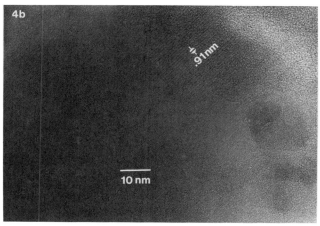

Fig. 4. (a) High resolution lattice fringe image of matrix clay, scale bar = 10 nm. Arrows mark areas in which subunit cell resolution is shown. (b) High resolution lattice fringe image of a large clay or mica mineral, believed to be talc. Scale bar = 10 nm.

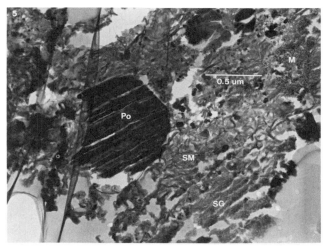

Fig. 5. Pyrrhotite grain in matrix. Po = pyrrhotite, SM = smectite replacing matrix, SG = smectite replacing grain, M = magnetite. Scale bar = 0.5 μm.

analysis were either high nickel pyrrhotite or low nickel pentlandite. Figure 5 shows a particular large pyrrhotite for which quantitative elemental data (shown in Table 4), and diffraction data were obtained. The irregular linear features are artifacts of the microtoming procedure and are not original features of the mineral. Quantitative results are shown in Table 4 for three other sulfides, including 10-20 nm pentlandite grains (analysis #4 in Table 4) that surround the pseudomorphically replaced grain shown in Fig. 3b. As pointed out by *Tomeoka and Buseck* (1984a), low-Ni pentlandite is a rare phase that is not known from terrestrial samples. It clearly appears to be a secondary phase in this IDP, and its presence may indicate a particular mode or mechanism of diagenesis. A number of other grains were analyzed qualitatively and contain major Fe and S, with minor Ni. Figure 6 shows some smaller pyrrhotite grains for which quantitative analyses were obtained (analyses 2-3, Table 4).

Magnetite. Another phase that is abundant in this IDP is magnetite. Magnetite, identified by both electron diffraction and semiquantitative elemental analysis, occurs as ~10 nm sized subhedral grains in 0.2-0.8 μm patches or clusters throughout the matrix. Typical clusters of magnetite are shown in Fig. 6 as well as Figs. 5 and 2.

Mg-Fe silicate phase. Several grains of a phase that contain primarily O, Mg, Fe, and Si are present in the IDP. The grains very in size from 0.5 μm to 0.1 μm, and appear to have no diffraction pattern. Quantitative analyses for three of the grains are shown in Table 5, compared with a quantitative analysis of a "tar ball" reported by *Bradley and Brownlee* (1987). The phase we believe to be an amorphous silica glass, since too little oxygen is present to balance with silicon, and the cation/silicon ratios do not correspond to any known silicates.

DISCUSSION AND CONCLUSIONS

Secondary Nature of Some Phases and Mechanism of Pseudomorphic Replacement

While a number of authors have alluded to the secondary nature of some phyllosilicates in hydrous IDPs, the IDP described here is the only one in which *in situ* phase relationships maintained throughout the sample preparation

TABLE 4. Composition of Fe-Ni sulfides in Ames-Dec86-11, given in atomic element %.

Element	Po*	Po†	Po‡	Pn§
Fe	42.14	45.6	43.14	50.65
Ni	2.80	4.08	7.44	6.10
S	55.06	50.3	49.42	43.29
Ni/Fe	0.07	0.09	0.17	0.12
Fe+Ni/S	0.82	0.99	1.02	1.17

*Large pyrrhotite, shown in Fig. 6.
†Diamond shaped grain in Fig. 6.
‡Grain G3.
§Small grains decorating replacement smectite in Fig. 3.

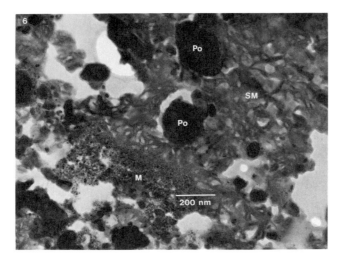

Fig. 6. Sulfide grains in smectite matrix. Po = pyrrhotite, Sm = matrix smectite, M = magnetite. Scale bar = 200 nm.

procedure demonstrate unequivocally the secondary nature of some phases. In addition to the grain replacement, smectite, pentlandite, and magnetite appear to be secondary based on their morphology, size, and phase relationships. The mechanism for emplacement of the matrix smectite is problematical. If one considers the matrix smectite to be secondary, it follows that the source of the material could either be external or internal. That is to say, the volume now occupied by the matrix could have originally been pore space, or it could have been a precursor phase of some kind. We favor invoking a precursor phase of some kind for the following reasons: (1) Texturally, there do not appear to be any alterations (flow structure, compaction, or the like), indicative of infilling matrix; and (2) the bulk composition of the IDP is low in Mg relative to chondritic or solar abundance ratios (Table 1). Since the matrix is Mg-rich relative to other phases, simply removing the matrix clay to render a pre-alteration bulk chemistry would make the material even less chondritic. Philosophically speaking, one would expect the more "primitive" material to be closer to chondritic abundances, not the opposite. The pre-alteration matrix material presumably would have a bulk chemistry similar

TABLE 5. Comparison of cation/silicon ratios from "Unknown Phase," to "Tar Ball" of *Bradley and Brownlee*, 1987.

Element	"Tar ball"	Unk 1	Unk 2	Unk 3
O	*	1.64	2.08	*
Mg	0.777	0.55	0.94	0.65
Al	0.102	0.01	n.d.	n.d.
Si	1.0	1.0	1.0	1.0
S	0.327	0.04	0.02	n.d.
Ca	0.38	0.11	n.d.	n.d.
Cr	0.011	0.02	n.d.	n.d.
Fe	0.481	0.20	0.05	0.04
Ni	0.032	n.d.	n.d.	n.d.

n.d. = not detected.
*Not measured.

to that of the matrix replacement smectite. It is most likely that the matrix smectite either originated as an alteration product of a pre-existing fine-grained crystalline phase, or as the product of devitrification of an original glassy matrix of similar composition.

The grain replacement smectite is clearly secondary, although the identify of the precursor phases is not evident. It is possible that the precursor phases were sulfides, in which case one must allow for a flux of elements into and out of the altered material. This is not a serious restriction in terms of mass balance, since replaced grains compose no more than a few percent of the total particle. Secondary pentlandite and magnetite in the matrix could be the result of such a process. Silicon, oxygen, and magnesium could have easily been provided by the surrounding matrix phase. The presence of euhedral and subhedral pyrrhotites in the matrix suggests, however, that sulfides were not subject to extensive dissolution during the history of the particle.

Comparison with Carbonaceous Chondrite Matrix

The hydrous IDP described here is dissimilar to carbonaceous chondrite matrix in terms of the phases present, texture, and elemental composition. By way of comparison, Fig. 7 shows low magnification and high magnification lattice fringe images of a microtomed section of Murchison matrix. The major phase in the IDP is smectite, which is rare in carbonaceous chondrites. Although smectite (montmorillonite) was tentatively identified in Orgeuil by *Bass* (1971) and *Caillere and Rautureau* (1974) by X-ray and TEM methods, the identifications were made on the basis of partial data, i.e., either structural or elemental data were obtained, but never both from the same sample. Mg- or Mg, Al-rich layer silicates in CM carbonaceous chondrites are typically serpentine, berthierine, and talc (*Barber*, 1985). Smectite may be present in extremely small amounts (*Barber*, 1981). The low abundance of smectite in carbonaceous chondrites and its great abundance in hydrous IDPs does not eliminate the possibility that IDP phyllosilicates and carbonaceous chondrites phyllosilicates share a similar origin. However, different formational conditions must have existed within the parent bodies of the two materials (e.g., temperature, kinetics, precursor grain size, composition, etc.). The IDP sulfides are likewise dissimilar to typical sulfides in CM meteorites, in terms of Fe/Ni contents, morphologies, and grain sizes. As *Tomeoka and Buseck* (1984a) point out, sulfides in the "LOW-CA" IDP are different from carbonaceous chondrites, and most sulfur in CM matrices is present in tochilinite, which is a complex alteration product (*Bunch and Chang*, 1980; *Zolensky and Mackinnon*, 1986; *Mackinnon and Zolensky*, 1984). Tochilinite has not been identified to date in hydrated IDPs. *Tomeoka and Buseck* (1984a) also suggest that the sulfide aggregations in the "LOW-CA" IDP resemble the assemblages of opaque phases in terrestrially serpentinized rocks. We agree with the analogy on the basis of similar sulfide textures in Ames-Dec86-011, together with the clumped nature of the magnetite (Figs. 2 and 6). The unknown Mg-Si-Fe-O phase in the particle described here has no known counterpart in carbonaceous chondrites.

Many TEM investigations of IDPs have yielded excellent images, electron diffraction data, and phase compositions. Bulk analyses of hydrated particles have not been previously made, however, nor have *in situ* grain-to-grain relationships been characterized and correlated with phase and bulk compositions. Such characterizations are necessary to make unambiguous interpretations. By comparing cation/silicon ratios of this IDP to CM elemental abundance ratios (Table 1), we see that Mg, Fe, and Ni in the particle are lower than for typical CM meteorites, about the same for Ca, K, and Al, and significantly higher for S and Na. There is little correspondence to chondritic abundance ratios, with the exception of Ca, Al, and Ni. We can assume from this that either the bulk composition of the particle was never chondritic or that the bulk composition was changed by alteration processes in a mobile aqueous medium. Since we have little understanding of the original phases, we know very little about the original composition of the particle and whether or not it approached chondritic composition.

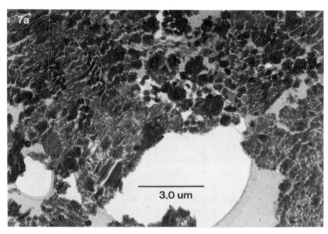

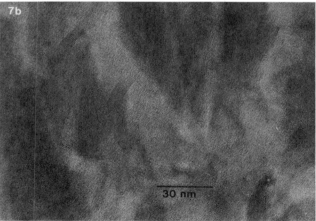

Fig. 7. (a) Low magnification image of microtomed thin section of Murchison carbonaceous chondrite. (b) High resolution lattice fringe image of phyllosilicate from (a) above. Major spacings are coincident with serpentine.

Bunch and Chang (1980) conclude that CM meteorite textures and chemistries are the result of aqueous alteration processes that operated within a limited open system on or in a parent body. CM meteorites are more complex than hydrous IDPs in terms of their texture, mineralogy, and composition, although some analogies can be made. The overall characteristics of the IDP (pseudomorphic replacement, interlocking clay matrix, aggregates of magnetite and sulfides of variable composition and texture) are consistent with *in situ* alteration.

The database for layer silicates in carbonaceous chondrites is extensive and provides support for the physical and chemical environment within which secondary minerals may have formed. The environmental constraints for smectite (and talc) formation, however, are not well understood. In terrestrial occurrences, these minerals commonly form from alteration of mafic and ultramafic rocks under water saturated conditions and by hydrothermal (<40°C) alterations of seafloor detrital sediments (*McMurtry et al.*, 1983). Amorphous silica and silica-rich phases may have formed cogenetically with smectite in the latter environment. The amorphous silica-rich phase in the hydrous IDP we describe may be a prealteration relic, or it may have formed cogenetically during multiple alteration events or during a single event of changing environmental conditions. Textural characteristics of the particle do not support alteration by topotactic solid state transformation (*Eggleton and Boland*, 1982). The presence of discrete clusters of fine-grained, (apparently diagenetic) magnetite throughout the matrix appears to require more element mobility than would be feasible via cryogenic alteration (summarized in *Rietmeijer*, 1985, and *Rietmeijer and Mackinnon*, 1985).

The rather simple mineralogy and nonchondritic composition of the particle do not leave much support for a direct kinship with known carbonaceous chondrites. The available data do strongly suggest that at least one aqueous alteration event took place in which fine-grained material, possibly glass, was transformed to smectite. This event appears to be unique to hydrous IDPs.

Origin of Hydrous IDPs

Hydrous IDPs appear to be unlike the matrix of carbonaceous chondrites, and unlike any other meteoritic asteroidal material analyzed to date. The secondary phyllosilicate phases present in the IDP described in this paper attest to the hydrous alteration of precursor phases by some mechanism that would allow for the migration of replacing and replaced elements over a distance of several microns. If hydrous IDPs are indeed fragments of cometary material, we are left with a number of possibilities: (1) Current models of comets are incorrect, and liquid water is or was present, as is proposed for the parent bodies of carbonaceous chondrites. (2) Hydrated IDPs result from hydrous alteration of the surface layer of comets during periodic transits through the inner solar system. (3) Hydrated IDPs originate from "burned out" comets or the rocky cores of extinct comets, where hydrous alteration takes place. (4) Some or all comets experienced a heat pulse at some time during their history which provided for the existence of liquid water [for example, by radiogenic heating from ^{26}Al in their early history (*Wallis*, 1980)].

A wealth of data support the view that comets are "icy snowballs" (*Whipple*, 1950, 1951). Some evidence for surface heating of comets during inner solar system transit has been presented by *Gombosi and Houpis* (1986), who reported surface temperatures for Halley as high as 80°C at 1 Astronomical Unit. We do not believe, however, that this surficial heating is capable of producing the textures and phase relationships observed in hydrous IDPs. The delicate replacement of pre-existing grains by smectite with sub-μm fabric preservation (e.g., Fig. 2), does not seem possible by such a short-lived and violent mechanism.

It is difficult to evaluate mechanisms (3) and (4) above since, in the former case, IDPs cannot with certainty be assigned to particular cometary parent bodies, and in the latter case, the extent of radiogenic heating by ^{26}Al depends critically on rates of accretion which are only imperfectly known. It is our belief, however, that any accretionary or diagenetic model which attempts to explain the origin of hydrous IDPs should allow for alteration mediated by liquid water at some time in the history of the parent body.

Acknowledgments. We thank Dr. Del Philpott for providing the facilities for the microtomy. This research was in part supported by a National Research Council associateship to DFB and a grant from the National Aeronautics and Space Administration, Exobiology Division, to TEB. Support from the U. S. Department of Energy under contract #DE-AC03-76SF00098 is gratefully acknowledged for work performed at the National Center for Electron Microscopy. We also thank Dr. I. D. R. Mackinnon and an anonymous reviewer for helpful and constructive comments concerning the manuscript.

REFERENCES

Anders E. and Ebihara M. (1982) Solar-system abundances of the elements. *Geochim. Cosmochim. Acta, 46*, 2363-2380.

Barber D. J. (1981) Matrix phyllosilicates and associated minerals in C2M carbonaceous chondrites. *Geochim. Cosmochim. Acta, 45*, 945-970.

Barber D. J. (1985) Phyllosilicate and other layer-structured materials in stony meteorites. *Clay Miner., 20*, 415-454.

Bass M. N. (1971) Montmorillonite and serpentine in Orgeuil meteorite. *Geochim. Cosmochim. Acta, 35*, 139-147.

Blake D. F., Bunch T. E., Philpott D. E., and Zeiger R. (1987) A simple device for the preparation of materials science specimens for ultramicrotomy. *J. Electron Microscopy Technique, 6*, 305-306.

Bradley J. P. and Brownlee D. E. (1986) Cometary Particles: Thin sectioning and Microbeam analysis. *Science, 231*, 1542-1544.

Bradley J. P., Sandford S., and Walker R. M. (1987) Interplanetary Dust Particles. In *Meteorites and the Early Solar System* (J. Kerridge, ed.), pp. University of Arizona, Tucson.

Brownlee D. E. (1978) Interplanetary Dust: Possible Implications for Comets and presolar interstellar grains. In *Protostars and Planets*, (T. Gehrels, ed.), pp. 134-150, University of Arizona, Tucson.

Brownlee D. E. (1979) Interplanetary Dust. *Rev. Geophys. Space Phys., 17*, 1735-1742.

Brownlee D. E., Tomandl D. A., Hodge P. W., and Hörz F. (1974) Elemental abundances in interplanetary dust. *Nature, 252*, 667-669.

Brownlee D. E., Olszewski E., and Wheelock M. (1982) A working taxonomy for micrometeorites (abstract). In *Lunar and Planetary Science XIII*, pp. 71-72. Lunar and Planetary Institute, Houston.

Bunch T. E. and Chang S. (1980) Carbonaceous Chondrites - II. Carbonaceous chondrite phyllosilicates and light element geochemistry as indicators of parent body processes and surface conditions. *Geochim. Cosmochim. Acta, 44*, 1543-1578.

Caillere S. and Rautureau M. (1974) Determination des silicates phylliteux des meteorites carbonees par microscopie et microdiffraction electroniques. *Canadian R. Acad. Sci., 279*, 539-542.

Eggleton R. A. and Boland J. N. (1982) Weathering of enstatite to talc through a sequence of transitional phases. *Clays and Clay Miner., 30*, 11-20.

Fraundorf P., Patel R. I., and Freeman J. J. (1981) Infrared spectroscopy of interplanetary dust in the laboratory. *Icarus, 47*, 368-380.

Gombosi T. I., and Houpis H. L. F. (1986) An icy-glue model for cometary nuclei. *Nature, 324*, 43-44.

McMurtry G. M., Wang C.-H., and Yeh H.-W. (1983) Chemical and isotopic investigations into the origin of clay minerals from the Galapagos hydrothermal mounds field. *Geochim. Cosmochim. Acta, 47*, 475-489.

Mackinnon I. D. R. and Rietmeijer F. J. M. (1983) Layer silicates and a bismuth phase in chondritic porous aggregate W7029*A. *Meteoritics, 18*, 343-344.

Mackinnon I. D. R. and Rietmeijer F. J. M. (1987) Mineralogy of chondritic interplanetary dust particles. *Rev. Geophys.*, in press.

Mackinnon I. D. R. and Zolenski M. E. (1984) Proposed structures for poorly characterized phases in C2M carbonaceous chondrite meteorites. *Nature, 309*, 240-242.

Mackinnon I. D. R., Lumpkin G. R., and Van Deusen S. B. (1986) Thin-film analyses of silicate standards at 200 kV: The effect of temperature on element loss. *Microbeam Analysis - 1986*, pp. 451-454.

Rietmeijer F. J. M. (1985) A model for diagenesis in protoplanetary bodies. *Nature, 313*, 293-294.

Rietmeijer F. J. M. and Mackinnon I. D. R. (1984) Diagenesis in interplanetary dust: chondritic porous aggregate W7029*A. *Meteoritics, 19*, 301.

Rietmeijer F. J. M. and Mackinnon I. D. R. (1985) Layer silicates in a chondritic porous interplanetary dust particle. *J. Geophys. Res., 90*, D149-D155.

Tomeoka K. and Buseck P. R. (1984a) Transmission electron microscopy of the "LOW-CA" hydrated interplanetary dust particle. *Earth Planet. Sci. Lett., 69*, 243-254.

Tomeoka K. and Buseck P. R. (1984b) A hydrated interplanetary dust particle containing calcium- and aluminum-rich pyroxene: possible relations to carbonaceous chondrites. *Meteoritics, 19*, 322-323.

Tomeoka K. and Buseck P. R. (1985) Hydrated interplanetary dust particle linked with carbonaceous chondrites? *Nature, 314*, 338-340.

Tomeoka K. and Buseck P. R. (1986) A carbonate-rich interplanetary dust particle: possible residue from protostellar clouds. *Science, 231*, 1544-1546.

Wallis M. K. (1980) Radiogenic melting of primordial comet interiors. *Nature, 284*, 431-433.

Whipple F. L. (1950) A comet model. I. The acceleration of Comet Encke. *Astrophys. J., 111*, 375-394.

Whipple F. L. (1951) A comet model. II. Physical relations for comets and meteors. *Astrophys. J., 113*, 464-474.

Williams D. B. (1984) *Practical Analytical Electron Microscopy*. Philips Electron Optics Publishing Group, Mahwah, New Jersey.

Zolensky M. E. and Mackinnon I. D. R. (1986) Microstructures of cylindrical tochilinites. *Am. Mineral., 71*, 1201-1209.

Identification of Two Populations of Extraterrestrial Particles in a Jurassic Hardground of the Southern Alps

C. Jéhanno[1], D. Boclet[2], Ph. Bonté, A. Castellarin[3], and R. Rocchia[2]

[1]Centre des Faibles Radioactivités, CEA-CNRS, Boîte Postale 1, 91190 Gif-sur-Yvette, France

[2]Service d' Astrophysique, Centre d'Etudes Nucléaires de Saclay, 91191 Gif-sur-Yvette, Cedex, France

[3]Istituto di Geologia e Paleontologia dell' Università, via Zamboni, Bologna, Italia

---

Two populations of extraterrestrial particles occur in a Lower-Middle Jurassic hardground. One is unambiguously identified as consisting of highly weathered iron-nickel cosmic spherules. This population is common in old sediments, especially in hardgrounds. The second and most abundant population is unusual, consisting of particles with irregular shapes. Their structures and chemical compositions indicate a probable derivation from chondritic material. However, their very unusual morphology, high magnetite content, and homogeneous iridium volume distribution preclude an origin from the steady influx of chondritic micrometeoroids. Their properties are more consistent with a short accretionary event and we suggest that this second population results from the ablation of a large meteoroid accreted by the Earth during hardground formation.

INTRODUCTION

We recently reported (*Rocchia et al.*, 1986) the existence of an iridium anomaly in an ~180-m.y.-old Lower-Middle Jurassic sequence. This sequence, described by *Sturani* (1964, 1971) and *Castellarin* (1966), crops out in the southern Alps near the village of Loppio, a few kilometers east of the lake of Garda, Italy.

The maximum iridium concentration (~3 ng g^{-1}) is found in a hardground 3-5 mm thick located on top of upper Liassic limestones (Fig. 1). This hardground is the result of the slow sedimentation rate that prevailed for several 10^5 years encompassing an undetermined fraction of the Aalenian (+Toarcian ?). A lower anomalous iridium concentration of 0.1-0.2 ng g^{-1} is observed in the 50 cm of Bajocian-Bathonian sediments (Ammonitico Rosso Inferiore formation) overlying

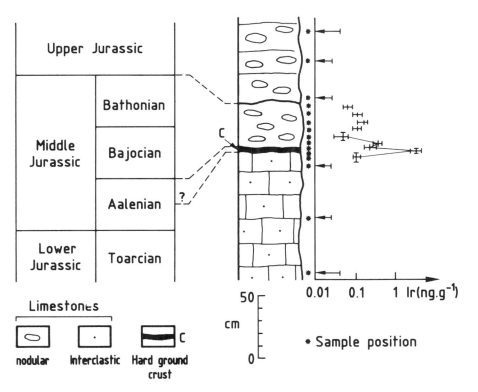

Fig. 1. Stratigraphy and iridium profile of the Loppio Jurassic section.

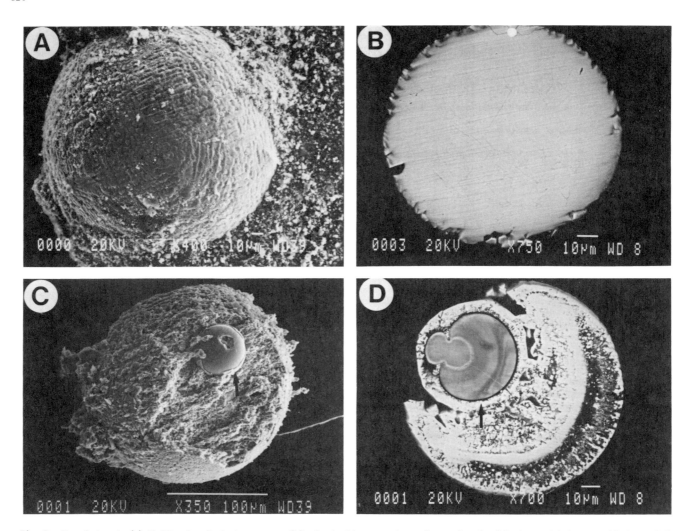

Fig. 2. Population 1. (a) Fe-Ni spherule in its gangue; (b) plantinoid nugget in an iron spherule; (c) ghost nickel core with its nickel-rich skin (arrow); (d) polished section of a nickel-core-bearing spherule.

the hardground. Above the Bajocian-Bathonian sediment and below the hardground, iridium was not detected (upper limit: 0.04 ng g^{-1}).

We have carefully analysed the iridium-enriched layers, with special attention to the basal layer (the crust of the hardground), and searched for particulate phases in which iridium might be concentrated. It is worth remembering that hardgrounds are geological settings resulting from extremely low sedimentation rates over long periods and thus are natural accumulation sites for the steady rain of cosmic dust and also for any infall produced by an exceptional event having occurred during the gap in sedimentation. From the basal layers we have extracted two populations of strongly magnetic particles that are the subject of this paper.

EXPERIMENTAL METHOD

The experimental procedure is similar to that described by *Bonté et al.* (1987). The main successive steps are as follows. First, samples are enclosed in ultra-pure quartz containers. They are irradiated in the OSIRIS reactor (35 hours; flux: 1.24 10^{14} n cm^{-2} s^{-1}) at Pierre Süe laboratory (Saclay). Standards, including USGS rocks (A1 and MAG1) and an aluminum wire with 1 ppm Ir, are irradiated in the same rabbit. The average iridium concentration in each population is measured with a HP-Ge detector (FWHM: 1.8 keV at 1332 keV). Then each magnetic particle is mounted in epoxy resin, and the gamma-rays of ^{192}Ir, ^{51}Cr, ^{59}Fe, ^{58}Co, ^{46}Sc, and ^{60}Co are counted to determine the weights of Ir, Cr, Fe, Ni, Sc, and Co, respectively.

Second, about 20% to 30% of each sample is ground off and the residual weights of each element are determined in a second count. Each polished section is then examined in a JEOL JSM-840 scanning electron microscope equipped with two solid-state detectors for Z-contrast image, and a LINK energy dispersive X-ray spectrometer for quantitative elemental analysis. Major element compositions are determined on large 50 μm × 50 μm areas. From the Fe concentration and using INAA results we calculate the weight of each sample and then all minor element concentrations. Some samples are next sequentially polished and the gamma ray measurements of Ir, Co, and Fe conducted after each polishing step give the "depth profile" of each element.

RESULTS

Population 1

Population 1 (P1) consists of nearly perfect black spheres (Fig. 2a). The average iridium concentration in 12 spherules is about 10 $\mu g\ g^{-1}$. This high value suggests that population 1 consists of ancient iron cosmic spherules, relict of the steady micrometeoroid influx. (In the following, "ancient" refers to cosmic spherules found in sediments more than a few million years old; "modern" refers to recently accreted cosmic spherules, which are found in Greenland (*Maurette et al.,* 1986) or in young deep sea sediments (*Brownlee,* 1981) and which have not yet been deeply weathered.) Individual spherules show a wide distribution of iridium but lower variations of other elements.

Two species have been identified: (1) spherules with a platinoid nugget, similar to those described by *Brownlee et al.* (1984), containing 2 to 20 $\mu g\ g^{-1}$ of iridium (Figs. 2a,b); (2) spherules with a "ghost" nickel core (Figs. 2c,d), the core being altered and replaced by iron hydroxides. Nickel curiously subsists in a thin skin wrapping the core (arrow on Fig. 2c). Their iridium content (~1 $\mu g\ g^{-1}$) is lower than in the equivalent population of modern spherules.

The "depth profiles" are similar to those observed in modern iron cosmic spherules (*Bonté et al.,* 1987): Iridium is preferentially concentrated in nuggets or in a Ni-enriched core but is never homogeneously dispersed in the host sphere. Conversely, Co, Fe, and often Ni steadily decrease upon polishing, indicating a nearly uniform volume distribution.

In both species the wustite, an unstable oxide phase associated with magnetite in modern Fe-Ni spherules, is absent, probably transformed since the Jurassic into magnetite and/or iron hydroxides, occurring with the relict magnetite as a nearly homogeneous brittle matrix. Alteration probably caused the rather low iridium content of Ni-core-bearing spherules. Conversely, platinoid nuggets have been perfectly preserved, as illustrated in Fig. 3, which shows a nugget (7 μm diameter) extracted from its gangue and exhibiting a fresh acicular texture. In this nugget, the platinoid concentrations are similar to those observed in modern Fe-Ni spherules but vary from one acicular feature to the other.

The identification of population 1 as ancient, weathered Fe-Ni cosmic spherules is obvious. Its presence in a Jurassic hardground is not surprising; these spherules have previously been reported in several hardgrounds of the southern Alps by *Castellarin et al.* (1974), *Del Monte et al.* (1976), and *Czajkowski* (1987).

Population 2

Population 2 (P2) is very unusual. It consists in rather large particles, from 100 to more than 600 μm in size, 10 or 20 times more abundant than population 1. These particles seem to be made of melted material and show a wide morphological diversity: spherical, ovoidal, cylindrical, more or less oblate, or simply irregular rounded bodies. The largest particles often exhibit globular outgrowths, suggesting either the aggregation or the coalescence of several smaller melted objects or, in some cases, swelling produced by heating (Figs. 4a-f).

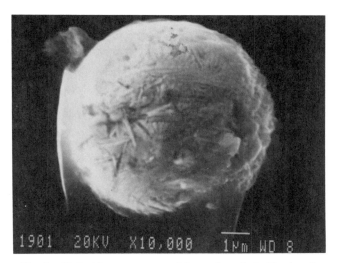

Fig. 3. Platinoid nugget extracted from a Jurassic Fe-Ni spherule.

The chemical composition of the particles is characterized by the presence of Fe, Cr, Co, Ni, and Ir. When compared to their ratios in chondritic meteorites and to the mean ratios in modern chondritic cosmic spherules (*Bonté et al.,* 1987), we note an enrichment of Fe and Cr, a small depletion of Ir and Ni, and a greater depletion of Co (Fig. 5).

Iridium concentrations lie between 100 and 500 ng g^{-1} with an average value around 300 ng g^{-1}. Depth profiles obtained from a stepwise polishing of irradiated samples indicate that iridium is uniformly distributed in volume (Fig. 6). Nuggets seem to be absent in population 2 although they are found in modern chondritic spherules (*Bonté et al.,* 1987).

The internal texture is characterized by high concentrations of massive or dendritic magnetite crystals (Figs. 7a,c), sometimes as large as 15 μm, nearly uniformly distributed in a matrix or groundmass. The matrix is made of iron hydroxides mixed with clays and sometimes abundant calcite. In some cases, magnetite crystals are more abundant near the external surface, forming a continuous coating or shield (Fig. 7b). Dendritic magnetite textures may be very fine and reveal small chondrules (Fig. 7d). The matrix is generally structureless; however, two organized structures, barred and porphyritic, can be observed and are similar to those frequently found in modern chondritic cosmic spherules. In modern spherules, bars and crystals consist in olivine separated by a residual glass in which micron-sized magnetite has crystallised. In Jurassic P2 particles, olivine and glass have been replaced by iron hydroxides, clays, and/or calcite (Figs. 7e,f). Figure 8a shows a magnified portion of the texture of the particle in Fig. 7e. It can be compared with the barred texture observed in a modern spherule (Fig. 8b): The morphological similarity is striking. However, we note that the rather large sizes of magnetite crystals in the Jurassic sample are not typical in cosmic spherules. When several particles have coalesced, both barred and porphyritic structures occur in the same composite grain.

Magnetite crystals contain Fe, Mg, Al, Ni, and Cr and appear to be the only phase in which the original composition has

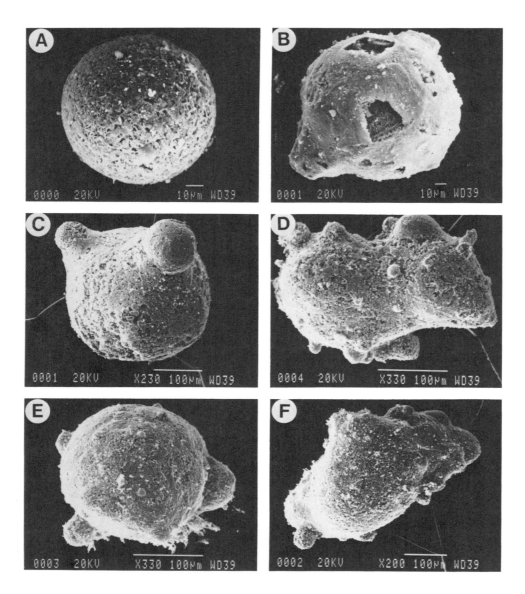

Fig. 4. (a-f) Population 2 particles.

been preserved. We have analysed 3 to 5 magnetites in 12 P2 particles. On the diagram MgO/Al$_2$O$_3$ (Fig. 9a), representative points of any particular particle are clustered in relatively small areas but the different clusters form an elongated figure representing the overall magnetite composition. We note that Al$_2$O$_3$ is relatively constant but MgO differs from one P2 particle to the other. On the diagram Cr$_2$O$_3$/NiO (Fig. 9b), the whole set of points has a lower dispersion. The Cr$_2$O$_3$ content varies from one crystal to the other inside the same particle. Moreover, a zoning occurs in the largest particles with concentrations varying by a factor 10 from the edge to the center: Cr replaces Fe while Mg, Al, and Ni are invariable. The sum of the four oxides (MgO, Al$_2$O$_3$, Cr$_2$O$_3$, and NiO) totals more than 20% of the magnetite weight; the remainder is FeO and Fe$_2$O$_3$. This uniform enrichment is never observed in the magnetites (always smaller in size) of modern chondritic cosmic spherules. P2 magnetite crystals are similar in size and composition to those found in the basal layer in the K-T section of Caravaca (*Bohor et al.*, 1985). If the magnetites of Caravaca can only be extracted as individual grains, due to the complete weathering of their carrier particle, they are still found in clumpy assemblages reminiscent of the original texture. In the Jurassic hardground, however, magnetites are still locked in their well-preserved host bodies.

Comparison with Extraterrestrial Spherules

The texture, magnetite composition, and iridium concentration of P2 particles can be compared to chondritic cosmic spherules like those described by *Brownlee et al.* (1981) and *Robin et al.* (1987). Several marked differences exist: (1) Coalescence outgrowths are never observed in typical cosmic

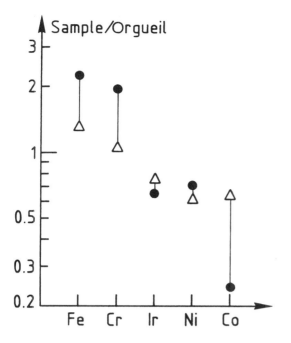

Fig. 5. Abundances of Fe, Cr, Co, Ni, and Ir normalised to the Orgueil chondrite in population 2 particles (●) and modern chondritic micrometeorites (△).

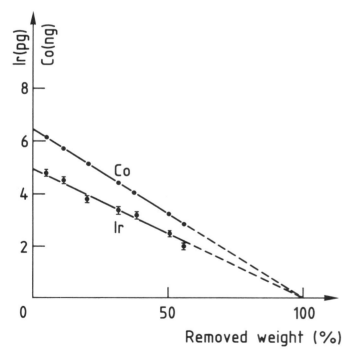

Fig. 6. Typical depth profiles of cobalt and iridium in a P2 particle.

spherules; (2) barred and porphyritic structures are observed in both objects, but the magnetite crystals are more abundant and/or bigger in population 2; (3) the concentrations of Mg, Al, Ni, and Cr are considerably higher than in the magnetites of modern spherules; (4) iridium is uniformly distributed in the volume of P2 particles but found in tiny nuggets in modern chondritic spherules: This dispersion probably does not result from alteration since nuggets are well-preserved in the companion P1 iron spherules found in the same sediment; and (5) P2 particles are, generally, larger than modern cosmic spherules: Indeed, 80% of them and only 10% of Greenland cosmic spherules (*Robin et al.,* 1987) are larger than 150 μm. All of these properties show that P2 particles and chondritic cosmic spherules are similar but distinctly different objects.

DISCUSSION

Origin of Population 2

The discovery of P2 particles in the ~180-m.y.-old hardground was unexpected; usually only Fe-Ni cosmic spherules are preserved in old sediments and the chondritic fraction is rapidly weathered out (*Kyte,* 1983). Tentatively, we propose the following scenario, which involves a special accretionary event.

Morphology

Coalescence can be explained only if the accretion mechanism produced a high volume density of melted fragments in a short time period. This situation occurs during the encounter of large meteoroids with the Earth. The lack of any terrestrial inclusions in the analysed samples suggests that coalescence operated during atmospheric ablation and not during impact cratering. The ablation might also cause boiling of the meteoroid skin and swelling of small torn-out fragments. Derivation of all the P2 particles from a single parent object is consistent with the homogeneity of the analysed samples. The relative richness of population 2 in iron and magnetite could be accounted for by an Fe-rich chondritic meteoroid or ablation-enrichment of Fe by evaporation of the silicates.

If the ablation hypothesis is correct, a comparison of population 2 with the ablation surface of a massive meteorite would be very instructive. *Blanchard and Kyte* (1978) have already studied natural and laboratory ablation products. They noticed the formation of many magnetite crystals in the fusion crust of the Murchison meteorite. We have analysed a polished section of the Saint Severin (LL6) meteorite and found abundant dendritic magnetite in the 150-μm-thick fusion crust (Fig. 10). The Saint Severin magnetite is rich in Ni and Cr, similar to the magnetite in some P2 particles (Figs. 6c,d). These observations are consistent with the proposed scenario.

Iridium Dispersion

As noted above, this uniform distribution of Ir throughout the P2 particles probably did not result from alteration. We

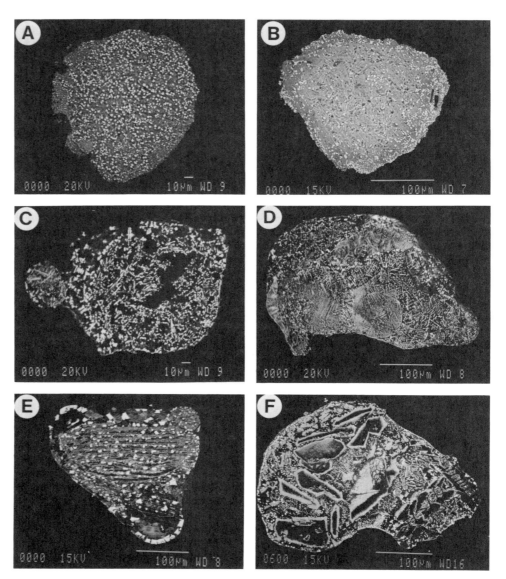

Fig. 7. (a-f) Polished sections of P2 particles.

suggest that it is more likely the consequence of the accretion mode, i.e., ablation from a large meteoroid. In the micrometeorite flux, objects are finely divided before their high-velocity atmospheric entry. The ablation of micrometeoroids produces Fe-Ni globules, nickel cores, or platinoid nuggets in both iron and chondritic cosmic spherules. Thermal conditions (temperature and duration of ablation) are probably quite different for the droplets ablated from the surface of a large meteoroid since fresh material is continually supplied to the ablation surface as the meteoroid passes through the atmosphere. Then temperatures would be sufficient to produce melting and coalescence, and an efficient cooling would prevent the droplets from reaching the high temperatures required for platinoid nugget formation (*Czajkowski*, 1987).

Conservation

The preservation of P2 grains is remarkable. *Kyte* (1984) showed that, in recent deep-sea sediments, chondritic cosmic spherules can weather rapidly and may not survive more than a few million years. This is consistent with the presence of only Fe-Ni spherules in two other younger hardgrounds of the Venetian region: Middle-Upper Jurassic and Lower Cretaceous (*Castellarin et al.*, 1974; *Czajkowski*, 1987; *Jéhanno*, unpublished data, 1987. In these settings, as in old deep-sea sediments, chondritic cosmic spherules are absent, probably removed by alteration, although relict micron-sized magnetite crystals and platinoid nuggets may exist but have not yet been detected. The problem in the Loppio hardground is that the normal chondritic cosmic spherules are absent but P2 particles have been well-preserved. We have no mechanism to propose for this apparent selective preservation. Less severe thermal conditions during ablation could explain the apparently exceptional resistance of population 2, but other parameters such as the exact nature of the parent body and, eventually, the special depositional conditions existing during the hardground formation might have some importance.

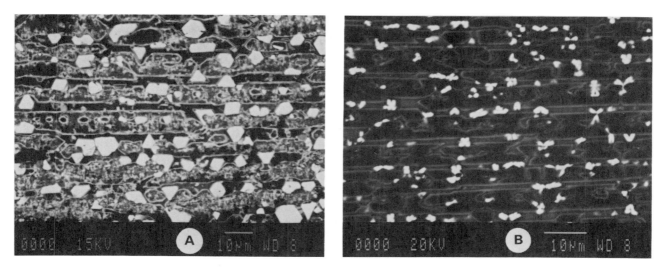

Fig. 8. (a) Barred structure in a magnified part of Fig. 6e; (b) barred structure in a modern spherule.

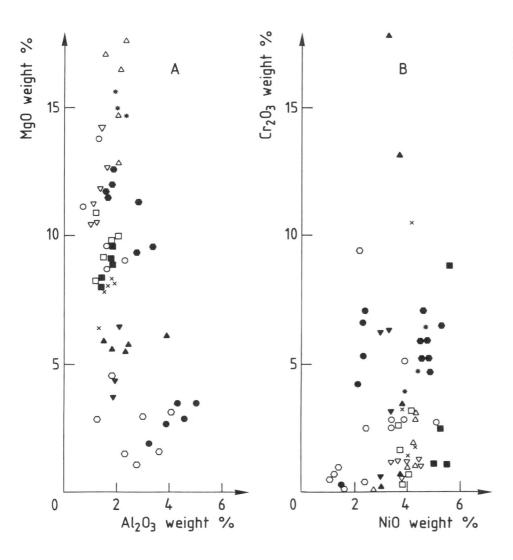

Fig. 9. MgO/Al_2O_3 and Cr_2O_3/NiO diagrams of magnetite crystals.

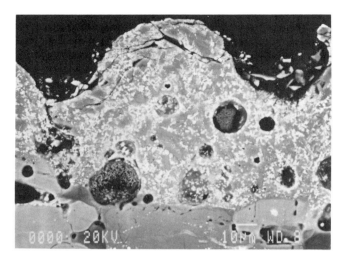

Fig. 10. Texture of the fusion crust in the Saint Severin meteorite.

CONCLUSIONS

Two different extraterrestrial particle populations occur in a Lower-Middle Jurassic hardground at Loppio. One population is the residue of the background influx of micrometeoroids. The second population appears to have resulted from an exceptional and sudden accretionary event, probably the atmospheric ablation of a large meteoroid. If this hypothesis is correct this second population should be found at least on a regional scale comparable to the ablation altitude (~100 km). Several hardgrounds at the same stratigraphical position and located a few tens of kilometers from Loppio will soon be studied. If P2 particles were distributed over 10^3 km^2 with the surface density observed at Loppio (~10^{-3}g cm^{-2}), the mass of ablation material would be 10^{10}-10^{11}g corresponding to an asteroid with a minimum diameter of 50 meters.

A general remark derived from this study is that when an iridium-enriched sediment horizon is associated with low sedimentation rates, two potential sources can explain the iridium anomaly: the steady micrometeoroid flux and any more of less exceptional accretionary event that occurred during the period of slow sedimentation. In both cases a part of iridium is deposited within particles; the rest, deriving from the most finely divided fraction, precipitated from seawater.

Acknowledgments. We thank F. Kyte and G. Blanford for their very helpful review of this manuscript. This work was funded by C.E.A. and C.N.R.S. This is C.F.R. contribution no. 871.

REFERENCES

Blanchard M. D. and Kyte F. T. (1978) Are the stratospheric dust particles meteor ablation debris or interplanetary dust? *NASA Tech. Mem. 78507.* NASA Ames Research Center, Moffett Field, CA.

Bohor B. F., Foord E. E., and Modreski P. J. (1985) Extraterrestrially derived magnesioferrite at the K-T boundary (abstract). In *Lunar and Planetary Science XVI*, pp. 77-78. Lunar and Planetary Institute, Houston.

Bonté Ph., Jéhanno C., Maurette M., and Brownlee D. E. (1987) Platinum metals and microstructures in magnetic deep-sea cosmic spherules. *Proc. Lunar Planet. Sci. Conf. 17th,* in *J. Geophys. Res., 92,* E641-E648.

Brownlee D. E. (1981) Extraterrestrial components in deep-sea sediments. In *The Sea* (C. Emiliani, ed.), pp. 733-762. Wiley, New York.

Brownlee D. E., Bates B. A., and Wheelock M. M. (1984) Extraterrestrial group nuggets platinum in deep-sea sediments. *Nature, 309,* 693-695.

Castellarin A. (1966) Filoni sedimentary nel Giurese di Loppio (Trentino Meridionale). *Giornale di Geologia, Bologna, 33,* 527-555.

Castellarin A., Del Monte M., and Frascari F. (1974) Cosmic fallout in the "hardgrounds" on the Venetian region (Southern Alps). *Giornale di Geologia, Bologna, 39,* 333-346.

Czajkowski J. (1987) Cosmo and geochemistry of the Jurassic hardgrounds. Ph.D. thesis, University of California, San Diego.

Del Monte M., Giovanelli G., Nanni T., and Tagliazucca M. (1976) Black magnetic spherules in condensed sediments from topographic highs. *Arch. Met. Geoph. Biokl., Ser. A, 25,* 151-157.

Kyte F. T. (1983) Analysis of extraterrestrial materials in terrestrial sediment s. Ph.D. thesis, University of California, Los Angeles.

Maurette M., Hammer C., Brownlee D. E., Reeh N., and Thomsen H. H. (1986) Placers of cosmic dust in the blue ice lakes of Greenland. *Science, 233,* 869-872.

Robin E., Jéhanno C., Maurette M., and Hammer C. (1987) A micrometeorite "spectrum" for the mass distribution of well-preserved Greenland cosmic dust grain (abstract). In *Lunar and Planetary Science XVIII*, pp. 844-845. Lunar and Planetary Institute, Houston.

Rocchia R., Boclet D., Bonté Ph., Castellarin A., and Jéhanno C. (1986) An iridium anomaly in the Lower-Middle Jurassic of the venetian region, northern Italy. *Proc. Lunar Planet. Sci. Conf. 17th,* in *J. Geophys. Res., 91,* E259-E262.

Sturani C. (1964) La sussessione delle faune ad Ammoniti nelle formazioni medio-giurassiche delle Prealppi venete occidentali. *Mem. Ist. Geol. Mineral. Univ. Padova, 24,* 1-63.

Sturani C. (1971) Ammonites and stratigraphy of the "Posidonia alpina" beds of the Venetian Alps. *Mem. Ist Geol. Mineral. Univ. Padova, 28,* 190 pp.

Trapping Ne, Ar, Kr, and Xe in Si$_2$O$_3$ Smokes

Joseph A. Nuth III
Code 691, NASA Goddard Space Flight Center, Greenbelt, MD 20771

Chad Olinger, Dan Garrison, and Charles Hohenberg
McDonnell Center for the Space Sciences, Washington University, St. Louis, MO 63130

Bertram Donn
Code 690, NASA Goddard Space Flight Center, Greenbelt, MD 20771

We have condensed simple Si$_2$O$_3$ smokes at both low (<750 K) and high (>1000 K) temperature at 35 torr H$_2$ pressure in the presence of 0, 10, 100, and 1000 microns of a noble gas mixture containing Ne, Ar, Kr, and Xe. In general, both Ne and Ar are quite loosely bound in the smokes (6.0 × 10^{-8} and 2.6 × 10^{-4} ccSTP/g, respectively) and are degassed at temperatures below 1200 K. Both Kr and Xe are somewhat more strongly bound at concentrations of 1.0 × 10^{-7} and 8.2 × 10^{-8} ccSTP/g, respectively, and in addition show a double release with a second component at a temperature of ~1875 K. With the exception that Si$_2$O$_3$ smokes appear to show a particular affinity for argon, possibly due to an anomalous absorption of atmospheric argon, none of the other noble gases are found in sufficient concentration to explain the gases observed in meteorites as primary circumstellar condensates. However, our data in conjunction with observations of *Honda et al.* (1979) do seem to show a degree of dependence between noble gas retention and chemical composition.

1. INTRODUCTION

Isotopically distinct rare gas components have been discovered in numerous forms of carbonaceous grains (*Lewis and Anders*, 1983; *Lewis et al.*, 1987; *Ott et al.*, 1981); none have yet been observed in silicates. Do silicates hold rare gas less tightly than do carbonaceous grains? Do silicates form in an environment depleted in rare gases or do carbonaceous condensates simply survive passage through the interstellar medium to a larger degree than do silicates? No measurements have yet been made of the efficiency with which silicates trap rare gas as they are condensed from the vapor. However, *Honda et al.* (1979) did measure the efficiency with which argon was trapped during the condensation of CdTe, Zn, Mg, and Fe$_3$O$_4$ grains and found that argon was loosely trapped in the smokes, roughly in proportion to the ambient argon pressure over the range 10-1000 torr. They also found that mechanical compaction of the powder greatly increased the retention of argon during stepped heating experiments.

Anders and coworkers have studied the sorption of noble gases by magnetite and carbon (*Yang et al.*, 1982), by chromite and carbon (*Yang and Anders*, 1982a), and by sulfides, spinels, and other substances (*Yang and Anders*, 1982b) and have concluded that the planetary gases could have been adsorbed in the early stages of the contraction of the solar nebula. More recent studies in the same laboratory (*Zadnik et al.* 1985; *Wacher et al.*, 1985) have examined the mechanism by which xenon is trapped in three types of carbonaceous matter and suggest that the xenon is held in at least two different sites; an adsorbed component is held near the grain surface, while a diffused component is held within a labyrinth of pores at a much deeper level within the grain. *Niemeyer and Marti* (1981) measured the noble gas distribution coefficients for carbon condensates produced via laser ablation of a graphite target. Although the elemental ratio of the gas fraction trapped in the grains was similar to the planetry noble gas ratios, the distribution coefficients were much too small to explain the amounts of gas observed to be trapped in carbonaceous meteorites.

We condensed Si$_2$O$_3$ smokes at Goddard Space Flight Center at both low and high temperatures in an ambient H$_2$ atmosphere into which 0, 10, 100, and 1000 microns of a rare gas mixture had been added. These smokes were then shipped to the McDonnell Center for the Space Sciences at Washington University for analysis of the isotopic and chemical composition of the trapped gas. We have previously argued (*Nuth and Donn*, 1984) that Si$_2$O$_3$ smokes, though simple, are good analogs to the condensates in oxygen-rich circumstellar environments. Even though our experiments were performed at relatively high pressure, we feel that our results can be extrapolated quite easily to the stellar regions in which such grains are likely to condense.

2. EXPERIMENTAL

Prior to the start of these experiments, a 30 liter mixture of rare gases was made in the following proportions: 10 torr Xe, 42 torr Kr, 217 torr Ne, and 500 torr Ar. This gas mixture was allowed to equilibrate for four days before the first experiment was run. Si$_2$O$_3$ smokes were made by the evaporation of SiO into an ambient H$_2$ atmosphere (at 35 torr) in a bell jar/furnace system described previously (*Nuth and Donn*, 1983). In the current experiments however a small quantity (0-1000 microns) of the noble gas mixture was introduced into the bell jar after bakeout and before the introduction of the H$_2$. Experiments were run at both high temperature (nucleation occurred at T ≥ 1000 K) and at low temperature (T ≤ 750 K) in order to check for decreased adsorption at higher temperatures. Samples were collected after the runs, placed in labeled vials and shipped to Washington University for noble gas analysis.

For noble gas analysis a few milligrams of Si_2O_3 smokes were loaded into platinum foil boats and placed into the high sensitivity, pulse counting mass spectrometer system (*Hohenberg*, 1980). The system was then closed and outgassed at 400°C to remove absorbed water and gas from the walls of the chamber and from the samples. During the analysis progressively higher temperatures were attained for 30 minutes each by the use of a tungsten resistance heater. Gas samples were purified by exposure to three films from titanium flash getters: Each noble gas was then analyzed separately by selective desorption from an activated charcoal trap.

3. RESULTS

The results of an analysis of a low temperature condensation run with 1000 microns of the noble gas mixture added to 35 torr of H_2 is shown in Table 1. For each gas (Ne, Ar, Kr, Xe) we show its partial pressure (in microns) during condensation, the quantity trapped in the grains (ccSTP/g), the individual ratios of Ne, Kr, and Xe to Ar in both the original atmosphere and trapped in the grains, and the relative trapping efficiency of Ne, Kr, and Xe as compared to argon. As can be seen, the trapping efficiency of argon is quite high when

TABLE 1.

	Ambient Gas (microns)	Trapped Gas ccSTP/g×10^8	Ambient Gas vs. Ar	Trapped Gas vs. Ar	Trapping Efficiency
Ne	282	6.0	0.43	2.3×10^{-4}	5.3×10^{-4}
Ar	650	26000	1.0	1.0	1.0
Kr	55	10.0	0.084	4.0×10^{-4}	4.8×10^{-3}
Xe	13	8.2	0.02	3.2×10^{-4}	1.6×10^{-2}

compared to that for Ne, Kr, and Xe. However, since the measured argon content of the blank smokes (those condensed at zero microns) was also high (within a factor of three of those condensed at 1000 microns), the high apparent argon trapping efficiency may be indicative of a different phenomena, post-formational trapping of atmospheric gases. The high abundance (about 1%) of argon in the terrestrial atmosphere, especially when compared with the other noble gases (neon, a thousand times less abundant; xenon, a million times less abundant), would make post-formational incorporation of the terrestrial atmosphere a substantial factor only for argon. In addition, it appears that both trapped argon and neon are quite easily lost from the grains (Figure 1a). Figure 1a shows the result of a stepped heating experiment for neon and argon trapped in Si_2O_3 smokes at low temperature with 1000 microns of noble gas mixture in the ambient gas. Figure 1b is the plot of similar data for krypton and xenon. It appears that the maximum in the release temperature shifts to higher values as the molar weight of the rare gas increases, from less than 600 C for Ne and Ar, to 700°C for Kr, to about 900°C for Xe. In addition, it appears that both Kr and Xe have a component trapped in a more retentive site that is released near 1600°C.

No significant isotopic fractionation either in the total trapped gas or in fractions released at any particular temperature were

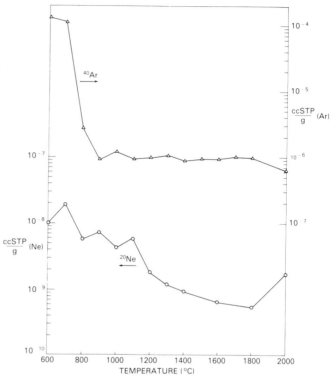

Fig. 1a. Quantity of Ne and Ar released during stepped heating experiment from Si_2O_3 smokes prepared in 35 torr of H_2 to which 282 microns of Ne and 650 microns of Ar had been added. Smokes were condensed at temperatures less than 750K.

noted. A very slight enrichment in ^{20}Ne was detectable in the gas released at low temperature, which was compensated for by a slight enrichment of ^{22}Ne at higher release temperatures. This effect could easily be the result of diffusion from the grains and, because the total neon released was isotopically normal, was probably not due to isotopic fractionation during condensation.

The amount of gas trapped in the grains appears to be independent of the condensation temperature, as grains condensed at high temperature contain as much of each gas as do those condensed at low temperature. The amount of gas trapped in the smokes does, however, depend on the partial pressure of the rare gas mix in the ambient atmosphere from which the grains condensed. The quantity of trapped Ne, Kr, and Xe gas is roughly proportional to its partial pressure (Table 2), as would be expected from the results of *Honda et al.* (1979). However, in our experiments, the quantity of Ar trapped in the Si_2O_3 smokes is independent of pressure. The

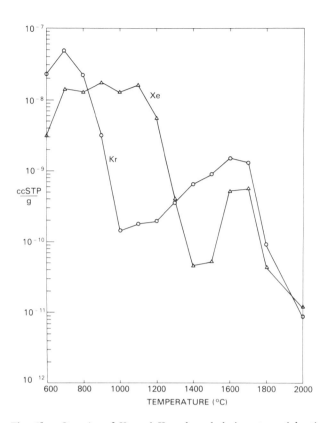

Fig. 1b. Quantity of Kr and Xe released during stepped heating experiment from Si_2O_3 smokes prepared in 35 torr of H_2 to which 55 microns of Kr and 13 microns of Xe had been added. Smokes were condensed at temperatures less than 750K.

quantity of Ar in smokes condensed at 0 microns of the noble gas mixture is, within a factor of 3, the same as that of smokes condensed in 1000 microns of noble gas.

4. DISCUSSION

If the rare gases (Ne, Kr, and Xe) are trapped as a surface layer and are incorporated into the grains as the particles grow, then one would expect that the higher the partial pressure of the ambient rare gas, the higher the surface concentration of the gas, and consequently the higher the concentration ultimately trapped in the smokes. This behavior is observed for Kr and Xe (see Table 2). As there appears to be no temperature dependence, the rare gases are most likely physically trapped by condensing silicate molecules as the grains grow. An explanation for the seemingly preferential trapping of argon could be that argon atoms fit nicely into the vacancies in the growing Si_2O_3 smokes. Neon obviously slips through the cracks whereas krypton and xenon may be less mobile on the surface (due to their mass and polarizability and, therefore, even though they may not readily fit in normal voids, krypton and xenon may be less likely to escape the growing surface before they are buried. An alternative explanation for the anomalously high relative concentration of argon in our smokes is that Si_2O_3 preferentially absorbs argon from the atmosphere. Again, this could be due to a size effect, where argon atoms easily slip into voids in the Si_2O_3 lattice. Typical particle sizes for grains prepared in our system are on the order of 20-40 nm in diameter and the particles are usually aggregated into a fairly open structure (*Rietmeijer et al.*, 1986). Such grains have very large surface to volume ratios and would therefore be expected to efficiently expose large numbers of trapping sites to atmospehric constituents.

Data contained in Table 2 allows us to make a preliminary estimate of the importance of atmospheric contamination for each of the samples for which gas concentrations were measured in this series of experiments. It is fairly obvious that the argon trapped in these samples was an atmospheric contaminant introduced after condensation of the smokes. It is not clear when the neon was introduced since almost all of the the neon was probably lost in the initial outgassing of the smokes ($T \leq 400°C$). Both Kr and Xe were present in the "0 micron" samples at a concentration of $\sim 2 \times 10^{-9}$ ccSTP/g, well above the System Blank Levels for these gases ($\leq 10^{-14}$ ccSTP/g. If this concentration is assumed to be "background" contamination from the atmosphere, then $\sim 5 \times 10^{-9}$ ccSTP/g of Kr is trapped when the Kr partial pressure is 0.55 microns, $\sim 7 \times 10^{-9}$ ccSTP/g is trapped at a Kr partial pressure of 6.3 microns and $\sim 135 \times 10^{-9}$ ccSTP/g is trapped at a Kr pressure of 55 microns. Similarly, if the atmospheric Xe component trapped in the smokes is $\sim 2 \times 10^{-9}$ ccSTP/g, then only a small amount of Xe ($\sim 1 \times 10^{-9}$ ccSTP/g) was trapped at a Xe partial pressure of 1.6 microns, while $\sim 57 \times 10^{-9}$ ccSTP/g was trapped at a Xe partial pressure of 13 microns. Experiments to accurately measure the atmospheric adsorption efficiency of noble gases in Si_2O_3 smokes which have equilibrated with atmospheric rare gases for ~ 5 years are planned.

How do these experiments relate to the condensation process either in the solar nebula or in circumstellar environments? Under astrophysically relevant conditions the concentration of the noble gas in the ambient environment is likely to be much lower than it was in our experiments. Therefore, with the possible exception of argon, silicates are not likely to carry significant quantities of noble gas. We do note that the amount of argon trapped in the study of *Honda et al.* (1979) appeared to be a function of the composition of the smoke and we also note that Si_2O_3 traps roughly 3-4 orders of magnitude more argon than would be expected on the basis of that earlier study. It therefore may be worthwhile to attempt some additional experiments in which the trapping efficiency of the various gases is measured as a function of grain composition and in which the adsorption efficiency of the smokes is measured directly.

The thermal retention of argon in Si_2O_3 is consistent with the results of *Honda et al.* (1979) for CdTe smokes. In both studies the vast majority of the argon is lost before the sample reaches 900°C. We did not try to compact the Si_2O_3 smokes to increase the gas retention since we are more interested in the retention of individual grains rather than in the properties of compact aggregates. In our studies, the majority of Ne, Ar, and Kr was lost relatively easily, whereas Xe was retained somewhat more efficiently. As grains in circumstellar shells typically encounter temperatures in excess of 500-1000 K for

TABLE 2.

	^{20}Ne (ccSTP/g)	^{40}Ar (ccSTP/g)	^{84}Kr (ccSTP/g)	^{132}Xe (ccSTP/g)
System Blank	$\leq 10^{-9}$	$\leq 10^{-9}$	$\leq 10^{-14}$	$\leq 10^{-14}$
0 micron, high T	-	5.2×10^{-5}	-	1.9×10^{-9}
0 micron, low T	1.7×10^{-7}	2.2×10^{-5}	2.1×10^{-9}	-
11 microns, high T	$\leq 10^{-9}$	4.1×10^{-5}	7.4×10^{-9}	-
10 microns, low T	$\leq 10^{-9}$	2.0×10^{-5}	6.3×10^{-9}	-
115 microns, high T	$\leq 10^{-9}$	2.9×10^{-5}	9.2×10^{-9}	-
115 microns, high T*	$\leq 10^{-9}$	4.1×10^{-5}	9.6×10^{-9}	2.9×10^{-9}
125 microns, low T	$\leq 10^{-9}$	2.3×10^{-6}	7.0×10^{-9}	3.7×10^{-9}
1000 microns, high T	2.5×10^{-6}	4.1×10^{-5}	1.39×10^{-7}	6.35×10^{-8}
1000 microns, high T*	4.1×10^{-7}	2.2×10^{-5}	1.35×10^{-7}	5.52×10^{-8}
1000 microns, low T	$\leq 10^{-9}$	0.9×10^{-5}	5.2×10^{-8}	5.70×10^{-8}

- Trapped gas component lost during the analysis.
*These samples were analyzed twice for noble gas content as tests of the degree of homogeneity within the Si_2O_3 sample.

time scales of up to several weeks, the bulk of Ne and Ar trapped during condensation in a circumstellar environment would probably be lost. It is possible, however, that some Kr and even more Xe would remain in the grains. Unfortunately, even if the grains were to condense at a total pressure of one atmosphere (highly unlikely), extrapolation of our data indicates that such grains would contain just a few times 10^{-12} ccSTP/g of Xe and only a bit more Kr. Such small quantities are trivial in comparison to the amounts actually observed in meteorites (e.g., see *Ott et al.*, 1981). It is interesting to note that *Niemeyer and Marti* (1981) concluded on the basis of their condensation experiments that carbon is also a less efficient trap for noble gases than one would need in order to explain the meteoritic observations. It appears that understanding the trapping process itself is at least as important as the measurement of the properties of the grains which serve as the traps.

5. CONCLUSIONS

We have measured the quantity of Ne, Ar, Kr, and Xe trapped in Si_2O_3 smokes during vapor phase condensation as a function of the pressure of the individual rare gases in the ambient atmosphere. Our data is in substantial agreement with previous work by *Honda et al.* (1979), which showed that the quantity of rare gas trapped during condensation was approximately proportional to its partial pressure and that the quantity of gas trapped is dependent on the chemical composition of the condensate. We find that the bulk of the Ne and Ar trapped during grain growth is lost upon heating to ~1000 K for prolonged times although a substantial portion of the trapped Kr and Xe would probably survive heating to such temperatures. Unfortunately, extrapolation of our measured trapping efficiencies to realistic condensation pressures predicts that only a trivial quantity (~10^{-12} ccSTP/g) of even Kr and Xe would be trapped in the first place by this process.

Because the trapping efficiency for argon was significantly greater in Si_2O_3 than would be expected from the results of *Honda et al.* (1979), we conclude that argon is efficiently trapped by Si_2O_3 smokes and that more efficient traps for the other noble gases may also exist. It is not apparent at this time if the argon is trapped during formation, where it would occur preferential to the other noble gases (Table 1), or if it occurred upon later exposure to atmospheric gases, in which case the other atmospheric species would not be readily observable. In either case, Si_2O_3 smokes represent substrates for noble gas incorporation. For this reason, we intend to investigate both the trapping and absorption efficiency of simple and mixed oxides of Ti, Al, Si, and Fe. *Fanale and Cannon* (1972) have already shown that adsorption of rare gas occurs efficiently on silicate minerals only at low temperature, while others (*Lancet and Anders*, 1973; *Zaikowski and Schaefer*, 1976) have shown that incorporation of rare gases during hydrothermal alteration yields abundance patterns that are inconsistent with those observed in meteorites. If we fail to observe a significant affinity for the rare gases by oxide condensates of Ti, Al, Si, and Fe, then we would be forced to conclude that significant rare gas enhancements will only be found in carbonaceous grains. This is consistent with the mechanism proposed by *Huss and Alexander* (1987) in which the bulk of the rare gases found in solar system materials is bound in the carbonaceous mantles of pre-solar grains that formed in the molecular cloud from which the nebula collapsed. Unfortunately, before we can draw definitive conclusions from our work, several additional experiments must be performed in order to characterize the noble gas absorption efficiency of mixed oxide smokes.

Acknowledgment. The authors gratefully acknowledge the constructive criticism of our reviewers, Drs. D. Bogard and R. Becker, and helpful discussions with Dr. S. Chang.

REFERENCES

Fanale F. P. and Cannon W. A. (1972) Origin of planetary primordial rare gas: The possible role of adsorption, *Geochim. Cosmochim. Acta, 36,* 319.

Hohenberg C. H. (1980) High sensitivity pulse-counting mass spectrometer system for noble gas analaysis, *Rev. Sci. Instrum. 51,* 1075-1082.

Honda M., Ozima M., Nakada Y., and Onaka T. (1979) Trapping of rare gases during the condensation of solids, *Earth Planet. Sci. Lett.*, 43, 197-200.

Huss G. R. and Alexander E. C. (1987) On the presolar origin of the "normal planetary" noble gas component in meteorites, *Proc. Lunar Planet. Sci. Conf. 17th*, in *J. Geophys. Res.*, 92, E710-E716.

Lancet M. S. and Anders E. (1973) Solubility of noble gases in magnetite: Implications for planetary gases in meteorites, *Geochim Cosmochim Acta*, 37, 1371.

Lewis R. and Anders E. (1983) Interstellar matter in meteorites, *Sci. Am.*, 249, 66-77.

Lewis R. S., Ming T., Wacker J. F., Anders E., and Steel E. (1987) Interstellar diamonds in meteorites, *Nature*, 326, 160-162.

Niemeyer S. and Marti K. (1981) Noble gas trapping by laboratory carbon condensates, *Proc. Lunar Planet. Sci. 12B*, 1177-1188.

Nuth J. A. and Donn B. D. (1983) Laboratory studies of the condensation and properties of amorphous silicate smokes, *Proc. Lunar Planet. Sci. Conf.*, in *J. Geophys. Res.*, 88 A847-852.

Nuth J. A. and Donn B. D. (1984) Thermal metamorphism of Si_2O_3 (a circumstellar dust analog), *Proc. Lunar Planet. Sci.* in *J. Geophys. Res.*, 89, B657-B661.

Ott U., Mack R., and Chang S. (1981) Noble-gas-rich separates from the Allende meteorite, *Geochim. Cosmochim. Acta*, 45, 1751-1788.

Rietmeijer F. J. M., Nuth J. A., and Mackinnon I. D. R. (1986) Analytical electron microscopy of Mg-SiO smokes: A comparison with infrared and XRD studies, *Icarus*, 66, 211-222.

Wacker J. F., Zandik M. G., and Anders E. (1985) Laboratory simulation of meteoritic noble gases. I. Sorption of xenon on carbon: Trapping experiments, *Geochim. Cosmochim. Acta*, 49, 1035-1048.

Yang J. and Anders E. (1982a) Sorption of noble gases by solids, with reference to meteorites. II. Chromite and carbon, *Geochim. Cosmochim. Acta*, 46, 861-875.

Yang J. and Anders E. (1982b) Sorption of noble gases by solids, with reference to meteorites. III. Sulfides, spinels and other substances; on the origin of planetary gases. *Geochim. Cosmochim. Acta*, 46, 879-892.

Yang J., Lewis R. S., and Anders E. (1982) Sorption of noble gases by solids, with reference to meteorites. I. Magnetite and carbon, *Geochim. Cosmochim. Acta*, 46, 841-860.

Zadnik M. G., Wacker J. F., and Lewis R. S. (1985) Laboratory simulation of meteoritic noble gases. II. Sorption of xenon on carbon: Etching and heating experiments, *Geochim. Cosmochim Acta*, 49, 1049-1059.

Zaikowski A. and Schaeffer O. A. (1976) Incorporation of noble gases during synthesis and equilibration of serpentine and implications for meteoritical noble gas abundances, *Meteoritics*, 11, 394.

Stochastic Histories of Refractory Interstellar Dust

Kurt Liffman and Donald D. Clayton
Department of Space Physics and Astronomy, Rice University, Houston, TX 77251

We calculate histories for refractory dust particles in the interstellar medium. The double purposes are to learn something of the properties of interstellar dust as a system and to evaluate with specific assumptions the cosmic chemical memory interpretation of a specific class of isotopic anomalies. We assemble the profile of a particle population from a large number of stochastic, or Monte Carlo, histories of single particles, which are necessarily taken to be independent with this approach. We specify probabilities for each of the events that may befall a given particle and unfold its history by a sequence of random numbers. We assume that refractory particles are created only by thermal condensation within stellar material during its ejection from stars, and that these refractory particles can be destroyed only by being sputtered to a size too small for stability or by being incorporated into the formation of new stars. In order to record chemical detail, we take each new refractory particle to consist of a superrefractory core plus a more massive refractory mantle. We demonstrate that these superrefractory cores have effective lifetimes much longer than the turnover time of dust mass against sputtering. As examples of cosmic chemical memory we evaluate the ^{16}O-richness of interstellar aluminum and mechanisms for the ^{48}Ca/^{50}Ti correlation. Several related consequences of this approach are discussed.

1. INTRODUCTION

One class of the isotopic anomalies found in meteorites appears to have previously resided in a refractory component of interstellar dust that carries an isotopic signature of initial condensation within the events of nucleosynthesis themselves. Because the major refractory elements were synthesized overwhelmingly in supernova explosions, dust condensation within the supernova interior during its expansion is implicated (*Clayton*, 1975a,b, 1978, 1979). The acronym SUNOCON describes these supernova condensates, and it is their subsequent history that is the topic of this paper, an abbreviated version of which was published in the Lunar and Planetary Science Conference abstract volumes (*Liffman and Clayton*, 1987). In order that the SUNOCONs be able to remember their events of nucleosynthesis and condensation, it is quantitatively necessary that they be highly persistent, with an age spectrum far older than that of the bulk of interstellar dust, which is probably finely grained carbonaceous material of recent cosmic vintage. The class of interpretations based on this tenacious presolar memory has come to be called "cosmic chemical memory" (e.g., *Clayton*, 1982 and references therein). It is to the foundations of that theoretical picture that this paper is addressed.

A basic distinction of relevance is the one between the lifetime of the total interstellar dust mass and the lifetimes of the individual particles as entities. If the total mass of interstellar dust is designated by M, then we define the mass lifetime τ_M by $\tau_M = M/(dM/dt)$. A large number of studies have evaluated τ_M in the range 10^8-10^9 yr. The lifetimes of particles as entities, however, are determined by the times required for their removal from existence. Similarly, the lifetime of a specific refractory structure within the cores of composite grains differs from τ_M. Without this distinction it is hard to appreciate how substantial amounts of a refractory isotopic tracer can survive a galactic history spanning 10^{10} yr if $\tau_M \approx$ few $\times 10^8$ yr. Two examples of great importance to meteoritic science can illustrate this enigma. Aluminum-rich solids are the most ^{16}O-rich specimens found, with roughly 5% excess ^{16}O in the calcium aluminum inclusions (CAI) (*Clayton et al.*, 1977). One of us has presented a persistent stream of arguments (*Clayton*, 1975a, 1977a,b, 1979, 1982, 1986a,b) that the cosmic chemical memory at work is that of ^{16}O-pure Al_2O_3 SUNOCON cores within composite interstellar grains.

Taking that view as correct for purposes of motivation, how can Al nucleosynthesis over 10^{10} yr still result in a 5% ^{16}O excess in macroscopic solar-system solids even after 10^{10} yr of galactic destruction and reprocessing and after the early solar-system chemistry required for the production of the CAI? Surely interstellar Al would have to be characterized by a ^{16}O-richness considerably greater than 5% in order that a 5% excess remain following the solar system chemical processing associated with production of large Al-rich minerals by a gathering together and fusing of the aluminous portions of countless microscopic grains. *Clayton* (1986b) has already shown how to formally include SUNOCON lifetimes in the theory of chemical evolution of the galaxy, with the intuitively appealing result that the SUNOCON remainder surviving is approximately $r_s \approx \tau/T_{galaxy}$, where τ is the lifetime of the SUNOCON mass. But the ^{16}O-richness of Al is too small if $\tau_M(Al) \approx \tau_M$ (all dust). Therefore one purpose of this investigation is to calculate, within the framework of a specific numerical model of the interstellar history of refractory dust, how much more persistent are the Al_2O_3 cores of refractory dust than are the bulk amounts of refractory dust. The fossil ^{26}Mg in refractory aluminum has also come under scrutiny (*Clayton*, 1986a).

A related example concerns the ^{50}Ti anomalies that are endemic in CAIs (*Niederer et al.*, 1980; *Niemeyer and Lugmair*, 1984). These variations reach 10% in magnitude and are of both algebraic signs (*Hinton et al.*, 1987; *Zinner et al.*, 1986) in hibonite inclusions. *Zinner et al.* (1986) also find a clear correlation with ^{48}Ca anomalies. All of these effects demand an explanation in the cosmic chemical memory picture. The numerical results we will report here illustrate how this history may operate, although our major objective is the understanding of complicated histories themselves rather than any specific isotopic applications.

The interstellar medium is complicated and still poorly understood. It is therefore necessary for us to make some simplifying assumptions about the ways in which dust particles cycle among its various physical environments (also called "phases"), how dust particles are destroyed in those environments, and how they grow by accreting gaseous atoms. It is also necessary to make specific assumptions about the birth of the refractory component and its initial structure. With such things fixed it is then possible to follow the cosmochemical history of dust particles in their migration through time. This last objective is our main goal, because it illuminates ways in which cosmic chemical memory can function. If these cosmochemical mechanisms can be established, their modifications by improved descriptions of the interstellar medium can more easily follow.

In section 2 we describe the specific assumptions that we make in this exploratory calculation. In section 3 we describe the techniques of the calculation and its results. Following this we will discuss our calculation critically and return to quantitative examples of interpretation of isotopic anomalies. Several special technical demonstrations are relegated to appendices.

2. MODEL ASSUMPTIONS

Most of the interstellar mass resides in molecular clouds (H_2) and in the diffuse clouds of neutral hydrogen (HI). Smaller quantities exist as warm (T ~ 7000 K) diluted (n ~ 0.1 cm^{-3}) neutral and ionized hydrogen, and still smaller amounts as coronal-type gas (T ~ 10^5 K, n ~ 10^{-3} cm^{-3}) that nonetheless occupies large volume. Refractory grains are destroyed by sputtering and by grain-grain collisions in the wakes of shock waves in these media, but primarily in the diffuse neutral-H phases. Refractory atoms are reaccreted by grains when they make return contact with grains, primarily, we think, when the sputtered diffuse media reform clouds and rejoin molecular clouds. These continuous and complicated processes are idealized by us as a two-phase interstellar medium: a diffuse part through which supernova shock waves and shock waves due to cloud collisions periodically propagate causing grain destruction, and a molecular-cloud phase in which stars are born astrating the dust. The bursts of star formation also return molecular cloud material to the diffuse phase. We take these two phases to have approximately equal masses and we assume that matter cycles continuously between the two phases with a turnover timescale τ_{cloud}, which we numerically take to be 10^8 yr in our study. The situation we envision is sketched in Fig. 1 and quantitatively analyzed in the Appendix.

Our general computational approach involves treating each particle independently of others in terms of probabilities for the various things that happen to it. This allows Monte Carlo type calculations of stochastic histories for each particle, a simplification that in turn allows us to follow the fate of a large number of particles (6 × 10^6 particles in our largest calculations). This picture imposes certain assumptions concerning how the transfer of dust particles between the two

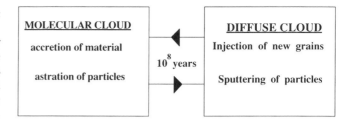

Fig. 1. Schematic of the basic structure of the program. Dust particles reside in two separate phases of the ISM and move from one phase to the other on an average timescale of 10^8 years.

phases occurs. Each particle is to have an equal probability for changing phases, independent of its neighbors and independent of its past history. This may be justified by the assumption that each phase is well-mixed on a timescale shorter than the transfer time. We adopt such an assumption without detailed justification. This mixing assumption also simplifies the prescription for the reaccretion of gaseous atoms by dust grains, which we take to occur during their transition from the diffuse phase to the molecular cloud phase. Each volume element of the diffuse phase is then regarded as containing the same composition of gaseous atoms and the same ratio of gas to dust. It then follows that when a randomly chosen volume element within the diffuse phase transfers to the molecular cloud phase, accreting all of the refractory gaseous atoms as it does so, each grain will add the same thickness Δr of accreted mantle independent of its size (e.g., *Clayton*, 1980). Because the gaseous refractory atoms are reaccreted in less than 10^6 yr in the molecular cloud, a time less than 1% of characteristic residence times, we take this reaccretion episode to be instantaneous as the parcel of matter joins the molecular cloud. This happy simplification allows the history of each particle to be calculated independently of the history of others.

For clarification, consider briefly the contrasting situation in which different portions of the diffuse medium have undergone different degrees of sputtering and therefore contain different gas/dust ratios. When these unmixed elements cool to join the molecular cloud, each particle would add a Δr that would in that case depend upon the past sputtering history of the particles within that volume element. If the total mass of refractory material is to be conserved, one would then find no migration in grain size if each particle reaccretes the same mass that was previously sputtered away from it, whereas in the independent-particle histories that we actually compute, each particle does a random walk in grain size by undergoing different amounts of sputtering (see below) before reaccreting the universally shared Δr.

We assume that the sputtering occurs only in the diffuse phase, and that it continues acting on each particle as long as that particle remains in the diffuse phase. By taking a constant probability per unit time $1/\tau_{cloud}$ for each particle to leave the diffuse medium, it is easy to choose residence times by a random number generator for each particle independently. Namely, the probability that the particle resides for a time t before being ejected during the subsequent interval dt is

$$p(t)\,dt = e^{-t/\tau}\,dt/\tau \qquad (1)$$

so that $\int_0^\infty p(t)\,dt = 1$ for transfer at some time. From the Monte Carlo equation, therefore, we set the residence time t during each residence by

$$\int_0^t p(t)\,dt = G \qquad (2)$$

where G is a random-number generator on the interval [0,1]. Inverting gives the residence time

$$t = \tau \ln(1-G)^{-1} \qquad (3)$$

This is the duration of the sputtering for each residence in the diffuse medium.

For the sputtering rate itself we have used two different prescriptions. The first envisions particles sitting in a hot postshock medium being sputtered away by thermal ions at a constant rate independent of size

$$\left(\frac{dr}{dt}\right)_{sputter} = -k \qquad (4)$$

where k is chosen such that a characteristic particle, which we envision as one having a radius $r = 0.1\ \mu m$, will have a mass sputtering lifetime $m(dm/dt)^{-1} = \tau_M$ equal to the mass sputtering lifetime 2×10^8 yr calculated by *Draine and Salpeter* (1979) for diffuse-medium dust. This gives $k = 0.02\ \mu m\ (10^8\ yr)^{-1}$.

In an alternate description of the sputtering rate we assume that the sputtering is dominated by the inertia of the grains causing different stopping times in the postshock flow (*Shull*, 1978). The larger particles are then sputtered at a greater rate, which we parameterize by

$$\left(\frac{dr}{dt}\right)_{sputter} = -\frac{r_i}{\tau_I} \qquad (5)$$

where r_i is the *initial radius* of the spherical grain at the time it enters the diffuse phase. In this calculation we have taken $\tau_I = 2.6 \times 10^8$ years. This value gives a mass sputtering lifetime $M/(dM/dt) = 8.5 \times 10^7$ years, which is consistent with the rate calculations by *Draine and Salpeter* (1979). These two parameterizations of the sputtering rate lead to interestingly different grain-size distributions that have similar cosmochemical fractionation patterns, however. These abundance fractionation patterns are a major objective of our modeling. The mechanism of sputtering and cycling, followed during solar system formation by grain-size sorting mechanisms, was advanced by *Clayton* (1980) in a paper that has greatly influenced our perceived need for the present study.

Grain-grain collisions were not included in our computer program, although the work of *Jura* (1976), *Shull* (1977, 1978) and *Seab and Shull* (1983) show that collisional processes are very important for the evolution of dust grains. It is our hope to include grain-grain collisions in a later study.

We have chosen to concentrate on the refractory SUNOCON component for this study. The major refractory cations, Mg, Al, Si, Ca and Ti, that appear in minerals of low vapor pressure near 1500 K all have their nucleosynthetic origins in supernova shells of carbon, oxygen, and silicon burning (*Woosley et al.*, 1973; *Clayton*, 1988). For the sake of calculation, we will adopt the otherwise debatable argument that *all* of these refractory cations initially condense in SUNOCONs during the expansion of the supernova interior, so that they are placed in the gas phase only by destruction mechanisms. This assumption is really not so restrictive, because our calculations are also relevant even if only a fixed fraction of them condense in this initial form. But with this assumption of complete initial condensation, we see that if the system has no additional sources or sinks of refractory atoms, the mass of refractory atoms will thus be conserved even after repeated sputtering and accretion episodes. We will refer to this as *mass conservation* in what follows, and make it one of our objectives. But even now it is clear that if only some fixed fraction condenses in the initial SUNOCONs, the total mass of refractory atoms at any later time must be equal to the total mass of SUNOCONs created up to that time divided by that initial-condensation fraction. We would represent this by a non-mass-conserving system, i.e., one in which the total mass of dust exceeds the mass of the injected SUNOCONS. Whether our calculation results in mass conservation or not will depend on the assumed thickness Δr of accretion mantle of refractory atoms that each grain adds during its random transfer from diffuse medium to molecular cloud. When we speak of mass conservation, however, we speak not of total galactic mass but rather of the mass of refractory atoms contained within a fixed mass of interstellar material; i.e., the mass is really a *concentration*. This allows the problem to be treated in the same spirit as the treatment of interstellar abundances (concentrations) in the chemical evolution of galaxies (*Clayton*, 1984, 1985).

This last point is sufficiently important that we ask the reader for a moment to consider the simpler problem of the abundance of aluminum atoms in the interstellar medium. That total abundance may decline owing simply to the conversion of interstellar material into stars; however, the associated concentration of interstellar Al remains unchanged. The processes of nucleosynthesis and destruction (if any) are what manifest themselves as a changing interstellar concentration. That this distinction is of considerable magnitude can be seen considering the exponential model of galactic evolution, in which each of these quantities declines in the same exponential way: the total mass of interstellar medium, the rate of star formation, and the rate of nucleosynthesis. However, in the same model the *concentration* of Al, defined as the mass of Al per unit mass of ISM, increases linearly with time despite the almost exponential decline in the total mass of Al, and, moreover, the age spectrum of interstellar Al nuclei is flat (*Clayton*, 1985) rather than having the profile of the rate of nucleosynthesis, which is exponential. It is just in this same sense that we mean the "mass of refractory atoms" to mean the "mass of refractory atoms within a fixed mass of interstellar medium."

Motivated by both this result and by computational simplicity, we inject the SUNOCONs into the interstellar medium at a constant rate. This makes their birthday spectrum flat, just as the age distribution of nuclei is flat in exponential galactic models. *Clayton* (1986b) has shown how to formally treat the SUNOCON age spectrum within his "standard analytic model" of chemical evolution (*Clayton*, 1985), but we feel that for the present exploratory study, a flat age spectrum is adequate for the problem of quantifying the major physical effects in the cosmic chemical memory of the SUNOCONs. Other approximations introduce larger errors or uncertainties than this one.

We assume that star formation occurs only in the molecular cloud phase. Dust there is incorporated into stars (astrated) at a constant rate equal to $(1.5 \times 10^9 \text{ yr})^{-1}$, so that for the ISM as a whole the lifetime against astration is 3×10^9 yr. During the Monte Carlo histories of individual particles, therefore, the probability of astration during a residence time t in the molecular cloud is again given by an equation like (1). This is easily handled. However, associated issues of interpretation must be included in our model assumptions. Roughly half the matter that enters stars returns to the ISM with its refractory metal concentrations unchanged, and a question arises about our treatment of that ejecta. In the first place, we assume that the bursts of star formation disrupt the molecular clouds at the point of star formation, and that this disruption is one of the major avenues by which molecular cloud dust is transferred to the diffuse medium. The material returned from the stars is therefore assumed to join the diffuse medium as well. But in what chemical form does it rejoin the diffuse medium? The answer is not known. *Clayton* (1982, 1986b) has argued that these refractory atoms recondense as refractory STARDUST. If that is so, we should reinject STARDUST into the diffuse phase at half the astration rate of refractory dust. In order to obtain a mass-conserving system in our calculations, however, we must return all of the astrated refractory elements either as gas or as refractory STARDUST, because our simple representation of galactic history suppresses the decrease in total mass owing to the stellar remnants remaining after astration.

We handle this in one of two ways, depending upon which extreme assumption we make for the old refractory atoms returned from stars. If they are totally gaseous upon return, then astration has the effect of reducing the total number of interstellar particles, and the Δr chosen to conserve mass will represent the reaccretion of the sum of atoms made gaseous both by sputtering and by astration. If, on the other hand, they are totally condensed as refractory STARDUST upon return, the number of particles will not decline owing to astration, and the Δr chosen to conserve mass will represent only the reaccretion of atoms made gaseous by sputtering. In this case, moreover, the cores of the interstellar grains will be of two types of thermal condensates—SUNOCONs and STARDUST—whereas the accretion mantles on both represent the average composition sputtered from the particles. In this case we can easily demonstrate that

(STARDUST injection rate)$=\omega_*\mathrm{t}\times$(SUNOCON injection rate)(6)

where $\omega_* = (3 \times 10^9 \text{ yr})^{-1}$ is the rate at which matter is turned into stars (the formation rate, or astration rate). For the galactic histories we calculate below, having age 6×10^9 yr and constant SUNOCON injection rate, the STARDUST injection rate then is initially zero (a result of the initial zero metallicity) and increases to a value twice the SUNOCON injection rate at $t = 6 \times 10^9$ yr; moreover, the total mass of STARDUST injection over the full galactic age is then equal to the total mass initially injected as SUNOCONs.

One of our assumptions is that the refractory atoms can be treated independently of the more volatile ones. This seems reasonable for the thermal condensates, the SUNOCONs and the STARDUST, but what of the mantles accreted when matter joins our molecular cloud phase? These atoms stick cold on the grains, causing the refractories to mix with much more abundant volatile elements. This could totally invalidate the radius histories of our particles and is a serious problem, especially for carbon, but we can partially justify our assumption as follows. There exist many transitional phases between molecular cloud interiors and diffuse medium shock waves. The volatile atoms can be sputtered much more readily by thermal sputtering at modest temperatures. We imagine a low level of thermal sputtering that is much more frequent than the strong (v > 50 km/sec) shocks needed to sputter refractory particles. We imagine, therefore, that as mantle laden molecular-cloud particles are returned by a nearby burst of star formation to the diffuse medium, their volatile mantles are selectively removed first. We do not defend this argument except as a rationalization of a simplifying assumption that allows this exploratory investigation to proceed in a more straightforward manner. In actual fact, many carbonaceous materials are also quite persistent, so that carbon can be regarded as volatile only with substantial peril. However, it should be noted that accreted carbon may be removed by chemically reacting with monatomic hydrogen (*Duley and Williams*, 1986). Nonetheless, our initial calculations treat the heavy refractory metals (Mg, Al, Si, Ca, Ti, and their oxides) as if they are the only condensible atoms. When this system is better understood, the interaction with volatile atoms can perhaps be included more intelligently.

Another assumption that we make is that the refractory elements can be divided into a superrefractory class that condenses first followed by a mantle class that condenses around the superrefractory nucleus. There is good precedent for this. The corundum mineral Al_2O_3 and related aluminum-rich minerals can condense about 200 K earlier in a falling temperature sequence than can MgO or Mg silicates (*Grossman and Larimer*, 1974). We take it that the relatively rapid expansion and cooling of stellar ejecta is slow enough to approximate this equilibrium sequence, yet fast enough that the superrefractory cores maintain their identity. In other words, our SUNOCONs and STARDUST are layered structures. When they are first injected into the diffuse phase of interstellar medium, the SUNOCONs are represented in our calculations by spheres, with the superrefractory core, phase A, having initial radius half that of the refractory mantle, phase B. We choose these radii so that phase A is initially one-eighth of the refractory SUNOCON mass both for definiteness and simplicity and also because the mass ratio Al_2O_3/MgO is not far from one-eighth with solar abundances. Furthermore, we take each SUNOCON to have that initial structure even when we inject them with a spectrum of initial sizes.

Another reason for choosing this layered structure is that it allows us to numerically evaluate aspects of cosmochemical memory mentioned in our introduction. Because these SUNOCONs are initially monoisotopic ^{16}O oxides, the association of ^{16}O-richness with SUNOCON cores generally and with aluminum can be remembered if those Al_2O_3 cores are substantially shielded from destruction. This would require that they be sputtered more slowly than the phase B mantles, as *Clayton* (1986b) has evaluated within analytic mathematical models of SUNOCON persistence. The layered structure also allows us to evaluate, at least within the assumptions of our calculation, the idea that the very high depletion factor of interstellar aluminum from the gas phase has its roots in this SUNOCON structure (e.g., *Clayton*, 1982). In such ways our assumption concerning initial structure is designed not only to be reasonable but also to facilitate system tests of aspects of cosmic chemical memory.

In those calculations wherein we assume that the refractory atoms returned from stars also condense, we give the STARDUST the same structure, a phase A core with phase B mantle. These STARDUST particles must be tagged to distinguish them from SUNOCONs, however, because they are no longer ^{16}O-rich, and except where explicitly argued to the contrary, can best be thought of as isotopically normal.

The atoms that have been sputtered from the ensemble of particles are reaccreted in a second mantle, which we call phase C. This material represents a well-mixed diffuse medium having the composition of the matter sputtered from the mixture of particles. The phase C mantles are identical in composition for SUNOCON cores and for STARDUST cores. That composition is in first approximation solar, both elementally and isotopically, but it is deficient in those species that have remained locked up within the cores. Simple mass balance considerations identify the composition of phase C. For example, it is slightly ^{16}O-poor because the remaining SUNOCON mass is pure ^{16}O and it is Al-poor because a significant fraction of Al is locked up in the phase A cores. Other examples can be added by further definitions of the injection structure. We envision the phase C mantles as being chemically homogeneous and structurally amorphous. For our calculations we will take all phases to have the same density, so that particle mass and particle volume can be used interchangeably.

Many of the features of our calculations are shown in Fig. 2, which traces a portion of the evolution of one particle. The duration of its residence in either diffuse medium or molecular cloud is determined by a random number G chosen for each residence according to equation (3). In the upper panel thermal sputtering erodes each particle in the diffuse phase at a constant rate according to equation (4), whereas in the lower panel the erosion rate is proportional to particle radius as in equation (5). The contrast in the slopes of the lines is evident in Fig. 2. In either panel, the particle accretes a phase C mantle of thickness Δr when it enters the molecular

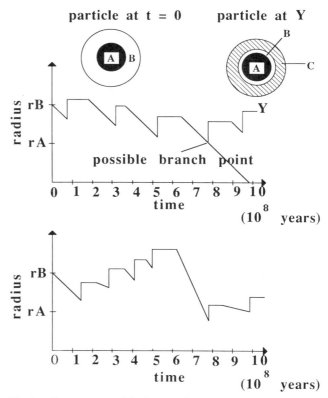

Fig. 2. Short segment of the history of one particle. At t = 0 when the dust particle is first created, the particle consists of an inner core (labelled A) and an outer core-mantle (labelled B). After a number of cycles between the diffuse and molecular phases of the ISM, the particle may have the cross-section as shown for the particle at Y; namely, most of the B layer has been eroded by sputtering and a new layer (C) has been formed by accretion of material during the particle's entries of molecular clouds. The top graph is a plot of grain radius versus time for sputtering erosion occuring at a constant rate (see equation (4)) while the particle is in diffuse clouds. The discontinuous jump in the particle size represents the material accreted when it transfers from a diffuse cloud to a molecular cloud. A possible branch point in the particle's history is shown, at which it could remain in the diffuse cloud until it is destroyed or it could be transferred to a molecular cloud, accreting material and surviving. The lower graph shows for comparison the particle size versus time for a sputtering erosion rate proportional to the size of the particle (see equation (5)).

cloud, where Δr is independent of particle size. Its magnitude is chosen to achieve a mass conserving system. The radius remains fixed during the subsequent molecular cloud residence, but another random number determines whether the particle will be astrated during that molecular cloud residence. The upper panel shows one example of a branch point in the history of that particle. The particle structure at two different times in the upper-panel history is shown above that panel. Although the initial sputtering after injection is of phase B material, it is interesting that most of the subsequent sputtering will be of the phase C material added during transfer episodes. This cartoon makes very clear the distinction between the lifetime of a particle as an entity and its mass sputtering lifetime. Each particle in our calculation takes a random walk through its possible histories. The particle vanishes only when it is sputtered below a minimum stable size, which we take to be 5 Å, or when it is astrated.

When we inject the initial SUNOCONs we choose an injection size spectrum. For a trial calculation we first injected all SUNOCONS at a = 0.1 μm, and noticed the subsequent spreading of the grain-size spectrum. Some were randomly walking downward in size and some upward. For the calculations reported here, however, we have taken the injection spectrum to be a power law in radius

$$n(a)\, da = c\, a^{-n}\, da \quad a_{min} < a < a_{max}$$
$$= 0 \text{ otherwise} \tag{7}$$

where we have used both n = 2 and n = 3. (It should be noted that we denote the initial radius by a. Otherwise the radius is denoted by r.) The latter spectrum distributes the injection mass uniformly over the grain sizes. Our standard calculations use a_{min} = 0.01 μm and a_{max} = 0.1 μm, but other calculations to be reported vary these limits by factors of two or three. When we inject STARDUST as well, we use the same size spectrum as for the SUNOCONs, although there is no reason to do this except that it seems an unbiased choice. The method of calculation can be employed for any specified distribution of sizes. Because of our intent to calculate the histories of a large number of particles, we choose each injection size by a random number appropriately matched to a probability and we choose the injection time by another random number that begins each new history at starting times distributed uniformly over six billion years. We choose a galactic age of 6 b.y. as being representative of the molecular cloud in which the sun formed rather than one today. The difference is that the particles today would be more heavily sputtered and more heavily astrated.

With this preamble we now turn to the details of the calculations and their results.

3. CALCULATIONS AND RESULTS

We set the parameters of the problem to define the nature of the injected dust and the various probabilities for the events that each particle may encounter, as described in the previous section. The history of a single particle is then computed in roughly 0.1 sec on a modest computer. The duration of a total computation depends upon the number of particle histories that we require to adequately define questions of interest. After some exploring with smaller populations, we report here separate calculations containing 6×10^6 particles in order to define adequately the population of each size bin. It is not feasible to record the exact final configuration of each particle, so at the end of the 6×10^9 yr galactic history, each particle is entered into a bin {r, Δr} to indicate that a particle of radius r in the interval Δr exists at the conclusion of its history. Because we are interested in composite particles, having phases A, B, and C as shown in Fig. 2, we also need a bin {r_A, Δr_A} to define the population of phase A cores, etc. We took a bin size Δr = 20 Å so that roughly 150 bins in outer radius are needed to record the range size from the smallest particles to those of about 0.3 μm. The phase A and phase B structures are limited to the maximum size with which they are injected, whereas the phase C mantles may stochastically grow. When both SUNOCONs and STARDUST are injected, another binary index is needed to identify the nature of the particle core. The phase C mantles are common material that has previously been, for whatever reason, in the gas phase. That reason is sputtering alone if all refractory atoms condense as matter is ejected from stars.

In the calculations we now report, we do assume that the refractory heavy elements condense entirely when ejected from stars, and we therefore seek an increment Δr to each phase C mantle upon cloud reentry that conserves the total mass of the dust system. As noted previously, this choice is not at all necessary in calculations of this type. If substantial refractory matter is ejected in gaseous form, we would instead choose a Δr that achieves a total grain mass greater than the total grain mass injected from the stars. However, it was our choice to examine the extreme system of total condensation or of no condensation for the STARDUST. We report calculations of four separate systems.

Case 1. No STARDUST particles are formed, so that all refractory grains possess SUNOCON cores. Sputtering in the diffuse medium erodes the particle radius at a constant rate $dr/dt = -0.02\, \mu m/10^8$ yr.

Case 2. No STARDUST particles are formed as in case 1, but the particle sputtering is proportional to its radius as in equation (5).

Case 3. SUNOCONS and STARDUST both condense 100% and sputtering occurs at a constant rate as in case 1.

Case 4. SUNOCONs and STARDUST both condense 100% and sputtering occurs proportional to radius as in case 2.

These four cases span an interesting range of behavior for the refractory component of interstellar dust. We recognize that the truth lies somewhere in between these extreme assumptions, but we judge such a situation to be appropriate for a first attempt at modelling such systems. Our results will be shown largely in tables and in figures. All four cases injected SUNOCONs uniformly over 6×10^9 yr. and astrated particles at the rate $(3 \times 10^9 \text{ yr})^{-1}$.

Case 1 and Case 2

In the tables of results we summarize a large amount of numerical information. The first of these, Table 1, pertains to cases 1 and 2, wherein SUNOCONs are the only source of ISM particles because no STARDUST condenses in the stellar ejecta. The first column in this table describes the quantities being tabulated, with the following conventions: physical units are shown in brackets, whereas abbreviations used are contained in braces. Because Tables 1-3 have similar nomenclature, we digress here to comment upon the meaning of the entries for Table 1 according to line number. An entry 2.0E7 means 2.0×10^7.

Lines 1-3 characterize the power law a^{-n} in initial radius with which the particles were injected between $a_{min} < a < a_{max}$ (lines 2, 3).

TABLE 1. Grain population characteristics: SUNOCONs only (cases 1 and 2).

	Quantity	Case 1a	Case 1b	Case 1c	Case 1d	Case 2
1	Injected particle distribution	a - 2	a - 3	a - 3	a - 3	a - 3
2	Minimum radius of injected particles (m)	1.00E - 08	1.00E - 08	3.00E - 08	3.00E - 08	1.00E - 08
3	Maximum radius of injected particles (m)	1.00E - 07	1.00E - 07	1.00E - 07	2.00E - 07	1.00E - 07
4						
5	Sputtering rate (μm/1e8 years)	0.02	0.02	0.02	0.02	variable
6	Number of particles injected {Ni}	600,000	6,000,000	600,000	600,000	2,500,000
7	Increase in size due to accretion {Δr} (m)	1.78E - 08	1.65E - 08	1.83E - 08	1.95E - 08	1.57E - 08
8	# destroyed in the diffuse cloud {NdDC}	409,965	4,656,931	276,442	240,223	1,138,412
9	# destroyed in the molecular cloud {NdMC}	115,045	828,808	192,620	211,079	826,109
10	# surviving in the diffuse cloud {NsDC}	38,663	265,423	67,589	76,837	276,377
11	# surviving in the molecular cloud {NsMC}	36,327	248,838	63,349	71,861	259,102
12	total # surviving {Nf}	74,990	514,261	130,938	148,698	535,479
13	Nf/Ni (%)	12.50%	8.57%	21.82%	24.78%	21.42%
14						
15	Total mass injected {M.injected} (kg)	1.38E - 13	4.57E - 13	3.48E - 13	7.87E - 13	1.91E - 13
16	Total mass sputtered {M.sputter} (kg)	1.28E - 12	5.28E - 12	2.99E - 12	5.24E - 12	3.60E - 12
17	Total mass astrated {M.astration} (kg)	1.16E - 13	3.83E - 13	2.87E - 13	6.59E - 13	1.56E - 13
18	Total mass accreted {M.accretion} (kg)	1.40E - 12	5.66E - 12	3.28E - 12	5.9E - 12	3.75E - 12
19	Total surviving mass {M.final} (kg)	1.38E - 13	4.59E - 13	3.47E - 13	7.88E - 13	1.92E - 13
20	M.final/M.injected	1.00	1.00	1.00	1.00	1.01
21	M.final/Nf (kg)	1.84E - 18	8.92E - 19	2.65E - 18	5.30E - 18	3.59E - 19
22	M.injected/Ni (kg)	2.30E - 19	7.62E - 20	5.80E - 19	1.31E - 18	7.62E - 20
23	M.sputter/M.accretion	0.92	0.93	0.91	0.89	0.96
24	M.sputter/M.final	9.29	11.51	8.62	6.65	18.73
25	M.astration/M.final	0.84	0.84	0.83	0.84	0.81
26	M.accretion/M.final	10.12	12.35	9.44	7.48	19.55
27	M.sputter/M.injected	9.28	11.54	8.60	6.65	18.88
28	M.astration/M.injected	0.84	0.84	0.82	0.84	0.82
29	M.accretion/M.injected	10.12	12.38	9.42	7.49	19.70
30						
31	Total mass of core injected {MA.injected} (kg)	1.73E - 14	5.72E -14	4.35E - 14	9.84E - 14	2.38E - 14
32	Total mass of core sputtered {MA.sputter} (kg)	8.79E - 15	3.50E -14	2.19E - 14	2.86E - 14	1.65E - 14
33	Total mass of core astrated {MA.astration} (kg)	5.08E - 15	1.35E - 14	1.29E - 14	4.03E - 14	4.57E - 15
34	Total mass of core surviving {MA.final} (kg)	3.41E - 15	8.68E - 15	8.71E - 15	2.95E - 14	2.70E - 15
35	MA.final/MA.injected (%)	19.76%	15.18%	20.02%	29.96%	11.33%
36	MA.sputter/MA.injected (%)	50.83%	61.28%	50.31%	29.04%	69.48%
37	MA.astration/MA.injected (%)	29.41%	23.54%	29.68%	40.99%	19.19%
38	MA.final/M.final (%)	2.47%	1.89%	2.51%	3.74%	1.41%
39	MA.injected/M.injected (%)	12.50%	12.50%	12.50%	12.50%	12.50%
40						
41	Total mass of core-mantle injected {MB.injected} (kg)	1.21E - 13	4.00E - 13	3.05E - 13	6.89E - 13	1.67E - 13
42	Total mass of core-mantle sputtered {MB.sputter} (kg)	9.84E - 14	3.43E - 13	2.47E - 13	4.71E - 13	1.51E - 13
43	Total mass of core-mantle astrated {MB.astration} (kg)	1.38E - 14	3.51E - 14	3.49E - 14	1.29E - 13	1.02E - 14
44	Total mass of core-mantle surviving {MB.final} (kg)	8.81E - 15	2.17E - 14	2.24E - 14	8.85E - 14	5.73E - 15
45	MB.final/MB.injected (%)	7.28%	5.42%	7.37%	12.86%	3.44%
46	MB.sputter/MB.injected (%)	81.31%	85.82%	81.18%	68.39%	90.47%
47	MB.astration/MB.injected (%)	11.40%	8.76%	11.46%	18.75%	6.10%
48	MB.final/M.final (%)	6.37%	4.73%	6.46%	11.24%	2.98%
49	MB.injected/M.injected (%)	87.50%	87.50%	87.50%	87.50%	87.50%
50						
51	MB.injected/MA.injected	7.00	7.00	7.00	7.00	7.00
52	MB.final/MA.final	2.58	2.50	2.58	3.00	2.12
53						
54	M/(dM/dt).sputter [for injected particles] (years)	9.17E + 07	6.51E + 07	9.69E + 07	1.49E + 07	8.53E + 07
55	M/(dM/dt).sputter [time averaged] (years)	3.23E + 08	2.61E + 08	3.48E + 08	4.51E + 08	1.60E + 08
56	M/(dM/dt).astration [time averaged] (years)	3.59E + 09	3.59E + 09	3.63E + 09	3.59E + 09	3.70E + 09

Line 5 denotes whether the sputtering reduced the particle radius at a constant rate (0.02 $\mu m/10^8$ yr) or at a rate proportional to radius (variable).

Line 6 gives the total number of particles injected during the history, and lines 10-12 give the final numbers remaining at the end, with line 13 the ratio.

Line 7 gives the increase Δr of each particle by accretion during molecular cloud entry. We choose this number to satisfy mass-balance considerations.

Lines 8 and 9 gives the total numbers of particles destroyed during the history, by sputtering in the diffuse phase and by astration in the molecular phase. For example, in case 1b the sputtering deaths exceed astration deaths by almost 6 to 1, with the fraction surviving at the end equal to 8.57%.

Lines 15-29 concern total masses and their ratios within the particle system: the total mass ever injected, total mass ever sputtered (including repeats), total mass ever astrated, and the total mass ever accreted during cloud entry. Considering again case 1b as an example, one sees 6×10^6 SUNOCONs injected bearing total mass 4.57×10^{-13} kg. Line 20 shows final mass (after 6×10^9 yr) equal to 1.00 times the injected mass, showing that mass conservation was exactly achieved by the value of Δr chosen in this calculation. Line 24 shows the interesting result that the total mass ever sputtered away from particles is 11.51 times the injection mass, confirming that the phase C mantles are sputtered and reaccreted many times over. Sputtering and reaccretion is the main mass exchange.

Lines 31-39 tabulate quantities pertaining to the phase A superrefractory cores. For example, line 36 shows that 61.28% of the phase A cores in case 1b have been sputtered at some time in history, a quantity directly relatable to the ^{16}O-richness of interstellar aluminum.

Lines 41-49 tabulate analagous quantities pertaining to the phase B refractory mantles (here called *core* mantles to distinguish from phase C).

Lines 51-52 show B/A ratios in the remaining SUNOCON cores. Note that this bulk ratio would be 7.0, the injection ratio, if the phase C mantles were to be included, because the total system is mass conserving. Because M_B/M_A is reduced in the ensemble to roughly one-third its initial value, the ratio of corresponding chemical elements $(B/A)_c$ in the phase C mantles must be larger than the injection ratio, and can be computed easily by mass balance. This would be accomplished with the aid of lines 35 and 45, which show, respectively, that in the case 1b example 15.18% of phase A remains unsputtered and unastrated whereas only 5.42% of phase B does so. The complements of these numbers reside in phase C mantles, so that $(B/A)_c = 7.00 (94.58/84.82) = 7.81$, a value 11.5% higher than the injection ratio. This calculation is typical of cosmochemical information that is not included in Table 1 but is easily derived from its entries.

Lines 54-56 give some interesting lifetimes for mass transfer in this system. For both sputtering and astration we define a *mass lifetime* $\tau_M = M/(dM/dt)$ for the total mass of interstellar dust, with the entered numbers being a grand average over the 6×10^9 yr of the system of particles. To evaluate it (dM/dt) is approximated by the ratio, for sputtering, say, of the total mass ever sputtered including repeats (line 16) to the total time, 6×10^9 yr, whereas M is replaced by its average value during the history, namely $M = M_f/2$, considering that the SUNOCONs were injected at a constant rate. Line 54 shows the value of τ_M for the injection spectrum itself, and its contrast with line 55 is so great as to warrant explanation. Consider again the case 1b example. For an a^{-3} injection spectrum between $a_{min} = 0.01$ μm and $a_{max} = 0.1$ μm, the value of τ_M with erosion rate 0.02 $\mu m/10^8$ yr is 0.651×10^8 yr, whereas the historical average $\tau_M = 2.61 \times 10^8$ yr. Two major features account for that fourfold discrepancy. The first is that only approximately half of the particles reside in the sputtering medium. The second is that the injection spectrum contains a much larger fraction of small particles than in the final evolved spectrum, owing primarily to the sputtering destruction of small particles as demonstrated in Fig. 3 below. Thus τ_M increases as this galaxy ages because the average size of the particles is increasing. That this must be so can be seen in line 13, where it is shown that only 8.57% of the number of particles injected survives to the end, but they still carry all of the mass so that they must be about 12 times more massive than the average in the injection spectrum. Now it is a property of constant erosion rate (equation (4)) that τ_M is larger for large particles than for small ones, being proportional to size. Note the difference in case 2, where an erosion rate proportional to size yields a τ_M independent of particle size, being 8.5×10^7 yr for the numerical values chosen. The average τ_M is not quite two times longer, owing in this case almost exclusively to having only half the particles in the sputtering medium.

In a similar vein, line 56 shows that the average mass lifetime against astration is somewhat greater than the value $\tau = 3 \times 10^9$ yr that is, by construction, exactly the expected lifetime of a particle against astration. This modest difference reflects two small differences between the particles in our diffuse medium and in the molecular clouds. The final number of particles in DC exceeds that in MC (lines 10 and 11) and this can be understood via the quantitative model given in the Appendix. Counteracting this somewhat is the greater mass of average MC particles, as shown in Fig. 4 below, which results in a mass lifetime only slightly greater than 3×10^9 yr.

One aspect of these lifetimes must be emphasized again owing to its importance for cosmic chemical memory. Even though the mass lifetime against sputtering is only a few $\times 10^8$ yr, the fraction of specific structures (e.g., phase A cores) that survive 6×10^9 yr of history indicates a much larger effective mass lifetime for those phase A cores (*Clayton*, 1986b).

Figures 3 and 4 show the distribution of particle and core sizes in the case 1b calculation. Figure 3 compares the numbers of particles in DC and in MC with the total number ever injected, and Fig. 4 makes the same comparison for the mass of particles in that size bin. Note in Fig. 4 that the injection spectrum is flat in mass for an a^{-3} number distribution. The total integral of the injection spectrum in Fig. 3 is 6×10^6 particles. The small solid squares show numbers (or masses) as a function of the appropriate radius of the final particle. Note that the final distribution falls off more or less exponentially up to r ~ 0.26 μm, even though $a_{max} = 0.1$ μm at injection. This is stochastic growth of those particles

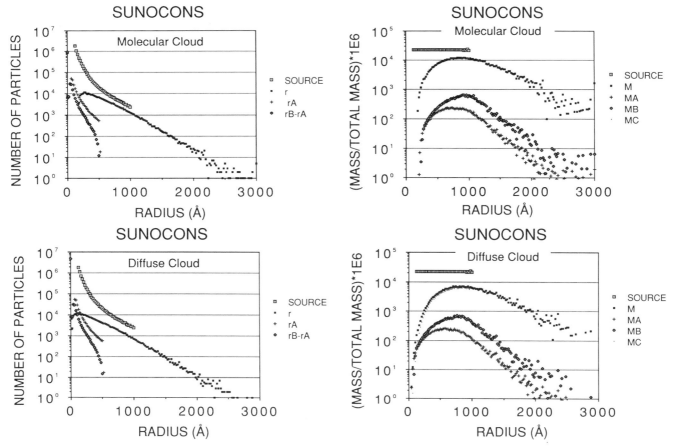

Fig. 3. The number of particles per bin versus size for case 1b. Bin sizes are 20 Å. Numbers injected over 6×10^9 yr history (SOURCE) and numbers surviving in both molecular and diffuse clouds are shown. SOURCE particles were initially injected into the diffuse-cloud phase with an assumed a^{-3} size distribution. Numbers are plotted versus total size (r), radius of phase A core (r_A), and shell thickness of phase B core mantle (r_B-r_A). Total radius can shrink or grow stochastically, but r_A and r_B-r_A can be smaller but never greater than their initial values. The diffuse-cloud distributions are virtually identical to those in molecular clouds, except total size (r) is shifted by $\Delta r = 165$ Å in molecular cloud. In case 1b, N(r) declines almost exponentially, reaching sizes up to three times the largest injected particles. Note that in all the figures where particle radii are shown, the unit of length is the angstrom. For Figs. 4, 5, 6, 9, 10, 11, 14, 15, 16, 18, 19, and 20 the radius refers to the total radius of the particles for all layers of the particle, while in Figs. 3, 8, 13, and 17 the radius can also be the radius of the core (r_A) or the width of the initial core mantle (r_B-r_A).

Fig. 4. Sums of particle masses per bin, normalized to the total mass of the system, for the same case 1b data as Fig. 3. Sum of total masses in bin for particle radius r = M(r); sum of phase A core masses of particles in same bin = M_A(r); sum of core-mantle masses = M_B(r); sum of mantle masses = M_C(r). $M = M_A + M_B + M_C$. Mantle mass M_C appears as simple points barely beneath (or hidden by) the square for M on this logarithmic plot, because those mantles dominate the total masses. Although molecular-cloud and diffuse-cloud distributions are virtually identical, M(r) and M_C(r) are greater in the molecular cloud because its particles have extra $\Delta r = 165$ Å mantle. The mass per bin is seen to remain within several percent of its maximum value even out to r = 3000 Å. In Figs. 4, 5, 9, 10, 14, 15, 18, and 19 the normalized mass is multiplied by a factor of 10^6. This factor has no physical significance and was used only because our computer graphics software cannot plot very small ordinate values.

that were lucky enough to have spent shorter than average times in the DC between repeated entries to the MC. Note that with $\Delta r = 0.01654$ μm with each MC entry (line 7), a minimum of 10 MC entries would be required to reach the largest sizes even if the particle suffered no intervening sputtering. Because the transition probability was chosen by us to be $(10^8 \text{ yr})^{-1}$, an average particle having age half the galactic age would have been expected to make 30 transitions from one phase to another. These ideas illuminate the random-walk nature of these histories. The distribution of core sizes as a function of phase A radius r_A and of r_B-r_A, both of which have the permanent upper limit 0.05 μm, is also shown in those figures. The greater steepness of the number distribution in r_B-r_A over that in r_A reflects the much greater erosion rate of the phase B core mantles, which must after all be completely removed before the phase A cores can be attacked. Evidently the phase A cores retain a distribution shape very similar to their injected distribution, except that their number has been

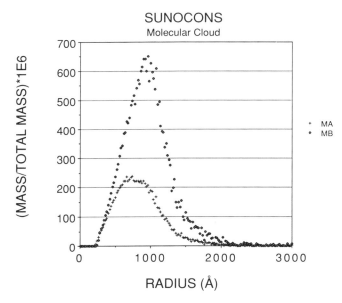

Fig. 5. Magnified portion of Fig. 4 showing sums $M_A(r)$ and $M_B(r)$ of SUNOCON cores on linear scale versus total-particle bin size. Case 1b.

reduced. [Remembering that *at injection* $n(r_A = 0.05\mu m) = n(a = 0.1 \mu m)$, etc.]

In Fig. 4 the format differs in the sense that the mass within phase A, for example, is shown as a function of the *total particle size*. Easily the dominant observation here is that the total mass (solid squares) at all particular sizes is much greater than that in either SUNOCON phase, so that the phase C mantles are dominating the mass (roughly 90% of the mass at the peak in the mass distribution near $r = 0.08 \mu m$). Another significant result is that the ratio M_B/M_A is near unity for the small particles and approaches a constant ratio near 4 in the large particles. [Recall that the injection ratio was $M_B/M_A = 7$ for all particular sizes.]

Figure 5 shows fractional masses $M_A(r)$ and $M_B(r)$ on an expanded linear scale for the MC particles and Fig. 6 shows the ratio $M_B(r)/M_A(r)$. The scatter for large r reflects the statistical variations owing to small populations of bin [r, Δr]. Thus as far as SUNOCON mass is concerned, a collection of small particles has a smaller M_B/M_A than will a macroscopic aggregate of large particles. This is a significant cosmochemical effect, showing that chemical fractionation can be achieved by macroscopic aggregates having different grain sizes (*Clayton*, 1980). One example of this effect of possible importance to isotopic studies of meteorites will be given. The isotopic composition of Mg in phase A, where it exists only in $MgAl_2O_4$, differs from that of other Mg bearing SUNOCONs in phase B. Thus macroscopic aggregates of different grain sizes will differ in Mg isotopic composition. This merits a separate study but is documented here for the sake of comprehension of the relevance of this study to cosmic chemical memory.

Another obvious conclusion from Figs. 3 and 4 is the near identity of the distributions in diffuse and molecular media. This is to be expected, because average MC particles are derived from average DC particles by accretion of phase C mantle increment Δr (= 0.0165 μm = 165 Å for case 1b), shoving the distribution uniformly toward larger r by that amount. Thus the MC particles are slightly more massive with extra phase C material. Having confirmed that our program runs in this regard according to expectation, we make all subsequent figures pertain to the molecular cloud only, that being the composition relevant to the origin of the solar system.

The large fatality rate for particles is dominated by the sputtering destruction of the very small particles in case 1. That the large particles live much longer can be seen in many ways, of which we call attention to three. First, Fig. 3 shows that in case 1b the final number of particles having $r = 1000$ Å is much closer to the a^{-3} injection spectrum than is the number of much smaller particles. Second, one can instructively compare the fraction that survive (line 13) for the four cases 1a, b, c, d, which differ only in the shape of the injection spectrum, which is normalized to contain the same number of injected particles. Thus going from case 1a to case 1b lowers the surviving fraction from 12.50% to 8.57% because the a^{-3} spectrum injects more of its particles at the small end than does the a^{-2} spectrum. Going from case 1b to case 1c increases the surviving fraction from 8.57% to 21.82% because the minimum radius a_{min} of the a^{-3} injection spectrum was increased from 100 Å to 300 Å. Going to case 1d increases survivors even more by increasing a_{max} to 0.2 μm. Third, the average ages of individual particles when they died in our calculations, as illustrated in Fig. 7 from the case 1b calculation, increases steadily with radius of the injected particle. *Large particles live longer if the sputtering erosion rate is constant.* As evident as this may be, many cosmochemical consequences and results of our calculations are explained by it. Many others that we do not address also follow, such as the larger $^{235}U/^{238}U$ isotopic ratio in the smaller particles.

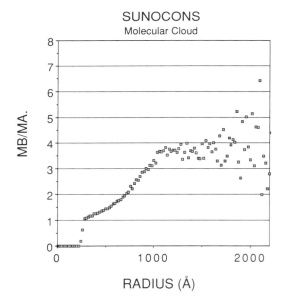

Fig. 6. Ratio of core-mantle sum M_B to core sum M_A within particles of total radius r. Note increase with particle size. At injection $M_B/M_A = 7$.

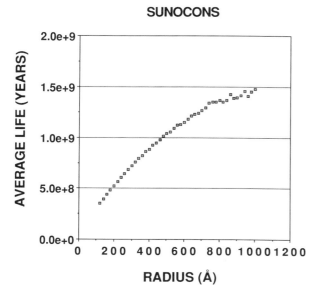

Fig. 7. Average lifetimes of particles as entities for those particles destroyed in case 1b as a function of their *initial* size ($a_{max} = 1000$ Å). Small-particle deaths are dominated by sputtering, but for largest initial particles astration becomes half the death rate ($\tau_{astration} = 3 \times 10^9$ yr).

Case 2

Case 2 is best compared with case 1b, because all aspects of the injection spectra are the same. However, in case 2 the sputtering rate is taken proportional to the radius of the particle (equation (5)), as might occur if grain inertia rather than

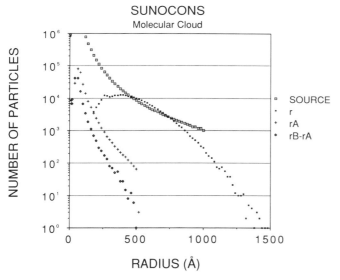

Fig. 8. Number of particles per bin versus size for case 2. Format is identical to Fig. 3, but only the molecular-cloud particles are shown in this case. Note that particle radii are very restricted in comparison with Fig. 3 because a sputtering rate proportional to radius accelerates the decline for $r > 1000$ Å. Note also more final particles near $r = 600$ Å than the number injected, a funneling of particles toward the median size with this sputtering law.

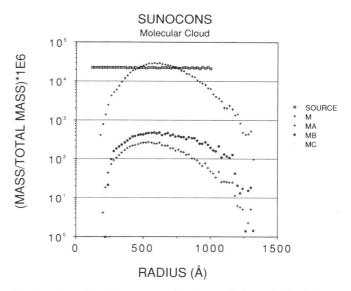

Fig. 9. Sums of particle masses per bin for case 2. Format is identical to Fig. 4, but only molecular-cloud particles are shown.

thermal ions dominates the sputtering. The result on the final grain size spectrum shown in Fig. 8 is dramatic. The particle population is now much more compressed in radius than it was in case 1b in Fig. 3. The number of large particles (larger than $a_{max} = 1000$ Å) is very much reduced, whereas the final number near 500 Å is as great as the total ever injected at that size. The mass distributions are in Figs. 9 and 10. The ratio M_B/M_A is more nearly uniform below 700 Å in Fig. 11 than in case 1b as well. Appreciation of these effects enables a readier understanding of the changes tabulated in Table 1. An example of interest compares lines 54 and 55 for these two calculations. The mass lifetime against sputtering in the

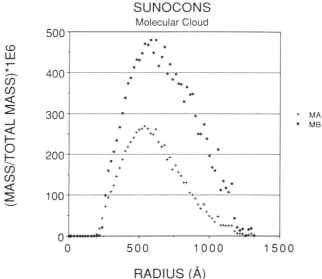

Fig. 10. Enlargement of M_B and M_A from Fig. 9 on linear scale. Compare with Fig. 5.

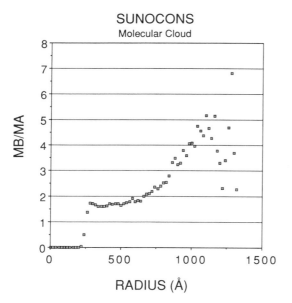

Fig. 11. Ratio of M_B/M_A for case 2, as in Fig. 6.

injection spectrum is slightly greater for case 2 than for case 1b, but the time averaged mass lifetime against sputtering (line 55) shows just the opposite change. The average mass lifetime is larger for case 1b because it builds up many more large particles than does case 2. The age variation with size that was so marked (Fig. 7) for case 1 calculations now virtually disappears, as shown in Fig. 12. Only for the smallest particles is the lifetime shortened, because small particles near the destruction threshold injected into the sputtering medium may

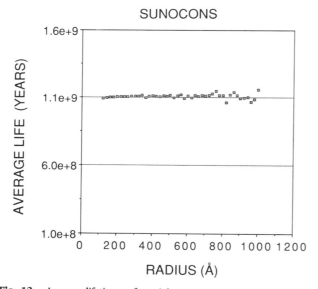

Fig. 12. Average lifetimes of particles as entities for those particles destroyed in case 2 as a function of initial size. This sputtering law gives constant lifetime. Compare with Fig. 7. Because $\tau_{astration} = 3 \times 10^9$ yr the lifetime against sputtering deaths is $\tau_{sputter} = 1.8 \times 10^9$ yr for all sizes (this is determined from $1/\tau = 1/\tau_{sputter} + 1/\tau_{astration}$ where τ is obtained from the figure).

be largely destroyed before transferring to a molecular cloud medium even once, but because we inject all particles with $a > a_{min} = 100$ Å, this effect does not occur at all for case 2 because of its small sputtering rate for small particles.

Cases 3 and 4: SUNOCONs and STARDUST

When material is incorporated into stars (astration), it consumes the bulk interstellar medium. The metallicity $Z(t)$ in that medium grows with time (linearly in the closed-box exponential model), so the mass of heavy elements astrated per unit mass of star formation also increases with time. Our equation (6) models the simplest chemical evolution behavior for the galaxy by returning stellar material (taking return fraction R = 1/2) at a linearly increasing rate while injecting SUNOCONs at a constant rate. Because we make the extreme choice that *all* refractory material condenses when returned, we injected that mass as STARDUST having the same injection structure and injection spectrum as the SUNOCONs. Note that the isotopic and even the chemical structures of phases A and B will be different for STARDUST than for SUNOCONs, but this difference is not explicitly incorporated into our calculations, which merely label the core structure as either SUNOCON or STARDUST. The phase C mantles have a common composition, coming as they do from the gas phase, and that phase C composition is as easily calculated by mass balance in this case as it was in case 1 and 2, where STARDUST did not exist.

Table 2 lists the case 3 results, computed with the constant sputtering erosion rate. The quantities tabulated are very similar to those in Table 1, but the columns list the values for SUNOCONs, for STARDUST, and for the sum. Note that the total number of particles injected, hence the total mass, is smaller than in case 1b. The injection spectrum can be seen to be the same as that for case 1b. So let us call to attention some of the differences that STARDUST introduces.

Although the numbers of particles of each type injected are equal (line 8), the fraction of STARDUST particles surviving at the end with STARDUST cores (12.44%, line 15) exceeds the fraction of SUNOCONs surviving at the end with SUNOCON cores (7.04%). The reason for this is the younger average age of STARDUST particles owing to the linear increase in their injection rate. For the same reason, SUNOCON-core particles have sputtered, astrated, and accreted more mass over galactic time than have STARDUST-core particles. Likewise both phase A and phase B are more heavily sputtered and astrated for SUNOCON cores. On a matter of format, note that percentages listed in the separate SUNOCON and STARDUST columns are percentages within that family of particles rather than percentages of the total dust population. The final column expresses some averages for the combined system. In this vein it is of interest to note explicitly that the existence of STARDUST particles renders the SUNOCONs more vulnerable to sputtering than they were in the analogous case 1b without STARDUST. For example, the final value of M_B is 4.53% of the total injected for SUNOCONs in case 3 (line 47), whereas it was 5.42% in case 1b. The reason for this increased vulnerability is the STARDUST's carrying of dust mass that would have otherwise

TABLE 2. Grain population characteristics: SUNOCONs and STARDUST (case 3).

1	Particle type	SUNOCONS	STARDUST	SUNOCONS & STARDUST
2				
3	Injected particle distribution	a - 3	a - 3	1 - 3
4	Minimum radius of injected particles (m)	1.00E - 08	1.00E - 08	1.00E - 08
5	Maximum radius of injected particles (m)	1.00E - 07	1.00E - 07	1.00E - 07
6				
7	Sputtering rate (μm/1e8 years)	0.02	0.02	0.02
8	Number of particles injected {Ni}	600,000	600,000	1,200,000
9	Increase in size due to accretion {Δr} (m)	1.448E - 08	1.448E - 08	1.448E - 08
10	# destroyed in the diffuse cloud {NdDC}	486,086	461,117	947,203
11	# destroyed in the molecular cloud {NdMC}	71,654	64,218	135,872
12	# surviving in the diffuse cloud {NsDC}	21,758	38,945	60,703
13	# surviving in the molecular cloud {NsMC}	20,502	35,720	56,222
14	Total # surviving {Nf}	42,260	74,665	116,925
15	Nf/Ni (%)	7.04%	12.44%	9.74%
16				
17	Total mass injected {M.injected} (kg)	4.57E - 14	4.57E - 14	4.57E - 14
18	Total mass sputtered {M.sputter} (kg)	2.95E - 13	2.47E - 13	
19	Total mass astrated {M.astration} (kg)	1.74E - 14	1.37E - 14	
20	Total mass accreted {M.accretion} (kg)	2.86E - 13	2.42E - 13	
21	Total surviving mass {M.final} (kg)	1.85E - 14	2.72E - 14	4.56E - 14
22	M.final/M.injected	0.40	0.59	1.00
23	M.final/Nf (kg)	4.37E - 19	3.64E - 19	
24	M.injected/Ni (kg)	7.62E - 20	7.62E - 20	
25	M.sputter/M.accretion	1.03	1.02	
26	M.sputter/M.final	16.00	9.10	
27	M.astration/M.final	0.94	0.50	
28	M.accretion/M.final	15.47	8.92	
29	M.sputter/M.injected	6.46	5.41	
30	M.astration/M.injected	0.38	0.30	
31	M.accretion/M.injected	6.25	5.30	
32				
33	Total mass of core injected {MA.injected} (kg)	5.72E - 15	5.72E - 15	
34	Total mass of core sputtered {MA.sputter} (kg)	3.78E - 15	3.44E - 15	
35	Total mass of core astrated {MA.astration} (kg)	1.20E - 15	1.05E - 15	
36	Total mass of core surviving {MA.final} (kg)	7.38E - 16	1.23E - 15	
37	MA.final/MA.injected (%)	12.92%	21.52%	
38	MA.sputter/MA.injected (%)	66.15%	60.14%	
39	MA.astration/MA.injected (%)	20.93%	18.34%	
40	MA.final/(M.final(sunocon & stardust)) (%)	1.62%	2.70%	
41	MA.injected/M.injected (%)	12.50%	12.50%	
42				
43	Total mass of core-mantle injected {MB.injected} (kg)	4.00E - 14	4.00E - 14	
44	Total mass of core-mantle sputtered {MB.sputter} (kg)	3.51E - 14	3.41E - 14	
45	Total mass of core-mantle astrated {MB.astration} (kg)	3.04E - 15	2.73E - 15	
46	Total mass of core-mantle surviving {MB.final} (kg)	1.81E - 15	3.14E - 15	
47	MB.final/MB.injected (%)	4.53%	7.85%	
48	MB.sputter/MB.injected (%)	87.87%	85.33%	
49	MB.astration/MB.injected (%)	7.60%	6.83%	
50	MB.final/(M.final(sunocon and stardust)) (%)	3.97%	6.88%	
51	MB.injected/M.injected (%)	87.47%	87.47%	
52				
53	MB.injected/MA.injected	7.00	7.00	
54	MB.final/MA.final	2.45	2.55	
55				
56	M/(dM/dt).sputter [for injected particles] (years)	6.51E + 07	6.51E + 07	
57	M/(dM/dt).sputter [time averaged] (years)	1.87E + 08	3.30E + 08	
58	M/(dM/dt).astration [time averaged] (years)	3.18E + 09	5.96E + 09	

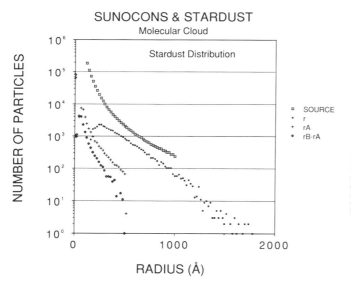

Fig. 13. Numbers of particles per bin versus size for case 3. Format is identical to Fig. 3. Only STARDUST-core particles are shown, and only in molecular cloud. SUNOCON-core particle distributions are visually indistinguishable on this logarithmic scale (but see Table 2 for differences). Comparing with otherwise identical case 1b in Fig. 3, introduction of 100% condensation of refractory STARDUST restricts growth of large particles by lowering average mass of particles. Note (Table 2) that total interstellar mass is ten times smaller in this calculation than in case 1b, but the rate of decline for $r > 1000$ Å is faster than in Fig. 3.

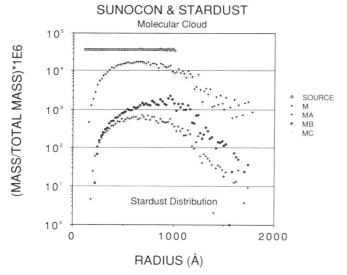

Fig. 14. Sums of STARDUST-core-particle masses per bin for case 3. SUNOCON-core distribution is visually indistinguishable. Format identical to Fig. 4. Note that M_A and M_B are much larger fractions of total mass than in Fig. 4, because STARDUST condensation reduces phase C mass.

been distributed in phase C mantles in case 1. The average masses of final particles (line 23) in case 3 are 4.37×10^{-19} kg for SUNOCON-core particles and 3.64×10^{-19} kg for STARDUST-core particles, whereas it was 8.92×10^{-19} kg in case 1b, so that the average mass of phase C mantle is

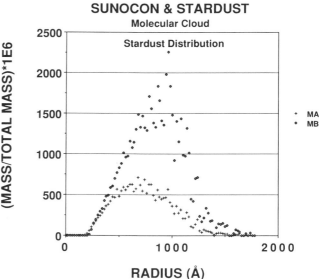

Fig. 15. Enlargement of STARDUST M_A and M_B from Fig. 14 on linear scale. SUNOCON distributions are very similar. Compare with Fig. 5 to see effect of condensation of STARDUST.

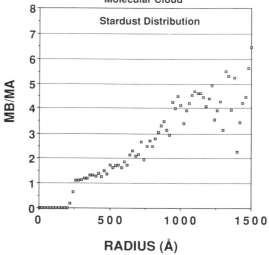

Fig. 16. Ratio of M_B/M_A for case 3, as in Fig. 6.

considerably smaller in case 3. Because the STARDUST injection maintains to some degree the total number of particles, the mass increment Δr of phase C mantle during each cloud reentry is smaller (0.0144 μm) in case 3 than it was in case 1b (0.0165 μm). The total final masses are equal in each case to the total SUNOCON mass ever injected. As a final remark on the format of Table 2, we caution that the simple prescription used for average mass lifetimes for sputtering and for astration (lines 57 and 58) is not as physically relevant as it was in the absence of STARDUST. Because the rate of injection of STARDUST increases linearly with the time (in contrast to the SUNOCON injection rate), the replacements of M by $M_F/2$ and of dM/dt

TABLE 3. Grain population characteristics: SUNOCONs and STARDUST (case 4).

Particle type	SUNOCONS	STARDUST	SUNOCONS & STARDUST
Injected particle distribution	a - 3	a - 3	a - 3
Minimum radius of injected particles (m)	1.00E-08	1.00E-08	1.00E-08
Maximum radius of injected particles (m)	1.00E - 07	1.00E - 07	1.00E - 07
Sputtering rate (μm/1e8 years)	variable	variable	variable
Number of particles injected {Ni}	600,000	600,000	1,200,000
Increase in size due to accretion {Δr} (m)	1.11E - 08	1.11E - 08	1.11E - 08
# destroyed in the diffuse cloud {NdDC}	275,641	228,838	504,479
# destroyed in the molecular cloud {NdMC}	197,030	165,867	362,89 7
# surviving in the diffuse cloud {NsDC}	65,825	107,340	173,165
# surviving in the molecular cloud {NsMC}	61,504	97,955	159,459
# surviving {Nf}	127,329	205,295	332,624
Nf/Ni(%)	21.22%	34.22%	27.72%
Total mass injected {M.injected} (kg)	4.57E - 14	4.57E - 14	4.57E - 14
Total mass sputtered {M.sputter} (kg)	3.61E - 13	2.96E - 13	
Total mass astrated {M.astration} (kg)	1.56E - 14	1.26E - 14	
Total mass accreted {M.accretion} (kg)	3.49E - 13	2.91E - 13	
Total surviving mass {M.final} (kg)	1.82E - 14	2.76E - 14	4.58E - 14
M.final/M.injected	0.40	0.60	1.00
M.final/Nf (kg)	1.43E - 19	1.35E - 19	
M.injected/Ni (kg)	7.62E - 20	7.62E - 20	
M.sputter/M.accretion	1.03	1.02	
M.sputter/M.final	19.86	10.71	
M.astration/M.final	0.86	0.46	
M.accretion/M.final	19.20	10.51	
M.sputter/M.injected	7.88	6.47	
M.astration/M.injected	0.34	0.28	
M.accretion/M.injected	7.62	6.35	
Total mass of core injected {MA.injected} (kg)	5.72E - 15	5.72E -15	
Total mass of core sputtered {MA.sputter} (kg)	4.12E - 15	3.76E -15	
Total mass of core astrated {MA.astration} (kg)	1.00E - 15	9.10E - 16	
Total mass of core surviving {MA.final} (kg)	5.88E - 16	1.05E - 15	
MA.final/MA.injected (%)	10.29%	18.38%	
MA.sputter/MA.injected (%)	72.12%	65.70%	
MA.astration/MA.injected (%)	17.58%	15.92%	
MA.final/(M.final(sunocon & stardust)) (%)	1.28%	2.29%	
MA.injected/M.injected (%)	12.50%	12.50%	
Total mass of core-mantle injected {MB.injected} (kg)	4.00E - 14	4.00E - 14	
Total mass of core-mantle sputtered {MB.sputter} (kg)	3.67E - 14	3.58E - 14	
Total mass of core-mantle astrated {MB.astration} (kg)	2.14E - 15	2.01E - 15	
Total mass of core-mantle surviving {MB.final} (kg)	1.20E - 15	2.24E -15	
MB.final/MB.injected (%)	2.99%	5.61%	
MB.sputter/MB.injected (%)	91.67%	89.38%	
MB.astration/MB.injected (%)	5.34%	5.01%	
MB.final/(M.final (sunocon & stardust)) (%)	2.61%	4.90%	
MB.injected/M.injected (%)	87.50%	87.50%	
MB.injected	7.00	7.00	
MB.final/MA.final	2.03	2.14	
M/(dM/dt).sputter [for injected particles] (years)	8.53E + 07	8.53E + 07	
M/(dM/dt).sputter [time averaged] (years)	1.51E + 08	2.80E + 08	
M/(dM/dt).astration [time averaged] (years)	3.49E + 09	6.56E + 09	

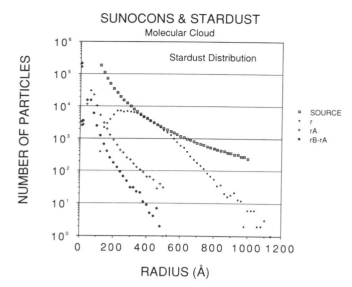

Fig. 17. Numbers of STARDUST-core particles per bin versus size for case 4. Format identical to Fig. 3. SUNOCON particles are indistinguishable. This case is severely restricted in size by combination of STARDUST condensation and a sputtering rate proportional to size. Total mass equal to that in Fig. 13 for case 3.

by (total mass sputtered)/(total time) are not really physically appropriate for STARDUST, with the result that the average τ_M for sputtering and for astration of STARDUST both exceed unphysically the corresponding average values for SUNOCONs. Furthermore, there has occurred a steady transfer of SUNOCON-particle mass to STARDUST-particle mass during the evolution (line 21), so that no simple global average has rigorous meaning. These entries in lines 57 and 58 are to be regarded as purely formed results of this averaging algorithm, therefore.

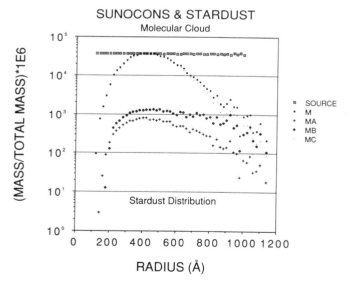

Fig. 18. Sums of STARDUST-core-particle masses per bin for case 4. Format identical to Fig. 4. Compare also to Fig. 14. Note relative flatness of M_B distribution.

In Figs. 13-16 we display analogous case 3 distributions. We show here only the STARDUST-core population in the molecular cloud, because the SUNOCON-core distributions are uninterestingly similar except for being slightly less abundant. The shapes are the same, however. We call explicitly to attention only the much smaller number of large particles in Fig. 13 than in corresponding Fig. 3 without STARDUST. The introduction of STARDUST has restricted the growth of large particles by severely limiting the number of gaseous atoms needed to grow phase C mantles.

Case 4

This is the same mixture of SUNOCON and STARDUST injection as in case 3, but the sputtering rate is taken to be proportional to grain size according to equation (5). Macroscopic characteristics of the results are listed in Table 3 in the same form as Table 2. These entries, as well as Figs. 17-20, show that the particles do not spread in radius as much as in case 3, nor do as many sputtering deaths occur as in that case, because when particles get small their sputtering rate decreases. These comparisons are analogous to the differences between cases 2 and 1b in the absence of STARDUST, but the introduction of STARDUST also narrows the distributions, making case 4 the narrowest of all studied. Despite the injection of several hundred particles per bin just below a = 1000 Å, Fig. 17 shows less than 20 final particles having r > 1000 Å and only a single particle (out of 600,000 STARDUST particles) reaching r > 1100 Å.

4. DISCUSSION

In this work we have numerically surveyed some ways in which the refractory component of interstellar dust evolves owing to sputtering and accretion cycles in the interstellar medium. Our purpose has been to raise awareness and understanding of the kinds of chemical fractionations that are affected by this type of situation. To illustrate the range of results consistent with our approach we have taken a grid of four extreme assumptions: (1) Dust erosion occurs at a constant radial rate in the sputtering medium or (2) it occurs at a rate proportional to the size of each particle, and (3) refractory matter returned from old stars condenses with 100% efficiency into STARDUST or (4) does not condense at all. We have presented extensive tables of numerical results for each calculation that will have to be studied by those readers needing a full physical grasp of the results.

We have simplified nature in many ways in order to produce this survey of effects, but two of the simplifications stand out in our minds as being physically the most serious in the sense of potentially invalidating our results: (1) We have ignored collisions between grains and (2) we have ignored more volatile elements. By omitting (1) we have room for neither high-speed interstellar collisions that fragment and/or vaporize particles nor aggregation into macroscopic bodies in stellar accretion disks that are subsequently returned (Oort clouds) to the interstellar medium. The former increases the mass rate of particle destruction but may actually increase the numbers of

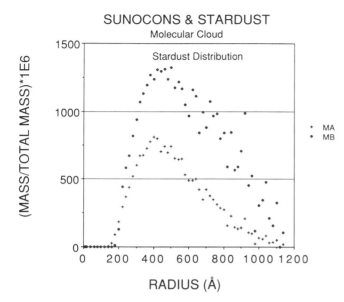

Fig. 19. Enlargement of STARDUST M_A and M_B on linear scale from Fig. 18.

particles, which in our calculation never exceeds the number injected from stars. By omitting (2) we have essentially assumed that more volatile elements (e.g., C, Na, P, S, K, and others) do not condense at all, nor do they accrete with the refractories on cloud entry. Were this to occur, the radii of particles and their subsequent sputtering rate of the *refractory* elements would require modification. We intend for these events to be the subject of continuing study. Even if our results are numerically invalidated by their eventual roles (which is by no means certain), they nonetheless offer insights of many kinds into such physical systems. The cosmochemical fractionations that we

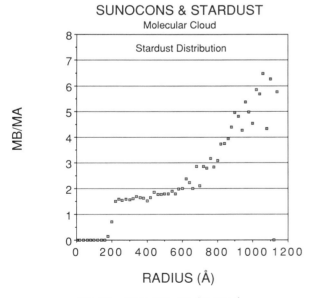

Fig. 20. Ratio of M_B/M_A for case 4.

have evaluated will have natural counterparts in the interstellar medium.

We have been able to derive many mathematical results of interest for this system as we defined it. Instead of presenting these here, we have displayed only the method of numerical computation and its results, preferring to publish mathematical results later as needed to clarify further results. What we have presented here is a numerical experiment as well as a physical model.

Astronomical Opacity

On the astronomical side it can be noted that none of these distributions produce anything like the $n(r) \propto r^{-3.5}$ (*Mathis et al.*, 1977) distribution of small 100 Å $<$ r $<$ 1000 Å particles that has been advanced to explain observed dependences of extinction and reddening upon wavelength. It seems unlikely, however, that the very refractory elements, which are the ones we are studying, are the sources of that abundant small opacity. Carbonaceous material in particular is probably at least an order of magnitude more abundant in the interstellar medium, so it need not immediately concern us that our spectra n(r) level off and even decrease below 500 Å. More to the point is the realization that some very special arguments may be needed to produce a sharp upturn in the population of small particles for that opacity examining Fig. 3 that one must here discuss the *diffuse cloud* distributions that are not shown for subsequent calculations, because, as Fig. 3b shows, the molecular cloud distributions all begin for r $\gtrsim$ 150 Å, which is the mantle we regard them as taking on during each molecular cloud entry. This translation by Δr can be taken into account on the later figures with confidence. Having said that, we nonetheless see that the small-particle spectrum is much flatter than the injection spectrum in models of this type. We therefore conclude that one (at least) of our assumptions, though reasonable for refractory elements perhaps, is not relevant to the major interstellar opacity. We would only conjecture, moreover, the following possible differences from our assumptions: (1) The injection spectrum has itself many small particles steeper than a^{-3}; (2) grain-grain collisions are fragmenting larger particles into many smaller ones; (3) the small particles are resistant to sputtering and collisions for some specific reasons; and/or (4) interstellar chemistry is able to create large numbers of small particles. Other reasons may exist. We, however, return to our topic—very refractory elements.

^{16}O-Richness of Interstellar Al

In the initial discussion of any exploratory calculation it becomes relevant to reexamine its impact on the ideas that motivated the calculation. In the present case the motivations were largely cosmochemical, driven by curiosity over the functioning of the cosmic-chemical-memory interpretation of isotopic anomalies in meteorites. Many of these applications have been addressed in a simpler computational framework as one of us has sought believable physical effects for that theory, as summarized for example by *Clayton* (1982). Let us examine how a few of those ideas work out quantitatively here.

A key thread is a putative association of ^{16}O-richness with the element aluminum. This and the associated preferential survival of phase A SUNOCON cores was evaluated analytically with simplifying assumptions by *Clayton* (1986b), who discussed the astrophysical requirements within the theory of the chemical evolution of the galaxy. These ideas can be instructively compared in two of our present calculations, case 1b having SUNOCON cores only, and case 4 having both SUNOCON and STARDUST cores and a different paradigm for sputtering as well. In the simpler case 1b all Al was initially injected in phase A cores that, for simplicity, we approximate as pure Al_2O_3, with the oxygen being pure ^{16}O (*Clayton*, 1977a,b, 1978). Line 35 tells that 15.18% of these phase A cores persist after the 6×10^9 yr evolution. This is itself noteworthy considering that an *a priori* mass lifetime against sputtering is $\tau_M = 6.51 \times 10^7$ yr (line 54). If mass were really destroyed at this rate, the fraction remaining after continuous injection for $T = 6 \times 10^9$ yr would be $T = 6 \times 10^9$ and

$$(1-e^{-T/\tau_M})/\left(\frac{T}{\tau_M}\right) = 1.1\%$$

which is 14 times smaller than our answer. This would appear to support the reduction by more than an order of magnitude in the Al sputtering rate called for by *Clayton* (1982, 1986b). The resolution of this difference is a key one for very refractory memory. To make further progress, take it that the phase B SUNOCONS, which we specify for simplicity as MgO, are also pure ^{16}O as befitting the products of carbon burning in stars, but take the remainder of oxygen to have just that complementary value needed to render the totality of oxygen isotopically normal. Only 5.42% of this phase B SUNOCON survived (line 45). Take the relative solar abundances (*Cameron*, 1982) to be $Al:Mg:^{16}O = 8.5 \times 10^4 : 1.06 \times 10^6 : 1.84 \times 10^7$, all per 10^6 Si. It follows that 15.18% of the Al plus 5.42% of the Mg remaining in the chemical forms specified carry 0.42% of all ^{16}O atoms. It then also follows that the entire remainder of oxygen, which we take to be well-mixed, has a 0.42% deficit in ^{16}O [a ^{16}O "ghost" according to definition by *Clayton* (1977b)]. This ghost characterizes not only the phase C mantles of all of the particles, but all of the considerably larger amounts of gaseous oxygen as well. This description makes it possible to calculate the ^{16}O-richness of the oxygen associated with total bulk aluminum, but one must decide how to define such a procedure. One way is to say that some process gathers together all of the Al_2O_3 units without exchanging their oxygen. This might be almost possible as a distillation residue, because although Al_2O_3 units in phase C would be exchanging oxygen during the sublimation of the less refractory silicates, it would be exchanging with oxygen identical to itself, whereas the pure ^{16}O in the phase A cores might be sufficiently refractory and shielded to not do so during evaporation of the mantles. In any case we can *calculate* the oxygen composition contained in the sum of interstellar Al_2O_3 units, whether nature can gather them without exchange or not. As we have seen, 15.18% of Al carries pure ^{16}O and 84.82% of Al carries oxygen deficient by 0.42% in ^{16}O. The weighted mean then shows bulk Al to be 17.6% enriched in ^{16}O. Inasmuch as 5% is the largest detected ^{16}O excess (*Clayton et al.*, 1977), this calculation confirms the candidacy of cosmic chemical memory among the explanations. Nor should one think of this particular method of calculation as being the only or even the preferred one within the cosmic chemical memory interpretation. Figures 4 and 5 illustrate that phase A is a considerably larger fraction of the total grain mass for $r < 750$ Å than it is for $r > 1000$ Å. Quite evidently, therefore, a macroscopic aggregate of small grains is more ^{16}O-rich than a macroscopic aggregate of large grains. This conclusion remains if we include phase B in the ^{16}O-rich material, as we logically should. We perform no integrals over size separates to make this point but are content to observe that at $r = 750$ Å phases A and B comprise 14% of the total particle mass whereas at $r = 1500$ Å they comprise only 2.5% of the total particle mass. Such aggregates differ by more than 10% in the bulk ^{16}O-richness. This calculation confirms the theoretical arguments (*Clayton*, 1980) that cycles of sputtering and redeposition followed by grain-size sorting could introduce chemical and isotopic fractionation into particles that did not possess such bulk properties initially—a nontrivial observation. Several other interpretations of the observed ^{16}O-richness could find relevance in these calculations.

One of the results of this paper is the demonstration that the evolution of the particle distributions are restricted both by the inclusion of STARDUST and by the utilization of a sputtering erosion rate proportional to grain size. We therefore jump to the extreme change of both choices, case 4, to examine the resultant quantitative change in the previous argument. Line 37 of Table 3 shows that the phase A SUNOCON mass at the end is now 10.29% of that ever injected rather than 15.18% above. This is not a large enough change to invalidate the substantial ^{16}O-richness of oxygen bonded to aluminum, so that argument proceeds as before. The major difference for the element aluminum is that 18.38% of it is now in STARDUST phase A cores, leaving 71.33% of it distributed through the phase C mantles of both particles.

Depletion from Interstellar Gas

Another astronomical observable is the relative abundances of atoms in the gas phase. Their abundance pattern is fractionated by the differential depletion from gas onto grains. One of us has long argued (e.g., *Clayton*, 1982) that these abundance patterns cannot be understood without fractionated structure within the dust itself—another variant of cosmic chemical memory. In particular, the observation that Mg is an order of magnitude more gaseous than Al has been attributed to the latter's sheltering within phase A cores, although both remain overwhelmingly within grains. The present calculation shows that this idea encounters quantitative difficulty, although it may be savable. The problem is that an insufficiently large fraction of Al is retained within these cores. Each of the cases studied have the same difficulty, so we consider case 4. Lines 37 and 47 of Table 3 show respectively that the fraction of Al within phase A cores is 28.67% and that the fraction of Mg within its initial phase B core mantle is 8.60%. If the complements are taken to be uniformly distributed within phase C mantles, the Mg/Al ratio within those mantles is enhanced by only the modest factor $(1 - .086)/(1 - .287) = 1.28$. If Mg and Al are sputtered with equal efficiency from the grain surfaces (phase C), this small

Mg/Al enhancement falls far short of the values needed for a satisfactory explanation of the observed Mg/Al fractionation. As attractive as this idea is, therefore, it must be given up in its simplest form. The most likely places to look for a repair are probably in the assumptions that Mg condenses completely upon ejection and that the remainder of the less refractory elements can be ignored.

Titanium Isotopic Anomalies

Niederer et al. (1980) and *Niemeyer and Lugmair* (1984) documented that titanium isotopic anomalies are, like those of oxygen, routinely present in carbonaceous meteorites. By far the largest variations occur at one isotope, ^{50}Ti, occuring with both algebraic signs (*Hinton et al.*, 1987) and in correlation with similar anomalies in ^{48}Ca (*Zinner et al.*, 1986). Each of these groups recognized a genuine nucleosynthetic anomaly arising, as the nucleosynthesis of these isotopes is known to do (*Clayton*, 1988), from a neutron-rich equilibrium process near supernova cores. Why those variations show up in meteoritic samples produces no such unanimity. Some favor a nonuniform admixture of those nuclei into the solar system from a neighboring supernova core, whereas others favor the chemical memory of average interstellar matter. Our present calculation can say nothing about the former possibility, but it will be applied to the latter possibility, for which *Clayton* (1981) advanced these theoretical grounds: Because these isotopes find themselves in expanding core material devoid of both oxygen and sulfur, they cannot initially appear, as other isotopes of those elements do, in oxidized or sulfidized SUNOCONs, but must instead condense within metallic droplets or not at all. Our calculations at this point cannot really address the possibility of ^{48}Ca, ^{50}Ti, and ^{54}Cr residing in interstellar metallic SUNOCONs, because we have addressed a simplified system involving only a single type of SUNOCON. What we do address is the possibility of ejection in gaseous form or in the form of very small metallic droplets that are overwhelmingly destroyed by subsequent sputtering.

Suppose the nuclei near the neutronized core cannot condense during the reexpansion. They, and only they, violate the condition of our calculation requiring 100% condensation within SUNOCONs. However, they will, like other gaseous refractory atoms, adhere to grain surfaces during transitions of their local portion of interstellar medium into the molecular cloud phase; that is, they will be quickly absorbed into phase C material. The evolution of this system will therefore be characterized by correlated isotopic excesses in phase C material and correlated absence from the SUNOCON cores. That chemical memory would then be developable in macroscopic samples by forming some of them from predominantly phase C material and others from SUNOCON-rich material. The easiest mechanical (as opposed to chemical) way to achieve this is through aggregates of differing grain size.

Consider this more quantitatively with the aid of the case 1b example. It would take a more elaborate SUNOCON description to assign the other isotopes of Ca and Ti to either phase A or phase B, so for simplicity take them to be initially uniformly distributed within the initial SUNOCONs. At the conclusion of the 6×10^9 yr galactic evolution the remaining SUNOCON mass is 6.63% (line 38 plus line 48) of the total dust mass. Thus 6.63% of other Ca and Ti isotopes remain in SUNOCON cores free of ^{48}Ca and ^{50}Ti. Clearly then, those phase C mantles contain an isotopic excess (anomaly) of 6.63%/.934 = 7.1% in ^{48}Ca and ^{50}Ti. The total system is isotopically normal in bulk, but correlated anomalies of ^{48}Ca and ^{50}Ti will characterize each grain. The large ones are phase-C-rich and thus ^{48}Ca-^{50}Ti-rich, whereas the small final grains are relatively poor in phase C. As described previously, the SUNOCON cores, which are 6.63% on average, are 14% of 750 Å grains and only 2.5% of 1500 Å grains, so a filtering by grain size can be quite efficient. This efficiency is certainly required, because the observed anomalies range up to 10% in macroscopic samples. Differential siting of the other isotopes between phases A and B will produce muted endemic variations in the other isotopes.

It is evident that a very similar argument follows if one assumes instead that the neutron-rich equilibrium zones condense, but do so only in small (say 100 Å) metallic droplets. As Fig. 3 shows, the final number of particles near $r = 100$ Å is less than 1% of the number injected near $a = 100$ Å, requiring that they have been overwhelmingly destroyed by sputtering. That destruction puts ^{48}Ca and ^{50}Ti temporarily in the gas phase so that they join phase C as before.

We have concluded this paper on interstellar dust with a discussion of isotopic anomalies in meteorites because they illustrate so clearly the ways in which cosmic chemical memory is hidden within the dust grains. It is our hope that these demonstrations will intensify both interest and scientific progress in this area of great importance to both astrophysics and meteoritics.

APPENDIX

To obtain some quantitative understanding of the system presented in Fig. 1 we consider the equations

$$\frac{dN_1}{dt} = c - \alpha N_1 - B_1 N_1 + B_2 N_2 \qquad (A1)$$

$$\frac{dN_2}{dt} = B_1 N_1 - \gamma N_2 - B_2 N_2 \qquad (A2)$$

where

N_1 = the number of particles in the diffuse cloud phase (DCP).

N_2 = the number of particles in the molecular cloud phase (MCP).

α = rate of destruction of particles in the DCP (assumed constant in equation (1), but actually a function of time).

γ = rate of destruction of particles in the MCP.

B_1 = rate of particle transfer from the DCP to the MCP.

B_2 = rate of particle transfer from the MCP to the DCP.

c = creation rate of particles in the DCP.

Equations (A1) and (A2) have the solutions:

$$N_1(t) = c \frac{(e^{\chi_+ t} - e^{\chi_- t})}{(\chi_- - \chi_+)} + c(\gamma + B_2) \left(\frac{(\chi_+ - \chi_-) + \chi_- e^{\chi_+ t} - \chi_+ e^{\chi_- t}}{\chi_+ \chi_- (\chi_+ - \chi_-)} \right)$$

and

$$N_2(t) = B_1 c \left(\frac{(\chi_+ - \chi_-) + \chi_- e^{\chi_+ t} - \chi_+ e^{\chi_- t}}{\chi_+ \chi_- (\chi_+ - \chi_-)} \right)$$

where

$$\chi_\pm = -\frac{(\gamma + \alpha + B_1 + B_2) \pm \sqrt{(\alpha + B_1 + B_2 + \gamma)^2 - 4[(\alpha + B_1)(\gamma + B_2) - B_1 B_2]}}{2}$$

and $\chi_\pm \leq 0$.

If we consider the case $\chi_\pm < 0$ then as $t \to \infty$

$$N_1(t) \to \frac{c(\gamma + B_2)}{\chi_+ \chi_-} \quad \text{and} \quad N_2(t) \to \frac{B_1 c}{\chi_+ \chi_-}$$

$$\Rightarrow \frac{N_1(t)}{N_2(t)} \to \frac{B_2}{B_1}\left(1 + \frac{\gamma}{B_2}\right)$$

In our program we have assumed average transfer times of 10^8 years

$$\Rightarrow B_1 = B_2 = \frac{1}{10^8} \text{ year}^{-1}$$

We have also assumed that the grains are destroyed in the MCP on an average timescale of 1.5×10^9 years

$$\Rightarrow \gamma = \frac{1}{1.5 \times 10^9} \text{ year}^{-1}$$

$$\Rightarrow \frac{N_1(t)}{N_2(t)} \to \frac{16}{15}$$

which is the result we obtain when we look at the number of particles surviving in the DCP and the MCP (see rows 10 and 11 in Table 1, rows 12 and 13 in Tables 2 and 3).

The fact that the two phases have equal masses is a consequence of taking $B_1 = B_2$. This can be seen from the equations

$$\left.\frac{dM_1}{dt}\right|_{\text{exchange}} = B_1 M_1 \quad (A3)$$

$$\left.\frac{dM_2}{dt}\right|_{\text{exchange}} = B_2 M_2 \quad (A4)$$

where
M_1 = total mass of gas and dust in the DCP.
M_2 = total mass of gas and dust in the MCP.
t = time.

Equations (A3) and (A4) express the rate of exchange of mass between the phases.

When the system has reached steady state then

$$\left.\frac{dM_1}{dt}\right|_{\text{exchange}} = \left.\frac{dM_2}{dt}\right|_{\text{exchange}}$$

$$\Rightarrow \frac{M_1}{M_2} = \frac{B_2}{B_1}$$

thus if $B_1 = B_2$ then $M_1 \approx M_2$.

Acknowledgments. This research was supported in part by the Robert A. Welch Foundation and in part by NASA grant NAG-9-100 BASIC.

REFERENCES

Cameron A. G. W. (1982) Elemental and nuclidic abundances in the solar system. In *Essays in Nuclear Astrophysics* (C. A. Barnes, D. D. Clayton, and D. N. Schramm, eds.), pp. 23-43. Cambridge University Press, Cambridge.

Clayton D. D. (1975a) Extinct radioactivities: trapped residuals of presolar grains. *Astrophys. J., 199,* 765-769.

Clayton D. D. (1975b) ^{22}Na, Ne-E, extinct radioactive anomalies and unsupported ^{40}Ar. *Nature, 257,* 36-37.

Clayton D. D. (1977a) Solar system isotopic anomalies: Supernova neighbor or presolar carriers. *Icarus, 32,* 255-269.

Clayton D. D. (1977b) Cosmoradiogenic ghosts and the origin of CaAl-rich inclusions. *Earth Planet. Sci. Lett., 35,* 398-410.

Clayton D. D. (1978) Precondensed matter: Key to the early solar system. *Moon and Planets, 19,* 109-137.

Clayton D. D. (1979) Supernovae and the origin of the solar system. *Space Sci. Rev., 24,* 147-226.

Clayton D. D. (1980) Chemical and isotopic fractionation by grain size separates. *Earth Planet. Sci. Lett., 47,* 199-210.

Clayton D. D. (1981) Some key issues in isotopic anomalies: Astrophysical history and aggregation. *Proc. Lunar Planet. Sci. 12B,* pp. 1781-1802.

Clayton D. D. (1982) Cosmic chemical memory: A new astronomy. *Quar. J. Royal Astron. Soc., 23,* 174-212.

Clayton D. D. (1984) Galactic chemical evolution and nucleocosmochronology: A standard model with terminated infall. *Astrophys. J., 285,* 411-425.

Clayton D. D. (1985) Galactic chemical evolution and nucleocosmochronology: A standard model. In *Nucleosynthesis* (W. D. Arnett and J. W. Truran, eds.), pp. 65-88. Univ. of Chicago, Chicago.

Clayton D. D. (1986a) Interstellar fossil ^{26}Mg and its possible relationship to excess meteoritic ^{26}Mg. *Astrophys. J., 310,* 490-498.

Clayton D. D. (1986b) Isotopic anomalies and SUNOCON survival during galactic evolution. In *Cosmogonical Processes* (W. D. Arnett,

C. Hansen, J. Truran, and S. Tsuruta, eds.), pp. 101-122. VNU Science, Utrecht.

Clayton D. D. (1988) Stellar nucleosynthesis and the chemical evolution of the solar neighborhood. In *Meteorites and the Origin of the Solar System* (J. Kerridge and M. Mathews, ed.), in press. University of Arizona, Tucson.

Clayton R. N., Onuma N., Grossman L., and Mayeda T. K. (1977) Distribution of the presolar component in Allende and other carbonaceous chondrites. *Earth Planet. Sci. Lett., 34*, 209-224.

Draine B. T. and Salpeter E. E. (1979) Destruction mechanisms for interstellar dust. *Astrophys. J., 231*, 438-455.

Duley W. W. and Williams D. A. (1986) PAH molecules and carbon dust in interstellar clouds. *Mon. Not. Roy. Astr. Soc., 219*, 859-864.

Grossman L. and Larimer J. W. (1974) Early chemical history of the solar system. *Rev. Geophys. Space Phys., 12*, 71-101.

Hinton R. W., Davis A. M., and Scatena-Wachel D. E. (1987) Large negative ^{50}Ti anomalies in refractory inclusions from the Murchison carbonaceous chondrite: Evidence for incomplete mixing of neutron-rich supernova ejecta into the solar system. *Astrophys. J., 313*, 420-428.

Jura M. (1976) Calcium abundance variations in diffuse interstellar clouds. *Astrophys. J., 206*, 691-698.

Liffman K. and Clayton D. D. (1987) Stochastic models of refractory interstellar dust (abstract). In *Lunar and Planetary Science XVIII*, pp. 552-553. Lunar and Planetary Institute, Houston.

Mathis J. S., Rumple W., and Nordsieck K. H. (1977) The size distribution of interstellar grains. *Astrophys. J., 217*, 425-433.

Niederer F. R., Papanastassiou D. A., and Wasserburg G. J. (1980) Endemic isotopic anomalies in titanium. *Astrophys. J. (Letters), 240*, L123-L128.

Niemeyer S. and Lugmair G. W. (1984) Titanium isotopic anomalies in meteorites. *Geochim. Cosmochim. Acta, 48*, 1401-1416.

Seab C. G. and Shull J. M. (1983) Shock processing of interstellar grains. *Astrophys. J., 275*, 652-660.

Shull J. M. (1977) Grain disruption in interstellar hydromagnetic shocks. *Astrophys. J., 215*, 805-811.

Shull J. M. (1978) Disruption and sputtering of grains in intermediate-velocity interstellar clouds. *Astrophys. J., 226*, 858-862.

Woosley S. E., Arnett W. D., and Clayton D. D. (1973) The explosive burning of oxygen and silicon. *Astrophys. J. Suppl., 26*, 231-312.

Zinner E. K., Fahey A. J., Goswami J. N., Ireland T. R., and McKeegan K. D. (1986) Large ^{48}Ca anomalies are associated with ^{50}Ti anomalies in Murchison and Murray hibonites. *Astrophys. J. (Letters), 311*, L103-L107.

Venus Lives!

Charles A. Wood

Code SN4, NASA Johnson Space Center, Houston TX 77058

Peter W. Francis

Lunar and Planetary Institute, 3303 NASA Road One, Houston, TX 77058

Observations of electrical noise and sulfur dioxide variations by the Pioneer Venus spacecraft have been interpreted as evidence for ongoing volcanic activity (*Scarf and Russell*, 1983; *Esposito*, 1984). Recently this interpretation has been challenged, with the general conclusion that there is no evidence for active volcanism on the planet (*Taylor and Cloutier*, 1986). This debate does not impinge on the abundant comparative planetological reasons for believing that Venus should be active (i.e., it is compositionally similar to and has nearly the same mass as the Earth), and direct observations of volcanic landforms on Venera 15 and 16 and Arecibo radar images. The 0.5-1.0 b.y. age of these volcanic materials, as originally determined by *Ivanov et al.* (1986), is likely to be an overestimate for a variety of reasons. Comparison with the duration of major volcanic systems on Earth (75-80 m.y. for the Hawaiian-Emperor chain) and Mars (perhaps 1 b.y. for Olympus Mons) suggests that volcanism is likely to continue on Venus today if it existed 0.5-1.0 b.y. ago. Unfortunately, there is little hope of remotely detecting active volcanism because of the masking effects of the pervasive and hot atmosphere of Venus. The Venera data confirm or strengthen various predictions about volcanism on Venus based upon a comparative planetological approach.

We know of no geological evidence for currently active volcanism on Venus.
... the available evidence appears to indicate that it is inactive.
—Taylor and Cloutier, 1986, p. 1093.

INTRODUCTION

Of all the extraterrestrial planets Venus is most like the Earth in terms of bulk properties, but it has vastly different surface conditions. The similarities led to early speculation that the planet might even be an abode of life. By contrast, the discovery of a CO_2-rich atmosphere at high temperature and pressure led to more recent speculation of an inert, Dantësque hell.

Recently, there have been some equally discordant speculation on the *geologic* state of the planet. On the basis of evidence that is now controversial, Venus has been claimed to be volcanically active with ongoing explosive eruptions monitored by the Pioneer-Venus spacecraft (*Scarf and Russell*, 1983; *Esposito*, 1984; *Ksanfomaliti*, 1985). Critics dismiss the evidence and see no evidence that Venus is active (*Taylor and Cloutier*, 1986). We suggest that these conflicting views are based on only a limited set of data about Venus and ignore the principal conclusions from two decades of comparative planetological studies of Venus and dozens of other planets and moons of the solar system. These studies strongly imply that Venus is almost certainly active volcanically and tectonically.

VOLCANIC LIGHTNING ON VENUS?

Electrical impulses recorded by the plasma wave instrument on Pioneer-Venus were interpreted as lightning "whistlers" by *Taylor et al.* (1979), *Scarf and Russell* (1984), *Singh and Russell* (1986), and others. The details of the interpretation are complex and beyond the scope of the present study; however, one aspect of the lightning hypothesis has been a major factor in the question of volcanic activity on Venus. *Scarf and Russell* (1983) attempted to trace the lightning down magnetic field lines back to source regions, and argued that the putative lightning is clustered near Beta and Phoebe Regios and perhaps near Atla Regio. Noting that lightning is often generated in volcanic eruption columns on Earth, Scarf and Russell speculated that the lightning postulated to exist on Venus was of the same origin, and quoted suggestions by *Masursky et al.* (1980) that these highland regios are probably young volcanoes.

Taylor and Cloutier (1986) carried out a detailed reanalysis of the electrical data, and convincingly concluded that the electric noise is related to solar wind interactions with the ionosphere of Venus, and is not correlated with topographic highs that could be volcanoes. These authors further extrapolated their negation of the proposed electric noise/lightning/volcano relation to claim that there was no evidence of any kind to indicate that Venus is currently active. *Singh and Russell* (1986) disagreed with this interpretation, and argued that lightning is still the best explanation of the data, but they also admitted that newly considered stochastic factors impair the correlation between lightning sources and Venus surface topography. Thus there is little evidence, regardless of the origin of the electrical noise, that lightning on Venus is associated with active volcanism. *Taylor and Cloutier* (1986) also concluded, based on physical evidence such as the lack of aerosols necessary for significant electrical charge buildup, that lightning of any origin is unlikely on Venus.

In any case, the original interpretation of active volcanism on Venus based on the detection of electrical noise was unrealistic for a number of reasons.

1. Virtually all lightning on Earth originates in thunderstorms (i.e., it is nonvolcanic in origin), with an average of 100 lightning strikes per second over the globe (*Navarra*, 1979). By contrast, the frequency of occurrence of volcanic-derived lightning apparently has not been studied. Most volcanology texts mention lightning only briefly. Although most eruption descriptions do not mention lightning at all [e.g., there is no entry in the index to *The Volcano Letter* (*Fiske et al.*, 1987), which reports 30 years of volcanic activity in Hawaii], it is apparently a relatively common feature of explosive eruptions. *Blong* (1984) reports that volcanically derived lightning is most likely to be associated with strongly explosive eruptions that loft ash particles through the atmosphere. Such explosive eruptions are likely to be rare on Venus because of the massive atmosphere (*Wood*, 1979; *Head and Wilson*, 1986), and the existing radar data also reveal little evidence for pyroclastic deposits from such eruptions (*Head and Wilson*, 1986).

2. Five hundred sixty-seven individual electrical noise events (interpreted as lightning strikes) were recorded by the plasma wave instrument during Pioneer-Venus' first 3.5 years in orbit around Venus (*Scarf and Russell*, 1983). If all of these widespread events were caused by volcanism, Venus would have to have an extraordinary number of large-scale explosive eruptions simultaneously occurring at many volcanoes. This seems unlikely based on the lack of evidence for pyroclastic deposits as mentioned above, and on the much lower occurrence rate of large scale volcanism on Earth [only six large eruptions have occurred during the first 80 years of this century (*Newhall and Self*, 1982)].

ATMOSPHERIC SO$_2$ VARIATIONS

During its first five years of operation the Pioneer-Venus ultraviolet spectrometer detected an order of magnitude decrease in sulfur dioxide abundance at the top of Venus' clouds and in the amount of submicron haze above the clouds (*Esposito*, 1984). Review of telescopic data revealed that a similar variation probably occurred during the late 1950s. Esposito suggested that these variations were due to episodic injections of sulfur dioxide into the atmosphere, possibly caused by episodic volcanic eruptions. *Esposito* (1984) calculated that the rise of a volcanic plume high enough to emplace SO$_2$ in the upper atmosphere of Venus requires an eruptive thermal energy release equal to that of the largest eruptions in historic times on Earth. Thus, Esposito suggested that a Krakatoa scale eruption might have occurred just before the arrival of the Pioneer-Venus spacecraft at Venus in 1978 and again in about 1959. On Earth such eruptions occur only about once per century.

This intriguing suggestion that volcanism may be vigorously active on Venus has also been challenged by *Taylor and Cloutier* (1986) who, in debunking the lightning/volcanism claim, vaguely proposed that the SO$_2$ variations might "result from as yet unexplained changes in the general circulation of the Venus atmosphere." Whereas there almost certainly are many unknown processes affecting Venus' atmosphere, Esposito's explanation generally agrees with terrestrial experience (e.g., the rapid formation and multiyear dissipation of the stratospheric SO$_2$ cloud), and should be considered a viable alternative until it is shown to be unlikely. The main argument against the proposal is that explosive volcanism is unlikely on Venus unless the exsolved magma volatile content exceeds 4 wt % (*Head and Wilson*, 1986). On Earth, large gas-rich explosive eruptions derive their volatiles from meteoric water, which enhances volatile concentrations in silicic magma chambers (*Sommer*, 1977); presumably this mechanism is absent on Venus.

VOLCANISM ON VENUS: THE COMPARATIVE PLANETOLOGICAL APPROACH

The two phenomena discussed above, lightning and SO$_2$ variations, are not the only or the best evidence for a volcanically active Venus. The most important arguments are inferential, deriving from cosmochemical and comparative planetological considerations, and observational (see next section). Because Venus is so similar to the Earth in diameter, density, and location in the solar system it has long been considered Earth's twin, and likely to be made of similar materials and to have had a similar thermal history (e.g., *Basaltic Volcanism Study Project*, 1981, Chapter 9). Although the atmospheric conditions of Venus differ greatly from those of the Earth, modeling indicates that the onset and continuation of greenhouse heating may be due to accidental rather than fundamental differences between the planets (*Donahue and Pollack*, 1983).

A number of petrological and morphological comparisons of volcanism in the solar system also imply that Venus should have experienced significant volcanic activity that may continue today. *Walker et al.* (1979) recognized that for planetary basaltic volcanism for which we have samples, the maximum

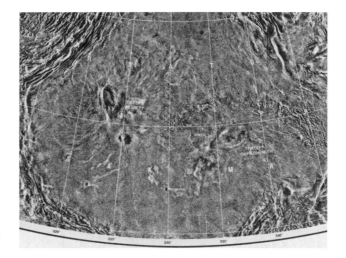

Fig. 1. The volcanic calderas Colette (80 × 120 km) and Sacajawea (140 × 280 km), with bright (= radar rough) radial flows up to 300 km long. Photograph from Venera 15/16 mosaic; dimensions from *Barsukov et al.* (1986).

eruption temperature, degree of fractionation, chemical diversity, and duration all increase from the eucrite meteorite parent body to the Moon to Earth, i.e., with increasing planet size. *Head et al.* (1977) defined a planetary evolution index, the ratio of volcanic terrain to impact cratered terrain, which they found to increase from the Moon to Mercury to Mars to Earth. Similarly, *Wood* (1984) found that the number of calderas on a planet increases from the Moon to Mars to Earth. All three of these statistical relations, which are based on observed details of petrology, volcanic morphology, and gross planetary surface characteristics, predict that Venus should have had, and probably continues to have, a complex and dynamic volcanic history. This result can readily be extended to tectonism, for the same sort of progression in diversity, magnitude, and duration of tectonic activity occurs from the Moon to Mercury to Mars and the Earth, based on photogeologic evidence. Thus, Venus should be geologically alive today.

VOLCANISM ON VENUS: OBSERVATIONS

There is significant evidence from spacecraft and Earth-based observations for volcanic landforms on Venus. The coarse resolution radar data from Pioneer-Venus and Goldstone Observatory revealed large mountains that were somewhat speculatively interpreted to be volcanic (*Masursky et al.*, 1980; *Malin and Saunders*, 1977). The increased resolution of the Venera 15/16 and Arecibo radar images leaves little doubt, however, that volcanic flows and mountains are common landforms on Venus. *Head and Wilson* (1986) reviewed the topographic and morphologic evidence that Rhea and Theia Montes, a massif south of Ishtar Terra, and Collette (Fig. 1) are volcanoes, each of which has radial lobes of radar-bright material, which they interpreted as lava flows. *Barsukov et al.* (1986) pointed out other likely calderas (including Sacajawea), as well as many other possible volcanic landforms. They also noted that many large areas of plains show similarities to lunar maria and other presumed basaltic plains on Mars. Preliminary mapping (*Barsukov et al.*, 1986) suggests that much of the northern quarter of Venus is formed of volcanic units of various types and ages.

Chemical support for these morphological conclusions is provided by analyses of samples by Venera and Vega landers. Analyses of rocks at five different landing sites on Venus are consistent with their interpretation as various kinds of basalts, similar to those found on Earth (*Surkov et al.*, 1986). Interestingly, *all* of the samples analysed by Venera and Vega landers on different terrain types are quite mafic: alkaline basalts, tholeiitic basalts, and olivine gabbronorite (*Surkov et al.*, 1986). The lack of any silicic rocks in this admittedly limited sample, and the lack of strong morphological evidence for silicic volcanic landforms (*Head and Wilson*, 1986), indicate a more restricted range of volcanic compositions and processes than on Earth.

WHAT IS THE AGE OF VENUS VOLCANISM?

Radar data demonstrate that Venus has a diversity of volcanic landforms over a substantial portion of its mapped surface, and *in situ* measurements document a variety of probable igneous rock types; the only uncertainty is the age range of this volcanic activity. Based on crater counts over the entire northern quarter of Venus, *Ivanov et al.* (1986) deduced an average crater retention age of 500-1000 m.y. A number of additional factors need to be considered, however.

1. W. K. Hartmann (personal communication, 1987), whose lunar crater count/absolute age relation was used by *Ivanov et al.* (1986), estimates uncertainties in crater counts and the applicability of the lunar crater relation to Venus might yield errors of factors of 3-4. Thus, the average age of the northern cap of Venus could be as young as about 100 m.y. or as old as 4 b.y.

2. *Schaber et al.* (1987) proposed that the surface of Venus is even younger than suggested by *Ivanov et al.* (1986), because (1) the Hartmann lunar cratering rate over the past 3.3 b.y. (corrected for Venus) is 2 to 3 times higher than the present rate for asteroid and comet impacts on Earth, and (2) this rate is essentially the same as that for Venus, based on recent calculations of the orbits of astronomically observed asteroids and comets (*Shoemaker and Wolfe*, 1987). Schaber et al. admitted that if the Venusian atmosphere inhibits the formation of all but the largest craters on Venus, then the surface age could be older than their estimate of 100-300 m.y., but they suggest that no physical evidence indicates an age as great

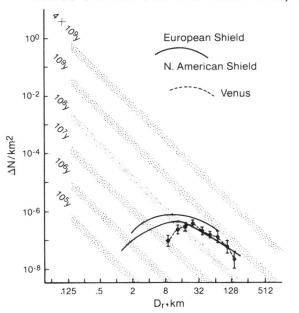

Fig. 2. Impact crater densities versus crater diameter for Venus and Earth, modified from Fig. 9 of *Ivanov et al.* (1986). The highlighted diagonal line represents a crater retention age of 100 m.y. The data for the European shield (top curved line) and the North American shield (bottom curved line) bracket the data for Venus. Earth data from Chap. 8 of the *BVSP* (1981).

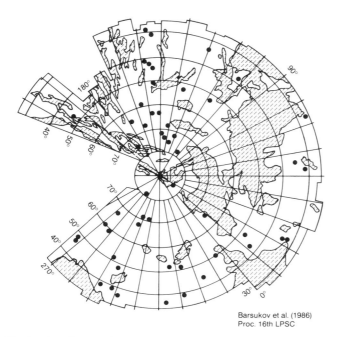

Fig. 3. Distribution of impact craters (dots) and terrains lacking them (shaded) based on mapping of the northern portion of Venus by *Barsukov et al.* (1986). Based on the lack of observed impact craters the indicated terrains must be considerably younger than the average age of this part of the planet. These young areas include ridged belted plains, ridged plains, and several other units that generally appear tectonized.

as 1 b.y. In any case, the intrinsic uncertainty in crater counting discussed above is still pertinent to the Schaber et al. estimate: The age could be as young as 45 m.y.!

3. No matter which crater retention age is accepted for the northern quarter of Venus, it is an *average* age; some units must be older and others younger, just as on Earth. According to *Cogley* (1984), the average age of the Earth's continental crust is 450 m.y. [which is also the crater retention age for the North American and European shields according to the *Basaltic Volcanism Study Project*, 1981, Chapter 8; note that these shield crater counts bracket the Venus data (Fig. 2)]. Clearly the Earth, with its average age at the lower range of the estimates for Venus by Ivanov et al., is a volcanically and tectonically active world, and thus Venus probably is also. Although only a small-scale geologic map of the northern portion of Venus is available (*Barsukov et al.,* 1986), it does show that some of the terrain units apparently have no superposed impact craters and thus must be significantly younger than the average age (Fig. 3).

CAN WE DETECT ACTIVE VOLCANISM ON VENUS?

Cosmochemical considerations and radar data suggest that volcanism probably continues as an active process on Venus today. How can we test this hypothesis? How can active volcanism, or recently deposited volcanic materials, be detected? Before addressing this question it is necessary to clarify what "active volcanism" encompasses. Although the term "active" is commonly used in studies of terrestrial volcanoes to denote those that have some historic eruption record, the term is best applied to those volcanoes that retain the potential for future activity. Clearly, this is difficult to determine, and as a matter of convenience, volcanologists generally accept as active those volcanoes that have erupted within the last 10,000 years, as for example in the catalog *Volcanoes of the World* (*Simkin et al.,* 1981). Vesuvius, which has not erupted since 1944, is listed as active, as are most of the Cascade volcanoes. There is a clear relationship between the explosiveness of an eruption and the repose interval between eruptions (*Simkin et al.,* 1981, Fig. 6). Thus, large terrestrial eruptions of the same magnitude as those hypothesized to emplace SO_2 in the upper atmosphere of Venus typically occur after repose periods of 100-10,000 years. In other words, since eruptions are generally of short duration, the most explosive volcanoes typically spend only 1-0.01% of their time actually in an eruptive state. On longer time scales, large volcanic systems such as that which produced Valles Caldera in New Mexico may have only a few major eruptions in the course of a 10-m.y. lifetime (*Luedke and Smith,* 1978). The major mantle anomaly that produced the Hawaiian-Emperor chain of volcanoes has been producing magma for at least 75-80 m.y. (*Clague and Dalrymple,* 1987). Thus, although individual volcanoes have life spans of 10^5-10^6 years, mantle source regions such as hot spots may be active 10^7 years. Evidence from crater counts on Olympus Mons and other giant shield volcanoes on Mars (*Blasius,* 1976; *Tanaka,* 1987) suggest that they have had lifetimes up to a billion years. Thus, "active" volcanism can span periods from tens of years to hundreds of millions of years; the failure to detect an eruption at any given instant certainly does not imply that a volcano is dead.

Detection of an eruption in progress on Venus would be very difficult. On Earth the most common method of detecting an explosive eruption (other than from reports from local citizens) is through imaging of an eruption plume by satellite or Shuttle photography (e.g., *Robock and Matson,* 1983; *Wood,* 1986). On Venus, an eruption cloud would not be visible within the already completely cloudy atmosphere. Similarly, the thermal anomalies that are sometimes remotely detected at terrestrial volcanoes (e.g., *Francis and Rothery,* 1987) would be undetectable through Venus' massive and hot atmosphere. Interestingly, some terrestrial eruptions have been detected by their SO_2 signatures in the upper atmosphere (e.g., the Mystery Cloud of 1981), exactly like the SO_2 variations *Esposito* (1984) reported on Venus. On Earth, the clear temporal and spatial association of atmospheric SO_2 concentrations with observed eruptions leaves no doubt of the physical link; for Venus, some doubt can still exist.

The discussion above concerns the detection of plumes from large explosive eruptions, which are not expected to be common on Venus, however (*Wood,* 1978; *Head and Wilson,* 1986). Effusive eruptions, which largely emplace lava flows, would be even more difficult to detect on Venus as they occur, but could be detected through comparison of high resolution radar images obtained at different times. For example, large lava flows, such as those around Colette, are obviously detectable, but recognition that a flow is new requires

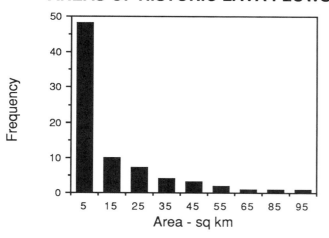

Fig. 4. Distribution of areas of the largest terrestrial lava flows in historic times based upon data from a variety of sources. All historic flows from Mauna Loa and Kilauea (Hawaii), Reunion and Paricutin (Mexico) are included. Examination of much volcanic literature and maps suggest that few flows are larger than these.

possession of an earlier image in which the flow does not exist. Venera 15/16 radar provides "before" images and the proposed Magellan spacecraft will, we hope, provide "after" images in the 1990s. If flows on Venus are comparable in size to those on Earth, flows that formed since 1986 are unlikely to be detected with the Venera/Magellan data pairs because the typical area of *larger* historic flows on Earth is only 15 km^2 (Fig. 4); probably only flows larger than 100 km^2 could be easily detected on Venus. Similarly, *Head and Wilson* (1986) calculate that radar detectable pyroclastic deposits would only extend 10-15 km downwind from explosive volcanic vents. Thus, unless very large lava flows or pyroclastic deposits are formed very frequently on Venus, there is little hope of detecting the recent volcanic activity that ought to be taking place on Venus with existing or planned spacecraft techniques.

CONCLUSIONS

There is no compelling direct evidence that Venus has been volcanically active during the last decade. There is however, excellent evidence for youthful volcanic mountains and lava flows. The surface of the northern quarter of Venus is probably younger than 1 b.y. old; some units are likely to be much younger. Considering the lifetimes of major volcanic systems on Earth and Mars, it is likely that Venus maintains its volcanic vigor today. It is unlikely that any direct evidence of eruptions will be detected with existing and planned spacecraft because of the small sizes of likely volcanic manifestations and the long intervals expected between eruptions. Perhaps the best evidence for active volcanism on Venus will come if future studies of the dynamics and chemical mixing of the planet's atmosphere cannot account for Esposito's variations of SO$_2$ by any process other than volcanism. Finally, it is interesting to note that three sets of previously published predictions, based on comparative planetological considerations, concerning volcanism on Venus—that it should be compositionally diverse and of long duration (*Walker et al.*, 1979), that volcanic materials should cover substantial portions of the planet's surface (*Head et al.*, 1977), and that there should be many calderas (*Wood*, 1984)—are all apparently correct.

REFERENCES

Barsukov V. L., Surkov Yu. A., Moskalyova L. P., Shcheglov O. P., Manvelyan O. S., and Kharyukova V. P. (1982) Geochemical studies of the Venusian surface from Veneras 13 and 14. *Geokhimiya*, 7, 899-919.

Barsukov V. L. et al. (1986) The geology and geomorphology of the Venus surface as revealed by the radar images obtained by Veneras 15 and 16. *Proc. Lunar Planet. Sci. Conf. 16th*, in *J. Geophys. Res.*, 91, D378-D398.

Basaltic Volcanism Study Project (1981) *Basaltic Volcanism on the Terrestrial Planets*. Pergamon, New York. 1286 pp.

Blasius K. R. (1976) The record of impaact cratering on the great volcanic shields of the Tharsis region of Mars. *Icarus*, 29, 343-361.

Blong R. J. (1984) *Volcanic Hazards—A Sourcebook on the Effects of Eruptions*. Academic, Sydney, Australia. 424 pp.

Clague D. A. and Dalrymple G. B. (1987) The Hawaiian-Emperor volcanic chain: Part 1—Geologic evolution. In *Volcanism in Hawaii* (R. W. Decker, T. L. Wright, and P. H. Stauffer, eds.), pp. 5-54. U.S. Geol. Survey Prof. Paper 1350.

Cogley J. G. (1984) Continental margins and the extent and number of the continents. *Rev. Geophys. Space Phys.*, 22, 101-122.

Donahue T. M. and Pollack J. B. (1983) Origin and evolution of the atmosphere of Venus. In *Venus* (D. M. Hunten, L. Colin, T. M. Donahue, and V. I. Moroz, eds.), pp. 1003-1036, University of Arizona, Tucson.

Esposito L. W. (1984) Sulfur dioxide: Episodic injection shows evidence for active Venus volcanism. *Science*, 223, 1072-1074.

Fiske R. S., Simkin T., and Nielsen E. A., eds. (1987) *The Volcano Letter*. Smithsonian Inst. Press, Washington, D.C.

Francis P. W. and Rothery D. A. (1987) Using the Landsat Thematic Mapper to detect and monitor active volcanoes: An example from the Lascar volcano, Northern Chile. *Geology*, 15, 614-617.

Head J. W. and Wilson L. (1986) Volcanic processes and landforms on Venus: Theory, predictions, and observations. *J. Geophys. Res.*, 91, 9407-9446.

Head J. W., Wood C. A., and Mutch T. A. (1977) Geologic evolution of the terrestrial planets. *Amer. Sci.*, 65, 21-29.

Ivanov B. A., Basilevsky A. T., Kryuchkov V. P., and Chernaya I. M. (1986) Impact craters of Venus: Analysis of Venera 15 and 16 data. *Proc. Lunar Planet. Sci. Conf. 16th*, in *J. Geophys. Res.*, 91, D413-D430.

Ksanfomaliti L. V. (1985) Signs of possible volcanism on Venus. *Solar System Res.*, 18, 200-207.

Luedke R. G. and Smith R. L. (1978) Map showing distribution, composition, and age of Late Cenozoic volcanic centers in Arizona and New Mexico. *U. S. Geol. Surv. Map I-1091-A*.

Malin M. C. and Saunders R. S. (1977) Surface of Venus: Evidence of diverse landforms from radar observations. *Science*, 196, 987-990.

Masursky H., Eliason E., Ford P. G., McGill G. E., Pettengill G. H., Schaber G. G., and Schubert G. (1980) Pioneer Venus radar results: Geology from images and altimetry. *J. Geophys. Res.*, 85, 8232-8260.

Navarra J. G. (1979) *Atmosphere, Weather and Climate*. Saunders, Philadelphia. 519 pp.

Newhall C. G. and Self S. (1982) The volcanic explosivity index (VEI): An estimate of explosive magnitude for historical volcanism. *J. Geophys. Res., 87*, 1231-1238.

Robock A. and Matson M. (1983) Circumglobal transport of the El Chichon volcanic dust cloud. *Science, 221*, 195.

Scarf F. L. and Russell C. T. (1983) Lightning measurements from the Pioneer Venus Orbiter. *Geophys. Res. Lett., 10*, 1192-1195.

Schaber G. G., Shoemaker E. M., and Kozak R. C. (1987) Is the Venusian surface really old? (abstract). In *Lunar and Planetary Science XVIII*, pp. 874-875. Lunar and Planetary Institute, Houston. A definitive, but less accessible paper by these authors is: (1987) The surface age of Venus: Use of the terrestrial cratering record, *Astronomicheskiiy Vestnik, 21* (in Russian), 144-150.

Simkin T., Siebert L., McClelland L., Bridge D., Newhall C., and Latter J. H. (1981) *Volcanoes of the World*. Hutchinson Ross, Stroudsburg, PA. 233 pp.

Singh R. N. and Russell C. T. (1986) Further evidence for lightning on Venus. *Geophys. Res. Lett., 13*, 1051-1054.

Sommer M. A. (1977) Volatiles H_2O, CO_2, and CO in silicate melt inclusions in quartz phenocrysts from the rhyolitic Bandelier airfall and ash-flow tuff, New Mexico. *J. Geol, 84*, 423-432.

Surkov Yu. A., Moskalyova L. P., Kharyukova A. D., Dudin A. D., Smirnov G. G., and Zaitseva S. Ye (1986) Venus rock composition at the Vega 2 landing site. *Proc. Lunar Planet. Sci. Conf. 17th*, in *J. Geophys. Res., 91*, E215-E218.

Tanaka K. L. (1987) The stratigraphic history of Mars. *Proc. Lunar Planet. Sci. Conf. 17th*, in *J. Geophys. Res., 91*, E139-E158.

Taylor H. A., Jr. and Cloutier P. A. (1986) Venus: Dead or Alive?, *Science, 234*, 1087-1093.

Taylor W. W. L., Scarf F. L., Russell C. T., and Brace L. H. (1979) Evidence for lightning on Venus. *Nature, 279*, 614.

Walker D., Stolper E. M., and Hays J. F. (1979) Basaltic volcanism: The importance of planet size. *Proc. Lunar Planet. Sci. Conf. 10th*, 1995-2015.

Wood C. A. (1979) Venus volcanism: Environmental effects on style and landforms. *NASA Tech. Memo. TM-80339*, 244-246.

Wood C. A. (1984) Calderas: A planetary perspective. *J. Geophys. Res., 89*, 8391-8406.

Wood C. A. (1986) Volcanoes: The Space Shuttle legacy (abstract). *Eos Trans. AGU, 67*, 1073.

Note added in proof:

Prinn (1985, The volcanoes and clouds of Venus: *Scientific American, 252*, 46-53.) has argued from chemical considerations that the amount of SO_2 in Venus' atmosphere is ten times that which could co-exist in equilibrium with likely surface materials. He argues that the presence of sulphurous gases in the atmosphere far in excess of their equilibrium proportions demonstrates that these gases must have been introduced into the atmosphere by geologically very recent processes, and that this is in itself good evidence for continuing volcanic activity on Venus.

The Resurfacing History of Mars: A Synthesis of Digitized, Viking-Based Geology

Kenneth L. Tanaka, Nancy K. Isbell, and David H. Scott
U.S. Geological Survey, 2255 N. Gemini Dr., Flagstaff, AZ 86001

Ronald Greeley
Department of Geology, Arizona State University, Tempe, AZ 85287

John E. Guest
University of London Observatory, London, NW7 2QS, England

A new global geologic map series of Mars has been digitized at high resolution (1.846 km^2/pixel), permitting accurate measurement of the areas of the map series' 90 geologic units. Relative ages of the units are assigned to one or more of eight stratigraphic epochs of Mars [which are subdivisions of the Noachian (oldest), Hesperian, and Amazonian Periods]. The types of resurfacing that the units represent or are modified by are classified among volcanic, eolian, fluvial, periglacial, and impact processes. We find that the surface of Mars (144 × 10^6 km^2) is dominantly volcanic (84 × 10^6 km^2). By estimating the total extent (including buried area) of the geologic units, a resurfacing history was constructed (the extents of exposed areas are given in parentheses). Eolian resurfacing was prevalent during the Late Amazonian Epoch (particularly at the poles), affecting 4.9 × 10^6 km^2; evidence for fluvial activity is most extensive in Late Noachian material (associated with valley-network formation), dissecting about 14.7 × 10^6 km^2 (5.4 × 10^6 km^2), and in Late Hesperian material (corresponding to most outflow-channel formation), modifying about 11.0 × 10^6 km^2 (7.2 × 10^6 km^2); and periglacial activity modified 13.3 × 10^6 km^2 (6.5 × 10^6 km^2) of Late Hesperian northern plains material (from the aftereffects of ponded flood waters from the outflow channels). The area of resurfacing by all processes generally decreases from 98 × 10^6 km^2 in the Middle Noachian Epoch to 11 × 10^6 km^2 in the Late Amazonian Epoch; cumulative resurfacing totals 313 × 10^6 km^2. Resurfacing rates vary according to absolute-age scheme used and generally decrease with time; resurfacing rates were about 1000 km^2/yr during the Middle Noachian Epoch, one hundred to several hundred km^2/yr during the Late Noachian to Late Hesperian Epochs, and tens of km^2/yr or less during the Amazonian Period.

INTRODUCTION

A general resurfacing history of a planet provides a global context in which to evaluate accretionary, geologic, and climatologic events of varying magnitude and duration. Hence, this history is of primary importance in understanding the evolution of the planet as a whole. The resurfacing history of Mars is of particular interest because of the diverse record, both in space and time, of its geologic phenomena.

The dataset acquired from the Viking mission includes tens of thousands of images of the surface taken from orbit. A new 1:15,000,000-scale global map series of Mars based on these data was finished recently (*Scott and Tanaka*, 1986; *Greeley and Guest*, 1987; *Tanaka and Scott*, 1987). This new map series has more than 90 geologic units and supersedes the Mariner 9-based map of Mars (*Scott and Carr*, 1978) that has 24 units. Additionally, crater-density boundaries were determined for the three stratigraphic systems of Mars (*Scott and Tanaka*, 1986), which were divided into eight stratigraphic series (*Tanaka*, 1986) on the basis of material referents (explained further below). The resulting geologic history of Mars indicated in these works is in general agreement with previous work (e.g., *Mutch et al.*, 1976, pp. 316-319; *Scott and Carr*, 1978; *Hartmann et al.*, 1981; *Neukum and Hiller*, 1981), but our work represents a more thorough and detailed analysis of geologic units and their stratigraphic relations.

We have produced a high-resolution digital database of these geologic maps that allows precise measurement of the areas of each geologic unit. We present a detailed resurfacing history of Mars that is based on the geologic interpretations of origin and stratigraphic position for each unit, and we show the change over time in extent of surfaces formed or modified by volcanic, impact, fluvial, eolian, and periglacial processes. We also estimate rates of resurfacing by these processes. This history is compared with previous analyses of resurfacing (*Arvidson et al.*, 1980; *Greeley and Spudis*, 1981; *Neukum and Hiller*, 1981; *Greeley*, 1987).

DIGITIZATION AND ORGANIZATION OF MAP UNITS

Geologic units of the four 1:15,000,000-scale maps of Mars (western and eastern equatorial and north and south polar regions) were digitized and converted to raster image files. Linear features, such as narrow channels, fault scarps, and so on, were not digitized. The map files were geometrically transformed to an equal-area sinusoidal projection with a resolution of 0.03215 degree/pixel, or 1.846 km^2/pixel (Plates 1 and 2). This projection conforms to a biaxial ellipsoid standard for the shape of Mars and allows direct determination of the areal extent of units.

We grouped the map units in a hierarchy based on geographic, geologic, and geomorphic associations. Because of

TABLE 1. Symbol, name, resurfacing processes, stratigraphic position, and area of exposure of map units and groups in digitized geologic maps of Mars.

Symbol	Name	Resurfacing processes	Stratigraphic position	Area (10^3 km^2)
	SURFICIAL MATERIALS			5,993
	Eolian and dune materials			832
Ae	Eolian deposits	eolian	UA	153
Ad	Dune material	eolian	UA	14
Adc	Crescentic dune material	eolian	UA	443
Adl	Linear dune material	eolian	UA	222
Am	*Mantle material*	eolian	UA	1,802
As	*Slide material*	periglacial	MA-UA	644
	Polar deposits			2,715
Api	Polar ice deposits-north	eolian	UA	837
	Polar ice deposits-south	eolian		88
Apl	Polar layered deposits-north	eolian	UH?-UA	395
	Polar layered deposits-south	eolian		1,395
	CHANNEL-SYSTEM MATERIALS			4,326
	Channel materials			1,563
Ach	Younger channel material	fluvial	UA	132
Hch	Older channel material	fluvial	UH	1,431
	Flood-plain materials			1,073
Achp	Younger flood-plain material	fluvial	LA-UA	42
Hchp	Older flood-plain material	fluvial	UH	1,031
Hcht	*Chaotic material*	fluvial	UH	829
Achu	Younger channel and flood-plain material, undivided	fluvial	UA	808
b	Channel bar	fluvial	UH	53
	LOWLAND TERRAIN MATERIALS			30,912
	Northern plains assemblage			25,450
	Arcadia Formation			6,623
Aa$_5$	Member 5	volcanic	UA	212
Aa$_4$	Member 4	volcanic	UA	641
Aa$_3$	Member 3	volcanic	MA	1,765
Aa$_2$	Member 2	volcanic	MA	142
Aa$_1$	Member 1	volcanic	LA	3,863
	Medusae Fossae Formation			2,487
Amu	Upper member	volcanic/eolian	UA	997
Amm	Middle member	volcanic/eolian	MA	891
Aml	Lower member	volcanic/eolian	MA	599
	Vastitas Borealis Formation			12,896
Hvm	Mottled member	periglacial/fluvial/volcanic	UH	3,418
Hvg	Grooved member	periglacial/fluvial/volcanic	UH	1,968
Hvr	Ridged member	periglacial/fluvial/volcanic	UH	1,245
Hvk	Knobby member	periglacial/fluvial/volcanic	UH	6,265
	Other materials			3,444
Aps	Smooth plains material	eolian/fluvial/volcanic	LA-UA	2,970
AHpe	Etched plains material	eolian/periglacial	UH-MA	474
	Eastern volcanic assemblage			5,454
	Elysium Formation			4,050
Ael$_4$	Member 4	volcanic/fluvial	LA	84
Ael$_3$	Member 3	volcanic	LA	1,293
Ael$_2$	Member 2	volcanic	LA	105
Ael$_1$	Member 1	volcanic	LA	2,568
AHat	Albor Tholus Formation	volcanic	UH-LA	19
Hhet	Hecates Tholus Formation	volcanic	UH	28
Hs	Syrtis Major Formation	volcanic	UH	1,357
d	dome	volcanic?	UH?	8

TABLE 1. (continued)

Symbol	Name	Resurfacing processes	Stratigraphic position	Area (10^3 km^2)
	HIGHLAND TERRAIN MATERIALS			81,570
	Western volcanic assemblage			14,102
	Olympus Mons Formation			1,746
Aop	Plains member	volcanic	UA	438
Aos	Shield member	volcanic	UA	388
Aoa$_4$	Aureole member 4	volcanic/periglacial	MA	258
Aoa$_3$	Aureole member 3	volcanic/periglacial	MA	140
Aoa$_2$	Aureole member 2	volcanic/periglacial	MA	74
Aoa$_1$	Aureole member 1	volcanic/periglacial	LA	448
	Tharsis Montes Formation			6,897
At$_6$	Member 6	volcanic	UA	603
At$_5$	Member 5	volcanic	MA	3,064
At$_4$	Member 4	volcanic	LA	1,206
AHt$_3$	Member 3	volcanic	UH-LA	1,210
Ht$_2$	Member 2	volcanic	UH	628
Ht$_1$	Member 1	volcanic	UH	186
	Alba Patera Formation			3,167
Aau	Upper member	volcanic	LA	170
Aam	Middle member	volcanic	LA	1,008
Hal	Lower member	volcanic	LH-UH	1,989
AHcf	Ceraunius Fossae Formation	volcanic	UH-LA	462
	Syria Planum Formation			1,830
Hsu	Upper member	volcanic	UH	1,222
Hsl	Lower member	volcanic	UH	608
	Hellas assemblage			4,255
Ah$_8$	Knobbly plains floor unit	eolian/volcanic	MA	79
Ah$_7$	Rugged floor unit	eolian	MA	108
Ah$_6$	Reticulate floor unit	volcanic/eolian	MA	20
Ah$_5$	Channeled plains rim unit	volcanic/eolian	LA	435
Ah$_4$	Lineated floor unit	volcanic/eolian	LA	40
Hh$_3$	Dissected floor unit	fluvial/volcanic	UH	833
Hh$_2$	Ridged plains floor unit	volcanic	LH	876
Nh$_1$	Basin-rim unit	impact/volcanic	LN	1,864
	Plateau and high-plains assemblage			63,213
	Valles Marineris interior deposits			324
Avf	Floor member	periglacial/eolian/volcanic/fluvial	LA-MA	216
Hvl	Layered member	volcanic/fluvial	UH	108
	Highland Paterae			745
AHt	Tyrrhena Patera Formation	volcanic	UH	44
AHa	Apollinaris Patera Formation	volcanic	UH	45
AHh	Hadriaca Patera Formation	volcanic	UH	97
	Amphitrites Formation			
Hap	patera member	volcanic	LH	119
Had	dissected member	volcanic/fluvial	LH	440
	Tempe Terra Formation			417
Htu	Upper member	volcanic	UH	183
Htm	Middle member	volcanic	UH	203
Htl	Lower member	volcanic	LH	31
	Dorsa Argentea Formation			1,255
Hdu	Upper member	volcanic	LH-UH	793
Hdl	Lower member	volcanic	LH-UH	462
	Plateau sequence			53,473
Hpl$_3$	Smooth unit	volcanic/eolian/fluvial	LH	3,505
Hplm	Mottled smooth plains unit	eolian/volcanic/fluvial	LH-UH	268
Npl$_2$	Subdued cratered unit	volcanic/eolian/fluvial	UN	8,270
Npl$_1$	Cratered unit	volcanic/impact	MN	18,866
Npld	Dissected unit	fluvial	MN-UN	

TABLE 1. (continued)

Symbol	Name	Resurfacing processes	Stratigraphic position	Area (10^3 km^2)
		volcanic/impact	MN	12,961
Npl	Etched unit	eolian	MN-UH	
		volcanic/impact	MN	2,557
Nplr	Ridged unit	volcanic	MN-UN	4,462
Nplh	Hilly unit	impact/volcanic	LN	2,584
	Highly deformed terrain materials			3,075
Hf	Younger fractured material	volcanic	LH-UH	1,306
Nf	Older fractured material	volcanic	MN-UN	1,499
Nb	Basement complex	impact/volcanic	LN	270
HNu	Undivided material	volcanic/impact/periglacial	LN-LH	3,351
v	Volcano (highland)	volcanic	MN-UH	323
Nm,m	Mountain material	volcanic/impact	LN-MN	250
	MATERIALS THROUGHOUT MAP AREA			
Apk	*Knobby plains material*	volcanic/eolian/periglacial	LA-MA	2,386
Hr	*Ridged plains material*	volcanic	LH	13,302
	Impact-crater materials			5,509
c$_s$	superposed	impact	UN-UA	3,443
c$_b$	partly buried	impact	LN-UN	1,040
s	smooth crater-floor	volcanic/eolian/fluvial	UN-UA	1,026
	TOTAL			143,998

Symbols, stratigraphic data, and origins adapted from *Scott and Tanaka* (1986), *Tanaka* (1986), *Greeley and Guest* (1987), and *Tanaka and Scott* (1987); N = Naochian, H = Hesperian, A = Amazonian, L = Lower, M = Middle, U = Upper.

several inconsistencies in map-unit organization among the four geologic maps in the new series, some of the organization has been modified (as discussed in the following), and thus a global organization of units was achieved (Plate 3, Table 1).

Surficial materials, including eolian (unit Ae), dune (units Ad, Adc, Ad1), mantle (unit Am), slide (unit As), and polar materials (units Api, Apl), generally form by mass wasting and atmospheric processes, producing thin and commonly transient deposits. Surficial deposits on the western equatorial map and polar deposits on the polar maps were previously shown as part of the plateau and high-plains assemblage because of their limited geographic occurrence. Dune and mantle materials were grouped with channel-system materials on the polar maps because no grouping of surficial deposits was used. However, because these materials are interpreted to originate by similar processes and because they form local surficial deposits in various geographic situations, we grouped them together.

Channel-system materials include channel-floor (units Ach, Hch), floodplain (units Achp, Hchp), chaotic terrain (unit Hcht), channel-bar (symbol "b"), and undivided channel (unit Achu) materials, a scheme generally consistent with the organization on the geologic maps.

Lowland terrain materials consist of the northern plains and eastern volcanic assemblages that fill the northern lowlands of Mars. Lowland-highland distinctions were not made on the polar maps because of the dominance of lowlands in the north polar region and highlands in the south polar region. The northern plains assemblage is composed of the Arcadia, Medusae Fossae, and Vastitas Borealis Formations and other smooth (unit Aps) and etched (unit AHpe) plains materials, in agreement with the maps. The eastern volcanic assemblage consists of various volcanic materials in the Elysium province and in Syrtis Major Planum (geographic locations in this report are from *U.S. Geological Survey*, 1985), and it occurs almost entirely within the eastern map region.

Highland terrain materials (western volcanic, Hellas, plateau and high-plains assemblages) occur primarily in the equatorial and southern latitudes of Mars, where the highlands dominate. The western volcanic assemblage includes lava flows and shields of the Tharsis province; this assemblage was emplaced mainly on older highland rocks. The Hellas assemblage consists of basin-rim material (unit Nh$_1$) and basin-fill deposits (units Hh$_2$-Ah$_8$) associated with the Hellas impact basin. The plateau and high-plains assemblage includes a variety of materials that occur locally in the highlands, such as Valles Marineris floor (unit Avf) and layered (unit Hvl) materials, highland paterae (in the eastern and south polar map regions), and Tempe Terra (lava flows and volcanic shields in eastern map) and Dorsa Argentea (flows in the south polar region) Formations. The assemblage also includes the widespread plateau sequence, which comprises the majority of highland rocks. This sequence consists of hilly (unit Nplh), ridged (unit Nplr), etched (unit Nple), dissected (unit Npld), cratered (unit Npl$_1$), subdued cratered (unit Npl$_2$), mottled (unit Hplm), and smooth (unit Hpl$_3$) units. Also scattered throughout the highlands are highly deformed terrain (units Nf and Hf, primarily in the Tharsis province), basement material (unit Nb), undivided highland material (unit HNu), volcanoes (symbol "v"), and isolated mountains (unit Nm, symbol "m").

A few materials occur thoughout the map region: knobby (unit Apk) and ridged (unit Hr) plains materials and superposed (symbol "c$_s$"), partly buried (symbol "c$_b$"), and interior-floor (symbol "s") impact-crater materials. The ridged plains material is included in the plateau and high-plains assemblage on the western and south polar maps, where it mostly occurs in the highlands. However, lowland occurrences of the unit are also common below the highland scarp.

ASSIGNMENT OF MAP-UNIT AREAS TO RESURFACING PROCESSES

The type(s) of resurfacing assigned to individual geologic units (Table 1) was based primarily on interpretations given on our 1:15,000,000-scale maps. We chose to place all resurfacing into the categories of volcanic, impact, periglacial (includes dominatly ice-related deformation and mass-wasting), fluvial (includes alluvial), and eolian processes. Volcanism brings forth new material to the surface, whereas the other processes produce materials resulting from modification of existing surface material. Many of the geologic units were mapped on the basis of morphologies resulting from these processes. Although the surfaces of all the map units were fitted into these process categories, designations for some of the surfaces are highly uncertain or complex, and may include more than one resurfacing process. Additionally, the precision of the geologic mapping was limited by scale (for example, many patches of units smaller than about 20 to 30 km across were not mapped). The following sections briefly explain our assignments of resurfacing processes; a more detailed explanation of the interpretation of each unit can be obtained from the geologic maps.

Volcanic Resurfacing

Resurfacing by volcanic activity is defined here as the areal extent of volcanic units produced for a given period. The volcanic units consist of lava flows, pyroclastic deposits, volcanoes, and possible lahars; in many cases, individual units are made up of sequences of flows and deposits representing multiple resurfacing events.

The morphologies of these volcanic features vary widely and diagnostic geomorphic evidence for volcanism commonly is circumstantial in older, eroded materials. Broad, smooth to rolling plains deformed by wrinkle ridges and of Noachian to Lower Hesperian stratigraphic position are believed to be thick, voluminous lava deposits by analogy with mare basalts on the Moon (*Greeley and Spudis*, 1981). Interpretation of this material on Noachian surfaces of Mars, however, is less certain because of modification by impact, eolian, and fluvial processes and the smaller size of the wrinkle ridges (except for those on unit Nplr). Lobate flows are much stronger evidence of volcanism; such flows make up most of the units of the western and eastern volcanic assemblages and the Arcadia and Apollinaris Patera Formations. Locally, a few flows can be seen on high-resolution images of the Vastitas Borealis Formation, indicating that parts of it are volcanic. Parts of several other plains units are probably lava as well.

Interpretation of pyroclastic deposits is more controversial. Most highland paterae appear to be circular depressions or low sheilds composed of sheets of friable material that has been heavily eroded, forming gullies that radiate away from the paterae. The latter paterae were thus interpreted to be composed of ash deposits (*Greeley and Spudis*, 1981), whereas the circular depressions were proposed to be calderas (*Tanaka and Scott*, 1987). Younger lobate and channeled deposits in the Elysium and Arcadia Formations, which are contemporaneous with local lava-flow emplacement, also appear to be composed of easily eroded material. Such deposits have been proposed to consist of lahars or pyroclastic flows (*Christiansen and Greeley*, 1981; *Tanaka and Scott*, 1986). Particularly controversial is the Medusae Fossae Formation, an extensive and thick sequence of highly eroded deposits along the highland-lowland scarp in southern Amazonis Planitia. *Schultz and Lutz-Garihan* (1981) proposed that the sequence is composed of paleopolar deposits resulting from polar wandering. *Scott and Tanaka* (1982), however, mapped the deposits into seven units and interpreted them to be welded and nonwelded ash flows and lava flows. *Squyres et al.* (1987) also found flow features in parts of the Medusae Fossae Formation, and they suggested an origin as lahars released from plateau materials and produced by ground-ice melting caused by shallow intrusion of dikes and sills. The preservation of joint patterns in Medusae Fossae deposits suggests some degree of induration; fractures are not found in present-day Martian polar materials. The varied stratigraphy, flow features, and induration appear to favor a largely pyroclastic origin for the formation. Similarly, the deeply eroded layered deposits in Valles Marineris (unit Hvl) may also be pyroclastic (*Peterson*, 1981).

The volcanoes on Mars fall into several morphologic classes (including shields, tholii, and highland paterae) that may reflect variance in lava compositon and eruption mechanics (e.g., *Plescia and Saunders*, 1979; *Greeley and Spudis*, 1981). Hesperian and Amazonian volcanoes were generally identified with confidence because they display lobate flows, rilles, pit chains, and summit craters. Noachian volcanoes, however, are more eroded and are covered or embayed by impact-crater material and younger deposits. Many such older volcanoes have been proposed because they show remnant volcanic features (*Scott and Tanaka*, 1981a).

Impact Resurfacing

A variety of impact-crater forms and related features have been identified on Mars. Morphologic analysis of large craters provides easy discrimination of impact from volcanic and tectonic origins; features such as secondary craters, ramparts in the ejecta, central pits and peaks, and thick, raised rims identify impact craters. Two crater units (c_s and c_b) consist solely of impact material; part of most Lower and Middle Noachian units, which are heavily cratered, are designated impact material. Although degradational states have been used to classify relative ages of craters on Mars, our crater units have only limited usefulness in the reconstruction of an impact resurfacing history, for the following reasons: (1) Only craters with rim diameters or ejecta blankets greater than 150 km across were mapped; (2) most of the older impact craters and basins are embayed by younger material, which obscures ejecta blankets; and (3) precise stratigraphic assignment of most of the craters was not possible. However, the inclusion of crater material in this study permits a qualitative view of its resurfacing effect and is a necessary component in the paleostratigraphic analysis below.

Fluvial Modification

The surface of Mars has been carved by a variety of channel forms over most of its history. Most of the channels that are assigned a fluvial origin are categorized as either outflow or runoff channels (also known as valley networks) (*Sharp and Malin*, 1975). The runoff channels dissected a substantial part of the highlands; where they are closely spaced, the highland material was mapped as the dissected unit of the plateau sequence (unit Npld). The channels formed mostly during the Middle and Late Noachian Epochs (*Tanaka*, 1986), and they commonly feed into Late Noachian subdued cratered (intercrater plains) material (unit Npl$_2$).

Chaotic terrain, outflow channels, and flood plains mainly of Late Hesperian age indicate that some surfaces were modified by catastrophic water release and runoff and fluvial sedimentation (channel-system materials in Table 1). Chaotic material (unit Hcht) is the source of many of the outflow channels and may be composed of alluvial aquifers (*Tanaka and MacKinnon*, 1987). The channels (unit Hch) may also be filled with fluvial material in places, but elsewhere they show scoured bottoms (*Carr and Clow*, 1981). Terraces in some channels, such as Kasei Valles, indicate several periods of fluvial activity (*Neukum and Hiller*, 1981). *Lucchitta et al.* (1986) and *McGill* (1986) proposed that plains material mapped as Vastitas Borealis Formation includes extensive sediments washed out through the outflow channels, and so part of the extent of these units is attributed to fluvial deposition. The extent of fluvial activity involved in the emplacement and modification of materials in Hellas Planitia is not well known and has been represented only in the dissected floor unit of the Hellas assemblage (unit Hh$_3$). Flood plains and channels in southern Elysium Planitia (unit Achu) have been dated as Late Amazonian (*Tanaka and Scott*, 1986).

Eolian Resurfacing

Thin eolian mantles are common on Mars, as observed on Viking Lander images and in very high resolution Orbiter images, and form large bright and dark patches on lower resolution Orbiter images. For this study, however, we are concerned only with materials tens of meters or more in thickness that form mappable units at small scale.

The present climate of Mars allows for a variety of eolian-atmospheric processes to occur, including sublimation and precipitation of CO_2 (and some H_2O) ices and wind transport of sand and dust. Such processes are particularly active at the poles, depositing and redistributing ice and dust mantles (*Kieffer et al.*, 1976; *Pollack et al.*, 1979; *Squyres*, 1979; *Carr*, 1982). Adjacent to, and possibly derived from, the polar deposits are fields of dunes, especially surrounding the north pole (*Breed et al.*, 1979; *Thomas*, 1982). All together, these deposits make up most of the surficial materials (Table 1). The Medusae Fossae Formation is also heavily reworked by the wind and possibly contains eolian material.

Other eolian materials include deposits and mantles that form smooth plains or that drape over underlying topography in Hellas Planitia (Amazonian members of the Hellas assemblage), Argyre Planitia (unit Nple), and Valles Marineris (unit Avf). Eolian deposits and mantles also make up smooth plains adjacent to the highland scarp in Utopia, Isidis, and Elysium Planitiae (units Aps and AHpe) and cover low areas in the highlands (units Nple and Npl$_2$) (see also Table 1). Pedestal craters in various areas also attest to possible extensive former mantles that may date back to the Hesperian Period (*Tanaka and Scott*, 1987). How much eolian material has been removed is uncertain, although remnants of relatively old, apparently eolian deposits occur. Therefore, how accurate our estimation of the extent of eolian resurfacing for older periods is also uncertain.

Periglacial Modification

Many surface features on Mars have been attributed to formational processes in which ground ice was involved (e.g., *Carr and Schaber*, 1977; *Lucchitta*, 1981); we apply the term "periglacial" to the processes and material units that result from cold temperatures and the presence of ground ice. Because the past climate of Mars restricted, if not precluded, rainfall, the Martin outflow channels attest to a large inventory of ground water and ice (*Carr*, 1986). Under present climate conditions, ground ice is stable near the surface at latitudes higher than about $\pm30°$ (*Squyres and Carr*, 1986). In equatorial areas, ice may occur below a depth of 100 m, but it may have been within tens of meters of the surface during early geologic time, depending on the pore structure of the subsurface (*Fanale et al.*, 1986).

Because of this probable abundance of ground ice, several high-latitude units were likely modified by periglacial activity: the Vastitas Borealis Formation (forming several types of patterned ground that have been attributed to removal of ground ice (*Lucchitta et al.*, 1986) and other ice-related phenomena), slide material (unit As, which forms ice-rich debris aprons), knobby (unit Apk) and etched (unit AHpe) plains materials (which are partly formed by ground-ice removal and resulting degradation of plateau and plains materials), and undivided material (unit HNu, consisting of remnant knobs and mesas of highland rocks surrounded by debris aprons). Because few debris aprons are exposed along the highland-lowland boundary, only a modest amount of mass wasting through ground-ice release is assigned to development of the highland knobs and fretted terrain as represented by the undivided material.

Several units occurring in the equatorial belt (below latitudes $\pm30°$) also apparently have origins that involved deeper sources of ground ice. These consist of (1) slide material (unit As) and floor material (unit Avf) derived from the possibly ice-rich walls of Valles Marineris (*Lucchitta*, 1979), (2) peculiar slidelike features on the flanks of Tharsis Montes and Olympus Mons (unit As) (*Lucchitta*, 1981), and (3) the aureole deposits of Olympus Mons (units Aoa_{1-4}) that may be gravity-spreading features lubricated by ground ice (*Lopes et al.*, 1982; *Tanaka*, 1985).

STRATIGRAPHIC ASSIGNMENTS

Resurfacing of geologic units commonly accompanied their formation. Stratigraphic positions (Table 1) of geologic units were drawn from the correlation charts of the 1:15,000,000-scale geologic maps; the charts are based on impact-crater densities and superposition relations. *Tanaka* (1986) assigned the units to positions according to a scheme of eight stratigraphic series (material units), which was based on an overlapping sequence of referents. These series and their referents are as follows: (1) Lower Noachian, basin-rim material (unit Nplh); (2) Middle Noachian, cratered terrain material (unit Npl_1); (3) Upper Noachian, intercrater plains material (unit Npl_2); (4) Lower Hesperian, ridged plains material (unit Hr); (5) Upper Hesperian; Vastitas Borealis Formation (northern plains material; units Hvk, Hvr, Hvg, and Hvm); (6) Lower Amazonian, smooth plains material of Acidalia Planitia (unit Aa_1); (7) Middle Amazonian, lava flows of Amazonis Planitia (units Aa_2 and Aa_3); and (8) Upper Amazonian, flood-plain material of southern Elysium Planitia (unit Ach). The period of formation of each of these series corresponds to an epoch, in which the adjective is changed to denote time instead of position (e.g., the Lower Noachian Series formed during the Early Noachian Epoch).

The relative ages for most units are well documented through the stratigraphic record. Exceptions include possible older surficial materials that have been reworked or removed, and undivided and impact materials. Impact materials occur throughout the history of Mars but, in most cases, are difficult to date precisely. Highly degraded and partly buried craters (unit c_b) generally predate Late Noachian fluvial and eolian erosion, whereas relatively fresh, superposed craters (unit c_s) and smooth crater-floor material (unit s) postdate that activity. Undivided material (unit HNu) makes up highly eroded highland remnants and scarps that expose a variety of indiscriminate materials emplaced throughout the Noachian and Hesperian Periods.

Stratigraphic-position assignments for Martian geologic units by other workers (*Scott and Carr*, 1978; *Greeley and Spudis*, 1981; *Hartmann et al.*, 1981; *Neukum and Hiller*, 1981; *Barlow*, 1987) generally agree well with our assignments. *Greeley and Spudis* (1981, Fig. 3) presented a detailed stratigraphy of volcanic rocks that was based on Mariner 9-based geologic mapping (*Scott and Carr*, 1978) and crater counting (*Condit*, 1978). The histories of *Hartmann et al.* (1981) and *Neukum and Hiller* (1981, Fig. 18) based on the crater counting of Viking images, include volcanic rocks, highland materials, channels, and mensae and tectonic terrain.

Resurfacing ages, however, do not always correspond to material ages determined by stratigraphic position and crater counts. In particular, fluvial and eolian erosional surfaces and mantles of the dissected (unit Npld) and etched (unit Nple) units of the plateau sequence did not correspond to material ages, and the lower stratigraphic age limit of polar layered deposits was loosely defined because of lack of stratigraphic relations and reworking of the surface. The resurfacing ages designated for these units (the only ones whose ages did not agree entirely with material ages) are described below.

The dissected unit of the plateau sequence (unit Npld) is essentially made up of areas of the cratered unit (unit Npl_1) that are heavily dissected by valley networks. Therefore, the unit is highly resurfaced by fluvial activity, which is dated as Late Noachian and older (*Tanaka*, 1986). The age of the dissected material (Middle Noachian) defines the lower limit of this activity. Similarly, the etched unit is composed of cratered unit material and is deeply furrowed by eolian processes. In many areas, remnants of a former mantle remain. The age limits of the resurfacing are uncertain. The lower limit is only restricted by the age of the cratered unit itself. The assigned Late Hesperian upper limit is estimated from the moderate density of superposed (often pedestal) craters on the mantle. The mantle is also similar to possibly Late Hesperian mantle remnants in the south polar region (*Tanaka and Scott*, 1987).

Presently, the polar layered deposits overlie the Late Hesperian Vastitas Borealis and Dorsa Argentea Formations. Parts of the latter formation, which occurs in the south polar region, are pitted, perhaps due to presence of volatiles in subsurface materials (*Sharp*, 1973). This observation, and the Late Hesperian age of the outflow channels (which exposed large volumes of H_2O to the atmosphere), suggest that polar layered deposits may have been forming since that epoch.

THE EXPOSED RESURFACING RECORD

All geologic units presently exposed were assigned to the category of resurfacing process(es) and stratigraphic position(s) that they represent. In cases where multiple resurfacing processes are inferred, we assigned a weight of twice the area to the first process relative to the second process; additional processes were assigned the same weight as the second process. Where a unit comprises more than one series, the area is divided equally among the series. This procedure was further complicated for units Npld and Nple in which multiple resurfacing processes occurred during different time periods. Calculations based on this procedure for all geologic units resulted in estimates of the exposed area on Mars resurfaced by a given process for each epoch, as shown in Table 2 and Fig. 1.

The degree of error in these area determinations varies according to the limitations, uncertainties, and subjectivity contained in our dataset. The assignments of origin in many cases are necessarily arbitrary without more detailed analysis that is beyond the scope of this study. We generally could determine the dominant process represented by the unit on the basis of geomorphic evidence and previous work (particularly our geologic maps). The accuracy of this procedure should be within 50% for most individual assignments of proportion of resurfacing processes. The

TABLE 2. Extent of resurfaced areas (in 10^3 km^2) exposed on Mars according to age and process.

Epoch	Volcanic	Eolian	Fluvial	Periglacial	Impact	Total
Late Amazonian	3280	4878	1245	322	574	10299
Middle Amazonian	7272	2076	326	1021	574	11269
Early Amazonian	12676	1569	354	394	574	15567
Late Hesperian	11503	983	7235	6501	574	26796
Early Hesperian	19267	1306	1099	209	783	22664
Late Noachian	7702	2432	5353	209	1130	16826
Middle Noachian	20019	320	3241	209	10766	34555
Early Noachian	2074	—	—	209	3,741	6024
Total	83793	13564	18853	9074	18716	144000

accuracy of the overall resurfacing measurements for each epoch also depends on the accuracy of the stratigraphic designations. In this regard, the greatest problem involved resurfaced areas that span more than one series. Again, we generally had little basis for estimating the relative amounts of resurfacing attributable to each series for these units. Fortunately, most of these units were of limited extent and duration (the units of greatest duration are HNu, c_s, c_b, and s).

Overall, our volcanic resurfacing history for exposed rocks is consistent with that of previous workers (*Arvidson et al.*, 1980; *Greeley and Spudis*, 1981; *Neukum and Hiller*, 1981), who show extensive areas of exposed volcanic rocks for all stratigraphic positions equivalent to Middle Noachian to Lower Amazonian and show smaller areas of exposure in younger rocks. *Arvidson et al.* (1980) suggested that most surfaces between latitudes ±30° were volcanic and largely greater than 3.5 b.y. old, according to the *Neukum and Wise* (1976) crater-flux model. This age range, for that crater-flux model, corresponds to the Noachian and Hesperian Periods (*Tanaka*, 1986). *Greeley and Spudis* (1981) used the same crater-count database [4- to 10-km-diameter counts by *Condit* (1978)] as Arvidson and colleagues, but they included all latitudes. With the inclusion of the northern plains material as volcanic, their history indicates constant resurfacing as a function of crater density for rocks of Noachian to Middle Amazonian position. Qualitatively, *Neukum and Hiller* (1981) indicated that volcanic and fluvial activity decreased dramatically during the equivalent of the Early Amazonian Epoch; this is consistent with our mapping.

Further analysis of Fig. 1 shows that the record of eolian resurfacing is prevalent in Amazonian materials. This is probably due to the preservation of young, transient polar deposits forming the ice caps, dust mantles, and dunes, whereas older deposits of this kind were probably removed and reworked. Fluvial activity was sporadic, gullying highland rocks during the Noachian Period, producing extensive outflow channels and flood plains during the Late Hesperian Epoch, and flooding a plain in southern Elysium Planitia during the Late Amazonian. The extent of time that fluvial activity occurred through existing channels is not known in most cases. Ice-related modification of surface material appears pronounced during two periods— the Late Hesperian activity arises from the modification of plains material in Vastitas Borealis, and the Middle Amazonian peak stems from the interpretation that the aureole deposits of Olympus Mons were formed by ice-lubricated gravity spreading (*Tanaka*, 1985). Stratigraphic constraints are not well developed for impact material, and the proportion of highland rocks made up by impact material is not known.

PALEOSTRATIGRAPHIC RESTORATION OF MAP UNITS

Estimation of the extent of burial of map units and the resurfacing that they underwent was complicated. It was necessary to identify and estimate the relative proportions of different buried map units by interpreting map information. This procedure involved observing the proportions of stratigraphically lower units in surrounding areas and the modes of occurrence of probable underlying units (i.e., some units are exposed only in polar areas, lowlands, near volcanic vents, etc.) as related to surface characteristics. Main occurrences of the geologic units already were assessed during mapping. Table 3 gives estimates of the relative proportion of buried units underlying each map unit; generally only buried units of considerable extent are included. This technique was commonly repeated several times to arrive at totals for areas of units covered by a sequence of units. Two examples are the south polar ice cap (unit Api) which may have beneath it units Apl, Hdu, Hdl, Npl$_2$, and Npl$_1$, and the flanks of the

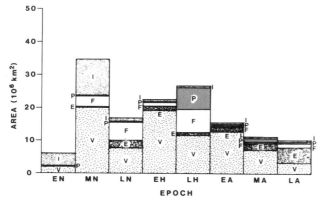

Fig. 1. Bar graph indicating exposed area of resurfacing for each of the Martian epochs (total height of bars) and divided into areas modified by volcanic (V), eolian (E), fluvial (F), periglacial (P), and impact (I) processes; data from Table 2.

TABLE 3. Estimated proportions of underlying units and total restored surface areas for map units on Mars.

Map unit	Underlying unit(s)	Total area of unit (10^3 km^2)
Surficial Materials		
Ae	Aoa$_4$	153
Ad	Hr-Hpl$_3$-Npl$_2$-Npl$_1$-Npld-Nple-Nplr-Nplh	14
Adc	Hvm-Hvg-Hvr-Hvk	443
Adl	Hvm-Hvg-Hvr-Hvk	222
Am	Hvm-Hvg-Hvr-Hvk	1802
As	At$_5$ (western map); 1/2 Hr, 1/2 Hpl$_3$-Npl$_2$-Npl$_1$-Npld-Nple-Nplr-Nplh (remainder)	644
Api	Apl	925
Apl	Hvm-Hvg-Hvr-Hvk (north pole)	1232
	1/2 Hdu, 1/2 Hpl$_3$-Npl$_2$-Npl$_1$-Npld-Nple-Nplr-Nplh (south pole)	1483
Channel-System Materials		
Ach	1/2 Aa$_1$, 1/2 Hr	132
Hch	1/2 Hr, 1/2 Npl$_2$	1677
Achp	Hr	42
Hchp	Hr	1064
Hcht	Npl$_2$	829
Achu	Aps	808
b	Hr	53
Lowland Terrain Materials		
Aa$_5$	1/3 Hvg, 1/3 Aa$_4$, 1/3 Aa$_3$	212
Aa$_4$	Aa$_3$	712
Aa$_3$	1/2 Aa$_1$, 1/4 Hvk, 1/4 Hr	2477
Aa$_2$	Aa$_1$	142
Aa$_1$	1/2 Hvm-Hvg-Hvr-Hvk, 1/4 Hr, 1/4 HNu	6671
Amu	Amm	997
Amm	1/2 Aa$_1$, 1/4 Aml, 1/8 Aoa$_1$, 1/8 AHt$_3$	1888
Aml	2/3 Hr, 1/3 Aa$_1$	821
Hvm	3/4 Hr, 1/4 HNu	6811
Hvg	3/4 Hr, 1/4 HNu	3998
Hvr	3/4 Hr, 1/4 HNu	2489
Hvk	1/2 Hr, 1/2 HNu	13108
Aps	2/3 Hvm-Hvg-Hvr-Hvk, 1/3 Hr	3064
AHpe	HNu	487
Ael$_4$	1/2 Hvm-Hvg-Hvr-Hvk, 1/4 Hr, 1/4 HNu	87
Ael$_3$	1/2 Hvm-Hvg-Hvr-Hvk, 1/4 Hr, 1/4 HNu	1334
Ael$_2$	1/2 Hvm-Hvg-Hvr-Hvk, 1/4 Hr, 1/4 HNu	105
Ael$_1$	1/2 Hvm-Hvg-Hvr-Hvk, 1/4 Hr, 1/4 HNu	2650
AHat	Hr	19
Hhet	HNu	28
Hs	1/2 Nple, 1/4 Npld, 1/4 Nplh	1400
d	Hr	8
Highland Terrain Materials		
Aop	1/2 Aos, 1/2 Aoa$_1$	438
Aos	Aoa$_2$	826
Aoa$_4$	3/4 Aoa$_1$, 1/4 Aoa$_3$	411
Aoa$_3$	Aoa$_2$	243
Aoa$_2$	Aoa$_1$	105
Aoa$_1$	1/2 Hf, 1/2 Hr	1080
At$_6$	At$_5$	603
At$_5$	13/20 At$_4$, 1/10 AHt$_3$, 1/10 Hf, 1/20 Hch, 1/20 Hsu, 1/20 Nf	4016
At$_4$	4/5 AHt$_3$, 1/10 Hf, 1/20 Hsu, 1/20 Nf	3854
AHt$_3$	7/10 Ht$_2$, 1/10 Hf, 1/10 Nf, 1/10 Hpl$_3$-Npl$_2$-Npl$_1$-Npld-Nple-Nplr-Nplh	4695
Ht$_2$	1/2 Ht$_1$, 3/10 Hpl$_3$-Npl$_2$-Npl$_1$-Npld-Nple-Nplr-Nplh, 1/10 Hf, 1/10 Nf	3935
Ht$_1$	1/3 Hf, 1/3 Nf, 1/3 Hpl$_3$-Npl$_2$-Npl$_1$-Npld-Nple-Nplr-Nplh	2153
Aau	Aam	170

TABLE 3. Estimated proportions of underlying units and total restored surface areas for map units on Mars.

Map unit	Underlying unit(s)	Total area of unit (10^3 km^2)
Aam	Hal	1178
Hal	1/2 Nf, 1/4 Hr, 1/4 HNu	3843
AHcf	Hal	462
Hsu	1/2 Hsl, 1/2 Hf	1655
Hsl	3/4 Hf, 1/4 Hr	1436
Ah_8	Hh_3	79
Ah_7	1/3 Hh_3, 1/3 Hh_2, 1/3 Nh_1	108
Ah_6	Hh_2	20
Ah_5	2/3 Nh_1, 1/3 Hpl_3	435
Ah_4	Hh_2	40
Hh_3	Hh_2	948
Hh_2	Nh_1	1072
Nh_1	basement	3951
Avf	1/2 Hvl, 1/2 HNu	216
Hvl	HNu	324
AHt	Hr	44
AHa	Hr	45
AHh	Hpl_3	97
Hap	Nh_1	119
Had	Nh_1	440
Htu	2/3 Htm, 1/3 Hal	183
Htm	1/2 Hal, 1/4 Npl_1, 1/4 Nf	305
Htl	Npl_1	31
Hdu	1/2 Hdl, 1/4 Npl_2, 1/4 Npl_1	1560
Hdl	1/2 Npl_2, 1/2 Npl_1	1257
Hpl_3	Npl_2	4016
Hplm	Npl_2	280
Npl_2	Npl_1-Npld-Nple-Nplr	23970
Npl_1	basement	51629
Npld	basement	34620
Nple	basement	6417
Nplh	basement	11199
Hf	3/4 Nf, 1/4 Npl_2-Npld-Npl_1-Nplh	5,414
Nf	basement	6879
Nb	basement	11962
HNu	basement	17953
v	1/3 Npl_1, 1/3 Nf, 1/3 Nplh	323
Nm, m	basement	250
Materials Throughout Area		
Apk	1/2 Hvm-Hvg-Hvr-Hvk, 1/2 Hr	2462
Hr	Npl_2-Npl_1-Npld-Nple-Nplr-Nplh-Nf-Nb	40023
c_s	many units	3443
c_b	Nh_1-Npl_2-Npl_1-Npld-Nple-Nplr-Nplh-Nf-Nb	1093
s	many units	1026

Areas of units dashed together are divided according to relative amounts of exposure (Table 1). Total areas are sums of exposed and esimated buried areas. Underlying "basement" is undivided Lower Noachian material.

Tharsis Montes, which may include the six members of the Tharsis Montes Formation, two units of fractured material, and some of the eight units of the plateau sequence. A similar exercise of reconstructing progressively older surfaces was done for the Tharsis region (*Scott and Tanaka*, 1981b) on the basis of lava-flow mapping at 1:2,000,000 scale (*Scott et al.*, 1981).

Most of the estimates of proportions of underlying units were fairly straightforward because of clear stratigraphic relations and our thorough knowledge of where the units occur. The northern plains, however, are so broad and thoroughly resurfaced by the Vastitas Borealis Formation that we can have only a sketchy idea of what underlies it. The formation embays ridged plains material (unit Hr) and knobs of undivided material (unit HNu) in Elysium Planitia and other places. Crater distributions that included partly buried crater rims in Utopia Planitia (*McGill*, 1986) indicate a buried surface having a stratigraphic age equivalent to ridged plains material. Based on these data and the relative abundance of units Hr and HNu

Plate 1. Digitized Lambert equal-area projection of the western hemisphere (longitudes 0° to 180°) of Mars showing Viking-based geologic map composited from work of *Scott and Tanaka* (1986) and *Tanaka and Scott* (1987) (shaded-relief base courtesy of U.S. Geological Survey). Interval of grid is 10° (except above latitudes ±60°, where interval of meridians is 30°); extra grid lines are at latitudes ±55° where polar geologic maps were joined to equatorial map. See Plate 3 for correlation of map units.

Plate 2. Digitized Lambert equal-area projection of the eastern hemisphere (longitudes 180° to 360°) of Mars showing Viking-based geologic map composited from work of *Greeley and Guest* (1987) and *Tanaka and Scott* (1987) (shaded-relief base courtesy of U.S. Geological Survey). Interval of grid is 10° (except above latitudes ±60°, where interval of meridians is 30°); extra grid lines are at latitudes ±55° where polar geologic maps were joined to equatorial map. See Plate 3 for correlation of map units.

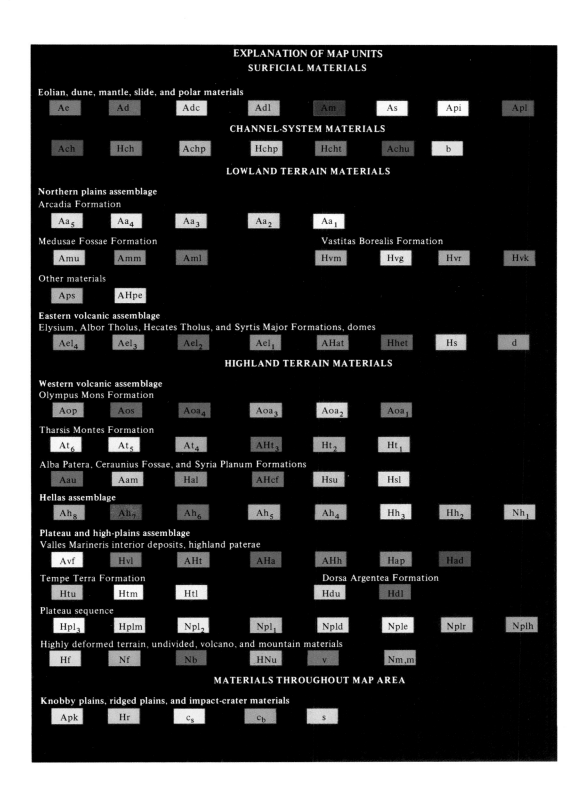

Plate 3. Correlation of map units shown in Plates 1 and 2. For names and detailed organization of map units, see Table 1 and text.

TABLE 4. Total areal extent (including buried surfaces) of resurfacing on Mars (in 10^3 km^2) according to age and process.

Epoch	Volcanic	Eolian	Fluvial	Periglacial	Impact	Total
Late Amazonian	3796	5380	1252	322	574	11324
Middle Amazonian	9965	2997	334	1340	574	15210
Early Amazonian	20621	2084	363	405	574	24047
Late Hesperian	27984	1702	11043	13258	574	54561
Early Hesperian	52761	1919	1229	1122	1696	58727
Late Noachian	25975	6838	14691	1122	2060	50686
Middle Noachian	58664	802	8655	1122	28997	98240
Total	199766	21722	37567	18691	35049	312795

exposed in the northern plains, we assigned three-fourths of the area beneath the Vastitas Borealis Formation to unit Hr, and a fourth to unit HNu. In turn, evidence indicating what type of material exists beneath unit Hr is lacking. Ridged plains material does embay unit HNu and in highland exposures unit Hr generally overlies units of the plateau sequence (see Table 1). Therefore, plateau sequence units are assigned to underlie unit Hr in the northern plains.

After the buried surface areas for units were determined, the total areas (exposed plus buried) of each unit were calculated (Table 3). The total areas for some units were much greater than exposed areas alone, especially for widespread and stratigraphically lower units (e.g., units Hr, Npl$_2$, Npl$_1$) and basal units of a local sequence (e.g., Tharsis Montes Formation and Hellas assemblage). Following the same procedure used to produce Table 2, we calculated, on the basis of these values, the amount of area resurfaced by each of the five processes for each stratigraphic interval (Table 4, Fig. 2). The Early Noachian Epoch was left out in this table because the primary origins of these rocks are unknown, and they include basement material that underlie the surface materials over the entire planet. We repeat here that the resurfacing estimates involve a maximum of about 50% uncertainty for exposed rocks; this uncertainty increases as estimates of buried areas are included. Also, the thickness to which resurfacing occurred varies widely among the geologic units and processes studied.

As expected, volcanism is the major resurfacing process, repeatedly covering many areas with volcanic material, adding up to an area exceeding the equivalent of the planet's surface (Table 4). Our estimates of total volcanic rocks is virtually the same as given by Greeley (1987) but much different in detail, particularly for the Middle Noachian in which nearly 30% of our volcanic rocks occur. Our estimates indicate that, following highland volcanism, an intermediate peak of volcanic activity occurred during the Early Hesperian, most of which is attributed to emplacement of ridged plains material. The western and eastern volcanic assemblages and some lowland materials account for areas resurfaced by volcanism during Late Hesperian and Amazonian time.

Eolian resurfacing generally increased with younger stratigraphic age, except for a pronounced peak in the Late Noachian. As discussed earlier, this trend may reflect the likelihood that most older deposits of eolian material do not become well indurated and are easily eroded away. The pronounced Late Noachian peak of eolian resurfacing is due to the large amount of intercrater plains material (unit Npl$_2$) estimated for this period, which may be partly eolian. If this modification were spread over the Hesperian Period as well (as for the young mantle on unit Nple), a more even degree of resurfacing would result for Late Noachian to Late Amazonian times. Constant resurfacing would be consistent with the hypothesis that the early peak in impact bombardment on Mars would have caused the comminution of large volumes of surface material into sand and dust, and only relatively minor volumes

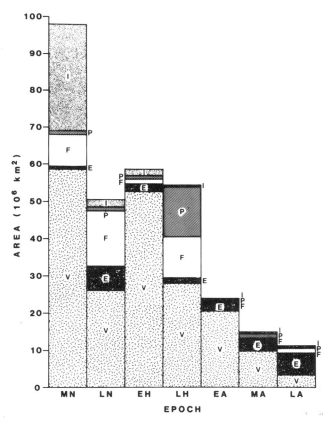

Fig. 2. Bar graph showing total areas of resurfacing (including inferred buried surfaces) for each of the Martian epochs (except Early Noachian) and divided into areas modified by volcanic (V), eolian (E), fluvial (F), periglacial (P), and impact (I) processes; data from Table 4.

of comminuted material by impact and other processes would be added later on.

Fluvial modification had major peaks in the Noachian Period, when the formation of valley networks was fairly widespread, and in the Late Hesperian Epoch, when most outflow-channel activity occurred. Coincident with Late Hesperian fluvial activity is the peak in periglacial modification associated with the northern plains, which was probably a consequence of the large volume of water which was released onto the plains (*Carr*, 1986), subsequently modifying the plains material because of freezing and desiccation (*Lucchitta et al.*, 1986; *McGill*, 1986). Overall, resurfacing of Mars was extensive during the Noachian and Hesperian, and gradually tapered off in the Amazonian.

RESURFACING RATES

Resurfacing rates are of great significance in analyzing the evolution of a planet's surface, atmosphere, and interior. The preceding summary of the resurfacing history of Mars is based on a relative-age stratigraphy, and so the peaks in resurfacing discussed there relate to areas resurfaced over an undefined time period. To ascertain rates of resurfacing activity, absolute ages are needed. Unfortunately, the absolute-age chronology of Mars is not as well established as is the stratigraphy. The chronologies that have been proposed for Mars are based on estimated crater fluxes derived from comparisons with lunar crater distributions and planetestimal distributions. Lack of returned samples, and hence, lack of radioisotopic ages (as available for the Moon) and various complicating astronomical and planetary factors necessitate the use of various assumptions in constructing a time scale. Because of this, several widely varying chronologies for Mars have been proposed.

Tanaka (1986, Fig. 2) took two of the more commonly used, yet greatly different chronologies (model 1 of *Neukum and Wise*, 1976; *Hartmann et al.*, 1981) and correlated them to the boundaries of eight epochs assigned to Mars. Tanaka compared crater distributions of heavily cratered terrain on the Moon and Mars and assigned age limits to the Middle Noachian epoch that are amenable to the Hartmann chronology. For the same epoch, absolute-age limits were not developed for the Neukum and Wise model. To calculate these, we used the Neukum and Wise standard crater curve as defined by *Neukum and Hiller* (1981, p. 3098) to determine the expected ratio of cumulative crater frequency at 1 km diameter to that at 16 km diameter. This ratio is 500, and therefore by this technique the crater-density boundaries of the Middle Noachian Series (and using the 16-km crater-density boundaries defined by *Tanaka*, 1986) are 50,000 and 100,000 craters larger than 1 km per 10^6 km^2. These crater densities indicate age boundaries for the Middle Noachian Epoch of about 4.3 and 4.2 b.y. Also, we changed the Middle Amazonian-Early Amazonian boundary from 2.30 to 2.50 b.y. to conform more accurately to the Neukum and Wise model.

On the basis of these two chronologies, rate histories for each major resurfacing process on Mars are shown in Fig. 3 (Hartmann-Tanaka chronology, hereafter "HT") and Fig. 4 (Neukum and Wise chronology, hereafter "NW"). The HT chronology produces long durations for the Late Hesperian

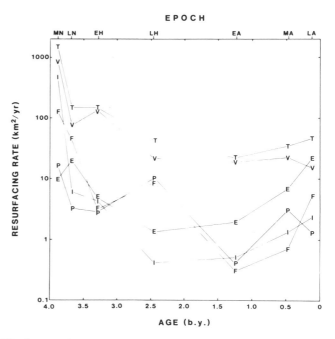

Fig. 3. Semi-log graph showing average rates of resurfacing for total (T), volcanic (V), eolian (E), fluvial (F), periglacial (P), and impact (I) processes for each Martian epoch (except Early Noachian); rates based on resurfacing areas given in Table 4 and on crater model by *Hartmann et al.* (1981) and chronostratigraphy by *Tanaka* (1986); data points are at mean ages of epochs.

and Early Amazonian Epochs, whereas the NW chronology predicts that the Noachian and Hesperian Periods covered only the first billion years of Mars' history. Some contrasts in resurfacing rates result because of these differences. Both show total and volcanic resurfacing rates of about 1,000 km^2/yr for the Middle Noachian. The HT chronology indicates a rapid decline to 100 km^2/yr for the Late Noachian and Early Hesperian, leveling off to a few tens of square kilometers per year for the remainder of geologic time; the NW chronology indicates a similar drop in rates to about 100 km^2/yr for the Late Noachin Epoch, but shows an increase in resurfacing rates to several hundred km^2/yr during the Hesperian and dropping to less than 20 for the Amazonian. The minor increase in total resurfacing rate for the Late Amazonian in both models is probably due to the high degree of preservation of younger surficial and fluvial materials and the effect of evenly dividing resurfacing areas for a given unit (e.g., unit Apl) among periods of decreasing length. The Late Amazonian Epoch is the only one in which eolian resurfacing, rather than volcanism, is the dominant resurfacing process. The analysis of *Arvidson et al.*, (1980), who use the NW chronology, indicates a rapidly decreasing total resurfacing rate for the equatorial zone to about 100 km^2/yr (equivalent to 0.7% area/b.y. in Arvidson et al.) by the end of the Hesperian, and decreasing to less than 10 km^2/yr for the Middle and Late Amazonian. These rates agree well with our results; our higher rates for Late Hesperian and Amazonian are accounted for by the resurfacing that occurred at high latitudes.

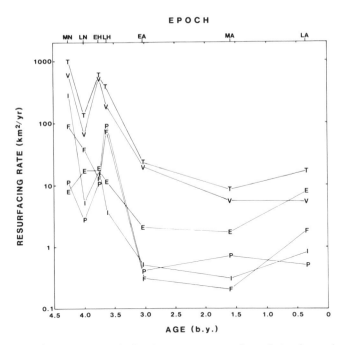

Fig. 4. Semi-log graph showing average rates of resurfacing for total (T) volcanic (V), eolian (E), fluvial (F), periglacial (P), and impact (I) processes for each Martian epoch (except Early Noachian); rates based on resurfacing areas given in Table 4 and on crater model by *Neukum and Wise* (1979) and chronostratigraphy by *Tanaka* (1986) and additional work (in text); data points are at mean ages of epochs.

For fluvial activity, the HT chronology suggests a major peak in the Noachian and a relatively minor pulse in the resurfacing rate during the Late Hesperian, whereas the NW model would predict a more substantial rate of resurfacing during the Late Hesperian.

CONCLUSIONS

The history presented in this paper is more detailed in areal, chronologic, and process information than is previous work, but it is also preliminary, and we expect it will undergo modification and improvement, perhaps in major ways, when new spacecraft data applicable to the chronology and geology of Mars become available. At this time, however, our results are generally consistent with qualitative and less detailed assessments made thus far regarding the surface history of Mars.

The following is a summary of our main results.

1. Ninety geologic units compiled on 1:15,000,000-scale maps of Mars were digitized for accurate measurements of areal extents. The unit identifications and interpretations are based on medium-resolution (130-300 m/pixel) images. According to these data and our inferences of unit origins (Table 1), we find that volcanic surfaces (84×10^6 km^2) cover more than half of Mars; surfaces affected by eolian (14×10^6 km^2), fluvial (19×10^6 km^2), periglacial (9×10^6 km^2), and impact (19×10^6 km^2) processes each cover a significant portion of the remainder (Table 2).

2. On the basis of our assignments of stratigraphic age for the geologic units and their resurfacing features (Table 1), we determined for each Martian epoch the proportion of exposed surface that has been covered or altered by geologic processes (Fig. 1, Table 2). Very little of the Early Noachian crust is exposed. Extensive volcanic, impact, and fluvial surfaces are prevalent in the Middle and Late Noachian Epochs. The Early Hesperian is dominated by volcanic ridged plains material; volcanic rocks continue to be prevalent throughout the remainder of the stratigraphic record. Fluvial and periglacial surfaces are abundant in the Late Hesperian as a result of the formation of outflow channel and flood-plain surfaces and northern plains material (Vastitas Borealis Formation). Eolian material forms extensive Late Amazonian cover in the polar regions. Because of poor preservation and dating of older eolian materials, their resurfacing history is not well known and conceivably was fairly uniform through geologic time.

3. Through various map relations, we inferred the buried resurfacing record (Table 3), enabling us to estimate the total areal extent of resurfacing on Mars by age and process (Fig. 2, Table 4). Generally, the same peaks determined for the exposed resurfacing as a function of age and process are shown. The magnitudes, however, are much greater when buried rocks are considered; Hesperian and Noachian surfaces were 2 to 3 times as extensive than what is exposed. For example, exposed volcanic rocks of Middle Noachian age cover about one-seventh of the planet's surface; our estimate of their total resurfacing extent is nearly three-sevenths of the planet. Overall, resurfacing since the Middle Noachian Epoch has modified the equivalent of more than twice the area of the entire surface of Mars, nearly two-thirds of which was caused by volcanism.

4. Overall resurfacing rates (Figs. 3 and 4) have decreased dramatically from about 1000 km^2/yr for the Middle Noachian Epoch. Depending on the absolute-age scheme used and assumptions in the dataset, resurfacing rates of volcanic, eolian, fluvial, and periglacial activity underwent sporadic changes during the Late Noachian to Late Hesperian; total resurfacing was on the order of one hundred to several hundred km^2/yr. For each of these geologic processes (particularly eolian), our data suggest that activity occurred at modest rates (between about 1 and 20 km^2/yr) through Amazonian time.

Acknowledgments. We thank Haig Morgan and Chris Isbell of the U.S. Geological Survey, Flagstaff, Arizona, for assistance in manipulating the massive digital datasets used in this study. Ray Arvidson, Phil Davis, Baerbel Lucchitta, and Jeff Plaut performed helpful reviews of the manuscript. Work performed for NASA contrct no. W-15,814.

REFERENCES

Arvidson R. E., Goettel K. A., and Hohenberg C. M. (1980) A post-Viking view of Martian geologic evolution. *Rev. Geophys. Space Phys.*, 18, 565-603.

Barlow N. G. (1987) A revised martian relative age chronology and some geologic implications (abstract). In *Lunar and Planetary Science XVIII*, pp. 44-45. Lunar and Planetary Institute, Houston.

Breed C. S., Grolier M. J., and McCauley J. F. (1979) Morphology and distribution of common 'sand' dunes on Mars: Comparison with the Earth. *J. Geophys. Res., 84*, 8183-8204.

Carr M. H. (1982) Periodic climate change on Mars: Review of evidence and effects on distribution of volatiles. *Icarus, 50*, 129-139.

Carr M. H. (1986) Mars: A water-rich planet? *Icarus, 68*, 187-216.

Carr M. H. and Clow G. D. (1981) Martian channels and valleys: Their characteristics, distribution, and age. *Icarus, 48*, 91-117.

Carr M. H. and Schaber G. G. (1977) Martian permafrost features. *J. Geophys. Res., 82*, 4039-4054.

Christiansen E. H. and Greeley R. (1981) Mega-lahars (?) in the Elysium region, Mars (abstract). In *Lunar and Planetary Science XII*, pp. 138-140. Lunar and Planetary Institute, Houston.

Condit C. D. (1978) Distribution and relations of 4- to 10-km-diameter craters to global geologic units of Mars. *Icarus, 34*, 465-478.

Fanale F. P., Salvail J. R., Zent A. P., and Postawko S. E. (1986) Global distribution and migration of subsurface ice on Mars. *Icarus, 67*, 1-18.

Greeley R. (1987) Release of juvenile water on Mars: Estimated amounts and timing associated with volcanism. *Science, 236*, 1653-1654.

Greeley R. and Guest J. E. (1987) Geologic map of the eastern equatorial region of Mars. *U. S. Geol. Survey Misc. Inv. Series Map I-1802-B*, in press.

Greeley R. and Spudis P. D. (1981) Volcanism on Mars. *Rev. Geophys. Space Phys., 19*, 13-41.

Hartmann W. K., Strom R. G., Weidenschilling S. J., Blasius K. R., Woronow A., Dence M. R., Grieve R. A. F., Diaz J., Chapman C. R., Shoemaker E. M., and Jones K. L. (1981) Chronology of planetary volcanism by comparative studies of planetary cratering. In *Basaltic Volcanism on the Terrestrial Planets*, Chap. 8, pp. 1049-1127. Pergamon, New York.

Keiffer H. H., Chase S. C., Jr., Martin T. Z., Miner E. D., and Palluconi F. D. (1976) Martian north pole summer temperatures: Dirty water ice. *Science, 194*, 1341-1344.

Lopes R. M. C., Guest J. E., Hiller K. H., and Neukum G. P. O. (1982) Further evidence for a mass movement origin of the Olympus Mons aureole. *J. Geophys. Res., 87*, 9917-9928.

Lucchitta B. K. (1979) Landslides in Valles Marineris, Mars. *J. Geophys. Res., 84*, 8097-8113.

Lucchitta B. K. (1981) Mars and Earth: Comparison of cold-climate features. *Icarus, 45*, 264-303.

Lucchitta B. K., Ferguson H. M., and Summers C. (1986) Sedimentary deposits in the northern lowland plains, Mars. *Proc. Lunar Planet. Sci. Conf. 17th*, in *J. Geophys. Res., 91*, E166-E174.

McGill G. E. (1986) The giant polygons of Utopia, northern martian plains. *Geophys. Res. Lett., 13*, 705-708.

Mutch T. A., Arvidson R. E., Head J. W., Jones K. L., and Saunders R. S. (1976) *The Geology of Mars*. Princeton University Press, Princeton, N. J.

Neukum G. and Hiller K. (1981) Martian ages. *J. Geophys. Res., 86*, 3097-3121.

Neukum G. and Wise D. U. (1976) Mars: A standard crater curve and possible new time scale. *Science, 194*, 1381-1387.

Peterson C. (1981) A secondary origin for the central plateau of Hebes Chasma. *Proc. Lunar Planet. Sci. 12B*, 1459-1471.

Plescia J. B. and Saunders R. S. (1979) The chronology of the Martian volcanoes. *Proc. Lunar Planet. Sci. Conf. 10th*, 2841-2859.

Pollack J. B., Colburn D. S., Flasar F. M., Kahn R., Carlston C. E., and Pidek D. (1979) Properties and effects of dust particles suspended in the Martian atmosphere. *J. Geophys. Res., 84*, 2929-2945.

Schultz P. H. and Lutz-Garihan A. B. (1981) Ancient polar locations on Mars: Evidence and implications (abstract). In *Papers Presented to the Third International Colloquium on Mars*, pp. 229-231. Lunar and Planetary Institute, Houston.

Scott D. H. and Carr M. H. (1978) Geologic map of Mars. *U. S. Geol. Survey Misc. Invest. Series Map I-1083*.

Scott D. H., Schaber G. G., Tanaka K. L., Horstman K. C., and Dial A. L. Jr. (1981) Map series showing lava-flow fronts in the Tharsis region of Mars. *U. S. Geol. Survey Misc. Invest. Series Maps I-1266 to I-1280*.

Scott D. H. and Tanaka K. L. (1981a) Mars: A large highland volcanic province revealed by Viking images. *Proc. Lunar Planet. Sci. 12B*, 1449-1458.

Scott D. H. and Tanaka K. L. (1981b) Paleostratigraphic restoration of buried surfaces in Tharsis Montes. *Icarus, 45*, 304-319.

Scott D. H. and Tanaka K. L. (1982) Ignimbrites of Amazonis Planitia region of Mars. *J. Geophys. Res., 87*, 1179-1190.

Scott D. H. and Tanaka K. L. (1986) Geologic map of the western equatorial region of Mars. *U. S. Geol. Survey Misc. Invest. Series Map I-1802-A*.

Sharp R. P. (1973) Mars: South polar pits and etched terrain. *J. Geophys. Res., 78*, 4222-4230.

Sharp R. P. and Malin M. C. (1975) Channels on Mars. *Geol. Soc. Amer. Bull., 86*, 593-609.

Squyres S. W. (1979) The evolution of dust deposits in the Martian north polar region. *Icarus, 40*, 244-261.

Squyres S. W. and Carr M. H. (1986) Geomorphic evidence for the distribution of ground ice on Mars. *Science, 231*, 249-252.

Squyres S. W., Wilhelms D. E., and Moosman A. C. (1987) Large-scale volcano-ground ice interactions on Mars. *Icarus, 70*, 385-408.

Tanaka K. L. (1985) Ice-lubricated gravity spreading of the Olympus Mons aureole deposits. *Icarus, 62*, 191-206.

Tanaka K. L. (1986) The stratigraphy of Mars. *Proc. Lunar Planet. Sci. Conf. 17th*, in *J. Geophys. Res., 91*, E139-E158.

Tanaka K. L. and MacKinnon D. J. (1987) Development of the Chryse hydrologic system, Mars (abstract). In *Lunar and Planetary Science XVIII*, pp. 996-997. Lunar and Planetary Institute, Houston.

Tanaka K. L. and Scott D. H. (1986) Channels of Mars' Elysium region: Extent, sources, and stratigraphy. In *Reports of the Planetary Geology Program—1985*, pp. 403-405. NASA TM 88383, NASA, Washington, DC.

Tanaka K. L. and Scott D. H. (1987) Geologic maps of the polar regions of Mars. *U. S. Geol. Survey Misc. Invest. Series Map I-1802-C*, in press.

Thomas P. (1982) Present wind activity on Mars: Relation to large latitudinally zoned sediment deposits. *J. Geophys. Res., 87*, 9999-10008.

U. S. Geological Survey (1985) Shaded relief maps of the eastern, western and polar regions of Mars. *U. S. Geol. Survey Misc. Invest. Series Map I-1618*.

A Widespread Common Age Resurfacing Event in the Highland-Lowland Transition Zone in Eastern Mars

Herbert Frey

Geophysics Branch, NASA Goddard Space Flight Center, Greenbelt, MD 20771

Ann Marie Semeniuk, Jo Ann Semeniuk, and Susan Tokarcik

Astronomy Program, University of Maryland, College Park, MD 20742

One approach to understanding the nature and origin of the fundamental crustal dichotomy on Mars is to attempt to unravel the complex history of the boundary transition zone between the cratered highlands and northern lowland plains. This is an area in which resurfacing events have played a major role, and the cumulative frequency curves for impact craters larger than 5 km diameter within the transition zone and for the adjacent cratered terrain and smooth plains all show evidence of such resurfacing: Toward smaller diameters these curves bend away from a "Mars standard production curve" (*Neukum and Hiller,* 1981). The points at which such bendaways occur can be used to isolate separate branches representing separate crater retention surfaces between which a resurfacing event has occurred. The oldest surface resolvable in the cumulative frequency curves is found only in the cratered terrain south of the transition zone, and its crater retention age $N(1)$ (cumulative number of craters with $D > 1$ km per 10^6 km^2) is about 210,000 both east and west of the Isidis Basin. No evidence for a surface of this age is found within the transition zone or the northern plains. A major resurfacing event that ceased at $N(1) \approx 86,000$ and that is probably responsible for the formation of the cratered plateau material occurred throughout the cratered terrain. Some relics of a surface of this age underlie portions of the transition zone. The most widespread resurfacing within the transition zone also affected the adjacent cratered terrain and ceased at $N(1) \approx 20,000$ to 30,000, which is equal to or somewhat earlier than the time of formation of ridged plains in Lunae Planum and elsewhere. That this event occurred throughout the transition zone is shown by analysis of cumulative frequency curves for individual terrain units within the transition zone: Each of the units shows a branch of this age. Even younger resurfacing occurred in the transition zone and in the adjacent northern plains, but ages that range from $N(1) \approx 5000$–10,000 vary from unit to unit. There is in both the Lunae Planum age resurfacing event and the younger events (for common units) evidence that the resurfacing was more efficient further from the cratered terrain in the sense that larger craters were removed by the event. Furthermore, the Lunae Planum age resurfacing appears to have ceased in different areas at different times, with the age of cessation being progressively younger west of the Isidis Basin further north from the cratered terrain through the transition zone to the northern plains. The *oldest* resurfacing event at $N(1) \approx 86,000$ was much more efficient at removing larger craters in what is today the transition zone than it was in the cratered terrain, implying that some difference in the (topographic?) nature of the crust had been established prior to this resurfacing event.

INTRODUCTION

One of the outstanding problems in Martian geological evolution is the origin of the fundamental dichotomy between the cratered highlands and the northern lowland plains. Recognized early as the major physiographic characteristic of the planet (*Masursky,* 1973; *Carr et al.,* 1973), this dichotomy is as basic to Mars as is the continental/oceanic crustal difference to the Earth. Explanations offered to account for the crustal dichotomy range from extreme endogenic to extreme exogenic processes: subcrustal erosion and subsidence in the north with underplating in the south due to mantle convection (*Wise et al.,* 1979) to a megaimpact event (*Wilhelms and Squyres,* 1984). The diversity of these ideas, both of which have severe problems matching the observational data (*Frey et al.,* 1986a), illustrates our current state of ignorance about the most basic characteristic of the Martian crust.

Over much of the eastern hemisphere of Mars and in restricted portions of western Mars there is an easily recognizable boundary between the cratered highlands and northern lowlands. This border is not a sharp line but a zone whose width and character change with longitude between 180° and 360° W. Characteristics of this zone include a locally abrupt change in topography, an abrupt decrease in the density of impact craters, and in places an abundance of broken or detached plateaus (mesas) and knobs. The history inferred from earlier geologic mapping of this zone is a complex one (*Lucchitta,* 1978; *Greeley and Guest,* 1978; *Meyer and Groilier,* 1977; *Hiller,* 1979; *Scott et al.,* 1978). Major uncertainties still exist, such as the nature of the knobs that are found throughout the region (*Greeley and Guest,* 1978). It is generally agreed that the highland/lowland boundary (and therefore the crustal dichotomy it marks) is an ancient feature of the Martian crust (*Wise et al.,* 1979; *Neukum and Hiller,* 1981; *Wilhelms and Squyres,* 1984; *Maxwell et al.,* 1984; *Tanaka,* 1986). To date there has been little agreement on the exact mode of origin, the true extent of the boundary (was it originally global in extent? was it originally a small circle?), or the duration of its formation and that of the subsequent processes that have modified it.

In this paper we address the last of these issues. Crater counts for the transition zone as a whole, for individual units within the transition zone, and for the adjacent cratered terrain and smooth plains units are used to place limits on the number and age of major resurfacing events that have played a role in the development of this region. We present evidence for a widespread, common-age resurfacing that affected the entire region and that may have been contemporaneous with other major events elsewhere on Mars. The sequence of major resurfacing events that these data imply are used to constrain the history of the region and some implications of these results for the origin of the crustal dichotomy are discussed.

HIGHLAND/LOWLAND BOUNDARY AND TRANSITION ZONE

The distribution of detached plateaus, knobby terrain, impact craters, and intervening smooth plains that characterize the transition zone have been mapped in detail (*Semeniuk and Frey*, 1984, 1985, 1986; *Frey and Semeniuk*, 1985, 1986). The detached plateaus are generally considered to be mesas isolated from the adjacent uplands of cratered plateau material (*Lucchitta*, 1978; *Greeley and Guest*, 1978; *Scott et al.*, 1978). The origin of the knobs is much less certain. As *Greeley and Guest* (1978) have shown, the geologic mapping allow several possibilities: The knobs could be exhumed heavily cratered crust underlying the cratered plateau material, or they could represent resistant material within the cratered plateau, or they could be igneous material intruded through the cratered plateau material. Whatever their true nature, the knobs and plateaus serve to identify the transition zone and their varying distribution can be used to subdivide the transition zone into individual morphological terrain units (*Frey et al.*, 1986b). Figure 1 shows the units discussed in this paper on two scales: the large-scale cratered terrain (CT), transition zone as a whole (TZ) and adjacent smooth plains (SP) in Fig. 1a, and the individual transition zone units in Fig. 1b. Table 1 lists the characteristics of these units as mapped by us from the 1:2 million controlled photomosaics. Reference to previous geologic mapping is also provided. Our mapping is based strictly on morphology; principal considerations are the abundance and distribution of detached plateaus, knobby terrain, and intervening plains. Crater abundance was not used as a criterion as has sometimes been the case with other mapping.

Two of the transition zone morphological units, TZ1 and TZ8, are found both east and west of the Isidis Basin. These two units represent extremes in terrain unit morphologies found within the transition zone (see Table 1). TZ1 consists almost exclusively of large, closely-spaced detached plateaus with almost no occurrence of knobs. The unit lies immediately adjacent to the cratered terrain and is almost certainly a fractured cratered plateau unit (*Lucchitta*, 1978; *Greeley and Guest*, 1978; *Scott et al.*, 1978). The widespread TZ8 unit contains almost no detached plateaus. It is dominated by plains-forming units with clusters of small (<5 km across), sometimes densely grouped, knobs. The plains are continuous with similar units lying to the north of the region we designate as the transition zone; it is the fall-off in occurrence of knobs that permits drawing the TZ/SP boundary seen in Fig. 1.

The other individual terrain units (TZO, TZ2, TZ3, oTZ, and sTZ) are unique to either the western or eastern portion of the transition zone, and are in part responsible for the difference in character of the TZ from west to east. In general the transition zone is wider and lies at higher latitudes west of Isidis. For these reasons we treat the entire area as consisting of two regions: ISMCAS (Ismenius-Casius) west of Isidis and AMAE (Amenthes-Aeolis) east of Isidis. This distinction is maintained throughout the cumulative frequency curves shown below.

Impact craters larger than 4 km in diameter were counted in each of the individual terrain units within the transition zone and in the cratered terrain and smooth plains lying on either side (see Fig. 1). Cumulative frequency curves for $5 \leq D \leq 100$ km were then plotted for each unit (large-scale and individual). For reasons described below, the data are binned at 18 points per decade. The large scale areas (CT, TZ, SP) are discussed first in order to understand the regional context of the transition zone.

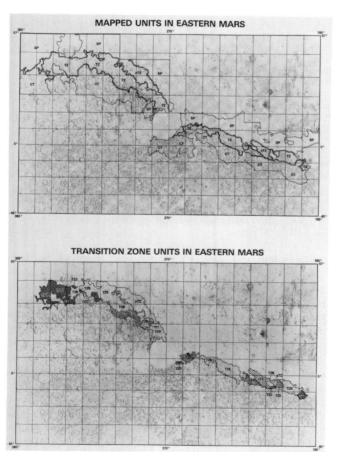

Fig. 1. Units for which cumulative frequency curves were derived and analyzed are shown superimposed on a shaded relief map of the eastern hemisphere of Mars. (a) Large-scale units are the cratered terrain (CT), transition zone (TZ), and smooth plains (SP). (b) Individual terrain units within the transition zone are designated TZ-. See text for details.

TABLE 1. Definitions of units shown in Fig. 1.

I. LARGE SCALE UNITS

Cratered Terrain (CT): Heavily cratered uplands lying immediately south of the transition zone. Mostly *Nplc* and *Nhc* of *Scott and Carr* (1978), but includes *HNbr* of Isidis Basin.

Transition Zone (TZ): Complex mixture of detached plateaus with both rough and smooth upper surfaces, knobs of large and small size occurring singly or in clusters, and intervening smooth plains-forming units. West of Isidis (ISMCAS) consists of *ANch*, *HNk*, and some *Nplc* (*Scott and Carr*, 1978); east of Isidis (AMAE) is mostly *HNk* with some *Apc*, *Nplc*, *Hpr*, and *Aps* (*Scott and Carr*, 1978). As mapped here includes units TZ0, TZ1, TZ2, TZ3, and TZ8, described below.

Smooth Plains (SP): Sparsely cratered plains of generally smooth or mottled appearance. West of Isidis (ISMCAS) consists mostly of *Apc* with some *Npm* units, but east of Isidis (AMAE) includes *Aps* and *Hpr* mixed with mostly *Apc* (*Scott and Carr*, 1978).

II. INDIVIDUAL UNITS WITHIN THE TRANSITION ZONE

A. West of the Isidis Basin (ISMCAS)*

TZ1: Consists of large, closely spaced plateaus of irregular outline and rough upper surfaces. Lies immediately adjacent to cratered terrain. Probably fractured plateau material. Previously mapped as units *h, k* (L); *h, k, p* (GG); *pf, k* (MG).

TZ2: Called "fretted terrain" by some. Consists of large, discrete, generally widely spaced plateaus with smooth upper surfaces. Separated by smooth plains-forming units. Previously mapped as *pl, pl+p, k* (L).

TZ8: Characterized by clusters of small knobs (less than 5 km across) often closely spaced. Clusters frequently well separated by intervening smooth plains. Occasional large knobs or highly degraded small plateaus may occur. Most extensive individual unit. Previously mapped as mixed *k, h, p,* and *pl* (L); *h, k, p, pv* (GG); *k, p* (MG).

oTZ: An "outer transition zone" consisting mostly of mottled and smooth plains with scattered small knobs. Knobs sometimes occur in small groups but overall density much less than in adjacent TZ8. Previously mapped as mostly *pv*, some *k, p, h* (GG).

B. East of the Isidis Basin (AMAE)†

TZ0: Eastern rim of Isidis Basin. Consists of large, rough blocks dissected by long fractures. Previously mapped as mostly *brd* (H).

TZ1: Same as in ISMCAS. Previously mapped as *pl* (SMW).

TZ3: Similar to TZ1 but plateaus are small, closely spaced with occasional interplateau plains. Upper surfaces of plateaus are rough. Knobs occur occasionally between plateaus. Previously mapped as *k, pr, pd* + some *ps* (SMW).

TZ8: Same as in ISMCAS. Previously mapped as *k* (H); *pr* with some *k* (SMW).

sTZ: Consists of rough, stripped unit within which occasional degraded plateaus of smooth appearance occur. Grooved and fluted terrain with no obvious counterpart elsewhere in the transition zone. Previously mapped as *pr, k* (SMW); *pr* (SA).

Key to previous mapping: (L) = *Lucchitta* (1978) (MC-5); (MG) = *Meyer and Grolier* (1977) (MC-13); (SA) = *Scott and Allingham* (1976) (MC-15); (GG) = *Greeley and Guest* (1978) (MC-6); (H) = *Hiller* (1979) (MC-14); (SMW) = *Scott et al.* (1978) (MC-23).

*ISMCAS average TZ = TZ1, TZ2, and TZ8 only.
†AMAE average TZ = TZ0, TZ1, TZ3, and TZ8 only.

CUMULATIVE FREQUENCY CURVES AND RESURFACING

Figure 2 shows cumulative frequency curves for cratered terrain, the transition zone, and adjacent smooth plains for the area west (ISMCAS) and east (AMAE) of the Isidis Basin, as marked in Fig. 1. The three regions are well separated in terms of their cumulative frequency curves, implying they can be treated as three separate temporal units with separate ages and cratering histories. The cratered terrain curves are nearly identical for ISMCAS and AMAE, as shown in Fig. 2c. This is not surprising since, in terms of the large scale units described here, CT is the most uniform between ISMCAS and AMAE (see Table 1). That is, the averaging that has been done to treat this part of Mars as a single unit has not included a wide variety of terrains with possibly different cratering histories. Likewise, the SP units are fairly similar averaged over the large areas shown in Fig. 1, although the AMAE unit is slightly younger.

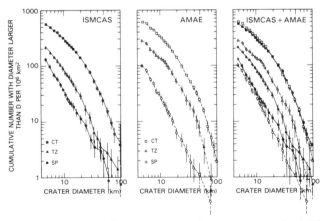

Fig. 2. Cumulative frequency curves for cratered terrain (CT), transition zone (TZ), and smooth plains (SP) are shown for the diameter range 5 to 100 km. Units west (ISMCAS) and east (AMAE) of the Isidis Basin are shown separately and then combined in the last panel. Note the generally clear separation of the TZ data from that for CT and SP in each region.

TABLE 2. Cumulative number of craters with D ≥ 5 and with D ≥ 16 per 10^6 km² and comparison with Martian stratigraphic series.

Region	Unit	Area (km²)	# Craters	N(5)	SS	N(16)	SS	N(50)	Notes
ISMCAS	CT	1,783,297	1037	582	mN	192	mN	24	
	TZ	1,415,218	311	219	uN	43	uN	3	
	tz1	207,891	53	253	uN	55	uN	-	
	tz2	370,855	52	140	lH	38	uN	-	lH/uN
	tz8	836,472	206	246	uN	42	uN	2	
	oTZ	348,843	64	184	lH	48	uN	9	lH/uN
	SP	2,312,313	299	129	lH	17	lH	2	
AMAE	CT	2,583,499	1587	614	mN	198	mN	25	
	TZ	1,131,769	322	284	uN	83	uN	8	
	tz0	105,623	36	341	uN	104	mN	10	uN/mN
	tz1	143,128	35	245	uN	56	uN	-	
	tz3	82,191	27	329	uN	79	uN	-	
	tz8	739,261	207	280	uN	86	uN	10	
	sTZ	263,749	31	118	uH	31	uN	-	uH/uN
	SP	2,698,042	274	102	uH	13	lH	1	uH/lH

SS = Stratigraphic series from *Tanaka* (1986) as follows: uH = upper Hesperian; lH = lower Hesperian; uN = upper Noachian; mN = middle Noachian.

The transition zone shows more variation from ISMCAS to AMAE, although in both areas it is clearly of an age intermediate between CT and SP. The average ISMCAS TZ (which consists of TZ1, TZ2, and TZ8 units) is younger than the AMAE TZ (made up of TZO, TZ1, TZ3, and TZ8 units), but the curves cross at diameters larger than 60 km. This implies a greater loss of smaller craters in ISMCAS, as described below. The somewhat greater complexity of the averaged TZ curves is consistent with the greater complexity of this region and the variation that occurs from west to east, as described above. It is because of this variation and complexity that the TZ was subdivided into individual terrain units (Table 1) within which crater counts were also done, as described below (see Fig. 7 and following figures).

Table 2 compares cumulative number of craters at D ≥ 5, 16, and 50 km (per 10^6 km²) for the three large-scale units and for the individual terrain units within the TZ. Unit symbols are the same as in Fig. 1. The total counting area of each unit and the total number of craters counted are shown. Note the large counting areas for the CT, TZ, and SP but the relatively small counting areas for the individual terrain units within the TZ (except for TZ8). There is a corresponding small number of craters for some of these units that provides some of the uncertainty associated with the interpretation of later cumulative frequency curves. The N(50) column shows that some large craters are preserved not only in the CT but also in some TZ units. Comparison with *Tanaka's* (1986) stratigraphy is also shown for the N(5) and N(16) columns. The cratered terrain is Middle Noachian, as *Tanaka* (1986) shows. The average TZ is Upper Noachian for both the ISMCAS

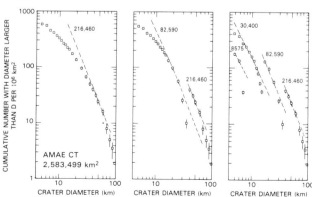

Fig. 3. Progressive splitting of the cumulative frequency curves for cratered terrain (CT) into separate branches according to the *Neukum and Hiller* (1981) technique. ISMCAS (a) and AMAE (b) data shown separately. First panel shows all the binned data from Fig. 2, with a production curve of the given N(1) crater retention age fit to the largest diameter points. The middle panel shows the effect of subtracting the last point lying on the production curve from the remaining smaller diameter points and replotting those points. A new, lower age production curve is fit to these points. Because the cumulative frequency curve bends away again from the production curve, the subtraction process is repeated. This produces a third branch that can be fit with a lower age production curve as shown. Note the similarity of ages derived for the three main branches for both the ISMCAS and AMAE data. Counting area shown in lower left.

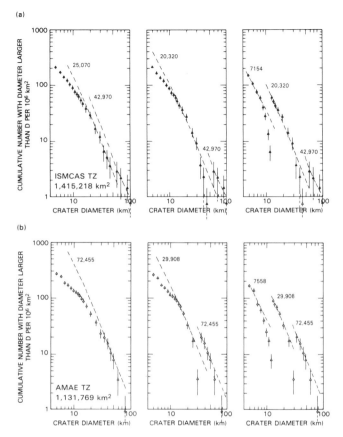

Fig. 4. Same as for Fig. 3, but for transition zone (TZ). See text for details.

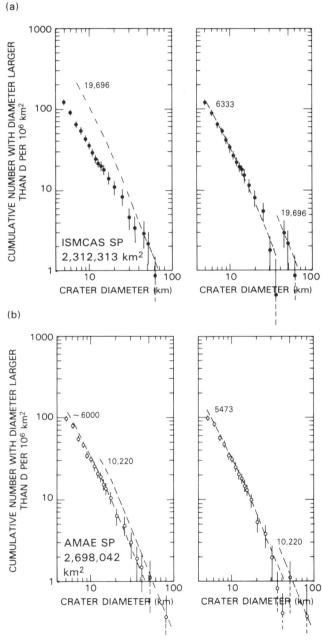

Fig. 5. Same as for Fig. 3, but for smooth plains (SP). See text for details.

in a cumulative frequency curve lying along a new standard Mars production curve of younger age. This new curve records the time (in cumulative crater number) since the resurfacing stopped.

In the *Neukum and Hiller* (1981; hereafter N&H) approach the cumulative frequency curve is split into two or more branches as follows: At the point where the curve departs from the standard Mars production curve fitted through the large diameter craters, the cumulative number for the last point remaining on the production curve is subtracted from the

and AMAE portions. Note, however, that this averaging smears out some potentially important information revealed by the individual TZ units. ISMCAS TZ2 and oTZ and AMAE sTZ all show a younger stratigraphic age (based on the cumulative crater density) at N(5) than at N(16). All three of these units appear to be Upper Noachian based on N(16). The two ISMCAS units (TZ2 and oTZ) would be classified as Lower Hesperian based on N(5) and the AMAE sTZ unit would be called *Upper* Hesperian based on N(5). The large average SP unit in AMAE also shares this complexity, appearing to be Lower Hesperian at N(16) but Upper Hesperian at N(5).

This difference in stratigraphic age at different crater diameters is due to the relative depopulation of craters at smaller diameters. For the CT and average TZ units in Fig. 2 and for the individual TZ units in Fig. 7, the cumulative frequency curves bend over toward smaller diameters, away from a simple power-law fit through the large diameters. *Neukum and Hiller* (1981) suggested treating this bending over in terms of one or more "resurfacing events." Larger craters lying along a "standard Mars production curve" represent an underlying surface that had up to some point accumulated craters consistent with that production curve. A resurfacing event is imagined to have removed craters smaller than some diameter. When this event ceased, the new surface again accumulates craters that after some period of time will result

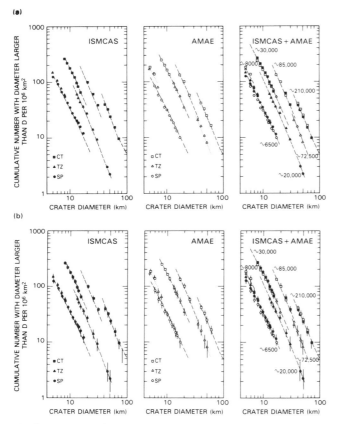

Fig. 6. Summary of the curve splitting for cratered terrain, transition zone, and smooth plains from Figs. 3-5. The branches derived in the previous figures are shown along with the points that define them, separately for ISMCAS and AMAE and then combined. **(a)** Points shown without error bars. **(b)** Points shown with error bars. Note that the derived branches in general are well separated from one another within their error bars even though in many cases the branches are defined by different units over different diameter ranges.

remaining smaller diameter points. These points are then replotted, producing two branches that both should lie along standard production curves. The newer branch shows the cumulative curve for the case of no survivors from before the resurfacing event; it records the "age" of the upper surface since resurfacing, whereas the older branch provides the age of the underlying, preresurfacing surface. Crater ages are determined by extrapolation of the production curves along which the two branches lie to a diameter of 1 km and are the cumulative number with diameter greater than or equal to 1 km per million square kilometers [N(1)].

The approach used by N&H is based on a specific and too-simplistic model of crater depopulation. It treats resurfacing as being composed of discrete events. While such events have undoubtably occurred on Mars (such as the emplacement of volcanic flows), many resurfacing processes may not be abrupt events efficient at removing all craters smaller than some limiting diameter. For example, degradation of landforms through mass-wasting may occur over long periods of time, be locally variable in its efficiency, and compete with crater production in such a way that a clean splitting of a cumulative frequency curve will not be possible. But extensive resurfacing by plains emplacement has been an important part of the modification history of the transition zone on Mars as well as for the adjacent cratered terrain and smooth plains (*Lucchitta*, 1978; *Greeley and Guest*, 1978; *Scott et al.*, 1978; *Hiller*, 1979; *Neukum and Hiller*, 1981; *Greeley and Spudis*, 1981; *Tanaka*, 1986). In the Late Noachian this produced the cratered plateau material and in the Early Hesperian the ridged plains appeared (*Tanaka*, 1986); these are both components of the transition zone and the adjacent cratered terrain and smooth plains (see Table 1 and below). Treating the cumulative frequency curves as the product of one or more resurfacing events may be a reasonable approach for this part of Mars and may provide insight into the sequence and number of major events that produced the complicated terrain of the transition zone.

The N&H method can be similar to that of counting separately superimposed and protruding craters in that it provides information on the upper surface and on the cumulative or aggregate lower surface. For a single resurfacing event (which would produce two branches after splitting the cumulative frequency curve) both methods should produce similar results, provided the standard production curve is a true representation of the crater production over the entire time craters were accumulated. In principle, however, the *Neukum and Hiller* (1981) technique could be more powerful for the following reason. Subtraction at the bendaway point and replotting may produce a second branch that itself bends away from the younger standard production curve at some still smaller diameters (see Figs. 3-5, below). The second bendaway can be treated as before: Subtraction from the remaining smaller diameter points and replotting of those points will produce a third branch that will also lie along a standard production curve of still younger age, if the split is a clean one (i.e., if the resurfacing event was efficient at removing all the craters smaller than some limiting diameter and then abruptly ceased). In such cases three different surface ages will be revealed, representing at least two resurfacing events. Many of the cumulative frequency curves shown below have this multiple branch character after splitting the curves by the N&H technique.

Besides being very model-dependent (in the assumed nature of the resurfacing), the *Neukum and Hiller* (1981) method has other limitations. In order to resolve multiple resurfacing events, the diameter range of craters counted (and the number of craters counted at medium and large diameters) must be large. This generally requires large counting areas. The identification of bendaways from a standard production curve will be extremely sensitive to the choice of production curve, about which there is little agreement for Mars. At large diameters the cumulative frequency curves can be approximated by a power law, but *Woronow* (1977) suggested that a log-normal curve was a more appropriate choice for the early Mars production curve. The Neukum and Hiller choice, based on that derived in *Neukum and Wise* (1976), is flatter than a power law with more structure. *Maxwell and McGill* (1987), in a more limited study of portions of these same

TABLE 3. Derived crater ages for end of resurfacing events in large scale units.

Region	Unit	$N(1)_{av}(n)$ h/l (diam. range)	$N(1)_{av}(n)$ h/l (diam. range)	$N(1)_{av}(n)$ h/l (diam. range)	$N(1)_{av}(n)$ h/l (diam. range)
ISMCAS	CT	$203,820(5)^{212,005}_{193,456}$(40-100)	$88,155(3)^{90,023}_{85,146}$(20-35)	$31,663(8)^{33,315}_{28,549}$(5-17)	
	TZ		$[42,970(4)^{53,955}_{34,442}$(60-100)]	$20,320(7)^{22,677}_{18,962}$(13-45)	$7,154(4)^{7,348}_{6,999}$(5-12)
	SP			$19,696(2)^{20,027}_{19,365}$(45-60)	$6,333(12)^{6,878}_{5,854}$(5-35)
AMAE	CT	$216,460(4)^{228,966}_{202,272}$(40-100)	$82,590(3)^{85,146}_{78,850}$(17-35)	$30,404(4)^{30,882}_{29,961}$(8-13)	$[8,575(2)^{8,972}_{8,178}$(5-7)]
	TZ		$72,455(4)^{77,822}_{70,419}$(35-80)	$29,908(5)^{31,287}_{28,663}$(13-30)	$7,479(3)^{7,933}_{7,237}$(5-12)
	SP			$[10,220(2)^{10,756}_{9,683}$(50-80)]	$5,473(11)^{5,701}_{5,062}$(5-40)

Crater age [N(1)] = cumulative number with D ≥ 1 km per 10^6 km^2.
$N(1)_{av}$ = average crater age from points falling along N&H production curve.
n = number of points from which $N(1)_{av}$ was calculated.
h/l = high and low value included in calculation of $N(1)_{av}$.
diam. range = range of crater diameters for which $N(1)_{av}$ applies.
[ ] = poorly determined or uncertain result.

areas, used a higher order polynomial with even more structure recommended by *Neukum* (1983) in their splitting of cumulative frequency curves. For a planet like Mars where erosional modification and resurfacing have clearly been an important part of its history, some caution is warranted in the use of apparent production curves with significant structure: These may be incorporating planet-wide depopulation processes in the curves themselves. Production curves with more structure will result in fewer branches when splitting the cumulative frequency curves because fewer bendaways will be apparent. By contrast, cumulative frequency curves such as those in Figs. 2 and 7 will appear to be "splittable" into several branches if the chosen production curve is more like a straight line. It is not only the number of derived branches that depends on the choice of standard curve, but also the diameter at which the bendaways appear to occur. This in turn affects the derived age of the upper surface, and therefore the derived age of cessation of the resurfacing event.

There is also the additional possible complication that the Mars production curve, whatever its true nature, may have changed over time due either to separate populations of impactors or to evolution of the impacting population (*Strom et al.*, 1981). *Tanaka* (1986) suggests that the Martian crater production curve may have changed from log-normal in Middle Noachian and earlier times to a power law distribution in later times, with the transition occurring in Late Noachian time. If true, this would be particularly unfortunate in that it is precisely this period of Martian history we are examining in this paper. *Neukum and Hiller* (1981) consider their standard Mars production curve to be time-invariant over the diameter range $0.1 \leq D \leq 20$ km (within a factor of 2), and *Neukum and Wise* (1976) showed that cumulative frequency curves closely followed the production curves to even larger diameters.

As shown in Fig. 3a, some heavily cratered areas have cumulative frequency curves that lie along the N&H production curve out to 100 km diameter. The general lack of agreement among previous workers as to the nature of the Martian crater production curve over time is a serious problem in that the results of curve splitting are so dependent on the choice of standard curve. Because we adopt their technique for analyzing the cumulative frequency curves for the TZ and its surroundings, we also choose to use the N&H production curve in order that our results may be compared more closely with theirs.

One final problem in the application of the N&H method is the tendency of the replotted points to initially lie low with respect to a production curve. This is a direct consequence of the subtraction process. At a bendaway the cumulative number is changing more slowly than it would if the curve continued to follow the production curve; differences between successive cumulative numbers are small and subtraction of the last point lying on the old branch from those falling off the production curve will initially produce very small values for the replotted points. If the break is clean in the sense of a resurfacing event that ceases abruptly, the replotted points will smoothly recover from the subtraction process and at smaller diameters lie again along a production curve. Failure of the points to closely follow a production curve after replotting may be an indication that the break was not a clean one; i.e., the resurfacing event did not abruptly cease but continued to compete for some time with crater production, or the resurfacing event did not cleanly remove all craters down to a limiting diameter (i.e., some "survivors" are contained in the replotted points). Examples of such branches where the split from the previous branch may not be clean are shown below (see Figs. 9a,c).

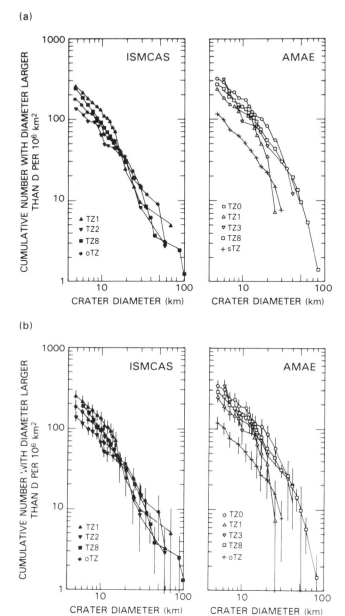

Fig. 7. Cumulative frequency curves for individual morphological units within the transition zone, as defined in Table 1. Data for ISMCAS and AMAE are shown separately, even though two of the units (TZ1 and TZ8) are common to both regions. (a) Data without error bars. (b) Data with error bars. Note that despite their different character and the fact that curves cross, all show a bend over toward smaller diameters.

CUMULATIVE FREQUENCY CURVE BRANCHES FOR LARGE-SCALE UNITS

Figures 3-5 show the results of applying the *Neukum and Hiller* (1981) technique to the cumulative frequency curves for the cratered terrain, the average transition zone, and the adjacent smooth plains as shown in Fig. 2. The figures are all in the same format, consisting of three panels that show the progressive splitting of the cumulative curve into branches. The cumulative number of craters larger than a given diameter per million square kilometers were binned, as recommended by N&H, into 18 bins per decade. For our counts with diameter range 5 to 100 km, this means plotting points at the following diameters: 5, 6, 7, 8, 9, 10, 11, 12, 13, 14, 15, 17, 20, 25, 30, 35, 40, 45, 50, 60, 70, 80, 90, 100 km. Only existing points were plotted. Gaps between bins (no change in cumulative number per million square kilometers) were retained as these often are due to removal of craters by resurfacing.

Points are plotted with error bars that represent +/- the square root of the cumulative *number* of craters (per unit area). When curves are split, the replotted points are shown with new error bars corresponding to the new (lower) number of craters after subtraction.

All the binned data are plotted in the first panel, along with a production curve from N&H fit to the largest diameter values. The break point where the curves bend away from the standard production curve is generally easy to recognize. The middle panel shows the results of subtracting the value of the last point lying on the production curve (the last point representing the older underlying surface) from the remaining (smaller diameter) points and replotting those points. The original branch is also replotted for comparison. A new production curve is fitted through the replotted points of the second branch. If these points also bend away from the second production curve at still smaller diameters (which is generally the case for Figs. 3 and 4), then the process is repeated and the result shown in the third panel. The crater ages shown by each production curve are the extrapolation of those curves to 1 km diameter, but are averages computed (not read from graphs) using the analytic expression provided by *Neukum and Hiller* (1981) from the points lying on the curve. These ages are N(1), the cumulative number of craters with D ≥ 1 km diameter per 10^6 km², which each of the surfaces would have had if no resurfacing had occurred, and represent crater retention age or time since the resurfacing stopped.

The Cratered Terrain

As shown in Fig. 1, the CT unit (Fig. 3) consists mostly of *Scott and Carr's* (1978) Nplc and Nhc units (Table 1). The portion west of the Isidis Basin has an average cumulative frequency curve that closely follows a N&H production curve with N(1) = 203,800 over the diameter range 40-100 km. At smaller diameters the points clearly bend away from this production curve. Replotting after subtraction yields a second branch which very closely follows a production curve with age N(1) = 88,200 from 35 km down to 20 km diameter, at which point the cumulative frequency curve bends away again. Repeating the subtraction process produces a third branch that lies along a production curve with age N(1) = 31,700 between 8 and 17 km (the point at 17 km being low due to subtraction, as was the point at 35 km for the middle branch). At diameters less than 8 km the curve seems to depart from the production line again, but replotting these few points produced no convincing fourth branch.

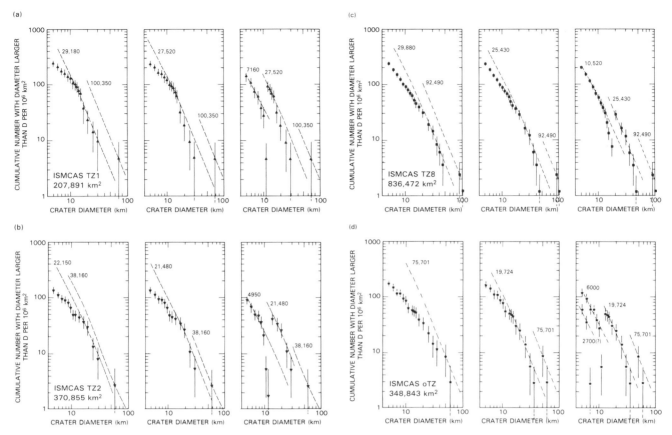

Fig. 8. Progressive splitting of the cumulative frequency curves for ISMCAS transition zone terrain units. Format the same as in Figs. 3-5; see Fig. 3 for details. (**a**) TZ1 unit. (**b**) TZ2 unit. (**c**) TZ8 unit. (**d**) oTZ unit.

Because the CT curves for ISMCAS and AMAE in Fig. 2 are so similar, we would expect the N&H method to produce similar branches for both data sets. Figure 3b shows the results for the AMAE portion. Three branches seem well derived by the curve splitting process. They have ages $N(1) = 216,500$, 82,600, and 30,400 which are identical to within the limitations of the method to the corresponding branches shown for ISMCAS in Fig. 3a. A possible fourth branch with $N(1) = 8600$ may be barely resolvable in the AMAE data, which also bends away from the third branch at diameters less than 8 km as did the ISMCAS CT data.

The three branches derived by the N&H technique would represent three surfaces of different ages separated (or caused by) two separate resurfacing events. It appears these two resurfacing events each occurred at about the same time east and west of the Isidis Basin. The older event ceased at crater number $N(1) = 82,000-88,000$ and the younger event stopped at $N(1) = 30,000-32,000$.

The Average Transition Zone

The counting area for the average TZ (Fig. 4) is less than that of the CT, but some large craters are included (see Table 2). The ISMCAS data show a gap in the cumulative frequency curve (Fig. 4a) at 50 km. Craters larger than this lie along a production curve with $N(1) \approx 43,000$, but the distribution of these four points (with a gap also occurring at 80 km diameter) is unusual in that the points lie first low (probably due to undersampling), then high, then low again with decreasing diameter. Because average TZ is an aggregate of several distinct units (Table 2), these points have to be treated with caution. They may well represent contributions from only one or two units within the western TZ (e.g., TZ8, see Figs. 7 and 8 below). The points at D < 50 km seem to lie on a production curve with age $N(1) = 25,000$. While it is tempting to consider the curve as shown in Fig. 4a (since the points cluster so well about the 25,000 production curve), the larger diameter points lying roughly along the 43,000 curve are probably survivors from an underlying surface (see Fig. 8). Even though it will produce a poorer fit for the first few replotted points, we follow the N&H approach and subtract at D = 45 km and smaller diameters. The resulting cumulative frequency curve now lies along a production curve with $N(1) = 20,300$, somewhat younger than the case before subtraction, and the first three points all lie low. The second branch also departs from the standard curve at D = 13 km, implying a third branch can be derived. This has an age of $N(1) = 7100-7200$ as shown in the third panel, and is reasonably well determined despite the first three points in the replotted data.

The AMAE portion of the average TZ shows better definition of an old branch over the diameter range 35-80 km where

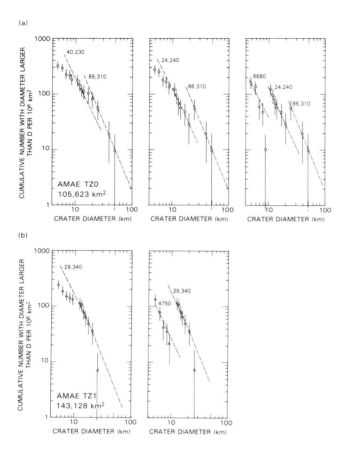

Fig. 9. Same as Fig. 8 but for the AMAE transition zone terrain units. **(a)** TZ0 unit. **(b)** TZ1 unit. **(c)** TZ3 unit. **(d)** TZ8 unit. **(e)** sTZ unit.

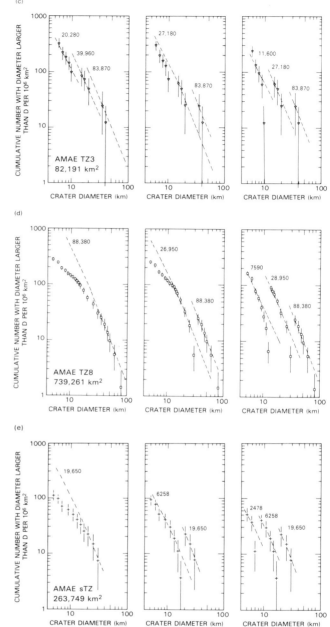

the points cluster tightly (for D = 35-50) about a production curve with N(1) = 72,500. Subtracting at the breakaway at 30 km yields a second branch with a reasonably well-determined N(1) = 29,900. A third branch with N(1) = 7500 can also be derived from the data, although this is not so well defined as the young 7150 branch in the ISMCAS data.

The Smooth Plains

The SP unit (Fig. 5) used here for comparison with the TZ is a mix of Npm and Apc in *Scott and Carr's* (1978) designation for the ISMCAS portion, and mixed Apc, Aps, and some Hpr east of the Isidis Basin. We are therefore averaging over stratigraphic boundaries. The ISMCAS data show a few large diameter points which lie approximately along a curve with N(1) = 19,700. The plotted points bend away at 35 km diameter, a fact made more obvious by the gap at 40 km. The replotted points after subtraction approximately fit a production curve with N(1) = 6000-6500, but some lie above this curve, perhaps due to the mixture of units that are lumped in SP. There does seem to be evidence for two branches, however, although the older one is defined by only a few (large

diameter) values. Note that the 19,700 branch is very similar (identical within the sensitivity of the technique) to the 20,300 branch found in ISMCAS for the average TZ.

Two widely separated points lie along a production curve with N(1) = 10,200 in Fig. 5b; the remaining AMAE SP data cluster along a N&H production curve with N(1) = 6000. The older branch is not well determined, but we subtract at 40 km diameter and replot to produce a somewhat younger branch with N(1) = 5500, for which the overall fit of the points is slightly better than in the original (prior to subtraction) plotting.

Summary for CT, TZ, and SP

The above results are summarized in Fig. 6 in which the final derived branches for each unit are replotted using different symbols to represent the CT, TZ, and SP. For ISMCAS it is the cratered terrain that defines the oldest three branches [with $N(1) = 210{,}000$, $85{,}000$, and $30{,}000$]. The $N(1) = 20{,}000$ branch is defined by the TZ data and a few points from SP; both TZ and SP replotted points lie along a production curve with a mean value of about 6500. In AMAE there is the same 210,000 branch defined by the CT data, two branches at 72,500-82,600 shown by replotted CT and oldest TZ values, a branch at about 30,000 again defined by replotted CT and TZ, and a branch with $N(1) = 5500$. The 8000 branch is defined (if at all) by only a few points from CT. The third panel shows the ISMCAS and AMAE results plotted together; note the clustering of the points especially for the $N(1) \sim 6500$ and $N(1) = 20{,}000\text{-}30{,}000$ branches. Note also that all the major branches about which these points cluster are well separated within their error bars (Fig. 6b). These data are summarized in Table 3, which shows for each area and unit the derived ages $N(1)$, the number of points that went into the average value of $N(1)$, the high and low values in that average, and the diameter range over which the branch is thought to apply. Poorly determined or marginal results based on only a few points or on points that show large scatter about a production curve are shown in brackets.

The overall resurfacing history implied by these results can be summarized as follows:

1. The oldest surface resolvable in the areas studied lies in the cratered terrain unit and has $N(1) \sim 210{,}000$ for both ISMCAS and AMAE. There is no evidence within the adjacent transition zone for any surviving buried surface of this age.

2. A resurfacing event within the cratered terrain removed craters smaller than about 40 km diameter in both ISMLAS and AMAE. This "event" ceased about $N(1) \sim 85{,}000$ in both portions of the CT. There may be a relic of this surface preserved in the average TZ in AMAE in the 72,500 branch, which is defined by craters larger than 35 km.

3. Resurfacing in the transition zone *and* in the adjacent cratered terrain followed, removing craters smaller than 17-20 km in CT but craters smaller than 35-45 km within the TZ. This resurfacing event produced a counting surface at about age $N(1) \sim 30{,}000$ in both the CT and TZ east of Isidis Basin and in the CT west of the Isidis Basin, but many have ended somewhat later in the TZ in ISMCAS at about 20,000. A relic of this surface may also survive in portions of the SP in ISMCAS, which show an age of about 20,000 based on a few craters in the 45-60 km diameter range.

4. One additional resurfacing event is recorded in some of the units. The average TZ in both ISMCAS and AMAE have a branch at $N(1) \sim 7100\text{-}7500$ and the SP units show ages of 5500-6500 for a large diameter range of 5-35 km. In ISMCAS this does indeed seem to be a resurfacing but in AMAE the evidence for an earlier, underlying surface is weak.

Later in this paper we discuss the overall implications of these results and their connection with events derived form other studies. Here it is important to note the 20,000-30,000 age surface that appears in the data from all three units in ISMCAS and in the CT and average TZ in AMAE. The close correspondence of the derived crater ages between units for this branch implies that the resurfacing event it represents was widespread and important in the overall development of the transition zone. But the transition zone data shown in Figs. 2, 4, and 6 are averaged over a complex collection of recognizably separate morphological terrain units. It is important to determine whether the 20,000-30,000 age "event" is an average of many events of different ages in different TZ units or whether it is in fact present throughout the transition zone (i.e., present in all different terrain units).

Below the cumulative frequency curves for individual terrain units within the transition zone in ISMCAS and AMAE are shown. These curves are analyzed as above in order to compare the resurfacing histories of the individual units with that of the large-scale areas.

CUMULATIVE FREQUENCY CURVE BRANCHES FOR TZ UNITS

The individual units within the transition zone are described above (see Table 1) and are also discussed in some detail below. As shown in Table 2 the counting areas for these individual terrain units are significantly smaller than the areas over which craters were counted for the large CT, average TZ, and SP units. This means the error bars will be significantly larger, and as shown in Figs. 8 and 9, frequently a single large (surviving?) crater will occur in the cumulative frequency curves that is well removed or separated from the other points. While it is not possible to infer a branch from these individual points, it is important to note their existence, as they may be clues to the underlying surfaces that have been nearly completely resurfaced over most observable crater diameters.

Figure 7 shows the cumulative frequency curves for all the transition zone units in ISMCAS and in AMAE. For clarity the curves are shown without (Fig. 7a) and with (Fig. 7b) error bars. In ISMCAS the average TZ discussed above consisted of TZ1, TZ2, and TZ8; the oTZ unit appears to be transitional between TZ8 and SP and is therefore treated independently (see Table 1 and discussion above). Likewise, the sTZ unit is so different from the rest of the AMAE TZ units that it was not included in the previous discussion of average TZ.

Although the error bars are large due to the smaller counting area and therefore smaller number of available craters, it is clear that all the curves show a great deal of structure, and all bend over at some point with decreasing diameter. Note the very low curve of the sTZ unit, implying a significantly younger age than for most of the other units. As shown in Table 2, most of the TZ units are of Upper Noachian to Lower Hesperian age, but some show discrepancies in the stratigraphic series implied by the cumulative number of craters at $N(5)$ and $N(16)$. Note also that especially in the ISMCAS data the curves converge and cross one another at about 17 km diameter. This implies that the depopulation (resurfacing?) throughout the transition zone was variable between the units, which probably accounts for their variable morphological character.

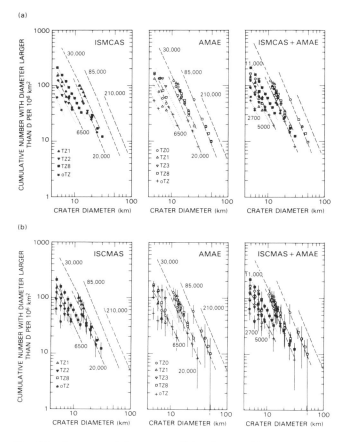

Fig. 10. Summary of the curve splitting for transition zone terrain units from Figs. 8 and 9. Branches and ages indicated by dashed lines are from Fig. 6, derived from the large-scale units. Dotted lines indicate branches and ages not seen in the large-scale areas. Format the same as in Fig. 6. Data shown separately for ISMCAS and AMAE and then combined. (a) Points shown without error bars. (b) Points shown with error bars. See text for details.

Each of the terrain units is treated as were the major units: The cumulative frequency curves are analyzed as in *Neukum and Hiller* (1981) in Figs. 8 and 9 below for ISMCAS and AMAE units, respectively. The branches resulting from splitting the curves are summarized in Fig. 10 and Table 4.

ISMCAS TZ Terrain Units

In Figs. 8a and 8b a single large crater of diameter 70 and 60 km (respectively) lies well removed from the remaining data points that occur at D ≤ 30 km. The smaller diameter values lie along N&H production curves with N(1) = 29,200 and N(1) = 22,200, respectively. The large craters are likely isolated survivors from an underlying surface that was nearly completely resurfaced; these probably contribute to the questionable branch with age 43,000 found in the average TZ curve in Fig. 4a. Although it will produce a poor fit for the first few points, we subtract the survivor and replot the smaller diameter points, which then lie along somewhat younger production curves of N(1) = 27,500 and N(1) = 21,500 for the TZ1 and TZ2 units, respectively. The TZ1 branch is better defined, in part because it extends over a larger diameter interval (12-30 km) than does the TZ2 branch (17-30 km) before the points again bend away from the production curve. Note how prominant is the bendaway for TZ2, implying a significant resurfacing that extended over an appreciable period of time. This may well be related to the different morphological characteristics of the two units (see Table 1). The youngest branches determined by a second round of subtraction (shown in the third panel) have N(1) = 7200 and N(1) = 5000 for TZ1 and TZ2 and are reasonably well determined within the error bars.

The TZ8 unit (Fig. 8c) has a larger counting area than the other units within the ISMCAS TZ. There is a large gap between 45 and 90 km diameter; the two large diameter points are probably due to two surviving craters, but the age implied by these is unreliable. Subtraction of these survivors reduces the age of the production curve about which the replotted points lie to 25,400 from the 29,900 value about which the points with D ≤ 45 km appeared to lie prior to subtraction. At D < 20 km the points bend away from the production curve; a third branch with N(1) = 10,500 seems to be well determined by the large number of points over the diameter range 5-13 km with the points from 14-17 km falling low due to subtraction.

The outer transition zone (oTZ; see Fig. 8d) that consists mostly of rolling plains (*Scott and Carr*, 1978), variegated plains (*Greeley and Guest*, 1978), or ridged plains (*Tanaka*, 1986) has a small gap between possible survivors at D ≥ 50 km and the remaining points that lie at D ≤ 35 km. When the survivors are subtracted, the remaining points lie along a production curve with age 19,700 down to 14-15 km diameter. Craters smaller than 12 km replot after subtraction along a production curve with age 6000. Although within their error bars all the points can be associated with this N(1) = 6000 curve, the bendover at 7-8 km can also be interpreted as providing another (poorly determined) branch with age about 2700, although this is based only on two points.

In summary, the individual units within the TZ in ISMCAS all show a branch in their split cumulative frequency curves that implies a resurfacing event terminated at crater number N(1) ~ 20,000-30,000. The individual units also all appear to have one or two large craters surviving from an older surface. In all the units the 20,000-30,000 age surface was again resurfaced, but the time at which this ceased varied from unit to unit.

AMAE TZ Terrain Units

In Fig. 9a the cumulative frequency curve for the TZ0 unit (the Isidis Basin dissected basin material unit brd of *Hiller* 1979) shows two distinct branches in that points lie along two distinct production curves even before subtraction. Despite the large error bars associated with the small number of large craters, and the gaps that occur between 25 and 40 km, the 86,300 production curve seems to be a good representation of the large diameter branch. The bendaway at 20 km diameter leads to the second branch, which *after subtraction* lies with some scatter around a production curve of $N(1) = 24,200$ down to a diameter of 11 km. There is a gap at 10 km diameter and the points at $D \leq 9$ km do bend away from the 24,200 curve. However, when replotted after subtraction the smaller diameter points do not cluster very well about a third production curve. Within their error bars the $D = 5-8$ km points could lie on a $N(1) = 6700$ production curve, but the fit is far worse than previous experience would lead us to expect. This may be a case where the break is not showing a clean resurfacing event and the resurfacing event may not have ceased abruptly. If continued removal of craters fitfully competed with crater production it is unlikely that the replotted points would cluster neatly along a single production line.

The TZ1 unit in AMAE consists mostly of plateau material (*Scott et al.*, 1978) and, as was the case for TZO, the counting area is small. Only two branches can be derived from the cumulative frequency curve (Fig. 9b), which includes craters of diameters only ≤ 25 km. Down to 12 km these craters show a good fit about a production curve with $N(1) = 29,000$. There is a clear bendaway at this point that includes a gap for $D = 10-11$ km. The replotted smaller diameter points again show more scatter about a mean production curve of age 4800 than was generally the case for the youngest branches in the ISMCAS TZ units, but within the error bars there does seem to be evidence that a resurfacing event did occur and that two branches can be derived from the data.

The very small counting area of TZ3 results in only 27 craters larger than 5 km diameter (Table 2). Gaps occur between 20-35 and 10-15 km. These alone suggest that some crater removal has occurred. The points in Fig. 9c seem to lie along three separate production curves of ages 20,300, 40,000, and 84,000, but the oldest of these is based only on two points with very large error bars. For consistency in approach the curves were broken at the obvious places; the resulting branches suffer from the process. The $N(1) = 27,200$ is a reasonable fit to a few points but the scatter about the 11,600 line is so large that it is questionable that a true termination of a resurfacing event can be defined. While it appears that better results would have been achieved without subtraction, the ages inferred would have been inconsistent because they would have included the cumulative effects of *all* underlying surfaces.

Figure 9d shows the results of the N&H approach for the much larger area of the highly modified TZ8 unit. An old branch with $N(1) = 84,400$ is clearly defined by the points with $D \geq 35$ km. Subtraction and replotting gives a well-determined second branch with a good fit about a production curve of age 29,000. The third branch at $N(1) = 7600$ is not so well defined but still clearly present. Note that this is the only unit within the TZ for which a well-defined old branch is unmistakably present, although the TZO unit also showed a possible branch at $N(1) = 86,300$. These are important because of their similarity to the CT branches at 82,600 and 88,200.

The highly modified stripped unit sTZ only has craters smaller than 35 km despite a respectable counting area. A 19,700 branch appears to be separable from a 6300 branch with some indication of a third very young branch at $N(1) = 2500$. This last branch is based only only two points, however.

Summary for Individual Transition Zone Units

Figure 10 summarizes the above analysis for the transition zone terrain units in the same format as Fig. 6. Points lying along production curves are shown in Fig. 10a without error bars and in Fig. 10b with error bars. The dashed lines are the production curves from Fig. 6, based on the large-scale area curves, for comparison with those previously derived branches. Details of the ages derived for branches for the individual transition zone terrain units are given in Table 4, which has the same format as the corresponding summary in Table 3 for the large-scale areas CT, TZ, and SP.

Despite the morphological differences between the individual units in the TZ and the obvious differences in the original cumulative frequency curves (Fig. 7), there is one overwhelming invariant shown in Table 4 and Fig. 10: *All the terrain units* show a branch with crater age $N(1) = 20,000$ to 30,000 (in round numbers). This is an important result because it shows that the 20,000-30,000 age resurfacing event apparently recorded in the overall TZ was not the result of averaging over many diverse units with different resurfacing histories. The inferred event occurred throughout the transition zone and in the adjacent cratered terrain (CT) as well. That it is shown to have occurred in each of the individual units makes this an extremely robust result.

The occurrence of this resurfacing event throughout the transition zone and adjacent cratered terrain (and the survival of an underlying surface of this age in the ISMCAS smooth plains) implies that the event was a major, widespread one that played a significant role in shaping the present-day structure of the highland/lowland boundary zone.

Other important conclusions can be drawn from Table 4 and Fig. 10. These include:

1. There is nowhere within the transition zone any evidence for a surviving surface as old as the oldest CT surface [$N(1) \approx 210,000$].

2. The oldest surviving surface within the transition zone (which underlies the 20,000-30,000 age surface) has an age of $\sim 86,000$. Even though this is well determined for only two of the terrain units, it is important because it is comparable to the resurfacing age of 82,600-88,200 found in the cratered terrain (Table 3).

3. Resurfacing events more recent than the 20,000-30,000 event show a wide variety of ages at which the events terminated, ranging from ≈ 5000 to over 10,000 for those that are well determined.

There are additional subtle points that may be important in the overall evolution of the transition zone. For the 20,000-30,000 age event, on average the ages determined for the ISMCAS units are less than those of the AMAE units, as was the case for the ages determined from the average TZ in ISMCAS and AMAE. This is especially true for the common units TZ1 and TZ8 (27,500 and 25,400 respectively in ISMCAS, but 29,300 and 29,000 respectively in AMAE). This could be, however, within the sensitivity of the technique. Note, however, that for the younger events the reverse is true: The ISMCAS events for the common units ceased earlier (at 7200 and 10,500 in ISMCAS compared with 4700 and 7600 in AMAE for TZ1 and

TABLE 4. Derived crater ages for end of resurfacing events in transition zone terrain units.

Region	Unit	$N(1)_{av}(n)$ h/l (diam. range)	$N(1)_{av}(n)$ h/l (diam. range)	$N(1)_{av}(n)$ h/l (diam. range)	$N(1)_{av}(n)$ h/l (diam. range)
ISMCAS	TZ1	$[100,351(1)^{100,351}_{100,351}(70)]$	$27,518(4)^{28,117}_{26,195}(12\text{-}30)$	$7,160(4)^{7,625}_{6,863}(5\text{-}11)$	
	TZ2	$?[38,161(1)^{38,161}_{38,161}(60)$	$21,475(2)^{22,762}_{20,181}(17\text{-}30)$	$4,950(5)^{5,632}_{4,369}(5\text{-}12)$	
	TZ8	$[92,494(1)^{92,494}_{92,494}(90\text{-}100)$	$25,431(3)^{27,771}_{24,195}(20\text{-}45)$	$10,517(9)^{10,966}_{9,960}(5\text{-}17)$	
	oTZ	$[75,701(1)^{75,701}_{75,701}(50\text{-}60)]$	$19,724(5)^{21,750}_{18,069}(14\text{-}35)$	$6,071(5)^{6,674}_{5,737}(8\text{-}12)$	$[2,692(2)^{2,921}_{2,462}(5\text{-}7)]$
AMAE	TZ0	$86,309(3)^{92,565}_{82,740}(25\text{-}50)$	$24,240(6)^{27,203}_{20,887}(11\text{-}20)$	$6,684(4)^{8,747}_{5,062}(5\text{-}9)$	
	TZ1		$29,342(6)^{33,066}_{27,208}(10\text{-}25)$	$4,747(4)^{6,141}_{3,734}(5\text{-}9)$	
	TZ3	$[83,867(1)^{83,867}_{83,867}(35\text{-}40)$	$27,176(2)^{28,009}_{26,343}(15\text{-}20)$	$11,600(2)^{11,925}_{11,275}(5\text{-}10)$	
	TZ8	$88,381(4)^{92,564}_{84,212}(35\text{-}80)$	$28,950(6)^{30,408}_{27,314}(13\text{-}30)$	$7,594(4)^{8,747}_{6,895}(5\text{-}12)$	
	sTZ		$19,650(3)^{22,142}_{17,588}(20\text{-}30)$	$6,258(3)^{6,700}_{5,963}(9\text{-}17)$	$[2,478(2)^{2,502}_{2,453}(5\text{-}7)]$

Crater age $[N(1)]$ = cumulative number with $D \geq 1$ km per 10^6 km^2.
$N(1)_{av}$ = average crater age from points falling along N&H production curve.
n = number of points from which $N(1)_{av}$ was calculated.
h/l = high and low value included in calculation of $N(1)_{av}$.
diam. range = range of crater diameters for which $N(1)_{av}$ applies.
[ ] = poorly determined or uncertain result.

TZ8, respectively). Note also for the TZ8 unit (the unit dominated by plains-forming material) that the diameter ranges over which the major 20,000-30,000 age unit is defined differ from ISMCAS (20-45 km) to AMAE (13-30 km). If the resurfacing event was the same in both regions (as perhaps is implied by both the common morphology and age of resurfacing), then the efficiency of this event at removing craters was greater in ISMCAS than in AMAE. Craters with diameters 35 to 80 km survive in AMAE as shown by the older branch with surface age 88,400. In ISMCAS no craters between 45 and 90 km survive, although two craters with diameter ≥ 90 km do indicate the presence of an underlying surface.

DISCUSSION

The *Neukum and Hiller* (1981) technique for breaking cumulative frequency curves into separate branches corresponding to different resurfacing events seems to be a viable way to study the cratering history of the transition zone and its surroundings. The generally good fit of the plotted and replotted points (allowing for the effects of subtraction) about production curves provides some confidence that those branches are well determined and that the breaks or bendaways from the previous production curves are due to resurfacing events. Perhaps the reason this approach works so well for the data presented here is because resurfacing (by plains-forming materials) has been an obvious and important part of the history of this region.

However, it must be emphasized that the number of derived branches and the crater ages inferred for the cessation of the resurfacing events implied are highly dependent on the choice of the production curve against which the cumulative frequency data are compared. As described above, a production curve with more structure (more bends) might result in fewer derived branches, while a straight-line power law production curve would make the bendovers at smaller diameters in the cumulative frequency curves that much more obvious. Our results are tied to the *Neukum and Hiller* (1981) curve, which was derived by *Neukum and Wise* (1976), and are therefore most easily compared to the results of others who have also used this standard.

Table 5 summarizes in simplified form the information from Tables 3 and 4 on the ages of derived resurfacing events and the diameter ranges over which these events branches are defined. Poorly determined or uncertain ages are shown in brackets; these may be due to either too few points (e.g., the old ages for ISMCAS transition zone units) or the fact that the replotted points never grouped tightly about a new production curve (e.g., the 11,600 age branch in AMAE TZ3). These last are of more interest because they may identify resurfacing events that competed with crater production (see discussion above). That is, while a bendaway from the previous production curve is obvious, the cause of that depopulation cannot be ascribed to a complete resurfacing in the sense that *Neukum and Hiller* (1981) idealized the process. In Table 5 the average TZ results are enclosed in the double parentheses to emphasize that these are average results from a complex area that is made up of a number of individual morphological terrain units (also shown in Table 5).

From Table 5 and the preceeding results described above we derive the following conclusions about the transition zone and its adjacent cratered terrain and smooth plains.

1. All the units show evidence for resurfacing in the N&H sense, and most appear to have experienced more than one resurfacing event.

2. *All* the units (except AMAE SP) had a resurfacing event (or show a surface that survives) with a crater age $N(1) = 20,000-30,000$. This marks the time that the resurfacing event stopped. This event is seen in the cratered terrain and in each terrain unit of the transition zone, and a surface of this age survives below the present day surface in some of the smooth plains west of the Isidis Basin (ISMCAS SP).

3. A previous resurfacing in the cratered terrain ceased at crater number $N(1) \approx 86,000$. Relics of this underlying surface may be present throughout the transition zone but are well defined from the cumulative frequency curves only in Amenthes-Aeolis in units TZO and TZ8.

4. The oldest surface recorded in this area is in the cratered terrain and has an age $N(1) \approx 210,000$. Nothing this old is preserved as an underlying surface in the TZ or SP units. Note that evidence for this underlying surface in the CT is preserved despite the two subsequent resurfacing events at 86,000 and $\sim$ 31,000.

5. In the TZ, its individual units and in the SP there is evidence for resurfacing at times younger than the common 20,000-30,000 event, but the crater ages derived from these vary from 5,000 to more than 10,000 from unit to unit.

The resurfacing history implied by the above conclusions derived from Table 5 can be summarized as follows:

1. A resurfacing event occurred throughout the highlands adjacent to the present-day transition zone, removing craters smaller than 40 km diameter from a surface with original crater retention age $\sim$ 210,000. This event terminated at crater age $\sim$ 86,000. Relics of this new surface may be present below (portions of) the transition zone.

2. A second major resurfacing of the cratered terrain and of (the underlying terrain in what is today) the transition zone occurred, terminating everywhere at crater number $\sim$ 20,000-30,000. In the cratered terrain this removed craters smaller than 17-20 km, but in the transition zone even larger craters were removed. West of the Isidis Basin craters smaller than 60 km were depopulated by this event, while east of Isidis the depopulation was efficient for D < 25-35 km. If this resurfacing was the same process throughout the entire area, then it was more efficient at removing craters west of Isidis. But in the transition zone, the event also terminated later west of Isidis, at $N(1) \sim 21,500$ in TZ2 compared with 27,200 to 29,300 for TZ3, TZ8, and TZ1 in AMAE. In the cratered terrain in *both* ISMCAS and AMAE the event was over early, by $\sim$ 31,000.

3. This major resurfacing event left a surface with age $\sim$ 19,000-20,000 in the outer transition zone (oTZ) and in the smooth plains in ISMCAS, and in the highly modified sTZ unit in AMAE.

4. Subsequently, resurfacing events occurred throughout the transition zone and adjacent smooth plains (but not, for the diameter range considered here, in the cratered terrain to the south). These events vary in the age at which they stopped (from $N(1) \approx 5000$ to over 10,000) and in their ability to remove craters. Within the transition zone craters down to 20 km diameter were removed in ISMCAS TZ2 and TZ8 and in AMAE sTZ. Elsewhere the diameter cutoff is smaller (10-15 km), but in the smooth plains the craters that survived

TABLE 5. Summary of derived crater ages for end or resurfacing events in different units.

Region	Unit	$N(1)_{av}$(diam. range)	$N(1)_{av}$(diam. range)	$N(1)_{av}$(diam. range)	$N(1)_{av}$(diam. range)	$N(1)_{av}$(diam. range)
ISMCAS	CT	203,800 (40-100)	88,200 (20-35)	31,700 (5-17)		
	((TZ$_{av}$		[43,000](60-100)	20,300 (13-45)	7,200 (5-12)))	
	tz1		[100,400] (70)	27,500 (12-30)	7,200 (5-11)	
	tz2		[38,200] (60)	21,500 (17-30)	5,000 (5-12)	
	tz8		[92,500](90-100)	25,400 (20-45)	10,500 (5-17)	
	oTZ		[75,700](50-60)	19,700 (14-35)	6,100 (8-12)	[2,700](5-7)
	SP			19,700 (45-60)	6,300 (5-35)	
AMAE	CT	216,600 (40-100)	82,600 (17-35)	30,400 (8-13)	[8,600] (5-7)	
	((TZ$_{av}$		72,500 (35-80)	29,900 (13-30)	7,500 (5-12)))	
	tz0		86,300 (25-50)	24,200 (11-20)	6,700 (5-9)	
	tz1			29,300 (10-25)	4,700 (5-9)	
	tz3		[83,900](35-40)	27,200 (15-20)	[11,600](5-10)	
	tz8		88,400 (35-80)	29,000 (13-30)	7,600 (5-12)	
	sTZ			19,700 (20-30)	6,300 (9-17)	[2,500](5-7)
	SP		[10,200](50-80)		5,500 (5-40)	

Crater age $[N(1)]$ = cumulative number with D $\geq$ 1 km per 10^6 km^2.
$N(1)_{av}$ = rounded average crater age from points along N&H production curve.
diam. range = range of crater diameters for which $N(1)_{av}$ applies.
[ ] = poorly determined or uncertain result.

from below are ≥ 45 km in diameter. These subsequent events were clearly able to bury larger craters at greater distances from the cratered terrain/transition zone boundary. One exception to this is the oTZ that preserves craters larger than 14 km on the underlying surface, while the TZB unit to the west lost craters down to 17 km.

The overall picture that emerges is one of multiple resurfacing events, some of which were common in timing, occurring throughout the area, with subsequent activity moving progressively northward (from the cratered terrain toward the northern lowlands) and with craters of larger diameter being more effectively removed *by a given resurfacing event* if they lay further from the cratered terrain. The 20,000-30,000 age event is most significant in that it appears to have affected the widest area and to have ceased at about the same time in all the units. Below we compare the sequence of events derived from this analysis with scenarios described by others based on other data.

Comparison with Other Evolutionary Chronologies

The sequence of resurfacing events described above is completely consistent with the scenario described by *Wise et al.* (1979): formation of the highland crust with intense cratering at N(1) = 400,000-100,000; resurfacing of the northern one-third of Mars with lowering of elevation, possible volcanism, and burial at N(1) = 100,000-50,000; oldest plains deposits and some fracturing at N(1) = 70,000-25,000; formation of Lunae Planum and related surfaces at N(1) = 25,000-15,000; and subsequent events consisting of faulting, plains deposits, major outflow channel formation, and construction of large volcanic cones at N(1) ≤ 15,000. The correspondence of these ages with our major resurfacing events at 210,000, 86,000, 20,000-30,000, and ≤ 10,000 is very good in part because they too use the *Neukum and Wise* (1976) production curve as a standard. *Wise et al.* (1979) set the establishment of the crustal dichotomy on Mars at 100,000 to 50,000, well after the formation of the ancient highland crust. Our first cratered terrain resurfacing event at 86,000, for which there appear to be some relic surfaces underlying parts of the transition zone, lies within these ages. The second major resurfacing event in the cratered terrain and the first that had widespread effect on the transition zone and portions of the adjacent smooth plains occurred at 20,000 to 30,000, which is the age that *Wise et al.* (1979) associate with the formation of Lunae Planum and related surfaces. *Neukum and Hiller* (1981) and *Neukum and Wise* (1976) give a similar age from their counts in Lunae Planum projecting along the *Neukum and Wise* (1976) production curve. We also have counted craters in Lunae Planum and find, for the older of the two resolvable branches, an age of N(1) = 22,000 (*Frey et al.*, 1986b).

The major widespread resurfacing that occurred at 20,000 to 30,000 throughout the transition zone (as recorded in each of the terrain units), in the adjacent cratered terrain and perhaps preserved in parts of the adjacent smooth plains, appears to have occurred at or slightly before the time of cessation of the Lunae Planum volcanics.

Tanaka (1986) gives the stratigraphic level of the ridged plains (such as Lunae Planum, Hesperian Planum, Syrtis Major Planum) as Lower Hesperian. Unfortunately, it is not possible to compare directly his cumulative crater values at N(5) and N(16) with those derived from N(1) in this paper. This is a consequence of the different production curves that apply to our analysis and his stratigraphy. Table 6 shows the problem. We show representative ages N(1) for many of the resurfacing events we find (from Table 5) and the N(5) and N(16) values these would have using the *Neukum and Hiller* (1981) production curve. The stratigraphic series implied by these cumulative numbers are taken from Table 3 of *Tanaka* (1986). Consider the 20,000 and 30,000 ages discussed above that correspond to the Lunae Planum resurfacing. The inferred stratigraphic level from N(5) is *Middle* Noachian and from N(16) is *Upper* Noachian. This inconsistency is due to the shape of the N&H production curve between 5 ≤ D ≤ 16 km, which is very close to a D^{-2} power law over this diameter range. The N(5) and N(16) values corresponding to the stratigraphic boundaries (e.g., Middle Noachian/Upper Noachian imply a much shallower slope over this diameter range. Furthermore, neither of these cumulative values produces an inferred stratigraphic age that is Lower Hesperian, as *Tanaka* (1986) shows, for the ridged plains. This does not mean that one of the two dating schemes is wrong (the ridged plains are the referrant for the Lower Hesperian in the stratigraphic series), only that they are incompatible in terms of crater number at specific diameters. Use of the N&H production curve (from *Neukum and Wise* 1976) generally results in higher cumulative crater numbers (and therefore older stratigraphic ages) than *Tanaka* (1986) would use.

This point is further demonstrated by treating the problem the other way: In the bottom portion of Table 6 we show the N&H values of N(1) that the *Tanaka* (1986) N(5) and N(16) would give, using the *Neukum and Wise* (1976)

TABLE 6. Cross comparison of resurfacing ages at N(5) and N(16) with *Tanaka's* (1986) stratigraphy.

Major Resurfacing Event Ages N(1)	N&H N(5)	Inferred SS	N&H N(16)	Inferred SS
5,000	108	uH	10	-
10,000	216	uN	20	1H
20,000	432	mN	41	uN
30,000	648	mN	61	uN
86,000	1858	-	175	mN
210,000	4537	-	428	1N

Straigraphic Series	N(5)	Inferred N&H N(1)	n(16)	Inferred N&H N(1)
uH	67-125	3100-5800		
1H	125-200	5800-9300	<25	<12,500
uN	200-400	9300-18,500	25-100	12,500-50,000
mN	>400	>18,500	100-200	50,000-100,000
1N			>200	>100,000

production curve. The Lower Hesperian N(5) would result in N(1) = 5800-9300 and the N(16) would result in N(1) < 12,500. Both of these are lower than the 15,000-25,000 value of N(1) inferred by *Wise et al.* (1976), *Neukum and Hiller* (1981), and us (*Frey et al.*, 1986b) for Lunae Planum (one of the surfaces used to define the Lower Hesperian).

However, even though the cumulative numbers at any given diameter can not be compared directly, *the sequence of major resurfacing events should be similar*. Our age of 210,000 for the (preresurfaced) cratered terrain (i.e., the survivors of the first resolvable resurfacing event) is in principle Lower Noachian in Tanaka's stratigraphy (see Table 6). In terms of resurfacing events, *Tanaka* (1986) lists a major intercrater plains resurfacing in the Late Noachian and embayment of the lowlands along the highland/lowland boundary by ridged plains in Lower Hesperian time. Between these two events intense "erosion" (not necessarily "resurfacing" in the Neukum & Hiller sense) of the cratered terrain and intercrater plains occurred, producing the mesas (detached plateaus) through development of fretted channels. In late Upper Hesperian time the northern plains were (again) resurfaced, with smooth plains formation continuing into the Amazonian epoch. In broad outline the same sequence of events is found from breaking the cumulative frequency curves (Table 5): resurfacing of the cratered terrain at N(1) ~ 86,000, further resurfacing that was extensive and approximately contemporaneous (in its cessation) in the transition zone *and* adjacent cratered terrain at N(1) ≈ 20,000 to 30,000, and later resurfacing in the transition zone and especially in the adjacent northern plains at N(1) ≲ 10,000.

If the formation of the ridged plains such as Lunae Planum is taken as a common reference and if the 20,000-30,000 age resurfacing corresponds to Lunae Planum time (*Wise et al.*, 1979; *Frey et al.*, 1986b), then the previous major event at N(1) ≈ 86,000 should correspond to *Tanaka's* (1986) Upper Noachian intercrater plains resurfacing, which produced the cratered plateau material (*Scott and Carr*, 1978). Remnants of this surface are preserved in the detached plateaus (mesas) that are found in the transition zone (*Lucchitta*, 1978; *Greeley and Guest*, 1978; *Hiller*, 1979; *Scott et al.*, 1978). That this event occurred throughout the cratered terrain and within portions of what is today the transition zone, and that it was able to remove craters smaller than 40 km diameter throughout the highlands, supports this association. From Fig. 3 it is clear that there was no obvious earlier event in the cratering record.

The association of *Tanaka's* (1986) ridged plains resurfacing (of Lower Hesperian age) with our 20,000-30,000 crater age event (Lunae Planum age) is strengthened by considering the outer transition zone unit (oTZ) of ISMCAS. The N(5) value from the original cumulative frequency curve (see Table 2) implies an overall stratigraphic age of Lower Hesperian. From Table 5 it is clear that this average age is a composite of two branches with N(1) = 19,700 and 6100. The older of these is comparable to Lunae Planum in age. The oTZ unit is composed mostly of rolling plains (see Table 1) of Hesperian age (*Scott and Carr*, 1978), which are similar or perhaps slightly younger than the ridged plains.

We conclude that, although the cumulative numbers cannot be directly compared, the sequence of events suggested by *Tanaka* (1986) is the same as found by us through breaking the cumulative frequency curves into separate branches that are due to resurfacing events.

Extent and Efficiency of Resurfacing in the Transition Zone

The *Neukum and Hiller* (1981) technique identifies individual crater retention surfaces that existed prior to and following major resurfacing events, and therefore provides detailed information on the age and extent of those surfaces. It is also possible to infer something of the nature and extent of the resurfacing from the survivors showing through the event (those craters that define the crater retention surface prior to resurfacing). If a given resurfacing event is found in a number of locations, then information on the timing of its termination and how that may have varied spatially can also be derived.

The N(1) ≈ 86,000 resurfacing event associated above with the formation of the cratered plateau material clearly had a global context. Within the area considered here the event terminated at essentially the same time both east and west of the Isidis Basin (there is no difference between 82,600 and 88,200 given the sensitivity of the technique used here and the uncertainties associated with it). The event appears to have occurred within what is today the transition zone: The identification of the detached plateaus with cratered plateau material (also called plateau material) provides geologic evidence for this (*Lucchitta*, 1978; *Greeley and Guest*, 1978; *Hiller*, 1979; *Scott et al.*, 1978). In the TZ0 and TZ8 units in AMAE there is evidence for this event in the oldest branch derived from the cumulative frequency curve, and the age is the same (86,300 to 88,400; see Table 5). The other units do not show a well-defined branch of this age, but most have one or two craters of large size (which may have been from this earlier time) surviving the next resurfacing event. The diameter range over which the 86,000 event is defined is different in the cratered terrain and in the two TZ units. In the cratered terrain the branch is marked by craters as large as 35 km, implying that craters of this size were initially depopulated by this event. In TZ8 the size range of craters defining the 86,000 age branch is 35-80, which means that craters of much larger size were removed by the resurfacing event within what is today the transition zone than in the cratered terrain immediately to the south.

Therefore, although the major period of scarp retreat and other events that produced the transition zone and helped to establish the crustal dichotomy may have occurred between the 86,000 and 20,000-30,000 age resurfacing events (*Wise et al.*, 1979; *Tanaka*, 1986), in some fashion there was already a difference between the region that would become the cratered highlands and the one that would become the transition zone, such that the size of craters affected by the older (86,000) resurfacing event was different between the two. Perhaps an elevation difference had already been established by the time the cratered plateau material resurfaced the old cratered terrain, permitting more efficient burial of craters in the lower-lying sections.

There is similar evidence for variations in the efficiency of the 20,000-30,000 (Lunae Planum age) resurfacing. In the cratered terrain this event is defined by craters in the 5-17 km and 8-13 km diameter range for ISMCAS and AMAE, respectively. This means that craters larger than 20 and about 17 km survived the resurfacing and are left to identify the previous crater retention surface with age ≈ 86,000 (see Table 5). By contrast the Lunae Planum age event is defined by craters with diameters extending to large values for the units within the transition zone (typically to about 30 km or larger). Also, comparison with the size of "survivors" (craters falling on older production curves) shows that the Lunae Planum age event appears to have removed craters as large as 45-80 km in ISMCAS. To the east of Isidis the event was less efficient and only depopulated craters with diameters less than about 35 km. If the formation of ridged plains was the resurfacing event in both areas, an obvious conclusion is that the thickness of the plains-forming materials was significantly greater to the west of the Isidis Basin. Further to the north, in what are today the SP units in Fig. 1, no craters larger than 30 km survive in AMAE but craters as large as 60 km help to define this Lunae Planum age surface at $N(1) = 19,700$ in ISMCAS. The resurfacing at this time was more efficient west of the Isidis Basin and progressively more efficient in ISMCAS the further north it occurred.

There is also in Table 5 indication that the timing of the 20,000-30,000 age event may have varied spatially in a systematic way. The actual range of $N(1)$ values for this event (19,700 to 31,700) is large by comparison with the range for the previous resurfacing event (82,600 to 88,400 for well-determined branches), even though in general older events should be more poorly determined because of the scarcity of craters. We believe that this technique is sensitive enough that the spread of values about an average of 25,700 (± 6000) represents a real spread in the crater retention age of the surface in the different units in which it is found. The ages in Table 5 can be grouped as follows: 19,500 to 21,500, 24,000 to 27,500, and 29,000 to 32,000, or as 20,500 (±1000), 25,750 (±1750), and 30,500 (±1500). Figure 11 shows the units in which these subdivisions of the 20,000-30,000 age resurfacing event occur.

The oldest age associated with the (Lunae Planum age) resurfacing event [$N(1) \geq 29,000$] occurs throughout the cratered terrain and in large portions of the transition zone east of the Isidis Basin (the TZ1 and TZ8 units). The middle group of ages for this resurfacing event are found in the units of the transition zone that abut the Isidis Basin and extend to the northwest (TZ0 in AMAE, TZ1 and TZ8 in ISMCAS) and in the TZ3 unit in AMAE. The youngest cessation of the resurfacing event is found in the TZ2 and oTZ units in ISMCAS and in the SP units northward of these, and in the unusual sTZ unit in AMAE. This picture is similar to that found for the size of craters removed by the resurfacing: The overall activity seems to shift westward and then northward (toward the smooth plains from the cratered terrain) with time.

Because the branches separated from the cumulative frequency curves only provide information on the end or termination of a resurfacing event, there is no way to know

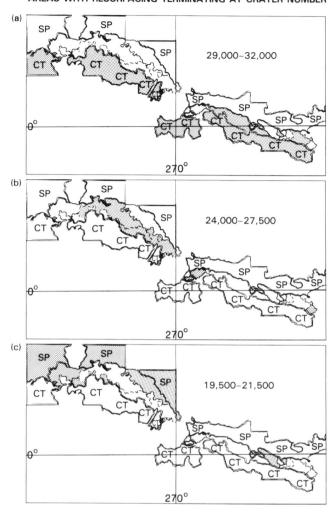

Fig. 11. The Lunae Planum age resurfacing event [$N(1) \approx 20,000$ to 30,000] is subdivided into three ages and units in which those ages occur are shown shaded. Separation of the transition zone into individual units is retained in the figure. (**a**) Units in which the resurfacing stopped at $N(1) = 29,000$ to 32,000. (**b**) Units in which the resurfacing stopped at $N(1) = 24,000$ to 27,500. (**c**) Units in which the resurfacing stopped at $N(1) = 19,500$ to 21,500. In general the age of cessation shifts westward and northward over time.

whether or not resurfacing was occurring throughout ISMCAS (for example) from before $N(1) = 32,000$. It does seem clear, however, that the event stopped earliest in the cratered terrain (and AMAE TZ units), and progressively later further from the CT.

In common TZ units the *youngest* resurfacing event shows a different pattern. In the TZ1 and TZ8 units the ages at which the resurfacing stopped are younger *east* of the Isidis Basin: 4700 and 7600 in AMAE compared with 7200 and 10,500 in ISMCAS, respectively. But for TZ8 the craters surviving this most recent event are larger in ISMCAS ($D \geq 20$ km) than in AMAE ($D \geq 13$ km), as found for the earlier resurfacing.

If the youngest resurfacing event was the same in both areas, it was more efficient again in ISMCAS, although in this case it stopped sooner.

From the above we conclude that in general the resurfacing events determined here (especially the Lunae Planum age event and those younger) were generally more efficient at removing craters to the *west* of the Isidis Basin, where the transition zone is generally at higher latitudes. We also believe that there is evidence that the 20,000-30,000 age event ceased later further from the cratered terrain as well as generally removing larger craters further from the cratered terrain.

Implications for the Origin of the Crustal Dichotomy

The data presented here provide information only on the resurfacing history of the transition zone and its surroundings in eastern Mars. There is in the technique we have used no direct information on tectonic or geologic events that do not in and of themselves cause a depopulation of impact craters over large areas that can be described as a resurfacing. For example, there is no information on when the fracturing and channeling that produced the detached plateaus actually occurred. We can only bracket that event between the two major resurfacing events at 86,000 and 20,000 to 30,000.

Some implications of these results for the nature and origin of the crustal dichotomy should be pointed out, however. Nowhere within the transition zone (or in the SP to the north) is there any surviving relic of the $\approx 210,000$ age we find for the oldest branch in the cratered terrain cumulative frequency curve. The oldest crater retention age found within the transition zone is the $N(1) \approx 86,000$ age associated with the formation of the cratered plateau material. *Wilhelms and Squyres* (1984) suggested that the rim of their Borealis Basin (which resulted from the megaimpact that caused the Martian crustal dichotomy) lay in part along the transition zone. Furthermore, this impact was supposed to have occurred very early in Martian history, prior to most other recognizable geologic events (e.g., formation of the cratered plateau material?). We find, using the N&H technique, that the oldest portions of the knobby terrain in Elysium-Amazonis (within the proposed Borealis Basin) have a crater retention age of 230,000 (*Frey et al.*, 1987), which is quite similar to that found for the cratered terrain south of the transition zone. Thus there is evidence for very ancient crust outside *and* inside the proposed Borealis Basin, but not along what is considered by *Wilhelms and Squyres* (1984) to be part of its rim. While it is possible that subsequent geologic processes have managed to remove most relics dating from the time of the impact or immediately thereafter, these results do cast some doubt on the identification of the whole transition zone as a part of the rim of the Borealis Basin. We have argued elsewhere (*Frey et al.*, 1986a) that there are other problems with this hypothesis for producing the crustal dichotomy on Mars, as attractive as it is. At the very least it would seem that most of the transition zone preserves a record considerably younger than the proposed megaimpact event suggested by *Wilhelms and Squyres* (1984).

There is, however, indirect evidence that some sort of (topographic?) dichotomy or difference may have existed from before the formation of the cratered plateau material in the resurfacing event at $N(1) \approx 86,000$. As described above, craters removed by this process were larger north of the cratered terrain/transition zone boundary than south of it. The difference in the "survivors" provides only indirect evidence for some spatial difference in the preresurfaced crust and does not in itself define what that difference might have been. While structural or topographic possibilities are attractive, other causes can also be imagined. For example, a prior resurfacing event just within what would become the transition zone could have made easier the removal of large craters by the 86,000 event and left no record if the craters accumulated between such events were all subsequently removed.

What is most important in the results presented here concerning the evolution of the transition zone is that the widespread, common (Lunae Planum) age resurfacing event recorded in each unit of the transition zone, in the adjacent cratered terrain to the south, and in parts of the smooth plains to the north was apparently associated with a global scale volcanic resurfacing at many locations around Mars. Formation of ridged plains at the $N(1) \approx 20,000$ to 30,000 time was not restricted to just Lunae Planum and the transition zone area in eastern Mars, but also occurred at Sinai, Syrtis Major, Hesperia, Malea Plana, Tempe, Hellas, and elsewhere (*Scott and Carr*, 1978; *Tanaka*, 1986). Although not generally mapped as ridged plains within the transition zone, a surface of this same age was apparently laid down in this area. Furthermore, the resurfacing that occurred at this time also affected the adjacent highlands, in that some craters smaller than about 20 km diameter were depopulated. It is therefore possible to infer from the cumulative frequency curves more about the extent of this resurfacing event than the surficial geology alone would allow. The major resurfacing of the highland/lowland boundary in eastern Mars occurred at the time of Lunae Planum and related ridged plains development and extended both north and south of the transition zone where it was extremely important in modifying the existing surface.

CONCLUSIONS

The *Neukum and Hiller* (1981) technique for breaking the cumulative frequency curves into separate branches provides a viable and useful way to determine the number and timing of major resurfacing events in the highland/lowland boundary zone.

From crater counts over the diameter range 5 to 100 km in the transition zone (TZ), its individual morphological terrain units, the cratered terrain (CT) to the south, and the smooth plains (SP) to the north, we find evidence for multiple resurfacing events with the following characteristics.

1. The oldest surface resolvable lies under the cratered terrain and has a crater retention age $N(1) \approx 210,000$. No underlying surface of comparable age exists within the (crater retention record of the) transition zone or smooth plains.

2. A major resurfacing of the CT adjacent to the transition zone occurred at $N(1) \approx 86,000$. This produced the cratered

plateau material, relics of which are preserved in some portions of the TZ as the oldest recognizable surface derived from the cumulative frequency curves.

3. A second major resurfacing of the cratered terrain and of *all* the transition zone (as shown by being recorded in each of the individual terrain units in the TZ) occurred, producing a surface of $N(1) \approx 20,000$ to $30,000$. This age surface also underlies portions of the smooth plains adjacent to the TZ in the area west of the Isidis Basin (ISMCAS). This resurfacing event occurred at the time of ridged plains formation in Lunae Planum and elsewhere.

4. In the transition zone and adjacent smooth plains there is evidence for still younger resurfacing, stopping at $N(1) \leq 10,000$ in the TZ and at 5500-6500 in the SP.

From the same type of analysis applied to the individual terrain units within the transition zone, we add the following conclusions.

5. The Lunae Planum age resurfacing event varied in a quasisystematic way in the timing of its cessation and the efficiency of its resurfacing throughout the transition zone. In general the resurfacing stopped first at $N(1) = 29,000$-$32,000$ in the cratered terrain to the south in both ISMCAS and AMAE and in large portions of the TZ in AMAE east of the Isidis Basin. The cessation occurred at $N(1) = 24,000$-$27,500$ within large portions of the transition zone in ISMCAS. In the more northerly oTZ unit and the smooth plains in ISMCAS the event stopped at $N(1) = 19,500$-$21,500$.

The resurfacing event also in general removed larger craters further from the cratered terrain. In the CT craters larger than about 20 km survived the resurfacing, but in the TZ it is craters larger than 35 km in AMAE and larger than 50 km in ISMCAS that survive from before the Lunae Planum age event. In the SP the event depopulated craters smaller than 60 km in ISMCAS and smaller than 80 km in AMAE.

6. The earlier $N(1) \approx 86,000$ resurfacing associated with the formation of the cratered plateau material also had greater efficiency in the TZ (where recorded than in the CT. Craters as large as 80 km were removed in TZ8 (AMAE) but only craters with $D \leq 40$ km were affected in the cratered terrain. This implies that some difference in the nature of the crust (topography?) may have already existed between what would become the cratered highlands and the adjacent transition zone prior to the resurfacing event.

7. Resurfacing events more recent than the major Lunae Planum age event stopped at different crater retention ages in different portions of the transition zone and smooth plains. For common TZ units the same tendency for greater efficiency at removing somewhat larger craters west of Isidis than east of Isidis persisted through these events.

Although these results are dependent on the use of the *Neukum and Hiller* (1981) technique and crater production curve, they are consistent with resurfacing scenarios derived by others from geologic mapping and stratigraphic relations. The value of the approach used here is that it provides more detailed information on the number of major resurfacing events, their spatial extent, and the timing (and how that timing may have varied spatially) of their cessation. For example, it would be hard to appreciate from geologic mapping alone that the Lunae Planum age resurfacing event (associated with the formation of ridged plains) was so widespread, also affected the cratered terrain adjacent to the transition zone, and varied spatially in a systematic way in its efficiency and the timing of its cessation in the highland/lowland boundary zone area.

Acknowledgments. We would like to thank two reviewers and an associate editor for provoking a much more careful and detailed version of this paper. Roberta Anderson typed the manuscript and Tammy Grant, Barbara Conboy, and Rick Evans helped in the preparation of the figures and text. This work was supported by the Planetary Geology and Geophysics Program of NASA Headquarters, in part through a grant to the University of Maryland (NAG5-245).

REFERENCES

Carr M. H., Masursky H., and Saunders R. S. (1973) A generalized geologic map of Mars. *J. Geophys. Res., 87,* 4031-4036.

Frey H. and Semeniuk A. M. (1985) The highland boundary on Mars: Distribution of characteristic structures (abstract). In *Lunar and Planetary Science XVI,* pp. 252-253. Lunar and Planetary Institute, Houston.

Frey H. and Semeniuk A. M. (1986) Characteristic structures of the highland boundary on Mars and comparison with topography. *Reports of the Planetary Geology and Geophysics Program - 1985,* pp. 430-432. NASA Tech. Mem. 88388, NASA, Washington, D.C.

Frey H., Schultz R. A., and Maxwell T. A. (1986a) The martian crustal dichotomy: Product of accretion and not a specific event? (abstract). In *Lunar and Planetary Science XVII,* pp. 241-242. Lunar and Planetary Institute, Houston.

Frey H., Semeniuk A. M., Semeniuk J. A., and Tokarcik S. (1986b) Crater counts in the transition zone in eastern Mars: Evidence for common resurfacing events (abstract). In *Lunar and Planetary Science XVII,* pp. 243-244. Lunar and Planetary Institute, Houston.

Frey H., Semeniuk J. A., and Tokarcik S. (1987) Common age resurfacing events in the Elysium-Amazonis knobby terrain on Mars (abstract). In *Lunar and Planetary Science XVIII,* pp. 304-305. Lunar and Planetary Institute, Houston.

Greeley R. and Guest J. E. 91978) Geologic map of the Casius quadrangle of mars. *USGS Atlas of Mars, 1:5 Geologic Series, Map I-1038* (MC 6).

Greeley R. and Spudis P. D. (1981) Volcanism on Mars. *Rev. Geophys. Space Phys., 19,* 13-41.

Hiller K. H. (1979) Geologic map of the Amenthes quadrangle of Mars. *USGS Atlas of Mars, 1:5 M Geologic Series, Map I-1110* (MC 14).

Lucchitta B. K. (1978) Geologic map of the Ismenius Lacus quadrange of Mars. *USGS Atlas of Mars, 1:5 M Geologic Series, Map I-1065* (MC 5).

Masursky H. (1973) An overview of geological results from Mariner 9. *J. Geophys. Res., 78,* 4009-4030.

Maxwell T. A. and McGill G. E. (1987) Ages of fracturing and resurfacing along the martian dichotomy boundary between Nepenthes and Nilosyrtis Mensae (abstract). In *Lunar and Planetary Science XVIII,* pp. 604-605. Lunar and Planetary Institute, Houston.

Maxwell T. A., Frey H., and Semeniuk A. M. (1984) Geomorphology and evolution of the ancient cratered terrain-smooth plains boundary on Mars. *Geol. Sco. Am. Agstr. with Programs, 16,* 586.

Meyer J. D. and Grolier M. J. (1977) Geologic map of the Syrtis Major aquandrangle of Mars. *USGS Atlas of Mars, 1:5 M Geologic Series, Map I-995* (MC 13).

Neukum G. (1983) *Habilitationsschrift* (as referenced by Maxwell and McGill, 1987).

Neukum G. and Hiller K. (1981) Martian ages. *J. Geophys. Res., 86*, 3097-3121.

Neukum G. and Wise D. U. (1976) Mars: A standard crater curve and possible new time scale. *Science, 194*, 1381-1387.

Scott D. H. and Allingham J. W. (1976) Geologic map of the Elysium quadrangle of Mars. *USGS Atlas of Mars, 1:5 M Geologic Series, Map I-935* (MC 15).

Scott D. H. and Carr M. H. (1978) Geologic map of Mars. *USGS Atlas of Mars, 1:25 M Geologic Series*, Map I-1083.

Scott D. H., Morris E. C., and West M. N. (1978) Geologic map of the Aeolis quadrangle of Mars. *USGS Atlas of Mars, 1:5 M Geologic Series, Map I-1111* (MC 23).

Semeniuk A. M. and Frey H. V. (1984) Distribution of characteristic features across the boundary scarp in Acidalia and Amazonis-Memnonia (abstract). In *Lunar and Planetary Science XV*, pp. 748-749. Lunar and Planetary Institute, Houston.

Semeniuk A. M. and Frey H. (1985) Outlier detached plateaus and the highland boundary on Mars (abstract). In *Lunar and Planetary Science XVI*, pp. 759-760. Lunar and Planetary Institute, Houston.

Semeniuk A. M. and Frey H. (1986) Characteristic structures of the highland boundary on Mars: Evidence against a single mega-impact event? (abstract). In *Lunar and Planetary Science XVII*, pp. 791-792. Lunar and Planetary Institute, Houston.

Strom R. G., Woronow A., and Gurnis M. (1981) Crater populations on Ganymede and Calisto. *J. Geophys. Res., 86*, 8659-8674.

Tanaka K. L. (1986) The stratigraphy of Mars. *Proc. Lunar Planet. Sci. Conf. 17th*, in *J. Geophys. Res., 91*, E139-E158.

Wilhelms D. E. and Squyres S. W. (1984) The martian hemispheric dichotomy may be due to a giant impact. *Nature, 309*, 138-140.

Wise D. U., Golombek M. P., and McGill G. E. (1979) Tectonic evolution of Mars. *J. Geophys. Res., 84*, 7934-7939.

Woronow A. (1977) A size frequency study of large martian craters. *J. Geophys. Res., 82*, 5807-5802.

Ages of Fracturing and Resurfacing in the Amenthes Region, Mars

Ted A. Maxwell

Center for Earth and Planetary Studies, National Air and Space Museum, Smithsonian Institution, Washington, DC 20560

George E. McGill

Department of Geology and Geography, University of Massachusetts, Amherst, MA 01003

The timing of resurfacing events and structural modification of outlier plateaus and mesas in Mars' eastern hemisphere provides a constraint on the history of tectonic events along the cratered terrain—northern plains boundary. Based on a combination of superposition, crosscutting relations, and stratigraphy, the crater ages of discrete geologic units that can be used to bracket structural events have been determined for the Isidis basin rim, on which circumferential faulting ceased by $N(5) = 235$; cratered plateau material, which experienced a resurfacing event at $N(5) = 230$; and Syrtis Major Planum, emplaced at $N(5) = 150$. North of the boundary ages were determined for knobby terrain [resurfaced at $N(5) = 190$] that marks the transition between the cratered plateau and the smooth plains and various types of plains units [$N(5) = 60$-190]. Age determinations using production curves fit to crater density distributions are highly dependent on the choice of production curve, as are ages derived by subtracting and replotting populations that deviate from such curves. In order to determine crater ages of structural events in the Amenthes region, separate counts were made for geologic units that exhibit independent evidence for resurfacing or crosscutting by fractures. The oldest event dated by such techniques is the formation of the circumferential grabens surrounding the Isidis basin; they ceased forming before the final emplacement of ridged plains on the adjacent northern lowlands [$N(5) = 190$]. The cratered plateau east of the Isidis basin retains two crater populations; stripping of the rims of craters (an old population) was complete before downfaulting of the transition zone between the cratered terrain and the northern plains, and a young population of craters on the plateau records the same age as the ridged plains units north of the boundary. Faulting in this region is thus constrained to a narrow time period at about $N(5) = 190$. Similar sequences of faulting, erosion, and infilling by younger plains are observed at the northern edge of Elysium Mons [faulting between $N(5) = 170$ and 105], and at the western edge of Isidis Planitia [faulting between $N(5) = 180$ and 68]. Although the sequence of events was the same, the timing of this activity was unique to each geographic region. All these events postdate the presumed formation of the Martian dichotomy, and suggest that the formation of presently observed structures are the result of a long and complex geologic history.

INTRODUCTION

The Martian crustal dichotomy is generally believed to be a very old feature, due to mantle convection (*Wise et al.*, 1979) or a giant single impact (*Wilhelms and Squyres*, 1984). Later erosion has been proposed to explain the present location of the boundary between the northern lowlands and southern highlands in some localities, and to account for the numerous erosional remnants north of the present boundary (*Scott*, 1978; *Hiller*, 1979; *Maxwell et al.*, 1985; *Frey et al.*, 1986). The overall model of crustal history implied by the "mantle overturn" and "megaimpact" hypotheses involves a single, probably catastrophic, early structural event that created the crustal dichotomy, followed by episodic or continuous erosion and deposition that have modified the position and appearance of the boundary, and partially filled the northern lowlands with volcanic rocks and/or sediments (*McGill*, 1986; *Lucchitta et al.*, 1986). Mapping and crater dating of surfaces and materials on both sides of the present dichotomy boundary allow us to place limits on the later history of erosion and deposition, and suggest that the structural history of the northern plains is complex; explaining the dichotomy by a unique early structural "megaevent" may be a significant oversimplification.

In order to look at the detailed timing of structural events at the transition zone between the ancient cratered terrain and the northern smooth plains, we are concentrating our studies of crater dating in the Amenthes region, where relatively old surfaces are not obscured by the younger volcanic plains related to Tharsis volcanism. In this region the old upland

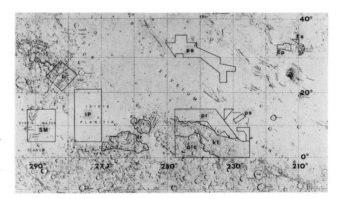

Fig. 1. Shaded relief map of the equatorial region of Mars' eastern hemisphere. Areas outlined were used for crater counts to determine age relations along the dichotomy. Ir = Isidis rim; plc = cratered plateau; SM = Syrtis Major plateau; kt = knobby terrain; pr = ridged plains; IP = Isidis Planitia; ps = smooth plains; Es = Elysium shield; Ep = Elysium plains. Base map is 1:15,000,000 Shaded Relief Map of the Eastern Region of Mars (*U.S. Geological Survey*, 1985).

surface immediately south of the northern plains boundary appears more modified than its pristine condition farther south. This modification is the result of one or more processes, including burial by lavas, planation, and pervasive fracturing followed by erosion and faulting. North of the dichotomy boundary, the modified upland is buried by younger materials, but remnants of older surfaces are present in many places in the form of knobs and mesas (*Scott*, 1978; *Semeniuk and Frey*, 1986; *McGill*, 1986). Geologic mapping based on Mariner 9 and early Viking Orbiter images allowed *Hiller* (1979) and *Scott et al.* (1978) to interpret the knobby terrain as remnants of the cratered plateau. *Tanaka* (1986) used crater densities read directly from cumulative distribution plots to infer that border faulting occurred between the Late Noachian and Early Hesperian Epochs. The exposure of faults and modified surface units provides the opportunity to place the various tectonic, depositional, and erosional events in chronological order, and to test crater counting methods for age determinations in an area where there is independent evidence for the timing of geologic events.

The rim material of the Isidis basin consists of faulted massifs marking a semicircular arc on the south side of the plains filled structure (Fig. 1). Additional ancient terrain deposits consist of the cratered plateau material (plc) south of the boundary, which also records resurfacing and reestablishment of a surface capable of retaining a crater population. Crater frequency counts were done on these units, as well as transitional units north of the boundary, and different plains units of the northern plains to bracket the timing of faulting and resurfacing. Counting results reported here also include areas of the shield and surrounding plains of Elysium, a large volcanic construct northeast of Amenthes.

The overall objective of this research is to determine whether there is any tectonic evidence in the relatively recent history of the boundary zone that will place constraints on the origin of the Martian dichotomy. Several individual questions are thus addressed: (1) Are the age relations determined by superposition, crosscutting faults, and visible resurfacing episodes consistent with those determined by crater counting methods that assume a specific Mars crater production curve (e.g., *Frey et al.*, 1987)? (2) What can the detailed age relations tell us about the evolutionary history in the Amenthes region itself? and (3) Is the sequence of "postdichotomy" events consistent with either the time of scarp modification or sequence of modification processes elsewhere in the eastern hemisphere, or are we looking at the results of an event (megaimpact or mantle overturn) unique to this particular region?

TABLE 1. Summary of surface ages in the Amenthes Region.

Region	n	Area (×1000 km^2)	N(1)*	N(1)†	N(5)‡	N(5)§	N(16)§
Elysium Region							
Surrounding Plains (Ep)	89	41	2000	3300	80	105	—¶
Northern Shield (Es)	84	23	4000	6800	140	170	—¶
Plains Units							
Isidis Planitia (Ip)	133	444	1600	3400	60	68	—¶
Smooth Plains (ps)	467	351	2100	4600	70	115	—¶
Syrtis Major (sm)	186	441	4100	8200	150	180	23
Ridged Plains (pr)	186	371	5100	12000	190	190	30
Transition Zone (kt)							
Postresurfacing	114	382	5000	13500	190	205	40
Old Surface	132	382	14000	62000	—¶	250	77
Upland Surfaces (plc)							
Postresurfacing	155	173	5800	11000	230	290	23
Cratered Plateau	272	173	19500	80000	790	790	175
Isidis Rim (Ir)							
Postfaulting (MC14-SW)	75	192	6000	27000	240	245	50
Postfaulting (MC13-NE)	54	117	3300	4900	120	120	—¶
Rim Material	102	192	19000	85000	—¶	390	135

All values given are number of craters greater than the given diameter/10^{6}km^2.
*Derived from *Neukum* (1983) crater production curve.
†Derived from *Neukum and Hiller* (1981) crater production curve.
‡Determined from 1983 production curve where it crosses 5 km diameter line.
§Determined directly from where crater distribution crosses the given diameter (*Tanaka*, 1986).
¶Ages are not given where craters were not counted in that diameter range, or for older units, where obvious effects of resurfacing have modified the crater distribution so that it is not fit by a production curve.

Methods

We have determined the ages of surface units and tectonic events in the Amenthes region (Fig. 1) by using traditional photogeologic mapping of material units, determining the ages of faults and fractures relative to erosional and depositional surfaces using superposition and cross-cutting relationships, measuring the crater size-frequency distributions of the surfaces, and determining the age of the surface with a variety of techniques (Table 1). In previous studies, *Wise et al.* (1979) used areas of "what appear to be homogeneous crater populations" to determine ages of surfaces, *Neukum and Hiller* (1981) used a Mars standard crater production curve fit to a variety of geologic units to determine relative and absolute ages, and *Tanaka* (1986) used a variety of crater sizes to determine relative and absolute ages for Mars time-stratigraphic system boundaries. Because we are investigating the relative timing of structural modification in a discrete area of Mars, we have not attempted to derive absolute ages. Instead, we are concentrating on the consistency of crater count methods for events in the boundary region and the implications for its structural evolution.

All resurfacing ages were determined by checking the superposition relationships for every crater used to derive an age. For upland surfaces in quadrangles MC-5 SW, MC-13 NE, and MC-14, we have determined a lower limit for the age of the old upland surface by counting all craters present, and the age of resurfacing by counting only those craters superposed on the eroded or partially buried upland surface.

Methods for reducing crater count data to standard frequencies of a given size crater per unit area have been reviewed extensively by *Neukum and Wise* (1976), *Neukum and Hiller* (1981), and *Tanaka* (1986). Briefly, the individual measurements for a given area are plotted as a cumulative distribution normalized to a unit area on logarithmic paper, and the number of craters greater than a given diameter [N(D)] is either read directly from the plot, or more commonly, determined from fitting a production curve to the data and reading values from the curve. *Neukum and Wise* (1976) and *Neukum and Hiller* (1981) used a unit area of 1 km^2 in calculation of their crater densities, but more recent work is based on number of craters per 10^6 km^2, and it is that value we adopt for this paper.

The key difference in absolute crater-density number (alias crater age or crater-retention age) determined by different workers is the choice of production curve used to fit the data. Although early work on Martian crater distributions suggested a D^{-2} power law distribution (*Hartmann*, 1973), more recent work by *Neukum and Wise* (1976) and *Neukum and Hiller* (1981) has suggested that a polynomial function best fits a variety of Martian terrains. These standard production curves were based on crater counts in the 0.1-20 km diameter range from Lunae Planum and other plains units. More recently, *Neukum* (1983) has added counts from additional areas on Mars and produced an even flatter standard production curve.

Additional complexities in surface-age determinations are due to resurfacing, the obliteration of craters smaller than a given diameter by erosion or deposition. Thus, *Neukum and Hiller* (1981) split their plots where they deviated from the production curve and replotted the smaller craters to obtain a resurfacing age. The results obtained with this method are acutely dependent on which production curve is used. The 1981 production curve flattens in the 1- to 10-km-diameter range, and thus incorporates a distribution that has a "bump" in the curve at 10 km. The 1983 production curve is even more irregular in the 2- to 20-km-diameter range, and using this curve eliminates several of the additional populations that would be inferred from the use of more linear production curves.

To illustrate that determining resurfacing ages is model dependent, we have determined resurfacing ages for an area of Isidis rim (Fig. 2) using two different production curves (Fig. 3a), and by separation of crater populations using superposition (Fig. 3b). To obtain a minimum formation age for the surface, all craters were counted on the rim material (location shown in Fig. 1) and plotted as a cumulative distribution in Fig. 3a. Both the *Neukum and Hiller* (1981) and the *Neukum* (1983) curves were fit visually to the frequency distribution of all craters (solid circles), and the 5-km values read from the curves. As shown in Fig. 3, neither curve fits the data distribution very well because of the resurfacing that has reduced the number of craters <20 km in diameter. Craters smaller than 20 km were replotted as a cumulative distribution as done by *Neukum and Hiller* (1981), and again fit to both production curves (Fig. 3a), suggesting a resurfacing age of N(5) = 260 (1983 fit) or N(5) = 390 (1981 fit). The difference between these ages is significant.

In order to determine whether either the 1981 or 1983 resurfacing age was consistent with that determined stratigraphically, the resurfacing age was determined by counting only those craters that are postresurfacing, based on

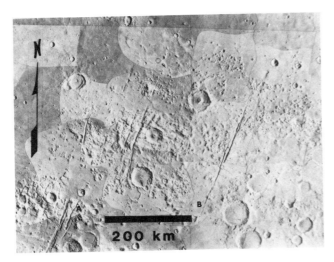

Fig. 2. Eastern rim of the Isidis basin. Two sets of northeast faults are evident; those flooded by plains (**a**) and those that cross cut plains units (**b**). Large discontinuous plains-flooded grabens are overlain by a younger population of craters that suggest stabilization of the surface by crater age N(5) = 240. (Viking Orbiter Photomosaic MC-14 SW).

superposition relationships (solid circles in Fig. 3b). The age of N(5) = 240 read from the 1983 curve on Fig. 3b agrees very well with the age determined by curve splitting using the 1983 curve, but this is not always true. In addition, the replotted smaller craters in Fig. 3a suggest the possibility of a second resurfacing if the 1981 curve is used (cf. *Frey et al.,* 1987), but not if the 1983 curve is used. For these reasons, we believe that it is better to use superposition to sort craters into pre- and postresurfacing populations than it is to depend on production curves to do the sorting.

Finally, we have found that the 1983 curve provides a better overall fit for ages of upland resurfacing and most northern plains units. Thus, for internal consistency, we have used this curve throughout. Because many of our cumulative diameter-frequency plots do not deviate from this curve near D = 5 km (except for normal data scatter), many of our N(5) ages are essentially the same as other direct N(5) ages determined by estimating the 5-km density by using a D^{-2} line (*Tanaka,* 1986) or by fitting a line to the half dozen or so points closest to 5 km on the plot. Table 1 provides other crater ages determined from our plots for comparison with ages determined by other workers.

STRUCTURAL CHRONOLOGY IN THE AMENTHES REGION

Upland Surfaces

The Isidis basin rim is one of the oldest exposed surfaces in the Amenthes region. Only rugged, massif-like rim materials still remain in the immediate vicinity of the basin, and no trace of an ejecta blanket exists in the surrounding terrain. According to the stratigraphic chronology proposed by *Tanaka* (1986), the basin is part of the Lower Noachian Series with a crater density N(16) = 120-300. The rim material is faulted by

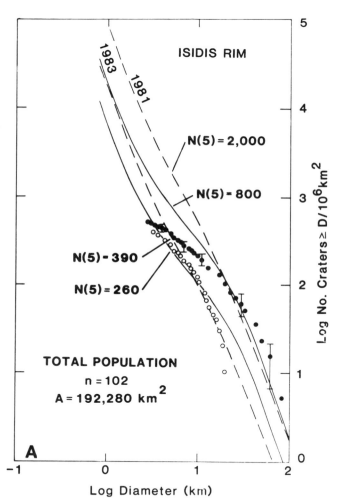

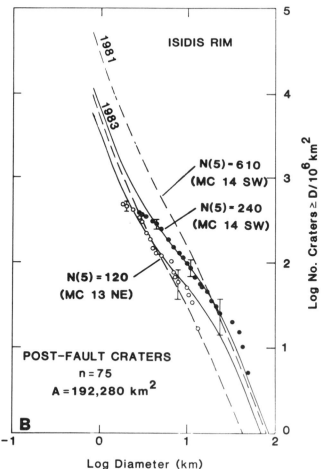

Fig. 3. Cumulative frequency plots of Isidis rim. (**a**) Total population fit to *Neukum and Hiller* (1981) production curve (dashed line) and *Neukum* (1983) curve (solid line). Solid circles are total population of craters, and open circles are remaining population after subtracting craters >20 km. (**b**) Postfaulting crater population on Isidis rim material (solid circles). This distribution is based on superposition rather than subtraction (Fig. 3a), and demonstrates the model dependence of the subtraction technique. Open circles are postfault craters on the northwest rim of the basin (from MC-13 NE).

Fig. 4. Rimless, plains-filled craters in the cratered plateau material south of the cratered terrain boundary. The etched and sculpted appearance of the terrain suggests erosional stripping of rim materials rather than deposition of material up to the rim crest. (Viking Orbiter Frame 381S49 in quadrangle MC-14 SW)).

grabens circumferential to the center of the basin that are well developed on the northwest and southeast sides and infilled by plains material. Additional small north/northeast-trending faults cut the rim material and a few craters on the southeast side of the basin (Fig. 2), although most of the major faults occurred before emplacement of the plains. For the young crater population shown in Fig. 3b, all craters that were faulted or plains filled were omitted from the count. An additional crater count on the northwest rim of the Isidis basin yields an age of N(5) = 120 for the age of termination of circumferential faulting (Fig. 3b). Consequently, by crater number N(5) = 120-240, the circumferential faulting of the Isidis basin had ceased. The larger faults had ceased forming earlier, but statistics on the few craters that are cut by these large faults are so poor that it is unreliable to determine that age.

Stratigraphically, the next youngest upland unit has been mapped as cratered plateau material (*Hiller,* 1979) and occurs to the east of the Isidis basin. The surface is characterized by scattered dendritic drainage patterns, numerous north/ northwest-oriented scarps, and has two morphologically distinct populations of craters. Numerous craters in the cratered plateau lack a raised rim and evidence for an ejecta blanket, and their interiors are commonly flooded with smooth plains material. As shown in Fig. 4, the surface surrounding these craters has an etched and lineated appearance implying erosional stripping of their rims and ejecta rather than deposition of material up to the rim crest. The maximum diameter for these modified craters is 40 km, which, based on Martian diameter/rim height curves (*Pike and Davis,* 1984), suggests removal (or deposition) of 300 to 600 m of material.

As would be expected from visual examination of the crater distribution on the cratered plateau, it is difficult to fit the frequency data by one curve (Fig. 5). The maximum age for the surface (all craters, Fig. 5) determine from the 1983 curve crossing is N(5) = 790, identical to that which would be obtained by reading a "direct" age from the crater distribution (Table 1). The "young" crater population yields an age of N(5) = 230, essentially identical to the postfaulting age determined for Isidis rim material (Fig. 5).

Structural deformation of the cratered plateau is present in two types of scarps. An older, highly degraded system of west-northwest oriented ridges and scarps occurs parallel to the cratered terrain boundary and radial to the Isidis basin. Fresher appearing scarps have orientations similar to the old ridges, but commonly deviate to more northerly trends (Fig. 6). These scarps are continuous across the old rimless craters, suggesting that structural modification continued to N(5) ~ 240, but did not affect the younger crater population.

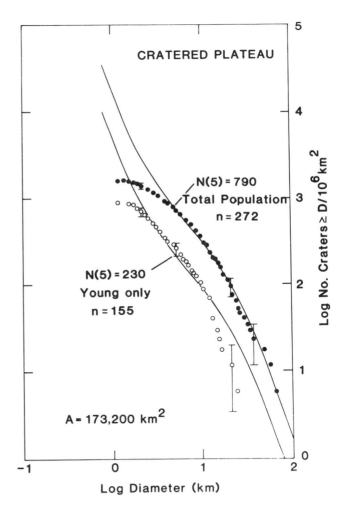

Fig. 5. Crater frequency plots for cratered plateau material south of the cratered terrain boundary in MC-14 SW. Total population of craters (solid circles) shows a fall-off at about 4 km, indicative of resurfacing of the unit. Young population of craters including only those with rims and ejecta (open circles) suggests stabilization of the surface by crater age N(5) = 230.

Fig. 6. Young west/northwest trending scarp in the cratered plateau material that crosscuts rimless crater. Faulting in the region occurred between N(16) = 135 and N(5) = 230, but did not continue into the time represented by younger craters in the region. (Viking Orbiter Frame 381S62 in quadrangle MC-22 NW).

Transition Zone

A 300-km-wide zone of scattered knobs and plateaus extends northward from the cratered terrain boundary into the northern lowlands where the density of knobs and mesas gradually decreases. Large mesas near the cratered plateau are most easily interpreted as remnants from backwasting of that unit, but the smaller isolated plateaus and knobby hills are subject to more uncertainty in interpretation. They have been interpreted as ancient terrain protruding through the younger plains material, resistant volcanic layers of once more extensive plains deposits, or as consolidated portions of a once widespread aeolian deposit (*Greeley and Guest,* 1978; *Scott,* 1978; *Hiller,* 1979). Like the crater counts of the Isidis rim and cratered plateau materials, our counts of this region were designed to determine whether the buried (?) unit north of the boundary was the same age as the cratered plateau, and hence, whether the knobs were remnants of that unit.

Within this zone (kt in Fig. 1) all possible craters, even vaguely circular arcs of knobs that may once have been a continuous crater rim, were counted to obtain the maximum age of the surface. As shown in Fig. 7, we have fit the production curve to the larger (>20 km) craters because of the deviation from the curve due to resurfacing by plains material. The maximum age thus derived is N(16) = 77; a 5-km age is not reliable because of the fall-off in crater frequency. The 16-km age read directly from the distribution is less than the maximum age of the cratered plateau, but within the range of the 1 sigma error associated with the crater diameters used to fit both curves. The large crater distribution (>20 km) of both units is indistinguishable when they are plotted on the same graph. Consequently, we believe the younger maximum age of the knobby terrain may be partially due to inherent difficulties in fitting the production curve to a crater distribution that has undergone resurfacing, although it is possible that some of the discrepancy is real if we are not seeing all of the craters on the buried surface because of their disruption by faulting.

The youngest age of materials within the transition zone is based on only those craters that are superposed on either the knobs and plateaus, or the intervening plains materials. The "young" crater distribution matches the 1983 production curve quite well (Fig. 7), resulting in an age of N(5) = 190 for stabilization of this surface.

Plains Materials

Four types of plains units occur both north and south of the cratered terrain boundary in the Amenthes region: Syrtis Major Planum (SM), ridged plains (pr), smooth plains (ps), and Isidis Planitia (IP). On the southwest edge of the Isidis basin, the Syrtis Major volcanic construct is surrounded by a

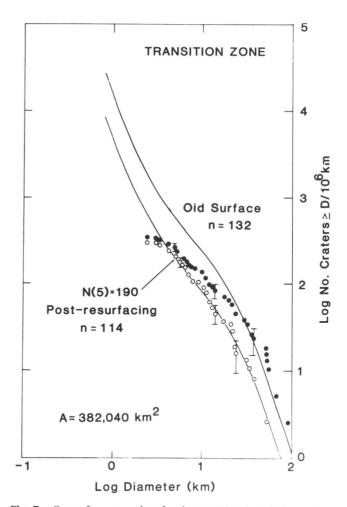

Fig. 7. Crater frequency plots for the transition zone between the cratered plateau and the northern smooth plains. Total population of craters including buried rims is shown by solid circles; open circles are craters superposed on knobs, mesas, and intervening plains units.

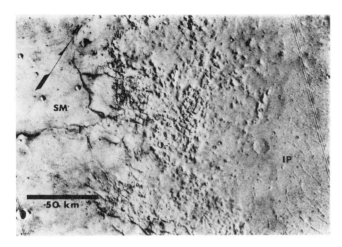

Fig. 8. Contact between plains of Syrtis Major Planum (SM) and Isidis Planitia (IP). Isolated plateaus and knobs extend beneath the smooth plains material of Isidis Planitia, and most erosional backwasting to produce mesas and knobs took place prior to emplacement of smooth plains that encircle the central part of the Isidis basin. (Viking Orbiter Frame 377S60 in quadrangle MC-13 SE).

complex of plains units that is elevated 3-4 km above the floor of Isidis Planitia (*Schaber*, 1982). The plains display prominent wrinkle ridges and scarps, and obscure the Isidis rim where they descend into the basin floor. Juxtaposed to the Syrtis Major plains units are the hummocky plains of the Isidis basin floor. These materials are characterized by small, semicircular arcuate ridges (not wrinkle ridges), and linear strings of isolated mounds that *Grizzaffi and Schultz* (1987) interpret as evidence for deposition and removal of a thick layer of material.

The contact between plains units of Syrtis Major and Isidis Planitia is marked by plateaus and intervening smooth plains that encircle the hummocky plains of the central Isidis basin. Individual plateaus of Syrtis Major Planum materials gradually are replaced by more subdued knobs and hills, and eventually grade into the hillocks and hummocky plains of the central Isidis basin (Fig. 8). The fracturing that preceded mesa development must be younger than the Syrtis Major plateau, which has a formation age of $N(5) = 150$ (Fig. 9). The young limit on both fracturing and backwasting leading to mesa development is the age of the Isidis Planitia plains, $N(5) = 60$.

Northern lowland plains units similarly provide a young limit on mesa development adjacent to the cratered plateau material. Both smooth plains and ridged plains occur intermixed with isolated knobs in the transition zone and within depressions in the cratered plateau south of the boundary. North of the transition zone, contacts between these two types of plains are gradational. Counts on the ridged plains in MC-14 SE north of the transition zone indicate an age of $N(5) = 190$ (Fig. 10), identical to the age determined for postresurfacing within the transition zone. The smooth plains yield an age of $N(5) = 70$, slightly older but virtually indistinguishable from that determined for Isidis Planitia. Both ages provide a young limit on faulting that contributed to initial plateau development, but debris aprons that are superposed on plains adjacent to plateaus indicate that erosion and cliff retreat continued past these times.

Elysium Region

On the north flank of Elysium Mons, photogeologic evidence for a sequence of faulting, plateau formation, and cliff retreat of isolated mesas is similar to that seen in the western part of the Isidis basin. The northern part of the volcanic shield is cut by curved faults concave to the northern plains, which enclose polygonal plateaus separated by straight fault segments. Away from the shield, the plateaus are more widely spaced, and are mixed with knobby hills 1-5 km in diameter (Fig. 11). Because the sequence of structural disruption and modification appears similar to that observed in the Isidis basin,

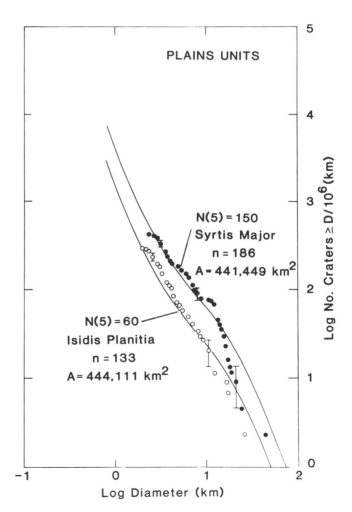

Fig. 9. Crater frequency plots for Syrtis Major Planum (solid circles) and the central part of Isidis Planitia (open circles).

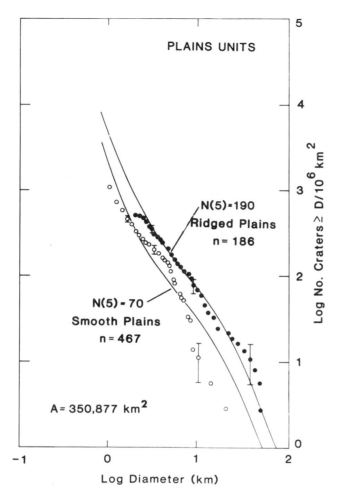

Fig. 10. Crater frequency plots for smooth and ridged plains of the lowlands north of the transition zone. Data for the ridged plains (solid circles) is well fit by the 1983 production curve, but that of the smooth plains (open circles) is not well fit, possibly due to a further resurfacing episode removing craters less than ~3-4 km (?).

we determined the age of the shield deposits and of the intervening plains surrounding the mesas to place upper and lower limits on the age of faulting.

Crater ages determined for the plains and shield materials (Fig. 12) are consistent with later infilling of plains units surrounding the shield. The crater frequency on volcanic flows on the shield itself are fit fairly well by the 1983 production curve, although both shield and surrounding plains counts show evidence for resurfacing by the fall-off in frequency at diameters <1.5 km. On the shield, this deviation could be attributed to resurfacing by younger flows, but there is no such similar evidence for resurfacing on the plains unit north of the shield. The age of the shield is $N(5) = 140$, essentially the same as that determined for Syrtis Major Planum [$N(5) = 150$], although the lower density of large craters alone suggests that the shield is younger than Syrtis Major, as found by *Tanaka* (1986). The age of the surrounding plains is $N(5) = 80$, which

is also similar to that of the smooth plains to the southwest in MC-14. This is interpreted to be the time at which most of the plateau development ceased, except for minor continued cliff retreat.

DISCUSSION

Cratered Terrain Tectonics and Resurfacing

The oldest surfaces we have dated in the Amenthes region, the Isidis rim and cratered plateau material, are cut by faults and scarps, remnants of the earliest observed tectonism along the highland side of the cratered terrain boundary (Fig. 13). The time of formation of large grabens circumferential to Isidis is constrained by infilling of the grabens by ridged plains material, but more closely by dating of the crater population that includes plains-filled highland craters and those craters that are superposed on the faults themselves. Formation of these grabens ceased by $N(5) = 240$ at about the same time as cessation of the process that removed the rims from craters on the cratered plateau. *Comer et al.* (1985) preferred a model for concentric graben formation in which the superisostatic

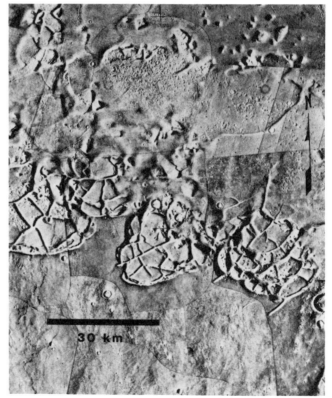

Fig. 11. Faulted plateaus and isolated outliers on the northern flank of Elysium Mons. The shield was faulted and the isolated plateaus were created by cliff retreat before emplacement of the intervening plains deposits that surround the shield. (Viking 1:500,000 photomosaic MTM 35212).

load responsible for faulting extended to the edge of the volcanic fill of Isidis. If this model is correct, our results imply that this portion of the fill is now buried by more recent plains.

The old population of rimless craters in the cratered plateau can be used to constrain highland tectonic events regardless of the process that removed the rims. As shown below, however, the interpretation of that event(s) may also allow more precise dating of boundary faulting. *Wilhelms* (1974) preferred a depositional origin for what was then termed the plateau plains, based on analogy with lunar plains and images available at that time from Mariner 9. However, rimless craters in the cratered plateau material are not in all cases surrounded by smooth plains. Such craters also interrupt the contact between gullied uplands and smoother plateau material, and may also be completely surrounded by rugged plateau material. Deposition of highland plains material may have helped to

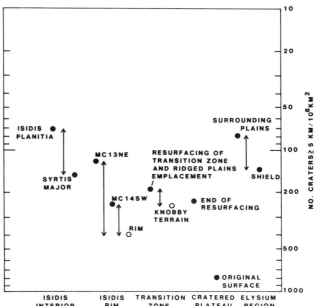

Fig. 13. Summary of resurfacing and tectonic events in the Isidis-Amenthes-Elysium region. In areas of volcanic plains emplacement, a similar sequence of faulting, erosion, and younger plains deposition is seen, but the timing of individual events in the sequence varies with location, as shown by double headed arrows indicating age ranges during which fracturing occurred. Open circles represent ages determined directly (see footnotes in Table 1).

subdue the rims but, because of the varied terrain exposed surrounding the craters, was not solely responsible for obscuring the rim. Consequently, we believe an erosional origin for rim and ejecta removal is supported by the geologic setting of these upland craters, and that the event(s) occurred well after the Isidis impact, since such craters are present on the Isidis rim.

Based on crater dating alone, it is not possible to tell whether the erosional process operated for a long time period (as would be expected from enhanced atmospheric conditions accelerating erosion), or was a function of the substrate (volatile-rich material that lost cohesion due to climate change). Our estimate of 300-600 m of removal is similar to the 300 m of removal estimated by *Grizzaffi and Schultz* (1987) for the interior of Isidis Planitia. A further suggestion of burial of this terrain was made by *Wilhelms and Baldwin* (1987), who proposed that both the uplands and preplains knobby terrain were covered by ice-rich deposits of varying thickness, which created the gullied uplands and interknob deposits when later eroded. If the erosion of this transient material was the same process that removed crater rims, then both deposition and erosion of such material occurred between $N(16) = 135$ and $N(5) = 230$, the postresurfacing age of the cratered plateau material.

Although the exact cause of planation remains unspecified, the timing of planation does place a constraint on the time of border faulting. If the faulting occurred after planation (as

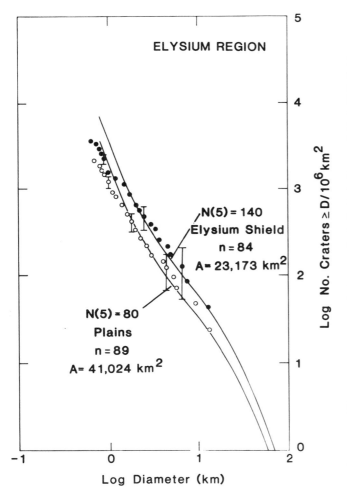

Fig. 12. Crater frequency plots for the northern shield of Elysium Mons (solid circles) and surrounding plains (open circles). Fall-off in shield crater population at diameters <1.5 km is probably due to resurfacing by multiple flows on the north flank of the volcano.

is suggested by the smooth-topped character of mesas adjacent to the uplands), then faulting had to occur after $N(5) = 230$, the resurfacing age, but before $N(5) = 190$, the age of the adjacent unfaulted plains.

Plateau Formation and Cliff Retreat

Based on age determinations reported here and additional dating of tectonic events in eastern Utopia Planitia, the same sequence of events occurred in MC-5 SW and MC-13 NE as in the Amenthes region (*McGill,* 1987). Dating indicates a rapid sequence of border zone formation that involved: (1) planation of the upland surface, (2) downfaulting of transition zones as separate blocks rather than continuous terraces, (3) erosion to produce mesas along the northern edge of the transition zone, and (4) flooding by ridged plains materials.

In the fretted terrain north of the flank of Elysium Mons, the complex plains surface has an age of $N(5) = 80$. The scattered mesas on these plains surfaces were present as discrete, isolated landforms before the intervening plains units formed, as indicated by two lines of evidence.

1. Wherever age relationships can be determined, the craters used to date the plains formed after the mesas existed in their present form; crater ejecta lies on plains surfaces between mesas, and clear evidence exists locally for obstruction of the flow of fluidized ejecta by mesas.
2. Plains lavas commonly terminate near groups of mesas so that it appears as if many of the mesas occur within or between younger lava flows. This relationship makes sense if the preplains surface near older mesas was slightly higher than in intervening areas where all of the older plateau material had been eroded away, and where younger plains lavas have ponded.

Thus, the lava capped plateau surface of Elysium was faulted and eroded in a relatively short time interval between $N(5) = 140$ and $N(5) = 80$, an interval almost certainly larger than the actual one because flow lobes younger than $N(5) = 140$ are faulted, and the $N(5) = 80$-age plains lavas need not have formed immediately after cessation of mesa erosion. These events occurred after the final volcanic eruptions of the north flank of Elysium, since there is no evidence for recent flows extending beyond or over the escarpment of the Elysium shield (*Mouginis-Mark et al.,* 1984).

The polygonal terrain material of Adamas Labyrinthus was deposited in the same time interval (*McGill,* 1986) as faulting and erosion of the cratered terrain boundary and the northern plateaus. Between Elysium Mons and Adamas Labyrinthus are numerous channels on a terrain characterized by mixed smooth and intricately eroded surfaces referred to as "disordered plains" by *Carr* (1981, pp. 82-83). It thus is not unreasonable to infer that materials eroded from the plateau peripheral to Elysium Mons were deposited as what is now polygonal terrain, an origin for these materials consistent with the mechanically most reasonable hypothesis for large-scale patterned fracturing (*McGill,* 1986), and consistent with large-scale patterns of channels and northern hemisphere topography (*Lucchitta et al.,* 1986).

CONCLUSIONS

1. Dating of surfaces and tectonic events in the eastern hemisphere of Mars, where there is independent stratigraphic control on periods of resurfacing, indicates that crater frequency ages determined by subtraction where crater frequency deviates from production curves are highly model dependent. Often they are not equivalent to ages determined from separate crater populations isolated by superposition and cross-cutting relationships.

2. The oldest structural event in the Amenthes region is the formation of grabens concentric to the Isidis basin (Amenthes Fossae), which ceased forming before final emplacement of the uppermost plains units of Isidis Planitia or the northern plains.

3. The planation process that removed crater rims from old craters in the upland cratered plateau ceased by $N(5) = 240$, suggesting that border faulting in the Amenthes region took place in a very short time period, post-planation of the cratered plateau, but pre-deposition of the plains at $N(5) = 190$. Relationships in the transition zone suggest that the true age of border faulting is very close to $N(5) = 190$.

4. Erosion of plateaus in the Elysium region took place at the same time as the deposition of the uppermost surface of the polygonal northern plains, consistent with the hypothesis of a sedimentary origin for that terrain.

5. Different areas of volcanic plains and upland surfaces have undergone similar sequences of tectonic detachment, faulting, and infilling by younger plains. Although the same sequence of events occurred at these different localities, the dating reported here indicates that faulting and plateau formation occurred at different times.

6. Both the "megaimpact" (*Wilhelms and Squyres,* 1984) and the "mantle overturn" (*Wise et al.,* 1979) models proposed to explain the Martian crustal dichotomy imply a single major structural disruption of the crust, and a "multiple impact" model (*Frey et al.,* 1986) requires numerous scattered disruptions early in Martian history. Our results indicate that fracturing and associated resurfacing in and around the Amenthes region was an ongoing process, and also apparently of different ages in different places. Although this result by itself is not strong enough to eliminate any of the models, it is not favorable to them, and suggests that a search for a more protracted origin for the dichotomy may be in order.

Acknowledgments. This paper has benefitted from the useful review comments of K. Tanaka and J. Zimbelman. Part of this work was done while one of us (GEM) was a Visiting Scientist at the Center for Earth and Planetary Studies, Smithsonian Institution. This research was supported by NASA grants NAGW-129 (at the Smithsonian Institution) and NGR-22-010-076 (at the University of Massachusetts). We are also indebted to the Smithsonian Office of Fellowships and Grants for support through the Short-Term Visit Program.

REFERENCES

Carr M. H. (1981) *The Surface of Mars.* Yale University Press, New Haven. 232 pp.

Comer R. P., Solomon S. C., and Head J. W. (1985) Mars: Thickness of the lithosphere from the tectonic response to volcanic loads. *Rev. Geophys., 23,* 61-92.

Frey H. V., Schultz R. A., and Maxwell T. A. (1986) The Martian crustal dichotomy: Product of accretion and not a specific event? (abstract). In *Lunar and Planetary Science XVII*, pp. 241-242. Lunar and Planetary Institute, Houston.

Frey H., Semeniuk J. A., and Tokarcik S. (1987) Common age resurfacing events in the Elysium-Amazonis knobby terrain on Mars (abstract). In *Lunar and Planetary Science XVIII*, pp. 304-305. Lunar and Planetary Institute, Houston.

Greeley R. and Guest J. (1978) Geologic Map of the Casius Quadrangle of Mars. *U. S. Geol. Surv. Misc. Invest. Map I-1038*, U.S. Geol. Survey, Reston.

Grizzaffi P. and Schultz P. (1987) Evidence for a thick transient layer in the Isidis impact basin (abstract). In *Lunar and Planetary Science XVIII*, pp. 370-371. Lunar and Planetary Institute, Houston.

Hartmann W. K. (1973) Martian cratering, 4, Mariner 9 initial analysis of cratering chronology. *J. Geophys. Res.*, 78, 4096-4116.

Hiller K. H. (1979) Geologic Map of the Amenthes Quadrangle of Mars. *U.S. Geol. Surv. Misc. Invest. Map I-1110*, U.S. Geol. Survey, Reston.

Lucchitta B. K., Ferguson H. M., and Summers C. (1986) Sedimentary deposits in the northern lowland-plains, Mars. *Proc. Lunar Planet. Sci. Conf. 17th*, in *J. Geophys. Res.*, 91, E166-E174.

Maxwell T. A., Leff C., Hooper D., and Doorn P. (1985) The Mars hemispheric dichotomy: Morphology and stratigraphic implications of Viking color data (abstract). *Geol. Soc. Amer. Abstracts with Program*, 17, 655.

McGill G. E. (1986) The gian polygons of Utopia, northern Martian plains. *Geophys. Res. Lett.*, 13, 705-708.

McGill G. E. (1987) Relative age of faulting, mesa development, and polygonal terrane, Eastern Utopia Planitia, Mars (abstract). In *Lunar and Planetary Science XVIII*, pp. 620-621. Lunar and Planetary Institute, Houston.

Mouginis-Mark P. J., Wilson L., Head J. W., Brown S. H., Hall J. L., and Sullivan K. D. (1984) Elysium Planitia, Mars: Regional geology, volcanology, and evidence for volcano-ground ice interactions. *Earth, Moon, and Planets*, 30, 149-173.

Neukum G. (1983) Meteoritenbombardment und Datierung Planetarer Oberflachen, Habilitationsschrift, Ludwig-Maximilians-Universitat, Munich. 186 pp.

Neukum G. and Hiller K. (1981) Martian ages. *J. Geophys. Res.*, 86, 3097-3121.

Neukum G. and Wise D. U. (1976) Mars: A standard crater curve and possible new time scale. *Science*, 194, 1381-1387.

Pike R. J. and Davis P. A. (1984) Toward a topographic model of martian craters from photoclinometry (abstract). In *Lunar and Planetary Science XV*, pp. 645-646. Lunar and Planetary Institute, Houston.

Schaber G. G. (1982) Syrtis Major: A low relief volcanic shield. *J. Geophys. Res.*, 87, 9852-9866.

Scott D. H. (1978) Mars, highlands-lowlands: Viking contributions to Mariner relative age studies. *Icarus*, 34, 479-485.

Scott D. H., Morris E. C., and West M. N. (1978) Geologic Map of the Aeolis quadrangle of Mars. *U.S. Geol. Surv. Misc. Invest. Map I-1111*, U.S. Geol. Survey, Reston.

Semeniuk A. M. and Frey H. (1986) Characteristic structures of the highland boundary on Mars: Evidence against a single mega-impact event? (abstract). In *Lunar and Planetary Science XVII*, pp. 791-792. Lunar and Planetary Institute, Houston.

Tanaka K. L. (1986) The stratigraphy of Mars. *Proc. Lunar Planet. Sci. Conf. 17th*, in *J. Geophys. Res.*, 91, E139-E158.

U.S. Geological Survey (1985) Shaded relief maps of the eastern, western, and polar regions of Mars. *U.S. Geol. Surv. Misc. Invest. Map I-1618*, U.S. Geol. Survey, Reston.

Wilhelms D. E. (1974) Comparison of martian and lunar geologic provinces. *J. Geophys. Res.*, 79, 3933-3041.

Wilhelms D. E. and Baldwin R. J. (1987) Uplands/knobby-terrain relation on Mars (abstract). In *Lunar and Planetary Science XVIII*, pp. 1084-1085. Lunar and Planetary Institute, Houston.

Wilhelms D. E. and Squyres S. U. (1984) The Martian hemispheric dichotomy may be due to a giant impact. *Nature*, 309, 138-140.

Wise D. U., Golombek M. P. and McGill G. E. (1979) Tectonic evolution of Mars. *J. Geophys. Res.*, 84, 7934-7939.

Gossans on Mars

Roger G. Burns

Department of Earth, Atmospheric, and Planetary Sciences, Massachusetts Institute of Technology, Cambridge, MA 02139

The appearance of rusty iron-rich oxidized cappings over sulfide-bearing rocks on Earth suggests that similar gossans may have formed on the surface of Mars. On the assumption that oxidative weathering of sulfides might be a significant process on Mars, reactions involving iron-rich ultramafic igneous rocks containing pyrrhotite-pentlandite-(pyrite-chalcopyrite) assemblages similar to those occurring in terrestrial komatiites are described. Electrochemical reactions causing oxidation of massive and disseminated Fe-Ni sulfide minerals are initiated by pyrite, which may be present either as a primary phase or be formed during deep weathering of pyrrhotite. Pyrite in contact with aerated groundwater generates strongly acidic sulfate-rich solutions that stabilize and mobilize simple and complex cations of iron. Below the water table, dissolved ferric ions generate secondary sulfides enriched in Fe, Ni, Cu, etc. Above the water table, mixtures of jarosite and hydrated ferric oxides ("limonite") are precipitated by the oxidation of pyrite and dissolved Fe^{2+} and the hydrolysis of complexed ferric ions, together with silica and clay silicates derived from acid-leached silicates in basaltic host-rock. Surface features on Mars indicative of the flow of water suggest that gossan deposits probably were formed there in the past in aqueous environments. However, incomplete wall-rock alteration reactions between groundwater and poorly fractured basalt may have resulted in present-day permafrost remaining acidic. Oxidative weathering of subsurface sulfides may still be occurring on Mars, but electrochemical reactions are restricted now by the limited accessibility of fluid electrolyte to the postulated subsurface pyrrhotite-pentlandite mineralization. Gossans on Mars could indicate subsurface enrichments of Ni, Cu, and other chalcophilic metals not sampled during the Viking Lander experiments. Remote-sensed visible spectra of Mars' surface are consistent with the presence of poorly-crystalline FeOOH (ferrihydrite), jarosite, silica, and clay silicates found in gossans, while incompletely weathered pyrrhotite may represent the magnetic material observed in Martian regolith.

"Gossans are signposts that point to what lies beneath the surface. They arrest attention and incite interest as to what they may mask. . . . The finding of one may herald the discovery of buried wealth."
—M. L. Jensen and A. M. Bateman, 1979.

INTRODUCTION

The term *gossan* was probably first applied to the iron-rich oxidized cappings over sulfide-bearing ore deposits in Cornwall (*Blain and Andrew*, 1977). In Cornish usage (*Pryce*, 1778), a gossan was defined as "a kind of imperfect iron ore, commonly of a tender rotten substance, and red or rusty iron colour." The surface of Mars, too, has a red or rusty iron color and the duricrusts there are pitted and friable. The question arises, therefore, whether there are any paragenetic relationships between Martian regolith and gossans that may indicate the presence of subsurface sulfide mineralization on Mars.

Mars is generally regarded as a volatile-rich planet (*Dreibus and Wanke*, 1985), but a substantial portion of its present-day water inventory is tied up in subsurface permafrost (*Clifford*, 1985). Geomorphological features, attributed to fluvial processes during diurnal, seasonal, and historical melting of the permafrost, indicate that groundwater levels have changed periodically near the surface of Mars (*Carr*, 1987). When the water table levels change on Earth, chemical weathering processes of surficial rocks are altered, particularly those involving oxidation reactions of sulfide minerals in crustal rocks, leading to the development of conspicuous gossans. Whether or not such leached iron-stained oxidized cappings over sulfide bodies on Earth are indicative of sulfide mineralization on Mars is the focus of this paper, which examines chemical weathering processes that may have produced localized and disseminated gossans on the surface of Mars. Earlier studies have examined reaction pathways of sulfur in the Martian lithosphere (*Clark and Baird*, 1979), thermodynamic stabilities of primary and secondary minerals in the present-day weathering environment of Mars (*Gooding*, 1978), and hydroxo ferric sulfate minerals that might occur on Mars (*Burns*, 1987). The present study discusses electrochemical processes and thermodynamic relationships linking acidity to oxidation-reduction reactions between primary sulfide minerals and their oxidative weathering products present in the regolith of Mars.

FEATURES OF TERRESTRIAL GOSSANS

The identification of red or rusty iron-rich oxidized cappings over sulfide-bearing lodes has led to the discovery of subsurface ore deposits during mineral exploration on Earth (*Jensen and Bateman*, 1979; *Rose et al.*, 1979). Indeed, the assessment of colors and textures of gossans is a routine and often essential guide to the economic potential of underlying mineralization (*Locke*, 1926; *Blanchard*, 1968; *Blain and Andrew*, 1977). Gossans represent an advanced stage of chemical weathering, during which sulfide minerals and their host rocks are extensively oxidized and leached of soluble ions leaving behind poorly crystalline hydrated iron oxides and sulfates ("limonite") and disseminated silica (jasper).

In the near-surface environment sulfide minerals are unstable in the presence of aqueous solutions containing dissolved O_2, CO_2, and ionic species, which percolate slowly through the zone of aeration to the water table. In transit, chemical weathering reactions cause the primary (hypogene) sulfides to form secondary (supergene) sulfide, sulphate, and oxide mineral assemblages near the water table, producing a vertically zoned weathered profile such as that depicted in Fig. 1. A sulfide deposit thus becomes oxidized and generally leached of many of its constituents to as great a depth as oxidation can proceed, and this level is often the water table. Beneath the water table, dissolved oxygen is normally absent, but other ionic species, notably dissolved ferric iron, may continue to oxidize the primary sulfides at depth and cause additional metal enrichments in secondary sulfide minerals there.

Because most sulfide minerals are coherent electronic conductors, weathering of them proceeds by electrochemical reactions involving coupled half-cell reactions, one cathodic and one anodic, which may extend from a few Angstroms in individual crystals to several hundreds of meters in massive ore bodies (*Thornber*, 1975a,b). Since oxidation involves loss of electrons, the sulfide acts as the anode and electrons are conducted from the sulfide through the groundwater electrolyte to the cathode reaction sites. The reaction pathways depend on the composition of the original sulfide minerals.

By way of illustration, consider the evolution of a gossan forming from a mineral assemblage containing predominantly pyrrhotite (Fe_7S_8), some pentlandite [$(Fe,Ni)_9S_8$], and accessory pyrite (FeS_2) or chalcopyrite ($CuFeS_2$), which may be associated with ultrabasic and mafic igneous rocks (*Nickel et al.*, 1974; *Naldrett*, 1981; *Guilbert and Park*, 1986) and be relevant therefore to Mars (*McGetchin and Smyth*, 1978; *Morgan and Anders*, 1979). The mechanism of chemical weathering of such sulfide mineralization has been studied extensively (*Thornber*, 1975a,b) and pertinent alteration zones and chemical reactions are summarized in Fig. 2.

At the surface, meteoric water containing dissolved oxygen constitutes the cathodic half-cell reactions

$$O_2 + 2H_2O + 4e^- = 4OH^- \quad (1a)$$

or

$$O_2 + 4H^+ + 4e^- = 2H_2O \quad (1b)$$

which participate in oxidation reactions above or adjacent to the water table. Another cathodic reaction important below the water table, where the concentration of dissolved oxygen may be very low, is

$$Fe^{3+} + e^- = Fe^{2+} \quad (2)$$

These cathodic reactions couple with a variety of anionic half-cell reactions, depending on the sulfide minerals present and their proximity to the water table. In the pyrrhotite (+ pentlandite ± pyrite) assemblage under discussion, less than 10% of the ore was originally pyrite (*Thornber*, 1975b). However, in deep-weathering reactions beneath the water table (Fig. 2), secondary pyrite may form from pyrrhotite

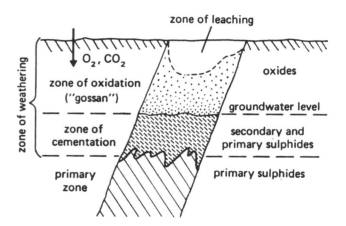

Fig. 1. Zones of weathering associated with gossan formation above sulfide mineralization (after *Bauman*, 1976). The scale of the sulfide vein may be a few microns to several meters in diameter.

$$Fe_7S_8 = 4FeS_2 + 3Fe^{2+} + 6e^- \quad (3)$$

This anodic reaction, when coupled with the cathodic reaction in equation (2), is represented by the overall reaction

$$Fe_7S_8 + 6Fe^{3+} = 4FeS_2 + 9Fe^{2+} \quad (4)$$

Pyrite, either as a primary or secondary mineral, is unique because it is the intermediary in many reactions generating gossans above most sulfide ore deposits. Thus pyrite (or its polymorph marcasite) yields the chief reagents in the chemical weathering reactions of underlying ore bodies and their host rocks, and generates the ultimate ferric oxide phases in gossans above them (*Blain and Andrew*, 1977). At or near the water table, oxidation of pyrite and dissolved Fe^{2+} occurs. For example, the anodic reaction

$$FeS_2 + 8H_2O = Fe^{2+} + 2SO_4^{2-} + 16H^+ + 14e^- \quad (5)$$

when coupled with the cathodic reaction in equation (1a) or (1b), is represented by the overall reaction

$$2FeS_2 + 2H_2O + 7O_2 = 2Fe^{2+} + 4SO_4^{2-} + 4H^+ \quad (6)$$

Further oxidation ultimately occurs above the water table

$$4Fe^{2+} + 6H_2O + O_2 = 4FeOOH + 8H^+ \quad (7)$$

$$4FeS_2 + 2H_2O + 15O_2 = 4Fe^{3+} + 8SO_4^{2-} + 4H^+ \quad (8)$$

The dissolution of pyrite is further promoted by Fe^{3+} ions

$$FeS_2 + 14Fe^{3+} + 8H_2O = 15Fe^{2+} + 2SO_4^{2-} + 16H^+ \quad (9)$$

The overall result is that strongly acidic and sulfate-rich aqueous solutions [typically pH 2 to 5, and $(SO_4^{2-}) = 10^{-2}$ m; *Sato*, 1960a; *Thornber*, 1975a,b] are produced by the oxidation of pyrite (*Garrels and Thompson*, 1965). These acidic solutions attack silicate minerals and facilitate the chemical weathering

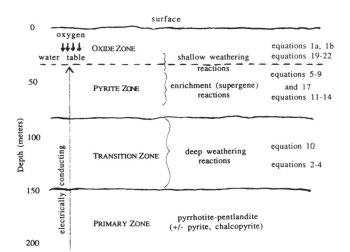

Fig. 2. Schematic representation of the zones of alteration associated with a primary pyrrhotite-pentlandite (± pyrite, chalcopyrite) ore body (modified from *Thornber*, 1975a). The equations relate to reactions formulated in the text.

of ultramafic and basaltic host rocks (*Siever and Woodford*, 1979).

Associated with pyrrhotite in mafic igneous rocks is pentlandite, $(Fe,Ni)_9S_8$, which also undergoes deep-weathering reactions. First, violarite (ideally $FeNi_2S_4$) is formed

$$(Fe,Ni)_9S_8 + 2Fe^{3+} = 2(Fe,Ni)Ni_2S_4 + 3Fe^{2+} \quad (10)$$

and subsequently reacts to violarite-polydymite ($FeNi_2S_4$-Ni_3S_4) solid-solutions by the supergene enrichment reaction

$$(FeNi)Ni_2S_4 + Ni^{2+} = (Ni,Fe)Ni_2S_4 + Fe^{2+} \quad (11)$$

Some violarite may also form from pyrrhotite (*Thornber*, 1975a). Near the water table, dissolution of violarite-polydymite phases occur

$$2(Ni,Fe)_3S_4 + 15O_2 + 2H_2O = 6(Ni^{2+},Fe^{2+}) + 8SO_4^{2-} + 4H^+ \quad (12)$$

The dissolved Ni^{2+} ions then become available for further supergene enrichment reactions below the water table (e.g., equation (11)). If chalcopyrite ($CuFeS_2$) is present in the original ore, supergene enrichment reactions lead to the formation of covellite (CuS) near the water table

$$CuFeS_2 + 4H_2O = CuS + Fe^{2+} + SO_4^{2-} + 8H^+ + 8e^- \quad (13)$$

which combined with equation (1b) gives

$$CuFeS_2 + 2O_2 = CuS + Fe^{2+} + SO_4^{2-} \quad (14)$$

The reactions formulated above, particularly equations (5)-(9), indicate that highly acidic sulphate-rich solutions containing dissolved ferrous and ferric iron enter the groundwater adjacent to weathering sulfide minerals, and become potent solvents for dissolving primary and secondary sulfides. Even in the absence of dissolved oxygen, ferric iron is a powerful oxidizing agent and lixiviant (*Dutrizac and MacDonald*, 1974). Numerous economically-important supergene enrichments in ore bodies owe their metal concentrations to the leaching properties of dissolved ferric salts. Ferric and sulfate ions are not only being continuously replenished from pyrite (e.g., equations (6) and (8)), but dissolved Fe^{3+} also generates secondary pyrite from pyrrhotite (equation (4)). Alternatively, the dissolved iron may be transported great distances away from the primary sulfide mineralization, particularly in strongly acidic groundwater, and ultimately be deposited as insoluble ferric oxides at the surface (e.g., equation (7)).

In summary, deep-weathering of primary igneous Fe-Ni-Cu sulfides is initiated by dissolved Fe^{3+} ions in percolating groundwater that may be acidic. Although dissolved oxygen is necessary to produce the ferric iron in the near-surface environment, oxygen is not essential in the formation of secondary sulfide minerals. These supergene sulfides may enrich high concentrations of Ni, Cu, etc., which might not be apparent in chemically analysed surface weathering products constituting gossans.

FORMATION OF JAROSITES

In the highly acidic sulfate solutions near the water table, the chemical weathering reactions are, in fact, more complex than those depicted by equations (5)-(9). This is because the soluble ferric iron produced by the dissolution or oxidation of pyrite exists predominantly as the complex ions $FeSO_4^+$, $Fe(SO_4)_2^-$, $FeOH^{2+}$ and $Fe(OH)_2^+$ (*Sapieszko et al.*, 1977; *Dousma et al.*, 1979). Such dissolved ferric sulfato and hydroxo complex ions may lead to the formation of a variety of hydroxo ferric sulfate minerals (*Burns*, 1987), most notably jarosites [$MFe_3(SO_4)_2(OH)_6$, where $M = H_3O^+$, K^+, Na^+, etc.]

$$2FeSO_4^+ + FeOH^{2+} + 5H_2O = [Fe_3(SO_4)_2(OH)_6]^- + 5H^+ \quad (15)$$

$$Fe(SO_4)_2^- + 2FeOH^{2+} + 4H_2O = [Fe_3(SO_4)_2(OH)_6]^- + 4H^+ \quad (16)$$

Jarosites may also form by direct oxidation of FeS_2 or dissolved Fe^{2+}

$$12FeS_2 + 45O_2 + 30H_2O + 4K^+ = 4KFe_3(SO_4)_2(OH)_6 + 16SO_4^{2-} + 36H^+ \quad (17)$$

$$6Fe^{2+} + 4SO_4^{2-} + 9H_2O + 3/2\,O_2 = 2[Fe_3(SO_4)_2(OH)_6]^- + 6H^+ \quad (18)$$

At low pH, hydronium jarosite (carphosiderite) mono-dispersed sols are stabilized (equations (15) and (16)), but in the presence of alkali metal cations (e.g., K+, Na+ derived from weathered feldspars) extremely insoluble jarosite minerals are precipitated (equation (17)). Significantly, however, such precipitates are inhibited by dissolved Ni^{2+} and Mg^{2+} (*Matijevic et al.*, 1975). These cations could be present in significant concentrations in groundwater derived from leached pentlandite and ferromagnesian silicates (olivine, pyroxenes)

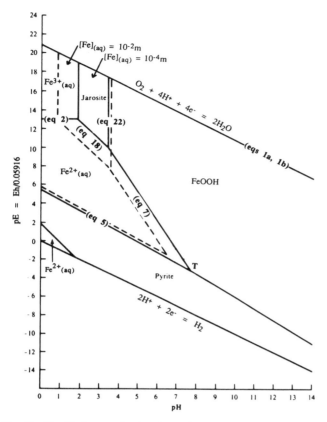

Fig. 3. Eh-pH diagram for jarosite and pyrite in equilibrium with FeOOH and dissolved iron at 25°C (modified from *Nordstrom and Munoz*, 1985, p. 303). Heavy and dashed lines correspond to total concentrations of dissolved Fe species of 10^{-4} m and 10^{-2} m, respectively, in equilibrium with 10^{-2} m dissolved S species and 10^{-4} m dissolved K^+. Equations refer to reactions cited in the text. Note that at 298.15°K, Eh = 0.05916× pE.

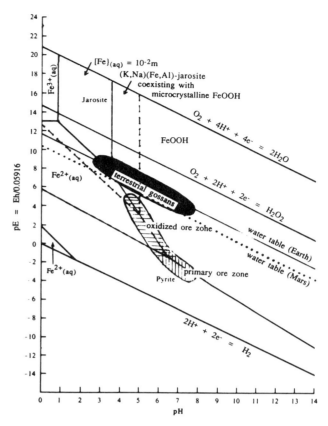

Fig. 4. Equilibrium diagram for pyrite and its oxidative products, including jarosite [$KFe_3(SO_4)_2(OH)_6$] and FeOOH, occurring in gossans at 25°C. The area bounded by dashed lines represents the approximate stability field of jarosite expanded by (K^+-Na^+) and (Fe^{3+}-Al^{3+}) solid solutions. Ranges of pE and pH measured in oxidized pyrrhotite-pentlandite assemblages are shown (data from *Thornber*, 1975b, and *Sato*, 1960a,b). Also shown are pE-pH curves for dissolved oxygen in equilibrium with the atmospheres of Earth and Mars.

in weathered mafic igneous rocks (*Siever and Woodford*, 1979).

In the less acidic environment above the water table, the dissolved ferric ions or monodispersed sols are hydrolysed to poorly crystalline FeOOH phases

$$Fe^{3+} + 2H_2O = FeOOH + 3H^+ \qquad (19)$$

$$FeOH^{2+} + H_2O = FeOOH + 2H^+ \qquad (20)$$

$$FeSO_{4+} + 2H_2O = FeOOH + SO_4^{2-} + 3H^+ \qquad (21)$$

$$[Fe_3(SO_4)_2(OH)_6]^- = 3FeOOH + 2SO_4^{2-} + 3H^+ \qquad (22)$$

The formation of FeOOH from jarosites (equation (22)) is sluggish, partially on account of its extreme insolubility (*Brown*, 1971; *Vlek et al.*, 1974), so that the two phases frequently coexist. These reactions, together with the oxidation of dissolved Fe^{2+} (equation (5)), account for the presence of jarosite and goethite in the oxidized zone above weathered sulfide bodies. It is this mixture of conspicuously colored FeOOH and basic ferric sulfate phases that characterizes "limonite" in many terrestrial gossans.

Eh-pH RELATIONSHIPS

The stability relationships of pyrite, jarosite, FeOOH (goethite or ferrihydrite), and dissolved Fe^{2+} may be shown graphically in plots of oxidation-reduction potential (Eh) versus acidity (pH) (*Garrels and Christ*, 1965, Ch. 7), such as that illustrated in Fig. 3 (*Nordstrom and Munoz*, 1985, p. 303). The relationship between Eh and pE at 25°C is given by

$$pE = Eh/0.05916 \qquad (23)$$

where Eh is the half-cell potential (volts) of such as those represented in equations (1a), (1b), (2), and (5) above and highlighted in Fig. 3. The pE-pH or "predominance-area" diagram shown in Fig. 3 outlines areas where the designated species make up more than 50% of the total concentration. Each line represents pE-pH conditions where two iron species are in thermodynamic equilibrium (e.g., pyrite and $Fe^{2+}_{(aq)}$ in equation (5)), and every intersection of three lines represents the pE-pH point where three species are in equilibrium (e.g., pyrite, FeOOH and $Fe^{2+}_{(aq)}$ at point T). The thermodynamic data summarized in Fig. 3 indicate that at 25°C pure jarosite coexists with equal concentrations of FeOOH and dissolved Fe^{2+} at pH and pE values approximating 3.3 and 10, respectively. When the activity of dissolved iron species increases from 10^{-4} m to 10^{-2} m, the stability area (or predominance field) of jarosite also increases (see Fig. 3). The jarosite-FeOOH equilibrium boundary could extend beyond pH 5 for very small grain size FeOOH (*Nordstrom and Munoz*, 1985, p. 299). Another noteworthy feature of Fig. 3 is the large stability range of dissolved Fe^{2+} ions in acidic solutions.

Plotted on Fig. 4 are pE-pH relationships that apply to groundwater in equilibrium with oxygen in Earth's atmosphere, as well as measured pE and pH values in gossan and supergene sulfide zones formed during weathering of pyrrhotite-pentlandite (+pyrite) deposits associated with mafic igneous rocks at Kambalda, Western Australia (*Thornber*, 1975a,b; *Blain and Andrew*, 1977). In this terrestrial ore deposit, the groundwater below the water table is sufficiently acidic to stabilize pure jarosite.

The stability field of jarosite may also be expanded as a result of atomic substitution. Many jarosites, including naturally-occurring and synthetic specimens, show extensive atomic substitution of K^+ and Fe^{3+} by other cations in the crystal structure. Thus, solid-solutions involving K^+-Na^+-H_3O^+ (and Rb^+, Ag^+, NH_4^+, Pb^{2+}, etc.,) and Fe^{3+}-Al^{3+} substitutions have been characterized (*Brophy et al.*, 1962; *Brophy and Sheridan*, 1965; *Leclerc*, 1980; *Ripmeester et al.*, 1986). In their calculations of stability diagrams at 298K for jarosite-type compounds in air, *Hladky and Slansky* (1981) showed that jarosite, natrojarosite, alunite, and natroalunite are stable to pH values of 3.2, 2.1, 7.5, and 6.9, respectively. A solid-solution jarosite phase such as $(K_{0.5}Na_{0.5})(Fe_{0.5}Al_{0.5})_3(SO_4)_2OH)_6$, therefore, might be expected to be stable in air beyond pH 5 (see Fig. 4).

KINETICS OF OXIDATION

The rates of oxidation of aqueous Fe^{2+} ions derived from the dissolution of sulfides in aerated groundwater have been summarized by *Stumm and Morgan* (1970, p. 540). They formulated the steps in the oxidation process as

$$FeS_{2\,(s)} + 7/2\,O_{2\,(aq)} + H_2O = Fe^{2+} + 2SO_4^{2-} + 2H^+ \quad (24)$$

$$Fe^{2+} + 1/4\,O_{2\,(aq)} + H^+ = Fe^{3+} + 1/2\,H_2O \quad (25)$$

$$Fe^{3+} + 3H_2O = Fe(OH)_{3\,(s)} + 3H^+ \quad (26)$$

$$FeS_{2\,(s)} + 14Fe^{3+} + 8H_2O = 15Fe^{3+} + 2SO_4^{2-} + 16H^+ \quad (27)$$

The importance of Fe^{3+} ions in the oxidation of pyrite (*Garrels and Thompson*, 1960) was further demonstrated by *Moses et al.* (1987) who showed that pyrite oxidation rates in acidic solutions (pH 2) saturated in ferric iron (equation (27)) are at least two orders of magnitude higher than solutions saturated with dissolved oxygen (equation (24)), indicating the low probability of a direct gas-solid reaction between atmospheric oxygen and pyrite. The rate-determining step is considered to be equation (25). One rate equation given by *Stumm and Morgan* (1970, p. 534) is

$$-d\ln[Fe(II)]/dt = k_H\,[O_{2\,(aq)}]\,/\,[H^+]^2 \quad (28)$$

This equation shows that in acidic solutions depleted in dissolved oxygen, the rate of oxidation of dissolved Fe^{2+} ions is extremely slow. Moreover, the rate decreases ten-fold for each 15°C fall in temperature. Such slow reactions are particularly relevant to the present-day surface of Mars.

INTERRUPTIONS DURING THE WEATHERING SEQUENCE

The oxidative power of atmospheric oxygen is the driving force in the weathering of sulfides in the absence of bacterial activity. The dissolution of oxygen in groundwater and its migration to cathode surfaces adjacent to sulfides involve diffusion processes and are probably rate-controlling steps. Other factors affecting the rate of abiotic oxidation are the nature and extent of fracturing and permeability of sulfide minerals and host rock, and whether or not the water table is rising or falling.

When the concentration of dissolved oxygen becomes very low and the water supply is limited, cathodic reduction reactions may become sluggish and involve hydrogen peroxide (*Sato*, 1960b)

$$O_2 + 2H_2O + 2e^- = H_2O_2 + 2OH^- \quad (29A)$$

or

$$O_2 + 2H^+ + 2e^- = H_2O_2 \quad (29B)$$

Such peroxide-forming reactions during incomplete oxidation of sulfide minerals may be significant in Martian regolith and could account for the presence of peroxides and superoxides suggested by the Viking Lander biology experiments (*Huguenin*, 1982). The H_2O_2 boundary is shown in Fig. 4.

Furthermore, ferric-bearing solutions may liberate elemental sulfur

$$FeS_2 + 2Fe^{3+} = 3Fe^{2+} + 2S \quad (30)$$

$$Fe_7S_8 + 14Fe^{3+} = 21Fe^{2+} + 8S \quad (31)$$

$$(Fe,Ni)_9S_8 + xFe^{3+} = yFe^{2+} + zNi^{2+} + 8S \quad (32)$$

$$CuFeS_2 + 4Fe^{3+} = Cu^{2+} + 5Fe^{2+} + 2S \quad (33)$$

Metastable sulfur has been observed in pyrite-jarosite-sulfur assemblages formed during the chemical weathering of ultramafic pyrrhotite-pentlandite ore deposits (*Nickel*, 1984).

Acidic solutions formed during weathering of sulfides, particularly minerals with low metal-sulfur ratios (e.g., pyrrhotite and pyrite), which aid in the transport of dissolved cations to supergene and deep-weathering zones, are generally neutralized by wall-rock alteration of silicate minerals in host rock. Thus, hydrous clay silicates (e.g., kaolinite, montmorillonite) are frequently associated with weathered ore bodies. However, rates of such acid-buffering reactions involving feldspars, pyroxenes, and olivine are probably minimized in massive, unfractured host rocks (*Blain and Andrew*, 1977). Silica released during chemical weathering of host-rock silicates is ultimately deposited as opal or jasper associated with "limonite" in gossans.

The consequence of a rising water table is to reduce access of oxygen to sites of cathodic reactions, thereby slowing down shallow weathering oxidative reactions. However, deep weathering processes involving Fe^{3+} ions transported in groundwater may still proceed leading to a build-up of dissolved Fe^{2+} ions (e.g., equations (4) and (10)). Freezing of the groundwater and formation of permafrost could limit the mobility of dissolved cations to and from reaction centers where deep weathering reactions commence, slow down supergene enrichment reactions, and interfere with the precipitation of insoluble oxides occurring in gossans. Such a situation, which may currently exist on the surface of Mars, is discussed later.

On the other hand, a receding water table, perhaps induced by increased fracture flow at depth, could strand gossan-coated sulfides above the water table. If the partial pressure of atmospheric oxygen is low, hydrogen peroxide formation (equations (29a) and (29b)) could be the rate-determining step. Again such a situation could exist on arid surfaces of Mars and lead to survival of oxide-coated magnetic pyrrhotite particles observed in the Viking magnetic properties experiments (*Hargraves et al.*, 1979).

THE CHEMICAL WEATHERING ENVIRONMENT ON MARS

The terrestrial gossan-forming reactions portrayed in equations (1)-(22) and summarized in Fig. 3 are applicable to the standard temperature and pressure (298.16°K, 1 atmos.) of terrestrial aqueous solutions. On Mars, present-day environmental parameters influencing chemical weathering reactions differ somewhat from those on Earth, particularly with respect to temperatures, partial pressure of oxygen in the atmosphere, fluidity of groundwater, and perhaps the bulk composition and permeability of parent crustal rocks. Each of these factors may have influenced gossan-forming processes on Mars.

Parent Host Rock and Sulfide Mineralization

Evidence summarized elsewhere (*Burns*, 1986) suggests that the mantle of Mars is considerably more iron-rich than the Earth's mantle (*McGetchin and Smyth*, 1978; *Morgan and Anders*, 1979), and that partial melting of the Martian mantle produced iron-rich basaltic magmas (*Bussod and McGetchin*, 1979). An iron-rich Martian mantle also allows for high nickel concentrations comparable to Earth's mantle (*Morgan and Anders*, 1979). Experimental studies (*Haughton et al.*, 1974; *Carroll and Rutherford*, 1985) have demonstrated that the sulfur solubility increases with FeO content in a wide range of silicate melt compositions. *Carroll and Rutherford* (1985) suggested that FeO-rich magmas on Mars could carry large amounts of dissolved magmatic sulfur to the planet's surface. Here, magmatic segregation and liquid immiscibility processes would produce massive and disseminated Fe-Ni sulfide mineralization, analogous to ore deposits associated with terrestrial ultramafic komatiites occurring in Precambrian cratonic terranes in Canada, Australia, South Africa, Antarctica, and Siberia (*Naldrett*, 1981; *Guilbert and Park*, 1986). Thus, pyrrhotite-pentlandite ($\pm$ pyrite and chalcopyrite) assemblages in iron-rich mafic and ultramafic host rocks might be expected to occur near the surface of Mars, and to be vulnerable to chemical weathering by aerated aqueous solutions.

On Earth, basaltic magma derived from the mantle has reached the surface predominantly along submarine spreading centers. Such seafloor basalts and gabbros have undergone extensive hydrothermal interaction with seawater, which is facilitated by fracturing of underlying oceanic crust. Thus, the composition of aqueous solutions, particularly the alkalinity (pH 8.2) of the oceans, has been buffered by such seawater-basalt interactions. On Mars, where plate tectonic activity appears to have been insignificant (*Carr*, 1973), vast volumes of basaltic magma have reached the surface of the planet via immense shield volcanoes in the Tharsus and Elysium regions. There also has been neglible recycling of Martian crust into the mantle along subduction zones, so that the ratio of sulfur to other volatiles (e.g., chlorine) has not been affected by subduction of sulfur-rich lithosphere into the mantle (*Clark and Baird*, 1979). The absence of spreading centers and associated fracturing on Mars has also minimized acid-buffering reactions of aqueous solutions by subsurface hydrothermal activity, with the result that acidity of groundwater (now largely permafrost) has probably been maintained during the evolution of Martian regolith.

Temperature

Gossan formation on Earth is facilitated in humid and semi-arid climates where temperatures may attain 40°C. The chemical weathering processes of underlying sulfides depend on the transport of dissolved oxygen and/or ions to reaction centers in fluid aqueous solutions. The present-day cold surface of Mars may interfere with these oxidative reactions, but such inhibiting conditions have probably not prevailed throughout Martian history. The immense channels and canyon on Mars attest to the flow of water and hence to the availability of an aqueous electrolyte phase so essential for oxidative weathering of sulfides. Thus, gossan formation undoubtedly occurred in the past on Mars.

Athough the *ranges* of present-day temperature extremes are probably comparable over the surfaces of Mars and Earth (*Clark*, 1979), the *average* latitudinal and seasonal values are perhaps 50°-60° lower on Mars than on Earth, and are generally below the freezing point of pure water except near the equator. Permafrost appears to prevail below the present-day Martian regolith, so that chemical weathering on Mars is now limited by the mobility of dissolved ionic species but not necessarily electron mobilities. Some diurnal and seasonal melting may occur, particularly near the equator, producing fluid electrolyte and enabling dissolved ions to continue developing gossans in equatorial regions. *Gibson et al.* (1983) noted that Dry Valley temperatures in Antarctica approach those measured at and near the surface of Mars. They observed chemical alteration and oxidation of igneous minerals above and within the permanently frozen zone, and pointed out that *Ugolini and Anderson* (1973) had demonstrated that ionic migration occurs even in frozen soils along liquid films on individual soil particles. These observations suggest that fluid aqueous electrolyte, so essential for promoting anionic oxidative reactions of sulfide, may exist to a limited extent in permafrost. Moreover, some brine compositions proposed for Mars' surface may have eutectic temperatures some 40° to 50° lower than the melting point of ice (*Brass*, 1980), thereby increasing the areal extent of a fluid aqueous electrolyte phase on Mars.

Gooding (1978), in his thermodynamic calculations of mineral stabilities toward chemical weathering, adopted 240°K as a typical surface temperature of Mars. He showed that the oxidation of FeS to Fe(II) sulfate should occur spontaneously by gas-solid reactions on Mars, although as discussed earlier (equation (28)), reaction rates may be slower on Mars. Gooding suggested that in the presence of liquid water the dissolved sulfate would free Fe^{2+}, which should be readily oxidized to Fe^{3+} and precipitate ultimately as FeOOH or Fe_2O_3. These results complement the electrochemical mechanism of oxidative weathering of sulfides described earlier.

Temperature also affects the thermodynamic data forming the basis of pE-pH diagrams, such as those represented in Figs. 3 and 4. Not only do predominance areas change, but also the slopes of the various boundaries. For example, there are indications that the stability field for jarosite increases to higher pH as the temperature is lowered. Thermodynamic data do not exist to construct an equilibrium pE-pH diagram appropriate to Mars at 240°K, a task made difficult by the metastability of liquid water at this temperature and by uncertainties about brine compositions there.

Atmospheric Oxygen on Mars

Not only is the abundance of oxygen lower in the atmosphere of Mars (0.13%) than Earth (20.95%), but the atmospheric pressure on Mars' surface is much smaller (0.00642 atm). As a result, the oxygen partial pressure in the Martian atmosphere is considerably lower (10^{-5} atm) than that on Earth (0.2 atm). This affects the concentration of oxygen dissolved in groundwater; at 273°K, $log_{10}[O_2]$ is -3.35 on Earth and -7.7 on Mars (*Gooding*, 1978). However, the lowering by almost four orders of magnitude of the oxygen partial pressure and the dissolved oxygen activity on Mars has only a small effect on the boundaries of the pE-pH curves shown in Fig. 4; that for the water table is lowered by less than one pE unit. The nature and type of gossans produced on Mars are likely to be very similar to those forming on Earth, particularly if reaction rates of oxidative weathering reactions were raised during times when the surface temperatures of Mars were higher than present-day ranges.

DISCUSSION

The foregoing descriptions of gossan formation on Earth and environmental parameters on Mars lead to the following scenario for surface-weathering processes that might have generated Martian regolith.

The iron-rich basaltic magmas containing significant Ni concentrations transported significant amounts of sulfides to the surface of Mars, depositing them as massive and disseminated pyrrhotite-pentlandite (± pyrite, chalcopyrite) assemblages. Aerated groundwater was sufficiently oxygenated to initiate chemical weathering of these sulfide minerals when the surface of Mars was warmer than it it now. Pyrite, either as an accessory mineral in the primary ores or as a major secondary phase formed by deep weathering of pyrrhotite, facilitated the oxidative weathering reactions of underlying sulfides. These proceeded by electrochemical processes and released relatively high concentrations of dissolved Fe^{2+}, H^+ and SO_4^{2-} ions into percolating groundwater. Oxidation of Fe^{2+} to simple and complex ferric sulfato- and hydroxo-ionic species generated aqueous ferric ions that induced deep-weathering well below the water table and enrichments of Fe, Ni, Cu, etc., in supergene sulfides nearer to the water table. The highly acidic groundwater attacked primary ferromagnesian silicates (olivine, pyroxenes) and feldspars in host mafic and ultramafic rocks, liberating dissolved silica and additional iron cations and producing secondary clay silicate minerals. Such wall-rock alteration processes were limited, however, by the low fracturing and permeability of the host rocks, so that groundwater on Mars has remained acidic, thereby aiding the solubility and transport of dissolved Fe^{2+} cations. Above the water table, poorly crystalline FeOOH (e.g., ferrihydrite) and jarosite phases were precipitated, either as a result of oxidation of remnant pyrite and dissolved Fe^{2+}, or due to hydrolysis of simple and complexed ferric ions. These insoluble ferric phases, together with precipitated silicate and clay silicate residues, resemble the "limonite"-jasper-clay deposits in terrestrial gossans, and constitute the present-day regolith of Mars. Such deposits, however, may obscure enrichments of Ni, Cu, etc., in supergene sulfides and the presence of primary sulfide minerals just below the surface of Mars. As a result, they were not sampled in the Viking XRF experiment (*Clark et al.*, 1982).

Although oxidative weathering of subsurface sulfides may be continuing below the present-day surface of Mars, the rates of the electrochemical processes are reduced by the low temperatures and are controlled by the accessibility of fluid electrolyte to reaction centers through the frozen permafrost.

Evidence in support of gossan formation on Mars stems from several sources. First, laboratory diffuse reflectance spectral measurements of ferrihydrite-silica gels (*Sherman et al.*, 1982) and jarosite-bearing clay mineral assemblages (*Hunt*, 1979; *Hunt and Ashley*, 1979; *Townsend*, 1987) match closely remote-sensed reflectance spectral profiles of bright regions of Mars (*Singer*, 1985; *Burns*, 1987). Second, jarosite, which is so characteristically an oxidative weathering product of iron sulfides, may be present in SNC meteorites believed to have originated from Mars, although uncertainties exist as to whether such oxidized Fe-S phases may have formed in the Antarctic environment (*Smith and Steele*, 1984; *Gooding*, 1986). Third, the Viking XRF experiment (*Clark et al.*, 1982) analysed adequate sulfur and iron to exist as jarosite in the regolith (*Burns*, 1987). However, the low potassium concentration would necessitate the presence of a hydronium- and/or sodium-rich jarosite phase that would be precipitated from acidic groundwater. Fourth, the failure to identify any carbonate minerals in the regolith (*Clark and Van Hart*, 1981) would be explained by the high acidity of the groundwater (now permafrost). However, neutralization of the acidity by some wall-rock alteration on Mars could account for the occurrence of calcite and sulfur-rich aluminosilicates in impact glass found in Antarctic shergottite EETA 79001 (*Gooding and Muenow*, 1986; *Gooding et al.*, 1987). Fifth, the magnetic phase detected in the Viking magnetic experiment (*Hargraves et al.*, 1979) may, in fact, be remnant pyrrhotite that was incompletely oxidized, particularly if the level of the water table has dropped on Mars. Finally, a low concentration of oxygen in a limited supply of water, which favors the formation of H_2O_2 and promotes the production of peroxide and superoxide phases, may account for the results obtained in the Viking biology experiments (*Huguenin*, 1982).

CONCLUSIONS

Mars appears to have had all of the essential ingredients for the generation of gossans by oxidative weathering of sulfide minerals on its surface. The iron-rich Martian mantle probably had adequate S and Ni concentrations to evolve primary Fe-Ni sulfides associated with mafic and ultramafic igneous rocks when they were erupted onto the surface of the planet. Degassing of the mantle yielded H_2O and eventually oxygen in the atmosphere, so that aerated groundwater was available to initiate oxidative weathering of near-surface igneous silicate and sulfide minerals, producing dissolved ferrous and ferric ions. Acidic groundwater transporting Fe^{3+} ions penetrated through tectonic fractures associated with volcanic intrusions and initiated deep weathering reactions of primary pyrrhotite and pentlandite to secondary pyrite and violarite-polydymite phases. They, in turn, released dissolved Fe^{2+}, Ni^{2+}, and SO_4^{2-} into highly acidic solutions, which facilitated limited wall-rock alteration of feldpars, pyroxenes, and olivine in basaltic, gabbroic, and ultramafic host-rocks to clay silicate and iron oxyhydroxide phases. Dissolved Na^+, Ca^{2+}, Mg^{2+}, Al^{3+}, silica, and Fe cations derived from these silicates were added to the groundwater, to be precipitated by hydrolysis and oxidation reactions as poorly crystalline clay silicate (smectite), silica (jasper), ferric sulfate (e.g., jarosite), and FeOOH (ferrihydrite) phases in gossans near the surface of Mars. The groundwater, now present-day permafrost, has probably remained acidic on Mars due to incomplete buffering by wall-rock alteration reactions with relatively unfractured crustal rocks.

Acknowledgments. Drs. James L. Gooding and W. Bourcier, who reviewed the original version of the manuscript, provided helpful suggestions that were embodied in the revised version. The research is supported by NASA grants NSG-7604 and NAGW-1078.

REFERENCES

Bauman L. (1976) *Introduction to Ore Deposits*. Scottish Academic Press, Edinburgh.

Blain C. F. and Andrew R. L. (1977) Sulphide weathering and the evaluation of gossans in mineral exploration. *Minerals Sci. Engng.*, 9, 119-150.

Blanchard R. (1968) Interpretation of leached outcrops. *Nevada State Bur. Mines, Bull.*, 66, xx-yy.

Brass G. W. (1980) Stability of brines on Mars. *Icarus*, 42, 20-28.

Brophy G. P. and Sheridan M. F. (1965) Sulfate studies IV: The jarosite-natrojarosite-hydronium jarosite solid solution series. *Amer. Mineral.*, 50, 1595-1607.

Brophy G. P., Scott E. S., and Snellgrove R. A. (1962) Sulfate studies II. Solid solution between alunite and jarosite. *Amer. Mineral.*, 47, 112-126.

Brown J. B. (1971) Jarosite-goethite stabilities at 25°C, 1 atm. *Mineral. Deposita*, 6, 245-252.

Burns R. G. (1986) Terrestrial analogues of the surface rocks of Mars? *Nature*, 320, 55-56.

Burns R. G. (1987) Ferric sulfates on Mars. *Proc. Lunar Planet Sci. Conf. 17th*, in *J. Geophys. Res.*, 92, E570-E574.

Bussod G. and McGetchin T. R. (1979) Martian lavas—reconnaissance experiments on a model ferro-picrite composition (abstract). In *Lunar and Planetary Science X*, pp. 172-174. Lunar and Planetary Institute, Houston.

Carr M. H. (1973) Volcanism on Mars. *J. Geophys. Res.*, 78, 4039-4054.

Carr M. H. (1987) Water on Mars. *Nature*, 326, 30-35.

Carroll M. R. and Rutherford M. J. (1985) Sulfide and sulfate saturation in hydrous silicate melts. *Proc. Lunar Planet. Sci. Conf. 15th*, in *J. Geophys. Res.*, 90, C601-C612.

Clark B. C. (1979) Chemical and physical microenvironments at the Viking Lander sites. *J. Molec. Evol.*, 14, 13-31.

Clark B. C. and Baird A. K. (1979) Is the martian lithosphere sulfur rich? *J. Geophys. Res.*, 84, 8395-8403.

Clark B. C. and Van Hart D. C. (1981) The salts of Mars. *Icarus*, 45, 370-378.

Clark B. C., Baird A. K., Weldon R. J., Tsusaki D.M., Schnabel L., and Candelaria M. P. (1982) Chemical composition of martian fines. *J. Geophys. Res.*, 87, 10059-10067.

Clifford S., ed. (1985) *Water on Mars*. LPI Tech. Rpt. 85-03, Lunar and Planetary Institute, Houston.

Dousma J., den Ottelander D., and de Bruyn P. L. (1979) The influence of sulfate ions on the formation of iron (III) oxides. *J. Inorg. Nucl. Chem.*, 41, 1565-1568.

Dreibus G. and Wanke H. (1985) Mars, a volatile-rich planet. *Meteoritics*, 20, 367-381.

Dutrizac J. E. and MacDonald R. J. C. (1974) Ferric ion as a leaching medium. *Minerals Sci. Engng*, 6, 59-100.

Garrels R. M. and Christ C. L. (1965) *Solutions, Minerals, and Equilibria*. Harper and Row, New York.

Garrels R. M. and Thompson M. E. (1960) Oxidation of pyrite by iron sulfate solutions. *Amer. J. Sci.*, 258A, 57-67.

Gibson E. K., Wentworth S. J., and McKay D.S. (1983) Chemical weathering and diagenesis of a cold desert soil from Wright Valley, Antarctica: an analog of martian weathering processes. *Proc. Lunar Planet. Sci. Conf., 13th*, in *J. Geophys. Res.*, 88, A912-A928.

Gooding J. L. (1978) Chemical weathering on Mars: thermodynamic stabilities of primary minerals (and their alteration products) from mafic igneous rocks. *Icarus*, 33, 483-513.

Gooding J. L. (1986) Clay-mineral weathering products in Antarctic meteorites. *Geochim. Cosmochim. Acta*, 50, 2215-2223.

Gooding J. L. and Muenow D. W. (1986) Martian volatiles in shergottite EETA 79001: New evidence from oxidized sulfur and sulfur-rich aluminosilicates. *Geochim. Cosmochim. Acta, 50,* 1049-1059.

Gooding J. L., Wentworth S. J., and Zolensky M. E. (1987) Calcium carbonate and sulfate of possible extraterrestrial origin in the EETA79001 meteorite. *Geochim. Cosmochim. Acta,* in press.

Guilbert J. M. and Park C. F., Jr. (1986) *The Geology of Ore Deposits.* W. H. Freeman and Co., New York.

Hargraves R. B., Collinson D. W., Arvidson R. E., and Cates P. M. (1979) Viking magnetic properties experiment: extended mission results. *J. Geophys. Res., 84,* 8379-8384.

Haughton D. R., Roeder P. L., and Skinner B. J. (1974) Solubility of sulfur in mafic magmas. *Econ. Geol., 69,* 451-462.

Hladky G. and Slansky E. (1981) Stability of alunite minerals in aqueous solutions at normal temperature and pressure. *Bull. Mineral., 104,* 468-477.

Hunt G. R. (1979) Near-infrared (1.3-2.4mm) spectra of alteration minerals -Potential for use in remote sensing. *Geophysics, 44,* 1974-1986.

Hunt G. R. and Ashley R. P. (1979) Spectra of altered rocks in the visible and near infrared. *Econ. Geol., 74,* 1613-1629.

Huguenin R. L. (1982) Chemical weathering and the Viking biology experiments on Mars. *J. Geophys. Res., 87,* 10069-10082.

Jensen M. L. and Bateman A. M. (1979) *Economic Mineral Deposits.* 3rd ed. Wiley and Sons, New York.

Leclerc A. (1980) Room temperature Mossbauer analysis of jarosite-type compounds. *Phys. Chem. Mineral., 6,* 327-334.

Locke A. (1926) *Leached Outcrops as Guides to Copper Ore.* William and Wilkins, Baltimore.

Matijevic E., Sapieszko R. S., and Melville J. B. (1975) Ferric hydrous oxide sols. 1. Mono-dispersed basic iron (III) sulfate particles. *J. Colloid. Interface Sci., 50,* 567-581.

McGetchin T. R. and Smyth J. R. (1978) The mantle of Mars: some possible geological implications of its high density. *Icarus, 34,* 512-536.

Morgan J. W. and Anders E. (1979) Chemical composition of Mars. *Geochim. Cosmochim. Acta, 43,* 1601-1610.

Moses C. O., Nordstrom D. K., Herman J. S., and Mills A. L. (1987) Aqueous pyrite oxidation by dissolved oxygen and by ferric iron. *Geochim. Cosmochim. Acta, 51,* 1561-1571.

Naldrett A. J. (1981) Nickel sulfide deposits: classification, composition, and genesis. *Econ. Geol., 75th Anniv. Vol.,* 628-625.

Nickel E. H. (1984) An unusual pyrite-sulphur-jarosite assemblage from Arkaroola, South Australia. *Mineral. Mag., 48,* 139-142.

Nickel E. H., Ross J. R., and Thornber M. R. (1974) The supergene alteration of pyrrhotite-pentlandite ore at Kambalda, Western Australia. *Econ. Geol., 69,* 93-107.

Nordstrom D. K. and Munoz J. L. (1985) *Geochemical Thermodynamics.* The Benjamin/Cummings Publ. Co., Inc., Menlo Park. 477 pp.

Pryce W. (1778) *Mineralogia Cornubiensis.* 1st ed. Pitman Press, Bath, reprinted 1972.

Ripmeester J. A., Ratcliffe C. I., Dutrizac J. E., and Jambor J. L. (1986) Hydronium ion in the alunite-jarosite group. *Can. Mineral., 24,* 435-447.

Rose A. W., Hawkes H. E., and Webb J. S. (1979) *Geochemistry in Mineral Exploration.* Academic Press, New York.

Sapieszko R. S., Patel R. C., and Matijevic E. (1977) Ferric hydrous oxide sols. 2. Thermodynamics of aqueous hydroxo and sulfato ferric complexes. *J. Phys. Chem., 81,* 1061-1068.

Sato M. (1960a) Oxidation of sulfide ore bodies, I. Geochemical environments in terms of Eh and pH. *Econ. Geol., 55,* 928-961.

Sato M. (1960b) Oxidation of sulfide ore bodies, II. Oxidation mechanisms of sulfide minerals at 25°C. *Econ. Geol., 55,* 1202-1231.

Sherman D. M., Burns R. G., and Burns V. M. (1982) Spectral characteristics of the iron oxides with application to the martian bright region mineralogy. *J. Geophys. Res., 87,* 10169-10180.

Siever R. and Woodford N. (1979) Dissolution kinetics and the weathering of mafic minerals. *Geochim. Cosmochim. Acta, 43,* 717-724.

Singer R. B. (1985) Spectroscopic observation of Mars. *Adv. Space Res., 5,* 59-68.

Smith J. V. and Steele I. M. (1984) Achondrite ALHA77005: Alteration of chromite and olivine. *Meteoritics, 19,* 121-133.

Stumm W. and Morgan J. J. (1970) *Aquatic Chemistry.* Wiley-Interscience, New York. 583 pp.

Thornber M. R. (1975a) Supergene alteration of sulphides, I. A chemical model based on massive nickel sulphide deposits at Kambalda, Western Australia. *Chem. Geol., 15,* 1-14.

Thornber M. R. (1975b) Supergene alteration of sulphides, II. A chemical study of the Kambalda nickel deposits. *Chem. Geol., 15,* 117-144.

Townsend T. E. (1987) Discrimination of iron alteration minerals in visible and near-infrared reflectance data. *J. Geophys. Res., 92,* 1441-1454.

Ugolini F. C. and Anderson D. M. (1973) Ionic migration and weathering in frozen Antarctic soils. *Soil Sci., 115,* 461-470.

Vlek P. L. G., Blom T. J. M., Beek J., and Lindsay W. L. (1974) Determination of the solubility product of various iron hydroxide and jarosite by the chelation method. *Soil Sci. Soc. Amer. J., 44,* 1089-1095.

Martian Mantle Primary Melts: An Experimental Study of Iron-Rich Garnet Lherzolite Minimum Melt Composition

Constance M. Bertka[1] and John R. Holloway[2]
Department of Geology, Arizona State University, Tempe, AZ 85287

[1] *Now at the Geophysical Laboratory, Carnegie Institute of Washington, Washington, D.C. 20008*
[2] *Also at Department of Chemistry, Arizona State University, Tempe, AZ 85287*

The minimum melt composition in equilibrium with an iron-rich garnet lherzolite assemblage has been determined from an experimental study of the liquidus relations of iron-rich basaltic compositions at 23 kb. The iron contents of the basaltic compositions studied are appropriate for model partial melts of an olivine-bearing, iron-rich Martian mantle. The convergence of experimental liquids toward a melt multiply saturated in garnet, clinopyroxene, orthopyroxene, and olivine was used to isolate the invariant phase assemblage. Sodium partitioning calculations were used to constrain the sodium content of the experimental minimum melt, thereby aiding a description of its normative mineralogy. Liquidus phases in equilibrium with the minimum melt composition were chosen from an evaluation of Fe/Mg distribution coefficients and mass balance calculations. As representatives of residual peridotite phases, the liquidus phases were used to model Martian mantle garnet lherzolite modal abundances and bulk density. Calculations using the experimentally determined liquidus phases indicate that recent model Martian mantle compositions are in agreement with a garnet lherzolite assemblage but the high calculated density of these assemblages suggests the model compositions may be too iron rich. The experimentally determined primary melt composition and its calculated sodium content indicate that Martian garnet lherzolite minimum melts are picritic alkali olivine basalts. Martian primary melts are more picritic than terrestrial garnet lherzolite primary melts. Olivine fractionation of Martian garnet lherzolite minimum melts en route to the surface can decrease their normative olivine content but their silica undersaturated character will be preserved.

INTRODUCTION

Planetary mantles are the source regions for basaltic magmas (*Yoder*, 1976). The nature of the lavas erupted at the surface of a planet are a reflection of the bulk composition, modal assemblage, volatile content, and oxygen fugacity of the planet's interior. The effects of variations in these parameters have been examined experimentally for terrestrial mantle compositions (e.g., *Wyllie*, 1979). In comparison to the terrestrial mantle the Martian mantle is characterized by its iron enrichment (*Ringwood*, 1966, 1981; *McGetchin and Smyth*, 1978; *Morgan and Anders*, 1979; *Goettel*, 1983; *Dreibus and Wanke*, 1985). Experimental work appropriate to understanding magma generation from iron-rich mantle source regions is lacking. The purpose of this study is to determine the composition of a partial melt in equilibrium with an iron-rich, anhydrous Martian mantle assemblage. These data will provide initial minimum melt composition constraints for one endmember out of a range of possible source region volatile contents and phase assemblages. Martian mantle bulk composition and phase assemblage models, along with interpretations of Martian lavas based on Viking Lander, remote sensing, and SNC meteorite data, are the basis of the compositional range and experimental conditions chosen for study in this investigation.

Bulk Composition of the Martian Mantle

Calculated mantle densities indicate that the Martian mantle has a higher density than the Earth's mantle (see Table 1). As suggested by several workers (e.g., *McGetchin and Smyth*, 1978), this difference is most likely due to differences in the amount of iron contained in the two mantles. The amount of iron enrichment in the Martian mantle and its bulk composition have been derived from petrologic and cosmochemical arguments. *McGetchin and Smyth* (1978) modeled the Martian mantle after a terrestrial pyrolite mantle enriched in iron. *Morgan and Anders* (1979), *Goettel* (1977, 1978, 1983), and *Ringwood* (1966, 1981) used cosmochemical arguments (chondritic, equilibrium condensation, and homogeneous accretion models, respectively) to derive their bulk compositions. Cosmochemical models have also used compositional data from meteorites with a hypothesized Martian origin: the SNCs (*Dreibus et al.*, 1982; *Burghele et al.*, 1983). If Mars is the parent body of SNC meteorites then these samples help constrain the mantle composition. *Burghele et al.* (1983) and *Dreibus et al.* (1982) used SNC bulk composition data and a two-component planetary mixing model to explore this possibility. Compositions proposed for the Martian mantle are given in Table 2.

Goettel's estimate is closest to the terrestrial mantle, but all show a significant iron enrichment. Differences in iron

TABLE 1. Published estimates of planetary mantle density.

Reference	Density (g/cc)
Mars	
Goettel (1981)	3.44 ± 0.06
Okal and Anderson (1978)	3.33-3.41
Johnson and Toksöz (1977)	3.47-3.58
Johnston et al. (1974)	3.71-3.74
Earth	
Ringwood (1969)	3.30-3.40

TABLE 2. Martian and Earth bulk mantle composition.

Wt % Oxide	Mars					Earth
	Ringwood (1981)	McGetchin-Smyth* (1978)	Morgan-Anders (1979)	Goettel (1983)	Dreibus-Wanke (1985)	Pyrolite; Green-Ringwood (1963)
SiO_2	36.8	40.04	41.60	45.07	44.4	45.16
TiO_2	0.2	0.63	0.33	-	0.14	0.71
Al_2O_3	3.1	3.14	6.39	3.26	3.02	3.54
Cr_2O_3	0.4	0.38	0.65	-	0.76	0.43
Fe_2O_3	-	0.41	-	-	-	0.46
FeO	26.8	18.48	15.85	15.07	17.9	8.04
MnO	0.1	0.12	0.15	-	0.46	0.14
NiO	-	0.18	-	-	-	0.20
MgO	29.9	33.22	29.78	32.08	30.2	37.47
CaO	2.4	2.73	5.16	3.03	2.45	3.08
Na_2O	0.2	0.51	0.10	1.40	0.50	0.57
K_2O	-	0.12	0.01	0.12	-	0.13
P_2O_5	-	0.05	-	-	0.16	0.06
mg#†	66.5	75.9	77.0	79.1	75.0	88.8

*Composition recalculated by *Morgan and Anders* (1979).
†mg# = atomic Mg/(Mg + Fe^{2+}).

enrichment between the McGetchin-Smyth, Morgan-Anders, Goettel, and Ringwood models stem from assumed mantle densities, Mg/Si ratios, or oxidation state. *Dreibus and Wanke* (1985) used element correlations between measured ratios in SNCs and chondritic abundances to predict iron enrichment. The aluminum and calcium content is notably higher in the Morgan-Anders composition than in the other estimates.

Phase Assemblages of a Martian Upper Mantle

Phase assemblages of the Martian mantle have been estimated from mantle norm algorithms (*McGetchin and Smyth*, 1978; *Morgan and Anders*, 1979), and projection of predicted bulk compositions into experimentally determined phase regions of the CaO-MgO-Al_2O_3-SiO_2 (CMAS) system (*Stolper*, 1980), as well as thermodynamic calculations in this system (*Patera*, 1982) and in the Na_2O-FeO-CMAS system (*Wood and Holloway*, 1982). These studies indicate that the transition from a spinel- to garnet-bearing mantle assemblage occurs at lower pressures in the iron-rich Martian mantle than in the Earth's mantle. This result is also supported by experimental work. Melting experiments on iron-rich mare basalts (e.g., *Walker et al.*, 1976) suggest a spinel to garnet transition occurs at 14 kb to 15 kb and work with the Morgan-Anders Martian mantle composition (*Patera and Holloway*, 1982) indicates that this transition occurs at 11 kb to 16 kb, depending on temperature. The transition pressure predicted by the theoretical studies varies from 6 kb to 25 kb, due in part to the choice of bulk mantle composition.

The bulk mantle composition also determines if the phase assemblages of the upper Martian mantle resemble those of the Earth's upper mantle: that is, a progression from plagioclase to spinel to garnet lherzolite with increasing pressure. This Earth-like progression is predicted for the Morgan-Anders bulk composition (*Stolper*, 1980; *Patera*, 1982; *Wood and Holloway*, 1982). Alternatively, as predicted for the McGetchin-Smyth composition, the Martian mantle may be dominated by wehrlites, i.e., orthopyroxene-deficient assemblages (*Morgan and Anders*, 1979; *Stolper*, 1980; *Patera*, 1982; *Wood and Holloway*, 1982). *Morgan and Anders* (1979) also predicted a wehrlite assemblage of their model mantle composition. The low (Mg + Fe)/Si ratios of the most recently proposed compositions (*Goettel*, 1983; *Dreibus and Wanke*, 1985) would favor a lherzolite assemblage.

Characterization of Martian Lavas from Viking Lander Data, Remote Sensing Data, and SNC Meteorites

Greeley and Spudis (1981) summarized the evidence for and the extent of volcanism on Mars. Morphologically the terrain is characterized by central volcanoes, located predominantly in the Tharsis region, and more widespread volcanic plains. The depth of magma source regions in the Tharsis plateau has been determined from analysis of gravity and topography data (*Blasius and Cutts*, 1976; *Carr*, 1976; *Thurber and Toksöz*, 1978). Models that assume uniform lithospheric thickness and use the height or surface load of shield volcanoes to calculate source region depths (*Carr*, 1976; *Thurber and Toksöz*, 1978) predict depths of 150 km to 200 km, corresponding to pressures of ~20 kb to ~30 kb. *Blasius and Cutts* (1976) preferred a nonuniform lithospheric thickness model. Source region depths in this model vary from 100 km, ~13 kb for shield volcanoes at the center of the Tharsis plateau to 140 km, ~20 kb for those at its margin.

Compositional data for the Martian surface, supplied by the X-ray fluorescence spectrometers on board the Viking landers, suggest that windblown sediments of the northern hemisphere volcanic plains are derived from basaltic parent materials, but classification of this basalt is model dependent (*Maderazzo and Huguenin*, 1977; *Baird and Clark*, 1981; *Clark et al.*, 1982). *Maderazzo and Huguenin* (1977) used Earth-based reflectance spectral data, Viking XRF data, and a photochemical weathering model to predict a silica undersaturated, picritic parent material. In contrast, *Baird and Clark* (1981) constructed model basalt analogs constrained by the normative mineralogy of the major oxide fraction of the Martian fines. Their model basaltic compositions range from slightly quartz normative to slightly olivine normative. Spectral reflectance information supplied by both Earth-based and orbiter data predict surface mineralogies consistent with slightly oxidized mafic igneous rocks and their weathering products (*Soderblom et al.*, 1978; *Singer et al.*, 1979; *Singer*, 1980). Contrasting with earlier interpretations (*Maderazzo and Huguenin*, 1977), recent data indicate a mineralogy dominated by pyroxene with only five percent modal olivine (*Singer*, 1980) and favor the igneous source model of *Baird and Clark* (1981).

Descriptions of the shergottites, nakhlites, and Chassigny achondrites, referred to as the SNC group, and the evidence suggestive of a Martian origin have been summarized elsewhere (*McSween*, 1985; *Laul et al.*, 1986). As likely Martian volcanic products, the SNCs provide possible clues to the Martian mantle oxygen fugacity and bulk composition. Calculations of oxygen fugacity from coexisting iron-titanium oxides in Shergotty and Nakhla yield fO_2 values at QFM, comparable to terrestrial basalt values (*Stolper and McSween*, 1979). Mantle compositions calculated from SNC bulk data (*Dreibus et al.*, 1982; *Burghele et al.*, 1983) resemble terrestrial values but are iron- and volatile-enriched. Information about parent body primary melt compositions from these meteorites is more elusive. The SNCs are cumulates with varying proportions of intercumulus liquid and cumulus phases (*Bunch and Reid*, 1975; *Floran et al.*, 1978; *Stolper and McSween*, 1979; *Treiman*, 1986). Predictions of magma compositions from such a limited number of cumulus samples is highly model dependent. Furthermore, the high Fe/Mg ratio of the SNC cumulus phases and the absence of olivine as a liquidus phase in any experimental shergottite intercumulus liquid compositions is evidence for the highly fractionated character of SNC parent magmas (*Stolper and McSween*, 1979; *McSween*, 1985). These factors inhibit a constraint of Martian mantle primary melt composition or source region mineralogy using SNC data. However, *Stolper and McSween* (1979) concluded from experiments at 1 bar on two members of the shergottite group that shergottite intercumulus liquids are fractionated derivatives of a silica saturated melt. *Stolper et al.* (1979) assumed the source region for shergottite primary melts was a low pressure, plagioclase peridotite. With this assumption the crystallization sequence of the shergottite parent liquids requires that plagioclase be depleted from the source region during the partial melting event (*Stolper and McSween*, 1979).

The assumption of a plagioclase peridotite source region for shergottite parental melts has been questioned, mainly on the basis of interpretations of REE distribution data (*Shih et al.*, 1982; *Smith et al.*, 1984; *McKay and Wagstaff*, 1985). *Smith et al.* (1984) favored a spinel lherzolite source, whereas *McKay and Wagstaff* (1985) used experimentally determined distribution coefficients to support the conclusions of *Shih et al.* (1982), that calculated REE patterns for shergottite melts suggest either a complex melting process or a garnet-bearing source region. *Treiman* (1986) also predicted a spinel- or garnet-bearing source region for model ultrabasic nakhlite parent magmas. He concluded that the primary melt from which the parental liquid was derived was enriched in normative olivine or augite and that the high CaO/Al_2O_3 ratio of the parental magma relative to a chondritic value indicated that aluminum was retained in the source region. The lack of a Eu anomaly and of plagioclase as an early crystallizing phase suggests a spinel- or garnet-bearing mantle assemblage (*Treiman*, 1986). Conversely, the low Al content for model Martian basalts and their similarity to komatiite compositions, as deduced from the igneous source model of *Baird and Clark* (1981), may imply high degrees of partial melting (*Longhi*, 1982) with no spinel or garnet left in the source region or an Al-depleted source with a variable degree of partial melting (Longhi, personal communication, 1987).

In summary, calculated partial melting depths imply spinel- or garnet-bearing source regions for Martian volcanic products in general agreement with models for SNC petrogenesis. However, surface data can also be interpreted as implying Al-depleted source regions. SNC data and surface compositional data suggest basaltic compositions for the volcanic products but more definitive descriptions are model dependent.

Previous Martian Mantle Primary Melt Estimates

McGetchin and Smyth (1978) estimated the minimum melt composition yielded by their proposed mantle mineralogy at 30 kb, garnet + olivine + clinopyroxene + oxide, by a consideration of experimental work in the diopside-forsterite-pyrope system (*Davis and Schairer*, 1965) and in the diopside-forsterite-periclase system (*Presnall*, 1966). The predicted primary melt is an olivine-rich alkalic basalt.

Stolper (1980) also referred to work with terrestrial peridotites and the CMAS system (*Presnall et al.*, 1978) to describe the melts in equilibrium with a proposed modal mineralogy. The modal mineralogy was calculated from a bulk mantle composition that is an average of model compositions given in BVSP (1981). This methodology yielded a mantle consisting of spinel-plagioclase wehrlite at <9 kb, spinel lherzolite at 9-25, and spinel-garnet wehrlite above 25 kb. Stolper concluded that primary melts produced from these average assemblages at less than 5 kb will be silica undersaturated, at 9 kb to 15 kb will be tholeiitic, and in a 15 kb to 25 kb region will be alkali olivine basalts. Above 25 kb the melts lack a terrestrial counterpart and so remain elusive.

Wood and Holloway (1982) organized a thermodynamic data base for the NFCMAS system in a free energy minimization program that calculates the stable phase assemblage, the phase compositions, and the mode for a chosen bulk composition at a given pressure and temperature. The free energy minimization method allowed for the calculation of the activity of all components in each phase. By using a model terrestrial partial melt composition and assuming a constant ratio of activity to melt mole fraction, Martian primary melt compositions were predicted at 30 kb for the McGetchin-Smyth and Morgan-Anders bulk composition. These model melt compositions are alkali olivine basalts and olivine tholeiites, respectively.

Longhi (1982) calculated partial melting sequences for proposed Martian mantle compositions at 20 kb. These calculations suggested that the minimum melts of all the proposed mantle compositions would be alkali olivine basalts.

EXPERIMENTAL APPROACH

Models that attempt to outline the petrogenetic history of a lava or account for the diversity of lavas on a planet benefit from knowledge of the characteristics of mantle-derived primary melts. Primary melt compositions are liquid compositions in equilibrium with their source region, that is, unmodified by fractionation processes. Terrestrial studies have shown that primary melt compositions are not directly comparable to the composition of surface basalts (*O'Hara*, 1965, 1968; *Green and Ringwood*, 1967), in contrast to earlier studies (*Kuno*, 1960), which did attribute basalt diversity solely to variations in primary melt composition with partial melting depth (for a review see *Wyllie*, 1971). Although later investigations agreed on the importance of fractionation processes subsequent to melt production in the source region, they disagreed on the nature of these fractionation processes as well as on the specific variation of primary melt composition with depth. Despite these differences, terrestrial models agreed that primary melt composition depends on the depth of partial melting because of variations in mantle mineralogy with depth and the degree of partial melting in the source region. Likewise, an initial study of Martian mantle primary melt composition is dependent on a choice of source region characteristic and degree of melting.

A review of previous work on Martian mantle petrology, volcanic features, and proposed surface basaltic compositions (see above) indicates depths of partial melting equivalent to 20 kb to 30 kb pressure in an iron-rich source region. In this pressure region, proposed mantle mineralogies consist of garnet or spinel lherzolite or wehrlite assemblages. We chose to begin our study of Martian mantle primary melt compositions with an anhydrous, iron-rich garnet lherzolite assemblage at 23 kb. Our experiments isolate a primary melt composition that represents a minimum melt; that is, a liquid in equilibrium with all four lherzolite phases, garnet, clinopyroxene, orthopyroxene, and olivine at the temperature where melting of the chosen phase assemblage and mg# begins. Terrestrial garnet lherzolite studies (*Harrison*, 1981) suggest that none of the four phases are exhausted as long as melt fractions are lower than 10%. If garnet lherzolite modal abundances and melting modes in the Martian mantle are similar to terrestrial values, then, by comparison, our minimum melt composition most likely represents a liquid produced at less than 10% partial melting.

Searching for the minimum melt composition, rather than primary melts representing greater proportions of partial melt, allows flexibility in choosing a starting composition. Assuming an isobaric invariant point in the peridotite system, the presence of all four crystalline phases, but not their modal abundance, fixes the melt composition. Therefore, the liquidus phases of this melt have the same composition as the source region residual phases. The mantle bulk composition determines the mode of the residual phases for a given assemblage, but the minimum melt composition is independent of this mode. The minimum melt is isolated in this study by searching for the basaltic composition multiply-saturated in peridotite phases at a chosen pressure.

The approach outlined above has been followed in terrestrial studies where the high pressure liquidus relations of natural basaltic compositions are determined in order to identify a multiply-saturated liquidus surface or to describe plausible high pressure fractionation paths (*Green and Ringwood*, 1967; *Green*, 1969; *Bultitude and Green*, 1971; *Thompson*, 1974). Conversely, terrestrial studies have also centered on analyzing interstitial melt areas produced in peridotite melting experiments (*Ito and Kennedy*, 1967; *Kushiro*, 1972a,b, 1973; *Jaques and Green*, 1979, 1980). Experimental difficulties encountered in both approaches are the loss of iron to noble metal sample containers (*Stern and Wyllie*, 1975) and the quench crystallization of melt at high pressures. In peridotite melting experiments, quench crystal overgrowth on stable crystals results in interstitial melt patches that are unrepresentative of primary melt composition (*Green*, 1973, 1976; *Jaques and Green*, 1979). A review of these problems and partial solutions are given by *Jaques and Green* (1979) and *Takahashi and Kushiro* (1983).

Iron loss from the sample is prevented by using noble metal capsules lined with graphite. *Green* (1973) and *Jaques and Green* (1979) used mass balance calculations to remove the quench overgrowth factor from interstitial melt analysis. *Takahashi and Kushiro* (1983), *Takahashi* (1985), and *Fujii and Scarfe* (1985) sandwiched charges of basaltic composition between peridotite phases creating a layer of melt free of

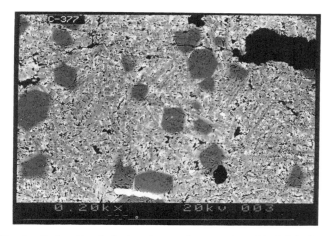

Fig. 1. Backscattered electron image of run product. The bright Pt inclusion, bottom-center, is 100 μm long. Melt does not quench to a glass in these iron-rich basaltic compositions. Darker Mg-rich, euhedral orthopyroxene crystals are surrounded by lighter Fe-rich quench crystals that are representative of the melt composition. Dark patch in upper right corner is a void space.

quench crystal modification in equilibrium with the volumetrically more abundant peridotite composition. Unfortunately, the iron-rich nature of the Martian compositions as well as the high pressure conditions of our experiments serve to enhance the difficulties encountered in terrestrial studies.

Attempts at locating interstitial melt areas produced in starting materials of iron-rich crystalline peridotite or isolating melt between layers of peridotite phases failed. Severe quench crystallization inhibits the formation of interstitial glass patches and iron-rich low viscosity melts are not retained in a layer between peridotite phases. Successful melt analysis requires a large proportion of melt to crystalline phases produced. However, the multiply-saturated basalt approach is not an escape from quench crystallization difficulties.

At our working pressure of 23 kb, well above the spinel-to-garnet transition for a Martian mantle composition (*Patera and Holloway*, 1982), iron-rich basaltic compositions do not quench to glasses. Run products are a mixture of stable and quench crystals with *no* interstitial glass whatsoever (see Fig. 1). Distinguishing quench from stable crystals is vital to the success of our approach since we use the liquidus relations of iron-rich basaltic compositions to isolate the minimum melt. Textural observations with both backscattered electron images and polarized light, as well as compositional criteria, are used to distinguish quench from stable crystals. Quench crystals occur as large optically continuous patches with anomalous extinction and indistinct outlines. Their stoichiometry does not match any single peridotite phase and compositionally they are more iron-rich than coexisting stable crystalline phases. Stable crystals have distinct outlines in backscattered images (often with ~3-μm-wide quench overgrowths), generally sharp extinction, and appropriate stoichiometry.

Our initial basalt composition was a terrestrial garnet lherzolite partial melt composition (*Takahashi*, 1985; Table 2, run #43) to which iron was added. The degree of iron enrichment was calculated from an olivine/liquid Fe/Mg distribution coefficient, $K_D = 0.3$ (*Roeder and Emslie*, 1970), assuming an olivine-dominated bulk mantle with a mg# = 78 (mg# = atomic Mg/Mg+Fe^{2+}). The chosen magnesium number is an average of two endmember magnesium numbers calculated from *Goettel's* (1983) proposed range of model Martian mantle Fe/Mg ratios.

After locating the liquidus temperature of this composition at 23 kb and identifying the liquidus phase, new basaltic compositions were created by adding the components necessary to saturate the liquid in all four peridotite phases: garnet, clinopyroxene, orthopyroxene, and olivine. Initially, for each basaltic composition created, a series of runs from subsolidus to liquidus temperatures were performed. The compositional and textural information gained from these runs assured an accurate distinction between quench and stable crystals and correct identification of liquidus phases. Details of the experimental and analytical technique are discussed below. The constraints placed on the minimum melt and residual phase composition with this approach is revealed in a consideration of the phase rule as it applies to the experimental system.

Ignoring the minor constituents Na$_2$O and Fe$_2$O$_3$ in the experimental system, the remaining number of components, SiO$_2$-Al$_2$O$_3$-CaO-MgO-FeO, at fixed pressure and temperature in the presence of five phases (cpx, opx, ol, gt, melt) defines an invariant assemblage. The bulk composition of coexisting phases at this (P, T) point for chosen mantle mg# is fixed in the CMASF system. Introduction of the minor components Fe$_2$O$_3$ and Na$_2$O results in a system with two degrees of freedom. However, the effect on the phase equilibria of the system with the addition of these components in minor amounts is expected to be negligible.

The ferric iron content of the system is dependent on its oxygen fugacity, which is not controlled in our experiments but is calculated from an expression for the equilibrium constant of the iron-quartz-fayalite buffer reaction

$$fO_2 = \frac{(X^{OL}_{Fe_2SiO_4})^2}{(K)(a^{system}_{SiO_2})(a^{Pt}_{Fe})^2}$$

where

$X^{OL}_{Fe_2SiO_4}$ = analyzed mole fraction of fayalite in olivine

K = equilibrium constant for the iron-quartz-fayalite buffer reaction (*Eugster and Wones*, 1962)

$a^{system}_{SiO_2}$ = calculated activity of SiO$_2$ in the system, which is constrained by the coexistence of olivine and orthopyroxene (*Nicholls*, 1977)

a^{Pt}_{Fe} = activity of iron in platinum, defined by the analyzed mole fraction of iron in platinum inclusions within the charge and with data provided by *Taylor and Muan* (1962) for the activities of iron in Fe-Pt alloys.

The calculated fO$_2$ falls 2.9 log 10 units below the quartz-fayalite-magnetite buffer with an uncertainty of ± one log unit due to analytical error and estimates of iron activity in Fe-Pt alloys. An upper limit on fO$_2$ in our experiments is also established by calculated equilibria, using the technique and data described by *Holloway* (1981), in the graphite-saturated C-O-H system, which constrains the maximum fO$_2$ to values 1.2 to 1.4 log units below QFM at 23 kb and 1400°-1500°C.

The sodium content of a primary melt in equilibrium with an mg# = 78 garnet lherzolite assemblage is not constrained by our experiments. Sodium in the subsolidus assemblage is restricted to clinopyroxene. The abundance of sodium in the melt depends on clinopyroxene/liquid partitioning, bulk mantle sodium content, melting mode, melting components, and degree of melting. Partitioning calculations that consider these variables are performed so that the character of the primary melt in terms of normative silica saturation can be explored.

In addition to yielding a minimum melt composition, our experiments also provide useful Martian mantle assemblage data. As noted previously, the crystalline phases in equilibrium with a minimum melt, the liquidus phases in our experiments, are equivalent to the residual phases in the source peridotite. At the small degree of partial melting associated with minimum

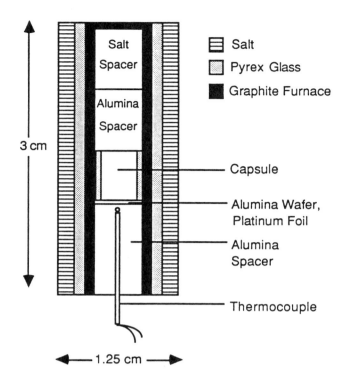

Fig. 2. Solid media cell assembly for the piston cylinder.

a pyrex glass and anouter NaCl sleeve. The sample container was surrounded by a pyrex glass sleeve and remaining space in the furnace was filled with powdered pyrex, crushable alumina and NaCl spacers. A W-Re (W5%Re, W26%Re) thermocouple, inside an alumina spacer topped by an alumina wafer and platinum foil, was used to record run temperatures. Thermocouple precision is estimated to be ±5°C. No correction for pressure effect on the emf output of the thermocouple was made. All assembly materials were dried prior to loading; salt and pyrex sleeves at 500°C and the graphite furnace and crushable alumina spacers at 800°C. A piston-out technique was employed. The pressure was initially raised 10% higher than the required run pressure and maintained while the sample was brought to run temperature. Then the pressure was reduced to the desired value. No correction for friction was made. Previous studies using this furnace assembly found that pressures are accurate to ±0.5 kb (*Esperanca and Holloway*, 1986).

Starting materials. Starting materials for these experiments were synthesized from reagent grade oxides and acid-leached Brazilian quartz. Prior to mixing, the SiO_2, Al_2O_3, MgO, and $CaSiO_3$ were dried at 1000°C for 24 hours, Fe_2O_3 at 150°C for 24 hours, and Na_2CO_3 at 400°C for two hours. Five-gram total oxide mixtures were then ground under ethanol in an agate ball mixer for one hour. The initial basalt composition was reduced in a gas mixing furnace at 1000°C with fO_2 controlled at one log unit below the quartz fayalite magnetite buffer in a $Pd_{60}Ag_{40}$ crucible for 24 hours. To ensure homogeneity the basalt was reground and melted under the same fO_2 conditions at 1250°C for 30 minutes in a $Pd_{60}Ag_{40}$ crucible presaturated in the same composition. The resulting glass was then ground in a carbide container to less than 5 μm.

A set of mineral compositions with mg# similar to that of the melt was made in order to conveniently change the starting glass composition in terms of the four crystalline phases. Olivine and orthopyroxene, with mg#s = 52, were synthesized in platinum crucibles in a 1 atm gas mixing furnace two log units below QFM at 1200°C for 4 hours. A mg# = 52 garnet composition was also sintered at one log unit below QFM at 1000°C for 24 hours. At several times during the synthesis the samples were removed from the furnace, reground in an agate mortar, and replaced in the furnace. The resulting olivine and orthopyroxene phases and the garnet composition mixture were also ground to less than 5 μm in a carbide container. Glasses of the initial basalt and mineral compositions were

melts, the composition of the residual phases will be largely identical to their subsolidus composition at the same pressure. Given that the mode of the assemblage is controlled by the bulk composition, mass balance calculations using our experimental mineral compositions and bulk composition models proposed for the Martian mantle will identify those models that are in agreement with a garnet lherzolite assemblage at 23 kb. Finally, estimates of bulk density can be derived from the calculated modes of these models and used to refine estimates of the mantle bulk composition.

Experimental Technique

Apparatus. A non-end-loaded 1.3 cm diameter piston cylinder (*Patera and Holloway*, 1978) with a solid media assembly was used to achieve mantle pressure and temperature conditions. The solid media cell assembly (see Fig. 2) consisted of a 3-cm-long, 0.6-cm-I.D. graphite tube furnace, encased in

TABLE 3. Starting basalt compositions.

Wt % Oxide	Fetak	TG1	TG2	TG3	TG4	TG5	TG6	TG7	TG8
SiO_2	45.2	43.2	43.9	42.7	42.2	43.5	39.7	41.2	42.5
Al_2O_3	10.6	15.0	13.6	12.2	10.4	7.8	7.1	9.5	11.5
FeO	19.5	19.1	19.2	21.9	24.1	24.1	29.4	25.6	22.4
MgO	11.8	11.6	11.7	13.6	15.2	15.2	17.8	15.5	13.6
CaO	11.8	10.3	10.8	8.9	7.5	8.7	5.6	7.6	9.2
Na_2O	1.1	0.7	0.9	0.7	0.6	0.8	0.5	0.6	0.8

TABLE 4. Experimental conditions and run products.

Starting Composition	Run #	Minutes	Temperature °C	Run Products*				Calculated Mode† Wt %	% FeO Lost	Corrected Bulk mg#
FeTak Sintered Oxides	Tak 2	180	1300	gt	cpx	ol				
	Tak 3	180	1350		cpx		m			
	Tak 4	180	1400		cpx		m	10.7 cpx	0	-
	Tak 5	180	1450				m			
FeTak Glass	Tak 8	315	1225	gt	cpx	ol				
	Tak 9	315	1290	gt	cpx	ol				
	Tak 10	325	1320	gt	cpx					
	Tak 11	185	1420		cpx		m			
TG1	Tak 12	300	1330	gt	cpx					
	Tak 13	300	1360	gt	cpx		m			
	Tak 14	300	1390	gt	cpx		m			
	Tak 15	300	1405	gt			m	9.1 gt	0	-
TG2	Tak 16	360	1375	gt	cpx		m			
	Tak 17	340	1400		cpx		m			
	Tak 18	300	1387	gt	cpx		m			
	Tak 29	180	1410	gt	cpx		m	4.9 cpx 1.7 gt	5.98	53.5
	Tak 19	300	1425				m			
TG3	Tak 20	180	1400				opx m	8.3 opx	10.82	55
	Tak 21	220	1380	gt	cpx	ol	opx m			
TG4	Tak 23	285	1400			ol	opx m			
	Tak 24	200	1435				opx m	13.9 opx	16.11	57.3
	Tak 25	180	1455				m			
TG5	Tak 26	190	1425			ol	opx m			
	Tak 27	180	1440				opx m	5.3 opx	11.47	55.9
TG6	Tak 28	180	1450			ol	m			
TG7	Tak 30	180	1440			ol	m	10.8 ol	6.46	53.7
TG8	Tak 31	180	1405	gt	cpx	ol	opx m	1 > gt 1 > cpx 10 opx	4.47	53.2

*gt = garnet; cpx = clinopyroxene; ol = olivine; opx = orthopyroxene; m = melt in quantities large enough to collect in pools within the charge.
†Remainder is liquid.

TABLE 5. Melt analyses.

Starting Composition	FeTak		TG1		TG2		TG3		TG4		TG5		TG7		TG8	
Run #	Tak 4		Tak 15		Tak 29		Tak 20		Tak 24		Tak 27		Tak 30		Tak 31	
	Wt %	±1σ	Wt %	±1σ	Wt %	±1σ	Wt %	±1σ	Wt %	±1σ	Wt %	±1σ	Wt %	±1σ	Wt %	±1σ
SiO_2	43.9	1.8	43.0	0.3	43.6	0.7	42.5	0.9	41.7	1.1	43.9	0.9	41.9	0.4	42.0	0.8
Al_2O_3	11.5	0.5	14.0	0.2	13.9	0.2	13.4	0.2	11.8	0.3	8.6	0.2	11.6	0.5	12.2	0.4
FeO	20.4	1.2	19.3	0.6	18.7	0.3	20.6	0.6	22.3	1.4	22.3	0.5	24.2	0.8	22.5	1.0
MgO	11.4	0.6	11.5	0.2	11.5	0.3	13.0	0.6	14.5	0.5	15.1	0.3	13.0	0.8	12.5	0.4
CaO	11.4	0.5	11.5	0.2	11.4	0.2	9.8	0.3	9.0	0.3	9.3	0.2	8.7	0.6	10.1	0.3
Na_2O	1.3	0.1	0.9	0.1	1.0	0.1	0.7	0.2	0.7	0.1	0.8	0.1	0.8	0.1	0.9	0.1
mg#	49.9		51.5		52.3		53.0		53.7		54.7		48.9		49.7	

Reported analyses are normalized averages of at least ten 40-μm × 40-μm areas. mg# = atomic $Mg/(Mg + Fe^{2+})$.
*Standard deviation is reported as absolute wt %.

TABLE 6. Mineral analyses.

Phase	cpx	cpx	cpx	cpx	cpx		opx	opx	opx	opx	opx	
Run #	Tak 4	Tak 14	Tak 21	Tak 29	Tak 31	$\pm 1\sigma$*	Tak 20	Tak 21	Tak 24	Tak 27	Tak 31	$\pm 1\sigma$*
Wt % Oxide												
SiO_2	52.0	50.9	52.4	52.2[a]	52.4	0.4 (a = 1.3)	52.4	52.0[c]	52.9	54.7	51.9[f]	0.5 (e = 1.3, f = 0)
Al_2O_3	5.1	7.7[b]	4.8	5.8[c]	5.2	0.4 (b = 1.1, c = 1)	6.4	5.2[g]	5.7	2.1	6.4	0.3 (g = 1.2)
FeO	10.2	9.7	13.1	11.3	13.4	0.5	12.7	13.88	12.7	13.2	14.2	0.3
MgO	17.6	17.0	21.3	20.5[d]	20.4	0.5 (d = 1.0)	26.7	26.23[h]	27.1	27.9	25.1	0.5 (h = 0.8)
CaO	14.7	14.3	8.1	9.9	8.3	0.7	1.8	2.53	1.6	2.1	2.5	0.3
Na_2O	0.5	0.4	0.3	0.3	0.3	0.05	0.1	0.09	0.1	0.1	0.1	0.05
mg#	75.5	75.7	74.4	76.4	73.0		79.0	77.1	79.2	79.0	76.0	
Coexisting phases†	m	gt,m	all	gt,m	all		m	all	m	m	all	

Numbers of atoms on the basis of six oxygens

Si	1.9027	1.8538	1.9048	1.8905	1.9092		1.8642	1.8705	1.8818	1.9542	1.8644	
Al	0.2180	0.3315	0.2073	0.2457	0.2214		0.2703	0.2208	0.2388	0.0872	0.2692	
Fe	0.3119	0.2952	0.3967	0.3434	0.4087		0.3775	0.4175	0.3772	0.3942	0.4252	
Mg	0.9603	0.9227	1.1546	1.1093	1.1055		1.4181	1.4063	1.4358	1.4831	1.3439	
Ca	0.5750	0.5581	0.3146	0.3844	0.3243		0.0677	0.0975	0.0618	0.0794	0.0943	
Na	0.0336	0.0295	0.0184	0.0234	0.0195		0.0040	0.0060	0.0034	0.0002	0.0055	

made with an iridium strip heater and analyzed to ensure the desired compositions were successfully synthesized. Experimental sample compositions were created from 0.1-g mixtures of the initial basalt and mineral compositions ground under ethanol in an agate mortar for fifteen minutes. One weight percent of natural clinopyroxene and garnet crystals were added as seeds to encourage nucleation. These starting compositions are listed in Table 3.

Encapsulation. Approximately 20 mgs of the starting basaltic compositions were sealed in graphite-line 6-mm-long

TABLE 6. (continued)

Phase	gt	gt	gt		ol	ol	
Run #	Tak 15	Tak 29	Tak 31	$\pm 1\sigma$	Tak 30	Tak 31	$\pm 1\sigma$
Wt % Oxide							
SiO_2	40.7	40.8	40.9	0.4	37.8	38.1	0.4
Al_2O_3	23.5	23.1	22.8	0.3	0.1	0.1	0.05
FeO	12.5	13.0[i]	14.1	0.5 (i = 0.8)	23.3	24.6	0.4
MgO	17.4	17.3	16.6	0.6	38.4	36.8	0.5
CaO	5.9	5.9	5.51	0.4	0.3	0.4	0.05
Na_2O	0.0	0.0	0.0	0.05	0.0	0.01	0.05
mg#	71.3	70.4	67.7		74.6	72.7	
Coexisting phases†	m	m,cpx	all		m	all	

	Numbers of atoms on the basis of twelve oxygens				*Numbers of atoms on the basis of four oxygens*		
Si	2.9464	2.9585	2.9809		0.9973	1.0090	
Al	2.0071	1.9718	1.9550		0.0046	0.0040	
Fe	0.7554	0.7880	0.8598		0.5095	0.5396	
Mg	1.8769	1.8721	1.8052		1.4787	1.4397	
Ca	0.4568	0.4555	0.4296		0.0092	0.0113	
Na	0.0014	0.0012	0.0034		0.0	0.0007	

Reported analyses are normalized averages of at least three grains.
*Largest absolute standard deviation of all runs except as noted by superscripts—values in parentheses.
†Coexisting phases: m = melt, gt = garnet, cpx = clinopyroxene, opx = orthopyroxene, ol = olivine, all = m + gt + cpx + opx + ol.

4-mm-O.D. platinum capsules. The graphite crucibles that lined the platinum to prevent iron loss had a wall thickness of 0.4 mm and were fired at 800°C prior to loading.

Analytical Technique

Longitudinal sections of the experimental charges were mounted and polished and phase relations determined with both polarized light and backscattered electron images. Quantitative analyses of major elements were obtained with a JEOL JXM-8600 electron microprobe equipped with both a wavelength and an energy dispersive system. Operating conditions for the wavelength dispersive analyses were 15 kv accelerating potential and 10 nA beam current. A 2-μm beam was used to analyze both stable mineral phases and quench material. The quench material was analyzed by rastering the beam across a 40-μm × 40-μm area. Standards for the analyses were natural oxides and silicates. The probe data were reduced using the *Bence and Albee* (1968) correction method.

RESULTS

Minimum Melt Composition

The experimental phase assemblages of the starting basaltic compositions at 23 kb over a range of near solidus to near liquidus temperatures are given in Table 4. Selected analyses of these phases are given in Tables 5 and 6. Each melt composition represents an average of at least 10 analyses of quench crystal areas 40 μm × 40 μm. The areas chosen for analysis were at least 100 μm from stable crystals so that the effect of quench overgrowth on these phases would be minimized in the analyzed melt compositions. The reported crystalline compositions, unless otherwise indicated, are averages of three to five core analyses.

Modes were calculated for each starting bulk composition utilizing run product analyses from the runs closest to the liquidus temperature. Mass balance calculations to express the starting compositions as proportions of the analyzed run product phases were performed with a least squares fit computer program. Mass balance calculations restricted to SiO_2, Al_2O_3, CaO, and MgO result in errors limited largely to deficiencies in total FeO. This iron oxide error is attributed to variable amounts of iron loss to the platinum containers through cracks in the graphite crucibles. Optical examination of the run products often revealed the presence of quench crystals between the platinum and graphite crucibles confirming that pathways existed for iron loss. The amount of iron loss is unpredictable and varies with individual runs. Mass balance calculations that are corrected for this loss in general have a remaining total error of less than one weight percent. For example, Table 4 lists the relative percentage of iron lost from the starting composition of Tak 29 as 6.0%. The least squares fit mass balance calculation for this run yielded 4.9 wt % clinopyroxene, 1.7 wt % garnet, and 93.4 wt % melt with a total oxide error equal to 2 wt %. A large portion of the total error, 1.2 wt %, is accounted for by the FeO fit. This suggests 1.2 wt % FeO (or 6.0% of the FeO in the starting

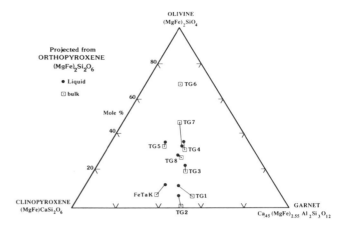

Fig. 3. Projection of bulk compositions and analyzed liquid compositions onto an olivine-clinopyroxene-garnet plane. Starting bulk compositions corrected for iron loss (open squares) and analyzed liquid compositions (filled squares) in mol % projected from an ideal orthopyroxene composition, $(MgFe)_2Si_2O_6$. The endmembers in this plane also have an ideal composition. Olivine = $(MgFe)_2SiO_4$; clinopyroxene = $(MgFe)CaSi_2O_6$; garnet = $Ca_{0.45}(MgFe)_{2.55}Al_2Si_3O_{12}$. Iron and Mg are treated as a single component, (FeO + MgO). Bulk compositions are labeled and connected by tie lines to their analyzed liquid compositions. Liquid compositions are all from 23-kb run products.

composition) was lost during the run. For each basaltic composition the charge closest to the liquidus temperature with the least percentage of total oxide error was chosen as representative of the basalt's liquidus relations. The calculated modal abundances, the percentage of iron loss, and bulk mg#s adjusted for iron loss are given in Table 4.

Iron loss is limited to less than 16% relative to the initial bulk iron content (and is usually much lower) and resulting melt mg#s range from 48.9 to 54.7 (see Table 5) corresponding to bulk mantle mg#s of 76 to 80, well within Martian mantle estimates. In a given charge, the change of the experimental melt mg# from the starting mg# of 52 is controlled not only by the amount of iron loss but also by the modal percentage of crystalline phases. In the charges used to isolate the minimum melt composition, modal calculations indicate that the crystalline liquidus phases are restricted to less than 14 wt %. In interpreting our experimental data we assume that only the mg# of these phases, not their stability, is affected by the small range in experimental melt mg#.

The near liquidus analyzed melt compositions for each sample are projected from orthopyroxene onto an olivine-clinopyroxene-garnet plane in Fig. 3. The endmembers used in this projection scheme have an ideal composition and serve only to show the convergence of melt compositions toward a multiply saturated invariant point. Starting bulk compositions corrected for iron loss are also plotted.

As shown in Fig. 4, where the phases stable with each melt are indicated, the liquidus data are used to locate the clinopyroxene-garnet cotectic and to isolate the minimum melt composition. Although the topology of the liquidus surface in this projection is not uniquely defined due to variation in bulk

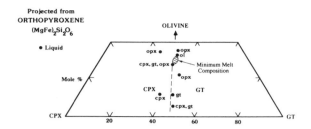

Fig. 4. Projection of analyzed liquid compositions onto an olivine-clinopyroxene-garnet plane and location of minimum melt composition. Projection scheme same as in Fig. 3. Crystalline phases stable with each analyzed liquid composition are indicated. The dashed line approximates the location of the clinopyroxene-garnet cotectic.

mg# and modal percentage of phases, in general there is a decrease in liquidus temperature as the invariant region is approached. Also, the range in temperature from solidus to liquidus for a given basalt composition decreases near this region. The solidus temperature at 23 kb for a Martian mantle yielding the minimum melt composition projected in Fig. 4 is estimated to be 1400° ± 15°C. Our best estimate of the melt composition is given in Table 7; it is an average of two experimental melt analyses, TG8 and TG7, which border the invariant region of the projection.

Crystalline Compositions

The liquidus phases of the minimum melt are equivalent to the residual phases of the source peridotite. Thus the experimental liquidus phases whose compositions are closest to equilibration with our minimum melt estimate provide data concerning potential Martian mantle assemblages. These compositions are chosen in part from an evaluation of the compositional variation of the crystals in the run products as starting bulk compositions approach the invariant region. Variation between runs is due to differences in crystal/liquid modal proportions and changes in the compositional buffering provided by coexisting phases. The compositions in equilibrium with the minimum melt are buffered by the coexistence of all four lherzolite phases.

TABLE 7. Estimate of minimum melt and liquidus phase composition.

Wt % Oxide	Minimum Melt* Estimate	Liquidus Phases*			
		Cpx	Opx	Gt	Ol
SiO_2	41.9	52.4	51.9	40.9	37.8
Al_2O_3	11.9	5.2	6.4	22.8	0.1
FeO	23.3	13.4	14.2	14.1	23.3
MgO	12.7	20.4	25.1	16.6	38.4
CaO	9.4	8.3	2.5	5.5	0.3
Na_2O	0.8	0.3	0.1	0.1	0
mg#	49.3	73.0	76.0	67.7	74.6

*Minimum melt estimate average of Tak 30 and Tak 31 melt analyses; cpx, opx, gt analyses from run # Tak 31, olivine from run # Tak 30.

Clinopyroxene. In most runs this phase is 10 to 15 μm in size with 2-to 3-μm quench rims. In any given run CaO and Al_2O_3 show the greatest variability between crystals—up to 2 wt %. Al_2O_3 contents of all analyzed clinopyroxenes range from 4 to 6 wt %, but CaO contents show a marked decrease from 14 to 16 wt % (in most runs) to 8 wt % in those runs whose liquid compositions approach the minimum melt. This decrease is attributed to the CaO content constraint imposed by the coexistence of orthopyroxene (*Lindsley*, 1983).

Orthopyroxene. Orthopyroxene occurs as large 60- to 200-μm grains with 3-μm quench rims. Al_2O_3 and CaO contents of analyzed grains from the same charge vary by less than 1 wt %. The Al_2O_3 content of orthopyroxenes in equilibrium with liquids near the minimum melt composition is 5 to 6.5 wt %, as compared to 2 to 3 wt% in other liquidus orthopyroxenes. Increasing orthopyroxene Al_2O_3 content is correlated to increasing Al_2O_3 in the coexisting melt. The CaO content is also 1 wt % greater in orthopyroxene coexisting with clinopyroxene than in orthopyroxene without coexisting clinopyroxene.

Garnet. Garnet grains range from 15 to 25 μm with 3-μm quench rims. Excluding Fe/Mg ratios, they have a similar core composition in all runs. The Al_2O_3 and CaO contents within a charge vary by less than 1 wt % for the most magnesium-rich crystals. Backscattered electron images reveal a zonation of garnets in many runs from iron-rich cores to magnesium-rich margins. In relation to bulk composition these cores are volumetrically insignificant.

Olivine. With the exception of one starting composition, olivine is limited to runs where it coexists with at least one other phase. The grains are 10 to 15 μm with 3-μm quench rims. The Al_2O_3 and CaO contents are both less than 0.5 wt % and similar in all charges. No attempt was made to precisely determine these values.

DISCUSSION

Approach to Equilibrium

A measure of equilibrium is the partitioning of elements between coexisting phases. Equilibrium Fe/Mg distribution coefficients have been determined by previous workers: olivine/liquid (*Roeder and Emslie*, 1970) olivine/orthopyroxene (*Mori and Green*, 1978), and garnet/clinopyroxene (*Raheim and Green*, 1974; *Ellis and Green*, 1979). These values can be compared to our experimental Fe/Mg distribution coefficients (Table 8), with the limitations discussed below, to evaluate the approach to equilibrium in our experiments.

A determination of Fe/Mg partitioning between phases in experimental run products must consider the unequal rates of adjustment of these phases to any iron loss from the system (*Green*, 1976). Previous workers (*Green*, 1973; *Stern and Wyllie*, 1975; *Jaques and Green*, 1979) have established rates of adjustment for peridotite phases as follows: liquid > olivine > pyroxene > garnet. A comparison of diffusion rates for selected elements in these phases highlights the scale of these relative differences. Using estimated maximum diffusion coefficients for Ca in melt, Fe and Mg in olivine, and Ca and

TABLE 8. Fe/Mg distribution coefficients.

ol/liq. Fe/Mg Run #	Experimental		Roeder and Emslie (1970)
		±1σ*	
Tak 31	0.37	0.02	0.3 ± 0.01
Tak 30	0.33	0.02	
Tak 26	0.36	0.04	

ol/opx Fe/Mg Run #	Experimental		Mori and Green (1978)
		±1σ	
Tak 31	1.18	0.04	1.1 ± 0.01
Tak 26	1.16	0.07	
Tak 21	1.40	0.13	

gt/cpx Fe/Mg Run #	Experimental		Ellis and Green (1979)
		±1σ	
Tak 31	1.29	0.08	1.36
Tak 29	1.37	0.12	1.37
Tak 14	1.38	0.04	1.45
Tak 13	1.46	0.08	1.54
Tak 12	1.60	0.09	1.65
Tak 10	1.60	0.07	1.64

*Standard deviation reported as absolute value.

Mg in pyroxenes (from data reviewed by *Takahashi and Kushiro*, 1983), characteristic diffusion distances during three-hour run times and a temperature range applicable to our experiments are as follows: Ca in melt, 3300 microns; Fe + Mg in olivine, 10 microns; Ca + Mg in pyroxene, 0.1 micron. Diffusion distances in garnet are probably comparable to those in clinopyroxene. If the bulk composition of the system changes during the run due to loss of iron to the platinum capsule, the melt experiences the largest and most homogeneous iron depletion. The remaining phases are slow to fully reequilibrate with the new melt composition and are zoned toward Mg-rich rims. Olivine reequilibrates with the new melt composition more quickly than the other phases.

With the exception of large 200-μm orthopyroxene grains in two runs where core to rim analyses were collected, all reported crystalline compositions are core analyses. The transverse analyses of the orthopyroxene grains reveal a minor zonation of these grains toward Mg-rich margins, with Fe-rich rims from quench overgrowth. Mass balance calculations for orthopyroxene-saturated liquidus runs, utilizing crystalline core analyses and analyzed liquid compositions, confirm iron depletion was limited largely to the melt.

Iron loss from the melts in our runs was not severe enough to stabilize new phases. This is evidenced by the internally consistent disappearance of phases with increasing temperature in the run products of a single basalt composition. The internal agreement is maintained with variations in starting material, sintered oxide mix versus glass, and run time (see Table 4, Fe Tak runs). The only exception to the above is for runs Tak29 and Tak17, but the Tak17 run is suspect due to poor mass balance results. The proposed minimum melt composition is dependent on a correct estimate of iron loss as this loss may result in silica-enriched melt compositions. Our minimum melt composition is derived from runs where the calculated iron loss is less than 7 wt % of the amount present; a negligible amount in terms of melt composition and normative mineralogy.

Olivine/orthopyroxene. Although our results agree with the assumption that crystals grew early from a liquid not depleted in Fe and were slow to reequilibrate as Fe was lost, the discrepancy between our experimental Fe/Mg olivine/orthopyroxene and olivine/liquid distribution coefficients and distribution coefficients previously determined cannot be accounted for by variation in reported iron and magnesium diffusion rates in these phases. Experimental olivine/orthopyroxene Fe/Mg K_Ds are higher than predicted values as are Fe/Mg olivine/liquid K_Ds (Table 8). The apparently high iron content of the liquidus olivine relative to the melt can be accounted for by preferential iron loss from the melt for run Tak30 and, if K_D values up to 0.33 are accepted as equilibrium, for run Tak26. The larger K_D value is suggested by experimental work with mare basalts (e.g., *Walker et al.*, 1976) whose mg#s are lower relative to terrestrial basalts. A recalculation of the Fe/Mg olivine/liquid K_D with the addition of the lost iron to the melt then yields an equilibrium value for Tak30 and Tak26. However, a similar calculation for Tak31 does not result in an equilibrium Fe/Mg olivine/liquid K_D. The olivines in Tak31 and Tak26 are also iron-enriched relative to coexisting crystalline phases. This may be due to poor analyses, resulting from the small size of the olivine grains and their large quench rims. A more likely possibility is that these crystals are individual quench crystals. Except for the starting composition in which olivine is the liquidus phase, Tak30, the modal abundance of olivine in our runs is minor and its exclusion from mass balance calculations consistently yields the best fit between starting compositions and analyzed run products.

Garnet/clinopyroxene. Our experimental garnet/clinopyroxene Fe/Mg K_Ds are within the error limits of those of *Ellis and Green* (1979). *Ellis and Green* (1979) compare garnet-clinopyroxene equilibration temperatures calculated with their expression for K_D to experimental data not used in the derivation of this expression. The uncertainty between calculated and experimental temperatures is ± 5%. Our experimental values fall well within the limits of this uncertainty. Furthermore, a classification of these grains as

TABLE 9. Calculated garnet lherzolite modes.

Phase	Dreibus-Wänke (1985)	Goettel (1983)
Olivine	49.5	49.7
Orthopyroxene	31.5	21.9
Clinopyroxene	18.8	26.3
Garnet	0.3	2.1

Values in wt %.

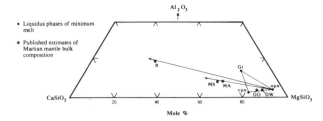

Fig. 5. Projection of liquidus phases of the minimum melt and published estimates of Martian mantle bulk composition from olivine, Mg_2SiO_4, onto the plane $CaSiO_3$-$MgSiO_3$-Al_2O_3. Iron and Mg are treated as a single component, (FeO + MgO). Phases: Gt, garnet; Cpx, clinopyroxene; Opx, orthopyroxene. Bulk composition: MS, McGetchin-Smyth; MA, Morgan-Anders; GO, Goettel; DW, Dreibus-Wänke; R, Ringwood.

quench crystals is not supported by textural or compositional evidence. The modal abundances of clinopyroxene and garnet are minor only in near-liquidus runs where they are still distinguishable from quench crystal. Also, clinopyroxene CaO contents are in general agreement with pyroxene solvus relations (*Lindsley*, 1983). Pyroxene aluminum coordination calculations after *Papike et al.* (1974) indicate that within the uncertainty of analytical error the clinopyroxenes have no ferric iron, consistent with the low fO_2 estimates for run conditions.

The above evaluation of the crystalline analytical data in light of iron loss and Fe/Mg partitioning indicates that olivine compositions may not represent equilibrium olivine at run conditions, except where olivine is the only liquidus phase (run # Tak30). Garnet, orthopyroxene, and clinopyroxene compositions are accepted as representing stable crystal analyses and those from run Tak31 are chosen as best representatives of our minimum melt liquidus phases (Table 7).

Mineral Proportions in Garnet Lherzolite Martian Mantles

The modal abundances of garnet lherzolite phases can be determined with mass balance calculations that use the crystalline liquidus phases of the minimum melt and model bulk mantle compositions. The mass balance calculations are performed on a molar basis by treating FeO + MgO as one component in order to compare bulk mantle assemblages independent of magnesium number. The *Goettel* (1983) and *Dreibus and Wanke* (1985) Martian mantle compositions can be expressed as proportions of the coexisting liquidus phases. Calculated modal abundances for these models in weight percent of the various phases are given in Table 9. The *Ringwood* (1966, 1981), *McGetchin and Smyth* (1978), and *Morgan and Anders* (1979) models yield orthopyroxene-deficient assemblages indicating they would form spinel-garnet wehrlite, as predicted by *Morgan and Anders* (1979). These relationships are illustrated in Fig. 5, where the liquidus phases of the minimum melt and published estimates of Martian mantle bulk composition are projected from olivine into the plane $CaSiO_3$-$MgSiO_3$-Al_2O_3. The Goettel and Dreibus-Wanke compositions fall within the garnet lherzolite field, which is defined by the experimental phases. The Ringwood, McGetchin-Smyth, and Morgan-Anders models project outside this field along negative extensions of trajectories from the orthopyroxene apex.

Early descriptions of Martian mantle bulk composition (*Ringwood*, 1966, 1981; *McGetchin and Smyth*, 1978; *Morgan and Anders*, 1979) have higher Fe + Mg/Si ratios than recent models (*Goettel*, 1983; *Dreibus and Wanke*, 1985). *Ringwood* (1966, 1981) accounts for a mantle rich in FeO with a high mantle oxidation state. *McGetchin and Smyth* (1978) and *Morgan and Anders* (1979) estimated, based on available data, a mantle density greater than the recent estimate of 3.44 ±0.06 g/cc (*Goettel*, 1981). Higher density estimates are incorporated into bulk composition models by the addition of iron, increasing the Fe + Mg/Si ratio and forming orthopyroxene-deficient assemblages. Our experimental results confirm that orthopyroxene is not restricted to minor proportions in all prospective Martian mantle garnet assemblages. Recent bulk composition models (*Goettel*, 1983; *Dreibus and Wanke*, 1985) can be defined with our experimentally-determined lherzolite phases and have minor modal proportions of garnet and abundant orthopyroxene resembling the modes of terrestrial garnet lherzolites.

Density of Garnet Lherzolite Martian Mantles

Mineral densities are calculated from the endmember densities listed in Table 10. Minerals are assumed to have no excess mixing volume and corrections for thermal expansion and compressibility are not made. The densities of the model garnet lherzolite assemblages are shown in Table 10. The modal proportions of the *Goettel* (1983) and *Dreibus and Wanke* (1985) models are similar and it follows that their calculated densities will be similar. However, the Fe/Mg ratios of the experimental minimum melt and liquidus phases are appropriate for equilibrium with a mantle mg# of 75-76, which is closer to the mantle mg# estimate of *Dreibus and Wanke* (1985) than that of *Goettel* (1983). Therefore, the density of the *Goettel* (1983) assemblage is recalculated choosing crystalline Fe/Mg ratios appropriate for a mg# = 78 mantle. The mg# = 75-76 mantle densities are higher than the *Goettel* (1981) upper Martian mantle density range, 3.44 ± 0.06 g/cc, whereas the mg# = 78 mantle density falls within the upper limits of this range.

Goettel (1981) provides a discussion of the factors influencing density estimates and argues in favor of the 3.44 ±0.06 g/cc estimate. That the calculated density of the experimental assemblages falls within or slightly above the upper limit of the 3.44 ± 0.06 g/cc geophysical estimate suggests the abundance of iron in the model compositions may be high. Regardless, the proximity of the calculated density of the garnet lherzolite assemblage of the *Dreibus and Wanke* (1985) mantle composition to the most recent estimate of Martian mantle density is noteworthy. The Dreibus-Wanke composition was derived independent of Martian mantle density estimates. Therefore, the calculated density supports the hypothesized Martian origin of the SNC meteorites.

TABLE 10. STP densities (g/cc) of lherzolite endmember minerals and calculated garnet lherzolite densities.

Mineral	Formula	Density g/cc
Olivine		
forsterite	Mg_2SiO_4	3.213
fayalite	Fe_2SiO_4	4.393
Pyroxene		
enstatite	$Mg_2Si_2O_6$	3.198
ferrosilite	$Fe_2Si_2O_6$	3.966
diopside	$CaMgSi_2O_6$	3.277
hedenbergite	$CaFeSi_2O_6$	3.550
CaTsh	$CaAl_2SiO_6$	3.435
MgTsh	$MgAl_2SiO_6$	3.436*
Garnet		
pyrope	$Mg_3Al_2Si_3O_{12}$	3.559
almandine	$Fe_3Al_2Si_3O_{12}$	4.318
grossular	$Ca_3Al_2Si_3O_{12}$	3.595
andradite	$Ca_3Fe_2Si_3O_{12}$	3.860

	Calculted lherzolite densities (g/cc)	
Mantle mg#	*Dreibus and Wänke* (1985)	*Goettel* (1983)
75-76	3.52	3.52
78	-	3.48

STP densities from *Robie et al.* (1978), unless otherwise noted.
Gasparik and Newton (1984).

The Sodium Content of Primary Martian Magmas

Identifying a reasonable range for the Na content of the minimum melt is crucial to establishing the relation of the experimental primary melt to silica saturation. We have estimated Na contents for a range of mantle bulk compositions with partitioning calculations based on trace element fractionation for a partial melting event in which the liquid remains in equilibrium with the residual solid (*Shaw*, 1970). In a garnet lherzolite subsolidus assemblage the bulk Na content is accounted for largely as a jadeite component in clinopyroxene. If Na is assumed to be restricted to clinopyroxene then its weight fraction, cl, in the experimental primary melt can be described as

$$cl = Co/[Wcpx\ Dcpx/l + F(1 - Pcpx\ Dcpx/1)]$$

where

Co	= weight fraction Na_2O in subsolidus bulk composition
Wcpx	= subsolidus weight fraction clinopyroxene
Dcpx/l	= distribution coefficient for Na_2O between clinopyroxene and melt
F	= melt fraction
Pcpx	= melting mode.

Primary melt compositions whose Na contents are calculated with the above expression for a chosen range of each variable are given in Table 11 along with their normative mineralogies. Input parameters for these calculations are chosen from a consideration of model Martian mantle compositions and their modal mineralogy as defined by our experimentally determined phase compositions. Also, our minimum melt definition requires the fraction of melt to be less than 10 wt %. The cpx/L distribution coefficient is varied from the value suggested by our experimental products to a maximum coefficient chosen from a comparison to previous experimental work (*Jakobsson*, 1984). Both equilibrium and nonequilibrium melting modes are considered. The calculated melt compositions yield nepheline normative, silica undersaturated, primary melts.

The largest variation in the calculated Na contents shown in Table 11 is due to changing Co from 0.5 to 1.5. The majority of Na contents of the proposed mantle compositions in Table 2 are closer to 0.5 than 1.5. The high value proposed by *Goettel* (1983) is not well constrained. Changes in Na concentration due to the range of Wcpx, Dcpx/L, and Pcpx are small. As the most realistic melt fraction value for the minimum melt probably lies between those chosen for our calculations, our best liquid estimate is bordered by the compositions listed in columns 1 and 5 of Table 11. A large uncertainty associated with the calculated compositions is caused by our lack of

TABLE 11. Calculated primary melt compositions and normative mineralogy.

*Input Parameters**						
Co	0.5	1.5				
Wcpx	0.26		0.19			
Dcpx/L	1/3			1/2		
F	0.10				0.01	
Pcpx	0.26					0.50
Oxide (wt %)						
SiO_2	41.07	38.69	40.89	41.28	40.05	41.01
Al_2O_3	11.65	10.97	11.60	11.71	11.36	11.63
FeO	22.84	21.51	22.74	22.96	22.27	22.80
MgO	12.44	11.72	12.39	12.51	12.13	12.42
CaO	9.20	8.67	9.16	9.25	8.97	9.19
Na_2O	2.81	8.43	3.23	2.30	5.22	2.94
Normative mineralogy (wt %)						
ab	0	0	0	1.5	0	0
an	19.2	0	17.2	21.6	7.6	18.5
di	21.9	6.5	21.0	20.3	16.5	21.7
ol	46.0	48.6	46.1	46.9	46.6	46.0
hy	0	0	0	0	0	0
ne	12.9	30.6	14.8	9.7	23.9	13.5
cs	0	10.8	1.0	0	5.3	0
ns	0	3.5	0	0	0	0

co = weight fraction Na_2O in subsolidus bulk composition; Wcpx = subsolidus weight fraction clinopyroxene; Dcpx/l = distribution coefficient for Na_2O between clinopyroxene and melt; F = melt fraction; Pcpx = melting mode.
*Input parameters varied individually; for each calculation the parameter that is different from those given in column one is indicated.

knowledge of the exchange reaction for Na between clinopyroxene and melt. Our calculations assume Na in the melt is proportional to Na in clinopyroxene. This assumption is valid only if the activities of CaO, Al_2O_3, SiO_2, and (MgO + FeO) are fixed by the coexistence of the garnet lherzolite plus liquid assemblage. Under these circumstances with small changes in Na concentration in the melt the concentration ratios of other major elements remain constant.

The Effects of Volatiles and High Degrees of Melting on Primary Martian Magma Compositions

Our minimum melt composition is representative of a melting product of an anhydrous iron-rich garnet lherzolite mantle. It is probable that the Martian mantle contains small amounts of C-O-H volatiles such as CO_2 and H_2O. These will partition strongly into the melt such that, at small degrees of partial melting, the melt will be enriched in these components and its composition may differ significantly from the one determined in this work (e.g., *Wyllie*, 1979). Alternatively, liquids formed by high degrees of melting will be less affected by volatiles because the volatile concentrations will be low. For larger degrees of partial melting, however, the melt composition will eventually be affected by the disappearance of a phase.

Judging from the minor modal abundance of garnet at 23 kb in our calculated assemblages it will be the first phase to disappear, probably at ≤5% melting. Higher degrees of melting will produce liquids in equilibrium with clinopyroxene, orthopyroxene, and olivine and so should follow terrestrial trends, modified by the effect of the lower mg#.

Comparison of Martian and Terrestrial Garnet Lherzolite Minimum Melt Compositions

As terrestrial source regions, modeled as dry, garnet-bearing lherzolites, produce silica undersaturated minimum melts (*Scarfe et al.*, 1979; *Takahashi and Kushiro*, 1983), it is reasonable that iron-enriched garnet lherzolite sources yield similar compositions. Although our results indicate iron-enriched 23 kb source regions produce low-mg# undersaturated minimum melts, these melts are also less calcic and silceous than 30 kb and 40 kb terrestrial model melts (*Takahashi*, 1985; *Davis and Schairer*, 1965).

O'Hara (1968) demonstrates that, in the iron-free diopside-garnet-clinopyroxene system, the olivine and clinopyroxene fields expand relative to garnet with decreasing pressure such that the invariant melts of garnet lherzolites become less picritic. Our experimental results indicate that the addition of iron to this system produces invariant melts at 23 kb that are more picritic than magnesium-rich invariant melts generated at higher pressure (compare Table 11 to Table 12). The validity of this conclusion depends on the assumption that olivine is not in a reaction relation at our experimentally-located invariant point. That olivine is not in a reaction relation at the experimentally-located invariant point is supported by the subliquidus precipitation of both orthopyroxene and olivine in starting compositions whose liquidus phases are orthopyroxene. Furthermore, work in the iron-free forsterite-

TABLE 12. Terrestrial minimum melt compositions and normative mineralogy.

Wt % Oxide	Davis and Schairer (1965)		Takahashi (1985)
SiO_2	50.18		47.71
Al_2O_3	11.89		11.19
FeO	-		7.93
MgO	25.77		19.53
CaO	12.17		12.41
Na_2O	-		1.22
mg#	100		81.4
Wt % normative mineralogy			
ab	—	ab	10.32
an	32.4	an	25.05
di	21.8	di	29.18
fo	19.9	ol	35.11
en	26.0	hy	0.32
ne	—	ne	0.0
cs	—	cs	0.0

diopside-silica system (*Kushiro*, 1969) indicates that at ≥7 kb forsterite does not react with liquids to yield orthopyroxene.

We conclude that on a relative scale the minimum melts of Martian mantle garnet lherzolites are more olivine normative than their terrestrial counterparts. A measure of this relative difference in pictritic character between our model 23-kb melt and model terrestrial 30-kb and 40-kb melts rests on the assumption that the reported terrestrial melt compositions are also representative of minimum melts, as higher degrees of partial melting would produce more olivine normative melts.

Relation of Garnet Lherzolite Minimum Melt Composition to Surface Basaltic Compositions

The minimum melt composition is not likely to be equivalent to any surface composition. This does not, however, rule out primary melts of anhydrous garnet lherzolite assemblages as potential sources of these magmas, as the surface basalts may be fractionated derivatives of primary melts. These primary melts may be minimum melts or melts generated by larger degrees of partial melting, in which case the modal proportions of phases exert a control on the partial melt composition. A likely differentiation process operative on a primary melt en route to the surface is olivine fractionation (*O'Hara*, 1968).

The compositions and normative mineralogies of liquids produced by 10 wt % perfect olivine fractional crystallization of two of the calculated primary melt compositions (columns 1 and 5 in Table 11) are given in Table 13. The fractionated liquids are less picritic than the minimum melt compositions but the normative Ne content of the silica undersaturated compositions increases. This trend continues up to 26 wt % fractionated olivine, suggesting the silica undersaturated character of a garnet lherzolite minimum melt will be retained in the surface basalt composition if the only differentiation process active is olivine fractionation.

The minimum melt composition and the olivine fractionated derivatives of this composition are not equivalent to shergottite

TABLE 13. Composition and normative mineralogy of olivine fractionated liquids.

Wt % Oxide	1*	2†
SiO_2	41.4	40.2
Al_2O_3	12.9	12.6
FeO	22.8	22.1
MgO	9.7	9.4
CaO	10.2	9.9
Na_2O	3.1	5.8
Wt % normative mineralogy		
ab	0	0
an	21.2	8.4
di	24.4	18.6
ol	40.1	40.8
hy	0	0
ne	14.3	26.4
cs	0	5.8

Liquids are derivatives of calculated primary melt compositions at 10 wt % perfect olivine fractional crystallization.
*Column 1, Table 11.
†Column 5, Table 11.

parent magmas or to Baird-Clark models for Martian basalt, both of which predict silica-saturated basalt compositions. The liquid compositions calculated by *Treiman* (1986) for Nakhla parent magmas share an ultrabasic, silica-undersaturated character with our minimum melt composition. However, in addition to lower mg#s, the Nakhla parent magma compositions differ in their much lower alumina and higher silica contents. Simple petrogenetic models, such as olivine fractional crystallization as the sole differentiation process, cannot relate our minimum melt composition to present-day estimates of the composition of Martian volcanics.

CONCLUSIONS

1. The bulk mantle compositions proposed by *Goettel* (1983) and *Dreibus and Wanke* (1985) will form lherzolite assemblages: spinel lherzolite at low pressures and garnet lherzolite at higher pressures.

2. Mantle densities calculated using the experimentally determined phase compositions are slightly higher than the geophysical estimates, implying higher Martian mantle mg#s than represented by existing bulk composition models.

3. Anhydrous equilibrium melting of an iron-rich garnet lherzolite source at 23 kb produces picritic alkali olivine minimum melts. With larger degrees of melting garnet will disappear from model Martian mantle assemblages and partial melt compositions will be constrained by equilibrium with olivine, orthopyroxene, and clinopyroxene.

4. Martian mantle garnet lherzolite source regions probably yield primary melts that are more picritic than terrestrial equivalents corresponding terrestrial primary melts.

5. Perfect olivine fractional crystallization of Martian mantle garnet lherzolite minimum melts will result in basalts that are less picritic than initially but that retain their silica undersaturated source region character.

Acknowledgments. We thank Dave Joyce, Hanna Nekvasil, Rick Hervig, reviewers J. Longhi and J. W. Morgan, and associate editor John Jones for their valuable comments on the manuscript. James Clark provided expert technical assistance with the electron microprobe. The electron microprobe facility was supported by Arizona State University and the National Science Foundation. This work was supported by NASA grant NAGW-182 Supp. #6.

REFERENCES

BSVP (Basaltic Volcanism Study Project) (1981) Geophysical and cosmochemical constraints on properties of mantles of the terrestrial planets. In *Basaltic Volcanism on the Terrestrial Planets*, pp. 633-699. Pergamon, New York.

Baird A. K. and Clark B. C. (1981) On the original igneous source of Martian fines. *Icarus, 45,* 113-123.

Bence A. E. and Albee A. E. (1968) Empirical correction factors for the electron microanalysis of silicates and oxides. *J. Geol., 76,* 382-403.

Blasius K. R. and Cutts J. A. (1976) Shield volcanism and lithospheric structure beneath the Tharsis plateau, Mars. *Proc. Lunar Sci. Conf. 7th,* 3561-3573.

Bultitude R. J. and Green D. H. (1971) Experimental study of crystal-liquid relationships at high pressure in olivine nephelinite and basanite compositions. *J. Petrol., 12,* 121-147.

Bunch T. E. and Reid A. M. (1975) The Nakhlites part 1: Petrography and mineral chemistry. *Meteoritics, 10,* 303-315.

Burghele A., Dreibus G., Palme H., Rammensee W., Spettel B., Weckwerth G., and Wanke H. (1983) Chemistry of Shergottites and the Shergotty parent body (SPB): Further evidence for the two component model of planet formation (abstract). In *Lunar and Planetary Science XIV,* pp. 80-81. Lunar and Planetary Institute, Houston.

Carr M. H. (1976) The volcanoes of Mars. *Sci. Amer., 234,* 32-43.

Clark B. C., Baird A. K., Weldon R. J., Tsusaki D. M., Schnabel L., and Candelaria M. P. (1982) Chemical composition of Martian fines. *J. Geophys. Res., 87,* 10,059-10,067.

Davis B. T. C. and Schairer J. F. (1965) Melting relations in the join diopside-forsterite-pyrope at 40kb and 1 atm. *Carnegie Inst. Wash. Yearbk., 63,* 123-126.

Dreibus G. and Wanke H. (1985) Mars: a volatile rich planet. *Meteoritics, 20,* 367-382.

Dreibus G., Palme H., Rammensee W., Spettel B., Weckwerth G., and Wanke H. (1982) Composition of the Shergotty parent body; Further evidence of a two component model of planet formation (abstract). In *Lunar and Planetary Science XIII,* pp. 186-187. Lunar and Planetary Institute, Houston.

Ellis D. J. and Green D. H. (1979) An experimental study of the effect of Ca upon garnet-clinopyroxene Fe-Mg exchange equilibria. *Contrib. Mineral. Petrol., 71,* 13-22.

Esperanca S. and Holloway J. R. (1986) The origin of high-K latites from Camp Creek, Arizona: Constraints from experiments with variable fO_2 and aH_2O. *Contrib. Mineral. Petrol., 93,* 504-512.

Eugster H. P. and Wones D. R. (1962) Stability relations of the ferruginous biotite, annite. *J. Petrol., 3,* 82-125.

Floran R. J., Prinz M., Hlava P. F., Keil K., Nehru C. E., and Hinthorne J. R. (1978) The Chassigny meteorite: A cumulate dunite with hydrous amphibole-bearing melt inclusions. *Geochim. Cosmochim. Acta, 42,* 1213-1229.

Fujii T. and Scarfe C. M. (1985) Composition of liquids coexisting with spinel lherzolite at 10 kb and the genesis of MORBs. *Contrib. Mineral. Petrol., 90,* 18-28.

Gasparik T. and Newton R. C. (1984) The reversed alumina contents of orthopyroxene in equilibrium with spinel and forsterite in the system $MgO-Al_2O_3-SiO_2$. *Contrib. Mineral. Petrol, 85,* 186-196.

Goettel K. A. (1977) Composition of the terrestrial planets. *NASA Tech. Memo. X-3511,* pp. 7-10.

Goettel K. A. (1978) The composition of Mars. *NASA Tech. Memo. 79729,* pp. 116-117b.

Goettel K. A. (1981) Density of the mantle of Mars. *Geophys. Res. Lett., 8,* 497-500.

Goettel K. A. (1983) Present constraints on the composition of the mantle of Mars. *Carnegie Inst. Wash. Yearbk., 82,* 363-366.

Greeley R. and Spudis P. D. (1981) Volcanism on Mars. *Rev. Geophys. Space Phys., 19,* 13-41.

Green D. H. (1969) The origin of basaltic and nephelinitic magmas in the earth's mantle. *Tectonophysics, 7,* 409-422.

Green D. H. (1973) Experimental melting studies on a model upper mantle composition at high pressure under water-saturated and water-undersaturated conditions. *Earth Planet. Sci. Lett., 19,* 37-53.

Green D. H. (1976) Experimental testing of equilibrium partial melting of peridotite under water-saturated high pressure conditions. *Can. Mineral., 14,* 255-268.

Green D. H. and Ringwood A. E. (1963) Mineral assemblages in a model mantle composition. *J. Geophys. Res., 68,* 937-945.

Green D. H. and Ringwood A. E. (1967) The genesis of basaltic magmas. *Contrib. Mineral. Petrol., 15,* 103-190.

Harrison W. J. (1981) Partitioning of REE between minerals and coexisting melts during partial melting of a garnet lherzolite. *Amer. Mineral., 66,* 242-259.

Holloway J. R. (1981) Volatile interactions in magmas. In *Thermodynamics of Minerals and Melts: Advances in Physical Geochemistry, Vol. I* (R. C. Newton, A. Navrotsky, and B. J. Wood, eds.), pp. 273-293. Springer Verlag, New York.

Ito K. and Kennedy G. C. (1967) Melting and phase relationships in a natural peridotite. *Amer. J. Sci., 265,* 519-538.

Jakobsson S. (1984) Melting experiments on basalts in equilibrium with a graphite-iron-wustite buffered C-O-H fluid. Ph.D. dissertation, Arizona State University, Tempe. 186 pp.

Jaques A. L. and Green D. H. (1979) Determination of liquid compositions in high pressure melting of peridotite. *Amer. Mineral., 64,* 1312-1321.

Jaques A. L. and Green D. H. (1980) Anhydrous melting of peridotite at 0-15 kb pressure and the genesis of tholeiitic basalts. *Contrib. Mineral. Petrol., 73,* 287-310.

Johnston D. H. and Toksöz M. N. (1977) Internal structure and properties of Mars. *Icarus, 32,* 73-84.

Johnston D. H., McGetchin T. R., and Toksöz M. N. (1974) The thermal state and internal structure of Mars. *J. Geophys. Res., 79,* 3959-3971.

Kuno H. (1960) High alumina basalt. *J. Petrol., 1,* 121-145.

Kushiro I. (1969) The system forsterite-diopside-silica with and without water at high pressures. *Amer. J. Sci., 267A,* 269-294.

Kushiro I. (1972a) Effect of water on the composition of magmas formed at high pressures. *J. Petrol., 13,* 311-334.

Kushiro I. (1972b) Partial melting of synthetic and natural peridotites at high pressures. *Carnegie Inst. Wash. Yearbk., 71,* 357-362.

Kushiro I. (1973) Partial melting of garnet lherzolites from kimberlites at high pressure. In *Lesotho Kimberlites* (Peter H. Nixon, ed.), pp. 294-299. Lesotho National Development Corporation, Maseru.

Kushiro I., Shimizu N., and Nakamura Y. (1972) Composition of coexisting liquid and solid phases formed upon partial melting of natural garnet and spinel lherzolites at high pressure. *Earth Planet. Sci. Lett., 14,* 19-25.

Laul J. C., Smith M. R., Wanke H., Jagoutz E., Dreibus G., Palme H., Spettel B., Burghele A., Lipschutz M. E., and Verkouteran R. M. (1986) Chemical systematics of the Shergotty meteorite and the composition of its parent body (Mars). *Geochim. Cosmochim. Acta, 50,* 909-926.

Lindsley D. H. (1983) Pyroxene thermometry. *Amer. Mineral., 68,* 477-493.

Longhi J. (1982) Modeling high pressure partial melting of the Martian mantle (abstract). In *Lunar and Planetary Science XIII,* pp. 445-446. Lunar and Planetary Institute, Houston.

Maderazzo M. and Huguenin R. (1977) Petrologic interpretation of Viking XRF analysis based on reflectance spectra and the photochemical weathering model. *Bull. Amer. Astron. Soc., 9,* 527-528.

McGetchin T. R. and Smyth J. R. (1978) The mantle of Mars: Some possible geological implications of its high density. *Icarus, 34,* 512-536.

McKay G. and Wagstaff J. (1985) Clinopyroxene REE distribution coefficients for shergottites: REE content of the Shergotty/Zagami melts (abstract). In *Lunar and Planetary Science XVI,* pp. 540-541. Lunar and Planetary Institute, Houston.

McSween H., Jr. (1985) SNC meteorites: Clues to Martian petrologic evolution. *Rev. Geophys., 23,* 391-416.

Morgan J. W. and Anders E. (1979) Chemical composition of Mars. *Geochim. Cosmochim. Acta, 43,* 1601-1610.

Mori T. and Green D. H. (1978) Laboratory duplication of phase equilibria observed in natural garnet lherzolites. *J. Geol., 86,* 83-97.

Nicholls J. (1977) The activities of components in natural silicate melts. In *Thermodynamics in Geology* (D. G. Fraser, ed.), pp. 327-348. D. Reidel, Boston.

O'Hara M. J. (1965) Primary magmas and the origin of basalts. *Scot. J. Geol., 1,* 19-40.

O'Hara M. J. (1968) The bearing of phase equilibria studies in synthetic and natural systems on the origin and evolution of basic and ultrabasic rocks. *Earth Sci. Rev., 4,* 69-133.

Okal E. A. and Anderson D. L. (1978) Theoretical models for Mars and their seismic properties. *Icarus, 33,* 514-528.

Papike J. J., Cameron K. L., and Baldwin K. (1974) Amphiboles and pyroxenes: characterization of other than quadrilateral components and estimates of ferric iron from microprobe data (abstract). *Geol. Soc. Amer. Abstracts with Program, 6,* 1053-1054.

Patera E. S. (1982) Phase equilibria of the upper Martian mantle: Theoretical calculations and experiments. Ph.D. dissertation, Arizona State University. 138 pp.

Patera E. S. and Holloway J. R. (1978) A non-end-loaded piston cylinder design for use to forty kilobars (abstract). *Eos Trans. AGU, 59,* 1217-1218.

Patera E. S. and Holloway J. R. (1982) Experimental determinations of the spinel-garnet boundary in a Martian mantle composition. *Proc. Lunar Planet. Sci. Conf. 14th,* in *J. Geophys. Res., 87,* A31-A36.

Presnall D. C. (1966) The join forsterite-diopside-iron oxide and its bearing on the crystallization of basaltic and ultramafic magmas. *Amer. J. Sci., 264,* 753-809.

Presnall D. C., Dixon S. A., Dixon J. R., O'Donnell T. H., Brenner N. L., Schrock R. L., and Dycus D. W. (1978) Liquidus phase relations on the join diopside-forsterite-anorthite from 1 atm to 20 kb: Their bearing on the generation and crystallization of basaltic magma. *Contrib. Mineral. Petrol., 66,* 203-220.

Raheim A. and Green D. H. (1974) Experimental determination of the temperature and pressure dependence of the Fe-Mg partition coefficient for coexisting garnet and clinopyroxene. *Contrib. Mineral. Petrol., 48,* 179-203.

Ringwood A. E. (1966) Chemical evolution of the terrestrial planets. *Geochim. Cosmochim. Acta, 30,* 41-104.

Ringwood A. E. (1969) Composition and evolution of the upper mantle. In *The Earth's Crust and Upper Mantle* (P. J. Hart, ed.), pp. 1-17. American Geophysical Union, Washington, DC.

Ringwood A. E. (1981) Geophysical and cosmochemical constraints on properties of mantles of the terrestrial planets. In *Basaltic Volcanism on the Terrestrial Planets*, pp. 633-699. Pergamon, New York.

Robie R. A., Hemingway B. S., and Fisher J. R. (1978) Thermodynamic properties of minerals and related substances at 298.15K and 1 bar pressure (10^5 pascals) and at higher temperature. *U.S. Geol. Surv. Bull.*, 1452, 456 pp.

Roeder P. L. and Emslie R. F. (1970) Olivine-liquid equilibrium. *Contrib. Mineral. Petrol.*, 29, 275-289.

Scarfe C. M., Mysen B. O., and Rai C. S. (1979) Invariant melting behavior of mantle materials: Partial melting of two lherzolite nodules. *Carnegie Inst. Wash. Yearbk.*, 78, 498-501.

Shih C.-Y., Nyquist L. E., Bogard D. D., McKay G. A., Wooden J. L., Bansal B. M., and Wiesmann H. (1982) Chronology and petrogenesis of young achondrites Shergotty, Zagami and ALHA 77005: Late magmatism on a geologically active planet. *Geochim. Cosmochim. Acta*, 46, 2323-2344.

Shaw D. M. (1970) Trace element fractionation during anatexis. *Geochim. Cosmochim. Acta*, 34, 237-243.

Singer R. B. (1980) The dark materials on Mars: II. New mineralogic interpretations from reflectance spectroscopy and petrologic implications (abstract). In *Lunar and Planetary Science XI*, pp. 1048-1050. Lunar and Planetary Institute, Houston.

Singer R. B., McCord T. B., and Clark R. N. (1979) Mars surface composition from reflectance spectroscopy: A summary. *J. Geophys. Res.*, 84, 8415-8426.

Smith M. R., Laul J. C., Ma M.-S., Huston T., Verkouteren R. M., Lipschutz M. E., and Schmitt R. A. (1984) Petrogenesis of the SNC (Shergottites, Nakhlites, Chassignites) meteorites: Implications for their origin from a large dynamic planet, possibly Mars. *Proc. Lunar Planet. Sci. Conf. 14th*, in *J. Geophys. Res.*, 89, B612-B630.

Soderblom L. A., Edwards K., Eliason E. M., Sanchez E. M., and Charette M. P. (1978) Global color variations on the Martian surface. *Icarus*, 34, 446-464.

Stern C. R. and Wyllie P. J. (1975) Effect of iron absorption by noble-metal capsules on phase boundaries in rock-melting experiments at 30 kbar. *Amer. Mineral.*, 60, 681-689.

Stolper E. (1980) Predictions of mineral assemblages in planetary interiors. *Proc. Lunar Planet. Sci. Conf. 11th*, 235-250.

Stolper E. and McSween H. Y., Jr. (1979) Petrology and origin of the Shergottite meteorites. *Geochim. Cosmochim. Acta*, 43, 1475-1498.

Stolper E., McSween H. Y., Jr., and Hays J. F. (1979) A petrogenetic model of the relationships among achondritic meteorites. *Geochim. Cosmochim. Acta*, 43, 589-602.

Takahashi E. (1985) Melting of a dry peridotite KLB 1 up to 14 Gpa: Implications on the origin of peridotitic upper mantle. *J. Geophys. Res.*, 91, 9367-9382.

Takahashi E. and Kushiro I. (1983) Melting of a dry peridotite at high pressures and basalt magma genesis. *Amer. Mineral.*, 68, 859-879.

Taylor R. W. and Muan A. (1962) Activities of iron in iron-platinum alloys at 1300°C. *Trans. Met. Soc. AIME*, 224, 500-502.

Thompson R. N. (1974) Primary basalts and magma genesis. *Contrib. Mineral. Petrol.*, 45, 317-341.

Thurber C. H. and Toksöz M. N. (1978) Martian lithospheric thickness from elastic flexure theory. *Geophys. Res. Lett.*, 5, 977-980.

Treiman A. H. (1986) The parental magma of the Nakhla achondrite: Ultrabasic volcanism on the Shergottite parent body. *Geochim. Cosmochim. Acta*, 50, 1061-1070.

Walker D., Kirkpatrick R. J., Longhi J., and Hays J. F. (1976) Crystallization history of lunar picritic basalt sample 12002: Phase-equilibria and cooling-rate studies. *Geol. Soc. Amer. Bull.*, 87, 646-656.

Wood B. J. and Holloway J. R. (1982) Theoretical prediction of phase relationships in planetary mantles. *Proc. Lunar Planet. Sci. Conf. 13th*, in *J. Geophys. Res.*, 87, A19-A30.

Wyllie P. J. (1971) *The Dynamic Earth: Textbook in Geosciences*. Wiley and Sons, New York. 416 pp.

Wyllie P. J. (1979) Petrogenesis and physics of the earth. In *The Evolution of Igneous Rocks: Fiftieth Anniversary Perspectives* (H. S. Yoder, Jr., ed.), pp. 483-520. Princeton University Press, Princeton, NY.

Yoder H. S., Jr. (1976) *Generation of Basaltic Magmas*. National Academy of Sciences, Washington, DC. 265 pp.

A Model of the Porous Structure of Icy Satellites

Janusz Eluszkiewicz and Jacek Leliwa-Kopystynski
Warsaw University, Institute of Geophysics, ul. Pasteura 7, 02-093 Warsaw, Poland

---

The effect of porosity on pressure distribution within a spherical, isothermal body consisting of a mixture of ice and rock is described by an ordinary nonlinear differential equation of second order. It is shown that this equation may be treated as a generalization of the Emden equation $\omega'' + 2/x\, \omega' + \omega^n = 0$ with the term ω^n replaced by $Q(\omega)$, where Q is the function describing the pressure dependence of porosity. For a plausible choice of Q this equation is solved numerically. It is shown that porosity can significantly influence the dimensionless moment of inertia γ of a medium-size icy satellite (e.g., Mimas). The combined effect of porosity and differentiation is also investigated. This effect is sufficient to explain the low value of γ quoted for Mimas by S. F. Dermott (personal communication, 1986). In addition, the effect of porosity on the shape of figure of a synchronously rotating satellite is evaluated. Results show that in the first-order theory porosity alone is not sufficient to explain the actual shape of Mimas.

INTRODUCTION

There is now strong evidence that the surface layers of airless bodies in the solar system are very porous (e.g., the review article by *Veverka et al.,* 1986). For example, on the Moon typical densities of the upper 1 cm are ~1.4 g cm^{-3}, well below the density of individual particles (~3.5 g cm^{-3}). Photometric studies of bright icy satellites suggest a very high porosity of the optically active layers (*Buratti,* 1985). For Io, a surface porosity of ~85% was found by *Matson and Nash* (1983) from thermal measurements that are sensitive to the topmost few centimeters. For the Moon, microwave observations (probing up to some tens of meters below the surface) indicate that the porosity at greater depths may be significantly lower (*Troitsky,* 1965; *Lindsay,* 1976). However, it seems possible that on smaller objects porosity decreases more gradually with depth. In particular, such may be the situation on the icy Saturnian and Uranian moons. As *Smoluchowski et al.* (1984) write, "There is no reason to suppose that ices in the solar system whether on satellites, in cometary nuclei, on interplanetary grains or on the ring particles of Saturn have no pores."

In this work the effect of porosity on the moment of inertia and the shape of figure of a satellite is investigated. Because laboratory data on porosity in ice-rock mixtures at low temperatures are lacking, a numerical model must be used. In spite of this we believe that the main conclusions of our work (as stated in the abstract) are indeed of general importance.

BASIC EQUATIONS

The only parameters of the satellites that are rather well known at present are their masses M, radii R, and therefore mean densities $\bar{\rho} = 3M/4\pi R^3$. The latter for the "icy" medium-size Saturnian and Uranian satellites are in the range from 1.1 g cm^{-3} (Mimas) to 1.6 g cm^{-3} (Titania). The surface features (e.g., albedo and spectroscopic data) obtained by Voyagers (*Smith et al.,* 1982, 1986) indicate that the main component of the surface is water ice ($\rho_i = 0.94$ g cm^{-3}). The fact that $\bar{\rho} > \rho_i$ indicates that the icy satellites should partly consist of a rocky component (commonly assumed density ρ_r of chondritic rock varies between 2.6 g cm^{-3} and 3.5 g cm^{-3} depending on whether the rock is hydrated or anhydrous). Because of the low pressure range (Fig. 1) we will assume that ice and rock are uncompressible. In calculations of the internal structure the satellite is assumed to be an isothermal sphere.

The starting equations of the model are those of hydrostatic equilibrium

$$\frac{dp}{dr} = -\frac{GM(r)\rho(r)}{r^2} \qquad (1)$$

$$M(r) = 4\pi \int_0^r \rho(r) r^2 dr \qquad (2)$$

An expression for $\rho(r)$ that takes into account the presence of voids between rock and ice grains can be obtained by writing the unit of volume V as the sum

$$V = V_{ice} + V_{rock} + V_{voids} \qquad (3)$$

The porosity q and the rock concentration C are defined as follows:

$$V_{voids} = q\, V \qquad (4)$$

$$C = \frac{m_r}{m_r + m_i} \qquad (5)$$

where m_r and m_i are the rock and ice masses respectively contained in V. Introducing the densities $\rho_r = m_r/V_{rock}$, $\rho_i = m_i/V_{ice}$, one obtains from equation (3) the following equation for the local density

$$\rho(r) = \frac{m_i + m_r}{V} = \frac{\rho_r \rho_i}{\rho_r(1-C) + \rho_i C}(1-q) \qquad (6)$$

This formula holds in the general case when ρ_i, ρ_r, q, and C all depend on r. In this work C is assumed to be constant

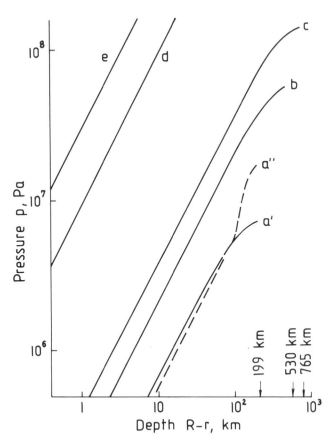

Fig. 1. Pressure versus depth within some of the satellites; pressure distribution in the Earth's crust and within a terrestrial glacier are shown for comparison. Curves (a') and (a'') correspond to Mimas (mean density $\rho = 1.137$ g cm^{-3}, radius R = 198.8 km): (a') uniform Mimas, and (a'') fully differentiated Mimas with an icy mantle (density of ice $\bar{\rho}_i = 0.94$ g cm^{-3}) and a rocky core (assumed rock density $\rho_r = 3$ g cm^{-3}). (b) Uniform Tethys, $\bar{\rho} = 1.21$ g cm^{-3}, R = 530 km. (c) uniform Rhea, $\bar{\rho} = 1.33$ g cm^{-3}, R = 765 km. (d) Glacier on Earth, $\rho_i = 0.94$ g cm^{-3}. (e) Earth's crust, $\rho_r = 3$ g cm^{-3}. The central pressure in uniform satellites is $p_c = 2\pi/3\ G\rho^2R^2$ (7.14 MPa, 57.5 MPa, and 145 MPa for Mimas, Tethys and Rhea, respectively). It is seen that the pressure range within the satellites corresponds to that in the outermost layers of the Earth.

throughout the satellite (differentiation is not taken into account). C is then equal to the total silicate mass fraction. According to the incompressibility assumption, ρ_i and ρ_r are taken to be constant also. Equation (6) may then be rewritten in the following form

$$\rho(r) = \rho_0[1 - q(p)] \equiv \rho_0 Q(p) \quad (7)$$

where

$$\rho_0 = \rho_0(C) \equiv \frac{\rho_i \rho_r}{\rho_r(1-C) + \rho_i C} \quad (8)$$

Equations (1), (2), and (7) together with an assumed (or known from experiment) relation

$$q = q(p) \quad (9)$$

determine pressure and density distributions within a spherical, porous, undifferentiated, and isothermal body. Unfortunately the porosity-pressure relation, equation (9), for ice-rock mixtures at satellite temperatures (T ≤ 100K) is unknown. There is some theoretical work concerning the compressibility of random dense packing of equal spheres that has been applied by *Arzt et al.* (1983) in constructing mechanism maps for hot-isostatic pressing of ice. In their model, densification of ice due to pressure is described by an algebraic relation followed (for porosites lower than 10%) by an exponential law. In the present work an exponential relation will be adopted for simplicity throughout the whole porosity range.

Substituting equations (2) and (7) into equation (1) and differentiating with respect to r one obtains

$$\frac{d^2p}{dr^2} - \frac{d\log Q}{dp}\left(\frac{dp}{dr}\right)^2 + \frac{2}{r}\frac{dp}{dr} + 4\pi G \rho_0^2 Q^2 = 0 \quad (10)$$

With this equation are associated the boundary conditions

$$p(r = R) = 0, \quad \left.\frac{dp}{dr}\right|_{r=0} = 0 \quad (11)$$

In dimensionless variables equations (10) and (11) transform into

$$\frac{d^2\tilde{p}}{dx^2} - \frac{d\log Q}{d\tilde{p}}\left(\frac{d\tilde{p}}{dx}\right)^2 + \frac{2}{x}\frac{d\tilde{p}}{dx} + Q^2 = 0 \quad (10')$$

$$\tilde{p}(x = 1) = 0, \quad \left.\frac{d\tilde{p}}{dx}\right|_{x=0} = 0 \quad (11')$$

$$x = r/R, \quad \tilde{p} = p/p_0, \quad p_0 = 4\pi G \rho_0^2 R^2 \quad (12)$$

In Appendix A it is shown that equation (10') may be treated as a generalization of the well-known Emden equation describing a polytropic self-gravitating sphere (*Chandrasekhar*, 1967).

The use of a temperature-independent relation, equation (9), can to some extent be justified as follows. As *Arzt et al.* (1983) have pointed out there are three basic mechanisms of porosity decrease under the action of pressure: the rapid temperature-independent densification, temperature-dependent creep-controlled densification, and the densification resulting from atomic diffusion (which also depends on temperature). However, according to the calculations carried out by *Smoluchowski and McWilliams* (1984), the temperature-dependent processes are negligible at the low temperatures expected in the interiors of the icy Saturnian and Uranian satellites (but the pore migration due to a temperature gradient can modify this picture).

PHYSICAL CONSEQUENCE

In this section the effect of pressure-dependent porosity on two gross properties of a satellite, moment of inertia and the shape of figure, is demonstrated by means of a numerical example.

Numerical Example

In the computations it was assumed that an increase of pressure from p to p + dp causes a change of porosity q → + dq according to the following linear law

$$dq/q = -\alpha \, dp, \quad \alpha = \text{const} \qquad (13)$$

After integration, equation (13) leads to an exponential dependence of porosity on pressure

$$q = q_0 \exp(-p/\alpha) = q_0 \exp(-\tilde{p}/\tilde{\alpha}) \qquad (13')$$

This form of the porosity-pressure relation not only proved to be easy to handle numerically but is also reasonable from a physical standpoint. According to equation (13') porosity decreases toward the center from its value q_0 at the surface of the satellite where p = 0. The parameter q_0 may thus be called "the surface porosity" ($0 < q_0 < 1$). The value of q_0 may be found either from photometric measurements (e.g., by means of Hapke's compaction parameter h; see *Buratti*, 1985) or from infrared observations (as was done for Io by *Matson and Nash*, 1983). The problem is, however, that the usually very high values of q_0 so obtained correspond to very thin surface layers and it is not clear how relevant they may be to the present model where no distinction is made between regolith and bedrock (this issue depends on the mode of formation, grain size distribution, and other factors). Since surface porosity should occur in any form of equation (8) it would be very interesting to know its relevant value. In sedimentary rocks on the Earth's surface the primary porosity (i.e., the porosity not modified by postdepositional events) is roughly 0.25 ÷ 0.4 for sandstones and 0.4 ÷ 0.7 for carbonates (*Choquette and Pray*, 1970). On the theoretical side it is known that for random dense packing of equal spheres, $q_0 = 0.36$ (*Finney*, 1970). To the best of our knowledge the porosity of a randomly packed system of spheres with an arbitrary distribution of radii f(r)dr has never been evaluated. It might be expected that a continuous form of f [corresponding to the mass spectrum of planetesimals N(m)] would lead to a complete filling of space, but this is apparently not the case; as *Omnés* (1985) has shown, the minimum of porosity is obtained for a discrete distribution of radii. From the above considerations it emerges that a surface porosity of about 0.4 is the most plausible. The parameter α in equation (13') describes how porosity decreases with increasing pressure: In the limit $\alpha \to 0$ there is, apart from a thin surface layer, no porosity in the interior of the satellite; whereas for $\alpha \to \infty$ the satellite is uniform ($q = q_0 = \text{const}$). In this work α was treated as a free parameter.

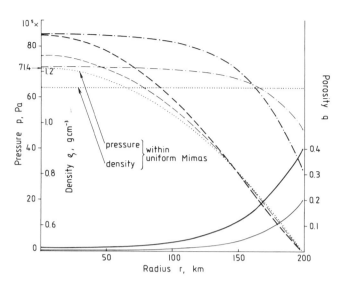

Fig. 2. Distribution of pressure (p(r) (dashed lines), density $\rho(r)$ (dashed/point lines), and porosity q(r) (continuous lines) within porous Mimas for two assumed values of the surface porosity: $q_0 = 0.4$ (heavy lines) and $q_0 = 0.2$ (thin lines). The parameter $\tilde{\alpha} = 0.04$.

In calculations the values R = 198.8 km, $\bar{\rho} = 1.137$ g cm^{-3} were adopted. They are quoted with high accuracy by Dermott (personal communication, 1986) for Mimas. Mimas appears to be very suitable as an example for which the model is employed because: (1) According to thermal history calculations carried out by *Ellsworth and Schubert* (1983) it may indeed be in an isothermal state. (2) Its medium size and therefore moderate pressure range indicate that porosity may be an important factor. For the largest satellites, porosity is probably significant only at small depths and its overall effect is not great, whereas for small bodies (e.g., cometary nuclei) porosity, although high, is presumably constant and therefore not important in the context of the present model where its decrease with pressure is explicitly taken into account. (3) For Mimas, in addition to M and R, the dimensionless moment of inertia is quoted with high accuracy: $\gamma = 5/2 \, I/MR^2 = 0.890 \pm 0.035$ (Dermott, personal communication, 1986). This fact will play an important role in the discussion of our results.

The solution of the boundary value problem equation (10') and (11') was obtained by means of the Goodman-Lance method together with the four step Runge-Kutta integration scheme for the auxiliary initial value problem. The distributions of p, q, and ρ for q = 0.2 and 0.4 and $\tilde{\alpha} = 0.04$ are shown in Fig. 2. This value of $\tilde{\alpha}$ is characteristic in two respects: It corresponds to a minimum of the dimensionless moment of inertia and also gives rise to a (local) minimum in the tidal distortion of a synchronously rotating satellite (see discussion that follows). It is seen from Fig. 2 that higher values of q_0 and therefore stronger mass concentration toward the center enhance central pressure (but not very significantly) and steepen the density gradient.

Moment of Inertia

For a spherical body the dimensionless moment of inertia γ (normalized to unity) is defined as follows

$$\gamma = \frac{1}{\frac{2}{5}MR^2} \frac{8\pi}{3} \int_0^R \rho(r) r^4 dr \qquad (14)$$

In dimensionless variables this equation may be rewritten as

$$\gamma \xi = 5 \int_0^1 Q[\tilde{p}(x)] x^4 dx \qquad (14')$$

where the function Q is defined in equation (7) and the dimensionless density factor ξ is defined as

$$\xi = \frac{\bar{\rho}}{\rho} = 3 \int_0^1 Q[\tilde{p}(x)] x^2 dx \qquad (15)$$

After $\tilde{p}(x)$ had been found from equation (10'), ξ and γ were calculated by means of equations (15) and (14'). With known density of ice ρ_i and an assumed density of rock ρ_r the silicate mass fraction C was calculated from equations (15) and (8). For fixed q_0 the relation $\gamma(C)$ was established in this way (see Fig. 3). From Fig. 3 some interesting observations may be made.

1. The rock concentration C takes on only values belonging to some interval $[C_{min}, C_{max}]$. The value of C_{min} (corresponding to $\alpha \to 0$) is independent of q_0; it decreases from $C_{min} = 0.265$ for $\rho_r = 2.6$ g cm^{-3} to $C_{min} = 0.225$ for $\rho_r = 3.8$ g cm^{-3}. There is some value of $q_0 = q_0^*$ such that $C_{max} = 1$ for $q_0 > q_0^*$. This fact is intuitively obvious: If the porosity is too great the satellite must consist of rock only in order to give the desired mass.

2. $\gamma = 1$ for $C = C_{min}$ and for $C = C_{max}$ if $C_{max} < 1$. This fact can also be easily explained: $C = C_{min}$ corresponds to $\alpha \to 0$ (virtually no porosity) and $C = C_{max} < 1$ corresponds to $\alpha \to \infty$ (uniform porosity).

3. For fixed C, the greater q_0, the lesser γ.

4. $\gamma(C)$ has a minimum for each q_0. This minimum is obtained for $\tilde{\alpha} = 0.03 \div 0.05$ with greater values of $\tilde{a}$ corresponding to lower values of q_0. γ_{min} tells how significant the effect of porosity on moment of inertia may be.

The fact that for a celestial body $\gamma < 1$ indicates that the body is not uniform. There might be several causes of deviations from uniformity: (1) full or partial differentiation; (2) self-compression of matter (in the first approximation negligible for small bodies; (3) the existence of a crust (rather unlikely for minor satellites). Small self-gravitation of minor satellites permits them still another possibility that is considered in the present work; (4) the presence of voids between grains.

When the departures from uniformity are small, the true moment of inertia may be written as the product

$$\gamma = \Pi \gamma_j \qquad (16)$$

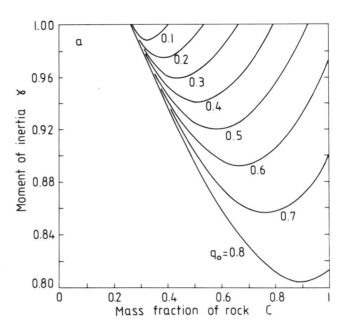

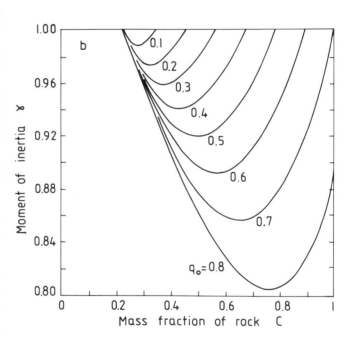

Fig. 3. The dimensionless moment of inertia γ versus rock concentration C, as calculated for the exponential porosity-pressure relation (equation 13')). The curves are parameterized through the value of surface porosity $q_0 = 0.1, \ldots, 0.8$. On each curve the parameter $\tilde{\alpha} = \infty$ (rightmost point). This figure corresponds to R = 198.8 km, $\bar{\rho} = 1.137$ g cm^{-3}, and to two different densities of the rocky component (a) $\rho_r = 2.6$ g cm^{-3} and (b) $\rho_r = 3.8$ g cm^{-3}. As the value of rock density is increased the point $C = C_{min}$ is shifted to the left but the minimum value of γ remains unchanged.

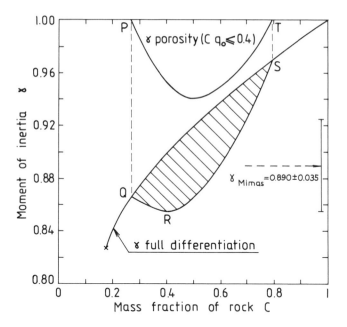

Fig. 4. The joint effect of porosity and differentiation on the dimensionless moment of inertia for Mimas. The shadowed area corresponds to the product of $\gamma_{porosity}(q_o \leq 0.4)$ and $\gamma_{full\ differentiation}$. If differentiation is only partial then the region of accessible γ's is enlarged to the interior of the curve PQRST. The points of intersection of the line $\gamma = 0.890$ with the boundary of the corresponding area determine the range of C for which the value of γ quoted for Mimas by Dermott (personal communication, 1986) can be recovered. Density of rock $\rho_r = 2.6$ g cm^{-3}.

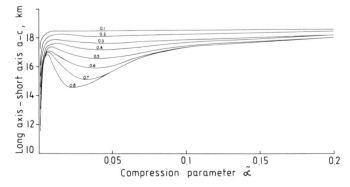

Fig. 5. The difference between the equatorial semiaxis directed toward Saturn, a, and the polar semiaxis, c, for tidally distorted Mimas as a function of the dimensionless compression parameter $\tilde{\alpha}$. The curves are labelled by the value of the surface porosity q_o.

where γ_j denotes the moment of inertia calculated for the jth case of nonuniformity (e.g., *Cook*, 1980). For a fully differentiated satellite with rocky core and icy mantle the dimensionless moment of inertia is equal to

$$\gamma_1(C,\bar{\rho},\rho_i) = \frac{1}{\rho}\left[(\bar{\rho} - \rho_i)\left(\frac{C\bar{\rho} - \bar{\rho} + \rho_i}{\rho_i}\right)^{2/3} + \rho_i\right] \quad (17)$$

The combined effect of full differentiation and porosity is illustrated in Fig. 4. It is seen that for $q_o \leq 0.4$ porosity alone cannot explain (in the present model) the low value of γ quoted for Mimas by Dermott (personal communication, 1986). Nevertheless, it can reduce γ significantly. If in addition to being porous the satellite is fully differentiated, then γ_{Mimas} can be recovered for each C between 0.35 and 0.6; curiously, this is roughly the range of C generally expected for the icy satellites. If allowance is made for only partial differentiation then the situation is even more favourable: In order to give the desired γ the silicate mass fraction can be as low as C_{min} discussed above (the upper limit 0.6 remains unchanged). This fact is interesting because for Enceladus (which is comparable in size and mass to Mimas) a low value of C is expected, whereas values greater than 0.6 are in general unlikely for the medium size icy satellites (*Schubert et al.*, 1986).

Shape of Figure

The effect of porosity on the figure of the satellite was evaluated in the first approximation by means of the method described by *Zharkov et al.* (1985) (see Appendix B). For this purpose it was assumed that the satellite rotates synchronously about its parent planet, can be regarded as a liquid body, and the density distribution within its interior is spherical. The deformation of the figure of the satellite is caused by the tide-forming potential and the centrifugal potential. For a uniform density distribution the difference between the equatorial semiaxis directed toward the planet, a, and the polar semiaxis, c, should be equal in the case of Mimas to about 18 km (Dermott, personal communication, 1986). From the control networks of Mimas, *Davies and Katayama* (1983) obtained the actual value a - c = 6 ± 3 km. In the present model the difference a - c was calculated as a function of the dimensionless compression parameter $\tilde{\alpha}$ for $0.1 \leq q_o \leq 0.8$ (recall that $\tilde{\alpha}$ has the physical meaning of a parameter describing the thickness of the porous layer). The results are shown in Fig. 5. If the porosity is confined to the uppermost layers (low $\tilde{\alpha}$) then a-c is low. However, the value of 6 (or even 9) km has not been recovered for any $\tilde{\alpha}$. Such low values of a-c could only be obtained for extremely low values of $\tilde{\alpha}$ ($\ll 10^{-5}$) that are both physically and numerically unreasonable. For plausible values of $\tilde{\alpha}$, a-c is greater than 14 km; we conclude that in the present model porosity alone is not sufficient to explain the actual shape of Mimas (at least in the first-order theory). We note that for each value of q_o, a-c has a local minimum at some $\tilde{\alpha} = \tilde{\alpha}^*$. This minimum is more pronounced for greater q_o. It is interesting that $\tilde{\alpha}^*$ coincides roughly with that value of $\tilde{\alpha}$ that minimizes γ. It is perhaps worth investigating if this is more than a numerical artifact.

APPENDIX A

The Generalized Emden Equation

The pressure distribution within a porous satellite obeys equation (10')

$$\frac{d^2\tilde{p}}{dx^2} - \frac{d\log Q}{d\tilde{p}}\left(\frac{d\tilde{p}}{dx}\right)^2 + \frac{2}{x}\frac{d\tilde{p}}{dx} + Q^2 = 0 \quad (A1)$$

Q is a specified function of $\tilde{p}$

$$Q = Q(\tilde{p}) \quad (A2)$$

With equation (A1) the following boundary conditions are associated

$$\tilde{p}(x = 1) = 0, \quad \frac{d\tilde{p}}{dx}(x = 0) = 0 \quad (A3)$$

We seek the solution of equation (A1) in the form

$$\tilde{p}=\tilde{p}[\omega(x)], \quad \frac{d\tilde{p}}{dx} = \tilde{p}_\omega \omega', \quad \frac{d^2\tilde{p}}{dx^2} = \tilde{p}_{\omega\omega}(\omega')^2 + \tilde{p}_\omega \omega'' \quad (A4)$$

where prime denotes differentiation with respect to x. Then equation (A1) transforms into

$$\tilde{p}_\omega \omega'' + \left(\tilde{p}_{\omega\omega} - \frac{d\log Q}{d\tilde{p}}\tilde{p}_\omega^2\right)(\omega')^2 + \frac{2}{x}\tilde{p}_\omega \omega' + Q^2 = 0 \quad (A5)$$

In order to eliminate the term containing $(\omega')^2$ we solve the following equation

$$\tilde{p}_{\omega\omega} - \frac{d\log Q}{d\tilde{p}}\tilde{p}_\omega^2 = 0 \quad (A6)$$

Equation (A6) can be reduced to an equation of first order by making the substitution

$$\tilde{p}_\omega = u = u(\tilde{p}), \quad \tilde{p}_{\omega\omega} = \frac{du}{d\tilde{p}}u \quad (A7)$$

which transforms it into the following

$$\frac{du}{d\tilde{p}}u - \frac{d\log Q}{d\tilde{p}}u^2 = 0 \quad (A8)$$

Since we are interested in the case of $u \neq 0$ then equation (A8) is equivalent to

$$d\log u = d\log Q \quad (A9)$$

We choose the solution

$$u = Q = \frac{d\tilde{p}}{d\omega} \quad (A10)$$

Then equation (A5) takes the form

$$Q\omega'' + \frac{2}{x}Q\omega' + Q^2 = 0 \quad (A11)$$

We are interested in the case of $Q \neq 0$. Then equation (A11) reduces to

$$\omega'' + \frac{2}{x}\omega' + Q[\tilde{p}(\omega)] = 0 \quad (A12)$$

Equation (A12) can be treated as a generalization of the Emden equation

$$\omega'' + \frac{2}{x}\omega' + \omega^n = 0 \quad (A13)$$

APPENDIX B

Figure of a Synchronously Rotating Porous Satellite

It is assumed that the satellite rotates synchronously about its parent planet, is liquid, and the density distribution within its interior is spherically symmetric. Then it may be shown that, to the first approximation, the semiaxes of figure of the satellite are given by the following formulas (e.g., *Zharkov et al.*, 1985)

$$a = R + \Delta a, \quad \Delta a = \frac{7}{6}\beta h_2 R,$$
$$b = R + \Delta b, \quad \Delta b = -\frac{1}{3}\beta h_2 R, \quad (B1)$$
$$c = R + \Delta c, \quad \Delta c = -\frac{5}{6}\beta h_2 R$$

where a and b are equatorial semiaxes directed toward the planet and along the satellite orbital motion, respectively, c is the polar semiaxis, and R is the radius of a sphere of equivalent volume. The coefficient β is equal to

$$\beta = \frac{M'}{M}\left(\frac{R}{R'}\right)^3 \quad (B2)$$

where M' is the planet's mass, M is the satellite mass, and R' is the radius of the satellite's orbit (considered circular). Using Dermott's data for Mimas one obtains $\beta = 0.0188$. This value is slightly greater than the value $\beta = 0.0180$ used by *Zharkov et al.* (1985) (they write α instead of β). The quantity

h_2 entering equation (B1) is the Love number of second order. It is equal to

$$h_2 = \frac{T(R)}{Rg_0} \quad (B3)$$

where g_0 is the gravitational acceleration on the surface of the satellite, and the function $T(r)$ satisfies the second order differential equation (*Molodensky*, 1953)

$$\frac{d^2T}{dr^2} + \frac{2}{r}\frac{dT}{dr} - \left[\frac{4\pi G}{g(r)}\frac{d\rho}{dr} + \frac{n(n+1)}{r^2}\right]T = 0 \quad (B4)$$

with $n = 2$. In equation (B4) r is the radial distance from the center of the satellite, ρ is the density, and g is the gravitational acceleration. With equation (B4) the following boundary conditions are associated

$$\left.\frac{dT}{dr}\right|_{r=R} = -\frac{n+1}{R}T(r=R) + \left.\frac{4\pi G\rho T}{g}\right|_{r=R} + (2n+1)g_0 \quad (B5)$$

$$T \text{ finite for } r \to 0$$

For a porous satellite by virtue of equations (1), (7), and (13′), equations (B4) and (B5) take the following form

$$\frac{d^2\tilde{T}}{dx^2} + \frac{2}{x}\frac{d\tilde{T}}{dx} + \left\{\frac{1}{\tilde{\alpha}}\,q(\tilde{p})\left[1-q(\tilde{p})\right] - \frac{6}{x^2}\right\}\tilde{T} = 0 \quad (B6)$$

and

$$\left\{\frac{d\tilde{T}}{dx} + 3\left[1 - \xi(1-q_0)\right]\tilde{T}\right\}\bigg|_{x=1} = 5 \quad (B7)$$

$$\tilde{T}(x=0) = 0$$

where the dimensionless variables are defined as follows

$$x = r/R, \quad \tilde{T} = T/Rg_0 \quad (B8)$$

The Love number of second order is given as

$$h_2 = \tilde{T}(x=1) \quad (B9)$$

REFERENCES

Arzt E., Ashby M. F., and Easterling K. E. (1983) Practical applications of hot isostatic pressing diagrams: Four case studies. *Metall. Trans. A, 14*, 211-221.

Buratti B. (1985) Application of a radiative transfer model to bright icy satellites. *Icarus, 61*, 208-217.

Chandrasekhar S. (1967) *An Introduction to the Study of Stellar Structure*, Chap. 4, Dover, New York.

Choquette P. W. and Pray L. C. (1970) Geologic nomenclature and classification of porosity in sedimentary carbonates. *Am. Assoc. Petrol. Geol. Bull., 54*, 207-250.

Cook A. H. (1980) *Interiors of the Planets*. Cambridge University Press, Cambridge, England.

Davies M. E. and Katayama E. Y. (1983) The control networks of Mimas and Enceladus. *Icarus, 53*, 332-340.

Ellsworth K. and Schubert G. (1983) Saturn's icy satellites: thermal and structural models. *Icarus, 54*, 490-510.

Finney J. L. (1970) Random packings and the structure of simple liquids — I. The geometry of random close packing. *Proc. Roy. Soc. Lond., A319*, 479-493.

Lindsay J. F. (1976) *Lunar Stratigraphy and Sedimentology*. Elsevier, New York.

Matson D. M. and Nash D. B. (1983) Io's atmosphere: Pressure control by subsurface regolith cold trapping. *J. Geophys. Res., 88*, 4771-4783.

Molodensky M. S. (1953) The elastic tides, free nutation and some problems of the Earth's constitution. *Tr. Geofiz. Inst. Akad. Nauk SSSR, 19*, 3-52.

Omnés R. (1985) Sur l'empilement désordonné d'un système de spheres de rayons différents. *J. Physique, 46*, 139-147.

Smith B. A., Soderblom R., Batson R., Bridges P., Inge J., Masursky H., Shoemaker E., Beebe R., Boyce J., Briggs G., Bunker A., Collins S. A., Hansen C. J., Johnson T. V., Mitchell J. L., Terrile R. J., Cook A. F., II, Cuzzi J., Pollack J. B., Danielson G. E., Morrison D., Owen T., Sagan C., Veverka J., Strom R., and Suomi V. E. (1982) A new look at the Saturn system: The Voyager 2 images. *Science, 215*, 504-537.

Smith B. A., Soderblom L. A., Beebe R., Bliss D., Boyce J. M., Brahic A., Briggs G. A., Brown R. H., Collins S. A., Cook A. F., II, Croft S. K., Cuzzi J. N., Danielson G. E., Davies M. E., Dowling T. E., Godfrey D., Hansen C. J., Harris C., Hunt G. E., Ingersoll A. P., Johnson T. V., Krauss R. J., Masursky H., Morrison D., Owen T., Plescia J., Pollack J. B., Porco C. B., Rages K., Sagan C., Shoemaker E. M., Sromovsky L. A., Stoker C., Strom R. G., Suomi V. E., Synnott S. P., Terrile R. J., Thomas P., Thompson W. R., and Veverka J. (1986) Voyager 2 in the Uranian system: Imaging science results. *Science, 233*, 43-64.

Smoluchowski R. and McWilliam A. (1984) Structure of ices on satellites. *Icarus, 58*, 282-287.

Smoluchowski R., Marie M., and McWilliam A. (1984) Evolution of density in solar system ices. *Earth, Moon and Planets, 30*, 281-288.

Schubert G., Spohn T., and Reynolds R. T. (1986) Thermal histories, compositions, and internal structures of the moons of the solar system. In *Satellites* (J. A. Burns and M. S. Matthews, eds.), pp. 224-292. University of Arizona, Tucson.

Troitsky V. S. (1965) Investigation of the surfaces of the Moon and planets by thermal radiation. *Radio Sci., 69*, 1585-1612.

Veverka J., Thomas P., Johnson T. V., Matson D., and Housen K. (1986) The physical characteristics of satellite surfaces. In *Satellites* (J. A. Burns and M. S. Matthews, eds.), pp. 342-402. University of Arizona, Tucson.

Zharkov V. N., Leontjev V. V., and Kozenko A. V. (1985) Models, figures, and gravitational moments of the Galilean satellites of Jupiter and icy satellites of Saturn. *Icarus, 61*, 92-100.

Author Index

Ahrens T. J. 435
Arnold J. R. 79

Basu A. 283
Becker R. H. 87
Bell J. F. 573
Beran A. 403
Bertka C. M. 723
Blake D. F. 615
Boclet D. 623
Bonté Ph. 623
Boslough M. B. 443
Brannon J. C. 555
Brearley A. J. 513
Britt D. T. 503
Bunch T. E. 615
Burns C. A. 455
Burns R. G. 713

Castellarin A. 623
Cintala M. J. 409
Clayton D. D. 637
Clayton R. N. 513, 545
Coombs C. R. 339
Craddock R. A. 331
Cygan R. T. 443

Delano J. W. 59
Donn B. 631

Echer C. J. 615
Ehlmann A. J. 545
Eluszkiewicz J. 741

Farrand W. H. 319
Fisenko A. V. 537
Flynn G. J. 607
Francis P. W. 659
Frey H. 679
Frick U. 87

Garrison D. 631
Gerke T. 283
Glass B. P. 455
Goldstein J. I. 493
Goodacre A. K. 375
Grady M. M. 513
Greeley R. 243, 331, 665
Grieve R. A. F. 375
Guest J. E. 665

Haskin L. A. 1
Hawke B. R. 155, 339, 355
Head J. W. III 307
Heymann D. 525
Hohenberg C. 631
Holloway J. R. 723

Honda M. 79
Hörz F. 409, 423

Ignatenko K. I. 537
Imamura M. 79
Isbell N. K. 665

Jéhanno C. 593, 623

Keil K. 513, 545, 573
Klöck W. 471
Kobayashi K. 79
Koeberl C. 403
Kohl C. P. 79
Kowalik J. A. 493
Kracher A. 485

Laul J. C. 187, 203
Laurenzi M. A. 299
Lavielle B. 565
Lavrukhina A. K. 537
Leliwa-Kopystyński J. 741
Liffman K. 637
Lindroth D. P. 365
Lindstrom D. J. 1
Lindstrom M. M. 11, 45, 121, 139, 169, 219, 255
Ljul A. Yu. 537
Longhi J. 459
Lucey P. G. 155, 355, 397
Lugmair G. W. 555

Mardinly A. J. 615
Marti K. 565
Marvin U. B. 169
Maurette M. 593
Maxwell T. A. 701
Mayeda T. K. 513, 545
McConville P. 299
McGee J. J. 21
McGill G. E. 701
McKay D. S. 67, 283
Miyamoto M. 33
Moore H. J. 383
Mori H. 33

Nagai H. 79
Neal C. R. 121, 139
Nishiizumi K. 79
Nuth J. A. III 631

Olinger C. 631

Palma R. L. 525
Palme H. 471
Pan V. 459

Papike J. J. 187, 203
Pepin R. O. 87
Pieters C. M. 503, 581
Pillinger C. T. 513
Podnieks E. R. 365
Podosek F. A. 555

Reedy R. C. 79
Robin E. 593
Rocchia R. 623
Roush T. L. 397
Rupert J. D. 375
Ryder G. 219, 273

Salpas P. A. 11
Schultz L. 513, 545
Scott D. H. 665
Scott E. R. D. 513, 545
See T. H. 423
Semeniuk A. M. 679
Semeniuk J. A. 679
Sharpton V. L. 307, 375
Shervais J. W. 45, 169, 255
Signer P. 513
Simon S. B. 187, 203
Spudis P. D. 155, 243
Steele A. 273
Sutton S. R. 607
Swann G. A. 243

Tagai T. 33
Takeda H. 33
Tanaka K. L. 665
Taylor L. A. 11, 45, 121, 139
Thompson T. W. 383
Tokarcik S. 679
Turner G. 299
Tyburczy J. A. 435

Vetter S. K. 169, 255

Warren P. H. 233
Webb S. 599
Weber H. W. 545
Wentworth S. J. 67
Wieler R. 513
Williams D. B. 493
Willis K. 219
Wilson L. 339
Wood C. A. 659

Yamashita H. 79
Yon S. A. 581
Yoshida K. 79

Zolensky M. E. 599

Subject Index

Accretion 435
Adcumulate 21
Agglutinates 409, 423
Analytical electron microscopy 615
Anorthosites 1, 169
  ferroan 11, 21, 187
  magnesium 187
  noritic 21
Antarctic ice samples 599
Apennine Bench Formation 243
Apennine Front 169, 187, 203, 219, 283
Apennine mountains 155
Apollo 14 155
Apollo 15 155, 169, 187, 203, 243, 255, 273, 339
Apollo 16 155
Apollo 17 11
Aristarchus 355
Assimilation 121, 139, 233
Asteroids 409, 503, 573
Atmosphere-surface equilibration 435

Basalt petrogenesis 121
Basalts 273, 459
  Fra Mauro 59
  highland 59
  KREEP 187, 219
  mare 139, 243, 255, 273
  olivine normative 255
  quartz normative 255
  regolith 255
  very high potassium 121
Basin ejecta 155
Basin rings 155
Bidirectional spectra 503
Breccia modeling 1,
Breccias 45, 169, 299, 573, 573
  feldspathic 21
  fragment modes 1
  fragmental 545
  gas-rich 573
  granulitic 33
  highlands 21
  meteorite 545
  poikilitic 33
  regolith 33, 59, 67, 545

Caldera 659
Carbon 513
Carbon isotopes 513
Carbonaceous chondrite 615
Cayley Formation 155
Chalcophiles 471

Chemical weathering 443
Chondrites 513, 545, 555, 573
  type 3 ordinary 513
Chondritic clasts 513
Chronology 555
Coarse fines 187, 219
Cometary particle 615
Cometary showers 375
Comminution 409
Cosmic chemical memory 637
Cosmic dust 631
Cosmic grains 593
Cosmic-ray exposure age 299
Cosmic-ray reactions 565
Cosmogenic nuclide 79
Crater ages 375
Crater dating 701
Cratering 409
Cratering mechanics 155
Craters 383
  identification 383
Cretaceous/Tertiary boundary 455

Dehydration kinetics 435
Differentiation 33
Diffusive fractionation 87
Diogenite 459
Dissolution 443

Ejecta 409
Electromagnetic energy 365
Eucrite 459
Extraterrestrial mining 365

Faults 307
Fra Mauro 59, 155, 203
Fractional crystallization 121, 139, 459
Fractionation 273, 409, 423

Galactic cosmic ray 79
Geology 339
Germanium 485
Glasses
  Apollo 16 67
  GASP 67
  highland basalt 67
  impact 403
  impact melt 299
  KREEP 67
  LKFM 67
  mare 67
  silica-poor 67
Gossans 713

Grain size 423
Granite 121
Granulites 11
Graphite 513

Hadley Rille 243, 283
Highland contamination 319
Highland gabbros 355
Howardite 459

Icy satellites 741
Imbrium basin 155, 355
Impact 409, 423, 435
Impact cratering 443
Impact glass 403, 455
Impact melt 21, 169, 203, 219, 573
Impact-induced devolatilization 435
Inert gases 525
Interplanetary dust particles 599, 615
Interstellar dust 637
Irghizites 403
Iridium 593, 623
Iridium anomaly 455
Iron-nickel meteorites 503
Irradiation hardness 565
Isotopic anomalies 637

KREEP 121, 139, 203, 233
KREEP basalts 219
Kamacite 493

Laser 365
Laser probe 299
Lava 243, 339
Lava tubes 243
Layer lattice silicate 615
Layered pluton 21
Liquid density 459
Luna 16 319
Lunar basins 155
Lunar crust 33
Lunar highlands 11
Lunar orbital geochemistry 319
Lunar resources 397
Lunar tectonics 307

Magma 459
Magma mixing 233
Magma ocean 233
Magmasphere 233
Magnetic separates 525
Mantle 723
Mare Fecunditatis 319

Mare Tsiolkovsky 331
Mare deposits 331
Mare ridges 307
Mare thickness 319, 331
Mare volcanism 243
Mare volume 331
Mars 679, 701, 713, 723
   crater ages 679
   eolian history 665
   fluvial history 665
   geology 665
   highland-lowland boundary 679
   mantle 723
   resurfacing 665
   stratigraphy 665
   surface 443
   volcanism 665
Maturity 283
Metamorphism 555
Meteorite parent bodies 573
Meteorites 503, 573, 593, 623
   Antarctic 33
   iron 493
   lunar 33
Mg-rich rocks 233
Microgabbros 273
Micrometeorites 593, 623
Microprobe analysis 485
Microtektites 455
Microtexture 33
Microtomy 615
Microvitrophyres 455
Microwaves 365
Mineral chemistry 283
Mixing model 283
Modal analysis 283
Moon
   cometary impact 397
   surface water 397
Mozart Rille 243

Near-infrared spectroscopy 155
Neodymium 555
Nitrogen 87
Noble gas retention 631
Noble gases 87, 513
Norite 169
   anorthositic 21
North Ray Crater 21

Oblique impacts 331
Olivine 273, 471
Orbital geochemistry 155
Oxygen isotopes 513

Partial differentiation 485
Partial melting 459
Partial melts 723
Periodic impacts 375
Petrogenesis 139
Phase boundary 459
Phosphates 555
Phyllosilicate 615
Picrites 273
Plagioclase 33
Planetary core formation 471
Plessite 493
Plutonic rocks 1
Polymict rock 21
Presolar grains 631
Primary melts 723
Pristine glass 59
Pristine rocks 233
Procellarum basin 155
Proportional growth 155
Pyroclastics 59
Pyroxene 33

Radar 339
Radar images 383
Radar resolution 383
Rare earth elements 11, 139, 187, 203
Rare gas trapping 631
Reflectance spectroscopy 355, 397, 573
Regoliths 219, 283, 409, 423, 503
   asteroid 545
Reiner Gamma 397
Remote sensing 339, 355, 397
Resurfacing 679
Rima Mozart 339
Rima Parry 59
Rock fragmentation 365

Secondary cratering 409
Shock 409, 435
Shock activation 443
Shock melting 423
Shock-enhanced dissolution 443
Shocked glass 33
Short-range unmixing 273
Siderophiles 11, 203, 375, 471
Sieve fractions 525
Sinuous rilles 243, 339
Solar abundances 87
Solar cosmic ray 79
Solar energy 365
Solar gas 525
Solar wind 87
Spallation krypton 565

Spectra 339, 503
Spectral alteration 573
Spherules
   clinopyroxene-bearing 455
   cpx 455
   early Archean 455
Spur Crater 299
Sputtering 637
Stepwise heating 525
Stratigraphy 339
Stratospheric particles 607
Strontium 555
Structural evolution 701
Sulfides 713
SUNOCON 637
Synchrotron X-ray fluorescence 607

Taenite 493
Tektites
   Muong Nong 403
   water content 403
Terrestrial cratering 375
Time-series analysis 375
Trace elements 607
   transition zone 679
Transmission electron microscopy 615
Troctolite 169
Tycho 355
Venus 659
Vertical mixing 319
Vertical profile 283
Volcanic eruption 339
Volcanoes 659
Weathering 435, 443
Widmanstatten 493
Winonaites 485
Zhamanshinites 403

Sample Index

10046 87	15445 155, 169, 219, 243	67955 33
10084 87, 365	15455 1, 155, 169, 219, 243, 299	67975 21, 169
	15459 169, 219	
12001 87	15465 169, 299	68415 21
12002 79	15466 299	68815 79
	15498 169	
14047 59	15531 87	71501 87
14049 59	15555 273, 459	
14072 459	15556 273	72275 11, 233
14149 87	15620 273	
14179 233	15623 273	73215 11
14259 87	15633 273	
14301 59	15634 273	74241 87
14303 121, 233	15641 273	74275 79
14305 121	15643 273	
14307 59	15651 273	76230 11, 33
14310 79, 187	15663 273	76335 1
14313 59	15668 273	76505 11
14318 59	15672 273	76535 1
14321 79, 139, 233		
	60010 67	77017 11
15003 233	60015 1	77545 33
15007 233, 283	60016 33, 67	
15008 283	60019 33	78155 11
15009 283	60639 233	78235 1
15010 283		
15011 283	61016 1	79035 87
15065 459	61156 33	79215 11, 33
15205 169, 243, 283	61175 67	
15223 187, 203	61516 67	
15240 219		
15243 219	62236 1, 219	
15245 219	62237 1, 219	
15250 219		
15255 219	63507 67	
15256 273	63588 67	
15263 187		
15271 219	65015 33	
15274 273	65095 67	
15299 273	65715 67	
15303 187, 203		
15304 187	66035 67	
15314 187, 203	66036 67	
15382 169, 233	66075 67	
15385 273	66095 397	
15386 169, 187, 233		
15387 273	67015 21	
15388 273	67016 21	
15405 169, 233, 355	67075 21	
15415 1, 11, 155, 169, 187	67215 1, 21	
15418 169, 219	67415 33	
15426 169	67455 21	
15434 187, 203	67601 87	
15435 169	67701 87	
15437 1	67915 21	

Meteorite Index

ALHA 76003 545
ALHA 77005 79
ALHA 77011 513
ALHA 77015 513
ALHA 77047 513
ALHA 77049 513
ALHA 77050 513
ALHA 77081 485
ALHA 77167 513
ALHA 77214 513
ALHA 77241 513
ALHA 77260 513
ALHA 77278 513
ALHA 77299 513
ALHA 78015 513
ALHA 78119 513
ALHA 81005 33
ALHA 81024 513
ALHA 81025 513
ALHA 81030 513
ALHA 81031 513
ALHA 81032 513
ALHA 81121 513
Alais 607
Allende 525, 607
Ambapur Nagla 555
Ames-Dec86-011 615

Barwell 545
Binda 459
Bjurbole 555
Bluff 545
Bovedy 545
Bremervorde 513
Bruderheim 565

Cangas de Onis 573
Canyon Diablo 493, 503
Cold Bokkeveld 87

Dayton 493
Dhajala 513
Dimmitt 513, 545, 573
Dubrovnik 573

EETA 79001 87
Elenovka 555

Farmington 545

Ghubara 545
Gibeon 503
Grady 513
Guarena 555

Ivuna 607

Jelica 555
Juvinas 459

Kapoeta 545
Kendleton 545
Khohar 513

Lissa 545
Lodran 573

Menow 555
Mezo-Madaras 513
Moama 459
Mocs 555
Murchison 87, 545, 615, 623
Murray 87, 615

Nadiabondi 545
Negrillos 593
Ngawi 545
Novo Urei 87
Nuevo Laredo 459

Odessa 493
Olivenza 573
Orgueil 87, 525, 607, 615
Orvinio 545

Peace River 299, 555
Pesyanoc 87
Piancaldoli 513
Plainview 513, 545

Renazzo 87
Roosevelt County 014 545
Rose City 545

Semarkona 513
Sharps 513
Sioux County 459
Soko Banja 555
St. Marks 555
St. Mesmin 545
St. Severin 555, 565, 623
Stannern 459
Steinbach 573

Tieschitz 513, 555
Tolvca 493

Vishnupur 545

W7029*A 615
Walters 545
Weston 87, 513

Yamato 75097 545
Yamato 791197 33
Yamato 82192 33
Yamato 82193 33